Carraher's
POLYMER
CHEMISTRY

Eighth Edition

Carraher's
POLYMER
CHEMISTRY

Eighth Edition

Charles E. Carraher, Jr.

CRC Press
Taylor & Francis Group
Boca Raton   London   New York

CRC Press is an imprint of the
Taylor & Francis Group, an **informa** business

CRC Press
Taylor & Francis Group
6000 Broken Sound Parkway NW, Suite 300
Boca Raton, FL 33487-2742

© 2011 by Taylor and Francis Group, LLC
CRC Press is an imprint of Taylor & Francis Group, an Informa business

No claim to original U.S. Government works

Printed in the United States of America on acid-free paper
10 9 8 7 6 5 4 3 2 1

International Standard Book Number: 978-1-4398-0955-6 (Hardback)

Library of Congress Cataloging-in-Publication Data

Carraher, Charles E.
　　Carraher's polymer chemistry. -- 8th ed. / Charles E. Carraher, Jr.
　　　　p. cm.
　　Summary: "Updated to reflect a growing focus on green chemistry in the scientific community and in compliance with the American Chemical Society's Committee on professional training guidelines, Carraher's polymer chemistry, eighth edition integrates the core areas that contribute to the growth of polymer science. It supplies the basic understanding of polymers essential to the training of science, biomedical, and engineering students. New in the eighth edition: updating of analytical, physical, and special characterization techniques. Increased emphasis on carbon nanotubes, tapes and glues, butyl rubber, polystyrene, polypropylene, polyethylene, poly(ethylene glycols), shear-thickening fluids, photo-chemistry and photophysics, dental materials, and aramids. New sections on copolymers, including fluoroelastomers, nitrile rubbers, acrylonitrile-butadiene-styrene terpolymers, and EPDM rubber. New units on spliceosomes, asphalt, and fly ash and aluminosilicates. Larger focus on the molecular behavior of materials, including nano-scale behavior, nanotechnology, and nanomaterials. Continuing the tradition of providing a user-friendly approach to the world of polymeric materials, the book allows students to integrate their chemical knowledge and establish a connection between fundamental and applied chemical information. It contains all of the elements of an introductory text with synthesis, property, application, and characterization. Special sections in each chapter contain definitions, learning objectives, questions, and additional reading, and case studies are woven into the text fabric. Symbols, trade names, websites, and other useful ancillaries appear in appendices to supplement the text"-- Provided by publisher.
　　Summary: "This new edition of Carraher's Polymer Chemistry maintains the scope and organization of its bestselling predecessors, while fulfilling the ACS-CPT advanced course requirement. Emphasizing the fundamental behavior of polymers in nature reflected in technological developments, it features new chapters on composites, fibers, and naturally occurring polymers in plants and animals. It also highlights nanoscale polymer applications with new sections on drug design, electronics, optical fibers, textiles, and adhesives. Each chapter includes study tips, glossaries, and exercises. This edition also expands listings for lab exercises, polymer structures, trade names, and Internet resources. Professors adopting the text will be aided by a test-bank available for their exclusive use"-- Provided by publisher.
　　Includes bibliographical references and index.
　　ISBN 978-1-4398-0955-6 (hardback)
　　1. Polymers. 2. Polymerization. I. Seymour, Raymond B. (Raymond Benedict), 1912-1991. Seymour/Carraher's polymer chemistry. II. Title. III. Title: Polymer chemistry.

QD381.S483 2010
547′.7--dc22
　　　　　　　　　　　　　　　　　　　　　　　　　　　　　　　　　　　　　　2010034105

Visit the Taylor & Francis Web site at
http://www.taylorandfrancis.com

and the CRC Press Web site at
http://www.crcpress.com

Foreword

Polymer science and technology have developed tremendously over the last few decades, and the production of polymers and plastics products has increased at a remarkable pace. By the end of 2000, nearly 200 million tons per year of plastic materials were produced worldwide (about 2% of the wood used, and nearly 5% of the oil harvested) to fulfill the ever-growing needs of the plastic age; in the industrialized world plastic materials are used at a rate of early 100 kg per person per year. Plastic materials with more than $250 billion per year contribute about 4% to the gross domestic product in the United States. Plastics have no counterpart in other materials in terms of weight, ease of fabrication, efficient utilization, and economics. It is no wonder that the demand and the need for teaching in polymer science and technology have increased rapidly. To teach polymer science, a readable and up-to-date introductory textbook is required that covers the entire field of polymer science, engineering, technology, and the commercial aspect of the field. This goal has been achieved in Carraher's textbook. It is eminently useful for teaching polymer science in departments of chemistry, chemical engineering, and material science, and also for teaching polymer science and technology in polymer science institutes, which concentrate entirely on the science and technologies of polymers.

This eighth edition addresses the important subject of polymer science and technology, with emphasis on making it understandable to students. The book is ideally suited not only for graduate courses but also for an undergraduate curriculum. It has not become more voluminous simply by the addition of information—in each edition less important subjects have been removed and more important issues introduced. Polymer science and technology is not only a fundamental science but also important from the industrial and commercial point of view. The author has interwoven discussion of these subjects with the basics in polymer science and technology. Testimony to the high acceptance of this book is that early demand required reprinting and updating of each of the previous editions. We see the result in this new significantly changed and improved edition.

Otto Vogl
Herman F. Mark Professor Emeritus
Department of Polymer Science and Engineering
University of Massachusetts
Amherst, Massachusetts

Preface

As with most science, and chemistry in particular, there is an explosive broadening and increase in the importance of the application of foundational principles of polymers. This broadening is seen in ever increasing vistas allowing the advancement of our increasingly technologically dependant society and solutions to society's most important problems in such areas as environment and medicine. Some of this broadening is the result of extended understanding and application of already known principles, but it also includes the development of basic principles and materials known to us hardly a decade ago. Most of the advancements in communication and computers, medicine, and air and water purity are tied to macromolecules and a fundamental understanding of the principles that govern their behavior. Much of this revolution is of a fundamental nature and is explored in this latest edition. This book deals with these basic principles and their application in real-life situations. Technology is the application of scientific principles. In polymers there is often little, if any, division between science and technology.

The importance of the environment and our interaction with it is becoming increasingly evident. Industries are increasingly emphasizing on green science and practices that are favorable to the environment. Polymer science also emphasizes on these practices and contributes critical components toward solutions. This text continues to emphasize these measures including special sections that deal directly with environmental issues as well as integrating green science appropriately woven within the fabric that is polymer chemistry. Consistent with the continued emphasis on green chemistry, new sections dealing with photochemistry and green materials have been added.

Polymers are found in the organic natural world as the building blocks for life itself. They are also found as inorganic building blocks that allow construction of homes, skyscrapers, and roads. Synthetic polymers serve as basic building blocks of society today and tomorrow. This text includes all three of these critical segments of polymeric materials.

A basic understanding of polymers is essential to the training of today's science, biomedical, and engineering students. *Carraher's Polymer Chemistry* complies with the American Chemical Society's Committee on Professional Training guidelines as an advanced or in-depth course. It naturally integrates and interweaves the important foundational areas since polymers are critical to all of the foundational areas with all of these foundational areas contributing to the growth of polymer science. Most of the fundamental principles of polymers are an integral part of the syllabi of the undergraduate and graduate training courses for students. This allows students to integrate their chemical knowledge and establish a connection between fundamental and applied chemical information. Thus, along with the theoretical information, application is integrated as an essential part of the information. As in other areas such as business and medicine, short case studies are presented as historical material.

While this book is primarily written as an introductory graduate-level course, it can also be used as an undergraduate course, or as the introductory undergraduate–graduate course. The topics are written so that the order and inclusion or exclusion of chapters or parts of chapters will still allow students an adequate understanding of the science of polymers. Most of the chapters begin with the theory part followed by the application portion. The most important topics are generally at the beginning of the chapter followed by important, but less critical, sections. Some would prefer to take up the synthesis-intense chapters first, some would prefer to take up the analytical/analysis/ properties chapters first, and others may simply prefer to take up the chapters as they appear in the text. The book contains all of the elements of an introductory text with synthesis, property, application, and characterization all present, allowing this to be the only polymer course taken by an individual or the first in a series of polymer-related courses taken by the student.

This edition continues in the "user-friendly" mode, with special sections in each chapter containing definitions, learning objectives, questions, and additional reading. Application and theory are integrated so that they reinforce one another. There is a continued emphasis on picturing, reinforcing, interweaving, and integrating basic concepts. The initial chapter is shorter, allowing students to become acclimated. Other chapters are written so they can be covered in about a week's time or less. Where possible, difficult topics are distributed and reinforced over several topics. Case studies are woven into the text fabric.

The basic principles that apply to synthetic polymers apply equally well to inorganic and biological polymers and are present in each of the chapters covering these important polymer groupings.

The updating of analytical, physical, and special characterization techniques continues. A number of topics have been increased and include carbon nanotubes, tapes and glues, butyl rubber, polystyrene, polypropylene, polyethylene, poly(ethylene glycols), shear-thickening fluids, photochemistry and photophysics, dental materials, and aramids. New sections on a number of copolymers, including fluoroelastomers, nitrile rubbers, acrylonitrile–butadiene–styrene terpolymers, and EPDM rubber, have been added. In addition, new units on spliceosomes, asphalt, and flyash and aluminosilicates have been included. There is more emphasis on the molecular behavior of materials, that is, nano-scale behavior, and on nanotechnology and nanomaterials.

Acknowledgments

The author gratefully acknowledges the contributions and assistance of the following in preparing this text: John Droske, Eli Pearce, Charles Pittman, Edward Kresge, Gerry Kirshenbaum, Sukumar Maiti, Alan MacDiarmid, Les Sperling, Eckhard Hellmuth, Mike Jaffe, Otto Vogel, Thomas Miranda, Murry Morello, and Graham Allan; and a number of our children who assisted in giving suggestions for the text—Charles Carraher III, Shawn Carraher, Colleen Carraher-Schwarz, Erin Carraher, and Cara Carraher—to Erin for discussions on materials, Cara for her help with the biomedical material, and to Shawn for his help in relating the business and industrial aspects. Special thanks to Gerry Kirshenbaum for his kind permission to utilize portions of articles by me that appeared in *Polymer News*. This book could not have been written if not for those who are ahead of us in this field, especially Raymond Seymour, Herman Mark, Charles Gebelein, Paul Flory, and Linus Pauling; all of these friends shepherded and helped me. My thanks to them.

I thank my wife Mary Carraher for her help in proofing and allowing this edition to be written.

Contents

Polymer Nomenclature

As with most disciplines, the system followed for naming or defining things is important. Here we will focus on naming polymers with emphasis on synthetic polymers. Short presentations on how to name proteins and nucleic acids are provided in Chapter 4 and those for nylons are provided in Chapter 5.

The fact that synthetic polymer science grew up in many venues before nomenclature groups were present to assist in standardization of the naming approach resulted in many popular polymers having several names including common names. Many polymer scientists have not yet accepted the guidelines given by the official naming committee of the International Union of Pure and Applied Chemistry, IUPAC, because the common names have gained such widespread acceptance. Although there is a wide diversity in the practice of naming polymers, we will concentrate on the most utilized systems.

COMMON NAMES

Little rhyme or reason is associated with many of the common names of polymers. Some names are derived from the place of origin of the material, such as *Hevea brasilliensis*— literally "rubber from Brazil"—for natural rubber. Other polymers were named after their discoverer, as is Bakelite, the three-dimensional polymer produced by condensation of phenol and formaldehyde, which was commercialized by Leo Baekeland in 1905.

For some important groups of polymers, special names and systems of nomenclature were developed. For instance, the nylons were named according to the number of carbons in the diamine and dicarboxylic acid reactants used in their synthesis. The nylon produced by the condensation of 1,6-hexamethylenediamine (six carbons) and adipic acid (six carbons) is called nylon-66. Even here, there is not a set standard as to how nylon-66 is to be written with alternatives including nylon 66 and nylon-6,6.

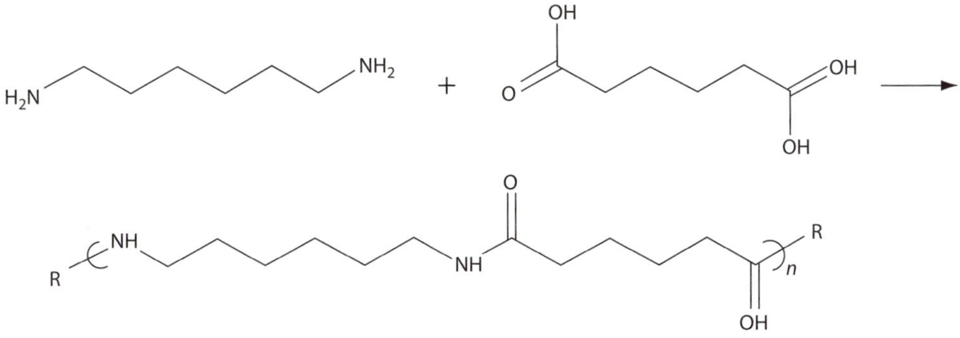

Nylon-6,6

SOURCE-BASED NAMES

Most common names are source based, that is, they are based on the common name of the reactant monomer, preceded by the prefix "poly." For example, polystyrene is the most frequently used name for the polymer derived from the monomer 1-phenylethene, which has the common name styrene.

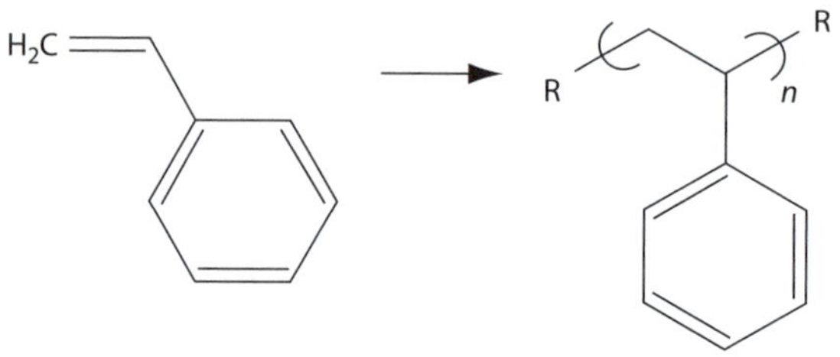

Styrene   Polystyrene

The vast majority of commercial polymers based on the vinyl group ($H_2C=CHX$) or the vinylidene group ($H_2C=CX_2$) as the repeat unit are known by their source-based names. Thus, polyethylene is the name of the polymer synthesized from the monomer ethylene; poly(vinyl chloride) from the monomer vinyl chloride, and poly(methyl methacrylate) from methyl methacrylate.

Many condensation polymers are also named in this manner. In the case of poly(ethylene terephthalate), the glycol portion of the name of the monomer, ethylene glycol, is used in constructing the polymer name, so that the name is actually a hybrid of a source-based and a structure-based name.

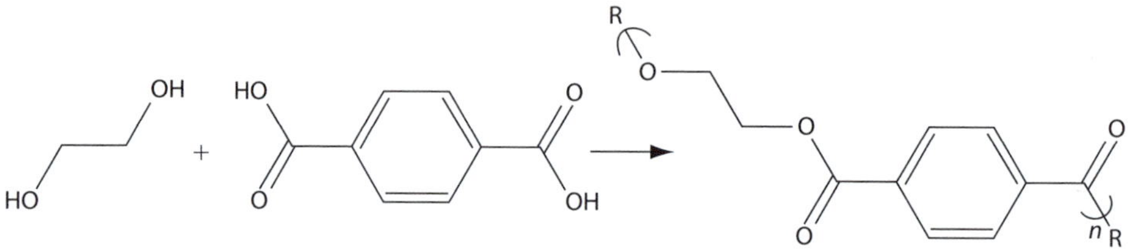

Poly(ethylene terephthalate)

This polymer is well known by a number of trade names, such as Dacron, its common grouping, polyester, and by an abbreviation PET or PETE.

Although it is often suggested that parentheses be used in naming polymers of more than one word [like poly(vinyl chloride)], but not for single word-polymers (like polyethylene), some authors entirely omit the use of parentheses for either case (like polyvinyl chloride) so even here there exists a variety of practices. We will employ the use of parentheses for naming polymers of more than one word.

Copolymers are composed of two or more monomers. Source-based names are conveniently employed to describe copolymers using an appropriate term between the names of the monomers. Any of the half dozen or so connecting terms may be used depending on what is known about the structure of the copolymer. When no information is known or intended to be conveyed the connective term "co" is employed in the general format poly(A-co-B), where A and B are the names of the two monomers. An unspecified copolymer of styrene and methyl methacrylate would be called poly[styrene-co(methyl methacrylate)].

Kraton, the yellow rubber-like material often found on the bottom of running shoes, is a copolymer whose structural information is known. It is formed from a group of styrene units, that is, a "block" of polystyrene attached to a group of butadiene units, or a block of polybutadiene, which is attached to another block of polystyrene forming a triblock copolymer. The general representation of such a block might be –AAAAAAAABBBBBBBBAAAAAAAA–, where each A and B represents an individual monomer unit. The proper source-based name for Kraton is polystyrene-block-polybutadiene-block-polystyrene, or poly-block-styrene-block-polybutadiene-block-polystyrene with the prefix "poly" being retained for each block. Again, some authors will omit the "poly" use, giving polystyrene-block-butadiene-block-styrene.

STRUCTURE-BASED NAMES

Although source-based names are generally employed for simple polymers, IUPAC has published a number of reports for naming polymers. These reports are being widely accepted for the naming of complex polymers. A listing of such reports is given in the references section. A listing of source- and structure-based names for some common polymers is given in Table 1.

LINKAGE-BASED NAMES

Many polymer "families" are referred to by the name of the particular linkage that connects the polymers (Table 2). The family name is "poly" followed by the linkage name. Thus, those polymers that contain an ester linkage are known as polyesters; those with an ether linkage are called polyethers, and so on.

TABLE 1
Source and Structure-Based Names

Source-Based Names	Structure-Based Names
Polyacrylonitrile	Poly(1-cyanoethylene)
Poly(ethylene oxide)	Polyoxyethylene
Poly(ethylene terephthalate)	Polyoxyethyleneoxyterephthaloyl
Polyisobutylene	Poly(1,1-dimethylethylene)
Poly(methyl methacrylate)	Poly[(1-methoxycarbonyl)-1-metylethylene]
Polypropylene	Poly(1methylethylene)
Polystyrene	Poly(1-phenylethylene)
Polytetrafluoroethylene	Polydifluoromethylene
Poly(vinylacetate)	Poly(1-acetoxyethylene)
Poly(vinyl alcohol)	Poly(1-hydroxyethylene)
Poly(vinyl chloride)	Poly(1-chloroethylene)
Poly(vinyl butyral)	Poly[(2-propyl-1,3-dioxane-4,6-diyl)methylene]

TABLE 2
Linkage-Based Names

Family Name	Linkage	Family Name	Linkage
Polyamide	$-N-\overset{\overset{O}{\|\|}}{C}-$	Polyvinyl	$-C-C-$
Polyester	$-O-\overset{\overset{O}{\|\|}}{C}-$	Polyanhydride	$-\overset{\overset{O}{\|\|}}{C}-O-\overset{\overset{O}{\|\|}}{C}-$
Polyurethane	$-O-\overset{\overset{O}{\|\|}}{C}-\overset{\overset{H}{\|}}{N}-$	Polyurea	$-\overset{\overset{H}{\|}}{N}-\overset{\overset{O}{\|\|}}{C}-\overset{\overset{H}{\|}}{N}-$
Polyether	$-O-$	Polycarbonate	$-O-\overset{\overset{O}{\|\|}}{C}-O-$
Polysiloxane	$-O-Si-$	Polysulfide	$-S-$

TRADE NAMES, BRAND NAMES, AND ABBREVIATIONS

Trade (and/or brand) names and abbreviations are often used to describe a particular material or a group of materials. They may be used to identify the product of a manufacturer, processor, or fabricator, and may be associated with a particular product or with a material or modified material, or a material grouping. Tradenames (or trade names) are used to describe specific groups of materials that are produced by a specific company or under license of that company. Bakelite is the tradename given for the phenol-formaldehyde condensation developed by Baekeland. A sweater whose contents are described as containing Orlon contains polyacrylonitrile fibers that are "protected" under the Orlon trademark and produced or licensed to be produced by the holder of the Orlon trademark. Carina, Cobex, Dacovin, Darvic, Elvic, Geon, Koroseal, Marvinol, Mipolam, Opalon, Pliofex, Rucon, Solvic, Trulon, Velon, Vinoflex, Vygen, and Vyram are all tradenames for poly(vinyl chloride) manufactured by different companies. Some polymers are better known by their tradename than their generic name. For instance, polytetrafluoroethylene is better known as Teflon, the tradename held by Dupont.

Abbreviations, generally initials in capital letters, are also employed to describe polymers. Table 3 contains a listing of some of the more widely used abbreviations and the polymer associated with the abbreviation.

CHEMICAL ABSTRACTS-BASED POLYMER NOMENCLATURE

The most complete indexing of any scientific discipline is done in chemistry and is done by chemical abstracts (CA). Almost all of the modern searching tools for chemicals and chemical information are based on CA for at least some of its information base. It is critical for polymer chemists to have some grasp of how CA names chemical compounds. The full description of the guidelines governing the naming of chemical compounds and related properties is given in Appendix IV at the end of the CA *Index Guide*. This description is about 200 pages. While small changes are made with each new edition, the main part has remained largely unchanged since about 1972. Today, there are computer programs, including that associated with SciFinder Scholar, that names materials once the structure is given. For small molecules this is straight forward, but for polymers care must be taken. Experiment with simple polymers before moving to more complex macromolecules. If the Chemical Abstracts Service (CAS) # is known, this can be entered and names investigated for appropriateness for your use.

TABLE 3
Abbreviations for Selected Polymeric Materials

Abbreviation	Polymer	Abbreviation	Polymer
ABS	Acrylonitrile–butadiene–styrene terpolymer	CA	Cellulose acetate
EP	Epoxy	HIPS	High-impact polystyrene
MF	Melamine-formaldehyde	PAA	Poly(acrylic acid)
PAN	Polyacrylonitrile	SBR	Butadiene–styrene copolymer
PBT	Poly(butylene terephthalate)	PC	Polycarbonate
PE	Polyethylene	PET	Poly(ethylene terephthalate)
PF	Phenyl-formaldehyde	PMMA	Poly(methyl methacrylate)
PP	Polypropylene	PPO	Poly(phenylene oxide)
PS	Polystyrene	PTFE	Polytetrafluoroethylene
PU	Polyurethane	PVA, PVAc	Poly(vinyl acetate)
PVA,PVAl	Poly(vinyl alcohol)	PVB	Poly(vinyl butyral)
PVC	Poly(vinyl chloride)	SAN	Styrene–acrylonitrile
UF	Urea-formaldehyde		

CA organizes the naming of materials into twelve major arrangements that tie together about 200 subtopics. These main topic headings are as follows:

A. Nomenclature systems and general principles
B. Molecular skeletons
C. Principle chemical groups
D. Compound classes
E. Stereochemistry and stereoparents
F. Specialized substances
G. Chemical substance names for retrospective searches
H. Illustrative list of substitute prefixes
J. Selective bibliography of nomenclature of chemical substances
K. Chemical prefixes
L. Chemical structural diagrams from CA Index Names
M. Index

The section dealing with polymers is subtopic 222. Polymers. The subsection dealing with polymers builds on the foundations given before and thus some of the guidelines appear to be confusing and counterproductive to the naming of polymers but the rules were developed for the naming of small molecules. Following is a description of the guidelines that are most important to polymer chemists. Additional descriptions are found in the CA Appendix IV itself and in articles given in the readings. The Appendix IV concentrates on linear polymers. A discussion of other more complex polymeric materials is also found in articles cited in the readings section.

General Rules

In searching the chemical literature, in particular systems based on CA, searches for particular polymers can be conducted using the Chemical Abstract Service Number, CAS # (where known), or by repeat unit. The IUPAC and CAS have agreed upon a set of guidelines for the identification, orientation, and naming of polymers based on the structural repeat unit (SRU). IUPAC names polymers as "poly(constitutional repeat unit)" while CAS utilizes a "poly(structural repeating unit)." These two approaches typically give similar results.

Here we will practice using the sequence "identification, orientation, and naming" first by giving some general principles and finally by using specific examples.

In the *identification* step, the structure is drawn, usually employing at least two repeat units. Next, in the *orientation* step, the guidelines are applied. Here we will concentrate on basic guidelines. Within these guidelines are subsets of guidelines that are beyond our scope.

Structures will be generally drawn in the order, from left to right, in which they are to be named.

Seniority

The starting point for the naming of a polymer unit involves determining seniority among the subunits.
A. This order is
   Heterocyclic rings>
   Greatest number of most preferred acyclic heteroatoms>
   Carbocyclic rings>
   Greatest number of multiple bonds>
   Lowest or closest route (or lowest locant) to these substituents>
   Chains containing only carbon atoms

with the symbol ">" indicating "is senior to."

This is illustrated below:

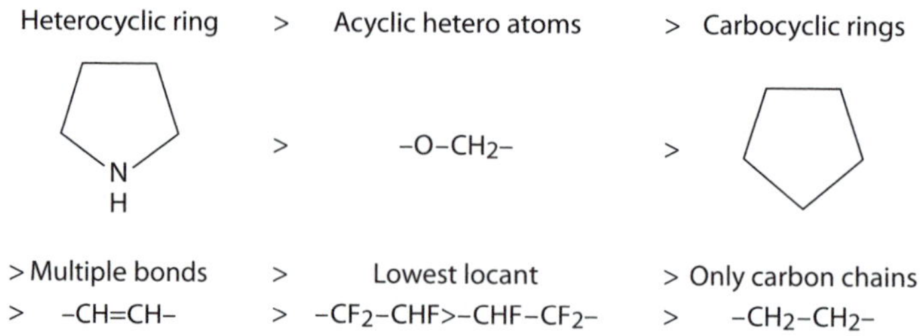

| Heterocyclic ring | > | Acyclic hetero atoms | > | Carbocyclic rings |

| | > | $-O-CH_2-$ | > | |

| > Multiple bonds | > | Lowest locant | > Only carbon chains |
| > $-CH=CH-$ | > | $-CF_2-CHF>-CHF-CF_2-$ | > $-CH_2-CH_2-$ |

This order is partially derived from guidelines found in other sections, such as section 133 Compound Radicals where the ordering is given as

Greatest number of acyclic hetero atoms>
Greatest number of skeletal atoms>
Greatest number of most preferred acyclic hetero atoms>
Greatest number of multiple bonds>
Lowest locants or shortest distance to nonsaturated carbons

The lowest locant or shortest distance refers to the number of atoms from one senior subunit to the next most senior subunit when there is only one occurrence of the senior subunit.

This order refers to the backbone and not to the substitutions. Thus, polystyrene and poly(vinyl chloride) are contained within the "chains containing only carbon atoms" grouping.

B. For ring systems the overall seniority is

Heterocyclic>
Carbocyclic

but within the rings there is also an ordering (Section 138) that is

Nitrogenous heterocyclic>
Heterocyclic>
Largest number of rings>
Cyclic system occurring earliest in the following list of systems; spiro, bridged fused, bridges nonfused, fused>
Largest individual ring (applies to fused carbocyclic systems)>
Greatest number of ring atoms

For example,

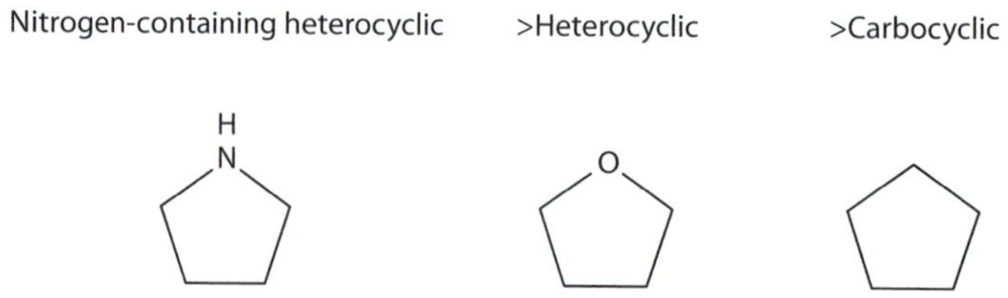

| Nitrogen-containing heterocyclic | >Heterocyclic | >Carbocyclic |

and

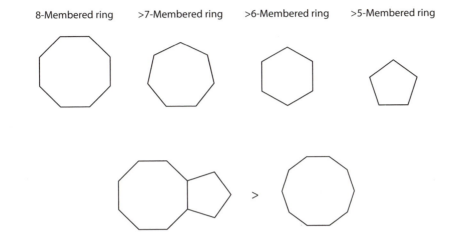

8-Membered ring   >7-Membered ring   >6-Membered ring   >5-Membered ring

and

> C. For hetero-atomed linear chains or cyclic rings, the order of seniority is O> S> Se> Te> N> P> As> Sb> Bi> Si> Ge> Sn> Pb> B> Hg

Thus, because $-O-CH_2-$ is senior to $-S-CH_2-$, it would be named first in a polymer that contained both $-O-CH_2-$ and $-S-CH_2-$ segments. Further, a polymer containing these alternating units would *not* be poly(thiomethyleneoxymethylene) but would be named poly(oxymethylenethiomethylene).

Another example,

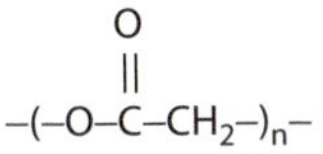

$$-(-O-\overset{\overset{\displaystyle O}{||}}{C}-CH_2-)_n-$$

is named poly[oxy(1-oxy-1,2-ethanediyl)] or less preferred poly[oxy(1-oxoethylene)] but not poly[(2-oxo-1,2-ethanediyl)oxy] or poly[(2-oxoethylene)oxy].

> D. Unsaturation is senior to saturation. The more it is unsaturated the more senior it is with all other items being equal. Thus 1,4-phenylene is senior to 2,5-cyclohexadiene-1,4diyl, which in turn is senior to 2-cyclohexene-1,4-diyl, which is senior to cyclohexane-1,4diyl. For linear chains $-CH=CH-CH=CH-$ is senior to $-CH=CH-CH_2-CH_2-$, which is in turn senior to the totally saturated chain segment.

ROUTE

> A. From the senior subunit determined from "Seniority" take the shortest path (smallest number of atoms) to another like or identical unit or to the next most preferred subunit. Thus, for the homo polymer poly(oxymethylene) it is simply going from one oxygen to the next oxygen and recognizing that this is the repeat unit. For a more complex ether this means going until the chain begins to repeat itself going in the shortest direction from the senior unit or atom to the next most senior unit or atom.

Thus, $-O-C-C-O-C-C-C-$ is named "oxy-1,2-ethanediyloxy-1,3propanedily" rather than "oxy-1,3-propanediyloxy-1,2-ethanediyl."

> B. Where the paths are equal, such as in some nylons, the repeat unit is named so that the heteroatom "N" is first named and then more highly substituted (carbonyl) unit appears next. Thus, nylon 3,3, with the structure

$$
\begin{array}{cc}
O & O \\
\parallel & \parallel \\
\end{array}
$$
$$-(-NH-C-CH_2-C-NH-CH_2-CH_2-CH_2-)_n-$$

is named poly[imino(1, 3-dioxo-1,3-propanediyl)imino-1,3-propanediyl].

C. In otherwise identical subunits, there are three items to be considered in decreasing order of importance:

   a. Maximum substitution: thus, 2,3,5-trichloro-*p*-phenylene is senior to 2,5-dichloro-*p*-phenylene, which in turn is senior to 2-chloro-*p*-phyenylene

   b. Lowest locants: thus, 2,3-dichloro-*p*-phenylene is senior to 2,5-dichloro-*p*-phenylene

   c. Earliest alphabetical order: thus, 2-bromo-*p*-phenylene is senior to 2-chloro-*p*-phenylene, which is senior to 2-iodo-*p*-phenylene.

D. Where there is no conflict with other guidelines, triple bonds are senior to double bonds, which in turn are senior to single bonds; multiple bonds should be assigned the lowest possible locants. Thus, the polymer from 1,3-butanediene polymerized in the so-called "1,4-" mode is usually drawn as $-(-C-C=C-C-)-$ but it is named as drawn as $-(-C=C-C-C-)-$ and named poly(1-butene-1,4-diyl) with the appropriate "cis-" or "trans-" designation. Polyisoprene, typically drawn as

$$-(-CH_2-C(CH_3)=CH-CH_2-)_n-$$

is frequently named poly(2-methyl-1,3-butadiene) but it is named as though its structure is

$$-(C(CH_3)=CH-CH_2-CH_2-)_n-$$

with the name poly(1-methyl-1-butene-1,4-diyl).

Substituents are named as one of several classes. The most important ones are dealt with here. For monoatomic radicals from borane, methane, silane (and other Group IVA elements) they are named by replacing the "ane" ending by "yl," "ylene," and "ylidyne" to denote the loss of one, two, or three hydrogen atoms, respectively.

H_2B- boryl H_3C- methyl $H_2C=$ methylene $HC\equiv$ methylidyne

Acyclic hydrocarbon radicals are named from the skeletons by replacing "ane," "ene," and "yne" suffixes by "yl," "enyl," and "ynyl," respectively.

CH_3-CH_2- ethyl $CH_3-CH_2-CH_2-$ propyl $-CH_2-CH_2-$ 1,2-ethanediyl

$-CH=CH-$ 1,2-ethenediyl $H_2C=CH-CH=$ 2-propenylidene

$$
\begin{array}{c}
\parallel \\
-CH_2-C-CH_2-1,3\text{-propanediyl-2-ylidene}
\end{array}
\qquad
\begin{array}{c}
\mid \\
-CH_2-CH-CH_2-1,2,3\text{-propanetriyl}
\end{array}
$$

Table 4 contains the names of selected bivalent radicals that may be of use to polymer chemists.

SEARCHING

Searching is made much simpler with computer systems such as SciFinder Scholar, where the name or CA# can be entered and references related to that compound are forthcoming. Teasing out the

TABLE 4
Names of Select Bivalent Radicals

"Common" or "Trivial" Name	CAS Name	Structure
Adipyl, adipoly	1,6-Dioxo-1,6-hexanediyl	$-CO-(CH_2)_4-CO-$
1,4-Butanediyl	1,4-Butanediyl	$-(CH_2)_4-$
Carbonyl	Carbonyl	$-CO-$
Diglycoloyl	Oxybis(1-oxo-2,1-ethanediyl)	$-CO-CH_2-O-CH_2-CO-$
Ethylene	1,2-Ethanediyl	$-CH_2-CH_2-$
Imino	Imino	$-NH-$
Iminodisulfonyl	Iminobis(sulfonyl)	$-SO_2-NH-SO_2-$
Methene, methylene	Methylene	$-CH_2-$
Oxybis(methylenecarbonylimino)	Oxybis[(1-oxo-2,1-ethanediyl)imino)]	$-NHCO-CH_2-O-CH_2-CO-N-$
Pentamethylene	1,5-Pentanediyl	$-(CH_2)_5-$
Phenylene	1,4-Phenylene	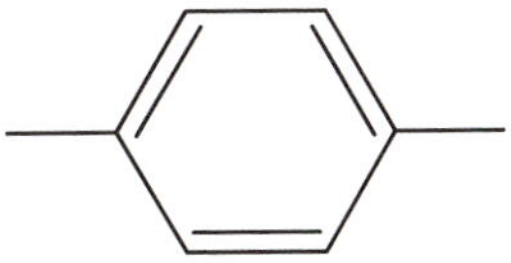
Phenylenedimethylene	1,4-Phenylenebis(methylene)	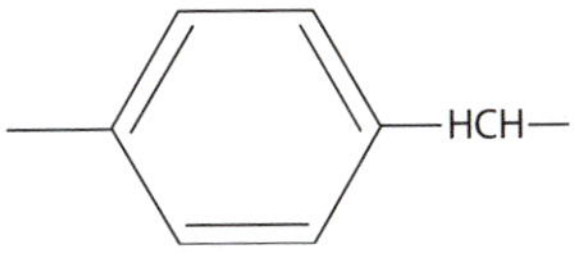
Phenylenedioxy	1,4-Phenylenebis(oxy)	
Sebacoyl	1,10-Dioxo-1,10-decanediyl	$-CO-(CH_2)_8CO-$
Styrenyl	1-Phenyl-1,2-ethanediyl	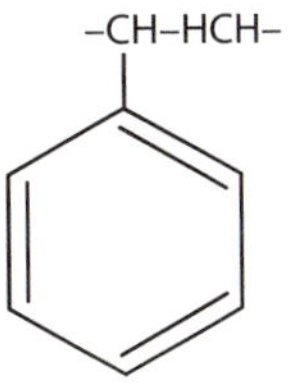 $-CH-HCH-$
Sulfonyl, sulfuryl	Sulfonyl	$-SO_2-$
Tartaroyl	2,3-Dihydroxy-1,4-dioxo-1,4-butanediyl	$-CO-CH(OH)-CH(OH)-CO-$
Terephthaloyl	1,4-Phenylenedicarbonyl	
Thio	Thio	$-S-$
Thionyl	Sulfinyl	$-SO-$
Ureylene	Carbonyldiimino	$-NH-CO-NH-$
Vinylene	1,2-Ethenediyl	$-CH=CH-$

particular information you want is more difficult and often requires thinking about associations that might yield the desired information. Thus, if you want to know the solubility of a given polymer, you might look up molecular weight determinations for that polymer since most molecular weight determinations require the polymer be dissolved.

Following is a longer version that can be used if hard copy or electronic version of CA is available.

In searching, polymers from a single monomer are indexed as the monomer name with the term "homopolymer" cited in the modification. Thus, polymers of 1-pentene are listed under the monomer

1-Pentene
    homopolymer

Polymers formed from two or more monomers, such as condensation polymers and copolymers, and homopolymers are indexed at each inverted monomer name with the modifying term "polymer with" followed by the other monomer names in uninverted alphabetical order. The preferential listing for identical heading parents is in the order: (a) maximum number of substituents, (b) lowest locants for substituents, (c) maximum number of occurrences of index heading parent, and (d) earliest index position of the index heading. Examples are

1-Pentene
    polymer with 1-hexene

2,5-Furandione
    polymer with 1,4-butanedisulfonic acid

Silane, dichlorodiethyl-
    polymer with dichlorodiphenylsilane

While the percentage composition of copolymers (i.e., the ratio of comonomers) is not given, copolymers with architecture other than random or statistical are identified as "alternating, block, graft, and so forth." Random or statistical copolymer are not so identified in the CA index. Oligomers with definite structure are noted as dimer, trimer, tetramer, and so on.

Often similar information is found at several sites. For instance, for copolymers of 1-butene and 1-hexene, information will be listed under both 1-butene and 1-hexene, but the listings are not identical so both entries should be consulted for completeness.

CA's policy for naming acetylenic, acrylic, methacrylic, ethylenic, and vinyl polymers is to use the source-based method, and source-based representation is used to depict the polymers graphically; thus a synonym for polyethylene is polyethylene and not poly(1,2-ethanediyl); a synonym polypropylene is polypropylene, and poly(vinyl alcohol) is named ethenol, homopolymer although ethenol does not exist. Thus, these polymers are named and represented structurally by the source-based method, not the structure-based method.

EXAMPLES

Following are examples that illustrate CAS guidelines of naming.

$-(CH_2-)_n-$	Poly(methylene)
$-(CH_2-CH_2-)_n-$	Poly(ethylene)
$-(CH=CH-)_n-$	Poly(1,2-ethenediyl)

$$\begin{array}{cc} O & O \\ \| & \| \\ \end{array}$$
$-(-C-C-CH_2-CH_2-)_n-$ Poly1,2-dioxo-1,4-butanediyl

$-(-CH=CH-CH-CH_2-)_n-$
                |
               CH_3 Poly(3-methyl-1-butene-1,4-diyl)

$$-(NH-\overset{\overset{\displaystyle O}{\|}}{C}-CH_2-CH_2-)_n-$$

Poly[imino(1-oxo-1,3-propanediyl)]

$$-(O-\overset{\overset{\displaystyle O}{\|}}{C}-O-CH_2-CH_2-)_n-$$

Poly[oxocarbonyloxy(1,2-ethanediyl)]

$$-(-CH_2-S-NH-CH_2-CH_2-O-\\CH_2-S-CH_2-NH-CH_2-)_n-$$

Poly(oxymethylenethioimino-1,2-ethaned iyloxymethylenethiomethyleneiminom- ethylene)

$$-(-CFH-CH_2-)_n-$$

Poly(1-fluoro-1,2ethanediyl)

$$-(-O-CH_2-CH_2-)_n-$$

Poly(oxy-1,2-ethanediyl)

$$-(-O-CH_2-)_n-$$

Poly(oxymethylene)

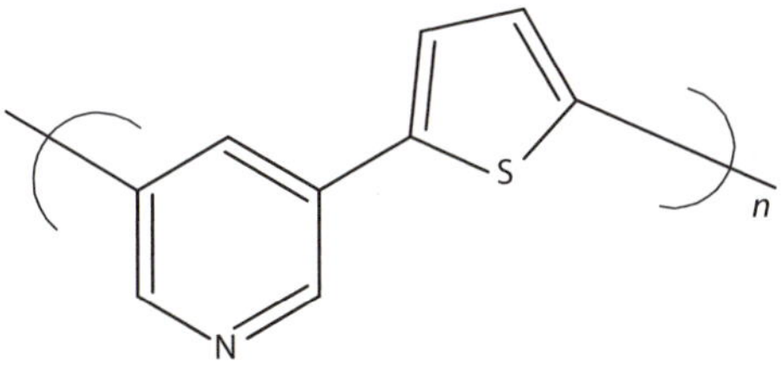

Poly(3,5-pyridinediyl-2,5-thiophenediyl)

$$-[-NH-\overset{\overset{\displaystyle O}{\|}}{C}-(CH_2)_4-\overset{\overset{\displaystyle O}{\|}}{C}-NH-(CH_2)_6-]_n-$$

Poly[imino(1,6-dioxo-1,6hexanediyl) imino-1,6-hexanediyl]

Poly(oxy-1,4-phenylene)

Poly(thio-1,4-phenylene)

In this text, we will typically employ the more "common" (semisystematic or trivial) names of polymers, but it is important in searching the literature using any CA-driven search engine that you are familiar with CA naming for both monomers and polymers.

SUMMARY

While there are several important approaches to the naming of polymers, in this book we will utilize common and source-based names because these are the names that are most commonly utilized by polymer chemists and the general public, and these names, in particular the source-based names, allow a better understanding of the basics of polymers as a function of polymer-structure relationships based on starting materials. Even so, those wishing to do further work in polymers must become proficient in the use of the guidelines used by chemical abstracts and IUPAC.

I wish to acknowledge the assistance of Edward S. Wilks for his help in preparing the section on Chemical Abstracts-Based Polymer Nomenclature.

ADDITIONAL READING

IUPAC. (1952): Report on nomenclature in the field of macromolecules. *J. Poly. Science*, 8:257–277.

IUPAC. (1966): Report on nomenclature dealing with steric regularity in high polymers. *Pure. Appl. Chem.*, 12:645–656 (1966); previously published as Huggins, M. L., Natta, G., Desreus, V., and Mark, H. (1962). *J. Poly. Science*, 56:153–161.

IUPAC. (1969): Recommendations for abbreviations of terms relating to plastics and elastomers. *Pure. Appl. Chem.*, 18:583–589.

IUPAC. (1976): Nomenclature of regular single-strand organic polymers, *Pure. Appl. Chem.*, 48:373–385.

IUPAC. (1981): Stereochemical definitions and notations relating to polymers, *Pure. Appl. Chem.*, 53:733–752.

IUPAC. (1985). Nomenclature for regular single-strand and quasi-single strand inorganic and coordination polymers, *Pure. Appl. Chem.*, 57:149–168.

IUPAC.(1985): Source-based nomenclature for copolymers, *Pure. Appl. Chem.*, 57:1427–1440.

IUPAC. (1987): Use of abbreviations for names of polymeric substances, *Pure. Appl. Chem.*, 59:691–693.

IUPAC. (1989): Definitions of terms relating to individual macromolecules, their assemblies, and dilute polymer solutions, *Pure. Appl. Chem.*, 61:211–241.

IUPAC. (1989): Definition of terms relating to crystalline polymers, *Pure. Appl. Chem.*, 61:769–785.

IUPAC-A. (1989): Classification of linear single-strand polymers, *Pure. Appl. Chem.*, 61:243–254.

IUPAC. (1991): *Compendium of Macromolecular Nomenclature*, pp. 171. Blackwell Scientific Pubs., Oxford, UK. (Collection of summaries).

IUPAC. (1993): Nomenclature of regular double-strand (ladder or spiro) organic polymers, *Pure. Appl. Chem.*, 65:1561–1580.

IUPAC. (1994): Graphic representations (chemical formulae) of macromolecules, *Pure. Appl. Chem.*, 66:2469–2482.

IUPAC. (1994): Structure-based nomenclature for irregular single-strand organic polymers, *Pure. Appl. Chem.*, 66:873–889.

Carraher, C., Hess, G., and Sperling, L. (1987): *J. Chem. Educ.*, 64:36–38.

Carraher, C. (2001): *J. Polym. Mater.,* 17(4):9–14.

Chemical Abstract Service. *Appendix IV; Chemical Abstracts Service*, Chemical Abstract Service, Columbus, OH.

Polymeric Materials: Science and Engineering: 68 (1993) 341; 69 (1993) 575; 72 (1995) 612; 74 (1996) 445; 78 (1998), Back Page; 79 (1998) Back Page; 80 (1999), Back Page; 81 (1999) 569.

Polymer Preprints: 32(1) (1991) 655; 33(2) (1992) 6; 34(1) (1993) 6; 34(2) (1993) 6; 35(1) (1994) 6; 36(1) (1995) 6; 36(2) (1995) 6; 37(1) (1996) 6; 39(1) (1998) 9; 39(2) (1998) 6; 40(1) (1999) 6; 41(1) (2000) 6a.

How to Study Polymers

Studying about polymers is similar to studying any science. Following are some ideas that may assist you in your study.

Much of science is abstract. While much of the study of polymers is abstract, it is easier to conceptualize, make mental pictures, of what a polymer is and how it should behave than many areas of science. For linear polymers, think of a string or rope. Long ropes get entangled with themselves and other ropes. In the same way, polymer chains entangle with themselves and with chains of other polymers that are brought into contact with them. Thus, create mental pictures of the polymer molecules as you study them.

Polymers are real and all about us. We can look at giant molecules on a micro or atomic level or on a macroscopic level. The PET bottles we have may be composed of long chains of poly(ethylene terephthate), PET, chains. The aramid tire cord is composed of aromatic polyamide chains. Our hair is made up of complex bundles of fibrous proteins, again polyamides. The polymers you study are related to the real world in which we live. We experience these "large molecules" at the macroscopic level everyday of our lives, and this macroscopic behavior is a direct consequence of the atomic-level structure and behavior. Make pictures in your mind that allow you to relate the atomic and macroscopic worlds.

At the introductory level we often examine only the primary factors that may cause particular giant molecule behavior. Other factors may become important under particular conditions. The polymer molecules you study at times examine only the primary factors that impact polymer behavior and structure. Even so, these primary factors form the basis for both complex and simple structure–property behavior.

The structure–property relationships you will be studying are based on well-known basic chemistry and physical relationships. Such relationships build upon one another and as such you need to study in an ongoing manner. Understand as you go along. Read the material BEFORE you go to class.

This course is an introductory-level course. Each chapter or topic emphasizes knowledge about one or more area. The science and excitement of polymers has its own language. It is a language that requires you to understand and memorize certain key concepts. Our memory can be short term or long term. Short-term memory may be considered as that used by an actor or actress for a TV drama. It really does not need to be totally understood, nor retained after the final "take." Long-term memory is required in studying about giant molecules since it will be used repeatedly and is used to understand other concepts (i.e., it is built upon).

In memorizing, learn how you do this best—time of day, setting, and so on. Use as many senses as necessary—*be active*—read your assignment, write out what is needed to be known, say it, listen to yourself say it. Also, look for patterns, create mnemonic devices, avoid cramming too much into too small a time, practice associations in all directions, and test yourself. Memorization is hard work.

While knowledge involves recalling memorized material, to really "know" something involves more than simple recall—it involves comprehension, application, evaluation, and integration of the knowledge. Comprehension is the interpretation of this knowledge—making predictions, applying it to different situations. Analysis involves evaluation of the information and comparing it with other information and synthesis has to do with integration of the information with other information.

In studying about giant molecules please consider doing the following:

- Skim the text BEFORE the lecture.
- Attend the lecture and take notes.
- Organize your notes and relate information.
- Read and study the assigned material.
- Study your notes and the assigned material, and then
- Review and self-test.

Learning takes time and effort. Study daily by skimming the text and other study material, think about it, visualize key points and concepts, write down important material, make outlines, take notes, study sample problems, and so on. All of these help—but some may help you more than others—so focus on these modes of learning but not at the exclusion of the other aspects.

In preparing for an exam consider the following:

- Accomplish the above—DO NOT wait until the day before the examination to begin studying; create good study habits.
- Study wisely—study how YOU study best—time of day, surroundings, and so on.
- Take care of yourself; get plenty of sleep the night before the examination.
- Attend to last-minute details—is your calculator working, is it the right kind, do I have the needed pencils, review the material once again, and so on.
- Know what kind of test it will be—if possible.
- Get copies of old examinations if possible; talk to others who might have already had the course.

During the test:

- Stay cool, do NOT PANIC.
- Read the directions; try to understand what is being asked for.
- In an essay or similar exam, work for full or at least partial credit; plan your answers.

The study of polymer molecules contains several types of content:

- *Facts*—the term *polymer* means "many" (poly) "units" (mers).
- *Concepts*—linear polymers are long molecules like a string.
- *Rules*—solutions containing polymer chains are more viscous, slower flowing, than solutions that do not contain polymers.
- *Problems*—what is the approximate molecular weight of a single polystyrene chain that has 1,000 styrene units in it?

These varied types of content are often integrated within any topic, but in this introduction to polymer molecules, the emphasis is often on concepts but all the aspects are important.

1 Introduction to Polymers

1.1 HISTORY OF POLYMERS

Since most materials are polymeric and most of the recent advances in science and technology involve polymers, some have called this the polymer age. Actually, we have always lived in a polymer age. The ancient Greeks classified all matter as animal, vegetable, and mineral. Minerals were emphasized by the alchemists, but medieval artisans emphasized animal and vegetable matter. All are largely polymeric and are important to life as we know it. Most chemists, biochemists, and chemical engineers are now involved in some phase of polymer science or technology.

The word *polymer* is derived from the Greek *poly* and *meros*, meaning many and parts, respectively. Some scientists prefer to use the word *macromolecule*, or large molecule, instead of polymer. Others maintain that naturally occurring polymers, or *biopolymers*, and synthetic polymer should be studied in different courses. Others name these large molecules simply "giant molecules." However, the same principles apply to all polymers. If one discounts the end uses, the differences between all polymers, including plastics, fibers, and elastomers or rubbers, are determined primarily by the intermolecular and intramolecular forces between the molecules and within the individual molecule, respectively, and by the functional groups present, and most of all, by their size allowing an accumulation of these forces.

In addition to being the basis of life itself, protein is used as a source of amino acids and energy. The ancients degraded or depolymerized the protein in meat by aging and cooking, and they denatured egg albumin by heating or adding vinegar to the eggs. Early humans learned how to process, dye, and weave the natural proteinaceous fibers of wool and silk and the carbohydrate fibers from flax and cotton. Early South American civilizations, such as the Aztecs, used natural rubber (*Hevea brasiliensis*) for making elastic articles and for waterproofing fabrics.

There has always been an abundance of natural fibers and elastomers but few plastics. Of course, early humans employed a crude plastic art in tanning the protein in animal skins to make leather and in heat-formed tortoise shells. They also used naturally occurring tars as caulking materials and extracted shellac from the excrement of small coccid insects (*Coccus lacca*).

Until Wohler synthesized urea from inorganic compounds in 1828, there had been little progress in organic chemistry since the alchemists emphasized the transmutation of base metals to gold and believed in a vital force theory. Despite this essential breakthrough, little progress was made in understanding organic compounds until the 1850s when Kekule developed the presently accepted technique for writing structural formulas. However, polymer scientists displayed a talent for making empirical discoveries before the science was developed.

Charles Goodyear grew up in poverty. He was a Connecticut Yankee born in 1800. He began work in his father's farm implement business. Later he moved to Philadelphia where he opened a retail hardware store that soon went bankrupt. Charles then turned to being an inventor. As a child he had noticed the magic material that formed a rubber bottle he had found. He visited the Roxbury India Rubber Company to try and interest them in his efforts to improve the properties of rubber. They assured him that there was no need to do so.

He started his experiments with a malodorous gum from South America in debtor's prison. In a small cottage on the grounds of the prison, he blended the gum, the raw rubber called hevea rubber with anything he could find—ink, soup, castor oil, and so on. While rubber-based products were available, they were either sticky or became sticky in the summer's heat. He found that treatment of the raw rubber with nitric acid allowed the material to resist heat and not to adhere to itself. This success attracted backers who helped form a rubber company. After some effort he obtained a contract to supply the U.S. Post Office with 150 rubber mailbags. He made the bags and stored them in a hot room while he and his family were away. When they returned they found the bags in a corner of the room, joined together as a single mass. The nitric acid treatment was sufficient to prevent surface stickiness, but the internal rubber remained tacky and susceptible to heat.

While doing experiments in 1839 at a Massachusetts rubber factory, Charles accidentally dropped a lump of rubber mixed with sulfur on the hot stove. The rubber did not melt, but rather charred. He had discovered vulcanization, the secret that was to make rubber a commercial success. While he had discovered vulcanization, it would take several years of ongoing experimentation before the process was really commercially useful. During this time he and his family were near penniless. While he patented the process, the process was too easily copied and pirated so that he was not able to fully profit from his invention and years of hard work. Even so, he was able to develop a number of items.

Charles Goodyear, and his brother Nelson, transformed natural rubber, hevea rubber, from a heat "softenable" thermoplastic to a less heat-sensitive product through the creation of cross-links between the individual polyisoprene chain-like molecules using sulfur as the cross-linking agent. *Thermoplastics* are two-dimensional molecules that may be softened by heat. *Thermosets* are materials that are three-dimensional networks that cannot be reshaped by heating. Rather than melting, thermosets degrade. As the amount of sulfur was increased, the rubber became harder, becoming a hard rubber-like (ebonite) material.

The spring of 1851 found the construction of a remarkable building on the lawns of London's Hyde Park. The building was designed by a maker of greenhouses so it was not unexpected that it had a "greenhouse-look." This Crystal Palace was to house almost 14,000 exhibitors from all over the world. It was the chance for exhibitors to show their wares. Charles Goodyear, then 50 years old, used this opportunity to show off his over two decades worth of rubber-related products. He decorated his Vulcanite Court with rubber walls, roof, furniture, buttons, toys, carpet, combs, and so on. Above it hung a giant six-foot rubber raft and assorted balloons. The European public was introduced to the world of new man-made materials.

Within a little more than a decade Charles Goodyear was dead. Within a year of his death, the American Civil War broke out. The Union military used about $27 million worth of rubber products by 1865, helping launch the American rubber industry.

In 1862, Queen Victoria, while in mourning for her recently departed husband Albert, opened the world's fair in London. One of the exhibitors was Alexander Parks. He was displeased with the limited colors available for rubber products—generally dull and dark. In his workshop in Birmingham, England, he was working with nitrocellulose, a material made from the treatment of cotton and nitric and sulfuric acids. Nitrocellulose solutions were made by dissolving the nitrocellulose in organic liquids such as ethanol and ether. Thin films and coatings were made by simply pouring the nitrocellulose solutions onto the desired item or surface and allowing the solvent to evaporate. He wanted to make solid objects from nitrocellulose. After years of work he developed a material he called Parkesine, from which he made buttons, combs, and in fact many of the items that were made of rubber—the difference being that his materials were brightly colored, clear, or made to shine like mother of pearl. At the London world's fair he advertised "PATENT PARKESINE of various colors: hard elastic, transparent, opaque, and waterproof." Even with his work he had not developed a material that could be "worked" or was stable, and even with his hype, the material never caught on except within exhibition halls.

About this time, John Wesley Hyatt, a printer from Albany, New York, was seeking a $10,000 prize for anyone who could come up with a material that was a substitute for ivory billiard balls, and had developed a material that was stable and could be "worked" from shellac and wood pulp. He turned to nitrocellulose, discovering that shredded nitrocellulose could be mixed with camphor, heated under pressure, to produce a tough white mass that retained its shape. This material, dubbed celluloid, could be made into the usual rubber-like products, but also solid pieces like boxes, wipe-clean linen, collars, cuffs, and ping-pong balls. Celluloid could also, like the shellac–wood pulp mixture, be worked—cut, drilled, and sawed. But celluloid was flammable, and did not stand up well in hot water. The wearers of celluloid dentures truly could have their "teeth curled" when drinking a hot cup of coffee. One of its best qualities was that it could be made to "look like" other materials—it could be dyed to look like marble, swirled to mimic tortoiseshell and mother of pearl, and even look and feel like ivory. It did not make good billiard balls. One account had billiard balls hitting and exploding like a shot that caused cowboys to draw their guns.

Both cellulose and cellulose nitrate are linear or two-dimensional polymers, but the former cannot be softened because of the presence of multitudinous hydrogen bonds between the chainlike molecules. When used as an explosive the cellulose nitrate is essentially completely nitrated, but the material used by Parks and Hyatt was a dinitrate, still potentially explosive, but less so. Parks added castor oil and Hyatt added camphor to plasticize—to reduce the effect of the hydrogen bonding—the cellulose nitrate, allowing it some flexibility.

Worldwide, rubber gained in importance with the invention of the air-filled or pneumatic tires by a Scotsman, John Dunlop, in 1888. He had a successful veterinarian practice in Belfast. In his off time he worked to improve the ride of his son's tricycle. His invention happened at the right time. The automobile industry was emerging, and the air-filled tires offered a gentler ride. Thus the tire industry came into being.

All of these inventions utilized natural materials as at least one ingredient. After years of work in his chemistry labs in Yonkers, New York, Leo Baekeland, in 1907, announced in an American Chemical Society meeting the synthesis of the first truly synthetic polymeric material latter dubbed Bakelite.

Baekeland was born in Belgium in 1863; he was the son of an illiterate shoe repairman and a maid. He was bright, and he received, with highest honors, his doctorate at the age of 20. He could have spent the remaining part of his life in academics in Europe, but following the words of Benjamin Franklin, he sailed to America. In the 1890s, he developed the first photographic paper, called Velox, which could be developed in synthetic light rather than sunlight. George Eastman saw the importance of this discovery and paid Bakeland $750,000 for the rights to use this invention.

It was generally recognized by the leading organic chemists of the nineteenth century that phenol would condense with formaldehyde. Since they did not recognize the concept of functionality, Baeyer, Michael, and Kleeberg produced useless cross-linked goos, gunks, and messes and then returned to their research on reactions of monofunctional reactants. However, by the use of a large excess of phenol, Smith, Luft, and Blumer were able to obtain a hard, but meltable thermoplastic material.

With his $750,000, Baekeland set up a lab next to his home. He then sought to solve the problem of making the hard material made from phenol and formaldehyde soluble. After many failures, he thought about circumventing the problem by placing the reactants in a mold of the desired shape and allowing them to form the intractable solid material. After much effort he found the conditions under which a hard, clear solid could be made—Bakelite was discovered. Bakelite could be worked: it was resistant to acids and organic liquids, stood up well to heat and electrical charge, and could be dyed to give colorful products. It was used to make bowling balls, phonograph records, telephone housings, gears, and cookware. His materials also made excellent billiard balls. Bakelite also acted as a binder for sawdust, textiles, and paper, forming a wide range of composites including Formica laminates, many of which are still used. It was also used as an adhesive giving us plywood.

While there is no evidence that Baekeland recognized what polymers were, he appeared to have a grasp on functionality and how to "use" it to produce thermoplastic materials that could later be converted to thermosets. Through control of the ratio of phenol to formaldehyde he was able to form a material that was a thermoplastic. He coined the term *A-stage resole resin* to describe this thermoplastic. This A-stage resole resin was converted to a thermoset cross-link, *C-stage Bakelite*, by additional heating. Baekeland also prepared thermoplastic resins called *novolacs* by the condensation of phenol with a lesser amount of formaldehyde under acidic conditions. The thermoplastic novolacs were converted to thermosets by addition of more formaldehyde. While other polymers had been synthesized in the laboratory, Bakelite was the first truly synthetic plastic. The "recipes" used today differ little from the ones developed by Baekeland and show his ingenuity and knowledge of the chemistry of the condensation of the trifunctional phenol and difunctional formaldehyde.

While poly(vinyl chloride) (PVC) was initially formed by Baumann in 1872, it awaited interest until 1926 when B. F. Goodrich discovered how to make sheets and adhesives from PVC—and the "vinyl-age" began. While polystyrene was probably first formed by Simon in 1839, it was almost 100 years latter, 1930, that the German giant company I. G. Farben placed polystyrene on the market. Polystyrene molded parts became commonplace. Rohm and Haas bought out Plexiglass from a British firm in 1935 and began the production of clear plastic parts and goods, including replacements for glass as camera lenses, aircraft windows, clock faces, and car taillights.

To this time, polymer science was largely empirical, instinctive, and intuitive. Before World War I, celluloid, shellac, Galalith (casein), Bakelite, and cellulose acetate plastics; hevea rubber, cotton, wool, silk rayon fibers; Glyptal polyester coatings; bitumen or asphalt and coumarone-indene and petroleum resins were all commercially available. However, as evidenced by the chronological data shown in Table 1.1, there was little additional development in polymers prior to World War II because of a general lack of fundamental knowledge of polymers. But the theoretical basis was being built. Only a few of many giants will be mentioned.

TABLE 1.1
Chronological Developments of Commercial Polymers (upto 1991)

Before 1800	Cotton, flax, wool, and silk fibers; bitumens caulking materials; glass and hydraulic cements; leather and cellulose sheet (paper); natural rubber (*Hevea brasiliensis*), gutta percha, balata, and shellac
1839	Vulcanization of rubber (Charles Goodyear)
1845	Cellulose esters (Schonbein)
1846	Nitration of cellulose (Schonbein)
1851	Ebonite (hard rubber; Nelson Goodyear)
1860	Molding of shellac and gutta percha
1868	Celluloid (plasticized cellulose nitrate; Hyatt)
1888	Pneumatic tires (Dunlop)
1889	Cellulose nitrate photographic films (Reinchenbach)
1890	Cuprammonia rayon fibers (Despeisses)
1892	Viscose rayon fibers (Cross, Bevan, and Beadle)
1903	First tubeless tire (Litchfield of Goodyear Tire Co.)
1897	Poly(phenylene sulfide)
1901	Glyptal polyesters
1907	Phenol-formaldehyde resins (Bakelite; Baekeland)
1908	Cellulose acetate photographic fibers
1912	Regenerated cellulose sheet (cellophane)
1913	Poly(vinyl acetate)
1914	Simultaneous interpenetrating network (SIN)
1920	Urea-formaldehyde resins
1923	Cellulose nitrate automobile lacquers
1924	Cellulose acetate fibers
1926	Alkyd polyester (Kienle)
1927	Poly(vinyl chloride) wall covering
1927	Cellulose acetate sheet and rods
1927	Graft copolymers
1928	Nylon (Carothers, Dupont)

TABLE 1.1 (continued)
Chronological Developments of Commercial Polymers (upto 1991)

1929	Polysulfide synthetic elastomer (Thiokol; Patrick)
1929	Urea-formaldehyde resins
1930	Polyethylene (Friedrich/Marvel)
1931	Poly(methyl methacrylate) (PMMA) plastics
1931	Polychloroprene elastomer (Neoprene; Carothers)
1934	Epoxy resins (Schlack)
1935	Ethylcellulose
1936	Poly(vinyl acetate)
1936	Poly(vinyl butyral) (safety glass)
1937	Polystyrene
1937	Styrene–butadiene (Buna-S) and styrene–acrylonitrile (Buna-N) copolymer elastomers
1939	Melamine-formaldehyde resins
1939	Nylon 6 (Schlack)
1939	Nitrile rubber (NR)
1940	Isobutylene-isoprene elastomer (butyl rubber; Sparks and Thomas)
1941	Low-density polyethylene (LDPE)
1941	Poly(ethylene terephthalate) (PET)
1942	Butyl rubber
1942	Unsaturated polyesters (Ellis and Rust)
1943	Fluorocarbon resins (Teflon; Plunket)
1943	Silicones
1945	SBR
1946	Polysulfide rubber (Thiokol)
1948	Copolymers of acrylonitrile, butadiene, and styrene (ABS)
1949	Cyanoacrylate (Goodrich)
1950	Polyester fibers (Winfield and Dickson)
1950	Polyacrylonitrile fibers
1952	Block copolymers
1953	High-impact polystyrene (HIPS)
1953	Polycarbonates (Whinfield & Dickson)
1956	Poly(phenylene ether); Poly(phenylene oxide) (GE)
1957	High-density polyethylene (HDPE)
1957	Polypropylene
1957	Polycarbonate
1958	Poly(dihydroxymethylcyclohexyl terephthate) (Kodel, Eastman Kodak)
1960	Ethylene–propylene copolymer elastomers
1961	Aromatic nylons (Aramids, Nomex, Dupont)
1962	Polyimide resins
1964	Poly(phenylene oxide)
1964	Ionomers
1965	Polysulfone
1965	Styrene–butadiene block copolymers
1966	Liquid crystals
1970	Poly(butylene terephthate)
1974	Polyacetylene
1982	Polyetherimide (GE)
1991	Carbon nanotubes (Iijima; NEC Lab)

Over a century ago, Graham coined the term *colloid* for aggregates with dimensions in the range of 10^{-9}–10^{-7} m. Unfortunately, the size of many macromolecules is in this range, but it is important to remember that unlike colloids, whose connective forces are ionic and/or secondary forces, polymers are individual molecules whose size cannot be reduced without breaking the covalent bonds that hold the atoms together. In 1860, an oligomer, a small polymer, was prepared from ethylene glycol and its structure was correctly given as $HO-(-OCH_2CH_2-)_n-OH$. But when poly(methacrylic acid) was made by Fittig and Engelhorn in 1880 it was incorrectly assigned a cyclic structure. Polymers were thought of as being colloids or cyclic compounds like cyclohexane. By use of the Raoult and van't Hoff concepts, several scientists obtained high molecular weight values for these materials and for a number of other polymeric materials. But since the idea of large molecules was not yet accepted they concluded that these techniques were not applicable to these molecules rather than accepting the presence of giant molecules.

The initial "tire-track in the sand" with respect to tires was the discovery of vulcanization of rubber by Charles Goodyear in 1844. The first rubber tires appear in the mid-1880s. These tires were solid rubber, with the rubber itself absorbing the bumps and potholes. John Dunlop invented the first practical pneumatic or inflatable tire with his patent granted in 1888. Andre Michelin was the first person to use the pneumatic tire for automobiles. The Michelin brothers, Andre and Edouard, equipped a racing car with pneumatic tires and drove it in the 1895 Paris–Bordeaux road race. They did not win, but it was sufficient advertising to begin interest in pneumatic tires for automobiles. Further, because they did not cause as much damage to the roads, pneumatic tires were favored by legislation. It is interesting to see that the names of these three pioneers still figure prominently in the tire industry. Even so, another inventor had actually been given the first patent for a vulcanized rubber pneumatic tire in 1845 but it did not take off. Thompson was a prolific inventor, also having patented a fountain pen in 1849 and a steam traction engine in 1867.

A number of the giant tire companies started at the turn of the century. In America, many of these companies centered around Akron, the capital of the rubber tire. In 1898, the Goodyear Tire and Rubber Company started. The Firestone Tire and Rubber Company was started by Harvey Firestone in 1900. Other tire companies followed shortly.

Hermann Staudinger studied the polymerization of isoprene as early as 1910. Intrigued by the difference between this synthetic material and natural rubber, he began to focus more of his studies on such materials. His turn toward these questionable materials, of interest to industry but surely not academically important, was viewed unkindly by his fellow academics. He was told by one of his fellow scientists, "Dear Colleague, Leave the concept of large molecules well alone... There can be no such thing as a macromolecule."

Staudinger systematically synthesized a variety of polymers. In the paper "Uber Polymerization" in 1920, he summarized his findings and correctly proposed linear structures for such important polymers as polyoxymethylene and polystyrene. X-ray studies of many natural and synthetic materials were used as structural proof that polymers existed. Foremost in these efforts were Herman Mark and Linus Pauling. Both of these giants contributed to other important areas of science. Pauling contributed to the fundamental understanding of bonding and the importance of vitamins. Mark helped found the academic and communication (journals, short courses, workshops) basis that would allow polymers to grow from its very diverse roots.

Probably the first effort aimed at basic or fundamental research in the chemical sciences was by DuPont. Their initial venture in artificial fibers was in 1920 when they purchased a 60% interest in Comptoir des Testiles Artificels, a French rayon company. The combined company was named the DuPont Fiber Company. DuPont spent considerable effort and money on expanding the properties of rayon. In 1926, Charles M. A. Stine, director of the chemical department, circulated a memo to DuPont's executive committee suggesting that the company move from investing in already existing materials to investigating new materials. This was a radical idea that a company supposedly focused on profit spends some of its effort on basic research. The executive approved much of Stine's proposal and gave him $25,000 a month for the venture, which allowed him to hire 25 chemists for the task. The initial hiring was difficult because academic chemists did not trust DuPont to allow them to do basic research. A year later he was able to make his central hiring, Wallace Hume Carothers.

Wallace Hume Carothers is the father of synthetic polymer science. History is often measured by the change in the flow of grains of sand in the hour glass of existence. Carothers is a granite boulder in this hourglass. Carothers was born, raised, and educated in the Midwest of the United States. In 1920, he left Tarkio College with his BS degree and entered the University of Illinois where he received his MA in 1921. He then taught at the University of South Dakota where he published his first paper. He returned to receive his PhD under Roger Adams in 1924. In 1926, he became an instructor in organic chemistry at Harvard.

In 1927, the DuPont Company reached a decision to begin a program of fundamental research "without any regard or reference to commercial objectives." This was a radical departure since the bottom line was previously products marketed and not papers published.

Charles Stine, director of DuPont's chemical department, was interested in pursuing fundamental research in the areas of colloid chemistry, catalysis, organic synthesis, and polymer formation and convinced the Board to hire the best chemists in each field to lead this research. Stine visited many in the academic community, including the then president of Harvard, one of the author's distant uncles, J. B. Conant, an outstanding chemist himself, who told him about Carothers. Carothers was persuaded to join the DuPont group and was attracted with a generous research budget and an approximate doubling of his academic salary to $6,000. This was the birth of the Experimental Station at Wilmington, Delaware.

Up to this point, it was considered that universities were where discoveries were made and industry was where they were put to some practical use. This separation between basic and applied work was quite prominent at this juncture and continues in many areas even today in some fields of work, though the difference has decreased. But in polymers, most of the basic research was done in industry having as its inception the decision by DuPont to bridge this "unnatural" gap between fundamental knowledge and application. In truth, they can be considered as the two hands of an individual, and to do manual work both hands are important.

Staudinger believed that large molecules were based on the jointing, through covalent bonding, of large numbers of atoms. Essentially he and fellow scientists like Karl Freudenberg, Herman Mark, Michael Polanyi, and Kurt Myer looked at already existing natural polymers. Carothers, however, looked at the construction of these giant molecules from small molecules forming synthetic polymers. His intention was to prepare molecules of known structure through the use of known organic chemistry and to "investigate how the properties of these substances depended on constitution." Early work included the study of polyester formation through reaction of diacids, with diols forming polyesters. But he could not achieve molecular weights greater than about 4,000 below the size where many of the interesting so-called polymeric properties appear.

DuPont was looking for a synthetic rubber. Carothers assigned Arnold Collins to this task. Collin's initial task was to produce pure divinylacetylene. While performing the distillation of an acetylene reaction, in 1930, he obtained a small amount of an unknown liquid that he set aside in a test tube. After several days the liquid turned to a solid. The solid bounced and eventually was shown to be a synthetic rubber polychloroprene whose properties were similar to those of vulcanized rubber but it was superior in its resistance to ozone, ordinary oxidation, and to most organic liquids. It was sold under its generic name "neoprene" and the trade name "Duprene."

Also in 1930, Carothers and Julian Hill designed a process to remove water that was formed during the esterification reaction. Essentially they simply froze the water as it was removed using another recent invention called a molecular still (basically a heating plate coupled to vacuum) allowing the formation of longer chains. In April, Hill synthesized a polyester using this approach and touched a glass stirring rod to the hot mass and then pulled the rod away, effectively forming strong fibers; the pulling helped in reorienting the mobile polyester chains. The polyester had a molecular weight of about 12,000. Additional strength was achieved by again pulling the cooled fibers. Further reorientation occurred. This process of "drawing" or pulling to produce stronger fibers is now known as "cold drawing" and is widely used in the formation of fibers today. The process of "cold drawing" was discovered by Carothers' group. Although interesting, the fibers were not considered to be of commercial use. Carothers and his group then moved to look at the reaction of diacids with diamines instead of diols. Again, fibers were formed but these initial materials were deemed not to be particularly interesting.

In 1934, Paul Flory was hired to work with Carothers to help gain a mathematical understanding of the polymerization process and relationships. Thus, there was an early association between theory and practice or structure–property relationships.

In 1934, Donald Coffman, a member of the Carothers team, pulled a fiber from an aminoethylester (polyamide) polymer. The fiber retained the elastic properties of the polyesters previously investigated but had a higher melting point, which allowed it to be laundered and ironed. The field of candidates for further investigation was narrowed to two—polyamide 5,10 made from pentamethylene diamine and sebacic acid, and polyamide 6,6 synthesized from hexamethylenediamine and adipic acid. Polyamide 6,6 won because the monomers could be made from benzene, a readily available feedstock from coal tar.

The polyamide fiber project was begun in earnest using the reaction of adipic acid with hexamethylenediamine. They called the polyamide fiber 66 because each carbon-containing unit had six carbons. It formed a strong, elastic, largely insoluble fiber with a relatively high melt temperature. DuPont chose this material for production. These polyamides were given the name "nylons." Thus was born nylon-6,6. It was the first synthetic material whose properties equaled or exceeded the natural analog, namely silk. (In reality, this may not be the truth, but at the time it was believed to be true.)

The researchers had several names for polyamide 6,6, including rayon 66, fiber 66, and Duparon derived from "Dupont pulls a rabbit out [of] the hat nitrogen/nature/nature/nozzle/naphtha." The original "official" name was "Nuron," which implied newness and also spelled "on run" backwards. This name was too close to other trademarked names and was renamed "Nirton" and eventually to what we know today as "Nylon."

As women's hem lines rose in the 1930s, silk stockings were in great demand but were very expensive. Nylon changed this. Nylon could be woven into sheer hosiery. The initial presentation of nylon hose to the public was by Stine at a forum of women's club members in New York City on October 24, 1938. Nearly 800,000 pairs were sold on May 15, 1940 alone—the first day they were on the market. By 1941 nylon hosiery held 30% of the market but by December 1941 nylon was diverted to make parachutes, and so on.

From these studies Carothers established several concepts. First, polymers could be formed by employing already known organic reactions but with reactants that had more than one reactive group per molecule. Second, the forces that bring together the individual polymer units are the same as those that hold together the starting materials, namely, primary covalent bonds. Much of the polymer chemistry names and ideas that permeate polymer science today were standardized through his efforts.

Representing the true multidisciplinary nature of polymers, early important contributions were also made by physicists, engineers, and those from biology, medicine, and mathematics, including W. H. Bragg, Peter Debye, Albert Einstein, and R. Simha.

World War II helped shape the future of polymers. Wartime demands and shortages encouraged scientists to seek substitutes and materials that even excelled currently available materials. Polycarbonate (Kevlar), which could stop a "speeding bullet," was developed, as was polytetrafluoroethylene (Teflon), which was super slick. New materials were developed spurred on by the needs of the military, electronics industry, food industry, and so on. The creation of new materials continues at an even accelerated pace brought on by the need for materials with specific properties and the growing ability to tailor make giant molecule—macromolecule—polymers.

TABLE 1.2
Commercialization of Selected Polymers

Polymer	Year	Company
Bakelite	1909	General Bakelite Corp.
Rayon	1910	American Viscose Company
Poly(vinyl chloride)	1927	Goodrich
Styrene–Butadiene copolymer	1929	I.G. Farben
Polystyrene	1929/1930	I.G. Farben & Dow
Neoprene	1931	Dupont
Poly(methyl methacrylate)	1936	Rhom and Haas
Nylon-66	1939/1940	Dupont
Polyethylene (LDPE)	1939	ICI
Poly(dimethyl siolxane)	1943	Dow Corning
Acrylic fiber	1950	Dupont
Poly(ethylene terephthalate), PET	1953/1954	Dupont/ICI
Polyurethane block copolymers (Spandex)	1959	Dupont
Poly(phenylene terephthalamide)	1960	Dupont

TABLE 1.3
Nobel Prize Winners for Their Work with Synthetic Polymers

Scientist(s)	Year	Area
Herman Staudinger	1953	Polymer hypothesis
Karl Ziegler and Giulio Natta	1963	Stereoregulation of polymer structure
Paul Flory	1974	Organization of polymer chains
Bruce Merrifield	1984	Synthesis on a solid matrix
Pierre de Gennes	1991	Polymer structure and control at interfaces
A.J. Heeger, Alan Mac Diarmid, and H. Shirakawa	2000	Conductive polymers

Unlike other areas of chemistry, most of the basic research has been done in industry so that there is often a close tie between discoveries and their commercialization. Table 1.2 lists some of the dates of commercialization for some important synthetic polymer discoveries.

A number of Nobel Prizes have been given for polymer work. Table 1.3 contains winners for advances in synthetic polymers. In truth, there are many more since most of the prizes given out in medicine and biology deal with giant molecules.

Throughout this text advances are placed in some historical setting. This adds some texture to the topics as well as case histories that are widely used in subject areas such as business and medicine.

1.2 WHY POLYMERS?

Polymers are all about us. They serve as the very basis of both plant and animal life as proteins, nucleic acids, and polysaccharides. In construction they serve as the concrete, insulation, and wooden and composite beams. At home they are found as the materials for our rugs, curtains, coatings, waste paper baskets, water pipes, window glass, ice cube trays, and pillows. In transportation they are present in ever increasing amounts in our air craft, automobiles, ships, and trucks. In communication they form critical components in our telephones, TVs, computers, CDs, newspaper, optical fibers, and cell phones. Plastics act as favorite materials for our toys such as toy soldiers, plastic models, toy cars, dolls, skip ropes, hula hoops, and corvettes. Our food is polymer intense as meat, vegetables, breads, and cookies. In history, polymers have been the vehicle for the Magna Carter, Torah, Bible, Koran, and our Declaration of Independence. Outside our homes they are present in our flowers, trees, soil, spider webs, and beaches. In fact, it is improbable that a polymer is not involved in your present activity—reading a paper book, holding a plastic-intense writing device, sitting on a cloth-covered chair or bed, and if your eyes need corrective vision, glasses of one variety or another.

Polymers gain their importance because of their size. Many polymers are made from inexpensive and readily available materials, allowing vast quantities of products to be made for a high increase in value, but they are typically inexpensive compared to nonpolymer alternatives. They also often have desirable physical and chemical properties. Some polymers are stronger on a weight basis than steel. Most are resistant to rapid degradation and rusting. You will learn more about these essential materials for life and living in this text.

Polymers are often divided according to whether they can be melted and reshaped through application of heat and pressure. These materials are called *thermoplastics.* The second general classification belongs to compounds that decompose before they can be melted or reshaped. These polymers are called *thermosets.* While both thermoset and thermoplastic polymers can be recycled, because thermoplastics can be reshaped simply through application of heat and pressure, recycling of thermoplastics is easier and more widespread.

In general groups, synthetic polymers are often described by their "use" and "appearance" as fibers, elastomers, plastics, adhesives, and coatings. A common toothbrush illustrates the three major physical forms of synthetic polymers—the rubbery (elastomeric) grips, plastic shaft, and fibrous brussels. The rubbery grips have a relatively soft touch; the plastic shaft is somewhat flexible and hard, and the brussels are highly flexible. Another illustration of the breath of polymers about us is given in Table 1.4 where polymers are divided according to the source.

To get an idea of the pervasiveness of polymers in our everyday life, we can look at containers. Most containers are polymeric—glass, paper, and synthetic polymer. It is relatively easy to identify each of these general categories. Even, within the synthetic polymer grouping, it has become relatively easy to identify the particular polymer used in some applications, such as with disposable containers. Most of these synthetic polymers are identified by an "identification code" that is imprinted somewhere on the plastic container, generally on their bottom. The numbers and letters are described in Figure 1.1. The recycling code was developed by the Society of Plastics Industry for use with containers. Today, the "chasing-arrows" triangle is being used more widely for recycling by the public. A colorless somewhat hazy water container has a "2" within the "chasing" arrows and underneath it "HDPE," indicating that the bottle is made of HDPE. The clear, less flexible soda bottle has a "1" and "PETE," both signifying that the container is made out of PET, a polyester. A brownish clear medicine container has a "5" and the letters "PP" on its bottom conveying the information that the bottle is made of polypropylene. Thus, ready identification of some common items is easy.

But, because of the use of many more complex combinations of polymers for many other items, such identification and identification schemes are not as straightforward. For some items, such as clothing and rugs, labels are present that tell us the major materials in the product. Thus, a T-shirt might have "cotton" on its label indicating that the T-shirt is largely made of cotton. A dress shirt's label may say 55% cotton and 45% polyester meaning it is made from two polymers. Some items are identified by trade names. Thus, a dress advertised as being made from Fortrel (where "Fortrel" is a trade name) means it is made largely of a polyester material, probably the same polyester, PET or PETE, which made our soda bottle. Some everyday items are a complex of many materials but only some or none are noted. This is true for many running shoes and tires. Tires will often be described as being polyester (again, probably the same PETE) or nylon (or aramid). This describes only the composition of the tire cord but does not tell us what other materials are included in the tire's composition. Yet, those that deal with tires generally know what is used in the manufacture of the tire in addition to the "stated ingredients." You will be introduced, gently, to the identification of the main

TABLE 1.4
Common Polymers

Material/Name	Typical Polymer	Chapter
Styrofoam	Polystyrene	6
PVC pipe	Poly(vinyl chloride)	6
Nylon stockings	Polyamide, Nylon 6,6	4
Concrete	Cement	12
Meat	Protein	10
Plexiglass	Poly(methyl methacrylate)	6
Automotive bumpers and side panels	Polyethylene and polyethylene/polypropylene blends	5,7
Potatoes	Starch	9
Compact discs (case)	Polycarbonate (polystyrene)	4
Hula hoop	Polypropylene, polyethylene	5
Diamond	Carbon	12
Silicon sealants	Polydimethylsiloxane	11
Bakelite	Phenol-formaldehyde cross-linked	4
Super glue	Poly(ethyl cyanoacrylate)	18
Cotton T-shirt	Cellulose	9
Fiberglass	Composite	8
Saran wrap	PVC copolymer	7
Velcro	Polyamide	4
Rubber band	Natural rubber	9
Soda bottle	Poly(ethylene terephthalate), PET	4
Teflon	Polytetrafluoroethylene	6
Orlon sweater	Polyacrylonitrile	6
Sand	Silicon dioxide	12
Pillow stuffing	Polyurethane	4
Wood, paper	Cellulose	9
Human genome	Nucleic acids	10

polymers that are present in many everyday items, either through looking at labels, researching on the web, simply knowing what certain items are generally composed of, through the feel and gross physical properties (such as flexibility and stiffness) of the material, and so on.

Further, the properties of essentially the same polymer can be varied through small structural changes giving materials with differing properties and uses. There is a match between desired properties and the particular material used. For instance, for plastic bags, strength and flexibility are needed. The bag material should be somewhat strong, inexpensive (since most bags are "throw-away" items), and readily available in large quantities. Increased strength is easily gained from increasing thickness. But, with increased thickness comes decreased flexibility, increased cost since more material is needed to make the thicker bags, and increased transportation (because of the additional weight) and storage costs. Thus, there is a balance between many competing factors. Plastic bags are typically made from three polymers, HDPE, low-density polyethylene (LDPE), and linear low-density polyethylene (LLDPE; actually a copolymer with largely ethylene units). These different polyethylene polymers are similar, differing only in the amount of branching that results in differing tendencies to form ordered (crystalline) and less ordered (amorphous) chain arrangements. You will learn more about them in Chapter 5. Grocery bags are generally made from HDPE which is a largely linear polymer that has a high degree of crystallinity. Here, in comparison to LDPE film with the same strength, the bags are thinner, allowing a decrease in cost of materials, transportation cost, and storage space. The thinness allows good flexibility. LDPE is used for dry cleaning garment bags where simply covering the garments is the main objective rather than strength. The LDPE is less

PETE

Poly(ethylene terephthalate)—**PET** or **PETE**
PET is the plastic used to package the majority of soft drinks. It is also used for some liquor bottles, peanut butter jars, and edible-oil bottles. About one-quarter of plastic bottles are PET. PET bottles can be clear; they are tough and hold carbon dioxide well.

HDPE

High-density polyethylene—**HDPE**
HDPE is a largely linear form of PE. It accounts for more than 50% of the plastic bottle market and is used to contain milk, juices, margarine, and some grocery snacks. It is easily formed through application of heat and pressure and is relatively rigid and low cost.

V

Poly(vinyl chloride)—**PVC** or **V**
PVC is used "pure" or as a blend to make a wide variety of products, including PVC pipes, food packaging film, and containers for window cleaners, edible oils, and solid detergents. It accounts for only 5% of the container market.

LDPE

Low-density polyethylene—**LDPE**
LDPE has branching and is less crystalline, more flexible, and not as strong as HDPE. The greater amount of amorphous character makes it more porous than HDPE, but it offers a good inert barrier to moisture. It is a major material for films from which trash bags and bread bags are made.

PP

Polypropylene—**PP**
PP has good chemical and fatigue resistance. Films and fibers are made from it. Few containers are made of PP. It is used to make some screw-on caps, lids, yogurt tubs, margarine cups, straws, and syrup bottles.

PS

Polystyrene—**PS**
PS is used to make a wide variety of containers, including those known as Styrofoam plates, dishes, cups, etc. Cups, yogurt containers, egg cartons, meat trays, and plates are made from PS.

OTHER

Other Plastics
A wide variety of other plastics are coming to the marketplace including copolymers, blends, and multilayered combinations.

FIGURE 1.1 The Society of Plastics Industry recycling codes utilizing the numbers 1–7 and bold, capital letters to designate the material utilized to construct the container.

crystalline and weaker but more flexible because of the presence of more branching in comparison to HDPE. The thicker glossy shopping bags from malls are LLDPE, which, like HDPE, is largely linear. This increased thickness results in the bags being less flexible. These bags can be used many times.

Thus, most of the common items about us are polymeric. Table 1.4 gives a brief listing of some of these materials along with the locations where they will be dealt within the book.

With the electronic age we can access the web to gather lots of general information about almost any topic, including polymers. This book allows you to have a greater appreciation and understanding of such information and the products about us, including our own bodies.

1.3 TODAY'S MARKETPLACE

As noted above, polymers are all around us. More than 100 billion pounds (50 million tons) of synthetic polymers is produced annually in the United States (Tables 1.5 through 1.8), and the growth of the industry is continuing at a fast rate. There is every reason to believe that this polymer age will continue as long as petroleum and other feedstocks are available and as long as consumers continue to enjoy the comfort, protection, and health benefits provided by elastomers, fibers, plastics, adhesives, and coatings. The 100

TABLE 1.5
U.S. Production of Plastics (Millions of Pounds)

Year →	1995	2000	2008
Thermosetting resins			
Epoxies	570	620	560
Ureas and Melamines	1,900	2,800	3,100
Phenolics	3,100	3,900	4,300
Thermoplastics			
Polyethylenes			
Low density	7,000	6,300	6,800
High density	10,000	12,600	14,700
Linear low density	4,700	7,200	10,900
Polypropylene	9,800	14,200	15,300
Polystyrene			
Polystyrene	5,300	6,200	4,700
Acrylonitrile-butadiene–styrene and Styrene–acrylonitrile	2,700	2,900	2,900
Polyamides, Nylons	900	1,100	1,100
Poly(vinyl chloride) and copolymers	11,000	13,000	11,600
Polyesters			2,400

Source: American Plastics Council.

TABLE 1.6
U.S. Production of Man-made Fibers (Millions of Pounds)

Year →	1995	2000	2008
Noncellulosic			
Acrylics	400	310	120
Nylons	2,400	2,400	1,500
Olefins	2,100	2,900	2,200
Polyesters	3,500	3,500	2,100
Cellulosic			
Acetate and Rayon	450	310	54

Source: Fiber Economics Bureau.

TABLE 1.7
U.S. Production of Paints and Coatings (Millions of Gallons)

Year →	1995	2000	2005
Architectural	620	650	860
Product	380	450	410
Special	195	180	175

Source: Department of Commerce.

TABLE 1.8
U.S. Production of Synthetic Rubber
(Millions of Pounds; 2008)

Styrene–butadiene	1,750
Polybutadiene	1,210
Nitrile	180
Ethylene–propylene	540
Other	1,100

Source: International Institute of Synthetic Rubber Producers.

TABLE 1.9
Polymer Classes—Natural and Synthetic

Polymeric Materials					
Inorganic		Organic/Inorganic	Organic		
Natural	Synthetic			Natural	Synthetic
Clays	Fibrous glass siloxanes			Proteins	Polyethylene
Cement	Poly(Sulfur nitride)	Polyphosphazenes		Nucleic acids	Polystyrene
Pottery	Poly(Boron nitride)	Polyphosphate esters		Lignins	Nylons
Bricks	Silicon carbide	Polysilanes		Polysaccharides	Polyesters
Sands		Sol-Gel networks		Melanins	Polyurethanes
Glasses				Polyisoprenes	Poly(methyl methacrylate)
Rock-like					Polytetrafluoroethylene
Agate					Polyurethane
Talc					Poly(vinyl chloride)
Zirconia					Polycarbonate
Mica					Polypropylene
Asbestos					Poly(vinyl alcohol)
Quartz					
Ceramics					
Graphite/Diamond					
Silicas					

billion pounds of synthetic polymers consumed each year in the United States translates to more than 300 pounds for every man, woman, and child in the United States. This does not include paper and wood-related products, natural polymers such as cotton and wool, or inorganic polymers (Table 1.9).

Polymers are all about us. The soils we grow our foods from are largely polymeric as are the foods we eat. The plants around us are largely polymeric. We are walking exhibits as to the widespread nature of polymers—from our hair and finger nails, our skin, bones, tendons, and muscles; our clothing—socks, shoes, glasses, undergarments; the morning newspaper; major amounts of our automobiles, airplanes, trucks, boats, space craft; our chairs, waste paper baskets, pencils, tables, pictures, coaches, curtains, glass windows; the roads we drive on, the houses we live in, and the buildings we work in; the tapes and CDs we listen to music on; packaging—all are either totally

TABLE 1.10
U. S. Chemical Trade—Imports and Exports (Millions of Dollars)

Year →	1985		1994		2008	
	Imports	**Exports**	**Imports**	**Exports**	**Imports**	**Exports**
Organic chemicals	6,000	4,600	12,800	10,800	34,800	47,700
Inorganic chemicals	2,000	2,000	4,100	4,100	13,100	16,800
Oils and perfumes	—	—	3,500	2,000	12,100	9,600
Dyes, colorants	—	—	2,300	1,900	6,400	3,100
Medicinals/pharmaceuticals	2,700	1,080	6,100	4,700	38,200	59,600
Fertilizers	2,160	1,000	2,700	1,300	7,500	8,400
Plastics and resins	3,800	1,600	8,500	3,300	40,100*	18,800*
Others	5,300	4,220	7,600	2,700	26,000	12,700
Total chemicals[†]	22,000	14,500	51,600	33,400	179,100	176,800
Total	213,000	345,000	502,800	669,100	1,855,000	1,954,000

*Includes both plastics in primary and nonprimary forms but does not include rubber and rubber products.
[†]Includes nonlisted chemicals.
Source: Department of Commerce.

polymeric or contain a large amount of polymeric materials. Table 1.9 lists some general groupings of important polymers. Welcome to the wonderful world of polymer science.

The number of professional chemists directly employed with polymers as part of their interest and assignment is estimated to be 40%–60% of all chemists. As the diversity of chemistry increases, the dispersion of those dealing with polymers increases. Polymer chemistry is a major tool applied in biomedical research, synthesis, manufacturing, chemical engineering, pharmaceutical efforts, the environment, communications, and so forth. "As it was in the beginning" polymers continue to draw strength from those with diverse training, allowing polymers to directly contribute to solutions for most of our technological problems including fuel and transportation, building and construction, communication, medicine and dentistry, and so on. Analytical chemistry is related to the analysis of materials; inorganic is related to the catalysts employed in the synthesis of natural and synthetic polymers; organic is related to the synthesis of diverse materials; physical chemistry describes the kinetics, thermodynamics, and properties of macromolecules; and biological chemistry deals with biopolymers.

Polymeric materials, along with several other chemical industrial products, contribute positively to the balance of trade (Table 1.10). In fact, plastics and resins show the greatest value increase of exports minus imports with more than $20 billion net favoring exports. The polymer-intense material numbers are higher than that noted in Table 1.10 since fiber and rubber materials are absent as a separate entry. Even so, the figures demonstrate the positive nature polymers play in our balance of trade situation.

Table 1.11 contains a listing of the major chemical producers in the United States and the world. These producers are involved directly and/or indirectly with some form of synthetic polymers. Essentially, all of the industrially advanced countries of the world have major chemical producers. Table 1.11 contains a partial listing of these companies.

Thus, polymers play a critical role in our everyday lives, actually forming the basis for both plant and animal life, and represent an area where chemists continue to make important contributions.

1.4 ENVIRONMENTAL ASSESSMENT

Daily, we are becoming more aware of the importance of environmental planning and good environmental actions (practices) and their effect on us. This emphasis is being driven by a number of pressures,

TABLE 1.11
Major Chemical Producers Based on (Net) Sales (>2,500 Million Dollars; in Millions of U.S. Dollars; 2008)

United States

Dow	57,500	ExxonMobile	38,400
Dupont	30,500	PPG	15,900
Chevron Phillips	12,600	Monsanto	11,400
Praxair	10,800	Air Products	10,400
Huntsman	10,100	Rhom and Haas	9,500
Celanese	6,800	Eastman Chemical	6,700
Hexion Specialty	6,100	Mosaic	5,700[*]
Lubrizol	5,000	Dow Corning	4,900[*]
Honeywell	4,900[*]	Occidental Petroleum	4,700[*]
Nalco	4,200	Chemtura	3,600
Cytec Ind.	3,600	Solutia	3,500[*]
Westlake	3,200[*]	W.R. Grace	3,300
Cabot	3,200	FMC	3,100
Georgia Gulf	2,900	Terra	2,900
Sunoco	2,800[*]	CF	2,800[*]
Rockwood Specialities	2,700[*]	Polyone	2,700
Momentive	2,500[*]		

Europe

BASF (Germany)	79,500	Ineos Group (UK)	47,000
Bayer (Germany)	44,400	Lyondell-Basell (Netherlands)	38,400
Linde (Germany)	16,900	Air Liquide (France)	16,300
Akzo Nobel (Netherlands)	14,000	Solvay (Belgium)	13,100
DSM (Netherlands)	12,200	Merck (Germany)	9,700
Lanxess (Germany)	9,100	Arkema (France)	7,800
Syngenta (Switzerland)	7,700	Clariant (Switzerland)	7,100
Rhodia (France)	7,000	Ciba (Switzerland)	5,400
Wacker (Germany)	5,200	Kemira (Finland)	3,900
Givaudan (Switzerland)	3,400		

Japan

Mitshuishi Chem. Inds.	20,200	Sumitomo Chemical	17,300
Asahi Kasei	15,000	Mitsui Chem.	14,400
Toray Inds.	14,200	Shin-Etsu	11,600
Showa Denko	9,700	Teijin	9,100
DIC	9,000	Taiyo Nippon Sanso	4,800
Kaneka	4,300	JSR Corp.	3,400

Canada

Agrium	10,300	Potash Corp.	9,400
Nova Corp.	7,400		

Other

SABIC (Saudi Arabia)	34,400	Sinopec (China)	33,800
Formosa Plastics Group (Formosa)	27,500	PetroChina (China)	16,000

TABLE 1.11 (continued)
Major Chemical Producers Based on (Net) Sales (>2,500 Million Dollars; in Millions of U.S. Dollars; 2008)

Yara (Norway)	15,800	LG Chem (S. Korea)	13,600
Reliance Industries (India)	12,200	Braskem (Brazil)	9,800
Sasol (South Africa)	8,300	National Petrochemical Co. (Iran)	7,800
Israel Chemicals (Israel)	6,900		

*2007 values.
Source: C & EN.

including federal and state laws and initiatives, industrial consciousness, reality, international efforts, individual actions, and so on. Chemistry, chemists, and the chemical industry have been directing much effort in this direction for over several decades with this effort magnified over the past several years. This textbook highlights some of these efforts in polymers. Here a number of terms related to these efforts will be introduced. While these terms are described individually, the environmental activity is a matrix of actions and activities, each one dependent on others to be successful. Many of these studies and assessments are governed somewhat by procedures described by the International Organization for Standardization (ISO). (More about ISO can be found on the web and in Appendix H.)

Environmental impact assessment—An environmental impact assessment (EIA) is simply an assessment of the possible impact that a project or material may have on the natural environment. This possible impact may be positive or negative, and often is a combination of positive and negative impacts. The intension of such EIAs is to identify where changes can and should be made and to make us aware of these instances. The International Association for Impact Assessment (IAIA) describes such impact assessments as a process for "identifying, predicting, evaluating and mitigating the biophysical, social, and other relevant effects of development proposals prior to major decisions being taken and commitments made." (Petts, J. (Ed.) (1999): *Handbook of Environmental Impact Assessment*, Vols. 1 and 2, Blackwell, Oxford.) The need and specifications for EIAs depends on the particular country. For the United States it originated as part of the National Environmental Policy Act of 1970. States may also have other requirements.

Ecological footprint—The ecological footprint is a measure of our demand on the Earth's ecosystems. It includes our demand on both natural resources and on the ability for these resources to be regenerated. In the past, the methods and items to be measured varied widely, often based on factors suiting the particular sector making the footprint assessment. Today, more homogeneous standards are emerging. It is intended to reflect a measure of the land, fresh water, and ocean area required to produce, for instance, a product. Such footprints are often calculated to reflect other average measures. One such measure is a per capita ecological footprint that compares consumption and lifestyles with the natural ability to provide this consumption.

Life cycle assessment—A life cycle assessment (LCA) investigates and valuates the environmental impact of a product or service. It is also referred to as an ecobalance analysis, cradle-to-grave analysis, and life cycle analysis. Such assessments are intended to measure the effects of the cascade of technologies related to products and services. Here we will restrict our comments to products. "Life cycle" can refer to a holistic assessment of raw material used in the production of a product, including energy consumption for procurement and transport of the material, manufacture, distribution, use, and finally disposal, including recycling, and if possible, environmental costs. Often guesses and averages are included in these studies. Thus, the cost of transportation could include some proportion of the truck or train construction, road or rail construction, and the deterioration and repair of the same.

According to ISO 14040 and 14044 standards, an LCA contains four distinct phases. These phases are goal and scope, inventory analysis, impact assessment, and interpretation of the first three phases. In the "goal and scope" phase the functional unit is described with effort focused on defining the boundaries of the system or product to be studied. It includes describing the methods used for assessing possible environmental impacts and which impact categories are to be included in the study. In the "inventory" phase data is collected, including the description and verification of the data along with various modeling programs to be used. In general, items considered include inputs such as quantities of materials, land usage, and energy and outputs such as air emission, solid waste, and water emissions. Software packages have been developed, and are being developed, to assist in such evaluations. The "impact" phase is intended to describe contributions to more global situations such as global warming and acidification. The final phase, "interpretation," brings together the other three phases, and conclusions are made.

There are varying types of LCA studies that act to limit or define the type of LCA study being made. The cradle-to-grave assessment is a full LCA study from manufacture (cradle) to the disposal (grave) of a product. A cradle-to-gate study looks at the life cycle from manufacture (cradle) to factory gate (before it is sent to the consumer). These assessments are often used as a basis for environmental product declarations. A cradle-to-cradle assessment involves products where the product is recycled so the study terminates when a new product is made from the recycled original product. A life cycle energy analysis looks at all energy inputs to a product and not solely direct energy inputs during manufacture, including energy necessary to produce components and materials needed for the manufacturing process.

1.5 SUMMARY

After reading this chapter, you should understand the following concepts:
1. Polymers or macromolecules are giant molecules with large structures and high molecular weights. In spite of their varieties they are governed by the same laws that apply to small molecules.
2. If we disregard metals and some inorganic compounds, practically everything else in this world is polymeric. Polymers form the basis for life itself and for our communications, transportation, buildings, food, and so on. Polymers include protein and nucleic acids in our bodies, the fibers (natural and synthetic) we use for clothing, the protein and starch we eat, the elastomers in our automotive tires, the paint, plastic wall and floor coverings, foam insulation, dishes, furniture, pipes, and so on.
3. There are some systems in place that allow us to readily identify the nature of many polymeric materials including clothing and containers.
4. Early developments in polymers were largely empirical because of a lack of knowledge of polymer science. Advancements in polymers were rapid in the 1930s and 1940s because of the theories developed by Staudinger, Carothers, Mark, and many other scientists.
5. This is truly the age of the macromolecule. Essentially every important problem and advance includes polymers, including synthetic (such as carbon nanotubes) and biological (such as the human genome and proteins). There are more chemists working with synthetic polymers than in all of the other areas of chemistry combined.
6. The environmental impact of materials with respect to health and the impact on the environment (today and in the future) is a critical factor as we move forward. Polymers are an essential part of this impact and the solutions to essentially all of the important environmental issues. We must be aware of these environmental issues contributing as individuals and as groups to promote and practice responsible "green science."

GLOSSARY

ABS: A polymer produced by the copolymerization of acrylonitrile, butadiene, and styrene.
Bakelite: A polymer produced by the condensation of phenol and formaldehyde.
Cellulose: A naturally occurring carbohydrate polymer.
Ecological footprint: A measure of our demand on the Earth's ecosystems.

Elastomer: A rubber.

Environmental impact assessment (EIA): An assessment of the possible impact that a project or material may have on the natural environment.

Filament: The individual extrudate emerging from the holes in a spinneret; forms fibers.

Functionality: The number of reactive groups.

Intermolecular forces: Secondary forces between macromolecules.

Intramolecular forces: Secondary forces within the same macromolecular chain.

Life Cycle Assessment (LCA): Investigation and evaluation of the environmental impact of a product or service. It is also referred to as an ecobalance analysis, cradle-to-grave analysis, and life cycle analysis.

Linear: A continuous chain.

Macromolecule: A polymer. Large chained molecular structure.

Natural rubber (NR): Polyisoprene obtained form rubber plants; *Hevea brasiliensis.*

Nylon-66: A polyamide produced form the condensation of adipic acid and 1,6-hexanediame.

Oligomer: Low molecular weight polymer with generally 2–10 repeat units.

Plasticizer: An additive that reduces intermolecular forces in polymers making it more flexible.

Polymer: A giant molecule, macromolecule, made up of multiple repeating units, where the backbone is connected by covalent bonds.

Protein: A natural polyamide composed of many amino acid-derived repeat units.

Rayon: Regenerated cellulose in the form of filaments.

Thermoplastic: A linear polymer that softens when heated.

Thermoset: A network polymer containing chemical cross-linking that does not soften when heated.

Vital force concept: A hypothesis that stated that organic compounds can be produced only by natural processes and not in the laboratory.

Vulcanization: Process where elastomers such as natural rubber are cross-linked by heating with sulfur.

EXERCISES

1. Name six polymers that you encounter daily.
2. Why are there more chemists that work with polymers than with other areas?
3. Why are there so many outstanding polymer chemists alive today?
4. Which of the following are polymeric or contain polymers as major components?
    (a) water, (b) wood, (c) meat, (d) cotton, (e) tires, and (f) paint
5. Name three inorganic polymers.
6. Name three synthetic polymers.
7. You look at the bottom of several containers. Identify what the bottles are made of from the following recycling codes found on their bottoms: (a) four with the letters LDPE, (b) six with the letters PS, and (c) five with the letters PP.
8. Why is there a time delay between discovering a new polymer and commercializing it?
9. What are some advantages of polymers over metals?
10. It has been said that we are walking exhibitions of the importance of polymers. Explain.
11. Why are polymer-intense industries often located in the same geographical area?
12. Why might simple identification codes such as those employed for containers fail for objects such as sneakers and tires?
13. How can (should) we become more environmentally responsible?

ADDITIONAL READING

Allcock, H. (2008): *Introduction to Materials Chemistry*, Wiley, Hoboken, NJ.

Allcock H., Lampe, F., Mark, J. E. (2003): *Contemporary Polymer Chemistry*, 3rd Ed., Prentice-Hall Upper Saddle River, NJ.

Bahadur, P., Sastry, N., (2006): *Principles of Polymer Science*, 2nd Ed., CRC Press, Boca Raton, FL.

Billmeyer, F. W. (1984): *Textbook of Polymer Science*, 3rd Ed., Wiley, NY.

Bower, D. (2002): *Introduction to Polymer Physics*, Cambridge University Press, Cambridge, UK.

Callister, W. D. (2000): *Materials Science and Engineering*, 5th Ed., Wiley, NY.

Campbell, I. M. (2000): *Introduction to Synthetic Polymers*, Oxford University Press, Oxford, UK.

Campo, E. A. (2007): *Industrial Polymers*, Hanser-Gardner, Cincinnati, OH.

Cardarelli, F. (2008): *Materials Handbook: A Concise Desktop Reference*, Springer, NY.

Carraher, C. (2004): *Giant Molecules*, Wiley, Hoboken, NJ.

Carraher, C. (2010): *Introduction to Polymer Chemistry*, Taylor and Francis, Boca Raton, FL.

Carraher, C., Gebelein, C. (1982): *Biological Activities of Polymers*, ACS, Washington, DC.

Challa, G. (2006): *Introduction to Polymer Science and Chemistry*, Taylor and Francis, Boca Raton, FL.

Cherdron, H. (2001): *Tailormade Polymers*, Wiley, NY.

Clark, H., Deswarte, F., Clarke, J. (2008): *The Introduction to Chemicals from Biomass*, Wiley, Hoboken, NJ.

Cowie, J., Arrighi, V. (2007): *Polymers: Chemistry and Physics of Modern Material*, CRC, Boca Raton, FL.

Craver, C., Carraher, C. (2000): *Applied Polymer Science*, Elsevier, NY.

Davis, F. J. (2004): *Polymer Chemistry*, Oxford University Press, NY.

Ebewele, R. O. (2000): *Polymer Science and Technology*, CRC Press, Boca Raton, FL.

Elias, H. G. (1997): *An Introduction to Polymers*, Wiley, NY.

Elias, H. G. (2008): *Macromolecules: Physical Structures and Properties*, Wiley, Hoboken, NJ.

Fischer, T. (2008): *Materials Science for Engineering Students*, Wiley, Hoboken, NJ.

Fried, J. R. (2003): *Polymer Science and Technology*, 2nd Ed., Prentice-Hall, Upper Saddle River, NJ.

Gnanou, Y., Fontanille, M. (2008): *Organic and Physical Chemistry of Polymers*, Wiley, Hoboken, NJ.

Gooch, J. (2007): *Encyclopedia Dictionary of Polymers: (2007),* Springer, NY.

Grosberg, A., Khokhlov, A. R. (1997): *Giant Molecules*, Academic Press, Orlando, FL.

Hiemenz, P., Lodge, T. (2007): *Polymer Chemistry*, 2nd Ed., CRC, Boca Raton, FL.

Hummel, R. E. (1998): *Understanding Materials Science*, Springer-Verlag, NY.

Kricheldorf, H., Nuyken, O., Swift, G. (2005): *Handbook of Polymer Synthesis*, Taylor and Francis, Boca Raton, FL.

Mark, H. (2004): *Encyclopedia of Polymer Science and Technology*, Wiley, Hoboken, NJ.

Mishra, A. (2008): *Polymer Science: A Test Book*, CRC, Boca Raton, FL.

Morawetz, H. (2003): *Polymers: The Origins and Growth of a Science*, Dover Publications, Mincola, NY.

Munk, P., Aminabhavi, T. M. (2002): *Introduction to Macromolecular Science*, 2nd Ed., Wiley, Hoboken, NJ.

Newell, J. (2008): *Modern Materials Science and Engineering*, Wiley, Hoboken, NJ.

Nicholson, J. W. (2006): *The Chemistry of Polymers*, Royal Society of Chemistry, London.

Osswald, T. (2003): *Materials Science of Polymers for Engineering*, Hanser-Gardner, Cinncinati, OH.

Ravve, A. (2000): *Principles of Polymer Chemistry*, 2nd Ed., Kluwer, NY.

Rodriguez, F. (1996): *Principles of Polymer Systems*, 4th Ed., Taylor and Francis, Philadelphia, PA.

Rosen, S. L. (1993): *Fundamental Principles of Polymeric Materials*, 2nd Ed., Wiley, NY.

Sandler, S., Karo, W., Bonesteel, J. Pearce, E. M. (1998): *Polymer Synthesis and Characterization: A Laboratory Manual*, Academic Press, Orlando, FL.

Shashoua, Y. (2008): *Conservation of Plastics*, Elsevier, NY.

Sorsenson, W., Sweeny, F., Campbell, T. (2001): *Preparative Methods in Polymer Chemistry*, Wiley, NY.

Sperling, L. H. (2006): *Introduction to Physical Polymer Science*, 4th Ed., Wiley, NY.

Stevens. M. P. (1998): *Polymer Chemistry*, 2nd Ed., Oxford University Press, Oxford, England.

Strobl, G. (2007): *The Physics of Polymers*, Springer, NY.

Tanaka, T. (1999): *Experimental Methods in Polymer Science*, Academic Press, NY.

Tonelli, A. (2001): *Polymers from the Inside Out*, Wiley, NY.

Walton, D. (2001): *Polymers*, Oxford, NY.

GENERAL ENCYCLOPEDIAS AND DICTIONARIES

Alger, M. (1997): *Polymer Science Dictionary*, 2ⁿᵈ Ed., Chapman and Hall, London, UK.

Brandrup, J., Immergut, E. H., Grulke, E. A. (1999): *Polymer Handbook*, 4ᵗʰ Ed., Wiley, NY.

Compendium of Macromolecular Nomenclature. (1991): *IUPAC*, CRC Press, Boca Raton, FL.

Gooch, J. W. (2007): *Encyclopedic Dictionary of Polymers*, Springer, NY.

Harper, C. A. (2002): *Handbook of Plastics, Elastomers, and Composites*, McGraw-Hill, NY.

Kaplan, W. A. (Published yearly): *Modern Plastic World Encyclopedia*, McGraw-Hill, NY.

Kroschwitz, J. I. (2004): *Encyclopedia of Polymer Science and Engineering*, 3ʳᵈ Ed., Wiley, NY.

Mark, J. E. (1996): *Physical Properties of Polymers Handbook*, Springer, NY.

Mark, J. E. (1999): *Polymer Data Handbook*, Oxford University Press, NY.

Olabisi, O. (1997): *Handbook of Thermoplastics*, Dekker, NY.

Salamone, J. C. (1996): *Polymer Materials Encyclopedia*, CRC Press, Boca Raton, FL.

Wilkes, E. S. (2001): *Industrial Polymers Handbook*, Wiley-VCH, Weinheim.

2 Polymer Structure (Morphology)

The size and shape of polymers are intimately connected to their properties. The shape of polymers is also intimately connected to the size of the various units that comprise the macromolecule and the various primary and secondary bonding forces that are present within the chain and between chains. This chapter covers the basic components that influence polymer shape or morphology.

We generally describe the structure of both synthetic and natural polymers in terms of four levels of structure. The *primary structure* describes the precise sequence of the individual atoms that compose the polymer chain. For polymers where there is only an average structure, such as proteins, polysaccharides, and nucleic acids, a representative chain structure is often given.

The structure can be given as a single repeat unit such that the full polymer structure can be obtained by simply repeating the repeat unit 100, 500, and 1,000 times depending on the precise number of repeat units in the polymer chain. For poly(vinyl chloride), PVC, this is

$$R—(—CH_2—CH—)—R \quad \text{or} \quad R—(—CH_2—CH(Cl)—)—R \quad \text{or} \quad —(CH_2—CH—) \tag{2.1}$$
$$\qquad\qquad | \qquad\qquad\qquad\qquad\qquad\qquad\qquad\qquad\qquad\qquad\qquad\qquad\qquad | $$
$$\qquad\qquad Cl \qquad\qquad\qquad\qquad\qquad\qquad\qquad\qquad\qquad\qquad\qquad\qquad\qquad Cl$$

Or some fuller description of the primary structure may be given such as that below for three repeat units of PVC where the particular geometry about each chiral carbon is given.

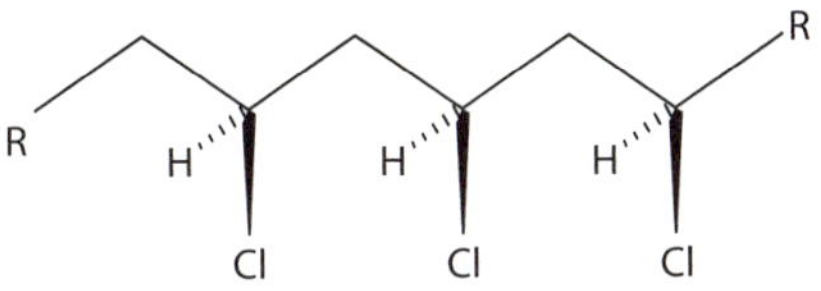

$$\tag{2.2}$$

The ends may or may not be given depending on whether they are important to the particular point being made. Thus, for the single PVC repeat unit given above the end groups may be as follows:

$$CH_3—CH—(—CH_2—CH—)_n—CH_2—CH_2Cl \tag{2.3}$$
$$\qquad\quad | \qquad\qquad\qquad | $$
$$\qquad\quad Cl \qquad\qquad\quad Cl$$

Natural polymers also have general repeat units such as for cellulose (2.4):

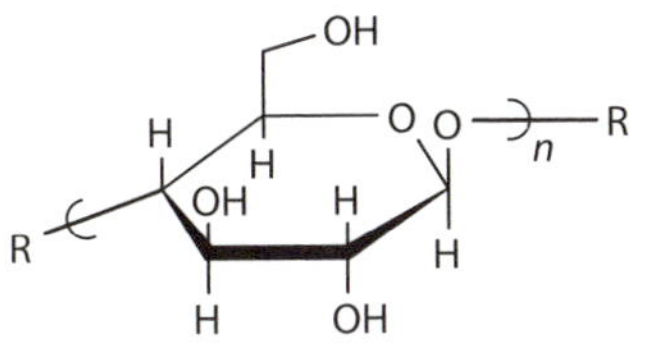

$$\tag{2.4}$$

The *secondary structure* describes the molecular shape or *conformation* of the polymer chain. For most linear polymers this shape approaches a helical or "pleated skirt" (or sheet) arrangement

23

depending on the nature of the polymer, the treatment, and the function. Examples of secondary structures appear in Figure 2.13.

The *tertiary structure* describes the shaping or folding of the polymer. Examples of this are given in Figures 2.16b, 2.17a, and 2.17c.

Finally, the *quaternary structure* represents the overall shape of groups of the tertiary structures, where the tertiary structures may be similar or different. Examples are found in Figures 2.15, 2.16a, and 2.17b.

2.1 STEREOCHEMISTRY OF POLYMERS

The terms "memory" and "to remember" are similar and used by polymer chemists in similar, but different ways. The first use of the terms "memory" and "to remember" involves reversible changes in the polymer structure usually associated with stress–strain deformation of a rubber material where the dislodged, moved polymer segments are connected to one another through chemical and physical cross-links so that once the particular stress–strain is removed the polymer returns to its original, prestress–strain arrangement of the particular polymer segments. Thus, the polymer "remembers" its initial segmental arrangement and returns to it through the guiding of the cross-links.

The second use involves nonreversible changes of polymer segments and whole chain movements also brought about through stress–strain actions or other means to effect nonreversible changes. These changes include any synthetic chain and segmental orientations as well as post-synthesis changes including fabrication effects. These changes involve "permanent" differences in chain and segmental orientation, and in some ways these changes represent the total history of the polymer materials from inception (synthesis) through the moment when a particular property or behavior is measured. These irreversible or nonreversible changes occur with both cross-linked and noncross-linked materials and are largely responsible for the change in polymer property as the material moves from being synthesized, processed, fabricated, and used in what ever capacity it finds itself. Thus, the polymeric material "remembers" its history with respect to changes and forces that influence chain and segmental chain changes. The ability of polymers to "remember" and have a "memory" is a direct consequence of their size.

We can get an idea of the influence of size in looking at the series of hydrocarbons as the number of carbon atoms increases. For hydrocarbons composed of low numbers of carbons, such as methane, ethane, propane, and butane, the materials are gases at room temperature. For the next grouping such as hexane and octane (Table 2.1) the materials are liquids. The individual hydrocarbon chains are held together by dispersion forces that are a sum of the individual methylene and end group forces. There is a gradual increase in boiling point and total dispersion forces for the individual chains until the materials become waxy solids such as that found in bee waxes and in birthday candles. Here, the total dispersion forces are sufficient to be greater than individual carbon–carbon bond strength so the chains decompose before their evaporation. These linear chains are sufficiently long to be crystalline waxy solids but not sufficiently long to allow the chains to interconnect various crystalline groupings. Thus, they are brittle solids with little physical strength. As the chains increase in length the chain lengths are finally sufficient to give tough solids we call linear polyethylene. It is interesting to note that many rocks and diamonds are very strong but brittle because they exhibit essentially no flexibility. Single, almost completely linear, polyethylene crystals can be grown; these are very strong but brittle. But, most linear polyethylene chains gain strength and some flexibility from the chains being sufficiently long to connect the various crystalline domains into larger groupings. This connecting allows applied forces to be distributed throughout the surrounding areas on a microlevel. Linear polyethylene generally contains some portions that are not completely regular or crystalline. These regions are referred to as amorphous and introduce into the polyethylene flexibility with sufficient free volume to allow some segmental mobility. Since most linear polyethylene are not completely linear but contain some branching, this branching prevents complete ordering of the chains, contributing to

TABLE 2.1
Typical Properties of Straight Chain Hydrocarbons

Average Number of Carbon Atoms	Boiling Range (°C)	Name	Physical State at Room Temp.	Typical Uses
1–4	<30	Gas	Gas	Heating
5–10	30–180	Gasoline	Liquid	Automotive fuel
11–12	180–230	Kerosene	Liquid	Jet fuel, heating
13–17	230–300	Light gas oil	Liquid	Diesel fuel, heating
18–25	305–400	Heavy gas oil	Viscous liquid	Heating
26–50	Decomposes	Wax	Waxy	Wax candles
50–1,000	Decomposes		Tough waxy to solid	Wax coatings of food containers
1,000–5,000	Decomposes	Polyethylene	Solid	Bottles, containers, films
>5,000	Decomposes	Polyethylene	Solid	Waste bags, ballistic wear, fibers, automotive parts, truck liners

the presence of amorphous regions. These amorphous regions include the portions of the chains that interconnect the various ordered regions as well as regions resulting from the branching and other related phenomena. Thus, synthetic polyethylene contains both *crystalline* regions, where the polymer chains are arranged in ordered lines and which impart strength to the material, and *amorphous* regions, where the chains are not arranged in as ordered lines and which contribute flexibility with the combination giving a strong material, which on the basis of weight is stronger than steel. The tensile strength/density for bulk steel is 500 while for ultrahigh molecular weight polyethylene (UHMPE) it is about 3,800. The strength of UHMPE is recognized in many applications, including acting as one of the materials employed in the construction of many ballistic resistant body armors (i.e., bullet-proof vests), where it acts to both blunt and distribute the energy of incoming projectiles.

As a side note, low molecular weight polyethylene with appreciable side branching has a melting range generally below 100°C whereas high molecular weight polyethylene with few branches has a melting range approaching the theoretical value of about 145°C.

In general, most polymers contain some combination of crystalline and amorphous regions providing a material that has a combination of flexibility and strength.

High-density polyethylene (HDPE), formerly called low-pressure polyethylene, $[H(CH_2CH_2)_nH]$, like other alkanes $[H(CH_2)_nH]$, may be used to illustrate a lot of polymer structure. As in introductory organic chemistry, we can understand the properties and chemical activities of many complex organic compounds if we understand their basic chemistry and geometry. HDPE, like decane $[H(CH_2)_8H]$ or paraffin $[H(CH_2)_{about\ 50}H]$, is a largely linear chain-like molecule consisting of catenated carbon atoms bonded covalently. The carbon atoms in all alkanes, including HDPE, are joined at characteristic tetrahedral bond angles of approximately 109.5°. While decane consists of 10-methylene groups, HDPE may contain more than 1,000 of these methylene units (Figure 2.1). While we use the term normal or linear to describe nonbranched chains, we know that because of the tetrahedral bond angles and ability for twisting that the chains are zigzag shaped with many possible structural variations.

The distance between the carbon atoms is 1.54 Å or 0.154 nanometers (nm). The apparent zigzag distance between carbon atoms in a chain of many carbon atoms is 0.126 nm. Thus, the length of an extended nonane chain is 8 units times 0.126 nm/units = 1.008 nm. For polyethylene, the repeat unit has two methylenes so that the apparent zigzag distance is 2 × 0.126 nm or 0.252 nm for

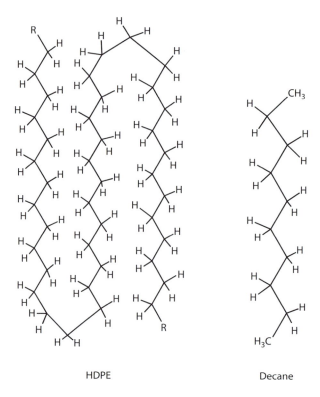

HDPE   Decane

FIGURE 2.1 Simulated structure of linear high-density polyethylene (HDPE) contrasted with the structural formula of linear or normal decane.

each "ethylene" unit. The zigzag or contour length of a HDPE chain 1,000 units long (effectively 2,000 methylenes), $[H(CH_2CH_2)_{1000}H]$ is 1,000 units times 0.252 nm/units or 252 nm. However, because of rotations about the carbon atoms, chains seldom extend to their full extended *contour length* but are present in many different shapes or *conformations*.

The full contour length of a polymer is obtained by multiplying the apparent repeat unit length (*l*), that is the length of each mer or unit, by the number of units (*n*); contour length = *nl*. Even so, it is important to know the average end-to-end distance of polymer chains. The statistical method for this determination, called the *random flight technique*, was developed by Lord Raleigh in 1919. The classical statistical approach may be used to show the distance traveled by a blindfolded person taking *n* number of steps of length *l* in a random walk or the distance flown by a confused moth, bird, or bee.

The distance traveled from start to finish is not the straight-line path measured as nl (the contour length) but it is the root-mean-square distance ($[r^2]^{1/2}$), which is equal to $ln^{1/2}$. Thus, the root-mean-square length of a flexible PE chain with 1,000 units is 0.252 nm times $(1,000)^{1/2}$ = 7.96 nm or about 3% of the contour length. Nobel laureate Paul Flory and others have introduced several corrections so that this random flight technique could be applied to polymer chains approaching a full contour length of *nl*; that is, rigid rod structures.

Each specific protein molecule has a specific chain length, like classical small molecules, and are said to be monodisperse with respect to chain length or molecular weight. However, most synthetic commercial polymers such as HDPE are composed of molecules of different lengths. Thus, the numerical value for the number of repeat units, *n*, or the *degree of polymerization* (DP) should be considered an average DP or average molecular weight. This average notion is often noted by a bar over the top of the DP or M. Accordingly, the average molecular weight of a *polydisperse* polymer will equal the product of the average DP times the molecular weight of the repeating unit or *mer*.

In organic chemistry, it is customary to call a nonlinear molecule, like isobutane, a branched compound. However, polymer scientists use the term *pendant group* to label any group present on the repeat unit. Thus, polypropylene (PP)

$$-(- CH_2—CH—)- \quad \overset{\displaystyle CH_3}{\underset{|}{}}$$

(2.5)

has a methyl group as a pendant unit, but PP is designated as a linear polymer. In contrast, low-density polyethylene (LDPE), formally called high-pressure polyethylene, is a branched polymer because chain extensions or branches of methylene units are present coming off of branch points along the typically linear backbone chain (Figure 2.2). For LDPE the frequency of this branching is about 1.5 per 20 methylene units to 1 per 2,000 methylene units. This branching, like branching in simple alkanes, increases the specific volume and thus reduces the density of the polymer. The linearity provides strength since it increases the opportunity of forming a regular crystalline structure while the branching provides flexibility and toughness since this encourages the formation of amorphous regions. Recently, low-pressure processes have been developed that produce linear low-density polyethylene (LLDPE) that is largely linear but with much less branching (Table 2.2).

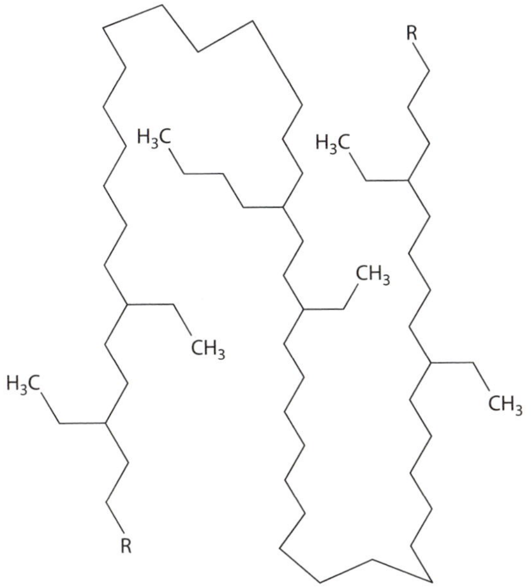

FIGURE 2.2 Simulated structural formula for branched low-density polyethylene (LDPE); compare with Figure 2.1 for HDPE.

TABLE 2.2
Types of Commercial Polyethylene

Type	General Structure	Crystallinity (%)	Density (g/cc)
LDPE	Linear with branching	50	0.92–0.94
LLDPE	Linear with less branching	50	0.92–0.94
HDPE	Linear with little branching	90	0.95

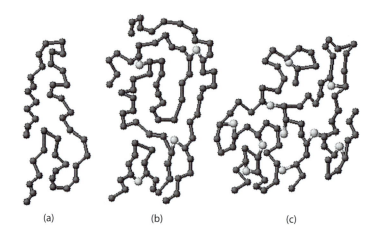

FIGURE 2.3 Skeletal structural formulas of a linear polymer (a) and a network (cross-linked) polymer with low cross-linking density (b) and high-density cross-linking (c). Cross-links are noted by the noncolored spheres.

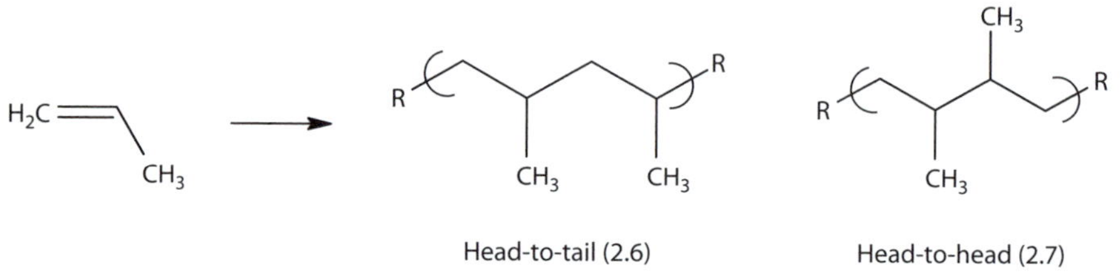

Head-to-tail (2.6)   Head-to-head (2.7)

FIGURE 2.4 Simulated structural formulas showing the usual head-to-tail and unusual head-to-head configurations of polypropylene.

Both linear and branched polymers are typically *thermoplastics* that melt when heated. However, cross-linked three-dimensional, or network, polymers are *thermoset polymers* that do not melt when heated but rather decompose before melting. The cross-link density can vary from low, such as found in a rubber band, to high as found in ebonite (Figure 2.3).

While there is only one possible segment arrangement for the repeat unit in HDPE, there are two possible repeat units in PP and most other polymers. The units can be connected using what is called a *head-to-tail* or through a head-to-head arrangement (Figure 2.4). The usual arrangement by far is the head-to-tail so that pendant methyl groups appear on alternate backbone carbons.

The polymerization of monosubstituted vinyl compounds that gives polymers like polystyrene (PS) and PP produces polymer chains that possess chiral sites on every other carbon in the polymer backbone. Thus, the number of possible arrangements within a polymer chain is staggering since the number of possible isomers is 2^n, where "n" is the number of chiral sites. For a relatively short chain containing 50 propylene units the number of isomers is about 1 times 10^{15}. While the presence of such sites in smaller molecules can be the cause of optical activity, these polymers are not optically active since the combined interactions with light are negated by similar, but not identical, other sites contained on the particular and other polymer chains. Further, it is quite possible that no two polymer chains made during a polymerization will be exactly identical because of chiral differences.

The particular combinations of like and mirror image units within a polymer chain influences the polymer properties on a molecular level. On the bulk level, the average individual chain structure influences properties because of chiral differences.

In the early 1950s, Nobel laureate, Giulio Natta used stereospecific coordination catalysts to produce stereospecific isomers of PP. Natta used the term *tacticity* to describe the different

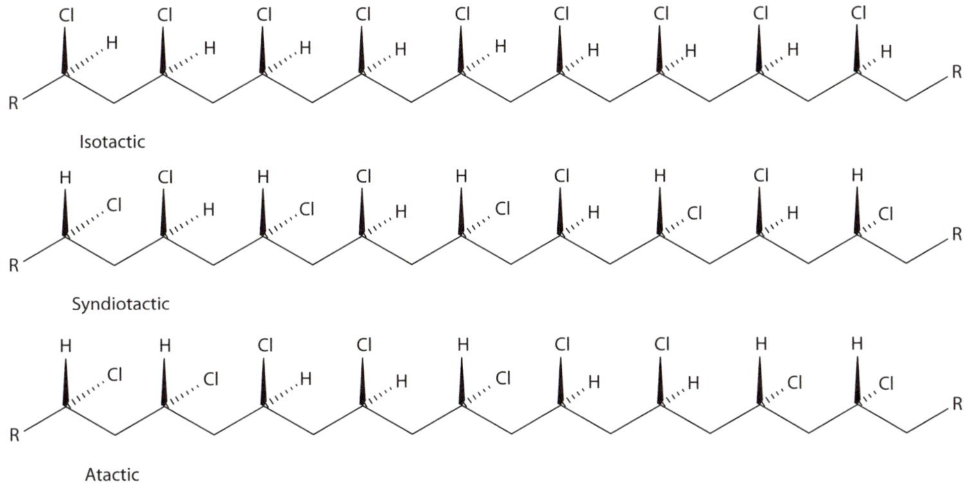

FIGURE 2.5　Skeletal formulas of isotactic, syndiotactic, and atactic of poly(vinyl chloride), PVC.

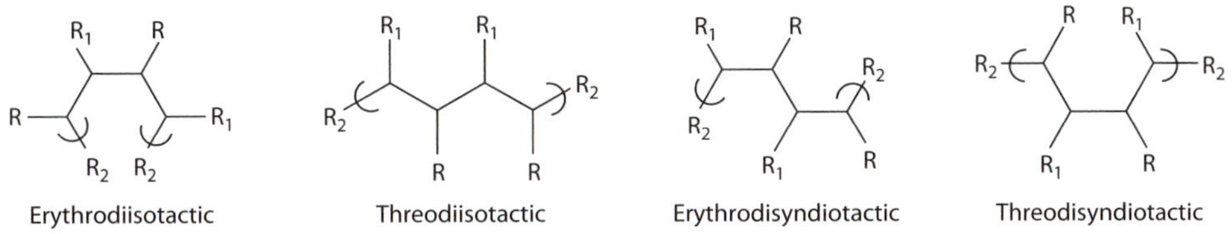

FIGURE 2.6　Simulated formulas of ditactic isomers where R_2 are chain extensions and R and R_1 are not hydrogen.

possible structures. As shown in Figure 2.5 the isomer corresponding to the arrangement DDDDDD or LLLLLL is called *isotactic* (same). The isomer corresponding to the DLDLDLDL alternating structural arrangement about carbon is called *syndiotactic* (alternating). The isomer arrangement that corresponds to some mix of stereo arrangements about the chiral carbons is called *atactic* (having nothing to do with). The differences in stereoregularity about the chiral carbon influence the physical properties of the polymers. Thus, those with isotactic or syndiotactic arrangements are more apt to form compact crystalline arrangements and those with atactic stereoregularity are more apt to form amorphous arrangements. Isotactic PP (iPP) has a MP of about 160°C and it is highly crystalline whereas atactic PP (aPP) melts at about 75°C and is amorphous. The term *eutactic* is used to describe either an isotactic or syndiotactic polymer or a mixture of both.

While most polymers contain only one chiral or asymmetrical center in each repeat unit, it is possible to have diisotacticity where two different substituents are present at chiral centers. These isomers are labeled erythro- and threodiisotactic and erythro- and threosyndiotactic isomers. This topic is further described in Appendix J (Figure 2.6).

The many different conformers resulting from rotation about the carbon–carbon bonds in simple molecules like ethane and *n*-butane may be shown by Newman projections (Figure 2.7). The most stable is the anti or trans projection where the steric hindrance is minimized. There are a number of eclipsed and gauche arrangements of which only one of each is shown in Figure 2.7. The energy difference between the anti and the eclipsed, the least stable form, is about 12 kJ/mol.

The ease in going from one conformer to the other conformer decreases as the pendant groups increase in size and in secondary bonding. Thus, poly(methyl methacrylate) (PMMA) is hard at

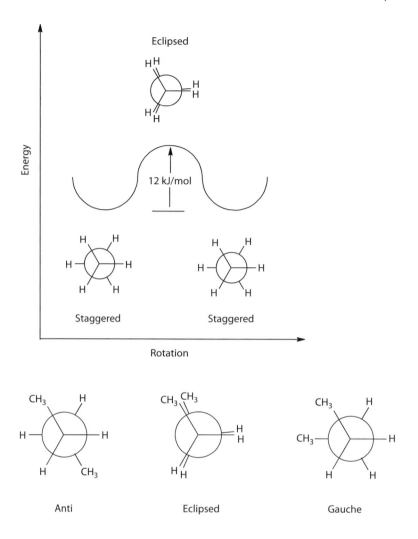

FIGURE 2.7 Top-potential energy profile illustrating the potential energy changes associated with rotation about a C–C bond of ethane and bottom—Newman projections of designated conformers of n-butane.

room temperature because of the polar groups and steric hindrance. By comparison, polyisobutylene, offering only a small amount of steric hindrance, is flexible at room temperature.

2.2   MOLECULAR INTERACTIONS

Forces in nature are often divided into primary forces (typically greater than 200 kJ/mol) and secondary forces (typically less than 40 kJ/mol). *Primary bonding* forces can further be subdivided into ionic (characterized by a lack of directional bonding; occurs between atoms of large differences in electronegative; normally not found in polymer backbones), metallic (the number of outer, valence electrons is too small to provide complete outer shells; often considered as charged atoms surrounded by a potentially fluid sea of electrons; lack of bonding direction), and covalent (including dative and coordinate) bonding (the major bonding in organic compounds and polymers; directional). The bonding length of primary bonds is generally about 0.1–0.22 nm. The carbon–carbon bond length is about 0.15–0.16 nm. Atoms in individual polymer chains are joined to one another by covalent bonds with bonding energies about 320–370 kJ/mol for single bonds.

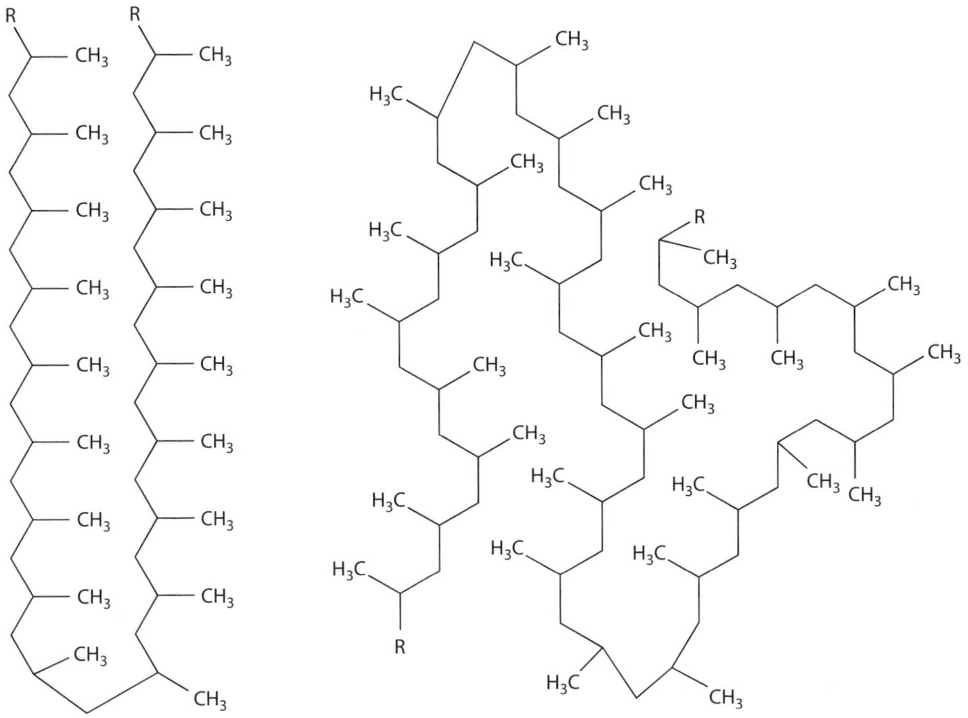

FIGURE 2.8 Representation of a crystalline portion from isotactic polypropylene (a), and an amorphous portion from atactic polypropylene (b).

Polymer molecules are also attracted to one another through secondary forces. *Secondary forces*, often called *van der Waals forces* because they are the forces responsible for the van der Waals corrections to the ideal gas relationships, are of longer distance in interaction in comparison to primary forces. Secondary bonding distances are generally on the order of 0.25–0.5 nm. The force of these interactions is inversely proportional to some power of the distance, generally 2 or greater (i.e., force is proportional to 1/[distance]^r). Thus, many physical properties are sensitive to the polymer *conformation* (arrangements related to rotation about single bonds) and *configuration* (arrangements related to the actual chemical bonding about a given atom), since both affect the proximity one chain can have relative to another. Thus, amorphous PP is more flexible than crystalline (generally isotactic or syndiotactic) PP because the crystalline PP has the units closer to one another, allowing the secondary bonding to be stronger (Figure 2.8).

These intermolecular forces are also responsible for the increase in boiling points within a homologous series, such as the alkanes, for the higher-than-expected boiling points of polar molecules such as alkyl chlorides, and for the abnormally high boiling points of water, alcohols, amines, and amides. While the forces responsible for these increases in boiling points are all van der Waals forces, the forces can be further subdivided in accordance to their source and intensity. Secondary forces include London dispersion forces, induced permanent forces, and dipolar forces, including hydrogen bonding.

All molecules, including nonpolar molecules such as heptane and polyethylene, are attracted to each other by weak *London* or *dispersion forces* that result from induced dipole- induced dipole interactions. The temporary or transient dipoles are due to instantaneous fluctuations in the electron cloud density. The energy range of these forces is fairly constant and about 8 kJ/mol. This force is independent of temperature and is the major force between chains in largely nonpolar polymers such as those in classical elastomers and soft plastics such as PE.

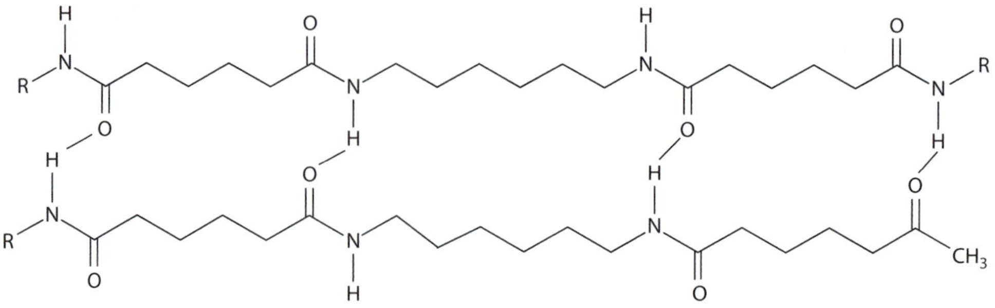

FIGURE 2.9 Typical hydrogen bonding (shown as "–" between hydrogen on nitrogen and oxygen for nylon-66.

It is of interest to note that methane, ethane, and ethylene are all gases; hexane, octane, and nonane are all liquids (at room conditions), while low molecular weight PE is a waxy solid. This trend is primarily due to an increase in the mass per molecule and to an increase in the London forces per polymer chain. The London force interaction between methylene units is about 8 kcal/mol. Thus, for methane molecules the attractive forces is 8 kJ/mol; for octane it is about 64 kJ/mol; and for polyethylene with 1,000 ethylene (or 2,000 methylenes) it is about 2,000 methylene units × 8 kJ/mol per methylene unit = 16,000 kJ/mol well sufficient to make PE a solid and to break backbone bonds before it boils. (Polymers do not boil because the energy necessary to make a chain volatile is greater than the primary backbone bond energy.)

Polar molecules such as ethyl chloride and PVC are attracted to each other by both the London forces, but also to *dipole–dipole interactions* resulting from the electrostatic attraction of a chlorine atom in one molecule to a hydrogen atom in another molecule. These dipole–dipole forces are of the order of 8–25 kJ/mol, generally greater than the London forces and they are temperature dependent. Hard plastics, such as PVC, have dipole–dipole attractive forces present between the chains.

Strongly polar molecules such as ethanol, poly(vinyl alcohol) (PVA), cellulose, and proteins are attracted to each other by a special type of dipole–dipole force called *hydrogen bonding*. Hydrogen bonding occurs when a hydrogen present on a highly electronegative element, such as nitrogen or oxygen, comes close to another highly electronegative element. This force is variable but for many molecules it is about 40 kJ/mol and for something like hydrogen fluoride (HF), which has particularly strong hydrogen bonding, it is almost as strong as primary bonding. Intermolecular hydrogen bonding is usually present in classical fibers such as cotton, wool, silk, nylon (Figure 2.9), polyacrylonitrile, polyesters, and polyurethanes. Intramolecular hydrogen bonds are responsible for the helices observed in starch and globular proteins.

It is important to note that the high melting point of nylon-66 (Figure 2.9; 265°C) is a result of a combination of dipole–dipole, London, and hydrogen-bonding forces. The relative amount of hydrogen bonding decreases as the number of methylene groups increases and a corresponding decrease is seen in the melting point for nylon-6,10 in comparison to nylon-66. Polyurethanes, polyacrylonitrile, and polyesters are characterized by the presence of strong hydrogen and polar bonding and form strong fibers. In contrast, iPP, which has no hydrogen bonding holding the PP chains together, is also a strong fiber but because of the ability of the similar chains to fit closely together. Thus, both the secondary bonding between chains and the ability to tightly fit together, steric factors are important factors in determining polymer properties.

In addition to the contribution of intermolecular forces, chain entanglement is also an important contributory factor to the physical properties of polymers. While paraffin wax and HDPE are homologs with relatively high molecular weights, the chain length of paraffin is too short to permit chain entanglement and hence it lacks the strength and many other physical characteristic properties of HDPE.

Chain entanglement allows long chains to act as though they were even longer because entanglement causes the entangled chains to act together. Many of the physical properties, such as tensile strength, of polymers increase dramatically when chain entanglement occurs. The *critical chain length* (z) required for the onset of entanglement is dependent on the polarity and shape of the polymer. The number of atoms in the critical chain lengths of PMMA, PS, and polyisobutylene are 208, 730, and 610, respectively.

Viscosity is a measure of the resistance to flow. The latter is a result of cooperative movement of the polymer segments from vacate location, hole, to another vacate location in a melted state. This movement is impeded by chain entanglement, high intermolecular forces, the presence of reinforcing agents, and cross-links. The *melt viscosity* (η) is often found to be proportional to the 3.4 power of the critical chain length, as shown in Equation 2.6, regardless of the polymer. The constant K is temperature dependent.

$$\log \eta = 3.4 \log z + \log K \tag{2.6}$$

The flexibility of amorphous polymers above the glassy state, where segmental mobility is possible, is governed by the same forces as melt viscosity and is dependent on a wriggling type of segment motion in the polymer chains. This flexibility is increased when many methylene groups ($-CH_2-$) or oxygen atoms ($-O-$) are present. Thus, the flexibility of aliphatic polyesters usually increases as "m" is increased (2.7).

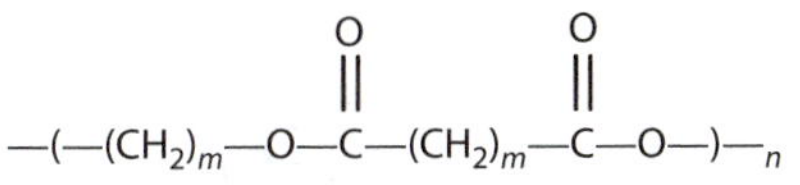

$$-(-(CH_2)_m-O-C-(CH_2)_m-C-O-)-_n \tag{2.7}$$

Aliphatic polyester

Flexibilizing groups include methylene and ethylene oxides, dimethylsiloxanes, and methylene groups.

In contrast, the flexibility of amorphous polymers above the glass state is decreased when stiffening groups such as

1,4-Phenylene (2.8) 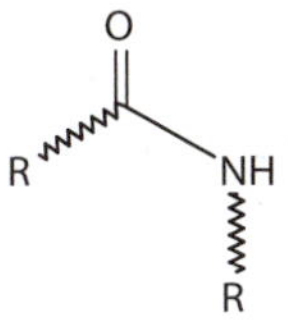   Amide (2.9) 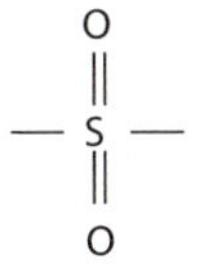   Sulfone (2.10) 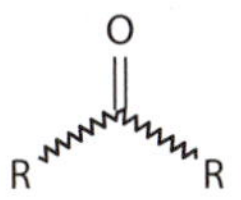   Carboxyl (2.11)

are present in the polymer backbone. Thus, poly(ethylene terephthalate) (PET; 2.12) is stiffer and with higher melting point than poly(ethylene adipate; 2.13), and the former is stiffer than poly(butylene terephthalate) because of the presence of fewer methylene groups between the stiffening groups.

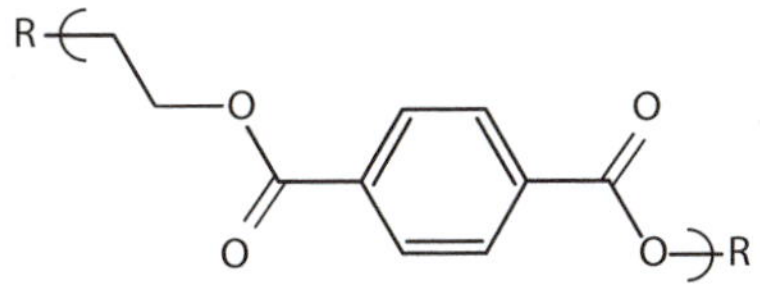

Poly(ethylene terephthalate) (2.12) 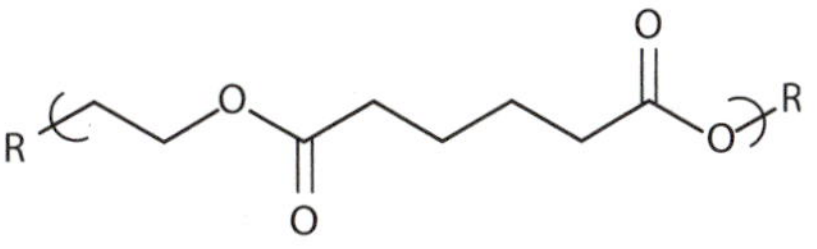   Poly(ethylene adipate) (2.13)

TABLE 2.3

Approximate Glass Transition Temperatures (T_g) for Selected Polymers

Polymer	T_g (K)	Polymer	T_g (K)
Cellulose acetate butyrate	323	Cellulose triacetate	430
Polyethylene (LDPE)	148	Polytetrafluoroethylene	160, 400*
a-Polypropylene	253	Poly(ethyl acrylate)	249
i-Polypropylene	373	Poly(methyl acrylate)	279
Polyacrylonitrile	378	a-Poly(butyl methacrylate)	339
Poly(vinyl acetate)	301	a-Poly(methyl acrylate)	378
Poly(vinyl alcohol)	358	Poly(vinyl chloride)	354
cis-Poly-1,3-butadiene	165	Nylon-66	330
trans-Poly-1,3-butadiene	255	Poly(ethylene adipate)	223
Polydimethylsiloxane	150	Poly(ethylene terephthalate)	342
PS	373		

*Two major transitions observed.

These groups act as stiffening units because the groups themselves are inflexible as in the case of 1,4-phenylene or because they form relatively strong bonding, such as hydrogen bonding between chains as is the case of the amide linkage.

Small molecules such as water can exist in three phases: solid, liquid, and gas. Polymers do not boil so this phase is missing for them but they do melt. But polymers undergo other transitions besides melting. The most important of these is called the glass transition, T_g, which will be discussed below. Before we turn to the glass transition it is important to note that polymers may undergo many other transitions. About 20 transitions have been reported for polyethylene. PS undergoes several transitions that have been identified. At about –230°C, the movement, often described as wagging or oscillation, of the phenyl groups begins. At about –140°C, movement of four-carbon groups in the PS backbone begins. At about 50°C, torsional vibration of the phenyl groups begins. At about 100°C, long-range chain movement begins corresponding to the reported T_g value for PS (Table 2.3). It is important to remember that while small molecules have a precise temperature associated with its transitions, such as 0°C for melting for water, polymer values, while often reported as a specific value, are a temperature range. This temperature range is the result of at least two features. First, there is a variety of polymer chain environments at the molecular level, each with its own energy-associated features. Second, transitions that require large segment or whole chain movement will also have a kinetic factor associated with them because it takes time for chains to untangle/tangle and rearrange themselves. Thus heating/cooling rate affect the temperatures required to effect the changes.

As noted before, small molecules can generally exist in three phases—liquid, solid, and gas, but polymers degrade before boiling, so they do not exist in the gas scale. Even so, polymers generally undergo several major thermal transitions. At low temperatures polymers are brittle, glassy since there is not sufficient energy present to encourage even local or segmental chain movement. As the temperature is increased, at some temperature there is sufficient energy available to allow some chain mobility. For a polymer containing both amorphous and crystalline portions or is only amorphous, the onset of this segmental chain mobility for the amorphous segments is called the *glass transition temperature*, T_g. Because there is unoccupied volume in the amorphous polymer structure some segmental chain movement occurs. This segmental chain movement is sometimes likened to a snake slithering "in place" within the grass. The localized chain movement causes a further increase in unoccupied volume, and larger segments are able to move eventually, allowing the snake further movement in the grass. As the temperature is increased, there is sufficient temperature to overcome the forces present in the crystalline portion of the polymer, allowing a breaking up of the crystalline portion. This temperature is often referred to as the crystalline transition temperature, T_c. This temperature is directly related to the melt transition temperature, T_m or melting point. Finally, as the temperature continues to rise there is sufficient energy available to overcome the primary bonds holding the polymer together and the polymer decomposes at the *decomposition temperature, T_d*. The T_g and T_m are described further below.

The flexibility of amorphous polymers is reduced drastically when they are cooled below a characteristic transition temperature called the *glass transition temperature (T_g)*. At temperatures below T_g there is no ready segmental motion, and any dimensional change in the polymer chain is the result of temporary distortions of the primary covalent bonds. Amorphous plastics perform best below T_g but elastomers must be used above the brittle point, T_g, or they will act as a glass and be brittle and break when bent. The importance of a material being above its T_g to offer some flexibility was demonstrated by the space shuttle Challenger disaster where the cool temperature at the launch pad resulted in the "O" ring not being flexible so that fuel escaped resulting in the subsequent explosion.

The *melting point* (also called the melt transition temperature) is the temperature range where total or whole polymer chain mobility occurs. The melting point (T_m) is called a first-order transition temperature, and T_g is sometimes referred to as a second-order transition. The values for T_m are usually 33%–100% greater than the T_g. Symmetrical polymers like HDPE exhibit the greatest difference between T_m and T_g. The T_g values are low for elastomers and flexible polymers, such as PE and dimethylsiloxane, are relatively high for hard amorphous plastics, such as polyacrylonitrile and PET (Table 2.3).

The T_g for isotactic PP is 373 K or 100°C, yet, because of its high degree of crystallinity, it does not flow to any great extent below its melting point of 438 K (165°C). In contrast, the highly amorphous polyisobutylene, which has a T_g value of 203 K (−70°C), flows at room temperature (Table 2.3). T_g decreases as the size of the ester groups increases in polyacrylates and polymethylacrylates. The effect of the phenylene stiffening groups is also demonstrated with the T_g of PET (2.12) being about 120° higher than that of poly(ethylene adipate) (2.13).

The main reasons why amorphous polymers go from a solid glassy state to a more flexible plastic state are the presence of sufficient energy and unoccupied volume. The energy is supplied by heating the sample and allows the polymer sufficient energy for the chain segments to become separated and to begin movement, which in turn creates free or unoccupied volume that allows the chain segments to slip past one another, resulting in the material being more flexible. For the chains to begin moving, the secondary forces that hold the chains together must be overcome. As movement begins, additional unoccupied volume is created and this expansion within a complex maze of intertwining chains creates additional free volume. A measure of this expansion is the *thermal coefficient of expansion*. The temperature range where the available free volume and energy necessary to overcome segmental chain interactions is available is called the *glass transition temperature, T_g*. Since the specific volume of polymers increases at T_g to accommodate the increased segmental chain motion, T_g values can be estimated from plots of the change in specific volume with temperature (Figure 2.10).

Below the T_g, the chains are "frozen" into place and the material acts as a brittle solid or glass, hence the name *glassy state*.

Other properties such as stiffness (modulus), refractive index, dielectric properties, gas permeability, X-ray adsorption, and heat capacity (Figure 2.11) all change at T_g and have been used to determine the T_g. As seen in Figure 2.11, both T_g and T_m are endothermic because energy is absorbed when segmental mobility or melting occur. This idealized plot of energy as a function of temperature is for a polymer with about a 50%–50% mixture of crystalline and amorphous regions. Bringing about wholesale mobility, and melting takes more energy than causing segmental mobility. This is illustrated in Figure 2.11, where the area under the T_g-associated peak is less than the area under the curve for the T_m-associated peak (the area under the peaks is a direct measure of the heat [energy]). While T_g values are the most often reported related to the onset of segmental motion in the principle polymer backbone, other secondary values may be observed for the onset of motion of large pendant groups, branches, and so on.

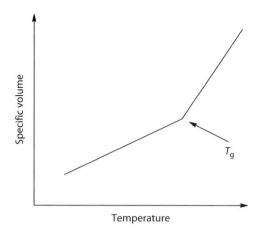

FIGURE 2.10 Determination of T_g by noting the abrupt change in specific volume.

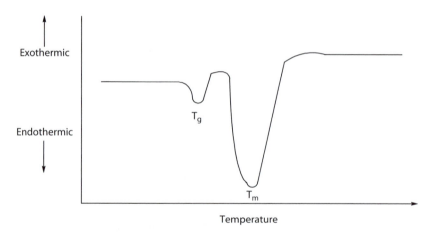

FIGURE 2.11 Typical DSC thermogram of a polymer.

The number of possible conformers increases with chain length and can be shown statistically to equal 2^{2n}, where n is the number of units. Thus, when $n = 1,000$, the number of possible conformers of HDPE is $2^{2,000}$ or 10^{600}, more than the grains of sand at all of our beaches combined. Four of these possible conformers are shown in Figure 2.12.

Because there are many possible ends in branched polymers, it is customary to use the radius of gyration (S) instead of the end-to-end distance for such polymers. The *radius of gyration* is actually the root-mean-square distance of a chain end from the polymer's center of gravity. S is less than the end-to-end distance (r), and for linear polymers $r^2 = 6S^2$.

In general, polymers (both natural and synthetic) "emphasize" two general shapes—helical and pleated (Figures 2.9 and 2.13). The intermolecular bonds in many polyamides, including natural polyamides such as β-keratin, where the steric requirements are low produce strong pleated sheets. Hair, fingernails and toenails, feather, and horns have a β-keratin structure. Helical structures are often found where there is high steric hindrance because helical structures allow the minimization of these steric factors by "radiating" them outward from a central (backbone) core distributing the steric groups about a helical circle. It is important that secondary bonding, generally hydrogen

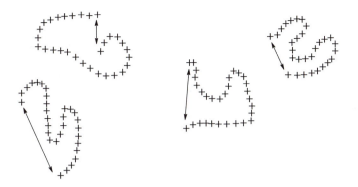

FIGURE 2.12   End-to-end distances for four 30-unit chains.

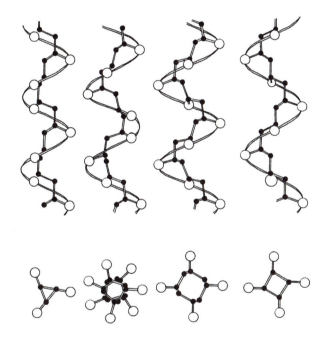

FIGURE 2.13   Helical conformation of isotactic vinyl polymers. From Gaylord, N. (1959): *Linear and Steroregular Addition Polymers*, N. Gaylord and H. Mark, eds., Wiley, NY.

bonding, occurs in both structures. With sheets, bonding occurs between chains and for helical structures bonding occurs within the same chain. Further, some compounds, such as α-keratin, form helical structures within sheet structures.

   α-Keratin (composed of parallel polypeptide α-helices) and most globular proteins are characterized by intramolecular bonds. More about the structures of the two types of keratins is discussed in Section 10.2. These, and many other polymers, including nucleic acids, form helices. Ribonucleic acid (RNA) exists as a single-stranded helix, while DNA exists as a double-stranded helix. For both natural and synthetic polymers, these helices vary with respect to the number of backbone carbons per complete cycle. Figure 2.13 contains helical conformations for isotactic vinyl polymers. The "R" groups are designated by the larger open circles. From left to right the nature of the R group varies as expected. For the furthest left

R = $-CH_3$, $-CH_2CH_3$, $-CH=CH_2$, $-OCH_3$, $-CH_2CH_2CH(CH_3)_2$, $-O-CH_2CH_2CH(CH_3)_2$ and cyclohexyl. For the next R = $-CH_2CH(CH_3)CH_2CH_3$, $-CH_2CH(CH_3)_2$. For the third from the left R = $-CH(CH_3)_2$, and for the extreme right R = a variety of substituted cyclohexyls, including 2-methylcyclohexyl and 4-fluorocyclohexyl.

2.3 POLYMER CRYSTALS

Before 1920, leading scientists not only stated that macromolecules were nonexistent, but they also believed, if they did exist, they could not exist as crystals. However, in the early 1920s Haworth used X-ray diffraction techniques to show that elongated cellulose was a crystalline polymer consisting of repeat units of cellobiose. In 1925, Katz in jest placed a stretched natural rubber band in an X-ray spectrometer and to his surprise observed an interference pattern typical of a crystalline substance. This phenomenon may be shown qualitatively by the development of opacity when a rubber band is stretched (try it yourself) and by the abnormal stiffening and whitening of unvulcanized rubber when it is stored for several days at 0°C. The opacity noted in stretched rubber and cold rubber is the result of the formation of crystallites or regions of crystallinity. The latter was first explained by a fringed micelle model that is now found not consistent with much of the current experimental findings (Figure 2.14).

Amorphous polymers with irregular bulky groups are seldom crystallizable, and unless special techniques are used even ordered polymers are seldom 100% crystalline. The combination of amorphous and crystalline structures varies with the structure of the polymer and the precise conditions that have been imposed on the material. For instance, rapid cooling often decreases the amount of crystallinity because there is not sufficient time to allow the long chains to organize themselves into more ordered structures before they become frozen in place. The reason linear ordered polymers fail to be almost totally crystalline is largely kinetic, resulting from an inability of the long chains to totally disentangle and perfectly align themselves during the time the polymer chain is cooling and mobile.

Mixtures of amorphous and mini-crystalline structures or regions may consist of somewhat random chains containing some chains that are parallel to one another forming short-range mini-crystalline regions. Crystalline regions may be formed from large-range ordered platelet-like structures, including polymer single crystals, or they may form even larger organizations such as spherulites as shown in Figures 2.15 and 2.16. Short- and long-range ordered structures can act as physical cross-links.

In general, linear polymers form a variety of single crystals when crystallized from dilute solutions. For instance, highly linear PE can form diamond-shaped single crystals with a thickness on the order of 11–14 nm when crystallized from dilute solution. The surface consists of "hairpin

FIGURE 2.14 Schematic two-dimensional representation of a modified micelle model of the crystalline-amorphous structure of polymers.

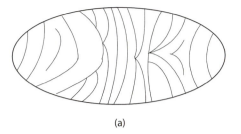 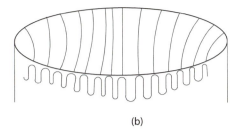

(a) (b)

FIGURE 2.15 Structure of a spherulite from the bulk. (a) shows a slice of a simple spherulite. As further growth occurs, filling in, branch points, and so on occur as shown above (b). The contour lines are simply the hairpin turning points for the folded chains.

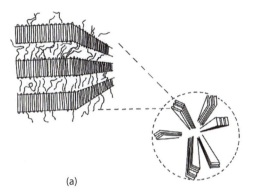 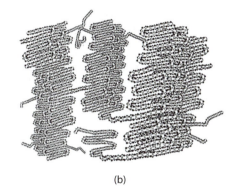

(a) (b)

FIGURE 2.16 Spherulite structure showing the molecular-level lamellar chain-folded platelets and tie and frayed chain arrangements (a) and a more complete model of two sets of three lamellar chain-folded platelets formed from polyethylene (b). Each platelet contains about 850 ethylene units as shown here.

turned" methylene units as pictured in Figures 2.15 and 2.16. The polymer chain axes are perpendicular to the large flat crystal faces. A single polymer chain with 1,000 ethylene (2,000 methylene) units might undergo on the order of 50 of these hairpin turns on the top surface and another 50 turns on the bottom face with about 20 ethylene units between the two surfaces.

Many polymers form more complex single crystals when crystallized from dilute solution, including hollow pyramids that often collapse on drying. As the polymer concentration increases, other structures occur, including twins, spirals, and multilayer dendritic structures, with the main structure being spherulites.

When polymer solids are produced from their melts, the most common structures are these spherulites that can be seen by the naked eye and can be viewed as Maltese cross-like structures with polarized light and crossed Nicol prisms under a microscope.

For linear PE, the initial structure formed is a single crystal with folded chain *lamellae*. These quickly lead to the formation of sheaf-like structures called axialites or hedrites. As growth proceeds, the lamellae develop on either side of a central reference point. The lamellae continue to fan out, occupying increasing volume sections through the formation of additional lamellae at appropriate branch points. The result is the formation of spherulites as pictured in Figures 2.15 and 2.16.

While the lamellar structures present in spherulites are similar to those present in polymer single crystals, the folding of chains in spherulites is less organized. Further, the structures that exist between these lamellar structures are generally occupied by amorphous structures, including atactic chain segments, low molecular weight chains, and impurities.

The individual spherulite lamellae are bound together by "tie" molecules that are present in more than one spherulite. Sometimes these tie segments form intercrystalline links, which are threadlike

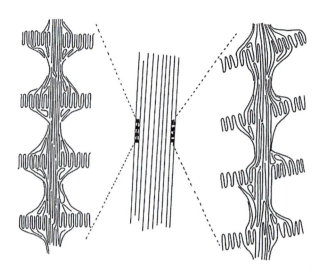

FIGURE 2.17 Crystalline polymer structures formed under applied tension including flow conditions. Middle shows the tertiary monofibrilar structure including platelets and the left shows these monofibrilar structures bundled together forming a quaternary structure fibril. Right shows the distorted shish kebab formed with more rapid flow.

structures that are important in developing the characteristic good toughness found in semicrystalline polymers. They act to tie together the entire assembly of spherulites into a more or less coherent "package."

Depending upon the particular conditions of crystallization, a number of secondary and tertiary structures can be formed. In most cases, crystalline polymers attempt to form crystalline platelets. Under little or no externally applied stress, these platelets organize themselves in spherulites as pictured in Figures 2.15 and 2.16. They start by a nucleating process and begin to radiate outward from the central nucleating site. Amorphous chain segments get trapped between the forming crystalline platelet combinations, giving a kind of fuzzy or frayed exterior. These platelets are generally either planar, as shown in Figure 2.17, or they can be helical or twisted. The platelets continue to grow until they butt up against other spherulites.

Under externally applied stress, including simple melt flow, the tertiary structure can approach a shish kebab arrangement where there are planes of platelets separated by areas where there exists both crystalline and amorphous regions as pictured in Figure 2.17, left. These shish kebab structures often organize into quaternary structures consisting of bundles of shish kebab single-strand filaments forming fibrils. Under more rapid flow conditions the shish kebab itself become distorted (Figure 2.17, right).

Interestingly, the amorphous regions within the spherulite confer onto the material some flexibility while the crystalline platelets give the material strength, just as in the case with largely amorphous materials. This theme of amorphous flexibility and crystalline strength (and brittleness) is a central idea in polymer structure–property relationships.

It must be remembered that the secondary structure of both the amorphous and crystalline regions typically tend toward a helical arrangement of the backbone for most polymers but not polyethylene, which forms a crank-shaft structure because of the lack of steric restraints (i.e., lack of pendant groups off the backbone).

The kind, amount, and distribution of polymer chain order/disorder (crystalline/amorphous) is driven by the processing (including pre- and post-) conditions, and thus it is possible to vary the polymer properties through a knowledge of and the ability to control the molecular-level structures. The crystalline regions may be disrupted by processing techniques such as thermoforming and extrusion of plastics and drawing of fibers. In the last process, which is descriptive of the others, the

TABLE 2.4
Avrami Values for Particular Crystallization Growth for Sporadic and Ordered (or Predetermined) Nucleation

Crystallization Growth Pattern	Sporadic Nucleation	Ordered Nucleation	Overall Dimensionality
Fiber/rod	2	1	One
Disc	3	2	Two
Spherulite	4		Three
Sheaf	6		

crystallites are ordered in the direction of the stress, the filament shrinks in diameter, and heat is evolved and reabsorbed as a result of additional orientation and crystallization.

Crystallization often occurs over a wide area/volume almost simultaneously. It is similar to raindrops or grains of sand falling into water and setting up waves that progress outward until they overlap with one another. Avrami and others have studied the rate of crystallization and have derived various relationships to describe and differentiate the various crystallizations. The rate of crystallization can be followed using dilatometry using the Avrami equation 2.14 that was developed to follow the crystallization of metals. Here, the quotient of the difference between the specific volume, V_t, at time t and the final specific volume, V_f, divided by the difference between the original specific volume, V_o, and the final volume is equal to an experimental expression where K is the kinetic constant related to the rate of nucleation and growth and n is an integer related to the nucleation and growth of crystals. In theory, the value of n is related to the dimensionality of the growing crystallinity (Table 2.4). The value of n has been calculated using several scenarios.

$$\frac{V_i - V_f}{V_o - V_f} = e^{-K_t^n} \tag{2.14}$$

Table 2.4 contains values for two of these scenarios. These values are valid for only the initial stages of crystallization.

Noninteger values for n are not uncommon. As noted before, depending on the particular conditions several crystalline formations are possible and are found for the same polymer. Sperling has collected a number of Avrami values for some common values given in literature. The range of values for polyethylene is 2.6–4.0; that for poly(decamethylene terephthalate) is 2.7–4.0; that for PP is 2.8–4.1; that for poly(ethylene oxide) is 2.0–4.0; and that for isotactic-polystyrene is 2.0–4.0.

The kind, amount, and distribution of polymer chain order/disorder (crystalline/amorphous) is driven by a number of factors, including structure and processing. With respect to processing, it is possible to influence polymer properties through a knowledge of and ability to control the molecular morphology (structure). Crystalline structures can be disrupted by processing techniques such as thermoforming and extrusion of plastics and drawings of films and fibers.

Crystallization of polymers containing bulky groups occurs more slowly than polymers that do not contain bulky substituents. In addition to crystallization of the backbone of polymers, crystallization may also occur in regularly spaced bulky groups even when an amorphous structure is maintained in the backbone. In general, the pendant group must contain at least 10 carbon atoms for this side-chain crystallization to occur. Ordered polymers with small pendant groups crystallize more readily than those with bulky groups. Rapid crystallization producing films with good transparency may be produced by addition of a crystalline nucleating agent such as benzoic acid and by rapid cooling.

While polymeric hydrocarbons have been used as illustrations for simplicity, it is important to note that the principles discussed apply to all polymers, organic as well as inorganic, and natural as well as synthetic, and to elastomers, plastics, and fibers.

2.4 AMORPHOUS BULK STATE

An amorphous bulk polymer contains chains that are arranged in something less than a well-ordered, crystalline manner. Physically, amorphous polymers exhibit a T_g but not a T_m, and do not give a clear X-ray diffraction pattern. Amorphous polymer chains have been compared to spaghetti strands in a pot of spaghetti, but in actuality the true extent of disorder that results in an amorphous polymer is still not fully understood.

Section 13.3 contains a discussion of a number of techniques employed in the search for the real structure of the amorphous bulk state. Briefly, evidence suggests that little order exists in the amorphous state, with the order being similar to that observed with low molecular weight hydrocarbons. There is evidence that there is some short-range order for long-range interactions and that the chains approximate a random coil with some portions paralleling one another. In 1953, Flory and Mark suggested a random coil model whereby the chains had conformations similar to those present if the polymer were in a theta solvent. In 1957, Kargin suggested that amorphous polymer chains exist as aggregates in parallel alignment. Models continue to be developed, but all contain the elements of disorder/order suggested by Flory and Mark and the elements of order suggested by Kargin.

2.5 POLYMER STRUCTURE–PROPERTY RELATIONSHIPS

Throughout the text we will relate polymer structure to the properties of the polymer. Polymer properties are related not only to the chemical nature of the polymer, but also to such factors as extent and distribution of crystallinity, distribution of polymer chain lengths, and nature and amount of additives, such as fillers, reinforcing agents, and plasticizers, to mention a few. These factors influence essentially all the polymeric properties to some extent, including hardness, flammability, weatherability, chemical stability, biological response, comfort, flex life, moisture retention, appearance, dyeability, softening point, and electrical properties.

Materials must be varied to perform the many tasks required of them in today's society. Often they must perform them repeatedly and in a "special" manner. We get an ideal of what materials can do by looking at some of the behavior of giant molecules in our body. While a plastic hinge must be able to work thousands of times, the human heart, a complex muscle largely composed of protein polymers (Section 10.6), provides about 2.5 billion beats within a lifetime moving oxygen (Section 16.8) throughout the approximately 144,000 km of the circulatory system with (some) blood vessels the thickness of hair and delivering about 8,000 L of blood every day with little deterioration of the cell walls. The master design allows nerve impulses to travel within the body at the rate of about 300 m/min; again polymers are the "enabling" material that allows this rapid and precise transfer of nerve impulses. Human bones, again largely polymeric, have a strength about five times that of steel (on a weight basis). Genes, again polymeric (Sections 10.10 and 10.11), appear to be about 99.9% the same between humans, with the 0.1% functioning to give individuals the variety of size, abilities, and so on, that confer uniqueness. In the synthetic realm, we are beginning to understand and mimic the complexities, strength, preciseness, and flexibility that are already present in natural polymers.

Here we will briefly deal with the chemical and physical nature of polymeric materials that permit their division into three broad divisions—elastomers or rubbers, fibers, and plastics. *Elastomers* are polymers possessing high chemical and/or physical cross-links. For industrial application the "use" temperatures must be above T_g (to allow for ready "chain" mobility), and its normal state (unextended) must be amorphous. The restoring force, after elongation, is largely due to entropy. On release of the applied force the chains tend to return to a more random state. Gross, actual mobility

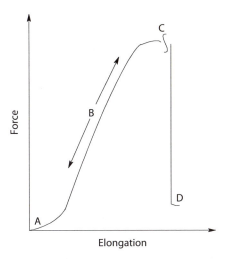

FIGURE 2.18 Elongation of an elastomer as a function of applied force, stress, where A is the original "relaxed" state, B is the movement to full extension, C is the point at which the elastomer "breaks," and D is the force necessary to pull two separate pieces of elastomer apart.

of chains must be low. The cohesive energy forces between chains should be low permitting rapid, easy expansion. In its extended state a chain should exhibit high tensile strength, whereas at low extensions it should have a low tensile strength. Cross-linked vinyl polymers often meet the desired property requirements. The material, after deformation, should return to its original shape because of the cross-linking. This property is often referred to as an elastic "memory." Figure 2.18 illustrates force versus elongation for a typical elastomer. As the elastomer is pulled, the largely random chain segments become "stretched out" forming microcrystalline domains resulting in a decreased entropy or increased order. Eventually, most of the chains are part of these microcrystalline domains resulting in further elongation requiring much increased force (stress). This microcrystallinization, physical cross-linking, also confers to the elastomer a greater brittleness, eventually resulting in the rubber breaking as additional stress is applied.

Fiber properties include high tensile strength and high modulus (high stress for small strains). These properties can be obtained from high molecular symmetry and high cohesive energies between chains, both requiring a fairly high degree of polymer crystallinity. Fibers are normally linear and drawn (oriented) in one direction, producing higher mechanical properties in that direction. Typical condensation polymers, such as polyesters and nylons, often exhibit these properties. If the fiber is to be ironed, its T_g should be above 200°C, and if it is to be drawn from the melt, its T_g should be below 300°C. Branching and cross-linking are undesirable since they disrupt crystalline formation, even though a small amount of cross-linking may increase some physical properties, if effected after the material is drawn and processed. Permanent press garments often have some cross-linking, ensuring a "remembrance" of the "permanent press."

Products with properties intermediate between elastomers and fibers are grouped together under the heading "*plastics*." Plastics typically have some flexibility and have dimensional stability, that is, they act as somewhat flexible solids. Many polymers can act as members of two of these three categories depending on the treatment of the material. Thus, nylon-66 is fibrous in behavior when it is treated so that the chains have good alignment and are stretched to both increase this alignment and to form fibers. Nylon-66 is plastic when it has less alignment, that is, is more amorphous, and used as a bulk material rather than as a fiber. Polyesters also can be either fibers or plastics under the same conditions as given for nylon-66. Other materials, such as PVC and siloxanes, can be processed to act as plastics or elastomers.

TABLE 2.5
Selected Property–Structure Relationships

Glass Transition Temperature
    Increases with the presence of
        Bulky pendant groups
        Stiffening groups as 1,4-phenylene
        Chain symmetry
        Polar groups
        Cross-linking
    Decreases with the presence of
        Additives like plasticizers
        Flexible main chain groups
        Nonpolar groups
        Dissymmetry
Solubility
    Favored by
        Longer chain lengths
        Low interchain forces
        Disorder and dissymmetry
        Increased temperature
        Compatible solvent
Crystallinity
    Favored by
        High interchain forces
        Regular structure; high symmetry
        Decrease in volume
        Increased stress
        Slow cooling from melt
        Homogeneous chain length

Selected property–structure relationships are summarized in Tables 2.5 and 2.6. As noted before, some polymers can be classified into two categories, with properties being greatly varied by varying molecular weight, end group, processing, cross-linking, plasticizer, and so on. Nylon-66 in its more crystalline form behaves as a fiber, whereas less crystalline forms of nylon-66 are generally classified as plastics.

There are some general guidelines with respect to a material's T_g and T_m, its general amorphous/crystalline structure, and the potential use area. Elastomers are cross-linked, amorphous polymers where the use temperature is above its T_g. An adhesive is a linear or branched amorphous polymer that is used above its T_g. Coatings are generally near their T_g when used so that some flexibility is present allowing the coatings material to withstand temperature changes without cracking and so coalescing occurs on drying. Plastics can be either amorphous or partially amorphous. Amorphous (or partially crystalline) plastics such as PP and PE should have a use temperature below the T_m but above the T_g. Fibers are composed of crystalline polymers where the use temperature is below the T_m.

2.6 CROSS-LINKING

Cross-linking is important because this is a major mechanism for retaining shape which, in turn, influences the physical properties, such as solubility, of polymers. There are three types of cross-linking present in synthetic and natural polymer. Two of the three types are physical and the third is chemical.

TABLE 2.6
General Property Performance–Structure Relationships

	Increased Crystallinity	Increased Cross-linking	Increased mol. wt.	Increased mol. wt. Distribution	Addition of Polar Back Bone Units	Addition of Backbone Stiffening Groups
Abrasion resistance	+	+	+	−	+	−
Brittleness	−	M	+	+	+	+
Chemical resistance	+	V	+	−	−	+
Hardness	+	+	+	+	+	+
T_g	+	+	+	−	+	+
Solubility	−	−	−	0	−	−
Tensile strength	+	M	+	−	+	+
Toughness	−	−	+	−	+	−
Yield	+	+	+	+	+	+

+, increase in property; 0, little or no effect; −, decrease in property; M, property passes through a maximum; V, variable results dependent on particular sample and temperature.

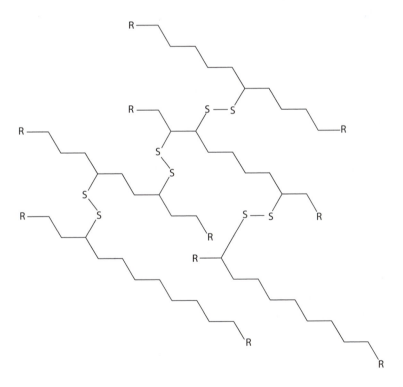

FIGURE 2.19   Chemical cross-linking of polyethylene chains through formation of disulfide linkages.

Chemical cross-linking is given many names depending on the particular area of application. For instance, for hair, the name "setting" is often associated with the breakage and subsequent reformation of thiol cross-links. For tires, the terms vulcanization and curing are associated with the formation of sulfur-associated chemical cross-links (Figure 2.19).

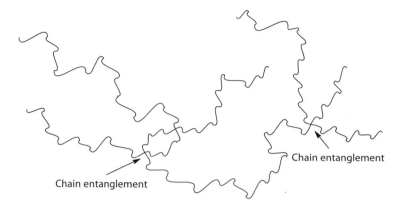

FIGURE 2.20 Physical cross-linking through chain entanglement.

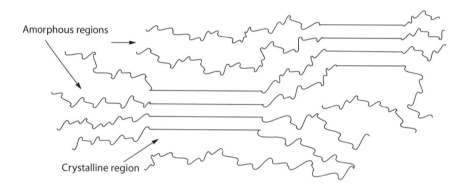

FIGURE 2.21 Crystalline portions that act as physical cross-links.

Chain entanglement is a physical means of forming cross-links (Figure 2.20). The incidence of chain entanglement is related to both the particular polymer (dependent on such factors as bond angles and substituents) and its length. The longer a polymer, the more apt that there is one or more chain entanglements. For most linear polymers, chain entanglements occur when chains of 100 units and more are present. Chain entanglements cause the material to act as if its molecular weight is much greater. For instance, if a chain of polyethylene of 100 units is connected to another polyethylene chain of 100 units, which is again connected to another chain of 100 units, then the effective chain length is about 300 units.

The third type of cross-linking involves formation of crystalline portions within an amorphous grouping (Figure 2.21). For vinyl polymers, the temperature that allows segmental chain mobility to occur is generally well below room temperature, but the temperature to disrupt crystalline formations is typically well above room temperature. Within mixtures of crystalline and amorphous structures below the T_m, the crystalline portions act to "tie-in" or connect the surrounding areas acting as cross-links.

In some situations, crystalline formation can occur because of addition of physical stress such as the stretching of a rubber band. As the rubber band is stretched, the amorphous random chains become aligned, forming small areas that are crystalline that oppose further stretching. Thus, rubber bands contain chemical cross-links, and on extension, crystalline domains form (Figure 2.22).

There are a number of consequences to the presence of cross-links. Cross-links impart to a material memory with the chains about and those chains involved with the cross-links locked into

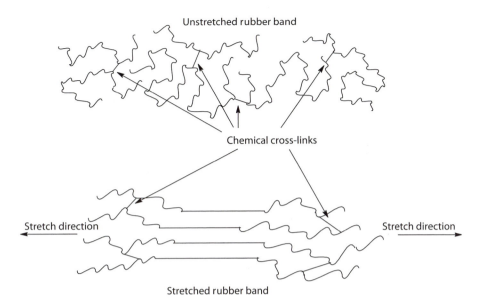

FIGURE 2.22 Chemically cross-linked rubber stretching resulting in the formation of physical crystalline cross-links.

particular environment. Chemically cross-linked materials are not soluble and they do not melt. Chemically cross-linked materials are also not easily environmentally recycled because melting and solubility are often involved in the ready recycling of polymeric materials.

The general name given to materials that do not melt when heated is *thermoset*. Thermoset materials generally contain chemical cross-links. Since they do not melt when heated, it is more difficult to recycle such materials. By comparison, materials that do melt when heated are called *thermoplastics*. Such materials do melt when heat is applied and so are more easily recycled. Examples of thermosets are phenolic and amino plastics and some elastomers. PS, PP, polyethylene, nylon, and polyesters are examples of thermoplastic materials. By weight and value, we use many more synthetic thermoplastic materials in comparison to thermosets. Even so, both thermoset and thermoplastic materials are important and can be recycled.

2.7 CRYSTALLINE AND AMORPHOUS COMBINATIONS

Most polymers consist of a combination of crystalline and amorphous regions. Even within polymer crystals such as spherulites (Figures 2.15–2.17), the regions between the ordered folded crystalline lamellae are less ordered, approximating amorphous regions. This combination of crystalline and amorphous regions is important for the formation of materials that have both good strength (contributed largely by the crystalline regions) and some flexibility or "softness" (derived from the amorphous portions). Figure 2.16 contains a space-filled model for polyethylene chains (a total of about 400 units with five branches, one longer and four shorter).

This model of polyethylene contains a mixture of amorphous and crystalline regions. Note the cavities within the amorphous regions with materials containing a majority of amorphous regions having a greater porosity and consequently a greater diffusion and greater susceptibility to chemical and natural attack. As noted before, materials that contain high amounts of the crystalline regions are referred to as being crystalline and are less flexible and stronger and offer better stability to nature and attack by acids and bases, oils, and so on. Also as noted before, amorphous regions give the material flexibility while the crystalline regions give the material strength. Thus, many materials contain both crystalline and amorphous region, giving the material a balance between strength

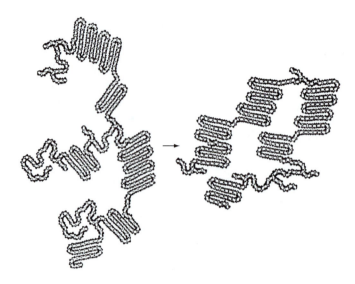

FIGURE 2.23   Idealized structure illustrating crystalline (ordered) and amorphous (nonordered) regions of lightly branched polyethylene chains for a prestressed and stressed orientation.

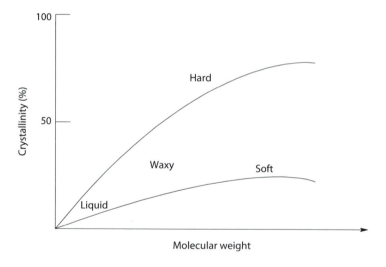

FIGURE 2.24   General physical states of materials as a function of crystallinity and molecular weight.

and flexibility. The final properties of a material are then dependent on the molecular structure of that material.

Through the use of specific treatment(s) the crystalline/amorphous regions can vary from being largely random to being preferentially oriented in one direction with a greater degree of "crystalline-type" structures when unidirectional stress is applied (Figure 2.23). Here the amount of free space or volume is less with the overall order greater and properties associated with these changes also changed. The material will be stronger, and have a greater ability to resist attack by acids, bases, oils, and other external agents, and the diffusion of gases and other agents through it is less. Polymers can be oriented (such as the pulling of fibers or films) in one or two directions. This preferential orientation results in the fiber or film material with anisotropic properties, with the material showing greater strength along the axis of pull.

Figure 2.24 shows the general relationship between material "hardness/softness" and the proportion that is crystalline for largely linear polymers.

2.8 SUMMARY

1. Polymers or macromolecules are high molecular weight compounds with chain lengths greater than the critical length required for the entanglement of these chains. There is an abrupt change in melt viscosity and other physical properties of high molecular weight polymers when the chain length exceeds the critical chain length.
2. While some naturally occurring polymers, such as proteins, are monodisperse, that is, all have the same molecular weight, other natural and synthetic polymers, such as cellulose and PE, are polydisperse, that is, they consist of a mixture of chains of differing numbers of units. Thus one uses the term average molecular weight when describing the molecular weight of these polydisperse materials.
3. Many polymers, such as cellulose and HDPE, are linear polymers containing long, continuous, covalently bonded atoms. Others may be branched or cross-linked. Both linear polymers and those with branching are generally thermoplastics that can be remolded by application of heat and pressure. Others that are cross-linked are thermosets that cannot be remolded by application of heat and pressure. Both groups of polymers can be recycled, but it is easier to recycle thermoplastic materials.
4. Functional groups in the polymer backbone, such as the methyl group in PP, are called pendant groups. Such polymers are formed giving a head-to-tail arrangement rather than a head-to-head arrangement.
5. The temperature at which local segmental mobility occurs is called the T_g and the temperature at which wholesale polymer chain mobility occurs is called the T_m.
6. The first-order transition or melting point (T_m) is energy wise larger than the T_g. Entirely crystalline polymers will have only a T_m, whereas a totally amorphous polymer will have only a T_g. Since most polymers are a combination of amorphous and crystalline regions, they have both a T_g and a T_m.
7. A polymer stretched out to its full contour length is only one of the myriad conformations possible for a polymer. The chain length is expressed statistically as the root-mean-square distance, which is only a fraction of the contour length.
8. Since branched chains like LDPE have many chain ends, it is customary to use the radius of gyration, S, as a measure of the distribution of polymer segments. The radius of gyration is the distance of a chain end from the polymer's center of gravity rather than a measure of the root-mean-square. Both measures are statistically related.
9. Fibers and stretched elastomers are translucent because of the presence of ordered crystallites or regions of crystallinity.
10. Crystalline regions of polymers can be represented as combinations of folded chains forming lamellar structures. Amorphous regions are less ordered than crystalline regions. Additional orientation of polymer chains occurs when stress is applied, resulting in increased strength in the direction of the applied stress. This results in increased strength in the order of the orientation.
11. The principal differences between elastomers, plastics, and fibers are the presence and absence of stiffening groups, molecular symmetry, and the strength of the intermolecular forces. Elastomers are typically characterized by the absence of stiffening groups, the presence of molecular asymmetry, low amount of crystallinity, and overall absence of strong intermolecular forces. In contrast, fibers are characterized by the presence of stiffening groups, molecular symmetry, high amount of crystallinity, and the presence of strong intermolecular forces. Fibers have a lack of branching and irregularly spaced pendant groups. Plastics have structures and properties that are between elastomers and fibers.

12. There are three types of cross-linking. These are chemical and two physical types of cross-linking. Physical cross-linking results from formation of crystalline regions within polymer structures and from chain entanglement. Cross-linked materials have good dimensional memory. Chemically cross-linked materials do not dissolve and do not melt.

GLOSSARY

Amorphous: Noncrystalline polymer or region in a polymer.

Atactic: Polymer in which there is a random arrangement of pendant groups on each side of the chain.

Backbone: Principle chain in a polymer.

Branched polymer: Polymer having extensions attached to the polymer backbone. Not pendant groups.

Bulky groups: Sterically large groups.

Chiral center: Asymmetric center such as a carbon atom with four different groups attached to it.

Cold drawing: Stretching a fiber under room temperature.

Configurations: Related chemical structures produced by the breaking and remaking of primary covalent bonds.

Conformations: Various shapes of polymer chains resulting from the rotation about single bonds in the polymer chain.

Conformer: Shape produced by a change in the conformation of a polymer.

Contour length: Fully extended length of a polymer chain; equal to the product of the length of a repeat unit times the number of units or mers.

Critical chain length (z): Minimum chain length required for entanglement of polymer chains.

Cross-linked density: Measure of the relative degree of cross-linking.

Crystalline polymer: Polymer with ordered structure.

Crystallites: Regions of crystallinity.

Degree of polymerization: Number of repeat units in a chain.

Dipole–dipole interactions: Moderate secondary forces between polar groups in different or the same polymer chain.

Dispersion forces: Low-energy secondary forces due to the creation of momentary induced dipoles; also known as London forces.

Glass transition temperature (T_g): Temperature range where a polymer gains local or segmental mobility.

Glassy state: Hard, brittle state; below T_g.

Gutta percha: Naturally occurring trans isomer of polyisoprene.

Head-to-tail configuration: Normal sequence of mers in which the pendant groups are regularly spaced; for PVC, the chlorine atom appears on every other carbon.

Intermolecular forces: Secondary forces between different molecules.

Intramolecular forces: Secondary forces within the same molecule.

Isotactic: Polymer where the geometry of the pendant groups are all on the same side of the polymer backbone.

Lamellar: Plate-like or planar (flat) in shape.

Linear polymer: Polymer without chains extending off the backbone.

Low-density polyethylene (LDPE): A branched form of PE produced at high pressure by the free–radical-initiated polymerization of ethylene.

Maltese cross: Cross with arms like arrowheads pointing inward.

Melting point (T_m): First-order transition when the solid and liquid phases are in equilibrium.

Mer: Repeat unit.

Modulus: Ratio of stress to strain, which is a measure of the stiffness of a polymer.

Monodisperse: Polymer mixture made up of molecules of one specific molecular weight.

Pendant groups: Groups attached to the main polymer backbone, like methyl groups in polypropylene.

Polydisperse: Polymer mixture containing chains of varying lengths.

Radius of gyration (S): Root-mean-square distance of a chain end to a polymer's center of gravity.

Root-mean-square distance ($[r^2]^{1/2}$): Average end-to-end distance of polymer chains; $l(n^{1/2})$.

Side-chain crystallization: Crystallization related to that of regularly spaced long pendant groups.

Single polymer crystals: Lamellar structure consisting of folded chains of a linear polymer.

Spherulites: Three-dimensional aggregates of polymer crystallites.

Stiffening groups: Groups in a polymer backbone that decrease the segmental motion of the polymer chain.

Syndiotactic: Polymer in which the pendant groups are arranged alternately on each side of the carbon backbone.

Tacticity: Arrangement of the pendant groups in space; that is, isotactic, syndiotactic, and atactic.

Thermoplastic: A linear polymer that softens when heated.

Thermoset: A network polymer containing chemical cross-linking that does not soften when heated.

van der Waals forces: Secondary forces based on the attraction between groups.

Viscosity: Measure of the resistance of a polymer or polymer solution to flow.

EXERCISES

(To answer some of these questions you may need to look at other parts in the book for structures and specific details.)

1. Make sketches or diagrams showing (a) a linear polymer, (b) a polymer with pendant groups, (c) a polymer with short branches, (d) a polymer with long branches, and cross-linked polymers with (e) low, and (f) high cross-linked density.

2. Which has (a) the greater volume for the same weight of material and (b) the lower softening point: HDPEC or LDPE?

3. What is the approximate bond length of the carbon atoms in (a) a linear and (b) a cross-linked polymer.

4. What is the approximate contour length of a HDPE chain with an average degree of polymerization (chain length) of $n = 2,000$ and of a PVC chain of the same number of repeating units?

5. Which of the following are monodisperse polymers with respect to chain length? (a) hevea rubber, (b) corn starch, (c) cellulose from cotton, (d) an enzyme, (e) HDPE, (f) PVC, (g) a specific DNA, (h) nylon-66, or (i) a specific RNA?

6. What is the average degree of polymerization of LDPE having an average molecular weight of 28,000?

7. What is the structure of the repeating unit in (a) polypropylene, (b) poly(vinyl chloride, and (c) hevea rubber?

8. Which of the following is a branched chain polymer: (a) HDPE, (b) Isotactic PP, (c) LDPE, or (d) amylose starch?

9. Which of the following is a thermoplastic: (a) ebonite, (b) Bakelite, (c) vulcanized rubber, (d) HDPE, (e) celluloid, (f) PVC, or (g) LDPE?

10. Which has the higher cross-linked density: (a) ebonite or (b) soft vulcanized rubber?

11. Do HDPE and LDPE differ in (a) configuration or (b) conformation?
12. Which is a trans isomer: (a) gutta percha or (b) hevea rubber?
13. Which will have the higher softening point: (a) gutta percha or (b) hevea rubber?
14. Show (a) a head-to-tail, and (b) a head-to-head configuration for PVC.
15. Show the structure of a typical portion of the chain of (a) s-PVC and (b) i-PVC.
16. Show Newman projections of the gauche forms of HDPE.
17. Name polymers whose intermolecular forces are principally (a) London forces, (b) dipole–dipole forces, and (c) hydrogen bonding.
18. Which will be more flexible: (a) poly(ethylene terephthalate) or (b) poly(butylene terephthalate)?
19. Which will have the higher glass transition temperature: (a) poly(methylene methacrylate) or (b) poly(butyl methacrylate)?
20. Which will have the higher T_g: (a) iPP or (b) aPP?
21. Which will be more permeable to a gas at room temperature: (a) iPP or (b) aPP?
22. Under what kind of physical conditions is a linear polymer more apt to form spherulites.
23. What is the full contour length of a molecule of HDPE with a DP of 1,500?
24. Which would be more flexible: (a) poly(methyl acrylate) or (b) poly(methyl methacrylate)?
25. Which would you expect to form "better" helical structures: (a) i-polypropylene or (b) a-polypropylene?
26. Which would you expect to have a higher melting point (a) nylon-66 or (b) an aramide?
27. What type of hydrogen bonds is present in the internal structure of a globular protein?
28. Which would have the greater tendency to "cold flow" at room temperature: (a) poly(vinyl acetate) ($T_g = 301$ K) or (b) PS ($T_g = 375$ K)?
29. Which would be least transparent: (a) combination of amorphous and crystalline PS, (b) entirely crystalline PS, or (c) entirely amorphous PS?
30. Which would be more apt to produce crystallites: (a) HDPE or (b) poly(butyl methacrylate)?
31. Which of the following would you expect to provide strong fibers (a) nylon-66, (b) a-polypropylene, or (c) wool.
32. Which would tend to be more crystalline when stretched: (a) unvulcanized rubber or (b) ebonite?
33. Which would be more apt to exhibit side-chain crystallization (a) poly(metnyl methacrylate) or (b) poly(dodecyl methacrylate)?
34. What must be present in order for movement to occur within a polymer at T_g or T_m?
35. What are the three major forms of cross-linking?

ADDITIONAL READING

Alfrey, T., Gurnee, E. F. (1956): *Dynamics of viscoelastic behavior, in Rheology-Theory and Applications*, F. R. Eirich, ed., Academic Press, NY.

Bicerano, J. (2002): *Prediction of Polymer Properties*, Dekker, NY.

Bower, D. (2002): *An Introduction to Polymer Physics*, Cambridge University Press, Ithaca.

Cheng, S. (2008): *Phase Transitions in Polymers: The Role of Metastable States*, Elsevier, NY.

Campbell, D., Pethrick, R., White, J. R. (2000): *Polymer Characterization: Physical Techniques*, 2nd Ed., Stanley Thornes, Cheltenham, UK.

Ciferri, A. (2005): *Supramolecular Polymer*, 2nd Ed., Taylor & Francis, Boca Raton, FL.

Ellis, B. (2008): *Polymers: A Property Database*, CRC, Boca Raton, FL.

Flory, P. J. (1953): *Principles of Polymer Chemistry*, Cornell University Press, Ithaca.

Geil, P. H. (1963): *Polymer Single Crystals*, Wiley, NY.

Henkel, M., Plemling, M., Sanctuary, R. (2007): *Ageing and the Glass Transition*, Springer, NY.

Howell, B., Zaikow, G. (2008): *Compounds and Materials with Specific Properties*, Nova, Hauppauge, NY.

Mandelkern, L (2004): *Crystallization of Polymers-Kinetics and Mechanisms*, Cambridge University Press, Ithaca.

Mark, J. (2009): *The Polymer Data Handbook*, Oxford University, Ithaca.

Natta, G. (1955): Stereospecific macromolecules, *J. Poly. Sci.,* 16:143.

Pauling, L., Corey, R., Branson, H. (1951): The structure of proteins, *Proc. Natl. Acad. Sci., USA*, 37:205.

Reiter, G. (2007): *Progress in Understanding of Polymer Crystallization*, Springer, NY.

Roe, R.-J. (2000): *Methods of S-Ray and Neutron Scattering in Polymer Chemistry*, Oxford University Press, NY.

Rotello, V., Thayumanavan, S. (2008): *Molecular Recognition and Polymers: Control of Polymer Structure and Self-assembly*, Wiley, Hoboken, NJ.

Schultz, J. (2001): *Polymer Crystallization*, Oxford University Press, Cary, NC.

Seymour, R. B., Carraher, C. E. (1984): *Structure-Property Relationships in Polymers*, Plenum, NY.

Viney, C. (2003): *Techniques for Polymer Organization and Morphology Characterization*, Wiley, Hoboken.

Wohlfarth, C. (2004): *Thermodynamic Data of Aqueous Polymer Solutions*, Taylor & Francis, Boca Raton, FL.

Wohlfarth, C. (2006): *Enthalpy Data of Polymer-Solvent Systems*, Taylor & Francis, Boca Raton, FL.

3 Molecular Weight of Polymers

3.1 INTRODUCTION

It is the size of macromolecules that give them their unique and useful properties. Size allows polymers to act more as a group so that when one polymer chain moves surrounding chains are affected by that movement. Size also allows polymers to be nonvolatile since the secondary attractive forces are cumulative (e.g., the London dispersion forces are about 8 kJ/mole per repeat unit), and, because of the shear size, the energy necessary to volatilize them is greater than the energy to degrade the polymer.

Generally, the larger the polymer, the higher is the molecular weight. The average molecular weight (M) of a polymer is the product of the average number of repeat units or mers expressed as the degree of polymerization, DP, times the molecular weight for the repeating unit. Thus, for polyethylene, PE, with an average DP of 100 the average molecular weight is simply 100 units times 28 daltons (Da)/unit = 2,800 Da. Note that amu and Da are often used interchangeably as units.

Polymerization reactions, which produce both synthetic and natural (but not for all natural materials, such as proteins and nucleic acids) polymers, lead to products with heterogeneous molecular weights, that is, polymer chains with different numbers of mers. Molecular weight distribution (MWD) may be rather broad (Figure 3.1), or relatively narrow, or may be mono-, bi-, tri-, or polymodal. A bimodal curve is often characteristic of a polymerization occurring under two different environments. Polymers consisting of chains of differing lengths are called *polydisperse* while polymers containing only one chain length, such as specific nucleic acids, are called *monodisperse*.

Some properties, such as heat capacity, refractive index, and density, are not particularly sensitive to molecular weight but many important properties are related to chain length. Figure 3.2 lists three of these. The melt viscosity is typically proportional to the 3.4 power of the average chain length; so η is proportional to $M^{3.4}$. Thus, the melt viscosity increases rapidly as the chain length increases, and more energy is required for the processing and fabrication of large molecules. This is due to chain entanglements that occur at higher chain lengths. However, there is a tradeoff between molecular weight–related properties and chain size such that there is a range where acceptable physical properties are present but the energy required to cause the polymers to flow is acceptable. This range is called the *commercial polymer range*. Many physical properties, such as tensile and impact strength (Figure 3.2), tend to level off at some point and increased chain lengths give little increase in that physical property. Most commercial polymer ranges include the beginning of this leveling off threshold.

While a value above the *threshold molecular weight value* (TMWV; lowest molecular weight where the desired property value is achieved) is essential for most practical applications, the additional cost of energy required for processing extremely high polymer molecular weights is seldom justified. Accordingly, it is customary to establish a commercial polymer range above the TMWV but below the extremely high molecular weight range. However, it should be noted that some properties, such as toughness, increases with chain length. Thus, extremely high molecular weight polymers, such as ultrahigh molecular weight polyethylene (UHMPE), are used for the production of tough articles such as waste barrels.

Oligomers and other low molecular weight polymers are not useful for applications where high strength is required. The word oligomer is derived from the Greek work *oligos*, meaning "a few."

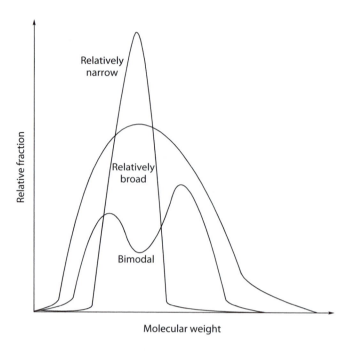

FIGURE 3.1   Relative differential weight distribution curves.

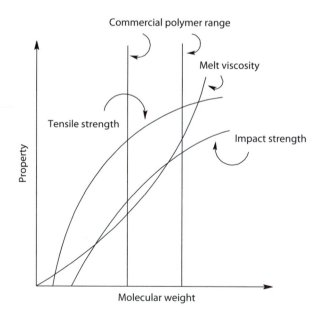

FIGURE 3.2   Relationship between molecular weight and polymer physical properties.

The value of TMWV is dependent on the cohesive energy density (CED) of amorphous polymers, the extent of crystallinity in crystalline polymers, and the effect of reinforcements in polymeric composites. Thus, while a low molecular weight amorphous polymer may be satisfactory for use as a coating or adhesive, chain lengths generally above 100 are often required if the polymer is to be used as an elastomer or plastic.

3.2   SOLUBILITY

Polymer mobility is an important aspect helping determine a polymer's physical, chemical, and biological behavior. Lack of mobility, either because of interactions that are too swift to allow the segments within the polymer chain some movement or because there is not sufficient energy (such as a high enough temperature) available to create mobility or because of a lack of available free volume, results in a brittle material. Many processing techniques require the polymer to have some mobility. This mobility can be achieved through application of heat and/or pressure and through dissolving the polymer. Because of its size, the usual driving force of entropy increase for the mixing and dissolving of materials (entropy) is smaller for polymers in comparison to small molecules. Traditional molecular weight determinations require that the polymer be dissolved. Here, we will focus on the general topic of polymer solubility and the factors that influence polymer solubility.

The first attempts at predicting solubility were largely empirical. Paint technologists employed various approaches. In one approach Kauri-butanol values were equal to the minimum volume of test solvent that produced turbidity when added to a standard solution of Kauri-Copal resin in 1-butanol. The aniline point is the lowest temperature where equal volumes of aniline and the test solvent are completely miscible. Both tests are really measures of the relative aromaticity of the test solvent.

Mixing can be described in terms of free energy. Free energy has two terms, an energy-related term and one related to order/disorder. The energy-related term is called enthalpy, H, and the order/disorder term called entropy, S.

For mixing to occur and for polymers to be dissolved it is essential that the change in free energy, ΔG, which is the driving force in the solution and mixing processes, decrease to below zero, that is, be negative. ΔH and ΔS are equal to the change in enthalpy and change in entropy and for constant temperature the relationship is the classical Gibbs equation.

$$\Delta G = \Delta H - T\,\Delta S \tag{3.1}$$

"Like-likes-like best of all" is a description that is useful at appropriate times in science. It is true of solubility. Thus, water-likes-water best of all and is infinitely soluble in itself. Hexane-likes-hexane best of all and is infinitely soluble in itself. Hexane and water are not soluble in one another because hexane is nonpolar and water is polar; thus they are not "like one another." In solubility, and in fact all of mixing, the ΔH term is always unfavorable when mixing or when solubility occurs. (Shortly, we will deal with attempts such as the CED and solubility parameter to minimize the unfavorable aspect of the ΔH term.) Thus, it is the ΔS term that allows mixing and solubility to occur. As seen in Figure 3.3 the amount of randomness or disorder gain is great when pure materials such as water and ethanol is changed from the ordered pure materials to the disordered mixture.

By comparison, the increase in randomness, ΔS, is much smaller if one of the materials is a polymer since the possible arrangement of the polymer chains is much more limited because the polymer units are attached to one another and not free to simply move about on their own. Figure 3.4 illustrates this with water and poly(ethylene glycol), PEG. We notice several aspects. First, as noted above, the number of arrangements of the PEG units is limited. Second, as in the case of an onion, each layer of PEG chains must be peeled back allowing water molecules to approach inner layers before entire solubility occurs and causes swelling. This results in polymer solubility often requiring a longer period of time, sometimes hours to week to months, in comparison to the solubility of smaller molecules where solubility can occur in seconds.

Polymer solubility, in comparison to small molecules, is

a. More limited with respect to the number of solvents as a result of the lower increase in randomness;
b. It is more limited with respect to the extent of solubility; and
c. Takes a longer time to occur.

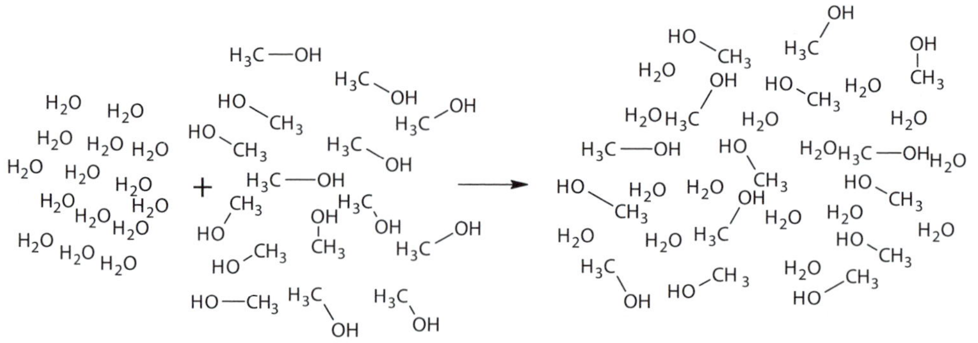

FIGURE 3.3 Illustration of the mixing of small molecules.

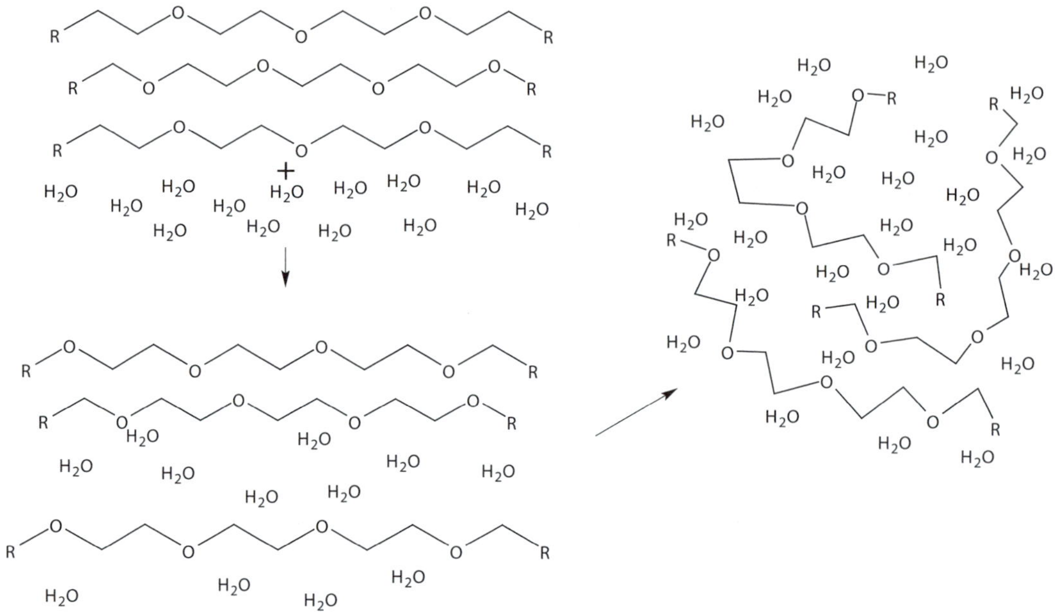

FIGURE 3.4 Illustration of water dissolving poly(ethylene glycol) (PEG).

There have been many attempts to describe the process of mixing/solubility of polymer molecules in thermodynamic terms. By assuming that the sizes of polymer segments were similar to those of solvent molecules, Flory and Huggins derived an expression for the partial molar Gibbs free energy of dilution that included the dimensionless Flory–Huggins interaction parameter, $X_1 = Z\Delta H/RT$, where Z is the lattice coordination number. It is now known that X_1 contains enthalpy and entropy contributions. The Flory–Huggins approach has been used to predict the equilibrium behavior between liquid phases containing an amorphous polymer. The theory has also been used to predict the cloud point, which is just below the critical solution temperature, T_c, where two phases coalesce. The Flory–Huggins interaction parameter has been used as a measure of solvent power. In general, the value of X_1 is 0.5 for poor solvents and decreases for good solvents.

Some limitations of the Flory–Huggins lattice theory were overcome by Flory and Krigbaum who assumed the presence of an excluded volume. The excluded volume is the volume occupied by a polymer chain including long-range intramolecular interactions. The long-range interactions were described in terms of free energy by introducing the enthalpy term, K_i, and entropy term, ψ_1. The entropy and enthalpy terms are equal when $\Delta G = 0$. The temperature where $\Delta G = 0$ is called the theta temperature. The *theta (θ) temperature* is the lowest temperature at which a polymer of infinite molecular weight is completely miscible with a specific solvent. The coil expands above the theta temperature and contracts at lower temperatures.

Physical properties of polymers, including solubility, are related to the strength of covalent bonds, the stiffness of the segments in the polymer backbone, the amount of crystallinity/amorphous, and the intermolecular forces between the polymer chains. The strength of the intermolecular forces is directly related to the CED, which is the molar energy of vaporization per unit volume. Since intermolecular attractions of solvent and solute must be overcome when a solute (here the polymer) dissolves, CED values may be used to predict solubility.

When a polymer dissolves, the first step is often a slow-swelling process called *solvation* in which the polymer molecules swell by a factor δ, which is related to CED. Linear and branched polymers dissolve in a second step, but network polymers remain in a swollen condition. In the dissolving process, external polymers are initially "dissolved" exposing additional polymer chains to the solvent, and so on eventually resulting in the entire polymer mass being dissolved. Thus, polymer solubility often takes considerably longer than the solubility of smaller molecules.

As early as 1926, Hildebrand showed a relationship between solubility and the internal pressure of the solvent and in 1931, Scatchard incorporated the CED concept into Hildebrand's equation. This led to the concept of a *solubility parameter, δ*, which is the square root of CED. Thus, as shown below, the solubility parameter, δ, for nonpolar solvents is equal to the square root of the heat of vaporization per unit volume.

$$\delta = \left(\frac{\Delta E}{V} \right)^{1/2} = (\text{CED})^{1/2} \quad \text{or} \quad \delta^2 = \text{CED} \tag{3.2}$$

According to Hildebrand, the heat of mixing a solute and a solvent is proportional to the square of the difference in solubility parameters, as shown below, where φ is the partial volume of each component, namely, solvent γ_1 and solute ϕ_2. Since, typically, the entropy term favors solution and the enthalpy term acts counter to the solution, the objective is to match solvent and solute so that the difference between their δ values is small, resulting in a small enthalpy acting against solubility occurring.

$$\Delta H_m = \phi_1 \phi_2 (\delta_1 - \delta_2)^2 \tag{3.3}$$

The solubility parameter concept predicts the heat of mixing liquids and amorphous polymers. It has been experimentally found that generally any nonpolar amorphous polymer will dissolve in a liquid or mixture of liquids having a solubility parameter that generally does not differ by more than ± 1.8 (cal/cc)$^{1/2}$. The Hildebrand with units of (cal/cc)$^{1/2}$ is preferred over these complex units giving for the previous expression ± 1.8 H.

The solubility parameter concept is based on obtaining a negative Gibbs' free energy. Thus, as the term ΔH_m approaches zero, ΔG will have the negative value required for solution to occur because the entropy term favors solution occurring. As noted before, the entropy (S) increases in the solution process hence the emphasis is on achieving low values of ΔH_m.

For nonpolar solvents, which were called regular solvents by Hildebrand, the solubility parameter is equal to the square root of the difference between the enthalpy of evaporation (H_v) and the product of the ideal gas constant (R) and the Kelvin (or Absolute) temperature (T) divided by the molar volume (V), as shown in the following equation:

$$\delta = \left(\frac{\Delta E}{V} \right)^{1/2} = \left(\frac{\Delta H_v - RT}{V} \right)^{1/2} \tag{3.4}$$

Since it is difficult to measure the molar volume, its equivalent the molecular weight (M) divided by density (D), is substituted for V as shown in the following equation:

$$\delta = D\left(\frac{\Delta H_v - RT}{M}\right)^{1/2}$$

(3.5)

Because the law of mixtures applies to the solubility parameter, it is possible to easily calculate the solubility parameter of blended liquids forming mixtures that can serve as solvents. For example, an equal molar mixture of n-pentane (δ = 7.1 H) and n-octant (δ = 7.6 H) will have a solubility parameter value of 7.35 H (simply [7.1 + 7.6]/2).

The solubility parameter of a polymer is generally determined by noting the extent of swelling or actual solution of small amounts of polymer in a series of solvents and comparing the solubility values of the ones that swell or dissolve the polymer and assigning the polymer a solubility parameter value that is close to the solvents that dissolve/swell the polymer. The solubility parameter can also be determined by adding a nonsolvent to a polymer solution and by noting the amount of nonsolvent needed to begin to precipitate the polymers.

Since polar forces are present in polar solvents and polar molecules, this must be considered when estimating solubilities with such "nonregular" solvents and polymers. Hydrogen bonding, a special case of secondary polar bonding, is also important for some solvents and polymers and again will influence the solubility parameters. Thus, special solubility values have been developed for polar and hydrogen-bonding solvents (Tables 3.1 and 3.2).

Plasticizers help the flexibility of polymers and are chosen so that they do not dissolve the polymer but rather allow segmental mobility to occur. Through experience, it is found that the solubility

TABLE 3.1
Solubility Parameters (δ) for Typical Solvents

Poor Hydrogen Bonding		Moderate Hydrogen Bonding		Strong Hydrogen Bonding	
Dimethylsiloxane	5.5	Diisopropyl ether	6.9	Diethylamine	8.0
Difluorodichloromethane	5.5	Diethylether	7.4	n-Amylamine	8.7
Neopentane	6.3	Isoamyl acetate	7.8	2-Ethylhexanol	9.5
Nitro-n-octane	7.0	Diisobutyl ketone	7.8	Isoamyl alcohol	10.0
n-Pentane	7.0	Di-n-propylether	7.8	Acetic acid	10.1
n-Octane	7.6	sec-Butyl acetate	8.2	m-Cresol	10.2
Turpentine	8.1	Isopropyl acetate	8.4	Analine	10.3
Cyclohexane	8.2	Methylamyl ketone	8.5	n-Octyl alcohol	10.3
Cymene	8.2	Ethyl acetate	9.0	t-Butyl alcohol	10.9
Carbon tetrachloride	8.6	Methyl ethyl ketone	9.3	n-Amyl alcohol	10.9
n-Propylbenzene	8.6	Butyl cellosolve	9.5	n-Butyl alcohol	11.4
p-Chlorotoluene	8.8	Methyl acetate	9.6	Isopropyl alcohol	11.5
Decalin	8.8	Dichloroethylether	9.8	Diethylene glycol	12.1
Xylene	8.8	Acetone	9.9	Furfuryl alcohol	12.5
Benzene	9.2	Dioxane	10.0	Ethanol	12.7
Styrene	9.3	Cyclopentanone	10.4	N-Ethylformamide	13.9
Tetraline	9.4	Cellosolve	10.5	Methanol	14.5
Chlorobenzene	9.5	N,N-Dimethylacetamide	10.8	Ethylene glycol	14.6
Ethylene dichloride	9.8	1,2-Propylene carbonate	13.3	Glycerol	16.5
p-Dichlorobenzene	10.0	Ethylene carbonate	14.7	Water	23.4
Nitroethane	11.1				
Acetonitrile	11.9				
Nitroethane	12.7				

TABLE 3.2
Approximate Solubility Parameter Values for Polymers

Polymer	Poorly H-Bonding	Moderately H-Bonding	Strongly H-Bonding
Polytetrafluoroethylene	5.8–6.4		
Poly(vinyl ethyl ether)	7.0–11.0	7.4–10.8	9.5–14.0
Poly(butyl acrylate)	7.0–12.5	7.4–11.5	
Poly(butyl methacrylate)	7.4–11.0	7.4–10.0	9.5–11.2
Polyisobutylene	7.5–8.0		
Polyethylene	7.7–8.2		
Poly(vinyl butyl ether)	7.8–10.6	7.5–10.0	9.5–11.2
Natural rubber	8.1–8.5		
Polystyrene	8.5–10.6	9.1–9.4	
Poly(vinyl acetate)	8.5–9.5		
Poly(vinyl chloride)	8.5–11.0	7.8–10.5	
Buna N	8.7–9.3		
Poly(methyl methacrylate)	8.9–12.7	8.5–13.3	
Poly(ethylene oxide)	8.9–12.7	8.5–14.5	9.5–14.5
Poly(ethylene sulfide)	9.0–10.0		
Polycarbonate	9.5–10.6	9.5–10.0	
Poly(ethylene terephthate)	9.5–10.8	9.3–9.9	
Polyurethane	9.8–10.3		
Polymethacrylonitrile		10.6–11.0	
Cellulose acetate	11.1–12.5	10.0–14.5	
Nitrocellulose	11.1–12.5	8.0–14.5	12.5–14.5
Polyacrylonitrile		12.0–14.0	
Poly(vinyl alcohol)			12.0–13.0
Nylon-6,6			13.5–15.0
Cellulose			14.5–16.5

parameter differences between the plasticizer and polymer should be less than 1.8 H for there to be compatibility between the plasticizer and polymer.

Because the heat of vaporization of a polymer is not readily obtained, small determined values for various components of a polymer chain that can be employed to calculate the solubility parameter. These values are called molar attraction constants and are additive and have been used for estimation of the solubility parameter for nonpolar polymers. In this approach $\delta = D\Sigma G/M$, where D is the density, G is the small molar attraction constants, and M is the molecular weight of the particular repeat unit. As expected, the more polar units have greater G values while the less polar units have smaller G values.

3.3 AVERAGE MOLECULAR WEIGHT VALUES

Small molecules, such as benzene and glucose, have precise structures such that each molecule of benzene has six carbon atoms and six hydrogen atoms. By comparison, each molecule of poly-1,4-phenylene may have a differing number of phenylene-derived units, while single molecules (single chains) of polyethylene may vary in the number of ethylene units, the extent and frequency of branching, the distribution of branching, and the length of branching. A few polymers, such as nucleic acids and many proteins, consist of molecules, individual polymer chains that must not vary, so they have a precise molecular weight.

While there are several statistically described averages, we will concentrate on the two that are most germane to polymers—number-average and weight-average. These are averages based

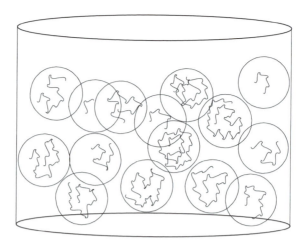

FIGURE 3.5 Jar with capsules, each capsule containing a single polymer chain where the capsule size is the same and independent of the chain size, illustrating the number-average dependence on molecular weight.

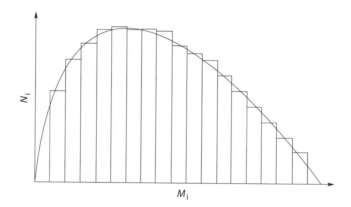

FIGURE 3.6 Molecular weight distribution for a polydisperse polymer sample constructed from "capsule-derived" data for the number-average situation such as given in Figure 3.5.

on statistical approaches that can be described mathematically and which correspond to physical measurements of specific values.

The number average value, corresponding to a measure of chain length average, is called the *number-average molecular weight*, $\overline{M}_n$. Physically, the number-average molecular weight can be measured using any technique that "counts" the molecules; that is, it is directly dependent on the number of chains. These techniques include vapor phase and membrane osmometry, freezing point lowering, boiling point elevation, and end-group analysis.

We can describe the number average using a jar filled with plastic capsules such as those used to contain tiny prizes (Figure 3.5). Here, each capsule contains one polymer chain. All the capsules are of the same size, regardless of the size of the polymer chain. Capsules are then withdrawn, opened, and the individual chain length measured and recorded. A graph such as that shown in Figure 3.6 can be constructed from the data with the maximum value being the number-average molecular weight. The probability of drawing a particular capsule is dependent on the *number* of each capsule and not on the size of the chain within the capsule.

The weight-average molecular weight is similarly described, except that the capsule size corresponds to the size of polymer chain contained within it (Figure 3.7). In this approach, the probability

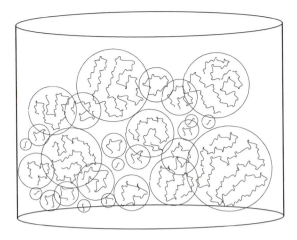

FIGURE 3.7 Jar with capsules, each containing a single polymer chain where the capsule size is directly related to the size of the polymer chain within the capsule.

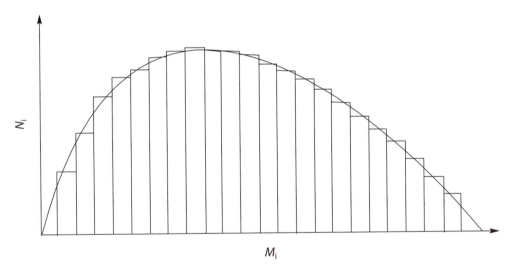

FIGURE 3.8 Molecular weight distribution for a polydisperse polymer sample constructed from "capsule-derived" data for the weight-average situation.

of drawing out a particular chain length is dependent on the *size* of the capsule. Larger chains have a greater probability (at least in this exercise) of being drawn out because they are larger and are contained within larger capsules. Again, a graph is constructed and the maximum value is the weight-average molecular weight (Figure 3.8).

Compare Figure 3.6 with Figure 3.8. Notice that the maximum occurs at a higher molecular weight for the weight-average situation. The area of the curve should be the same and the M_i ordinate is longer reflecting the extension of the molecular weight for the weight-average situation.

Several mathematical moments (about a mean) can be described using the differential or frequency distribution curve, and these can be described by equations. The first moment is the number-average molecular weight, $\overline{M}_n$. Any measurement that leads to the number of molecules, functional groups, end groups, or particles that are present in a given weight of sample allows the calculation of $\overline{M}_n$. The $\overline{M}_n$ is calculated like any other numerical average by dividing the sum of the individual molecular weight values by the number of molecules. Thus, $\overline{M}_n$ for three molecules having molecular weights of 1.00×10^5, 2.00×10^5, and 3.00×10^5 is $(6.00 \times 10^5)/3 = 2.00 \times 10^5$. Recalling that $W = \Sigma W_i = \Sigma M_i N_i$ the general solution is shown mathematically as

$$\overline{M}_n = \frac{\text{total weight of sample}}{\text{number of molecules of } N_i} = \frac{W}{\Sigma N_i} = \frac{\Sigma M_i N_i}{\Sigma N_i} \tag{3.6}$$

Most thermodynamic properties are related to the number of particles present and thus are dependent on $\overline{M}_n$.

Colligative properties are dependent on the number of particles present and are thus related to $\overline{M}_n$. $\overline{M}_n$ values are independent of molecular size and are highly sensitive to small molecules present in the mixture. Values of $\overline{M}_n$ are determined by Raoult's techniques that are dependent on colligative properties such as ebulliometry (boiling point elevation), cryometry (freezing point depression), osmometry, and end-group analysis.

Weight-average molecular weight, $\overline{M}_w$, is determined from experiments in which each molecule or chain makes a contribution to the measured result relative to its size. This average is more dependent on the number of longer chains than is the number-average molecular weight, which is dependent simply on the total number of each chain.

The $\overline{M}_w$ is the second moment average and is shown mathematically as

$$\overline{M}_w = \frac{\Sigma W_i M_i}{\Sigma W_i} = \frac{\Sigma M_i^2 N_i}{\Sigma M_i N_i} \tag{3.7}$$

Thus, the $\overline{M}_w$ of the three chains cited above is

$$\frac{(1.00 \times 10^{10}) + (4.00 \times 10^{10}) + (9.00 \times 10^{10})}{(6.00 \times 10^5)} = 2.33 \times 10^5$$

Bulk properties associated with large deformations, such as viscosity and toughness, are most closely associated with $\overline{M}_w$. $\overline{M}_w$ values are most often determined by light-scattering photometry.

However, melt elasticity is more closely related to the third moment known as the z-average molecular weight, $\overline{M}_z$. The $\overline{M}_z$ is most often determined using either light-scattering photometry or ultracentrifugation. It is shown mathematically as

$$\overline{M}_z = \frac{\Sigma M_i^3 N_i}{\Sigma M_i^2 N_i} \tag{3.8}$$

The $\overline{M}_z$ value for the three polymer chains cited above is 2.57×10^5:

$$\frac{(1 \times 10^{15}) + (8 \times 10^{15}) + (27 \times 10^{15})}{[(1 \times 10^{10}) + (4 \times 10^{10}) + (9 \times 10^{10})]} = 2.57 \times 10^5$$

While $Z + 1$ and higher average molecular weight values can be calculated, the major interests are in $\overline{M}_n$, $\overline{M}_v$, $\overline{M}_w$, and $\overline{M}_z$, which is the order of increasing size for a heterodisperse polymer sample as shown in Figure 3.9. Thus, for heterogeneous molecular weigh systems $\overline{M}_z > \overline{M}_w > \overline{M}_n$. The ratio of $\overline{M}_w/\overline{M}_n$ is called the *polydispersity index*. The most probable polydispersity index for polymers produced by the condensation technique with respect to molecular weight is 2. As the heterogeneity decreases, the various molecular weight values converge until $\overline{M}_z = \overline{M}_w = \overline{M}_n$.

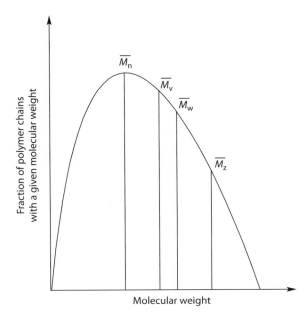

FIGURE 3.9   Molecular weight distributions.

Typical techniques for determining molecular weight are given in Table 3.3. The most popular techniques will be considered briefly. All classical molecular weight techniques require dilute solutions, generally 0.01 g/mL or 1% (1 g/100 mL) solutions. To further minimize solute interactions, extrapolation of the measurements to infinite dilution is normal practice.

For polydisperse polymer samples, measurements that lead directly to the determination of the molecular weight, such as light-scattering photometry and membrane osmometry, are referred to as "absolute molecular weight" methods. Techniques such as viscometry are not absolute molecular weight methods because they require calibration using an absolute molecular weight technique.

3.4   FRACTIONATION OF POLYDISPERSE SYSTEMS

The data plotted in Figure 3.9 was obtained by the fractionation of a polydisperse polymer sample. Polydisperse polymers can be fractionated by a number of techniques. The most widely used technique is chromatography. Other methods include addition of a nonsolvent to a polymer solution, cooling a polymer solution, solvent evaporation, extraction, diffusion, or centrifugation. The molecular weight of the fractions may be determined using any of the classic techniques given in Table 3.3.

Fractional precipitation is dependent on the slight change in the solubility with molecular weight. When a small amount of miscible nonsolvent is added to a polymer solution the product with the highest molecular weight precipitates first. The procedure is repeated after the precipitate is removed. Molecular weights are run for each fraction and a curve developed that is similar to Figure 3.9.

3.5   CHROMATOGRAPHY

As noted before, certain techniques such as colligative methods, light-scattering photometry, special mass spectral (MS) techniques, and ultracentrifugation allow the calculation of specific or absolute molecular weights. Under certain conditions, some of these also allow the calculation of the MWD.

There are a wide variety of chromatography techniques, including paper and column techniques. Chromatographic techniques involve passing a solution containing the to-be-tested sample through

TABLE 3.3
Typical Molecular Weight Determination Methods

Method	Type of Molecular Weight Average	Applicable Weight Range	Other Information
Light scattering	$\overline{M}_w$	To ∞	Can give other molecular weights and shape
Membrane osmometry	$\overline{M}_n$	10^4–10^6	
Vapor phase osmometry	$\overline{M}_n$	To 4×10^4	
Electron and X-ray microscopy	$\overline{M}_{n,w,z}$	10^2 to ∞	
Isopiestic method	$\overline{M}_n$	To 2×10^4	
Ebulliometry (BP elevation)	$\overline{M}_n$	To 4×10^4	
Cryoscopy (MP depression)	$\overline{M}_n$	To 5×10^4	
End-group analysis	$\overline{M}_n$	To 2×10^4	
Osmodialysis	$\overline{M}_n$	500–2.5×10^4	
Centrifugation			
Sedimentation equilibrium	$\overline{M}_z$	To ∞	
Archibald mod.	$\overline{M}_{z,w}$	To ∞	
Trautman's method	$\overline{M}_w$	To ∞	
Sedimentation velocity gives real M for only monodisperse systems			
Chromatography	Calibrated	To ∞	Gives molecular weight distribution
SAXS	$\overline{M}_w$		
Mass Spectrometry MALDI		To 10^7	Molecular weight distribution
Viscometry	Calibrated		
Coupled chromatography—LS		To ∞	Molecular weight distribution, shape, $\overline{M}_{n,w,z}$

"To ∞" means that the molecular weight of the largest particles soluble in a suitable solvent can be, in theory, determined.

a medium that shows selective absorption for the different components in the solution. *Ion-exchange chromatography* separates molecules on the basis of their electrical charge. Ion-exchange resins are either polyanions or polycations. For a polycation resin, those particles that are least attracted to the resin will flow more rapidly through the column and be emitted from the column first. This technique is most useful for polymers that contain changed moieties.

In *affinity chromatography*, the resin contains molecules that are especially selected and will interact with the particular polymer(s) that is being studied. Thus, for a particular protein, the resin may be modified to contain a molecule that interacts with that protein type. The solution containing the mixture is passed through the column and the modified resin preferentially associates with the desired protein allowing it to be preferentially removed from the solution. Later, the protein is washed through the column by addition of a salt solution and collected for further evaluation.

In *high-performance liquid chromatography* (HPLC), pressure is applied to the column that causes the solution to rapidly pass through the column allowing procedures to be completed in a fraction of the time in comparison to regular chromatography.

When an electric field is applied to a solution, polymers containing a charge will move toward either the cathode (positively charged species) or toward the anode (negatively charged species).

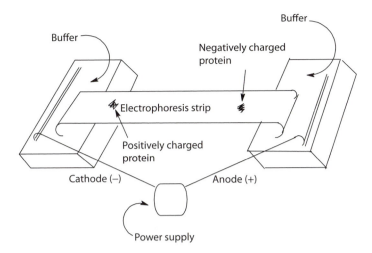

FIGURE 3.10 Basic components of an electrophoresis apparatus.

This migration is called *electrophoresis*. The velocity at which molecules move is mainly dependant upon the electric field and charge on the polymer driving the molecule toward one of the electrodes, and a frictional force dependant on the size and structure of the macromolecule that opposes the movement. In general, the larger and more bulky the macromolecule the greater the resistance to movement and the greater the applied field and charge on the molecule the more rapid the movement. While electrophoresis can be conducted on solutions, it is customary to use a supporting medium of a paper or gel. For a given system, it is possible to calibrate the rate of flow with the molecular weight and/or size of the molecule. Here, the flow characteristics of the calibration material must be similar to those of the unknown.

Generally though, electrophoresis is often employed in the separation of complex molecules such as proteins where the primary factor in the separation is the charge on the species. Some amino acids such as aspartic acid and glutamic acid contain an "additional" acid functional group, while amino acids such as lysine, arginine, and histidine contain "additional" basic groups. The presence of these units will confer to the protein tendencies to move toward the anode or cathode. The rate of movement is dependent on a number of factors, including the relative abundance and accessibility of these acid and base functional groups.

Figure 3.10 contains an illustration of the basic components of a typical electrophoresis apparatus. The troughs at either end contain an electrolyte buffer solution. The sample to be separated is placed in the approximate center of the electrophoresis strip.

Gel permeation chromatography (GPC) is a form of chromatography that is based on separation by molecular size rather than chemical properties. GPC or *Size exclusion chromatography* (SEC) is widely used for molecular weight and MWD determination. In itself, SEC does not give an absolute molecular weight and must be calibrated against polymer samples whose molecular weight has been determined by a technique that does give an absolute molecular weight.

Size exclusion chromatography is an HPLC technique whereby the polymer chains are separated according to differences in hydrodynamic volume. This separation is made possible by the use of special packing material in the column. The packing material is usually polymeric porous spheres often composed of polystyrene cross-linked by addition of varying amounts of divinylbenzene. Retention in the column is mainly governed by the partitioning (or exchanging) of polymer chains between the mobile (or eluent) phase flowing through the column and the stagnate liquid phase that is present in the interior of the packing material. The column-packing approaches being spherical. In reality, the packing material has various clefts that ensnare the passing polymer chains such that the progress down the column of smaller chains is preferentially slowed (Figure 3.11).

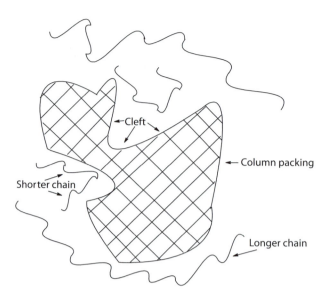

FIGURE 3.11 Illustration of column packing and chains.

Through control of the amount of cross-linking, nature of the packing material and specific processing procedures, spheres of widely varying porosity are available. The motion in and out of the stationary phase is dependent on a number of factors, including Brownian motion, chain size, and conformation. The latter two are related to the polymer chain's hydrodynamic volume—the real, excluded volume occupied by the polymer chain. Since smaller chains preferentially permeate the gel particles, the largest chains are eluted first. As noted above, the fractions are separated on the basis of size.

The resulting chromatogram is then a reflection of the molecular size distribution. The relationship between molecular size and molecular weight is dependant on the conformation of the polymer in solution. As long as the polymer conformation remains constant, which is generally the case, molecular size increases with increase in molecular weight. The precise relationship between molecular size and molecular weight is conformation dependant. For random coils, molecular size as measured by the polymer's radius of gyration, R, and molecular weight, M, R is proportional to M^b, where "b" is a constant dependent on the solvent, polymer concentration, and temperature. Such values are known and appear in the literature for many polymers allowing the ready conversion of molecular size data collected by SEC into molecular weight and MWD.

Figure 3.12 contains the results of a polymer separation using SEC. Here, two different polymer samples are initially added (far left). One polymer sample is a relatively low molecular weight sample with a fairly homogeneous chain size distribution. The second polymer contains longer chains with a broader MWD. As time elapses (second from the left) the two different samples separate with the sample containing the shorter chains moving more rapidly down the column. This separation occurs between the samples and within each polymer sample. Finally, the polymer chains containing the lower molecular weight sample emerge and are recorded as a relatively sharp band (third column). Then the sample containing the longer chains emerges giving a relatively broader band (fourth column).

There is a wide variety of instrumentation ranging from simple manually operated devices to completely automated systems. Briefly, the polymer-containing solution and solvent alone are introduced into the system and pumped through separate columns at a specific rate. The differences in refractive index between the solvent itself and polymer solution are determined using a differential refractometer. This allows calculation of the amount of polymer present as the solution passes out of the column.

The unautomated procedure was first used to separate protein oligomers using Sephadex gels. Today, there are a wide variety of specialized and general gels used as column packing. The efficiency of these packed columns can be determined by calculating the height in feet equivalent to a

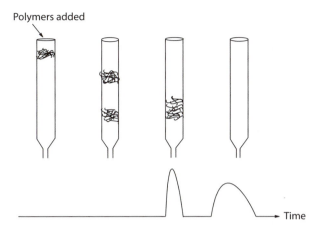

FIGURE 3.12 Illustration of a size-exclusion experiment. Column separations are given above with the SEC output given below.

theoretical plate (HETP), which is the reciprocal of the plate count per feet (*P*). *P* is directly proportional to the square of the elution volume (*V_c*) and inversely proportional to the height of the column in feet and the square of the baseline (*D*) as follows:

$$p = \left(\frac{16}{f} \right) \left[\left(\frac{V_c}{D} \right)^2 \right]$$ (3.9)

Conversion of retention volume for a given column to molecular weight can be accomplished using several approaches, including peak position, universal calibration, broad standard, and actual molecular weight determination by coupling the SEC to an instrument that gives absolute molecular weight.

In the *peak position* approach, well-characterized narrow fraction samples of known molecular weight are used to calibrate the column and retention times determined. A plot of log *M* versus retention is made and used for the determination of samples of unknown molecular weight. Unless properly treated, such molecular weights are subject to error. The best results are obtained when the structures of the samples used in the calibration and those of the test polymers are the same.

The *universal calibration* approach is based on the product of the limiting viscosity number (LVN) and molecular weight being proportional to the hydrodynamic volume. Benoit showed that for different polymers elution volume plotted again the log LVN times molecular weight gave a common line. In one approach, molecular weight is determined by constructing a "universal calibration line" through plotting the product of log LVN for polymer fractions with narrow MWD as a function of the retention of these standard polymer samples for a given column. Molecular weight is then found from retention time of the polymer sample using the calibration line.

Probably the most accurate approach is to directly connect, couple, the SEC to a device, such as a light-scattering photometer, that directly measures the molecular weight for each elution fraction. Here both molecular weight and MWD are accurately determined.

3.6 COLLIGATIVE MOLECULAR WEIGHTS

3.6.1 OSMOMETRY

A measure of any of the colligative properties involves counting solute (polymer) molecules in a given amount of solvent. The most common technique for polymers is membrane osmometry. The technique is based on the use of a semipermeable membrane through which solvent molecules freely pass, but though which the large polymer molecules are unable to pass. Existing membranes only approximate ideal semipermeability, the chief limitation being the passage of low molecular weight chains through the membrane.

There is a thermodynamic drive toward dilution of the polymer-containing solution with a net flow of solvent toward the cell containing the polymer. This results in an increase in liquid in that cell causing a rise in the liquid level in the corresponding measuring tube. This rise in liquid level is opposed and balanced by a hydrostatic pressure resulting in a difference in the liquid levels of the two measuring tubes. The difference is directly related to the osmotic pressure of the polymer-containing solution. Thus, solvent molecules pass through the semipermeable membrane reaching a "static" equilibrium.

Since osmotic pressure is dependent on the number of particles present, the measurement of this osmotic pressure can be used to determine the $\overline{M}_n$ of the dissolved polymer. The difference in height (Δh) of the liquids in the columns is converted to osmotic pressure (π) by multiplying the gravity (g) and the density of the solution (ρ), that is, $\pi = \Delta h \rho g$. In the old static osmometers, it might take weeks to months for equilibrium to become established allowing excessive passage of polymer chains through the membrane. Today, automated osmometers allow molecular-weigh measurements to occur in minutes with a minimal of passage of polymer chains through the membrane. The relationship between molecular weight and osmotic pressure is given in the following van't Hoff equation:

$$\pi = \frac{RTC}{\overline{M}_n} + BC^2 \tag{3.10}$$

Thus, the reciprocal of $\overline{M}_n$ is the intercept when data for π/RTC versus C are extrapolated to zero concentration (Figure 3.13).

The slop of the lines in Figure 3.13, that is, the virial constant B, is related to the CED. The value for B would be zero at the theta temperature. Since this slope increases as the solvency increases, it is advantageous to use a dilute solution consisting of a polymer and a poor solvent to minimize extrapolation errors.

In the vapor phase osmometry technique (VPO), drops of solvent and solution are placed in an insulated chamber in proximity to thermistor probes. Since the solvent molecules evaporate more rapidly from the solvent than from the polymer solution, a temperature difference results that is related to the molarity of the polymer (M) can be determined if the heat of vaporization per gram of solvent (λ) is known using the following relationship.

$$\Delta T = \frac{RT^2 M}{\lambda 100} \tag{3.11}$$

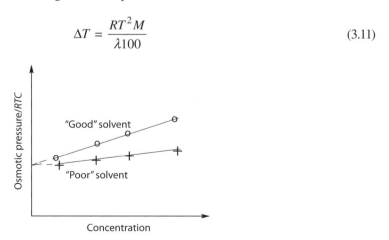

FIGURE 3.13 Plots of osmotic pressure, π, divided by RTC used to determine $1/\overline{M}_n$ in osmometry.

3.6.2 END-GROUP ANALYSIS

In cases where the end groups are known and their concentration can be determined, knowledge of their abundance allows a determination of $\overline{M}_n$. The sensitivity of this method decreases and the chain length becomes greater. Some end groups can be determined using spectroscopic techniques and other through titration.

3.6.3 EUBLLIOMETRY AND CRYOMETRY

Raoult's law works for small polymers as well as small molecules. Determination of $\overline{M}_n$ is based for both eublliometry (boiling point elevation) and cryometry (freezing point lowering) on the Clausius–Clapeyron equation:

$$\overline{M}_n = \left(\frac{RT^2V}{\Delta H} \right) \left(\frac{C}{\Delta T} \right)_{C \to 0} \tag{3.12}$$

By use of sensitive thermocouples and care, molecular weights to about 50,000 Da can be determined.

3.7 LIGHT-SCATTERING PHOTOMETRY

Ever watch a dog or young child chase moonbeams? The illumination of dust particles is an illustration of light scattering, not of reflection. Reflection is the deviation of incident light through one particular angle such that the angle of incidence is equal to the angle of reflection. Scattering is the radiation of light in all directions. Thus, in observing the moonbeam, the dust particle directs a beam toward you regardless of your angle in relation to the scattering particle. The energy scattered per second (scattered flux) is related to the size and shape of the scattering particle and to the scattering angle.

Scattering of light is all about us—the fact that the sky above us appears blue, the clouds white, and the sunset in shades of reds and oranges is a consequence of preferential scattering of light from air molecules, water droplets, and dust particles. This scattered light caries messages about the scattering objects.

The measurement of light scattering is the most widely used approach for the determination of $\overline{M}_w$. This technique is based on the optical heterogeneity of polymer solutions and was developed by Nobel Laureate Peter Debye in 1944.

Today, modern instruments utilize lasers as the radiation source because they provide a monochromatic, intense, and well-defined light source. Depending upon the size of the scattering object, the intensity of light can be essentially the same or vary greatly with respect to the direction of the oncoming radiation. For small particles the light is scattered equally independent of the angle the observer is to the incoming light. For larger particles the intensity of scattered light varies with respect to the angle of the observer to the incoming light. For small molecules at low concentrations this scattering is described in terms of the Raleigh ratio.

In 1871, Rayleigh showed that induced oscillatory dipoles were developed when light passed through gases and that the amount (intensity) of scattered light (τ) was inversely proportional to the fourth power of the wavelength of light. This investigation was extended to liquids by Einstein and Smoluchowski in 1908. These oscillations reradiate the light energy producing turbidity, that is, the Tyndall effect. Other sources of energy, such as X-rays or laser beams, may be used in place of visible light sources.

For light-scattering measurements, the total amount of the scattered light is deduced from the decrease in intensity of the incident beam, I_0, as it passes through a polymer sample. This can be described in terms of Beer's law for the absorption of light as follows:

$$\frac{I}{I_0} = e^{-tl} \tag{3.13}$$

where τ is the measure of the decrease of the incident-beam intensity per unit length (l) of a given solution and is called the turbidity of the solution.

The intensity of scattered light or turbidity (τ) is proportional to the square of the difference between the index of refraction (n) of the polymer solution and of the solvent (n_o), to the molecular weight of the polymer ($\overline{M}_w$) and to the inverse fourth power of the wavelength of light used (λ). Thus,

$$\frac{Hc}{\tau} = \frac{1}{\overline{M}_w P_\theta}\left(1 + 2Bc + Cc^2 + \cdots\right) \tag{3.14}$$

where the expression for the constant H and for τ is as follows:

$$H = \left[\frac{32\pi^2}{3}\right]\left[\frac{n_0^2\left(\dfrac{dn}{dc}\right)^2}{\lambda^4 N}\right] \quad \text{and} \quad \tau = K'n^2\left(\frac{i_{90}}{i_0}\right) \tag{3.15}$$

where n_0 = index of refraction of the solvent, n = index of refraction of the solution, c = polymer concentration, the viral constants B, C, and so on are related to the interaction of the solvent, P_θ is the particle-scattering factor, and N is Avogadro's number. The expression dn/dc is the specific refractive increment and is determined by taking the slope of the refractive index readings as a function of polymer concentration.

In the determination of $\overline{M}_w$, the intensity of scattered light is measured at different concentrations and at different angles (θ). The incident light sends out a scattering envelope that has four equal quadrants (Figure 3.14(a)) for small particles. The ratio of scattering at 45° compared with that for 135° is called the *dissymmetry factor* or dissymmetry ratio Z. The reduced dissymmetry factor Z_0 is the intercept of the plot of Z as a function of concentration extrapolated to zero concentration.

For polymer solutions containing polymers of moderate to low molecular weight, P_θ is 1 giving Equation 3.16. At low-polymer concentrations Equation 3.16 reduces to Equation 3.17 an equation for a straight line ($y = b + mx$) where the "c"-containing terms beyond the $2Bc$ term are small:

$$\frac{Hc}{\tau} = \frac{1}{\overline{M}_w}\left(1 + 2Bc + Cc^2 + \cdots\right) \tag{3.16}$$

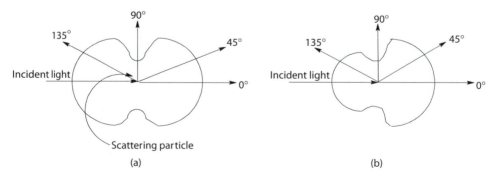

FIGURE 3.14 Light-scattering envelopes. Distance for the scattering particle to the boundaries of the envelope represents an equal magnitude of scattered light as a function of angle for a small-scattering particle (a) and a large-scattering particle (b).

$$\frac{Hc}{\tau} = \frac{1}{\overline{M}_w} + \frac{2Bc}{\overline{M}_w} \qquad (3.17)$$

Several expressions are generally used in describing the relationship between values measured by light-scattering photometry and molecular weight. One is given in Equation 3.14 and the others, such as Equation 3.18, are exactly analogous except that constants have been rearranged.

$$\frac{Kc}{R} = \frac{1}{\overline{M}_w}\left(1 + 2Bc + Cc^2 + \cdots\right) \qquad (3.18)$$

When the ratio of the concentration, c, to the turbidity, τ, (tau; related to the intensity of scattering at $0°$ and $90°$) multiplied by the constant H is plotted against concentration (Figure 3.15), the intercept of the extrapolated line is the reciprocal of $\overline{M}_w$ and the slope contains the viral constant B. Z_0 is directly related to the particle scattering factor, and both are related to both the size and shape of the scattering particle. As the size of the scattering particle, the individual polymer chain approaches about one-twentieth the wavelength of the incident light, scattering interference occurs giving a scattering envelope that is no longer symmetrical (Figure 3.14(b)). Here the scattering dependency on molecular weight reverts back to the relationship given in Equation 3.14.

The molecular weight for dilute polymer solutions is typically found using one of two techniques. The first technique is called the dissymmetrical method or approach because it utilizes the determination of Z_0 as a function of the particle-scattering factor as a function of polymer shape. $\overline{M}_w$ is determined from the intercept through substitution of the determined particle-scattering factor. The weakness in this approach is the necessity of having to assume a shape for the polymer in a particular solution. For small Z_0 values, choosing an incorrect polymer shape results in a small error, but for larger Z_0 values, the error becomes significant.

The second approach uses multiple detectors (Figure 3.16), allowing a double extrapolation to zero concentration and zero angle with the data forming what is called a *Zimm plot* (Figure 3.17). The extrapolation to zero angle corrects for finite particle size effects. The radius of gyration, related to polymer shape and size, can also be determined from this plot. The second extrapolation to zero concentration corrects for concentration factors. The intercepts of both plots is equal to $1/\overline{M}_w$.

The Zimm plot approach does not require knowing or having to assume a particular shape for the polymer in solution.

Related to the Zimm plot is the Debye plot. In the Zimm approach, different concentrations of the polymer solution are used. In the Debye approach, one low-concentration sample is used with $1/\overline{M}_w$ plotted against $\sin^2(\theta/2)$, essentially one-half of the Zimm plot.

Low-angle laser light-scattering photometry (LALLS) and multiangle low-angle laser light-scattering photometry (MALS) takes advantage of the fact that at low or small angles the scattering particle factor becomes one reducing Equation 3.14 to Equation 3.16 and at low concentrations to Equation 3.17.

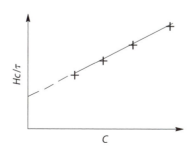

FIGURE 3.15 Typical simple plot used to determine $1/\overline{M}_w$ from scattering data.

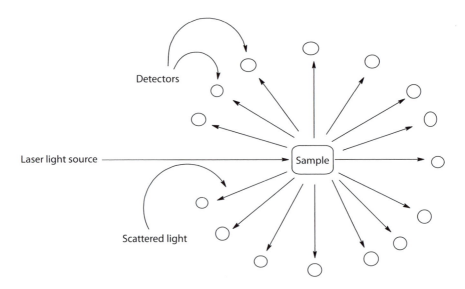

FIGURE 3.16 Multiple-detector arrangement showing a sample surrounded by an array of detectors.

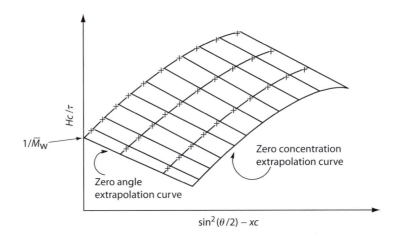

FIGURE 3.17 Zimm plot for a polymer scaled with a negative concentration coefficient (x) to improve data aesthetics and accessibility.

A number of automated systems exist with varying capabilities. Some internally carry out dilutions and refractive index measurements allowing molecular weight to be directly determined without additional sample treatment. The correct determination of dn/dc is very important since any error in its determination is magnified because it appears as the squared value in the expression relating light scattering and molecular weight.

Low-angle and multiangle light-scattering photometers are available that allow not only the determination of $\overline{M}_w$, but also additional values under appropriate conditions. For instance, a Zimm plot as shown in Figure 3.17 allows both $\overline{M}_w$ and $\overline{M}_n$ to be determined as well as the mean radius independent of the molecular conformation and branching.

These systems may also allow the determination of molecular conformation matching the radius and molecular weight to graphs showing the change in the root mean square radius of gyration (RMS) and molecular weight for different shaped molecules (Figure 3.18). The expression for the mean square radius of gyration is given as

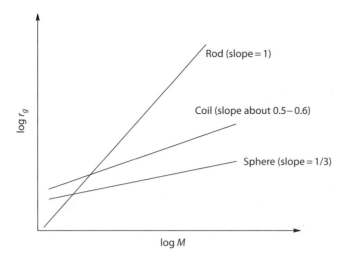

FIGURE 3.18 Standard plot of the log of the mean radium of gyration versus log smolecular weight for different-shaped macromolecules. Essentially, for a sphere the radius is proportion to the root-mean-square radius, rms radius, and $\overline{M}^{1/3}$ with a slope in the log r_g versus log $\overline{M}$ of 1/3; for rod-shaped polymers, length is proportional to rms radius and $\overline{M}$ with a slope of 1; and for random coils the end-to-end distance is proportional to the rms radius and $\overline{M}^{1/2}$ with a slope of about 0.5–0.6.

$$r_g^2 = \frac{\sum r_i^2 m_i}{\sum m_i} \tag{3.19}$$

One of the most important advances in polymer molecular weight determination is the "coupling" of SEC and light-scattering photometry, specifically LALLS or MALS. As noted in Section 3.5, SEC allows the determination of the MWD. In its usual operational configuration, it does not itself allow the calculation of an absolute molecular weight but relies on calibration with polymers of known molecular weight. By coupling HPLC and light-scattering photometry, the molecular weight of each fraction can be determined giving an MWD and various molecular weight values ($\overline{M}_w$, $\overline{M}_z$, $\overline{M}_n$).

The LALLS or MALS detector measures τ-related values, a differential refractive index (DRI) detector is used to measure concentration, and the SEC supplies samples containing "fractionated" polymer solutions allowing both molecular weight and MWD to be determined. Further, polymer shape can be determined. This combination represents the most powerful, based on ease of operation, variety of samples readily used, cost, means to determine polymer size, shape, and MWD available today.

A general assembly for a SEC-MALS instrument is given in Figure 3.19. A typical three-dimensional plot obtained from such an assembly is shown as Figure 3.20.

Dynamic light scattering is similar in principle to typical light scattering. When several particles are hit by oncoming laser light, a spotted pattern appears, the spots originating from the interference between the scattered light from each particle giving a collection of dark (from destructive interference) and light (from constructive interference) spots. This pattern of spots varies with time because of the Brownian motion of the individual scattering particles. The rate of change in the pattern of spots is dependant on a number of features, including particle size. In general, the larger the particle the slower the Brownian motion and, consequently, the slower the change in the pattern. Measurement of these intensity fluctuations with time allows the calculation of the translational diffusion constant of the scattering particles. The technique for making these measurements is given several names, including DLS emphasizing the fact that it is the difference in the scattered light with

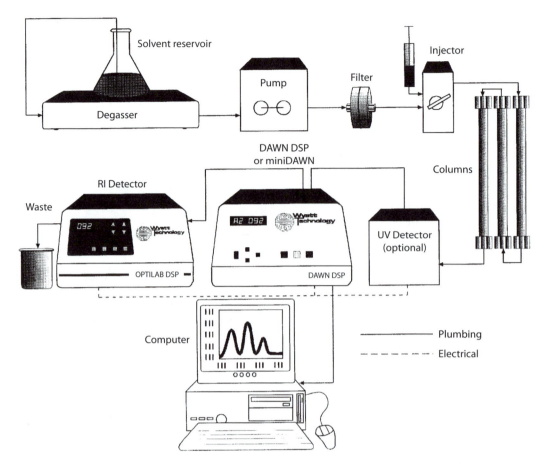

FIGURE 3.19 Typical SEC-MALS setup, including refractive-index refractometer. (Used with permission of Wyatt Technology Corporation, Santa Barbara, CA 93117; wyatt@wyatt.com)

time that is being measured; photon correlation spectroscopy (PCS) with the name emphasizing the particular mathematical technique employed to analyze the light-scattering data; and quasielastic-light scattering (QELS) with name emphasizing the fact that no energy is lost between the collision between the particle and the light photon.

The effect of subtle particle changes as a function of temperature, sample preparation, time, solvent, and other changes can be measured using DLS. Such changes can then be related to performance variations eventually interrelating structure shape and biological/physical property.

Another variation that is useful employing coupled light scattering is referred to as a *triple detection set*. The set consists of three detectors—a light-scattering detector, a differential-refractometer detector, and a capillary-differential viscometer. It functions in concert with a GPC and light-scattering source. The GPC separates the polymer mixture into molecular weight fractions. According to the Einstein equation the intrinsic viscosity times the molecular weight is equal to the hydrodynamic volume or size of polymers in solution. Thus, the molecular weight is determined using light-scattering photometry; viscometry gives the intrinsic viscosity, and the equation is solved for size.

In general, light-scattering photometry has a limit to determining molecular size with the lower limit being about 10 nm. The addition of the viscometer allows molecular sizes to be determined for oligomeric materials to about 1 nm. The assembly allows an independent measure of size and molecular weight as well as additional conformational and aggregation information,

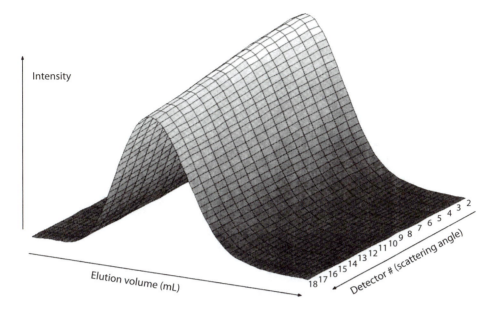

FIGURE 3.20 Three-dimensional plot of scattering intensity as a function of scattering angle and elution volume for a broad MWD polystyrene (NITS standard reference 706). (Used with permission of Wyatt Technology Corporation, Santa Barbara, CA 93117; wyatt@wyatt.com)

including small conformational changes. The assembly also allows good molecular determination to occur even when there are small dn/dc values, low molecular weight fractions, absorbing and fluorescent polymers, copolymers with varying dn/dc values, and chiral polymers that depolarize the incident beam.

Figure 3.21 contains data on myoglobin obtained using a triple detection setup. A molecular weight of 21,100 is found with a viscosity of 0.0247 dl/g and from this a hydrodynamic radius of 2.06 nm that is essentially the same as the Stokes value of 2.0 nm reported for myoglobin.

3.8 OTHER TECHNIQUES

3.8.1 Ultracentrifugation

Since the kinetic energy of solvent molecules is greater than the sedimentation force of gravity, polymer molecules remain suspended in solution. However, this gravitational field, which permits Brownian motion, may be overcome by increasing this force by use of high centrifugal forces, such as the ultracentrifugal forces developed by Nobel Laureate The Svedberg in 1925.

Both $\overline{M}_w$ and $\overline{M}_z$ may be determined by subjecting dilute polymer solutions to high centrifugal forces. Solvents with densities and indices of refraction different from the polymers are chosen to ensure polymer motion and optical detection of this motion. In sedimentation velocity experiments, the ultracentrifuge is operated at extremely high rotational speeds up to more than 70,000 rpm to transport the denser polymer molecules through the less-dense solvent to the cell bottom or to the top if the density of the solvent is greater than the density of the polymer. The boundary movement during ultracentrifugation can be followed using optical measurement to monitor the sharp change in refractive index (n) between the solvent and the solution.

The rate of sedimentation is defined by the sedimentation constant s, which is directly proportional to the polymer mass m, solution density ρ, and specific volume of the polymer V, and inversely proportional to the square of the angular velocity of rotation ω, the distance from the center of rotation to the point of observation in the cell r, and the fractional coefficient f, which is inversely related

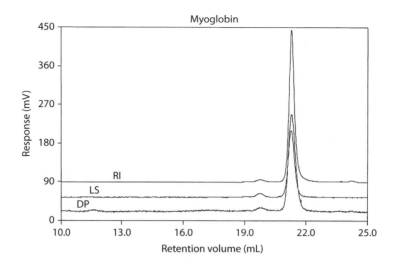

FIGURE 3.21 Response to selected detectors as a function of retention volume for myoglobin (dissolved in PBS buffer at a pH of 6.9). The three detectors are the RI = refractive index signal, LS = light-scattering signal, and DP = differential pressure transducer (viscosity signal). (Used with permission of Viscotek, Houston, TX.)

to the diffusion coefficient D extrapolated to infinite dilution. These relationships are shown in the following equations in which $(1-V_p)$ is called the buoyancy factor since it determines the direction of macromolecular transport in the cell.

$$s = \left(\frac{1}{\omega^2 r} \right) \frac{dr}{dt} = \frac{m(1-V_p)}{f} \tag{3.20}$$

$$D = \frac{RT}{Nf} \quad \text{and} \quad mN = M \tag{3.21}$$

$$\frac{D}{s} = \frac{RT}{\overline{M}_W \left(1-V_p\right)} \tag{3.22}$$

The sedimentation velocity determination is dynamic and can be completed in a short period of time. The sedimentation equilibrium method gives quantitative results, but long periods of time are required for centrifugation at relatively low velocities to establish equilibrium between sedimentation and diffusion.

The $\overline{M}_w$ is directly proportional to the temperature T and the ln of the ratio of concentration c_2/c_1 at distances r_1 and r_2 from the center of rotation and the point of observation in the cell and inversely proportional to the buoyancy factor, the square of the angular velocity of rotation and the difference between the squares of the distances r_1 and r_2 as shown below:

$$\overline{M} = \frac{2RT \left(\ln \frac{c_2}{c_1} \right)}{(1-V_p)\omega^2 (r_2^2 - r_1^2)} \tag{3.23}$$

TABLE 3.4

Mass Spectrometry Approaches Used in the Determination of Molecular Weights of Oligomeric and Polymeric Materials

MS Type	(Typical) Upper Molecular Weight Range (Da)
(Usual) Electron impact (EI)	To 2,000
Fast atom bombardment (FAB)	To 2,000
Direct laser desorption (Direct LD)	To 10^4
Matrix-assisted laser desorption/ionization (MALDI)	To 10^7

3.8.2 MASS SPECTROMETRY

Certain MS procedures allow the determination of the molecular weight or molecular mass of oligomeric to polymeric materials (Table 3.4).

In matrix-assisted laser desorption/ionization mass spectrometry (MALDI MS), the polymer is dissolved, along with a "matrix chemical," and the solution deposited onto a sample probe. The solution is dried. MALDI MS depends on the sample having a strong UV absorption at the wavelength of the laser used. This helps minimize fragmentation since it is the matrix UV-absorbing material that absorbs most of the laser energy. Often employed UV-matrix materials are 2,5-dihydroxybenzoic acid, sinnapinic acid, picplinic acids, and alpha-cyano-4-hydroxy cinnamic acid. The high energy of the laser allows both the matrix material and the test sample to be volatilized. Such techniques are referred to as "soft" since the test sample is not subjected to (much) ionizing radiation and hence little fragmentation occurs.

Mass accuracy on the order of a few parts per million are obtained. Thus, chain content can be determined for copolymers and other chains with unlike repeat units. Polymer MWD can also be determined using MALDI MS and related MS techniques. More about MALDI MS and other MS techniques is described in Section 13.4.

MALDI MS was developed for the analysis of nonvolatile samples and was heralded as an exciting new MS technique for the identification of materials with special use in the identification of polymers. It has fulfilled this promise to only a limited extent. While it has become a well used and essential tool for biochemists in exploring mainly nucleic acids and proteins, it has been only sparsely employed by synthetic polymer chemists. This is because of the lack of congruency between the requirements of MALDI MS and most synthetic polymers. MALDI MS requires that the test material, polymer, be soluble in a relatively volatile solvent. Proteins, nucleic acids, and poly(ethylene glycols) are water soluble so allow the use of MALDI MS for analysis of their chain length. Most polymers are not readily soluble in such solvents so MALDI MS is of little use in the analysis of these materials. Carraher and coworkers have developed an approach that is applicable to most materials that focuses on analyzing the fragmentation of the polymers. While this technique is not able to give molecular weights and MWDs, it allows the identification of ion fragments and ion fragment clusters up to several thousand daltons.

Recently, MS combinations have been available, including the TG-MS combination developed by Carraher that allows the continuous characterization of evolved materials as a polymer undergoes controlled thermal degradation.

3.9 VISCOMETRY

Viscosity is a measure of the resistance to flow of a material, mixture, or solution. Here we will consider the viscosity of solutions containing small, generally 1 g/100 cc (called *1% solutions*) and less, amounts of polymer. The study of such dilute polymer solutions allows a determination of a "relative" molecular weight. The molecular weight is referred to as "relative" since viscosity measurements have not been directly related, through rigorous mathematical relationships, to a specific molecular weight. By comparison, measurements made using light-scattering photometry and the other methods covered before are relatable to specific molecular weight values and these techniques are said to give us "absolute" molecular weights.

The relationship between the force "f" necessary to move a plane of area "A" relative to another plane a distance "d" from the initial plane (Figure 3.22) is described in Equation 3.24.

$$f \propto \frac{A}{d} \tag{3.24}$$

To make this a direct relationship a proportionality factor is introduced. This factor is called the *coefficient of shear viscosity* or simply *viscosity*.

$$f = \eta \left(\frac{A}{d} \right) \tag{3.25}$$

Viscosity is then a measure of the resistance of a material to flow. In fact, the inverse of viscosity is given the name *fluidity*. As a material's resistance to flow increases, its viscosity increases. Viscosities have been reported using a number of different names. The CGS unit of viscosity is called the *poise*, which is a dyne seconds per square centimeter. Another widely used unit is the pascal (or Pa.), which is Newton seconds per square centimeter. In fact, 1 Pa = 10 poise.

Table 3.5 contains the general magnitude of viscosity for some common materials. It is important to note the wide variety of viscosities of materials from gases such as air to viscoelastic solids as glass.

In polymer science we typically do not measure viscosity directly, but rather we look at relative viscosity measures by determining the flow rate of one material relative to that of a second material. Viscosity is one of the most widely used methods for the characterization of polymer molecular weight because it provides the easiest and most rapid means of obtaining molecular weight-related data that requires minimal instrumentation. A most obvious characteristic of polymer solutions is their high viscosity, even when the amount of added polymer is small. This is because polymers reside in several flow planes (Figure 3.22b) acting to resist the flow of one plane relative to another flow plane.

The ratio of the viscosity of a polymer solution to that of the solvent is called the *relative viscosity* (η_r). This value minus 1 is called the *specific viscosity* (η_{sp}), and the *reduced viscosity* (η_{red}) or *viscosity number* is obtained by dividing η_{sp} by the polymer concentration c; that is, η_{sp}/c. The *intrinsic viscosity* or *LVN* is obtained by extrapolating η_{sp}/c to zero polymer concentration. These relationships are given in Table 3.6 and a typical plot of η_{sp}/c and In η_r/c is given in Figure 3.23.

Staudinger showed that the intrinsic viscosity of a solution ([η]) or LVN is related to the molecular weight of the polymer. The present form of this relationship was developed by Mark–Houwink

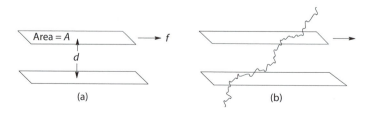

FIGURE 3.22 (a) Representation of Equation 3.24 (b) Illustration of a polymer chain between two flow planes.

TABLE 3.5
Viscosities of Selected Common Materials

Substances	General Viscosity (MPa)
Air	0.00001
Water	0.001
Polymer latexes/paints	0.01
PVC plastols	0.1
Glycerol	10
Polymer resins and "pancake" syrups	100
Liquid polyurethanes	1,000
Polymer "melts"	10,000
Pitch	100,000,000
Glass	1,000,000,000,000,000,000,000

TABLE 3.6
Commonly Used Viscosity Terms

Common Name	Recommended Name (IUPAC)	Definition	Symbol
Relative viscosity	Viscosity ratio	η/η_o	$\eta_{rel} = \eta_r$
Specific viscosity	—	$(\eta/\eta_o) - 1$ or $(\eta - \eta_o)/\eta_o$	η_{sp}
Reduced viscosity	Viscosity number	η_{sp}/c	η_{red} or η_{sp}/c
Inherent viscosity	Logarithmic viscosity number)	$\ln(\eta_r/c)$	η_{inh} or $\ln(\eta_r/c)$
Intrinsic viscosity	Limiting viscosity number	Limit $(\eta_{sp}/c)_{c\to 0}$ or limit $\ln(\eta_r/c)_{c\to 0}$	LVN

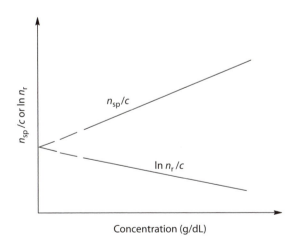

FIGURE 3.23   Reduced- and inherent-viscosity concentration lines for a dilute polymer solution.

(and is known as the *Mark–Houwink equation*) in which the proportionality constant "K" is characteristic of the polymer and solvent and the exponential "a" is a function of the shape of the polymer in a solution. For theta solvents, the value of "a" is 0.5. This value, which is actually a measure of the interaction of the solvent and polymer, increases as the coil expands and the value is between 1.8 and

2.0 for rigid polymer chains extended to their full contour length and 0 for spheres. When "a" is 1.0, the Mark–Houwink equation (3.26) becomes the Staudinger viscosity equation.

$$LVN = K\overline{M}^a \qquad (3.26)$$

Values of "a" and "K" have been determined and complied in several polymer handbooks and are dispersed throughout the literature. Typical values are given in Table 3.7. With known "a" and "K" values, molecular weight can be calculated using Equation 3.26. As noted before, viscosity is unable to give absolute molecular weight values and must be calibrated, that is, values of "a" and "K" determined using polymer samples where their molecular weights have been determined using some absolute molecular weight method such as light-scattering photometry. It is customary in determining the "a" and "K" values to make a plot of log LVN versus log $\overline{M}$ since the log of Equation 3.26, that is, Equation 3.27, is a straight line relationship where the slope is "a" and intercept "K." In reality, "a" is determined from the slope but "K" is determined by simply selecting a known LVN–M couple and using the determined "a" value to calculate the "K" value.

$$\log LVN = a \log \overline{M} + \log K \qquad (3.27)$$

The intrinsic viscosity or LVN, like melt viscosity, is temperature-dependent and decreases as temperature increases as shown in Equation 3.28.

$$LVN = Ae^{E/RT} \qquad (3.28)$$

However, if the original temperature is below the theta temperature, the viscosity will increase when the mixture of polymer and solvent is heated to a temperature slightly above the theta temperature.

Viscosity measurements of dilute polymer solutions are carried out using a viscometer, such as any of those pictured in Figure 3.24. The viscometer is placed in a constant temperature bath and the time taken to flow through a space measured.

Flory, Debye, and Kirkwood showed that [η] is directly proportional to the effective hydrodynamic volume of the polymer in solution and inversely proportional to the molecular weight, $\overline{M}$. The effective hydrodynamic volume is the cube of the root-mean-square end-to-end distance, $(r^2)^{3/2}$. This proportionality constant, N, in the Flory equation for hydrodynamic volume, Equation 3.29, has been considered to be a universal constant independent of solvent, polymer, temperature, and molecular weight.

TABLE 3.7
Typical "K" Values for the Mark–Houwink Equation

Polymer	Solvent	Temperature (K)	$K \times 10^5$ dL/g
Low-density polyethylene	Decalin	343	39
High-density polyethylene	Decalin	408	68
i-Polypropylene	Decalin	408	11
Polystyrene	Decalin	373	16
Poly(vinyl chloride)	Chlorobenzene	303	71
Poly(vinyl acetate)	Acetone	298	11
Poly(methyl acrylate)	Acetone	298	6
Polyacrylonitrile	Dimethylformamide	298	17
Poly(methyl methacrylate)	Acetone	298	10
Poly(ethylene terephthalate)	m-Cresol	298	1
Nylon-66	90% aqueous formic acid	298	110

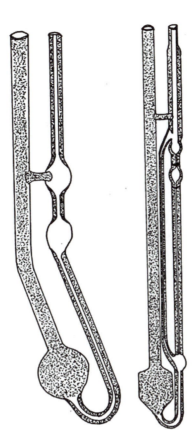

FIGURE 3.24 Common solution viscometers.

$$[\eta] = \frac{N(r^2)^{3/2}}{\overline{M}} \qquad (3.29)$$

The actual average end-to-end distance, r, is related to the nonsolvent expanded average end-to-end distance r_o using the Flory expansion factor, α, as follows:

$$r = \alpha\, r_o \qquad (3.30)$$

Substitution of Equation 3.30 into Equation 3.29 and rearrangement gives

$$[\eta]\overline{M} = N\, (r_o^2)^{3/2}\, \alpha^2 \qquad (3.31)$$

Values for α vary for Flory theta solvents from 0.5 to about 3 for polymers in good solvents.

In Equation 3.26, "a" values for random coils range from 0.5 for theta solvents to 0.8 for good solvents, 0 for hard spheres, about 1 for semicoils, and 2 for rigid rods.

The theta temperature corresponds to the Boyle point in an imperfect gas and is the range in which the virial coefficient B in the expanded gas law becomes zero. This same concept applies to the modification of the gas law ($PV = nRT$) used to determine the osmotic pressure of a polymer solution and is simply the van't Hoff equation that reduces to Equation 3.32 when $B = 0$.

$$\pi = \frac{RTC}{\overline{M}_n} \qquad (3.32)$$

For linear polymers at their theta temperature, that is, the temperature where the chain attains unperturbed dimensions, the Flory equation resembles the Mark–Houwink equation, where α is equal to 1.0 as shown below.

$$[\eta] = K\overline{M}^{1/2}\, \alpha^3 = K'\,\overline{M}^{1/2} \tag{3.33}$$

The intrinsic viscosity of a solution, like the melt viscosity, is temperature dependent and decreases as temperature increases.

$$[\eta] = Ae^{E/RT} \tag{3.34}$$

While the description of viscosity is complex, the relative viscosity is directly related to the flow-through times using the same viscometer as shown in Equation 3.29, where t and t_o are the flow times for the polymer solution and solvent, respectively, and the density of the solution (ρ) and solvent (ρ_o) are related as in Equation 3.35.

$$\frac{\eta}{\eta_0} = \frac{\rho t}{\rho_0 t_0} = \eta_r \tag{3.35}$$

Since the densities of the dilute solution and solvent are almost the same, they are normally cancelled giving Equation 3.36.

$$\frac{\eta}{\eta_0} = \frac{t}{t_0} = \eta_r \tag{3.36}$$

Thus, the relative viscosity is simply a ratio of flow times for the polymer solution and solvent. Reduced viscosity is related to the LVN by a virial equation 3.37

$$\frac{\eta_{sp}}{c} = [\eta] + k_1[\eta]^2 c + k'[\eta]^3 c^2 +\cdots \tag{3.37}$$

For most solutions, Equation 3.37 reduces to the *Huggins viscosity relationship*, Equation 3.38,

$$\frac{\eta_{sp}}{c} = [\eta] + k_1[\eta]^2 c \tag{3.38}$$

which allows $[\eta]$ to be determined from the intercept of the plot of η_{sp}/c versus c and is the basis for the top plot given in Figure 3.17.

Another relationship often used in determining $[\eta]$ is called the *inherent viscosity equation* and is given in Equation 3.39.

$$\ln\frac{\eta_r}{c} = [\eta] - k_2[\eta]^2 c \tag{3.39}$$

Again, a plot of In η_r/c versus c gives a straight line with the intercept $[\eta]$ (or LVN) after extrapolation to zero polymer concentration. This is the basis of the lower plot in Figure 3.17. While k_1 and k_2 are mathematically such that

$$k_1 + k_2 = 0.5 \tag{3.40}$$

many systems appear not to follow this relationship.

We will now turn our attention from the viscosity of dilute solutions and look at the viscosity of melted polymers. The viscosity of melted polymers is important in transferring resins and in polymer processing such as determining the correct conditions to have a specific flow rate for injection processing and in determining the optimum conditions to get the necessary dimensions of extruded shapes. Fillers, plasticizers, temperature, solvents, and molecular weight are just some of the variables that influence the viscosity of polymer melts. Here we will look at the dependence of melt viscosity on polymer molecular weight. Polymer melts have viscosities on the order of 10,000 MPas (1 centiposes is equal to 0.001 Pa/s).

For largely linear polymers, such as polystyrene, where particularly bulky side chains are not present the viscosity or the flow is mainly dependent on the chain length. In most polymers, the melt viscosity-chain length relationship has two distinct regions where the region division occurs when the chain length reaches some length called the *critical entanglement chain length*, Z, (or simply critical chain length) where intermolecular entanglement occurs. This intermolecular entanglement causes the individual chains in the melt to act as being much more massive because of the entanglement. Thus, the resistance to flow is a combination of the friction and entanglement between chains as they slide past one another. Below the critical entanglement length, where only the friction part is important, the melt viscosity, η, is related to the weight average molecular weight by

$$\eta = K_1 \overline{M}_w^{1.0} \tag{3.41}$$

And above the critical chain length, where both the friction and entanglement are important, the relationship is

$$\eta = K_h \overline{M}_w^{3.4} \tag{3.42}$$

where K_1 is a constant for the precritical entanglement chain length and K_h is for the situation above Z and where both K values are temperature dependant. The first power dependence is due to the simple increase in molecular weight as chain length increases, but the 3.4 power relationship is due to a complex relationship between chain movement as related to entanglement and diffusion and chain length.

The critical chain length is often the onset of "strength" related properties and is generally considered the lower end for useful mechanical properties. The Z value for polymers varies but is typically between about 200 and 1,000 units in length. For instance, the Z value for polystyrene is about 700; for polyisobutylene about 600; for poly(decamethylene sebacate) about 300; for poly(methyl methacrylate) about 200; and for poly(dimethyl siloxane) about 1,000.

A number of techniques have been developed to measure melt viscosity. Some of these are listed in Table 3.8. Rotational viscometers are of varied structures. The Couette cup and bob viscometer consists of a stationary inner cylinder, the bob, and an outer cylinder, cup that is rotated.

TABLE 3.8
Viscosity Measuring Techniques and Their Usual Range

Technique	Typical Range (Poise)
Capillary pipette	0.01–1,000
Falling sphere	1–100,000
Parallel plate	10,000–10^9
Falling coaxial cylinder	100,000–10^{11}
Stress relaxation	1,000–10^{10}
Rotating cylinder	1–10^{12}
Tensile creep	100,000–greater than 10^{12}

Shear stress is measured in terms of the required torque needed to achieve a fixed-rotation rate for a specific radius differential between the radius of the bob and cup. The Brookfield viscometer is a bob and cup viscometer. The Mooney viscometer, often used in the rubber industry, measures the torque needed to revolve a rotor at a specified rate. In the cone and plate assemblies, the melt is sheared between a flat plat and a broad cone whose apex contacts the plate containing the melt.

A number of capillary viscometers or rheometers have been employed to measure melt viscosity. In some sense these operate on a principle similar to the simple observation of a trapped bubble moving from the bottom of a shampoo bottle when it is turned upside down. The more viscous the shampoo, the longer it takes for the bubble to move through the shampoo.

3.10   SUMMARY

1. Some naturally occurring polymers such as certain proteins and nucleic acids consist of molecules with a specific molecular weight and are called *monodisperse*. However, many other natural polymers, such as cellulose and natural rubber, and most synthetic polymers consist of molecules with different molecular weights and are called *polydisperse*. Many properties of polymers are dependent on their chain length. Since the melt viscosity increases exponentially with chain length, the high-energy costs of processing high molecular weight polymers are not often justified.

2. The distribution of chain lengths in a polydisperse system may be represented on a typical probability-like curve. The $\overline{M}_n$ is the smallest in magnitude of the typically obtained molecular weights and is a simple arithmetic mean that can be determined using any technique base on colligative properties, such as osmotic pressure, boiling point elevation, freezing point depression, and end-group determination. $\overline{M}_w$ is larger than $\overline{M}_n$ and is referred to as the second power relationship for disperse polymer chains. This value is most often determined by light-scattering photometry. Light-scattering photometry and the colligative-related values are referred to as absolute molecular weight values because there is a direct mathematical connection between molecular weight and the particular property used to determine molecular weight.

3. For monodisperse samples, $\overline{M}_n = \overline{M}_w$. For polydisperse samples the ratio of $\overline{M}_w/\overline{M}_n$ is a measure of the polydisparity and is given the name polydispersity index. The viscosity molecular weight must be calibrated using samples whose molecular weight has been determined using an absolute molecular weight determination technique, thus it is not an absolute molecular weight determining technique, but it requires simple equipment, and is easy to measure. The Mark–Houwink relationship, $LVN = K\overline{M}^a$ is used to relate molecular weight and viscosity.

4. The number-average molecular weight is dependent on the number of polymer chains, while the weight-average molecular weight is dependent on the size of the chains. Thus, there is a correlation between the way the molecular weight is obtained and the type of molecular weight obtained.

5. MWD is most often measured using some form of chromatography. In GPC, cross-linked polymers are used in a column and act as a sieve allowing the larger molecules to elute first. After calibration, the molecular weight of the various fractions of the polymer can be determined. Combinations such as chromatography coupled with light-scattering photometry are used to obtain the molecular weight of the various fractions in a continuous manner.

6. While some techniques such as membrane osmometry and light-scattering photometry give absolute molecular weight, other techniques such as viscometry give only relative molecular weights unless calibrated employing a technique that gives absolute molecular weight. After calibration between viscometry values and chain length through some absolute molecular weight method, viscometry is a fast, inexpensive, and simple method to monitor molecular weight.

7. In general, polymers are soluble in less solvents and to a lower concentration than similar smaller molecules. This is because entropy is the driving force for solubility and smaller molecules have

larger entropy values when solubility is achieved in comparison to polymers. Polymers also take longer to dissolve since it takes time for the solvent molecules to penetrate the polymer matrix.

8. Flory and Huggins developed an interaction parameter that may be used as a measure of the solvent power of solvents for amorphous polymers. Flory and Krigbaum introduced the idea of a theta temperature, which is the temperature at which an infinitely long polymer chain exists as a statistical coil in a solvent.

9. Hildebrand developed solubility parameters to predict the solubility of nonpolar polymers in nonpolar solvents. The solubility parameter is the square root of the CED. For polar solvents, special solvent-polymer interactions can be incorporated into the solubility parameter approach.

GLOSSARY

Affinity chromatography: Chromatography in which the resin is designed to contain moieties that interact with particular molecules and/or units within a polymer chain.

Bingham plastic: Plastic that does not flow until the external stress exceeds a critical threshold value.

Brownian motion: Movement of larger molecules in a liquid that results from a bombardment of smaller molecules.

Buoyancy factor: In ultracentrifugation experiments, it determines the direction of polymer transport under the effect of centrifugal forces in the cell.

Chromatography: Family of separation techniques based on the use of a medium that shows selective absorption.

Colligative properties: Properties of a solution that are dependent on the number of solute molecules present.

Cloud point: Temperature at which a polymer starts to precipitate when the temperature is lowered.

Cohesive energy density (CED): Heat of vaporization per unit volume.

Commercial polymer range: Molecular weight range high enough to have good physical properties but not too high for economical processing.

Cryometry: Measurement of number-average molecular weight from freezing point depression.

Ebulliometry: Measurement of number-average molecular weight from boiling-point elevation.

Effective hydrodynamic volume: Cube of the root-mean-square end-to-end distance of a polymer chain.

Electrophoresis: Form of chromatography that uses an electric field to separate molecules.

End group analysis: Determination of number-average molecular weight by determination of end groups.

Flory–Huggins theory: Theory used to predict the equilibrium behavior between liquid phases containing polymer.

Fractional precipitation: Fractionation of polydisperse systems by addition of small amounts of nonsolvent to a solution of polymer.

Fractionation of polymers: Separation of a polydisperse polymer into fractions of similar molecular weight.

Gel permeation chromatography: Type of liquid–solid elution chromatography, which separates solutions of polydisperse polymer solutions into fractions containing more homogeneous chain sizes by means of a sieving action of a swollen cross-linked polymeric gel. Also called SEC.

High-performance liquid chromatography (HPLC): Chromatography in which pressure is applied that causes the solution to pass more rapidly through the column.

Hildebrand (H): Unit used for solubility parameter values.

Ion-exchange chromatography: Chromatography that separates molecules on the basis of their electrical charge employing polyanionic or polycationic resins.

Kauri-Butanol values: Measure of the aromaticity of a solvent.

Low-angle laser light-scattering photometry (LALLS): Light scattering that employs low-angle measurements minimizing the effect of polymer shape on the scattering.

Mark–Houwink equation: Relates limiting viscosity number to molecular weight; LVN = $K\overline{M}^a$.

Matrix-assisted laser desorption/ionization mass spectrometry (MALDI MS): Mass spectrometry in which the sample is placed in a matrix that contains a strong UV absorber chosen to match the UV absorption of the laser, which allows the molecules to become volatilized with minimal fragmentation.

Melt index: Measure of the flow related inversely to melt viscosity.

Monodisperse: System containing molecules of only one chain length.

Multiangle low-angle laser light-scattering photometry (MALS): Similar to LALLS except where the necessary angle ratios are made together; employs low-angle measurements minimizing the effect of polymer shape on the scattered light.

Number-average molecular weight: Arithmetical mean value obtained by dividing the sum of the molecular weights by the number of molecules.

Oligomer: Polymers with 2–10 repeat units. *Oligos* means "few."

Osmometry: Gives number-average molecular weight from osmotic pressure measurements.

Polydisperse: Mixture of polymer chains of different lengths.

Raoult's law: States that the vapor pressure of a solvent in equilibrium with a solution is equal to the product of the mole fraction of the solvent and the vapor pressure of the pure solvent. This relationship is used in obtaining number-average molecular weights.

SEC-MALS and SEC-LALLS: Coupled chromatography and light-scattering photometry that allows the determination of a number of important values along with chain length distribution.

Sedimentation equilibrium experiment: Ultracentrifugation technique that allows chain length information to be determined.

Semipermeable membrane: Membrane that permits the diffusion of solvent molecules but not large molecules.

Size exclusion chromatography (SEC): Chromatography in which separation is by molecular size or differences in hydrodynamic volume: also called *gel permeation chromatography* (GPC); can use the universal calibration approach to obtain molecular weight.

Solubility parameter: A numerical value equal to the square root of the CED, which is used to predict polymer solubility.

Theta solvent: Solvent in which the polymer chain exists as a statistical coil.

Theta temperature: Temperature at which a polymer of infinite molecular weight begins to precipitate.

Ultracentrifuge: Centrifuge that increases the force of gravity by as much as 100,000 times causing a distribution of materials in a solution to separate in accordance with chain length.

Vapor pressure osmometry: Technique for determining number-average molecular weight by measuring the relative heats of evaporation of a solvent form a solution and pure solvent.

Viscosity: Resistance to flow.

Intrinsic viscosity: The limiting viscosity number obtained by extrapolation of the reduced viscosity to zero concentration.

Reduced viscosity: Specific viscosity divided by the polymer concentration.

Relative viscosity: Ratio of the viscosities of a solution and its solvent.

Specific viscosity: Difference between the relative viscosity and 1.

Weight-average molecular weight: Second power average of molecular weight; dependent on the size of the particular chains.

Zimm plot: Type of double extrapolation used to determine the weight-average molecular weight in light-scattering photometry.

EXERCISES

1. Which of the following is polydisperse with respect to chain length: (a) casein, (b) commercial polystyrene, (c) paraffin wax, (d) cellulose, or (e) *Hevea brasiliensis*?
2. If the number average molecular weight for LDPE is 1.4 million, what is the corresponding average chain length?
3. What are the number- and weight-average molecular weights for a mixture of five molecules each having the following molecular weights: 1.25×10^6; 1.35×10^6; 1.5×10^6; 1.75×10^6; and 2.00×10^6?
4. What is the most probable value for the polydispersity index for (a) a monodisperse polymer and (b) a polydisperse polymer synthesized by a condensation technique?
5. List in increasing values: $\overline{M}_z$, $\overline{M}_n$, $\overline{M}_w$, and $\overline{M}_v$.
6. Which of the following provides an absolute measure of the molecular weight of polymers: (a) viscometry, (b) cryometry, (c) osmometry, (d) light-scattering photometry, and (e) GCP?
7. What is the relationship between the intrinsic viscosity or limiting viscosity number and average molecular weight?
8. What molecular weight determination techniques can be used to fractionate polydisperse polymers?
9. Which of the following techniques yield a number-average molecular weight: (a) viscometry, (b) light-scattering photometry, (c) ultracentrifugation, (e) osmometry, (e) ebulliometry, or (f) cryometry?
10. What kind of molecular weight do you generally get from light-scattering photometry?
11. What is the value of the exponent a in the Mark–Houwink equation for polymers in theta solvents?
12. How many amino groups are present in each molecule of nylon-66 made from an excess of hexamethylenediamine?
13. What is the value of the exponent in the Mark–Houwink equation for a rigid rod?
14. If the value of K and a in the Mark–Houwink equation are 1×10^{-2} cm^3/g and 0.5, respectively, what is the average molecular weight of a polymer whose solution has an intrinsic viscosity of 150 cc/g?
15. Which polymer of ethylene will have the highest molecular weight: (a) a trimer, (b) an oligomer, or (c) a UHMWPE?
16. What is a Zimm plot?
17. What type of molecular weight average, $\overline{M}_n$ or $\overline{M}_w$, is based on colligative properties?
18. What principle is used in the determination of molecular weight by vapor pressure osmometry?
19. Why does the melt viscosity increase faster with molecular weight increase than with other properties such as tensile strength?
20. In spite of the high cost of processing, ultrahigh molecular weight polyethylene is used for making trash cans and other durable goods. Why?
21. Under what conditions are the weight- and number-average molecular weight the same.
22. What is the driving force for polymer solubility?
23. What are the colligative methods for measuring molecular weight and what kind of molecular weight do you get?
24. What is the advantage of using viscometry to measure molecular weight?

25. Which will yield the higher apparent molecular weight values in the light-scattering method: (a) a dust-free system or (b) one in which dust particles are present?
26. Does HPLC need to be calibrated before it can give absolute molecular weights?
27. Which of the following does modern LC allow the calculation of (a) weight average molecular weight, (b) radius of gyration, (c) number-average molecular weight, (d) MWD, or (e) polydispersity index?
28. What is the significance of the virial constant B in osmometry and light-scattering equations?
29. According to Hildebrand, what is a regular solvent?
30. Which of the two steps that occur in the solution process, (a) swelling and (b) dispersion of the polymer particles, can be accelerated by agitation?
31. Define CED.
32. For solution to occur, the change in Gibbs free energy must be: (a) 0, (b) <0, or (c) >0.
33. Will a polymer swollen by a solvent have higher or lower entropy than the solid polymer?
34. Define the change in entropy in the Gibbs free-energy equation.
35. Is a liquid that has a value of 0.3 for its interaction parameter a good or a poor solvent?
36. What is the value of the Gibbs free energy change at the theta temperature?
37. What term is used to describe the temperature at which a polymer of infinite molecular weight precipitates from a dilute solution?
38. At which temperature will the polymer coil be larger in a poor solvent: (a) at the theta temperature, (b) below the theta temperature, or (c) above the theta temperature?
39. If the solubility parameter for water is 23.4 H, what is the CED for water?
40. What is the heat of mixing of two solvents having identical solubility parameters?
41. If the density of a polymer is 0.85 g/cc and the molar volume is 1,176,470 cc, what is the molecular weight?
42. Name some steps that occur when a polymer is dissolved.
43. Why is it important to determine polymer chain length?
44. Why do δ values decrease as the molecular weight increases in a homologous series of aliphatic polar solvents?
45. Which would be a better solvent for polystyrene: (a) n-pentane, (b) benzene, or (c) acetonitrile?
46. Which will have the higher or greater slop when its reduced viscosity or viscosity number is plotted against concentration: a solution of polystyrene (a) in benzene or (b) in noctane?
47. What are general typical values for *a* in the viscosity relationship to molecular weight?
48. When is the Flory Equation similar to the Mark–Houwink equation?
49. What is the term used for the cube root of the hydrodynamic volume?
50. Explain why the viscosity of a polymer solution decreases as the temperature increases.
51. Is MALDI MS restricted to use for natural polymers such as proteins and nucleic acids?
52. Which of the following would you expect to be most soluble in water—ethanol, hexane, or benzene? Why?
53. If entropy is the driving force for mixing and solubility why is there such a focus on enthalpy through the various approaches with solubility parameters and other similar values?
54. If MALDI MS is of such great use to biopolymer chemists why is it not more widely used by synthetic polymer chemists?

ADDITIONAL READING

Albertsson, A. (2008): *Chromatography for Sustainable Polymeric Materials: Renewable, Degradable, and Recyclable*, Springer, NY.

Brown, W. (1996): *Light Scattering Principles and Development*, Verlag, NY.

Carraher, C., Sabir, T, Carraher, C. L.(2009): *Inorganic and Organometallic Macromolecules*, Springer, NY.

Chabra, R., Richardson, J: (2008): *Non-Newtonian Flow and Applied Rheology: Engineering Applications*, Elsevier, NY.

Debye, P. J. (1944): Light scattering analysis, *J. Appl. Phys.*, 15:338.

Debye, P. J., Bueche, A. M. (1948): Intrinsic viscosity, diffusion, and sedimentation rates of polymers in solution, *J. Chem. Phys.*, 16:573.

Einstein, A. (1910): Theory of the opalescence of homogeneous liquids and liquid mixtures in the neighborhood of the critical state, *Ann. Physik.*, 33:1275.

Hansen, C. (2007): *Hansen Solubility Parameters: A User's Handbook*, CRC, Boca Raton, FL.

Huggins, M. L. (1942): The viscosity of dilute solutions of long-chain molecules. IV. Dependence on concentration, *J. Amer. Chem. Soc.*, 64:2716.

Jenekhe, S. A., Kiserow, D. (2004): *Chromogenic Phenomena in Polymers*, Oxford University Press, NY.

Krigbaum, W. R., Flory, P. J. (1952): Treatment of osmotic pressure data, *J. Polym. Sci.*, 9:503.

Kulichke, W. (2004): *Viscometry of Polymers and Polyelectrolytes*, Springer, NY.

Mac, W., Borger, L. (2006): *Analytical Ultracentrifugation of Polymers and Nanoparticles*, Springer, NY.

Mark, H., Whitby, G. S. (1940): *Collected Papers of Wallace Hume Carothers on High Polymeric Substances*, Interscience, NY.

Montaudo, G., Lattimer, R. (2002): *Mass Spectrometry of Polymers*, CRC Press, Boca Raton, FL.

Oliver, R. (1998): *HPLC of Macromolecules*, 2nd Ed., Oxford University Press, Cary, NY.

Pasch, H. (2003): *MALDI-TOF Mass Spectrometry of Synthetic Polymers*, Springer, NY.

Pasch, H., Schrepp, W. (2003): *MALDI-TOF of Polymers*, Springer-Verlag, Berlin.

Pasch, H., Trathnigg, B. (1999): *HPLC of Polymers*, Springer-Verlag, NY.

Rayleigh, J. W. S. (Lord) (1871): On the light from the sky, its polarization and color, *Phil. Mag.*, 41:107.

Schartl, W. (2007): *Light Scattering from Polymer Solutions and Nanoparticle Dispersions*, Springer, NY.

Staudinger, H. (1928): *Ber. Bunsenges Phys. Chem.*, 61:2427.

Svedberg, T., Pederson, K. O. (1940): *The Ultracentrifuge*, Clarendon, Oxford, UK.

Yau, W. (2009): *Size-Exclusion Chromatography*, Wiley, Hoboken, NJ.

Zimm, B. H. (1948): Apparatus and methods for measurement and interpretation of the angular variation of light scattering: preliminary results on polystyrene solutions, *J. Chem. Phys.*, 37:19.

Zimm, B. H. (1948): The scattering of light and the radical distribution function of high polymer solutions, *J. Chem. Phys.*, 16:1093.

4 Polycondensation Polymers
(Step-Reaction Polymerization)

In this chapter, we will emphasize condensation polymers. Since most of these are formed from stepwise kinetics, we will also focus on this kinetic process in this chapter. Even so, some of these polymers can be synthesized employing ring-opening polymerizations (ROPs), which employ the chain polymerization process for synthesis rather than the stepwise process for formation. The bulk of discussion for chain polymerization kinetics is contained in the following several chapters rather than here. We will begin with a comparison between the polymer types and kinetics of polymerization.

4.1 COMPARISON BETWEEN POLYMER TYPE AND KINETICS OF POLYMERIZATION

There is a large, but not total, overlap between the terms condensation polymers and stepwise kinetics and the terms addition (or vinyl) polymers and chain kinetics. In this section we will look at each of these four terms and illustrate the similarities and differences between them.

The terms addition and condensation polymers were first proposed by Carothers and are based on whether the repeat unit of the polymer contains the same atoms as the monomer. An *addition polymer* has the same atoms as the monomer in its repeat unit. Since many of the important addition polymers are formed from vinyl reactants, many addition polymers are also referred to as *vinyl polymers*.

$$
\begin{array}{cc}
X & X \\
| & | \\
H_2C=CH & \rightarrow \quad -(-CH_2-CH-)_n-
\end{array}
\tag{4.1}
$$

The atoms in the backbone of addition polymers are almost always only carbon.

Condensation polymers generally contain fewer atoms in the polymer backbone than in the reactants because of the formation of byproducts during the polymerization process and the backbone contains noncarbon atoms.

$$
X-A-R-A-X + Y-B-R'-B-Y \rightarrow -(-A-R-A-B-R'-B-)- + XY
\tag{4.2}
$$

where A-X can be most Lewis bases such as $-NH_2$, $-SH$, $-OH$ and B-Y can be Lewis acids such as

$$
\begin{array}{cccc}
O & O & O & O \\
|| & || & || & || \\
-C-OH, & -C-Cl, & -P-Cl, & -S-Cl \\
 & & | & || \\
 & & R & O
\end{array}
$$

The corresponding polymerizations to form these polymers are, typically, chain polymerizations to form addition polymers and stepwise polymerizations to form condensation polymerizations.

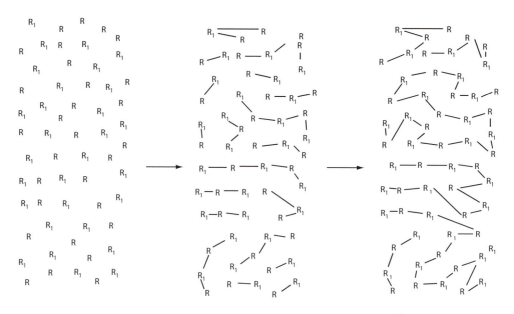

FIGURE 4.1 Depiction of stepwise chain growth for monomers R and R_1 as the polymerization begins (far left) and progresses toward the right.

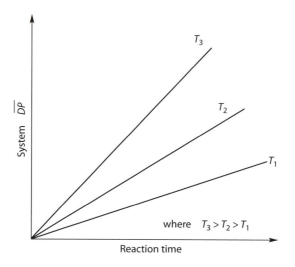

FIGURE 4.2 System molecular weight for stepwise kinetics as a function of reaction time and reaction temperature, T.

The term *stepwise kinetics*, or step-growth kinetics, refers to polymerizations in which the polymer's molecular weight increases in a slow, stepwise manner as reaction time increases. Figure 4.1 contains a depiction of the stepwise polymerization process. Initially, there is only monomer R and R_1 present (far left). After some time, all the monomer has reacted, yet no high molecular weight chains are found with the greatest degree of polymerization (DP) being 2.5 (middle). As polymerization continues, longer chains grow with a high DP of 10 (far right) for this depiction. Eventually, all the chains will connect finally giving a polymer. Figure 4.2 contains a representative plot of DP as a function of time for this process.

The formation of polyesters from a dialcohol (diol) and a dicarboxylic acid (diacid) will be used to illustrate the stepwise kinetic process. Polymer formation begins with one diol molecule reacting with one diacid forming one repeat unit of the eventual polyester (4.3).

$$\text{HO-C-R-C-OH + HO-R'-OH} \rightleftharpoons \text{HO-C-R-C-O-R'-OH + H}_2\text{O} \tag{4.3}$$

This ester-containing unit can now react with either an alcohol or an acid group producing chains ending with either two active alcohol functional groups or two active acid groups (4.4)

$$+ \text{HO-R'-OH} \rightleftharpoons \text{HO-R'-O-C(O)-R-C(O)-O-R'-OH + H}_2\text{O}$$
$$\text{HO-C(O)-R-C(O)-O-R'-OH}$$

$$+ \text{HO-C(O)-R-C(O)-OH} \rightleftharpoons \text{HO-C(O)-R-C(O)-O-R'-O-C(O)-}$$
$$\text{RC(O)-OH + H}_2\text{O} \tag{4.4}$$

The chain with two alcohol ends can now condense with a molecule containing an acid end and the chain with two acid ends can now condense with a molecule containing an alcohol group resulting in molecules that contain one acid and one alcohol active group. This reaction continues through the reaction matrix whenever molecules with the correct functionality, necessary energy of activation, and correct geometry collide. The net effect is the formation of dimers, trimers, and so on, until polymer is formed. The monomer concentration is low to nonexistent during most of the polymerization.

$$\text{HO-R'-O-C(O)-R-C(O)-O-R'-OH + HO-C(O)-R-C(O)-OH} \rightleftharpoons$$
$$\text{HO-R'-O-C(O)-R-C(O)-O-R'-O-C(O)-R-C(O)-OH + H}_2\text{O} \tag{4.5}$$

$$\text{HO-C(O)-R-C(O)-R'-O-C(O)-RC(O)-OH + HO-R'-OH} \rightleftharpoons$$
$$\text{HO-C(O)-R-C(O)-R'-O-C(O)-RC(O)-O-R'-OH + H}_2\text{O} \tag{4.6}$$

The reactants are consumed with few long chains formed until the reaction progresses toward total reaction of the chains with themselves. Thus, polymer formation occurs one step at a time, hence the name "stepwise" kinetics.

Chain-growth reactions require initiation to begin chain growth. We will look at the free radical growth forming polystyrene to illustrate chain growth. Here, the initiation of a styrene molecule, R, will illustrate the chain-growth process. The initiator reacts with a styrene monomer creating a free radical active chain end. This free radical chain end then reacts with another styrene monomer that, in turn, reacts with another styrene monomer, and so on until termination stops chain growth with the formation of a polystyrene polymer chain (Figure 4.3). Polystyrene chains are formed from the beginning of the polymerization process with growth occurring through formation of polymer chains. As the first chain is formed (Figure 4.3; middle), only polymer and monomer are present. As additional polymer chains are formed (Figure 4.3; right), the polymer mix still consists of only polymer chains and unreacted monomer. This is shown in the following sequences.

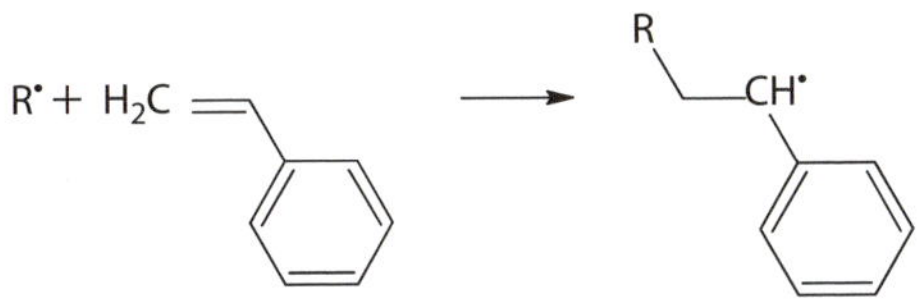

$$\text{R}^{\bullet} + \text{H}_2\text{C} = \quad \longrightarrow \quad \text{CH}^{\bullet} \tag{4.7}$$

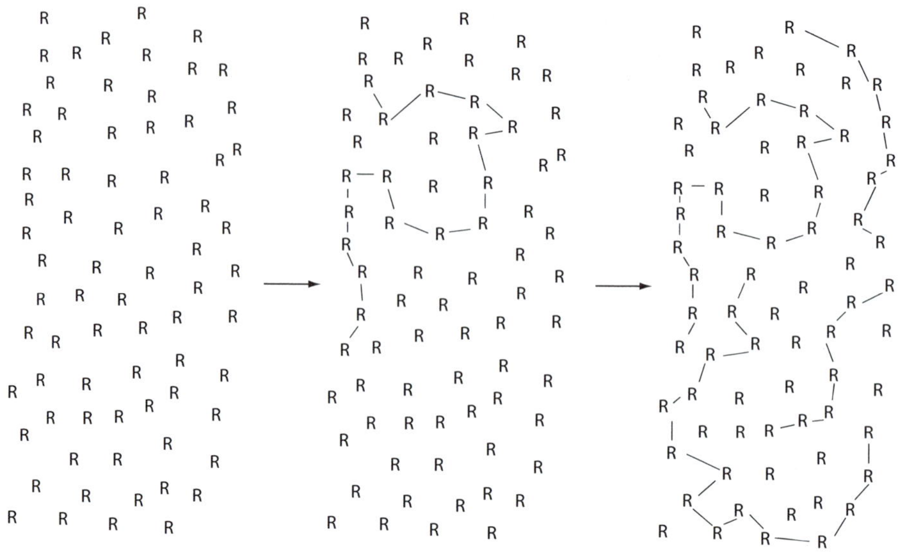

FIGURE 4.3 Molecular weight for chain-growth kinetics as a function of reaction time beginning with only monomer (R; far left) with reaction progressing toward the right.

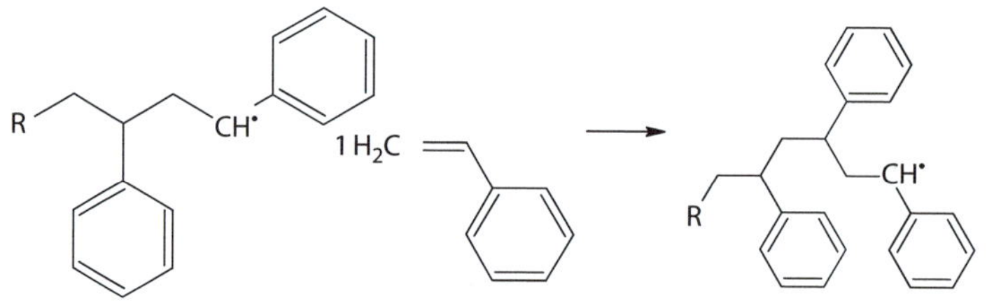

(4.8)

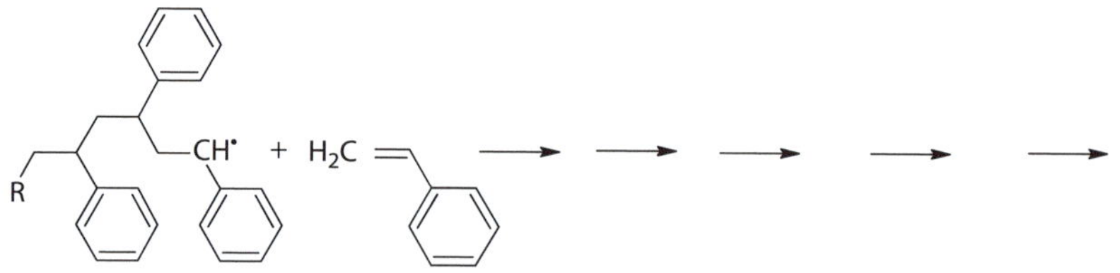

(4.9)

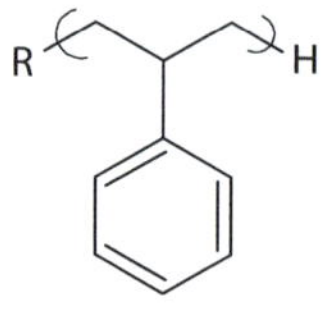

(4.10)

The average DP for the entire system, neglecting unreacted monomer, does not markedly change as additional polymer is formed (Figure 4.4). Also, unlike the stepwise process, average chain length decreases with increase in reaction temperature. More about this topic in discussed in Chapters 5 and 6.

Both the stepwise and chain-wise polymerizations produce polymers that are polydisperse with respect to chain lengths.

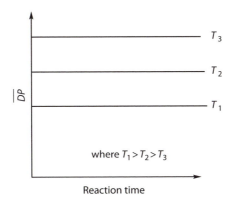

where $T_1 > T_2 > T_3$

Reaction time

FIGURE 4.4 Idealized average molecular weight of formed polymer as a function of reaction time and temperature, T, for chain-type polymerization.

TABLE 4.1
Comparison between Stepwise and Chain-wise Polymerizations

Chain	Step
Growth occurs by addition of one unit at a time to the active growing chain end	Any two unlike molecular units can react
Monomer concentrations decrease steadily throughout the polymerization	Monomer disappears early in the reaction
Polymer chains are formed from the beginning of the polymerization and throughout the process	Polymer chain length increases steadily during the polymerization
Average chain length for reacted species remains approximately constant throughout the polymerization	Average molecular weight for the reaction (for the reacted species) increases throughout the polymerization
As reaction time increases polymer yield increases, but molecular weight remains about the same	High "extents" of reaction are required to obtain high chain lengths
Reaction mixture contains almost only unreacted monomer, polymer, and very little growing polymer chains	Reaction system contains various stages, chain lengths, of product present in a calculable distribution

Most addition polymers are formed from polymerizations exhibiting chain-growth kinetics. This includes the typical polymerizations, via free radical or some ionic mode, of the vast majority of vinyl monomers such as vinyl chloride, ethylene, styrene, propylene, methyl methacrylate, and vinyl acetate. By comparison, most condensation polymers are formed from systems exhibiting stepwise kinetics. Industrially, this includes the formation of polyesters and polyamides (nylons). Thus, there exists a large overlap between the terms stepwise kinetics and condensation polymers and chain-wise kinetics and addition (or vinyl) polymers. A comparison of the two types of systems is given in Table 4.1.

Even so, there is not a total overlap between the various characteristics of vinyl-chain kinetics and condensation-step kinetics. Following are examples illustrating the lack of adherence to this overlap.

1. The formation of polyurethanes (PU) and polyureas typically occur through stepwise kinetics with the polymer backbones clearly containing noncarbon atoms. Yet, no byproduct is released through the condensation process because condensation occurs through an internal rearrangement and shift of the hydrogen—neither steps resulting in expulsion of a byproduct.

$$\text{OCN–R–NCO + HO–R'–OH} \longrightarrow -(-\overset{\overset{\displaystyle O}{\|}}{C}\text{–NH–R–NH–}\overset{\overset{\displaystyle O}{\|}}{C}\text{–O–R'–O–})- \qquad (4.11)$$

Polyurethane (PU)

2. Internal esters (lactones) and internal amides (lactams) are readily polymerized through a chain-wise kinetic process forming polyesters and polyamides, clearly condensation polymers with respect to having noncarbons in the backbone, but without expulsion of a byproduct.

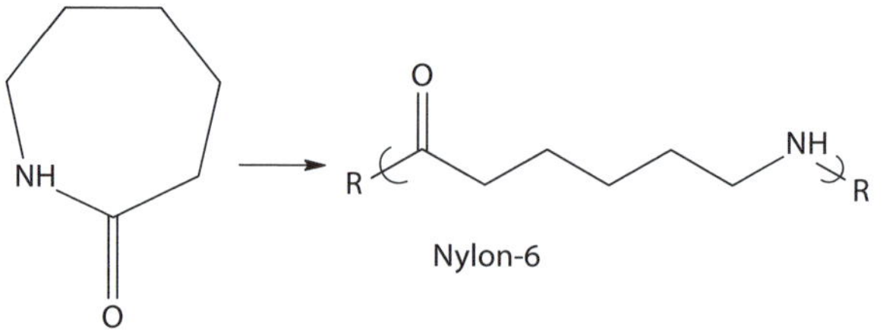

Nylon-6   (4.12)

A similar ROP with ethylene oxide forming poly(ethylene oxide) (PEO) forms a noncarbon backbone via a chain-wise kinetic process.

Poly(ethylene oxide)   (4.13)

These polymerizations are referred to as ring opening polymerizations, *ROPs*.

3. Interfacially formed condensation polymers such as polyesters, polycarbonates (PCs), nylons, and PU are typically formed on a microscopic level in a chain-growth manner due largely because of the highly reactive nature of the reactants employed for such interfacial polycondensations.

4.2 INTRODUCTION

While condensation polymers account for only a modest fraction of all synthetic polymers, most natural polymers are of the condensation type. The first all-synthetic polymer, Bakelite, was produced by the stepwise polycondensation of phenol and formaldehyde.

As shown by Carothers in the 1930s, the chemistry of condensation polymerizations is essentially the same as classic condensation reactions leading to the formation of monomeric esters, amides, and so on. The principle difference is that the reactions used for polymer formation are bifunctional instead of monofunctional.

Table 4.2 contains a listing of a number of industrially important synthetic condensation polymers.

4.3 STEPWISE KINETICS

The kinetics for stepwise polycondensation reactions and the kinetics for monofunctional aminations and esterifications, for example, are similar. Experimentally, both kinetic approaches are essentially identical. Usual activation energies (120–240 kJ/mol) require only about one collision in 10^{12} to 10^{15} to be effective in producing product at 100°C, whereas for the vinyl reactions, the activation energies are much smaller (8–20 kJ/mol), with most collisions of proper orientation being effective in lengthening the chain. This is in agreement with the slowness of the stepwise process in comparison to chain polymerizations.

TABLE 4.2
Properties and Uses of Some Important Synthetic Condensation-Type Polymers

Type (Common Name)	Typical Properties	Typical Uses
Polyamides (nylons)	Good balance of properties; high strength, good elasticity and abrasion resistance, good toughness favorable solvent resistance, outdoor weathering, moisture resistance	Fibers—about ½ of all nylon fiber goes into tire cord; rope, cord, belting, fiber cloths, thread, hose, undergarments, dresses; plastics—used as an engineering material, substitute for metal bearings, bearings, cams, gears, rollers, jackets on electrical wire
Polyurethanes	Elastomers—good abrasion resistance, hardness, resistance to grease, elasticity; fibers—high elasticity, excellent rebound; coatings—good resistance to solvents and abrasion; good flexibility, impact resistance; foams—good strength per weight, good rebound, high impact strength	Four major forms used—fibers, elastomers, coatings, foams; elastomers—industrial wheels, heel lifts; fibers—swimsuits, foundation garments; coatings—floors where impact and abrasion resistance are required, bowling pins; foams—pillows, cushions
Polyureas	High T_g, fair resistance to greases, oils, solvents	Not widely used
Polyesters	High T_g and T_m; good mechanical properties, resistance to solvents and chemicals; good rebound low moisture absorption, high modulus; film—high tensile strength (about that of steel), stiff, high resistance to failure on repeated flexing, high impact strength, fair tear strength	Fibers—garments, permanent press and "wash & wear" garments, felts, tire cord, film—magnetic recording tape, high-grade film
Polyethers	Good thermoplastic behavior, water solubility; moderate strength and stiffness	Sizing for cotton and synthetic fibers; stabilizers for adhesives, binders, and film formers in pharmaceuticals; thickeners; production of films
Polycarbonates	Crystalline with good mechanical properties, high impact strength; good thermal and oxidative stability, transparent, self-extinguishing, low moisture absorption	Machinery and business
Phenol-formaldehyde resins	Good heat resistance, dimensional stability, resistance to cold flow, solvent, dielectric properties	Used in molding applications; appliances, TVs, automotive parts, filler, impregnating paper, varnishes, decorative laminates, electrical parts, countertops, toilet seats; adhesive for plywood, sandpaper, brake linings, abrasive wheels
Polyanhydrides	Intermediate physical properties; medium to poor T_g and T_m	No large industrial applications
Polysulfides	Outstanding oil and solvent resistance; good gas impermeability, resistance to aging, ozone, bad odors, low tensile strength, poor heat resistance	Gasoline hoses, tanks, gaskets, diaphragms
Polysiloxanes	Available in a wide range of physical states—from liquids to greases, to waxes, to resins, to elastomers; excellent high and moderate low temperature physical properties; resistant to weathering and lubricating oils	Fluids—cooling and dielectric fluids, in waxes & polishes; as antifoam and mold release, for paper and textile treatment: elastomers—gaskets, seals, cable, wire insulation, hot liquids and gas movement, surgical and prosthetic devices, sealing compounds; resins-varnishes, paints, encapsulating, and impregnating agents
Polyphosphate and poly-phosphonate esters	Good fire resistance, fair adhesion, moderate moisture stability, fair temperature stability	Additives promoting flame retardance; adhesive for glass (since it has a similar refractive index), pharmaceuticals, surfactant

While more complicated situations can occur, we will consider only the kinetics of simple polyesterification. The kinetics of most other common polycondensations follows an analogous pathway.

For uncatalyzed reactions where the diacid and diol are present in equimolar amounts, one diacid is experimentally found to act as a catalyst. The experimental expression dependencies are described in the usual manner as follows:

$$\text{Rate of polycondensation} = -\frac{d[A]}{dt} = k[A]^2[D] \tag{4.14}$$

where $[A]$ is the concentration of diacid and $[D]$ is the diol concentration. Since $[A] = [D]$, we can write

$$-\frac{d[A]}{dt} = k[A]^3 \tag{4.15}$$

Rearrangement gives

$$-\frac{d[A]}{A^3} = kdt \tag{4.16}$$

Integration of Equation 4.16 over the limits of $A = A_o$ to $A = A_t$ and $t = 0$ to $t = t$ gives

$$2kt = -\frac{1}{[A_0]^2} - \frac{1}{[A_0]^2} = \frac{1}{[A_t]^2} + \text{constant} \tag{4.17}$$

It is usual to express Equation 4.17 in terms of the extent of reaction, p, where p is defined as the fraction of functional groups that have reacted at time t. Thus, $1 - p$ is the fraction of groups unreacted. A_t is in turn $A_o (1 - p)$, that is,

$$A_t = A_o (1 - p) \tag{4.18}$$

Substitution of the expression for A_t from Equation 4.18 into Equation 4.17 and rearrangement gives

$$2A_0^2 kt = \frac{1}{(1-p)^2} + \text{constant} \tag{4.19}$$

which is the equation of a straight line, that is,

$$mx = y + b$$

Where $m = 2A_0^2 k$, $x = t$, $y = 1/(1-p)^2$, and b is the constant. A plot of $1/(1 - p)^2$ as a function of time should be linear with a slope $2A_o^2 k$ from which k is determined. Determination of k at different temperatures allows the calculation of activation energy. Thus, one definition of the specific rate constant is $k = Ae^{-E_a/kt}$. The log of both sides gives $\log k = \log A - E_a/kt$, which again is the equation of a straight line where $y = \log k$, $b = \log A$, $m = -E_a/k$ and $x = 1/t$, where E_a is the activation energy, A is a constant, k is the specific rate constant, and t is the temperature.

The number-average DP can be expressed as

$$\overline{DP}_n = \frac{\text{number of original molecules}}{\text{number of molecules at time } t} = \frac{N_0}{N} = \frac{A_0}{A_t} \tag{4.20}$$

Thus,

$$\overline{DP}_n = \frac{A_0}{A_t} = \frac{A_0}{A_0(1-p)} = \frac{1}{1-p} \tag{4.21}$$

The relationship given in Equation 4.21 is called the *Carothers equation* because it was first found by Carothers while working with the synthesis of polyamides (nylons). For an essentially quantitative synthesis of polyamides where p is 0.9999, the $\overline{DP}_n$ is approximately 10,000, the value calculated using Equation 4.21.

$$\overline{DP}_n = \frac{1}{1-p} = \frac{1}{1-0.0000} = \frac{1}{0.0001} = 10,000 \tag{4.22}$$

Thus, the Carothers equation allows calculation of maximum DP as a function of extent of polymerization, and the purity of reactants. This value is sufficient to produce polyesters that will give strong fibers. The high value of p is decreased, as is the DP, if impurities are present or if some competing reaction, such as cyclization, occurred. Since the values of k at any temperature can be determined from the slope $(2kA_o^2)$ of when $1/(1-p)^2$ is plotted against t, $\overline{DP}_n$ at any time t can be determined from the expression

$$(\overline{DP}_n)^2 = 2kt[A_0]^2 + \text{constant} \tag{4.23}$$

Much longer times are required to effect formation of high polymer polyesters in uncatalyzed esterifications than for acid or base-catalyzed systems. For catalyzed systems, since the added acid or base is a catalyst, its apparent concentration does not change with time; thus, it is not included in the kinetic rate expression. In such cases the reaction follows the rate expression

$$\text{Rate of polycondensation} = -\frac{d[A]}{dt} = k[A][B] \tag{4.24}$$

For $[A] = [B]$ we have

$$-\frac{d[A]}{dt} = k[A]^2 \tag{4.25}$$

and rearrangement gives

$$-\frac{d[A]}{[A]^2} = kt \tag{4.26}$$

which on integration and subsequent substitution gives

$$kt = \frac{1}{A_t} - \frac{1}{A_0} = \frac{1}{A_0(1-p)} - \frac{1}{A_0} \tag{4.27}$$

Rearrangement gives

$$A_0kt = \frac{1}{(1-p)} - 1 = \overline{DP}_n - 1 \tag{4.28}$$

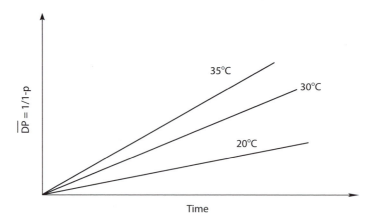

FIGURE 4.5   Plot of average chain length, $\overline{DP}_n$, as a function of reaction time for the acid-catalyzed condensation of ethylene glycol with terephthalic acid producing PET.

which predicts a linear relationship of $1/1 - p$ with reaction time. This is shown in Figure 4.5.

Useful polymers are not obtained unless the value for the fractional conversion p is at least 0.990, that is, a *DP* greater than 100.

We can also calculate the weight-average chain length as follows. We have that

$$\overline{DP}_w = \frac{1+p}{1-p} \tag{4.29}$$

The molecular weight distribution is defined as M_w/M_n or $\overline{DP}_w/\overline{DP}_n$ and

$$\frac{\overline{DP}_w}{\overline{DP}_n} = \frac{1+p}{1-p} \text{ divided by } \frac{1}{1-p} = 1+p \tag{4.30}$$

When *p* is near or equal to 1, that is, for a high molecular weight situation,

$$\frac{\overline{DP}_w}{\overline{DP}_n} = 1+1 = 2 \tag{4.31}$$

for condensation polymerizations.

Of note, the molecular weight distribution for condensation polymers formed from stepwise kinetics is generally small in comparison to vinyl polymers formed from the chain-wise kinetics, where $\overline{DP}_w/\overline{DP}_n$ values of 4–20 are not unusual.

It is important to note that the rate constant k for reactions for similar monofunctional compounds is essentially the same as for the difunctional compounds used in the formation of polycondensation polymers. Likewise, as in the case of reactions of small molecules, the rate constant k increases with temperature in accordance with the Arrhenius Equation 4.32.

$$k = Ae^{-Ea/kt} \tag{4.32}$$

As noted in Chapter 3, solution and bulk viscosity, resistance to flow increases as molecular weight increases. Energy and time are required to process polymers into useful items. Increases in energy and time are reflected in an increase in the price of the item and decrease in the lifetime of the equipment used to process the polymers. There is need to produce polymeric materials that exhibit the needed physical properties and yet have a reasonable viscosity to minimize the energy and time required to produce the particular product. Thus, molecular weight control is important.

We have already looked the major indicator of molecular weight control. It is the Carothers' relationship $\overline{DP}_n = 1/1 - p$ so that chain length can be controlled through extent of reaction and purity of reactants. This relationship was initially one of the least understood relationships and, early in the commercial production of polymers, costed industry a lot of time and money because of a lack of understanding its importance. Its importance is now well understood. Thus, the extent of reaction can be controlled through simply stopping the reaction by cooling it. More typically, it is controlled through the use of an excess of one or the other reactant or the introduction of some monofunctional reactant that will halt chain growth when it is added to a growing polymer chain.

For the situation when an excess of one reactant is employed, chain growth continues until all of the reactant that is present in the smaller molar amount reacts leaving chains with end groups derived from the reactant that is present in excess. For a reaction where the number of moles of "B" are in excess, $N_B > N_A$, the extent of reaction for A is p and for B it is "rp" where r is the ratio of N_A/N_B. The following relationship can be derived.

$$\overline{DP}_n = \frac{1+r}{1+r-2rp} \tag{4.33}$$

For an excess in N_B of one mole percent we have that $r = 100/101 = 0.9901$ and

$$\overline{DP}_n = \frac{1+0.9901}{1+0.9901-2(0.9901)(1)} = 201 \tag{4.34}$$

4.4 POLYCONDENSATION MECHANISMS

Proposed mechanisms for polycondensations are essentially the same as those proposed in the organic chemistry of smaller molecules. Here, we will briefly consider several examples to illustrate this similarity between reaction mechanisms for small molecules and those forming polymers. For instance, the synthesis of polyamides (nylons) is envisioned as a simple S_N2 type Lewis acid–base reaction with the Lewis base nucleophilic amine attacking the electron-poor, electrophilic carbonyl site followed by loss of a proton.

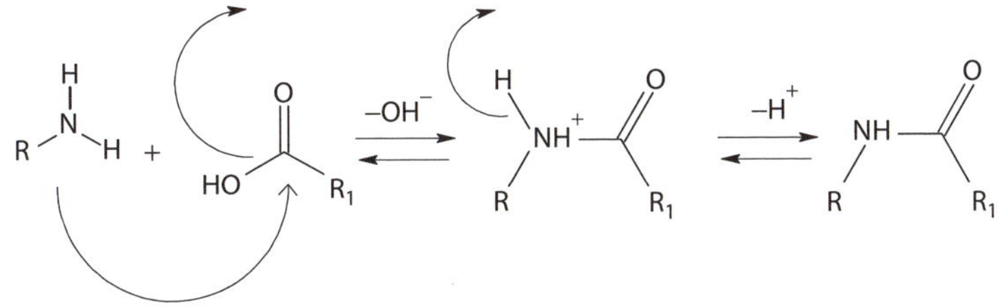

$$(4.35)$$

A similar mechanism is proposed for most polyesterifications (Equations 4.36 and 4.37).

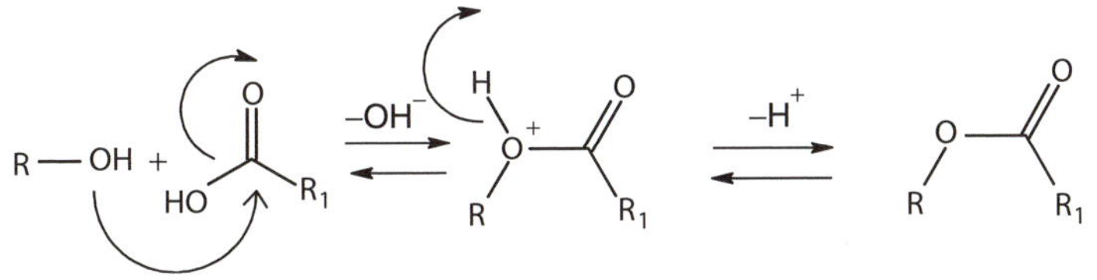

$$(4.36)$$

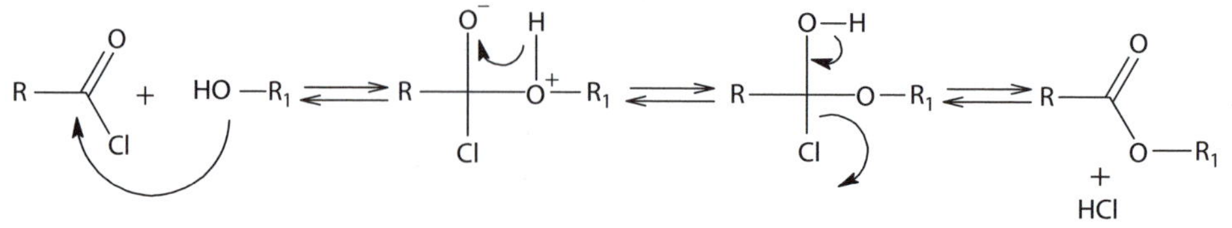

$$(4.37)$$

Below are a number of resonance forms for the isocyanate moiety. Because it is surrounded on both sides by atoms, N and O, that have greater electronegativities than the carbon atom, the carbon is electron poor and is the site for attack by amines (polyureas) and alcohols (PU).

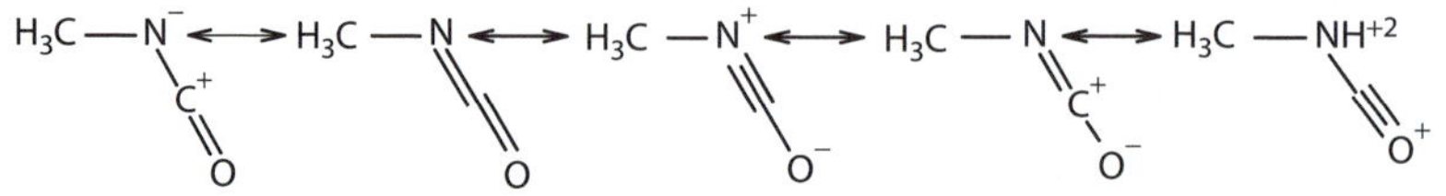

$$(4.38)$$

Polyurethane formation occurs with attack of the nucleophilic alcohol at the electron-poor isocyanate carbon with a proton shift followed by rearrangement to the urethane structure.

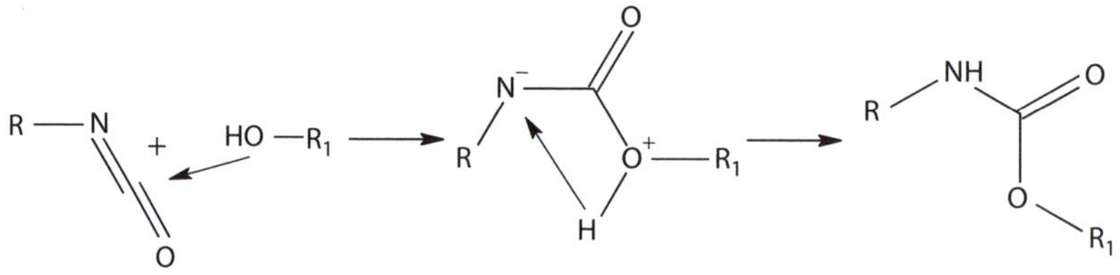

$$(4.39)$$

Polyether formation from ring opening forming PEO occurs through acid or base catalysis as shown in Equations 4.40 and 4.41.

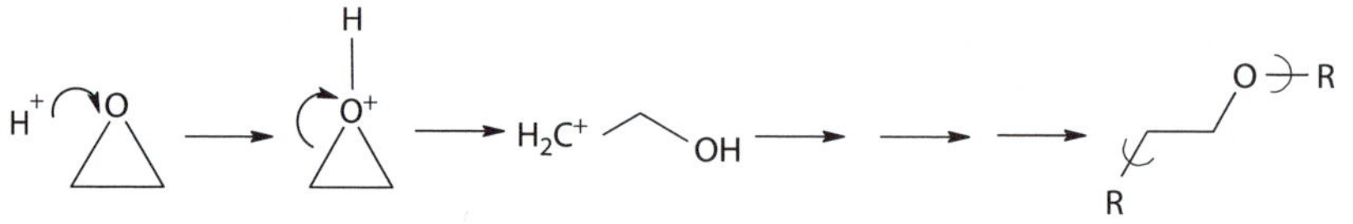

$$(4.40)$$

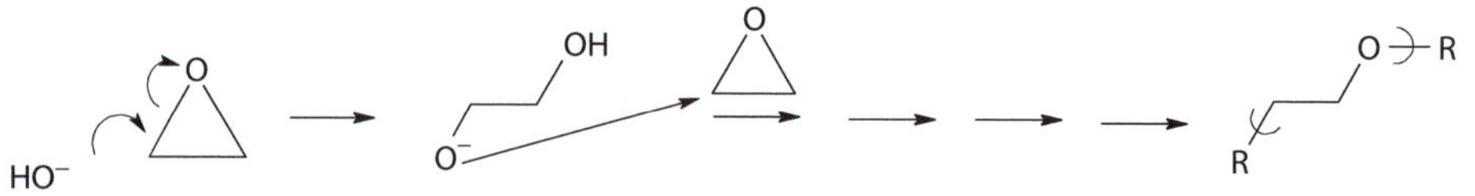

$$(4.41)$$

Ethylene oxide is a polar molecule with an excess of electron density on the oxygen making it the site for acid attack, whereas the ethylene moiety is electron poor and is the site for nucleophillic attack by the base.

Again, what we have learned from smaller molecule chemistry applies to polymer reactions typically with only modifications that consider the larger size of the polymer.

4.5 POLYESTERS

Carothers and his research group at DuPont began to investigate the formation of polymers from the reaction of aliphatic diacids with diols, generally adipic acid and ethylene glycol (derived from reaction of ethylene oxide with water; major ingredient in most antifreeze), in search of materials that would give them fibers. They were only able to form syrupy mixtures. This is because, unlike reactions with diamines (Section 4.7), the equilibrium reaction greatly disfavors ester formation. Further, the ability to have almost equal amounts of functional groups is easily achieved with the amines through formation of salts as shown in Equation 4.53, but diols do not form such salts. The critical need to have the reactants present in equal molar amounts for equilibrium determined reactions is clearly seen in Equation 4.21. Carothers' group understood the principle of "driving" an equilibrium reaction so sought to remove water thus forcing the reaction toward ester formation. For this they developed a so-called "molecular still" which was simply heating the mixture and applying a vacuum coupled with a "cold-finger" that allowed evacuated water to condense and be removed from the reaction system. Since the fractional conversion (p) was only 0.95, the average chain length of these polyesters was less than 20.

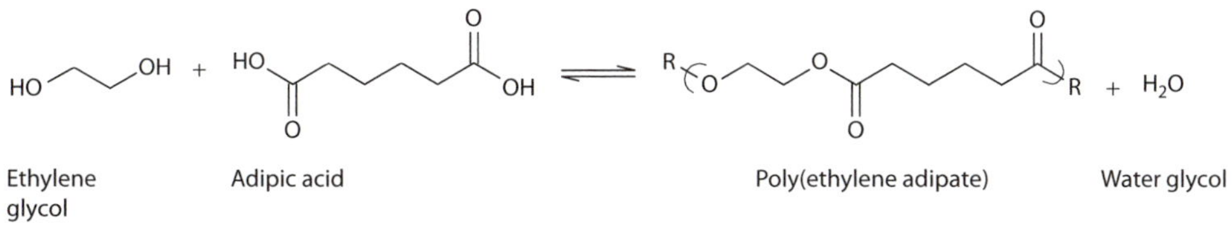

$$(4.42)$$

Ethylene glycol   Adipic acid   Poly(ethylene adipate)   Water glycol

The DuPont research team turned from the synthesis of polyesters to tackle, more successfully, the synthesis of the first synthetic fiber material, nylon, which approached and in some cases exceeded the physical properties of natural analogs (next section). The initial experience with polyesters was put to good use in the nylon venture.

Today, we know that Carothers would have had greater success in producing high molecular weight polyesters had his group employed transesterification (Equations 4.43, 4.52),

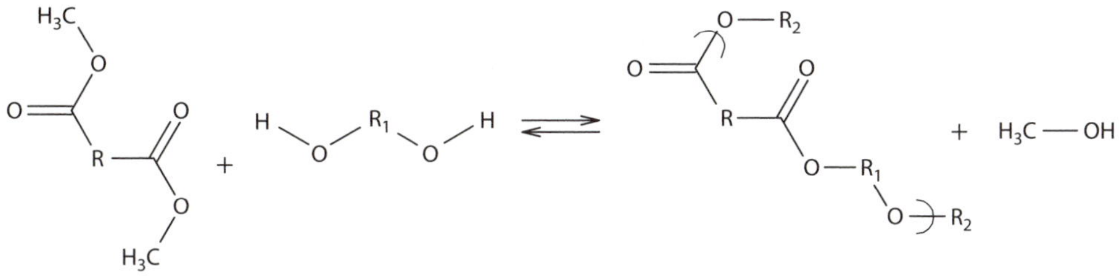

$$(4.43)$$

ring opening of an internal ester (lactone; Equation 4.44),

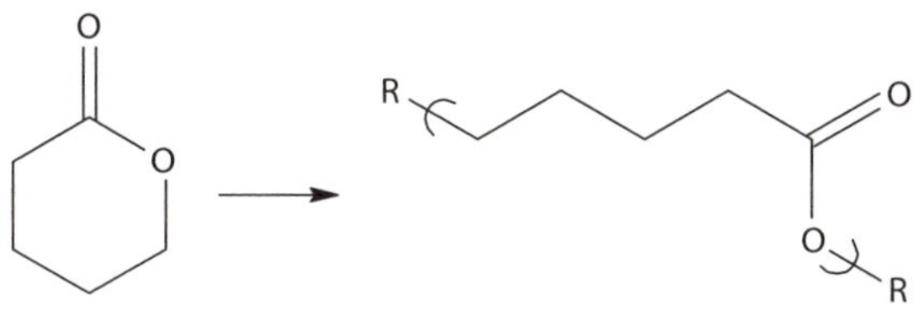

$$(4.44)$$

ring opening of internal ethers (Equations 4.45, 4.53),

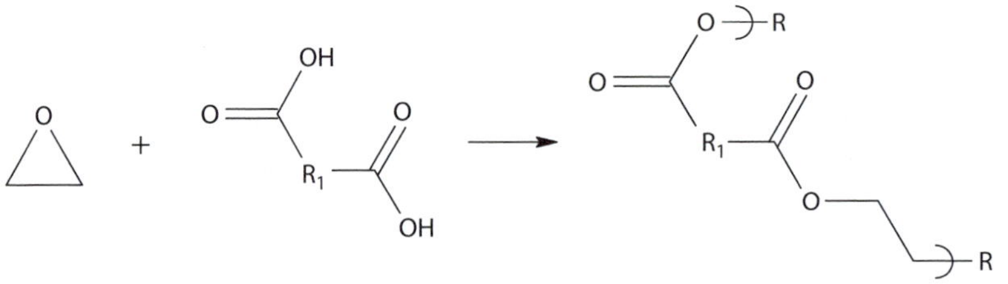

(4.45)

reaction of diols with acyl dichlorides (Schotten–Baumann reaction; Equations 4.37, 4.46)

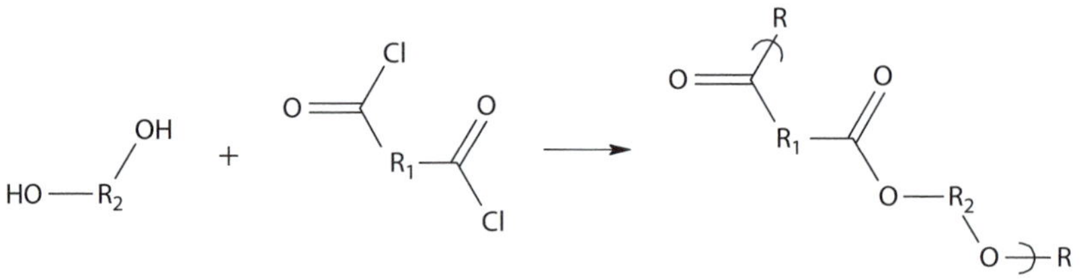

(4.46)

or diols with anhydrides (such as Equations 4.47, 4.50, and 4.51).

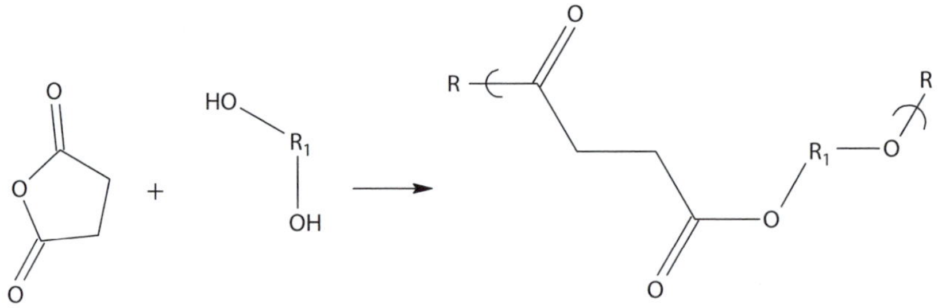

(4.47)

These techniques then compose, along with the simple reaction of diacids with diols (Sections 4.5 and 4.6), the major techniques employed in the synthesis of polyesters. Each of these reactions involves the nucleophillic addition to the carbonyl group as shown in Section 4.5 and is illustrated in Equation 4.37. Focusing on the carbonyl-containing reactant, polyester formation employing direct esterification (reaction between an acid and alcohol) and transesterification is relatively slow with each step reversible. Reaction rates are increased through the use of acid catalysts that coordinate the carbonyl oxygen enhancing the electrophilic nature of the carbonyl carbon (Equation 4.48).

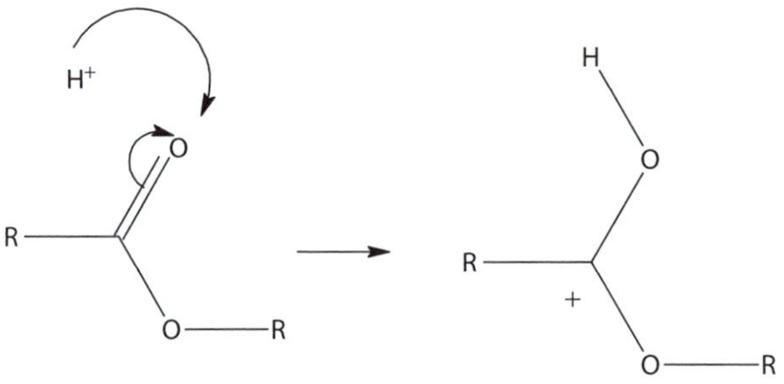

(4.48)

Basic catalysts are often employed in transesterification reactions probably to increase the nucleophilicity of the alcohol through formation of secondary bonding to the alcohol-proton resulting in the alcohol being more nucleophillic (Equation 4.49).

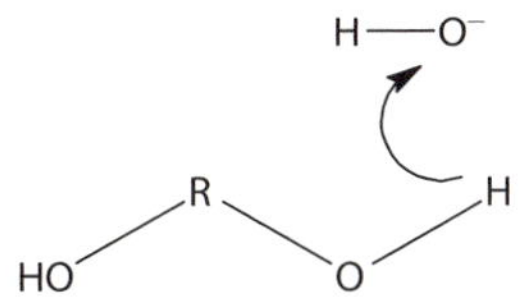

$$(4.49)$$

Reaction with anhydrides and acid chlorides are more rapid and can occur in an essentially nonreversible fashion. But, anhydrides and acid chlorides are considered so-called "high energy" reactants since they often involve additional energy-requiring steps in their production and thus are less suitable for large-scale production of materials. The activity energies for direct esterification and transesterification are on the order of 30 kcal/mole (120 kJ/mole) while the activation energies for anhydride and acid chloride reaction with alcohols are on the order of 15–20 kcal/mole (60–80 kJ/mole).

The initial polyester formation actually occurred early and is attributed to Gay Lussac and Pelouze in 1833 and Berzelius in 1847. These polyesters are called glyptals and alkyds, and they are useful as coatings materials and not for fiber production. While these reactions had low fractional conversions, they formed high molecular weight materials because they had funtionalities (i.e., number of reactive groups on a single reactant) greater than two resulting in cross-linking.

When the average functionality is greater than two, cross-linking occurs. Glyptal polyesters were produced in 1901 by heating glycerol and phthalic anhydride. Since the secondary hydroxyl is less active than the terminal primary hydroxyls in glycerol, the first product formed at conversions of less than 70% is a largely linear polymer. A cross-linked product is formed by further heating because the functionality of glycerol is three (Equation 4.50).

Alkyds were synthesized by Kienle in the 1920s from trifunctional alcohols and dicarboxylic acids. Unsaturated oils, called drying oils, were transesterified with the phthalic anhydride in the reaction so that an unsaturated polymer was obtained that could later be reacted producing a cross-linked product.

The term alkyd is sometimes used to describe all polyesters produced from polybasic acids and a polyhydric alcohol. The terms saturated and unsaturated polyesters have been used to distinguish between those alkyd polyesters that are saturated from those that contain unsaturation.

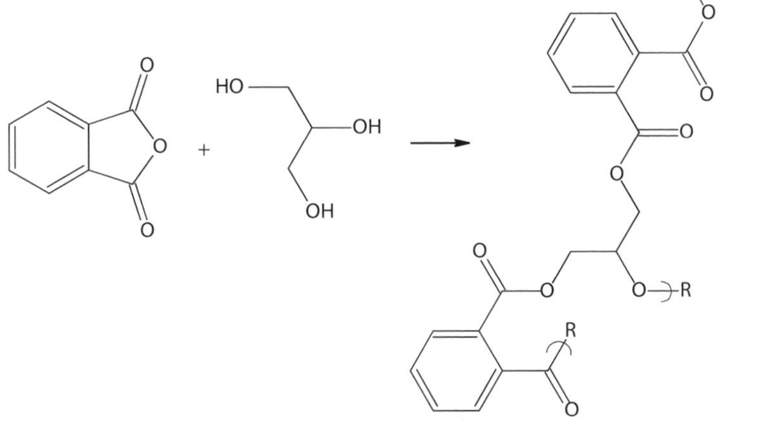

$$(4.50)$$

Phthalic anhydride + Glycerol $\rightarrow$ Cross-linked polyester

Unsaturated polyesters have been produced from reaction of ethylene glycol, phthalic anahy-dride, or maleic anhydride (Equation 4.51). These polyesters may be dissolved in organic solvents and is used as cross-linking resins for the production of fibrous glass-reinforced composites.

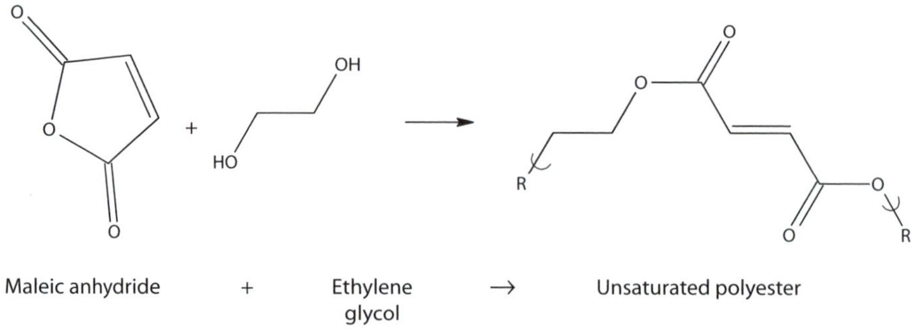

$$\text{(4.51)}$$

Maleic anhydride   +   Ethylene   →   Unsaturated polyester
                                              glycol

Aromatic polyesters had been successfully synthesized from reaction of ethylene glycol and vari-ous aromatic diacids but commercialization awaited a ready inexpensive source aromatic diacids. An inexpensive process was discovered for the separation of the various xylene isomers by crystallization. The availability of inexpensive xylene isomers allowed the formation of terephthalic acid through the air oxidation of the *p*-xylene isomer. DuPont, in 1953, produced polyester fibers from melt spinning, but it was not until the 1970s that DuPont-produced polyester fibers became commercially available.

Expanding on the work of Carothers and Hill on polyesters, Whinfield and Dickson, in England, overcame the problems of Carothers and coworkers by employing an ester interchange reaction between ethylene glycol and the methyl ester of terephthalic acid forming the polyester poly(ethylene terephthalate) (PET) with the first plant coming on line in 1953. This classic reaction producing Dacron, Kodel, and Terylene fibers—shown in Equation 4.52.

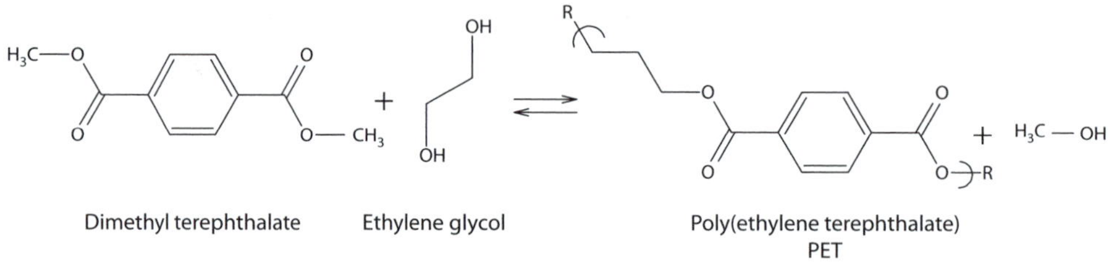

$$\text{(4.52)}$$

Dimethyl terephthalate   Ethylene glycol   Poly(ethylene terephthalate)
                                                                                          PET

While PET is normally made as described in Equation 4.45, it can also be made from the ring-opening reaction with ethylene glycol as shown in Equation 4.53.

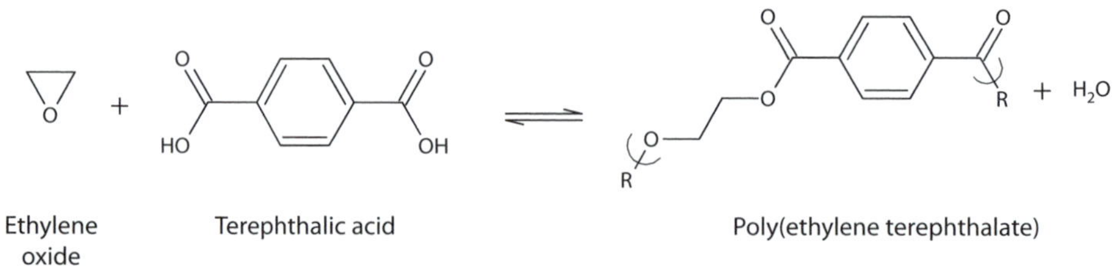

$$\text{(4.53)}$$

Ethylene   Terephthalic acid   Poly(ethylene terephthalate)
oxide

Polyester fibers (PET; Equations 4.52 and 4.53) are the world's leading synthetic fibers produced at an annual rate of over 1.5 million tons in the United States. Fibers are produced if the product is pushed through a small hole. As the polyester emerges from the hole, tension is applied assisting the polymer chains to align, giving the fiber additional strength in the direction of pulling. Crystallization of polyester resin can be achieved through heating to about 190°C followed by slow cooling. Rapid cooling, called *quenching*, produces a more amorphous material from which transparent film is made. Film strength is increased through application of heat and pulling of the film. Biaxially oriented PET film is one of the strongest films available. Thicker film, because of its low carbon dioxide permeability, is used in the manufacture of containers for carbonated drinks through injection molding. It is also used as magnetic film in X-ray and other photographic applications. Thinner film is used in such food applications as food packaging, including the boil-in-a-bag foods.

PET is difficult to mold because of its high melting, T_m 240°C, point. PET also crystallizes relatively slowly so that extra care must be exercised to insure that the PET molded products are fully crystallized or the partial crystallized portions will be preferred sites for cracking, crazing, shrinkage, and so on. Thus, nucleating agents and crystallization accelerators have been used to improve the crystallization rate. Postannealing has been used where appropriate.

Poly(butylene terephthalate) (PBT), because of the addition of two methylene units in the diol-derived portion, has lower melting point than PET with a T_g of about 170°C (Equation 4.54). Moldability of aryl polyesters has been improved through the use of PBT instead of PET or by use of blends of PET and PBT. These aryl polyesters are used for bicycle wheels, springs, and blow-molded containers.

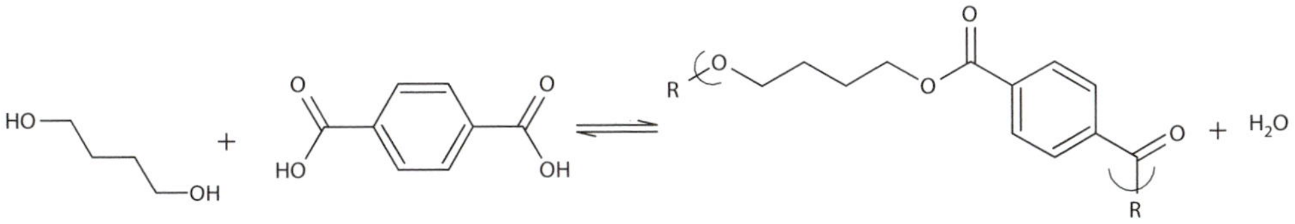

$$ (4.54) $$

Poly(butylene terephthalate), PBT

By comparison to PET, PBT melts at a lower temperature and crystallizes more rapidly and is often employed as a molding compound. PBT offers a balance of properties between nylons and acetals with low moisture absorption, good fatigue resistance, good solvent resistance, extremely good self-lubrication, and good maintenance of physical properties even at relatively high use temperatures. Fiber-reinforced PBT molding compound is sold under the trade name Celanex. A PBT molding compound was first sold under the trade name Valox. Today, there are many PBT molding compounds available.

Table 4.3 contains selected physical properties of PET and PBT.

TABLE 4.3
General Physical Properties of PET and PBT

	PET	PBT
Heat-deflection temperature (1,820 kPa; °C)	100	65
Maximum resistance to continuous heat (°C)	100	60
Crystalline melting point (°C)		
Coefficient of linear expansion (cm/cm-°C, 10^{-5})	6.5	7.0
Compressive strength (kPa)	8.6×10^4	7.5×10^4
Flexural strength (kPa)	1.1×10^5	9.6×10^4
Impact strength (Izod: cm-N/cm of notch)	26	53
Tensile strength (kPa)	6.2×10^4	5.5×10^4
Ultimate elongation (%)	100	100
Density (g/mL)	1.35	1.35

Poly(dihydroxymethylcyclohexyl terephthalate) (Equation 4.55) was introduced by Eastman Kodak as Kodel in 1958. In comparison to PET and PBT, the insertion of the cyclohexyl moiety gives a more hydrophobic material as well as a more moldable product that can be injected molded. The sole raw material is again dimethyl terephthalate. Reduction of the dimethyl terephthalate gives the dialcohol cyclohexanedimethanol. (Notice the reoccurring theme of reusing or using in a modified form the same readily available and inexpensive materials.) This material, along with blends and mixtures, is often extruded into film and sheeting for packaging. Kodel-type materials are used to package hardware and other heavy items as well as blow molded to produce packaging for shampoos, liquid detergents, and so on.

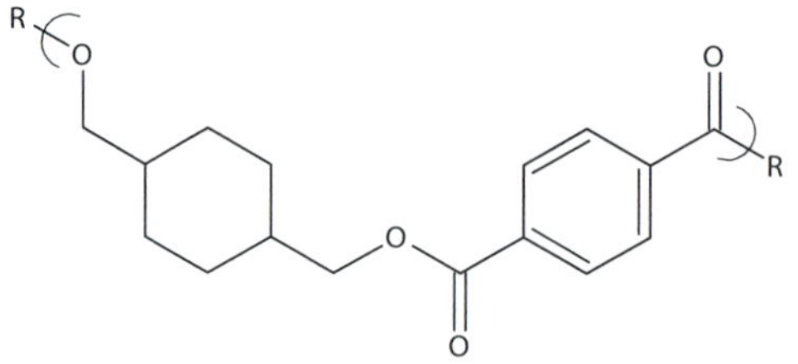

$$(4.55)$$

Poly(dihydroxymethylcyclohexyl terephthalate)

The hard–soft block copolymer approach employed to produce segmental PU (Section 7.6) has also been used with polyesters with the hard block formed from 1,4-butadienediol and terephthalic acid, while the soft block is provided from oligomeric (approximate molecular weight of 2,000 Da) poly(tetramethylene glycol) and is sold under the trade name of Hytrel.

Along with nylons, polyester fibers approach and exceed common natural fibers such as cotton and wool in heat stability, wash-and-wear properties, and wrinkle resistance. Blended textiles from polyester and cotton and wool also can be made to be permanent press and wrinkle resistant. The fibers are typically formed from melt or solvent spinning. Chemical and physical modification is often employed to produce differing fiber appearances from the same basic fiber material. Self-crimping textiles are made by combining materials with differing shrinkage properties. Different shaped dyes produce materials with varying contours and properties, including hollow fibers.

Several "wholly" aromatic polyesters are available. As expected, they are more difficult to process and stiffer, less soluble, but are employed because of their good high-thermal performance. Ekonol is the homopolymer formed from p-hydroxybenzoic acid ester (Equation 4.56). Ekonol has a T_g in excess of 500°C. It is highly crystalline and offers good strength.

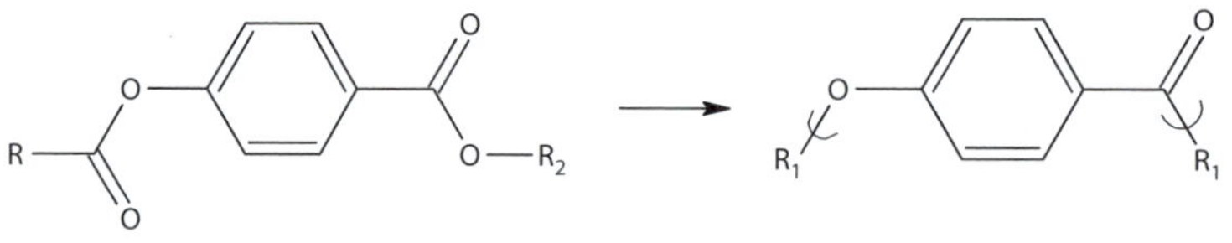

$$(4.56)$$

p-Hydroxybenzoic acid ester   Poly-p-benzoate

It is not unexpected that such aromatic polyesters have properties similar to PC because of their structural similarities.

4.6 POLYCARBONATES

Polycarbonates were originally produced from the interfacial polymerization reaction of phosgene with bisphenol A (BPA) [2,2-bis(4-hydroxyphenyl) propane] (Equation 4.57). The BPA is dissolved in a NaOH aqueous solution, resulting in the bisphenol losing the two "phenol" protons creating a more effective Lewis base or nucleophile. Phosgene is added typically via a chlorinated solvent and tertiary amine. The product is an unusually tough and transparent plastic available under the trade names of Lexan (General Electric) and Merlon (Mobay).

Polycarbonates can also be made from the ester interchange between diphenyl carbonate and BPA. The melting point of PCs is decreased from 225°C to 195°C when the methyl pendant groups are replaced by propyl groups.

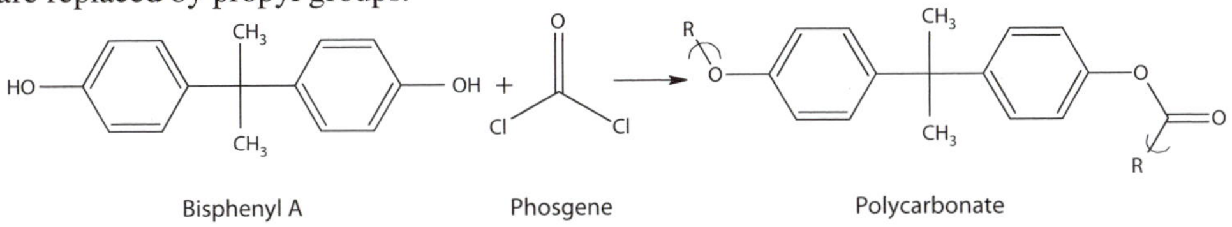

(4.57)

Bisphenyl A   Phosgene   Polycarbonate

Polycarbonates and PC-polyester copolymers are used for glazing, sealed beam headlights, door seals, popcorn cookers, solar heat collectors, and appliance housings.

Essentially all compact discs (CD), and related audio and video storing devices, have similar components. Here, we will focus on the composition of CDs that are purchased already containing the desired information and CDs that can be recorded on, CD/Rs. The major material of all of these storing devices is a PC base. Thus, these devices are PC laid over with thin layers of other materials. Of the less than 20 g CD weight, over 95% is the PC.

The aromatic rings contribute to PCs high glass-transition temperature and stiffness (Table 4.4). The aliphatic groups temper this tendency giving PC a decent solubility. The two methyl groups also contribute to the stiffness because they take up space somewhat hindering free rotation about the aliphatic central carbon moiety. Factors contributing to PC chain association are interaction between the aromatic rings of different parts of the same or different PC chain segments and the permanent dipole present within the carbonyl group. The lack of "hydrogen-bonding" hydrogens on PC means that this type of association is not present. The associations between PC segments contribute to a general lack of mobility of individual chains. This results in PC having a relatively high viscosity, which ultimately leads to a low melt flow during processing. The moderate inflexibility, lack of ready mobility, and nonlinear structure contribute to PC having a relatively long time constant for crystallization. Cooling is allowed to be relatively rapid so that most PC products possess a large degree of amorphous nature and accounts for PC having a high impact strength that is important in its use to blunt high impacts and important to CDs to provide a semirigid disc that can be dropped and not readily shatter. Thus, control of the rate of flow and cooling is an important factor in producing CD-quality PC material. A high degree of amorphous nature also contributes to the needed optical transparency with amorphous PC having a transparency near that of window glass.

TABLE 4.4
General Physical Properties of a Polycarbonate

Heat-deflection temperature (1,820 kPa; °C)	130
Maximum resistance to continuous heat (°C)	115
Crystalline melting point (°C)	225
Coefficient of linear expansion (cm/cm-°C, 10^{-5})	6.8
Compressive strength (kPa)	8.6×10^4
Flexural strength (kPa)	9.3×10^4
Impact strength (Izod: cm-N/cm of notch)	530
Tensile strength (kPa)	7.2×10^4
Ultimate elongation (%)	110
Density (g/mL)	1.2

While PC has the necessary desirable qualities as the basic material for information storage, it also has some debits. First, PC is relatively expensive in comparison with many polymers. Its superior combination of properties and ability for a large cost markup allows it to be an economically feasible material for specific commercial uses. Second, the polar backbone is susceptible to long-term hydrolysis so that water must be ruthlessly purged. The drying process, generally 4 hours, is often achieved by the placement of PC chips in an oven at 120°C with a dewpoint of −18°C.

The PC utilized for information storage has strict requirements including a high purity, greater than 87% spectral light transmission based on a 4 mm thick sample, a yellowness index less than 2, and light scattering less than 0.3 cd/(m²-lx). Sources of PC are mainly two, virgin and recycled. Virgin PC has an index of yellowing of 1.8 but the first reground PC has a yellow index to about 3.5. Thus, CDs employ only virgin PC.

Requirements for CD-quality material requires PC with low levels of chemical impurities, low particle levels, thermal stability, excellent mold release, excellent clarity, and constant flow and constant mechanical behavior (for reproducibility). There exists a time/cost balance. High molecular weight PC offers a little increase in physical property but the flow rate is slow making rapid production of CDs difficult. The molecular weight where good mechanical strength and reasonable flow occurs, and that allows for short cycles, is in the range of 16,000–28,000 Da.

Injection molding requires the barrel temperature to be about 350°C with a barrel pressure in excess of 138 MPa. The mold is maintained at 110°C to insure uniform flow and high definition, and to discourage an uneven index of refraction, birefringence. The CD is about four one-hundredths of an inch (0.5 mm) thick. For prerecorded CDs, the PC is compression molded on a stamper imprinted with the recorder information. This takes about 4 seconds. Once the clear piece of PC is formed, a thin, reflective aluminum layer is sputtered onto the disc. Then, a thin acrylic layer is sprayed over the aluminum to protect it. The label is then printed onto the acrylic surface and the CD is complete. This process is described in greater detail shortly.

The construction of the recordable CD or CD/R is more complex. Standard CDs contain four or five (a label layer) layers. These four (five) layers are:

- PC base
- Dye surface
- Reflective layer
- Lacquer layer
- Label

We must remember that the information is closest to the label side of the CD not the clear plastic side the data is read from. It is fairly easy to scratch the top surface of the CD rendering it unusable. Some CDs have a special hard top coating that helps resist surface damage. Dust, minor scratches, and finger prints are generally not harmful because the laser assemblies are designed to focus beyond the disc surface.

The "play only" CD contains a series of pits and lands generated during the mastering process based specifically on the data provided. The PC layer for the CD/R does not contain these pits and lands but rather contains a shallow groove or pregroove used for timing and tracking. A CD/R writes, records, information by using its lasers to physically "burn" pits into the organic dye layer. When heated beyond a certain temperature, the area that was "burned" becomes opaque and reflects a different amount of light in comparison to areas that have not been "burned." These "burned" and "unburned" areas correspond to the pits and lands of the prerecorded CD allowing it to be "read" by a regular CD player.

CD/Rs are Write Once/Read Many (WORM) storage discs. They cannot be erased once they have been written on.

The dye coating is the most important, expensive, and complex part of the CD manufacturing process. The dye layer serves to inhibit or permit laser radiation from piercing to the PC. There are three competing organic dye polymers used to manufacture CD/Rs. The original dye polymer used was the cyanine (and metal-stabilized cyanine) materials identified by a greenish tint with a gold reflective layer or blue when a silver reflective layer is present. The employed cyanine dye is itself blue. The suggested age for storage on cyanine dye discs is about 20 years and for metal-stabilized cyanine it is about 100 years.

The second dye system is based on an almost clear yellow–green phthalocyanine dye. The estimated lifetime of a disk based on the phthalocyanine dye is 100 years. The third dye system is based on azo dyes. The azo dye used in the process has a deep blue color partially caused by its unique silver alloy reflective layer. Again, a projected lifetime of 100 is sited.

The two most widely employed reflective layers are 24K gold and a silver alloy. The layers are thin enough, about 50–100 nm thick, to allow us to see through them.

The first step in the manufacturing of a blank CD is the creation of a glass master with a laser beam recorder (LBR). For the prerecorded CDs, pits and lands are etched into the photoresist or nonphotoresist on the glass. For the CD/R, the LBR records a shallow groove as a continuous spiral called the pregroove. This pregroove is not a perfect spiral, but it is "wobbled." On the recorded disc, the timing information necessary to control the disc's rate of spin is included as data. But for the CD/R disc, the CD recorder needs to have a way to guide the recording laser and the speed of the blank disc. The wobbled pregroove provides the tracking and timing information for the recording laser. The wobble is a slight sinusoidal wave that is about 0.03 mm from the center of the track path. This wobble corresponds to about one-thousandth of the length of one complete waveform so it is so small that it is not seen by the naked eye, but it is "seen" by the recoding laser. This slight diversion, from a completely spiral pathway provides the timing information and guides the recorder. This information is called the absolute time in pregroove or ATIP. It ensures that data is recorded at a constant rate. The resulting data trail obliterates the wobble pregroove, leaving recorded data in its place.

Once the glass master has been written, it is tested, silvered, and ready to be electroplated. It is electrically conductive and is placed in a reservoir with an electrolyte solution containing a nickel salt. Current is applied to the glass master eventually resulting in a metal layer on the glass after about 2 hours. The nickel copy of the glass master is called the metal "father" and is used as a stamper to produce CD/R discs. If the replication run is large, the metal "father" is returned to the electroplating process for creation of metal "mothers" that are used to make metal "sons" identical to the fathers. These "sons" are then employed in the manufacture of "blank" CD/R discs.

Injection molding is employed to stamp out copies of the master discs from PC. The molded PC discs are cooled and hardened quickly, about 4–6 seconds, and evenly. The dye layer is then applied. The applied dyes are often proprietary and continually modified in an attempt to get a "better" dye. The dye must be compatible with the system and adhere to the PC base. It is applied by spin coating; that is, the disc is spun and the dye is sprayed onto the surface. The dye is then dried and cured.

A metal layer is now added since the disc surface must reflect laser light. Aluminum is the most cost effective and widely used for prerecorded discs but most CD/R discs use gold or silver because of their greater reflectivity. The gold or silver is applied using sputtering or vacuum deposition. A thin acrylic plastic layer is then applied by spin coating and cured in ultraviolet (UV) light to aid in the protection of the disc. An optional label layer can now be added. I find that CD/Rs with this additional layer are stronger and appear to record better. The CD/Rs are now packaged, generally in packages of 10, 25, 50, and 100. The packaging is also plastic. My latest group of CD/Rs has a black polystyrene base and a clear polypropylene cone. The bulk purchase price is about 5–10 cents per disc, on sale. In a glass case, the individual CD/Rs often cost about a dollar. Thus, for all the high technology required to produce a CD/R, it is inexpensive.

A laser is used to encode information through creation of physical features sometimes referred to as "pits and lands" of different reflectivity at the PC-metal interface. As noted above, recordable CDs contain an organic dye between the PC and metal film. A laser creates areas of differing reflectiveness in the dye layer through photochemical reactions.

A beam from a semiconductor diode laser "interrogates" the undersides of both recordable and recorded CDs seeking out areas of reflected, corresponding to the binary "one," and unreflected, corresponding to the binary "zero" light. The ability to "read" information is dependent on the wavelength of the laser. Today, most of the CD players use a near-infrared laser because of the stability of such lasers. Efforts are underway to develop stable and inexpensive lasers of shorter wavelengths that will allow the holding of more information within the same space.

There is concern that BPA may cause neural and behavioral changes in infants and children. A major line of exposure to infants is in baby bottles. In our modern age, the composition of baby bottles has evolved from glass to polyethylene to PC and now back to glass and other newer materials such as Tritan, which is a proprietary copolyester developed by Eastman with properties similar to PC. While the use of most PC will continue, the "better to be safe than sorry" motto is best used for products that come in intimate contact with infants that are more susceptible than adults to various agents.

4.7 SYNTHETIC POLYAMIDES

Wallace Hume Carothers was brought to DuPont because his fellow researchers at Harvard and the University of Illinois called him the best synthetic chemist they knew. He started a program aimed at understanding the composition of natural polymers such as silk, cellulose, and rubber. Many of his efforts related to condensation polymers were based on his belief that if a monofunctional reactant reacted in a certain manner forming a small molecule (Equation 4.58) and that similar reactions except employing reactants with two reactive groups would form polymers (Equation 4.59).

$$R-OH + HOOC-R' \rightleftharpoons R-O-\overset{\overset{\displaystyle O}{\|}}{C}-R' + HOH \qquad (4.58)$$

Small ester

$$HO-R-OH + HOOC-R'-COOH \rightleftharpoons -(-O-R-O-\overset{\overset{\displaystyle O}{\|}}{C}-R'-\overset{\overset{\displaystyle O}{\|}}{C}-)- + H_2O \qquad (4.59)$$

Polyester

While the Carothers' group had made both polyesters and polyamides, they initially emphasized work on the polyesters since they were more soluble and easier to work with. One of Carothers' coworkers, Julian Hill, noticed that he could form fibers if he took a soft polyester material on a glass stirring rod and pulled some of it away from the clump. Because the polyesters had too low softening points for use as textiles, the group returned to work with the polyamides. They found that fibers could also be formed by the polyamides similar to those formed by the polyesters. These polyamides allowed the formation of fibers that approached, and in some cases surpassed, the strength of natural fibers. This new miracle fiber was introduced at the 1939 New York World's Fair in an exhibit that announced the synthesis of this wonder fiber from "coal, air, and water"—an exaggeration—but nevertheless eye catching. When the polyamides, nylons were first offered for sale in New York City, on May 15, 1940, nearly a million pairs were sold in the first few hours. Nylon sales took a large drop when it was noted that nylon was needed to produce the parachute material so critical to World War II.

The first polyesters produced by Carothers had relatively low molecular weights because of low fractional conversions. Carothers was successful in producing higher molecular weight polymers by shifting the equilibrium by removing the water produced. Equation 4.59 is an equilibrium process with the removal of water driving it toward polymer formation. However, these aliphatic polyesters, which he called "super polymers," lacked stiffening groups in the chain and thus had melting points that were too low for laundering and ironing.

Carothers next step was to move from polyesters to nylons and to increase the fractional conversion (p) by making salts by the equivalent reaction of 1,6-hexanediamine (hexamethylenediamine) and adipic acid. These salts were recrystallizable from ethanol giving essentially a true 1:1 ratio of reactants. Thus, a high molecular weight polyamide, generally known as simply a nylon, in this case nylon-66, was produced from the thermal decomposition of this equimolar salt as shown in Equation 4.60. This product has a melting point of 265°C.

Since the molecular weight of this nylon-66, produced by Carothers, was higher than he desired, 1% acetic acid (a monofunctional acid that acted to terminate chain growth) was added producing a product with a lower molecular weight and a melting point of about 200°C.

Molded nylon-66 is used for lawnmower blades, bicycle wheels, tractor hood extensions, skis for snowmobilies, skate wheels, motorcycle crank cases, bearings, and electrical connections. Fiber nylon-66 is used in clothing, fabrics, and rugs.

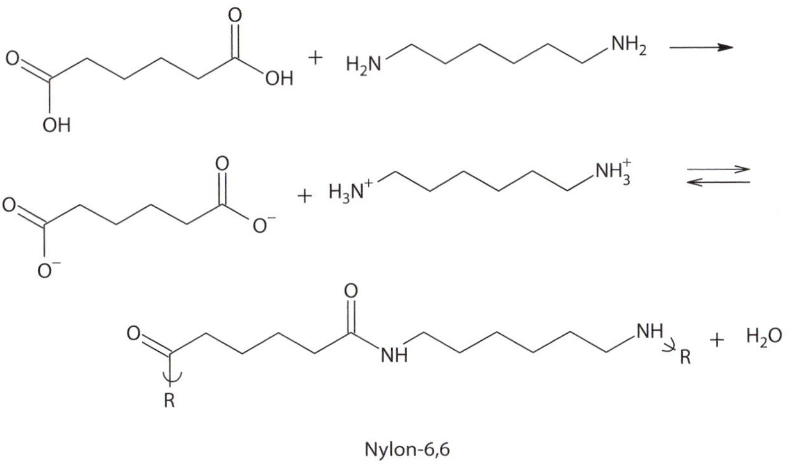

(4.60)

Nylon-6,6

In the early 1950s, George deMestral was walking in the Swiss countryside. When he got home he noticed that his jacket had lots of cockleburs on it. For some reason he examined the cockleburs and noticed that they had a lot of tiny "hooks." His cotton jacket had loops that "held" the cockleburs. He began putting into practice his observations, making combinations of materials with rigid hooks with materials that had flexible loops or eyes. The initial hook and eye for commercial use was made in France. Today, Velcro, the name given to the *hook-and-eye* combination, is often based on nylon as both the hook-and-eye material. Remember that nylon can be made to behave as both a fiber and as a plastic. Polyester is blended with the nylon to make it stronger. Polyesters have also been employed to make hook-and-eye material. The hook-and-eye material is used to fasten shoes, close space suits, and in many other applications.

The general structure for aliphatic nylons for naming purposes is

$$-(-NH-(CH_2)_a-NH-\overset{O}{\overset{||}{C}}-CH_2)_b-\overset{O}{\overset{||}{C}}-)- \tag{4.61}$$

where a is the number of carbons derived from the amine-associated portion, and b is the number of carbons, including the carbonyl carbon atoms, associated with the diacid. Thus, nylon-66 has six carbons derived from hexamethylenediamine, a is 6, and six carbons derived from adipic acid.

Nylon-610 is derived from hexamethylenediamine and sebacic acid (dodeconic acid) and is more resistant to moisture and more ductile than nylon-66 because of the presence of the additional flexible nonpolar methylenes.

TABLE 4.5
General Physical Properties of Nylon-66 and 6

	Nylon-66	Nylon 6
Heat-deflection temperature (1,820 kPa; °C)	75	80
Maximum resistance to continuous heat (°C)	120	125
Crystalline melting point (°C)	265	225
Coefficient of linear expansion (cm/cm-°C, 10^{-5})	8.0	8.0
Compressive strength (kPa)	1×10^5	9.7×10^4
Flexural strength (kPa)	1×10^5	9.7×10^4
Impact strength (Izod: cm-N/cm of notch)	80	160
Tensile strength (kPa)	8.3×10^4	6.2×10^4
Ultimate elongation (%)	30	–
Density (g/mL)	1.2	1.15

Nylon-6, structurally quite similar to nylon-66, was initially produced in Germany by the ring-opening polymerization of caprolactam partly as a way to avoid the patents established by DuPont a decade before. The copolymer of nylon-6 and nylon-66 has a smoother surface than either of the homopolymers.

$$-(-NH-(CH_2)_5-\overset{\overset{\displaystyle O}{\|}}{C}-)- \qquad (4.62)$$

Nylon-6

Nylon-6,6 is the dominant (sales wise) nylon in the United States, while nylon-6 is the dominant nylon in Europe.

Table 4.5 contains general physical properties of nylon-66 and nylon-6. As expected, they are similar.

The properties of polyamides are improved by the formation of polyether blocks (NBC) and by blending with thermoplastics such as EPDM, PET, PBT, and TPE (rubber-toughened nylons). NBC (Nyrim) is more expensive than reaction–injection molded (RIM) PU, but it can be heated to 200°C without melting. NBC moldings are produced from the RIM of poly(propylene glycol) (and other elastomeric materials) and caprolactam. Nylon and the elastomeric materials are incompatible but they are chemically combined in the RIM process giving a semicrystalline material with nylon and elastomeric blocks. *Nyrim* is referred to as a rubber-toughened recyclable nylon thermoplastic. Nyrim materials are used as dozer pads, skew blades, gears, half tracks, skew propellers, and as tracks for swamp-going vehicles. The tendency of these moldings to swell in water is reduced by reinforcing them with fibrous glass. They differ from toughened nylon, which are blends and not chemically connected. The ability to form strong hydrogen bonding is reduced, and the flexibility increased, by placing bulky methoxymethyl pendant onto nylons (Equation 4.63).

$$-(-NH-(CH_2)_6-\underset{\underset{\displaystyle CH_2-O-CH_3}{|}}{N}-\overset{\overset{\displaystyle O}{\|}}{C}-CH_2)_4-\overset{\overset{\displaystyle O}{\|}}{C}-)- \qquad (4.63)$$

While aliphatic-containing polyamides are given the name nylons, those where at least 85% of the amide groups are attached to an aromatic compound are called *aramids*. Aramids are stronger

TABLE 4.6
General Physical Properties of a Typical Aramid

Heat-deflection temperature (1,820 kPa; ºC)	260
Maximum resistance to continuous heat (ºC)	150
Crystalline melting point (ºC)	>370
Coefficient of linear expansion (cm/cm-ºC, 10^{-5})	2.6
Compressive strength (kPa)	2×10^5
Flexural strength (kPa)	1.7×10^5
Impact strength (Izod: cm-N/cm of notch)	75
Tensile strength (kPa)	1.2×10^5
Ultimate elongation (%)	5
Density (g/mL)	1.2

and tougher than nylons but they are also more difficult to solubilize and fabricate. Because the presence of the aromatic groups causes the aramids to be stiff, they often form liquid crystals (LC) that are present in a nematic liquid crystal state in concentrated solution.

Aramids are generally prepared by the solution or interfacial polycondensation of meta- and para-substituted diacid chlorides and/or diamines. In some systems, synthesis is achieved under rapidly stirred conditions where the polymer is quasi-soluble in the system. The polymer mixture is forced through a small opening into a nonsolvent forming a fiber without the need to dissolve the polymer before fiber formation. General properties of an amide such as Nomex are given in Table 4.6. Notice the generally higher "strength" values and the greater stiffness (less elongation) in comparison to nylon-66.

Poly(m-phenylene isophthalamide) (Equation 4.64), sold under the trade name of Nomex, exhibits good thermal stability decomposing above 370ºC. It is used in flame-resistant clothing. It is also used in the form of thin pads to protect sintered silica-fiber mats from stress and vibrations during the flight of the space shuttle.

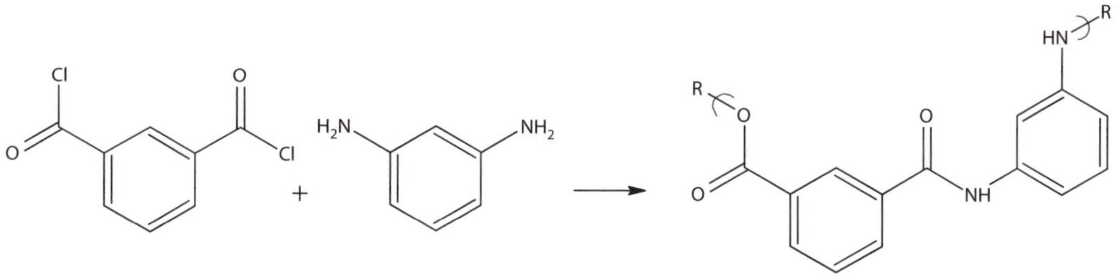

Isophthaloyl chloride + m-Phenylenediamine → Poly(m-phenylene isophthalamide)

(4.64)

Nomex is sold as both a fiber and a sheet. The sheet material is often a laminate with paper and used in electrical applications such as circuit boards and transformer cores. Firefighter gear generally includes a Nomex hood to protect the face not protected by the helmet. Race car drivers are required to wear flame-resistant underwear, balaclava, socks, shoes, and gloves that are typically made from Nomex fiber. Military pilots also wear flight suits that are often Nomex.

Because of the meta orientations of the reactants, Nomex chains do not realign on fiber formation, resulting is a material that is weaker than the corresponding para-oriented polymer, Kevlar.

The corresponding aramid produced using the para reactant in place of the meta gives poly(p-phenylene terephthalamide) (PPT) produced under the trademark of Kevlar and developed by DuPont by Stephanie Kwolek and Roberto Berendt in 1965 (Equation 4.65). Like Nomex, Kevlar exhibits good thermal stability decomposing above about 500°C. By weight it has higher strength and modulus than steel and is used in the manufacture of the so-called "bullet-proof" clothing. Because of its outstanding strength, it was used as the skin covering of the Gossamer Albatross, which was flown using only human-power across the English Channel. Aramids are also used as fiber reenforcement in composites and as tire cord.

(4.65)

Terephthaloyl chloride (PPT) *p*-Phenylenediamine Poly(*p*-phenylene terephthalamide)

Kevlar fiber has a high tensile strength (about 3,000 MPa) and a relatively low density (about 1.4 g/mL) leading to the often used statement that Kevlar is five times stronger than steel (on an equal weight basis). It gains part of its strength from the interchain hydrogen bonds formed between the carbonyl groups and hydrogen atoms on amines on neighboring chains and on the partial stacking of the phenylene rings allowing pi–pi interactions between the members of the stacking units. The molecularly rigid chains tend to form sheet-like structures similar to those of silk.

Kevlar is used in the manufacture of so-called "bullet-proof" clothing used by the military, policemen, and SWAT teams. In truth, most bullet-proof clothing is bullet-resistant and unless quite bulky not able to "stop" most rifle bullets or high caliber hand guns. It is also used in the construction of bullet-resistant facemasks by the military and motorcycle riders and to protect against abrasion by motorcycle riders.

There are many other commercial uses for Kevlar. It is used as the inner liner for some bicycle tires to protect against puncture. It is also being used for bow strings in archery. Drumheads have been made from Kevlar. Kevlar is widely employed as the protective outer sheath for fiber optic cable and as the reinforcing layer in rubber bellow expansion joints and hoses for high-temperature applications.

Aramid materials are also employed in the U.S. space program along with other "space-age" materials. The thermal Micrometeroid Garment on the Extravehicular Mobility Unit, Advanced Crew Escape Suit, thermal blankets, and the fuselage, bay doors, upper wind surfaces of the Space Shuttle Orbiter employs Nomex, Kevlar, and Gore-Tex materials. The airbags for the Mars Pathfinder and MER rovers, the Galileo atmospheric probe and the new Crew Exploration Vehicle all have aramid materials. Aramid materials are also used in the form of thin pads to protect sintered silica-fiber mats from stress and vibrations during the flight of the space shuttle.

Aramid fibers are widely employed in the construction of composites where high strength is required. The continuous phase is often an epoxy resin. Applications include cricket bats, helicopter rotor blades, bodies for formula one race cars, kayaks, tennis rackets, lacrosse sticks, and ice hockey sticks.

Several so-called semiaromatic nylons have been produced. Nylon-6,T is produced from condensation of terephthalic acid and 1,6-hexanediamine (4.66). Both reactants are readily available and inexpensive and the resulting materials offer greater strength than simply wholly aliphatic nylons such as nylon-6,6. Nylon-6,T has a very high T_m of 370°C and a T_g of 180°C. The high T_m results in the need for a high temperature to be employed in processing so that a third reactant is often introduced to lower the T_m and the processing temperature. "Third reactants" often used are adipic acid, caprolactam, isophthalic acid, and 1,5-hexyldiamine. These materials are sold under the trade names of Zytel HTN, Ultramid T, and Amodel R.

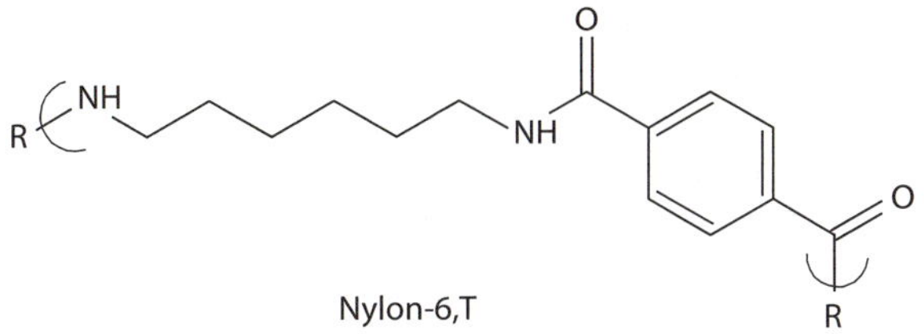

Nylon-6,T

(4.66)

Nylons offered new challenges to the chemical industry. Because of the presence of polar groups the attractive forces between chains was high in comparison to vinyl polymers. Nylons are generally semicrystalline, meaning they have a good amount of order. Thus, while they have a T_g, the main physical transition is the T_m so that they undergo a sharper transition from solid to melt in comparison to many of the vinyl polymers discussed in the next three chapters. Thus, the processing temperature window is narrower. If melt flow is required for processing, then the temperature must be sufficient to allow for ready flow but low enough so as not to break primary bonds within the processed material. Even so, processing techniques have been developed that allow nylons to be readily processed using most of the standard techniques.

Since chains of nylons having an even number of carbon atoms between the amide and acid groups pack better, some of their melting points are higher than comparable nylons with odd numbers of carbon atoms (Table 4.7). Further, the melting points decrease and the water resistance increases as the number of methylene groups between the amide and acid groups increases.

The presence of the polar groups result in materials with relatively high T_g and T_m values so that, unlike most vinyl polymers that must be above their T_g to allow needed flexibility, nylons and many condensation polymers function best where strength, and not flexibility, is the desired behavior. Because of the presence of these polar groups, which also allow for hydrogen bonding, nylons and most condensation polymers are stronger, more rigid and brittle, and tougher in comparison to most vinyl polymers. Nylons are also "lubrication-free" meaning they do not need a lubricant for easy mobility so that they can be used as mechanical bearings and gears without the need for periodic lubrication.

TABLE 4.7
Melting Point of Selected Polyamides

Nylon	Melting Point (°C)	Nylon	Melting Point (°C)
Aliphatic Nylons			
3	320	11	190
4	265	12	185
5	270	46	275
6	225	56	225
7	227	66	265
8	195	410	240
9	200	510	190
10	185	610	230

Aromatic (Terephthalamides) Nylons

Diamine	Melting Point (°C)	Diamine	Melting Point (°C)
1,2-Ethylene	460	1,4-Tetramethylene	440
1,3-Trimethylene	400	1,5-Pentamethylene	350
		1,6-Hexamethylene	370

In general, more crystalline nylons are used as fibers, while less crystalline nylon materials are more used as plastics. The amount of crystallinity is controlled through a variety of means, including introduction of bulky groups and asymmetric units, rapid cooling of nonaligned melts, and introduction of plasticizing materials. The theme of using asymmetric units was used by Grace and Company in developing Trogamid T, an amorphous transparent nylon, from the condensation of terephthalic acid with a mixture of 2,2,4- and 2,4,4-trimethylhexamethylene diamines.

4.8 POLYIMIDES

The initial announcement for the commercial preparation of polyetherimides (PEIs) was made by General Electric in 1982 under the trade name of Ultem. The final reaction involves the imidization of a diacid anhydride through reaction with a diamine, here m-phenylenediamine (4.67). The "ether" portion of the polymer backbone results from the presence of ether linkages within the diacid anhydride.

Ultem is used in medical and chemical instrumentation. Products made from Ultem are high melting, offer good stiffness, transparency, impact and mechanical strength, high flame resistance, low smoke generation, and broad chemical resistance. Some of these properties are expected. The high flame resistance is at least in part derived from the presence of already partially or largely oxidized atoms in the product. The low smoke generation is partially derived from the largely cyclic structure with other cyclic structures predictable from the product structure if exposed to sufficient heat. These cyclic structures often give products that are not evolved with good char formation when the material is exposed to ordinary flame conditions.

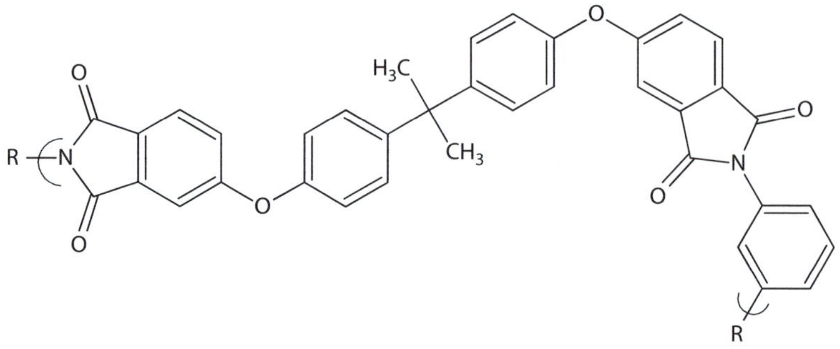

(4.67)

Polyetherimide

The general good mechanical properties are a result of the presence of strong double bonds present within polycyclic structures composing the polymer backbone plus the presence of strongly polar bonding units that allow the formation of good interactions between chains. Further, the structure is largely rigid with good dimensional stability along the polymer backbone. Any flexibility is gained largely from the presence of the ether linkages for the PEI and the presence of methylene units for the polyimides. These products offer good stable melt viscosities even after recycling several times. They can be processed using a variety of techniques including formation of sections as thin as 5 mil.

While Ultem is an example of thermoplastic polyimides, there are a number of thermoset polyimides. These are known for good chemical resistance and mechanical properties as well as good thermal stability. They are employed in electronic applications as insulating film on magnetic wire, for flexible cables, and for medical tubing. On a laptop computer, the cable connecting the main logic board to the display is often a polyimide with copper conductors. Polyimides are also used as high-temperature adhesives in the semiconductor sector.

4.9 POLYBENZIMIDAZOLES AND RELATED POLYMERS

Many heterocyclic polymers have been produced in an attempt to develop high–temperature-resistant polymers for aerospace applications. Among these are the polybenzimidazoles (PBIs) which are prepared from aromatic tetramines and esters of dicarboxylic acids, here terephthalic acid (Equation 4.68). In standardized procedures, the reactants are heated to below 300°C forming soluble prepolymer, which is converted to the final insoluble polymer by further heating.

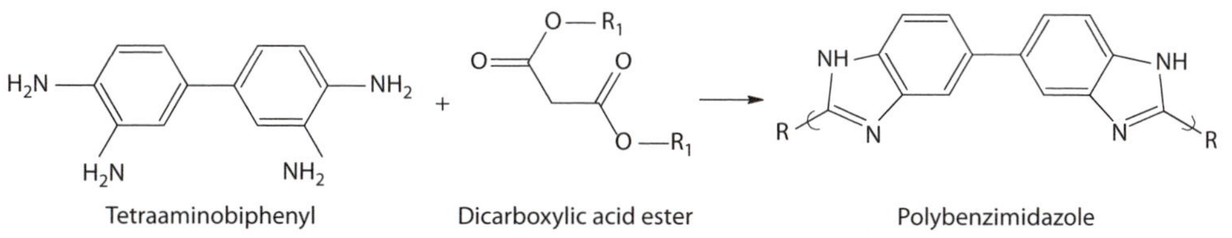

Tetraaminobiphenyl   Dicarboxylic acid ester   Polybenzimidazole   (4.68)

Polybenzimidazole fiber has an extremely high melting point and does not ignite. General properties appear in Table 4.8. They were initially made under the supervision of Carl Shipp "Speed" Marvel before 1983 as part of an overall project for the U.S. Air Force to create aerospace material. Celanese first produced PBI fibers in 1983 from reaction between tetra-aminobiphenyl and diphenyl isophthalate, Equation 4.69. The resulting structure is analogous to that shown in Equation 4.68.

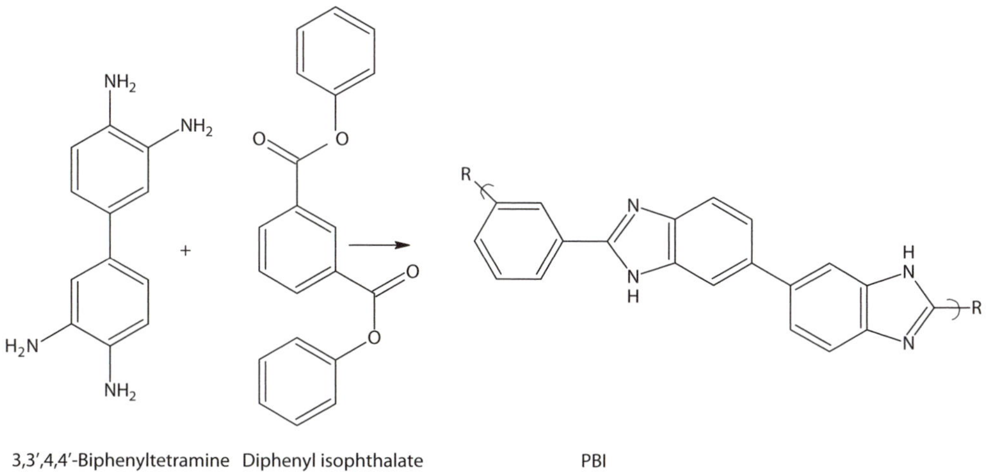

3,3′,4,4′-Biphenyltetramine Diphenyl isophthalate   PBI   (4.69)

PBI fibers are employed in the production of firefighter coats and suits, high-temperature gloves, welders' gloves, astronaut space suites, and race driver suits.

Polymers, such as PBI, have a weak link in them since a single covalent bond connects the phenyl rings. This weakness is overcome by the synthesis of ladder polymers, such as polyquinoxaline (4.70), polyquinoxalines (4.71), and polydithiones (4.72) that have two covalent bonds throughout the chain. Thus, the integrity of the polymer is maintained even if one bond is broken.

TABLE 4.8
General Physical Properties of PBI

Heat-deflection temperature (0.45 MPa; °C)	100	435
Coefficient of linear expansion (cm/cm-°C, 10^{-6})	23	
Compressive strength (MPa)	400	
Impact strength (Izod: J/m unnotched)	590	
Tensile strength (MPa)	160	
Poisson's ratio	0.3	
Ultimate elongation (%)	3	
Density (g/mL)	1.3	
Dielectric constant (at 1 MHz)	3.2	
Bulk resistivity (Ohm-cm)	10^{-15}	

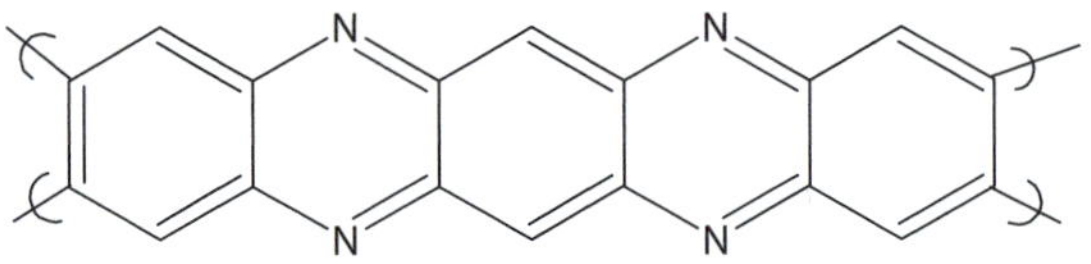

Polyquinoxaline (4.70)

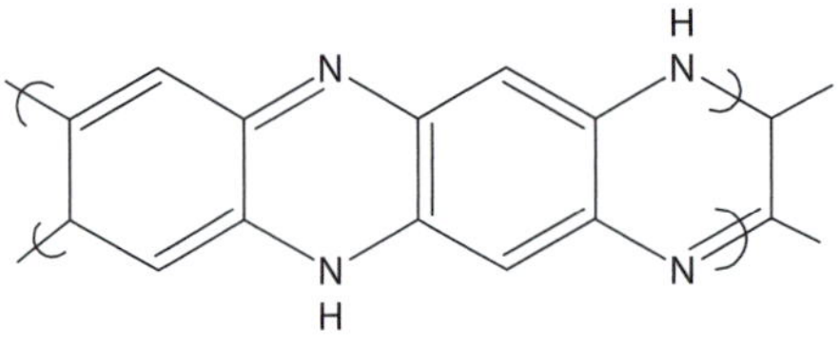

Polyquinoxaline (4.71)

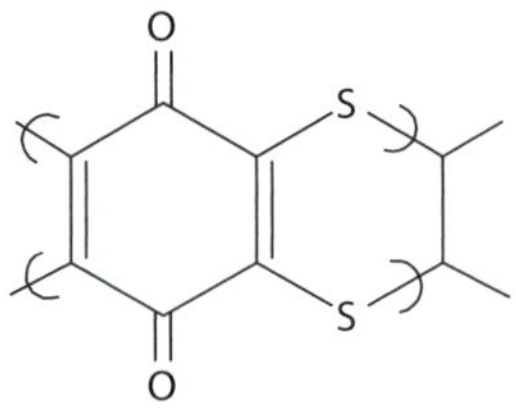

Polydithione (4.72)

4.10 POLYURETHANES AND POLYUREAS

Polyurethanes (PUs), or polycarbamates, were first made by Bayer and coworkers in 1937 by reacting diols and diisocyanates. This monomer combination avoided conflict with existing patents by Carothers and DuPont related to polyester production. The development of PU was stunted because of their use as aircraft coatings in World War II. In 1952, they became commercially available. PU are generally sold as flexible foams, rigid foams, and elastomers. Elastomer PUs materials are used as adhesives, coatings, and sealants. The basic reactants for these materials are diisocyanates and HO-containing reactants, including macroglycols called *polyols*. As expected, more flexible materials are made as the distance between the diol is larger and occupied by methylene and alkylene oxide moieties. Typical diisocyanates are tolyene diisocyanate (TDI; mixture of two isomers; (4.73), methylenediphenyl isocyanate (MDI), and polymeric isocyanate (PMDI) mixtures formed from phosgenating polyamines derived from the reaction of aniline with formaldehyde (Figure 4.6)). The aromatic unit provides stiffness to the polymer chain.

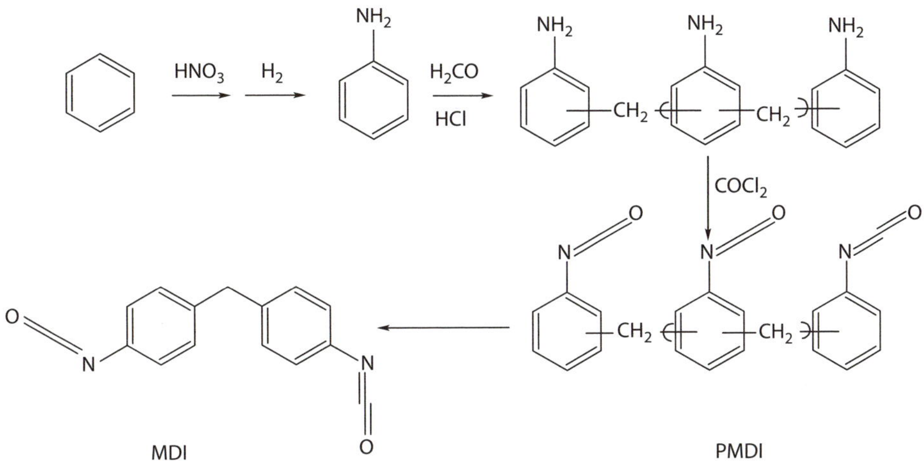

FIGURE 4.6 Synthesis of PMDI and MDI.

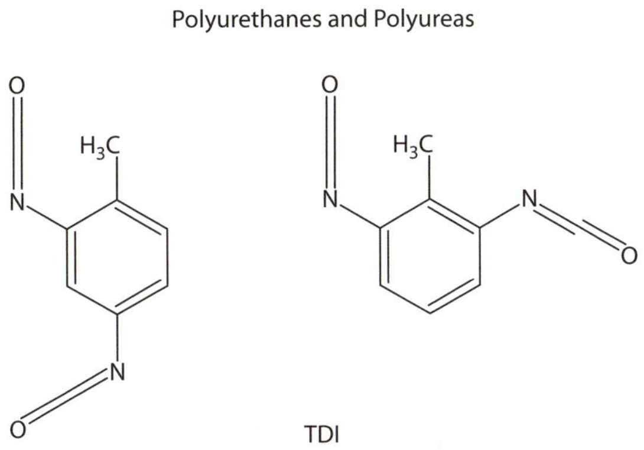

Polyurethanes and Polyureas

(4.73)

MDI and PMDI are both formed from the same reaction as shown in Figure 4.6

The hydroxyl-containing reactants are polyesters, polyethers, polycaprolactones, and diols. The manufacture of the polyether and polycaprolactone macroglycols is given in Figure 4.7.

The formation of polyurethanes from reaction with polyols can be catalyzed by addition of tertiary amines as shown in Figure 4.8

Production of foam was discovered when water was accidentally introduced into the reaction mixture producing what the scientists called "imitation Swiss cheese." Flexible foams are generally made from TDI and longer-chained polyether triols. The use of trifunctional reactants is needed to produce a three-dimensional product. These foams are generally water-blown, meaning that the water added reacts with isocyanate end groups producing carbon dioxide gas giving open-celled foam products. Isocyanates react with water producing unstable carbamic acids that decompose forming diamines and carbon dioxide. Low-density flexible foams are used in bedding applications and furniture while higher-density foams are used in automotive seating and semiflexible foams are employed in automotive interior padding. Some of the flexible foams are used as carpet underlay material.

Most rigid foam is made from PMDI and difunctional polyether polyols. The PMDI provides the needed additional functionality offering an average functionality of 2.7 resulting in the formation

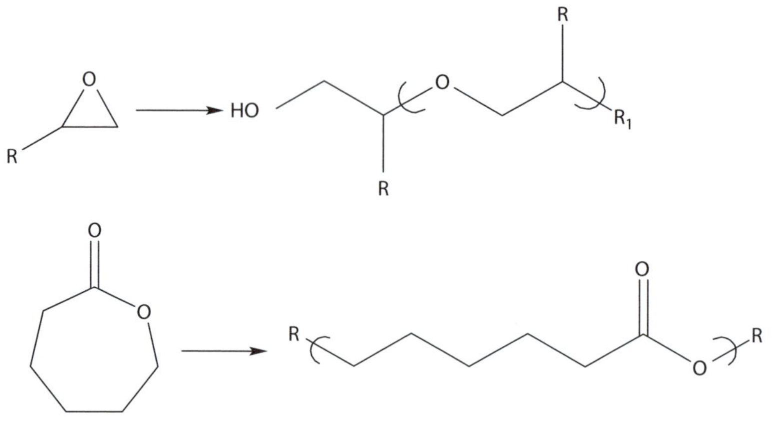

FIGURE 4.7 Synthesis of polyether and polycaprolactone macroglycols.

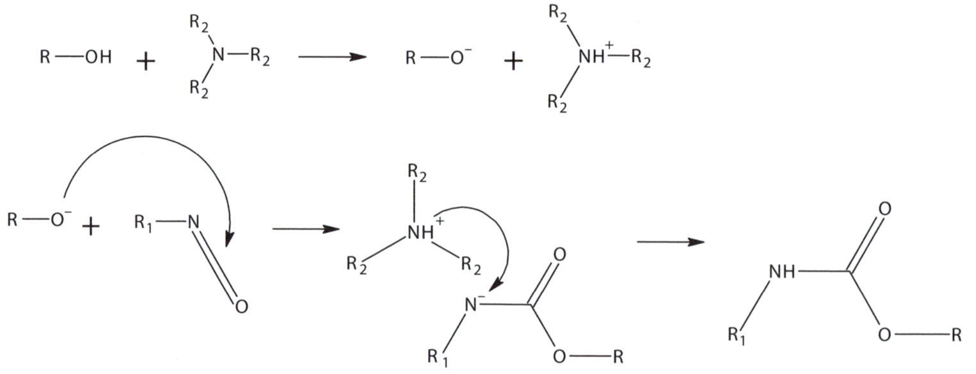

FIGURE 4.8 Tertiary amine catalyzed reaction between an alcohol and isocyanate.

of three-dimensional products. At times, the overall functionality is increased by introduction of polyols with a functionality greater than two. The resulting foams are closed celled with outstanding insulation properties. These foams are used as commercial insulations in the construction and transportation industries. They can be used as a spray-on product insulating pipes and tanks and are used in construction being sprayed on where insulation is needed.

Notice that open-celled products are employed where ready flexibility, softness, is needed while closed-cell products are used when insulation is needed. Closed-celled materials retain heat/cold while open-celled products allowed rapid heat exchange to occur.

Cross-linked PU coatings, elastomers, and foams can be produced using an excess of the diisocyanate, which reacts with the urethane hydrogen producing an allophanate, or by incorporating polyols such as glycerol in the reaction mixture. Larger polyols called macroglycols are used to form segmented products, including Spandex used in clothing and undergarments. These products are one basis for segmented elastomers, where aromatic PUs that form hydrogen bonding act as the "hard" segment, and the polyols such as poly(ethylene glycol) (PEG) form the "soft" segment. These segmented PUs can be thermoplastics or thermosets depending on whether cross-linking is introduced.

Reaction–injection molding is increasing in importance and emphasizes the production of thermoset PU. Here, liquid monomers are mixed together under high pressure before injection into the mold. Polymerization occurs within the mold. Most automotive dash panels are RIM produced.

Notice the tough surface and semiflexible underbelly of the dash. The amount of "foam" formation is controlled to give the finished product.

PUs are also widely used as coating materials sold as finished polymers, two-part systems, and prepolymer systems. Water-based PU systems are now available allowing easy home use. Aromatic diisocyanate-derived coatings generally offer poor external light stability while aliphatic-derived systems offer good light stability.

The nonaromatic PU shown in Equation 4.74 is sold under the trade name Perlon U. As in the case with nylons and polyesters, higher melting products are formed when the number of carbon atoms is even since this allows a closer packing of the chains.

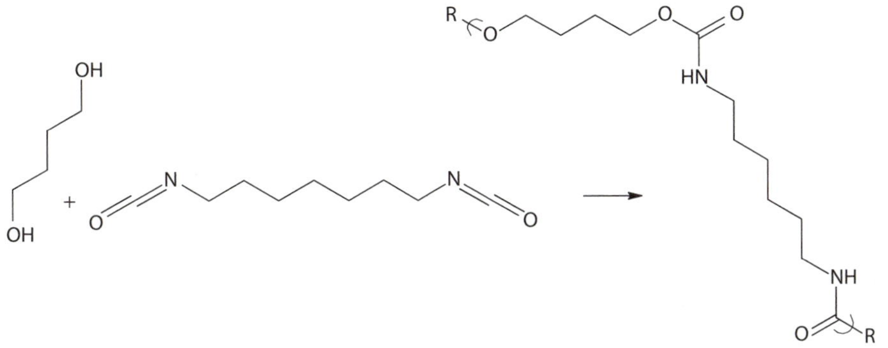

(4.74)

Polyureas (Equation 4.75) are made similar to PUs, except that diamines are employed in place of alcohols.

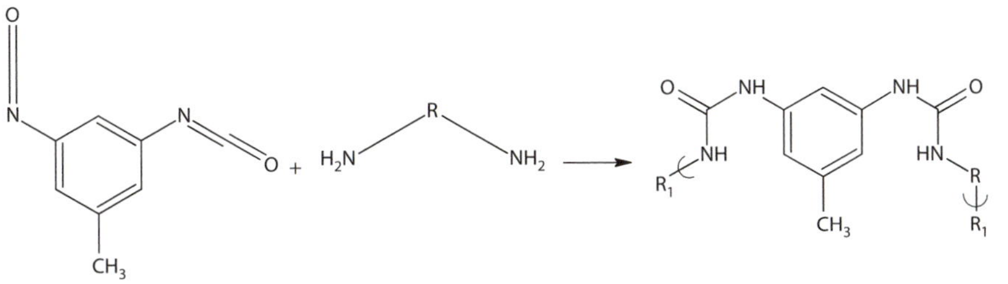

(4.75)

Polyureas are used where toughness is needed, such as in truck bed liners. Truck bed liners can also be sprayed onto the bed surface. Bridges and pipes are increasingly being coated with polyureas since these last longer than simply painting these surfaces. Tanks, sewers, manholes, and roofs are often coated with polyureas. Railcars are increasingly being coated with polyureas for abrasion protection. In all of these applications, corrosion protection is essential and polyurethanes offer good stability to weather and many chemicals. It is also finding application where water retention is needed such as in wastewater treatment and landscape and water containment. It is also being sprayed over foam creating architectural designs. Polyureas are used for flooring and parking decks because they can be rapidly applied and offer good protection when the floors are cleaned. Some automotive objects are formed from polyurea RIM systems replacing other molded pieces.

4.11 POLYSULFIDES

Thiokol (Equation 4.76), which was the first synthetic elastomer, was synthesized by Patrick in the 1920s, by the condensation of alkylene dichlorides and sodium polysulfides. These solvent-resistant elastomers have limited uses because of their foul order. They can be reduced to liquid polymers that can be reoxidized to solid elastomers used in caulking material and some rocket propellant formulations.

$$\text{Cl} \diagdown \diagup \text{O} \diagdown \diagup \text{Cl} + \text{Na}_2\text{S}_x \longrightarrow R \diagdown \diagup \text{O} \diagdown \diagup \text{S}_x R \qquad (4.76)$$

Because natural sulfur has eight sulfur atoms contained within each molecule, the number of sulfur atoms is generally variable being 1–8.

Poly(phenylene sulfide) (PPS; Ryton) is a solvent-resistant plastic that is useful in high-temperature services (Equation 4.77). PPS is used for pumps, sleeve bearing, cookware, quartz halogen lamp parts, and electrical appliance housings.

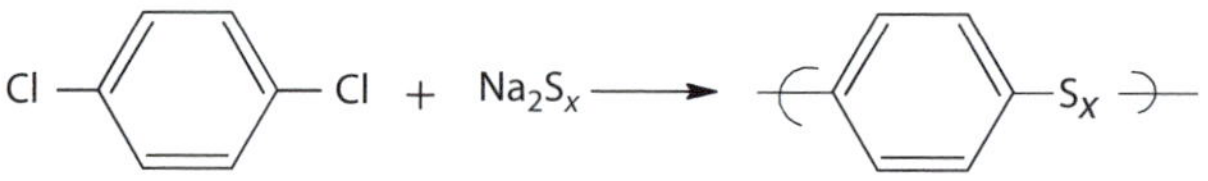

$$\text{Cl} - \bigcirc - \text{Cl} + \text{Na}_2\text{S}_x \longrightarrow \left(\bigcirc - \text{S}_x \right) \qquad (4.77)$$

4.12 POLYETHERS AND EPOXYS

Hay, in 1956, discovered an oxidative-coupling catalyst that allowed the production of polymeric aromatic ethers. The hope was to make polymers from readily available starting materials, mainly phenol. The main aromatic polyether available today is derived not from phenol but rather from the catalytic coupling of 2,6-dimethylphenol. The resulting polymer is referred to as poly(phenylene ether) (PPE), or poly(phenylene oxide) (PPO); Equation 4.78. While called PPO, it is actually composed of phenylene oxide units containing methyl groups in the 2 and 6 positions of the phenylene moiety. PPO is made by a room temperature oxidation brought about by bubbling oxygen through a solution of the phenol in the presence of copper (I) chloride and pyridine. Initially, there was not a ready, inexpensive source of 2,6-dimethylphenol but because of early found positive properties of PPO, an inexpensive source of 2,6-dimethylphenol was found and both the monomer and polymer became commercial in 1964.

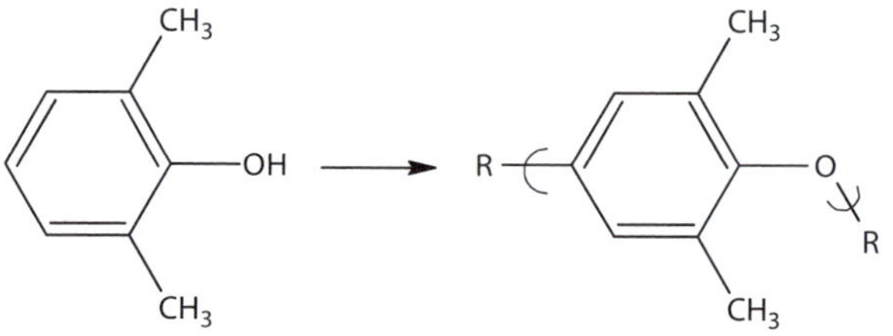

$$ (4.78) $$

PPO has a very high T_g of 215°C, a T_m of about 270°C, and exhibits good hydrolytic stability, but it has a very high melt viscosity and a tendency to oxidize and gel at processing temperatures (Table 4.9). In spite of these negative processing features, PPO showed good compatibility with polystyrene and so has found a place in the marketplace. While the methyl groups discouraged good interactions between PPO chains, the aromatic character positively interacts with the phenyl group on the polystyrene and the methyl groups interact positively with the aliphatic polystyrene backbone. The Noryl trade name covers a variety of related PPOs. PPO resins are the most important materials for forming blends and alloys with polystyrene and styrene derivatives. These blends and alloys with polystyrene raise the heat distortion temperature to over 100°C allowing production of materials that can be boiled. Combinations with PS are more easily processed and the PPO imparts needed flame resistance. The combinations also offer good hydrolytic stabilities and electrical properties, and they are relatively light weight. They can also be modified by addition to glass and other mineral fillers and are especially adaptable to metallizing. PPO-extruded sheet is being used for solar energy collectors, lifeguards on broadcasting towers, airline beverage cases, and window frames.

TABLE 4.9
General Physical Properties of PPO

Heat-deflection temperature (1,820 kPa; °C)	100
Maximum resistance to continuous heat (°C)	80
Crystalline melting point (°C)	215
Coefficient of linear expansion (cm/cm-°C, 10^{-5})	5.0
Compressive strength (kPa)	9.6×10^4
Flexural strength (kPa)	8.9×10^4
Impact strength (Izod: cm-N/cm of notch)	270
Tensile strength (kPa)	5.5×10^4
Ultimate elongation (%)	50
Density (g/mL)	1.1

The Noryl trade name covers a variety of PPO-intense materials, the most important being with polystyrene. While the methyl groups discouraged good interactions between PPO chains, the aromatic character positively interacts with the phenyl group on the polystyrene, and the methyl groups interact positively with the aliphatic polystyrene backbone. Unlike most polymer mixtures, the combination of PPO and polystyrene forms a miscible blend. These blends and alloys with polystyrene raise the heat distortion temperature to over 100°C, allowing production of materials that can be boiled. Combinations with PS, in particular high impact polystyrene (HIPS), are more easily processed and the PPO imparts needed flame resistance. The PS adds flexibility and lowers the melting point. The combinations also offer good hydrolytic stabilities and electrical properties, and they are relatively light weight. Noryl is widely used in switch boxes because of its good electrical resistance. It can also be modified by addition to glass and other mineral fillers and are especially adaptable to metallizing.

Aliphatic polyethers are also referred to as polyacetals. Polyoxymethylene (POM), precipitates spontaneously from uninhibited aqueous solutions of formaldehyde and was isolated by Butlerov in 1859. Staudinger, in the 1920s and 1930s, experimented with the polymerization of formaldehyde but failed to produce chains of sufficient length to be useful. While pure formaldehyde readily polymerized, it also spontaneously depolymerizes, un-zips. In 1947, DuPont began a program to make useful polymers from formaldehyde since formaldehyde is inexpensive and readily available. After 12 years, they announced the commercialization of the polymer from formaldehyde, POM, under the trade name of Delrin. The "secret" was capping the end groups by acetylation of the hydroxyl end groups, thus preventing the ready unzipping of the polymer chain (Equation 4.79). POM has a T_g of −75°C and a T_m of 180°C. General physical properties are given in Table 4.10.

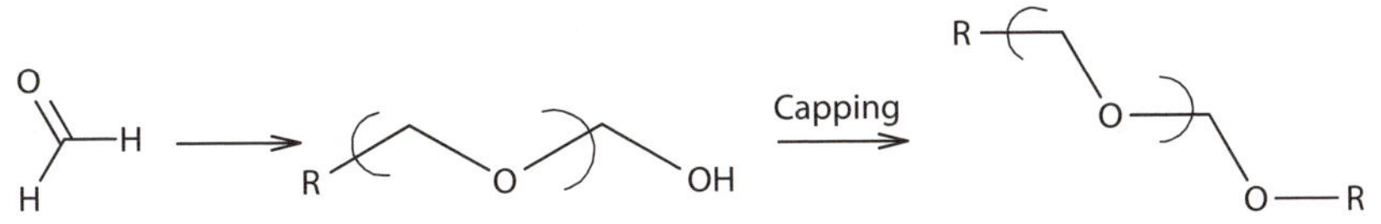

$$\text{(4.79)}$$

Celanese came out a year latter with a similar product under the trademark of Celcon. Celanese circumvented DuPont's patent on the basis of employing a copolymer variation that allowed enhanced stabilization against thermal depolymerization (Equation 4.80). The copolymer has a T_m of 170°C.

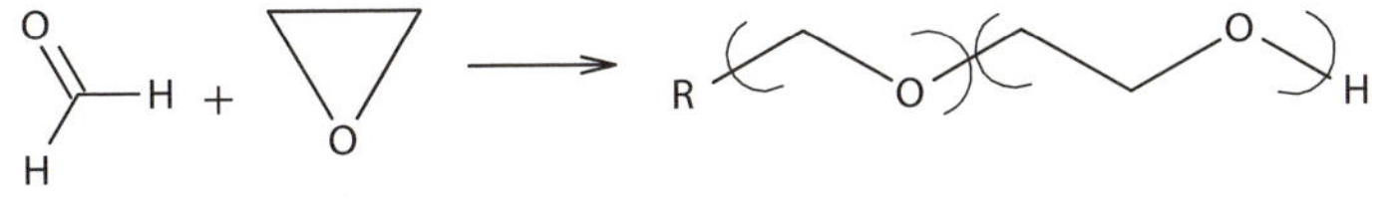

$$\text{(4.80)}$$

TABLE 4.10
General Physical Properties of POM and Epoxy Resins

	POM	Epoxy Resins
Heat-deflection temperature (1,820 kPa; °C)	125	140
Maximum resistance to continuous heat (°C)	100	120
Crystalline melting point (°C)	180	--
Coefficient of linear expansion (cm/cm-°C, 10^{-5})	10.0	2.5
Compressive strength (kPa)	1.1×10^5	1.2×10^5
Flexural strength (kPa)	9.7×10^4	1.2×10^5
Impact strength (Izod: cm-N/cm of notch)	80	50
Tensile strength (kPa)	6.9×10^4	5.1×10^4
Ultimate elongation (%)	30	5
Density (g/mL)	1.4	1.2

Polyacetals and other engineering plastics cost about half that of cast metals so are used as replacements for cast metal-intense applications. They have been approved by the Food and Drug Administration for contact with foods. Some of the uses of molded polyacetals are as valves, faucets, bearings, appliance parts, springs, automotive window brackets, hose clamps, hinges, video cassettes, tea kettles, chains, flush toilet float arms, gears, shower heads, pipe fittings, pasta machines, desktop staplers, and air gun parts.

Polyoxymethylenes are also employed in plumbing and irrigation because they resist scale accumulation, have good thread strength, torque retention, and creep resistance. POMs are also employed to assist in the flow of water in fire hoses, water displays, and for some large ships it is "squirted" from the front of the ship to cut down on friction because they help align the water allowing increased fuel efficiency.

Polyoxymethylene are resistant to many solvents, including aqueous salt and alkaline solution, and weak acids.

Epoxy resins are really polyethers but are named epoxies because of the presence of epoxide groups in the starting material. They were initially synthesized from epichlorohydrin and BPA in the 1940s. General properties are listed in Table 4.10.

The reaction is generally run in the presence of a base such as sodium hydroxide. BPA is a phenol and as such is a weak acid. The generated RO^- reacts with the electron-poor chlorine-containing carbon on epichlorohydrin creating a cyclic ether end group. The phenoxy moiety can also react with the cyclic ether eventually forming the polyether structure. This sequence is described in Figure 4.9.

High molecular weight thermoplastics called phenoxy resins are formed by the hydrolysis of the epoxy resins so that no epoxy groups are present. These transparent resins can be further reacted forming cross-linked material through reaction of the hydroxyl pendant groups with diisocyanates or cyclic anhydrides.

$\overline{DP}$ is dependent on the ratio of reactants. In general, an excess of the epichlorohydrin is used producing cyclic ether end groups. Epoxies are formed from reaction of diamines with low molecular weight epoxy resins that retain their cyclic ether end groups. Figure 4.10 shows the formation of this process. These materials are often sold as two-part or two-pot epoxy adhesives. Most "use-at-home" epoxy packages contain a part A or epoxy resin and a part B or hardener (typically a diamine). These two are mixed as directed and applied. They dry so they can be handled in 5–10 minutes. Full strength occurs after 5–7 days.

These cross-linked, cured epoxies have outstanding resistance to chemicals, durability, and toughness making them good coating materials. They are easily poured, before hardening, without

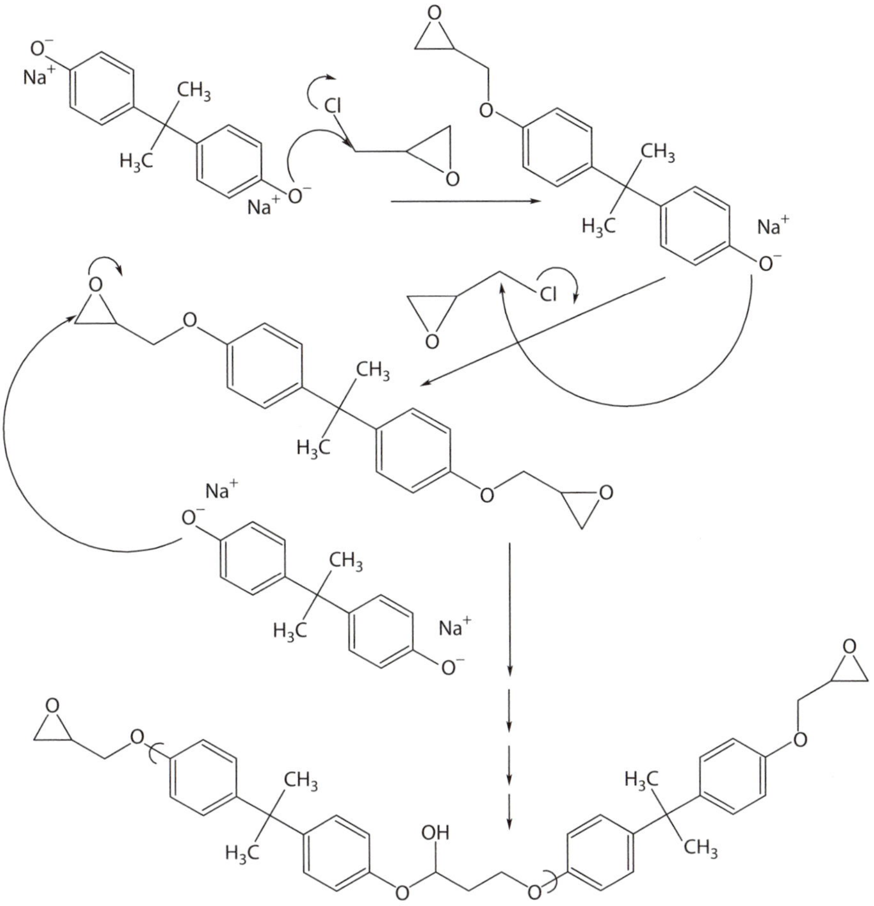

FIGURE 4.9 Mechanistic outline for the formation of epoxy resins.

bubble formation and are used for encapsulating electrical wires and electrical components for TVs and computers. As adhesives epoxies have good metal adhesion so are used to join automotive hoods and doors. They are also used in aerospace composites, batteries, as well as sporting and hardware goods.

There are also one-part adhesives based on low molecular weight epoxy units that contain the unopened cyclic ether end groups. These materials can be cured when heated to about 200°C for 30–60 minutes.

Commercially, the most important polyether is known as PEG, which is also known as PEO or polyoxyethylene (POE). While these names are synonymous, historically PEG was used for lower chained products, PEO for longer chains, and POE for both. Here we will simply use the term PEG. PEG comes in a variety of chain lengths from low that are liquids to high that are solids.

Poly(ethylene glycol) is generally synthesized from the ROP of poly(ethylene oxide) in water (Equation 4.81). Along with water, ethylene glycol and ethylene glycol oligomers are often present in the polymerization system.

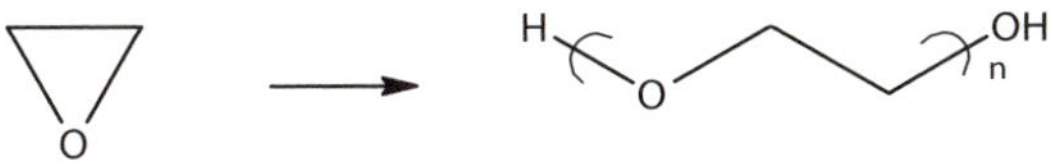

(4.81)

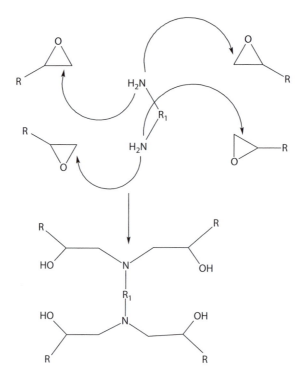

FIGURE 4.10 Curing of two-part epoxy resin adhesives.

Poly(ethylene glycols) are generally designated as to some average molecular weight. Thus, PEG4000 corresponds to n = about 90 (that is 4,000/44 amu/unit).

Poly(ethylene glycols) come with a variety of end groups. Thus, the monofunctional methyl ether of PEG, methoxypoly(ethylene glycol) has one of the end groups a methoxy rather than the usual hydroxyl. They also come in a variety of geometries. Star PEGs have 10–100 PEG units radiating from a central core. Branched PEGs have 3–10 PEG chains radiating from a central core. Comb PEGs have multiple PEG chains grafted to a polymer backbone.

Poly(ethylene glycols) have a low toxicity and are employed in a number of products. It forms the basis for a number of laxatives such as MiraLax, GlycoLax, and Movicol. It is used in a number of sexual lubricants and skin creams. Bowel irrigations often employ PEG with some electrolytes to be used before surgery, colonoscopy, or in drug overdoses. It is sold under trade names as Colyte, TriLyte, Fortrans, GoLately, and GlycoLax for this purpose. It is attached to selected protein drugs to allow a longer-acting medicinal effect. PEG-interferon alpha is used to treat hepatitis C and PEG–filgrastim is used to treat neutropenia.

It is employed as the soft segment in PU. PEG is used in toothpaste as a dispersant and as a lubricant in eye drops. It is also being used in certain body armor. Dr. Pepper adds PEG as an anti-foaming agent. It has been used to help preserve wood replacing water giving the wood increasing dimensional stability. Certain gene therapy vectors such as viruses can be coated with PEG to protect them from inactivation by immune systems and to de-target them from organs, where they could build up causing a toxic effect.

Because of their ready water solubility, PEGs are often used to impart water solubility to a polymer. This approach has been used to make electrically conductive polymers water soluble as well as a number of drugs water soluble.

4.13 POLYSULFONES

Polysulfones exhibit excellent thermal oxidative resistance and resistance to hydrolysis and other industrial solvents and creep. The initial commercial polysulfones were synthesized by the nucleophillic replacement of the chloride on bis(p-chlorophenyl) sulfone by the anhydrous sodium salt of BPA. It became commercially available in 1966 under the trade name Udel. It exhibits a reasonably high T_g of 190°C.

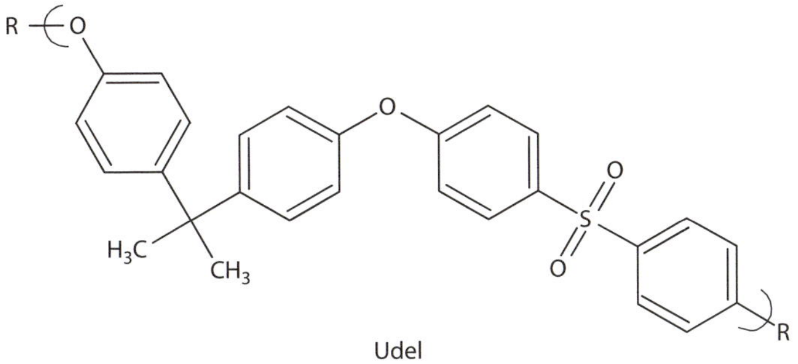

(4.82)

Udel

Union Carbide, in 1976 made available a second-generation polysulfone under the trade name of Radel. Radel was formed from the reaction of a bisphenol and bis(p-chlorophenyl) sulfone (4.83). This polysulfone exhibites greater chemical/solvent resistance, a greater T_g of 220°C, greater oxidative stability, and good toughness in comparison to Udel.

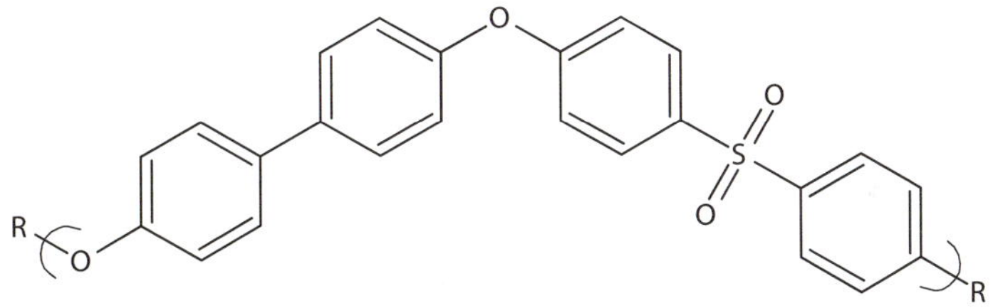

(4.83)

Radel

Polysulfones can also be made from the Friedel–Crafts condensation of sulfonyl chlorides.

Polysulfones are used for ignition components, hair dryers, cookware, and structural foams. Because of their good hydrolytic stability, good mechanical properties, and high thermal endurance, they are good candidate materials for hot water and food handling equipment, alkaline battery cases, surgical and laboratory equipment, life support parts, autoclavable trays, tissue culture bottles, and surgical hollow shapes, and film for hot transparencies. Membranes for use in hemodialysis, gas separation, food and beverage processing, and wastewater recovery have also been produced from polysulfones. Their low flammability and smoke production, again because of their tendency for polycyclic formation on thermolysis and presence of moieties that are already partially oxidized, makes them useful as materials for the aircraft and automotive industries. General properties of PPS are given in Table 4.11.

4.14 POLY(ETHER ETHER KETONE) AND POLYKETONES

Aromatic polyketones are semicrystalline materials that contain both ketone groups generally flanked by aromatic units. Many also have included within them ether moieties that allow for some flexibility and better processing. They have good thermal stabilities, also well as offering good mechanical properties, flame resistance, impact resistance, and resistance to the environment.

TABLE 4.11
General Physical Properties of PPS

Heat-deflection temperature (1,820 kPa; °C)	135
Maximum resistance to continuous heat (°C)	110
Crystalline melting point (°C)	190
Coefficient of linear expansion (cm/cm-°C, 10^{-5})	5.0
Compressive strength (kPa)	1.1×10^5
Flexural strength (kPa)	9.6×10^5
Impact strength (Izod: cm-N/cm of notch)	21
Tensile strength (kPa)	7.4×10^4
Ultimate elongation (%)	1.1
Density (g/mL)	1.3

Polyetherketone (PEK) was introduced by Raychem in the 1970s (Equation 4.84). It is made by the Friedel–Crafts reaction requiring good solvents or an excess of aluminum chloride to keep the polymer in solution allowing polymer growth to occur. Most polymerizations require that the reactants remain mobile, through solution or being melted, so that the individual units involved in the reaction can get together. Rapid precipitation of growing polymer chains often results in the formation of only oligomeric to small chains.

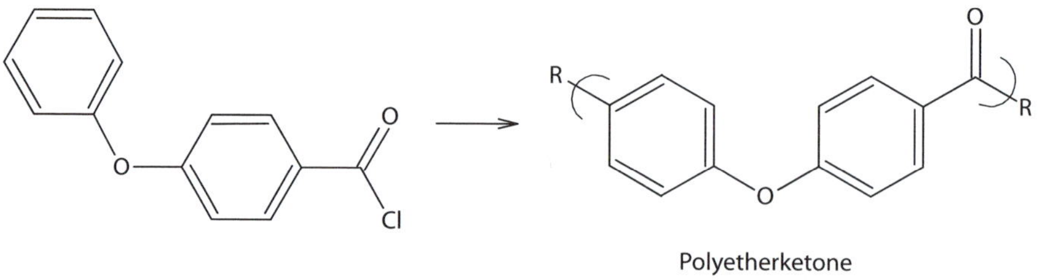

(4.84)

Polyetherketone

Imperial Chemical Industries (ICI) has introduced a new crystalline poly(ether ether ketone) (PEEK) (4.85). Applications include compressor plates, valve seats, thrust washers, bearing cages, and pump impellers. In the aerospace industry they are employed as aircraft fairings, fuel valves, and ducting. They are also used in the electrical industry as wire coating and semiconductor wafer carriers.

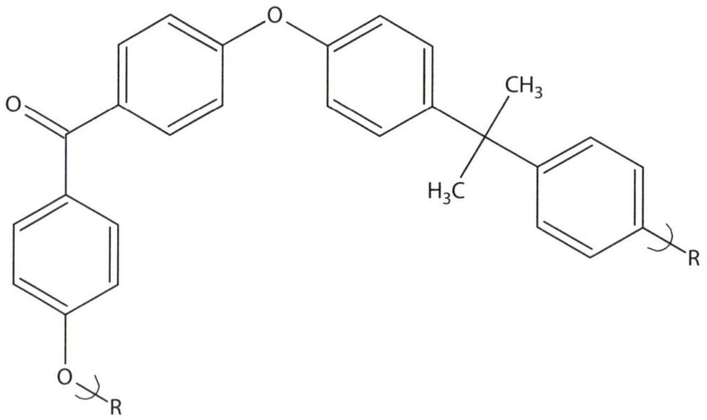

(4.85)

Aliphatic polyketones are made from the reaction of olefin monomers and carbon monoxide using a variety of catalysts. Shell commercialized a terpolymer of carbon monoxide, ethylene, and a small amount of propylene in 1996 under the trade name of Carilon (Equation 4.86). They have a useful range between the T_g (15°C) and T_m (200°C) that corresponds to the general useful range of use temperatures for most industrial applications. The presence of polar groups causes the materials to be tough with the starting materials readily available.

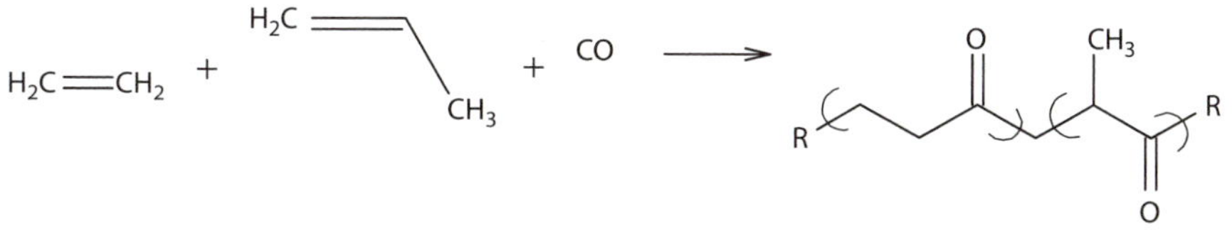

(4.86)

Aromatic polyketone

4.15 PHENOLIC AND AMINO PLASTICS

Baekeland found that a relatively stable resole prepolymer could be obtained by the controlled condensation of phenol and formaldehyde under alkaline conditions. These linear polymers (PF) may be converted to infusible cross-linked polymers called *resistes* by heating or by the addition of mineral acids. As shown in Equation 4.87, the initial products produced when formaldehyde is condensed with phenol are hydroxybenzyl alcohols. The linear resole polymer is called an *A-stage resin*, and the cross-linked resite is called a *C-state resin*.

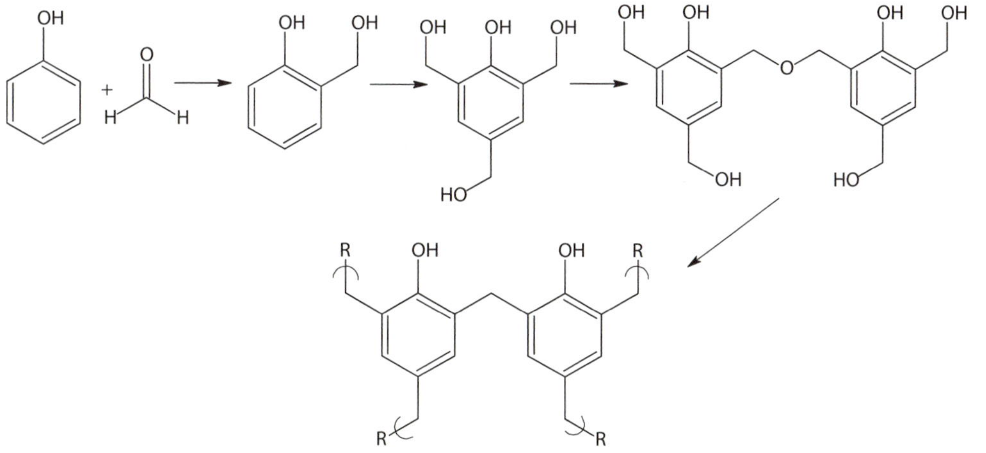

(4.87)

C-stage (resite resin)

Baekeland recognized that the trifunctional phenol would produce network polymers and thus used difunctional ortho- or para-substituted phenols to produce linear paint resins. Linear thermoplastic products are formed by alkaline or acid condensation of formaldehyde with phenol derivatives such as *p*-cresol (Equation 4.88).

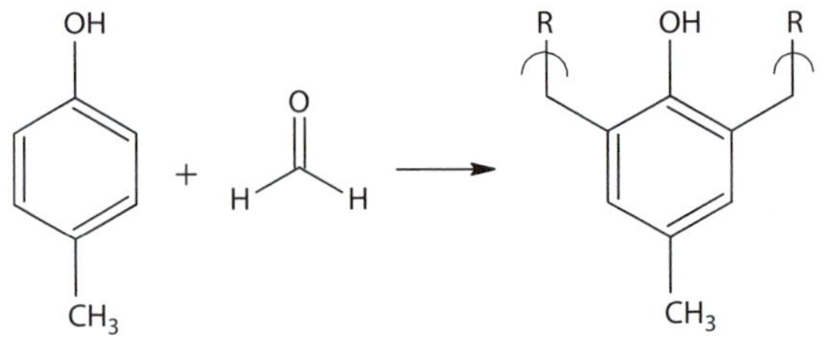

$$(4.88)$$

Since the acid condensation of one mol of phenol with 1.5 mol of formaldehyde produced thermoset C-stage products, Baekeland reduced the relative amount of formaldehyde used and made useful novolac resins in a two-step process. Thus, stable A-stage novolac resin is produced by heating one mol of phenol with 0.8 mol of formaldehyde in the presence of acid. After the removal of water by vacuum distillation, the A-stage resin produced is cooled and then pulverized. The additional formaldehyde required to convert this linear polymer into a thermoset resin is supplied by hexamethylenetetramine. The latter, which is admixed with the pulverized A-stage resin, is produced by the condensation of formaldehyde and ammonia. Other ingredients such as filler, pigments, and lubricants are mixed in with the resin and hexamethylenetetramine. The A-stage resin is further polymerized. The term phenol molding compound is applied to the granulated B-stage novolac resin.

While the condensation of urea and formaldehyde was described in 1884, urea-formaldehyde (UF) resins were not patented until 1918. Comparable products, based on the condensation of formaldehyde and melamine (2,4,6-triamino-1,3,5-triazine), were not patented until 1939. The term melamine-formaldehyde (MF; Equations 4.89, 4.90) is used to describe these products.

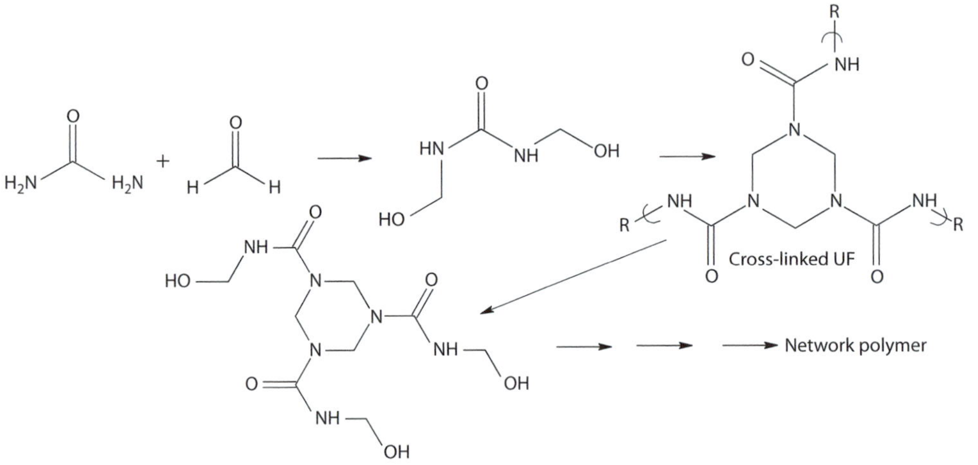

$$(4.89)$$

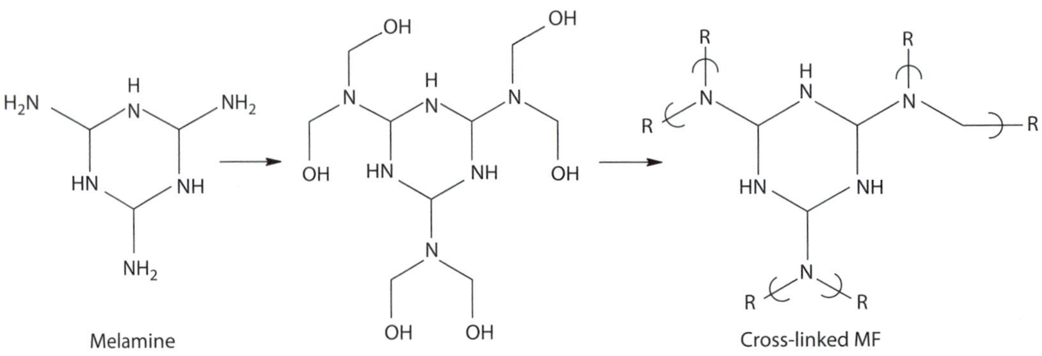

$$(4.90)$$

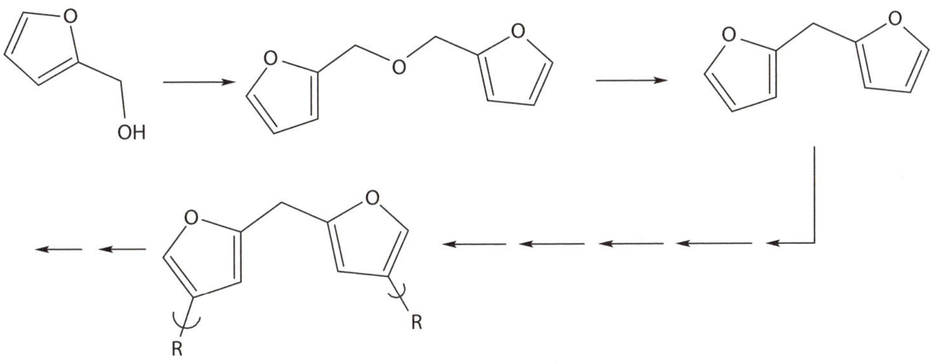

FIGURE 4.11 Synthesis of furan resins from furfuryl alcohol.

Urea and melamine are tetra- and hexa-functional molecules. However, the formation of a network polymer is prevented by adding alcohols such as n-butanol and by condensing with formaldehyde at low temperatures under basic conditions. While phenol resins have better moisture and weather resistance than urea resins, the latter are preferred for light-colored objects. For example, the interior layers of laminated countertops are bonded together by phenolic resins, but either urea or melamine resins are used for the decorative surface. Melamine plastics are more resistant to heat and moisture than UF and thus, are used for decorative surfaces and for dinnerware.

4.16 FURAN RESINS

Furan resins are produced by the polymerization of furfural or furfuryl alcohol in the presence of acids. Furan resins are deep brown colored and have a relatively low heat-deflection temperature (80°C) and good mechanical properties. They are used as jointing materials for brick and tile. They have excellent resistance to nonoxidizing acids, bases, and salts but are degraded by the presence of oxidizing acids such as nitric acid. They are resistant to nonpolar and most polar solvents at room temperature. Many of these are further reacted giving cross-linked thermoset materials. Since the furane starting materials are generally derived from vegetable matter, they are classified as a green chemistry polymer (Figure 4.11).

4.17 SYNTHETIC ROUTES

The previous sections describe the synthesis of a number of important condensation polymers. Here, we will briefly consider the three main synthetic techniques utilized in the synthesis of these polymers.

The *melt technique* is also called other names to describe the same or similar processes. These names include high melt, bulk melt, and simply bulk or neat. The melt process is an equilibrium-controlled process in which polymer is formed by driving the reaction toward completion, usually by removal of the byproduct. For polyesterifications involving the formation of hydrogen chloride or water, the driving force is the removal of the hydrogen chloride or water. Reactants are introduced along with any added catalyst to the reaction vessel. Heat is applied to melt the reactants, allowing them to condense together. The heat is maintained or increased above this melt temperature. Pressure is reduced to remove the condensate. Typical melt polymerizations take several hours to several days before the desired polymer is produced. Yields are necessarily high.

Solution condensations are also equilibrium processes with the reaction often driven by removal of the byproduct by distillation or by salt formation or precipitation. Many solution condensations are run near room temperature. Solvent entrapment is a problem, but since a reaction may occur under considerably reduced temperatures, compared to the melt process, thermally induced side reactions are minimized. Side reactions with the solvent may be a problem. Because the reactants must be energetic, many condensations are not suitable for the solution technique.

The *interfacial technique* (IF), while old, gained popularity with the work of Morgan and Carraher in the 1960s and 1970s. Many of the reactions can be carried out under essentially non-equilibrium conditions. The technique is heterophasic, with two fast-reacting reactants dissolved in a pair of immiscible liquids, one of which is usually water. The aqueous phase typically contains the Lewis base such as diol, diamine, or dithiol. The organic phase contains the Lewis acid, generally an acid halide, dissolved in a suitable organic solvent such as hexane. Reaction occurs near the interface, hence the name. With all the potential that the interfacial system offers, it has not attracted wide industrial use because of the high cost of the necessarily reactive monomers and cost of solvent removal. One commercial use for the IF system is the production of PC. Another involves the synthesis of aramids. Morgan and others noted that some polymers formed with rapid stirring would remain in solution for awhile before they precipitated. The problem with aramids was the need to form fibers from their solutions. Thus, the aromatic nylons had to be redissolved after formation. Today, aramids are synthesized using rapidly stirred systems where the polymer solution is sent through a small hole into a nonsolvent. This allows fibers to be produced without needing to redissolve the polymer.

Table 4.12 contains a comparison of these three major polycondensation processes.

4.18   LIQUID CRYSTALS

Everyday of our lives we "run across" LCs. They are commonly found in computer monitors, digital clocks, TV screens, and other "read-out" devices, and so on.

Reintzer, in 1888, first reported "liquid crystal" behavior. In working with chloesteryl esters, he observed that the esters formed opaque liquids, which on heating turned clear. We now know, that as a general rule, that many materials are clear if they are anisotropic, random or if the materials are composed of ordered molecules or segments of molecules, whereas they are opaque if there exists a mixture of ordered and disordered regions. Lehmann interpreted this behavior as evidence of a "third" phase that exists "between" the solid and isotropic liquid states. This new phase was named by Lehmann as the liquid crystal phase. Friedel called this phase the mesophase after the

TABLE 4.12
Comparison of Requirements for Different Polycondensation Techniques

Requirement	Melt	Solution	Interfacial
Temperature	High	Limited only by the MP and BP of the solvent used generally about room temperature	
Stability to heat	Necessary	Unnecessary	Unnecessary
Kinetics	Equilibrium, stepwise	Equilibrium, stepwise	Generally nonequilibrium, chain-wise
Reaction time	1hour to several days	Several minutes to 1 hour	Several seconds to 1 hour
Yield	Necessarily high	Less necessary high	Low to high
Stoichiometric equivalence	Necessarily high	Less necessary high	Less necessary
Purity of reactants	Necessary	Less necessary	Less necessary
Equipment	Specialized, often sealed	Simple, open	Simple to complex, can be open
Pressure	High, low	Atmospheric	Atmospheric

Greek word "mesos" meaning intermediate. The initial molecules investigated as LCs were large monomeric molecules.

Flory, in 1956, predicted that solutions of rod-like polymers could also exhibit liquid crystal behavior. The initial synthetic polymers found to exhibit liquid crystal behavior were concentrated solutions of poly(gamma-benzyl glutamate) and poly(gamma-methyl glutamate). These polymers exist in a helical form that can be oriented in one direction into "ordered groupings" giving materials with anisotropic properties.

Liquid crystals, LCs, are materials that undergo physical reorganization where at least one of the rearranged structures involve molecular alignment along a preferred direction causing the material to exhibit nonisotropic behavior and associated birefringent properties; that is, molecular asymmetry.

Liquid crystalline materials can be divided into two large groupings—thermotropic and lyotropic. Thermotropic LCs are formed when "pure" molecules such as cholesteryl form ordered structures upon heating. When LCs occurs through mixing with solvents they are called lyotropic LCs.

Thermotropic LCs can be further divided into (a) enantiotropic materials where the liquid crystal phases are formed on both heating and cooling cycles and (b) mesotropic materials where the LCs are stable only on supercooling from the isotropic melt. The mesotropic LCs have been further divided into three groupings as follows (Figure 4.12).

- Smectic meaning "soap"
- Nematic meaning "thread"
- Cholesteric derived from molecules with a chiral center

Liquid crystal polymers are typically composed of materials that are rigid and rod-like with a high length to breadth ratio or materials that have a disc shape. The smaller groupings that give the material LC behavior are called "mesogens." These mesogens are simply portions of the overall polymer that are responsible for forming the anisotropic LC and that, in fact, form the LC segments. Such mesogens can be composed of only segments from the backbone of the polymer, segments from the side chain or segments from both the backbone and side chain.

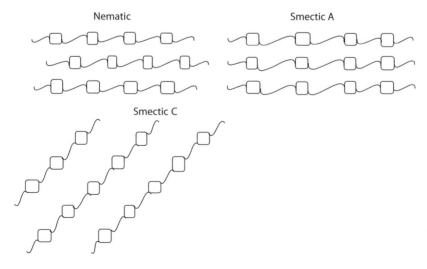

FIGURE 4.12 Different mesophasic structures, where the mesogenic unit is designated by a rounded square.

The mesogens form the ordered structures necessary to give the overall material anisotropic behavior. There have been identified a number of different mesogen groupings. Chains arranged so that the mesogen portions are aligned in one preferred direction with the ordering occurring in a three-dimensional layered fashion compose one group of arrangements called *smectic arrangements*. Here, the lateral forces between the mesogen portions are relatively higher than the lateral forces between the nonmesogen portions allowing a combination of segments that permit "flowing" (the passage of nonmesogen portions) and segments that retain contact (mesogen portions) as the material flows imparting a "memory"-type behavior of the material. A number of different "packings" of the mesogens have been identified. The most ordered of the mesogenic groupings is called "smectic B," which is a hexagonally, close-packed structure present in a three-dimensional arrangement. A much less ordered grouping is called the "Smectic A" phase. Here there is a somewhat random distribution of the mesogens between the layers.

Nematic LCs offer much less order in comparison to smectic arrangements. Here, the directional ordering of the mesogen portions along one axis is retained, but the centers of gravity of the mesogen portions are no longer "coupled." Thus, the forces between the chains are less resulting in a generally greater fluidity for nematic LCs in comparison with smectic structures. Nematic LCs also offers nonlinear behavior.

The chiral nematic assembly is formed by materials that have chiral centers and that form a nematic phase. Here, a "chiral-imposed twist" is imparted to the linear chains composing each layer resulting in a three-dimensional helical arrangement.

The molecular asymmetry typically occurs not because of intermolecular interaction, but because two molecules cannot occupy the same space at the same time. Molecular chains can exist in a random arrangement until a given concentration is exceeded causing the molecules to rearrange in a more ordered fashion to accommodate the larger number of molecules within the same volume. Often, this occurs such that there is an ordered phase and a more random phase. As the concentration of polymer increases, the ordered phase becomes larger at the expense of the disordered phase. This increase in polymer concentration can occur via several routes such as addition of more polymer, addition of a solution containing a higher concentration of polymer, and evaporation of the solvent.

For crystalline polymer systems, transition from the crystalline structure to a mesosphere structure occurs, whereas from amorphous polymer systems, the mesophase occurs after the T_g has occurred. Some polymer LC systems form several mesophases. Mesophases can be detected using differential scanning calorimetry (DSC), X-ray diffraction, and polarizing microscopy.

Introduction of flexible "spacer" units such as methylene, methylene oxide, and dimethylsiloxane groups lowers the melting point and increase the temperature range within which the mesophase is stable. Often these spacer units are introduced by copolymerization. Thus, preformed p-acetoxybenzoic acid is reacted with PET introducing a mesogenic unit in a polymer that has flexible spacer units (from the ethylene glycol) in it.

Poly(ethylene terephthalate), PET (4.91)   Mesogenic unit (4.92)

Along with the mesogen units contained within the polymer backbone, the mesogen units can occur as side chains. These mesogen units can be introduced either through reaction with monomers that contain the mesogen unit or through introduction with already formed polymers.

Liquid crystal materials have also been employed as films, plastics, and resins. Poly(1,4-benzoate) has been marketed under the name Ekonol. It decomposes before it melts, hence it does not form

LC melts. Copolymerization with 4,4'-biphenol and terephthalic acid gives Ekkcel. Ekkcel does melt before it decomposes. Certain forms can be compression molded and others injected molded. Reaction of poly(1,4-benzoate) with PET gives a material that can be injected molded. These LCs are chemical resistant, with high tensile strength. LC films with mesogenic side chains can be used in information storage devices.

In general, because of the high order present in LCs, especially within their ordered state, they have low void densities and as such exhibit good stability to most chemicals, including acids, bleaches, common liquids, and so on; low gas permeability; relatively high densities; they are strong and stiff with tensile moduli of the order of 10–25 GPs and tensile strengths in the range of 120–260 MPa.

Today, there exists a number of routes for processing that have been modified for LC materials. In general, they offer low shrinkage and warpage during molding, good repeatability from part to part and low heats of fusion, allowing rapid melting and cooling during processing.

Their low melt viscosity permits molding of thin sections and complex shapes. Counter, their tendency to form ordered structures causes LC materials to be particularly susceptible to molecular orientation effects during processing.

We can look at one illustration of the use of LC materials (Figure 4.13). A typical LC display, LCD, may contain thin layers of LC molecules sandwiched between glass sheets. The glass sheets have been rubbed in different directions and then layered with transparent electrode strips. The outside of each glass sheet is coated with a polarizer material oriented parallel to the rubbing direction. One of the sheets is further coated with a material to make it a reflecting mirror. The liquid crystal-line molecules preferentially align along the direction that the two glass surfaces have been rubbed. Because the two glass surfaces are put at 90° to one another, the liquid crystal orientation changes as one goes from one glass surface to the other creating a gradual twist of 90°.

Ordinary light consists of electromagnetic waves vibrating in various planes perpendicular to the direction of travel. As light hits the first polarizer material, only light that vibrates in a single plane is allowed to pass through. This plane-polarized light then passes through the layers of LC that effectively twists the plane of the light 90° allowing the "twisted" light to pass through the second polarized surface, striking the mirrored surface, and "bouncing back" being seen as a white background.

The LCD image is formed as voltage is applied to an appropriate pattern of tiny electrodes that causes reorientation in the LC. Orientated LC no longer are at 90° to one another and thus are unable to transmit light through to the mirrored surface and thus appear as dark areas. This combination of dark and light surfaces then creates the LCD image.

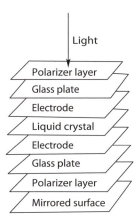

FIGURE 4.13 Composition of a typical LCD.

Assemblies similar to the above have been employed to create images in clocks and watches and other similar LCD image-containing products.

4.19 MICROFIBERS

Microfibers are not a new chemically distinct fiber, but rather the term refers to fibers that have smaller diameters. DuPont first introduced microfibers in 1989. Microfibers have diameters that are less than typical fibers. Microfibers are half the diameter of fine silk fiber, one-quarter the diameter of fine wool, and one hundred times finer than human hair. Denier, the weight in grams of 9,000-meter length of a fiber, is the term used to define the diameter or fineness of a fiber. While the definition for the thickness of microfibers is varied, a typical definition is that microfibers have a denier that is 0.9 denier or less. For comparison, the nylon stocking is knit from 10 to 15 denier fiber.

Microfibers allow a fabric to be woven that is lightweight and strong. Microfibers can be tightly woven so that wind, rain, and cold do not easily penetrate. Rainwear manufacturers use microfibers for this reason. They also have the ability to allow perspiration to pass through them. Thus, so-called microfiber athletic-wear is becoming more common place. Microfibers are also very flexible because the small fibers can easily slide back and forth on one another. The first fabric made from microfiber was Ultrasuede™ where short polyester microfibers were imbedded into a PU base. Today microfibers are made mainly from polyesters, nylon, acrylic, and rayon fibers.

The use of the term microfibers is now extended to glass and Teflon-related materials. Here we will restrict ourselves to only fabric applications.

In 1970, Toray Industries scientist Miyoshi Okamoto created the first microfiber. A few months later his colleague Toyohiko Hikota developed a process that allowed the production of fabric that was later trademarked as Ultrasuede™. Ultrasuede was produced from PET polyester fibers so thin that a pound of them laid end-to-end would reach from the earth to the moon and back. Ultrasuede is soft and supple, resistant to stains and discoloration, and machine washable and dry cleanable.

Because of progress made in spinning and fiber processing, smaller fibers can be routinely made with varying configurations, coatings, and so on. Microfiber production is mainly achieved using three techniques. The first technique will be illustrated using the processing technique employed to manufacture Ultrasuede.

The processing to form Ultrasuede is complex. First, ultramicrofibers are spun that are so light and fine that a single 50-miles long strand would weight less than one gram. These ultramicrofibers are then extruded through spinnerets creating a configuration that is similar to that present in tendons and hair bundles. These strands are then transformed by ironing, curling, cutting, and needle-punching into a felt-like material that is then impregnated with a special adhesive binder. The material is then formed into the desired contours and the protective coating is dissolved with a solvent and the material undergoes further processing creating the nonwoven fabric.

Microfibers are also made by simple extrusion through a spinneret with a smaller hole than normally employed for fiber production. The third method involves spinning a bicomponent fiber and using a solution to split the fiber into smaller pieces. Initially, bicomponent fibers in the range of 2–4 denier are spun after which the fibers are split into microfibers. If a 32-segment pie of nylon/polyester fiber is used, the final denier is in the range of 0.1 denier. Brushing and other techniques can be used to enhance the effects. Splittable, hollow fibers are also being used to achieve fiber splitting. For instance, for polyester/polypropylene fiber combinations, "natural" splitting occurs after passage through the spinneret. After mechanical drawing, the yarn has a denier of about 0.2 well within that described for microfibers. More recently, tipped fibers are being studied. Here, a bicomponent fiber is made except the second polymer is placed on the tip of the fiber. After spinning, the fibers are twisted and then wet heat is applied resulting in the tips of the fibers breaking apart into microfibers with a denier of about 0.2 because the two different polymers have different heats of elongation, and so on the physical changes cause the two polymer components to break apart.

The shape of the microfiber influences the end properties. For instance, Tomen has developed Technofine™, which is a polyester microfiber with a "W" shape cross-section. The increased surface allows a more rapid transport of water away from the skin and also increases the evaporation rate making garments made from it quicker drying and more adsorbent.

Most textiles have limited elongation but many of the microfiber textiles have elongations of 100%.

Currently, no industry regulations exist that describe the amount of microfiber that must be present to call it a microfiber material. Even so, typically industrial use calls for a material to be at least 35%–40% microfiber to be called a microfiber product. Microfibers are most commonly used in dress and blouse garments. They are also used to make hosiery, ties, scarves, intimate apparel, curtains, draperies, sheets, towels, rainware, swimwear, blankets, tents, sleeping bags, track and jogging-wear, as well as many other items. The greater surface area results in a fabric with deeper, richer, and brighter colors possible.

The care of microfiber products is similar to that of the normal-fiber materials made from the same polymer. One caution is heat sensitivity. Because the fibers are so fine, heat penetrates more quickly causing them to scorch or glaze more quickly than "normal" fibers if too much heat is applied or heat is applied over too long a period. Typically, microfibers are wrinkle resistant, but if ironing is done, then it should be accomplished using lower temperatures and only as directed.

4.20 GENERAL STEPWISE POLYMERIZATION

From an equation, such as Equation 4.93, it is possible to derive expressions describing the molecular weight distribution of stepwise polymerizations at any extent of polymerization. This relationship is more easily derived from a statistical standpoint. The following statistical treatment assumes the reaction to be independent of chain length.

We can write a general equation for the formation of a linear polymer formed from the reaction of bifunctional reactants A and B as follows:

$$nA + nB \rightarrow A(BA)_{n-1}B \tag{4.93}$$

The probability of finding a repeating unit AB in the polymer is "p," and the probability of finding "$n-1$" of these repeat units in the polymer chain is p^{n-1}. Likewise, the probability of finding an unreacted molecule of A or B is $1-p$. Thus, the probability (p_n) of finding a chain with n repeat units $(BA)_n$ is

$$P_n = (1 - p)\, p^{n-1} \tag{4.94}$$

N_n is the probability of choosing (at random) a molecule with $(AB)_n$ repeat units, where N is the total number of molecules and is given as follows:

$$N_n = N\,(1 - p)\, p^{n-1} \tag{4.95}$$

Since

$$\frac{N_0}{N} = \frac{1}{1-p} \text{ or } N = N_0(1 - p) \tag{4.96}$$

where N_o is the total number of structural units present, and is described by

$$N_n = N_o\,(1 - p)^2\, p^{n-1} \tag{4.97}$$

The corresponding weight-average molecular weight distribution, W_n, can be calculated from the relationship

$$W_n = \frac{nN_n}{N_o} \tag{4.98}$$

as follows

$$W_n = \frac{nN_o(1-p)^2 p^{n-1}}{N_o} = n(1-p)^2 p^{n-1} \tag{4.99}$$

The relationship shown in Equations 4.98 and 4.99 shows that high values of p (0.99) are essential in producing high N_n and W_n values. The number-average molecular weight M_n calculated from Equations 4.96 and 4.97 are as follows:

$$\overline{M}_n = \frac{mN_o}{N} = \frac{m}{1-p} \tag{4.100}$$

Where m = the molecular weight of the mer and

$$\overline{M}_w = \frac{m(1+p)}{1-p} = \overline{M}_n(1+p) \tag{4.101}$$

Thus, the index of polydispersity, $\overline{M}_w/\overline{M}_n$ becomes $1 + p$, as shown below:

$$\frac{\overline{M}_w}{\overline{M}_n} = \frac{m(1+p)/(1-p)}{m/(1-p)} = 1+p \tag{4.102}$$

Thus, when $p = 1$, the index of polydispersibility for the most probable distribution for stepwise polymerizations is 2.

Because the value of p is essentially 1 in some stepwise polymerizations employing very pure monomers, the products obtained under normal conditions will have very high molecular weights and are found to be difficult to process. The value of p can be reduced by using a slight excess of one of the reactants or by quenching (stopping) the reaction before completion. Thus, if a reaction is quenched when the fractional conversion p is 0.995, the average DP becomes 200.

When more than 1 mol of B is used with 1 mol of A, the ratio of A/B or "r" may be substituted in the modified Carothers' equation as follows:

$$\overline{DP} = \frac{\text{total } nA \text{ at } p}{\text{total } nA \text{ at } rp} = \frac{n[(1+1/r)]}{2n[1-p+(1-rp/r)]/2} = \frac{[1+(1/r)]}{1-p+[(1-rp)/r]} \tag{4.103}$$

Multiplying the top and bottom by r gives

$$\overline{DP} = \frac{r+1}{r(1-p)+(1-rp)} = \frac{(1+r)}{(1+r)-2rp} \tag{4.104}$$

For the formation of nylon-66, if $r = 0.97$ and p is about 1, the $\overline{DP}$ is equal to

$$\overline{DP} = 1 + r(1+r) - 2rp = \frac{1+0.97}{1+0.97-2(0.97)} = \frac{1.97}{0.03} = 66 \tag{4.105}$$

The $\overline{DP}$ of 66 is above the threshold limit of 50 required for nylon-66 fibers.

Since quenching the reaction or adding a stoichiometric excess of one reactant is seldom economical, the commercial practice is to add a specific amount of a monofunctional reactant in the synthesis of polyesters, nylons, and other similar polymers. In these cases, a functionality factor, f, is used that is equal to the average number of functional groups present per reactive molecule. While the value of f in the preceding examples has been 2.0, it may be reduced to lower values and used in the following modified Carothers' equation.

$$\overline{DP} = \frac{A_o}{A_o[1-(pf/2)]} = \frac{2}{2-pf} \tag{4.106}$$

Thus, if 0.01 mol of monofunctional acetic acid is used with 0.99 mol of two difunctional reactants, the average functionality or functional factor, f, is calculated as follows:

$$f = \frac{\text{mol of each reactant} \times \text{functionality}}{\text{total number of moles}} = \frac{0.99 \text{ mol} \times 2 + 0.99 \text{ mol} \times 2 + 0.01 \text{ mol} \times 1}{1.99 \text{ mol}} = 1.99 \tag{4.107}$$

Substitution of $f = 1.99$ and $p = 1.00$ in Equation 4.106 gives a $\overline{DP}$ of 200 representing an upper limit for chain size often employed commercially for nylon-66. The same calculation but employing $p = .95$ gives a $\overline{DP}$ of only 20, below the lower desired value for nylon-66.

Since the average molecular weight increases with conversion, useful high molecular weight linear polymers may be obtained by the step-reaction polymerization when the fractional conversion, p, is high (>.99). The concentration of reactants decreases rapidly in the early stages of polymerization, and differing chain lengths will be present in the final product. The requirement for a linear polymer is a functionality of 2. Network polymers are typically formed when the functionality is greater than 2.

4.21 SUMMARY

1. Many naturally occurring and some synthetic polymers are produced by condensation reactions many of which are described kinetically by the term stepwise polymerization. A high-fractional conversion is required to form linear polymers, such as polyesters, nylons, polysulfides, PUs, PC, polysulfones, polyimides, PBI, and polyethers. But a high-fractional conversion is not required for the production of network, cross-linked, products, such as epoxy, phenol, urea, formaldehyde, and melamine resins. One major exception to the production of condensation polymers through the stepwise kinetic process is the use of the interfacial reaction system employing reactive reactants that follows a chainwise kinetic process. The interfacial system is employed to produce PC and some aramids. The remaining condensation polymers are generally produced using the melt and solution techniques.

2. The rate expressions and values, mechanisms, and the activation energies for the condensation reactions forming polymers are similar to those of small molecule reactions. Reaction rate increases with temperature in accordance with the Arrhenius equation. Average DP also increases as the reaction temperature increases to the ceiling temperature, where polymer degradation occurs. Long chains are only formed at the conclusion of classical polycondensation processes.

3. The $\overline{DP}_n$ for formation of linear condensation polymers can be calculated using the Carothers' equation, $\overline{DP}_n = 1/(1-p)$.

4. Cross-linked products are formed when the functionality of either reactant is greater than two. Linear products are formed when the functionality of both reactants is two.

5. Condensation polymers tend to exist below their T_g at room temperature. They typically form fairly ordered structures with lots of strong interactions between the various chains giving strong materials with some, but not much, elongation when stretched. They are normally used as fibers and plastics. They have high stress–strain ratios.

6. Condensation polymers include many of the materials referred to as synthetic or man-made fibers, including polyesters (especially poly(ethylene terephthalate) known as PET) and a variety of nylons (mainly nylon-66 and nylon 6).

7. A number of plastics are condensation polymers and include polyesters and nylons that are not as highly oriented as the same materials in fiber form. Other plastics have been developed that have outstanding heat stability, strength, or other properties that allows their wide use. These plastics include PC, polyimides, PBI, polysulfides, polyethers, polysulfones, and polyketones.

8. PUs derived from reaction of diisocyanates and a variety of hydroxyl-containing materials form the basis for foamed products and a variety of elastomeric materials.

9. Highly cross-linked condensation materials form the basis for a number of important adhesives and bulk materials especially phenolic and amino plastics. Most of these products have formaldehyde as one of their starting reactants. These materials are thermosets that decompose before melting so are more difficult to recycle than most condensation polymers that are thermoplastics and do melt before decomposition.

10. Microfibers are simply fibers that are much thinner than typical fibers. They may be derived from any fiber-producing material that is suitably treated to give thin fibers.

GLOSSARY

Alkyds: Term originally used to describe oil-modified polyester, but now used for many polyester plastics and coatings.

Allophanates: Reaction product of a urethane and an isocyanate.

Amino resins: Urea and MF resins.

Aramids: Aromatic polyamides.

A-stage: Linear prepolymer of phenol and formaldehyde.

Bakelite: Polymer produced by condensation of phenol and formaldehyde first by Leo Baekeland.

Bifunctional: Molecule with two active functional groups.

Bisphenol A: 2,2'-Bis(4-hydroxphenol)propane.

B-stage: Advanced A-stage resin.

Carbamate: A urethane.

Carbamic acids: Unstable compounds that decompose spontaneously giving amines and carbon dioxide.

Carothers, W. H.: Inventor of nylon who also standardized much of the polymer nomenclature we use today.

Carothers equation: $\overline{DP}_n = 1/(1-p)$

Condensation reaction: Reaction in which two molecules react producing a third molecule and a byproduct such as water.

Cyclization: Ring formation.

Dacron: Trade name for PET fiber.

Drying: Cross-linking of an unsaturated polymer generally in the presence of oxygen.

Drying oil: An unsaturated oil like tung oil.

Engineering plastic: Plastics whose physical properties are good enough to permit their use as structural materials; generally they can be cut, sawn, and drilled.

Epoxy resin: Polymer produced by the condensation of epichlorohydrin and a dihydric alcohol or by the epoxidation of an unsaturated molecule.

Ester interchange: Reaction between an ester of a volatile alcohol and a less volatile alcohol in which the lower boiling alcohol is removed by distillation.

Filament: Extrudate when a polymer melts or when a solution is forced through a hole in a spinneret.

Functionality: Number of active functional groups present in a molecule.

Functionality factor: Average number of functional groups present per reactive molecule in a mixture of reactants.

Furan resin: Resin produced from furfuryl alcohol or furfural.

Gel point: Point at which cross-linking begins to produce polymer insolubility.

Glyptals: Polyesters, usually cross-linked by heating.

Incipient gelation: Point where the DP reaches infinity.

Interfacial polymerization: One in which the polymerization reaction occurs at the interface of two immiscible liquids.

Kodel: Trade name for a PET fiber.

Ladder polymer: Double-chained temperature-resistant polymer.

Laminate: Layers of sheets or paper or wood or other material adhered by resins and pressed together like plywood.

Long oil alkyd: One obtained in the presence of 65%–80% of unsaturated oil.

Medium oil alkyd: Alkyd obtained in the presence of 50%–65% of unsaturated oil.

Melamine-formaldehyde resin: Resin produced by the condensation of melamine and formaldehyde.

Microfibers: Small diameter fibers.

Molding compound: Name given to describe a mixture of a resin and essential additives.

Nonoil alkyd: An oil-free alkyd containing no unsaturated oils.

Novolac: Polymers prepared by the condensation of phenol and formaldehyde under acidic conditions.

Nylon: Synthetic polyamide.

Oil length: Term used to indicate the relative percentage of unsaturated oils used in the production of alkyds.

Phenoxy resin: Polymer with hydroxyl pendant groups resembling an epoxy resin without epoxy groups.

Poly(ethylene terephthalate) (PET): Linear polyester used to produce fibers and for blow-molding preparation of soft drink bottles; produced from terephthalic acid and ethylene glycol.

Prepolymer: Low molecular weight material (oligomer) capable of further polymerization.

Resite: Cross-linked resole.

Resole: Linear polymer prepared by condensation of phenol and formaldehyde under alkaline conditions.

Schotten–Baumann reaction: Traditionally the reaction between an acid chloride and a Lewis base.

Short oil alkyd: An alkyd obtained in the presence of 30%–-50% of unsaturated oil.

Step-reaction polymerization: Polymerization in which polyfunctional reactants react to produce larger units in a continuous stepwise manner.

Thiokol: Trade name for a polysulfide elastomer.

Unsaturated polyester: Term used to describe alkyds with unsaturated chains, particularly those produced by the condensation of maleic anhydride and ethylene glycol.

Urea-formaldehyde resin: Resin produced by condensation of urea and formaldehyde.

Wasted loops: Formation of cyclic compounds instead of polymer chains.

EXERCISES

1. Which of the following will give a polymer when condensed with adipic acid: (a) ethanol, (b) ethylene glycol, (c) glycerol, (d) aniline, or (e) ethylenediamine?
2. Could Carothers have produced strong polyester fibers by ester interchange or Schotten–Baumann reactions using aliphatic reactants?
3. Which would be useful as a fiber: (a) PET or (b) poly(hexylene terephthalate)?
4. If the fractional conversion in an ester interchange reaction is 0.99999, what would be the average DP of the polyester produced?
5. Use the logarithmic form of the Arrhenius equation to show that the value of the rate constant k increases as the temperature increases.
6. What is the first product produced when a molecule of sebacyl chloride reacts with a molecule of ethylene glycol?
7. What is the next product formed in question 6?
8. How would you improve the strength of the filament produced in the nylon rope trick without changing the reactants?
9. Name the product produced by the condensation of adipic acid and 1,4-tetramethylenediamine?
10. In which reaction would you expect the more "wasted loop": the reaction of oxalyl chloride with (a) ethylenediamine or with (b) 1,6-hexanediamine?
11. Which system would be more apt to produce "wasted loops: (a) a dilute solution or (b) a concentrated solution?
12. If the values of A_o and k are 10 mol/L and 10^{-3} L mol/s, respectively, how long would it take to obtain an average DP of 37?
13. Which will give the lower index of polydispersity: (a) $p = .000$ or (b) $p = .90$?
14. If you used a 2% molar excess of BPA with TDI, what would be the maximum average DP obtainable assuming $p = 1$?
15. Why would you predict that the product obtained in question 14 be a useful fiber assuming an average DP of 100?
16. Name the product formed from the reaction of a phenol with formaldehyde.
17. Give the Carother's equation. What is its significance?
18. What is the product of polymerized formaldehyde? Is it stable?
19. What is the product of a diol and diisocyanate? What are general uses of this product?
20. Which would be the better or stronger fiber: one made from an ester of (a) terephthalic acid or of (b) phthalic acid?
21. What would be the deficiency of a nylon film that was stretched in one direction only?
22. Which would be more flexible: (a) PBT or (b) poly(hexylene terephthalate)?
23. Which would be more apt to deteriorate in the presence of moisture: (a) Lexan molding powder or (b) Lexan sheet?
24. What reactants are typically employed to make a PC?
25. How would you prepare a nylon with greater moisture resistance than nylon-66?
26. How would you prepare a nylon that would be less "clammy" when used as clothing?
27. Which would be higher melting: (a) a polyamine or (b) a polyester with similar members of methylene groups in the repeat unit?
28. Why is a methoxymethylated nylon more flexible than nylon?
29. Why Bakelite not used in forming molded objects?
30. Isn't it wasteful to decompose a diisocyanate by hydrolysis to produce foams?
31. How would you prepare a hydroxyl-terminated polyester?
32. For the acid-catalyzed condensation of ethylene glycol with terephthalic acid, what is the relationship between DP and extent of reaction?

33. For stepwise kinetics what happens to chain length as time increases?
34. Why do PUs and epoxy resins have good adhesive properties?
35. Why are furan resins relatively inexpensive?
36. To product cross-linked polyesters what is normally present?
37. Could you produce a soluble novolac resin from resorcinol?
38. Can you explain why there are so many terms used, such as novolac, resole, and so on, in phenolic resin technology?
39. Why isn't Bakelite used for dinnerware?
40. Which of the following could be a nonpetrochemical plastic: (a) Bakelite, (b) urea plastics, or (c) melamine plastics?
41. Which would produce the better fiber: the reaction product of phthalic acid and (a) 1,4-butanediol, or (b) 2-hydroxylbutanol?
42. What are some properties of microfibers?
43. What are the main uses for PU?
44. What are the advantages of open-celled foams and how do they differ from closed-cell foams?
45. What is the most important property that LCs have?
46. Name three uses of LCs.
47. Why is PEG widely used in drug delivery?
48. What are aramides?
49. What are the general properties of aramids?

ADDITIONAL READING

Amato, I. (1997): *The Materials the World is Made of*, Harper Collins, NY.
Bayer, A. (1878): Phenol-formaldehyde condensates, *Ber. Bunsenges. Phys. Chem.*, 5:280, 1094.
Binder, W. (2007): *Hydrogen Bonded Polymers*, Springer, NY.
Brunelle, D. J., Korn, M. (2005): *Advances in Polycarbonates*, Oxfore University Press, NY.
Carothers, W. H. (1929): An introduction to the general theory of condensation polymers, *J. Amer. Chem. Soc.*, 51:2548.
Carothers, W. H. (1938): *US Patent* 2,130,947.
Carraher, C., Swift, G. (2002): *Functional Condensation Polymers*, Kluwer, NY.
Carraher, Swift, G., Bowman, C. (1997): *Polymer Modification*, Plenum, NY.
Craver, C., Carraher, C. (2000): *Applied Polymer Science*, Elsevier, NY.
Carraher, C., Preston, J. (1982): *Interfacial Synthesis*, Vol. 3, Dekker, NY.
Collins, P. J., Hird, M. (1997): *Introduction to Liquid Crystals*, Taylor and Francis, London.
D'Amore, A. (2006): *Monomers and Polymers: Reactions and Properties*, Nova, Commack, NY.
Donald, A. (2005): *Liquid Crystalline Polymers*, Cambridge University Press.
Hill, J. W., Carothers, W. H. (1932, 1933): Polyanhydrides, *J. Amer. Chem. Soc.*, 54:1569, 55:5023.
Imrie, I. (2004): *Introduction to Liquid Crystalline Polymer*, Taylor & Francis, NY.
Kadolph, S., Langford, A. (2001): *Testiles*, Prentice Hall, Englewood Falls, NJ.
Lee, S-T., Park, C. (2006): *Polymeric Foams*, Taylor and Francis, Boca Raton, FL.
Martin, S. M., Patrick, J. C. (1936): Thiokol, *Ind. Eng. Chem.*, 28:1144.
Marvel, C. S. (1959): *An Introduction to the Organic Chemistry of High Polymers*, Wiley, NY.
Millich, F. Carraher, C. (1977): *Interfacial Synthesis*, Vols. 1 and 2, Dekker, NY.
Morgan, P. W. (1965): *Condensation Polymers by Interface and Solution Methods*, Wiley, NY.
Morgan, P. W., Kwolek, S. L. (1959): The nylon rope trick, *J. Chem. Ed.*, 36:182.
Pascault, J., Sautereau, H., Verdu, J. Williams, R. (2002): *Thermosetting Polymers*, Dekker, NY.
Reneker, D. H., Fong, H. (2006): *Polymeric Nanofibers*, Oxford University Press, NY.
Rogers, M. (2003): *Synthetic Methods in Step-Growth Polymers*, Wiley, Hoboken, NJ.
Sadler, S. R., Karo, W. (1998): *Polymer Synthesis* (three volumes), Academic Press, Orlando, FL.
Seavey, K., Liu, Y. (2008): *Step-Growth Polymerization Process Modeling and Product Design*, Wiley, Hoboken, NJ.
Yamashito, H., Nakano, Y. (2008): *Polyester: Properties, Preparation and Applications*, Nova, Hauppauge, NY.

5 Ionic Chain-Reaction and Complex Coordination Polymerization

In contrast to the relatively slow step-reaction polymerizations discussed in Chapter 4, chain polymerizations are usually rapid, and the initiation species continues to propagate until termination. Thus, in the extreme case, a single initiation species could produce one high molecular chain, leaving all of the other monomer molecules unchanged. In any case, the concentration of monomer, which is often a substituted vinyl compound, decreases continuously throughout the reaction. In contrast to stepwise polymerization, the first species produced in chain polymerizations is a high molecular polymer.

A kinetic chain reaction usually consists of at least three steps, namely (1) initiation, (2) propagation, and (3) termination. The initiator may be an anion, a cation, a free radical, or a coordination catalyst. While coordination catalysts are the most important commercially, the ionic initiators will be discussed first in an attempt to simplify the discussion of chain-reaction polymerization.

In almost all of the polymerizations described in this chapter, there is a sensitive and critical balance between the activity of the catalyst and polymerization. For instance, if the catalyst is too active, it may bind at unwanted sites, including the solvent. If the catalyst complex is not sufficiently active, then ready initiation does not occur. The choice of the solvent is also important. Some solvents will react with the catalysts binding it rather than allowing the catalyst to initiate the desired polymerization. Others may "hold" the catalyst complex together rather than allowing the catalyst to initiate polymerization. Still others may not allow the catalyst complex to form. As in much of science, the precise ingredients and conditions were developed through a combination of intuition, science, art, and research. This process continues.

While many vinyl monomers undergo free radical polymerization (Chapter 6), a smaller number undergo ionic polymerization. Cationic polymerizations require monomers that have electron releasing groups such as an alkoxy, phenyl, or vinyl group. Anionic polymerization occurs with monomers containing electron-withdrawing groups such as carboxyl, nitrile, or halide. This selectivity is due to the strict requirements for stabilization of anionic and cationic species.

Compared to free radical polymerizations, the kinetics of ionic polymerizations are not as well defined. Reactions can use heterogeneous initiators, and they are usually quite sensitive to the presence of impurities. Thus, kinetic studies are difficult and the results sensitive to the particular reaction conditions. Further, the rates of polymer formation are more rapid.

Cationic and anionic polymerizations are similar. Both involve the formation and propagation of ionic species. While high energy, low stability, ions would be expected to react with most double bonds, ionic species that are stable enough to propagate are difficult to form and are easily destroyed. The "energenic-window" that allows the formation of such charged species that promote polymer formation is narrow. While polar solvents might be desirable to solvate the ions, and hence help stabilize them, they often cannot be used. Some polar solvents, such as water and alcohols, react with and destroy most ionic initiators. Other polar solvents, such as ketones, prevent initiation because of the formation of stable complexes with the initiators. Ionic polymerizations are therefore conducted in low or moderately polar solvents, such as hexane and ethylene dichloride.

By bulk, almost all vinyl polymers are made by four processes (Table 5.1)—free radical (about 50%), complex coordinate (about 20%), anionic (10%–15%), and cationic (8%–12%). Three of these techniques are covered in this chapter.

TABLE 5.1
Major Technique Used in the Production of Important Vinyl Polymers

Free Radical	Low-density polyethylene (LDPE)
	Poly(vinyl chloride)
	Poly(vinyl acetate)
	Polyacrylonitrile and acrylic fibers
	Poly(methyl methacrylate)
	Polyacrylamide
	Polychloroprene
	Poly(vinyl pyridine)
	Styrene–acrylonitrile copolymers, SAN
	Polytetrafluoroethylene
	Poly(vinylene fluoride)
	Acrylonitrile–butadiene–styrene copolymers (ABS)
	Ethylene-methacrylic acid copolymers
	Styrene–butadiene copolymers (SBR)
	Nitrile rubber (NBR)
	Polystyrene
Cationic	Polyisobutylene
	Butyl rubber
	Polyacetal
Anionic	Thermoplastic Olefin Elastomers (Copolymers of butadiene, isoprene, and styrene)
	Polyacetal
Complex	High-density polyethylene (HDPE)
	Polypropylene
	Polybutadiene
	Polyisoprene
	Ethylene–propylene elastomers

5.1 CHAIN GROWTH POLYMERIZATION—GENERAL

The next three chapters will deal with polymers formed from chain growth polymerization. Chain growth polymerization is also called addition polymerization and is based on free radical, cationic, anionic, and coordination reactions, where a single initiating species causes the growth of a polymer chain.

The kinetic chain reaction typically consists of three steps—initiation, propagation, and termination. The initiators for free radical, anionic, and cationic polymerizations are organic radicals, carbanions, and carboniums. Chain growth is exothermic with the polymerization mainly controlled by the steric and resonance factors associated with the monomer. Generally, the less resonance stabilization the more heat is given off during the reaction. Also, the greater the steric factors the less heat is given off during the polymerization.

5.2 CATIONIC POLYMERIZATION

The art of cationic polymerization, like that of many other types of polymerization, is at least a century old. However, the mechanisms for these reactions have only recently become better understood.

The first species produced in cationic polymerizations are carbocations, and these were unknown as such before World War II. It is now known that pure Lewis acids, such as boron trifluoride and aluminum chloride, are not effective as initiators. A trace of a proton-containing Lewis base, such as water, is also required. The Lewis base coordinates with the electrophilic Lewis acid, and the proton is the actual initiator. Since cations cannot exist alone, they are accompanied by a *counterion*, also called a *gegenion* (Equation 5.1).

$$BF_3 \quad + \quad H_2O \quad \rightleftharpoons \quad H^+, BF_3OH^-$$ (5.1)

Lewis acid   Lewis base   Catalyst–cocatalyst
Catalyst   Cocatalyst   complex

Since the required activation energy for ionic polymerization is small, these reactions may occur at very low temperatures. The carbocations, including the macrocarbocations, repel one another; hence, chain termination does not generally occur by combination but is usually the result of reaction with impurities.

Both the initiation step and the propagation step are dependent on the stability of the carbocations. Isobutylene (the first monomer to be commercially polymerized by ionic initiators), vinyl ethers, and styrene have been polymerized by this technique. The order of activity for olefins is $Me_2C=CH_2 > MeCH=CH_2 > CH_2=CH_2$, and for para-substituted styrenes the order for the substituents is Me-O > Me > H > Cl. The mechanism is also dependent on the solvent as well as the electrophilicity of the monomer and the nucleophilicity of the gegenion. Rearrangements may occur in ionic polymerizations.

The rate of initiation (R_i) for typical cationic reactions is proportional to the concentration of the monomer [M] and the concentration of the catalyst–cocatalyst complex [C] as follows:

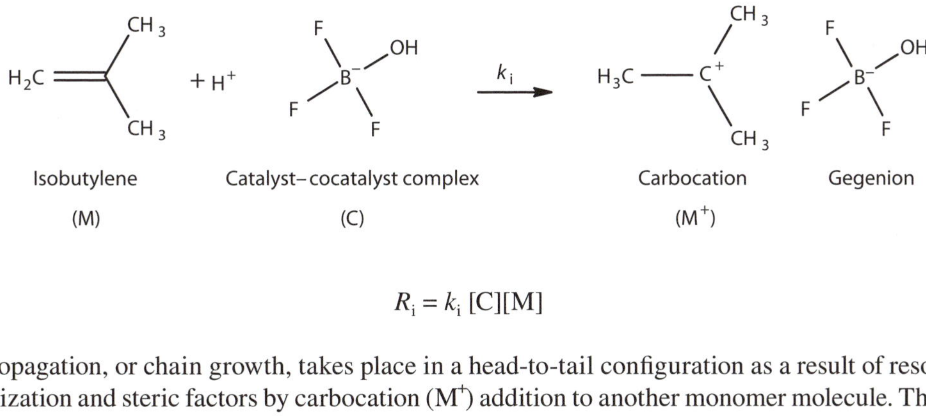

Isobutylene   Catalyst–cocatalyst complex   Carbocation   Gegenion

(M)   (C)   (M$^+$)

(5.2)

$$R_i = k_i\,[C][M]$$ (5.3)

Propagation, or chain growth, takes place in a head-to-tail configuration as a result of resonance stabilization and steric factors by carbocation (M$^+$) addition to another monomer molecule. The head stabilizes the cation best so it is the growing site, while the least sterically hindered site is the site for attack by the cation resulting in the typical head-to-tail arrangement. The rate constant for growth is essentially independent of chain length so is the same for all propagation steps and is influenced by the dielectric constant of the solvent. The rate is fastest in solvents with high-dielectric constants, promoting separation of the carbocation–gegenion pairs. The chemical and kinetic equations for this are as follows:

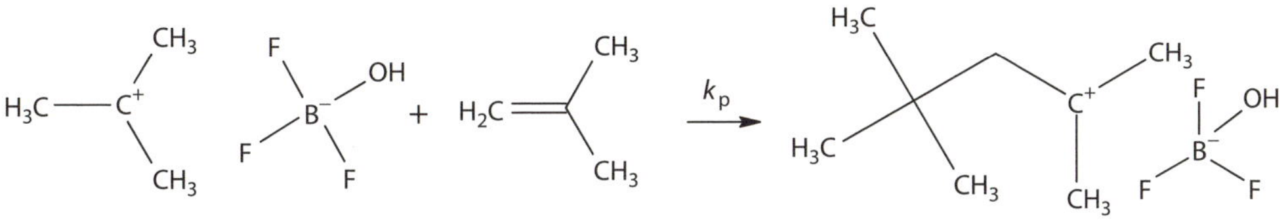

Carbocation (Gegenion)   Isobutylene   Macrocarbocation   Gegenion

(M$^+$)   (M)   (M$^+$)

(5.4)

$$R_p = k_p [M][M^+] \tag{5.5}$$

The termination rate R_t, assumed to be a first-order process, is simply the dissociation of the macrocarbocation–gegenion complex here forming BF_3 and H_2O and the now neutral "dead" polymer chain. This is expressed as follows:

$$R_t = k_t [M^+] \tag{5.6}$$

Termination may also occur by chain transfer, where a proton is transferred to a monomer molecule, leaving a cation that can serve as an initiator. The $\overline{DP}$ is equal to the kinetic chain length (ν) when chain transfer occurs. The chemical and kinetic equations for termination via chain transfer are as follows:

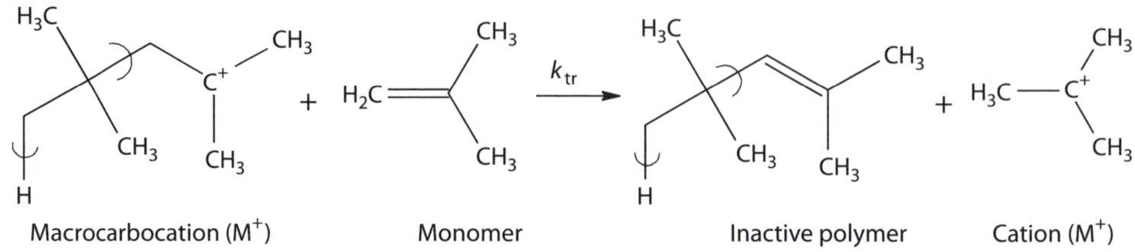

| Macrocarbocation (M^+) | Monomer | Inactive polymer | Cation (M^+) |

$$R_{tr} = k_{tr} [M][M^+] \tag{5.8}$$

It is experimentally found that the rate of initiation equals the rate of termination, and since the propagation step is so rapid, the number of growing chains is constant. Since it is difficult to determine values for some members of the kinetic expressions, including $[M^+]$ the following approach is normally taken to eliminate the need for determining $[M^+]$. Since there is a steady state of growing chains, the rate of initiation is equal to the rate of termination, giving $R_i = R_t$, and solving for $[M^+]$ gives the following:

$$k_i[C][M] = k_t[M^+], \tag{5.9}$$

therefore

$$[M^+] = \frac{k_i[C][M]}{k_t} \tag{5.10}$$

This expression of $[M^+]$ is substituted into the propagation rate expression Equation 5.5 giving Equation 5.11.

$$R_p = k_p[M][M^+] = \frac{k_p k_i[C][M]^2}{k_t} = k'[C][M]^2 \tag{5.11}$$

For termination by chain transfer we have

$$k_i[C][M] = k_t[M][M^+] \tag{5.12}$$

and

$$[M^+] = \frac{k_i[C][M]}{k_t[M]} = \frac{k_i[C]}{k_t} \tag{5.13}$$

giving

$$R_p = k_p[M][M^+] = \frac{k_p k_i[C][M]}{k_t} = k'[C][M] \tag{5.14}$$

The $\overline{DP}$ can also be described when internal dissociation is the dominant termination step as follows:

$$\overline{DP} = \frac{R_p}{R_t} = \frac{k_p[M][M^+]}{k_t[M^+]} = k''[M]$$
(5.15)

But if chain transfer is the dominant termination step then

$$\overline{DP} = \frac{R_p}{R_{tr}} = \frac{k_p[M][M^+]}{k_{tr}[M][M^+]} = \frac{k_p}{k_{tr}} = k''$$
(5.16)

It is important to note that regardless of how termination occurs the molecular weight is independent of the concentration of the initiator. However, the rate of ionic-chain polymerization is dependent on the dielectric constant of the solvent, the resonance stability of the carbonium ion, the stability of the gegenion, and the electropositivity of the initiator.

The rates of all single step reactions increase as the temperature increases. This may not be true for the "net effect" for multistep reactions such as those involved with multistep polymerizations, here the cationic polymerization. For cationic polymerizations the activation energies are generally of the order $E_{tr} > E_i > E_p$. Remembering that the description of the specific rate constant is

$$k = A\, e^{-Ea/RT}$$
(5.17)

the overall or "net" activation energy for chain growth from Equation 5.11 is

$$E_{(overall)} = E_p + E_i - E_{tr}$$
(5.18)

and for chain length from Equation 5.16 it is

$$E_{(overall)} = E_p - E_{tr}$$
(5.19)

For many cationic polymerizations, the net activation is negative, using the relationships given in Equation 5.11, so that the overall rate of polymerization decreases, for these cases, as the temperature is increased. Further, using Equation 5.16 and since $E_{tr} > E_p$, the overall degree of polymerization does decrease as the temperature is increased. This is pictured in Figure 4.4.

Butyl rubber (IIR) is widely used for inner tubes and as a sealant. It is produced using the cationic polymerization with the copolymerization of isobutylene in the presence of a small amount (10%) of isoprene. Thus, the random copolymer chain contains a low concentration of widely spaced isolated double bonds, from the isoprene, that are later cross-linked when the butyl rubber is cured. A representative structure is shown in 5.20, where the number of units derived from isobutylene units greatly outnumbers the number the units derived from the isoprene monomer. The steric requirements of the isobutylene-derived units cause the chains to remain apart giving it a low stress/strain value and a low T_g.

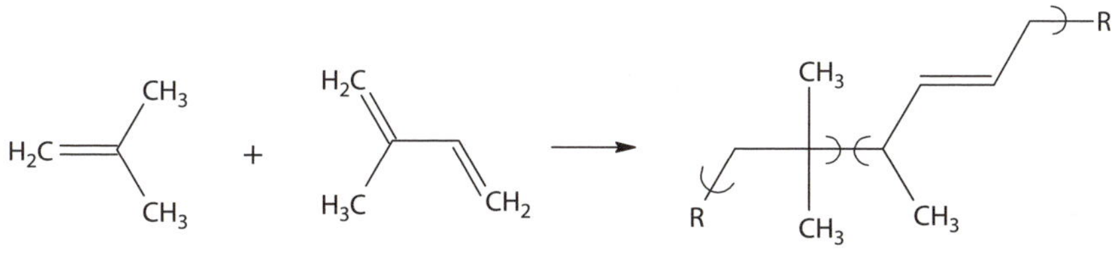

(5.20)

The cationic polymerization of vinyl isobutyl ether at −40°C produces stereoregular polymers (5.21). The carbocations of vinyl alkyl ethers are stabilized by the delocalization of "p" valence electrons in the oxygen atom, and thus these monomers are readily polymerized by cationic initiators. *Poly(vinyl isobutyl ether)* has a low T_g because of the steric hindrance offered by the isobutyl group. It is used as an adhesive and as an impregnating resin.

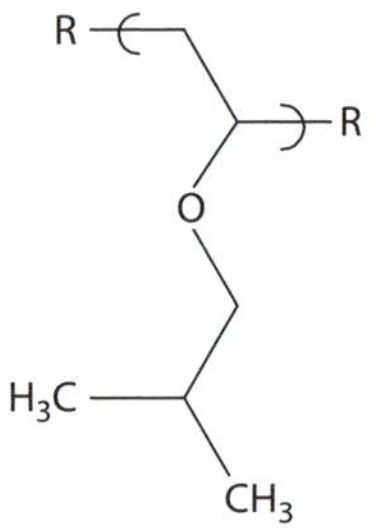

(5.21)

This production of stereoregular structures has been known for some time and is especially strong for vinyl ethers. Several general observations have been noted. First, the amount of stereoregularity is dependent on the nature of the initiator. Second, steroregularity increases with a decrease in temperature. Third, the amount and type (isotactic or syndiotactic) is dependent on the polarity of the solvent. For instance, the isotactic form is preferred in nonpolar solvents, but the syndiotactic form is preferred in polar solvents.

Commercial polymers of formaldehyde are also produced using cationic polymerization. The polymer is produced by ring opening of trioxane. Since the polyacetal, polyoxymethylene (POM), is not thermally stable, the hydroxyl groups are esterified (capped) by acetic anhydride (Equation 5.22). These polymers are also called *poly(methylene oxides)*. The commercial polymer is a strong engineering plastic. Engineering plastics typically have higher modulus and higher heat resistance than general-purpose plastics.

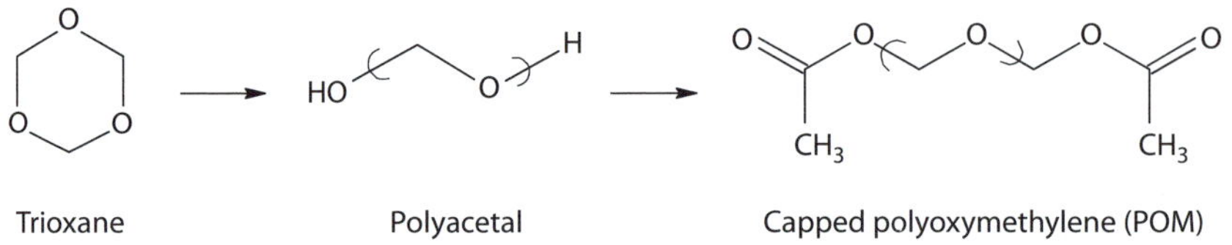

| Trioxane | Polyacetal | Capped polyoxymethylene (POM) | (5.22) |

Another stable *polyacetal* (POM; Celcon) is produced by the cationic copolymerization of a mixture of trioxane and dioxolane (Equation 5.23).

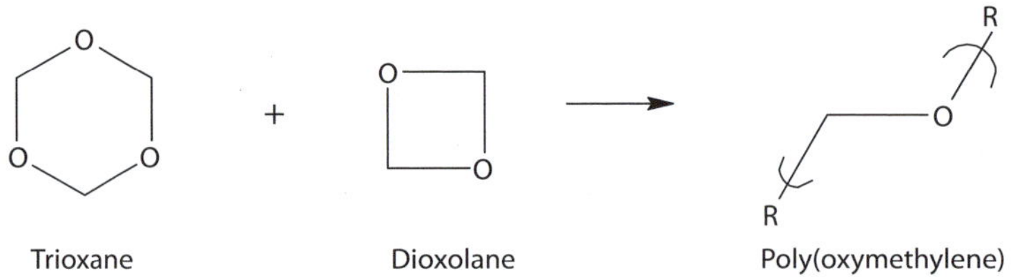

| Trioxane | Dioxolane | Poly(oxymethylene) | (5.23) |

Many other cyclic ethers have been polymerized using cationic polymerization. Ethylene oxide (also called oxirane) polymerizes forming *poly(ethylene oxide)* (PEO; also known as *poly(ethylene glycol)* (Equation 5.24) in the presence of acids such as sulfuric acid, producing a wide range of chain sized polymers sold under various trade names, including Carbowax and Polyox. PEO is also used in cosmetics and pharmaceuticals (as water-soluble pill coatings and capsules).

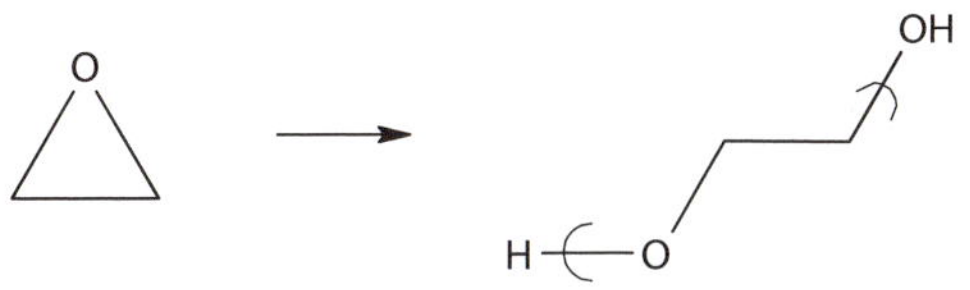

(5.24)

Polymer production proceeds as described below (Equation 5.25). An initiator, such as sulfuric acid, produces an oxonium ion and a gegnion. The oxonium ion then adds to the oxirane, ethylene oxide, producing a macrooxonium ion with growth eventually terminated by chain transfer with water.

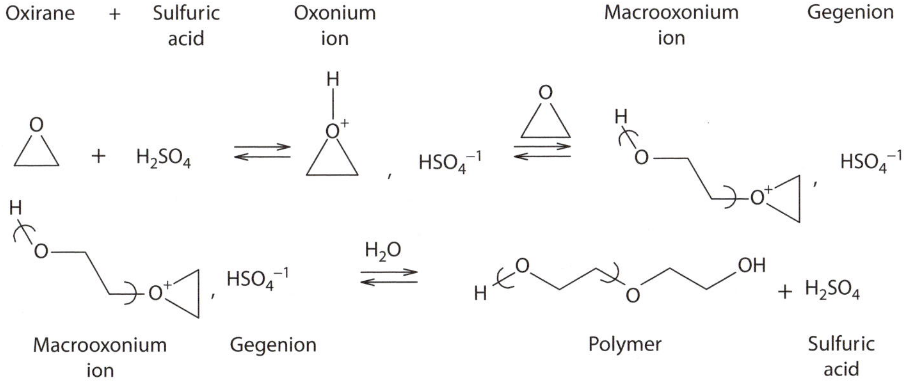

| Oxirane | + | Sulfuric acid | | Oxonium ion | | Macrooxonium ion | Gegenion |

(5.25)

| Macrooxonium ion | Gegenion | | Polymer | Sulfuric acid |

Propagation is an equilibrium reaction that limits polymerization. The highest chains are produced in mildly polar solvents, such as methylene chloride, at low temperatures (–20°C–100°C).

In addition to the production of polyacetals by the ring-opening polyomerization (ROM) of trioxane, Staundinger investigated ring opening of other ethers such as ethylene oxide (oxirane), as described above. Other homologous cyclic ethers, such as tetrahydrofurane, have been polymerized by cationic ROP reactions. The tendency for ring opening decreases as the size of the ring increases and some oxirane is generally added as a promoter for the polymerization of tetrahydrofurane. The six-member ring oxacyclohexane is so stable that it does not undergo cationic ROP even in the presence of a promoter.

The oxacyclobutane derivative, 3,3-bischloromethyloxacyclobutane, is polymerized (5.26) using cationic ROM, giving a water-insoluble, crystalline, corrosion-resistant polymer sold under the trade name of Penton.

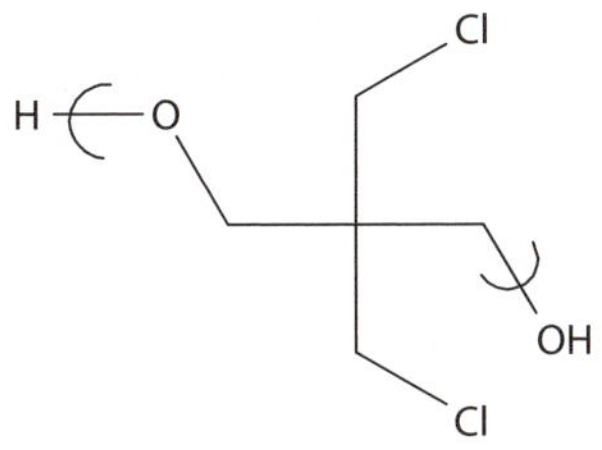

(5.26)

Polychloral (Equation 5.27) does not exist above its ceiling temperature of 58°C. Above this temperature the polymer decomposes. The production of polychloral is carried out by cationic polymerization through the introduction of a hot mixture of trichloroacetaldehyde and initiator into a mold and allowing polymerization to occur in situ as the mixture cools below the ceiling temperature. Even though polychloral has a low-ceiling temperature, it is used as a flame-resistant material because of the presence of the chlorine atoms.

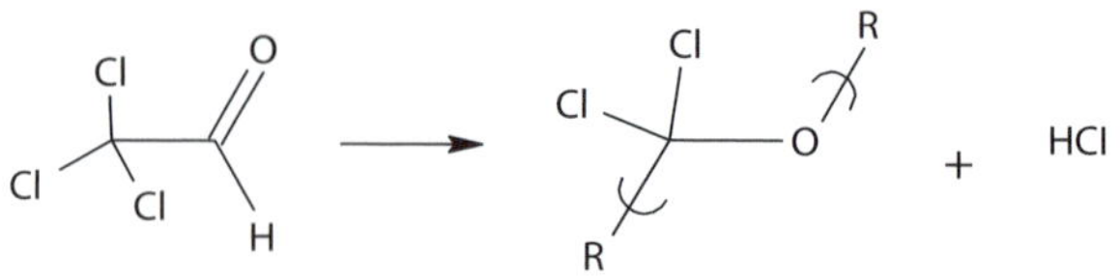

$$\text{(5.27)}$$

An acid-soluble polymer, Montrek, has been produced by the ROM of ethyleneimine, aziridine (Equation 5.28). The monomer is a carcinogen so care is taken to remove unreacted monomer.

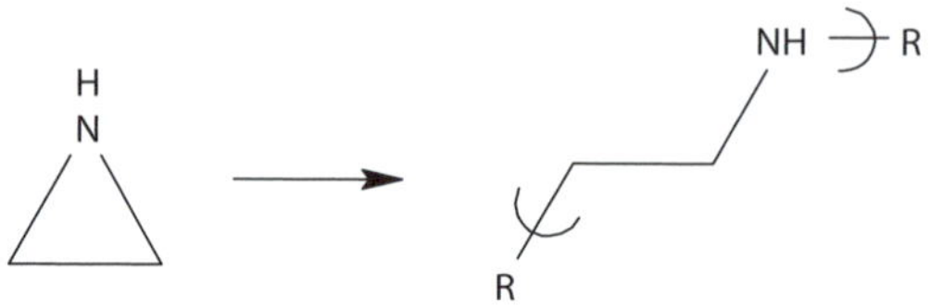

$$\text{(5.28)}$$

While lactams are often polymerized by anionic ROMs, N-carboxyl-alpha-amino acid anhydrides, called Leuchs' anhydrides, can be polymerized by either cationic or anionic techniques (Equation 5.29). These polypeptide products, which are called nylon-2 products, were initially produced by Leuchs in 1908. The synthetic approach can be used to produce homopolypeptides that are used as model compounds for proteins. Carbon dioxide is eliminated in each step of the propagation reaction.

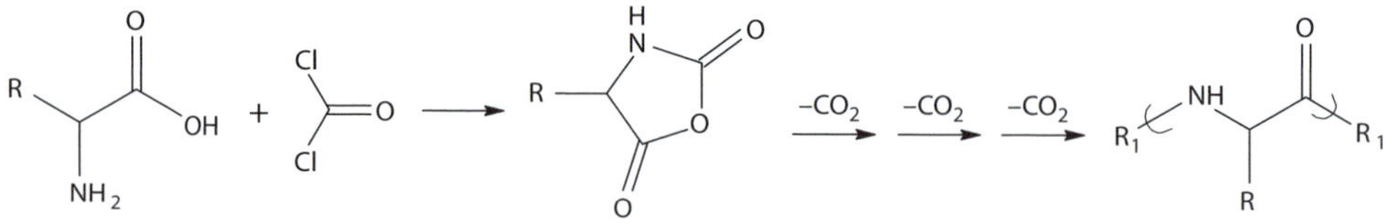

$$\text{(5.29)}$$

Polyterpenes, coumarone-indene resins, and so-called petroleum resins are produced commercially using cationic polymerization. These are used as additives for rubber, coatings, floor coverings, and adhesives.

5.3 ANIONIC POLYMERIZATION

Anionic polymerization was used to produce synthetic elastomers from butadiene at the beginning of the twentieth century. Initially, alkali metals in liquid ammonia were used as initiators, but by the 1940s they were replaced by metal alkyls such as n-butyllithium. In contrast to vinyl monomers with electron-donating groups polymerized by cationic initiators, vinyl monomers with electron-withdrawing groups are more readily polymerized using anionic polymerization. Accordingly, the order of activity using an amide ion initiator is acetonitrile> methyl methacrylate> styrene> butadiene.

The chemical and kinetic relationships for the anionic polymerization of acrylonitrile follow the same three major steps found for cationic polymerizations—namely initiation, propagation, and termination:

Amide ion   Acrylonitrile   Carboanion

$$\text{(5.30)}$$

$$R_i = k_i\,[C][M] \tag{5.31}$$

where C = :NH$_2^-$

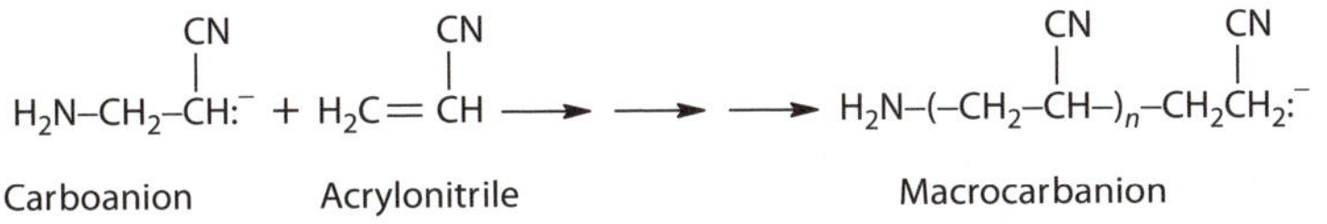

$$H_2N-CH_2-\overset{\overset{\displaystyle CN}{|}}{CH}{:}^- + H_2C= \overset{\overset{\displaystyle CN}{|}}{CH} \longrightarrow \longrightarrow \longrightarrow H_2N-(-CH_2-\overset{\overset{\displaystyle CN}{|}}{CH}-)_n-CH_2CH_2{:}^-$$ (5.32)

Carboanion   Acrylonitrile   Macrocarbanion

$$R_p = k_p\,[M][M^-]$$ (5.33)

with termination occurring through solvent transfer gives:

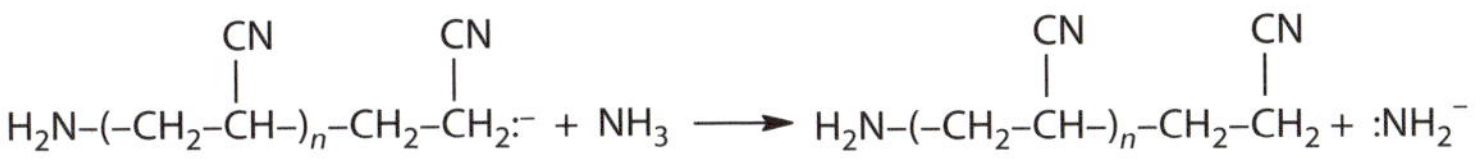

$$H_2N-(-CH_2-\overset{\overset{\displaystyle CN}{|}}{CH}-)_n-CH_2-CH_2{:}^- + NH_3 \longrightarrow H_2N-(-CH_2-\overset{\overset{\displaystyle CN}{|}}{CH}-)_n-CH_2-CH_2 + {:}NH_2^-$$ (5.34)

$$R_{tr} = k_{tr}\,[NH_3][M^-]$$ (5.35)

As in the case with cationic polymerizations, the number of growing chains is constant so that a steady state exists such as the $R_i = R_{tr}$. This is useful because it is difficult to determine the concentration of $[M^-]$ so that it can be eliminated as follows:

$$k_i[C][M] = k_{tr}\,[NH_3][M^-]$$ (5.36)

and

$$[M^-] = \frac{k_i[C][M]}{k_{tr}\,[NH_3]}$$ (5.37)

Substitution into Equation 5.33 gives

$$R_p = k_p\,[M][M^-] = \frac{k_p[M]k_i[C][M]}{k_{tr}[NH_3]} = \frac{k'[M]^2[C]}{[NH_3]}$$ (5.38)

$\overline{DP}$ can be described as follows:

$$\overline{DP} = \frac{R_p}{R_{tr}} = \frac{k_p[M][M^-]}{k_{tr}[NH_3][M^-]} = \frac{k'[M]}{[NH_3]}$$ (5.39)

Using the same approach we did with the cationic polymerization, we have for the rate of propagation

$$E_{(overall)} = E_p + E_i - E_{tr}$$ (5.40)

and for the dependence of chain length

$$E_{(overall)} = E_p - E_{tr}$$ (5.41)

Thus, the rate of propagation and the molecular weight are both inversely related to the concentration of ammonia. The activation energy for chain transfer is larger than the activation energy for propagation. The overall activation energy is about 160 kJ/mol. The reaction rate increases and molecular weight decreases as the temperature is increased as shown in Figure 4.4. The reaction rate is dependent on the dielectric constant of the solvent and the degree of solvation of the gegenion. Weakly polar initiators, such as Grignard's reagent, may be used when strong electron-withdrawing groups are present on the monomer, but monomers with less electron-withdrawing groups require more highly polar initiators such as *n*-butyllithium.

Synthetic *cis-1,4-polyisoprene* (Equation 5.42) is produced at an annual rate of about 100,000 tons by the anionic polymerization of isoprene when a low-dielectric solvent, such as hexane, and n-butyllithium is used. But, when a stronger dielectric solvent, such as diethylether, is used along with n-butyllithium, equal molar amounts of *trans*-1,4-polyisoprene and *cis*-3,4-polyisoprene units are produced. It is believed that an intermediate cisoid conformation assures the formation of a cis product. An outline describing the formation of *cis*-1,4-polyisoprene is given in Equation 5.42.

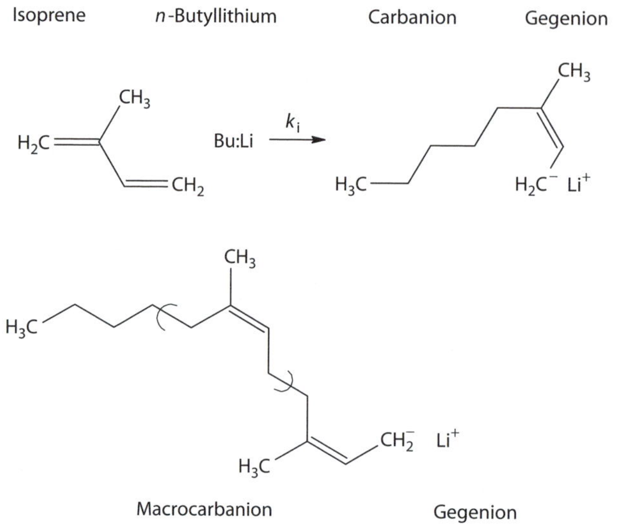

$$(5.42)$$

Macrocarbanion   Gegenion

No formal termination is given in Equation 5.42 since in the absence of contaminants the product is a stable macroanion. Szwarz named such stable active species "living polymers." These macroanions or macrocarbanions have been used to produce block copolymers such as Kraton. Kraton is an ABA block copolymer of styrene (A) and butadiene (B) (5.43). Termination is brought about by addition of water, ethanol, carbon dioxide, or oxygen.

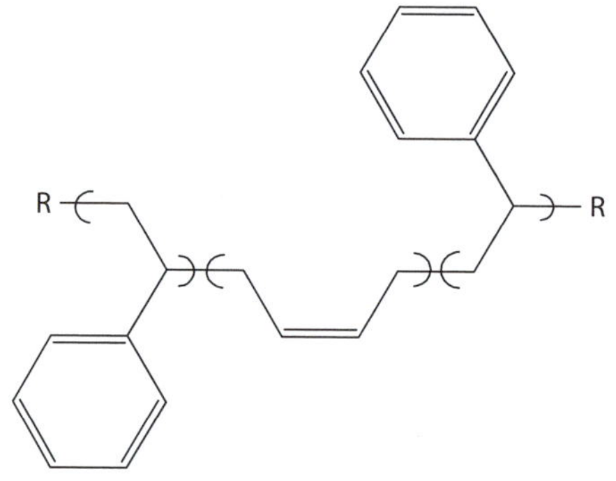

$$(5.43)$$

Block ABA copolymer of styrene and butadiene

Living polymers are generally characterized by (a) an initiation rate that is much larger than the polymerization rate; (b) polymer molecular weight is related to [monomer]/[initiator]; (c) linear molecular weight-conversion relationship; (d) narrow molecular weight range; and (e) stabilization of the living end groups allowing the formation of telechelics, macromers, block copolymers, and star polymers.

Group-transfer polymerizations make use of a silicon-mediated Michael addition reaction. They allow the synthesis of isolatable, well-characterized living polymers whose reactive end groups can be converted into other functional groups. It allows the polymerization of alpha-, beta-unsaturated esters, ketones, amides, or nitriles through the use of silyl ketenes in the presence of suitable nucleophilic catalysts such as soluble Lewis acids, fluorides, cyanides, azides, and bifluorides (HF_2^-).

As the polymerization occurs, the reactive ketene silyl acetal group is transferred to the head of each new monomer as it is added to the growing chain (Equation 5.44). Similar to anionic polymerization, the molecular weight is controlled by the ratio of the concentration of monomer to initiator. Reactions are generally carried out at low temperatures (about 0°C–50°C) in organic liquids such as THF. Compounds with "active" hydrogen atoms such as water and alcohols will stop the polymerization and their presence will curtain polymer chain length. Under the right conditions, polymerization will continue until all the monomer has been used up.

The trimethylsiloxy end group is a "living" end that continues to add units as long as the monomer is available or until it is neutralized.

In addition to the use of salt combinations to produce nylons described in the previous chapter, nylons may also be produced by the anionic ROM of lactams. In fact, this method was largely developed to overcome patent rights held by DuPont based on the work of Carothers and his group. This is the preferred method for the production of nylon-6, structurally analogous to nylon-66, and is widely practiced in Europe.

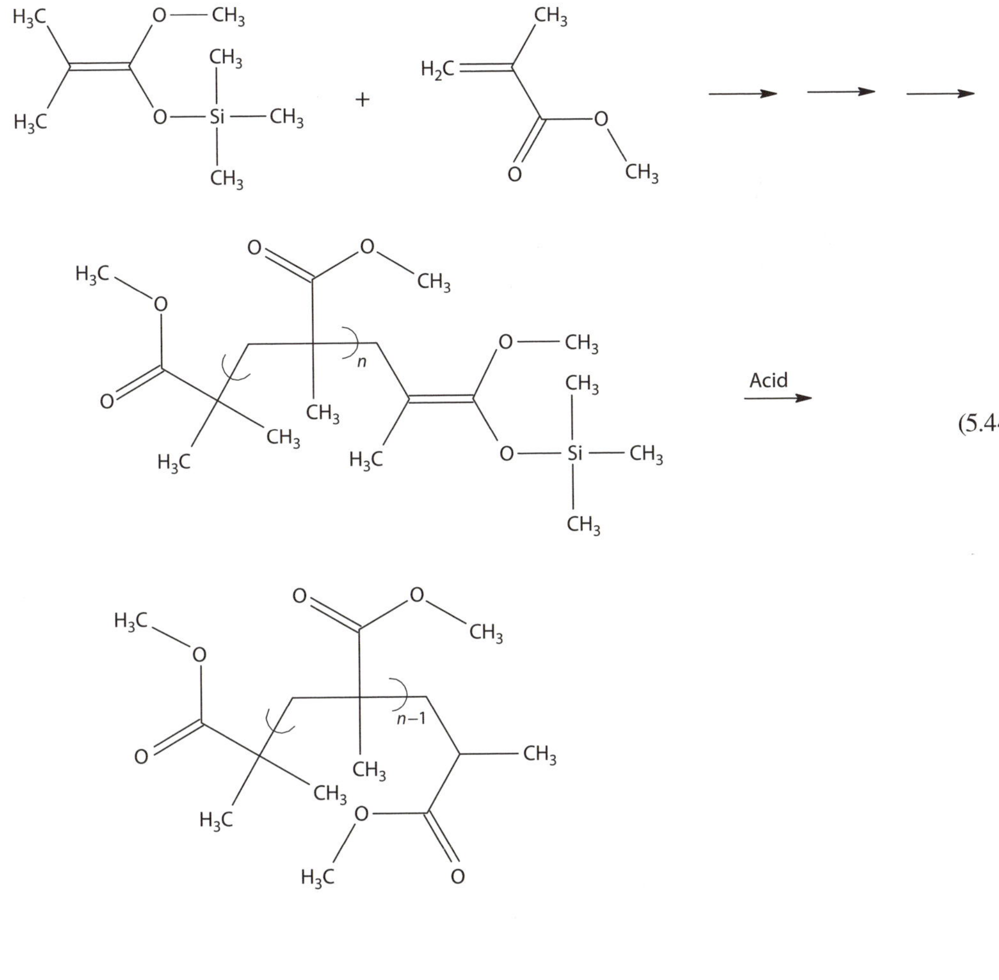

(5.44)

Nylon-6 is produced from the ROM of caprolactam (Equation 5.45). Since each unit contains six carbon atoms, it is named nylon-6 concurring to the naming of nylons. Along with having a structure analogous, it is not surprising that its physical and chemical properties are very similar to those of nylon-66, and they can generally be used to replace one another with almost no change in physical behavior. Below, Equation 5.45 describes the synthesis of nylon-6 employing sodium methoxide as the initiator.

The term monadic is used to describe nylons such as nylon-6 that have been produced from one reactant. The term dyadic is used to describe nylons such as nylon-66 that have been produced from two reactants.

The initiation period in the lactam-opening polymerization may be shortened by addition of an activator, such as acetyl chloride. Nylon-4, nylon-8, and nylon-12 are commercially available and are used as fibers and as coatings.

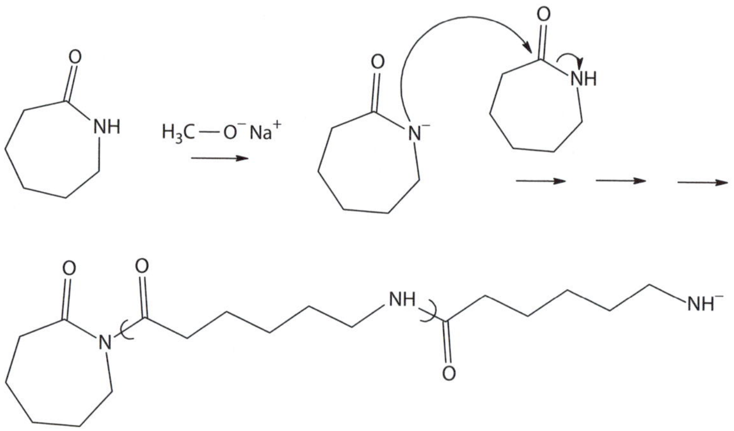

(5.45)

Lactomes may also be polymerized by ring-opening anionic polymerization techniques. While the five-membered ring is not readily cleaved, the smaller rings polymerize easily producing linear polyesters (Equation 5.46). These polymers are used commercially as biodegradable plastics and in polyurethane foams.

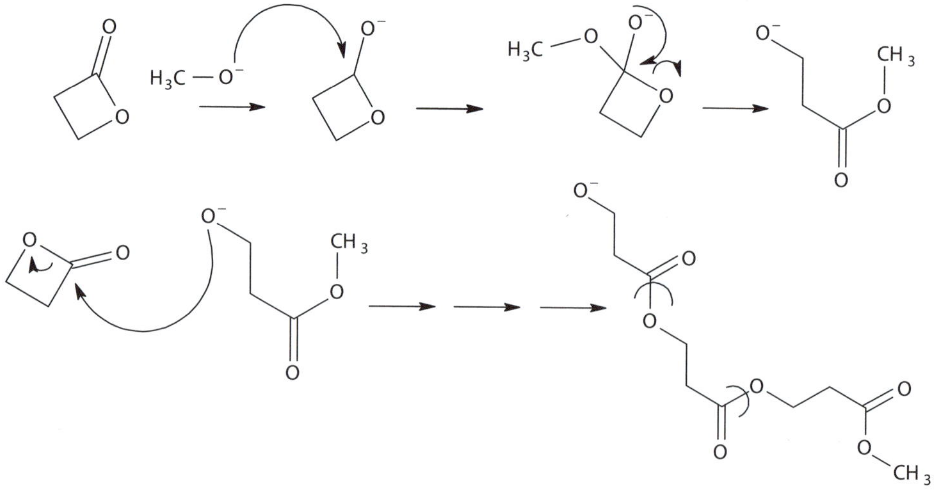

(5.46)

As seen, the anionic and cationic polymerizations are analogous differing mainly on the nature of the active species. The stereochemistry associated with anionic polymerization is also similar to that observed with cationic polymerization. For soluble anionic initiators at low temperatures, syndiotactic formation is favored in polar solvents, whereas isotactic formation is favored in nonpolar solvents. Thus, the stereochemistry of anionic polymerizations appears to be largely dependent on the amount of association the growing chain has with the counterion, analogous with the cationic polymerizations.

The stereochemistry of diene polymerizations is also affected by solvent polarity. For instance, the proportion of *cis*-1,4 units is increased by using organolithium or lithium itself as the initiator in the polymerization of isoprene or 1,3-butadiene in nonpolar solvents. A polymer similar to natural hevea rubber is obtained using the anionic polymerization of isoprene under these conditions. In more polar solvents employing sodium and potassium initiators, the amount of *cis*-1,4 units decreases and *trans*-1,4- and *trans*-3,4 units predominate.

5.4 STEREOREGULARITY

As noted in Chapter 2, there exists stereogeometry and stereoregularity in polymers. These differences have profound effects on the physical and, to a lesser degree, on the chemical properties of the polymers produced from the same monomer. Three possible units can be formed from the polymerization of butadiene as shown in Equation 6.36.

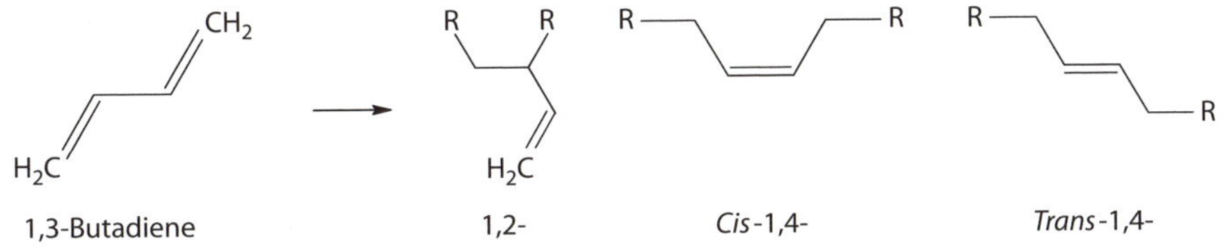

$$\text{(5.47)}$$

| 1,3-Butadiene | 1,2- | Cis-1,4- | Trans-1,4- |

For isoprene, there are four possible units formed (Equation 5.48).

As shown in Figure 2.5, there are three possible stereoregular forms for mono-substituted vinyl polymers. These are isotactic—all of the pendent groups are on one side of the chiral carbon; syndiotactic—the pendent groups appear on alternate sides of the chiral carbon; and atactic—some mixture of geometries about the chiral carbon.

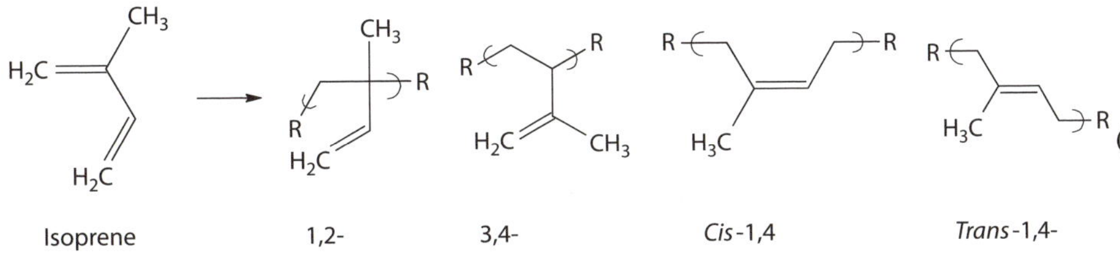

$$\text{(5.48)}$$

| Isoprene | 1,2- | 3,4- | Cis-1,4 | Trans-1,4- |

It is important to realize that polymer configuration and conformation are related. Thus, there is a great tendency for iostactic polymers (configuration) to form helical structures (conformation) in an effort to minimize steric constrains brought about because of the isotactic geometry.

5.5 POLYMERIZATION WITH COMPLEX COORDINATION CATALYSTS

Before 1950, the only commercial polymer of ethylene was a highly branched polymer called *high-pressure polyethylene,* where extremely high pressures were used in the polymerization process. The technique for making linear polyethylene (PE) was discovered by Marvel and Hogan and Banks in the 1940s and 1950s and by Nobel Laureate Karl Ziegler in the early 1950s. Ziegler prepared high-density polyethylene (HDPE) by polymerizing ethylene at low pressure and ambient temperatures using mixtures of triethylaluminum and titanium tetrachloride. Another Nobel Laureate, Giulio Natta, used Ziegler's complex coordination catalyst to produce crystalline, stereoregular polypropylene (PP). These catalysts are now known as Ziegler-Natta (or Natta-Ziegler) catalysts.

In general, a Ziegler-Natta catalyst is a combination of a transition-metal compound from Groups IVB (4) to VIIIB (10) and an organometallic compound of a metal from Groups IA (1) to IIIA (13) in the periodic table. It is customary to refer to the transition-metal compounds as the catalyst (because reaction occurs at the transition-metal atom site) and the organometallic compound as the cocatalyst.

Here we will use titanium to illustrate the coordination polymerization process. Several exchange reactions between catalyst and cocatalyst occur with Ti(IV) reduced to Ti(III). The extent and kind of stereoregulation can be controlled through a choice of reaction conditions and catalyst/cocatalyst. The titanium salt is present as a solid. The precise mechanism probably varies a little depending on the catalyst/cocatalyst and reaction conditions. Here we will look at the polymerization of propylene using titanium chloride and triethylaluminum. In general, a monomeric molecule is inserted between the titanium atom and the terminal carbon atom in a growing chain. Propagation occurs at the solid titanium salt surface—probably at defects, corners, and edges. The monomer is always the terminal group on the chain. Triethylaluminum reacts with the titanium-containing unit producing ethyltitanium chloride as the active site for polymerization.

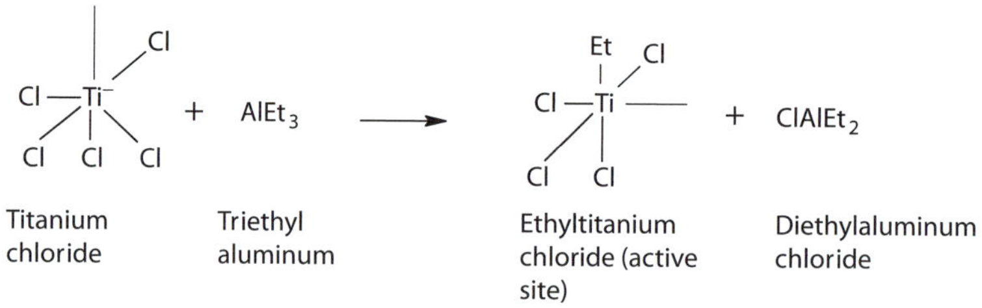

$$(5.49)$$

The propylene forms a pi-complex with the vacant d-orbital of titanium as shown in Equation 5.50.

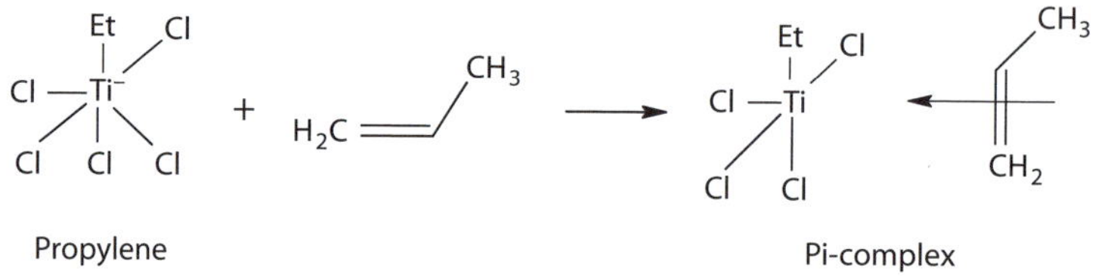

$$(5.50)$$

The ethyl groups transfer to the propylene opening up a new active site. The growing chain transfers to the site vacated by the ethyl group, creating a new active site that attracts, through pi-interactions, another propylene monomer. This sequence is shown in Equation 5.51. The edges of the solid titanium salt are believed to help provide the contour necessary to form the stereoregular chains.

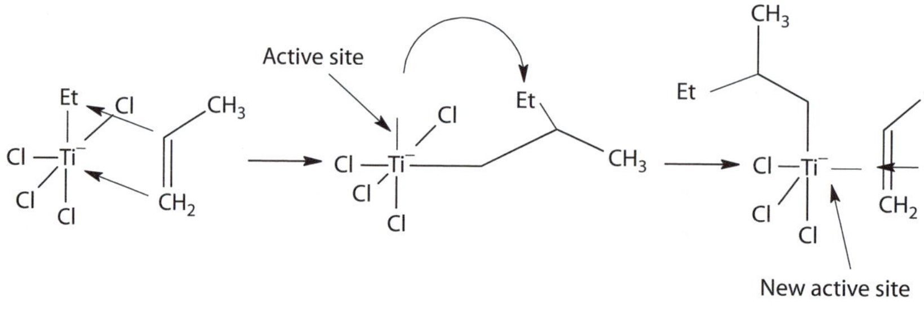

$$(5.51)$$

Most vinyl monomers give a predominance of the isotactic product. Typically, the more exposed the catalytic site, the less the stereoregularity of the resulting polymer. Isotactic-PP is produced using this technique as is HDPE.

The versatility of such stereoregulating systems is demonstrated in the polymerization of 1,3-butadiene, where all four of the potential structures- isotactic-1,2-, syndiotactic-1,2-, *trans*-1,4-, and *cis*-1,4, can be synthesized in relatively pure form using different catalysts systems.

Molecular weight is regulated to some degree by control of the chain transfer with monomer and with the cocatalyst, plus internal hydride transfer. However, hydrogen is added in the commercial processes to terminate the reaction because many systems tend to form longer chains beyond the acceptable balance between desired processing conditions and chain size.

The stereochemistry of the products is often controlled through control of the reaction temperature. For instance, use of low temperatures, where the alkyl shift and migration is retarded, favors formation of syndiotactic-PP. Commercial isotactic-PP is produced at room temperatures.

High-density polyethylene is typically produced using some stereoregulating catalysts. Much of it is produced using a Phillips catalyst system such as chromia catalyst supported on silica. Some HDPE and PP are commercially produced employing a Ziegler-Natta catalyst. This initiator is also employed for the production of polybutene and poly(-4-methyl-pentene-1) (TPX). TPX has somewhat high-melting point of about 300°C but because of the presence of the bulky butyl groups, a relatively low-specific gravity of 0.83. The percentage of polymer that is not soluble in n-hexane is called the *isotactic index* for some polymers, where the atactic and syndiotactic forms are hexane soluble.

5.6   SOLUBLE STEREOREGULATING CATALYSIS

The 1940s was a time of studying the kinetics and mechanism of production of vinyl polymers that took "center stage" in the 1950s. The 1950s incubated the solid-state stereoregulating catalysis that spawned a chemical revolution with the synthesis of stereoregular vinyl polymers in the 1960s. The 1980s and early 1990s served as a foundational time for soluble stereoregulating catalysis spawning another revolution related to the production of vinyl polymers with enhanced properties.

The solid-state stereoregulating catalysts "suffered" from at least three problems. First, while stereoregular polymers were formed with good control of the stereogeometry, polymer properties still fell short of predicted (upper limit) values. This was probably due to the presence of the associated solid-catalyst structure that accompanies the active catalytic site. This "excess baggage" restricts the motion of the growing chains so that while stereoregular control was good, the tendency to form good secondary structures was interrupted.

Second, in many cases the solid-state catalysis were incorporated, as contaminants, within the growing polymer, making an additional purification step necessary in the polymer processing to rid the polymer of this undesired material.

Third, many solid-state catalysts offered several "active polymerization sites" due to differences in the precise structure at and about the active sites. This resulted in an average stereoregular product being formed.

The new soluble catalysts offer a solution to these three problems. First, the "smaller" size of the active site, and associated molecules, allows the growing chains to "take advantage" of a natural tendency for the growing polymer chain to form a regular helical structure (in comparison to polymers formed from solid-state catalysts).

Second, the solution catalysts allow the synthesis of polymers that contain little or no catalytic agents, allowing the elimination of the typical additional "clean-up" steps necessary for polymers produced from solid-state catalysts.

Third, the newer soluble catalytic sites are homogeneous, offering the same electronic and stereostructure and allowing the synthesis of more stereoregular-homogeneous polymers.

The new soluble stereoregulating polymerization catalysts require the following three features:

- A metal atom (active) site
- A cocatalyst or counterion
- A ligand system

While the major metal site is zirconium, other metals have been successfully used including Ti, Hf, Sc, Th, and rare earths (such as Nd, Yb, Y, Lu, and Sm). Cyclopentadienyls (Cp) have been the most commonly used ligands, though a number of others have been successfully employed, including substituted Cp and bridged Cp. The most widely used metal-ligand grouping is zirconocene dichloride (zirconocene dichloride has a distorted tetrahedral geometry about Zr).

Methylalumoxane (MAO) (Equation 5.52) is the most widely utilized counterion. MAO is an oligomeric material with the following approximate structure.

$$
\begin{array}{c}
H_3C \qquad\quad CH_3 \quad\; CH_3 \\
\backslash \qquad\qquad | \qquad\; / \\
Al-O-[-Al-O-]_n-Al \\
/ \qquad\qquad\qquad\; \backslash \qquad\text{where } n = 4\text{--}20 \\
H_3C \qquad\qquad\qquad CH_3
\end{array}
\tag{5.52}
$$

Methylalumoxane

It is believed that MAO plays several roles. MAO maintains the catalyst complex as a cation, but doing so without strongly coordinating to the active site. It also alkylates the metallocene chloride, replacing one of the chloride atoms with an alkyl group and removing the second chlorine, thus creating a coordinately unsaturated cation complex, Cp_2MR^+. As an olefin approaches the ion pair containing the active metal, a metallocene-alkyl-olefin complex forms. This complex is the intermediate stage for the insertion of the monomeric olefin into a growing polymer chain.

The structure of the catalyst complex controls activity, stereoselectivity, and selectivity toward monomers. The catalyst structure is sensitive to Lewis bases such as water and alcohols encouraging the use of strongly oxyphilic molecules, such as MAO, to discourage the inactivation (poisoning) of the catalyst.

These soluble catalysts are able to give vinyl polymers that have increased stereogeometry with respect to tacticity as well as allowing the growing chains to form more precise helical structures. Further, the homogeneity of the catalytic sites also allows for the production of polymers with narrow molecular weight "spreads."

The summation of these affects is the production of polymers with increased strength and tensile properties. For PE, the use of these soluble catalysts allows the synthesis of PE chains with less branching compared to those produced using solid-state catalysts such as the Ziegler-Natta catalysts (ZNCs). PE produced employing soluble catalysts also show increased properties in comparison to PE produced by solid catalysts. Table 5.2 gives some comparisons of the PEs produced using the ZNCs with those produced with soluble catalysts.

Values of $\overline{M}_w/\overline{M}_n$ of 2 or less are common for the soluble catalyst systems, whereas values of 4–8 are usual for ZNC systems. The soluble catalyst systems also are able to polymerize a larger number and greater variety of vinyl monomers to form homogeneous polymers and copolymers in comparison to solid-catalyst systems.

The active site is a cationic metallocene-alkyl generated by reaction of a neutral metallocene formed from reaction with excess MAO or other suitable cocatalysts such as a borane Lewis acid. This sequence is shown in Figure 6.1 employing MAO with ethylene to form PE. Initiation and propagation occur through precoordination and insertion of the ethylene into the alkyl group-polymer chain. Here termination occurs through beta-hydride elimination producing a zirconium hydride and a long-chain alpha olefin. These long-chain alpha olefins can form linear HDPE or can

TABLE 5.2
Comparison of Properties of Polyethylene, Using Solid (ZNC) and Soluble Catalysts

Property	Unit	Soluble	ZNC
Density	g/cc	0.967	0.964
Melt index		1.3	1.1
Haze		4.2	10.5
Tensile yield	Psi	800	750
Tensile brake	Psi	9,400	7,300
Elongation break	%	630	670

From C. F. Pain, Proceedings Worldwide Metallocene Conference (Met Con '93), Catalyst Consultant Inc., Houston, TX, May 26–28.

be used as comonomers with monomers such as 1-propylene, 1-hexene, or 1,5-hexadiene to give a variety of branched and linear products. These Group IV B metallocene catalysts are very active, producing yields in excess of one ton of PE per gram of catalyst per hour with a total efficiency on the order of 25 tons of PE per gram of catalyst.

These catalysis systems are also used to form other hydrocarbon polymers such as a variety of PPs.

A major limitation of such Group IV B metallocene catalysts is that they are very air and moisture sensitive and not tolerant of heteroatom-containing monomers. In the case of heteroatom-containing monomers, the unbonded electron pairs on the heteroatom, such as oxygen, preferentially coordinate to the Lewis acid metal center in place of the carbon–carbon double bond. Some so-called middle- and late-transition metal organometallics are more tolerant to the presence of such heteroatoms and can be used as effective cocatalysts. These include some palladium, iron, cobalt, and nickel initiators.

The use of transition and selected main group metal catalysis is increasing with the ability to design special catalytic systems for special polymer architecture and property production. These catalysis systems involve the transition metal as a site for active polymer growth. The new soluble stereoregulating catalysts are one example of these systems. These growing sites may be more or less ionic/covalent depending upon the catalyst used and such sites are not generally appreciably dissociated as is the case in classical cationic and anionic systems. The metal's ligands can provide both electronic and steric structural control and are generally more robust in comparison to the anionic/cationic systems. Along with many advantages, there are some challenges. Because of their very nature, transition-metal initiators can be very complex requiring several synthetic steps; they may be expensive and/or require costly cocatalysts; and control of the particular reaction conditions is very important since small, seemingly subtle changes can be magnified into larger polymer structural changes (Figure 5.1).

There are an increasingly large number of metal-catalyzed polymerizations, including olefin metathesis reactions including *ring-opening metathesis polymerizations (ROMPs)*, formation of polyketones from the copolymerization of carbon monoxide, group-transfer polymerizations, and step-growth addition/elimination (coupling) polymerizations. The study of metal catalytic sites is a vigorous area of ongoing research.

Polymers produced from single-site catalysts are increasingly being used in the marketplace. As noted above, the strength of the materials is increased because of the greater order in the individual polymer chains. For PE this means the number of branches is less and for substituted polymers such as polyproplyene this means that the order about the substituted-carbon is increased, allowing for a denser, tighter fit of the individual polymer chain segments resulting in increased overall polymer strengths and less permeability for materials.

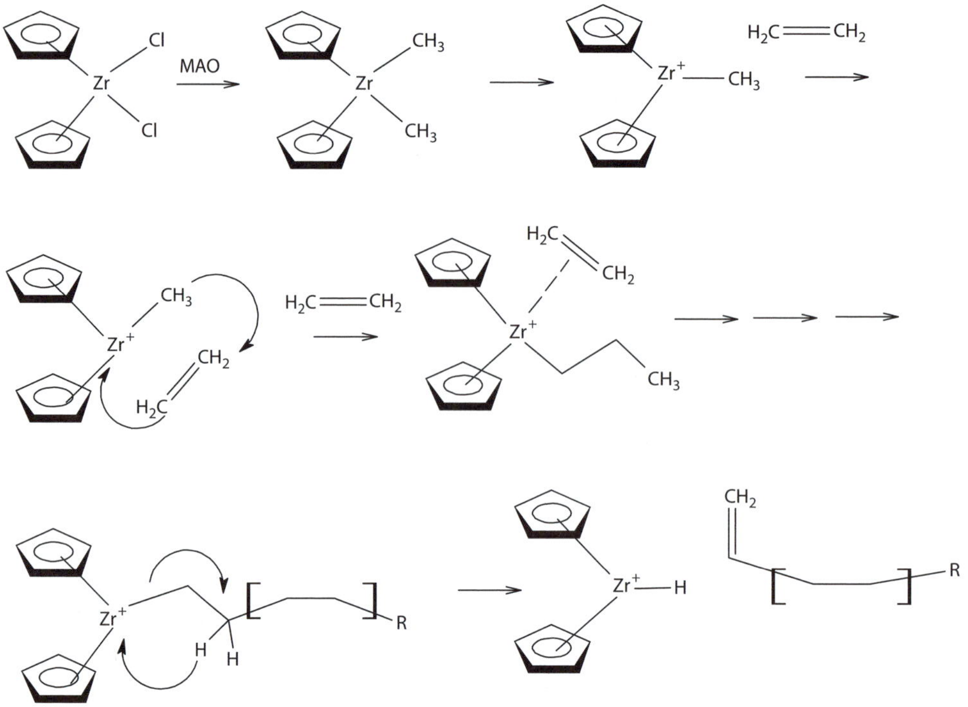

FIGURE 5.1 Proposed mechanism for soluble stereoregulating catalyst polymerizations.

Use of materials produced from single-site catalysts in areas employing thin films is increasing. For instance, bananas are generally produced at one location and shipped and stored to other locations for sale. Even when picked green, they ripen rapidly when exposed to oxygen. Regular LLDPE is generally employed as a thin film to protect bananas for shipment and storage. Regular LLDPE permits some transfer of oxygen, and because of the somewhat pointed nature of bananas the film may be punctured. Single-site metallocene-based LLDPE is less permeable and less apt to tear and is now replacing regular LLDPE in this use. Its use is also increasing in the containment of heavier materials such as topsoil and water-purification salt utilizing thicker films. In both cases, thinner films, and consequently less film material, is necessary to give an equal or better job performance. Single-site produced materials also offer better clarity, toughness, and easy sealability.

5.7 POLYETHYLENES

Tupperware was the idea of Earl Silas Tupper, a New Hampshire tree surgeon and plastics innovator. He began experimenting with PE during the early part of World War II. In 1947, he designed and patented the famous "Tupper seal" that "sealed in" freshness. To close the container it had to be "burped" to remove air. Tupperware was also bug proof, spill proof, did not rot or rust, and did not break when dropped. Even with all of these advantages, few were sold. Enter Brownie Wise, a divorced single mother from Detroit who desperately needed to supplement her income as a secretary. Her idea—"Tupperware Parties." By 1951, Tupper had withdrawn all of the Tupperware from the stores and turned over their sales to Brownie Wise with the only source of the ware being through the Tupperware Parties.

The initial synthesis of PE is attributed to many. In 1898, Han von Pechmann prepared it by accident while heating diazomethane. His colleagues Eugen Bamberger and Friedrich Tschirner characterized the white solid as containing methylene units and called it polymethylene. PE, from the ethylene monomer, was probably initially synthesized by M. E. P. Friedrich while a graduate student working for Carl S. Marvel in 1930 when it was an unwanted by-product from the reaction of ethylene and a lithium alkyl compound. In 1932, British scientists at the Imperial Chemical Industries, ICI, accidentally made PE while they were looking at what products could be produced from the high-pressure reaction of ethylene with various compounds. On March 1933, they found the formation of a white solid when they combined ethylene and benzaldehyde under high pressure (about 1,400 atmospheres pressure). They correctly identified the solid as PE. They attempted the reaction again, but with ethylene alone. Instead of again getting the waxy white solid, they got a violent reaction and the decomposition of the ethylene. They delayed their work until December 1935 when they had better high-pressure equipment. At 180°C, the pressure inside of the reaction vessel containing the ethylene decreased consistently with the formation of a solid. Because they wanted to retain the high pressure, they pumped in more ethylene. The observed pressure drop could not be totally due to the formation of PE, but something else was contributing to the pressure loss. Eventually, they found that the pressure loss was also due to the presence of a small leak that allowed small amounts of oxygen to enter into the reaction vessel. The small amounts of oxygen turned out to be the right amount needed to catalyze the reaction of the additional ethylene that was pumped in subsequent to the initial pressure loss (another "accidental" discovery). The ICI scientists saw no real use for the new material. By chance, J. N. Dean of the British Telegraph Construction and Maintenance Company heard about the new polymer. He had needed a material to encompass underwater cables. He reasoned that PE would be water resistant and suitable to coat the wire protecting it from the corrosion caused by the salt water in the ocean. In July of 1939, enough PE was made to coat one nautical mile of cable. Before it could be widely used, Germany invaded Poland and PE production was diverted to making flexible high-frequency insulated cable for ground and airborne radar equipment. PE was produced, at this time, by ICI and by DuPont and Union Carbide for the United States.

Polyethylene did not receive much commercial use until after the war when it was used in the manufacture of film and molded objects. PE film displaced cellophane in many applications being used for packaging produce, textiles, and frozen and perishable foods, and so on. This PE was branched and had a relatively low-softening temperature, below 100°C, preventing its use for materials where boiling water was needed for sterilization.

Karl Ziegler, director of the Max Planck Institute for Coal Research in Muelheim, Germany, was extending early work on PE, attempting to get ethylene to form PE at lower pressures and temperatures. His group found that certain organometallics prevented the polymerization of ethylene. He then experimented with a number of other organometallic materials that inhibited PE formation. Along with finding compounds that inhibited PE formation, they found compounds that allowed the formation of PE under much lower pressures and temperatures. Further, these compounds produced a PE that had fewer branches and higher softening temperatures.

The branched PE is called low-density, high-pressure PE because of the high pressures usually employed for its production and because of the presence of the branches, the chains are not able to closely pack, leaving voids and subsequently producing a material that had a lower density in comparison to low branched PE.

Giulio Natta, a consultant for the Montecatini company of Milan, Italy, applied the Zeigler catalysts to other vinyl monomers such as propylene and found that the polymers were higher density, higher melting, and more linear than those produced by the then classical techniques such as free radical initiated polymerization. Ziegler and Natta shared the Nobel Prize in 1963 for their efforts in the production of vinyl polymers using what we know today as solid-state steroregulating catalysts.

While many credit Natta and Ziegler as first having produced so-called high-density PE and stereoregular polyolefins, Phillips' scientists first developed the conditions for producing stereospecific olefin polymers and high-density PE. In 1952, J. Paul Hogan and Robert Banks discovered that ethylene and propylene polymerized into what we today know as high-density PE and stereoregular PP. As with many other advancements, their initial studies involved other efforts to improve fuel yields by investigating catalysts that converted ethylene and propylene to higher molecular weight

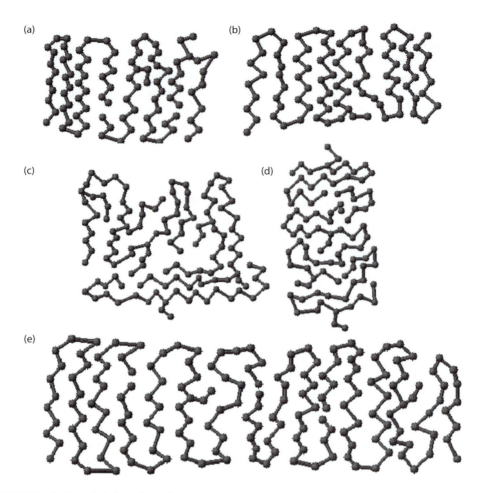

FIGURE 5.2 Ball-and-stick models of HDPE (a), UHMWPE (b), LDPE (c), LLDPE (d), and ULPE (e).

products. They found that chromium trioxide supported on a silica–alumina catalyst produced a hard solid rather than the usual waxy-like PE. They quickly looked at other olefins and soon discovered a crystalline PP, namely a stereoregular PP, specifically iPP.

While the common name for the monomer is ethylene, the official name is ethane so that what we know as PE is often referred to as *polyethene* or *polythene*. Even so, since the name PE is so entrenched in our common vocabulary, it will be employed here. PE can be produced employing radical, anionic, cationic, and ion-coordination polymerization. This is a result of the lack of substituents on the ethylene monomer. Each of these different polymerizations result in a different type of PE. Today there exists a wide variety of "polyethylenes" that vary in the extent and length of branching as well as molecular weight and molecular weight distribution and amount of crystallinity. Some of these are pictured in Figure 5.2. Commercial *low-density PE* (LDPE) typically has between 40 and 150 short-alkyl branches for every 1,000 ethylene units. It is produced employing high pressure (15,000–50,000 psi) and temperatures (to 350°C). It has a density of about 0.912–0.935. Because of the branching, the LDPE is amorphous (about 50%) and sheets can allow the flow through of liquids and gasses. Because of the branching and low amount of crystallinity, LDPE has a low-melting point of about 100°C, making it unsuitable for uses requiring sterilization through use of boiling water. LDPE has a combination of short to long branches, with long branches occurring at a rate of about ten short branches to every long branch.

High-density polyethylene, produced using organometallic catalysts such as the Ziegler-Natta or Phillips catalysts, have less than 15 (normally within the range of 1–6) short-alkyl branches (essentially no long branches) for 1,000 ethylene units. Because of the regular structure of the ethylene units themselves and the low extent of branching, HDPE chains can pack more efficiently, resulting in a material with greater crystallinity (generally up to 90%), higher density (0.96), with increased chemical resistance, hardness, stiffness, barrier properties, melting point (about 130ºC), and tensile strength. Low molecular weight (chain lengths in the hundreds) HDPE is a "wax" while "typical" HDPE is a tough plastic.

Linear low-density polyethylene (LLDPE) can be produced with less than 300 psi and at about 100ºC. It has a density between 0.915 and 0.925 g/mL. It is actually a copolymer of ethylene with about 8%–10% of an alpha olefin such as 1-butene, 1-pentene, 1-hexene, or 1-octene. Through control of the nature and amount of alpha olefin, we are able to produce materials with densities and properties between those of LDPE and HDPE. LLDPE does not contain the long branches found in LDPE. Because of its toughness, transparency, and flexibility it is used in film applications as packaging for cables, toys, pipes, and containers.

Very low-density polyethylene, (VLDPE) has a density range between 0.88 and 0.915 g/mL. It is largely linear chains with a high amount of short-chain branching generally made by copolymerization of the ethylene with short-chain alpha olefins such as 1-butene, 1-hexene, and 1-octene so that it is structurally similar to LLDPE except the alpha olefins are generally longer resulting is a lower density. VLDPE is generally made employing metallocene catalysts.

Ultrahigh molecular weight polyethylene, UHMWPE, is a high-density PE with chain lengths over 100,000 ethylene units. Because of the great length of the chains, they "inter-tangle" causing physical cross-links, increasing the tensile strength and related properties of these materials. (By comparison, HDPE rarely is longer than 2,000 ethylene units.) UHMWPE is about 45% crystalline and offers outstanding resistance to corrosion and environmental stress cracking, outstanding abrasion resistance and impact toughness, and good resistance to cyclical fatigue and radiation failure, and with a low-surface friction. It is produced utilizing catalysts systems similar to those employed for the production of HDPE (i.e., Ziegler-Natta and Phillips catalysts). It has a density of about 0.93.

Ultralinear polyethylene (ULPE) has recently become available through the use of soluble stereoregulating catalysts. Along with a decreased amount of short-chained alkyl branching, ULPE has a narrower molecular weight spread.

At extremely high chain lengths, fibers can be made of largely linear PE producing a material that is extremely strong and resistant to punctures and cutting even by scissors. One such product is "Spectra," which is about 70,000 units long, with the PE fibers stretched about 100% to align the chains. Spectra is used in the construction of bullet-resistant vests and surgical gloves where nicking and cutting of surgeon's hands by the sharp instruments is unwelcomed. Spectra is said to be 10 times stronger than steel on a weight basis and 35% stronger than aramid fibers such as Kevlar. Like other woven materials, these ultrahigh chain length PE gloves are vulnerable to punctures by sharp objects such as needles.

Cross-linked polyethylene, PEX or XLPE, is a medium to high-density PE that contains crosslinkes resulting in a thermoset material. The cross-linking causes a reduced flow for the material and permeability and increased chemical resistance. Because of the good retention and stability toward water, some water-pumping systems use PEX tubing since its cross-linking causes the tubing, once expanded over a metal nipple, to return to its original shape resulting in a good connection.

Medium-density polyethylene (MDPE) has a density between about 9.3 and 9.4 g/mL. It is mainly produced employing ion-coordination polymerization. It has good shock and drop resistance.

Polymethylene can be produced through several routes, including the use of diazomethane or a mixture of carbon monoxide and hydrogen. This polymer has only a little branching.

It is well accepted that the history of a polymer, including polymer processing, influences polymer behavior. Some of these influences are just becoming known. Interestingly, as in much of

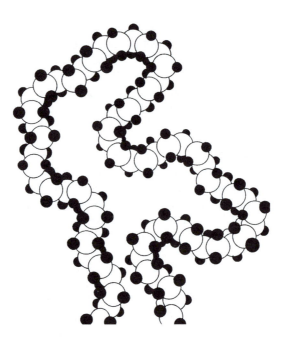

FIGURE 5.3 Space-filling structure of a portion of a linear amorphous polyethylene region.

science, once the critical parameters are known on a macrolevel, we are able to better understand them on a molecular and conceptual level. An example that illustrates this involves the processing of annealed PE. In general, for most linear PE, micelles and associated spherulites are formed when it is melted and then slowly cooled. If no force is applied during the annealing process (simple melt crystallization), a high amount of force and large deformation is required to breakdown the initial spherulite structures with reformation occurring along the axis of the pull when high-strength PE rod, film, and sheet is produced. However, if the PE is crystallized under pressure applied in one direction, less energy and lower deformation is required to align the PE spherulites since the spherulites are already partly aligned. In both cases, stretching of the molecular network is required. For the simple melt crystallized PE, the original spherulite structure is destroyed during the deformation followed by the formation of new fibrillar structures. For the pressure-associated annealing process, elongated micelles are formed that largely remain after the deformation process.

LDPE films are nearly clear even though they contain a mixture of crystalline and amorphous regions. This is because the crystalline portions are space filling and not isolated spherulites allowing a largely homogeneous structure with respect to refractive index resulting in a material that is transparent. In fact, the major reason that LDPE films appear hazy or not completely transparent is because of the roughness of the surface and is not due to the light scattering of the interior material.

Space-filling models of amorphous and crystalline linear PE are given in Figures 5.3 and 5.4.

Typical uses of the various PEs include the following:

1. UHMWPE-battery separators, light-weight fibers, permanent solid lubricant materials in railcar manufacture, automobile parts, truck liners; liners to hoppers, bins, and chutes; farm machinery as sprockets, idlers, wear plates, and wear shoes; moving parts of weaving machines and can and bottle handling machines, artificial joints, including hip and knee replacements, gears, butcher's chopping boards, sewage-treatment-bearings, sprockets, wear shoes; lumbering-chute, sluice, and chain-drag liners; neutron shields; also as components in bullet-resistant wear.

2. "Typical" HDPE-blow-molded products—bottles, cans, trays, drums, tanks, and pails; injection-molded products—housewares, toys, food containers, cases, pails, and crates; films,

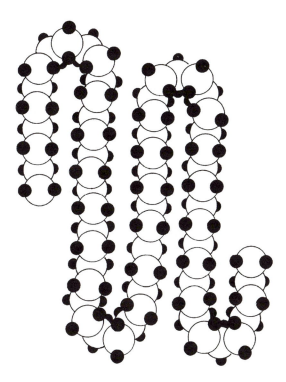

FIGURE 5.4 Space-filling structure of a portion of a linear crystalline polyethylene region.

pipes, bags, conduit, wire and cable coating, foam, insulation for coaxial and communication cables.

3. Low molecular weight HDPE—spray coatings, emulsions, prinking inks, wax polishes, and crayons.

4. LDPE—packaging products, bags, industrial sheeting, and piping and tubing, films, garbage cans, industrial containers, household items.

5. LLDPE—telephone jacketing, wire and cable insulation, piping and tubing, drum liners, bottles, films.

6. MDPE—shrink film, packaging film, gas pipes and fittings.

7. VLDPE—frozen food and ice bags, food packaging, stretch wrap.

Recently, efforts are under way to form ethylene, and hence PE, from green chemistry feed-stocks—sugarcane-derived ethanol and other natural feedstocks.

Plastomers is the name given to copolymers of ethylene that have a little crystallinity, but are largely amorphous. They are also called *very low-density PE* (VLDPE). They are more elastic than LLDPE but less stiff. They are used as a sealing layer in film applications and controlled permeation packaging for vegetables and fruits.

Low-density polyethylene, PP, and TPX are more susceptible to oxidation in comparison to HDPE because the presence of tertiary carbon atoms in the former. The degradation of LDPE, PP, and TPX is retarded through the use of antioxidants.

General properties of some of the important PEs are given in Table 5.3.

Most shopping bags are made from PE, PP, or paper. Today, there is a move to remove the so-called nongreen plastic bags. A brief comparison of the two is in order. In comparison to paper bags, plastic bags are lighter, stronger, low cost, and water and chemical resistant. They cost less energy to produce (about 70%), offer less of a carbon dioxide foot print (about 50%), transport and recycle. When disposed of improperly, plastic bags are unsightly and represent a hazard to wildlife and

TABLE 5.3
General Physical Properties of Selected Polyethylenes

Polyethylene →	LDPE	HDPE	UHMWPE
Heat deflection temperature (1,820 kPa; °C)	40	50	85
Maximum resistance to continuous heat (°C)	40	80	80
Coefficient of linear expansion (cm/cm-°C, 10^{-5})	10	12	12
Compressive strength (kPa)	—	2×10^4	—
Impact strength (Izod: cm-N/cm of notch)	No break	30	No break
Tensile strength (kPa)	5×10^3	3×10^4	6×10^4
Density (g/mL)	0.91	0.96	0.93

infants. Neither paper (because of lack of oxygen) nor plastic bags decompose in landfills. Both can be recycled. Paper is made from a renewable resource, cellulose from trees. With the move toward PLA bags, long-term disposal and degradation will become less of a problem. Thus, the answer is not apparent but what is apparent is that we must increase our efforts to recycle and properly dispose of both plastic and paper bags.

5.8 POLYPROPYLENE

Polypropylene is one of the three most heavily produced polymers (Section 1.3). The abundance of PP is the result of the variety of PP produced, its versatility allowing a wide variety of products to be produced, availability of a large feedstock, and its inexpensiveness. Today, PP is used in such diverse applications as a *film* in disposable diapers and hospital gowns to geotextile liners; *plastic* applications as disposable food containers and automotive components; and *fiber* applications such as in carpets, furniture fabrics, and twine.

While PP was produced in 1951 by Karl Rehn, it only became commercially available in the late 1950s with the production by Natta and coworkers at Phillips of somewhat stereoregular PP. The first PP was not highly crystalline because the tacticity, a measure of the stereoregularity, was only approximate. But, with the use of the Natta-Zeigler and Phillips catalysts systems PP with greater stereoregularity was produced giving PP with enhanced physical properties such as increased stiffness, better clarity, and a higher distortion temperature. Today, with better catalysts, including the soluble metallocene catalysts, the tacticity has been increased so that PP with 99% isotacticity can be produced. The more traditional Natta-Zeigler catalysts have high catalysts efficiencies with one gram of catalysts producing 1kg of PP. This high-catalytic efficiency eliminates the need for catalyst removal. Most iPP is made using bulk propylene, either as a gas or liquid.

A brief side trip allows some insight into industrial workings. Phillips Petroleum Company was busily working on ways to improve refinery processes because they were in the 1950s almost solely a fuel company. Two young chemists, J. Paul Hogan and Robert L. Banks, were trying to develop catalysts that would act as high-performance gasoline additives when one of the catalyst, mixed with the petroleum propylene present in a pipe, plugged up the pipe with a whitish, taffy-like material. While many companies might have told Hogan and Banks to get back to their original efforts that were the "bread-and-butter" of the company, instead they were told to investigate the formation of this off-white material. At this time, most of the known plastics were either too brittle or softened at too low a temperature for most practical uses. This off-white material, produced from propylene, could be hardened giving a flexible material that melted above 100°C. After some effort, it was discovered that this catalyst could also be used to give a PE that was superior to the "old" PE. Eventually, the "old" PE would be given the name of LDPE and the new PE the name of HDPE.

In 1953, a patent was applied for covering both the synthesis of PP and PE under the trade name of Marlex. Even after a material is discovered many steps are needed before a product becomes available to the general public. Management was getting glowing reports about the potential for these new materials. On the basis of these assessments, $50 million was committed for the construction of a large-scale PE plant at Adams Terminal.

Several things were in operation that is not immediately apparent. First, the great investment of money and time needed to bring a product into the marketplace. At that time $50 million was a lot of money, on the order of half a billion dollars today. Second, up to that time, technicians with little training were used in much of the chemical industry. The "educated" chemists were kept behind the research benches discovering "new" things. Most of the PhD chemists were in fact in universities rather than in industry. (Today, only about 15% of the professional chemists are teaching, the remainder are in industry and government.) Because of the sensitivity of the catalyst system used to produce the new polymers, substantially greater training had to be given to workers who dealt with the production of these new polymers. Third, this was a venture into a new arena by Phillips, up to now solely a gasoline-producing company. Finally, while not initially apparent, the catalysts used to produce these new polymers were part of a new group of catalysts being investigated internationally by many companies that allowed the production of so-called stereoregular polymers.

As noted previously, PP and other alpha olefins, can reside in three main stereoregular forms differentiated from one another because of the precise geometry of the methyl group as the polymer is formed. When the methyl groups reside on the same side as one looks down a barrow of the chain, as below, it is called isotactic-PP or simply iPP.

$$(5.53)$$

The methyl group can also exist in alternate positions as shown below. This form is called *syndiotactic-PP* or simply sPP.

$$(5.54)$$

The third structure consists of mixtures of the syndiotactic and isotactic structures favoring neither structure. This mixture of structures is called the *atactic form* (aPP) the "a" meaning having nothing to do with. Space-filling modes of all these three different tactic forms are given in Figure 5.5.

Each particular tactic form has its own physical properties. Table 5.4 contains representative values for iPP. The syndiotactic and isotactic forms are referred to as stereoregular forms and allow the chains to better fit together giving a material that is crystalline while the aPP is amorphous. Being crystalline causes the polymer to be stronger, denser, less porous to small molecules like water and oxygen, and to have a higher melting temperature. The old PP is of the atactic form and the form synthesized by Hogan and Banks is iPP. Only recently, using soluble stereoregulating systems has sPP become commercially available.

Other groups were working on developing these stereoregulating catalysts. In fact, more research dollars and effort was spent during the 1950s developing these stereoregulating catalysts than that spent on cancer research during that time. The competition was fierce and monetary stakes high. I was present at one of the initial presentations of the results from a number of different groups. Each group believed their catalyst system gave the best results and each believed that the product they obtained was the same product that other companies obtained. There were several times during the presentations where the groups would yell at one another and physically wrench the microphone

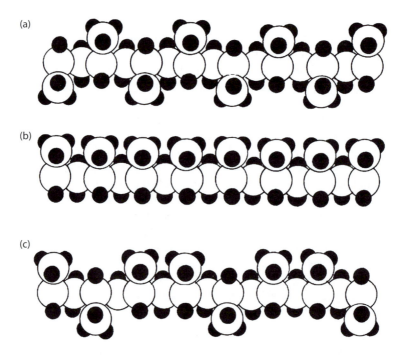

FIGURE 5.5 Space-filling models of syndiotactic (a), isotactic (b), and atactic (c) forms of polypropylene. (My children and now grandchildren believe that the isotactic polypropylene is really a chorus-line of dancing teddy bears.)

TABLE 5.4
General Physical Properties of iPP

Heat deflection temperature (1,820 kPa; °C)	55
Maximum resistance to continuous heat (°C)	100
Coefficient of linear expansion (cm/cm-°C, 10^{-5})	9
Flexural strength (kPa)	5×10^4
Impact strength (Izod: cm-N/cm of notch)	27
Tensile strength (kPa)	3.5×10^4
Ultimate elongation (%)	100
Density (g/mL)	0.90

from another speaker calling one another various unkind words. It was later learned that each of the groups had different catalytic systems that gave different products. Thus, the presenters were most probably giving accurate information but based on slightly differing catalyst systems.

Jumping from a laboratory-scale production to mass production is difficult and is made especially difficult because of the need to control the structure of the catalyst because only one form of the catalyst gave the desired iPP and PE while other forms would give mixtures of PP and PE structures. The first Marlex pellets came off the production line at the Adams plant and they were varied in color and size, and off specification. It was hard for the sales staff to convince buyers that this was the miracle material that they had promised. The realization that the catalysts form was so important became painfully evident. The material was better than the old PE and PP but it was not as good as that obtained in the laboratory. Warehouse after warehouse of somewhat inferior material was produced with few buyers.

Relief was spelled, not "R-E-L-I-E-F," but rather "H-U-L-A H-O-O-P" which was reinvented by Richard Knerr and Arthur "Spud" Melin who founded the Wham-O company that also reinvented another old toy, the frisbee. American children fell in love with the hula hoop and the somewhat inferior Phillips material was good enough to give good hula hoops. Demand for these plastic rings was sufficient to take the Adams plant output for a half year and turn it into hula hoops. Phillips president, Paul Endacott, was so pleased that he kept a hula hoop in his office to remind him of his "savior." By the time that particular wave of hula hoop mania wound down in 1959, the problems in the product line at the Adam terminal were overcome and Marlex had found new markets. One of the first markets was the use of PE and PP baby bottles in hospitals to replace the old glass bottles. The Marlex bottles were less expensive, could be sterilized (remember, both the new PE and PP could be heated to boiling, 100°C, without melting), and would not shatter when dropped. The manufacture of Chiffon household liquid detergent took a chance on using plastic bottles rather than glass to hold their detergent. Their success caused other manufacturers to change to the lighter, nonbreakable, and less-expensive plastic containers. Today, PP and PE hold the major share of the container market.

Copying the tubes used to make the hula hoop, plastic tubing for many varied applications, such as connecting air conditioners and ice makers, took off. Today, the two most used synthetic polymers are PE and PP. New catalysts and production procedures has allowed the physical properties and varied uses of these "big two" synthetic polymers to be continually increased.

While it was possible to produce syndiotactic-PP (sPP) employing the Ziegler-Natta solid systems, commercial sPP has only recently become commercial through the use of the soluble metallocene catalysts. These materials have a similar T_g as iPP but they have a different balance between stiffness and toughness.

Atactic or amorphous forms of PP are also used. Initially, atactic PP (aPP) was obtained as a by-product of the production of iPP. As an inexpensive by-product, it is used as a modifier for asphalt for roofing and in adhesives. As the effectiveness of catalyst systems becomes better, less aPP is available so that today some aPP is intentionally made for these applications.

While PP is often synthesized employing Ziegler-Natta catalyst, today there exists many other catalysts systems, including the soluble catalysts systems described in 5.6 "Polyethylenes." Many of these were developed by Walter Kaminsky and coworkers in 1980. These catalysts contain a Group IVB metallocene coupled with methylaluminoxane (MAO). By varying the structure of the metallocene cocatalyst, it is possible to tailor the tacticity of the product. Thus, for PP, the atactic product is formed when the simple zirconocene and hafnocene dichlorides are employed; syndiotactic-PP is produced when a bridged catalyst is employed; and isotactic-PP is produced when another bridged catalyst is employed.

Stretching a film tends to orientate the polymer chains. Biaxially stretched or oriented PP, BOPP is strong and clear, resulting in its wide use and producing clear bags and packaging materials. PP is often used in the construction of cold-weather clothing under-layers, such as long-sleeved shirts and underwear. It is also used in warm-weather wear because of its ability to remove perspiration from the skin. It is used in ropes that can float because of the low density of PP. In low-ventilation areas, it often replaces PVC because it emits less smoke and no dangerous chlorine-containing fumes when burned. It serves as the suture material in Prolene. Because of its ability to retain color and to form thin sheets, it is being used to make stationary folders, storage boxes, and the cube stickers for the Rubik's cubes. It is also used to construct trading card holders (such as baseball cards) with each sheet containing pockets for the trading cards. In medicine, it is also used in hernia repair operations with a small sheet or film of PP placed over the spot of the hernia to prevent reformation of the surgical hernia.

Expanded PP, EPP, foam is used for radio-controlled model aircraft, automobiles, and trucks by hobbyists because of the ability of the foam to absorb impacts and its low density.

5.9 POLYMERS FROM 1,4-DIENES

There are three important 1,4-dienes employed to produce commercially important polymers. These monomers possess a conjugated pi-bond sequence of -C=C-C=C- that readily forms polymers with

TABLE 5.5
General Physical Properties of Extrusion Grade ABS

Heat deflection temperature (1,820 kPa; °C)	90
Maximum resistance to continuous heat (°C)	90
Coefficient of linear expansion (cm/cm-°C, 10^{-5})	9.5
Compressive strength (kPa)	4.8×10^4
Flexural strength (kPa)	6.2×10^4
Impact strength (Izod: cm-N/cm of notch)	320
Tensile strength (kPa)	3.4×10^4
Ultimate elongation (%)	60
Density (g/mL)	1.0

an average energy release of about 20 kcal/mol (80 kJ/mol) with the conversion of one of the double bonds into lower (potential energy wise; generally more stable) energy single bond. For all of these products, cross-linking and grafting sites are available through the remaining double bond.

1,4-Butadiene can form three repeat units as described in Equation 5.47, the 1,2; *cis*-1,4; and *trans*-1,4. Commercial polybutadiene is mainly composed of the 1,4-*cis* isomer and is known as *butadiene rubber* (BR). In general, butadiene is polymerized using stereoregulating catalysts. The composition of the resulting polybutadiene is quite dependent on the nature of the catalyst such that almost total *trans*-1,4 units, or *cis*-1,4 units, or 1,2 units can be formed as well as almost any combination of these units. The most important single application of polybutadiene polymers is its use in automotive tires where over 10^7 tons are used yearly in the U.S. manufacture of automobile tires. BR is usually blended with natural rubber, NR, or *styrene–butadiene rubber* (SBR), to improve tire tread performance, particularly wear resistance.

A second use of butadiene is in the manufacture of *ABS* copolymers where the stereogeometry is also important. A polybutadiene composition of about 60% *trans*-1,4; 20% *cis*-1,4; and 20% 1,2 configuration is generally employed in the production of ABS. The good low-temperature impact strength is achieved in part because of the low T_g values for the compositions. For instance, the T_g for *trans*-1,4-polybutadiene is about −14°C while the T_g for *cis*-1,4-polybutadiene is about −108°C. Most of the ABS rubber is made employing an emulsion process where the butadiene is initially polymerized, forming submicron particles. The styrene–acrylonitrile copolymer is then grafted onto the outside of the BR rubber particles. ABS rubbers are generally tougher than HIPS rubbers but are more difficult to process. ABS rubbers are used in a number of appliances, including luggage, power tool housings, vacuum cleaner housings, toys, household piping, and automotive components such as interior trim. Table 5.5 contains representative data for extrusion grade ABS.

Another major use of butadiene polymer is in the manufacture of high-impact polystyrene (*HIPS*). Most HIPS has about 4%–12% polybutadiene in it so that HIPS is mainly a polystyrene intense material. Here the polybutadiene polymer is dissolved in a liquid along with styrene monomer. The polymerization process is unusual in that both a matrix composition of polystyrene and polybutadiene is formed as well as a graft between the growing polystyrene onto the polybutadiene is formed. The grafting provides the needed compatibility between the matrix phase and the rubber phase. The grafting is also important in determining the structure and size of rubber particles that are formed. The grafting reaction occurs primarily by hydrogen abstraction from the polybutadiene backbone by growing either polystyrene chains or alkoxy radicals if peroxide initiators are employed.

Interestingly, isoprene, 2-methyl-1,3-butadiene, exists as an equilibrium mixture of cis and trans isomers.

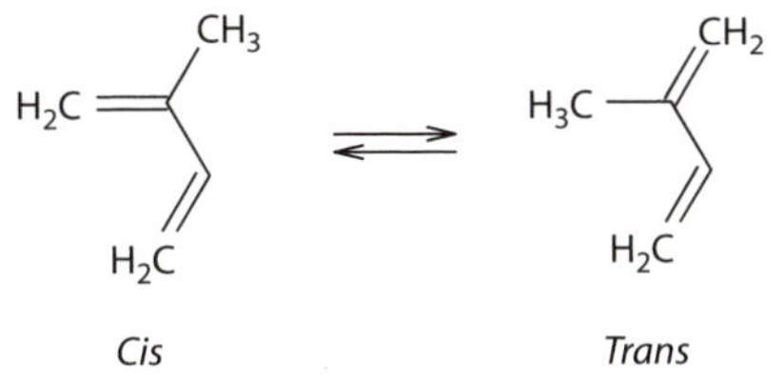

$$(5.55)$$

Cis   *Trans*

Polyisoprene is composed of four structures as shown in Equation 5.48. As in the case of polybutadiene, it is the *cis*-1,4 structure that is emphasized commercially. The *cis*-1,4-polyisoprene is similar to the *cis*-1,4- polybutadiene material except it is lighter in color, more uniform, and less expensive to process. Polyisoprene is composition wise analogous to natural rubber. The complete *cis*-1,4 product has a T_g of about −71°C. Interestingly, isomer mixtures generally have higher T_g values. Thus, an equal molar product containing *cis*-1,4; *trans*-1,4; and 3,4 units has a T_g of about −40°C.

As with many polymers, polyisoprene exhibits non-Newtonian flow behavior at shear rates normally used for processing. The double bond can undergo most of the typical reactions such as carbene additions, hydrogenation, epoxidation, ozonolysis, hydrohalogenation, and halogenation. As with the case of the other 1,4-diene monomers, many copolymers are derived from polyisoprene or isoprene itself.

Polyisoprene rubbers are used in the construction of passenger, truck, and bus tires and inner liners as well as sealants and caulking compounds, sporting goods, gaskets, hoses, rubber sheeting, gloves, belts, and footwear.

Polychloroprene was the first commercially successful synthetic elastomer introduced in 1932 under the trade names of DuPrene and Neoprene by DuPont. It was discovered by Carothers and coworkers. Because of its early discovery, good synthetic routes were worked out before the advent of good steroregulating catalytic systems. Thus, polychloroprene is largely manufactured by emulsion polymerization using both batch and continuous systems. Free radical products contain mainly 1,4-trans units. Along with the four "main" structural units analogous to those of polyisoprene, sequence distributions are available for both polyisoprene and polychloroprene. Polymerization can occur with the growing end being the four end or the one end (5.56). Generally, the 1,4-polymerization sequence is favored with the growing end being carbon 4.

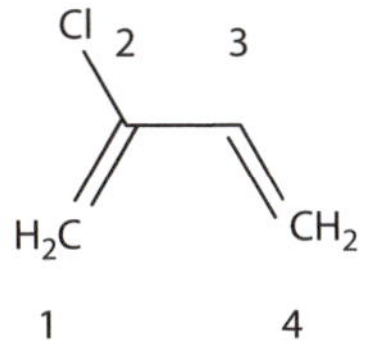

$$(5.56)$$

Structural regularity for inclusion of the 1,4-trans unit is inversely proportional to temperature. Thus, at 90°C the product contains about 85% of the *trans*-1,4 units while this increases to almost 100% at −150°C. Both uncured and cured polychloroprene exists as largely crystalline materials because of the high degree of stereoregularity. Cured polychloroprene has good high tensile strength because of this and application of stress to the material, either before or after curing, increases the tensile strength. The *trans*-1,4-polychloroprene has a T_g of about −49°C while 1,4-*cis*-polychloroprene has a T_g of about −20°C.

Compounding of polychloroprene is similar to that of natural rubber. Vulcanizing is achieved using a variety of agents, including accelerators. Because of its durability, polychloroprene rubber is often used where deteriorating effects are present. It offers good resistance to oils, ozone, heat, oxygen, and flame (the latter because of the presence of the chlorine atom). Automotive uses include as hoses, V-belts, and weather-stripping. Rubber goods include gaskets, diaphragms, hoses, seals, conveyer belts, and gaskets. It is also used in construction for highway joint seals, bridge mounts and expansion joints, and for soil-pipe gaskets. Further, it is used for wet laminating and contact-bond adhesives, in coatings and dipped goods, as modifiers in elasticized bitumens and cements, and in fiber binders.

As noted above, the presence of the chlorine atom in polychloroprene makes it less apt to burn in comparison to materials, such as hydrocarbon-only elastomers. Thus, it is used in fire doors, some combat-related attire, such as gloves and face masks and other similar applications. It is also used in the construction of objects that come into contact with water, such as diving suits and fishing wader boots.

5.10 POLYISOBUTYLENE

Polyisobutylene (PIB); Equation 6.46, was initially synthesized in the 1920s but was developed by William Sparks and Robert Thomas at Standard Oil's (to become Exxon) Linden, NJ laboratory. It is one of the few examples of the use of cationic catalysis to produce commercial-scale polymers. PIB and various copolymers are also called **butyl rubber** (IIR). Low molecular weight (about 5,000 Da) PIB can be produced at room temperature but large chains (over 1,000,000 Da) are made at low temperatures where transfer reactions are suppressed.

$$\tag{5.57}$$

As noted above, PIB and various copolymers are called butyl rubber. Butyl rubbers have lower permeability and higher damping than other elastomers, making them ideal materials for tire inner-liners and engine mounts.

Because of the symmetry of the monomer, it might be expected that the materials would be quite crystalline like linear PE. PIB does crystallize under stress but not under nonstressed conditions. This is because the geminal dimethyl groups on alternating carbons in the backbone cause the bond angles to be distorted from about the usual tetrahedral bond angle of 109.5 degrees to 123 degrees forcing the chain to straighten out. As a consequence of this the chains pack efficiently giving a relatively high-density material (density of 0.917 g/cc compared to densities of about 0.85 g/cc for many amorphous polymers) even when amorphous. This close packing reduces the incentive for crystallization, accounts for its low permeability, and produces an unusually low T_g of −60°C for such a dense material.

Because PIB is fully saturated, it is cured as a thermoset elastomer through inclusion of about 1%–2% isoprene that supplies the needed double bonds used in the curing process. Other materials, including brominated paramethyl styrene, are replacing isoprene for this use. PIB is also used in sealing applications and medical closures and sealants.

Polyisobutylene is often produced as a copolymer containing a small amount (1%–10%) of isoprene. Thus, the random copolymer chain contains a low concentration of widely spaced isolated double bonds that are later cross-linked when the butyl rubber is cured.

Butyl rubbers have lower permeability (including being essentially impermeable to air) and higher damping than other elastomers, making them ideal materials for tire innerliners, basketball inside coatings, and engine mounts. It is used as an adhesive, caulking compound, chewing gum base, and an oil additive. Its use as an oil additive is related to its change in shape with increasing temperature. Since lubricating oil is not a good solvent for PIB, PIB is present as a coil at room temperature. However, as the temperature increases, it begins to uncoil acting to counteract the decrease in viscosity of the oil as the temperature is increased. PIB is also used in sealing applications such as in roof repairs.

It is being used in a number of green chemistry applications. PIB (in the form of PIB succinimide) is added in small amounts to lubricating oils; this reduces the creation of an oil mist, thus lowering the workers, exposure to the oil mist. It is part of Elastol employed to clean up water-borne oil spills. It increases the crude oil's viscoelasticity, resulting in an increased ability of the oil to remain together as it is vacuumed from the water's surface.

Polyisobutylene acts as a detergent and is used as a fuel additive in diesel fuel reducing fuel injector fouling, resulting in increased mileage and lowered unwanted emissions. It is also employed as part of a number of detergent packages that are blended into gasoline and diesel fuels.

5.11 METATHESIS REACTIONS

Chauvin, Grubbs, and Schrock won the 2005 Nobel Prize for developing metathesis reactions. Olefin metathesis is a catalytically induced reaction wherein olefins, such as cyclobutene and cyclopentene, undergo bond reorganization, resulting in the formation of so-called polyalkenamers. Because the resulting polymers contain double bonds that can be subsequently used to introduce cross-linking, these materials have been used to produce elastomeric materials as well as plastics. Transition-metal catalysts are required for these reactions. Catalysts include typical Natta-Ziegler types and other similar catalysts–cocatalysts combinations. The reactions can be run at room temperature and the stereoregularity controlled through choice of reaction conditions and catalysts. For instance, the use of a molybdenum-based catalyst with cyclopeantene gives the cis product whereas the use of a tungsten-based catalyst gives the trans product.

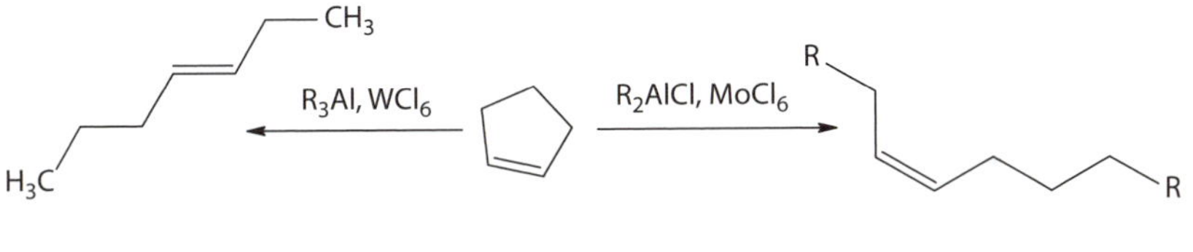

Trans-polypentenamer (5.58)   *Cis*-polypentenamer (5.59)

As expected, the metathesis polymerization of more strained cycloalkenes, such as cyclobutene, occurs more rapidly than less-strained structures such as cyclopentene.

It is believed that polymerization occurs via a chain polymerization where ring opening occurs via complete scission of the carbon–carbon double bond through the reaction with metal carbene precursors giving an active carbene species (Equation 6.49).

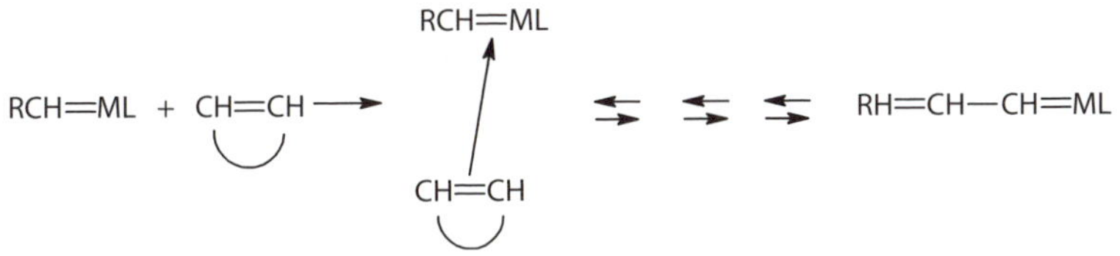

$$(5.60)$$

where "L" is the ligand attached to the metal.

5.12 ZWITTERIONIC POLYMERIZATION

While most polymerizations require an initiator, catalyst or some other form of activation, zwitterionic copolymerizations do not. These copolymerizations require a specific combination of one monomer that is nucleophilic and a second that is electrophilic in nature. The interaction of these two comonomers gives a zwitterion that is responsible for both the initiation and the propagation.

Initiation $\qquad$ $MN + ME \rightarrow {^+MN}{-ME^-}$ $\qquad$ (5.61)

Propagation $\qquad$ ${^+MN}{-ME^-} + {^+MN}{-ME^-} \rightarrow {^+MN}{-ME}{-MN}{-ME^-} \rightarrow \rightarrow \rightarrow \rightarrow$
$${^+MN}{-(-ME}{-MN-)}{-ME^-} \qquad (5.62)$$

If growth involves only addition and condensation reactions then an alternating copolymer is formed. Sometimes a lateral reaction occurs where the zwitterion interacts with one of the monomers giving a product that is a statistical copolymer.

5.13 ISOMERIZATION POLYMERIZATION

Isomerization polymerizations are polyaddition reactions where the propagating species rearranges to energetically preferred structures before subsequent chain growth.

$$nA \rightarrow -(-B-)_n- \qquad (5.63)$$

In 1962, Kennedy reported the first isomerization polymerization using 3-methyl-1-butene to give a 1,1-dimethyl PP as below.

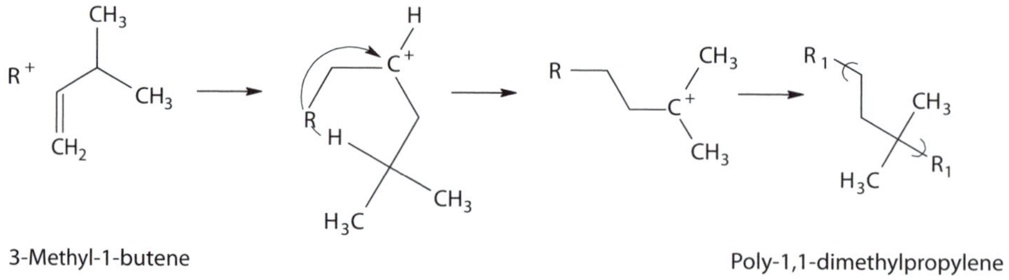

3-Methyl-1-butene $\qquad\qquad\qquad\qquad\qquad\qquad$ Poly-1,1-dimethylpropylene

$$(5.64)$$

Isomerization polymerizations can be associated with coordination catalysts systems, ionic catalysts systems, and free radical systems. The cationic isomerization polymerization of 4-methyl-1-pentene is of interest because the product can be viewed as an alternating copolymer of ethylene and isobutylene. This structure cannot be obtained by conventional approaches.

In the presence of certain ZNCs, an equilibrium exists between *cis*- and *trans*-1,3-pentadiene. Here, the *cis*-1,4-polypentadiene is formed from *trans*-1,3-pentadiene or from a mixture of the cis and trans isomers.

Such isomerizations are sometimes desired and sometimes are the cause of or explanation for unwanted structures. In the cationic polymerization forming poly(1-butene), nine different structural units have been found. Classical 1,2-hydride and 1,2-methide shifts, hydride transfer, and proton elimination account for these structures.

Unwanted branching of many polymers probably occurs through such isomerizations. PP, formed using cationic polymerization, has methyl, ethyl, n-propyl, n-butyl, isopropyl, gem-dimethyl, isobutyl, and t-butyl groups connected to the main chain.

5.14 PRECIPITATION POLYMERIZATION

Precipitation polymerization, also called slurry polymerization, is a variety of solution polymerization where the monomer is soluble but the polymer precipitates as a fine flock. The formation of olefin polymers via coordination polymerization occurs by a slurry process. Here the catalyst is prepared and polymerization is carried out under pressure and at low temperatures, generally less than 100°C. The polymer forms viscous slurries. Care is taken so that the polymer does not cake up on the sides and on the stirrer.

5.15 SUMMARY

1. Chain reactions, including ionic-chain polymerization reactions, consist of at least three steps: initiation, propagation, and termination. Termination generally occurs through chain transfer, producing a new ion and the "dead" polymer.

2. Cationic polymerizations occur with vinyl compounds that contain electron-donating groups using Lewis acids along with a cocatalyst as the initiators. Polymerizations generally occur at low temperatures in solvents with high-dielectric constants. The DP is proportional to the concentration of monomer, and overall rate of polymerization is proportional to the square of the monomer concentration. In general, the rate of polymerization is dependent on the dielectric constant of the solvent, resonance stability of the carbocation, the degree of solvation of the gegenion, and electropositivity of the initiator.

3. Monomers with electron-withdrawing groups can undergo anionic polymerization in the presence of anionic initiators. The rate of polymerization is dependent on the dielectric constant of the solvent, stability of the carbanion, electronegativity of the initiator, degree of solvation of the gegenion, and strength of the electron-withdrawing substituents.

4. Stereoregular polymers can be formed. These polymers can be divided into three general stereoregular unit combinations. When the pendent groups, such as methyl for PP, are all on one side of the polymer chain, the polymer combination is isotactic; when the methyl groups alternate from one side to the other, the polymer combination is syndiotactic; and when the position of the methyl group is somewhat random, it is atactic. The tacticity of polymers influences the physical properties of the products. In general, polymers with greater tacticity (order) have higher glass transition and melting temperatures, have a greater tendency to form crystalline products, and are stronger and denser. Stereoregular polymers are produced at low temperatures in solvents that favor formation of ion pairs between the carbocation and the gegenion. One of the most widely used stereoregulating systems is called the Ziegler-Natta catalyst system that generally employs a transition-metal salt such as titanium chloride and a cocatalyst such as alkylaluminum. A proposed mechanism involves a reaction on the surface of $TiCl_3$, activated by the addition of an alkyl group from the cocatalyst. The monomer adds to this active site producing a pi complex, which forms a new active center by insertion of the monomer between the titanium and carbon atoms. This step is repeated in the propagation reactions in which the alkyl group from the cocatalyst is the terminal group. Stereospecific polymers are also produced using the alfin and the chromia on silica initiators. The alfin system consists of allyl sodium, sodium isopropoxide, and sodium chloride.

5. Soluble stereoregulating systems have been developed using an organometallic transition complex such as Cp_2TiCl_2 and a cocatalyst often methylalumoxane. This system has advantages over the Ziegler-Natta and similar systems in that the polymers produced are more stereoregular, a wider range of monomers can be used, and little or no catalyst incorporated into the polymers, allowing the polymer to be directly used without having to undergo a procedure to remove the catalyst.

6. Two of the highest-volume polymers are made using ionic polymerization—HDPE and iPP. There exist a number of commercially available PEs that vary in extent and kind of branching, chain length, and amount of crystallinity.

7. A number of polymers have been made using ROMs. Nylon-6, similar to structure and properties to nylon-6,6, is made from the ring opening of the lactam caprolactam. Poly(ethylene oxide) is made from the ring opening of ethylene oxide and is made more stable by capping the ends preventing ready depolymerization.

GLOSSARY

Alfin catalyst: Complex catalyst system consisting of allyl sodium, sodium isopropoxide, and sodium chloride.

Anionic polymerization: A polymerization initiated by an anion.

Butyl rubber (IIR): Copolymer of isobutylene and isoprene.

Capping: Reacting the end groups to produce a stable polymer.

Carbanion: Negatively charged organic ion.

Carbocation: Positively charged organic ion, that is, one lacking an electron pair on a carbon atom.

Cationic polymerization: Polymerization initiated by a cation and propagated by a carbonium ion.

Ceiling temperature: Threshold temperature above which a specific polymer is unstable and decomposes.

Celcon: Trade name of copolymer of formaldehyde and dioxolane.

Chain-reaction polymerization: A rapid polymerization based on initiation, propagation, and termination steps.

Chain transfer: Process in which a growing chain becomes a dead polymer by abstracting a group from some other compounds, thereby generating another active site.

Copolymer: Polymer chain composed of units from more than one monomer.

Copolymerization: Polymerization of a mixture of more than one monomer.

Coupling: Joining of two active species.

Gegenion: A counterion.

Initiation: Start of a polymerization.

Isomerization polymerization: Polyaddition reaction in which the propagation species rearranges to energetically preferred structures before chain growth.

Kraton: Trade name for an ABA block copolymer of styrene–butadiene–styrene.

Lactom: Heterocyclic amide with one nitrogen atom in the ring.

Leuchs' anhydride: Cyclic anhydride that decomposes to carbon dioxide and an amino acid.

Living polymers: Macroanions or macrocarbanions.

Macroions: Charged polymer molecules.

Metathesis reaction: Catalytically induced reaction wherein olefins undergo bond reorganization resulting in the formation of polyalkenamers.

Monadic: Polyamide produced from one reactant.

Natta, Giulio: Discoverer of stereospecific polymers.

Oxirane: Ethylene oxide.

Polyacetal: Polyoxymethylene.

Polychloral: Polymer of trichloroacetaldehyde.

Promoter: Strained cyclic ethers that are readily cleaved.

Propagation: Continuous successive chain extension in a chain reaction.

Soluble stereoregulating catalyst: Soluble catalysts requiring a metal active site, cocatalyst or counterion, and a ligand system; capable of producing polymers with high stereoregularity and a minimum of branching.

Termination: Destruction of active growing chain in a chain reaction.

Trioxane: Treimer of formaldehyde.

Ziegler, Karl: Discoverer of complex coordination catalysts.

Ziegler-Natta (or Natta-Ziegler) catalyst: Able to produce stereoregular polymers.

Zwitterionic polymerization: Copolymerization between nucleophilic and electrophilic comonomers.

EXERCISES

1. Describe the contents of the reaction flask 10 minutes after the polymerization of (a) reactants in a stepwise polymerization such as dimethyl terephthalate and ethylene glycol and (b) monomer in chain reactions, such as isobutylene.
2. What is the initiator in the polymerization of isobutylene?
3. What is the general name of the product produced by cationic initiation?
4. What reactant besides the monomer is present in cationic chain propagation reactions?
5. What name is used to describe the negatively charged counterion in cationic chain-reaction polymerizations?
6. Is a Lewis acid (a) an electrophile or (b) a nucleophile?
7. Is a Lewis base (a) an electrophile or (b) a nucleophile?
8. Why isn't coupling a preferred termination step in the cationic chain polymerization of pure monomer?
9. Is the usual configuration of polymers produced by ionic-chain polymerization (a) head-to-tail or (b) head-to-head.
10. Which condition would be more apt to produce stereoregular polymers in ionic-chain polymerizations: (a) high temperature or (b) low temperatures?
11. Name (a) a thermoplastic, (b) an elastomer, and (c) a fiber that is produced commercially by ionic-chain polymerization.
12. Which technique would you choose for producing a polymer of isobutyl vinyl ether: (a) cationic or (b) anionic.
13. Which technique would you choose for producing a polymer of acrylonitrile: (a) cationic or (b) anionic?
14. Which of the following could be used to initiate the polymerization of isobutylene: (a) sulfuric acid, (b) boron trifluoride etherate, (c) water, or (d) butyllithium?
15. Which of the following could be polymerized by cationic chain polymerization?

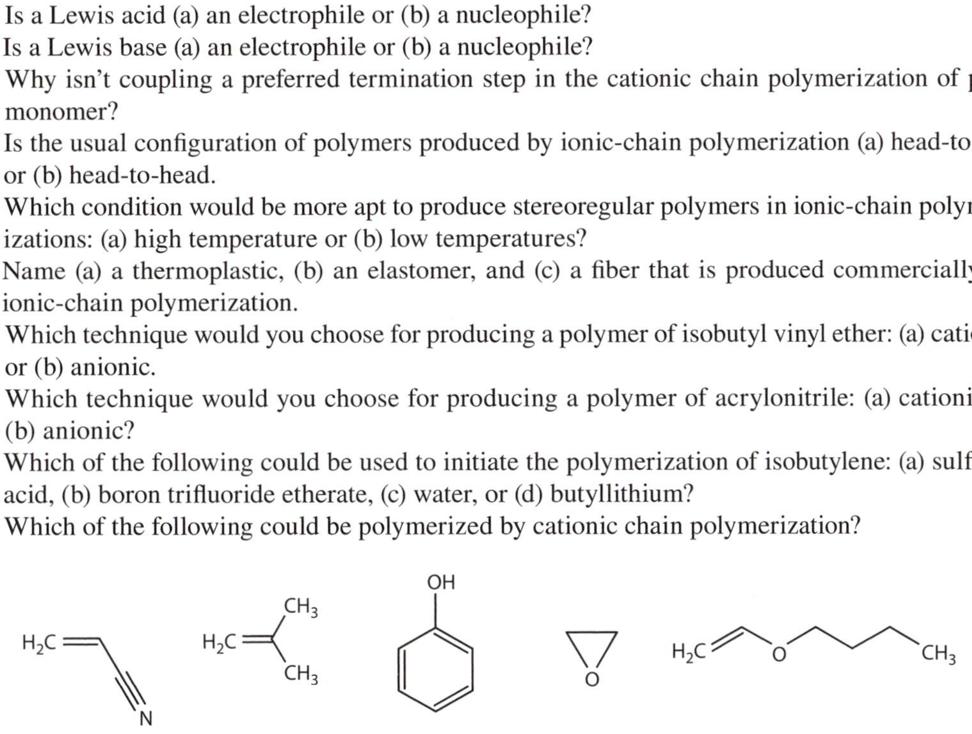

(a)   (b)   (c)   (d)   (e)

16. Which polymer is more susceptible to oxidation: (a) HDPE or (b) PP?
17. When termination is by chain transfer, what is the relationship of average DP and the kinetic chain length?
18. What would be the composition of the product obtained by the cationic low-temperature polymerization of a solution of isobutylene in ethylene?
19. What is the relationship between the rate of initiation to the monomer concentration in ionic-chain polymerization?
20. What effect will the use of a solvent with a higher dielectric constant have on the rate of propagation in ionic-chain polymerization?
21. How does the rate constant k_p change as the yield of polymer increases?
22. Which will have the higher T_g value: (a) polystyrene or (b) PIB?
23. Which of the following could serve as an initiator for an anionic chain polymerization? (a) $AlCl_3$-H_2O, (b) BF_3-H_2O, (c) butyllithium, or (d) sodium metal?
24. What species, in addition to a dead polymer, is produced in a chain transfer reaction with a macrocarbocation in cationic chain polymerization?

25. What is the relationship between R_i and R_t under steady-state conditions?
26. What is the relationship between average $\overline{DP}$ and R_p and R_t?
27. Draw a structure of what iPP looks like.
28. What percentage of polymer is usually found when a polymer produced by chain-reaction polymerization is heated above its ceiling temperature?
29. What is the relationship between the average degree of polymerization and initiator concentration in cationic chain polymerization.
30. Can the polymers found in the bottom of a bottle of insolubilized formaldehyde solution be useful?
31. How would you prepare stable polymers from formaldehyde?
32. Why is the thermal decomposition of polymers of formaldehyde called unzipping?
33. Discuss advantages of the soluble stereoregulating catalysts in comparison to the Natta-Zeigler catalysts.
34. Why are there so many widely used PEs.
35. How would you increase the flow rate of water in a fire hose?
36. Why is poly-3,3-bischloromethyloxybutylene crystalline?
37. Why are PP and PE the most widely used polymers?
38. What kind of polymers are made from lactones?
39. How could you remove unsaturated hydrocarbons from petroleum or coal tar distillates?
40. What species is produced by the reaction of an anionic chain polymerization initiator and the monomer?
41. What are the propagation species in anionic chain polymerizations?
42. Why are polymers produced by the anionic polymerization of pure monomers called "living polymers?"
43. Using the symbols A and B for repeating units in the polymer chain, which of the following is a block copolymer: (a) ABAABABBABAAB, (b) ABABABABAB, or (c) AAAAAABBBBBB?
44. What is the most widely used monadic nylon?
45. What is the repeating unit for nylon-4?
46. What is the catalyst and cocatalyst in the most widely used Ziegler-Natta catalyst?
47. Name two structures that are possible from the polymerization of 1,3-butadiene.
48. What is the principle difference between propagation reactions with butyllithium and a Ziegler-Natta catalyst?
49. What are some physical properties that iPP would have in comparison to aPP.
50. Show the skeletal structures of *cis*- and *trans*-polyisoprene.
51. Write formulas for repeating units in the chains of (a) poly-1,4-isoprene and (b) poly-1,2-isoprene.
52. What is the most widely used catalyst for the production of HDPE?
53. What elastomer is produced by anionic chain polymerization?
54. What elastomer is produced by use of a Ziegler-Natta catalyst?
55. What are difficulties associated with the use of solid-state catalysts in the production of polymers?
56. How are the difficulties associated with the use of solid-state catalysts overcome by the use of soluble stereoregulating catalysts.
57. What are some considerations in determining which PE is used to produce bags for a lawn and garden store?
58. How might you produce a polymer with good flexibility from crystalline HDPE that is strong but does not have the desired flexibility?
59. One table lists the ultimate elongation of a material as 60% and another lists the supposedly same general material with an ultimate elongation of 6,000. How can you account for this large difference?

60. Would you expect a copolymer rubber formed from styrene and 1,4-butadiene to be easily cross-linkable? Why?
61. Why are there so many different kinds of PE?

ADDITIONAL READING

Dragutan, V., Dragutan, I. (2009): *Polymers from Cycloolefins*, Taylor & Francis, Boca Raton, FL.

Dubois, P., Coulembier, O., Raquez, J., Degee, P. (2009): *Modern Arylation Methods*, Wiley, Hoboklen, NJ.

Gaylord, N. G., Mark, H. F. (1959): *Linear and Stereoregular Addition Polymers*, Interscience, NY.

Giarrusso, A., Ricci, G., Tritto, I. (2004): *Stereospecific Polymerization and Stereoregular Polymers*, Wiley, Hoboken. Hogan, J. P., Banks, R. L. (1955) (Philips Process): US Patent 2,717,888.

Jagur-Grodzinski, J. (2005): *Living and Controlled Polymerization: Synthesis, Characterization, and Properties of the Respective Polymers and Copolymers*, Nova, Commack, NY.

Moore, E. P. (1998): *The Rebirth of Polypropylene*, Hanser Gardner, Cincinnati, OH.

Morton, A., Bolton, F. H. (1952): Alfin catalysis, *Ind. Eng. Chem.*, 40:2876.

Natta, G, Danusso, F. (1967): *Stereoregular Polymers and Stereospecific Polymerization*, Pergamon, NY.

Oyama, S. T. (2008): *Mechanisms in Homogeneous and Heterogeneous Epoxidation Catalysis*, Elsevier, NJ.

Peacock, A. J. (2000): *Handbook of Polyethylene*, Taylor and Francis, Boca Raton, FL.

Scheirs, J. (2000): *Metallocene-Based Polyolefins*, Wiley, NY.

Severn, J., Chadwick, J. (2008): *Tailor-Made Polymers: Via Immobilization of Alpha-Olefin Polymerization Catalysts*, Wiley, Hoboken, NJ.

Szwarc, M. (1968): *Carbanions, Living Polymers and Electron-Transfer Processes*, Wiley, NY.

Ziegler, K. (1952, 1955): Ziegler catalysts, *Angew Chem.*, 64:323; 67:541.

6 Free Radical Chain Polymerization
(Addition Polymerization)

Since many synthetic plastics and elastomers and some fibers are prepared by free radical polymerization, this method is important. Table 6.1 contains a listing of commercially important addition polymers, including those that will be emphasized in this chapter because they are prepared using the free radical process.

As with other chain reactions, free radical polymerization is a rapid reaction that consists of the characteristic steps of initiation, propagation, and termination. Free radical initiators are produced by the homolytic cleavage of covalent bonds as well as numerous radiation-associated methods.

6.1 INITIATORS FOR FREE RADICAL CHAIN POLYMERIZATION

Free radical initiation can occur through application of heat (thermal), ultraviolet (UV) and visible light (photochemical), ionizing light, redox reagents, electricity (electrochemical), and so on, that is, any process that creates the essential free radicals.

Light in the UV and visible range can disrupt selected bonds forming free radicals. Such disruption occurs as we are exposed to sunlight. Suntan treatments often contain certain compounds that can accept this damaging radiation. Related compounds are also used in foods to give them longer shelf life. They are generally known as antioxidants. Synthetic antioxidants include benzophenones, benzils, and certain organic ketones. Thus, diphenylketone decomposes on exposure to UV radiation of the appropriate wavelength forming two free radicals (Equation 6.1).

$$
\underset{\text{Ph–C–Ph}}{\overset{\text{O}}{\|}} \xrightarrow{\text{UV}} \underset{\text{Ph–C}^{\bullet}}{\overset{\text{O}}{\|}} + \text{Ph}^{\bullet} \tag{6.1}
$$

The advantage of using such photochemical initiations is that polymerization can be carried out at room temperature.

When molecules are exposed to light of higher energy, shorter wavelength, or higher frequency, electrons can be removed or added depending on the specific conditions. Usual forms of ionizing radiation employed industrially and experimentally include neutrons, X-rays, protons, and alpha and beta particles. Oxidation–reduction, redox, reactions are also often employed to initiate free radical polymerizations in solution or heterogeneous organic-aqueous systems. Free radicals can be created by passing a current through a reaction system sufficient to initiate free radical polymerizations.

While application of heat or some other method can rupture the pi bond in the vinyl monomer causing the formation of a two-headed free radical that can act as a free-radical initiator, peroxides and dinitriles are generally employed as initiators. This is a consequence of the general bond dissociation energy trend of C–H>C–C>C–N>O–O. Dinitrile or azo compounds such as 2,2′-azo-bis-isobutyronitrile (AIBN) require temperatures of about 70°C–80°C to produce decomposition with free radical formation. Peroxides such as benzoyl peroxide (BPO) require temperatures higher than 60°C for decomposition and free radical formation. While the dissociation bond energy for C–N is generally greater than for O–O, the formation of a stable N_2 molecule is the thermodynamic driving force due to an entropy effect, allowing dissociation to occur at typically lower temperatures.

TABLE 6.1
Industrially Important Addition Polymers

Polymer Name	Typical Properties	Typical Uses
Polyacrylonitrile	High strength; good stiffness; tough, abrasion resistant; good flex life; good resistance to moisture, stains, fungi, chemicals, insects; good weatherability	Carpeting, sweaters, skirts, socks, slacks, baby garments
Poly(vinyl acetate)	Water sensitive with respect to physical properties such as adhesion and strength; good weatherability, fair adhesion	Lower molecular weight used in chewing gum, intermediate in production of poly(vinyl alcohol); water-based emulsion paints
Poly(vinyl alcohol)	Water soluble; unstable in acid and base system; fair adhesion	Thickening agent for various suspension and emulsion systems; packaging film, wet-strength adhesive
Poly(vinyl butyral)	Good adhesion to glass; tough, good stability to sunlight; good clarity; insensitive to moisture	Automotive safety glass as the interlayer
Poly(vinyl chloride) and poly(vinyidene chloride); (called "the vinyls or vinyl resins")	Relatively unstable to heat and light; resistant to fire, insects, fungi, moisture calendered products such as film, sheets, floor coverings; shower curtains, food covers, rainwear, handbags, coated fabrics, insulation for electrical cable and wire; old records	
Polytetrafluoroethylene	Insoluble in most solvents, chemically inert, low-dielectric loss, high-dielectric strength, uniquely nonadhesive, low friction, constant electrical and mechanical properties from 20°C to 250°C; high-impact strength	Coatings for frying pans, wire, cable; insulation for motors, oils, transformers, generators; gaskets; pump and valve packings; nonlubricated bearings, biomedical
Polyethylene (LDPE)	Good toughness and pliability over wide temperature range; outstanding electrical properties; good transparency in thin films; resistant to chemicals, acids, bases; ages poorly in sunlight and oxygen; low density, flexible, resilient, high-tear strength, moisture resistant	Films, sheeting used as bags, textile materials, pouches, frozen foods, produce wrapping, and so on; drapes, table cloths, covers for ponds, greenhouses, trash can liners, and so on; electrical wire and cable insulator; coating for foils, papers, other films, squeeze bottles
Polypropylene	Lightest major plastic; i-PP is the major form sold; high tensile strength, stiffness, hardness, resistance to marring; good gloss, high T_g allows it to be sterilized; good electrical properties, chemical inertness, moisture resistance	Filament—rope, webbing, cordage, carpeting, injection molding applications in appliances, small housewares, and automotive fields
Polyisoprene (cis-1,4-polyisoprene)	Structurally close to natural rubber; properties similar to those of natural rubber; good elasticity, rebound	Replacement of natural rubber; often preferred because of greater uniformity and cleanliness
SBR (styrene–butadiene rubber)	Random copolymer; generally slightly poorer physical properties than those of NR	Tire treads for cars; inferior to NR with respect to heat buildup and resilience, thus, not used for truck tires; belting, molded goods, gum, flooring, rubber shoe soles, hoses, electrical insulation
Butyl rubber (copolymer of isobutyl-ene	Amorphous isoprene-largely 1,4 isomer; good chemical inertness, low gas permeability, high viscoelastic response to stresses, less sensitive to oxidative aging than most isoprene rubbers; better ozone stability than NR; good solvent resistance	About 60%–70% used for inner tubes for tires

TABLE 6.1 (continued)
Industrially Important Addition Polymers

Polymer Name	Typical Properties	Typical Uses
Polychloroprene (mostly 1,4 isomer)	Outstanding oil and chemical resistance; high tensile strength; outstanding resistance to oxidative degradation, aging; good ozone and weathering response; dynamic properties same or better than most synthetic rubbers	Can replace NR in most applications; gloves, coated fabrics, cable and wire coatings, hoses, belts, shoe heels, solid tires
Polystyrene	Clear, easily colored; easily fabricated; transparent; fair mechanical and thermal properties; good resistance to acids, bases, oxidizing and reducing agents; readily attacked by many organic solvents; good electrical insulator	Production of ion-exchange resins, heat- and impact-resistant copolymer, ABS, resins, and so on, foams, toys, plastic optical components, lighting fixtures, housewares, packaging, home furnishings
Poly(methyl methacrylate)	Clear, transparent, colorless, good weatherability, good impact strength, resistant to dilute basic and acidic solutions; easily colored; good mechanical and thermal properties; good fabricability; poor abrasion resistance compared to glass	Used in cast sheets, rods, tubes, molding, extrusion compositions, tail- and signal-light lenses, dials, medallions, brush backs, jewelry, signs, lenses, skylight "glass"; generally used where good light transmission is needed

While initiation can occur via a number of routes, we will emphasize the use of chemical initiators for the formation of the free radicals necessary to begin the free radical polymerization process.

The rate of decomposition of initiators usually follows first-order kinetics and is dependent on the solvent present and the temperature of polymerization. The rate is usually expressed as a half-life time $(t_{1/2})$, where $t_{1/2} = \ln 2/k_d = 0.693/k_d$. The rate constant (k_d) changes with temperature in accordance with the Arrhenius equation as shown in Equation 6.2.

$$k_d = Ae^{-E_a/RT} \tag{6.2}$$

Typical equations for the dissociation of AIBN and BPO are shown below. It should be pointed out that because of recombination, which is solvent dependant, and other side reactions of the created free radical (R), the initiator efficiency is seldom 100%. Hence, an efficiency factor (f) is employed to show the fraction of effective free radicals produced.

The decomposition of AIBN and BPO to form radicals is given below:

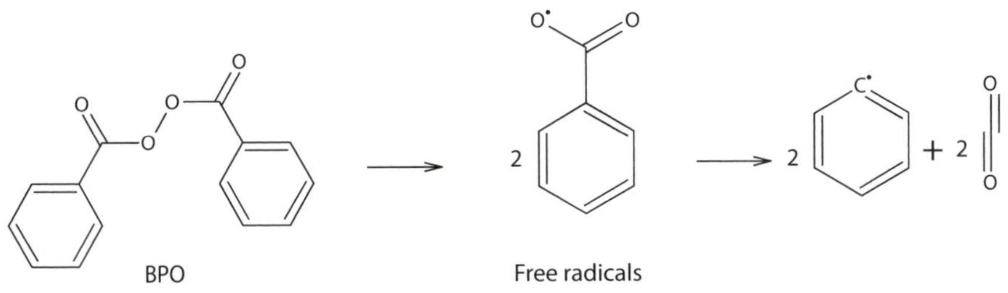

$$\tag{6.3}$$

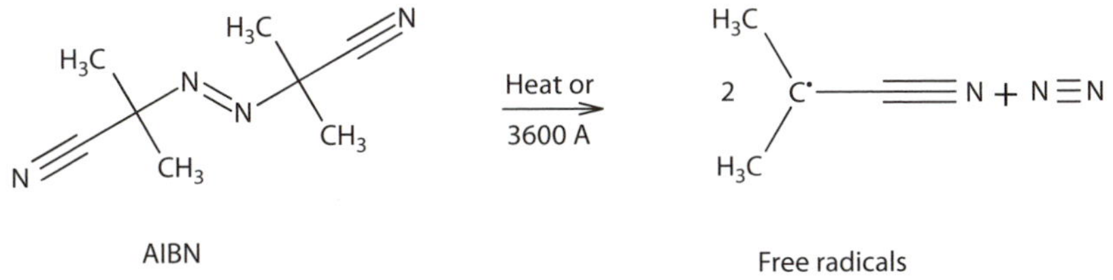

BPO   Free radicals

$$\tag{6.4}$$

AIBN   Free radicals

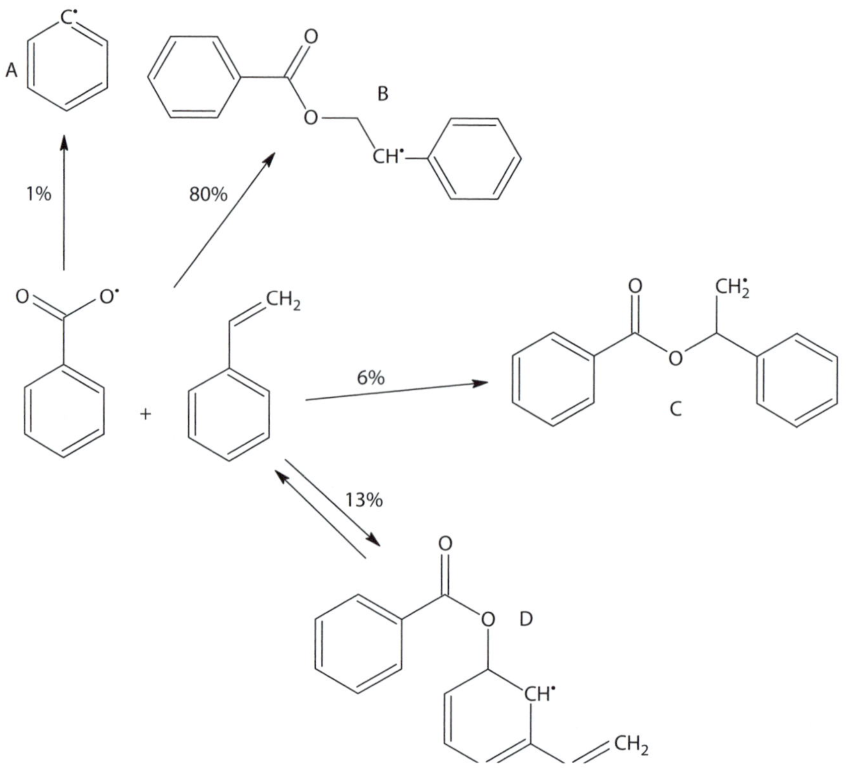

FIGURE 6.1   Initial reaction products of benzoyl peroxide and styrene monomer.

The precise structure of the initiating agent and initial addition to the monomer varies according to the reaction conditions, monomer, and initiator. For illustration, we will look at the products
formed from BPO. BPO is generally reported to break forming two benzoyl free radicals (Equation
6.3). In reality, the initial step is complex and involves the formation of a number of products as
shown in Figure 6.1 for styrene. The major reaction involves direct addition of the benzoyl radical
to the tail end of the styrene monomer creating a molecule where the radical resides at the head or
at a more sterically demanding and more radical stabilizing site (B). A lesser amount adds to the
head end (C). Some adds to the ring forming a variety of compounds including D. A small amount
decomposes forming the phenyl radical and carbon dioxide (A). For simplicity, we will employ
structure A as the initiating structure but all initiating structures give similar products, differing
only in the nature of the end group.

The BPO decomposes with a specific rate constant of about 10^{-8} s^{-1}, an Arrhenius constant, A, of
about 10^{16}, and an activation energy of about 28 kcal/mol (about 115 kJ/mol). As noted before, not
all radicals initiate new chains. Some terminate before initiation, forming inactive products mainly
phenyl benzoate (Equation 6.5). Thus, as noted before, an efficiency constant, f, is used that reflects
the ratio of BPO that actually form chains.

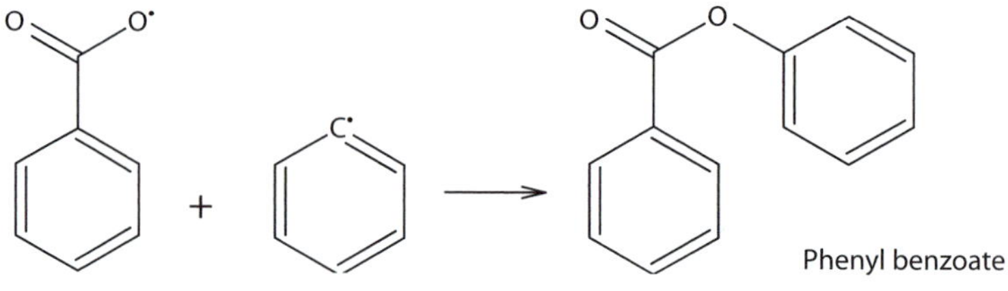

(6.5)

6.2 MECHANISM FOR FREE RADICAL CHAIN POLYMERIZATION

In general, the decomposition of the initiator (I) may be expressed by the following equation in which k_d is the specific rate or decay constant:

$$I \rightarrow 2R^\bullet \tag{6.6}$$

$$R_d = -\frac{d[I]}{dt} = k_d[I] \tag{6.7}$$

where R_d is the rate of decomposition.

Initiation of a free radical chain takes place by addition of a free radical ($R^\bullet$) to a vinyl monomer (Equation 6.8). Polystyrene (PS) will be used to illustrate the typical reaction sequences. (Styrene, like many aromatic compounds, is toxic and concentrations that come into contact with us should be severely limited.) It is important to note that the free radical ($R^\bullet$) is a companion of all polymerizing species and is part of the polymer chain acting as an end group, and hence should not be called a catalyst even though it is often referred to as such. It is most properly referred to as an initiator.

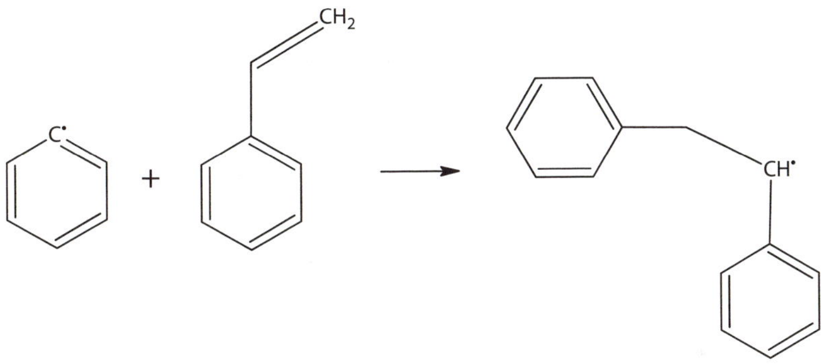

$$\tag{6.8}$$

$$R^\bullet + M \rightarrow RM^\bullet \tag{6.9}$$

$$R_i = \frac{d[RM^\bullet]}{dt} = k_i[R^\bullet][M] \tag{6.10}$$

where R_i is the rate of initiation, and $R^\bullet$ is the free radical from BPO.

The rate of decomposition of the initiator (I) (Equation 6.6) is the rate-controlling step in the free radical polymerization as well as formation of growing chains. Thus, the overall expression describing the rate of initiation can be given as

$$R_i = 2k_d f[I] \tag{6.11}$$

where "f" is the *efficiency factor* (a measure of the fraction of initiator radicals that produce growing radical chains, that is, are able to react with monomer).

A "2" is inserted in Equation 6.11 because, in this presentation, for each initiator molecule that decomposes two radicals are formed. The "2" is omitted from Equation 6.6 because this rate expression describes the rate of decomposition of the initiator but not the rate of formation of free radicals $R^\bullet$. (Similarly, in Equations 6.18 and 6.20, each termination results in the loss of two growing chains, thus a "2" appears in the descriptions.)

Propagation is a bimolecular reaction (Equations 6.12 and 6.13), occurring through addition of a new free radical (RM$^{\bullet}$) to another molecule of monomer (M). This step is repeated many times, resulting in the formation of the polymer chain. It is experimentally found that there may be slight changes in the propagation rate constant (k_p) in the first few steps, but the rate constant is generally considered to be independent of chain length. Hence, the symbols M$^{\bullet}$, RM$^{\bullet}$, and RM$_n$M$^{\bullet}$ may be considered equivalent in rate equations for free radical polymerization.

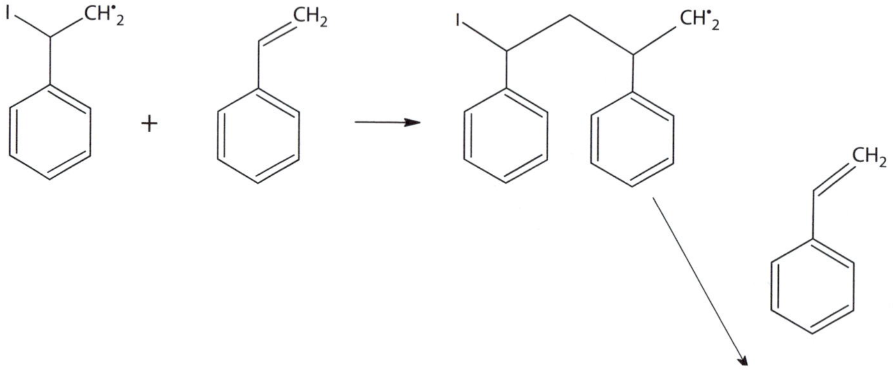

(6.12)

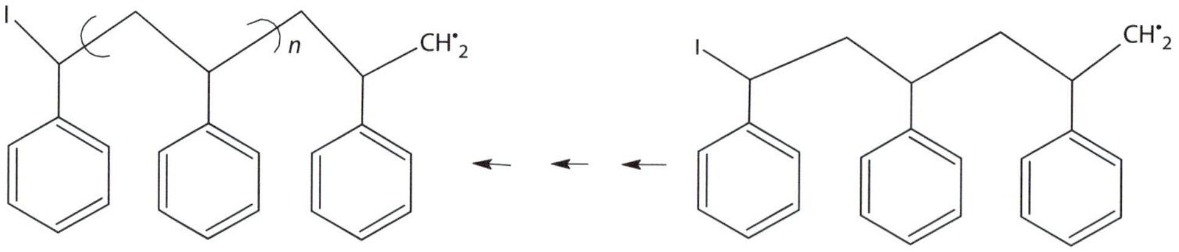

(6.13)

Styrene macroradical

Since the specific rate constants are approximately independent of the length of the growing chain, one specific rate constant is used to represent all of the propagation steps, k_p.

The rate of demise of monomer with time is described as

$$-\frac{d[M]}{dt} = k_p [M^{\bullet}][M] + k_i [R^{\bullet}][M] \qquad (6.14)$$

that is, monomer consumption only occurs in reactions described by Equations 6.8 and 6.13.

For long chains, the consumption of monomer by the initiation step (Equation 6.13) is small and can be neglected, allowing Equation 6.14 to be rewritten as

$$-\frac{d[M]}{dt} = k_p[M^{\bullet}][M] \qquad (6.15)$$

and

$$R_p = -\frac{d[M]}{dt} = k_p[M^{\bullet}][M] \qquad (6.16)$$

The polarity of the functional group in the monomers polymerized by free radical chain polymerization is between the positively inclined monomers characterized by undergoing cationic polymerization, and the negatively inclined monomers characterized by undergoing anionic polymerization. Figure 6.2 contains a list of the addition polymerization routes taken by various monomers. As is

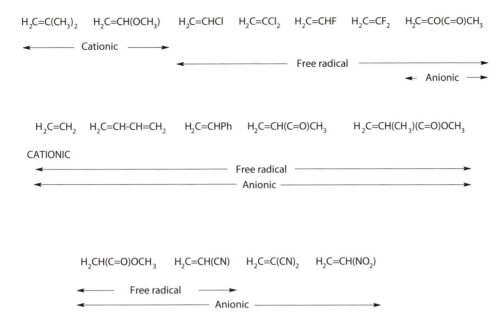

FIGURE 6.2 Type of chain initiation for some common monomers in order of general decrease in electron density associated with the double bond and their tendency to undergo chain polymerization.

true for the configuration of ionic growing chains, free radical polymers are also formed so that addition gives a head-to-tail configuration because functional groups on the vinyl monomers are better at stabilizing the free radical than are hydrogen atoms and because this balances the steric requirements present as addition occurs.

Unlike ionic polymerizations, the termination of the growing free radical chains usually occurs by the coupling of two macroradicals. Thus, the kinetic chain length (v) is equal to DP/2. The chemical and kinetic equations for bimolecular termination are shown below (Equations 6.17 and 6.18):

$$(6.17)$$

Termination is a head-to-head configuration at the juncture of the two macroradicals. The extend of coupling termination can then be obtained by determining the extent of head-to-head configuration in the product. The kinetic equation for coupling termination is shown in Equation 6.18.

$$R_t = -\frac{d[M^\bullet]}{dt} = 2k_t[M^\bullet][M] = 2k_t[M^\bullet]$$

$$(6.18)$$

In some situations, termination occurs by disproportionation. This termination process involves chain transfer to a hydrogen atom from one chain end to the free radical chain end of another growing chain, resulting in one of the "dead" polymer chains having an unsaturated chain end (Equations 6.19 and 6.20).

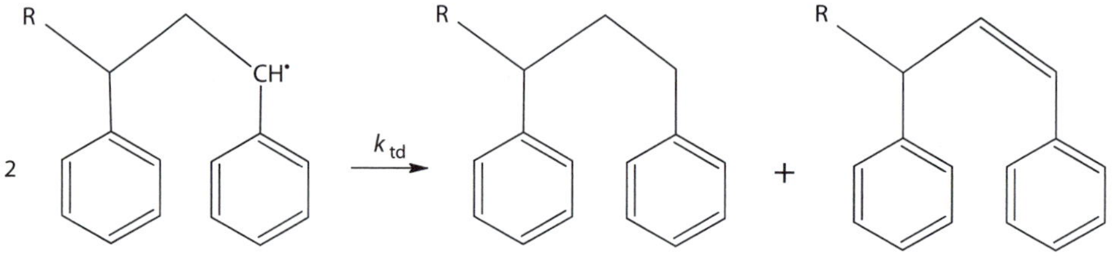

$$(6.19)$$

$$R_{td} = 2k_{td}[M^\bullet]^2 \qquad (6.20)$$

The kinetic chain length for termination by disproportionation is $\overline{DP} = v$ (compared with the relationship for coupling of $\overline{DP} = 2v$. The extent of the two types of termination is experimentally found by determining the number of head-to-head sites [coupling] and unsaturated end groups [disproportionation]).

The mode of termination varies with the monomer and the reaction conditions. While styrene macroradicals typically terminate by coupling, methyl methacrylate macroradicals terminate by coupling at temperatures below 60°C, but by disproportionation at higher temperatures.

The *kinetic chain length*, v, is described in Equation 6.21.

$$v = \frac{R_p}{R_i} = \frac{R_p}{R_{td}} = \frac{k_p[M][M^\bullet]}{2k_{td}[M^\bullet]^2} = \frac{k_p[M]}{k_{td}[M^\bullet]} = \frac{k'[M]}{[M^\bullet]} \qquad (6.21)$$

Because it is experimentally found that the number of growing chains is constant, there exists a steady state in $M^\bullet$ so that $R_i = R_{td}$. (A similar scenario can be used to develop rate expressions for coupling.)

While equations such as 6.15, 6.18, 6.20, and 6.21 are theoretically important, they contain $[M^\bullet]$, which is difficult to experimentally determine and are thus, practically of little use. The following is an approach to render such equations more useful by generating a description of $[M^\bullet]$ that involves more easily experimentally accessible terms.

The rate of monomer-radical *change* is described by

$$\frac{d[M^\bullet]}{dt} = [\text{monomer-radical formed}] - [\text{monomer-radical utilized}] = k_i[R^\bullet][M] - 2k_t[M^\bullet]^2 \qquad (6.22)$$

As noted before, it is experimentally found that the number of growing chains is approximately constant over a large extent of reaction. As noted before, this situation is referred to as a "steady state." For Equation 6.22, this results in $d[M^\bullet]/dt = 0$ and

$$k_i[R^\bullet][M] = 2k_t[M^\bullet]^2 \qquad (6.23)$$

Additionally, a steady-state value for the concentration of R· exists giving

$$\frac{d[R^\bullet]}{dt} = 2k_d f[I] - k_i[R^\bullet][M] = 0 \qquad (6.24)$$

Solving for [M] from Equation 6.23 gives

$$[M^\bullet] = \left(\frac{k_i[R^\bullet][M]}{2k_t} \right)^{1/2} \tag{6.25}$$

and [R] from Equation 6.24 gives

$$[R^\bullet] = \frac{2k_d f[I]}{k_i[M]} \tag{6.26}$$

Substituting in Equation 6.25 the expression for [R$^\bullet$], we obtain an expression for [M$^\bullet$], Equation 6.27, which contains more easily experimentally determinable terms.

$$[M^\bullet] = \left(\frac{k_d f[I]}{k_t} \right)^{1/2} \tag{6.27}$$

Using this relationship for [M] we get more useful rate (Equation 6.28) and kinetic chain length (Equation 6.30) equations that are free from the hard to measure [M].

$$R_p = k_p[M][M^\bullet] = k_p[M]\left(\frac{k_d f[I]}{k_t} \right)^{1/2} = \left(\frac{k_p^2 k_d f}{k_t} \right)^{1/2} [M][I]^{1/2} = k''[M][I]^{1/2} \tag{6.28}$$

Thus, the rate of propagation, or polymerization, is directly proportional to the concentration of the monomer and square root concentration of initiator.

In preparation of describing the kinetic chain length, we can also describe the rate of termination using the new description for [M].

$$R_t = 2k_t[M^\bullet]^2 = \frac{2k_t k_d f[I]}{k_t} = 2k_d f[I] \tag{6.29}$$

$$\overline{DP} = \frac{R_p}{R_i} = k_p[M]\left(\frac{(k_d f[I]/k_t)^{1/2}}{2k_t f[I]} \right) = \frac{k_p[M]}{2(k_d k_t f[I])^{1/2}} = \frac{k'[M]}{[I]^{1/2}} \tag{6.30}$$

Thus, chain length is directly proportional to monomer concentration and inversely proportional to the square root of initiator concentration.

Typical energies of activation for propagation and termination are given in Table 6.2 and typical free radical kinetic values in Table 6.3.

As done in Chapter 5, the effect of temperature can be determined using average activation of the various steps. Again, the rates of all single step reactions increase as the temperature increases but the overall result may be different for complex reactions. For free radical polymerizations, the activation energies are generally of the order $E_d > E_i \cong E_p > E_t$. Remembering that the description of the specific rate constant is

$$k = A \, e^{-Ea/RT} \tag{6.31}$$

the overall or "net" activation energy is

$$E_{(overall)} = E_t + E_i + E_p + E_d \tag{6.32}$$

TABLE 6.2
Energies of Activation for Propagation (E_p) and
Termination (E_t) in Free Radical Chain Polymerization

Monomer	E_p (kJ/mol)	E_t (kJ/mol)
Methyl acrylate	30	22
Acrylonitrile	17	23
Butadiene	39	–
Ethylene	34	–
Methyl methacrylate	26	12
Styrene	33	12
Vinyl acetate	31	22
Vinyl chloride	15	18

TABLE 6.3
Typical Free Radical Kinetic Values

Specific Rate Constant	Activation Energies (kJ/mol)		
k_d	10^{-3} s^{-1}	E_d	80–160
k_i	10^3 L/mol-s	E_i	20–30
k_p	10^3 L/mol-s	E_p	20–40
k_t	10^7 L/mol-s	E_t	0–20

Using only the specific rate constants involved with propagation from Equation 6.28 we have

$$R_p \, \alpha \, \frac{k_p^2 k_d}{k_t}$$

(6.33)

so that the overall activation energy using average values is a positive value (Equation 6.34) so the overall rate of polymerization increases as temperature increases.

$$E_{p(overall)} \, \alpha \, 2E_p + E_d - E_t = 2 \times 30 + 120 - 10 = 170$$

(6.34)

For chain length, from Equation 6.30 we have

$$\overline{DP} \, \alpha \, \frac{k_p}{2k_d k_t}$$

(6.35)

so that the overall activation energy using average values is

$$\overline{DP} \text{ (overall)} \, \alpha \, E_p - E_d - E_t = 30 - 120 - 10 = -100$$

(6.36)

so that $\overline{DP}$ decreases at temperature increases as pictured in Figure 4.4.

The Gibbs free-energy relationship for a reversible process at constant temperature for polymerization is described by

$$\Delta G_p = \Delta H_p - T\Delta S_p$$

(6.37)

where ΔH_p is the heat of polymerization defined by

$$\Delta H_p = E_p - E_{dp} \tag{6.38}$$

where E_p is the activation energy for propagation and E_{dp} is the activation energy for depolymerization.

The entropy term is negative so that it is the enthalpy or energy term that "drives" the polymerization. At low temperatures the enthalpy term is larger than the $T\Delta S_p$ term so that polymer growth occurs. At some temperature, called the *ceiling temperature*, the enthalpy term and entropy term are the same and $\Delta G_p = 0$. Above this temperature, depolymerization occurs more rapidly than polymer formation so that polymer formation does not occur. At the ceiling temperature depolymerization and polymerization rates are equal. The ceiling temperature is then defined as

$$T_c = \frac{\Delta H_p}{\Delta S_p} \tag{6.39}$$

since $\Delta G_p = 0$.

The ceiling temperature for styrene is about 310°C, for ethylene it is 400°C, for propylene it is 300°C, for methyl methacrylate it is 220°C, for tetrafluoroethylene it is 580°C, and for alpha-methylstyrene it is only 61°C.

It is interesting to note that due to their industrial importance, free radical polymerizations are the most studied reactions in chemistry. Furthermore, the kinetic approaches taken in this chapter are experimentally verified for essentially all typical free radical vinyl polymerizations.

There is some tendency for the formation of stereoregular sequences, particularly at low temperatures, but ionic and coordination catalysts are far superior in this aspect and are used to create stereoregular macromolecules.

Several additional comments are appropriate concerning chain type polymerization and termination. First, since the slow step is the initiation step and the other steps, and especially termination, have very low energy of activations and so are very fast, how does polymer form? Consider the relative concentrations of the various species. The concentration of polymer monomer is very high relative to the concentration of the free radical species favoring the formation of chain growth. The rate of termination is proportional to the square of the concentration of growing chain for both coupling (Equation 6.18) and disproportionation (Equation 7.20), and as a consequence of the concentration of growing chain being relatively quite small, polymer is allowed to form. Thus, chain growth resulting in polymer formation is the consequence of the high concentration of monomer and low concentration of growing chains. Second, in general, as the steric hindrance increases the tendency for termination occurring through disproportionation increases since coupling requires the approach of the ends of two growing chains. This approach of growing chain ends becomes, in general, less favorable as the steric hindrance increases.

6.3   CHAIN TRANSFER

Transfer of the free radical to another molecule serves as one of the termination steps for general polymer growth. Thus, transfer of a hydrogen atom at one end of the chain to a free radical end of another chain is a chain-transfer process we dealt within Section 6.2 under termination via disproportionation. When abstraction occurs intramolecularly or intermolecularly by a hydrogen some distance from the chain end, branching results. Each chain-transfer process causes the termination of one macroradical and produces another macroradical. The new radical sites serve as branch points for chain extension or branching. As noted above, such chain transfer can occur within the same chain as shown below:

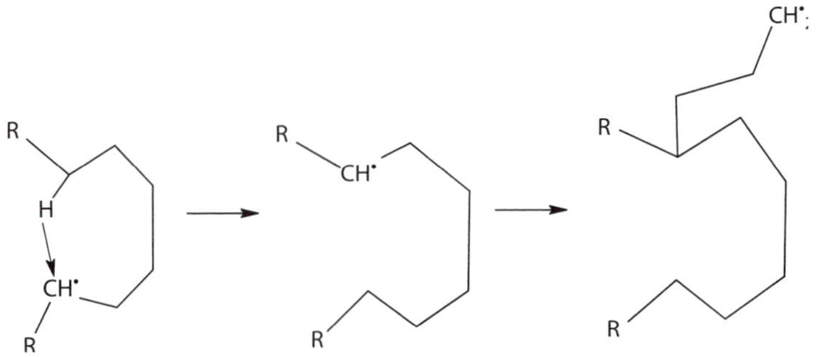

(6.40)

Chain transfer can also occur between chains.

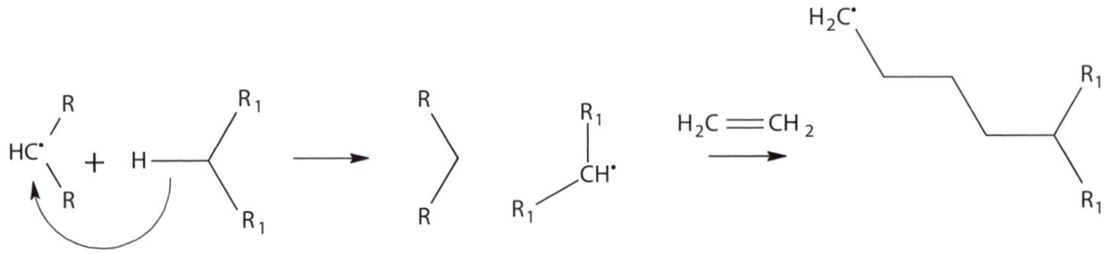

(6.41)

Chain transfer can also occur with initiator, impurity, solvent, or other additive present in the polymerization system. While the average chain length is equal to R_p divided by the sum of all termination reactions, it is customary to control all termination steps except the one that is being studied. Chain transfer to all other molecules, except solvent or some special additive, is typically negligible.

The chain-transfer reaction decreases the average chain length. This decrease in chain length increases as the concentration of the chain-transfer agent (S) increases and as the tendency of the chain-transfer agent to "chain-transfer" is increased. The resulting DP is equal to that which would have been obtained without the solvent or additive plus a factor related to the product of the ratio of the rate of propagation (R_p) and the rate of chain transfer (R_{tr}) and the ratio of the concentration of the monomer [M] to the concentration of chain-transfer agent [S].

The Mayo equation, Equation 6.42, which gives positive slops when the data is plotted (such as Figure 6.3), is the reciprocal relationship derived from the previously cited expression. The ratio of the rate of cessation or termination by transfer to the rate of propagation is called the *chain-transfer constant* (C_s).

$$\frac{1}{\overline{DP}} = \frac{1}{\overline{DP}_o} + C_s \frac{[S]}{[M]}$$

(6.42)

The chain-transfer constant is given as

$$C_S = \frac{k_{tr}}{k_p}$$

(6.43)

As shown in Figure 6.3, the molecular weight of PS is reduced when it is polymerized in solvents, and the reduction or increase in slope is related to the chain-transfer efficiency of the solvent. The slopes in this figure are equal to C_s.

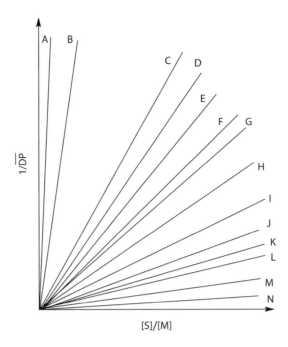

FIGURE 6.3 Molecular weight of polystyrene as a function of solvent and solvent concentration: A = n-butylmercaptan, B = carbon tetrabromide, C = carbon tetrachloride, D = o-cresol, E= p-cresol, F = m-cresol, G = phenol, H = sec-butylbenzene, I = cumene, J = ethylbenzene, K = chloroform, L = n-heptant, M = toluene, N = benzene where [S] = concentration of chain-transfer agent and [M] = concentration of styrene monomer.

TABLE 6.4
Chain-Transfer Constants of Solvent to Styrene in Free Radical Chain Polymerization at 60°C

Transfer Agent	$C_s \times 10^4$	Transfer Agent	$C_s \times 10^4$
Acetic acid	2.0	1-Dodecanethiol	148,000
Benzene	0.01	Hexane	0.9
Butyl alcohol	0.06	N,N-Dimethylaniline	12
t-Butyl alcohol	6.7	1-Naphthalenethiol	1,500
Butyl disulfide	0.24	1-Octanethiol	190,000
Carbon tetrabromide	18,000	p-Methoxyphenol	260
Carbon tetrachloride	84	Phenol	8.1
Chloroform	0.5	Triethylamine	1.4
o-Chlorophenol	6.0	Toluene	0.105
2,6-Ditert-butylphenol	49	Water	0

Chain-transfer constants of various solvents, including those given in Figure 6.3, are given in Table 6.4.

Chain-transfer agents have been called and employed as regulators (of molecular weight). When used in large proportions, they are called *telogens*, since they produce low molecular weight polymers (*telomers*).

6.4   POLYMERIZATION TECHNIQUES

The principle free radical polymerization techniques are bulk, solution, suspension, and emulsion. Tables 6.5 and 6.6 briefly describe these techniques.

Bulk Polymerization: Bulk polymerization of a liquid monomer such as methyl methacrylate is relatively simple in the absence of oxygen where small bottles or test tubes can be used as the reaction vessel. The monomer can be heated in the presence of an initiator giving a clear plastic shaped like the container, but a little smaller because of shrinkage. The volume of the monomers is generally larger than the final polymers, thus, the density of the polymer is greater than that of the original monomer.

The rate of bulk polymerization can be followed by monitoring the change in volume or increase in viscosity. When the viscosity is high, the termination reaction is hindered since the macroradicals are unable to diffuse readily in the viscous medium. Thus, the number of growing chains increases. This *autoacceleration*, called the Norris–Trommsdorff, Tranmsdorff, or gel effect, causes the formation of unusually high molecular weight chains. Since vinyl polymerizations are exothermic, there is a buildup of heat which further causes an additional autoacceleration of the reaction. If the temperature buildup is not controlled, it is possible that an explosion will occur. While the temperature can be easily controlled within a small test tube, it is more difficult in a large batch process. Stirring and external cooling are employed to control the polymerization process.

Suspension Polymerization: Water-insoluble monomers such as vinyl chloride may be polymerized as suspended droplets (10–1000 nm in diameter) in a process called *suspension* (pearl) *polymerization*. Coalescence of droplets is prevented by the use of small amounts of water-soluble polymers, such as PVA. The suspension process is characterized by good heat control and ease of removal of the discrete polymer particles.

Since PVC is insoluble in its monomer, it precipitates as formed in the droplet. This is actually advantageous since it permits ready removal of any residual carcinogenic monomer from the solid beads by stripping under reduced pressure.

Solution Polymerization: Monomers may also be polymerized in solution using good or poor solvents for homogeneous and heterogeneous systems, respectively. In solution polymerizations, solvents with low-chain transfer constants are used to minimize reduction in chain length.
Poly(vinyl acetate) (PVAc) may be produced by the polymerization of vinyl acetate (Equation 6.47). The viscosity of the solution continues to increase until the reaction is complete. Dilute polymer solutions are used to prevent the onset of autoacceleration because of the gel effect.

Poly(vinyl acetate) is used in adhesives and coatings and is hydrolyzed producing water-soluble PVA; Equation 6.44. The PVA may be reacted with butyraldehyde to produce poly(vinyl butyral) used as the inner lining of safety glass.

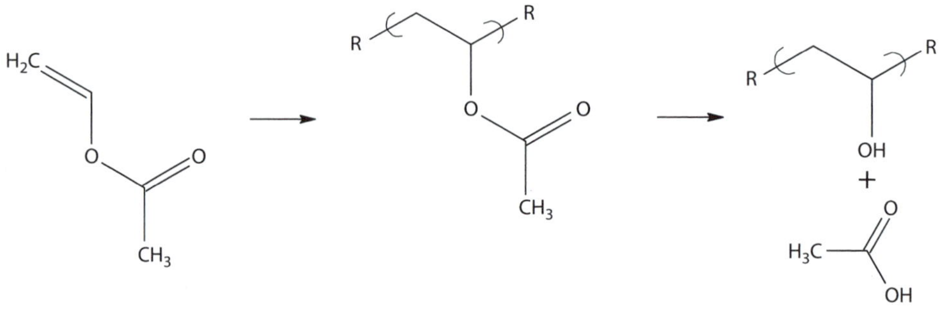

$$(6.44)$$

When a monomer such as acrylonitrile is polymerized in a poor solvent, macroradicals precipitate as they are formed. Since these are "living polymers," the polymerization continues as more

TABLE 6.5
Types of Polymerization Systems

Monomer–Polymer Phase Relationship	Monomer Location	
	Continuous	**Dispersed**
Homogeneous (same phase)	Bulk, solid state, solution	Suspension
Heterogeneous (different phase)	Bulk with polymer Precipitating	Emulsion; suspension with polymer precipitating

TABLE 6.6
Summary of Popular Polymerization Techniques

Bulk

Simplest of the techniques requiring only monomer and monomer-soluble initiator, and perhaps a chain-transfer agent for molecular weight control. Characterized, on the positive side, by high-polymer yield per volume of reaction, easy polymer recovery. Difficulty of removing unreacted monomer and heat control are negative features. Examples of polymers produced by bulk polymerization include poly(methyl methacrylate), PS, and low-density (high pressure) polyethylene

Solution

Monomer and initiator must be soluble in the liquid and the solvent must have the desired chain-transfer characteristics, boiling point (above the temperature necessary to carry out the polymerization and low enough to allow for ready removal if the polymer is recovered by solvent evaporation). The presence of the solvent assists in heat removal and control (as it also does for suspension and emulsion polymerization systems). Polymer yield per reaction volume is lower than for bulk reactions. Also, solvent recovery and removal (from the polymer) is necessary. Many free radical and ionic polymerizations are carried out utilizing solution polymerization, including water-soluble polymers prepared in aqueous solution (namely poly(acrylic acid), polyacrylamide, and poly(N-vinylpyrrolidinone). PS, poly(methyl methacrylate), poly(vinyl chloride) (PVC), and polybutadiene are prepared from organic solution polymerizations

Suspension

A water-insoluble monomer and initiator are used. Again, a chain-transfer agent may be used to control chain size. Stirring is usual. Droplets of monomer-containing initiator and chain-transfer agent are formed. A protective colloidal agent, often poly(vinyl alcohol) (PVA), is added to prevent coalescence of the droplets. Near the end, the particles become hard and are recovered by filtration. Because the liquid is water based, solvent recovery and treatment problems are minimal. The products may contain a number of impurities, including any of the agents added to assist in the polymerization process. Polymers produced by suspension polymerization include poly(vinyl chloride), PS resins, and copolymers such as poly(styrene-coacrylonitrile), SAN, and poly(vinyl chloride-co-vinylidene chloride)

Emulsion

The system usually contains a water-soluble initiator (in contrast to the requirement that the initiator must not be water soluble in suspension polymerizations), chain-transfer agent, and a surfactant. The hydrophobic monomer forms large droplets that are stabilized by the surfactant. At a certain surfactant concentration, the surfactant molecules form micelles that contain 50–100 surfactant molecules. During the polymerization, the monomer, that has a small but real water solubility, migrate from the monomer droplets through the water and into these micelles. Polymerization begins when the water-soluble initiator enters into the monomer-containing micelle. Because the concentration of micelles (about 10^{21}/L) is high compared with the concentration of monomer droplets (about 10^{13}/L), the initiator is more likely to enter a micelle than a monomer droplet. As polymerization continues, monomer is transferred to the growing micelles. At about 50%–80% conversion the monomer droplets disappear and the micells become large polymer-containing droplets. This suspension is called a *latex*. The latex is stable and can be used as is or the polymer recovered by coagulation. In inverse emulsion polymerization, the monomer, which is hydrophillic, is dispersed in an organic liquid. Here, the monomer is usually contained in an aqueous solution.

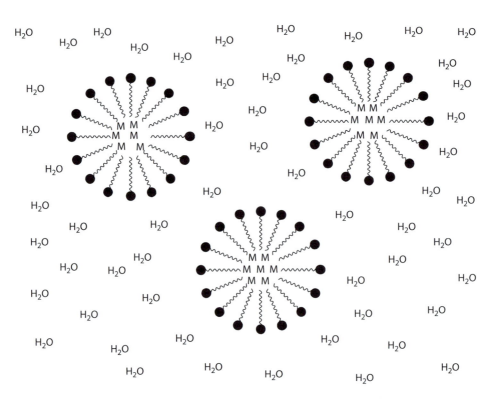

FIGURE 6.4 Micelles swollen with solubilized monomer. As the concentration increases the micelles change from spherical, as shown, to more rod-like in appearance.

acrylonitrile diffuses into the precipitated particles. This heterogeneous solution polymerization has been called *precipitation polymerization.*

Emulsion Polymerization: Many water-soluble vinyl monomers may be polymerized by the *emulsion polymerization* technique. This technique, which differs from suspension polymerization in the size of the suspended particles and in the mechanism, is widely used for the production of a number of commercial plastics and elastomers. While the particles in the suspension range from 10 to 1,000 nm, those in the emulsion process range from 0.05 to 5 nm in diameter. The small beads produced in the suspension process may be separated by filtering, but the latex produced in emulsion polymerization is a stable system in which the charged particles cannot be recovered by ordinary separation procedures.

Since relatively stable macroradicals are produced in the emulsion process, the termination rate decreases and a high molecular weight product is rapidly produced. It is customary to use a water-soluble initiator such as potassium persulfate, an anionic surfactant such as sodium sterate, and to stir the aqueous mixture of monomer, initiator, and surfactant in the absence of oxygen at 40°C–70°C. When the concentration of soap exceeds the critical micelle concentration (CMC), the molecules are present as micelles in which the hydrophillic carboxylic acid ends are oriented toward the water–micelle interface, and the lyophilic hydrocarbon ends are oriented toward the center of the micelle. The micelles are present as spheres with a diameter of 5–10 nm when the soap concentration is less than 2%. However, with the higher concentrations typically employed, the micelles resemble aggregates of rods which are 100–300 nm in length.

As shown in Figure 6.4, the water-insoluble monomer, M, is attracted to the lyophilic ends in the micelles, causing the micelles to swell. The number of swollen micelles per milliliter of water is on the order of 10^{18}. However, at the initial stages of polymerization (phase I) most of the monomer is present as globules that resemble those observed in suspension polymerization.

Since the initiation of polymerization takes place in the aqueous phase, essentially no polymerization occurs in the globules. Thus, they serve primarily as a reservoir of monomer supplied to the micelles to replace monomer converted to polymer. The number of droplets per milliliter of water is on the order of 10^{11}. Hence, since there are 10 million times as many micelles as droplets, the chance of initiation of monomer in a droplet is very small and the chance that more than one growing chain occurs within the same droplet is very very small.

The persulfate ion undergoes homolytic cleavage, producing two sulfate ion radicals.

$$S_2O_8^{-2} \rightarrow 2SO_4^{\bullet -} \tag{6.45}$$

The sulfate ion radical then initiates polymerization, here with a styrene monomer eventually forming a PS oligomer radical and eventually a PS radical (Equation 6.46).

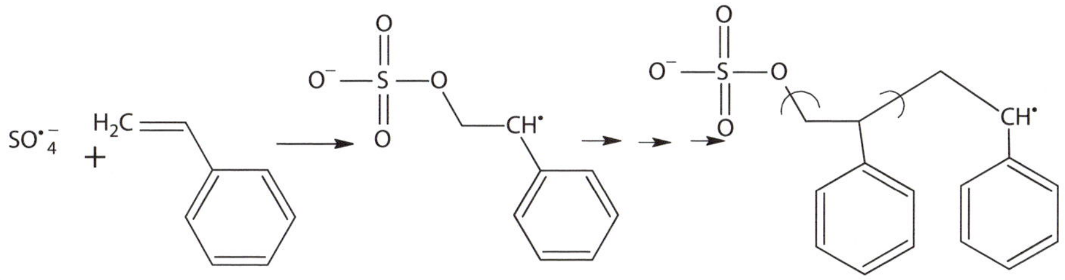

$$\tag{6.46}$$

According to a theory proposed by Harkins and refined by Smith and Ewart, the first stages of propagation in an emulsion system also take place in the aqueous phase to produce a more lyophilic surface oligoradical. When the degree of polymerization (DP) of the PS oligoradical is 3–5, its solubility is much similar to that of styrene, and it migrates to the swollen micelle where propagation continues with the styrene molecules already present. According to the accepted theories, each micelle can accommodate only one free radical, as noted before, and until a second one enters and terminates the propagation reaction through coupling, propagation continues to take place in the micelles. From a statistical point of view, only one-half of the micelles (N/2) will contain growing chains at any one time. It should also be noted that since propagation occurs in the micelles, the rate of polymerization will be proportional to the number of micelles present, that is, the rate is proportional to the soap concentration.

As the micelles grow by absorption of more monomer and formation of polymer, they become relatively large particles that absorb soap from micelles that have not been inoculated or stung by oligoradicals. Thus, in stage II, when about 20% of the monomer has been converted to polymer, the micelles disappear and are replaced by large, but fewer, monomer–polymer particles.

Polymerization continues in stage II and monomer continues to be supplied to the particles by the droplets in the aqueous phase. These droplets disappear when about 30% of the monomer has been converted to polymer. Polymerization continues in stage III after about 60% conversion, but all monomer must now be supplied to the macroradicals by a diffusion process in the micelles.

The rate of sulfate decomposition is described as

$$R_d = k_d[S_2O_8^{-2}] \tag{6.47}$$

The rate of initiation is then

$$R_i = k_i[SO_4^{\bullet -}][M] = 2k_d f[S_2O_8^{-2}] \tag{6.48}$$

The rate of propagation in the micelles is similar to that described for other free radical chain growth, but since the free radical concentration is equal to the number of active micelles, the value of N/2 is used instead of [M']. Thus, the rate of propagation is dependent on the number of micelles present.

$$R_p = k_p[M][M^\bullet] = k_p[M] \, (N/2) \tag{6.49}$$

The rate of production of free radicals at 50°C is about 10^{13} radicals/mL per 1 second. Thus, since there are 10^5 micelles for every free radical produced in one second, inoculation of any of the 10^{18} micelles/mL is infrequent. Hence, since propagation is a very fast reaction, long chains are produced before termination by coupling, which takes place as the result of the entrance of a new oligoradical in the active micelle. The $\overline{DP}$ is also proportional to the number of active micelles (N/2).

$$\overline{DP} = \frac{R_p}{R_i} = \frac{k_p(N/2)}{k_i[SO_4^{\bullet-}]} \tag{6.50}$$

6.5   FLUORINE-CONTAINING POLYMERS

Polytetrafluoroethylene (PTFE), better known by its trade name of Teflon, was accidentally discovered by Roy J. Plunkett, a DuPont chemist who had just received his PhD from Ohio State 2 years before. He was part of a group searching for nontoxic refrigerant gases. On April 6, 1938, he and his assistant, Jack Rebok, had filled a tank with tetrafluoroethylene. After some time, they opened the value but no gas came out. The tank weight indicated that there was no weight loss—so what happened to the tetrafluoroethylene. Using a hacksaw, they cut the cylinder in half and found a waxy white powder. He correctly surmised that the tetrafluoroethylene had polymerized. The waxy white powder had some interesting properties. It was quite inert toward strong acids, bases, and heat and was not soluble in any attempted liquid. It appeared to be quite "slippery."

Little was done with this new material until the military, working on the atomic bomb, needed a special material for gaskets that would resist the corrosive gas uranium hexafluoride that was one of the materials being used to make the atomic bomb. General Leslie Groves, responsible for the U.S. Army's part in the atomic bomb project, had learned of DuPont's new inert polymer and had DuPont manufacture it for them.

Teflon was introduced to the public in 1960 when the first Teflon-coated muffin pans and frying pans were sold. Like many new materials, problems were encountered. Bonding to the surfaces was uncertain at best. Eventually, the bonding problem was solved. Teflon is now used for many other applications, including acting as a biomedical material in artificial corneas, substitute bones for nose, skull, hip, nose, and knees; ear parts, heart valves, tendons, sutures, dentures, and artificial tracheas. It has also been used in the nose cones and heat shield for space vehicles and for their fuel tanks.

Over a half million vascular graft replacements are performed every year. Most of these grafts are made of poly(ethyleneterephthalate) (PET) and PTFE. These relatively large-diameter grafts work when blood flow is rapid, but they generally fail for smaller vessels.

Polytetrafluoroethylene is produced by the free radical polymerization process. While it has outstanding thermal and corrosive resistance, it is a marginal engineering material because it is not easily machinable. It has low tensile strength, resistance to wear, and it has low-creep resistance. Molding powders are processed by press and sinter methods used in powder metallurgy. It can also be extruded using ram extruder techniques.

PTFE is a crystalline polymer with melting typically occurring above 327°C. Because it is highly crystalline, it does not generally exhibit a noticeable T_g.

The C–F bond is one of the strongest single bonds known with a bond energy of 485 kJ/mol. While it is structurally similar to linear polyethylene (PE), it has marked differences. Because of the small size of hydrogen, PE exists as a crank-shaft backbone structure. Fluorine is a little larger (atomic radius of F = 71 pm and for H = 37 pm) than hydrogen, causing the teflon backbone to be helical and forming a complete twist every 13 carbon atoms. The size of the fluorine is sufficient to form a smooth "protective" sheath around the carbon backbone. The concentration of F end groups is low in ultrahigh molecular weight PTFE, contributing to its tendency to form crystals.

The electron density of PE and PTFE are also different. The electronegativity value for C is 2.5, F = 4.0, and for H = 2.1. Thus, the electron density on the fluorine surface of PTFE is greater than that for PE.

For high molecular weight linear PE, the repeat unit length is about 0.254 nm forming crystalline portions with a characteristic thickness of about 10 nm. The chain length for tough solids from PE is about 4.5 times the crystalline thickness. Thus, tough solids occur at molecular weights greater than 5,000 g/mol or chain lengths greater than about 45 nm. In comparison, the repeat unit length for PTFE is about 0.259 nm. The crystalline thicknesses for PTFE are about 100–200 nm or much thicker than for PE. Chain lengths for tough solids are about 4.5 times the crystalline thickness. Thus, much greater chain sizes, about 200,000–400,000 Da, are required to produce tough solids. The greater size of the crystalline portions also probably contributes to its higher T_m and greater difficulty in processing. The crystal thickness of PTFE is about 10–20 times the crystal thickness found for most other semicrystalline polymers such as PE.

At low molecular weights, PTFE is waxy and brittle. To achieve good mechanical properties ultrahigh molecular weights on the order of 10 million dalton is usually needed. These long chains disrupt crystal formation because they are longer than a single crystal. But the long chain lengths connect the crystals together adding to their strength. But these long chains result in extremely high viscosities so that ultrahigh molecular weight PTFE does not flow when melted and is thus, not melt processable. Form restrictive and costly methods are used to produce products from PTFE.

While vinyl fluoride was prepared in about 1900, it was believed resistant to typical "vinyl" polymerization. German scientists prepared vinyl fluoride through reaction of acetylene with hydrogen fluoride in the presence of catalysts in 1933 (Equation 6.51).

$$ H-F \ + \ HC \equiv CH \ \longrightarrow \ H_2C = \backslash_F \tag{6.51} $$

It was not until 1958 that DuPont scientists announced the polymerization of vinyl fluoride forming poly(vinyl fluoride) (PVF); Equation 6.52. Polymerization is accomplished using peroxide catalysts in water solutions under high pressure.

$$ H_2C = \backslash_F \ \longrightarrow \tag{6.52} $$

In comparison to PTFE, PVF is easily processable using a variety of techniques used for most thermoplastic materials. It offers good flame retardancy, presumably due to the formation of HF that assists in the control of the fire. Thermally induced formation of HF is also a negative factor because of its toxicity. As in the case of PVC, elimination of the hydrogen halide (HF) promotes formation of aromatic polycyclic products that themselves are toxic.

The difference in electronegativity between the adjacent carbons because of the differing electronegativities of H and F results in the C–F bond being particularly polar, resulting in it being susceptible to attack by strong acids. The alternating bond polarities on the PVF chain gives a tight structure, resulting in PVF films having a low permeability. This tight structure also results in good resistance, resistance to cracking, and resistance to fading.

Friction and wear are important related characteristics. If a material has a high friction then it will generally have a shorter wear time because water or other friction event chemicals pass over the material with the higher friction causing greater wear. The friction eventually "wears" away polymer chains layer-by-layer. The engineering laws of sliding friction are simple. According to Amontons' laws, the friction F between a body (rain drop, wind, or board rubbing against the material) and a plane surface (the polymeric material) is proportional to the load L and independent to the area of contact A. The friction of moving bodies is generally less than that of a static body. The kinetic friction is considered independent of the velocity. The coefficient of friction is defined as F/L. Polymers show a wide range of coefficients of friction so that rubbers exhibit relatively high values (BR = 0.4–1.5 and SBR = 0.5–3.0) whereas some polymers such as PTFE (0.04–0.15) and PVF (0.10–0.30) have low values.

The low coefficient of friction for PVF results in materials coated with it remaining somewhat free of dirt and other typical contaminants, allowing PVF-coated materials to be less frequently cleaned. It is essentially self-cleaning as rain carries away dust and other particulates, including bird droppings, acid rain, and graffiti. The low friction also results in longer lifetimes for materials coated with the PVF and for the PVF coating itself.

PVF has a T_g of about –20°C remaining flexible over a wide-temperature range (from about –20°C to 150°C), even under cold temperatures. Because of its low coefficient of friction and tightly bound structure, it retains good strength as it weathers. Films, in Florida, retain much of their thickness even after about a decade losing less than 20% of their thickness. To increase their useful lifetimes, relatively thick films, such as 1 mil, are generally employed. The "slickness" also acts to give the material a "natural" mildew resistance.

Unlike PVC that requires plasticizers to be flexible, PVF contains no plasticizers and does not "dry out" like PVC. PVF, because of its higher cost in comparison to PE and PP, is used as a coating and selected "high end" bulk applications such as films. Films are sold by DuPont under the trade name of Tedlar. Tedlar is used in awnings, outdoor signs, roofing, highway sound barriers, commercial building panels, and solar collectors. It is used as a fabric coating, protecting the fabric from the elements. PVF is resistant to UV-related degradation and unlike PVC, it is inherently flexible. While transparent, pigments can be added to give films and coatings with varying colors. Protective coatings are used on plywood, automotive parts, metal siding, lawn mower housings, house shutters, gutters, electrical insulation, and in packaging of corrosive chemicals. PVF has pizeoelectric properties generating a current when compressed.

The processability of fluorine-containing polymers is improved by replacement of one or more of the F atoms. Replacing one of eight fluorine atoms with a trifluoromethyl group gives a product called FEP or Viton, actually a copolymer of tetrafluoroethylene and hexafluoropropylene (6.53). Polytrifluoromonochloroethylene (PCTFE, Kel F; 6.54), in which one fluorine atom has been replaced by a chlorine atom has a less regular structure and is thus more easily processed. Poly(vinylidene fluoride) (PVDF; Kynar; Equation 6.55) is also more easily processable but less resistant to solvents and corrosives.

(6.53) (6.54)

Gore-Tex is named for its inventors W. L. Gore and R. W. Gore and the nature of the material, a textile. Gore-Tex materials are based on expanded PTFE and other fluoropolymers. They are used in a variety of areas, including gaskets, sealants, insulation for wires and cables, filter media, and medical implants and their most widely used area is as high-performance fabrics. Gore-Tex is a thin, highly porous fluoropolymer membrane bonded generally to a nylon or polyester fabric. The membrane has about 9 billion pores per square inch. Each of these pores are about 20,000 times smaller than a drop of water, preventing liquid water from penetrating but allowing individual water molecules to pass through. This results in a material that is breathable but also waterproof and wind resistant. This is in contrast to most raincoats that are waterproof but not breathable. Along with garments, Gore-Tex is also used to make tents and other outdoors goods.

TABLE 6.7
General Physical Properties of Selected Fluorine-Containing Polymers

Polymer→	PTFE	PCTFE	PVDF	PVF
Heat deflection temperature (1,820 kPa; °C)	100	100	80	90
Maximum resistance to continuous heat (°C)	250	200	150	125
Coefficient of linear expansion (cm/cm⁻°C, 10^{-5})	10	14	8.5	10
Compressive strength (kPa)	2.7×10^4	3.8×10^4	–	–
Flexural strength (kPa)	–	6×10^4	–	–
Impact strength (Izod: cm-N/cm of notch)	160	130	–	–
Tensile strength (kPa)	2.4×10^4	3.4×10^4	5.5×10^4	–
Ultimate elongation (%)	200	100	200	–
Density (g/mL)	2.16	2.1	1.76	1.4

A slice of quartz develops a net positive charge on one side and a net negative charge on the other side when pressure is applied. The same effect is found when pressure is applied by means of an alternating electric field. This effect is known as the piezoelectric effect and is used for quartz watches and clocks, for TVs, hearing aids, and so forth.

Several polymers are also effective piezoelectric materials. The best known of these is PVDF (6.55). PVDF is employed in loud speakers, fire and burglar alarm systems, earphones, and microphones.

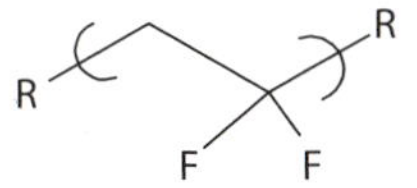

$$(6.55)$$

Table 6.7 contains physical properties of selected fluorine-containing polymers.

Nylon 11 (Equation 6.56) is also a piezoelectric material that can be aligned when placed in a strong electromagnetic field giving films used in infrared-sensitive cameras, underwater detection devices, and in electronic devices since it can be overlaid with printed circuits.

$$H_2N-(CH_2)_{10}-COOH \qquad \rightarrow -(-HN-(CH_2)_{10}-COO-)- \qquad (6.56)$$

11-Aminoundecanoic acid   Nylon 11

Ethylene tetrafluorethylene (ETFE) (6.57) was developed by DuPont in the 1970s to be an aeronautical insulation. ETFE was made from a waste product of lead and tin mining. ETFE is being used in a variety of applications. It is used as a covering for electrical wring, inflated in pillows as a building material, and as glass-like sheets that have an equal to and better light transmission compared with glass. In the nuclear industry, it is used for tie and cable wraps. Recently, it has become a material in innovative building exteriors, including use in the Eden Project, Cornwall; Kansas City Power and Light District, Kansas City; whale-shaped aquarium on New York's Coney Island; and most prominently displayed in Beijing's National Stadium also called the Bird's Nest. The Bird's Nest appears to be composed of loosely woven twigs. "Pillows" made from ETFE fill the spaces between and above the "twigs." It is also used to make the pneumatic panels that cover the outside of the soccer stadium Allianz Arena in Munich and the Beijing National Aquatics Center, also called the Water Cube, for the 2008 Olympics. ETFE panels are also employed as a dual laminate bonded with fiber-reinforced polymer composites that are used as corrosive protective liners in pipes, vessels, and tanks.

$$R \xrightarrow{\quad} \overset{\overset{F \quad F}{\diagdown \diagup}}{C} \xrightarrow{\quad} R \qquad (6.57)$$

ETFE has high-corrosion resistance and good strength over a wide range of temperatures of approximately –150°C to 150°C. Compared to glass, ETFE film is 1% of its weight, transmits more light, and is less expensive to install. It is highly resilient, able to bear 400 times its own weight. It is recyclable and self-cleaning because of the slickness brought about because of the presence of the tetrafluoroethylene units. On the negative side, as with most fluorine-containing polymers, combustion results in the release of highly corrosive HF.

6.6 POLYSTYRENE

Styrene monomer was discovered by Newman in 1786. The initial formation of PS was by Simon in 1839. While PS was formed almost 175 years ago, the mechanism of formation, described in Sections 6.1–6.3, was not discovered until the early twentieth century. Staudinger, using styrene as the principle model, identified the general free radical polymerization process in 1920. Initially commercialization of PS, as in many cases, awaited the ready availability of the monomer. While there was available ethyl benzene, it underwent thermal cracking rather than dehydrogenation until the appropriate conditions and catalysts were discovered. Dow first successfully commercialized PS formation in 1938. While most commercial PS has only a low degree of stereoregularity, it is rigid and brittle because of the resistance of the more bulky phenyl-containing units to move in comparison, for example, to the methyl-containing units of polypropylene. This is reflected in a relatively high T_g of about 100°C for PS. It is transparent because of the low degree of crystalline formation.

While PS is largely commercially produced using free radical polymerization, it can be produced by all four of the major techniques—anionic, cationic, free radical, and coordination-type systems. All of the tactic forms can be formed employing these systems. The most important of the tactic forms is syndiotactic PS (sPS). Metallocene-produced sPS is a semicrystalline material with a T_m of 270°C. It was initially produced by Dow in 1997 under the tradename of Questra. It has good chemical and solvent resistance in contrast to "regular" PS, which has generally poor chemical and solvent resistance because of the presence of voids that are exploited by the solvents and chemicals.

Physical properties of PS are dependent on the molecular weight and the presence of additives. General properties of PS are given in Table 6.8. While higher molecular weight PS offers better strength and toughness, it also offers poorer processability. Low molecular weight PS allows good processability but poorer strength and toughness. Generally, a balance is sought where intermediate chain lengths are used. Typically employed chain lengths are on the order of 1,500–3,500 with standard molecular weight distributions of about 2.2–3.5. Small amounts of plasticizers are often used to improve processability.

Styrene is employed in the formation of a number of co- and terpolymers. The best known is the terpolymer ABS.

Major uses of PS are in packaging and containers, toys and recreational equipment, insulation, disposable food containers, electrical and electronics, housewares, and appliance parts. Expandable PS is used to package electronic equipment such as TVs, computers, and stereo equipments.

PS is produced in three forms—extruded PS, expanded PS foam, and extruded PS foam (XPS). Expanded PS (EPS) was developed by the Koppers Company in Pittsburgh, PA in 1959. XPS insulation was developed by Dow Chemical and sold under the trade name Styrofoam. This term is often used for many other expanded PS materials. EPS and XPS are similar and both generally contain a

TABLE 6.8
General Properties of Polystyrene

Density (g/mL)	1.05
Density EPS (g/mL)	0.025–0.200
Dielectric constant	2.4–2.7
Electrical conductivity (S/m)	10^{-16}
Glass transition temp. (°C)	95
Melting point (°C)	240
Coefficient of linear expansion (1/K)	10^{-6}
Tensile strength (MPa)	45–60
Young's modulus (MPa)	3,000–3,600
Elongation at break (%)	3–4

mixture of 90%–95% PS and 5%–10% gas (also referred to as a blowing agent), generally pentane, nitrogen, or carbon dioxide. The solid plastic is heated generally using steam followed by introduction of the gas. EPS is produced from PS beads containing the entrapped blowing agents. When heated, the blowing agent turns to a gas expanding, resulting in the individual beads expanding and fusing together. XPS is similarly formed except the blowing agents become gaseous as the mixture emerges from an extruder giving a more continuous product in comparison to EPS.

Extruded PS foam is often known by its trade name Styrofoam. XPS has air inclusions giving the material some flexibility, low-thermal conductivity, and a low density. It is widely used as commercial and residential insulation. In construction, it is also used to make ornamental pillars that are subsequently coated with a harder material. Under roads and buildings, it is employed to prevent soil disturbance due to weathering. Life rafts are generally made from XPS but can be made from EPS. Styrofoam insulation is estimated to reduce energy costs by more than $10 billion annually.

About 3%–4% of landfill waste comes from old roofs. On the average, roofs need to be replaced every 7–10 years. Recently, Styrofoam is being employed in forming protected membrane roofs (PMRs) that insulate as well as waterproofs the roof area. Unlike many roof systems, the Styrofoam is placed over the PMR. It is believed that such systems increase the life expectancy of the roof to 40 years, thus reducing the landfill component from roofs.

Styrofoam structural insulation sheathing (SIS) has recently been introduced by Dow. SIS is a wall system that combines structural and water-resistance with insulation properties. It is made with about 80% recycled material and believed to reduce energy consumption by about 10%.

Expanded PS foam is incorrectly best known to the general public as Styrofoam cups, coolers, and egg containers. EPS can be generally distinguished from Styrofoam by the presence of small beads or spheres that are present in the EPS and are missing in the XPS. It is also used to make a number of other products such as inexpensive surfboards and other water and pool floating devices. It is used to make packing peanuts and molded packing material for cushioning items such as TVs, computers, and stereo equipment.

Extruded PS is used for objects where a somewhat rigid inexpensive plastic material is needed such as plastic Petri dishes, plastic test tubes, plastic model kits, CD jewel cases, toys, house wares, and appliance parts. Medical products are sterilized after the product is made using irradiation or treatment with ethylene oxide.

Legislation was put in place in some states to insure the recycling of PS. Interestingly, some of this legislation was written such that all PS had to be recycled within some period of time such as a year. This legislation was changed to reflect the real concern of fast-food containers when it was pointed out that less than 10% PS is used in this manner and that well over twice as much was used as house insulation that should not be recycled every year or so.

6.7   POLY(VINYL CHLORIDE)

Poly(vinyl chloride) is one of the three most abundantly produced synthetic polymers. PVC is one of the earliest produced polymers. In 1835, Justus von Liebig and his research student Victor Regnault reacted ethylene dichloride with alcoholic potash forming the monomer vinyl chloride. Latter Regnault believed he polymerized vinyl chloride but latter studies showed it to be poly(vinylidene chloride). In 1872, E. Baumann exposed vinyl chloride sealed in a tube to sunlight and produced a solid, PVC. Klasse, in Germany, found that vinyl chloride could be made by addition of hydrogen chloride to acetylene in a system that could be scaled up for commercial production. (Today, most vinyl chloride is made from the oxychlorination reaction with ethylene.) By World War I, Germany was producing a number of flexible and rigid PVC products. During World War I, Germany used PVC as a replacement for corrosion-prone metals.

Today, PVC is made from the polymerization of vinyl chloride as shown in Equation 6.58.

$$\tag{6.58}$$

Waldo Semon was responsible for bringing many of the PVC products to market. As a young scientist at BF Goodrich, he worked on ways to synthesize rubber and to bind the rubber to metal. In his spare time he discovered that PVC, when mixed with certain liquids, gave a elastic-like, pliable material that was rain proof, fire resistant, and that did not conduct electricity. Under the trade name Koroseal, the rubbery material came into the marketplace, beginning about 1926, as shower curtains, raincoats, and umbrellas. During World War II, it became the material of choice to protest electrical wires for the Air Force and the Navy. Another of his inventions was the synthetic rubber patented under the name Ameripol that was dubbed "liberty rubber" since it replaced natural rubber in the production of tires, gas masks, and other military equipment. Ameripol was a butadiene-type material.

Because of its versatility, some unique performance characteristics, ready availability, and low cost PVC is now the third largest produced synthetic polymer behind polyethylene and polypropylene.

As a side note, today there is a debate concerning the use of chlorine-containing materials and their effect on the atmosphere. This is a real concern and one that is being addressed by industry. PVC, and other chloride-containing materials have in the past been simply disposed of through combustion that often created unwanted hydrogen chloride. This practice has largely been stopped but care should be continued to see that such materials are disposed of properly. Further, simply outlawing of all chloride-containing materials is not possible or practical. For instance, we need common table salt for life and common table salt is sodium chloride. Chlorine is widely used as a water disinfectant both commercially (for our drinking water) as well as for pools. Further, PVC is an important material that is not easily replaced. Finally, the amounts of chloride-containing residue that is introduced into the atmosphere naturally is large in comparison to that introduced by PVC. Even so, we must exercise care as we want to leave a better world for our children and grandchildren, so a knowledge-based approach must be taken. As with most significant problems, it is better to error in the direction of caution.

Another health concern is the presence of certain plasticizers. As noted below, PVC employs a large amount of additives, including a variety of plasticizers. We do not live in a risk-free society but some risks should be eliminated or minimized. One group of plasticizers of interest is the phthalate plasticizers. As with other health concerns infant care is of greatest concern. Vinyl IV bags are often used in neonatal intensive care units. Food and Drugs Association (FDA) has requested that manufacturers eliminate the use of questionable plasticizers for the production of these bags. Another answer is simply eliminating vinyl IV bags using another material for construction of these bags. Further, baby and children toys should be constructed from materials that do not contain these plasticizers. Another group that is at particular risk is the critically ill or injured patients. Again, special care should be exercised when dealing with special groups of people. As we are dealing with toxic materials, we need to remember that all materials can be harmful if present in the wrong concentrations and/or locations. Even so, where we can we might minimize exposure to known potentially toxic materials.

Poly(vinyl chloride) materials are often defined to contain 50% or more by weight vinyl chloride units. PVC is generally a mixture of a number of additives and often other units such as ethylene, propylene, vinylidene chloride, and vinyl acetate. Structurally similar products, but with differing properties, are made from the chlorination of PE but almost all PVC is made from the polymerization of vinyl chloride. Typical homopolymers are about 400–1,000 units long.

Poly(vinyl chloride) is commercially produced by a number of techniques but mainly suspension, emulsion, bulk, and solution polymerization. Typically, product properties and form can be tailored through the use of a particular synthetic process and conditions. Particulate architecture is then controlled to achieve materials with specific sizes and distributions for specific uses and applications. Because of the tendency of PVC to split off hydrogen chloride, forming materials with high char at relatively low-general processing temperatures, special care is taken with respect to temperature control and particulate architecture that allows ready processing of PVC by most of the common processing techniques.

Tacticity of the PVC varies according to the particular reaction conditions, but generally manufactures favor a syndiotactic form with many PVC materials being about 50% sPVC. The reported amount of crystallinity is in the range of 5%–10%. This allows for a material with some strength, but one with sufficient amorphous regions to retain good flexibility.

Poly(vinyl chloride), in comparison to many other polymers, employs an especially wide variety of additives. For instance, a sample recipe or formulation for common stiff PVC pipe, such as that used in housing and irrigation applications, may contain in addition to the PVC resin, tin stabilizer, acrylic processing aid, acrylic lubricant-processing aid, acrylic impact modifier, calcium carbonate, titanium dioxide, calcium sterate, and paraffin wax. Such formulations vary according to the intended processing and end use. In such nonflexible PVC materials, the weight amount of additive is on the order of 5%–10%.

Plasticizers weaken the intermolecular forces in the PVC reducing crystallinity. A relatively stable suspension, called a *plastisol*, of finely divided PVC in a liquid plasticizer can be poured into a mod and heated to about 175°C, producing a solid flexible plastic as a result of fusion of the plasticizer in the PVC.

As noted before, there is a tendency for PVC to undergo elimination of hydrogen chloride when heated. The most labile chlorine atoms are those at tertiary or terminal sites. Once the initial chloride is eliminated, continued unzipping occurs with the formation of unsaturated backbone sites and the evolution of hydrogen chloride. The purpose of the stabilizer is to cap unzipping sites by substitution of more stable groups for the evolved chloride as depicted in Equation 6.59.

$$-----C–Cl + Sn–R \rightarrow -----C–R + Sn–Cl \tag{6.59}$$

Some of the tin stabilizers are based on oligomeric materials first made by Carraher and coworkers. These oligomeric materials are essentially "non-migratable."

A sample formula for a flexible upholstery fabric covering might contain PVC resin, medium-molecular weight polymeric plasticizer, stearic acid lubricant, calcium carbonate, pigment, antimony oxide, linear phthalate ester, epoxidized soy bean oil, and linear phthalate ester. Here the weight amount of additive is in the range of 40%–70% by weight with the plasticizer often being on the order of about 60%.

Poly(vinyl chloride) has a built in advantage over many other polymers in that it is itself flame resistant. About 50% of PVC is used as rigid pipe. About 70% of the water pipes in the United States are PVC. About 75% of the sewer pipes are PVC. Other uses of rigid PVC are as pipe fittings, electrical outlet boxes, and automotive parts. Uses of flexible PVC include gasoline-resistant hose, hospital sheeting, shoe soles, electrical tape, stretch film, pool liners, vinyl-coated fabrics, roof coatings, refrigerator gaskets, floor sheeting, and electrical insulation and jacketing. A wide number of vinyl chloride copolymers are commercially used. Many vinyl floor tiles are copolymers of PVC.

Many flat sheet signs are made of PVC. Films are also formed from PVC. Many of these signs and films are simply referred to as vinyl. These films and sheets form many of our commercial signs and markings on vehicles. Unplasticized or rigid PVC is used in the construction industry as a siding simply known as vinyl siding. It is also used to repair window frames and sills and fascia. It is also widely used in the construction of plastic gutters, downpipes, and drainpipes.

The flame resistance of PVC is a mixed blessing. In a fire, the PVC emits hydrogen chloride with the chlorine scavenging free radicals helping eliminate the fire. Hydrogen chloride fumes present a health concern when we breathe them. In fire, moisture helps dilute the hydrogen chloride causing it to settle on the cooler surfaces rather than remaining air borne. Even so, in closed structures such as tunnels alternative materials are advised.

Table 6.9 contains general physical properties of PVC. Because of the variety of additives, the values for the plasticized PVC are approximate.

Poly(vinyl chloride) is flexibilized by addition of plasticizers as already noted. It is also made more flexible through blending it with elastomers that act as impact modifiers. These blends are used when impact resistance is essential.

A number of copolymers are formed employing vinyl chloride. Because of the presence of the comonomer, these copolymers are more flexible than PVC itself. Vinylite (6.60) is a random copolymer of vinyl chloride (87%) and vinyl acetate. While Vinylite is not as strong as PVC, it is more easily processed.

(6.60)

TABLE 6.9
General Physical Properties of PVC

	Rigid PVC	Plasticized PVC
Heat deflection temperature (1,820 kPa; °C)	75	–
Maximum resistance to continuous heat (°C)	60	35
Crystalline melting point (°C)	170	–
Coefficient of linear expansion (cm/cm-°C, 10^{-5})	6	12
Compressive strength (kPa)	6.8×10^4	6×10^3
Flexural strength (kPa)	9×10^4	–
Impact strength (Izod: cm-N/cm of notch)	27	–
Tensile strength (kPa)	4.4×10^4	1×10^4
Ultimate elongation (%)	50	200
Density (g/mL)	1.4	1.3

Copolymers of vinyl chloride with vinyl acetate and maleic anhydride are made more adhesive to metals through hydrolysis of the ester and anhydride units.

Copolymers of vinyl chloride and vinylidene chloride are widely used (6.61). Two extremes are employed. The lesser known has a high vinyl chloride content compared with the vinylidene chloride comonomer. These copolymers are more easily dissolved and have greater flexibility. The most widely known copolymer has about 90% vinylidene chloride and 10% vinyl chloride and is known as Saran. This copolymer has low permeability to gases and vapors and is transparent. Films are sold as Saran Wrap. Because of potential health-related issues with Saran Wrap, it has been largely replaced.

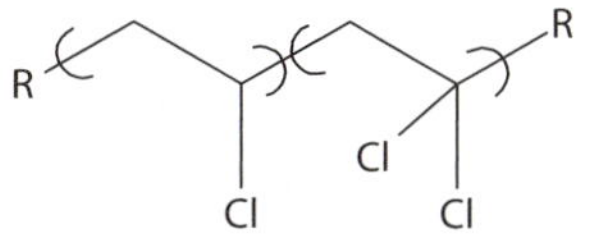

$$\text{(6.61)}$$

6.8 POLY(METHYL METHACRYLATE)

Poly(methyl methacrylate) (Equation 6.62) is the most widely employed of the alkyl methacrylates. For the most part, commercial PMMA is sold as an atactic amorphous polymer that has good light transparency (92%) and gives transparent moldings and films. General properties are given in Table 6.10. The presence of two substituents on every alternate carbon atom restricts chain mobility so PMMA is less flexible than the corresponding poly(alkyl acrylates). The presence of the alpha methyl group increases the stability of PMMA toward light-associated and chemical degradation.

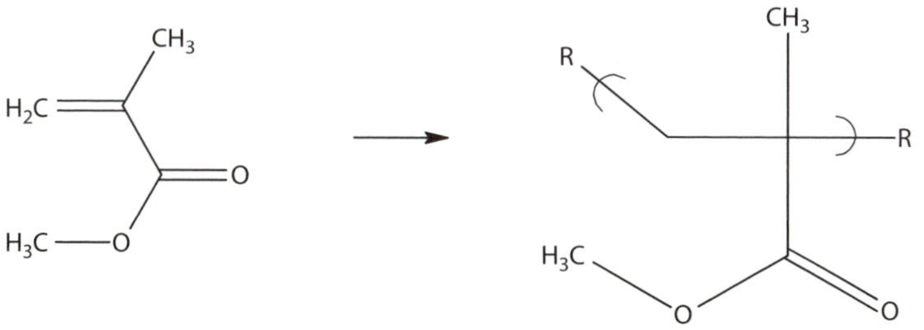

$$\text{(6.62)}$$

Unlike the poly(alkyl methacrylates) that degrade by random chain scission, PMMA undergoes degradation through unzipping when heated.

TABLE 6.10
General Physical Properties of PMMA

Heat deflection temperature (1,820 kPa; °C)	95
Maximum resistance to continuous heat (°C)	75
Coefficient of linear expansion (cm/cm-°C, 10^{-5})	7.0
Compressive strength (kPa)	1×10^5
Flexural strength (kPa)	9.6×10^4
Impact strength (Izod: cm-N/cm of notch)	21
Tensile strength (kPa)	6.5×10^4
Ultimate elongation (%)	4
Density (g/mL)	1.2

The most widely used acrylic plastics are PMMA (such as Lucite) and copolymers of methyl methacrylate that contain small amounts (2%–18%) of methyl or ethyl acrylate (such as Plexiglas). These materials are available as sheets and molding powders. PMMA is more resistant to impact than PS or glass but its scratch resistance is inferior to glass. PMMA is widely used in the creation of automotive (light) lenses for tail lights and some front light casings.

Poly(methyl methacrylate) is resistant to nonoxidizing acids, bases, and salts at ordinary temperatures but is attacked by oxidizing acids at room temperature. It is resistant to highly polar solvents such as ethanol but is soluble in less-polar solvents such as toluene.

Polymers such as the methacrylates play an essential role as *photoresists*. To make a photoresist, the methacrylate polymer is deposited onto silicon dioxide. A mask, which shields specific regions from subsequent exposure to light, is placed over the methacrylate polymer resist. The combination is exposed to light of such a strength as to induce methacrylate polymer bond breakage producing a somewhat degraded methacrylate product that is more soluble in organic liquids, allowing the preferential removal of the exposed photoresist. These methacrylate polymers are especially designed to allow both easy degradation and subsequent easy removal. One of these is a copolymer with adamantane and lactone-containing units (6.63).

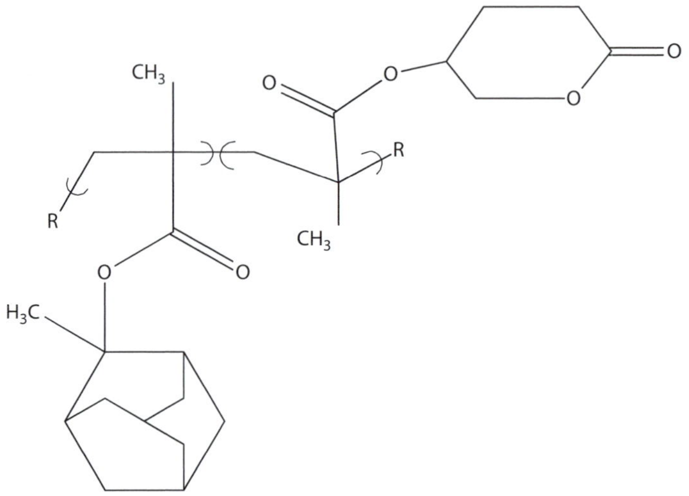

(6.63)

Modified methacrylate copolymer

This type of photoresist is called a "positive photoresist" since it is the exposed area that is removed.

Negative photoresists are formed from polymers that undergo reactions that decrease their solubility when exposed to radiation. Thus, polymers such as *cis*-1,4-polyisoprene (6.64) are used that crosslink when exposed to the appropriate radiation giving insoluble products.

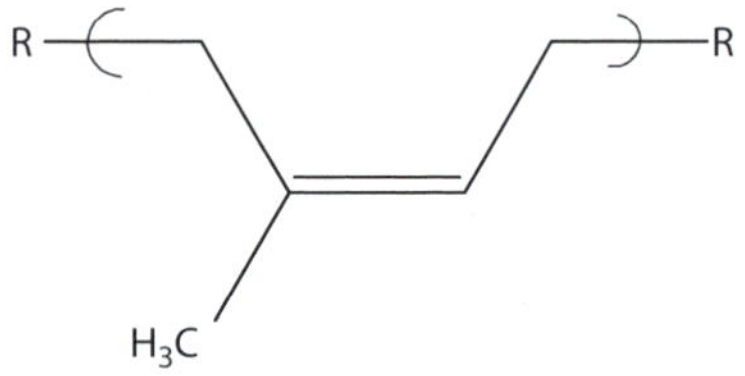

(6.64)

Since the 1950s, synthetic polymers have been used as art binders. The most common groups are referred to as the vinyls and acrylics by artists. Both groupings represent a wide variety of polymers and copolymers and are inaccurately, or at best not accurately, named. For polymer scientists, the term vinyl generally refers to PVC, but to artists it may refer to many other materials. Even so, most of the synthetic paint market today is based on acrylics. Acrylic paints are typically water emulsions of synthetic polymers. In art, the term acrylic is used to describe a wide variety of polymers and copolymers that can be considered as derivatives of acrylic acids. Most acrylics used in art binders are poly(methyl methacrylate) (Equation 6.62), PVAc (Equation 6.65), poly(*n*-butyl methacrylate) (6.66), and copolymers such as poly(ethyl acrylate-co-methacrylate).

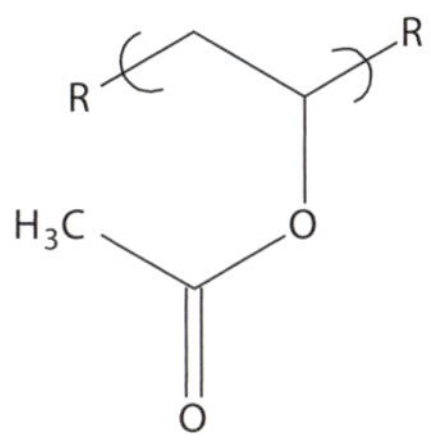

(6.65) 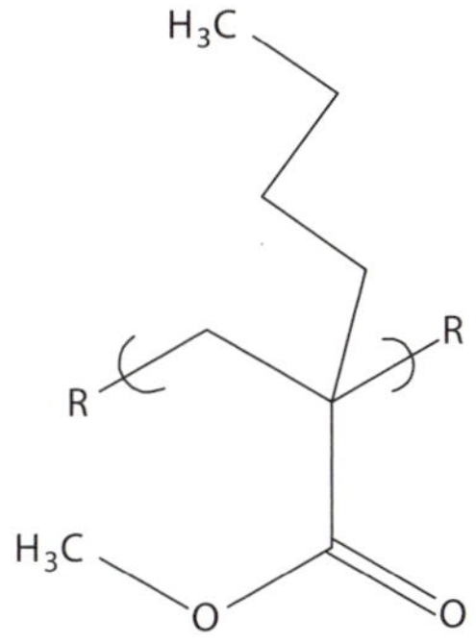   (6.66)

Acrylic emulsions tend to flow nicely "leveling out" rather than giving a three-dimensional effect sometimes offered in various oils. In the absence of pigment, acrylic emulsions give a milky white appearance. As the water evaporates, the binder particles coalesce forming a tight film. When dried, the film is clear and becomes water insoluble.

As noted above, acrylics are colorless when hardened so that the color comes from the addition of the coloring agents, pigments. When thick, acrylics form plastics, but when applied to give thin coatings, they give flexible films. Unlike many of the natural coatings, such as oil-based paints that crack and are not flexible, acrylic coatings are more flexible and do not crack as easily. Counter, the permanence of the acrylic coatings has not been field tested since they are only about half a century old, whereas oil-based paintings have remained for almost half a millennium. Even so, many artists claim that acrylic paints are more permanent than the natural-based paints. Only the test of time will truly give us the answer.

Acrylic paints have compositions similar to water-based house paints. Since they are synthetic, manufactures can design paints that fulfill specific requirements and that are the same from tube to tube. Because water is the main vehicle, such paints are largely odorless. They are not completely odorless because while the overwhelming majority of the vehicle is water, there are often minute amounts, generally 1%–10%, of organic liquid added to help the particles of polymer remain sufficiently flexible so that they will "lay down" a thin smooth coating.

Unlike oils that take days and weeks to dry "to the touch," acrylics dry in minutes to several hours. They are also considered nonyellowing.

Acrylics are also the most used paint for baked polymer clay.

6.9   POLY(VINYL ALCOHOL) AND POLY(VINYL ACETATE)

Poly(vinyl alcohol) is produced by the hydrolysis of PVAc (Equation 6.44). Because of the ability to hydrogen bond and small size of the hydroxyl grouping, PVA is a crystalline atactic material. Because of the formation of strong internal hydrogen bonds, completely hydrolyzed PVAc is not water soluble. Thus, hydrolysis of PVAc is stopped, giving a material that is about 88% hydrolyzed and water soluble. PVA is resistant to most organic solvents such as gasoline. PVA fibers (Kuralon) are strong and insoluble in water because of a surface treatment with formaldehyde that reacts with the surface hydroxyl groups, producing poly(vinyl formal) on the polymer surface.

Poly(vinyl alcohol) is used in the treatment of textiles and paper. PVA also acts as the starting material for the synthesis of a number of poly(vinyl acetals) with the general structure 6.67. The acetal rings on these random amorphous polymer chains restrict flexibility and increase the heat deflection temperature to a value higher than that of PVAc.

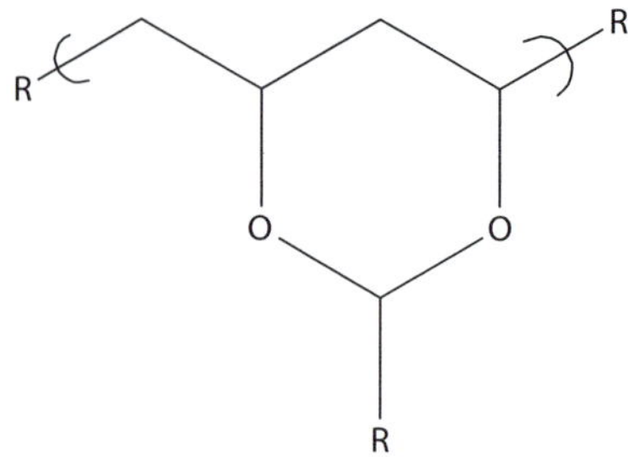

$$(6.67)$$

The deflection temperature of poly(vinyl formal) is about 90°C because of the presence of residual hydroxyl groups, commercial poly(vinyl formal) has a water absorption of about 1%. Poly(vinyl formal) (6.68) has a T_g of about 105°C and it is soluble in moderately polar solvents such as acetone.

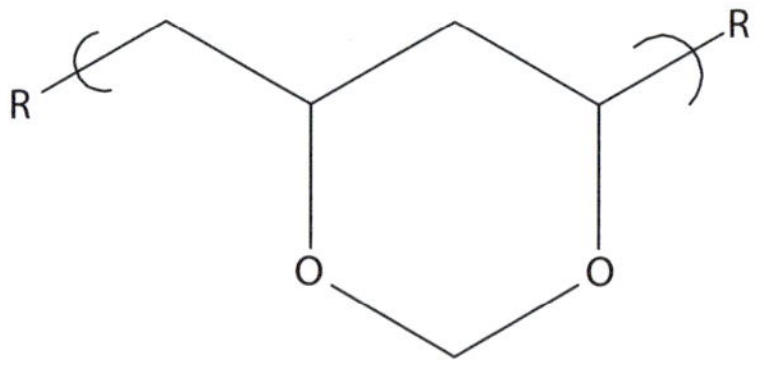

$$(6.68)$$

The most widely used poly(vinyl acetal) is poly(vinyl butyral) discussed in Section 6.4. As already noted, this plastic is used as the inner lining of safety windshield glass. Because of the presence of unreacted hydroxyl groups, poly(vinyl butyral) has excellent adhesion to glass.

6.10   POLYACRYLONITRILE

Polyacrylonitrile (PAN) (6.69), forms the basis for a number of fibers and copolymers. As fibers, they are referred to as acrylics or acrylic fibers. The development of acrylic fibers began in the early 1930s in Germany but they were first commercially produced in the United States by DuPont (Orlon) and Monsanto (Acrilan), about 1950.

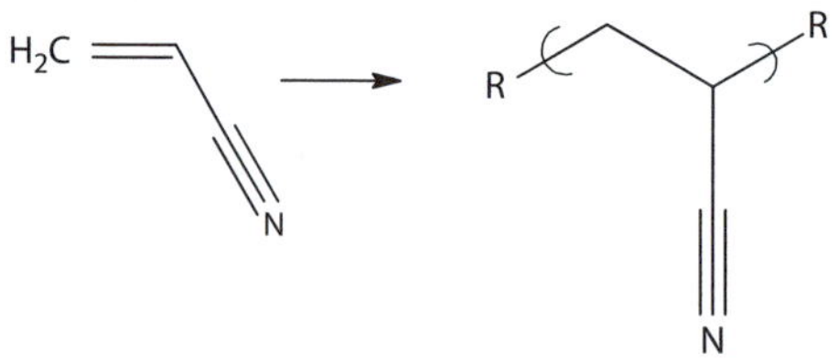

$$(6.69)$$

Because of the repulsion of the cyanide groups, the polymer backbone assumes a rod-like conformation. The fibers derive their basic properties from this stiff structure of PAN where the nitrile groups are randomly distributed about the backbone rod. Because of strong hydrogen bonding between the chains, they tend to form bundles. Most acrylic fibers actually contain small amounts of other monomers, such as methyl acrylate and methyl methacrylate. Because they are difficult to dye, small amounts of ionic monomers, such as sodium styrene sulfonate, are often added to improve their dyeability. Other monomers are also employed to improve dyeability. These include small amounts (about 4%) of more hydrophilic monomers such as *N*-vinyl-2-pyrrolidone (6.70), methacrylic acid, or 2-vinylpyridine (6.71).

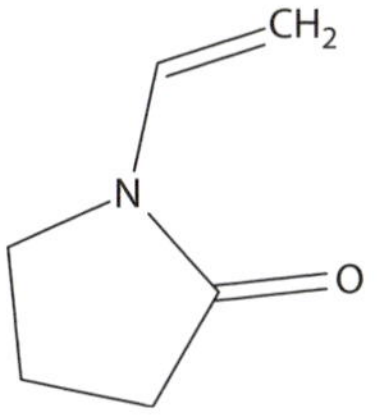

(6.70) 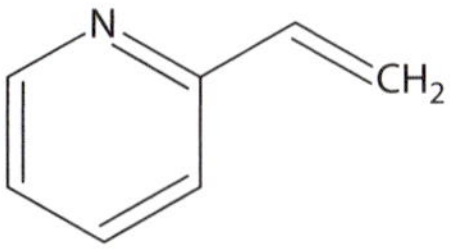   (6.71)

Acrylic fibers are used as an alternative to wool for sweaters. PAN is also used in the production of blouses, blankets, rugs, curtains, shirts, craft yarns, and pile fabrics used to simulate fur.

At temperatures above 160°C, PAN begins forming cyclic imines that dehydrogenate forming dark-colored heat-resistant fused ring polymers with conjugated C=C and C=N bonding. This is described in Section 6.4.

Fibers with more than 85% acrylonitrile units are called *acrylic fibers* but those containing 35%–85% acrylonitrile units are referred to as modacrylic fibers. The remainder of the modacrylic fibers are derived from comonomers such as vinyl chloride or vinylidene chloride, which are specifically added to improve flame resistance.

6.11 SOLID-STATE IRRADIATION POLYMERIZATION

There are numerous examples of solid-state polymerizations. Here we will briefly describe examples based on addition polymers. Generally, the crystalline monomer is irradiated with electrons or some form of high-energy radiation such as gamma or X-rays. Since many monomers are solids only below room temperature, it is customary to begin irradiation at lower temperatures with the temperature only raised after the initial polymerization occurs. (Some reactions are carried to completion at the lower temperature.) After polymerization, monomer is removed. Table 6.11 contains a list of some of the common monomers that undergo solid-state irradiation polymerization.

This approach can offer several advantages. First, polymers can be formed from monomers that do not give polymer under more typical reaction conditions. Second, under some cases, the crystalline structure acts as a template giving order that might be difficult to otherwise achieve. Third, removal and interference by solvent or additives is eliminated since they are not present. Fourth, the polymers produced by this technique are often different from those from the same monomer except that they are produced using typical reaction techniques.

TABLE 6.11
Monomers that Undergo Solid-State Irradiation Polymerization

Styrene	Formaldehyde
acetaldehyde	Acrylic acid (and salts)
Methacrylic acid (and salts)	Trioxane
1,3-Butadiene	3,3-Bischloromethylcyclooxabutane
Isoprene	Acrylonitrile
Acrylamide	Beta-Propiolactone
Diacetylenes	

6.12 PLASMA POLYMERIZATIONS

Organic and inorganic molecules can be placed in the vapor state either through heating, low pressure, simply spraying, or some combination of these. These molecules are then subjected to some ionizing energy that forms active species that react with one another, eventually depositing themselves on a surface. Often the products are polymeric with complex structures. The term plasma polymerization is generally used to describe the process resulting in surface film formation while the term deposition is generally used to describe the deposition of powdery particles formed in the gas phase. Others describe plasma polymerization as that polymerization that occurs at high rates in the gas phase, resulting in powder formation and deposition as any sorption occurring on the surface. In truth, it is difficult to separate the two reaction sequences because active molecules can react both in the gaseous phase upon collision and on the surface.

Plasma environments are often created using plasma jets, ion beams, glow discharges, corona discharges, laser induced plasmas, and electron beams. Low-temperature plasmas can also be created using radio frequency, audio frequency, microwave, or direct-current energy sources. In general terms, the molecules enter the reactor as neutral species. They become reactive species as electronic energy is transferred to them. The reactive species can be ions, free radicals, or excited molecules. Reaction can occur in the gaseous phase and/or at the solid surface. Commercially, reactors often consist of a low pressure glow discharge of reactive species. Because a small amount of electromagnetic radiation is emitted in the visible region, the term glow discharge was derived.

This approach allows the deposition of thin films at low temperatures. By comparison, polymer deposition generally requires very high temperatures. For instance, the chemical vapor deposition of silicon nitride requires a temperature of about 900°C whereas the plasma chemical deposition requires a temperature of only 350°C.

A number of typical polymer-forming monomers have been polymerized using plasma polymerization, including tetrafluoroethylene, styrene, acrylic acid, methyl methacrylate, isoprene, and ethylene. Polymerization of many nontypical monomers has also occurred, including toluene, benzene, and simple hydrocarbons.

Plasma films are usually highly cross-linked, resistant to higher temperatures, resistance to abrasion and chemical attack, and adhesion to the surface is high. Adhesion to the surface is generally high both because the growing polymer complex can fit the surface contour and thus "lock-itself in" (physical adhesion) and because in many instances, the species are active enough to chemically react with the surface molecules to chemically bond to the surface. The surface can be prepared so that the chemical reaction is enhanced.

Plasma surface treatment of many polymers, including fabrics, plastics, and composites, often occurs. The production of ultrathin films via plasma deposition is important in microelectronics, biomaterials, corrosion protection, permeation control, and for adhesion control. Plasma coatings are often on the order of 1–100 nm thick.

6.13 SUMMARY

1. For classical free radical polymerizations the rate of propagation is proportional to the concentration of monomer and the square root of the initiator concentration. Termination usually occurs through a coupling of two live radical chains but can occur through disproportionation. The rate of termination for coupling is directly proportional to initiator concentration. The $\overline{DP}$ is directly proportional to monomer concentration and inversely proportional to the square root of the initiator concentration.

2. The first chains produced are high molecular weight products. Within the polymerizing system, the most abundant species are the monomer and polymers chains.

3. Increasing the temperature increases the concentration of free radicals, thus decreasing the chain length. Increasing the temperature increases the rate of polymer formation.

4. The rate-controlling step is the rate of initiation.
5. There is a steady-state concentration of growing chains.
6. The kinetic chain length, v, is equal to $\overline{DP}$ for disproportionation termination, but $\overline{DP} = 2v$ for termination by coupling.
7. Chain-transfer reactions almost always decrease $\overline{DP}$ and will often introduce branching as it occurs within or between polymer chains. Solvent also can act as effective chain-transfer agents lowering chain length.
8. Vinyl monomers can be polymerized using solution, bulk, suspension, and emulsion techniques. Each have their own characteristic strengths and weaknesses.
9. When polymerizations become viscous, termination slows, allowing an increase in the number of growing chains and rate of polymerization. This is known as the gel of Trommsdorff effect. If such reactions are allowed to continue without cooling, explosions are possible.
10. Monomers may be polymerized using a water-soluble initiator while dispersed, by agitation, in a concentrated soap solution. In this emulsion system, initiation occurs in the aqueous phase and propagation occurs in the soap micelles. Since the growing macroradicals are not terminated until a new free radical enters the micelle, high molecular weight products are rapidly obtained. The rate of polymerization and $\overline{DP}$ are proportional to the number of activated micelles.
11. Polyfluorocarbons are resistant to heat, solvent, and corrosives. The resistance is greatest in the regularly structured PTFE and decreases as the geometry is upset by substitution of other atoms for fluorine.
12. PVC is an important polymer used for many commercial uses, including as pipes for water-delivery systems within homes.

GLOSSARY

Backbiting: Hydrogen atom abstraction that occurs when a chain end of a macroradical doubles back on itself forming a more stable hexagonal conformation.

Branch point: Point on a polymer chain where additional chain extension occurs producing a branch.

Bulk polymerization: Polymerization of monomer without added solvents or water; also called neat.

Ceiling temperature (T_c): Characteristic temperature above which polymerization occurs but the polymer decomposes before it is recovered.

Chain stopper: Chain-transfer agent that produces inactive free radicals.

Chain transfer: Process in which a free radical abstracts an atom or group of atoms from a solvent, telogen, or polymer.

Chain-transfer constant (C_s): Ratio of cessation or termination of transfer to the rate of propagation.

Critical micelle concentration: Minimum concentration of soap in water that will produce micelles.

Dead polymer: Polymer in which chain growth has been terminated.

Disproportionation: Process by which termination occurs as a result of chain transfer between two macroradicals yielding dead polymers.

Half-life time: Time required for half the reactants to be consumed (generally for a first-order reaction).

Heterolytic cleavage: Cleavage of a covalent bond that leaves one electron with each of the two atoms. The products are free radicals.

Homopolymer: Polymer composed of only one repeating unit.

Kel F: Trade name for polytrifluorochloroethylene.

Kinetic chain length: Length of the polymer chain initiated by one free radical.

Macroradicals: Electron-deficient polymers having a free radical present on the chain.

Micelles: Ordered groups of soap molecules in water.

Oligoradical: Low molecular weight macroradical.

Piezoelectric: Conversion of mechanical force, such as pressure, into electrical energy.

Plasticizer: High-boiling compatible liquid that lowers T_g and flexibilizes the polymer.

Retarder: Additive that acts as a chain-transfer agent producing less active free radicals.

Saran: Trade name for copolymers of vinyl chloride and vinylidene chloride.

SBR: Rubbery copolymer of styrene and butadiene.

Suspension polymerization: Process in which liquid monomers are polymerized in liquid droplets suspended in water.

Telogen: Additive that readily undergoes chain transfer with a macroradical.

Telomer: Low molecular weight polymer resulting form chain transfer of a macroradical with a telogen.

Telomerization: Process in which telomers are produced by chain-transfer reactions.

Trommsdorff effect: Decrease in termination rate in viscous media that results in higher molecular weight polymers being formed.

EXERCISES

1. Use a slanted line to show the cleavage of (a) boron trifluoride-water, (b) sodamide, and AIBN in cationic, anionic, and free radical initiations, respectively.
2. Which type of chain-reaction polymerization is most likely to terminate by coupling?
3. If an initiator has a half-life of 4 h, what percentage of this initiator will remain after 12 hours?
4. If some heat-to-head configuration is detected in a polymer chain know to propagate by head-to-tail addition, what type of termination has occurred?
5. Which is the better way to increase polymer production rates: (a) increasing the temperature or (b) increasing the initiator concentration?
6. Name three widely used thermoplastics produced by free radical chain polymerization.
7. What effect does the increase of polarity of the solvent have on free radical polymerization?
8. Show the repeat units for (a) PS, (b) PVC, and (c) PMMA.
9. Can you think of any advantage of the Trommsdorff effect?
10. What is the limiting step in free radical chain polymerization?
11. In general, which is more rapid: (a) free radical chain reactions or (b) step reaction polymerizations?
12. If one obtained a yield of 10% polymer after 10 min of polymerizing styrene by a free-radical mechanism, what would be the composition of the other 90%.
13. Why is $t_{1/2}$ for all first-order reactions equal to $0.693/k_d$?
14. How could you follow the rate of decomposition of AIBN without directly measuring the rate of polymerization?
15. What is the usual value for the energy of activation of free radical initiation?
16. What is the advantage of producing free radicals by UV radiation?
17. Why is PVC considered environmentally negative by some?
18. If [M·] is equal to 1×10^{-11} mol/L under steady-state conditions, what will [M] equal after (a) 30, (b) 60, and (c) 90 min?
19. In general, what is the activation energy in free radical chain propagation of polymer chains?
20. What is the relationship between the rate of propagation and the concentration of initiators [I]?
21. When chain transfer with solvent occurs, what effect does this have on the average degree of polymerization?
22. In the free radical polymerization name two steady-state assumptions.

23. What monomer is used to produce PVA?
24. Name one advantage and one disadvantage for the bulk-batch polymerization of PS.
25. Does k_p increase or decrease when the average DP for a single growing chain goes from 10 to 10^4?
26. Why is ethylene more readily polymerized by free radical chain polymerization than isobutylene?
27. What is the termination mechanism in free radical polymerization is the average $\overline{DP} = v$?
28. The value of v increases as the polymerization temperature of a specific monomer is increased. What does this tell you about the termination process?
29. In general, what is the activation energy of termination?
30. Why would not you recommend the use of poly-α-methylstyrene for the handle of a cooking utensil?
31. Why is polyfluoroethylene generally known by the public as Teflon?
32. Which would you expect to have the higher chain-transfer constant: (a) carbon tetrafluoride or (b) carbon tetrachloride?
33. While the addition of dodecyl mercaptan to styrene causes a reduction in $\overline{DP}$, the rate of polymerization is essentially unchanged. Explain.
34. Would it be safe to polymerize styrene by bulk polymerization in a 55-gal drum?
35. How does the kinetics of polymerization differ in the bulk and suspension polymerization methods?
36. Since the monomers are carcinogenic, should the polymerization of styrene, acrylonitrile, and vinyl chloride be banned?
37. What are some unusual properties of polyfluoroethylene?
38. Why does not polymerization take place in the droplets instead of in the micelles in emulsion polymerization?
39. Why doesn't initiation occur in the micelles in emulsion polymerization?
40. What would happen if one added a small amount of an inhibitor to styrene before bulk polymerization?
41. Name several places you might "run across" PS daily.
42. Why is the T_g of PTFE higher than that of FEP?
43. Why does an increase in soap concentration increase the average DP and R_p in emulsion polymerizations?
44. Which will have the higher specific gravity (density): (a) PVC or (b) PVDC?
45. By bulk, why are vinyl-derived polymers used in greater amounts than synthetic condensation polymers?
46. How might the kinetics change in an initiator if an E_a of 30 kJ/mol were employed?
47. Why were free radical polymerizations studied so well?
48. What is an engineering plastic and is Teflon an engineering plastic?
49. Describe how a plastisol is made.
50. How is PAN made so it can be colored?
51. Compare the properties of PMMA and glass.
52. Why is there a concern over the use of certain plasticizers?

ADDITIONAL READING

Bamford, C. H., Barb, W. G., Jenkins, A. D., Onyon, P. E. (1958): *Kinetics of Vinyl Polymerization by Radical Mechanism*, Butterworths, London.

Bhowmick, A., Stephens, H. (2000): *Handbook of Elastomers*, Dekker, NY.

Buback, M., van Herk, A. (2007): *Radical Polymerization: Kinetics and Mechanism*, Wiley, Hoboken, NJ.

Cazacu, M. (2008): *Advances in Functional Heterochain Polymers*, Nova, Hauppauge, NY.

Chern, C. (2008): *Principles and Applications of Emulsion Polymerization*, Wiley, Hoboken, NJ.

Dubois, P. (2000): *Advances in Ring Opening Polymerizations*, Wiley, NY.

Gaylord, N. G., Mark, H. F. (1958): *Linear and Stereoaddition Polymers*, Interscience, NY.

Harkins, W. D. (1947, 1950): Emulsion polymerization, *J. Amer. Chem. Soc.*, 69:1429; *J. Polymer Sci.*, 5:217.

Kissin, Y., Centi, G. (2008): *Alkene Polymerization Reactions with Transition Metal Catalysts*, Elsevier, NY.

Korolyov, G., Mogilevich, M. (2009): *Three-Dimensional Free-Radical Polymerization, Process, and Technology*, Springer, NY.

Mayo, F. R. (1943): Chain transfer, *J. Amer. Chem. Soc.*, 65:2324.

Mishra, M., Yagic, Y. (2008): *Handbook of Vinyl Polymers: Radical Polymerization, Process, and Technology*, CRC, Boca Raton, FL.

Niyazi, F. (2007): *Photochemical Reactions in Heterochain Polymers*, Nova, Hauppauge, NY.

Papaspyrides, C., Vouviouka, S. (2009): *Solid-State Polymerization*, Wiley, Hoboken, NJ.

Peacock, A. (2000): *Handbook of Polyethylenes*, Dekker, NY.

Plunkett, R. J. (1941): *Polytetrafluoroethylene*, US Patent 2,230,654.

Scheirs, J., Priddy, D. (2003): *Modern Styrenic Polymers*, Wiley, Hoboken, NY.

Smith, D. (2008): *Handbook of Fluoropolymer Science and Technology*, Wiley, Hoboken, NJ.

Smith, W. V., Ewart, R. H. (1948): Emulsion polymerization, *J. Chem. Phys.*, 16:592.

Trommsdorff, E., Kohle, H., Langally, P. (1948): Viscous polymerization, *Makromol. Chem.*, 1:169.

Vasile, C. (2000): *Handbook of Polyolefins*, Dekker, NY.

Walling, C. (1957): *Free Radicals in Solution*, Wiley, NY.

Scheirs, J. (2003): *Modern Styrenic Polymers*, Wiley, Hoboken.

7 Copolymerization

While the mechanism of copolymerization is similar to that discussed for the polymerization of one reactant (homopolymerization), the reactivities of monomers may differ when more than one is present in the feed, that is, reaction mixture. Copolymers may be produced by step reaction or by chain-reaction polymerization. It is important to note that if the reactant species are M_1 and M_2 then the composition of the copolymer is not a physical mixture or blend, though the topic of blends will be dealt within this chapter.

Many naturally occurring polymers are largely homopolymers, but proteins and nucleic acids are copolymers composed of a number of different mers. While many synthetic polymers are homopolymers, the most widely used synthetic rubber (SBR) is a copolymer of styrene (S) and butadiene (B), with the "R" representing "rubber." There are many other important copolymers. Here we will restrict ourselves to vinyl-derived copolymers.

Copolymers may be *random* in the placement of units,

$$-M_1M_2M_1M_2M_2M_1M_1M_2M_1M_1M_2M_1M_2M_2M_1M_1M_2M_2M_2M_1M_1M_2M_1M_1- \qquad (7.1)$$

alternating in the placement of units,

$$-M_1M_2M_1M_2M_1M_2M_1M_2M_1M_2M_1M_2M_1M_2M_1M_2M_1M_2M_1M_2M_1M_2- \qquad (7.2)$$

block in either or both of the different units in which there are long sequences of the same repeating unit in the chain,

$$-M_1M_1M_1M_1M_1M_1M_1M_2M_2M_2M_2M_2M_2M_2M_2M_2M_1M_1M_1M_1M_1M_1M_1M_1M_1M_2M_2M_2M_2- \qquad (7.3)$$

or *graft* copolymers in which the chain extension of the second monomer is present as branches.

$$-M_1M_1M_1M_1M_1M_1M_1M_1M_1M_1M_1M_1M_1M_1M_1M_1M_1M_1M_1M_1- \qquad (7.4)$$

M_2	M_2	M_2
M_2	M_2	M_2
M_2	M_2	M_2
M_2	M_2	M_2
M_2	M_2	
	M_2	

Each of these types of copolymers offers different physical properties for a particular copolymer combination. It is interesting to note that block copolymers may be produced from one monomer only if the arrangement around the chiral carbon atom changes sequentially. These copolymers are called *stereoblock* copolymers.

7.1 KINETICS OF COPOLYMERIZATION

Because of a difference in the reactivity of monomers, expressed as reactivity ratios (r), the composition of the copolymer (n) may be different from that of the reactant mixture or feed (x). When x equals n, the product is said to be an *azeotropic copolymer.*

In the early 1930s, Nobel Laureate Staundinger analyzed the product obtained from the copolymerization of equimolar quantities of vinyl chloride (VC) and vinyl acetate (VAc). He found that the first product produced was high in VC, but as the composition of the reactant mixture changed because of a preferential depletion of VC, the product was becoming higher in VAc. This phenomenon is called the *composition drift*.

Wall studied the composition drift and derived what is now called the *Wall equation,* where n was equal to rx when the reactivity ratio r was equal to the ratio of the propagation rate constants. Thus, r was the slope of the line obtained when the ratio of monomers in the copolymer (M_1/M_2) was plotted against the ratio of monomers in the feed (m_1/m_2). The Wall equation is not general.

$$n = \frac{M_1}{M_2} = r\left(\frac{m_1}{m_2}\right) = rx$$

(7.5)

The copolymer equation that is now accepted was developed in the late 1930s by a group of investigators, including Wall, Dostal, Lewis, Alfrey, Simha, and Mayo. These workers considered the four possible extension reactions when monomers M_1 and M_2 were present in the feed. As shown below, two of these reactions are homopolymerizations or self-propagating steps (Equations 7.6 and 7.8), and the other two are heteropolymerizations or cross-propagating steps (Equations 7.7 and 7.9). The ratio of the propagating rate constants are expressed as *monomer reactivity ratios* (or simply reactivity ratios), where $r_1 = k_{11}/k_{12}$ and $r_2 = k_{22}/k_{21}$. $M_1^\bullet$ and $M_2^\bullet$ are used as symbols for the macroradicals with M_1 and M_2 terminal groups, respectively.

Reaction	Rate constant	Rate expression	
$M_1^\bullet + M_1 \rightarrow M_1 M_1^\bullet$	k_{11}	$R_{11} = k_{11}[M_1^\bullet][M_1]$	(7.6)
$M_1^\bullet + M_2 \rightarrow M_1 M_2^\bullet$	k_{12}	$R_{12} = k_{12}[M_1^\bullet][M_2]$	(7.7)
$M_2^\bullet + M_2 \rightarrow M_2 M_2^\bullet$	k_{22}	$R_{22} = k_{22}[M_2^\bullet][M_2]$	(7.8)
$M_2^\bullet + M_1 \rightarrow M_2 M_1^\bullet$	k_{21}	$R_{21} = k_{21}[M_2^\bullet][M_1]$	(7.9)

Experimentally, it is found that the specific rate constants for the various reaction steps described above are essentially independent of chain length, with the rate of monomer addition primarily dependent only on the adding monomer unit and the growing end. Thus, the four reactions between two comonomers can be described using only these four equations.

As is the case with the other chain processes, determining the concentration of the active species is difficult so that expressions that do not contain the concentration of the active species are derived. The change in monomer concentration, that is the rate of addition of monomer to growing copolymer chains, is described as follows:

$$\text{Disappearance of } M_1: -\frac{d[M_1]}{dt} = k_{11}[M_1^\bullet][M_1] + k_{21}[M_2^\bullet][M_1]$$

(7.10)

$$\text{Disappearance of } M_2: -\frac{d[M_2]}{dt} = k_{22}[M_2^\bullet][M_2] + k_{12}[M_1^\bullet][M_2]$$

(7.11)

Since it is experimentally observed that the number of growing chains remains approximately constant throughout the duration of most copolymerizations (i.e., a steady state in the number of growing chains), the concentrations of $M_1^\bullet$ and $M_2^\bullet$ is constant, and the rate of conversion of $M_1^\bullet$ to $M_2^\bullet$ is equal to the conversion of $M_2^\bullet$ to $M_1^\bullet$; that is, $k_{12}[M_1^\bullet][M_2] = k_{21}[M_2^\bullet][M_1]$. Solving for $M_1^\bullet$ gives

$$[M_1^{\bullet}] = \frac{k_{21}[M_2^{\bullet}][M_1]}{k_{12}[M_2]} \tag{7.12}$$

The ratio of disappearance of monomers M_1/M_2 is described by Equation 7.13 from Equations 7.10 and 7.11. Remember, that this is also the average composition of the growing chains and the resulting polymer.

$$\frac{d[M_1]}{d[M_2]} = \frac{k_{11}[M_1^{\bullet}][M_1] + k_{21}[M_2^{\bullet}][M_1]}{k_{22}[M_2^{\bullet}][M_2] + k_{12}[M_1^{\bullet}][M_2]} = \frac{[M_1](k_{11}[M_1^{\bullet}] + k_{21}[M_2^{\bullet}])}{[M_2^{\bullet}](k_{22}[M_2^{\bullet}] + k_{12}[M_1^{\bullet}])} \tag{7.13}$$

Substitution of $[M_1^{\bullet}]$ into Equation 7.13 gives

$$\frac{d[M_1]}{d[M_2]} = \frac{[M_1]}{[M_2]} \left\{ \frac{(k_{11}k_{21}[M_2^{\bullet}][M_1]/k_{21}[M_2]) + k_{21}[M_2^{\bullet}]}{(k_{12}k_{21}[M_2^{\bullet}][M_1]/k_{21}[M_2]) + k_{22}[M_2^{\bullet}]} \right\} \tag{7.14}$$

Division by k_{12} and cancellation of the appropriate k's gives

$$\frac{d[M_1]}{d[M_2]} = \frac{[M_1]}{[M_2]} \left\{ \frac{(k_{11}[M_2^{\bullet}][M_1]/k_{12}[M_2]) + [M_2^{\bullet}]}{([M_2^{\bullet}][M_1]/[M_2]) + k_{22}[M_2^{\bullet}]/k_{21}} \right\} \tag{7.15}$$

Substitution of $r_1 = k_{11}/k_{12}$ and $r_2 = k_{22}/k_{21}$ and cancellation of $[M_2^{\bullet}]$ gives

$$\frac{d[M_1]}{d[M_2]} = \frac{[M_1]}{[M_2]} \left\{ \frac{(r_1[M_1]/[M_2]) + 1}{([M_1]/[M_2]) + r_2} \right\} \tag{7.16}$$

Multiplication by $[M_2]$ gives what are generally referred to as the "copolymerization equations," Equations 7.17 and 7.19, which gives the copolymer composition without the need to know any free radical concentration, and which gives the composition of the growing polymer as a function of monomer feed (Equation 7.19).

$$n = \frac{d[M_1]}{d[M_2]} = \frac{[M_1](r_1[M_1] + [M_2])}{[M_2]([M_1] + r_2[M_2])} \tag{7.17}$$

This equation, (Equation 7.17), is also presented in another form that allows greater ease of seeing the relationship between the monomer feed, x, and copolymer composition. This is achieved by the following steps.

Multiplying through by $[M_1]$ and $[M_2^{\bullet}]$ gives

$$n = \frac{d[M_1]}{d[M_2]} = \frac{[M_1]r_1[M_1] + [M_1][M_2]}{[M_2][M_1] + [M_2]r_2[M_2]} \tag{7.18}$$

Then division of both the top and bottom by $[M_1][M_2]$ gives the second form, Equation 7.19, of the copolymerization equation but in terms of the composition of the feed (x) on the composition of the copolymer (n) as shown below:

$$n = \frac{d[M_1]}{d[M_2]} = \frac{r_1([M_1]/[M_2]) + 1}{r_2([M_2]/[M_1]) + 1} = \frac{r_1 x + 1}{r_2/x + 1} \tag{7.19}$$

The reactivity ratios are determined by an analysis of the change in the composition of the feed during the early stages of polymerization. Typical free radical chain copolymerization reactivity ratios are given in Table 7.1. The copolymer composition and type can be predicted by looking at the values of r_1 and r_2. If the value of r_1 is greater than 1, then M_1 tends to react with itself producing homopolymers, or block copolymers in M_1. When r_1 is greater than 1 and r_2 less than 1, then a copolymer is formed with blocks in M_1 (A; Equation 7.20).

AAAAAABAAAAAAABAAAAAABAAAAAAABAAAAAAAABABAAAAAAA   (7.20)

If both r_1 and r_2 are greater than 1, then a copolymer is formed with blocks of M_1 (A) and blocks of M_2 (B) (Equation 7.21).

AAAAAABBBBBAAAAAABBBBBBBBAAAAAABBBBBBAAAAABBBAAAA   (7.21)

The length of each block will be some statistical average with the statistical length dependent on how large the reactivity ratios are. In general, the larger the value of the reactivity ratio the larger the block will be.

Preference for reaction with the unlike monomer occurs when r_1 is less than one. When r_1 and r_2 are approximately equal to 1, the conditions are said to be ideal, with a random (not alternating) copolymer produced, in accordance with the Wall equation. Thus, a random copolymer (ideal copolymer) would be produced when chlorotrifluoroethylene is copolymerized with tetrafluoroethylene (Table 7.1).

When r_1 and r_2 are both approximately zero, as in the case with the copolymerization of maleic anhydride and styrene, an alternating copolymer is produced (Equation 7.22). In general, there will be a tendency toward alternation when the product r_1r_2 approaches zero.

ABABABABABABABABABABABABABABABABABABABABA   (7.22)

In contrast, as noted above, if the values of r_1 and r_2 are similar and the product r_1r_2 approaches 1, tendency will be to produce random copolymers (Equation 7.23).

ABABBABAABABBAABABBBABAABAABBAABABBAABABBABAAABABBA   (7.23)

In batch, reactions where the yield is high and no additional monomer is added at the end of the reaction, the average content of each monomer unit in the total copolymer formed will necessarily be directly related to the initial ratio of the two monomers. Thus, the copolymer content will vary depending on when the particular chains are formed but the final overall product will have an average of units in it reflective of the initial monomer concentrations.

The values for r_1 and r_2 are used to help guide the ratio of reactants necessary to achieve a specific copolymer composition. When a desired copolymer composition is required and either reactivity ratio is unlike and not near zero, continuous flow processes or batch processes, where monomer is continuously added, are employed. Considering the reaction between VAc and VC, r_1 is 0.23 meaning growing chains with VAc radicals at their ends have a greater tendency to react with a VC monomer than a VAc monomer. The value for r_2 is 1.68 meaning that there is a tendency for growing chains with VC radical ends to add to VC monomer rather than VAc monomer. Thus, given both monomers present in equal molar amounts, the beginning polymer formed will be rich in VC-derived units. As the amount of VC is depleted, the forming copolymer chains will have greater amounts of VAc until all of the VC is almost used up and chains with almost only VAc units will be formed. If a copolymer composition that favors a copolymer rich in VC is desired, then the concentration of VC will have to be maintained sufficient to give

TABLE 7.1
Typical Free Radical Chain Copolymerization Reactivity Ratios at 60°C

M_1	M_2	r_1	r_2	r_1r_2
Acrylamide	Acrylic acid	1.38	0.36	0.5
	Methyl acrylate	1.30	0.05	0.07
	Vinylidene chloride	4.9	0.15	0.74
Acrylic acid	Acrylonitrile (50°C)	1.15	0.35	0.40
	Styrene	0.25	0.50	0.04
	Vinyl acetate (70°C)	2	0.1	0.2
Acrylonitrile	Butadiene	0.25	0.33	0.08
	Ethyl acetate (50°C)	1.17	0.67	0.78
	Maleic anhydride	6	0	0
	Methyl methacrylate	0.13	1.16	0.15
	Styrene	0.04	0.41	0.16
	Vinyl acetate	4	0.06	0.24
	Vinyl chloride	3.3	0.02	0.07
Butadiene	Methyl methacrylate	0.70	0.32	0.22
	Styrene	1.4	0.78	1.1
Isoprene	Styrene	2	0.44	0.88
Maleic anhydride	Methyl acrylate	0	2.5	0
	Methyl methacrylate	0.03	3.5	0.11
	Styrene	0	0.02	0
	Vinyl acetate (70°C)	0.003	0.055	0.0002
Methyl acrylate	Acrylonitrile	0.67	1.3	0.84
	Styrene	0.18	0.75	0.14
	Vinyl acetate	9.0	0.10	0.90
	Vinyl chloride	5	0	0
Methyl methacrylate	Styrene	0.50	0.50	0.25
	Vinyl acetate	20	0.015	0.3
	Vinyl chloride	12.5	0	0
Styrene	p-Chlorostyrene	0.74	1.03	0.76
	p-Methoxystyrene	1.2	0.82	0.95
	Vinyl acetate	55	0.01	0.55
	Vinyl chloride	17	0.02	0.34
	2-Vinylpyridine	0.56	0.9	0.50
Vinyl acetate	Vinyl chloride	0.23	1.7	0.39
	Vinyl laurate	1.4	0.7	0.98
Vinyl chloride	Dimethyl maleate	0.77	0.009	0.007

Temperatures other than 60°C are shown in parentheses.
Source: Data from Brandrup and Immergut, 1975.

copolymers that have high VC-derived unit content. This is done by simply adding VC as the reaction proceeds.

The resonance stability of the macroradical is an important factor in polymer propagation. Thus, for free radical polymerization, a conjugated monomer such as styrene is at least 30 times as apt to form a resonance-stabilized macroradical as VAc, resulting in a copolymer being rich in styrene-derived units when these two are copolymerized.

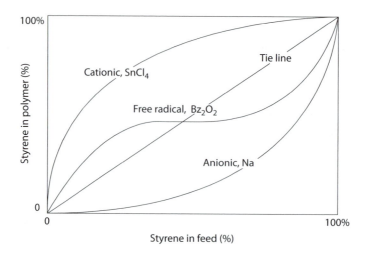

FIGURE 7.1 Instantaneous copolymer composition as a function of monomer composition and initiator employed for the comonomer system of styrene and methyl methacrylate using different modes of initiation. (Adapted from Landler, Y. (1950): *Comptes Rendus*, 230:539; with permission of Elsevier.)

Strongly electrophilic or nucleophilic monomers will polymerize exclusively by anionic or cationic mechanisms. However, monomers that are neither strongly electrophilic nor nucleophilic generally polymerize by ionic and free radical processes. The contrast between anionic, cationic, and free radical methods of addition copolymerization is clearly illustrated by the results of copolymerization utilizing the three modes of initiation (Figure 7.1). Such results illustrate the variations of reactivities and copolymer composition that are possible from employing the different initiation modes. The free radical "tieline" resides near the middle since free radical polymerizations are less dependent on the electronic nature of the comonomers relative to the ionic modes of chain propagation.

As noted before, the copolymerization can be controlled by control of the monomer feed in accordance of Equations such as 7.17 and 7.19.

7.2 Q–e SCHEME

A useful scheme for predicting r_1 and r_2 values for free radical copolymerizations was developed by Alfrey and Price in 1947. The Alfrey–Price Q–e scheme is similar to the Hammett equation approach except that it is not primarily limited to substituted aromatic compounds. In the semiempirical Q–e scheme, the reactivities or resonance effects of the monomers and macroradicals are evaluated empirically by Q and P values. The polar properties of both monomers and macroradicals are designated by arbitrary "e" values. Thus, as shown in Table 7.2, Q_1 and Q_2 are related to the reactivity, and e_1 and e_2 are related to the polarity of monomers M_1 and M_2, respectively. Styrene is assigned the Q value of 1 and an e value of –0.80. Higher Q values indicate greater resonance stability or reactivity, and higher e values (less negative) indicate greater electron-withdrawing power of the α-substituents on the vinyl monomer (in comparison to the phenyl substituent in styrene).

The Q–e scheme neglects steric factors, but it is a useful guide when data for r_1 and r_2 are not available. Following is an approach that relates the reactivity ratios to the Alfrey–Price e values.

$$k_{11} = P_1 Q_1 e^{-(e_1)^2} \tag{7.24}$$

$$k_{12} = P_1 Q_2 e^{-e_1 e_2} \tag{7.25a}$$

and,

TABLE 7.2
Typical Q and e Values for Monomers

Monomer	Q	e
Benzyl methacrylate	3.64	0.36
Methacrylic acid	2.34	0.65
2-Fluoro-1,3-butadiene	2.08	−0.43
P-Cyanostyrene	1.86	−0.21
P-Nitrostyrene	1.63	0.39
2,5-Dichlorostyrene	1.60	0.09
Methacrylamide	1.46	2.24
p-Methoxystyrene	1.36	−1.11
2-Vinylpyridine	1.30	−0.50
p-Methylstyrene	1.27	−0.98
Methacrylonitrile	1.12	0.81
p-Bromostyrene	1.04	−0.32
Styrene	1.00	−0.80
m-Methylstyrene	0.91	−0.72
Methyl methacrylate	0.74	0.40
Acrylonitrile	0.60	1.20
Methyl acrylate	0.42	0.60
Vinylidene chloride	0.23	0.36
Vinyl chloride	0.044	0.20
Vinyl acetate	0.026	−0.22

$$r_1 = \frac{k_{11}}{k_{12}} = \left[\frac{Q_1}{Q_2} \right] e^{-e_1(e_1 - e_2)}$$

(7.25b)

Similarly,

$$k_{22} = P_2 Q_2 e^{-(e_2)^2}$$

(7.26)

$$k_{21} = P_2 Q_1 e^{-e_1 e_2}$$

(7.27)

and,

$$r_1 = \frac{k_{22}}{k_{21}} = \left[\frac{Q_2}{Q_1} \right] e^{-e_2(e_2 - e_1)}$$

(7.28)

Therefore,

$$r_1 r_2 = e^{-(e_1 - e_2)^2} \quad \text{or} \quad r_1 r_2 = \exp[-(e_1 - e_2)^2]$$

(7.29)

It is important that while the reactivity is related to resonance stability of the macroradical M_1, the composition of the copolymer is related to the relative polarity of the two monomers $M_1^\bullet$ and $M_2^\bullet$

7.3 COMMERCIAL COPOLYMERS

One of the first commercial copolymers, introduced in 1928, was made of VC (87%) and VAc (13%) (Vinylite). Because the presence of the VAc mers disrupted the regular structure of PVC, the copolymer was more flexible and more soluble than PVC itself.

Copolymers of VC and vinylidene chloride were introduced in the 1930s. The copolymer with very high VC content is used as a plastic film (Pliovic), and the copolymer with high vinylidene chloride content is used as a film and filament (Saran).

Polybutadiene, produced in emulsion polymerization, is not useful as an elastomer. However, the copolymers with styrene (SBR) and acrylonitrile (Buna-N) are widely used as elastomers.

Ethylene–propylene copolymers (EPMs) show good resistance to ozone, heat, and oxygen and are used in blends to make today's external automotive panels. Two general types of ethylene–propylene copolymers are commercially available. EPMs are saturated and require vulcanization if used as a rubber. They are used in a variety of automotive applications, including as body and chassis parts, bumpers, radiator and heater hoses, seals, mats, and weather strips. EPMs are produced using Ziegler–Natta catalysts.

The second type of EPM is the ethylene–propylene–diene terpolymers (EPDMs). These are made by polymerizing ethylene, propylene, and a small amount (3–10 mol %) of nonconjugated diolefine employing Ziegler–Natta catalysts. The side chains allow vulcanization with sulfur. They are employed in the production of appliance parts, wire and cable insulation, coated fabrics, gaskets, hoses, seals, and high-impact polypropylene.

7.4 BLOCK COPOLYMERS

While block copolymers do not occur naturally, synthetic block copolymers have been prepared by all known classical polymerization techniques. The first commercial block copolymer was a surfactant (Pluronics) prepared by the addition of propylene oxide to polycarbanions of ethylene oxide. While neither water-soluble poly(ethylene oxide) nor water-insoluble poly(propylene oxide) exhibits surface activity, the ABA block copolymer consisting of hydrophilic and lyophilic segments, is an excellent surfactant. Each block has 20 plus repeat units of that variety (7.30).

$$(7.30)$$

The ethylene oxide block is hydrophilic, while the propylene oxide block is (relatively) hydrophobic. The copolymer forms micelles in aqueous solutions with the hydrophilic portions pointing outward, interacting with the water, while the hydrophobic portions form the inner core, shielded from the water by the ethylene oxide derived block. A micelle is also formed in organic liquids, but here the hydrophobic propylene oxide block "faces" outward while the ethylene oxide block acts as the inner core.

Polymers that contain two reactive end groups are referred to as *telechelic polymers*. Joseph Shivers, a DuPont chemist, invented Spandex in 1959 after about a 10-year search. It was first named Fiber K but DuPont chose the more appealing, smooth-sounding trade name of Lycra.

Elastomeric polyurethane fibers consisting of at least 85% segmented polyurethane are commercially available under the name Spandex. They are block copolymers and are among the first products specifically designed using the concept of soft and hard segments. The soft or flexible segment is composed of poly(ethylene oxide) chains that contain two hydroxyl end groups (one at each end of the poly(ethylene oxide) chain. The hard or stiff segment is derived from the reaction of the diisocyanate with the hydroxyl end groups forming polar urethane linkages that connect the poly(ethylene oxide) polymer segments through the urethane linkages. Such products are often referred to as segmented-polyurethane fibers (Equation 7.31). These products have found wide use in cloths including bras. Similar products have also been formed using hydroxyl-terminated polyethylene in place of the poly(ethylene oxide).

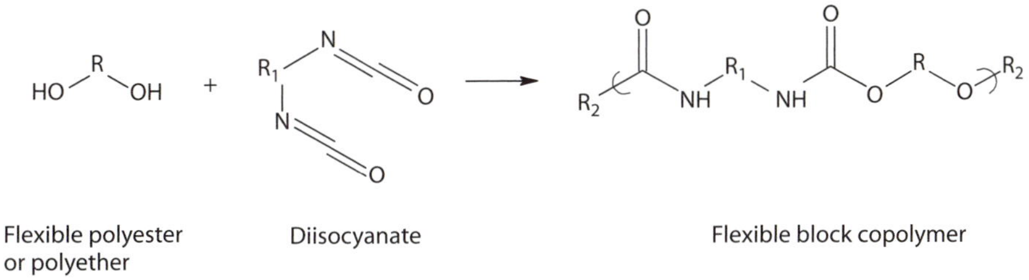

$$(7.31)$$

Flexible polyester   Diisocyanate   Flexible block copolymer
or polyether

A related product is formed from the analogous reaction using hydroxyl-terminated poly(ethylene oxide) with aromatic diacids to form a segmented aromatic polyester block copolymer that is sold under the trade name of Hytrel.

Poly(ethylene oxide) itself is made from ethylene oxide through a base catalyzed ring-opening polymerization. While the major use of poly(ethylene oxide) is in the manufacture of polyurethane products, including foams, it can be synthesized giving viscous liquids to solids depending on the molecular weight. It is used as surfactants, functional fluids, thickeners, lubricants and as plasticizers. Further, the circuit boards that hold the electronic components of computers are made from poly(ethylene oxide).

The most widely used chain-reaction block copolymers are those prepared by the addition of a new monomer to a macroanion. AB and ABA block copolymers called Soprene and Kraton, respectively, are produced by the addition of butadiene to styryl macroanions or macrocarbanions (7.32). This copolymer is normally hydrogenated (7.33).

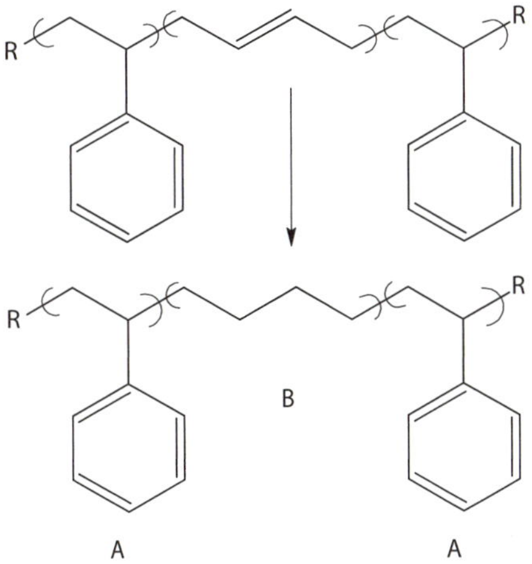

$$(7.32)$$

$$(7.33)$$

Today, new copolymers are making use of the hard-soft block strategy where the hard segment is a block portion as polyethylene that readily crystallizes forming a physical crosslink. The soft segment consists of blocks formed from α-olefins such as 1-butene, 1-hexene, and 1-octane where the substituted alkane-arm discourages crystallization.

A number of siloxane ABA block polymers have been formed. Copolymers of dimethylsiloxane (A)-diphenylsiloxane (B)-dimethylsiloxane (A) have been synthesized by the sequential polymerization of reactants. The diphenylsiloxane block acts as a hard block, and the more flexible dimethylsiloxane blocks act as soft blocks.

Hard segments have also been introduced using *p-bis*(dimethylhydroxysilyl)benzene with the soft block again being dimethylsiloxane.

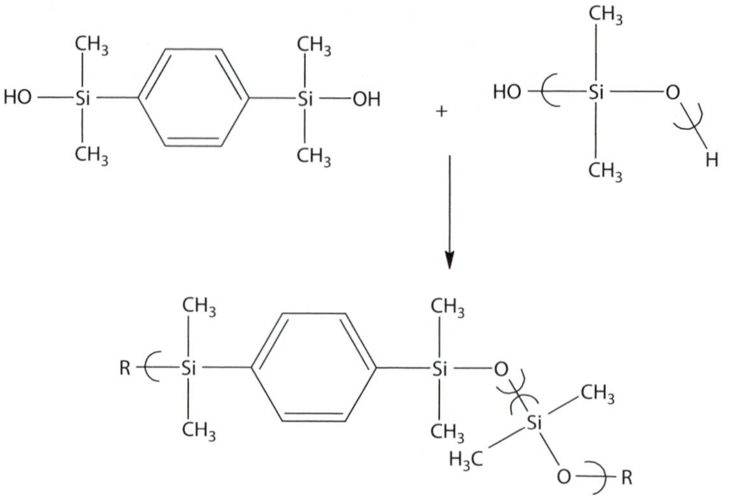

(7.34)

   Block copolymers, with segments or domains of random length, have been produced by the mechanical or ultrasonic degradation of a mixture of two or more polymers such as hevea rubber and poly(methyl methacrylate) (PMMA) (Heveaplus).

   Thus, there have been prepared a number of different block copolymers, including a number that utilize the hard-soft strategy.

7.5   GRAFT COPOLYMERS

The major difference between block and graft copolymers is the position of the second kind of unit. Thus, information that applies to block copolymers can often be applied to graft copolymers. So, domains where physical cross-linking occurs via crystallization can occur in either block components or within graft copolymers where the necessary symmetry occurs.

   Graft copolymers of nylon, protein, cellulose, or starch, or copolymers of vinyl alcohol have been prepared by reaction of ethylene oxide with these polymers. Graft copolymers are also produced when styrene is polymerized by Lewis acids in the presence of poly-p-methoxystyrene. The Merrifield synthesis of polypeptides is also based on graft copolymers formed from chloromethaylated polystyrene. Thus, the variety of graft copolymers is great.

   The most widely used graft copolymer is the styrene-unsaturated polyester copolymer (Equation 7.35). This copolymer, which is usually reinforced by fibrous glass, is prepared by the free radical chain polymerization of a styrene solution of unsaturated polyester.

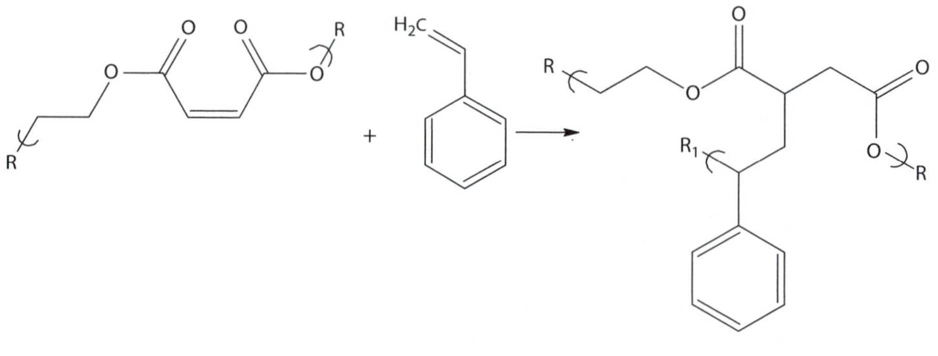

(7.35)

The graft copolymers of acrylamide, acrylic acid, and cellulose or starch are used as water absorbents and in enhanced oil recovery systems.

7.6 ELASTOMERS

Elastomers typically contain chemical and/or physical cross-links. While there are thermoplastic elastomers such as styrene–butadiene–styrene (SBS) and block copolymers, most elastomers are thermosets. Elastomers are characterized by a disorganized (high-entropy) structure in the resting or nonstressed state. Application of stress is accompanied by a ready distortion requiring (relative to plastics and fibers) little stress to effect the distortion. This distortion brings about an aligning of the chains forming a structure with greater order. The driving force for such a material to return to its original shape is largely a return to the original less-organized state. While entropy is the primary driving force for elastomers to return to the original resting state, the cross-links allow the material to return to its original shape giving the materials a type of memory. Materials that allow easy distortion generally have minimal interactions between the same or different chains. This qualification is fulfilled by materials that do not bond through the use of dipolar (or polar) or hydrogen bonding. Thus, the intermolecular and intramolecular forces of attraction are small relative to those present in fibers and plastics. Hydrocarbon-intense polymers are examples of materials that meet his qualification. Production generally requires the initial production of linear polymers, followed by the insertion of cross-links through a process called *vulcanization* or *curing*. Addition of fillers and other additives such as carbon black also is typical. Table 7.3 contains a listing of important elastomers.

The introduction of cross-links to inhibit chain slippage was discovered by Goodyear in 1839. He accomplished this through addition of sulfur to natural rubber (NR). Shortly after this, an accelerator, zinc (II) oxide, was used to speed up the process. Other additives were discovered often through observation and trial-and-error so that today's elastomers often have a number of important additives that allow them to perform demanding tasks.

Around 1915, Mote found that a superior abrasion-resistant elastomer was produced through the use of carbon black. Today, it is recognized that factors such as surface area, structure, and aggregate size are important features in the production of superior elastomers. For instance, high surface areas (small particle size) increase the reenforcement and consequently the tensile strength and improve the resistance to tearing and abrasion. Large aggregates give elastomers that have improved strength before curing, high modulus, and an improved extrusion behavior.

Rubbers typically have low hysteresis. *Hysteresis* is a measure of the energy absorbed when the elastomer is deformed. A rubber that absorbs a great amount of energy as it is deformed (such as a tire hitting bumps on the roadway) is said to have a high hysteresis. The absorbed energy is equivalent to the reciprocal of resilience such that a material with a low hysteresis has a high resilience. Rubbers with a particularly high hysteresis are used where heat buildup is desirable such as in tires to give the tread a better grip on the road and the tire a smoother ride. Thus, tires with high hysteresis are often preferred in drag-race tires where heat buildup allows better grip of the track.

The hard/soft segment scenario is utilized in the formation of a number of industrially important thermoplastic elastomers. Representative examples of such thermoplastic elastomers based on block copolymers are given in Table 7.4.

Such hard/soft scenarios can also be achieved through employing grafts (Table 7.5), where the pendant group typically acts as the hard segment with the backbone acting as the soft segment. Below, 7.36, is a representation of a typical graft copolymer chain. For an effective network to be formed each "A" chain needs to have at least two "B" grafts to allow for formation of a continuous

TABLE 7.3
Common Elastomers and Their Uses

Common Name (Chemical Composition)	Abbreviation	Uses and Properties
Acrylonitrile–butadiene–styrene (terpolymer)	ABS	Oil hoses, fuel tanks, gaskets, pipe and fittings, appliance and automotive housings. Good resistance to oils and gas
Butadiene rubber	BR	Tire tread, hose, belts. Very low hysteresis; high rebound
Butyl rubber (from isobutene and 0.5%–3% isoprene)	IIR	Innertubes, cable sheathing, tank liners, roofing, Seals, coated fabrics. Very low rebound; high hysteresis
Chloroprene rubber (polychloroprene)	CR	Wire and cable insulation, hose footwear, mechanical automotive products. Good resistance to oil and fire, good weatherability
Epichlorohydrin (epoxy copolymers)		Seals, gaskets, wire and cable insulation. Good resistance to chemicals
Ethylene–propylene rubbers (random copolymers with 60%–80% ethylene)	EP or EPM	Cable insulation, window strips. Outstanding insulative properties
Ethylene–propylene–diene (random terpolymers)	EPDM	Good resistance to weathering, resistant to ozone attack
Fluoroelastomers (fluorine-containing copolymers)		Wire and cable insulation, aerospace applications. Outstanding resistance to continuous exposure to high temperatures, chemicals, and fluids
Ionomers (largely copolymers of ethylene and acid-containing monomers with metal ions)		Golf ball covers, shoe soles, weather stripping Tough, flame-resistant, good clarity, good electrical properties, abrasion-resistant
Natural rubber (polyisoprene)	NR	General-purpose tires, bushings, and couplings, seals, footwear, belting. Good resilience
Nitrile rubber (random copolymer of butadiene and acrylonitrile)	NBR	Seals, automotive parts that are in contact with oils and gas, footwear, hose Outstanding resistance to oils and gas, little swelling in organic liquids
Polysulfide		Adhesive, sealants, hose binders Outstanding resistance to oil and organic solvents
Polyurethanes	PU	Sealing and joining, printing rollers, fibers, industrial tires, footwear, wire and cable coverings
Silicones (generally polydimethylsiloxane)		Medical applications, flexible molds, gaskets, seals. Extreme-use temperature range
Styrene–butadiene rubber (random copolymer)	SBR	Tire tread, footwear, wire and cable covering, adhesives. High hysteresis

interlinked network. While there has been a lot of research done with such graft materials they have not yet become very important commercially.

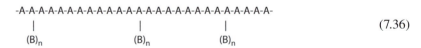

$$\text{-A-A-A-A-A-A-A-A-A-A-A-A-A-A-A-A-A-A-A-A-A-A-A-A-A-A-}$$

(7.36)

Thermoplastic elastomers can also be achieved through physical mixing of hard and soft segments (Table 7.6). These are fine dispersions of a hard thermoplastic polymer and an elastomer. The two materials generally form interdispersed cocontinuous phases. Often the physical combining is achieved through intense mechanical mixing but in some cases, such as with polyporpylene and

TABLE 7.4
Thermoplastic Elastomeric Block Copolymers

Hard Segment	Soft or Elastomeric Segment	General Type(s)
Polystyrene	Butadiene and polyisoprene*	Triblock, Branched
Polystyrene	Poly(ethylene-cobutylene)†	Triblock
Polystyrene	Polyisobutylene	Triblock, Branched
Poly(alpha-methylstyrene)	Polybutadiene	Triblock
Poly(alpha-mentylstyrene)	Polyisoprene	Triblock
Poly(alpha-mentylstyrene)	Poly(propylene sulfide)	Triblock
Polystyrene	Polydimethylsiloxane	Triblock, Multiblock
iPolypropylene	Poly(alpha-olefins)	Mixed
iPropylene	aPolypropylene	Mixed
Polyethylene	Poly(alpha-olefins)‡	Multiblock
Polyethylene	Poly(ethylene-cobutylene)	Triblock
Polyethylene	Poly(ethylene-copropylene)	Triblock
Polyurethane	Polydiacetylenes	Multiblock
Polyurethane	Polyester or Polyether§	Multiblock
Poly(methyl methacrylate)	Poly(alkyl acrylates)	Triblock, Branched
Polysulfone	Polydimethylsiloxane	Multiblock
Polyetherimide	Polydimethylsiloxane	Multiblock
Polycarbonate	Polyether	Multiblock
Polycarbonate	Polydimethylsiloxane	Multiblock
Polyamide	Polyester or Polyether**	Multiblock
Polyester	Polyether††	Multiblock

Sample trade names of block copolymers: *Kraton D, Cariflex TR, Taipol, Vector, Tufprene, Asaprene, Calprene, Europrene Sol T, Stearon, Flexprene, Quintac, Finaprene, Coperbo, Solprene, Stearon, and K-Resin. These products may be branched or linear with varying contents of the various components; †Elexar, C-Flex, Tekron, Hercuprene, Kraton G, Septon, Dynaflex, and Multi-Flex; ‡Flexomer, Exact, Affinity, and Engage; §Texin, Desmopan, Estane, Morthane, Elastollan, and Pellethane; **Montac, Pebax, Orevac, Vestamide, Grilamid, and Grilon; ††Arnitel, Hytrel, Ecdel, Lomod, Riteflex, and Urafil.

TABLE 7.5
Thermoplastic Elastomers Based on Graft Copolymers

Hard Pendant Segment	Soft Backbone
Polystyrene	Poly(butyl acrylate)
Polystyrene	Poly(ethyl-co-butyl acrylate)
Poly(4-chlorostyrene)	Polyisobutylene
Polystyrene and Poly(alpha-methylstyrene)	Polybutadiene
Polystyrene and Poly(alpha-methylstyrene)	Polyisobutylene
Polystyrene and Poly(alpha-methylstyrene)	Poly(ethylene-copropylene)
Poly(methyl methacrylate)	Poly(butyl acrylate)
Polyindene	Polyisobutylene
Polyindene	Poly(ethylene-copropylene)
Polyindene	Polybutadiene
Polyacenapthylene	Polyisobutylene

TABLE 7.6
Thermoplastic Elastomers Based on Hard-Soft Combinations

Hard Polymer	Soft/Elastomeric	Polymer Structure
Polypropylene	Ethylene–propylene Copolymer*	Blend
Polypropylene	Natural rubber†	Dynamic vulcanizate
Polypropylene	Ethylene–propylene–diene monomer‡	Dynamic vulcanizate
Polypropylene	Butyl rubber§	Dynamic vulcanizate
Polypropylene	Nitrile rubber**	Dynamic vulcanizate
Polypropylene	Poly(propylene-co-1-hexene)	Blend
Polypropylene	Poly(ethylene-covinyl acetate)	Blend
Polypropylene	Styrene–ethylene–butylene-styrene + oil	Blend
Polystyrene	Styrene–butadiene–styrene + oil	Blend
Nylon	Nitrile rubber	Dynamic vulcanizate
Poly(vinyl chloride)	Nitrile rubber + diluent††	Blend, dynamic vulcanizate
Chlorinated polyolefin	Ethylene interpolymer‡‡	Blend

Sample trade names: *,‡Flexothene, Ferroflex, Hifax, Polytrope, Ren-Flex, Telcar; †Geolast; ‡Hifax MXL, Santoprene, Sarlink 3000 and 4000, Uniprene; §Sarlink 2000, Trefsin; **Geolast, Vyram; ††Apex N, Chemigum, Sarlink 1000; ‡‡Alcryn.

EPMs the effect of blending is achieved through polymerizing the finely dispersed elastomer phase (EPM) simultaneously with the hard polypropylene.

At times, the phases are cross-linked during the mechanical mixing. This process is referred to as "dynamic vulcanization" and produces a finely dispersed discontinuous cross-linked elastomer phase. The products are referred to as thermoplastic vulcanizates or dynamic vulcanizates. The products of this process have an insoluble elastomer phase giving the material greater oil and solvent resistance. The cross-linking also reduces or eliminates the flow of this phase at high temperatures and/or under high stress. This allows the material better resistance to compression set.

Typical thermosetting elastomers are difficult to recycle because their cross-linking prevents them from being easily solubilized and reformed through application of pressure and heat. Recycling can be accomplished through the particalizing (grinding into small particles) of the elastomeric material followed by a softening-up by application of a suitable liquid and/or heat and, finally addition of a binder that physically or chemically allows the particles to bind together in the desired shape.

7.7 THERMOPLASTIC ELASTOMERS

A number of thermoplastic elastomers have been developed since the mid-1960s. The initial thermoplastic elastomers were derived from plasticized PVC and are called *plastisols*. *Plastisols* are formed from the fusing together of PVC with a compatible plasticizer through heating. The plasticizer acts to lower the T_g to below room temperature (RT). Conceptually, this can be thought of as the plasticizer acting to put additional distance between the PVC chains thus lowering the inter- and intrachain forces as well as helping solubilize chain segments. The resulting materials are used in a number of areas, including construction of boot soles.

The hard/soft segment scenario is utilized in the formation of a number of industrially important thermoplastic elastomers. Thermoplastic elastomers contain two or more distinct phases and their properties depend on these phases being intimately mixed and small. These phases may be chemically or physically connected. In order that the material be a thermoplastic elastomer, at least one phase must be soft or flexible under the operating conditions and at least one phase is hard with the hard phase(s) becoming soft (or fluid) at higher temperatures. Often the hard segments or phases are crystalline thermoplastics while the soft segments or phases are amorphous. In continuous chains containing blocks of hard and soft segments, the molecular arrangement normally contains crystalline regions where there is sufficient length in the hard segment to form the crystalline regions or phases where the soft segments form amorphous regions.

Such hard/soft scenarios can also be achieved through employing grafts where the pendant group typically acts as the hard segment with the backbone acting as the soft segment. Below, 7.37 is a representation of a typical graft copolymer chain. For an effective network to be formed, each "A" chain needs to have at least two "B" grafts to allow for formation of a continuous interlinked network. While there has been a lot of research done with such graft materials, they have not yet become very important commercially.

$$-A-A-A-A-A-A-A-A-A-A-A-A-A-A-A-A-A-A-A-A-A-A-A-A-A-A-A-$$
$$| \qquad\qquad | \qquad\qquad | \qquad\qquad\qquad (7.37)$$
$$(B)_n \qquad\qquad (B)_n \qquad\qquad (B)_n$$

Thermoplastic elastomers can also be achieved through physical mixing of hard and soft segments. These are fine dispersions of a hard thermoplastic polymer and an elastomer. The two materials generally form interdispersed cocontinuous phases. Often the physical combining is achieved through intense mechanical mixing but in some cases, such as with polypropylene and EPMs the effect of blending is achieved through polymerizing the finely dispersed elastomer phase (EPM) simultaneously with the hard polypropylene.

At times, the phases are cross-linked during the mechanical mixing. This process is referred to as "dynamic vulcanization" and produces a finely dispersed discontinuous cross-linked elastomer phase. The products are referred to as thermoplastic vulcanizates or dynamic vulcanizates. The products of this process have an insoluble elastomer phase giving the material greater oil and solvent resistance. The cross-linking also reduces or eliminates the flow of this phase at high temperatures and/or under high stress. This allows the material better resistance to compression set.

Styrene–butadiene–styrene block copolymers differ structurally from the random copolymer of styrene and butadiene (SBR). Because styrene and butadiene blocks are incompatible, they form separate phases joined at the junctions where the various blocks are connected. This gives an elastomeric material where the butadiene blocks form the soft segments and the styrene blocks form hard blocks.

The block copolymer made from connecting blocks of polystyrene with blocks of polybutadiene illustrates another use of soft and rigid or hard domains in thermoplastic elastomers. The polystyrene blocks give rigidity to the polymer while the polybutadiene blocks act as the soft or flexible portion. The polystyrene portions also form semicrystalline domains that add to the strength of the copolymer and these domains also act as "physical crosslinks" allowing the soft portions to respond in an "elastomeric" manner, while the semicrystalline domains give the material the "elastomeric" memory. The polybutadiene blocks act as the continuous phase while the polystyrene blocks act as the discontinuous phase. Heating the material above the T_m of the polystyrene domains allows whole chain mobility, allowing processability of the virgin material and subsequent reprocessability of used material. Upon cooling, the rigid domains reform. Block polystyrene–polybutadiene copolymers are used in the soles of many of the athletic shoes.

Worldwide sales of thermoplastic elastomers are on the order of one and a half million tons with a value of about $5 billion.

7.8 BLENDS

There is an ongoing search for new materials and materials that exhibit needed properties. Blends are one of the major avenues of achieving these new materials without actually synthesizing new polymers. *Polymer blends* are physical mixtures of two or more polymers though sometimes the various phases are chemically bonded together. These blended mixtures may offer distinct properties, one set of properties related to one member of the blend, and another set of properties related to the second member of the blend. The blended mixtures may also offer some averaging of properties. The property mix of polymeric blends is dependent on a number of factors, one of the major being the miscibility of the polymers in one another. This miscibility is in turn dependent on the nature of the polymers composing the blend and the amount of each component in the blend. Here polymer blends will be divided into miscible and immiscible polymer blends.

Extent of mixing is related to time since mixing requires sufficient time to allow the polymer chains to mix. Thus, for miscible blends particular structures can be "frozen-in" by rapid cooling when the desired mixing is achieved. Here, micelles of particular structures can cause the mixture to perform in one manner governed by the particular grouping that may not occur if more total mixing occurs.

Miscibility/immiscibility can be described in simple thermodynamic terms as follows, at constant temperature. Mixing occurs if the free energy of mixing is negative.

$$\Delta G_{mixing} = \Delta H_{mixing} - T\Delta S_{mixing} \tag{7.38}$$

Mixing is exactly analogous with polymer solubility. The driving force for mixing and solubility is the entropy or random-related term. The entropy-related term must overcome the opposing enthalpy energy term. In a more complete treatment, temperature and volume fraction must be considered.

7.8.1 Immiscible Blends

Immiscible combinations are all about us. Oil and water is an immiscible combination; as is the lava in the so-called lava-lamps; and chicken broth in chicken soup. *Immiscible blends* are actually a mis-naming at the molecular level since they are not truly mixed together. But at the macrolevel they appear mixed, so the name immiscible blends.

Immiscible blends are said to be phase separated, that is, there are different phases mixed together. Both phases are solid in behavior.

Because PS is brittle with little impact resistance under normal operating conditions, early work was done to impart impact resistance. The best known material from this work is called *high-impact polystyrene* or *HIPS*. HIPS is produced by dispersing small particles of butadiene rubber in with the styrene monomer. Bulk or mass polymerization of the styrene is begun producing what is referred to as prepolymerization material. During the prepolymerization stage styrene begins to polymerize with itself forming droplets of polystyrene with phase separation. When nearly equal phase volumes of polybutadiene rubber particles and polystyrene are obtained, phase inversion occurs and the droplets of polystyrene act as the continuous phase within which the butadiene rubber particles are dispersed. The completion of the polymerization generally occurs employing either bulk or aqueous suspension conditions.

Most HIPS has about 4%–12% polybutadiene in it so that HIPS is mainly a polystyrene intense material. The polymerization process is unusual in that both a matrix composition of polystyrene and polybutadienes is formed as well as a graft between the growing polystyrene onto the polybutadiene is formed. Grafting provides the needed compatibility between the matrix phase and the rubber phase. Grafting is also important in determining the structure and size of rubber particles that are formed. The grafting reaction occurs primarily by hydrogen abstraction from the polybutadiene backbone either by growing polystyrene chains or alkoxy radicals if peroxide initiators are employed.

High-impact polystyrene is an immiscible blend that is used in many applications and used to be employed as the material for many of the automotive bumpers. The polystyrene portion is strong and inflexible while the polybutadiene particles are flexible, allowing an impact to be distributed over a larger area. The polybutadiene rubbery portion allows the bumper to bend and indent and protects the PS from fracturing while the PS phase resists further deformation. This combination gives a strong flexible material.

In general, the morphology on a molecular level varies with the fraction of each component in the mixture. In general terms, we can talk about a continuous and discontinuous phase. For a combination of polymers A and B, such as PS and polybutadiene, at low amounts of A, polymer A will typically act as a discontinuous phase surrounded by B. Thus, at low amounts of PS, the PS congregates as small particles in a "sea" or continuous phase of B, the polybutadiene. As the fraction of polymer A increases, the spheres eventually become so large as to join together forming a continuous phase, and so two continuous phases are present. As the fraction of A continues to increase, polymer B becomes the discontinuous phase, being surrounded by the continues phase polymer A.

The discontinuous phase generally takes the rough shape of a sphere to minimize surface area exposure to the other phase. The size of the spheres influences the overall properties and varies with concentration. In general, because of the affinity of like polymer chains, spheres tend to grow. Larger sphere sizes are promoted because they give less relative contact area with the other phase.

As noted above, immiscible blends can exhibit different properties. If the domains are of sufficient size, they may exhibit their own T_g and T_m values. Many commercially used immiscible blends have two separate T_g and/or T_m values.

Blends can also offer variable physical properties as already noted for HIPS. Consider a blend of polymer PS and butadiene where butadiene is the major component. PS is the stronger material with the blend weaker than PS itself. In some cases, the blend can be stronger than the individual polymers. Heat and pressure can result in the change of the discontinuous phase becoming flattened out when pressed against a mold. The spheres can also be caused to elongate forming rod-like structures with the resulting structure similar to composites where the rod-like structures strengthen the overall structure.

The strength can also be increased by using about the same amounts of the two polymers so that they form two continuous phases. Here, both phases can assist the blend to be strong. Another approach is to use compatibilizers. *Compatibilizers* are materials that help bind together the phases allowing stress–strain to be shared between the two phases. Many compatibilizers are block copolymers where one block is derived from polymers of one phase and the second block composed of units derived from polymers of the second phase. The two blocks get "locked" into the structures of the like phases and thus serve to connect the two phases.

Graft copolymers are also used as compatibilizers to tie together different phases. HIPS contain polystyrene grafted onto polybutadiene backbones. This allows stress/strain to be transferred from the PS to the polybutadiene phase transferring energy that might break the brittle PS to the more flexible polybutadiene phase. That is why HIPS is stronger than PS itself.

Compatibilizers also act to modify the tendency to form large spheres. The formation of large spheres is a result of the two polymer components trying to segregate. The compatibilizer causes the two phases to come together minimizing the tendency to form large spheres. For instance, for a mixture of 20:80 PS:polybutadiene the sphere size is about 5–10 μm, whereas addition of about 9% PS-polyethylene (polyethylene is enough like polybutadiene to be incorporated into the polybutadiene phases) block copolymer results in PS spheres of about 1 micron. This increases the interface between the two phases resulting in better mechanical properties because stress–strain can be more effectively transferred from one phase to the other.

7.8.2 MISCIBLE BLENDS

Miscible blends are not as easy to achieve as immiscible blends. As noted above, entropy is the major driving force in causing materials to mix. Because polymer chains are already in a state of relatively high order, increases in randomness are not easily achieved so that immiscible blends are often more easily formed. To make matters worse, for amorphous polymers the amount of disorder in the unmixed polymer is often higher than for blends that tend to arrange the polymer chains in a more ordered fashion.

Sometimes special attractions allow polymers to mix. This is true for the mixture of PS and poly(phenylene oxide), PPO, where the interaction between the phenyl groups allow the miscibility of the two polymers. For many combinations, such preferential associations are not present.

Another approach is the use of copolymers. There are a number of variations to this. In some situations, polymer–copolymer combinations are used where the adage "the enemy of my enemy is my friend" comes into play. Thus, the random copolymer of styrene and acrylonitrile forms a miscible blend with PMMA. The copolymer is composed of nonpolar styrene units and polar acrylonitrile units that are incompatible with one another. These units will blend with PMMA to avoid one another.

Another approach is to use copolymers where the structures of the copolymer are similar to that of the other phase. This is, what occurs for PE and copolymers of ethylene and propylene.

Generally, miscible blends will have properties somewhere between those of the unblended polymers. These properties will be dependent on the ratio of the two polymers and this ratio is often used to obtain a particular property. These properties include mechanical, chemical, thermal, weathering, and so on. For instance, PPO is thermally stable with a high T_g, about 210°C. While this is good for some applications, it is considered too high for easy processing. Thus, PS, with a T_g of about 100°C, is added to allow a lower processing temperature. Noryl is PPO blended with a second polymer which is generally PS or HIPS. Noryl is used in the construction of internal appliance components, brackets and structural components in office products, large computer and printer housings, automotive wheel covers, and high-tolerance electrical switch boxes and connectors.

Industrial companies have long-term strategies. For example, Exxon (now ExxonMobile) is the third largest chemical company in the United States. Some time ago, they made the decision to emphasize the ethylene and propylene monomers that are obtained from the petrochemical interests of ExxonMobile. Thus, ExxonMobile has a research emphasis on the commercialization of products from these monomers. The major materials made from ethylene and propylene are polymeric, either homopolymers or copolymers. Efforts include developing catalysts that allow the formation of polymeric materials from the ethylene and propylene monomers and the use of these catalysts to synthesize polymeric materials that have varying properties, allowing their application in different marketplaces in society.

A driving force for conversion of gasoline to polymeric materials is increased value in the products made from the polymers. The general trail is gasoline → ethylene, propylene monomers → raw polymers and copolymers → finished products.

As noted above, one major use of HIPS was in automotive bumpers. These have been largely replaced by another blend, but here it is a miscible blend of PE and PP. HIPS bumpers have a more rubbery feel to them while the PE–PP materials are more plastic in their behavior and feel. These PE and PP intense plastics are not only made into automotive bumpers but also as side and bottom panels. There are several processes employed to produce the raw materials used in the production of these automotive parts. One process developed by Exxon begins with the production of PP using a catalyst developed by Exxon. This catalyst system produces isotactic stereoregular polypropylene (i-PP) that is stronger and denser than atactic nonstereoregular polypropylene (a-PP). At some time during this polymerization, some of the liquid polypropylene monomer is removed and ethylene monomer is added to the mix. Because of the continued presence of the catalyst that can also polymerize the ethylene monomer, copolymer containing ethylene and i-PP units is produced. This product can be roughly pictured as being formed from i-PP particles that have unreacted propylene monomer removed creating open-celled sponge-like particles. These i-PP particles then become impregnated with ethylene monomer eventually resulting in the formation of a copolymer that contains both i-PP and PE units with a i-PP outer shell. This product is eventually mixed with an ethylene(60%)–i-PP(40%) copolymer giving a blended material that has two continuous phases with a final i-PP content of 70%. It is this material that is used in molding the automotive bumpers and panels. The i-PP units contribute stiffness and the PE units contribute flexibility to the overall product. The inner PE units in the impregnated particles allow stress to be rapidly distributed to the stiffer outer i-PP shell. The closeness of structure between the two components, the i-PP impregnated particles and the PE/i-PP copolymer allows miscible mixing of the two phases.

The actual conditions and concentrations and ratios of monomer were developed by research scientists such as Edward Kresge over a long period of time.

7.9 FLUOROELASTOMERS

Fluoroelastomers are specialty copolymers containing a high amount of fluorine-containing units. These copolymers can be divided into four groups. Viton A (Equation 7.39) is a copolymer of vinylidene difluoride and hexafluoropropylene. It is used as a general-purpose sealing material in the automotive industry and with aerospace fuels and lubricants.

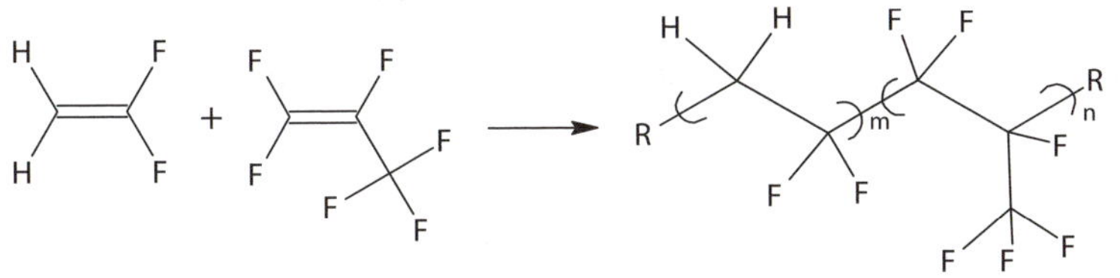

(7.39)

Viton B (Equation 7.40) is a terpolymer produced from the polymerization of tetrafluoroethylene, vinylidene fluoride, and hexafluoropropylene. It is used in power-utility seals and gaskets.

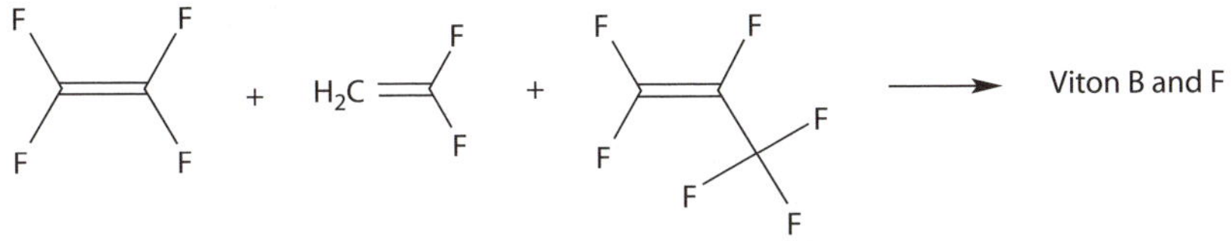

(7.40)

Viton F is also a terpolymer of tetrafluoroethylene, vinylidene fluoride, and hexafluoropropylene. It is used as seals for concentrated aqueous solutions of inorganic acids and for oxygenated automotive fuels. Finally, Viton Extreme is a copolymer of tetrafluoroethylene and propylene and the terpolymer of ethylene, tetrafluoroethylene, and perfluoromethylvinylether.

Viton O-rings are used in SCUBA diving when gas blends called Nitrox are employed. Viton has a low flammability even in the presence of high amounts of oxygen often found in Nitrox. It also stands up well in such increased oxygen conditions. As part of the move toward green fuels, Viton-lined hoses are often employed when biodiesel fuels are used.

These copolymers are often employed as mixtures with other polymers. They are compatible with many hydrocarbons but not compatible with ketones and organic acids.

7.10 NITRILE RUBBER

The copolymer of acrylonitrile and butadiene is referred to by a number of names, including nitrile rubber, Buna-N, and nitrile butadiene rubber (NBR). The composition of the copolymer varies depending on what the intended end product is. Since butadiene is employed, the resulting copolymer has sites of unsaturation that are often subsequently reacted forming thermosets. The greater the amount of cross-linking the more rigid the product along with the corresponding decreased porosity and diffusion properties. The greater the proportion of acrylonitrile incorporated into the copolymer, the greater the resistance to oils, fuels, and other chemicals and the stiffer the material. Also, the form of the butadiene monomer can vary.

Typically, NBR has a wide-operational temperature range of about −40°C to 120°C, making it useful for extreme automotive applications as well as cooling units. Its good resistance to oils and other chemicals allows its use around ketones, hydrocarbon liquids, esters, and aldehydes. In the lab, many of the gloves are made of nitrile rubber. These gloves are also used in home and industrial cleaning and medically as examination and disposable gloves. Nitrile gloves have greater puncture resistance compared to "rubber gloves." NBRs ability to withstand extreme temperatures and resistance to oils encourages their automotive uses as hoses, seals, belts, oil seals, and grommets. NBR is also used as adhesives, expanded foams, floor mats, and surface treatment of paper, synthetic leather, and footwear.

NBR is often synthesized using a radical initiating agent. Thus, emulsifier, acrylonitrile, and butadiene, catalysts, and radical initiators are added to the reaction vessel. The vessel is heated to 30°C–40°C for the formation of so-called hot NBR aiding in the polymerization process and promoting branch formation. Reaction continues for about 5–12 h to about 70% conversions when a terminating agent such as dimethyldithiocarbamate or diethylhydroxylamine is added. The production of cold NBR is similar except the polymerization temperature is in the range of 5°C–15°C. Lower-temperature NBR contains less branching generally producing a somewhat stiffer product. Unreacted monomer is removed and reused in a subsequent reaction. The latex is filtered, removing unwanted solids, and then sent to blending tanks where an antioxidant is added. The resulting latex is coagulated by addition of aluminum sulfate, calcium chloride, or other coagulating compound. The latex is dried giving a flaky crumb like product that is then used to produce the desired product. Because of the variation in reactants and conditions of polymerization, there exists a variation in product structure.

7.11   ACRYLONITRILE–BUTADIENE–STYRENE TERPOLYMERS

The terpolymer formed from reaction of acrylonitrile, butadiene, and styrene is referred to by the first letters of the three monomers, ABS. It is made by polymerizing acrylonitrile and styrene in the presence of polybutadiene. The proportions of reactants vary widely but are generally in the range of 15%–35% acrylonitrile, 5%–30% butadiene, and 40%–60% styrene. The resulting product contains long butadiene chains captured by shorter chains of poly(acrylonitrile-co-styrene). The polar nitrile groups attract one another binding the mixture together, resulting in a material that is stronger than simply polystyrene. The "plastic" styrene contributes a shiny product with surface that is largely impervious. The "rubbery" polybutadiene contributes resilience and a wide range of operating temperatures (OT) between −25°C and 60°C. This combination results in what is referred to as "rubber toughening" where the rubbery polybutadiene is dispersed within a plastic or a more rigid styrene matrix bound together by the acrylonitrile units. Typically, impacts are transferred from the more rigid styrene-rich portions to the rubbery butadiene-rich regions that are able to help absorb the impact through segmental chain movement. Impact resistance can be increased by increasing the amount of polybutadiene to a limit. Aging is also dependant on the ABS composition and is generally dependent on the amount of butadiene since the unsaturation is typically responsible for limiting the use time because of the increased cross-linking, and consequently increased brittleness, with time. Thus, it is customary to add an antioxidant to curtail aging.

The properties of the end product are dependant on the processing conditions. For instance, processing the ABS at higher temperatures increases the gloss and heat resistance of the material while lower processing temperatures result in increased impact resistance and strength.

The electrical properties of ABS are relatively independent over a wide range of applied frequencies, making it a good material where varied electrical frequencies are present.

ABS is used to make rigid-molded products such as piping and fittings, fuel tanks, automotive body parts, toys, wheel covers, enclosures, and where good shock absorbance is needed such as golf club heads and protective head gear. In fact, our ever-present Lego building blocks are made from ABS plastic.

7.12   EPDM RUBBER

Ethylene–propylene–diene terpolymers rubber's name is derived from E for ethylene, P for propylene, D for diene, and M for the ASTM rubber classification. The dienes currently employed are dicyclopendadiene (Equation 7.41), ethylidene norbornene (Equation 7.42), and vinyl norbornene (Equation 7.43). Along with supplying the needed double bond for subsequent cross-linking (curing), the dienes also supply steric hindrance encouraging amorphous formation.

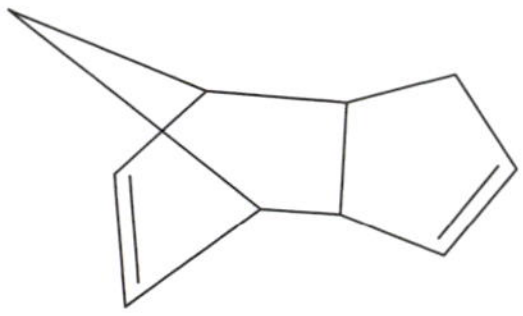

Dicyclopentadiene (7.41) 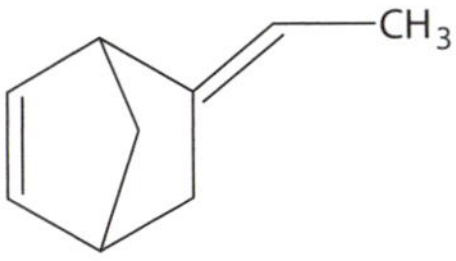   Ethylidene Norbornene (7.42) 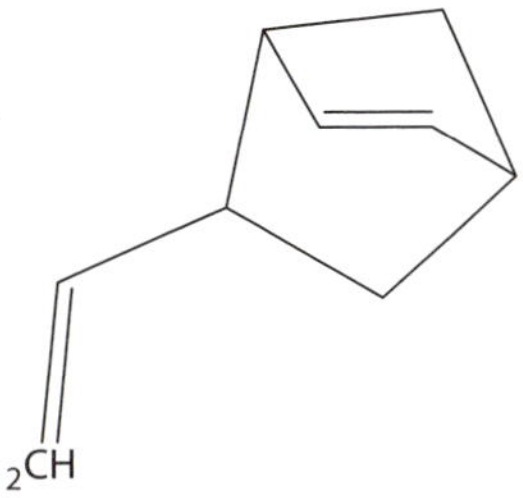   5-Vinyl-2-norbornene (7.43)

The ethylene content is about 45%–75%, diene content is about 2.5%–12%, with the remainder being propylene. The greater the ethylene content the higher the loading capability and better the mixing and the extrusion of the material. The polymers are cured employing a peroxide agent.

EPDM rubbers have excellent ozone, weather, and heat properties. While it is compatible with fireproofing hydraulic fluids, water, and bases, it is not compatible with most gasoline and other hydrocarbon liquids, concentrated acids, and halogen-intense solvents. The automotive industry uses EPDM rubbers as weather seals, including door, trunk, hood, and window seals. The noise in automobiles may be the result of the friction between the EPDM seals and the mated surface. This is normally corrected for by using a special coating that also increases the chemical resistance of the rubber. EPDM is also used to waterproof roofs. As a green material application, it allows the harvesting of rain water for other uses since it does not introduce pollutants into the rain water.

EPDM rubber is also used to produce garden hoses, washers, belts, tubing, electrical insulation, mechanical goods, motor oil additive, pond liner, RV roofs, and as an impact modifier.

7.13   NETWORKS—GENERAL

Polymer networks are generally one of two types. Thermoset elastomers have chemical cross-links where "un-linking" requires rupture of primary bonds. Thermoplastic elastomers have physical cross-links that include chain entanglement and formation of crystalline or ordered domains that act as hard segments. Here "un-linking" can be accomplished without rupture of primary bonds. For both types, stiffness and brittleness increase as the amount of cross-linking increases. Networks can also be divided according to flexibility. Highly cross-linked systems such as phenol-formaldehyde resins are network systems but they are very strong and brittle.

For elastomeric materials, high flexibility and mobility are required so that flexible nonpolar chain units are generally required. Such materials typically have low T_g values that allow ready segmental mobility to occur. Some materials can be "tricked" into being flexible below their T_g through introduction of appropriate flexibilizing agents, including plasticizers (diluents). Others can be made to become elastomeric through heating above the T_g of the crystalline areas.

Many of the copolymers are designed to act as elastomeric materials containing complex networks. In general, most elastomers undergo ready extension up to a point where there is an abrupt increase in the modulus. The cause of this limited extensibility was for many years believed to be due to the molecular extent of uncoiling of the polymer segments composing the elastomeric material. Today we know that while such ultimate chain elongations may contribute to this rapid increase in modulus, the primary reason for many elastomeric materials involves strain-induced crystallization. As a general observation, when the large increase in modulus is mainly due to limiting chain extension the limit will not be primarily dependent on temperature and presence/absence of a diluent. Conversely, when the abrupt increase in modulus is dependent on temperature and diluent, the limiting factor is probably stress-induced crystallization.

Ultimate properties of toughness (energy to rupture), tensile strength, and maximum extensibility are all affected by strain-induced crystallization. In general, the higher the temperature the lower the extent of crystallization and consequently the lower these stress–strain-related properties. There is also a parallel result brought about by the presence of increased amounts of diluent since this also discourages stress-related crystallization.

There are some so-called noncrystallizable networks where stress–strain behavior is fairly independent of diluent and temperature. One such system is formed from reaction of hydroxyl-terminated polydimethylsiloxane (PDMS), through reaction with tetraethyl orthosilicate. While these materials are not particularly important, commercially they allow a testing of various effects in particular the so-called "weakest-link" theory, where it is believed that rupture is initiated by the shortest chains because of their limiting extensibility. From such studies, it was found that at long extensions that short chains that should be the "weakest-link" or more aptly put, the "shortest-link," was not the limiting factor but rather the system "shared" the distortion throughout the network distributing the strain. This redistributing continues until no further distributing is possible at which case stress-induced rupture occurs. Introduction of short or limiting chains has a positive affect on the modulus related properties because shorter chains are better at distributing induced strain.

We are now able to construct so-called bimodal network systems composed of short and long (average length between cross-links) chains and multimodel systems containing chains of predetermined differing lengths. It has been found that short chains are better at reapportioning applied stress–strain than longer chains so that greater elongation is required to bring about the "upturn" in modulus. In general, the stress–strain curve for short chains is steeper than that for long-chain networks as expected. Interestingly, a combination of short- and long-chained networks gives a stress–strain curve that is between the one found for short-chained networks and the one for long-chained networks such that the area under the stress–strain curve, toughness, is much greater than for either of the monomodal systems. Products are being developed to take advantage of this finding.

Mismatching is important for some applications. Thus, moderately polar materials such as polyisoprene are ideal materials for hose construction where the liquid is nonpolar such as gasoline, flight fuel, lubricants, oils, and so on, while nonpolar materials such as polyethylene-intense copolymers would be suitable for use with transport and containment of polar liquids such as water.

Fillers are often added to polymeric networks. In particular, those elastomers that do not undergo strain-induced crystallization generally have some reenforcing filler added. The important cases involve carbon black that is added to many materials, including NR and silica that is added to siloxane elastomers. These fillers generally give the materials increased modulus, tear and abrasion resistance, extensibility, tensile strength, and resilience. Counter, they often create other effects such as giving the materials generally a higher hysteresis (heat build up when the material is exposed to repeated cycles of stress–strain) and compression set (permanent deformation).

7.14 POLYMER MIXTURES

Today, there exist a number of polymer mixtures, including blends noted in the previous section. Here we will briefly look at two more mixtures. A plastic or polymer *alloy* is a physical mixture of two or more polymers in a melt and is often structurally similar to blends, in fact the terms blends and alloys are sometimes used to describe the same materials. While some cross-linking may occur,

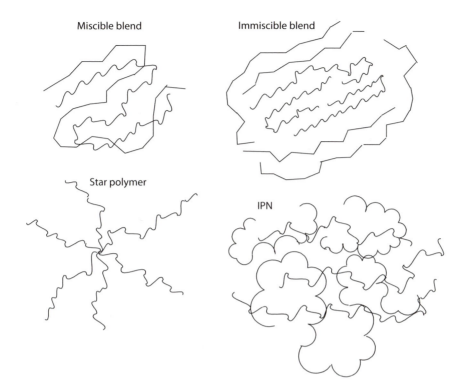

FIGURE 7.2 Polymer mixtures and special shapes.

such materials are generally not viewed as copolymers, though they are covered in this chapter along with blends and dendrites. Many alloys contain a matrix that is a mixture of polymer types with many of them containing ABS. ABS thermoplastics have a two-phase morphology consisting of elastic particles dispersed in a styrene-acrylonitrile (SAN) copolymer matrix.

Interpenetrating polymer networks (IPNs) are described by Sperling to be an intimate combination of two polymers, both in network form, where at least one of the two polymers is synthesized and/or cross-linked in the immediate presence of the other. This is similar to taking a sponge cake; soaking in it warm ice cream, and refreezing the ice cream, resulting in a dessert (or a mess) that has both spongy and stiff portions. Such IPNs, grafts, blocks, and blends can offer synergistic properties that are being widely exploited. Some of these special polymer mixtures and special shapes are given in Figure 7.2.

7.15 DENDRITES

Along with the varying structures given in the previous sections, there exist other structurally complex molecules called *dendrites* developed by a number of scientist, including Tomali and Frechet. These molecules can act as "spacers," "ball bearings," and as building blocks for other structures. Usually, they are either wholly organic or they may contain metal atoms. They may or may not be copolymers depending on the particular synthetic rout employed in their synthesis.

While some make a distinction between dendrimers and hyperbranched polymers, we will not do so here. In essence, hyperbranched polymers are formed under conditions that give a variety of related but different structures while dendrimers are formed one step at a time giving a fairly homogeneous product.

Dendrites are highly branched, usually curved, structures. The name comes from the Greek name for tree "dendron." Another term often associated with these structures is "dendrimers" describing the oligomeric nature of many dendrites. Because of the structure, dendrites can contain

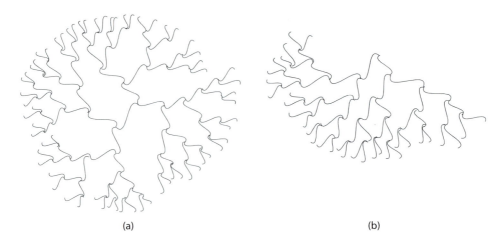

(a)   (b)

FIGURE 7.3 Dendrite structure derived from the reaction of 1,4-diaminobutane and acrylonitrile (a), and derived from "bent" difunctional reactants (b).

many terminal functional groups for each molecule that can be further reacted. Also, most dendrites contain "lots" of unoccupied space that can be used to "carry" drugs, fragrances, adhesives, diagnostic molecules, cosmetics, catalysts, herbicides, and other molecules.

The dendrite structure is determined largely by the "functionality" of the reactants. The dendrite pictured in Figure 7.3(a), can be considered as being derived from a tetra-functional monomer formed from the reaction of 1,4-diaminobutane and acrylonitrile. The resulting polypropylenimine dendrimer has terminal nitrile groups that can be further reacted, extending the dendrimer or terminating further dendrimer formation. The resulting molecule is circular with some three-dimensional structure. The dendrimer shown in Figure 7.3(b), is derived from difunctional reactants that are "bent" so as to encourage "fan-like" expansion rather than the typical linear growth.

Numerous approaches have been taken in the synthesis of dendrites or dendrimers. These approaches can be divided into two groupings. In divergent dendrimer growth, growth occurs outward from an inner core molecule. In convergent dendrimer growth, developed by Frechet and others, various parts of the ultimate dendrimer are separately synthesized and then they are brought together to form the final dendrimer.

The somewhat spherical shape of dendrimers gives them some different properties in comparison to more linear macromolecules. On a macroscopic level, dendrimers act as ball bearings rather than strings. In solution, viscosity increases as molecular weight increases for linear polymers. With dendrimers, viscosity also increases with molecular weight up to a point after which viscosity decreases as molecular weight continues to increase.

Dendrimers are being used as host molecules, catalysts, self-assembling nanostructures; analogues of proteins, enzymes, and viruses; and in analytical applications, including in ion-exchange displacement chromatography and electrokinetic chromatography.

We are continuing to recognize that polymer shape is important in determining material property. Another group of structurally complex shapes is referred to as stars. There are a number of synthetic routes to star polymers. Fetters and coworkers developed a number of star polymers based on chlorosilanes. For instance 3-, 12-, and 18-arm star polymers can be formed. These arms are now reacted with other reactants such as living polystyrene or polybutadiene giving now the star polymers with the silicon-containing inner core and polymer outer core. Through control of the length of the grafted polystyrene or other reactant, the size of the "star" can be controlled.

For dendrimers made using flexible arms, the core is mobile and depending upon the situation spends some time near the outer layer of the dendrimer sphere. Counter, stiff, rigid arms produce a dendrimer that "holds" its core within the interior of the dendrimer.

The term "generation" describes the number of times "arms" have been extended. The nature of each generation can be varied so that mixtures of steric requirements and hydrophobic/hydrophilic character can be introduced, offering materials with varying structures and properties. By varying the hydrophobic and hydrophilic interactions and steric nature of the arms, secondary and tertiary structural preferences can be imposed on the dendrimer.

The dendrite structure can be used as a synthetic tool to craft a particular property into an overall structure. For instance, structures have been formed that are a conducting rigid rod backbone with dendritic structures radiating from the rigid rod. The backbone collects photons of visible light in the range of 300–450 nm. The dendritic envelope collects light in the ultraviolet region from 220 to 300 nm and transfers it to the backbone that then fluoresces blue light at 454 nm. Unprotected poly(phenyleneethynylene), the nondendritic backbone alone, does emit light but suffers from both being brittle and from collisional quenching. The dendrimer product allows both flexibility and discourages collisional quenching by forcing a separation between the backbone rods. This shielding increases with each successive generation of the dendritic wedge.

7.16 IONOMERS

Ionomers are ion-containing copolymers typically containing more than 90% (by number) ethylene units with the remaining being ion-containing units such as acrylic acid. These "ionic" sites are connected through metal atoms. Ionomers are often referred to as processable thermosets. They are thermosets because of the cross-linking introduced through the interaction of the ionic sites with metal ions. They are processable or exhibit thermoplastic behavior because they can be reformed through application of heat and pressure.

As with all polymers, the ultimate properties are dependant upon the various processing and synthetic procedures that the material is exposed to. This is especially true for ionomers where the location, amount, nature, and distribution of the metal sites strongly determines the properties. Many of the industrial ionomers are made where a significant fraction of the ionomer is unionized and where the metal-containing reactants are simply added to the preionomer followed by heating and agitation of the mixture. These products often offer superior properties to ionomers produced from fully dissolved preionomers (Figure 7.4).

For commercial ionomers, bonding sites are believed to be of two different grouping densities. One of these groupings involves only a few or individual bonding between the acid groups and the metal atoms as shown in Figure 7.4. The second bonding type consists of large concentrations of acid groups with multiple metal atoms (clusters) as shown in Figure 7.5. This metal-acid group bonding (salt formation) constitutes sites of cross-linking. It is believed that the "processability" is a result of the combination of the movement of the ethylene units and the metal atoms acting as "ball bearings." The "sliding" and "rolling" is believed to be a result of the metallic nature of the acid-metal atom bonding. (Remember that most metallic salts are believed to have a high degree of ionic, nondirectional bonding as compared with typical organic bonds where there exists a high amount of covalent, directional bonding.) Recently, Carraher and coworkers have shown that the ethylene portions alone are sufficient to allow ionomers to be processed through application of heat and pressure.

Ionomers are generally tough and offer good stiffness and abrasion resistance. They offer good visual clarity, high-melt viscosities, superior tensile properties, oil resistance, and are flame retarders. They are used in the automotive industry in the formation of exterior trim and bumper pads, in the sporting goods industry as bowling pin coatings, golf ball covers, in the manufacture of roller skate wheels, and ski boots. Surlyn (DuPont; poly(ethylene-comethacrylic acid)) is used in vacuum packaging for meats, in skin packaging for hardware and electronic items (such as seal layers and as foil coatings of multiwall bags), and in shoe soles.

Sulfonated ethylene–propylene–diene terpolymers (EPDM) are formulated to form a number of rubbery products, including adhesives for footwear, garden hoses, and in the formation of calendered sheets. Perfluorinated ionomers marketed as Nafion (DuPont) are used for membrane applications, including

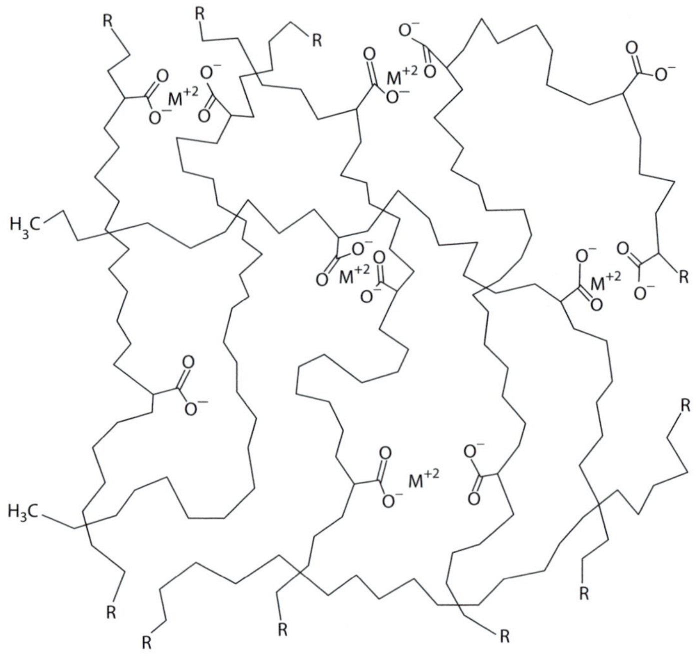

FIGURE 7.4 Ionomer structure showing individualized randomly dispersed bonding sites.

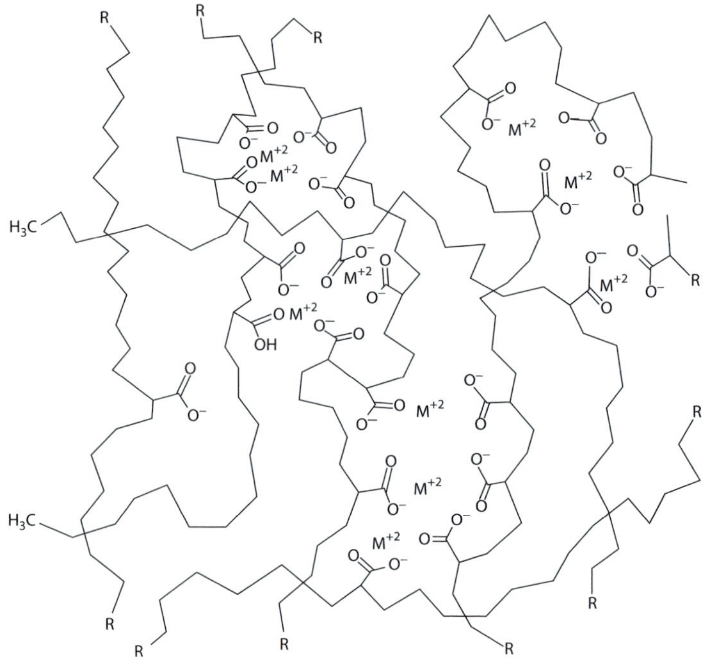

FIGURE 7.5 Ionomer bonding showing cluster bonding sites.

chemical processing separations, spent acid regeneration, electrochemical fuel cells, ion-selective separations, electrodialysis, and in the production of chlorine. It is also employed as a "solid"-state catalyst in chemical synthesis and processing. Ionomers are also used in blends with other polymers.

7.17 VISCOSITY MODIFIERS

Different mechanisms can be in operation to cause viscosity changes. As noted before, one general polymer property is that the addition of even small amounts of polymer to a solution can result in a relatively large increase in viscosity. This increased viscosity is related to the large size of the polymer chains causing them to be present in several flow planes resulting in what is referred to as viscous drag. Factors that influence the apparent size of a polymer are reflected in their viscosity both in solution and bulk.

For many polymers, as one moves from a good solvent to poorer solvent the extent of close coiling increases, resulting in a decreased viscosity because the more tightly coiled polymer chains now resides in fewer flow planes (Figure 7.6). A similar phenomena can occur as the temperature is decreased (above T_g) since the available energy for chain extension becomes less.

For some polymers additional specific mechanisms are in action. Most commercial motor oils are composed of oligomeric, highly branched chains containing from 12 to 20 carbons. These "oils" are derived from the usual fractionation of native oils with processing to remove the aromatic and other unwanted materials.

Synthetic motor oils also contain oligomeric materials containing 12–20 carbons formed from alpha-olefins generally containing 5, 6, and 7 carbon-containing units such as 7.44 for a C_5 unit.

$$(7.44)$$

Synthetic motor oils outperform normal oils in offering better stability because of the lack of sites of unsaturation that are found in native motor oils, and they offer a better wide temperature lubricity because of their designed lower T_g in comparison to native oils.

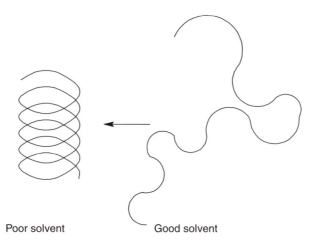

Poor solvent   Good solvent

FIGURE 7.6 Polymer-chain extensions in different solvents.

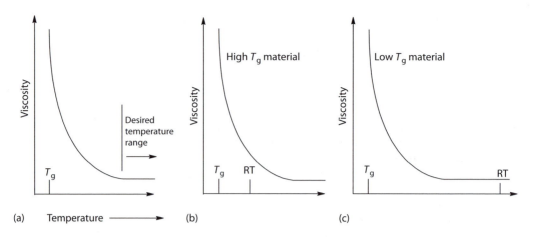

FIGURE 7.7 Plots of viscosity as a function of temperature. For from left to right: (a) General plot showing the desired temperature range where viscosity is approximately constant as temperature is varied, (b) plot for a high T_g material, and (c) plot for a low T_g material where RT = room temperature.

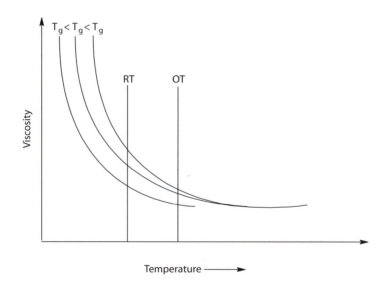

FIGURE 7.8 Viscosity as a function of molecular weight of a low molecular weight oil (extreme left), high molecular weight oil (extreme right), and of the low molecular weight oil to which an appropriate viscosity modifier has been added (middle).

In general, for solutions and bulk polymeric materials viscosity follows the general relationship shown in Figure 7.7 where viscosity is constant over much of the "fluid" range but sharply increases near T_g.

The purpose of a multiviscosity motor oil is to offer good engine protection over the operating range of the engine from a "cold start" to the normal operation temperature. Addition of a low T_g polymer allows for a wider range where viscosity is approximately constant.

Figure 7.8 is a similar plot to Figure 7.7 except where the influence of adding a low T_g material illustrates this phenomena. Here, the extreme left curve is the profile of a low molecular weight (low T_g) oil and extreme right curve is the profile of a high molecular weight (high T_g) oil. The high molecular weight oil offers the needed lubrication protection at the OT but at RT it is

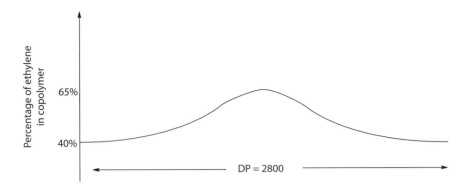

FIGURE 7.9 Polymer composition along a single polymer chain for the ethylene–propylene copolymer employed in most motor oils as the viscosity modifier.

too near its T_g so that the viscosity is high. The low molecular weight oil offers a good viscosity at RT but not at the OT. Addition of an appropriate viscosity modifier allows added lubrication protection and a somewhat constant viscosity over the needed temperature range. Today's motor oils are of the low molecular weight, low T_g variety with a polymer viscosity modifier added.

Addition of about 1% of an EPM is used today to achieve most of our multiviscosity oils. These copolymers were developed as part of Exxon's program that Ed Kresge and fellow researchers were involved with that also produced the miscible blends used to make automotive bumpers and panels (Section 7.8).

Exxon developed a single-site soluble catalyst from the reaction of vanadium tetrachloride and $Et_3Al_2Cl_2$ in hexane. This catalyst has special properties that allowed a tailoring of the chain size and chain composition within each chain. Unlike many catalysts where individual chain formation is rapid, this catalyst system creates individual chains in about 4 minutes, allowing growing chain composition to be controlled through varying the concentration of monomer feed as the chains are growing (see Section 7.1). They discovered that the chain extension could be varied by varying the amount of ethylene in the backbone. Figure 7.9 shows the average percentage of ethylene within each of the chains for today's multiviscosity modifiers. The polymer chains are all about 2,800 units long. Each end is rich in propylene-derived units while the middle is rich in ethylene-derived units. As the percentage of ethylene units moves toward 65%, there is an increased tendency for the ethylene units to form crystalline micelles, causing the polymer to collapse in size at lower temperatures, such as room temperature. As the temperature increases, the crystalline micelles melt, allowing the chain length of the copolymer to increase increasing the viscosity compensating for the effect of the increased temperature on viscosity. Thus the polymer allows an approximately constant viscosity over the needed temperature range.

7.18 SUMMARY

1. Unlike homopolymers, which consist of chains with identical repeating units, copolymers are macromolecules that contain two or more unlike units. These repeat units may be randomly arranged, or may alternate in the chain. Block copolymers are linear polymers that consist of long sequences of repeating units in the chain, and graft copolymers are branch polymers in which the branches consist of sequences of repeating units that differ from those present in the backbone. These different copolymers give materials with differing properties, even when synthesized using the same monomers.

2. The copolymerization between two different monomers can be described using only four reactions, two homopolymerizations and two cross-polymerization additions. Through appropriate

arrangements, equations that allow copolymer composition to be determined from the monomer-feed ratio are developed.

3. The product of the reactivity ratios can be used to estimate the copolymer structure. When the product of the reactivity ratios is near one, the copolymer arrangement is random; when the product is near zero, the arrangement is alternating; when one of the reactivity ratios is large, blocks corresponding to that monomer addition will occur.

4. Block and graft copolymers may differ from mixtures by having properties derived from each component. Block copolymers can be used as thermoplastic elastomers and graft copolymers with flexible backbones can be used for high-impact plastics. Block and graft copolymers can be produced by step and chain-reaction polymerization. The principle block copolymers are thermoplastic elastomers and elastic fibers such as ABS. Principal graft copolymers are grafted starch and cellulose.

5. Blends are physical mixtures of polymers. Depending on the extent and type of blend, the properties may be characteristic of each blend member or may be some "blend" of properties. Immiscible blends are phase separated with the phases sometimes chemically connected. They are generally composed of a continuous and discontinuous phase. HIPS is an example of an immiscible blend. Miscible blends occur when the two blended materials are compatible. Often the properties are a mixture of the two blended materials. The plastic automotive panels and bumpers are generally made from a miscible blend of polyethylene and a copolymer of polyethylene and polypropylene.

6. Dendrites are complex molecules formed from building up of the dendrite by individual steps (divergent growth) or from bringing together the units already formed (convergent growth).

7. Ionomers are often referred to as processable thermosets. They are cross-linked through the use of metal ions such that application of heat and pressure allow them to be reformed.

8. There are a number of applications of copolymers. One of these is the important area of viscosity modifiers that recognizes the influence that small amounts of polymers generally have on increasing the viscosity of solutions to which they have been added. One of these applications is to allow the viscosity of motor oils to remain approximately constant over an engine's temperature operating range. Here, the polymer additive increases its effective chain distance as temperature increases compensating for the temperature decrease caused by increased temperature.

GLOSSARY

AB: Block copolymer with two separate mers.

ABA: Block copolymer with three sequences of mers in the order shown.

ABS: Three-component copolymer of acrylonitrile, butadiene, and styrene.

Alloy: Rubber-toughened materials in which the matrix can be a mixture of polymer types.

Alternation copolymer: Ordered copolymer in which other building is a different mer.

Azeotropic copolymer: Copolymer in which the feet and composition of the copolymer are the same.

Blends: Mixtures of different polymers on a molecular level; may exist in one or two phases.

Block copolymer: Copolymer that contains long sequences or runs of one mer or both mers.

Buna-N: Elastomeric copolymer of butadiene and acrylonitrile.

Buna-S: Elastomeric copolymer of butadiene and styrene.

Butyl rubber: Elastomeric copolymer of isobutylene and isoprene.

Charge-transfer complex: Complex consisting of an electron donor (D) and an electron acceptor (A) in which an electron has been transferred form D to A resulting in the charge-transfer agent $D^+ A^-$.

Composites: Mixtures of different polymers, one forming a continuous phase (the matrix) and one the discontinuous phase (often a fiber).

Compositional drift: Change in composition of a copolymer that occurs as copolymerization takes place with monomers of different reactivities.

Copolymer: Macromolecule containing more than one type of mer in the backbone.

Dendrites: Complex molecules that are highly branched.

Domains: Sequences in a block copolymer or polymer.

Graft copolymer: Branched copolymer in which the backbone and the branches consist of different mers.

Homopolymer: Macromolecule consisting of only one type of mer.

Hytrel: Trade name for a commercial TEP.

Ideal copolymer: Random copolymer.

Interpenetrating polymer network (IPN): Intimate combination of two polymers, both in network form, where at least one of the two polymers is synthesized and or cross-linked in the immediate presence of the other.

Ionomers: Copolymers typically containing mostly ethylene units, with the remaining units being ion containing that are neutralized through reaction with metals.

Kraton: Trade name for an ABA block copolymer of styrene and butadiene.

Living polymers: Macrocarbanions.

Polyallomers: Block copolymers of ethylene and propylene.

Q-e scheme: A semiempirical method for predicting reactivity ratios.

Random copolymer: Copolymer in which there is no definite order for the sequence of the different mers.

Reactivity ratio: Relative reactivity of one monomer compared to another monomer.

Saran: Trade name for copolymer of VC and vinylidene chloride.

SBR: Elastomer copolymer of styrene and butadiene.

Spandex: Elastic fiber consisting of a block copolymer of a polyurethane, hard segment, and a polyester or polyether, soft segment.

Viscosity modifier: A polymer added to a solution that expands its effective length as temperature increases compensating for the effect that increased temperature has on decreasing the viscosity.

Wall equation: Predecessor to the copolymer equation.

EXERCISES

1. Draw representative structures for (a) homopolymers, (b) alternation copolymers, (c) random copolymers, (d) AB block copolymers, and (e) graft copolymers of styrene and acrylonitrile.
2. If equimolar quantities of M_1 and M_2 are used in an azeotropic copolymerization, what is the composition of the feed after 50% of the copolymer has formed?
3. Define r_1 and r_2 in terms of rate constants.
4. Do the r_1 and r_2 values increase or decrease during copolymerization?
5. What is the effect of temperature on r_1 and r_2?
6. What will be the composition of copolymers produced in the first part of the polymerization of equimolar quantities of vinylidene chloride and VC?
7. What monomer may be polymerized by anionic, cationic, and free radical chain techniques?
8. Which chain-polymerization technique would you select to polymerize (a) isobutylene, (b) acrylonitrile, and (c) propylene?
9. If r_1r_2 is about zero, what type of copolymer would be formed?
10. Show a structure for an AB block copolymer.
11. What is the value of r_1r_2 for an ideal random copolymer?
12. Which would polymerize more readily?

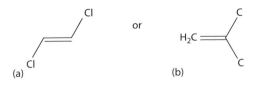

(a)   or   (b)

13. Why are most elastomers difficult to recycle?
14. What is the composition of the first copolymer chains produced by the copolymerization of equimolar quantities of styrene and methyl methacrylate in (a) free radical, (b) cationic, and (c) anionic copolymerization?
15. What is the composition of the first copolymer butyl rubber chains produced from equimolar quantities of the two monomers?
16. What is the composition of the first copolymer butyl rubber chains produced from a feed containing 9 mol of isobutylene and 1 mol of isoprene?
17. How would you ensure that production of butyl rubber of uniform composition occurs in question 16.
18. For a graft copolymer with a backbone derived from monomer A and the graft B what often occurs when B is long?
19. What is the composition of the first polymer chains produced by the copolymerization of equimolar quantities of VC and VAc?
20. What are the advantages, if any, of the VC–VAc copolymer over PVC itself?
21. Why are ionomers superior to LDPE?
22. What is the difference between buna-S, GRS, and SBR?
23. How can you connect two different phases in polymer blends?
24. Which sequence in the ABA block copolymer of ethylene oxide and propylene oxide is more lyophilic?
25. When will a blend exhibit two distinct properties.
26. Are dendrites really polymeric?
27. What product is obtained if 1.5 mol of styrene is copolymerized with 1 mol of maleic anhydride in benzene?
28. Why are ionomers referred to as processable thermosets?
29. Of what use is the copolymerization equation?
30. How could you use graft-copolymerization techniques to reduce the water solubility of starch?
31. What is the end group when azobiscyanopentanoic acid is used as an initiator?
32. What are some advantages of ionomers?
33. In the synthesis of dendrites, why must some of the reactants have a functionality greater than two?
34. Why do companies often emphasize only selected materials and/or processes?
35. How might the properties of two polymers containing the same amounts of monomer A and B differ if one polymer is a simple random copolymer and the other polymer is a block copolymer?
36. Name one important property of an elastomer.
37. What are potential useful properties of dendrites?

ADDITIONAL READING

Alfrey, T., Bohrer, J. Mark, H. (1952): *Copolymerization*, Interscience, NY.
Bhattacharya, A., Ray, P., Rawlins, J. (2008): *Polymer Grafting and Crosslinking*, Wiley, Hoboken, NJ.
Brandrup, J., Immergut, E. (1965): *Polymer Handbook*, Wiley, NY.
Calleja, F., Roslaniec, Z. (2000): *Block Copolymers*, Dekker, NY.
Carraher, C., Moore, J. A. (1984): *Modification of Polymers*, Plenum, NY.
Ciferri, A. (2005): *Supramolecular Polymers*, 2nd Ed., Taylor and Francis, Boca Raton, FL.
Dragan, E. S. (2006): *New Trends in Nonionic (Co)polymers and Hybrids*, Nova, Commack, NJ.
Eisenberg, A., Kim, J. S. (1998): *Introduction to Ionomers*, Wiley, NY.
Frechet, J. (2002): *Dendrimers and Other Dendritic Polymers*, Wiley, NY.
Frechet, J., Tomalia, D. (2002): *Dendrimers and Other Dendritic Polymers*, Wiley, NY.
Gupta, R. K. (2000): *Polymer and Composite Rheology*, 2nd Ed., Dekker, NY.

Hadjichristidis, N., Pispar, S., Floudas, G. (2007): *Block Copolymers: Synthetic Strategies, Physical Properties and Applications*, Wiley, Hoboken, NJ.

Ham, G. E. (1964): *Copolymerization*, Interscience, NY.

Hamely, I. (2008): *Block Copolymers in Solution: Fundamentals and Applications*, Wiley, Hoboklen, NJ.

Hamley, I. (2004): *Developments in Block Copolymer Science and Technology*, Wiley, Hoboken.

Harada, A., Hashidzume, A., Takashima, Y. (2006): *Supramolecular Polymers, Oligomers, Polymeric Betains*, Springer, NY.

Harrats, C., Groeninckx, G., Thomas, S. (2006): *Micro- and Nanostructured Multiphase Polymer Blend Systems*, Taylor and Francis, Boca Raton, FL.

Holden, G., Kricheldorf, H. R., Quirk, R. P. (2004): *Developments in Block Copolymer Science and Technology*, 3rd Ed., Hanser, Munich.

Klempner, D., Sperling, L. H., Ultracki, L. A. (1994): *Interpenetrating Polymer Networks*, American Chemistry Society, DC.

Lipatov, Y., Alekseeva, T. (2007): *Phase-Separated Interpenetrating Polymer Networks*, Springer, NY.

Mark, J., Etman, B., bokobza, L. (2007): *Rubberlike Elasticity*, Cambridge University, Ithaca, NY.

Mishra, M., Kobayashi, S. (1999): *Star and Hyperbranched Polymer*, Dekker, NY.

Newkome, G., Moorefield, C., Vogtle, F. (2001): *Dendrimers and Dendrons*, Wiley, NY.

Nwabunma, D. (2008): *Polyolefin Blends and Composites*, Wiley, Hoboken, NJ.

Pascaullt, J. (2003): *Polymer Blends*, Wiley, Hoboken.

Paul, D, Bucknall, C. B. (2000): *Polymer Blends: Formulation and Performance*, Vols. I and II, Wiley, NY.

Robeson, L. (2007): *Polymer Blends: An Introduction*, Carl Hanser GmbH, NY.

Shonaike, G., Simon, G. (1999): *Polymer Blends and Alloys*, Dekker, NY.

Simon, G. P. (2003): *Polymer Characterization Techniques and Their Application to Blends*, American Chemical Society, Washington, DC.

Sperling, L. H. (1997): *Polymeric Multicomponent Materials: An Introduction*, Wiley, NY.

Stann, M. (2000): *Blends*, Wiley, NY.

Stauart, G., Spruston, N., Hausser, M. (2007): *Dendrites*, Oxford University, Ithaca, NY.

Tomali, D. (2002): *Dendrimers and Other Dendric Polymers*, Wiley, NY.

Utracki, L. (2007): *Polymer Blends Handbook*, Springer, NY.

Wall, F. T. (1944): The structure of copolymers, *J. Amer. Chem. Soc.*, 66:2050.

Walling, C. (1957): *Free Radicals in Solution*, Wiley, NY.

Yan, D., Frey, H., Gao, C. (2008): *Hyperbranched Polymers: Synthesis, Properties, and Applications*, Wiley, Hoboken, NJ.

8 Composites and Fillers

Today, one of the fastest growing areas is composites. Composites offer good strength, corrosion resistance, and are light weight so they are continually replacing metals. Much of the replacement of metals is due to desired weight reductions as the cost of fuel continues to increase. Composites are generally composed of two phases, one called the *continuous* or *matrix phase* that surrounds the *discontinuous* or *dispersed phase*.

While many fillers approach being spherical in geometry, some are fiber like. When the length of these fibrous materials approach 100 times their thickness many combinations offer great increases in the strength-related properties of the materials that contain them. If these fibers are contained within a continuous phase, these materials are generally described as being traditional composites. Because of their importance, much of this chapter will deal with such traditional "long-fiber" composites.

8.1 FILLERS

According to the American Society for Testing and Materials standard (ASTM) D-883, a filler is a relatively inert material added to a plastic to modify its strength, permanence, working properties, or other qualities or to lower costs, while a reinforced plastic is one with some strength properties greatly superior to those of the base resin, resulting from the presence of high-strength fillers embedded in the composition. The word extender is sometimes used for fillers. Also, the notion that fillers simply "fill" without adding some needed property is not always appropriate. Some fillers are more expensive than the polymer resin and do contribute positively to the overall properties.

Many materials tend to approach a spherical geometry to reduce the interface the materials has with its surroundings. Further, a spherical geometry is favored because of the old adage that like-likes-like-the-best and a spherical geometry offer the best association of a material with itself.

The behavior of many fillers can be treated roughly as though they were spheres. Current theories describing the action of these spherical-acting fillers in polymers are based on the Einstein equation 8.1. Einstein showed that the viscosity of a viscous Newtonian fluid (η_o) was increased when small, rigid, noninteracting spheres were suspended in a liquid. According to the Einstein equation, the viscosity of the mixture (η) is related to the fractional volume (c) occupied by the spheres, and that was independent of the size of the spheres or the polarity of the liquid.

$$\eta = \eta_o (1 + 2.5c) \quad \text{and} \quad \frac{\eta_{sp}}{c} = 2.5 \tag{8.1}$$

Providing that c is less than 0.1, good agreement with the Einstein equation is found when glass spheres are suspended in ethylene glycol. The Einstein equation has been modified by including a hydrodynamics or crowding factor (β). The modified Mooney equation 8.2 resembles the Einstein equation when $\beta = 0$.

$$\eta = \eta_0 \left(\frac{2.5c}{(1 - \beta c)} \right) \tag{8.2}$$

Many other empirical modifications of the Einstein equation have been made to predict actual viscosities. Since the modulus (M) is related to viscosity, these empirical equations, such as the

Einstein–Guth–Gold (EGG) equation (8.3), have been used to predict changes in modulus when spherical fillers are added.

$$M = M_o(1 + 2.5c - 14.1c^2) \qquad (8.3)$$

Since carbon black and amorphous silica tend to form clusters of spheres (grasping effect), an additional modification of the Einstein equation was made to account for the nonspherical shape or aspect ratio (L/D). This factor (f) is equal to the ratio of the length (L) to the diameter (D) of the nonspherical particles ($f = L/D$).

$$\eta = \eta_o (1 + 0.67 fc + 1.62 f^2c^2) \qquad (8.4)$$

Both natural and synthetic fillers are used. Examples of such fillers are given in Table 8.1.

TABLE 8.1
Types of Fillers for Polymers

I. Organic materials

A. Cellulosic Products
  1. Wood products
    a. Kraft paper
    b. Chips
    c. Course flour
    d. Ground flour
  2. Comminuted cellulose products
    a. Chopped paper
    b. Diced resin board
    c. Crepe paper
    d. Pulp preforms
  3. Fibers
    a. α-cellulose
    b. Pulp preforms
    c. Cotton flock
    d. Textile byproducts
    e. Jute
    f. Rayon
    g. Sisal
B. Lignin-type products-processed lignin
C. Synthetic fibers
  1. Polyamides (nylons)
  2. Polyesters
  3. Polyacrylonitrile
D. Carbon
  1. Carbon black
    a. Channel black
    b. Furnace black

B. Silicates
  1. Minerals
    a. Asbestos
    b. Karolinite (China clay)
    c. Mica
    d. Talc
    e. Wollastonite
  2. Synthetic products
    a. Calcium silicate
    b. Aluminum silicate
C. Glass
  1. Glass flakes
  2. Solid and hollow glass spheres
  3. Milled glass fibers
  4. Fibrous glass
    a. Filament
    b. Rovings and woven rovings
    c. Yarn
    d. Mat
    e. Fabric
D. Metals
E. Boron fibers
F. Metallic oxides
  1. Ground material
    a. Zinc oxide
    b. Alumina
    c. Magnesia
    d. Titania

TABLE 8.1 (continued)
Types of Fillers for Polymers

2. Ground petroleum coke	2. Whiskers (including nonoxide)
3. Graphite filaments	a. Aluminum oxide
4. Graphite whiskers	b. Beryllium oxide
E. Polyfluorocarbons	
II. Inorganic materials	c. Zirconium oxide
A. Silica products	d. Aluminum nitride
1. Minerals	e. Boron carbide
a. Sand	f. Silicon carbide & nitride
b. Quarts	g. Tungsten carbide
c. Tripoli	h. Beryllium carbide
2. Synthetic materials	G. Calcium carbonate
a. wet-processed silica	1. Calk
b. Pyrogenic silica	2. Limestone
c. Silica aerogel	3. Precipitated calcium carbonate
	H. Barium ferrite and sulfate

Among the naturally occurring filler materials are cellulosics such as wood flour, α-cellulose, shell flour, and starch, and proteinaceous fillers such as soybean residues. Approximately 40,000 tons of cellulosic fillers are used annually by the U.S. polymer industry. Wood flour, which is produced by the attrition grinding of wood wastes, is used as filler for phenolic resins, urea resins, polyolefins, and poly(vinyl chloride) (PVC). Shell flour, which lacks the fibrous structure of wood flour, has been used as a replacement for wood flour for some applications.

α-Cellulose, which is more fibrous than wood flour, is used as filler for urea and melamine plastics. Melamine dishware is a laminated structure consisting of molded resin-impregnated paper. Presumably, the formaldehyde in these thermosetting resins react with the hydroxyl groups in cellulose, producing a more compatible composite. Starch and soybean fillers have been used to make biodegradable composites and other materials.

Carbon black, which was produced by the smoke impingement process by the Chinese more than 1,000 years ago, is now the most widely used filler for polymers. Much of the 1.5 million tons produced annually is used for the reinforcement of elastomers. The most widely used carbon black is furnace carbon black.

Conductive composites are obtained when powered metal fillers, metal flakes, or metal-plated fillers are added to resins. These composites have been used to produce forming tools for the aircraft industry and to overcome electromagnetic interference in office machines.

Calcium carbonate is available as ground natural limestone and as a synthetic chalk. It is widely used in paints, plastics, and elastomers. The volume relationship of calcium carbonate to resin or the pigment volume required to fill voids in the resin composite is called the pigment-volume concentration (PIVC).

8.2 TYPES OF COMPOSITES

There are a variety of polymer-intense composites that can be classified as shown in Figure 8.1. Many of these composite groups are used in combination with other materials, including different types of composites. Many naturally occurring materials such as wood are reinforced composites consisting of a resinous continuous phase and a discontinuous fibrous reinforcing phase.

Structural composites include laminas that can be sandwich or laminate. At times there is confusion between which materials are sandwich or laminate laminas. Even so, here we will consider

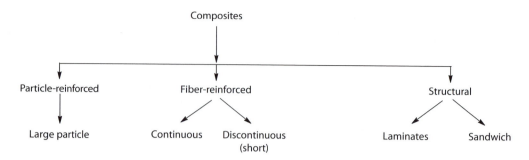

FIGURE 8.1 Classification of polymer-intense composites.

laminate composites as containing layers of material (generally considered the reinforcing agent) bound on one or both sides by an adhesive material. In general, there are a number of layers of reinforcing material present with the distance between the layers being small compared with those present in a *sandwich composites*. Plywood is an example of a sandwich laminar where layers or plies of wood are bound together using an adhesive such as one of the formaldehyde-related resins. Formica is an example of a laminate where paper, cloth, or other material is impregnated with the continuous-phase material. Sandwich laminas are widely used in the textile industry where foam, plastic, and fabric are bound together into new textiles. Many electronic boards are laminas. Examples of laminates include surfaces for countertops and wall paneling.

Particulate composites consist of the reinforcing materials being dispersed throughout the resin. Unlike fibrous composites, the reinforcing material is more bulky and not fibrous in nature.

Some materials to which fillers have been added can be considered as composites. These include a number of the so-called cements, including concrete (Section 12.2). As long as the added particles are relatively small, of roughly the same size, and evenly distributed throughout the mixture, there can be a reinforcing effect. The major materials in Portland cement concrete are the Portland cement, a fine aggregate (sand), a course aggregate (gravel and small rocks), and water. The aggregate particles act as inexpensive fillers. The water is also inexpensive. The relatively expensive material is the Portland cement. Good strength is gained by having a mixture of these such that there is a dense packing of the aggregates and good interfacial contact, both achieved by having a mixture of aggregate sizes—thus the use of large gravel and small sand. The sand helps fill the voids between the various larger gravel particles. Mixing and contact is achieved with the correct amount of water. Sufficient water must be present to allow a wetting of the surfaces to occur along with providing some of the reactants for the setting up of the cement. Too much water creates large voids and weakens the concrete.

At times the difference between fillers and particulate composites is small. For particulate composites, the emphasis is on the added strength due to the presence of the particulates within the resin. Particle board is often used as an example of particulate composites where the particulates are pieces of wood or wood-like or derived material. Unlike plywood, particle board generally uses flakes or chips of wood rather than layers of wood.

8.3 LONG-FIBER COMPOSITES—THEORY

Because of the use of new fibers and technology, most of the composites discussed here are referred to as "space-age" and "advanced materials" composites.

According to the ASTM definition, fillers are relatively inert while reinforcements improve the properties of the materials to which they are added. Actually, few fillers are used that do not improve properties, but reinforcing fibers produce dramatic improvements to the physical properties of the material to which it is added, generally to form composites.

The transverse modulus (M_T) and many other properties of a long fiber–resin composite may be estimated from the law of mixtures. The longitudinal modulus (M_L) may be estimated from the Kelly–Tyson equation (Equation 8.5), where the longitudinal modulus is proportional to the sum of the fiber modulus (M_F) and the resin matrix modulus (M_M). Each modulus is based on a fractional volume (c). The constant k is equal to 1 for parallel continuous filaments and decreases for more randomly arranged shorter filaments.

$$M_L = kM_Fc_F + M_Mc_M \tag{8.5}$$

Since the contribution of the resin matrix is small in a strong composite, the second term in the Kelly–Tyson term can be disregarded. Thus, the longitudinal modulus is dependent on the reinforcement modulus, which is independent of the diameter of the reinforcing fiber.

As noted before, for the most part, the resulting materials from the use of reinforcements are composites. Composites are materials that contain strong fibers embedded in a continuous phase. The fibers are called "*reinforcement*" and the continuous phase is called the *matrix*. While the continuous phase can be a metallic alloy or inorganic material, the continuous phase is typically an organic polymer that is termed a "resin." Composites can be fabricated into almost any shape and after hardening, they can be machined, painted, and so on as desired.

While there is a lot of science and "space-age technology" involved in the construction of composites, many composites have been initially formulated through a combination of this science and "trial-and-error" giving "recipes" that contain the nature and form of the fiber and matrix materials, amounts, additives, and processing conditions.

Composites have high tensile strengths (on the order of thousands of MPa), high Young's modulus (on the order of hundreds of GPa) and good resistance to weathering exceeding the bulk properties of most metals. The resinous matrix, by itself, is typically not particularly strong relative to the composite. Further, the overall strength of a single fiber is low. In combination, the matrix–fiber composite becomes strong. The resin acts as a transfer agent, transferring and distributing applied stresses to the fibers. In general, the fibers should have aspect ratios (ratio of length to diameter) exceeding 100, often much larger. Most fibers are thin (less than 20 um thick, about a tenth the thickness of a human hair). Fibers should have a high tensile strength and most have a high stiffness, that is, low strain for high stress or little elongation as high forces are applied.

There exists a relationship between the "ideal" length of a fiber and the amount of adhesion between the matrix and the fiber. For instance, assume that only the tip, one end, of a fiber is placed in a resin (Figure 8.2—left). The fiber is pulled. The adhesion is insufficient to "hold" the fiber and it is pulled from the resin (Figure 8.2—top right). The experiment is repeated until the fiber is broken (outside the matrix) rather than being pulled (without breaking) from the resin (Figure 8.2—right bottom). Somewhere between the two extremes, there is a length where there exists a balance between the strength of the fiber and the adhesion between the fiber and matrix. Most modern composites utilize fiber/matrix combinations that exploit this "balance."

Mathematically the critical fiber length necessary for effective strengthening and stiffening can be described as follows:

Critical fiber length = (Ultimate or tensile strength times fiber diameter/2) times the fiber–matrix bond strength OR the shear yield strength of the matrix—whichever is smaller.

Fibers where the fiber length is greater than this critical fiber length are called *continuous fibers* while those that are less than this critical length are called *discontinuous* or *short fibers*. Little transference of stress and thus little reinforcement is achieved for short fibers. Thus, fibers whose lengths exceed the critical fiber length are typically used.

Fibers can be divided according to their diameters. Whiskers are very thin single crystals that have large length to diameter ratios. They have a high degree of crystalline perfection and are essentially flaw free. They are some of the strongest materials know. Whisker materials include graphite, silicon carbide, aluminum oxide, and silicon nitride. Fine wires of tungsten, steel, and

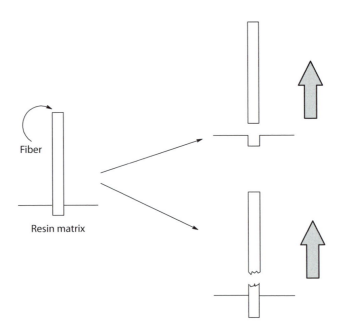

FIGURE 8.2 Tensile-loading experiments performed on single fibers embedded in a matrix. The illustration on the left is preapplication of the tensile loading and those on the right are postapplication of the tensile loading.

molybdenum are also used but here, even though they are fine relatively to other metal wires, they have large diameters. The most used fibers are "fibers," which are either crystalline or amorphous or semicrystalline with small diameters.

Fiber failure is usually of the "catastrophic" type where the failure is sudden. This is typical in polymeric materials where the material is broken at the "weak-link" and where the strength-related property is related to a combination of individual links (chains).

8.4   FIBERS AND RESINS

Tables 8.2 and 8.3 contain a partial listing of the main materials employed in the fabrication of composites. It is important to note that many of the entries given in Tables 8.1 and 8.2 represent whole families of materials. Thus, there are a large number of possible combinations, but not all combinations perform in a satisfactory manner. In general, good adhesion between the matrix and fiber is needed. Table 8.4 contains a listing of some of the more utilized combinations.

About 98% of the fibers employed in composites are glass (Sections 12.5 and 12.6), carbon (graphite, carbon fibers, etc.; Section 12.17), and aromatic nylons (often referred to as aramids; Section 4.8). New composites are emerging that employ carbon nanotubes and the fibers (Section 12.17). Asbestos (Section 12.13), a major fiber choice years ago, holds less than 1% of the market today because of the medical concerns linked to it.

Glass fibers are described using several terms, including fiberglass, glass fibers, and fibrous glass. Glass fibers are mainly composed of silicon dioxide glass. The glass fibers are "pulled" from the glass melt, forming fibers that range from 2 to 25 μm in diameter. The pulling action helps orientate the overall three-dimensional structure giving a material with greater strength and stiffness along the axis of the pull. The ability to pull fibers from molten glass was known for centuries, but fibrous glass was not produced commercially until the 1930s.

Table 8.5 contains a brief description of the most important glass fiber types.

TABLE 8.2
Frequently Employed Fibers in Composites

Alumina (Aluminum oxide)	Polyolefin
Aromatic nylons	Silicon nitride (Si_3N_4)
Boron	Titanium carbide (TiC)
Carbon and graphite	Tungsten carbide (WC)
Glass	Zirconia (ZrO_2)

TABLE 8.3
Polymer Resins Employed in the Fabrication of Composites

Thermosets	Thermoplastics
Epoxys	Nylons
Melamine-Formaldehyde	Polycarbonates
Phenol-Formaldehyde	Poly(ether ether ketone)
Polybenzimidazoles	Poly(ether ketone)
Polyimides	Poly(ether sulfones)
Polyesters	Poly(phenylene sulfide)
Silicones	Polyethylene
	Acetal
	Polyolefins

TABLE 8.4
Typically Employed Fiber/Resin Pairs

Alumina/	Carbon/nylon/epoxy
Alumina/polyimide	Carbon/polyimides
Boron/carbon/epoxy	Glass/epoxy
Boron/epoxy	Glass/carbon/polyester
Boron/polyimide	Glass/polyester
Boron/carbon/epoxy	Glass/polyimide
Carbon/acrylic	Glass/silicon
Carbon/epoxy	Nylon/epoxy

TABLE 8.5
Types of Glass Fibers

Designation	General Properties
C-Glass	Chemical resistant
E-Glass	"Typical" glass fiber
R-Glass and S-Glass	Stiffer and stronger than E-Glass

As with many polymers, the limits of strength are due to the presence of voids. For glass fibers, these voids generally occur on the surface, thus care is taken to protect these surfaces through surface treatments with methacrylatochromic chloride, vinyl trichlorosilanes, and other silanes. These surface agents chemically react with the fiber surface acting to repel and protect the surface from harmful agents such as moisture.

A number of kinds of carbon-intense fibers are used, the most common being carbon and graphite fibers and carbon black. As in the case of fibrous glass, surface voids are present. Carbon-intense fibers are often surface-treated with agents such as low molecular weight epoxy resins. Such surface treatments also aim at increasing the fiber–matrix adhesion.

Two general varieties of aromatic nylons are often employed. A less-stiff variety is employed when some flexibility is important, while a stiffer variety is used for applications where greater strength is required. While good adhesion with the resin is often desired, poor adhesion is sometimes an advantage such as in the construction of body armor where "delamination" is a useful mode for absorbing an impact.

As we understand materials better, we are able to utilize them for additional applications. It is known that "elongational" flow through orifices can result in the stretching and reorientation of polymer chains giving a stronger fiber in the direction of pull. Some polymers become entangled and the flow gives additional orientation. Finally, polymer solutions may be stable at rest, but under high rates of extrusion they may be removed from solution, forming a gel phase. These observations have allowed the production of a number of new polyolefin fibers, including ultrahigh-modulus polyethylene fibers that have low density but relatively high tensile strength with an elongation at break over two times greater than glass and aromatic nylon fibers.

Both thermoset and thermoplastic resin systems are employed in the construction of composites (Table 8.3). The most common thermoset resins are polyimides, unsaturated polyesters, epoxys, phenol-formaldehydes, and amino-formaldehydes. A wide variety of thermoplastic resins have been developed.

8.5 LONG-FIBER COMPOSITES—APPLICATIONS

Many of the applications for composite materials involve their (relative) light weight, resistance to weathering and chemicals, and their ability to be easily fabricated and machined. Bulk applications employ composites that are relatively inexpensive. Combinations of rigorous specifications, low volume, specific machining and fabrication specifications, and comparative price to alternative materials and solutions allow more expensive specialized composites to be developed and utilized.

Applications are increasing. Following is a brief description of some of these. One of the largest and oldest applications of composites is the construction of water-going vessels from rowboats, sailboats, racing boats, and motor craft to large seagoing ships. The use of fresh water and salt water resistant composites allowed the boating industry to grow and today includes a range from individually operated backyard construction to the use of large boatyards producing craft on an assembly line. Most of these craft are composed of fiberglass and fiberglass/carbon-combination composites.

Compositions are also important in the construction of objects to both propel material into and material to exist in outer space. Because of the large amount of fuel required to propel spacecraft into outer space, weight reduction, offered by composites, is essential. The polymeric nature of composites also makes it an ideal material to resist degradation caused by the vacuum of outer space.

Many biomaterials are composites. Bone and skin are relatively light compared to metals. Composite structures can approach the densities of bone and skin and offer the necessary inertness and strength to act as body-part substitutes.

Power-assisted arms been made by placing hot-form strips of closed-cell PE foam over the cast of an arm. Grooves are cut into these strips before application and carbon/resin are added to the grooves. The resulting product is strong, light, and the cushioned PE strips soften the attachment site of the arm to the living body. Artificial legs can be fashioned in glass/polyester and filled with polyurethane foam adding strength to the thin-shelled glass/polyester shell. Artificial legs are also made from carbon/epoxy composite materials. Some of these contain a strong interior core with a soft, flexible "skin."

Carbon/epoxy "plates" are now used in bone surgery replacing the titanium plates that had previously been employed. Usually, a layer of connective tissue forms about the composite plate.

Rejection of composite materials typically does not occur, but as is the case of all biomaterials, compatibility is a major criterion. Often, lack of biocompatibility has been found to be the result of impurities (often additives) found in the materials. Removal of these impurities allows the use of the purified materials.

Carbon and carbon/glass composites are being used to make "advanced-material" fishing rods, bicycle frames, golf clubs, baseball bats, racquets, skis and ski poles, basketball backboards, and so on. These come in one color—black—because the carbon fibers are black. Even so, they can be coated with about any color desired.

Composites are being employed in a number of automotive applications. These include racing car bodies as well as "regular" automobiles. Most automobiles have the lower exterior panels composed of rubbery, plastic blends, and/or composite materials. Corvettes have composite bodies that allow a light-weight vehicle with decent fuel economy and they do not rust. Other parts such as drive shafts and leaf springs in private cars and heavy trucks, antennas, and bumpers are being made from composite materials.

Industrial storage vessels, pipes, reaction vessels, and pumps are now made from composite materials. They offer needed resistance to corrosion, acids and bases, oils and gases, salt solutions, and the necessary strength and ease of fabrication to allow their continued adoption as a major industrial "building" material.

The Gulf War spot lighted the use of composite materials in the new-age aircraft. The bodies of both the Stealth fighter and bomber are mainly carbon composites. The versatility is apparent when one realizes that the Gossamer Albatross, the first plane to cross the English Channel with human power, was largely composite materials, including a carbon/epoxy and aromatic nylon composite body, propeller containing a carbon composite core, and so on.

The growth of composite materials in the aerospace industry is generally due to their outstanding strength and resistance to weathering and friction and their light weight, allowing fuel reduction savings. Its growth in commercial aircraft is seen in the increased use of fiber glass composite material in succeeding families of Boeing aircraft from about 20 sq yards for the 707, to 200 sq yd for the 727, to 300 sq yd for the 737, and more than 1,000 sq yd for the 747. This amount is increased in the Boeing 767 and includes other structural applications of other space-age composites. Thus, the Boeing 767 uses carbon/aromatic nylon/epoxy landing gear doors and wing-to-body fairings.

Until the late 1960s, almost all tactical aircraft were largely titanium. While titanium is relatively light, it is costly and has demanding production requirements so that its use was limited to moderate-temperature aircraft applications. Today, most tactical aircraft have a sizable component that is polymeric, mainly composite. The Boeing F/A 18E/F and Lockheed F/A-22 have about 25%, by weight, composite material. It is projected that future military aircraft will have more than 35% composite materials.

Composites have displaced more conventional materials because they are lighter with greater strength and stiffness, allowing them to carry a greater payload further. Composites are also relatively insensitive to flaws. In comparison to metals, fatigue testing of composites shows that they have a high resistance to cracking and fracture propagation. They are stable and are not subject to corrosion. However, in the design process, particular care must be taken with respect to the metal-composite interface because galvanic action of some metals will corrode when in contact with certain composites such as the carbon graphite/resin laminates.

The Stealth Bomber, more accurately known as the Northrop Grumman B-2 Spirit Stealth Bomber, is cited as the largest composite structure produced with more than 30% of the weight being carbon graphite epoxy composites. The Stealth Bomber was originally slated to hunt Russian mobile missiles that were built in the 1980s. It was deployed in 1993 and has a wingspan of more than 52 m, a length of 21 m, and is more than 5 m high. The reported cost of a single B-2 is varied ranging from about $2.4 billion to a low of $500 million. In any case, it is expensive and only about 21 have been produced with no plans to build more. The technology associated with the B-2 has been continuously upgraded, allowing it to perform tasks with even greater accuracy and stealthiness.

The word stealth comes from ancient roots meaning "to steal." There are a number of stealth aircraft in operation today with others coming on line. To be "stealthy," a plane should

- Be difficult to see with the eye
- Make little or no noise (achieved by muffling the engines)
- Display little heat from engines and other moving parts
- Minimize a production of contrails and other signs, and
- Absorb and scatter radar beams

The most common mode of aircraft detection is radar. Essentially, radar is the detection of radio waves that have been "thrown out" and which bounce off objects returning to the site of origin. Today's radar, if properly used, can help identify the location, speed, and identity of the aircraft. The radar cross-section (RCS) of an aircraft is how much echo the plane sends from radar. Birds have an RCS of about 0.01 m^2. The Stealth Bomber has a RCS of 0.75 m^2. The Stealth Bomber, and many stealth aircraft, gains their stealth character mainly from both the shape of the aircraft and the presence of radar absorbing material (RAM). The RAM is made to absorb and eliminate radio waves rather than reflecting them. Most of the RAM materials are polymeric.

The Stealth Bomber has a covering of RAM, skin. The basic elements of the composite cover consist of graphite carbon fibers embedded in an epoxy resin. The carbon fibers dissipate some of the absorbed radiation by ohmic heating. The material also contains ferrite particles that are important because they are good "lousy" radar absorbers. The good absorbency of radar signals is dependent on a low dielectric constant. In principle, the thicker the coating the greater the absorption of signals. Thick coatings increase cost and weight and decrease the aerodynamics of the aircraft. (In truth, the flying wing is aerodynamically unstable and is flown with a quadruple-redundant fly-by-wire [FBW] system.) The thickness of the layers needed to provide the necessary radar signature can be reduced through use of materials with intermediate absorption behavior, ferrite particles, to smooth the interference between the aircraft and the surrounding air since extreme differences cause strong radar reflections.

The graphite resin layers are laid in a manner similar to that employed in making the hull of a fiberglass boat hull. This layering, over a frame or mold, is accomplished using computer guided assistance. Thin layers of gold foil are also applied to the canopy to give radar reflecting characteristics similar to the rest of the aircraft.

The Stealth Bomber is shaped like a wafer-thin Batman boomerang with smooth, knife-like curved surfaces that lack right angles. The reflective nature of the design can be understood by considering looking straight on at a mirror. Your image will be reflected back at you. Tilt the mirror 45 degrees and it will reflect your image upward. The bomber design is a large system of triangles with flat surfaces angled to deflect radar waves away from the radar emitter. It scrambles the radar all about but not back to the receiver. This shape is also intended to offer a minimum of disruption to the air, that is, the shape brings about a minimum amount of air resistance, resulting in a minimization of radar difference between the aircraft and its surroundings.

The use of composites allow there to be few seams, thus reducing radar reflections. For years, seams were believed necessary for large aircraft. For instance, the wings of the large bombers actually "flap" in the wind, contributing to the ability of the aircraft to remain in the air for long periods of time. Without seams and some flexibility these large metal-based aircraft would literally snap apart. Composite assemblies have a minimum of flexibility. To discourage reflections in unwanted directions, the electrical continuity is maintained over the entire surface of the bomber. Thus, all moving and major body parts are tight fitting. The composite nature of the outer skin allows this to be closely achieved.

The engines are buried deep within the aircrafts interior, reducing noise and "heat signature." The heat signature is further reduced by mixing the jet's exhaust with cooled air and exiting it through slit-like vents.

Stealth aircraft are often painted a medium gray or bluish gray that matches the sky so that is does not stand out. Coatings are composed of RAM that penetrates into cracks, reducing the number of minor repairs and use of special tape for repair. Every screw, panel, seam, and gap is covered with a special tape or the RAM coating adding to the bomber's stealthiness.

Because most traditional polymers have low dielectric constants, they can be used as RAMs and many have been used in various applications in military and civil applications. While the B-2 largely depends on a graphite carbon-impregnated composite for its integrity and stealthiness, many other polymers are used as RAMs. Many of these are various elastomeric materials. Neoprene is widely used in naval applications as a RAM because of its good weather resistance. Nitrile is used for fuel and oil resistance and fluoroelastomers are used where wide temperature ranges are encountered. Often, thin flexible sheets or layers of the elastomer are adhered to a metal substrate. While the polysiloxanes offer good RAM behavior, they are difficult to adhere to many metal structures and thus, are not as widely employed in comparison to neoprene and nitrile rubbers. Products are also made using various molding methods. To improve weather resistance, the absorber is often painted with another RAM, an epoxy or urethane-based coating.

The modern ski is a good example of the use of composites to make a product with unique properties (Figure 8.3). The top and sides are composed of ABS polymer that has a low T_g, allowing it to remain flexible even at low temperatures. It is employed for cosmetic and containment purposes. Polyurethane forms the core and a damping layer that acts as filler and to improve chatter resistance. The base is a carbon-impregnated matrix composite that is hard, strong, and with good abrasion resistance. There are numerous layers of fibrous glass that are a mixture of bidirectional layers to

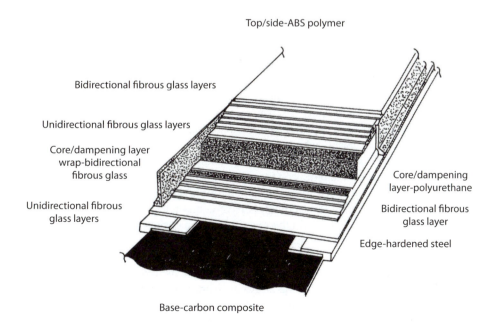

FIGURE 8.3 Cut-away illustration of a modern ski.

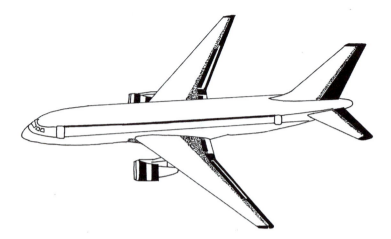

FIGURE 8.4 Use of graphite (solid) and graphite/Kevlar (TM) (dotted) composites in the exterior of the Boeing 767 passenger jet. Sites include wing tips, fixed trailing edge panels, inboard and outboard spoilers, and inboard and outboard ailerons for the large wings and for the tail wings the fin tip, rudder, elevators, stabilizer tips, and fin fixed trailing edge panels.

provide torsional stiffness, unidirectional layers that provide longitudinal stiffness with bidirectional layers of fiberglass acting as outer layers to the polyurethane layers composing a torsional box. The only major noncomposite material is the hardened steel edge that assists in turning by cutting into the ice. This combination works together to give a light, flexible, shock absorbing, tough ski.

Composites are also used extensively where light but very strong materials are needed such as in the construction of the new Boeing 767, where composites play a critical role in the construction of the exterior (Figure 8.4). The next generation of commercial airliners such as the 787 Boeing Dreamliner and the Airbus A350-XWB will use composites in about one-half of their structural components. The 787 has a one-piece all composite fuselage.

8.6 NANOCOMPOSITES

Nature has employed nanomaterials since the beginning of time. Much of the inorganic part of our soil is a nanomaterial with the ability to filter out particles often on a molecular or nanolevel. The driving force toward many of the nanomaterials is that they can offer new properties or enhanced properties unobtainable with so-called traditional bulk materials. Along with light weight, high strength to weight features, and small size, new properties may emerge because of the very high surface area to mass where surface atomic and molecular interactions become critical. The nanoworld is often defined for materials where some dimension is on the order of 1–100 nm. In a real way, single linear polymers are nanomaterials since the diameter of the single chain is within this range. The carbon–carbon bond length is on the order of 0.15 nm or the average zigzag bond length is about 0.13 nm. While some short- to moderate-length vinyl polymers have contour lengths less than 100 nm, higher molecular weight polymers have contour lengths that far exceed this. Even so, individual polymer chains, fall within the realm of nanomaterials when they act independently. Since the cumulative attractive forces between chains are large, polymer chains generally act in concert with other polymer chains leading to properties that are dependent on the bulk material. Chain folding, inexact coupling, chain branching are some of the reasons that bulk properties fall short of theoretical properties, but with the ability to work with chains individually, strength and related properties are approaching theoretical values. Much of the nanorevolution with materials involves how to synthesize and treat materials on an individual basis as well as visualizing uses for these materials.

The ultimate strength and properties of many materials is dependent upon the intimate contact between the various members. Thus, for ceramics, nanosized particles allow a more homogeneous structure, resulting in stronger ceramic materials.

As noted before, nanocomposites have also been with us since almost the beginning of time. Our bones are examples of nanocomposites. The reinforcement material is plate-like crystals of hydroxyapatite, $Ca_{10}(PO_4)_6(OH)_2$, with a continuous phase of collagen fibers. The shell of a mollusk is microlaminated containing as the reinforcement aragonite (a crystalline form of calcium carbonate) and the matrix is a rubbery material. Allowing nature to be a source of ideas is a continuing theme in synthetic polymer science, including modification of natural polymers. Much of the renewed interest in nanocomposite materials is the direct result of the availability of new nano-building blocks.

Within a composite material, much of the ultimate strength comes from the intimate contact the fiber has with the matrix material. Nanofibers allow more contact between the fiber (on a weight basis) and the matrix, resulting in a stronger composite because of an increased fiber surface–matrix interface.

A number of inorganic/organic nanocomposites have been made. These include nanoinorganics including nanofibers from silicon nitride, titanium (IV) oxide, zirconia (ZrO_2), alumina (Al_2O_3), titanium carbide (TiC), and tungsten carbide (WC). It also includes the use of special clays (layered silicates) mixed with nylons to form nanocomposites. The clay layers are separated giving platelets about 1 nm thick. These nylon–clay microcomposites are used to make the air intake cover of some of the Toyota automobiles. These individual clay platelets have also been used to form nearly single-layer polymer chain sheets similar to lignin. The interaction with the silicate surface encourages the polymer chains to take different arrangements. To be effective, the hydrophilic silicate surface is generally modified to accommodate the more hydrophobic common monomers and polymers.

While carbon fiber (thickness on the order of 1,000 nm) composites offer very strong materials, carbon nanotubes make even stronger composites. These carbon nanotubes have aspect ratios of more than 1,000 (ratio of length to diameter). Further, because some carbon nanotubes are electrically conductive, composites containing them can be made to be conductive. A number of carbon nanotube matrixes have been made, including using a number of engineering resins such as polyesters, nylons, polycarbonates, and poly(phenylene ether).

Individual polymer chains can be more flexible than groups of chains (bulk) even when the polymer is generally considered to be rigid. This is presumably because single chains have less torsional strain imparted by near neighbors and various chain entanglements and associations are not present. Compared with carbon fibers, carbon nanotubes imbedded within a polymer matrix can withstand much greater deformations before they break. Further, nanomaterials are generally more efficient in transferring applied load from the nanomaterial to the matrix. These factors contribute to the greater strength of carbon nanotube composites.

As noted before, adhesion between the reinforcing agent and matrix is important. Some matrix materials do not adhere well with certain fibers. This is partially overcome through the introduction of defects or functional groups onto the nanomaterials that act as hooks to anchor them to the matrix material.

Research with tires continues to be active with the end goals of reducing tire weight and increasing tire lifetime. For example, truck tires are now capable of running 750,000 miles with the carcass married to four tire threads. The truck tire liners weight about 9 pounds. A decrease in weight of 50% would translate into a significant increase in mileage of 3–5 mpg. Solutions should be inexpensive employing readily available and abundant materials.

Nanocomposites are being used in tires, in particular tire inner liners. Here, less permeable inner layers are achieved by the introduction of "clad" layers that allow the use of a thinner inner liner, resulting in an overall lighter tire. Tire inner layers are typically derived from butyl rubber, often halogenated butyl rubber.

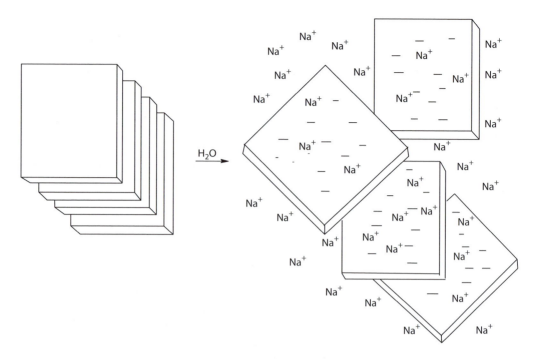

FIGURE 8.5   Formation of individual silicate sheets from exposure to water.

Exfoliate means to open, such as to open the pages of a book. New sheet-like nanocomposites are being investigated in the production of tire inner layers. These approaches focus on the use of two different materials, silicates and graphite. Most silicates and graphite exist in layered groupings.

Layered-clay silicates, generally from the intermediate-grained montmorillonite kaolin clay, are often used as filler in plastics and in the production of pottery and other ceramic items. These silicates consist of the silicate sheets held together mostly by the sodium cation with lesser amounts of other metal ions such as iron, copper, nickel, and so on. There are several approaches to open these silicate layers.

The layers of kaolin clay open up when treated with water. The water acts to remove the cations that hold together the clay matrix. The resulting sheets are negatively charged since the positively charged sodium ion is removed. These negatively charged sheets repel one another, resulting in the formation of nanosheets as shown in Figure 8.5.

Toyota works with kaolin clay that has about 10 nm between the sheets. The clay is treated with caprolactam monomer that impregnates the clay sheets, increasing the distance between sheets to 30–50 nm. The caprolactam monomer impregnating the silicate sheets is then polymerized giving nylon-silicate sheets. This sequence is pictured in Figure 8.6.

Another approach utilizes highly oriented pyrolytic graphite that is separated into individual graphite sheets as the inner layer material. In each case, the nanoclay or graphite material is present in about 2% by weight in the tire inner liner.

As the inner liner is processed, the sheets are caused to align forming a clad that reduces the ability for air to flow from the tire.

8.7   FABRICATION

Fabrication of composites can be divided into three general approaches—fibrous, structural materials, including laminates and particulates. Following is a brief description of each.

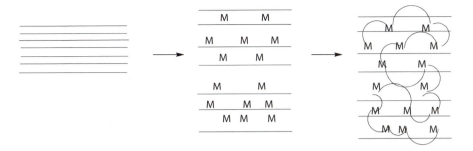

FIGURE 8.6 Silicate sheets impregnated with caprolactam monomer, M, which are eventually polymerized.

8.7.1 Processing of Fiber-Reinforced Composites

These exists a wide variety of particular operations but briefly they can be described in terms of filament winding, preimpregnation of the fiber with the partially cured resin, and pultrusion. Pultrusion is used to produce rods, tubes, and beams employing continuous fibers that have a constant cross-sectional shape. The fiber (as a continuous fiber bundle, weave or tow) is impregnated with a thermosetting resin and pulled through a die that shapes and establishes the fiber to resin ratio. This stock is then pulled through a curing die that can machine or cut, producing the final shape such as filled and hollow tubes and sheets.

The term used for continuous fiber reinforcement *preimpregnation* with a polymer resin that is only partially cured is "prepreg." Prepreg material is generally delivered to the customer in the form of a tape. The customer then molds and forms the tape material into the desired shape finally curing the material without having to add any additional resin. Preparation of the prepreg can be carried out using a calendaring process. Briefly, fiber from many spools are sandwiched and pressed between sheets of heated resin with the resin heated to allow impregnation but not so high as to be very fluid.

Thus, the fiber is impregnated in the partially cured resin. Depending upon the assembly the fiber is usually unidirectional, but can be made so that the fibers are bidirectional or are in some other combination. The process of fitting the prepreg into, generally onto, the mold is called "lay-up." In general, a number of layers of prepreg are used. The lay-up may be done by hand, called hand lay-up, or done automatically, or by some combination of automatic and hand lay-up. As expected, hand lay-up is more costly but is needed where one-of-a-kind products are produced.

In *filament winding,* the fiber is wound to form a desired pattern, usually but not necessarily hollow and cylindrical. The fiber is passed through the resin and then spun onto a mandrel. After the desired number of layers of fiber is added, it is cured. Prepregs can be filament wound. With the advent of new machinery, complex shapes and designs of the filament can be readily wound.

8.7.2 Structural Composites

Structural composites can be combinations of homogeneous and composite materials. Laminar composites are composed of two-dimensional sheets that generally have a preferred high-strength direction. The layers are stacked so that the preferred high-strength directions are different, generally at right angles to one another. The composition is held together by a resin. This resin can be applied as simply an adhesive to the various surfaces of the individual sheets or the sheet can be soaked in the resin before laying the sheets together. In either case, the bonding is usually of a physical type. Plywood is an example of a laminar composite. Laminar fibrous glass sheets are included as part of the modern ski construction. These fibrous glass sheets are fiber-reinforced composites used together as laminar composites.

Laminar materials are produced by a variety of techniques. Coextrusion blow molding produces a number of common food containers that consist of multilayers such as layers consisting of polypropylene/adhesive/poly(vinyl alcohol)/adhesive/adhesive/polypropylene.

Sandwich composites are combinations where a central core(s) is surrounded generally by stronger outer layers. Sandwich composites are present in the modern ski and as high-temperature stable materials used in the space program. Some cores are very light acting something like filler with respect to high strength, with the strength provided by the outer panels. Simple corrugated cardboard is an example of a honeycomb core sandwich structure except that the outer paper-intense layers are not particularly strong. Even in the case of similar polyethylene and polypropylene corrugated structures, the outer layers are not appreciatively stronger than the inner layer. In these cases the combination acts to give a light weight somewhat strong combination, but they are not truly composites but simply exploit a common construction.

8.7.3 Laminating

Laminating is a simple binding together of different layers of materials. The binding materials are often thermosetting plastics and resins. The materials to be bound together can be paper, cloth, wood, or fibrous glass. These are often referred to as the reinforcing materials. Typically, sheets, impregnated by a binding material, are stacked between highly polished metal plates, subjected to high pressure and heat in a hydraulic press producing a bonded product, which may be subsequently treated, depending on its final use. The end product may be flat, rod-shaped, tubular, rounded, or some other formed shape.

Reinforced plastics differ from high pressure laminates in that little or no pressure is employed. For instance, in making formed shapes, impregnated reinforcing material is cut to a desired shape, the various layers are added to a mold, and the molding is completed by heating the mold. This process is favored over the high pressure process because of the use of a simpler, lower-cost mold and the production of strain-free products.

8.7.4 Particulate

Particulate composites consist of the reinforcing materials being dispersed throughout the resin. Unlike fibrous composites, the reinforcing material in the particulate composites is more bulky and is not fibrous in nature. These particulates can be of many shapes and relative sizes. One of the employed shapes is that of a flat rounded rectangular "disk-like" shape where the disk-like particles can form overlapping layers, thus reinforcing one another (Figure 8.7). Particle board is an example of particulate composites. At times, there is little to differentiate between fillers and reinforcing agents. The emphasis on reinforcing agents is one where increased strength results from the presence of the reinforcing agent, whereas fillers fulfill the role of simply increasing the bulk of the material. Further, fillers are generally smaller than reinforcing agents.

8.8 METAL-MATRIX COMPOSITES

There were driving forces for the development of composites in the 1950s and 1960s. These included the need for materials that offered greater strength/density ratios. Another driving force was the promise that the marriage between two materials would produce materials with especially high-strength and elastic moduli. There was also the need for new materials to meet the demands of the aerospace industries that were rapidly developing. Applications for these materials rapidly spilled over from the aerospace and military into the general public with advancements made in one sector fueling advances in another sector. Many of the first advanced materials were of the clad or laminate type of composite. Here we will not be concerned with clad or laminate type of composites but rather those where the two phases are mixed together.

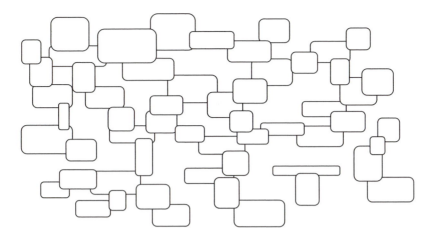

FIGURE 8.7 Composite formed from the reinforcing material being present as disk-like rounded rectangular particles.

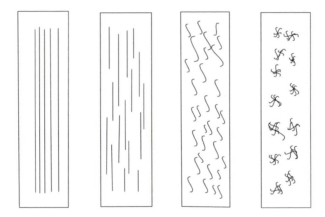

FIGURE 8.8 General forms for discontinuous-phase materials, from left, are continuous or long fibers, short fibers, whiskers, and particulates.

Space-age or advanced composites were front stage in the early 1960s with the development of high-modulus whiskers and filaments. Whiskers were easily made but at that time the science and engineering for the development of good composites had not been established. Counter, high-modulus fibers from boron were successfully impregnated into epoxy forming high-strength and modulus composites.

Often there is a borrowing of terms between metal-intense materials science and polymer-intense materials science where there is really little relationship between the two topics. This is not the case with metal-matrix composites (MMCs). While the materials are often different, there are a number of similarities. For polymer-intense composites, the matrix materials are organic polymers. For MMCs the matrix materials are typically a metal or less likely an alloy. Popular metals include aluminum, copper, copper-alloys, magnesium, titanium, and so-called superalloys.

In polymer-matrix composites, the noncontinuous phase or reinforcement material is a fiber such as glass, carbon fibers (graphite), aromatic nylons, and a number of inorganic fibers including WC, TiC, ZrO_2, and Al_2O_3. For the MMCs, the discontinuous phase generally exists as fibers, wires, whiskers, and particulates (Figure 8.8). A few of the discontinuous-phase materials overlap with the listing given for polymer-intense composites, including WC, TiC, Al_2O_3, and graphite,

TABLE 8.6
Typical Metal-Composite Matches

| Matrix Material | Discontinuous Phase | |
	Form	Material
Aluminum	Fibers	Boron
		Alumina
		Graphite
		Alumina-silica
		Silicon carbide
	Whiskers	Silicon carbide
	Particulates	Silicon carbide
		Boron carbide
Titanium	Fibers	Boron (coated)
		Silicon carbide
	Particulates	Titanium carbide
Copper	Fibers	Graphite
		Silicon carbide
	Particulates	Boron carbide
		Silicon carbide
		Titanium carbide
	Wires	Niobium-titanium
		Niobium-tin
Magnesium	Fibers	Alumina
		Graphite
	Particulates	Boron carbide
		Silicon carbide
	Whiskers	Silicon carbide
Superalloys	Wires	Tungsten

but there are a number of other materials whose fibers and whiskers are employed, including silicon carbide, boron carbide, coated boron, and wire materials such as niobium-titanium and niobium-tin and tungsten. By volume, the amount of whiskers, wires, and particulates is greater for the MMCs.

As in the case with polymer-intense composites, the matrix and fiber must be matched for decent properties. Table 8.6 contains a listing of typical matrix/fiber mixes.

In some cases, alloy formation is possible. The term MMC is restricted to materials where such alloy formation does not occur so that there is a phase separation between the matrix and reinforcing material.

In comparison to single-metal materials such as aluminum, copper, and iron, MMCs generally have

A higher strength-to-density ratio
A better fatigue and wear resistance
A better high-temperature strength
A lower creep related to lower coefficients of thermal expansion and they are stiffer

In comparison to polymer-intense composites, MMCs offer

No moisture absorption
Greater fire resistance
Higher use temperatures
Greater radiation resistance
Greater stiffness and strength, and
Higher thermal and electrical conductivities

MMCs also have some disadvantages in comparison to polymer-matrix composites. These include

Higher cost
Newer and less-developed technology and scientific understanding
Less known long-term information
Generally more complex fabrication, and
Greater weight

As noted above, the range of fibers employed does not precisely overlap with those employed for organic composites. Because the formation of the MMCs generally requires melting of the metal matrix the fibers need to have some stability to relatively high temperatures. Such fibers include graphite, silicon carbide, boron, alumina-silica, and alumina fibers. Most of these are available as continuous and discontinuous fibers. It also includes a number of thin metal wires made from tungsten, titanium, molybdenum, and beryllium.

As with organic-matrix composites, the orientation of the reinforcing material determines whether the properties will be isotropic or oriented in a preferential direction so that the strength and stiffness are greater in the direction of the fiber orientation.

There are a number of differences between organic and MMCs. These include the following: First, the metal matrix can also have considerable strength themselves so their contribution to the overall strength is more important than for organic-matrix composites. A second difference more often encountered for MMCs is a greater difference between the coefficient of expansions between the reinforcing material and metal matrix. Because of the greater use temperature differences often required for MMCs, these differences become more important. Such differences can result in large residual stresses in MMCs that may result in yielding. A third difference is related to the relative lower flexibility of MMCs. This leads to a greater need to be concerned with the marrying or joining of such composite parts. Many methods of joining these composite parts have been developed. A fourth difference is the possible greater reactivity between the matrix and fiber for MMCs. This limits combinations but has been overcome in many situations. One major approach is to place a barrier coating onto the reinforcement. For example, boron carbide is applied as a barrier coating to boron fibers, allowing their use to reinforce titanium. Because these coatings can be "rubbed" away from usage, composites made from these coated reinforcements should be monitored more closely and more often.

Some effort has gone into working with copper wire containing niobium and tin looking for how to make niobium-tin alloys which is one of the better low-temperature superconductors. Such alloys are brittle so copper was added in the hopes of creating a less-brittle material. It was noticed that the copper-niobium-tin, and also simply copper-niobium combination gave a material with greater than expected strength. Because copper and niobium are not miscible at low niobium concentrations, it formed dendrites within the copper matrix allowing strong, ductile wires and rods to be drawn. As the size of the rods became smaller, niobium filaments, about only 10 nm, about 30 atoms wide, across formed. If cast, the thickness is 5 microns thick, much thicker than the niobium filaments formed from the drawing down of the copper-niobium rod. This is then not only a MMC, but a metal–metal composite.

As with organic-matrix composites, advances with MMCs are continually being made to answer the call for better materials to meet the ever increasingly restrictive requirements and the ever broadening applications.

8.9 SUMMARY

1. Fillers are relatively inert materials that usually add bulk, but when well chosen they can enhance physical and chemical properties. Many natural and synthetic materials are used as fillers today. These include polysaccharides (cellulosics), lignin, carbon-based materials, glass, and other inorganic materials.
2. The most important reinforced materials are composites. They contain a continuous phase and a discontinuous phase. There are a variety of composites. These include particulate composites, structural composites, and fiber composites.
3. Fiber-reinforced composites contain strong fibers embedded in a continuous phase. They form the basis of many of the advanced and space-age products. They are important because they offer strength without weight and good resistance to weathering. Typical fibers are fibrous glass, carbon-based, aromatic nylons, and polyolefins. Typical resins are polyimides, polyesters, epoxys, phenol-formaldehyde, and many synthetic polymers. Applications include biomedical, boating, aerospace and outer space, sports, automotive, and industry.
4. Nanocomposites employ nanofibers that allow for much greater fiber–resin surface contact per mass of fiber, and consequently they generally offer greater strength in comparison to similar nonnanocomposites.
5. Concrete and particle board are examples of particulate composites. Plywood and Formica are laminate composites.
6. Metal-matrix composites offer a number of possible advantages over classical organic composites, including greater radiation resistance, greater strength, greater fire resistance greater stiffness, and higher thermal and electrical conductivities.

GLOSSARY

Aspect radio: Ratio of length to diameter.
Comminuted: Finely divided.
Composites: Materials that often contain strong fibers embedded in a continuous phase called a matrix or resin,
Diatomaceous earth: Siliceous skeletons of diatoms.
Discontinuous phase: Discrete filler additive, such as fibers, in a composite.
Extender: Term sometimes applied to inexpensive filler.
Fiberglass: Trade name for fibrous glass.
Fibrous filler: Fiber with an aspect ratio of at least 150:1.
Fibrous glass: Filaments made from molten glass.
Filament winding: Process in which resin-impregnated continuous filaments are sound on a mandrel and the composite is cured.
Filler: Usually a relatively inert material used as the discontinuous additive; generally inexpensive.
Fuller's earth: Diatomaceous earth.
Graphite fibers: Fibers made by the pyrolysis of polyacrylonitrile fibers.
Lamellar: Sheet like.
Laminate: Composite consisting of layers adhered by a resin.
Pultrusion: Process in which bundles of resin-impregnated filaments are passed through an orifice and cured.

Reinforced plastic: Composite whose additional strength is dependent on a fibrous additive.

Roving: Bundle of untwisted strands.

Sheet molding compound (SMC): Resin-impregnated mat.

Strand: Bundle of filaments.

Syntactic foam: Composite of resin and hollow spheres.

Whiskers: Single crystals used as reinforcement; extremely strong.

Wood flour: Attrition ground, slightly fibrous wood particles.

EXERCISES

1. Name three unfilled polymers.
2. What is the continuous phase in wood?
3. What filler is used in Bakelite?
4. Name three laminated plastics.
5. How would you change a glass sphere from an extender to a reinforcing filler?
6. If one stirs a 5 mL volume of glass beads in 1 L of glycerol, which will have the higher viscosity, small or large beads?
7. When used in equal volume, which will have the higher viscosity: (a) a suspension of loosely packed spheres or (b) a suspension of tightly packed spheres?
8. Why is the segmental mobility of a polymer reduced by the presence of filler?
9. What effect does filler have on the T_g?
10. Which would yield the stronger composite: (a) peanut shell flour or (b) wood flour?
11. What is the advantage and disadvantage, if any, of α-cellulose over wood flour?
12. What filler is used in decorative laminates such as Formica table tops?
13. What are potential advantages of natural fillers?
14. What is the disadvantage in using chipped glass as a filler in comparison to glass spheres?
15. What is the advantage of using BMC and SMC over hand lay-up techniques such as those used to make some boat hulls?
16. Was carbon black always used as a reinforcing filler for tires?
17. Why does a tire filled with oxygen lose pressure faster than a tire filled with nitrogen gas.
18. You discovered a new fiber and want to see if it might make a good composite. How might you do this rapidly to a first approximation?
19. What are some natural sources of fibers?
20. Why is phase separation an important consideration in the formation of composite materials?
21. What would you expect with respect to the density of metal composites and regular organic composites?
22. What are some advantages of MMCs in comparison to classical organic composites?

ADDITIONAL READING

Bart, J. (2005): *Additives in Polymers*, Wiley, Hoboken.

Bolgar, M., Hubball, J., Meronek, S. (2008): *Handbook for the Chemical Analysis of Plastic and Polymer Additives*, Taylor and Francis, Boca Raton, FL.

Craver, C., Carraher, C. E. (2000): *Applied Polymer Science*, Elsevier, NY.

Datta, S., Lohse, D. (1996): *Polymeric Compatibilizers*, Hanser Gardner, Cincinnati, OH.

Dubois, P., Groeninckx, G., Jerome, R., Legras, R. (2006): *Fillers, Filled Polymers, and Polymer Blends*, Wiley, Hoboken.

Gay, D., Hoa, S. (2007): *Composite Materials*, Taylor and Francis, Boca Raton, FL.

Gibson, R. (2007): *Principles of Composite Material Mechanics*, 2nd Ed., Taylor and Francis, Boca Raton, FL.

Gupta, R., Kennel, E., Kim, K-J. (2009): *Polymer Nanocomposites Handbook*, Taylor and Francis, Boca Raton, FL.

Harrats, C. (2009): *Multiphase Polymer-Based Materials*, Taylor and Francis, Boca Raton, FL.

Harrats, C., Groeninckx, G., Thomas, S. (2006): *Micro- and Nanostructured Multiphase Polymer Blend Systems*, Taylor and Francis, Boca Raton, FL.

Hatakeyama, T., Hatakeyama, H. (2005): *Thermal Properties of Green Polymers and Biocomposites*, Springer, NY.

Lutz, J. Grossman, R. (2000): *Polymer Modifiers and Additives*, Dekker, NY.

Mallick, P. K. (2008): *Fiber-reinforced Composites*, Taylor and Francis, Boca Raton, FL.

Marosi, G., Czigany, T. (2006): Wiley, Hoboken, NJ.

Mikitaev, A., Ligidov, M. (2006): *Polymers, Polymer Blends, Polymer Composites, and Filled Polymers*, Nova, Commack, NY.

Powell, G. M., Brister, J. E. (1900): *Plastigels*, US Patent 2,427,507.

Tuttle, M. E. (2004): *Structural Analysis of Polymeric Composite Materials*, Taylor and Francis, Boca Raton, FL.

9 Naturally Occurring Polymers—Plants

There may be many ways to divide nature. One way is to divide it according to being alive, that is, self-replicating and the ability to be dead, and nonliving such as rocks and sand. Again, we have two further main divisions—plant and animal. In this, and the following chapter, we will deal with materials that are alive or once were alive. With the exception of lignin, the general chemical groups that constitute animals and plants are the same on the macrolevel. Both base their ability to replicate on nucleic acids that form genes containing the blueprint for their life. But, the two classifications do differ in the major building material. Animals have as their basic building material proteins while plants have as their basic building material polysaccharides. All of the basic building and genetic materials are polymeric.

Industrially, we are undergoing a reemergence of the use of natural polymers as feedstocks and materials in many old and new areas of application. Since natural polymers are typically regeneratable or renewable resources, nature continues to synthesize them as we harvest them. Many natural polymers are available in large quantities. For instance, cellulose makes up about one-third of the bulk of the entire vegetable kingdom, being present in corn stocks, tree leaves, grass, and so on. With the realization that we must conserve and regulate our chemical resources comes the awareness that we must find substitutes for resources that are not self-renewing, thus, the reason for the increased emphasis in polymer chemistry toward the use and modification of natural, renewable polymers by industry.

Natural feedstocks must serve many human purposes. Carbohydrates as raw materials are valuable due to their actual or potential value. For example, commercial plants are already utilizing rapidly reproducing reengineered bacteria that metabolize cellulose wastes converting it to more protein-rich bacteria that is harvested and then used as a protein source feed meal for animals. Further, natural materials can be used in applications now reserved largely for only synthetic polymers. There is available sufficient natural materials to supply both food and polymer needs.

When plant or animal tissues are extracted with nonpolar solvents, a portion of the material dissolves. The components of this soluble fraction are called *lipids* and include fatty acids, triacylglycerols, waxes, terpenes, prostagladins, and steroids. The insoluble portion contains the more polar plant components, including carbohydrates, lignin, proteins, and nucleic acids.

Many renewable feedstocks are currently summarily destroyed (through leaving them to rot or burning) or utilized in a noneconomical manner. Thus, leaves are "ritualistically" burned each fall. A number of these seemingly useless natural materials have already been utilized as feedstock sources for industrial products with more becoming available.

Just to bring natural polymers a little closer to home, the following might be our breakfast from the viewpoint of natural polymers. This breakfast includes polymers derived from both the animal and plant kingdoms.

Milk	Coffee Cake
Proteins	Gluten
Fruit	Starches
Starches	Dextrins
Cellulose	Scrambled Eggs
Pectin	Ovalbumin

continued

Meat	Conalbumin
Bacon	Ovomucoid
	Mucins
	Globins

9.1 POLYSACCHARIDES

Carbohydrates are the most abundant organic compounds, constituting three-fourths of the dry weight of the plant world. They represent a great storehouse of energy as a food for humans and animals. About 400 billion tons of carbohydrates are produced annually through photosynthesis, dwarfing the production of other natural polymers, with the exception of lignin. Much of this synthesis occurs in the oceans, pointing to the importance of harnessing this untapped food, energy, and renewable feedstocks storehouse.

The potential complexity of even the simple aldohexose monosaccharides is indicated by the presence of five different chiral centers, giving rise to 2^5 or 32 possible steroisomeric forms of the basic structure, two of which are glucose and mannose. While these sugars differ in specific biological activity, their gross chemical reactivities are almost identical, permitting one to often employ mixtures within chemical reactions without regard to actual structure. Their physical properties are also almost the same, again allowing for the mixing of these structures with little loss in physical behavior.

Carbohydrates are diverse with respect to occurrence and size. Familiar mono and disaccharides include glucose, fructose, sucrose (table sugar), cellobiose, and mannose. Familiar polysaccharides are listed in Table 9.1 along with their source, purity, and molecular weight range.

The most important polysaccharides are cellulose and starch. These may be hydrolyzed to lower molecular weight carbohydrates (oligosaccharides) and finally to D-glucose. Glucose is the building block for many carbohydrate polymers. It is called a monosaccharide since it cannot be further hydrolyzed while retaining the ring. Three major types of representations are used to reflect saccharide structures. These are given in Figure 9.1. Here, we will mainly use the Boeseken-Haworth planer hexagonal rings to represent polysaccharide structures.

Simple sugars exist in both cyclic and linear forms. Intramolecular nucleophillic reactions occur creating equilibrium combinations of these linear and cyclic forms. Monosaccharides are classified according to certain characteristics. Here we will deal only with the cyclic designations since the polysaccharides are cyclic in nature. The only exception is the nature of the end groups that may be cyclic or linear. For our purposes these characteristics are the stereo placement of the alcohol groups on the ring; the size of the ring; and the placement of the ether linkage. Figure 9.2 contains cyclic representations of glucose, which contains six atoms in the ring so it is a hexose. With the exception of carbons 1 and 6, the carbons are stereocenters or stereogenic, making D-glucose one of $2^4 = 16$ possible stereoisomers.

Of particular importance is the placement of the hydroxyl group at carbon 1 since many of the most important complex or polysaccharides are linked through ether formation at this hydroxyl group. Assuming the bond between carbons 5 and 6 is coming toward you (out of the page), the linkage at carbon 1 is called the α form since it is going away from us or into the page; that is, it is on the opposite side of the bond between carbons 5 and 6; while β is the other form where the hydroxyl at carbon 1 is on the same side as the bond between carbons 5 and 6. In some ways, considering only the placement of carbon 6 relative to the hydroxyl on carbon 1, the α form is trans while the β form is *cis*.

Kobayashi and others have pioneered in the synthesis of synthetic polysaccharides. Polysaccharide synthesis is particularly difficult because of the need to control the stereochemistry of the anomeric carbon and regioselectivity of the many hydroxyl groups with similar reactivities. This problem was initially solved employing enzymatic polymerizations with the synthesis of cellulose in 1991 when the

TABLE 9.1
Naturally Occurring Polysaccharides

Polysaccharide	Source	Monomeric Sugar Unit(s)	Structure	Molecular Weight
Amylopectin	Corn, potatoes	D-Glucose	Branched	10^6–10^7
Amylose	Plants	D-Glucose	Linear	10^4–10^6
Chitin	Animals	2-Acetamidoglucose		
Glucogen	Animals (muscles)	D-Glucose	Branched	$>10^8$
Inulin	Arichokes	D-Fructose	(Largely) Linear	10^3–10^6
Mannan	Yeast	D-Mannose	Linear	–
Cellulose	Plants	D-Glucose	Linear (2D)	10^6
Xylan	Plants	D-Xylose	(Largely) Linear	–
Lichenan	Iceland moss	D-Glucose	Linear	10^5
Galactan	Plants	D-Galactose	Branched	10^4
Arabinoxylan	Cereal grains	L-Arabinofuranose linked to xylose chain	Branched	$>10^4$
Galactomannans	Seed mucilages	D-Mannopyrnose chains with D-Galactose side chains	(Largely) Linear	10^5
Arabinogalactan	Lupin, soybean, coffee beans	D-Galactopyranose chain, side chain galactose, arabinose	Branched	10^5
Carrageenan	Seaweeds	Complex	Linear	10^5–10^6
Agar	Red seaweeds	Complex	Linear	–
Alginic	Brown seaweeds	β-D-Mannuronic acid and α-L-Guluronic acid	Linear	–

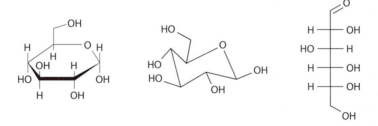

FIGURE 9.1 Beta-D-glucose using the Boeseken–Haworth, chair, and linear (from left to right) representations.

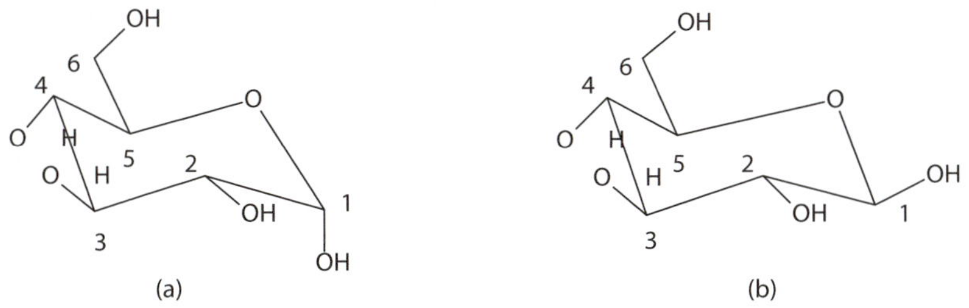

(a)

(b)

FIGURE 9.2 The α (a) and β (b) anomers of glucose.

cellobiosyl fluoride monomer was subjected to polycondensation catalyzed by celulase, a hydrolysis enzyme of cellulose. Before this, cellulose was formed only in living cells as a high-order material with parallel-chain alignment. The synthesis of other natural-like polysaccharides followed. Chitin, the most abundant animal-derived polysaccharide, was synthesized using a ring-opening polyaddition of an oxazoline monomer catalyzed by the chitinase enzyme. Chondroitin and hyaluronan, both glycosaminoglycan heteropolysaccharides have been synthesized using hyaluronidase enzymes. In these reactions, the hydrolase acted to catalyze bond formation, producing polymers but not the hydrolysis to break the bonds.

Such enzymatic catalyzed polycondensations have allowed the synthesis of a number of "natural" polysaccharide, but has also allowed the production of "non-natural" polysaccharides such as cellulose-xylan hybrids and functionalized hyaluronan, chondroitin sulfate, and chondroitin. Such work illustrates the ever narrowing bridge between natural and synthetic polymers and polymer syntheses.

9.2   CELLULOSE

Cellulose was originally "discovered" by Payen in 1838. For thousands of years impure cellulose formed the basis of much of our fuel and construction systems in the form of wood, lumber (cut wood), and dried plant material; served as the vehicle for the retention and conveying of knowledge and information in the form of paper; and clothing in the form of cotton, ramie, and flax. Much of the earliest research was aimed at developing stronger materials with greater resistance to the natural elements (including cleaning) and to improve dyeability so that the color of choice by common people for their clothing material could be other than a drab white. In fact, the dyeing of textile materials, mainly cotton, was a major driving force in the expansion of the chemical industry in the latter part of the nineteenth century.

Cellulose is a polydisperse polymer with an average degree of polymerization (DP) in the general range of 3,500–36,000. Native cellulose is widely distributed in nature and is the principle constituent of cotton, kapok, flax, hemp, jute, ramie, and wood. Cellulose comprises more than one-third of all vegetable matter and is the world's largest renewable resource. Approximately 50 billion tons is produced annually by land plants, which absorb 4×10^{20} cal of solar energy. Natural cotton fibers, which are the seed hairs from *Gossypium*, are about 1–2 cm in length and about 5–20 μm in diameter. The molecules in native cellulose are present in thread-like strands or bundles called *fibrils*. Cellulose is not found in a pure form, but rather it is associated with other materials such as lignin and hemicelluloses. Cotton contains the purest form of cellulose. Wood, in its dry state, contains 40%–55% cellulose, 15%–35% lignin, and 25%–40% hemicellulose. Plant pulp is the major source of commercial cellulose. The extraction of cellulose from plants is called *pulping*. Pulping is achieved using thermomechanical, chemical, or mechanical approaches. Plant pulp, from wood, is the major source of nontextile fibers while cotton is the major source of textile fibers.

Cellulose is used in the textile industry in cloths, cartons, carpets, blankets, and sheets. Paper is made from cellulose. Cellulosic fibers are also used as filter materials in artificial kidneys and reverse osmosis though today most kidney dialysis units use cuprammonium tubular films derived from cellulose rather than cellulose itself.

While the celluloses are often largely linear, they are not soluble in water because of the presence of strong intermolecular hydrogen bonding and sometimes the presence of a small amount of cross-linking. Highly ordered crystalline cellulose has a density as high as 1.63 g/cc, while amorphous cellulose has a density as low as 1.47 g/cc. High molecular weight native cellulose, which is insoluble in 17.5% aqueous sodium hydroxide, is called *α-cellulose*. The fraction that is soluble in 17.5% sodium hydroxide but insoluble in 8% solution is called *β-cellulose*, and that which is soluble in 8% sodium hydroxide is called *γ-cellulose*.

Strong basic solutions, such as sodium hydroxide, penetrate the crystalline lattice of α-cellulose, producing an alkoxide called *alkali* or *soda cellulose*. Mercerized cotton is produced by aqueous extraction of the sodium hydroxide.

Most linear celluloses may be dissolved in solvents capable of breaking the strong hydrogen bonds. These solutions include aqueous solutions of inorganic acids, calcium thiocyanate, zinc chloride, lithium chloride, ammonium hydroxide, iron sodium tartrate, and cadmium or copper ammonium hydroxide (Schweitzer's reagent). The product precipitated by the addition of a nonsolvent to these solutions is highly amorphous regenerated cellulose.

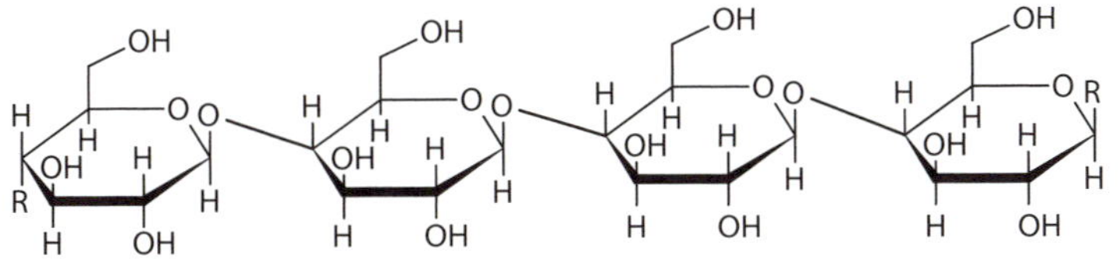

Cellulose

(9.1)

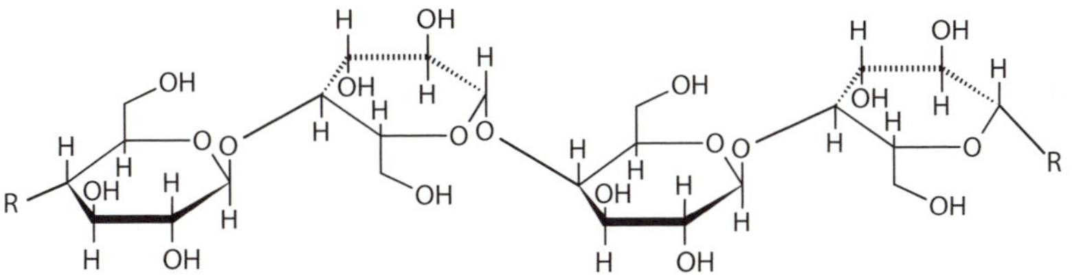

Cellulose 3D structure

(9.2)

Structure 9.1 is most commonly employed as a description of the repeat unit of cellulose but the lower structure (9.2) more nearly represents the actual three-dimensional structure with each D-glucosyl unit rotated 180 degrees. We will employ a combination of these two structural representations. Numbering is shown in (9.3) and the type of linkage is written as 1→4 since the units are connected through oxygen atoms contained on carbon 1 and 4 as below:

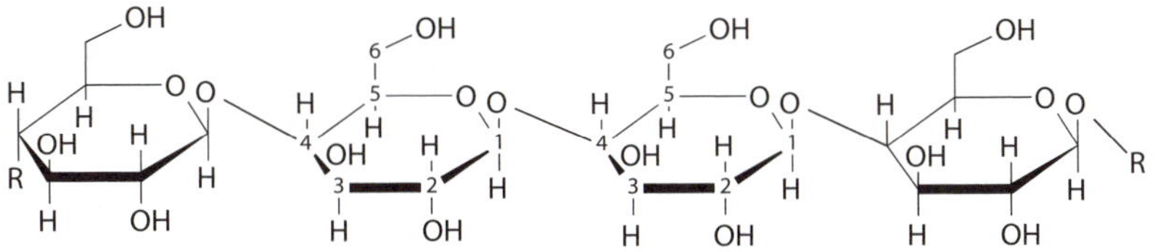

(9.3)

We also call the linkage, by agreement with the anometic nature of the particular carbons involved in linking together the glucosyl units, a beta or β-linkage. Thus, this linkage is a β 1→4 linkage. The other similar 1→4 linkage found in starch is called an *alpha* or α-*linkage*. The geometric consequence of this difference is great. The linear arrangement of cellulose with the β-linkage gives an arrangement where the OH groups reside somewhat uniformly on the outside of the chain, allowing close contact and ready hydrogen bond formation between chains. This arrangement results in a tough, insoluble, rigid, and fibrous material that is well suited as cell wall material for plants. By comparison, the α linkage of starch (namely amylose) results in a helical structure where the hydrogen bonding is both interior and exterior to the chain, allowing better wettability. This difference in bonding also results in one material being a "meal" for humans (the α-linkage), whereas the other is a meal for termites. The reason for this is the difference in the composition of enzymes present in the two species—humans with the enzyme capability to lyse or break α-linkages and for cows and termites and other species that contain symbiotic bacteria in their digestive systems that furnish the enzymes capable to digest or break the β-glucoside linkages.

While the stomach acid in humans and most animals can degrade polysaccharides to the energy-giving monomeric units, this is not efficient unless there is a specific enzyme, normally in the gut, which allows the ready and rapid degradation of polysaccharide. Since these enzymes are somewhat specific, their ability to degrade is polysaccharide specific.

The various crystalline modifications have different physical properties and chemical reactivities. These variations are a consequence of the properties varying according to plant source, location in the plant, plant age, season, seasonal conditions, treatment, and so on. Thus, in general, bulk properties of polysaccharides are generally measured with average values and tendencies given. These variations are not important for most physical applications but possibly are important for specific biological applications where the polysaccharide is employed as a drug within a drug delivery system or as a biomaterial within the body.

9.3 PAPER

It is believed that paper was invented by Ts'ai in China around the second century AD. The original paper was a mixture of bark and hemp. Paper was first produced in the United States by William Rittenhouse in Germantown, PA, in 1690 and this paper was made from rags. Paper was named after the papyrus plant, *Cyperus papyrus*.

Paper comes in many forms with many uses. The book you are reading is made from paper, we have paper plates, paper napkins, newspapers and magazines, cardboard boxes, in fact the amount of paper items is probably more than twice, by weight, that of all the synthetic polymers combined. About 30% paper is writing and printing paper. The rest is mainly used for tissues, toweling, and packaging. If you rip a piece of ordinary paper, not your book page please, you will see that it consists of small fibers. Most of these cellulosic fibers are randomly oriented, but a small percentage are preferentially oriented in one direction because the paper is made from a cellulose-derived watery slurry with the water largely removed through use of heated rollers that somewhat orient the fibers.

Modern paper is made from wood pulp, largely cellulose, which is obtained by the removal of lignin from debarked wood chips by use of chemical treatments with sodium hydroxide, sodium sulfite, or sodium sulfate. Newsprint and paperboard, which is thicker than paper, often contains a greater amount of residual lignin.

Wood is almost entirely composed of cellulose and lignin. In the simplest paper making scheme, the wood is chopped, actually torn, into smaller fibrous particles as it is pressed against a rapidly moving pulpstone. A stream of water washes the fibers away dissolving much of the water-soluble lignin. The insoluble cellulosic fibers are concentrated into a paste called *pulp*. The pulp is layered into thin sheets and rollers are used to both squeeze out much of the water and to assist in achieving paper of uniform thickness. This paper is not very white. It is also not very strong. The remaining lignin is somewhat acidic (lignin contains acidic phenolic groups that hydrolyze to give a weakly acidic aqueous solution) that causes the hydrolytic breakdown of the cellulose. Most of the newsprint is of this type or it is regenerated, reused paper.

Pulping processes are designed to remove the nonsaccharide lignin portion of wood which constitutes about 25% of the dry weight. The remaining is mostly cellulose with about 25% hemicellulose (noncellulose cell wall polysaccharides that are easily extracted by dilute aqueous base solutions). Pulping procedures can be generally classified as semichemical, chemical, and semimechanical. In semimechanical pulping, the wood is treated with water or sulfate, bisulfite, or bicarbonate solution that softens the lignin. The pulp is then ground or shredded to remove much of the lignin giving purified or enriched cellulose content. The semichemical process is similar but digestion times are longer and digesting solutions more concentrated giving a product with less lignin but the overall yield of cellulose-intense material is lowered by 70%–80%. Further, some degradation of the cellulose occurs.

Most paper is produced by the chemical process where chemicals are employed to solubilize and remove most of the lignin. While overall yields are lower than the other two main processes, the product gives good quality writing and printing paper. Three main chemical processes are used. In the soda process extracting solutions containing about 25% sodium hydroxide and 2.4% sodium carbonate are used. In the sulfite process the extracting solution contains a mixture of calcium dihydrogen sulfite and sulfur dioxide. The sulfide process utilizes sodium hydroxide, sodium monosulfide, and sodium carbonate in the extracting solution.

After the chemical treatment, the pulped wood is removed, washed, and screened. Unbleached, brown-colored paper is made directly for this material. Most whiten or bleached paper is made from treatment of the pulp with chlorine, chlorine dioxide, hypochlorite, and/or alkaline extraction. In general, sulfate pulped paper is darker and requires more bleaching and alkaline extraction to give a "white" pulp.

The *sulfide process*, also called the *kraft* process (the term "kraft" comes from the Swedish word for strong since stronger paper is produced), is more commonly used. The kraft process is favored over the sulfite treatment of the paper because of environmental considerations. The sulfite process employs more chemicals that must be disposed of-particularly mercaptans, RSHs, which are quite odorous. Research continues on reclaiming and recycling pulping chemicals.

If pure cellulose was solely used to make paper, the fiber mat would be somewhat water soluble with only particle surface polar groups and internal hydrogen bonding acting to hold the fibers together. White pigments such as clay and titanium dioxide are added to help "cement" the fibers together and to fill voids producing a firm, white writing surface. This often occurs as part of an overall coating process.

Most paper is coated to provide added strength and smoothness. The coating is basically an inexpensive paint that contains a pigment and a small amount of polymeric binder. Unlike most painted surfaces, most paper products are manufactured with a short lifetime in mind with moderate performance requirements. Typical pigments are inexpensive low-refractive index materials such as plate-like clay and ground natural calcium carbonate. Titanium dioxide is used only when high opacity is required. The binder may be a starch or latex or a combination of these. The latexes are usually copolymers of styrene, butadiene, acrylic, and vinyl acetate. Other additives and coloring agents may also be added for special performance papers. Resins in the form of surface coating agents and other special surface treatments (such as coating with polypropylene and polyethylene) are used for paper products intended for special uses such as milk cartons, ice cream cartons, light building materials, and drinking cups. The cellulose supplies the majority of the weight (typically about 90%) and strength with the special additives and coatings, providing special properties needed for the intended use.

A better understanding of the nature of paper and films made from synthetic polymers such as polyethylene can be seen when considering why authorities worry about anthrax escaping from a paper envelope yet confining anthrax in a plastic container with no fear of it escaping. When you hold good paper up to the light or tear it you will observe the tiny fibers that compose paper. Even when these fibers appear tiny, they are very large in comparison to individual polymer chains, be they cellulose or polyethylene. The web of chains for cellulose compose eventually the cellulose fibers that are put together physically, forming paper with lots of "unoccupied" spaces between the fibers with the spaces of sufficient size to allow the escape of the anthrax. By comparison, polyethylene film has little "unoccupied" spaces between the individual chains with the entire film composed of these individual chains with no large spaces so that the anthrax cannot escape.

Recycling of paper continues. Today, up to about one half of our paper products are recycled and this fraction is increasing as we do a better job of collecting and recycling paper products.

9.4 CELLULOSE-REGENERATING PROCESSES

Cellulose is sometimes used in its original or native form as fibers for textile and paper, but often it is modified through dissolving and reprecipitation or through chemical reaction. The *xanthate viscose process,* which is used for the production of rayon and cellophane, is the most widely used regeneration process. The cellulose obtained by the removal of lignin from wood pulp is converted to alkali cellulose. The addition of carbon disulfide to the latter produces cellulose xanthate.

While terminal hydroxyl and aldehyde groups, such as present in cellobiose, are also present in cellulose, they are not significant because they are only present at the ends of very long chains.

The hydroxyl groups are not equivalent. For instance, the pKa values of the two ring hydroxyl groups are about 10 and 12, which is about the same as the hydroxyl groups on hydroquinone and the first value about the same as the hydroxyl on phenol. The pKa value of the nonring or methylene hydroxyl group is about 14, same as found for typical aliphatic hydroxyl groups.

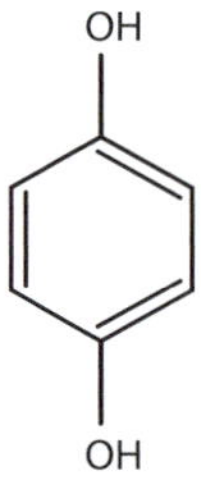

Hydroquinone (9.4) 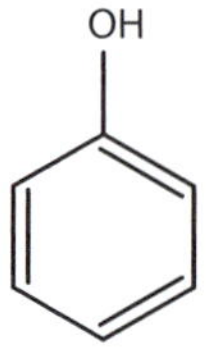   Phenol (9.5)

In the cellulose-regenerating process, sodium hydroxide is initially added such that approximately one hydrogen, believed to be predominately a mixture of the hydroxyl groups on carbons 2 and 3, is replaced by the sodium ion. This is followed by treatment with carbon disulfide forming cellulose xanthate, which is eventually rechanged back again, regenerated, to cellulose. This sequence is depicted below:

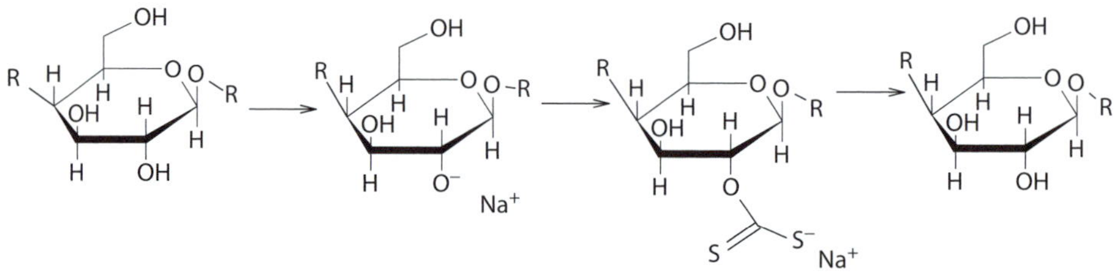

(9.6)

Cellulose ⟶ Sodium salt ⟶ Cellulose xanthate ⟶ Regenerated
                                                     cellulose–rayon or
                                                     cellophane

The orange-colored xanthate solution, or viscose, is allowed to age and is then extruded as a filament through holes in a spinneret. The filament is converted to cellulose when it is immersed in a solution of sodium bisulfite, zinc II sulfate, and dilute sulfuric acid. The tenacity, or tensile strength, of this regenerated cellulose is increased by a stretching process that reorients the molecules so that the amorphous polymer becomes more crystalline. Cellophane is produced by passing the viscose solution through a slit die into an acid bath. Important noncellulosic textile fibers are given in Table 9.2, and a listing of important cellulosic textile fibers is given in Table 9.3.

Since an average of only one hydroxyl group in each repeating glucose unit in cellulose reacts with carbon disulfide, the xanthate product is said to have a degree of substitution (DS) of 1 out of a potential DS of 3.

TABLE 9.2
Noncellulosic Textile Fibers

Fiber Name	Definition	Properties	Typical Uses	Patent Names (Sample)
Acrylic	Acrylonitrile units, 85% or more	Warm, light weight, shape retentive, resilient, quick drying, resistant to sunlight	Carpeting, sweaters, skirts, baby cloths, socks, slacks, blankets, draperies	Orlon
Modacrylic	Acrylonitrile units, 35%–85%	Resilient, softenable at low temperatures, easy to dye, abrasion resistant, quick drying, shape retentive, resistant to acids, bases	Simulated fur, scatter rugs, stuffed toys, paint rollers, carpets, hairpieces, wigs fleece fabrics	Verel, Dynel
Polyester	Dihydric acid–terephthalic acid ester, 85% or more	Strong, resistant to stretching and shrinking, easy to dye, quick drying, resistant to most chemicals, easily washed, abrasion resistant; retains heat set (permanent press)	Permanent press ware, skirts, slacks, underwear, blouses, rope, fish nets, tire cord, sails, thread	Kodel, Fortrel, Chemstrand Dacron
Spandex	Segmented polyurethane, 85% or more	Light, soft, smooth, resistant to body oils, can be stretched often, retain original form, abrasion resistant	Girdles, bras, slacks bathing suits, pillows	Lycra
Nylon	Reoccurring amide groups	Very strong, elastic, lustrous, easy to wash, abrasion resistant, smooth, resilient, low moisture absorbency, recovers quickly from extensions	Carpeting, upholstery, blouses, tents, sails, hosiery, suits, tire cord, fabrics, rope, nets	Caprolan, Astroturf, Celanese polyester

TABLE 9.3
Cellulosic Fibers

Fiber Name	Definition	Properties	Typical Uses	Patent Names (Sample)
Rayon	Regenerated cellulose with less than 15% OH substituted	Highly absorbent, soft, comfortable, easy to dye, good drapability	Dresses, suits, slacks, blouses, coats, tire cord, ties, curtains, blankets	Avril, Cuprel, Zantel
Acetate	Not less than 92% OH groups acetylated	Fast drying, supple, wide range of dyeability, shrink resistant	Dresses, shirts, slacks, draperies, upholstery, cigarette filters	Estron, Celanese acetate
Triacetate	Derived from cellulose by combining with acetic acid and/or acetic anhydride	Resistant to shrinking, wrinkling, and fading; easily washed	Skirts, dresses, sportswear	Arnel

Partially degraded cellulose is called *hydrocellulose* or *oxycellulose* depending on the agent used for degradation. The term holocellulose is used to describe the residue after lignin has been removed from wood pulp.

Control of the regeneration conditions, together with a wide variety of modification, allows the production of a wide variety of products, including high wet modulus fibers, hollow fibers, crimped fibers, and flame-resistant fibers. While almost all rayon is produced using the viscose process, some fibers are still produced using the cuprammonium process whereby cellulose is dissolved in an ammonium-copper II-alkaline solution.

9.5   ESTERS AND ETHERS OF CELLULOSE

It must be remembered that the three hydroxyl groups on the individual cellulose rings are not equivalent. The two ring hydroxyls are acidic with pKas similar to hydroquinone while the third nonring hydroxyl is similar to an aliphatic hydroxyl in acidity. Thus, in an aqueous sodium hydroxide solution the two ring hydroxyls will be deprotonated at high pHs. In theory, all three hydroxyls can undergo reaction, but in actuality less than three undergo reaction either because of reactivity restrictions and/or because of steric limitations. With many of the electrophilic/nucleophilic reactions it is the ring hydroxyls that are favored to react initially. The average number of hydroxyl groups that are reacted are often given as the *degree of substitution* or DS.

9.5.1   INORGANIC ESTERS

The most widely used so-called inorganic ester of cellulose is *cellulose nitrate* (CN), also called *nitrocellulose* and *gun cotton*. Celluloid is produced from a mixture of cellulose nitrate and camphor. Cellulose nitrate was first made in about 1833 when cellulose-containing linen, paper, or sawdust was reacted with concentrated nitric acid. It was the first "synthetic" cellulose recognized product. Initially, CN was used as a military explosive and improvements allowed the manufacture of smokeless powder. A representation of CN is given below:

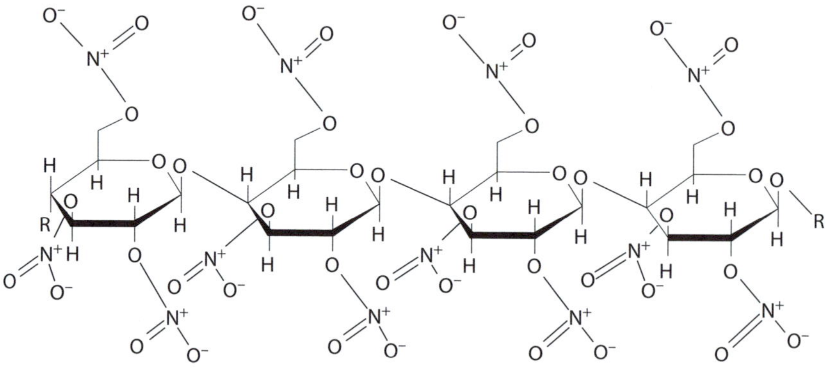

Cellulose nitrate

(9.7)

   The development of solvents and plasticizing agents for cellulose nitrate led to the production of many new and useful nonexplosive products. Celluloid was produced in 1870 from a mixture of CN and camphor. Films were cast from solution and served as the basis for the original still and motion pictures. After World War I, the development of stable CN solutions allowed the production of fast-drying lacquer coatings.
   While CN played an important role in the development of technology, its importance today is greatly diminished. It is still used as a protective and decorative lacquer coating, in gravure inks, in water-based emulsions as coatings, and to a lesser extent in plastics and films.
   Cellulose phosphate esters are produced from reaction with phosphoric acid and urea. The products are used to treat hypercalciuria because of its ability to bind calcium. It has also been used for the treatment of kidney stones.

9.5.2   ORGANIC ESTERS

The most important cellulose ester is cellulose acetate because of its use in fibers and plastics. They were first made in 1865 by heating cotton with acetic anhydride. During World War I, a cellulose acetate replaced the highly flammable CN coating on airplane wings and fuselage fabrics.

Varying properties are achieved by varying the amount of substitution. The melting point generally decreases with decreasing acetylation. Lower acetylation gives products with greater solubility in polar solvents and corresponding decreased moisture resistance. Cellulose acetate is made using heterogeneous solutions containing the cellulose, sulfuric acid as the catalyst, and acetic anhydride in acetic acid. Reaction occurs beginning with the surface or outermost layer and continues on layer by layer as new areas are exposed. When more homogeneous modification is desired, preswelling of the cellulose in water, aqueous acetic acid solutions, or in glacial acetic acid is carried out.

Reaction occurs differently since there are two "types" of hydroxyl groups (as noted before), the two ring hydroxyls and the methylene hydroxyl. In the typical formation of esters, such as cellulose acetate, the ring hydroxyl groups are acetylated initially (9.8) before the C-6 exocyclic hydroxyl. Under the appropriate reaction conditions, reaction continues to almost completion with all three of the hydroxyl groups esterified (9.9). In triacetate products, only small amounts (on the order of 1%) of the hydroxyls remain free and of these generally about 80% are the C-6 hydroxyl.

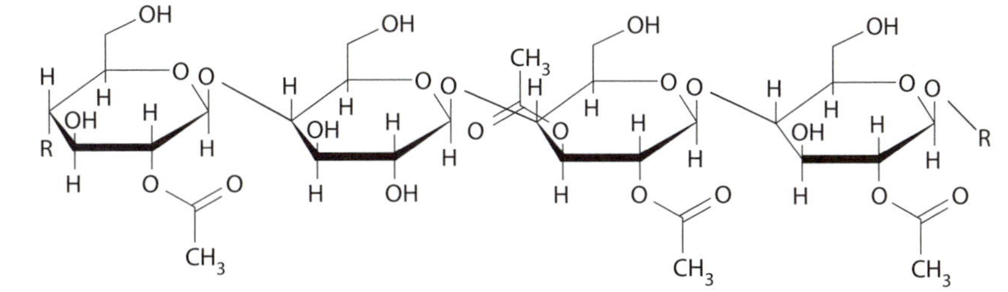

(9.8)

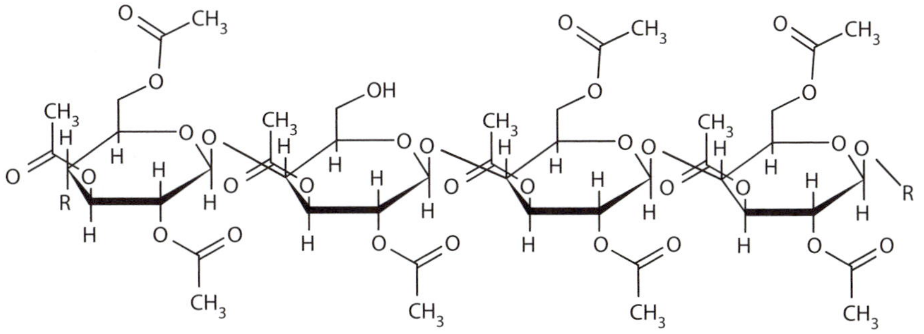

(9.9)

The most common commercial products are the triacetate (DS approaching 3) and the secondary acetate (DS about 2.45).

While other organic esters are commercially available, namely cellulose butyrate and cellulose propionate, by far the most widely used is cellulose acetate. Cellulose acetate is available as plastics, in films, sheets, fibers, and in lacquers. Cellulose acetate is used in the manufacture of display packaging and as extruded film for decorative signs, and to coat a variety of fibers. Injected molded products include toothbrush handles, combs, brushes. It is also used in lacquers and protective coatings for metal, glass, and paper. Cellulose acetate films are used in reverse osmosis to purify blood, fruit juices, and brackish water. Some eyeglass frames are made of cellulose acetate. Biodegradable film, sponges, and microencapsulation of drugs for control release also utilize cellulose acetate. Cellulose triacetate is used for photographic film bases. Numerous continuous filament yarns, tows, staples, and fibers are made from cellulose acetate. The precise form of filament produced is controlled by a number of factors, including the shape of the die.

As in all large-scale industrial processes, the formation of the cellulose esters involves recovery of materials. Acetic anhydride is generally employed. After reaction, acetic acid and solvent is recovered. The recovered acetic acid is employed in the production of additional acetic anhydride. The recovered solvent is also reintroduced after treatment.

Cellulose esters are used as plastics for the formation by extrusion of films and sheets and by injection molding of parts. They are thermoplastics and can be fabricated employing most of the usual techniques of (largely compression and injection) molding, extrusion, and casting. Cellulose esters plastics are noted for their toughness, smoothness, clarity, and surface gloss.

Acetate fiber is the generic name of a fiber that is partially acetylated cellulose. They are also known as cellulose acetate and triacetate fibers. They are nontoxic and generally nonallergic so are ideal from this aspect as clothing material.

While acetate and triacetate differ only moderately in the degree of acetylation, this small difference accounts for differences in the physical and chemical behavior for these two fiber materials. Triacetate fiber is hydrophobic and application of heat can bring about a high degree of crystallinity that is employed to "lock-in" desired shapes (such as permanent press). Cellulose acetate fibers have a low degree of crystallinity and orientation even after heat treatment. Both readily develop static charge and thus antistatic surfaces are typically employed to clothing made from them.

For clothing applications, there are a number of important performance properties that depend on the form of the textile. These properties include wrinkle resistance, drape, strength, and flexibility. These properties are determined using ASTM tests that often involve stress–strain behavior. Thus, the ability of a textile to resist deformation under an applied tensile stress is measured in terms of its modulus of elasticity or Young's modulus. As with any area of materials, specialty tests are developed to measure certain properties. Some of these are more standard tests, like the aforementioned Young's modulus, while others are specific to the desired property measured for a specific application. For instance, resistance to slightly acidic and basic conditions is important for textiles that are to be laundered. Again, these are tested employing standard test procedures. In general, triacetate materials are more resistant than acetate textiles to basic conditions. Both are resistant to mild acid solutions but degrade when exposed to strong mineral acids. Further, behavior to various dry cleaning agents is important. As the nature of dry cleaning agents change, additional testing and modification in the fabric treatments are undertaken to offer a textile that stands up well to the currently employed cleaning procedures. Again, both are stable to perchloroethylene dry cleaning solvents but can soften when exposed to trichloroethylene for extended treatment. Their stability to light is dependent upon the wavelength, humidity present, and so on. In general, they offer a comparable stability to light as that offered by cotton and rayon.

While cellulose acetates are the most important cellulose ester, they suffer by their relatively poor moisture sensitivity, limited compatibility with other synthetic resins, and a relatively high processing temperature.

9.5.3 Organic Ethers

Reaction with an epoxide such as ethylene oxide under alkaline conditions gives *hydroxyethylcellulose* (HEC).

$$\text{Cellulose–OH, NaOH} \;+\; \underset{\substack{\text{Ethylene}\\\text{oxide}}}{\text{H}_2\text{C–CH}_2} \overset{\text{O}}{\overbrace{}} \longrightarrow \underset{\text{Hydroxyethylcellulose}}{\text{Cellulose–(–O–CH}_2\text{–CH}_2\text{–)}_n\text{–OH}} + \text{NaOH} \tag{9.10}$$

This is an S_N2 reaction with the reaction proportional to the concentration of the epoxide and alkali cellulose, but since the base is regenerated, it is first order in epoxide.

$$\text{Rate} = k\,[\text{Epoxide}] \tag{9.11}$$

Industrially, HECs with DS values below 2 are used. Low DS materials (to about 0.5) are soluble only in basic solutions while those with DS values of about 1.5 are water soluble. Concentrated solutions of HEC are pseudoplastic with their apparent viscosities decreasing with increased rates of shear. Dilute solutions approach being Newtonian in their flow properties even under a wide range of shear rates.

HEC is used as a protective colloid in latex coatings and pharmaceutical emulsions; as a film former for fabric finishes, fibrous glass, and in aerosol starches; thickener for adhesives, latex coatings, toothpaste, shampoos and hair dressings, cosmetic creams and lotions, inks, and joint cements; lubricant for wallpaper adhesives and in pharmaceutical gels; and as a water binding for cements, plastics, texture coatings, ceramic glazes, and in printing inks.

Sodium carboxymethylcellulose is formed by the reaction of sodium chloroacetate with basic cellulose solutions. The sodium form of carboxymethylcellulose is known as CMC or as a food grade product as cellulose gum. It is soluble in both hot and cold water.

$$\text{Cellulose–OH, NaOH} + \text{Cl–CH}_2\text{COONa} \rightarrow \text{Cellulose–O–CH}_2\text{–COONa} \qquad (9.12)$$
$$\text{Sodium carboxymethylcellulose}$$

The most widely used cellulose gums have DS values about 0.65–1.0. CMCs are used as thickening, binding, stabilizing, and film-forming agents.

Carboxymethylhydroxyethylcellulose, CMHEC, is synthesized from the reaction of hydroxethylcellulose with sodium chloroacetate. The product is a mixed ether. It has properties similar to both CMC and HEC. Like CMC it exhibits a high water binding ability and good flocculating action on suspended solids but it is more compatible than CMC with salts. It forms ionic cross-links in the presence of salt solutions containing multivalent cations, allowing its viscosity to be greatly increased by the presence of such cations. Solutions can be gelled by addition of solutions of aluminum and iron. It is a water-soluble material used in oil recovery and in hydraulic fracturing fluids.

Methyl and hydroxyalkylmethylcelluloses are nonionic polymers soluble in cool water. Methylcellulose (MC), hydroxyethylmethylcellulose (HEMC), and hydroxypropylmethylcellulose (HPMC) do not interact with cations forming insoluble salts, but electrolytes that compete with MC for water can cause precipitation.

Hydroxylpropylcellulose (HPC) is a thermoplastic nonionic cellulose ester that is soluble in both water and a number of organic liquids. It is synthesized through reaction of the basic cellulose slurried with propylene oxide.

$$\text{Cellulose–OH, NaOH} + \text{H}_3\text{C–CH–CH}_2 \overset{\displaystyle O}{\overbrace{}} \rightarrow \text{Cellulose–(–O–CH}_2\text{–CH(CH}_3\text{)–)}_n\text{–OH} \qquad (9.13)$$
$$\text{Hydroxypropylcellulose}$$

Methylcellulose is formed from basic cellulose and its reaction with chloromethane.

$$\text{Cellulose–OH, NaOH} + \text{CH}_3\text{Cl} \rightarrow \text{Cellulose–O–CH}_3 \qquad (9.14)$$
$$\text{Methylcellulose}$$

Methylcellulose is used as an adhesive; in ceramics to provide water retention and lubricity; in cosmetics to control rheological properties and in the stabilization of foams; in foods as a binder, emulsifier, stabilizer, thickener, and suspending agent; in paints, paper products, plywood as a rheology control for the adhesive; in inks, and in textiles as a binder, and for coatings.

Ethylhydroxyethylcellulose (EHEC) is a nonionic mixed ether available in a wide variety of substitutions with corresponding variations in aqueous and organic liquid solubilities. It is compatible with many oils, resins, and plasticizers along with other polymers such as nitrocellulose. EHEC is synthesized through a two-step process, beginning with the formation of the HEC-like product through reaction between the basic cellulose and ethylene oxide. The second step involves further reaction with ethyl chloride.

$$\text{Cellulose-OH, NaOH} + \overset{\displaystyle O}{\overset{\displaystyle / \backslash}{H_2C-CH_2}} \rightarrow \text{Cellulose-(-O-CH}_2\text{CH}_2\text{-)-CH}_2\text{CH}_3 \qquad (9.15)$$

Ethylhydroxyethylcellulose

Uses for the water-soluble EHEC include water-borne paints, pastes, polymer dispersions, ceramics and cosmetics, and pharmaceuticals. Uses for organic soluble EHEC include inks, lacquers, and as coatings.

Cellulose undergoes reaction with activated ethylenic compounds such as acrylonitrile, giving cyanoethylcellulose via a Michael addition.

$$\text{Cellulose-OH} + H_2C{=}CHCN \rightarrow \text{Cellulose-O-CH}_2\text{-CH}_2\text{-CN} \qquad (9.16)$$

9.6 STARCH

While cellulose is the major structural polysaccharide, plant energy storage and regulation utilizes a combination of similar polysaccharides that combined are referred to as starch. Starch can be divided into two general structures, largely linear *amylase* (9.17) and branched *amylopectin* (9.18).

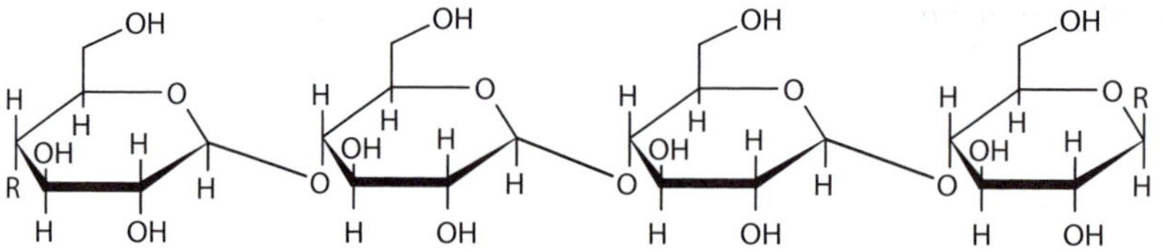

$$(9.17)$$

Linear amylose

Most starches contain about 10%–20% amylose and 80%–90% amylopectin thought the ratio can vary greatly. While cellulose can be considered a highly regular polymer of D-glucose with units linked through a β-1,4 linkage, amylose is a linear polysaccharide with glucose units linked in an α-1,4-fashion while amylopectin contains glucose units with chains of α -1,4 glucopyranosyl units but with branching occurring every 20–30 units, with the chain-branch occurring from the 6 position. While this difference in orientation is how the glucose units are connected appears small, it causes great differences in the physical and biological properties of cellulose and starch. As noted before, people contain enzymes that degrade the α-glucose units but we are unable to digest β units. Thus, starch is a food source for us, but cellulose is not. Also, the individual units of cellulose can exist in the chair conformation with all of the substituents equatorial, yet amylose must either have the glycosyl substituent at the 1 position in an axial orientation or exist in a nonchair conformation.

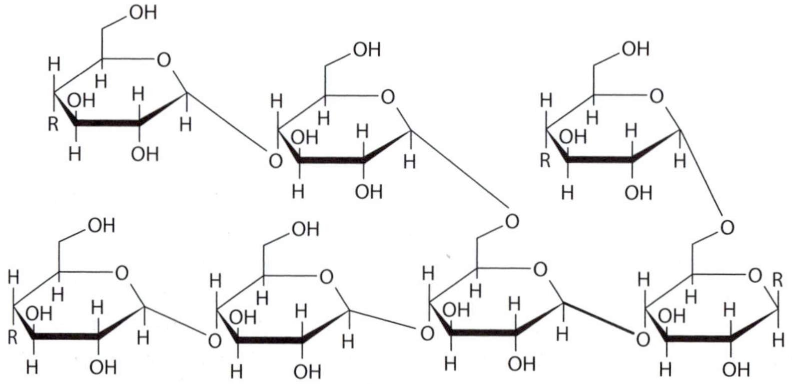

$$(9.18)$$

Branched amylopectin

Amylose typically consists of more than 1,000 D-glucopyranoside units. Amylopectin is a larger molecule containing about 6,000–1,000,000 hexose rings essentially connected with branching occurring at intervals of 20–30 glucose units. Branches also occur on these branches giving amylopectin a fan or tree-like structure similar to that of glycogen. Thus, amylopectin is a highly structurally complex material. Unlike nucleic acids and proteins where specificity and being identical are trademarks, most complex polysaccharides can boast of having the "mold broken" once a particular chain was made so that the chances of finding two exact molecules is very low.

Commercially, starch is prepared from corn, white potatoes, wheat, rice, barely, millet, cassava, tapioca, and sorghum. The fraction of amylose and amylopectin varies between plant species and even within the same plant varies depending on location, weather, age, and soil conditions. Amylose serves as a protective colloid. Mixtures of amylose and amylopectin, found combined in nature, form suspensions when placed in cold water. Starch granules are insoluble in cold water but swell in hot water, first reversibly until gelationization occurs at which point the swelling is irreversible. At this point, the starch loses its birefringence, the granules burst, and some starch material is leached into solution. As the water temperature continues to increase to near 100°C, a starch dispersion is obtained. Oxygen must be avoided during heating or oxidative degradation occurs. Both amylose and amylopectin are then water soluble at elevated temperatures. Amylose chains tend to assume a helical arrangement giving it a compact structure. Each turn contains six glucose units.

The flexibility of amylose and its ability to take on different conformations are responsible for the "retrogradation" and gelation of dispersions of starch. Slow cooling allows the chains to align to take advantage of inter- and intrachain hydrogen bonding squeezing out the water molecules, leading to precipitation of the starch. This process gives retrograded starch, either in the presence of amylose alone or combined in native starch, which is more difficult to redisperse. Rapid cooling of starch allows some inter- and intrachain hydrogen bonding, but also allows water molecules to be captured within the precipitating starch, allowing it to be more easily redispersed (Figure 9.3).

Most uses of starch make use of the high viscosity of its solutions and its gelling characteristics. Modification of starch through reaction with the hydroxyl groups lowers the gelation tendencies, decreasing the tendency for retrogradation. Starch is the major source of corn syrup and corn sugar (dextrose or glucose). In addition to its use as a food, starch is used as an adhesive for paper and as a textile-sizing agent.

Oligomeric materials called *cyclodextrins* are formed when starch is treated with *Bacillus macerans*. These oligomeric derivatives generally consist of six, seven, eight, and greater numbers of glucose units joined through 1,4-α linkages to form rings. These rings are doughnut-like with the hydroxyl groups pointing upward and downward along the rim. Like crown ethers used in phase-transfer reactions, cyclodextrins can act as "host" to "guest" molecules. In contrast to most phase-transfer agents, cyclodextrins have a polar exterior and nonpolar interior. The polar exterior allows guest molecules to be water soluble. The nonpolar interior allows nonpolar molecules to also be guest molecules. Cyclodextrins are being used as enzyme models since they can first bind a substrate and through substituent groups and act on the guest molecule—similar to the sequence carried out by enzymes.

A major commercial effort is the free radical grafting of various styrenic, vinylic, and acrylic monomers onto cellulose, starch, dextran, and chitosan. The grafting has been achieved using a wide variety of approaches, including ionizing and ultraviolet (UV)/visible radiation, charge-transfer agents, and various redox systems. Much of this effort is aimed at modifying the native properties such as tensil-related (abrasion resistance and strength) and care-related (crease resistance and increased soil and stain release) properties, increased flame resistance, and modified water absorption. One area of emphasis has been the modification of cotton and starch in the production of superabsorbent material through grafting. These materials are competing with all synthetic cross-linked

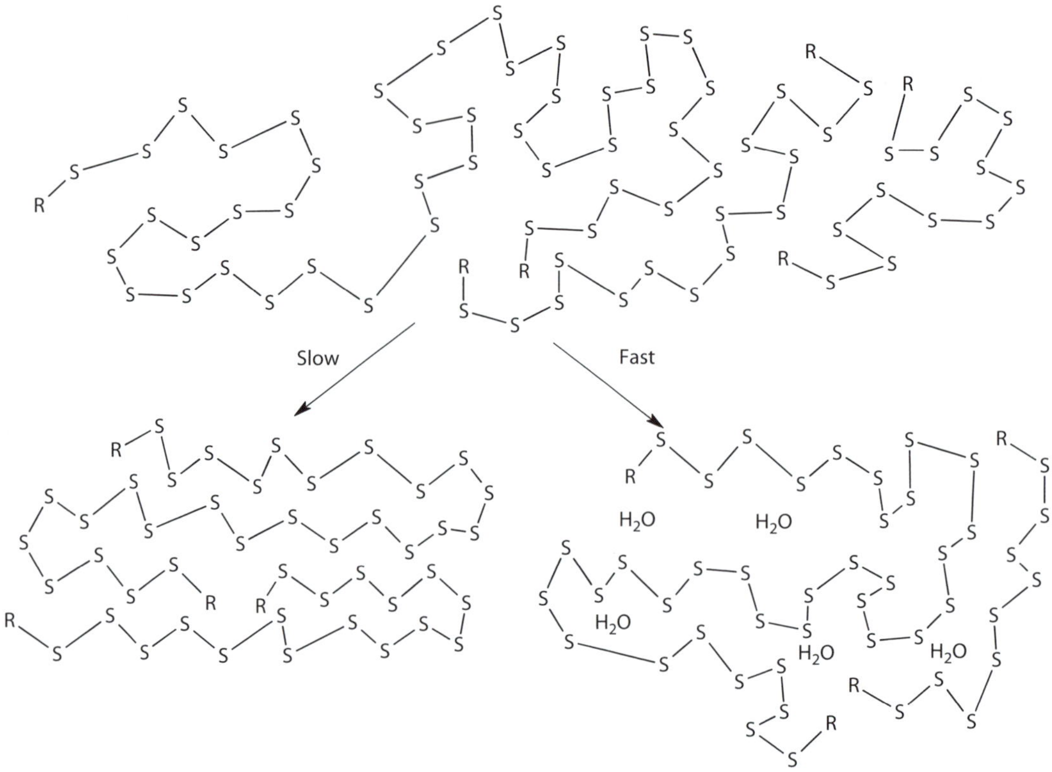

FIGURE 9.3 Behavior of amylose in a concentrated aqueous solution as a function of cooling rate.

acrylate materials that are finding use in diapers, feminine hygiene products, wound dressings, and sanitary undergarments.

9.7 HOMOPOLYSACCHARIDES

The best know homopolysaccharides are derived from D-glucose and known as glucans. Glucose has a number of reactive sites and a wide variety of polymers formed utilizing combinations of these reactive sites are found in nature. We have already visited the two most well-known members of this group—cellulose and starch containing amylose and amylopectin. Here we will visit some other important members.

Glycogen is a very highly branched glucan or polysaccharide formed from glucose. It is structurally similar to amylopectin though more highly branched. This greater branching gives glycogen a greater water solubility. Glycogens are the principle carbohydrate food reserve materials in animals. They are found in both invertebrates and vertebrates and likely found in all animal cells. The highest concentration of glycogen is found in muscle with the greatest amount found in our liver, the tissue from which it is most often isolated.

Glycogen is an amorphous polymer of high molecular weight, generally 10^6–10^9 Da. In spite of its high molecular weight, it has good water solubility because, as noted above, of its highly, but loosely branched character. It is polydisperse with respect to molecular weight as are other polysaccharides. The particular molecular weight and molecular weight distribution varies within and between cells and metabolic need. It stores D-glucose units until needed as an energy source. It also serves as a buffering agent helping control the amount of glucose in the blood. It is stored in tissues as spherical particles called β *particles*.

The average distance between branch points is only about 10–15 ring units in. comparison to amylopectin with about 20–30 units between branch points. Many glycogen particles contain small amounts of protein to which the polysaccharide chains are covalently bonded. Glycogen reacts weakly with iodine giving a yellow–orange color. It is believed that about 50 linear glucose units is required to form the blue complex found for amylose and because of the high degree of branching, few "runs" near 50 linear glucose units are found in glycogen.

Starch and glycogen are produced when the amount of glucose is high and are readily degraded back to glucose when energy is needed. In plants, this degradation occurs mainly through the action of two enzymes known as alpha- and beta-amylase. Interestingly, while the alpha-amylase can degrade starch and glycogen completely to glucose, beta-amylase is not able to degrade the branch points.

In animals, glycogen degradation to give the glucose needed as an energy source or to increase the blood sugar concentration begins with the action of phosphorylase. Phosphorylase occurs in active, a, and inactive, b, forms. Phosphorylase b is converted into phosphorylase a by phosphorylation that occurs at the end of a series of events initiated by an increased intercellular concentration of cyclic adenosine monophosphatase (cAMP) and activation of the protein kinase (Figure 10.8). This is reversed by a phosphoprotein phosphatase whose activity is hormonally regulated. Thus, phosphorylatin initiated by increased intracellular concentrations of cAMP inactivates glycogen synthetase and activates phosphorylase. This is an example of the complex steps that are moment-by-moment carried out in our bodies. Here, enzymes that are responsible for the glycogen metabolism do not act directly on glycogen but regulate the activity of other enzymes.

Skeletal muscle glycogen delivers glucose primarily as a response to contractile stress. Regulation occurs though both modification of the enzyme phosphorylase, primarily by the action of epinephrine-adrenaline-and allosteric regulation of phosphorylase related to a demand for adenosine triphosphate (ATP).

Glycogen found in the liver seldom is utilized as a source of energy but rather is employed to regulate blood sugar levels. Some tissues, such as nerve and brain tissue, rely solely of glucose as their energy source so that a steady supply of sugar is essential to their well-being. It is also found in some fungi and yeasts. Even some plants such as sweet corn synthesize a polysaccharide that is similar to glycogen.

(9.19)

Glycogen

Dextrans are a high molecular weight, branched extracellular polysaccharide synthesized by bacteria. These bacteria are found in many places, including the human mouth where they flourish on sucrose-containing food, which become trapped between our teeth. The generated dextrans become part of the dental plaque and thus are involved in tooth decay. Dextran-causing bacteria can also infect sugar cane and sugar beet after harvest and act to not only decrease the yield of sucrose but also interfere with sugar refining, clogging filters and pipelines. These bacteria can also contaminate fruit juices and wines, in fact any ready source of glucose or sucrose.

On the positive side, dextran itself has been refined and employed as a therapeutic agent in restoring blood volume for mass casualties. Natural dextrans are very high molecular weight (on the order of 10^8–10^9 Da) and are found to be unsuitable as a blood-plasma substitute. Lower molecular weight (about 10^6 Da) dextran is suitable and is often referred to as clinical dextran.

Dextran gels are commercially used. The gel formed from reaction with epichlorohydrin gives a cross-linked material used as a molecular sieve. Commercial cross-linked dextran is know as Sephadex(TM). Sephadex is formed in bead form from dissolving dextran in sodium hydroxide solution followed by dispersion in an immiscible organic liquid such as toluene-containing poly(vinyl acetate) and finally added to epichlorohydrin. Different series of Sephadex are used industrially and in research. Ionic groups are often incorporated to give anionic and cationic dextrans and ion-exchange molecular sieves. Sulfate esters of dextran are also used in separations.

Below illustrates some typical units that compose dextrans.

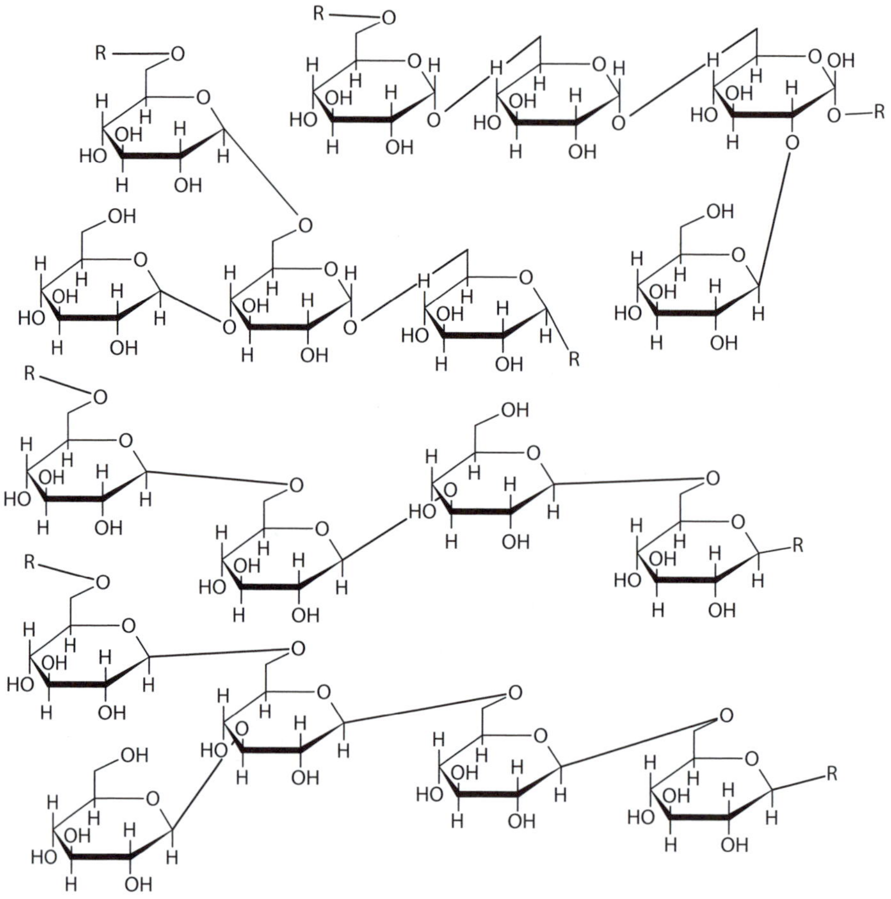

(9.20)

Representative dextran structures appear in 9.20—top left—1→6 linked glucose units with a 1→4 branch; top right—linear 1→6 linked glucose units with a 1→2 branch; middle—linear chain with both 1→6 and 1→3 linkages; bottom—linear chain of 1→6 linked glucose units with a 1→3 branch. All links are alpha linkages.

9.7.1 FRUCTANS

Fructans are polysaccharides composed of D-fructofuranose units. They are important in short-term energy reserves for grasses and some plants. Inulin, found in dahlias, and levans from grasses are examples of fructans. Levans are short linear polysaccharides composed of β 2→1 linked fructose units as pictured below:

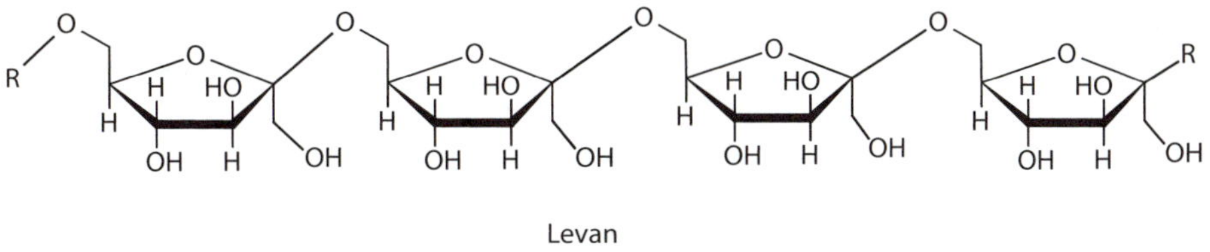

Levan (9.21)

9.7.2 CHITIN AND CHITOSAN

Chitin is generally a homopolymer of 2-acetamido-2-deoxy-D-glucose (*n*-acetylglucosamine) 1→4 linked in a β configuration; it is thus an amino sugar analog of cellulose. While it is widely distributed in bacteria and fungi, the major source is crustaceans. In fact, chitin is the most abundant organic skeletal component of invertebrates. It is believed to be the most widely distributed polysaccharide with the Copepoda alone synthesizing on the order of 10^9 tons each year. It is an important structural material often replacing cellulose in cell walls of lower plants. It is generally found covalently bonded to protein. Invertebrate exoskeletons often contain chitin that provides strength with some flexibility along with inorganic salts such as calcium carbonate that provide strength. In a real sense this is a composite where the chitin holds together the calcium carbonate domains.

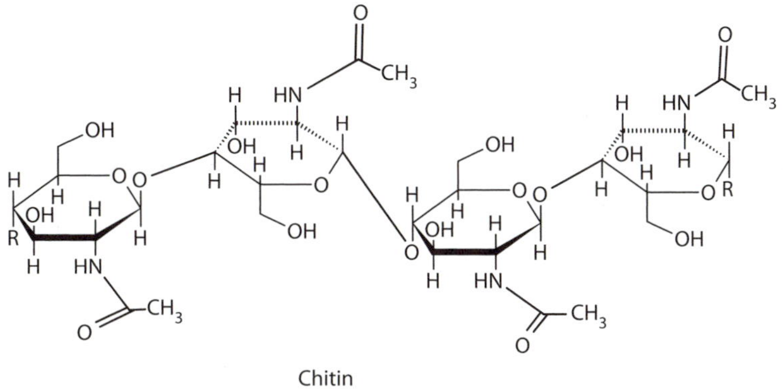

Chitin (9.22)

Chitosan is produced from the deacetylation of chitin. Chitosan is employed in the food industry. It is a hemostatic from which blood anticoagulants and antithrombogenic have been formed. It is often sold as a body fat reducing agent or to be taken along with eating to encapsulate fat particles.

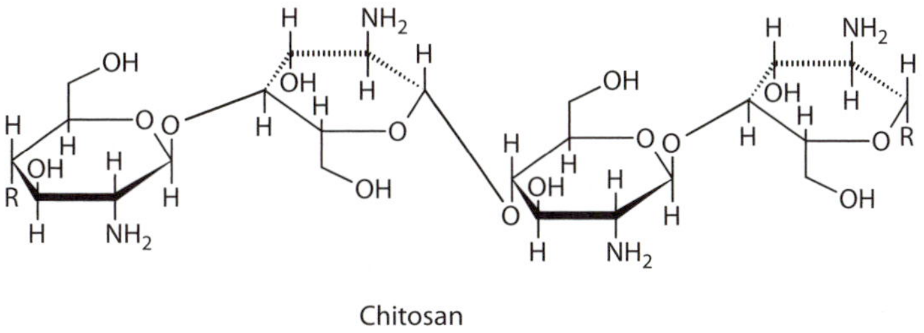

Chitosan (9.23)

Both chitosan and chitin are greatly underused readily available abundant materials that deserve additional study as commercial materials and feedstocks. Chitin itself is not antigenic to human tissue and can be inserted under the skin or in contact with bodily fluids generally without harm. In the body, chitin is slowly hydrolyzed by lysozyme and absorbed. Chitin and chitosan can be safely ingested by us and often we eat some since mushrooms, crabs, shrimp, many breads, and beer contain some chitin. Chitin and chitosan are believed to accelerate wound healing. Chitosan is also reported to exhibit bactericidal and fungicidal properties. Chitosan solutions are reported to be effective against topical fungal infections such as athlete's foot.

A continuing problem related to the introduction of bioengineering materials into our bodies is their incompatibility with blood. Many materials cause blood to clot (thrombosis) on the surfaces of the introduced material. Heparin (9.24) is an anticoagulant, nontoxic material that prevents clot formation when coated on vascular implants. While chitosan is a hemostatic material (stops bleeding by enhancing clotting), chitosan sulfate has the same anticoagulant behavior as heparin.

(9.24)

Representative structure of heparin

Cardiovascular disease is the second—to cancer—the leading cause of death in the United States. A contribution factor to cardiovascular disease is serum cholesterol. When ingested, chitosan exhibits hypocholesterolemic activity. Chitosan dissolves in the low pH found in the stomach and reprecipitates in the more alkaline intestinal fluid entrapping cholic acid as an ionic salt, preventing its absorption by the liver. The cholic acid is then digested by bacteria in the large intestine. Chitosan may also act to increase the ratio of high-density lipoprotein to total cholesterol. Chitosan has been studied in the formation of films, including membrane-gels that immobilize enzymes and other materials because of the mild conditions under which they form. Chitosan has been used as a flocculate in wastewater treatment. The presence of the amine gives coacervation with negatively charged materials such as negatively charged proteins allowing removal of unwanted protein waste. The amine groups also capture metal ions, in particular polyvalent and heavy metal ions, such as iron, lead, mercury, and uranium. Carraher, Francis, and Louda have also used chitosan to chelate with platinum salts to form materials with structures similar to the anticancer drug *cis*-dichlrodiamineplatinum II chloride. The amine and hydroxyl groups can be modified through use of a wide range of reactions, including formation of amides and esters.

Thus, there exists sufficient reason to consider these abundant materials in dietary, biomedical, cosmetic, and so forth applications.

9.7.3 OTHERS

Mannans are found in plants, particularly some seeds, and in some microorganisms such as algae and yeasts. Xylans are an important component of "hemicellulose," the base soluble materials closely associated with cellulose that are present in the secondary cell walls of higher plants. They are generally composed of β 1→4 linked D-xylopyranose units, thus the name. Arabinans are also plant material being present as a component of cell walls. Most arabinose-containing polysaccharides are actually combinations containing various saccharide units though there are some that contain largely only arabinose units.

9.8 HETEROPOLYSACCHARIDES

Heteropolysaccharides contain two or more different monosaccharides. Glycosaminoglycans are polysaccharides that contain aminosugar units. Most are of animal origin.

Above is a representative structure of *heparin* (9.24) that is complex containing D-glucuronic acid, L-iduronic acid, and D-glucosamine units. The glucosamine units may be N-acetylated or N-sulfonated. It is found in the lung, liver, and arterial walls of mammals. It is also found in intracellular granules of mast cells that line arterial walls and is released through injury. The glucuronic acid and iduronic acid units are not randomly present but occur in blocks. Heparin is found as the free polysaccharide and bonded to protein. Heparin acts as an anticoagulant, an inhibitor of blood clotting, and is widely used for this in medicine. In nature its purpose appears to be to prevent uncontrolled clotting.

Hyaluronic acid is found in connective tissues, umbilical cord, skin, and it is the synovial fluid of joints. It can have very large molecular weights, to 10^7 Daltons making solutions of hyaluronic acid quite viscous. They can form gels. As a synovial fluid in joints it acts as a lubricant and in the cartilage it may also act, along with chondroitin sulfates, as a shock absorber. In some diseases such as osteoarthritis the hyaluronic acid of the joints is partially degraded resulting in a loss of elasticity of the area. The molecules can adopt a helical structure.

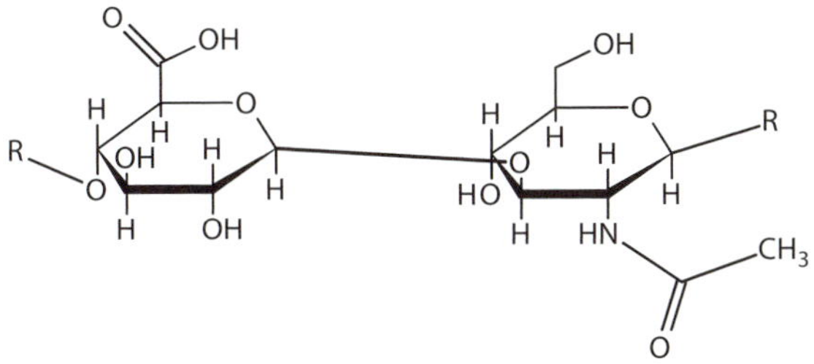

(9.25)

Hyaluronic acid

Chondroitin sulfates are found in bone, skin, and cartilage, but not as a free polysaccharide. Rather it exists as proteoglycan complexes where the polysaccharide is covalently bonded to a protein. The proteoglycan of cartilage contains about 10% protein, keratan sulfate (below), and chondroitin sulfate, mainly the 4-sulfate in humans. The chondroitin sulfate chains have a weight-average molecular weigh of about 50,000 Da but the complex has a molecular weight of several million. Again, chondroitin sulfates can adopt a helical conformation. The function of proteoglycan in cartilage is similar to that of noncellulosic polysaccharides and protein in plant cell walls. In cartilage collagen fibers provide the necessary strength that is provided in plants by cellulose fibers. Thus, cartilage proteoglycan is an important part of the matrix that surrounds the collagen fibers giving it rigidity and incompressibility. This network can also act as a shock absorber since on compression the water is squeezed out to a near by uncompressed region acting to "share the load" by distributing a shock or stress–strain.

Chondroitin sulfate is sold as a health aid to "maintain healthy mobile joints and cartilage."

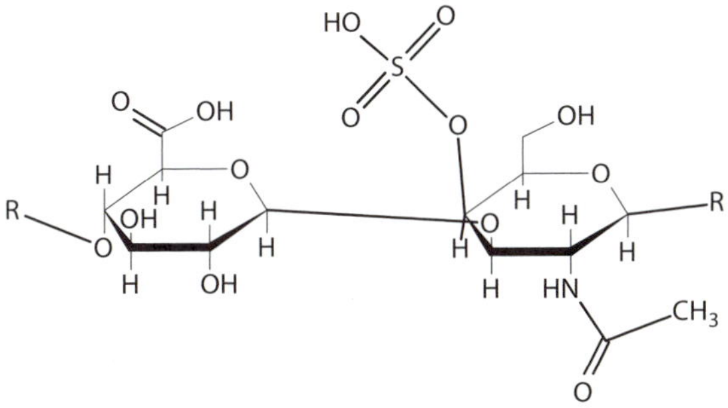

(9.26)

Chondroitin 4-sulfate

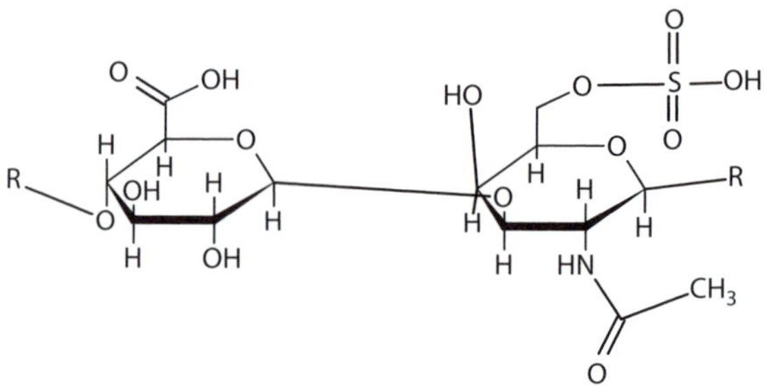

(9.27)

Chondroitin 6-sulfate

The second polysaccharide present in cartilage proteogylcan is keratan sulfate. It is generally found in shorter chains than chondroitin sulfate with a weight-average molecular weight to about 20,000 Da. It is also found in the cornea of the eye.

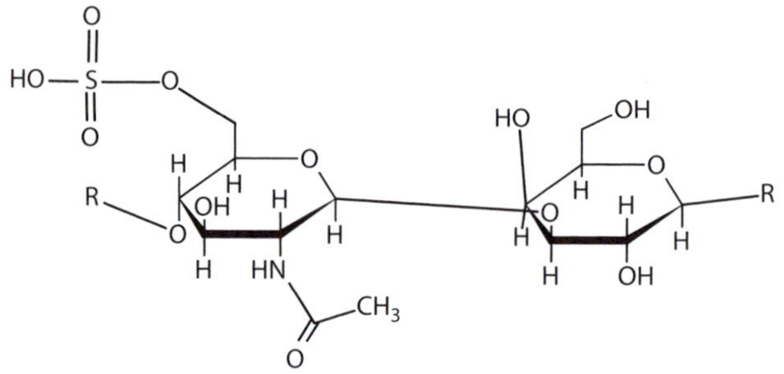

(9.28)

Keratan sulfate

Dermatan sulfate is found in the skin, arterial walls, and tendon, where it is a part of another proteoglycan complex. It is about the same size as chondroitin sulfate and also able to form helical conformations.

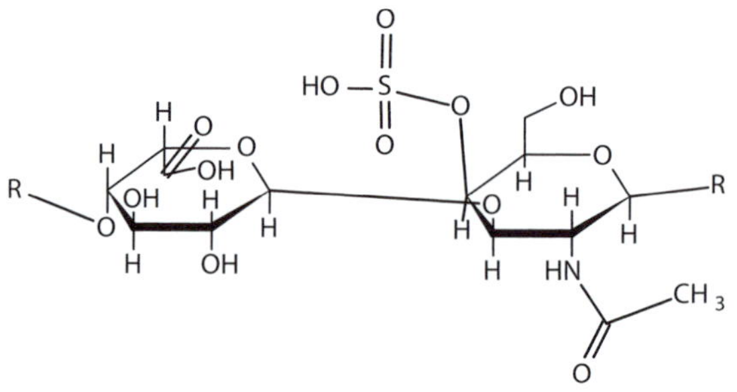

(9.29)

Dermatan sulfate

There are two main divisions of polysaccharides that contain unmodified galactose groups-arabinogalactans that contain many plant gums and carrageenas and agar. Seaweeds represent a source of many polysaccharides, including alginic acid, agar, and carrageenin. Alginic acid is a polymer of D-mannuronic acid and L-guluronic acid that may be arranged in a somewhat random fashion or in blocks. It is used as a stabilizer for ice cream, in paper coating, in the manufacture of explosives, and in latex emulsions.

The *carrageenans* and *agar* are generally linear galactans where the monomeric units are joined by alternating 1→4 and 1→3 bonds consisting then of disaccharide units. Carrageenan is the name given to a number of sulfated polysaccharides found in many red seaweeds where they play a structural role. The approximate repeat units for two industrially important carrageenans are given below. Both are able to form double helices containing two parallel staggered chains creating a gel. There are three disaccharide units per helix turn. The sulfate units are located on the outside of the helix with the helical structure stabilized by internal hydrogen bonds. In nature, red seaweeds contain an enzyme that converts the galactose-6-sulfate of the k-carrageenan to 3,6-anhydrogalactose that causes a stiffening of the helix. It has been found that red seaweeds found where there is strong wave action contain a high proportion of anhydrogalactose. Thus, it appears that the seaweed is able to control its structure in response to external stimuli to minimize shredding by the increased wave action.

Because of its gelling ability, carrageenan is widely used as food thickeners and emulsion stabilizers in the food industry and is present in many dairy products, including less-expensive ice cream and other desert products, providing a smooth, creamy texture. It is used as a stabilizer in foods such as chocolate milk.

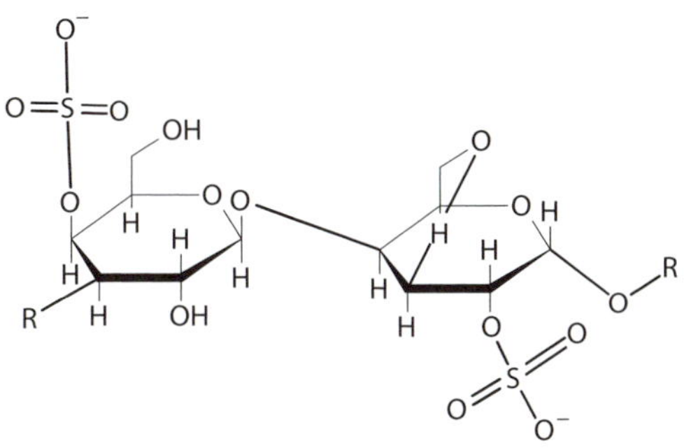

(9.30a)

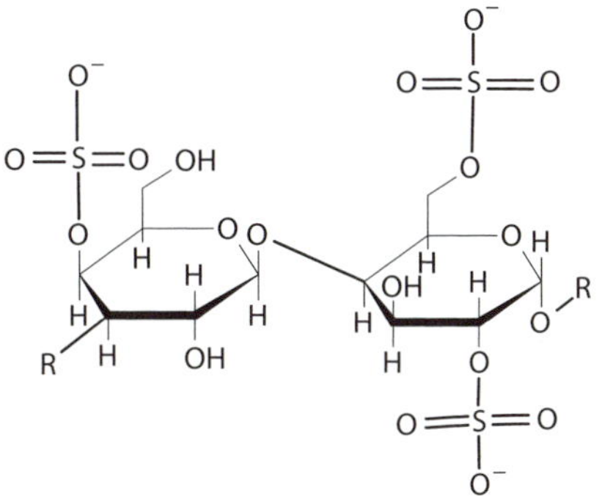

(9.30b)

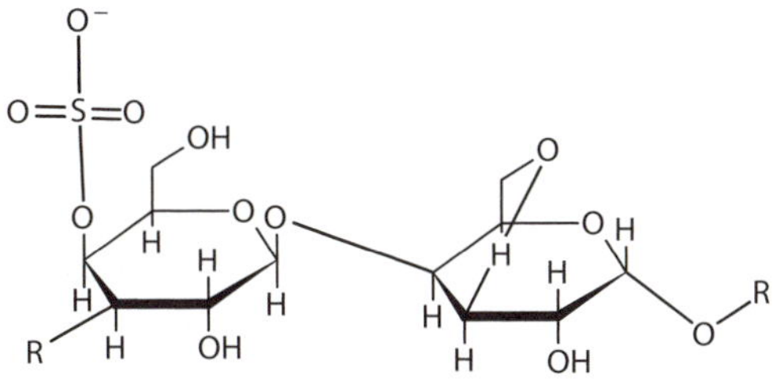

(9.31a)

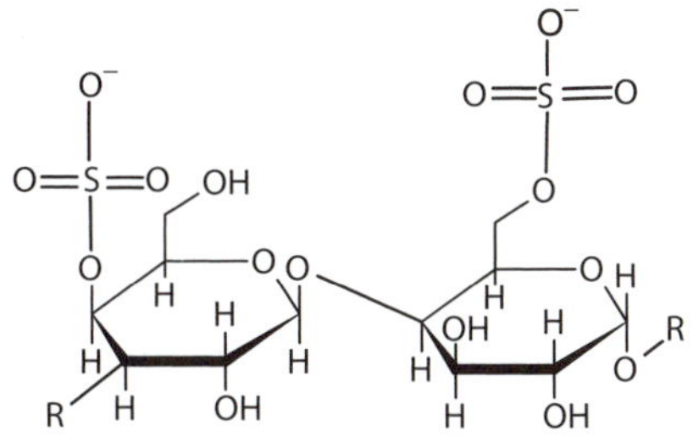

(9.31b)

The name agar refers to a family of polysaccharides that contain alternating β-D-galactopyranose and 3,6-anhydro-α-L-galactopyranose units and is thus similar to a carrangeenin where the anhydro-L-galactose is substituted for the anhydro-D-galactose. It is employed as the basis of many microbiological media and in canned food because it can be sterilized. The latter is an advantage over gelatin that is not able to withstand sterilization.

Agarose is the agar polysaccharide with the greatest gelling tendency. It contains no glucuronic acid units. It can form a compact double helix with the two chains being parallel and staggered, as in the case of carrageenan, forming a gel. Agarose gels are employed in gel-permeation chromatography (GPC) and gel electrophoresis.

Agarose

(9.32)

Glycoproteins contain both saccharide and protein moieties with the protein being the major component, but both portions are involved in the overall biological activities.

9.9 SYNTHETIC RUBBERS

Natural rubber (NR) has been known for more than one thousand years. The Aztecs played a game using a "rubber" ball. NR was used by the Mayan civilization in Central and South America before the twentieth century. In addition to using the latex from the ule tree for waterproofing of clothing, they played a game called "tlachtli" with large hevea rubber balls. The object of the game was to insert the ball into a tight-fitting stone hole in a vertical wall using only the shoulder or thigh. The game ended once a goal was scored, and the members of the losing team could be sacrificed to the gods.

Columbus, on his second voyage to America is reported to have seen the Indians of Haiti using rubber balls. By the eighteenth century Europeans and Americans used NR to "rub out" marks made by lead pencils. The "rubbing out" use led to the name "rubber." Because of the association of NR with the American Indian, it was also called "Indian rubber."

Early progress toward its use in Europe is attributed to Charles Macintosh and Thomas Hancock. NR was dissolved in relatively expensive solvents such as turpentine and camphene. The earliest applications were made by pouring these solutions containing the NR onto objects to be "rubberized." Later, other less-expensive solvents were discovered, including the use of coal-tar naphtha. Macintosh poured naphtha solutions containing the NR onto layers of cloth producing "waterproof" material, which was the origin of the Macintosh raincoat, misspelled by the English as "Mackintoshes." The layering of the NR not only produced a material that was waterproof, but also got around the problem that NR was sticky, becoming more sticky on hot days. NR also had an unpleasant odor that was somewhat captured and prevented from smelling up the place by placement between the pieces of cloth. Hancock, an associate of Macintosh, worked to develop other useful rubbers from NR. One of his first was rubber thread derived from cutting strips of NR and applied to cloths and footwear. He had lots of scrapes and found that by heating the scrapes he could reform sheets of the NR from which he could cut more strips. He also developed a crude mixing machine that allowed him to mix other materials, additives, into the rubber.

The development of rubber technology then shifted to North America in the early to mid-1800s. Mills were developed that allowed additives to be added to rubber and allowed rubber to be formed into sheets and small particles. Uses for NR were largely waterproof cloth items in Britain and waterproof boots in the United States.

But, the problem of stickiness remained until an accidental discovery by Charles Goodyear in 1839. As a young man Goodyear started a lifelong affair with NR to "tame it" for use. He recorded thousands of experiments with NR mixing materials of the day with it and observing what happened. When working with sulfur some of his rubber got mixed with sulfur and fire with the resulting mixture no longer sticky. After some effort, he worked with mixtures of NR, sulfur, and lead producing a "fireproof" gum that was later called "vulcanized rubber" after the god of fire Vulcan. About this same time, Hancock found a piece of Goodyear's vulcanized rubber and applied for a patent citing as the important ingredients heat, sulfur, and NR. Goodyear's combination included lead that allowed vulcanization to occur at lower temperatures. Goodyear applied for a British patent on January 30,

1844, 8 weeks before Hancock's application. Though in ill health and poverty, Goodyear battled to see that his contributions were recognized. Others entered the fight with related claims but history recognizes Goodyear for his scientific insight and Hancock for his applications of NR producing a number of products. (Chapter 1 contains more about the ventures of Charles Goodyear.)

Goodyear had trouble defending his patent, piling up huge debts before he died in 1860. Daniel Webster defended him in one of his patent infringement cases. By 1858, the yearly value of rubber goods produced in the United States was about $5 million. The major rubber producing plants clustered about Akron, Ohio, which the Goodyear Company founded in 1870.

Vulcanized rubber had many applications but one of the greatest is probably the "rubber tire." Early tires were made of solid NR and gave a stiff ride. John B. Dunlap rediscovered the air-filled tire, pneumatic tire, in 1887. At this time, the automobile was just beginning its ownership of the roads. Bicycles made by various small shops, including those made by the Wright Brothers of Dayton, Ohio, were being viewed as a poor man's horse. It did not need food nor did you need to tend to it or to its droppings. As roads, even gravel and dirt roads, become more common place in the cities, bicycles began to be the vehicle of choice here in America and in Europe. The air-filled tire gave a softer ride and became the tire of choice for cyclists. It also increased the value and need for NR. As cars and then trucks entered as major conveyers of people and goods, the need for greater amounts of rubber increased.

During this time, seeds from the Amazon valley were planted in British colonies in the east where plantations were founded using native labor, often enslaved, to harvest the NR. Of the various NR-producing trees, *Hevea Brasiliensis* was the best. With World War I, the need for rubber by the British and American forces was critical. To cope with this need several avenues were explored. Alternative rubber producing plants were explored emphasizing plants that would be producing within a year or less and which could survive under milder climates, milder with respect to temperature and need for water. While the exploration of such plants continues today, this avenue was not successful in supplying the greatly increased rubber needs (Figure 9.4).

The second avenue involved developing a synthetic rubber (SR). English scientist Michael Faraday helped set the stage by analyzing NR finding that it was generally composed of repeat units originally described by Faraday as $C_{10}H_{16}$, which was latter rendered $(C_5H_8)_n$. NR was later separated into three parts, an oil, tar, and "spirit." The most volatile "spirit" portion was evaluated and found to contain largely the isoprene unit. The task then was to produce isoprene synthetically, and then to convert it to rubber. By 1900, the synthesis of isoprene had been achieved, though not easily. Often, isoprene samples spontaneously underwent addition forming a material similar to NR.

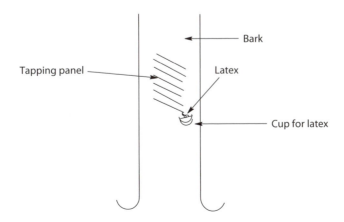

FIGURE 9.4 Harvesting of rubber latex from a rubber tree.

During this time other materials were found that gave rubber-like materials. In 1901 I. Kondakov, a Russian, discovered that dimethyl butadiene when heated with potash formed a rubber-like material. In 1910 S. V. Lebedev, another Russian, reacted butadiene forming a rubber-like material (Equation 9.33).

$$H_2C=CH–CH=CH_2 \rightarrow \text{Rubber-like product} \qquad (9.33)$$

During World War I, Germany was cut off from the Far East sources of NR and so developed their own alternative rubbers. They focused on the Kondakov process making two rubber-like materials-dubbed methyl "H," which was a hard rubber and methyl "W," which was a softer rubber. Before large-scale production started, the war ended and Germany returned to natural sources of NR. Even so, Germany continued some research toward the production of rubber. By the 1930s, they produced three "buna" rubbers. The name "buna" comes from the first two letters of the main ingredients, *bu*tadiene and *na*trium (sodium), the catalyst. They also discovered another of today's rubbers, buna-S, the "S" standing for styrene. Buna-S is derived from butadiene and styrene. The British referred to buna-S as SBR (styrene–butadiene rubber), while the United States referred to this as GR–S (Government rubber–styrene). Many of these SRs when vulcanized were superior to earlier synthesized rubbers and often to NR for some applications.

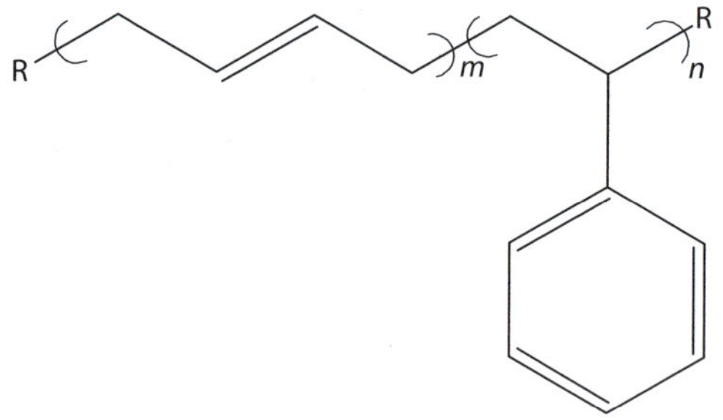

$$(9.34)$$

In 1937, Germany also developed buna-N (N = nitrile; 9.35) which offered good resistance to oil. Most of these rubbers are still in use today with small modifications.

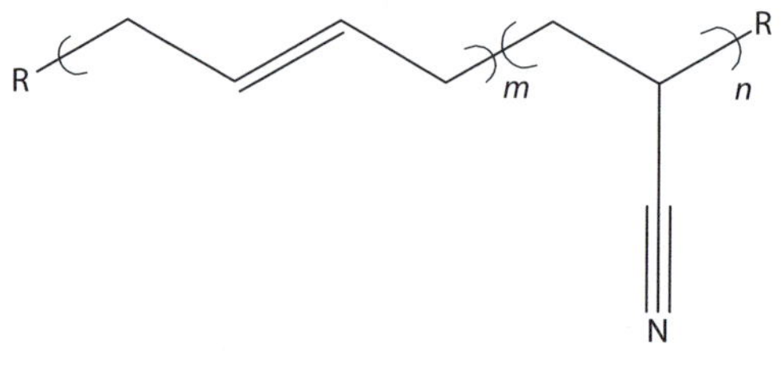

$$(9.35)$$

The USSR was producing SRs during this time using potatoes and limestone as starting materials. Thus, the move toward natural materials is not new.

As World War II approached, Germany and USSR worked frantically to be free from outside need of rubber. The United States by comparison were conducting research in a number of areas that would eventually prove useful, but had made no real push for self-reliance in the issue of rubber. U.S. experiments with SR before 1939 lead to the discovery of neoprene, Thiokol, and butyl rubber.

As noted above, chemists learned about the structure of rubber by degrading it through heating and analyzing the evolved products. One of the evolved products was isoprene, a five carbon hydrocarbon containing a double bond. Isoprene is a basic building block in nature serving as the "repeat" unit in rubber and also the building block of steroids such as cholesterol.

$$H_2C=C-CH_2=CH_3$$
$$|$$
$$CH_3$$

(9.36)

Isoprene

Again, as noted above, with knowledge that NR had isoprene units, chemists worked to duplicate the synthesis of rubber except using synthetic monomers. These attempts failed until two factors were realized. First, after much effort it was discovered that the methyl groups were present in a "cis" arrangement. Second, it was not until the discovery of stereoregular catalysts that the chemists had the ability to form NR-like material from butadiene.

The search for the synthesis of a purely SR, structurally similar to NR, continued and involved a number of scientists building upon one another's work—along with a little creativity. Nieuwland, a Catholic priest, President of Notre Dame University, and a chemist did extensive work on acetylene. He found that acetylene could be made to add to itself forming dimers and trimers (Equation 9.37).

$$HC\equiv CH \xrightarrow{} \quad \xrightarrow{}$$

(9.37)

Acetylene　　　Vinylacetylene　　　Divinylacetylene

Calcott, a DuPont chemist, attempted to make polymers from acetylene, reasoning that if acetylene formed dimers and trimers, conditions could be found to produce polymers. He failed, but went to Carothers who had one of his chemists, Arnold Collins, work on the project. Collins ran the reaction described by Nieuwland, purifying the reaction mixture. He found a small amount of material that was not vinylacetylene or divinylacetylene. He set the liquid aside. When he came back, the liquid had solidified giving a material that seemed rubbery and even bounced. They analyzed the rubbery material and found that it was not a hydrocarbon, but it had chlorine in it. The chlorine had come from HCl that was used in Nieuwland's procedure to make the dimers and trimers. The hydrogen chloride added to the vinylacetylene forming chloroprene.

This new rubber was given the name *Neoprene* (Equation 9.38). Neoprene had outstanding resistance to gasoline, ozone, and oil in contrast to NR. Today, Neoprene is used in a variety of applications, including as electrical cable jacketing, window gaskets, shoe soles, industrial hose, and heavy duty drive belts.

$$H_2C=C-CH=CH_2 \rightarrow -(-CH_2-C=CH-CH_2-)-$$
$$\quad\quad\; |\quad\quad\quad\quad\quad\quad\quad\quad\quad |$$
$$\quad\quad\; Cl\quad\quad\quad\quad\quad\quad\quad\quad\; Cl$$

Chloroprene　　　　Polychloroprene
　　　　　　　　　(Neoprene)

(9.38)

Thiokol was developed by J. C. Patrick in 1926. Joseph Cecil Patrick was born in 1892 in Jefferson County, MO. He was a physician rather than a chemist. Before completing his medical school studies, World War I began and he tried to join the U.S. Air Corps. Because of poor health, he could not meet the standards for combat service but was accepted in the U.S. Medical Corps and was sent to serve in France. While there, he was part of the influenza epidemic and was found in a coma surrounded by corpses awaiting shipment for interment. His escape from burial may have been the result of the administering of a large dose of quinine by a French nurse.

On returning to the United States he continued his medical education and accepted a position as a public health inspector in Kansas City, MO. Within the year he took a better paying job as an analytical chemist for Armor Packing Company in Buenos Aires, Argentina. He returned to the United States again, and finally completed his medical training from the Kansas City College of Medicine and Surgery in 1922.

Rather than practicing medicine he helped establish the Industrial Testing Laboratory, Inc. in Kansas City by the mid-1920s. He developed a process for the production of pectin as a precipitate from apples from a vinegar plant.

Another project undertaken by the testing laboratory was an attempt to improve the synthesis of ethylene glycol, the major ingredient in antifreeze, from the hydrolysis of ethylene chlorohydrin. They used sodium sulfide in the process. Instead of getting the desired antifreeze, he got a gunk. Instead of throwing away the gunk, which was often done during this time, he studied it. The material was foul smelling, much like rotting eggs, and it was stable in most solvents it was placed in. He called this material "Thiokol" after the Greek for "sulfur" and "gum" (Kommi). He got funds from Standard Oil Company of Indiana to do further work with Thiokol to try to eliminate its odor.

Sulfur naturally comes in packages of eight forming an octagon. Through varying the reaction conditions a variety of thiokols can be formed with the Thiokol chains having varying averages of sulfurs. Thiokols are formed from the reaction of ethylene with chlorine forming ethylene dichloride that is reacted with sodium polysulfide.

$$H_2C=CH_2 + Cl_2 \rightarrow H_2C(Cl)-C(Cl)H_2 \tag{9.39}$$

$$H_2C(Cl)-C(Cl)H_2 + NaS_x \rightarrow -(-CH_2-CH_2-S-S-)- + NaCl \tag{9.40}$$

Butyl rubber was discovered by R. M. Thomas and W. J. Sparks in 1937 and was developed by the Standard Oil Company (NJ) and was part of an exchange of information between the German chemical giant I. G. Farbenindustrie AG and Standard Oil.

Butyl rubber is a copolymer of 1-butene, $(CH_3)_2C=CH_2$, and small amounts (about 2%–3%) of isoprene or other unsaturated compounds. The unsaturation allows subsequent cross-linking of the material.

As World War II loomed with America pushed into the war by the Pearl Harbor attack, the need for an independent rubber supply was critical. The government instituted several nation-wide efforts, including the Manhattan and Synthetic Rubber Projects. Carl "Speed" Marvel was one of the chemists involved in the Synthetic Rubber Project.

When I first met "Speed" Marvel, there was little "speedy" about his gate but he was speedy with his mind and pleasant personality. As with many of the polymer pioneers, he was friendly and commented on the importance of my work before I could talk about my admiration of him and his efforts. Now that is a way to turn the tables on a young (now older) researcher. I had read about "Speed" and his part in "saving the war" and was anxious to simply meet him.

Carl Shipp "Speed" Marvel was born in 1894 on a farm in Waynesville, IL, and spent much of his 93 years furthering synthetic polymer chemistry. He acquired his name because when he registered for his first semester as a graduate student, in 1915, at the University of Illinois, it was believed he lacked some of the basic chemistry background so was given extra courses to take. He worked late into the night in the laboratory and studying but wanted to get to the breakfast table before the dining room door closed at 7:30 AM. His fellow students commented that was the only time he ever hurried and gave him the name "Speed" because of it.

During this time, chemistry was prospering in Europe but not in America so there was an effort by American chemists to establish routes to important chemicals, such as dyes, in academia and industry. There was a flow of chemists from academia to industry and from one academic institution to another. Roger Adams, one of the foremost organic chemists of the time, moved from Harvard to the University of Illinois at this time. This was a time of rapid growth in the chemical industry because before World War I much of the chemical industry centered in Germany and because of the war, this was changing. The American Society membership grew from 7,400 in 1915 to more than 10,600 in 1917. (Today, the ACS has a membership of about 160,000 and is the largest single discipline organization.) The shortage of chemicals was so severe that Adams had graduate students such as Speed Marvel actually work on the production of special chemicals related to the new chemical warfare that was introduced by the Germans. It was quite a task to make these dangerous chemicals because of the poor ventilating hoods used by the students.

America entered the war in 1917 and many university chemists were transferred to Washington, DC, to work on chemical warfare problems. Adams became a Major in a unit that emphasized chemical warfare problems and maintained contact with the group at Illinois asking them to furnish chemicals on a rush basis. Speed also helped in the preparation of steel furnishing steel makers with a chemical needed to analyze nickel steel.

At the end of the war there was a lessening in the need for chemists as industry reorganized itself for peace. At the end of the war, 1921, Speed was able to complete his graduate work at Illinois in organic chemistry. He had wanted to work in industry, but industrial jobs were hard to get at this time so he stayed on as an instructor. The academic hierarchy was similar to that present in places in Europe where junior faculty had bosses that were also part of the academic regime. Speed's boss was Oliver Kamm who left to become research director at Parke Davis so Speed was promoted. During this time, interest in chemistry was increasing so there was an increase in the number of graduate students in chemistry. In the fall of 1920, one of Speed's students was Wallace Carothers who went on to become one of the fathers of polymer chemistry. By 1924, Carothers became part of the organic faculty. During this time, there were evening seminars where the faculty discussed the latest trends and problems in chemistry. In 1927, Carothers left to join the Harvard faculty and eventually the DuPont Experimental Station.

In 1928, Speed became a consultant to the DuPont Experimental Station due to the nudging of Roger Adams. Chemistry was on the rise with the invention of the first synthetic rubber (neoprene) and first truly synthetic fiber (nylon) both discovered by Carothers. The plastics industry developed rapidly with chemistry becoming truly an industrial giant, mainly on the back of giant molecules.

War broke out in Europe and it was obvious that America would become part of the conflict. The Office of Scientific Research and Development and the National Research and Development Committee were established in late 1941. Various universities furnished the people for these agencies and most university labs accepted contracts related to studying problems of interest to the Department of Defense. Adams was important in this effort and so included Speed as a participant.

When Japan attacked Hawaii in December 1941, the most critical chemical problem was the solution to the rubber shortage because the far Eastern suppliers were cut off. Because Speed had "become" a polymer chemist during the years of 1930–1940, he became part of this rubber program. The major rubber companies, many universities, and other chemical companies put aside their competition and pooled their research efforts and in about 1 year's time developed a usable SR that was manufactured and used in smaller types of tires.

Speed helped organize, before the war years, a rubber program on the synthesis of rubber for the National Defense Research Committee under Adams, which was eventually responsible for the production of rubber during the war. He was also drafted and helped on a research program for the Committee on Medical Research dealing with malaria research.

In the 1950s, Speed was part of a large effort headed out of Wright Patterson Air Force Base aimed at developing thermally stable materials for a number of purposes, including use for outer space craft. This effort acted as an early focal point for the synthesis of metal-containing polymers. These metal-containing polymers lost out to the honey combed ceramic tiles now used on our space craft.

By 1953, Karl Ziegler and Giuilo Natta discovered a family of catalysts that allowed the introduction of monomer units onto growing polymer chains in an ordered manner. This allowed the synthesis of rubber-like polymers with greater strength and chemical stability in comparison to similar polymers made without the use of these steroregulating catalysts.

9.10　NATURALLY OCCURRING POLYISOPRENES

Polyisoprenes occur in nature as hard plastics called *gutta percha* and balata and as an elastomer or rubber known as *H. brasiliensis* or NR. Approximately 50% of the 500 tons of gutta percha produced annually is produced from trees grown on plantations in Java and Malaya. Balata and about 50% of the gutta percha used are obtained from trees in the jungles of South America and the East Indies. The first gutta-insulated submarine cable was installed between England and France

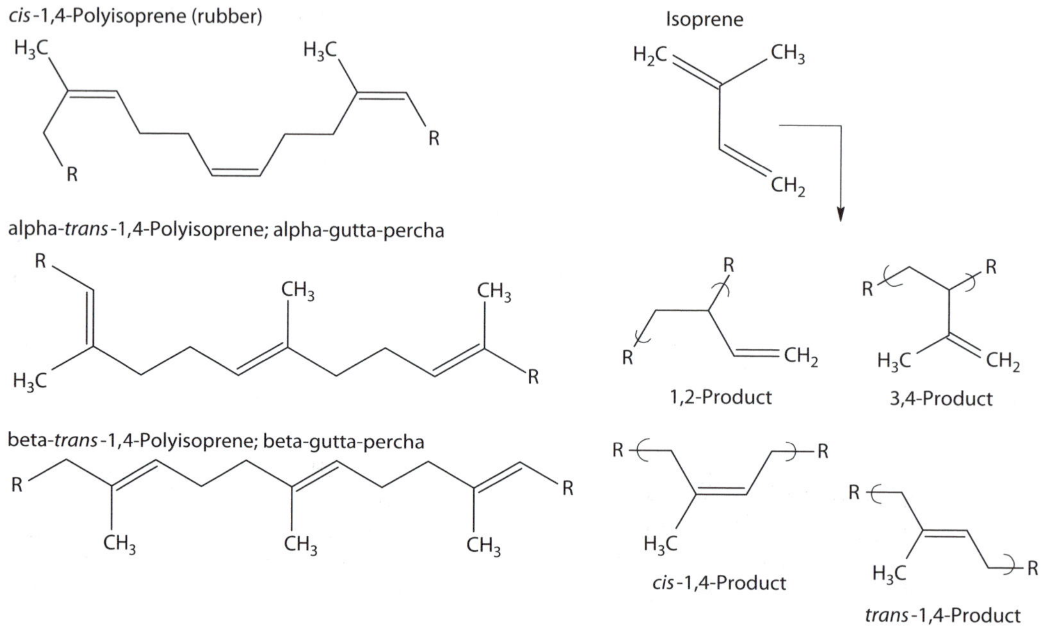

cis-1,4-Polyisoprene (rubber)

Isoprene

alpha-*trans*-1,4-Polyisoprene; alpha-gutta-percha

1,2-Product

3,4-Product

beta-*trans*-1,4-Polyisoprene; beta-gutta-percha

cis-1,4-Product

trans-1,4-Product

FIGURE 9.5 Abbreviated structural formulas for polyisoprenes.

in 1859. Gutta percha (*Palaquium oblongifolium*) continues to be used for wire insulation, and polyisoprene, balata (*Mimusops globosa*), and gutta percha are used as covers for some golf balls.

$$-(-CH_2-\overset{\overset{\displaystyle CH_3}{|}}{C}=CH-CH_2-)- \tag{9.41}$$

The hardness of some of the polydisperse naturally occurring crystalline polymers is the result of a trans configuration in 1,4-polyisoprene (Figures 9.5 and 9.6). This geometry facilitates good fit between chains and results in inflexibility in the chains. The hevea or NR is, by comparison, soft because it is a cis configuration (Figures 9.5 and 9.6) that does not allow a close fit between chains resulting in flexibility and a "rubbery" behavior. Both trans and cis isomers of polyisoprene are present in chicle, obtained from the *Achras sapota* tree in Central America. Both trans and cis isomers are also synthesized commercially.

Before the discovery of the vulcanization or cross-linking of hevea rubber with sulfur by Charles Goodyear in 1838, Faraday has shown that the empirical formula of this elastomer is C_5H_8, making it a member of the terpene family. The product obtained by pyrolsyis of rubber was named isoprene by Williams in 1860 and converted to a solid (polymerized) by Bouchardat in 1879.

Natural rubber crystallized when stretched in a reversible process. However, the sample remains in its stretched form (racked rubber) if it is cooled below its T_g. The racked rubber will snap back and approach its original form when it is heated above its T_g. The delay in returning to the original form is called *hysteresis*. These and other elastic properties of NR and other elastomers above the T_g are based on long-range elasticity. Stretching causes an uncoiling of the polymer chains, but these chains assume a more ordered microstructure creating crystalline domains acting to oppose further stretching. If the rubber is held in the stretched state for some time, slippage of chains occurs and precise return to its original shape does not occur when the stress is released.

The absence of strong intermolecular forces, presence of pendant methyl groups discouraging close association, and crankshaft action associated with the cis isomer all contribute to the flexibility of NR. The introduction of a few cross-links by vulcanization with sulfur reduces slippage of chains but still permits good flexibility.

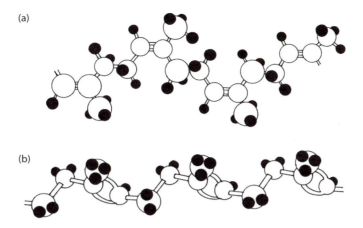

(a)

(b)

FIGURE 9.6 Ball-and-stick models of "soft" hevea rubber (*cis*-1,4-polyisoprene) (a) and "hard" gutta percha (*trans*-1,4-polyisoprene) (b).

When a strip of NR or SR is stretched at a constant rate, the tensile strength required for stretching (stress, s) increases slowly until elongation (strain) of several hundred percent is observed. This initial process is associated with an uncoiling of the polymer chains in the uncross-linked regions. Considerably more stress is required for greater elongation to about 800%. This rapid increase in modulus (G) is associated with better alignment of the polymer chains along the axis of elongation, crystallization, and decrease in entropy (ΔS). The work done in the stretching process (W_1) is equal to the product of the retractile force (F) and the change in length (dl). Therefore, the force is equal to the work per change in length.

$$W_1 = f \, dl \quad \text{or} \quad f = \frac{W_1}{dl} \tag{9.42}$$

W_1 is equal to the change in Gibbs free energy (dG), which under the conditions of constant pressure is equal to the change in internal energy (dE) minus the product of the change in entropy and the Kelvin temperature as follows:

$$f = \frac{W_1}{dl} = \frac{dG}{dl} = \frac{dE}{dl} - T\left(\frac{dS}{dl}\right) \tag{9.43}$$

The first term (dE/dl) in Equation 9.43 is important in the initial low-modulus stretching process, and the second term [$T(dS/dl)$] predominates in the second high modulus stretching process. For an ideal rubber, only the second term is involved.

As observed by Gough in 1805 and verified by Joule in 1859, the temperature of rubber increases as it is stretched, and the stretched sample cools as it snaps back to its original condition. (This can be easily confirmed by you by rapidly stretching a rubber band and placing it to your lips, noting that heating has occurred, and then rapidly releasing the tension and again placing the rubber band to your lips.) This effect was expressed mathematically by Kelvin and Clausius in the 1850s. The ratio of the rate of change of the retractive force (df) to the change in Kelvin temperature (dT) in an adiabatic process is equal to the specific heat of the elastomer (C_p) per degree temperature (T) times the change in temperature (dT) times the change in length (dl).

$$\frac{df}{dT} = \left(\frac{C_p}{T}\right)\left(\frac{dT}{dl}\right) \tag{9.44}$$

Equation 9.44 may be transformed to Equation 9.45.

$$\frac{dT}{df} = -\left(\frac{T}{C_p}\right)\left(\frac{dl}{dT}\right) \tag{9.45}$$

Unlike most solids, NR and other elastomers contract when heated.

9.11   RESINS

Shellac, which was used by Edison for molding his first photograph records and is still used as an alcoholic solution (spirit varnish) for coating wood, is a cross-linked polymer consisting largely of derivatives of aleuritic acid (9,10,16-triphydroxyhexadecanoic acid). Shellac is excreted by small coccid insects (*Coccus lacca*) which feed on the twigs of trees in Southeast Asia. More than 2 million insects must be dissolved in ethanol to produce 1 kg of shellac.

$$\text{HOCH}_2-(\text{CH}_2)_5-\text{CH(OH)}-\text{CH(OH)}-(\text{CH}_2)_7-\text{COOH} \tag{9.46}$$

Rosin, the residue left in the manufacture of turpentine by distillation, is a mixture of the diterpene, abietic acid, and its anhydride. It is nonpolymeric but is used in the manufacture of synthetic resins and varnishes.

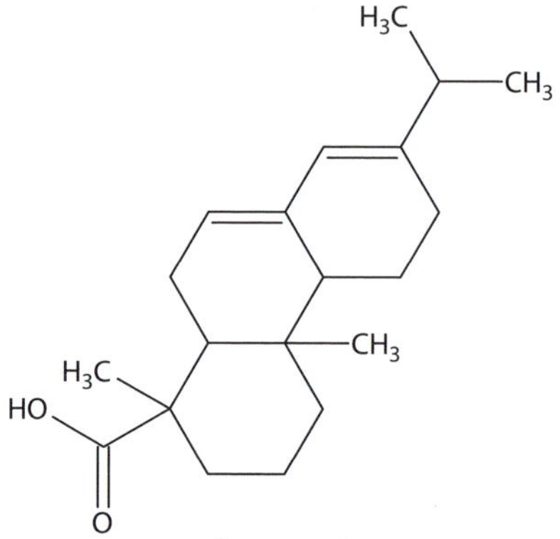

(9.47)

Abietic acid

Easter gum, a cross-linked ester, is obtained by the esterfication of glycerol or pentaerythritol with rosin.

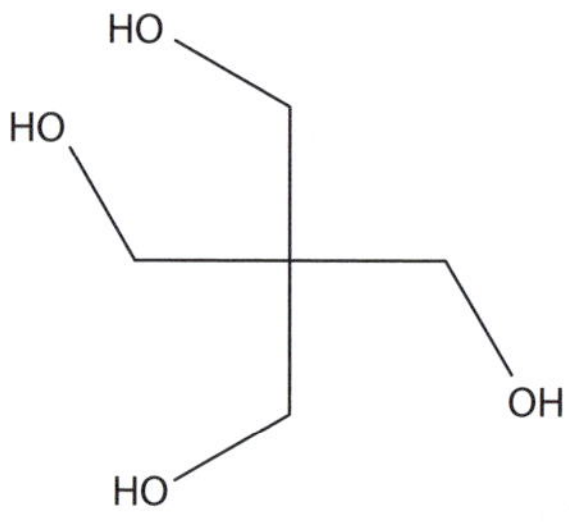

(9.48)

Pentaerythritol

Many natural resins are fossil resins exuded from trees thousands of years ago. Recent exudates are called *recent resins*, and those obtained from dead trees are called *semifossil resins*. Humic acid is a fossil resin found with peat, coal, or lignite deposits throughout the world. It is used as a soil conditioner, as a component of oil drilling muds, and as a scavenger for heavy metals. Amber is a fossil resin found in the Baltic Sea regions, and sandarac and copals are found in Morocco and Oceania, respectively. Casein, a protein from milk, under the name of Galalith, has been used as a molding resin and as an adhesive.

9.12 BALLOONS

Balloons have been part of our everyday lives from birthdays to graduations. As with many simple objects, their origin is unknown. Balloon-like objects have been part of ancient stories and probably were initially used in sport. These early balloons were made of animal bladders and intestines, both being protein-based, and hence polymeric. European jesters inflated the animal entrails using them to entertain others. Galileo inflated a pig's bladder to help measure the weight of air.

While there are also new world mentions of similar balloon construction, it was not until rubber arrived on the scene that the manufacture of balloons, as balloons we know today, began.

In 1824, Michael Faraday used NR to produce balloons for his hydrogen experiments at the Royal Institution in London. He writes "…The caoutchouc is exceedingly elastic….Bags made of it … have been expanded by having air forced into them, until the caoutchouc was quite transparent, and when expanded by hydrogen they were so light as to form balloons with considerable ascending power…" Faraday constructed his balloons by cutting two layered sheets of rubber and then pressing the edges together. Flour was applied to the inside of the balloons because the native rubber was tacky and would adhere preventing it from being inflated.

Toy balloons were introduced by Thomas Hancock in 1825 as a do-it-yourself kit that consisted of a rubber solution and a syringe. Vulcanized toy balloons were initially manufactured by J. G. Ingram of London in 1847. The vulcanizing caused the balloons to be nontacky and not susceptible to becoming excessively tacky on hot days. Montgomery Ward had balloons in their catalog by 1889.

In America, the story of the balloon coincided with the story of the tire, both requiring an understanding and use of rubber. The initial manufacture of balloons in the United States was in 1907 by the Anderson Rubber Company in Akron, Ohio. In 1912, Harry Rose Gill, founder of the National Latex Rubber Products of Ashland, Ohio, made a nonspherical cigar-shaped balloon. Gill also began packaging balloons in packs, the initial sanitary balloon package.

Neil Tillotson, at 16, began work at the Hood Rubber Company in Boston. After a 2-year period in the Seventh Cavalry during World War I, he returned to the Hood Rubber Company. Of about 25 Hood chemists, Neil was the only one without a college degree. But, he was inventive and with the first load of raw rubber latex in the 1920s, he started a lifelong romance with rubber. The company's initial efforts with the raw latex were unsuccessful and the company largely dropped its efforts to produce balloons. Tillotson persevered working out of his own home. Eventually, he managed to "tame" the latex sufficiently to make stylistic balloons using techniques similar to what is employed today. They cut a cat-like face from cardboard. He dipped the cat-like mold in latex and allowed it to dry. Inflated, it did resemble a cat. His first sale was an order for 15 gross for use at the annual Patriots Day Parade on April 19, 1931. Next, he formed a family business, incorporating later in 1931, using his family as an assembly line in the production of balloons. During the great depression, Neil traveled by bus around the country to sell his balloons. The Tillotson Rubber Company is still in operation with the balloon division named Dipco, a nice reminder of the way balloons are still made.

Rubber latex is sensitive to the influence of external forces or chemical agents. Thus, the so-called alligator balloon is formed when solid colored balloons are dipped in an acid bath, causing the balloon to turn into two shades, supposedly like an alligator. The radium balloon is produced by taking a solid colored balloon and dipping it a second time into another color.

Commercially inflated balloons held hydrogen and helium. Hydrogen was preferred because of its somewhat, about 10%, greater lifting ability. Hydrogen balloons were first produced by Faraday, as noted before. But, along with its greater lifting power, hydrogen offers a danger. On ignition, hydrogen burns rapidly forming water and energy, generally producing a flame. As early as 1914, firemen tried to curtail the use of hydrogen-filled balloons. In 1922, New York City banded the use of hydrogen-filled balloons because a city official was badly burned because of a fire caused by hydrogen-filled balloons.

There is an experiment that is often used in polymer demonstrations. Pointed objects such as long pins and sticks are gently pushed into a blow-up balloon without exploding the balloon. This results because the rubber polymer chains form about the impacting object to seal up the "hole" created by the object. Thus, even in its seemingly solid state, the rubber balloon's polymer chains are sufficiently mobile to allow it to "heal" itself.

Balloons today are still made from the rubber latex from the rubber tree using molds. Color is added as desired.

Foil balloons are derived from the effort of NASA Space Command with the concept and technology to metalize plastic sheeting. While the balloons are often referred to as Mylar or silver Mylars, they are not made from Mylar, a trade name for certain polyester film. They should be referred to as simply foil balloons. They are made from nylon film coated on one side with polyethylene and metallized on the other. The nylon film gives the balloon some strength and the polyethylene gives it some flexibility and helps retard the release of the held gas.

9.13   LIGNIN

Lignin is the second most widely produced organic material after the saccharides. It is found in essentially all living plants and is the major noncellulosic constituent of wood. It is produced at an annual rate of about 2×10^{10} tons with the biosphere containing a total of about 3×10^{11} tons. It contains a variety of structural units, including those pictured in Figure 9.7.

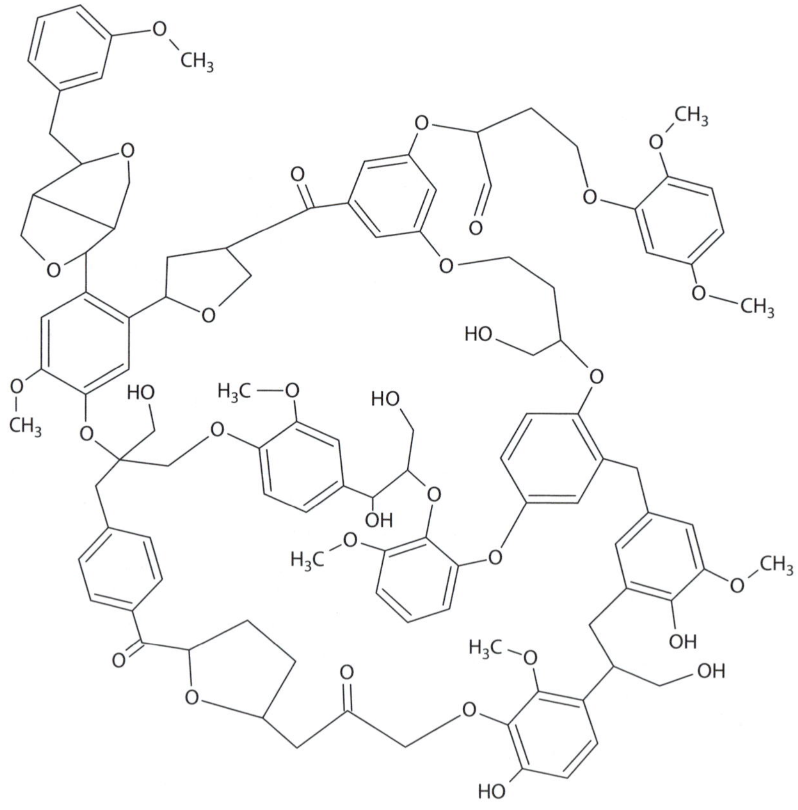

FIGURE 9.7   Representative structure of lignin.

Lignin has a complex structure that varies with the source, growing conditions, and so on. This complex and varied structure is typical of many plant-derived macromolecules. Lignin is generally considered as being formed from three different phenylpropanoid alcohols—coniferyl, coumaryl, and sinapyl alcohols, which are synthesized from phenylalanine via various cinnamic acid derivatives and is sometimes commercially treated as being composed of a C_9 repeat unit where the superstructure contains aromatic and aliphatic alcohols and ethers, and aliphatic aldehydes and vinyl units.

Lignin is found in plant cell walls of supporting and conducting tissue, mostly the tracheids and vessel parts of the xylem. It is largely found in the thickened secondary wall but can occur elsewhere close to the celluloses and hemicelluloses.

The presence of rigid aromatic groups and hydrogen bonding by the alcohol, aldehyde, and ether groups give a fairly rigid material that strengthens stems and vascular tissue in plants allowing upward growth. It also allows water and minerals to be conducted through the xylem under negative pressure without collapse of the plant. This structure can be flexibilized through introduction of a plasticizer, in nature mainly water. The presence of the hydrophillic aromatic groups helps ward off excessive amounts of water, allowing the material to have a variable flexibility but to maintain some strength. This type of balance between flexibility and strength is also utilized by polymer chemists as they work to modify lignin as well as synthetic polymers to give strong, semiflexible materials. Without the presence of water, lignin is brittle, but with the presence of water, the tough lignin provides plants with some degree of protection from animals.

Its chemical durability also makes it indigestible to plant eaters and its bonding to cellulose and protein material in the plant also discourages plant eaters from eating it. Formation of lignin also helps block the growth of pathogens and is often the response to partial plant destruction.

The role of transferring water by lignin makes the ability to produce lignin critical to any plants survival and permits even primitive plants to colonize dry land.

During the synthesis of plant cell walls, polysaccharides are generally initially laid down. This is followed by the biosynthesis of lignin that fills the spaces between the polysaccharide fibers acting to cement them together. The filling of cell wall spaces results in at least some of the lignin having a somewhat two-dimensional structure similar to a sheet of paper rather than the typical three-dimensional structure for most polysaccharides and other natural macromolecules. The lignin sheets act as a barrier toward the outside elements, including marauding pests as noted above.

The lignin is generally considered as being insoluble but it can be solubilized utilizing special systems such as strong basic solutions that disrupt the internal hydrogen bonding. It is not clear even now if the undegraded lignin is solubilized or if degradations accompany lignin solubilization. Weight-average molecular weights for some lignin samples vary depending on the type of solvent employed consistent with degradation occurring for at least some of these systems. Molecular weight values to about 10^7 have been reported for some alkali-extracted fractions.

Solubilized lignin solutions are easily oxidized and the presence of the aromatic unit containing electron-donating ether and alcohol moieties makes it available for electrophilic substitution reactions such as nitration, halogenation, hydroxylation, and so on.

Since its removal is the major step in the production of paper pulp, vast amounts of lignin are available as a byproduct of paper manufacture. Its industrial use is just beginning and it remains one of the major underused materials. Its sulfonic acid derivative is used as an extender for phenolic resins, as a wetting agent for oil drilling muds, and for the production of vanillin. Combined, this accounts for less than 1% of this important "byproduct." In the enlightened age of green chemistry, greater emphasis on the use of lignin itself and in products derived from it must occur.

9.14 MELANINS

Light is continuous ranging from wavelengths smaller than 10^{-12} cm (Gamma radiation) to greater than 10^8 cm. Radiation serves as the basis for the synthesis of many natural macromolecules via

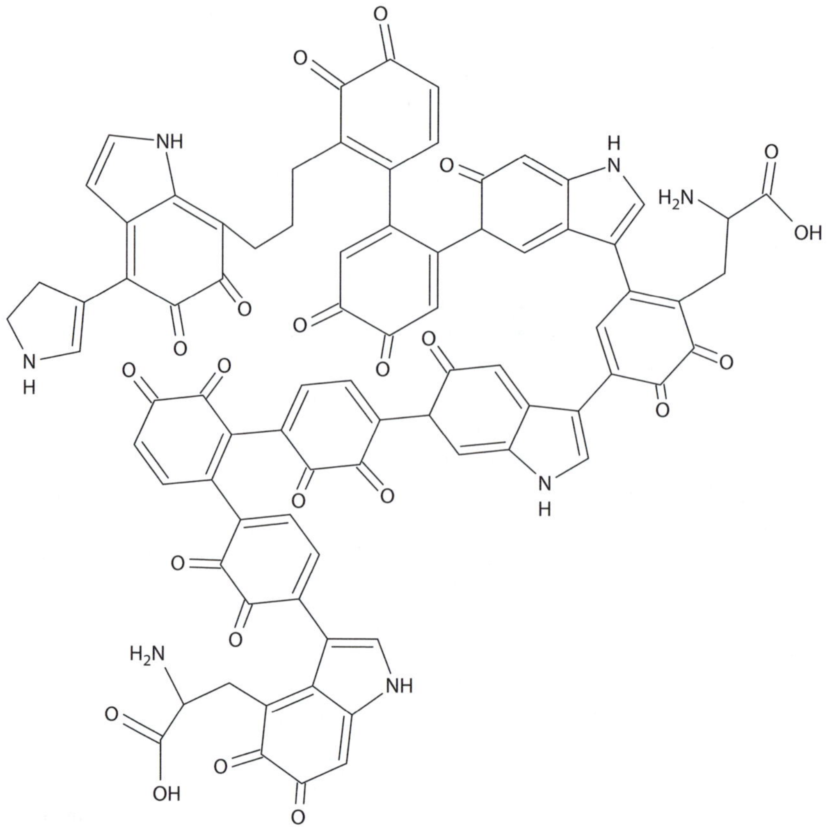

FIGURE 9.8   Representative structure of melanin.

photosynthesis. Radiation is used commercially to increase the wood pulp yield through cross-linking and grafting of lignin and other wood-components onto cellulosic chains. Radiation is also used in the synthesis and cross-linking of many synthetic polymers.

Radiation is also important in the synthesis and rearrangement of important "surface" macromolecules. Tanning of human skin involves the activation of the polypeptide hormone beta melanocyte-stimulating hormone (MSH) that in turn eventually leads to the phenomena of tanning. Exposure to higher energy light from about 297 to 315 nm results in both tanning and burning, whereas exposure to light within the 315–330 nm region results in mainly only tanning.

Ultraviolet radiation activates enzymes that modify the amino acid tyrosine in pigment-producing cells, the melanocytes. The enzyme tyrosinase, a copper-containing oxygenase, catalyzes the initial step, which is the hydroxylation of tyrosine to 3,4-dihydroxyphenylalanine that is oxidized to dopaquinone subsequently forming the melanins (Figure 9.8). The concentration of tyrosine is relatively high in skin protein. These modified tyrosine molecules undergo condensation forming macromolecules known as melanins. Melanins have extended chain resonance where the pi electrons are associated with the growing melamine structure. As the melanine structure grows, it becomes more colored giving various shades of brown color to our skin. This brown coloration acts to help protect deeper skin elements from being damaged by the UV radiation. The absence of the enzyme tyrosinase that converts tyrosine to melanin can lead to albinism.

At least two colored melanins are formed—a series of black melanins and a series of so-called red melanins. Our skin pigmentation is determined by the relative amounts of these red and black melanins in our skin.

The concentration of melanine also contributes to the color of our hair (except for redheads where the iron-rich pigment trichosiderin dominates). The bleaching of hair, generally achieved through exposure to hydrogen peroxide, is a partial result of the oxidation of the melanine. A side reaction of bleaching is the formation of more sulfur cross-links leading to bleached hair being more brittle because of the increased cross-linking leading to a decrease in hair flexibility.

Melanine also provides a dark background in our eye's iris, is involved in animal color changes (such as the octopus and chameleon), is formed when fruit is bruised, and is partially responsible for the coloration of tea.

9.15 ASPHALT

Nearly 100 billion pounds of asphalt are used annually in the United States. In crude oil, it is the most viscous component that is not easily distilled as are move volatile fractions such as gasoline and kerosene. They are also found in natural deposits such as the La Bera Tar Pits in California. It is sometimes referred to as bitumen, covered in Section 12.18. Here, we will cover additional material related to asphalt.

About 80% of asphalt is employed along with rocks to form our roadways. The asphalt acts as the binder for the rock aggregates similar to concrete. Asphalt concrete is the most widely used recycled material, on a weight and percentage basis, in the United States. More than 80% of asphalt from road surfaces removed from resurfacing projects is reused as a component of the new construction. Roofing shingles are also a major use of asphalt.

Asphalt is a viscoelastic material and this viscoelasticity must be present over the temperature range where the material is being used. Unfortunately, asphalt often softens during summer days, allowing rutting or permanent deformation to occur. During the winter, asphalt chains tend to become more organized, resulting in a more brittle material that can undergo thermal and fatigue cracking. With time, more volatile, shorter chains leave the mixture causing the material to lose some of its flexibility. This process is called aging.

Asphalt emulsions are made by mixing the asphalt, generally with petroleum solvents, but some formulations include the more environmentally acceptable water. These emulsions are often employed to resurface concrete and asphalt surfaces such as driveways.

Along with short-chained polymers with up to 150 carbons, asphalt contains a variety of other elements. The composition varies as to the source and treatment of the crude oils. Crude oil from Venezuela contains particularly high amount of metals, such as iron, nickel, and vanadium, in comparison to crude oil from the Middle East. Some of the asphalts contain polar functional groups such as alcohols, amines, thiols, and carboxylic-containing units. This allows the asphalt chains to aggregate similar to many detergents with the aliphatic units reaching outward from a polar core. For rocks with a mildly polar surface, this results in some binding between the rock surfaces and the asphalt, resulting in a more secure material.

9.16 SUMMARY

1. Polysaccharides are the most abundant, on a weight basis, naturally occurring organic polymer. They are truly complex molecules with most structures representable on only an "average" basis. They are diverse with respect to both occurrence and size. The most important, on a knowledge, weight, and use basis are cellulose and starch. Cellulose is composed largely of D-glucose units that are associated with other materials such as lignin. Cotton is one of the purest forms of cellulose, though we get most of our cellulose from wood pulp. Cellulose becomes soluble in aqueous solutions only when the hydrogen bonding within the cellulose is broken.
2. While cellulose is used extensively in the form of paper and paper-based products, it also forms the basis of a number of synthetic, including a variety of acetylated products such as rayon.

3. Starch is composed of two major components—amylopectin and amylose. It is the second most abundant polysaccharide and found largely within plants.
4. Polyisoprenes form the basis of many natural polymers, including the two most widely known natural plastics gutta percha and balata and an elastomer known as *H. brasiliensis*, or NR. The hard plastics balata and gutta percha are *trans* isomers while NR is the *cis* isomer of 1,4-polyisoprene. The polymer chain of amorphous NR and other elastomers uncoils during stretching and returns to its original low-entropy form when the material is above its T_g. Chain slippage is minimized through the presence of cross-links. Unlike other solids, stretched rubber contracts when heated. The long-range elasticity is dependent on the absence of strong intermolecular forces. Eventually, NR formed the basis of the rubber industry through gaining knowledge of such factors as cross-linking that allowed the material to remain coherent but elastomeric even when exposed to moderate temperatures.
5. Melamines are polymeric agents that play many roles in nature, including forming our skin pigmentation and with its growth through exposure to sunlight, acting to protect us from the harmful effects of the sunlight.
6. Lignin is a noncellulosic resinous component of wood. It is the second most abundant renewable natural resource. It has alcohol and ether units with many aromatic units. Much of lignin is sheet like in structure.

GLOSSARY

Accelerator: Catalyst for the vulcanization of rubber.

Alkali cellulose: Cellulose that have been treated with a strong basic solution.

α-Cellulose: Cellulose that is not soluble in 17.5% basic solution.

α Helix: Right-handed helical conformation.

Amylopectin: Highly branched starch polymer with branches or chain extensions on carbon 6 of the anhydroglucose repeating unit.

Amylose: Linear starch polymer.

Antioxidant: Compound that retards polymer degradation.

Balata: Naturally occurring *trans*-1,4-polyisoprene.

β Arrangement: Pleated sheet-like conformation.

β-Cellulose: Cellulose soluble in 17.5% basic solution but not soluble in 8% caustic solution.

Boeseken-Haworth projections: Planar hexagonal rings used for simplicity instead of staggered chain forms.

Carbohydrate: Organic compound often with an empirical formula CH_2O; sugars, starch, cellulose, are carbohydrates.

Carbon black: Finely divided carbon used for the reinforcement of rubber.

Carboxymethylhydroxyethylcellulose (CMHEC): Made from the reaction of sodium chloroacetate and hydroxyethylcellulose.

Carrageenin: Mixture of several polysaccharides containing D-galactose units; obtained from seaweed.

Casein: Milk protein.

Cellobiose: Repeat unit in cellulose.

Cellophane: Sheet of cellulose regenerated by the acidification of an alkaline solution of cellulose xanthate.

Celluloid: Product from a mixture of cellulose nitrate and camphor.

Cellulose: Linear polysaccharide consisting of many anhydroglucose units joined by beta-acetal linkages.

Cellulose acetate: Product from the acetylation of cellulose.

Cellulose nitrate: Made from the reaction of cellulose and concentrated nitric acid; also known as gun cotton.

Cellulose xanthate: Product of soda cellulose and carbon disulfide.

Chitin: Polymer of acetylated glucosamine present in the exoskeletons of shellfish.

Collagen: Protein present in connective tissue.

Compounding: Processing of adding essential ingredients to a polymer such as rubber.

Cyclodextrins: Oligomeric cyclic products formed from the reaction of starch treated with a certain enzyme.

Degree of substitution (DS): Number that designates the average number of reacted hydroxyl groups in each anhydroglucose unit in cellulose or starch.

Dextran: Branched polysaccharide synthesized from sucrose by bacteria.

Drying: Jargon used to describe the cross-linking of unsaturated polymers in the presence of air and a heavy metal catalyst (drier).

Ester gum: Ester of rosin and glycerol.

Ethylhydroxyethylcellulose (EHEC): Nonionic mixed ether formed form HEC and ethyl chloride.

Fibrils: Thread-like strands or bundles of fibers.

Fossil resins: Resins obtained from the exudate of prehistoric trees.

Galalith: Commercial casein plastics.

γ-Cellulose: Cellulose soluble in 8% caustic solution.

Glycogen: Highly branched polysaccharide that serves as the reserve carbohydrate in animals.

Guayule: Shrub that produces *cis*-1,4-polyisoprene rubber.

Gutta percha: Naturally occurring *trans*-1,4-polyisoprene.

Hevea Brasiliensis: Natural rubber, NR.

Humic acid: Polymeric aromatic carboxylic acid found in lignite.

Hydrocellulose: Cellulose degraded by hydrolysis.

Hydroxyethylcellulose (HEC): Produced from alkaline cellulose and ethylene oxide.

Hydroxypropylcellulose: Thermoplastic cellulose ether formed from alkaline cellulose and propylene oxide.

Latex: Stable dispersion of polymer particles in water.

Lignin: Noncellulosic resinous component of wood.

Mercerized cotton: Cotton fiber that has been immersed in caustic solution, usually under tension, and washed with water removing the excess caustic.

Methylcellulose (MC): Formed from alkaline cellulose and chloromethane.

Native cellulose: Naturally occurring cellulose; like cotton.

Oligosaccharide: Low molecular weight polysaccharide.

Racked rubber: Stretched rubber cooled below its T_g.

Rayon: Cellulose regenerated by acidification of a cellulose xanthate (viscose) solution.

Recent resins: Resins obtained from exudates of living trees.

Regenerated cellulose: Cellulose obtained by precipitation from solution.

Retrogradation: Process whereby irreversible gel is produced by the aging of aqueous solutions of amylose starch.

Shellac: Natural polymer obtained from the excreta of insects in Southeast Asia.

Starch: Linear or branched polysaccharides of many anhydroglucose units joined by an alpha linkage; amylose is the linear fraction and amylopectin is the branched fraction.

Tenacity: Term used for the tensile strength of a polymer.

Terpene: Class of hydrocarbons having the empirical formula C_5H_8.

Viscose: Alkaline solution of cellulose xanthate.

EXERCISES

1. Why is starch digestible by humans? Why is cellulose not digestible by humans?
2. How does cellobiose differ from maltose?

3. Why is cellulose stronger than amylose?
4. How does the monosaccharide hydrolytic product of celllulose differ from the hydrolytic product of starch?
5. What is paper made from?
6. How many hydroxyl groups are present on each anhydroglucose unit in cellulose?
7. Which would be more polar—tertiary or secondary cellulose acetate?
8. Why would you expect chitin to be soluble in hydrochloric acid?
9. Which is more apt to form a helix: (a) amylose or (b) amylopectin?
10. Why is amylopectin soluble in water?
11. How do the configurations differ for (a) gutta percha and (b) NR?
12. The formation of what polymer is responsible for tanning?
13. Will the tensile force required to stretch rubber increase or decrease as the temperature is increased?
14. Does a stretched rubber band expand or contract when heated?
15. List three requirements for an elastomer.
16. Why is there an interest in the cultivation of guayule?
17. Are the polymerization processes for synthesis and natural cis-polyisoprene (a) similar or (b) different?
18. What does the presence of C_5H_8 units in NR indicate?
19. Why does a rubber band become opaque when stretched?
20. What is the most important contribution to retractile forces in highly elongated rubber?
21. What is present in so-called vulcanized rubber compounds?
22. When a rubber band is stretched, what happens to its temperature?
23. Why are natural plastics not used more?
24. What type of solvent would you choose for shellac?
25. Why is lignin sometimes referred to as being a two-dimensional polymer?
26. Which is a polymer: (a) rosin or (b) ester gum?
27. If the annual production of paper is more than 100 million tons, how much lignin is discarded from paper production annually?
28. Might an article molded from Galalith be valuable?
29. What are some of the obstacles in using polymer-intensive plants as feedstocks for the preparation of fuels such as ethanol?
30. Since there are many plants that give a rubber-like latex, what are the impediments to their use as replacements for NR?
31. If you were beginning an industrial research project aiming at obtaining useful products from lignin, what might be some of the first areas of research you might investigate?
32. How is melanin related to race?
33. It is interesting that while a number of sun tan lotion ingredients are used, that one of the more recent ones that offers broad-range protection is titanium dioxide. What other major use of titanium dioxide is there?

ADDITIONAL READING

Carraher, C., Sperling, L. H. (1983): *Polymer Applications of Renewable-Resource Materials*, Plenum, NY.
Chiellini, E. (2001): *Biomedical Polymers and Polymer Therapeutics*, Kluwer, NY.
Dumitriu, S. (2001): *Polymeric Biomaterials*, Dekker, NY.
Gar, H., Cowman, M., Hales, C. (2008): *Carbohydrate Chemistry: Biology and Medical Applications*, Elsevier, NY.
Gebelein, C. Carraher, C. (1995): *Industrial Biotechnological Polymers*, Technomic, Lancaster, PA.
Hecht, S. M. (1998): *Bioorganic Chemistry: Carbohydrates*, Oxford University Press, Cary, NC.
Kennedy, J., Mitchell, J., Sandford, P. (1995): *Carbohydrate Polymers*, Elsevier, NY.
Paulsen, B. (2000): *Bioactive Carbohydrate Polymers*, Kluwer, NY.

Scheirs, J., Narayan, R. (2010): *Biodegradable Plastics: A Practical Approach*, Wiley, Hoboken, NY.

Scholz, C., Gross, R. (2000): *Polymers from Renewable Resources*, ACS, Washington, DC.

Steinbuckel, A. (2001): *Lignin, Humic, and Coal*, Wiley, NY.

Steinbuchel, A. (2001): *Polyisoprenoides*, Wiley, NY.

Steinbuchel, A. (2003): *Biopolymers, Miscellaneous Biopolymers and Biodegradation of Synthetic Polymers*, Wiley, Hoboken.

Yu., L. (2008): *Biodegradable Polymer Blends and Composites from Renewable Resources*, Wiley, Hoboken, NJ.

10 Naturally Occurring Polymers—Animals

One of the strongest, most rapidly growing areas of science and polymers is that involving natural polymers. Our bodies are largely composed of polymers: DNA, ribonucleic acid (RNA), proteins, and polycarbonates. Aging, awareness, mobility, strength, and so on, that is, all the characteristics of being alive, are related to the "health" of these polymers. Many medical, health, and biological projects focus on polymers. There is an increasing emphasis on such natural polymers. The emphasis on the human genome and relationships between genes, proteins, and our health underlies much of this movement. Thus, an understanding of polymeric principles is advantageous to those desiring to pursue a career related to their natural environment, be it medicine, biomedical, biological, bioengineering, and so forth.

Physically, there is no difference in the behavior, study, or testing of natural and synthetic polymers. Techniques suitable for application to synthetic polymers are equally applicable to the study and behavior of natural polymers.

Proteins and nucleic acids typically act as individual units, the nanoworld in action, while many other natural polymers and synthetic polymers act in concert with one another. This is not entirely true since proteins and nucleic acids, while acting as individual units, act with other essential biologically important units to carry out their tasks. Synthetic polymers generally act as groups of chains through chain entanglement and connected crystalline units giving the overall aggregate such desired properties as strength. In a real sense, the behavior of branched natural polymers such as amylopectin is similar to the branched low-density polyethylene while the behavior of linear amylose is similar to that of the largely linear high-density polyethylene.

While the specific chemistry and physics dealing with synthetic polymer is complicated, the chemistry and physics of natural polymers is even more complex because of a number of related factors, including (1) the fact that many natural polymers are composed of different, often similar but not identical, repeat units; (2) a greater dependency on the exact natural polymer environment; (3) the question of real structure of many natural polymers in their natural environment is still not well known for many natural polymers; and (4) the fact that polymer shape and size are even more important and complex in natural polymers than in synthetic polymers.

Biological polymers represent successful strategies that are being studied by scientists as avenues to different and better polymers and polymer structure control. Sample "design rules" and approaches that are emerging include the following:

- Identification of mer sequences that give materials with particular properties
- Identification of mer sequences that cause certain structural changes
- Formation of a broad range of materials with a wide variety of general/specific properties and function (such as proteins/enzymes) through a controlled sequence assembly from a fixed number of feedstock molecules (proteins—about 20 different amino acids; five bases for nucleic acids and two sugar units)
- Integrated, in situ (in cells) polymer productions with precise nanoscale control
- Repetitive use of proven strategies with seemingly minor structural differences but resulting in quite divergent results (protein for skin, hair, and muscle)
- Control of polymerizing conditions that allow steady-state production far from equilibrium

There often occurs a difference in "mindset" between the nucleic acid and protein biopolymers covered in this chapter and other biopolymers and synthetic polymers covered in other chapters. Nucleic acids and proteins are structure specific with one conformation. In general, if a molecule differs in structure or geometry from the specific macromolecule needed it is discarded. Nucleic acids and proteins are not a statistical average, but rather a specific material with a specific chain length and conformation. By comparison, synthetic and many other biopolymers are statistical averages of chain lengths and conformations. All of these distributions are often kinetic/thermodynamic driven.

This difference between the two divisions of biologically important polymers is also reflected in the likelihood that there are two molecules with the exact same structure. For molecules such as polysaccharides and those based on terpene-like structures, the precise structures of individual molecules vary, but for proteins and nucleic acids the structures are identical from molecule to molecule. This can be considered a consequence of the general function of the macromolecule. For polysaccharides, the major, though not the sole function, are energy and structure. For proteins and nucleic acids, main functions include memory and replication, in additional to proteins also serving a structural function.

Another difference between proteins and nucleic acids and other biopolymers and synthetic polymers involves the influence of stress–strain activities on the materials properties. Thus, application of stress on many synthetic polymers and some biopolymers encourages realignment of polymer chains and regions often resulting in a material with greater order and strength. Counter, application of stress to certain biopolymers, such as proteins and nucleic acids, causes a decrease in performance (through denaturation, etc.) and strength. For proteins and nucleic acids, this is a result of the biopolymer already existing in a compact and "energy favored" form and already existing in the "appropriate" form for the desired performance. The performance requirements for the two classifications of polymers are different. For one set, including most synthetic and some biopolymers, performance behavior involves response to stress–strain application with respect to certain responses such as chemical resistance, absorption enhancement, and other physical properties. By comparison, the most cited performances for nucleic acids and proteins involve selected biological responses requiring specific interactions occurring within a highly structured environment with specific shape and electronic requirements.

10.1 PROTEINS

The many different monodisperse polymers of amino acids, which are essential components of plants and animals, are called *proteins*. This word is derived from the Greek *porteios*, "of chief importance." The 20 different α-amino acids are joined together by peptide linkages (Table 10.1).

$$-(-C-NH-CH-)-$$

with O double-bonded to C and R on the CH carbon.   (10.1)

and are also called polyamides or *polypeptides*. The latter term is often used by biochemists to denote oligomers or relatively low molecular weight proteins. (Note the structural similarities and differences between proteins and polyamides-nylons [Section 4.7].)

All α-amino acids found in proteins are of the general structure

$$R-CH-COOH$$

with NH_2 on the CH carbon.   (10.2)

except glycine

$$H_2C-COOH$$

with NH_2 on the carbon.   (10.3)

TABLE 10.1
Structures of the 20 Common Amino Acids

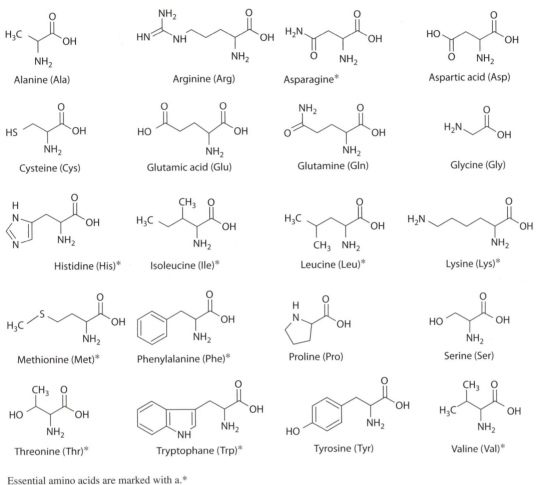

Alanine (Ala) Arginine (Arg) Asparagine* Aspartic acid (Asp)

Cysteine (Cys) Glutamic acid (Glu) Glutamine (Gln) Glycine (Gly)

Histidine (His)* Isoleucine (Ile)* Leucine (Leu)* Lysine (Lys)*

Methionine (Met)* Phenylalanine (Phe)* Proline (Pro) Serine (Ser)

Threonine (Thr)* Tryptophane (Trp)* Tyrosine (Tyr) Valine (Val)*

Essential amino acids are marked with a.*

and they contain a chiral carbon atom and are L-amino acids. The net ionic charge of an amino acid varies with the changes in the solution pH. The pH at which an amino acid is electrically neutral is called the *isoelectric point*. For simple amino acids (containing only one acid and one amine), this occurs at a pH of about 6 with the formation of a dipolar or *zwitterion* as shown below:

$$\overset{\quad\quad\;\; R}{\underset{H_3\overset{+}{N}-CH-COO^-}{|}} \tag{10.4}$$

Hence, α-amino acids, like other salts, are water-soluble, high-melting, polar compounds that migrate toward an electrode at pH values other than that of the isoelectric point in a process called *electrophoresis*.

In writing out sequences for polypeptides, it is usual to use a three-letter abbreviation or a one-letter abbreviation starting with the N-terminus to the left and going to the C-O terminus to the right. Thus the trimer

$$
\begin{array}{cccc}
\text{H} & \text{O} & \text{O} & \text{O} \\
| & || & || & || \\
\text{HN--CH}_2\text{--C--O--NH--CH--C--O--NH--CH--C--O--H} \\
& | & | \\
& \text{CH}_3 & \text{CH}_2 \\
& & | \\
& & \text{OH}
\end{array}
\qquad \text{becomes} \qquad (10.5)
$$

Gly–Ala–Ser or GlyAlaSer or EGAS, where the E signals the N-terminus.

It is important to remember that all proteins are polypeptides composed of a mixture of the 20 amino acids given in Table 10.1. The size of proteins is generally noted in either the number of amino acids units or more likely the molecular weight in kilodaltons (kDa).

Proteins are synthesized in nature from the N-terminus to the C-terminus. While all of the amino acids have only one variation with that variation being a lone substituent on the alpha carbon, this variation is sufficient to produce the variety of proteins critical to life. Some of these contain an "excess" of acid groups such as aspartic acid, while others contain one or more additional basic groups such as arginine. Still others possess sulfur-containing moieties, namely cysteine and methionine, which account for the need for sulfur to be present in our diets and eventually end up in our fossil fuels producing sulfur oxides on burning. All possess nitrogen-containing moieties and these also eventually end up in our fossil fuels and produce nitrogen oxides on burning. The presence of these two elements is then connected to acid-rain production.

The amino acids may be neutral, acidic, or basic, in accordance with the relative number of amino and carboxylic acid groups present. Cations can be formed with amino acids like tryptophane, lysine, histidine, and arginine, which have additional amine groups, while others that contain additional acid groups can be hydrolyzed to form anions like aspartic acid and glutamic acid. The presence of varying amounts of these amino acid moieties within a protein are primary driving forces for the separation of proteins using electrophoresis and result in polypeptides having different isoelectric points. If there are a number of acidic and/or basic groups on the polypeptide, the molecule is said to be a *polyampholyte* or if they contain only positive or negative charges, they are called *polyelectrolytes*. The behavior of these charged polypeptides is similar to the behavior of other charged polymers. Thus, a fully hydrolyzed poly(acrylic acid) acts as a rod because the negative sites repeal one another while a polypeptide with a large number of negative sites will also be elongated. The spacing and number of these charged sites helps determine the tertiary structure of such polypeptides.

Even though the atoms within a peptide bond are coplanar they can exist in two possible configurations, *cis* and *trans*.

$$
\begin{array}{ccc}
\text{O} \quad \text{C} & & \text{C} \quad \text{C} \\
|| \quad /\backslash & & /\backslash \quad /\backslash \\
\text{C--N} & \text{or} & \text{C--N} \\
\backslash / \quad \backslash & & || \quad \backslash \\
\text{C} \qquad \text{H} & & \text{O} \quad \text{H} \\
\textit{trans} & & \textit{cis}
\end{array}
\qquad (10.6)
$$

The trans form is usually favored whenever there is a bulky group on the adjacent alpha carbon(s) because bulky groups will interfere more in the cis structure.

While humans synthesize about a dozen of the 20 amino acids needed for good health, the other eight are obtained from outside our bodies, generally from eating foods that supply these *essential amino acids* (Table 10.1). Different foods are good sources of different amino acids. Cereals are generally deficient in lysine. Thus, diets that emphasize cereals will also have other foods that can supply lysine. In the orient, the combination of soybean and rice supply the essential amino acids while in Central America bean and corn are used.

Almost all of the sulfur needed for healthy bodies is found in amino acids as cysteine and methionine. Sulfur serves several important roles, including as a cross-linking agent similar to that served by sulfur in the cross-linking, vulcanization, of rubber. This cross-linking allows the various chains, which are connected by these cross-links, to "remember" where they are relative to one another. This cross-linking allows natural macromolecules to retain critical shapes to perform necessary roles.

The most widely used technique for producing polypeptides with specific sequences is the solid phase technique developed by Nobel Prize winner Bruce Merrifield in which all reactions take place on the surface of cross-linked polystyrene beads. The process begins with the attachment of the C-terminal amino acid to the chloromethylated polymer. Nucleophilic substitution by the carboxylate anion of an N-blocked protected C-terminal amino acid displaces chloride from the chloromethyl group forming an ester, protecting the C site while attaching it to the solid support. The blocking group is removed by addition of acid, and the polymer containing the nonprotected N-terminus is washed to remove unwanted byproducts. A peptide bond is formed by condensation to an N-blocked, protected amino acid. Again, the solid phase system is washed removing byproducts. The blocking group is removed and the site is ready for attachment by another amino acid. This cycle is repeated eventually producing the desired polypeptide without isolation of intermediate products. These steps are outlined below:

$$
\text{PS–Ph–CH}_2\text{–Cl} + \text{HOOC–CH–NH–BG} \xrightarrow{\ -\text{HCl}\ } \text{PS–Ph–CH}_2\text{–O–C–CH–NH–BG} \xrightarrow{\ \text{H}^+\ }
$$

(10.7)

$$
\text{PS–Ph–CH}_2\text{–O–C–CH–NH} \rightarrow \rightarrow \rightarrow \rightarrow
$$

where PS = polystyrene bead, Ph = phenyl group, and BG = blocking group.

10.2 LEVELS OF PROTEIN STRUCTURE

The shapes of macromolecules, both synthetic and natural, can be described in terms of primary, secondary, tertiary, and quaternary structure (Figure 10.1). Protein structure will be used to illustrate these structures.

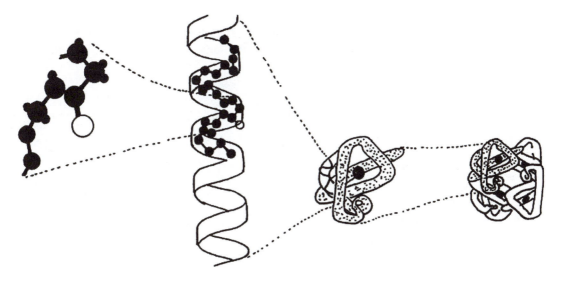

FIGURE 10.1 Four levels of structure elucidation. From left to right: primary, secondary, tertiary, and quaternary structures illustrate using a globular protein segment.

10.2.1 PRIMARY STRUCTURE

The term primary structure is used to describe the sequence of amino acid units (configuration) in a polypeptide chain. Thus, Equation 10.5 describes a primary structure.

10.2.2 SECONDARY STRUCTURE

The term *secondary structure* is used to describe the molecular shape or conformation of a molecule. The most important factor in determining the secondary structure of materials is its precise structure. For proteins, it is the amino acid sequence. Hydrogen bonding is also an important factor in determining the secondary structures of natural materials and those synthetic materials that can hydrogen bond. In fact, for proteins, secondary structures are generally those that allow a maximum amount of hydrogen bonding. This hydrogen bonding also acts to stabilize the secondary structure while cross-linking acts to lock-in a structure.

In nature, the two most common secondary structures are helices and sheets. In nature, extended helical conformations appear to be utilized in two major ways: to provide linear systems for the storage, duplication, and transmission of information (DNA, RNA), and to provide inelastic fibers for the generation and transmission of forces (F-actin, myosin, and collagen). Examples of the various helical forms found in nature are single helix (messenger and ribosomal DNA), double helix (DNA), triple helix (collagen fibrils), and complex multiple helices (myosin). Generally, these single and double helices are readily soluble in dilute aqueous solution. Often solubility is only achieved after the inter- and intrahydrogen bonding is broken.

There are a variety of examples in which linear or helical polypeptide chains are arranged in parallel rows. The two major forms that exist for proteins are illustrated in Figures 10.2 through 10.5. The chains can have the N→C directions running parallel making a parallel beta sheet (Figure 10.4), or they can have the N→C directions running antiparallel giving antiparallel beta sheets (Figure 10.5).

The structure of proteins generally fall into three groupings—fibers, membrane, and globular. The structural proteins such as the keratines, collagen, and the elastin are largely fibrous. A reoccurring theme with respect to conformation is that the preferential secondary structures of fibrous synthetic and natural polymers approximates that of a pleated sheet (or skirt) or helix. The pleated sheet structures in proteins are referred to as beta arrangements (Figure 10.5). In general, proteins with bulky groups take on a helical secondary structure while those with less bulky groups exist as beta sheets.

Helices can be described by the number of amino acid residues in a complete "turn." In order to fit into a "good" helix, the amino acids must have the same configuration. For proteins, that

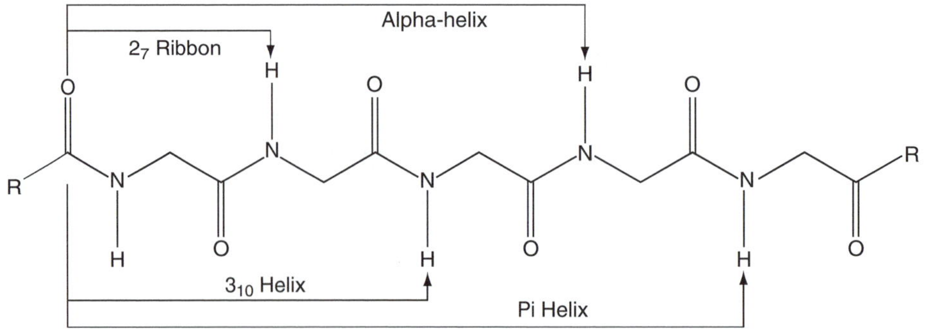

FIGURE 10.2 Commonly occurring repetitive helical patterns for polypeptides. (Source: Coates and Carraher, *Polymer News*, 9(3):77 (1983)).

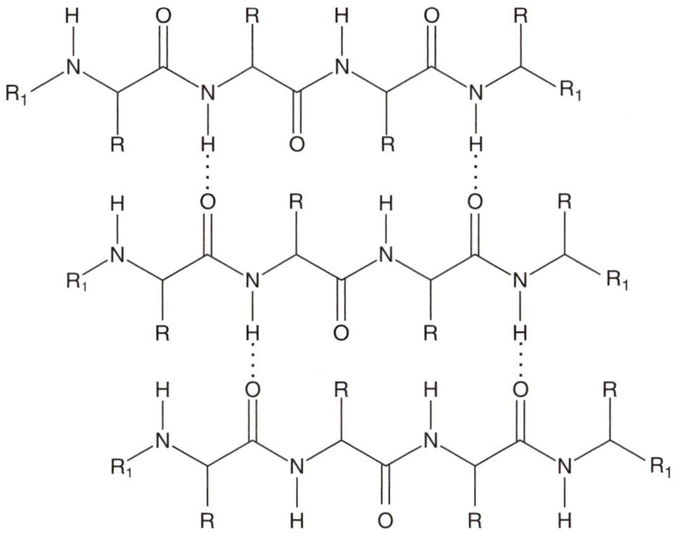

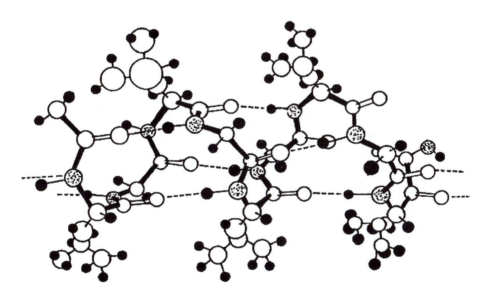

FIGURE 10.3 A beta arrangement or pleated-sheet conformation of proteins.

FIGURE 10.4 α-Keratine helix for the copolymer derived from glycine and leucine.

configuration is described as an L-configuration, with the helix being a "right-handed" helix. This right-handed helix is referred to as an alpha helix.

10.2.3 KERATINES

Keratines are structural proteins. As noted above, two basic "ordered" secondary structures predominate in synthetic and natural polymers. These are the helices and the pleated sheet structures. These two structures are illustrated by the group of proteins called the keratines. It is important to

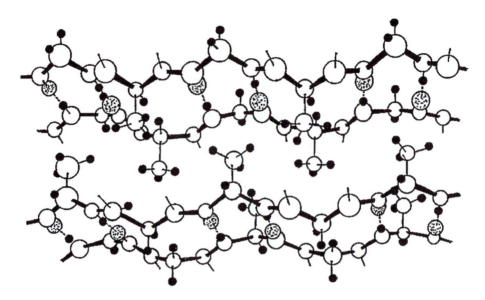

FIGURE 10.5 β-Keratin sheet for the copolymer derived from glycine and alanine.

remember that hydrogen bonding is critical in both structures. For helices, the hydrogen bonding occurs within a single strand, whereas in the sheets, the hydrogen bonding occurs between adjacent chains (Figure 10.3).

Helices are often described in terms of a repeat distance—the distance parallel to the axis in which the structure repeats itself; pitch—the distance parallel to the helix axis in which the helix makes one turn; and rise—the distance parallel to the axis from the level of one repeat unit to the next.

Helices generally do not have an integral number of repeat units or residues per turn. The alpha-helix repeats after 18 amino acid residues, taking five turns to repeat. Thus, the number of residues per turn is 18/5 = 3.6 residues/turn. For polypeptides, each carbonyl oxygen is hydrogen bonded to the amino proton on the fourth residue giving a "hydrogen-bonded loop" with thirteen atoms. Helices are often described in terms of this number, n. Thus, the alpha-helix is described as a 3.6_{13} helix. Because hydrogen bonds tend to be linear, the hydrogen bonding in proteins approximate this with the $-N-H \cdots O=C$ in a straight line. The "rise," h, of alpha-keratin (Figure 10.4) is found by X-ray spectroscopy to be about 0.15 nm for each amino acid residue. The pitch, p, of a helix is given by $p = nh$. For alpha-keratin, $p = 3.6$ amino acid residues/turn × 0.15 nm/amino acid residue = 0.54 nm/turn.

Hair and wool are composed of alpha-keratin. A single hair on our head is composed of many strands of keratine. Coiled, alpha-helices, chains of alpha-keratin intertwine to form protofibrils that in turn are clustered with other protofibrils forming a microfibril. Hundreds of these microfibrils, in turn, are embedded in a protein matrix giving a macrofibril that in turn combines giving a human hair. While combing will align the various hairs in a desired shape, after a while, the hair will return to its "natural" shape through the action of the sulfur cross-links pulling the hair back to its original shape.

The major secondary bonding is involved in forming the helical structures allowing the various bundles of alpha-keratine to be connected by weak secondary interactions that in turn allow them to readily slide past one another. This sliding or slippage along with the "unscrewing" of the helices allows our hair to be flexible. Some coloring agents and most permanent waving of our hair involves breakage of the sulfur cross-links and a reforming of the sulfur cross-links at new sites to "lock in" the desired hair shape.

Our fingernails are also composed of alpha-keratin, but keratin with a greater amount of sulfur cross-links giving a more rigid material. In general, for both synthetic and natural polymers, increased cross-linking leads to increased rigidity.

The other major structural feature is pleated sheets. Two kinds of pleated sheets are found. When the chains have their N→C directions running parallel they are called *parallel beta sheets*. The N→C directions can run opposite to one another giving what is called an *antiparallel beta sheet*. The beta-keratin (Figure 10.5) that occurs in silk produced by insects and spiders is of the antiparallel variety. While alpha-keratin is especially rich in glycine and leucine, beta-keratine is mostly composed of glycine and alanine with smaller amounts of other amino acids, including serine and tyrosine. Size-wise, leucine offers a much larger grouping attached to the alpha carbon than does alanine. The larger size of the leucine causes the alpha-keratine to form a helical structure to minimize steric factors. By comparison, the smaller size of the alanine allows the beta-keratine to form sheets. This sheet structure is partially responsible for the "softness" felt when we touch silk. While silk is not easily elongated because the protein chains are almost fully extended, beta keratin is flexible because of the low secondary bonding between sheets, allowing the sheets to flow past one another.

10.2.3.1   Silk

Silk is a protein fiber that is woven into fiber from which textiles are made, including clothing and high-end rugs. It is obtained from the cocoon of silkworm larvae. While most silk is harvested from commercially grown silkworms, some is still obtained from less well-established sources.

Silk was first developed as early as 6,000 BC. The Chinese Empress Xi Ling-Shi developed the process of retrieving the silk filaments by floating the cocoons on warm water. This process and the silkworm itself were monopolized by China until about 550 AD when two missionaries smuggled silkworm eggs and mulberry seeds from China to Constantinople (Istanbul). First reserved to use by the Emperors of China, its use eventually spread to the Middle East, Europe, and North America, but now its use is worldwide. The history of silk and the silk trade is interesting and can be obtained at http://en.wikipedia.org/wiki/Silk.

The early work focused on a particular silkworm, *Bombyx mori*, which lives on mulberry bushes. There are other silkworms each with its on special properties, but in general, most silk is still derived from the original strain of silkworm. Crystalline silk fiber is about four times stronger than steel on a weight basis.

In the silk fibroin structure, almost every other residue is glycine with either alanine or serine between them, allowing the sheets to fit closely together (Table 10.1). While most of the fibroin exists as beta sheets, regions that contain more bulky amino acid residues interrupt the ordered beta structure. Such disordered regions allow some elongation of the silk. Thus, in the crystalline segments of silk fibroin, there exists directional segregation using three types of bonding: covalent bonding in the first dimension, hydrogen bonding in the second dimension, and hydrophobic bonding in the third dimension. The polypeptide chains are virtually fully extended; there is a little puckering to allow for optimum hydrogen bonding. Thus, the structure is not extensible in the direction of the polypeptide chains. By comparison, the less specific hydrophobic (dispersive) forces between the sheets produce considerable flexibility. The crystalline regions in the polymers are interspersed with amorphous regions in which glycine and alanine are replaced by amino acids with bulkier pendent groups that prevent the ordered arrangements to occur. Furthermore, different silk worm species spin silks with differing amino acid compositions and thus, with differing degrees of crystallinity. The correlation between the extent of crystallinity and the extension at the break point is given in Table 10.2.

Another natural key that illustrates the tight relationship between structure and property is found in spider webs. The composition within a spider web is not all the same. We can look briefly at two of the general types of threads. One is known as the network or frame threads, also called the dragline fabric. It is generally stiff and strong. The second variety is the catching or the capture threads that

TABLE 10.2
Selected Properties as a Function of Silk Worm Species

Silk Worm Species	Approximate Crystallinity (%)	Extension at Break Point (%)
Anaphe moloneyi	95	12.5
Bombyx mori	60	24
Antherea mylitta	30	flows then extends to 35

Source: Coates and Carraher, *Polymer News*, 9(3):77 (1983). Used with permission.

are made of viscid silk that is strong, stretchy, and is covered with droplets of glue. The frame threads are about as stiff as nylon-6,6 thread and on a weight basis stronger than steel cable. Capture thread is not stiff but is more elastomeric-like and on a weight basis about one-third as strong as frame thread. While there are synthetic materials that can match the silks in both stiffness and strength there are few that come near the silk threads in toughness and their ability to withstand a sudden impact without breaking. Kevlar, which is used in bullet-resistant clothing, has less energy-absorbing capacity in comparison to either frame or capture threads. In fact, when weight is dropped onto frame silk, it adsorbs up to ten times more energy than Kevlar. On impact with frame thread, most of the kinetic energy dissipates as heat, which, according to a hungry spider, is better than transforming it into elastic energy, which might simply act to "bounce" the pray out of the web.

The frame threads are composed of two major components—highly organized microcrystals compose about one-quarter of the mass and the other three quarters are composed of amorphous spaghetti-like tangles. The amorphous chains connect the stronger crystalline portions. The amorphous tangles are dry, glassy-like, acting as a material below its T_g. The amorphous chains are largely oriented along the thread length as are the microcrystals giving the material good longitudinal strength. As the frame threads are stretched, the tangles straighten out allowing it to stretch without breaking. Because of the extent of the tangling, there is a lessening in the tendency to form micro-ordered domains as the material is stretched, though some micro-order domain are formed on stretching. Frame thread can be reversibly stretched to about 5%. Greater stretching causes permanent creep. Thread rupture does not occur until greater extension, such as 30%. By comparison, Kevlar fibers break when extended only 3%.

The capture threads are also composed of the same kinds of components but here the microcrystals compose less than 5% of the thread with both the amorphous and microcrystalline portions arranged in a more random fashion within the thread. Hydrated glue, which coats the thread, acts as a plasticizer imparting to the chains greater mobility and flexibility. It stretches several times its length when pulled and is able to withstand numerous shocks and pulls appropriate to contain the prey as it attempts to escape. Further, most threads are spun as two lines so that the resulting thread has a kind of build in redundancy. The spinning of each type of thread comes from a different emission site on the spider, and the spider leaves little to waste, using unwanted and used web parts as another source of protein.

Cloning of certain spider genes have been included in goats to specify the production of proteins that call for the production of silk-like fibroin threads that allow the production and subsequent capture of spider-like threads as part of the goat's milk.

The beta-keratin structure is also found in the feathers and scales of birds and reptiles.

10.2.3.2 Wool

Wool, while naturally existing in the helical form, forms a pleated skirt sheet-like structure when stretched. If subjected to tension in the direction of the helix axes, the hydrogen bonds parallel to the axes are broken and the structure can be irreversibly elongated to an extent of about 100%.

Wool is the fiber from members of the Caprinae family that includes sheep and goats. It is produced as an outer coat with the inner coat being more hair-like. Since it is removed without damaging the sheep's skin, it is renewable without sacrificing the animal. The fiber has two distinguishing characteristics that separate it from fur and hair. It has scales that overlap like shingles on a roof and it is curly, with many fibers having more than 20 bends inch. This crimp has both advantages and disadvantages. The disadvantage is that the fiber tends to mat and gather cockleburs when on the sheep. The curly nature gives the fiber lots of bulk and associated insulation ability. It retains air assisting the wearer to maintain their temperature. Thus, in the cold, the body retains needed heat, while in hot weather; the body can remain somewhat cool with respect to the surroundings. Thus, wool is a staple material for extreme temperatures.

The shingle nature results in a somewhat rough surface that helps the fibers to remain together. In turn, this allows the fibers to be easily formed into yarn.

As noted above, the inner coat is more hair like with little scale or crimp. This hair is called *kemp* while the fiber portion is the wool. Combined they are referred to as the fleece.

A common theme with natural products is the variability of the fiber with the particular plant or animal and the particular conditions under which they were grown. The amount of kemp to wool varies from breed to breed as does the number of crimps per inch. Thus, wool from different breeds raised in different parts of the world will have different properties and contribution to different properties of the fiber. Generally, the wool and kemp are physically separated. In fact, after shearing, the wool is divided into five groups—fleece (which is the largest), pieces, bellies, crutchings, and locks. The last four are packaged and sold separately. The desired part, the fleece, is further classified.

As the wool is removed from the sheep, it contains a high amount of grease that contains valuable lanoline. Natives make yarns out of this wool. But, for most commercial use, the lanolin is removed by washing with detergents and basic solutions.

A number of different wools are special and sold under different names such as cashmere wool. These wools have their own special characteristics that increase their cost and value. It is interesting that many animals have developed their own special coats to overcome their native environment. Thus, because cashmere goats live in a cold mountain temperature, their wool is an effective material to protect against the cold.

As an aside, wool is self-extinguishing so that flame retardants are not needed in wool rugs and suits. Further, we need not worry about herds of flammable sheep running about the hillsides.

10.2.4 COLLAGEN

Collagen is the most abundant single protein in vertebrates making up to one-third of the total protein mass. Collagen fibers form the matrix or cement material in our bones where mineral materials precipitate. Collagen fibers constitute a major part of our tendons and act as a major part of our skin. Hence, it is collagen that is largely responsible for holding us together.

The basic building block of collagen is a triple helix of three polypeptide chains called the *tropocollagen unit*. Each chain is about 1,000 residues long. The individual collagen chains form left-handed helices with about 3.3 residues per turn. To form this triple-stranded helix, every third residue must be glycine because glycine offers a minimum of bulk. Another interesting theme in collagen is the additional hydrogen bonding that occurs because of the presence of hydroxyproline derived from the conversion of proline to hydroxproline. The conversion of proline to hydroxproline involves vitamin C. Interestingly, scurvy, the consequence of a lack of vitamin C, is a weakening of collagen fibers giving way to lesions in the skin and gums and weakened blood vessels. Collagen fibers are strong. In tendons, the collagen fibers have a strength similar to that of hard-drawn copper wire.

Much of the toughness of collagen is the result of the cross-linking of the tropocollagen units to one another through a reaction involving lysine side chains. Lysine side chains are oxidized to aldehydes that react with either a lysine residue or with one another through an aldol condensation and dehydration resulting in a cross-link. This process continues throughout our life, resulting in our bones and tendons becoming less elastic and more brittle. Again, a little cross-linking is essential, but more cross-linking leads to increased fracture and brittleness.

Collagen is a major ingredient in some "gelation" materials. Here, collagen forms a triple helix for some of its structure while other parts are more randomly flowing single-collagen chain segments. The bundled triple-helical structure acts as the rigid part of the polymer, while the less ordered amorphous chains act as a soft part of the chain. The triple helix also acts as a noncovalently bonded cross-link.

10.2.5 ELASTIN

Collagen is found where strength is needed, but some tissues, such as arterial blood vessels and ligaments need materials that are elastic. Elastin is the protein of choice for such applications. Elastin is rich in glycine, alanine, and valine and it is easily extended and flexible. Its conformation approaches that of a random coil so that secondary forces are relatively weak, allowing elastin to be readily extended as tension is applied. The structure also contains some lysine side chains that are involved in cross-linking. The cross-linking is accomplished when four lysine side chains are combined to form a desmosine cross-link. This cross-link prevents the elastin chains from being fully extended and causes the extended fiber to return to its original dimensions when tension is removed. One of the areas of current research is the synthesis of polymers with desired properties based on natural analogues. Thus, elastin-like materials have been synthesized using glycine, alanine, and valine and some cross-linking. These materials approach elastin in its elasticity.

We are beginning to understand better how we can utilize the secondary structure of polymers as tools of synthesis. One area where this is being applied is "folded oligomers." Here, the secondary structure of the oligomer can be controlled through its primary structure and the use of solvents. Once the preferred structure is achieved, the oligomers are incorporated into larger chains eventually forming synthetic polymers with several precise structures "embedded" within them. The secondary structure of the larger polymers can also be influenced by the nature of the solvent allowing further structural variety. Further, other species, such as metal ions, can be added to assist in locking in certain desired structures and they can also be used to drive further structure modifications.

10.2.6 TERTIARY STRUCTURE

The term *tertiary structure* is used to describe the shaping or folding of macromolecules. These larger structures generally contain elements of the secondary structures. Often, hydrogen bonding, salt bridges, posttranslational modifications, and disulfide cross-linking lock in such structures (Figure 10.6). As noted above, proteins can be divided into three broad groups—fibrous or fibrillar proteins, membrane proteins, and globular proteins.

10.2.7 GLOBULAR PROTEINS

There is a wide variety of so-called globular proteins. Almost all globular proteins are soluble in acidic, basic, or neutral aqueous solutions. Many globular proteins are enzymes. Many of these have varieties of alpha- and beta-structures imbedded within the overall globular structure. Beta sheets are often twisted or wrapped into a "barrel-like" structure. They contain portions that are beta-sheet structures and portions that are in an alpha-conformation. Further, some portions of the globular protein may not be conveniently classified as either an alpha or beta structure.

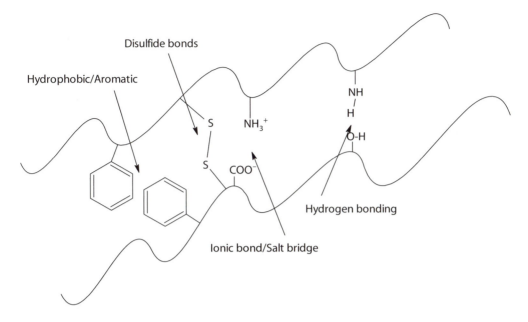

FIGURE 10.6 Chemical forces that help maintain the tertiary structure of proteins.

These proteins are often globular in shape so as to offer a different "look" or polar nature to its outside. Hydrophobic residues are generally found in the interior while hydrophilic residues are found on the surface interacting with the hydrophilic water-intense external environment. (This theme is often found for synthetic polymers that contain both polar and nonpolar portions. Thus, when polymers are formed or reformed in a regular water-filled atmosphere, many polymers will favor the presence of polar moieties on their surface.)

Folding depends on a number of interrelated factors. The chain folding process involves going from a system of random beta- and alpha-conformations to a single-folded structure. This change involves a decrease in the entropy or randomness. For folding to occur, this increase in order must be overcome by energy-related factors to allow the overall free energy to be favorable. These energy factors include charge–charge interactions, hydrogen bonding, van der Waals interactions, and hydrophilic–hydrophobic effects.

Within a particular globular polymer there may be one or more polypeptide chain folded backward and forward forming quite distinct structural domains. Each domain is characterized by a particular style of coiling or "sheeting" which may be nonrepetitive with respect to its peptide chain geometry or may be repetitive, conforming to one of several now well-recognized patterns. The specific chain conformations are determined by the side-chain interactions of the amino acids superimposed on intrapeptide hydrogen bonding along the chain. The form of chain folding is thus ultimately determined by the amino acid sequence and the polymeric nature of the polypeptide chains and is fundamental to the specific geometry of the given protein.

Protein units can be either negatively or positively charged. Attractions between unlike charges are important as are the repulsions between like changed units. As expected, these associations are pH dependent, and control of pH is one route for conformation changes to occur. The ability to hydrogen bond is also an important factor with respect to the internal folding scheme. Because the proteins are tightly packed, the weak van der Waals interactions can also play an important role in determining chain folding. The tendency for polarity-like segments to congregate can also be an important factor in chain folding. Thus, hydrophilic groupings generally are clustered to the outside of the globular protein allowing them to take advantage of hydrogen bonding and other polar bonding opportunities with the water-rich environment while hydrophobic clusters of amino acid units occupy the internal regions of the protein taking advantage of hydrophobic interactions.

Globular proteins act in maintenance and regulatory roles—functions that often require mobility and thus some solubility. Included within the globular grouping are enzymes, most hormones, hemoglobin, and fibrinogen that is changed into an insoluble fibrous protein fibrin that causes blood clotting.

Denaturation is the irreversible precipitation of proteins caused by heating, such as the coagulation of egg white as an egg is cooked, or by addition of strong acids, bases, or other chemicals. This denaturation causes permanent changes in the overall structure of the protein, and, because of the ease with which proteins are denatured, it makes it difficult to study "natural" protein structure. Nucleic acids also undergo denaturation.

Small changes in the primary structure of proteins can result in large changes in the secondary structure. For instance, researchers have interchanged the positions of two adjacent amino acid residues of a globular protein portion resulting in a beta strand becoming a right-handed helix.

Molecular recognition is one of the keys to life. Scientists are discovering ways to both modify molecular recognition sites and to "copy" such sites. One approach to modifying molecular recognition sites, namely enzymatic protein sites, is through what is referred to as *directed evolution*. Arnold and coworkers have employed the combinatorial approach by taking an enzyme with a desired catalytic activity and encouraging it to undergo mutation; selecting out those mutations that perform in the desired manner; and repeating this cycle until the new enzymes perform as desired. Ratner and coworkers have taken another approach whereby templates containing the desired catalytic sites are made. First, the protein is mounted on a mica support. The target protein is coated with a sugar monolayer that allows for specific recognition. A fluoropolymer plasma film is deposited over the sugar monolayer. The fluoropolymer reverse image is attached to a support surface using an epoxy resin. Solvents are then added to etch away the mica, sugar, and original protein leaving behind a "nano-pit" template that conforms to the shape of the original protein.

10.2.8 Fibrous Proteins

Fibrous proteins are long macromolecules that are attached through either inter or intrahydrogen bonding of the individual residues within the chain. Solubility, partial or total, occurs when these hydrogen bonds are broken. In general, they confer stiffness and rigidity to biological systems that are themselves fluid.

Fibrous proteins are found in animals. They appear as filaments and are often formed from a limited number of amino acid units. They are generally insoluble in water with hydrophobic portions protruding from the central core. Examples are collagen, elastin, and keratin. Actin and tubulin are globular soluble monomers that polymerize forming long stiff structures that make up the cytoskeleton that allows cells to maintain their shape. Structural proteins are used to construct tendons, bone matrices, connective tissues, and muscle fiber. They can also be used for storage. They are not as easily denatured as globular proteins.

Some structural proteins, such as kinesin, dynein, and myosin, serve as so-called motor proteins that can generate mechanical forces as muscles and are involved in allowing the movement of cells such as the sperm cell involved in sexual reproduction in multicellular organisms.

10.2.9 MEMBRANE PROTEINS

Membrane proteins are attached to or associated with the membrane of a cell. More than half of the proteins interact with these membranes. Membrane proteins are generally divided according to their attachment to a membrane. Transmembrane proteins span the entire membrane. Integral proteins are permanently attached to only one side of membranes. Peripheral membranes proteins are temporarily attached to integral proteins or lipid bilayers through combinations of noncovalent bonding such as hydrophobic and electrostatic bonding. These membranes often act as receptors or provide channels for charged or polar molecules to pass through them.

10.2.10 QUATERNARY STRUCTURE

The term quaternary structure is employed to describe the overall shape of groups of chains of proteins, or other molecular arrangements. For instance, hemoglobin is composed of four distinct but different myoglobin units, each with its own tertiary structure that comes together giving the hemoglobin structure. Silk, spider webs, and wool, already described briefly, possess their special properties because of the quaternary structure of their particular structural proteins.

Both synthetic and natural polymers have superstructures that influence/dictate the properties of the material. Many of these primary, secondary, tertiary, and quaternary structures are influenced in a similar manner. Thus, primary structure is a driving force for secondary structure. Allowed and preferred primary and secondary bonding influence structure. For most natural and synthetic polymers, hydrophobic and hydrophilic domains tend to cluster. Thus, most helical structures will have either a somewhat hydrophobic/hydrophilic inner core and the opposite outer core resulting from a balance between secondary and primary bonding factors and steric and bond angle constraints. Nature has used these differences in domain character to create the world about us.

As noted before, some proteins are linear with inner and intrachain associations largely occurring because of hydrogen bonding. Influences on globular protein structures are more complex, but again, the same forces and features are at work. Globular proteins have irregular three-dimensional structures that are compact but which when brought together form quaternary structures that approach being spherical. While the overall structure is spherical, the surface is irregular, with the irregularity allowing the proteins to perform different specific biological functions.

The preferred folding confirmation is again influenced by the same factors of bonding type, polarity, size, flexibility, and preferred bond angles. The folded conformations are possible because of the flexibility of the primary bonding present within proteins. Thus, polar portions, namely the amine and carbonyl moieties, are more fixed, but the carbon between them is more flexible. Again, the folding characteristic conformations are driven by secondary bonding. Some folding is chemically "fixed" through use of cross-links. In hair, these cross-links are often disulfides, $-S-S-$.

As previously noted, the flexibility of proteins allows them to carry out a wide variety of tasks. Our cells often build about 60,000 different kinds of proteins. A bacterial cell will synthesize only a little more than 1,000 different kinds of proteins.

When a protein contains roughly more than about 200 amino acid groups, it often assumes two or more somewhat spherical tertiary structural units. These units are often referred to as domains. Thus, hemoglobin is a combination of four myoglobin units with each of the four units influenced by the other three, and where each unit contains a site to interact with oxygen.

Enzymes act to lower the activation energy through a combination of holding the reactants in the correct geometry and making the number of "hits" or connections needed for reaction to be greatly decreased. The increases in reaction rate are generally huge. In the case of the enzyme

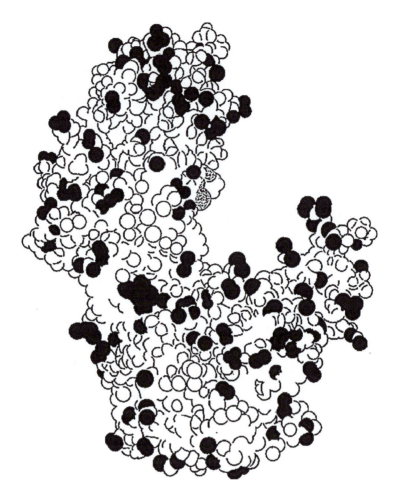

FIGURE 10.7 Protein enzyme phosphoglycerate kinase involved in the formation of adenosine-5′-triphosphate (ATP) from adenosine-5'-diphosphate (ADP).

orotate decarboxylase, the increase is 10^{17}, allowing a reaction that might take 80 million years without the enzyme to occur in about 20 ms in the presence of the enzyme. There are about 4,000 reactions known to be catalyzed by enzymes within the human body. While the size of the enzyme is large, the active site is generally only 3–4 amino acid units long. Enzymes are involved with much of our metabolism and are involved in the synthesis and repair of DNA and in transcription.

The specificity of enzymatic catalytic activity is dependent on tertiary structure. Phosphotglycerate kinase (Figure 10.7) is a protein composed of 415 amino acids. The protein chain is folded back and forward forming a claw-like structure that has two rigid domains divided by a more flexible hinge region. Phosphoglycerate kinase binds to phosphoglycerate, a fragment of glucose, transfering the phosphate forming adenosine triphosphate (ATP). If the enzyme binds to the phosphoglycerate and water, the phosphate could react with the water rather than forming ATP. The protein is designed to circumvent this by having the essential parts present in both halves of the claw. As the enzyme binds the phosphoglycerate and the adenosine diphosphate (ADP), the claw closes bringing together the two necessary active parts, excluding water and other nonessential compounds. In Figure 10.7, the active site phosphoglycerate is indicated by the "dotted" area about half way up one of the two sections of the claw.

TABLE 10.3
Shapes of Selected Biologically Important Proteins

Protein	Shape	Molecular Weight	Comments
Myoglobin	3D, oblate spheroid	1.7×10^4	Temporary oxygen storage in muscles
Hemoglobin	3D, more spherical than myoglobin	6.4×10^4	Oxygen transport through body
Cytochrome C	3D, prolate spheroid	$1.2–1.3 \times 10^4$	Heme-cont. protein, transports electrons rather than oxygen
Lysozyme	3D, short α-helical portions, region of antiparallel pleated sheets	1.46×10^4	Well studied, good illustration of structure-activity
Chymotrypsin and trypsin	3D, extensive β-structure	–	Hydrolysis of peptide bonds on the carboxyl side of certain amino acids
Insulin	3D, two α-helical section in A chain, B chain has α-helix and remainder is extended linear central core	6×10^3	Regulation of fat, carbohydrate, and amino acid metabolism
Somatotropin (human)	3D, 50% α-helix	2.2×10^4	Pituitary hormone
Collagen			
Keratin	Varies with source, 3D or 2D; most contain α-helix sections	$10^4–10^5$	Most abundant protein; major part of skin, teeth, bones, cartilage, and tendon
Fibroin	Varies with source, fibrous-linear with cross-links; crystalline regions contain antiparallel, pleated sheets	3.65×10^5	Major constituent of silk
Elastin	Varies with source; cross-linked, mostly random coil with some α-helix	$>7 \times 10^4$	Many properties similar to rubber; gives elasticity to arterial walls and ligaments

The structure given in Figure 10.7 also illustrates the usual arrangement whereby hydrophilic areas, noted as darkened areas, radiate outward from the enzyme surface while the less hydrophilic and hydrophobic areas tend to reside within the enzyme.

Table 10.3 contains a listing of some important proteins. Protein purification must be done under conditions where conformational and configurational changes are minimal. Such purification is most often carried out using varieties of chromatography, including affinity chromatography and electrophoresis. Somewhat common features of enzymes are the following:

a. α-Helix content not as high as myoglobin, but areas of β-sheeting are not unusual
b. Water-soluble enzymes have a large number of charged groups on the surface and those not on the surface are involved in the active site. Large parts of the interior are hydrophobic
c. The active site is found either as a cleft in the macromolecule or shallow depression on its surface

While enzymes are effective catalysts inside in the body, we have developed techniques for "capturing" some of this activity by immobilizing enzymes. The activity of many enzymes continues in this immobilized condition. In one approach, the enzyme is isolated and coupled to solvent-swellable gels using polyacrylamide copolymers that contain *N*-acryloxysuccinimide repeat units. The pendent groups react with the amino "ends" of the enzyme effectively coupling or immobilizing the enzyme. Modifications of this procedure have been used to immobilize a wide variety of enzymes. For instance, the particular reactive or anchoring group on the gel can be especially modified for a particular enzyme. Spacers and variations in the active coupling end are often employed. Amine groups on gels are easily modified to give other functional groups, including alcohols, acids, nitriles, and acids. Recently, other entities, such as fungi, bacteria, and cells have been immobilized successfully. This technique can be used for the continuous synthesis of specific molecules.

10.3   NUCLEIC ACIDS

Nucleoproteins, which are conjugated proteins, may be separated into nucleic acids and proteins. The name "nuclein," which was coined by Miescher in 1869 to describe products isolated form nuclei in pus, was later changed to nucleic acid. Somewhat pure nucleic acid was isolated by Levene in the early 1900s. He showed that either D-ribose or D-deoxyribose (Figure 10.8) was present in what are now known as RNA and DNA. They consist of two sugars that are identical except that the deoxyribose contains a hydrogen on carbon 2 rather than a hydroxyl (thus the name deoxy or without one "oxy" or hydroxyl [Figure 10.8]). These specific compounds were originally obtained from yeast (DNA) and the thymus gland (RNA).

In 1944, Avery showed that DNA was able to change one strain of bacteria to another. It is now known that nucleic acids direct the synthesis of proteins. Thus, our modern knowledge of heredity and molecular biology is based on our knowledge of nucleic acids. Recently, it was announced that the human genome was decoded. This is one of the most important events to date. The human genome is composed of natures most complex, exacting, and important macromolecule. It is composed of nucleic acids that appear complex in comparison to simpler molecules such as methane and ethylene, but simple in comparison to their result on the human body. Each unit is essentially the same containing a phosphate, and a deoxyribose sugar and one of four bases (Figure 10.7) with each base typically represented by the capital of the first letter of their name, G, C, A, and T. In fact, the complexity is less than having four separate and independent bases because the bases come in matched sets, they are paired. The mimetic Gee CAT allows an easy way to remember this pairing (G-C and A-T). The base, sugar, and phosphate combine forming nucleotides such as adenylic acid, adenosine-3′-phosphate shown below and represented by the symbols A, dA, and dAMP.

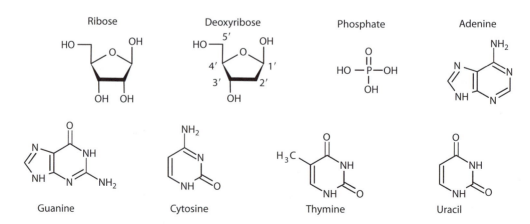

FIGURE 10.8 Components of nucleic acids.

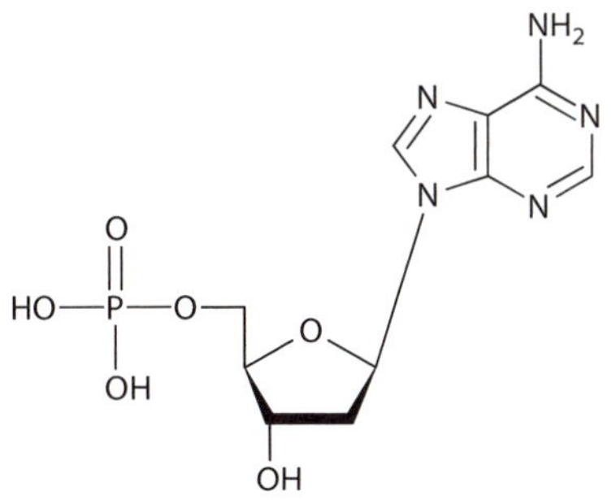

Adenylic acid, adenosine-3'-phosphate (10.8)

The backbone of nucleic acids is connected through the 3' and 5' sites on the sugar with the base attached at the 1' site. Because the sugar molecule is not symmetrical, each unit can be connected differently but there is order (also called *sense* or *directionality*) in the sequence of this connection so that phosphodiester linkage between units is between the 3' carbon of one unit and the 5' carbon of the next unit. Thus, nucleic acids consist of units connected so that the repeat unit is a 3'–5' (by agreement we consider the start to occur at the 3' and end at the 5' though we could just as easily describe this repeat as being 5'-3') linkage. Thus, the two ends are not identical—one contains an unreacted 3'-hydroxyl and the other an unreacted 5'-hydroxyl.

A shorthand is used to describe sequences. Following is a trimer containing in order the bases cytosine, adenine, and thymine.

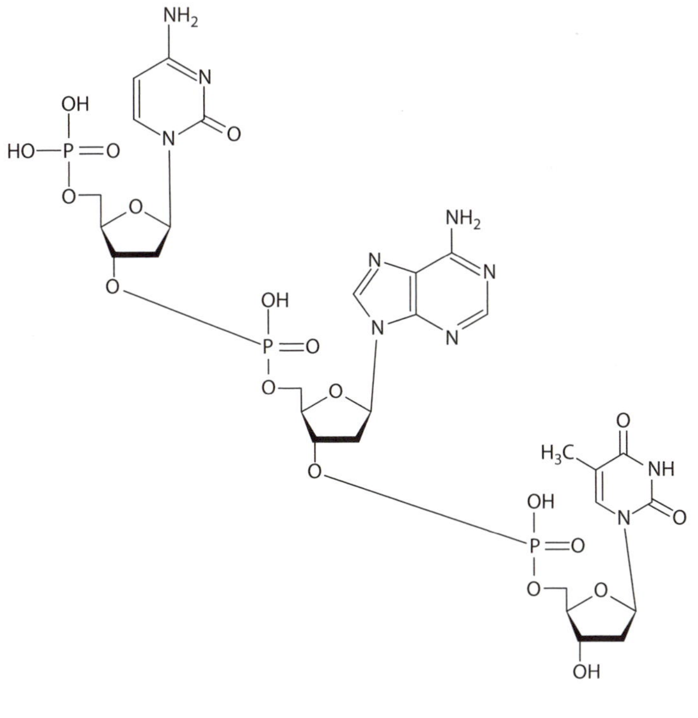

(10.9)

This sequence is described as

p-5'-C-3'-p-5'-A-3'-p-5'-T-3' or pCpApT or usually as simply CAT.

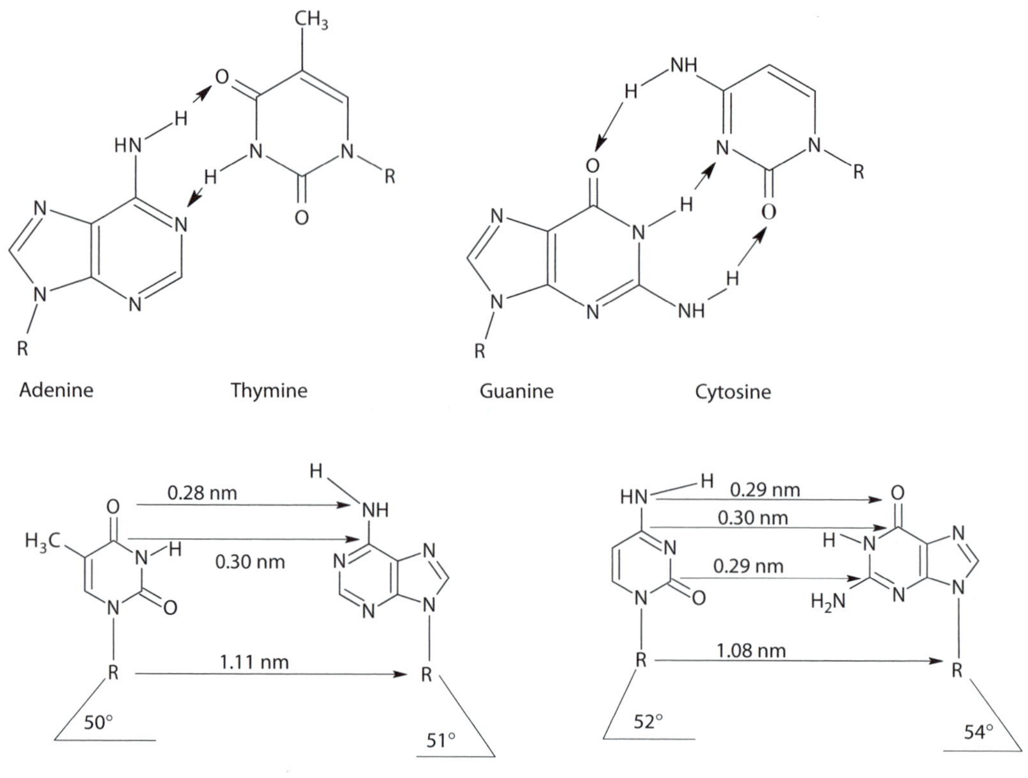

FIGURE 10.9 Hydrogen bonding between preferred base pairs in DNA. Top illustrating the number of hydrogen bonds and the bottom the bond distances between preferred base pairs.

Nobel Laureates Watson and Crick correctly deduced that DNA consisted of a double-stranded helix in which a pyrimidine base on one chain or strand was hydrogen bonded to a purine base on the other chain (Figure 10.9). The bonding distances are not the same with the GC paring more compact. This uneven pairing distances results in a DNA with a characteristic twisting giving unique structures. It is this twisting, and the particular base sequence, that eventually results in the varying chemical and subsequently biological activities for various combinations.

The stability of the DNA is due to both internal and external hydrogen bonding as well as ionic and other bonding. First, the internal hydrogen bonding is between the complementary purine–pyrimidine base pairs. Second, the external hydrogen bonding occurs between the polar sites along exterior sugar and phosphate moieties and water molecules. Third, ionic bonding occurs between the negatively charged phosphate groups situated on the exterior surface of the DNA and electrolyte cations such as Mg^{+2}. Fourth, the core consists of the base pairs, which, along with being hydrogen bonded, stack together through hydrophobic interactions and van der Waals forces. To take good advantage of pi-electron cloud interactions, the bases stack with the flat "sides" over one another so that they are approximately perpendicular to the long axis.

The AT and CG base pairs are oriented in such a manner so that the sugar-phosphate backbones of the two twined chains are in opposite or antiparallel directions with one end starting at the 5′ and ending at the 3′ and the starting end of the other across from the 5′ end being a 3′ end and opposite the other 3′ end is a 5′ end. Thus, the two chains "run" in opposite directions.

The glucose bonds holding the bases onto the backbone are not directly across the helix from one another. Thus, the sugar-phosphate repeat units are not the same. This dislocation creates structures referred to as major and minor grooves as pictured in Figure 10.8. It is known that at least some proteins that bind to DNA recognize the specific nucleotide sequences by "reading" the hydrogen bonding pattern presented by the edges of these grooves.

In solution, DNA is a dynamic, flexible molecule. It undergoes elastic motions on a nanosecond time scale most closely related to changes in the rotational angles of the bonds within the DNA backbone. The net result of these bendings and twistings is that DNA assumes a roughly globular or spherical tertiary shape. The overall structure of the DNA surface is not that of a reoccurring "barber pole" but rather because of the particular base sequence composition each sequence will have its own characteristic features of hills, valleys, bumps, and so on.

As the two strands in a double helix separate, they act as a template for the construction of a complementary strand. This process occurs enzymatically with each nucleotide being introduced into the growing chain through matching it with its complementary base on the existing chain. Thus, two identical strands are produced when one double-helix combination replicates.

DNA chains can contain 1 million subunits with an end-to-end contour length of about 1 mm. Even with the complexity of these large macromolecules, synthesis of new chains generally occurs without any change in the molecule. Even when changes occur, these giant machines have built into them "correcting" mechanisms that recorrect when mistakes occur.

The transcription product of DNA is always single-stranded RNA. The single strand generally assumes a right-handed helical conformation mainly caused by base-stacking interactions also present in the DNA. The order of interaction is purine–purine >> purine–pyrimidine > pyrimidine–pyrimidine. The purine–purine interaction is so strong that a pyrimidine separating two purines is often displaced from the stacking order to allow the interaction between the two purines to occur. Base paring is similar to that of the DNA except that uracil generally replaces thymine. For coupled RNA the two strands are antiparallel as in DNA. Where complementary sequences are present, the predominant double-stranded structure is an A form right-handed double helix. Many RNAs are combinations of complementary two-stranded helices, single-stranded segments, as well as other complex structures. Hairpin curves are the most common type of more complex structure in RNA. Specific sequences, such as UUCG, are generally found at the ends of RNA hairpin curves. Such sequences can act as starting points for the folding of a RNA into its precise three-dimensional structure. The tertiary structures for RNAs are complex with combinations being present. For instance, the tertiary structure for the transcription ribonucleic acid (tRNA) of yeast for phenylalanine consists of a cloverleaf, including three loops formed by hairpin curves and double-helix regions stabilized by hydrogen bonding. Hydrogen bonding sites that are not significant in the DNA structures are important. Thus, the free hydroxyl on the ribose sugar moiety can hydrogen bond with other units.

There are four major kinds of RNA. *Messenger RNA* (mRNA) varies greatly in size from about 75 units to more than 3,000 nucleotide units giving a molecular weight of 25,000 to one million. It is present at a percentage of about 2% of the total RNA in a cell. tRNA has about 73–94 nucleotides with a corresponding molecular weight range of 23,000–30,000. It is present in the cell at a level of about 16%. The most abundant RNA, 82%, is the ribose RNA (rRNA), which has several groupings of molecular weight with the major ones being about 35,000 (about 120 nucleotide units), 550,000 (about 1,550 units) and 1,100,000 (about 2,900 units). Eukaryotic cells contain an additional type called *small nuclear RNA* (snRNA).

Transfer RNAs generally contain 73–94 nucleotides in a single chain with a majority of the bases hydrogen bonded to one another. Hairpin curves promote complementary stretches of base bonding giving regions where helical double stranding occurs. The usual overall structure can be represented as a cloverleaf with each cloverleaf containing four of these helical double-stranded units. One of the loops acts as the acceptor stem that serves as the amino acid-donating moiety in protein synthesis.

Ribosomal RNA (rRNA) is a part of the protein synthesizing machinery of cells, ribosomes. Ribosomes contain two subunits called "small" and "large" with rRNAs being part of both of these units. rRNAs contain a large amount of intrastrand complementary sequences and are generally highly folded. Interestingly, there is a similarity between the folded structures of rRNA from many different sources even though the primary structure, base sequence, is quite varied. Thus, there appears to be preferred folding patterns for rRNAs.

Messenger RNA is the carrier of messages that are encoded in genes to the sites of protein synthesis in the cell where this message is translated into a polypeptide sequence. Because mRNAs are transcribed copies of the genetic unit, they are sometimes referred to as being the "DNA-like RNA". mRNA is made during transcription, an enzymatic sequence in which a specific RNA sequence is "copied" from a gene site. rRNA and tRNA are also made by transcription of DNA sequences but unlike mRNA, they are not subsequently translated to form proteins.

Actual reproduction steps involving DNA and RNA often occur in concert with protein where the protein can act as a clamp or vice holding the various important members involved with the particular reproduction step in place. Thus, the protein complex acts as an assembly line tunnel or doughnut with the reactants present within the interior.

There are two types of cells. *Prokaryote cells* lack a cell nucleus or any other membrane-bound organelles. Most are single celled. Bacteria and archaea are examples of organisms with prokaryote cells. *Eukaryote cells* possess enclosed membranes and form the basis of animals, plants, and fungi. In prokaryote cells the mRNA can be used immediately after it is produced or it may be bound to a ribosome. In comparison, in eukaryote cells mRNA is made in the cell nucleus and it is moved across the nuclear membrane after it is synthesized into the cytoplasm where protein synthesis occurs.

10.4   FLOW OF BIOLOGICAL INFORMATION

Nucleic acids, proteins, some carbohydrates and hormones are informational molecules. They carry directions for the control of biological processes. With the exception of hormones, these are macromolecules. In all these interactions, secondary forces such as hydrogen bonding and van der Waals forces, and ionic bonds, and hydrophobic/hydrophilic character play critical roles. *Molecular recognition* is the term used to describe the ability of molecules to recognize and interact (bond) specifically with other molecules. This molecular recognition is based on a combination of these interactions just cited and on structure.

Molecular recognition interactions have several common characteristics. First, the forces that are involved in these interactions are relatively weak and they are noncovalent. They are on the order of about 1–8 kcal/mol (4–30 kJ/mol) compared to covalent bonds of the order of about 80 kcal/mol (300 kJ/mol) for a C–C sigma bond. A single secondary bond is generally not capable of holding molecules together for any length of time. But for macromolecules, there is a cumulative effect so that the forces are not singular but are multiplied by the number of such interactions that are occurring within the particular domain. Second, these interactions are reversible. Initial contact occurs as the molecules come into contact with one another often through simple diffusion or movement of the molecules or segments of the molecules. These initial contacts are often not sufficient to cause the needed binding though some transitory interactions to occur. Even so, in some cases the cumulative bonding is sufficient to allow a transient but significant interaction to occur. This complex can then begin a specific biological process. Eventually, thermal motions and geometrical changes cause the complex to dissociate. Ready reversibility is an important key that allows a relatively few "signaling" molecules to carry out their mission. Third, bonding between the particular molecular sites is specific. There must exist a combination of complementary bonding, hydrophobic/hydrophilic sites, ionic charge, and geometry that allow effective long-term (generally no more than several seconds) interactions to occur.

In general, the flow of biological information can be mapped as follows:

$$DNA \rightarrow RNA \rightarrow Protein \rightarrow Cell\ structure\ and\ function$$

The total genetic information for each cell, called the *genome*, exists in the coded two-stranded DNA. This genetic information is expressed or processed either through duplication of the DNA so it can be transferred during cell division to a daughter cell or it can be transferred to manufactured RNA that in turn transfers the information to proteins that carry out the activities of the cell.

Duplication of double-stranded DNA is self-directed. The DNA, along with accessory proteins, directs the *replication* or construction of two complementary strands forming a new, exact replicate of the original DNA template. As each base site on the DNA becomes available through the unraveling of the double-stranded helix, a new nucleotide is brought into the process held in place by hydrogen bonding and van der Waals forces so that the bases are complementary. It is then covalently bonded through the action of an enzyme called *DNA polymerase*. After duplication, each DNA contains one DNA strand from the original double-stranded helix and one newly formed DNA strand. This is called *semiconservative replication* and increases the chance that if an error occurs, that the original base sequence will be retained.

How is DNA suitable as a carrier of genetic information? While we do not entirely understand several features are present in DNA. First, because of the double-stranded nature and mode of replication, retention is enhanced. Second, DNA is particularly stable within both cellular and extracellular environments, including a good stability to hydrolysis within an aqueous environment. Plant and animal DNA have survived thousands of years. Using polymerase chain reactions (PCR) we can reconstruct DNA segments allowing comparisons to modern DNA.

Transcription is the term used to describe the transfer of information from the DNA to RNA. The genome is quite large, on the order of a millimeter in length if unraveled, but within it exists coding regions called *genes*. Transcription is similar to DNA replication except ribonucleotides are the building units instead of deoxyribonucleotides; the base thymine is replaced by uracil; the DNA:RNA duplex unravels releasing the DNA to again form its double-stranded helix and the single-stranded RNA; and the enzyme linking the ribonucleotides together is called *RNA polymerase*.

Many viruses and retroviruses have genomes that are single-stranded RNA instead of DNA. These include the AIDS virus and some retroviruses that cause cancer. Here, an enzyme called *reverse transcriptase* converts the RNA genome of the virus into the DNA of the host cell genome thus infecting the host.

The transcription of the DNA gives three kinds of RNA—ribosomal, messenger, and transfer. The most abundant RNA is rRNA. Most rRNA is large and is found in combination with proteins in the ribonucleoprotein complexes called *ribosomes*. Ribosomes are subcellular sites for protein synthesis.

Transfer RNA is the smallest of the RNAs being less than 100 nucleotides long. tRNA combines with an amino acid incorporating it into a growing protein. There is at least one tRNA for each of the 20 amino acids used in protein synthesis. mRNA is varied in size but each carries the message found in a single gene or group of genes. The sequence of bases in mRNA is complementary to the sequence of DNA bases. mRNA is unstable and short-lived so that its message for protein synthesis must be rapidly decoded. The message is decoded by the ribosomes that make several copies of the protein from each mRNA.

The ultimate purpose of DNA expression is protein synthesis. mRNA serves as the intermediate carrier of the DNA genetic information for protein synthesis. The DNA message is carried in the form of base sequences that are transferred to RNA also in terms of base sequences and finally these are transferred into amino acid sequences through a translation process based on the genetic code. This process of information from the RNA to the protein is called *translation*.

A set of coding rules are in action as in the translation process. Briefly, these are as follows. First, a set of three adjacent nucleotides compose the code for each amino acid. A single amino acid can have several triplet codes or *codons*. Since there are four different nucleotides (or four different bases) in DNA and RNA there exists 4^3, or 64 trinucleotide combinations. For instance, using U as a symbol for uracil, present in RNA, the triplet or code or codon UUU is specific for phenylalanine.

Second, the code is nonoverlapping so that every three nucleotides code for an amino acid and the next three code for a second amino acid and the third set code for a third amino acid, and so on. Third, the sets of nucleotides are read sequentially without punctuation. Fourth, the code is nearly universal. Fifth, there are codes for other than amino acids, including stop or terminate UAG, and start or initiate AUG.

In essence, tRNA has two active sites—one that is specific for a given amino acid and the second that is specific for a given set of three bases. The tRNA "collects" an appropriate amino acid and brings it to the growing polypeptide chain inserting it as directed by the mRNA. There is then a collinear relationship between the nucleotide base sequence of a gene and the amino acid sequence in the protein.

The amount, presence, or absence of a particular protein is generally controlled by the DNA in the cell. Protein synthesis can be signaled external to the cell or within the cell. Growth factors and hormones form part of this secondary messenger service.

The translation and transcription of DNA information is polymer synthesis and behavior, and the particular governing factors and features that control these reactions are present in the synthesis and behavior of other macromolecules—synthetic and biological.

For the human genome there exists so-called coding or active regions called *exons* and noncoding regions called *introns*. The average size of an exon is about 120–150 nucleotide units long or coding for about 40–50 amino acids. Introns vary widely in size from about 50 to more than 20,000 units. About 5% of the genome is used for coding. It was thought that the other 95% was silent or junk DNA. We are finding that the introns regions play essential roles. Interestingly introns are absent in the most basic prokaryotes, only occasionally found in eukaryotes, but common in animals.

10.5   RNA INTERFERENCE

RNA interface (RNAi) is a somewhat newly discovered part of our body's natural immune system. RNAi is not another form of RNA, but rather it is a sequence involving protein enzymes that blocks the action of certain foreign RNA, thus its name RNA interference or simply RNAi. Andrew Fire and Craig Mello won the 2006 Nobel Prize in Medicine for their work with RNAi.

Following is a general discussion of what RNAi is and does, including some potential applications from its use.

While DNA is the depositary of our genetic information, it is "held captive" or "protected" by the nuclear envelope remaining within this envelope. Yet the information held by the double-stranded DNA is transferred throughout the cell with results felt throughout the body. Briefly, polymerase transcribes the information on the DNA into single-stranded mRNA. The mRNA single strands move from the cell nucleus into the cell cytoplasm through openings in the nuclear envelope called nuclear pore complexes.

In the cytoplasm, ribosome, another protein enzyme, translates the information on the mRNA into protein fragments eventually giving entire proteins. This "normal" sequence is depicted in Figure 10.10.

Viruses are genetic materials enclosed in a protein "coat." Viruses show a very high specificity for a particular host cell, infecting and multiplying only within those cells. Viral genetic material can be either DNA or RNA but is almost always double stranded. Viral attacks generally result in a virus infecting a cell by depositing its own genetic material, here RNA into the cell's cytoplasm. The purpose of depositing this viral-derived RNA is to have it replicate as rapidly as possible in an attempt to hijack the cell into producing viral-directed protein. We need to remember that almost all of the viral-derived RNAs are double stranded, rather than the cells single-stranded RNA.

By accident, it was discovered about a decade ago that the cell had a new weapon in its auto-immune system. Thought there had been prior hints, the first reasonable evidence that something was there was the work of Jorgensen, Cluster, English, Que, and Napoli (*Plant. Mol. Biol.* 31:957 [1996]). Their effort to darken the purple color of petunias, working toward a so-called black color,

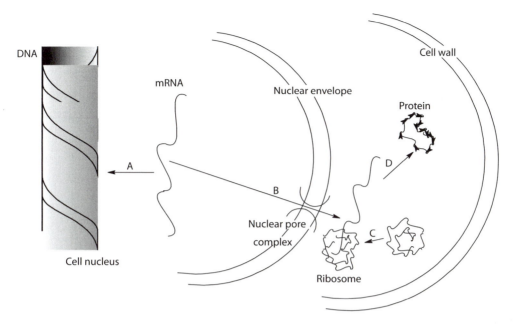

FIGURE 10.10 "Normal" protein formation sequence. First, A—mRNA is formed from the cell's DNA. Next, B—the mRNA enters into the cells cytoplasm. Third, C—ribosome forms about the mRNA resulting in the formation of the desired protein, D.

showed that some sort of suppression was at work giving variegated to white petunias instead of a darkened purple variety. They called this phenomenon cosuppression since the expression of both the introduced gene and the seeds own color gene was suppressed. Evidence for cosuppression was not new and had been found for fungi.

Eventually, it was discovered that the cell has its own ability to combat viral attacks. The initial step in the RNAi process is the use of an enzyme, nicknamed Dicer, which cuts the double-stranded viral-derived RNA into small, about 20 units, strands. The second part of the RNAi system involves the unwinding of these short sections using again a protein enzyme called *RNA-induced silencing complex* (RISC). This results in short chains of single-stranded RNA being present as a complex with the RISC. This complex is referred to as a RISC/RNA complex. This complex floats about the cell looking for complementary RNA segments in the much longer cellular mRNA. When such a fit is found, the complex deposits its viral-derived segment onto the mRNA rendering it either incapable or much less capable of creating viral-related protein. Thus, the cell has built into it, its own antiviral weaponry. This sequence is given in Figure 10.11.

The discovery of the RNAi sequence has led to researchers employing it to identify the activity of various gene segments as well as working on cures for a number of diseases. More about this later.

RNAi activity is most often induced using relatively small, generally 21–23 nucleotides long, segments of RNA that interfere with the activity (silences) of an endogenous gene. This silencing is called *posttranscriptional gene silencing* or PTGS. The interfering RNA is called *siRNA*. Thus, gene specific siRNAs are prepared and their effect monitored "downstream."

RNA interface is being used as a tool to help decipher the gene. Eventually, this may allow us to better understand the purpose for the various segments contained within specific genes. The basis of this is the ability of RNAi to curtail the activity of the cell to supply specific proteins. This loss of function then identifies the role that the particular RNAi has rendered inactive. Briefly, various double-stranded RNAs (dsRNA) are created. Each of these is tested to see where on the normal RNA they become attached and which function is impaired. This then tells us about the particular

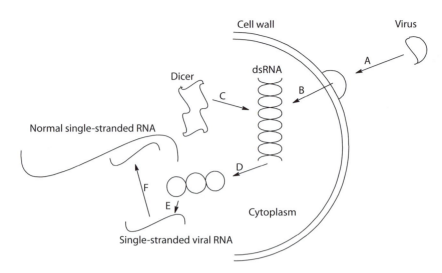

FIGURE 10.11 RNAi sequence. First the virus approaches and attaches itself to the cell wall, A—The virus injects double-stranded RNA, B—into the cell cytoplasm. The dicer attacks the dsRNA breaking it into smaller units, C and D. These smaller units are then acted on by RISC forming single-stranded viral RNA, E, which are rendered incapable of forming its own protein by attachment to a complement contained within the cells own single-stranded RNA, F.

function of that particular part of the normal RNA, which is then traced back to the cell's DNA and its location in a particular gene.

10.6 POLYMER STRUCTURE

In 1954, Linus Pauling received the Nobel Prize for his insights into the structure of materials, mainly proteins. Pauling showed that only certain conformations are preferred because of intermolecular and intermolecular hydrogen bonding. While we know much about the structures of natural macromolecules, there is still much to become known.

Two major secondary structures are found in nature—the helix and the sheet. These two structures are also major secondary structures found in synthetic polymers. The helix takes advantage of both the formation of intermolecular secondary bonding and relief of steric constraints. Some materials utilize a combination of helix and sheet structures such as wool that consists of helical protein chains connected to give a "pleated" skeet.

The Watson and Crick model for DNA as a double helix is only a generalized model to describe much more complex structures. Along with the typical double helix there exist structural elements such as supercoils, kinks, cruciforms, bends, loops, and triple strands and major and minor groves. Each of these structural elements can vary in length, shape, location, and frequency. Even the "simple" DNA double helix can vary in pitch (number of bases per helical turn), sugar pucker conformation, and helical sense (is the helix left- or right-handed).

Electron microscopy shows that DNA consists of either linear or circular structures. The chromosomal DNA in bacteria is a closed circle, a result of covalent joining of the two ends of the double helix (Figure 10.12). Note the presence of supercoils, branch points, intersections, and the generally thin and open structure. The chromosomal DNA in eukaryotic cells, like ours, is believed to be linear.

The most important of the secondary structures is supercoiling. *Supercoiling* simply is the coiling of a coil or in this case a coiling of the already helical DNA. The typical DNA structure is the thermally stable form. Two divergent mechanisms are believed responsible for supercoiling. The first, and less prevalent, is illustrated by a telephone cord. The telephone cord is typically coiled and

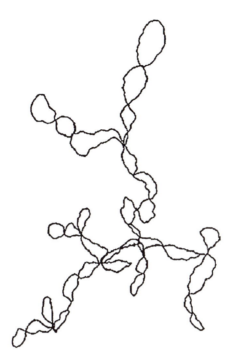

FIGURE 10.12 Description of a typical bacterial chromosome.

represents the "at rest" or "unstressed" coupled DNA. As I answer the telephone, I have a tendency to twist it in one direction and after answering and hanging up the telephone for awhile it begins forming additional coils. Thus, additional coiling tends to result in supercoiling. The second, and more common form, involves the presence of less than normal coiling. This can be illustrated by taking a rubber band, breaking one end and then attaching it about a stationary object. Begin bradding the two ends until just before a bunching or formation of supercoiling through over coiling. Then separate the two ends pulling them apart. The resulting strain produces supercoiling and illustrates supercoiling resulting in undercoiling or underwinding. Thus, underwinding occurs when there are fewer helical turns than would be expected. Purified DNA is rarely relaxed.

Supercoiling with bacterial DNA gives a largely open, extended, and narrow, rather than compacted, multibranched structure. By comparison, the DNA in eukaryotic cells is present in very compacted packages. Supercoiling forms the basis for the basic folding pattern in eukaryotic cells that eventually results in this very compacted structure. Subjection of chromosomes to treatments that partially unfold them show a structure where the DNA is tightly wound about "beads of proteins" forming a necklace-like arrangement where the protein beads represent precious stones imbedded within the necklace fabric (Figure 10.13). This combination forms the nucleosome, the fundamental unit of organization upon which higher-order packing occurs. The bead of each nucleosome contains eight histone proteins. Histone proteins are small basic proteins with molecular weights between 11,000 and 21,000 and specified by names such as H1, H2, and so on. H1 is especially important and its structure varies to a good degree from species to species whereas some of the other histones, such as H3 and H4, are very similar. Histones are rich in the amino acid residues from arginine and lysine.

Wrapping of DNA about a nucleosome core compacts the DNA length about seven fold. The overall compacting though is about 10,000 fold. Additional compacting of about 100 fold is gained from formation of so-called 30 nm fibers. These fibers contain one H1 within the nucleosome core. This organization does not occur over the entire chromosome but rather is punctuated by areas

30 nm

Histone core

DNA linker and "wrapper" →

FIGURE 10.13.   Illustration of regularly spaced nucleosomes consisting of histone protein bound to super-coiled DNA with DNA links between the histone bound units forming a 30 nm higher-order fiber.

containing sequence-specific (nonhistone containing) DNA-binding proteins. The name "30 nm fibers" occurs because the overall shape is of a fiber with a 30 nm thickness. The additional modes of compaction are just beginning to be understood but may involve scaffold assisting. Thus, certain DNA regions are separated by loops of DNA with about 20,000–100,000 base pairs, with each loop possibly containing sets of related genes.

The scaffold contains several proteins, especially H1 in the core and topoisomerase II. Both appear important to the compaction of the chromosome. In fact, the relationship between topoisomerase II and chromosome folding is so vital that inhibitors of this enzyme can kill rapidly dividing cells and several drugs used in the treatment of cancer are topoisomerase II inhibitors.

The central theme concerning the major secondary structures found in nature is also illustrated with the two major polysaccharides derived from sucrose, that is, cellulose and a major component of starch—amylose. Glucose exists in one of two forms—an alpha and a beta form where the terms alpha and beta refer to the geometry of the oxygen connecting the glucose ring to the fructose ring.

Cellulose is a largely linear homosaccharide of the beta-D-glucose. Because of the geometry of the beta linkage, individual cellulose chains are generally found to exist as sheets, the individual chains connected through hydrogen bonding. The sheet structure gives cellulose-containing materials good mechanical strength allowing them to act as structural units in plants. Amylose, by comparison, is a linear poly(alpha-D-glucose). Its usual conformation is as a helix with six units per turn (Figure 10.14). Amylose is a major energy source occurring in plants as granules.

A number of repeating features occur in nature. Along with the ones noted before, another depends on the buildup of structure from individual polymer chains to structures that are seen by our human eye. One of these assemblies is illustrated in Figure 10.15 where individual alpha-helix chains form slowly curving protofibril bundles. These bundles in turn form microfibril structures which in turn form macrofibril and finally the cortical cell of a single wool fiber. The bundling and further bundling eventually gives the wool fiber. A similar sequence can be described in the formation of our hair as well as our muscles. The difference for the muscle is that the basic structure is not the alpha helix but rather a sheet-like actin filament connected together by myosin filaments. In all cases, flexibility is achieved through several features with one of the major modes being the simple

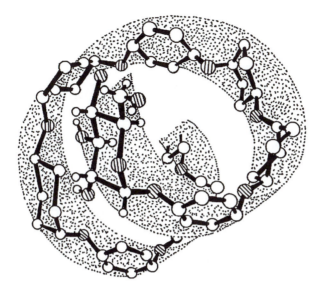

FIGURE 10.14 Helical structural arrangement of amylose derived from α-D-glucose units.

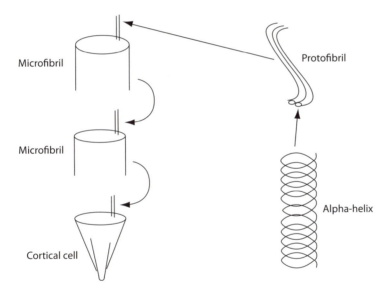

FIGURE 10.15 Buildup of wool hair.

sliding past one another the various bundles of protein. For the muscle, the sliding of the various sheets is also a factor.

These assemblies of more simple structures to form more complex structures is also found in the plant world. Thus, plant fibers often contain various cellulosic-intense layers that are not placed on top of one another, but rather are at some angle to one another resulting in increased strength and flexibility for the fiber.

10.7 PROTEIN FOLDING

Just as the functioning of nucleic acids depends in part on its overall structure, so also does the activity of proteins depend on overall structure. Protein folding is one of the "hot" areas today in science. To the synthetic polymer chemist, understanding the influences, the basics or fundamentals, which produce protein chain folding will allow the creation of new synthetic polymers that possess specifically desired properties. For biochemists, understanding these factors allows us to better understand other factors and to combat particular diseases related to chain folding.

The particular shape of protein chains is known as a *fold* and the process is called *chain folding*. While chain folding is often referred to as a self-assembly process, it also involves other specialized separate proteins that assist in this chain folding so it is not truly self-assembly. Chain folding occurs rapidly in the time scale of 1–10 millionths of a second or tens of microseconds.

Chain folding depends on the primary and secondary structures of materials. Thus, the particular atomic composition of a chain dictates, under equilibrium conditions, its tertiary and quaternary structures.

Proteins act as structural and "enzyme-type" materials. Much of the present discussion focuses on the folding of enzyme-type materials, but it is equally applicable to structural proteins. While the human genome has less than about 30,000 genes, these genes account for as many as one million proteins most of these being nonstructural in nature. Proteins also undergo maturing, or to be more truthful, aging, as it carries out its function(s). As with much of nature, after extended use and associated damage related to chemical and partial unfolding, the protein is degraded and its parts often becoming part of a newly synthesized protein.

As already noted, protein folding is dependent on the primary chemical structure. Chain folding begins even as it is being synthesized by ribosomes. The assembly area is crowed with this crowding increasing the risk of nonspecific association and aggregation. A primary driving force for aggregation is simply the push to bury hydrophobic portions of the molecule away from the influences of the hydrophilic, water-rich, birthing surrounding. Thus, left to itself, folded proteins have a hydrophilic surface that contains within it the more hydrophobic portions. But this tendency must be moderated or simple aggregation of the hydrophobic and hydrophilic portions, ribbons, occurs and even worse, unwanted aggregation of the proteins themselves occurs.

Crowding is one factor that is not easily achieved away from the native environment. Lack of crowding is one of the major reasons why in vivo and in vitro syntheses often result in proteins with different activities.

While we characterize nonstructural proteins as being globular in shape, these shapes vary considerably according to their use. Again, this overall shape is governed by a combination of factors, including primary and secondary structure.

A family of proteins assists in the folding process of proteins in general. These proteins are called *molecular chaperones*. As in the case of a date, the chaperones help guide appropriate interactions and discourage unwanted associations. Chaperones are found in all of our cells. Many of them are designated by the acronym Hsp for heat shock protein. They are also designated by the relative mass in kilodaltons so that a Hsp70 means it is a chaperone molecule that is about 70 kDa in mass. The main chaperones are Hsp60 (chaperonins), Hsp70, and Hsp90.

Typically, a series of steps is involved in the work carried out by chaperone molecules. Hsp70 operates on the protein as it is being formed on the ribosome. It recognizes extended or exposed protein chain regions that are more hydrophobic and acts to discourage unwanted association of these parts. It also acts to maintain the growing protein in a somewhat unfolded state.

Hsp70 hands off the protein to another class of chaperones known as Hsp60 or simply chaperonins. *Chaperonins* create a protected environment sometimes known as an "Anfinsen cage" because it creates an enclosed environment where the protein segments spontaneously fold, free from aggregating with other proteins and somewhat free from aqueous influences. These are large proteins that are somewhat cylindrical in shape. Chaperonins are composed of two major units, stacked rings.

There are two different classes of chaperonins. Class I includes eukaryotic cells and class II includes certain prokaryotic cells. Eukaryotic cells have nuclei and include our cells. Prokaryotic cells do not contain nuclei and other organelles. We will focus on eukaryotic cells. Much of the information we have on chaperonins is derived from studying bacteria such as *Escherichia coli*. The chaperonin in *E. coli* is given the designation GroES–GroEL. The GroEL is composed of two stacked 7-memebered rings of 60 kDa mass that form a cylinder about 15 nm high and 14 nm wide, with a 5 nm central cavity capable of holding proteins to 60 kDa. GroES is a cochaperonin that acts as a dome or cap for one end of the GroEL portion. It is composed of a single 7-membered ring of about 10 kD mass. Thus, much is know concerning the structures of at least a few of molecules involved in the folding process. Much is also known concerning the function of these molecules, but this is beyond the present scope.

Our cells contain about 1%–2% Hsp90 of the total cytosolic proteins. This is a huge percentage in comparison to most other proteins and signals their importance. Their action depends on the cyclic binding and hydrolysis of ATP. Hsp90 is involved with conformational regulation of signal transduction molecules.

Along with the guiding of chaperon molecules there are some "native" or natural tendencies with respect to chain folding, at least for smaller proteins. One somewhat common sense one but one only recently described involves the closeness of various segments and folding rate. Briefly, protein segments that are located close to one another and in a generally reasonably close orientation to the final folding location promote rapid chain folding. This is reasonable since it takes more time to organize protein segments that are further from one another relative to organization of structures that after folding will be close to one another. Even so, not all of what we are finding is the most "reasonable." For instance, interatomic interactions are not as important as previously believed. This suggests that local sequences, the primary structures, are most important in determining the final folded structures and rates at which these structures are achieved. These findings have been found for smaller groups to about 150 units in length. It is not known if they will be true for larger proteins. At times one tendency may be influenced by other factors. Thus, for beta-lactoglobulin, the local structure favors a helical structure but this tendency is overcome by tertiary interactions.

What is known is that we have not yet mastered the art of chain folding. We are aware that several groups of factors are involved with the folding and more will probably become apparent. We also know that there is a balance between these factors and again an understanding of factors influencing this balance will also become better know.

An added importance to understanding and being able to influence chain folding involves the number of diseases that are related to misfolds. For instance, misfolding can result in aggregation of the proteins, which is a symptom of Mad Cow, Creutzfeldt–Jakob, and Alzheimer's diseases.

It has been suggested that about 50% of cancers involve some chain misfolding. A key protein in this is the p53 protein, which exerts tumor suppression. A single DNA strand break can result in uncontrolled cell division but it normally activates p53 activity, which promotes the production of other proteins that block cell division or bring about cell death, and thus the cessation of the precancer activity. Mutation of a single nucleotide in p53 is believed to result in a misfold resulting in the protein's inability to recognize when it is needed or a failure to act correctly.

Diabetes can involve misfolding of proteins that formed in the endoplasmic reticulum (ER). The ER secretes certain hormones, enzymes, and antibodies and so is a key player in our health. In some cases, the misfolded proteins interfere with carbohydrate metabolism leading to diabetes.

Other protein misfolding-associated diseases include lung diseases, including cystic fibrosis and hereditary emphysema, blood coagulation, certain infectious diseases, and liver diseases. Thus, a better understanding of chain folding is important to the health of many of us.

10.8 GENETIC ENGINEERING

Genetic engineering is the alteration of an organism's genetic material. The aim is to introduce into the organism's genetic material some desirable trait that is otherwise absent. Alternation of genetic material entails the use of polymer chemistry on a molecular (or nano) level making use of somewhat straightforward chemical reactions, many of the reactions employing biological entities, such as enzymes, to carry out these reactions.

Essentially, gene segments are replaced to inject into the altered microorganism genetic material that expresses the desired trait. Today, routine gene alteration is taught in undergraduate laboratories. Even so, specific gene alteration requires extensive planning and is conducted in major research laboratories.

In the broadest sense, genetic engineering refers to any artificial process that alters the genetic composition of an organism. Such alterations can be carried out indirectly through chemical methods, through radiation, or through selective breeding. Today, the term usually refers to the process whereby genes or portions of chromosomes are chemically altered.

After the alteration of a single, or few, genes, the altered genes reproduce giving much larger numbers of genes with the alternation incorporated in the their genome. The term "**clone**" comes from the Greek work *klon*, meaning a cutting used to propagate a plant. Cell cloning is the production of identical cells from a single cell. In like manner, gene cloning is the production of identical genes from a single gene, introduced into a host cell. Today, the term cloning refers to one special type of genetic engineering.

Genes are a chromosomal portion that codes for a single polypeptide or RNA. Gene splicing is currently practiced as the enzymatic attachment of one gene or gene segment to another gene or gene segment. Genes are composed of DNA, which can be considered as a specialized polyphosphate polymer. The manipulation of DNA can occur for many reasons. One of these is the production of recombinant DNA. Here we will focus on the production of recombinant DNA. DNA cannot be directly transferred from one organism, the donor, to another recipient organism, the host. Instead, the donor DNA segment is cut and then recombined with a DNA from a host. *E. coli* is typically employed as the host cell since it is itself a harmless bacterium that reproduces rapidly. (But under the wrong conditions *E. coli* is responsible for many food poisonings.) The *E. coli* then acts as a "factory" that reproduces bacteria that contain the desired modification.

Enzymes, specialized proteins, are used as designing tools for the genetic engineering. One of these enzyme tools consists of *restriction endonucleases* that recognize a specific series of base pairs. They split the DNA at these specific points. This splitting is called "lysing," which in reality is simply the hydrolysis of DNA units as shown below:

$$R\diagdown O-P(=O)(O^-)-O\diagdown R \xrightarrow{\text{Restriction endonuclease}} R\diagdown O-P(=O)(O^-)-OH + HO-R \qquad (10.10)$$

Organisms produce restriction endonucleases that are specific for that organism. Certain restriction endonucleases cut double-stranded DNA asymmetrically in regions called *palindromes*, that is regions that "read" (have identical sequences) the same way from left to right on one strand as right to left on the other strand. This produces what is referred to as "sticky ends" that form not only a "cleft" for attachment but also a single-stranded end that has the ability to pair with another complimentary single-stranded strand end. Both strands of the original donor twin strand have a tendency to recombine with complementary strands of DNA from a host that has been treated to produce the complementary strands. The sticky ends, when mixed under the proper conditions in the presence of another enzyme, DNA-ligase, combine. The hydrogen bonding between complementary sticky ends reenforce the recombination reaction. The resulting recombination reaction results in a variety of products, including the desired recombination of host and donor DNA as well as the combination of the original donor strands and uncombined DNA. The mixture is often treated in one of two manners. The simplest case requires a chemical-resistant gene that is resistant to the employed chemical agent, such as tetracycline. The desired recombinant genes survive and are then transferred into the host organism so the new gene can express itself.

In some cases, such as the synthesis of insulin, the recombination mixture is added to a host organism, here *E. coli*. This infected mixture is then plated out and the individual colonies tested for insulin production. Those colonies that produce insulin are further plated out and grown for mass insulin production. Cells that accept the recombinant DNA are called *transformed*. More specialized sequences have been developed to increase the probability of gene incorporation and its successful reproduction.

A second tool employed by the genetic engineer is the enzyme terminal transferase that adds deoxyribonuclease resides to the 3′ end of DNA strands creating 3′ tails of a single type of residue.

Special modified *plasmid* DNA's, called *vectors* or carriers, are used as host or targets for gene modification. These circularly shaped vectors reproduce autonomously in host cells. Plasmids have two other important properties. First, they can pass from one cell to another allowing a single "modified" bacterial cell to inject neighboring bacterial cells with this "modification." Second, gene material from other cells can be easily formed into plasmids, allowing ready construction of modified carriers.

The steps involved in gene splicing, emphasizing the chemical nature of the individual steps, are as follows:

1. Lysing (which is really simply the hydrolysis of DNA units as shown above)
2. Construction of staggered, sticky, ends
3. Recombination or lysation, the reverse of lysing, chemically formation of a phosphate ester as below connecting the desired segment to the DNA of the host cell

$$R_1 - OH \;+\; \underset{\substack{| \\ O^-}}{\overset{\substack{R \\ \backslash \\ O - P - OH \\ \| \\ O}}{}} \;\underset{R}{\longrightarrow}\; \underset{\substack{| \\ O^-}}{\overset{\substack{R \\ \backslash \\ O - P - O \\ \| \\ O}}{}} \underset{R_1}{\backslash} \;+\; H_2O \qquad (10.11)$$

4. Chemical recombination of vector-insertion into the host cell; recombining plasmid genes into the host genetic complement
5. Replication of host cell

There are many uses of recombinant DNA. As noted above, one technique that produces recombinant DNA is called *cloning*. In one cloning technique used for the production of the sheep Dolly in 1996, the DNA nucleus from a female's egg is replaced with a nucleus from another sheep. The egg is placed in the uterus of a third animal, known as the surrogate mother. Dolly is nearly genetically identical to the animal from which the nucleus was obtained but not genetically related to the surrogate mother.

Recombinant DNA has been used in a variety of ways. The growth hormone gene of rainbow trout has been transferred into carp eggs resulting in the transgenic carp producing larger fish. The milk production of dairy cows has been increased by cloning and introducing into the cows the cattle growth hormone bovine somatotropin.

Transgenic strawberry and potato plants have been produced that are frost-resistant. Cotton, corn, soybean plants have been produced with increased resistance to herbicides allowing herbicide use without killing the transgenic crop-producing plants. Larger and smaller varieties of other food-producing plants have been produced using recombinant DNA as have plants that produce certain amino acids needed for our nutrition.

Transgenic bacteria have been produced that can metabolize petroleum products, including certain synthetic polymers.

Along with the production of insulin, many other medical uses have been achieved for recombinant DNA. This includes the production of *erythropoetin*, a hormone used to stimulate production of red blood cells in anemic people; tissue *plasminogen activator* an enzyme that dissolves blood clots in heart attack victims; and *antihemophillic human factor VIII*, used to prevent and control bleeding for hemophilia people. These three important genetically engineered proteins were all cloned in hamster cell cultures.

Gene engineering is the basis of gene therapy where genes are removed, replaced, or altered producing new proteins for the treatment of such diseases as muscular dystrophy, some cancers, adenosine deaminase deficiency, cystic fibrosis, and emphysema.

10.9 DNA PROFILING

DNA profiling is also referred to as DNA fingerprinting and DNA typing. It is used in paternity identification, classification of plants, criminal cases, identification of victims, heredity (of living, recently deceased, and anciently deceased), and so forth. DNA profiling is a tool that allows a comparison of DNA samples.

While about 99.9% of our DNA is alike, the 0.1% is what makes us individuals, and it is this 0.1% that allows for the identification of us as individuals. Of interest, it is not the within the gene portions that makeup our different physical and mental characteristics, but the DNA profiling employs DNA taken from what is referred to as the "junk DNA." Identification generally occurs because of the formation of different lengths of this junk DNA after appropriate treatment. This junk DNA contains the same sequence of base pairs, but in different repeat numbers. Thus, the sequence ATTCGG may appear four times, five times, six times, and so on. There are typically some statistical number of repeats. Other sequences such as GGCATCC and AATGCAAT also appear in some statistical number of repeats. While each of us have these different run sequences, individually we have unique run lengths of these different run sequences. These run sequences are called variable number of tandem repeats or VNTRs. The repeat runs used for identification are generally from specific locations within a chromosome. Enzymes "cut" the associated DNA at specific locations leading to decreases in DNA molecular weight. In fact, these DNA chain length decreases are apparent as bandshifts in DNA gels. The combination of the differences in decreased DNA chain lengths becomes unique as results are obtained from different enzymes are accumulated. These changes in the movement of DNA segments are then compared with results from different individuals and identification as to whether the individuals are the same or different. The identity results are often given as some percentage or ratio.

There are two basic types of DNA profiling: one that uses PCR enzymes and the second that employs the restriction fragment length polymorphism (RFLP) enzymes. The PCR approach utilizes a sort of molecular copying process where a specific region is selected for investigation. The PCR approach requires only a few nanograms of DNA. The DNA polymerase makes copies of DNA strands in a process that mimics the way DNA replicates naturally within the cell. Segments are selected for special study and the results used to identify the DNA pattern.

With the exception of identical twins, each individual has a DNA profile that is unique. As previously noted, in excess of 99.9% of the more than 3 billion nucleotides in human DNA are the same. But, for every 1,000 nucleotides there is an average of one site of variation or polymorphism. These DNA polymorphisms change the length of the DNA fragments produced by certain restriction enzymes. The resulting fragments are called *RFLP*. Gel electrophoresis is typically employed to separate the sizes and thus create a pattern of RFLLPs. The number and size of the fragments is used to create the DNA profile.

Several steps are involved in creating the genetic fingerprint. First, a sample of cells is obtained from a person's blood, bone, semen, hair roots, or saliva. The individual cells from the sample are split open and DNA isolated. The DNA is treated with restriction enzymes that cleave the DNA strands at specific locations, creating fragments of varying lengths and composition. The resulting fragments undergo electrophoresis using a gel that allows the separation of the fragmented DNA. Because the gel is fragile, a thin nylon membrane, covered by a towel is laid over the gel. As moisture is drawn to the towel from the electrophoresis gel, the DNA fragments are transferred to the nylon membrane. This process is called *blotting*. The DNA bands are visible to the eye but they are too numerous to be useful. Thus, a radioactive solution is washed over the nylon membrane that binds to select fragments, generally to only 6–20 of the DNA clusters. A sheet of photographic film is placed on top of the nylon membrane that records these cluster sites. The film is then developed producing a pattern of thick-and-thin bands. This pattern is the genetic pattern for that particular sample. This process can take a month or more at commercial labs for routine analysis, but when needed, the analysis can be made in only a day or two.

There are different restriction enzymes that cut DNA at different sites. The previous sequence can be repeated several times for the same DNA sample. From a study of each restriction enzyme, a probability that another person will have the same profile is assigned. Thus, one restriction enzyme may have the possibility that another person has the same match of 1 in 100 or 1%. A second restriction enzyme may have the probability of 1 in 1,000 or 0.1%. A third restriction enzyme may have a probability for a match being 1 in 500 or 0.2%. If there is a match with all three restriction enzymes, the probability would be $0.01 \times 0.001 \times 0.002$ or 0.00000002 or 0.000002% or 1 part in 50,000,000. There is a caution to using the "multiplication rule," in that DNA sequences are not totally random. In fact, DNA sequence agreements generally diverge as one's ancestors are less closely related.

The RFLP method requires a sample about 100 times larger than that required for the PCR approach, but with repeated sequences using different restriction enzymes, RFLP is more precise.

It must be noted that factors leading to DNA degradation, such as moisture, chemicals, bacteria, heat, and sunlight will impact negatively on DNA profiling since the precise sequences and length of the DNA and DNA fragments may be changed. While DNA, in general, is robust and can exist "alive" more than thousands of years (such as the germination of seeds found in the pyramids of Egypt), DNA degradation decreases the probability of precise matches. Also, DNA contamination by addition of DNA from another source greatly confuses the final results.

DNA sequencing has found importance in a wide range of areas. It is being used to identify individuals at greater risk for having certain diseases such as breast cancer. It is used for the screening of certain diseases such as the presence of the sickle-cell gene.

The initial VNTRs were several hundred nucleotide units long requiring long lab periods for the various segments to separate on the gel. Today, most tests employ shorter, 3–5 nucleotides long, VNTRs that allow for more rapid movement on the gel resulting in faster and less costly results. It also allows for the production of a greater number of sequences that are looked at and hence, a greater ability to match/not match the results. These shorter sequences are called short tandem repeats (STRs).

While DNA is more robust than often depicted in movies, age and extreme conditions such as a fire can substantially degrade it. In such cases mitochondrial DNA (mtDNA) is best used. Unlike nuclear DNA, mitochondrial genome exists in thousands of copies and is less apt to degrade and it is inherited only from the mother. Here, STRs are not analyzed, but rather the focus is on variable regions of the mitochondrial genome. Such analyses take much longer but are used for situations where time is not essential.

This type of DNA profiling has allowed taxonomists to determine evolutionary relationships among plants, animals, and other life forms. Currently it is a basis for the so-called Eve theory that says all of us are related to a common women, called Eve after the Biblical Eve. It is also being used to trace the (ancient) movement of people about Earth.

DNA profiling was used to determine whether bones unearthed that were said to be from Jesse James were in fact his. DNA samples were taken from grandchildren and compared to those obtained from the bone material and shown to be similar, so that while it cannot be absolutely said the bones were or were not from Jesse James, DNA evidence was consistent with them being his bones. DNA profiling has also been used in the identification of 9/11 victims, and a number of mass graves throughout the world.

In 1998, the Combined DNA Index System (CODIS) was begun by the FBI. It is an automated forensic data bank and contains DNA profile data related to most of the recent major crimes. It is also connected with state systems as well as similar worldwide data bases.

10.10 HUMAN GENOME—GENERAL

The unraveling of much of the human genome is one of the most important advances made since our dawning. An online tour of the *human genome* is found at a number of web sites allowing access to some of this valuable information.

As noted before, there are two general kinds of cells, those having a membrane-bound nucleus called *eukaryotic cells*, and those without a nuclear envelope called *prokaryotic cells*. Humans have eukaryotic cells. Other than blood cells, eukaryotic cells contain a nucleus that contains the genome, the complete set of genes. Unless noted otherwise our discussion will be restricted to eukaryotic cells.

The human genome and other mammalian cells contain about 600 times as much DNA as *E. coli*. But many plants and amphibians contain an even greater amount. While eukaryotic cells contains more DNA than do bacterial cells (prokaryotic cells), the gene density of bacterial cells is greater. For instance, human DNA contains about 50 genes mm while *E. coli* contains in excess of 2,500 genes mm. As will be noted, the human genome, while it contains a lot of nongene material, this nongene material appears to be active in the organization of the chromosome structure playing a number of roles, including supplying hyperfine contours to assist in the replication, protection, and selectivity of the sites. The contour length, the stretched out helical length, of the human genome material in one cell is about 2 m in comparison with about 1.7 m for *E. coli*. An average human body has about 10^{14} cells giving a total length that is equivalent in length to traveling to and from the earth and sun about 500 times or 1,000 one way trips.

Replication occurs with a remarkably high degree of fidelity such that errors occurs only once per about 1,000 to 10,000 replications or an average single missed base for every 10^9–10^{10} bases added. This highly accurate reproduction occurs because of a number of reasons, including probably some that are as yet unknown. As noted before, the GC group has three hydrogen bonds while the AT has two. In vitro studies have found that DNA polymerases inserts one incorrect base for every 10^4–10^5 correct ones. Thus, other features are in place that assist in this process. Some mistakes are identified and then corrected. One mechanism intrinsic to virtually all DNA polymerases is a separate $3'$–$5'$ exonuclease activity that double-checks each nucleotide after it has been added. This process is very precise. If a wrong base has been added this enzyme prevents addition of the next nucleotide removing the mispaired nucleotide and then allowing the polymerization to continue. This activity is called *proofreading* and it is believed to increase the accuracy another 10^2–10^3 fold. Combining the accuracy factors results in one net error for every 10^6–10^8 base pairs, still short of what is found. Thus, other factors are at work.

In general, replication occurs simultaneously as both strands are unwound. It is bidirectional with both ends of the loop having preferentially active starting points or sites. A new strand is synthesized in the $5'$ to $3'$ direction with the free $3'$ hydroxyl being the point at which the DNA is elongated. Because the two DNA strands are antiparallel, the strand serving as the template is read from its $3'$ end toward its $5'$ end. If DNA replication always occurs in the $5'$ to $3'$ direction then how can it occur simultaneously? The answer is that one of the strands is synthesized in relatively short segments. The leading strand, or the strand that "naturally" is going in the correct direction, replicates somewhat faster than the so-called lagging strand that is synthesized in a discontinuous matter with the required direction occurs at the opposite end of the particular segment, consistent with the observation that all new strands, and here strand segments, are synthesized in a $5'$ to $3'$ manner.

As a scientist, along with knowledge comes questions that in turn point to gathering more information, and so forth. We have already noted that the precise folding and compaction for chromosomal

DNA is not yet fully know (Section 10.10). Further, as noted above we are not fully aware of how replication occurs in such a precise manner.

A third major area of evolving knowledge involves how so much information is packed into the relatively small number of genes we believe we now have. Shortly ago we believed that the number of human genes was on the order of 100,000, a number that appeared appropriate even after the first two animal genomes were deciphered. The roundworm, sequenced in 1998 has 19,098 genes, and in 2000 the fruit fly was found to have 13,601 genes. Currently the number of human genes is believed by many, but not all, to be on the order of 20,000–40,000, less than half of the original number. This means that the genes are probably more complex than originally believed. It is now believed that the reason why we are able to function with so few genes is that our genes carry out a variety of activities. This ability is a consequence of several features. One involves the coordinated interactions between genes, proteins, and groups of proteins with variations of the interactions changing with time and on different levels. There is then a complex network with its own dynamics. It is a network that is probably largely absent in lower species such as the roundworm. The roundworm is a little tubed animal with a body composed of only 959 cells of which 302 are neurons in what passes for a brain. Humans, by comparison have 100 trillion cells including 100 billion brain cells. While protein domains exist in primitive animals such as the roundworm, they are not as "creative" as those found in more advanced animals. These domains that we have allow the "creation" of more complex proteins.

It appears that another way to gain complexity is the division of genes into different segments and by using them in different combinations increasing the possible complexity. These protein coding sequences are known as *exons* and the DNA in between them as introns. The initial transcript of a gene is processed by a spliceosome that strips out the introns and joins the exons together into different groupings governed by other active agents in the overall process. This ability to make different proteins from the same gene is called *alternative splicing*. Alternative splicing is more common with the higher species. Related to this is the ability of our immune system to cut and paste together varying genetic segments that allow the immune system to be effective against unwanted invaders.

In *eukaryotic* cells transcription and translation occur in two distinct temporal and spacial events whereas in *prokaryotic* cells it occurs in one step. Humans have eukaryotic cells so we will look at this process. Transcription occurs on DNA in the nucleus and translation occurs on *ribosomes* in the cytoplasm.

Our genes are split into coding or exon and noncoding or intron regions. The introns are removed from the primary transcript when it is made into a so-called mature or completed RNA—namely mRNA, tRNA, rRNA, and so on.

Such split genes occur in a wide variety of sizes and interruptions. Even so, the transcription must be precise. Several features are worth noting about this process. First, the order of the exons is fixed as is the size and order of the introns. Also, the order of the exons on the mature RNA is the same as in the original DNA. Second, each gene has the same pattern and size of exons and introns in all tissues and cells of the organism and, with the exception of the immune response and the major histocompatibility complex, no cell-specific arrangements exist. Third, many introns have nonsense codons in all three reading frames so nuclear introns are nontranslatable. Introns are found in the genes of mitochondria and chromoplasts as well as in nuclear genes.

We must remember that each of these steps consist of simple, thought complex when considered as a whole, chemical reactions.

Another source of increased complexity involves the fact that human proteins often have sugars and other chemical groups attached to them allowing subtle, and possibly not so subtle, changes in behavior to occur.

Also, it has been found that at least some, about 75%, of the DNA sequences in our genome, is apparently nonactive material. The coding regions may occupy only about 1%–1.5% of the genome. (It must be remembered that while only a small amount contain coding regions, that the structure about these regions is also important and that these structures are also important to the overall activity of the gene.) These so-called active regions are not evenly distributed across the cell's 23 pairs of *chromosomes* but are arranged in patches or regions, some being gene-rich and others gene-poor or deprived. They appear to be sticky, liking to associate with one another. It is similar to the United States where most of the people occupy a small fraction of the land area with large areas having only a low population. There are even preferred base sequences for these different regions. The populated regions tend to be high in C and G sequences whereas nonpopulated areas, regions were there are few active areas, have higher amounts of A and T sequences. These differences in preferential sequencing actually help account for the banding found in chromosome patterns. The light bands are rich in C and G and the dark ones in A and T.

Gene expression simply refers to its transcription resulting subsequently, in most cases, the synthesis of a protein or protein part. The flow of information typically is DNA → RNA → protein → cell structure and function. Transcription is the term used to describe the transfer of information from the DNA to RNA; the flow of information from the RNA to the protein is called *translation*. Genes whose product is needed essentially all the time are present at a constant amount in virtually every cell. Genes for enzymes of the central metabolic pathways are of this type and are often called *housekeeping genes*. Unvarying expression of a gene is called *constitutive gene expression*.

The cellular levels of some gene products vary with time in response to molecular signals. This is called *regulated gene expression*. Gene products that increase in concentration are called *inducible* and the process of increasing their expression is called *induction*. Conversely, gene products that decrease in concentration in response to a molecular signal are said to be repressible and the process called *repression*. Transcription is mediated and regulated by protein–DNA interactions. Thus, while we will focus on the DNA, protein interactions are critical to the operation, expression of genes.

We are beginning to understand some of the language of the genes. We are already aware of the sequences that code for particular amino acids. We are also becoming more aware of the meaning of other sequences. Many of these sequences are involved with transcription regulation. Promoters are DNA sites where the RNA polymerase can bind leading to initiation of transcription. They are generally located nearby the gene. There are a number of these sequences. The CAAT box has a consensus sequence of GGCCAATCT and its presence indicates a strong promoter site. One or more copies of the sequence GGGCGG, called the *GC box*, are often found upstream from transcription start sites of housekeeping genes. The TATA box has a sequence of TATAAAA.

Enhancers are DNA sequences that assist the expression of a given gene and may be located several hundred or thousand base pairs from the gene. They are also called *upstream activation sequences* because they exist somewhat removed from transcription start site. Their location varies between genes. Such sequences are bidirectional occurring the same in both directions.

Response elements are promoter modules in genes responsive to common regulation. Examples include the heat shock element (HSE) with a sequence CNNGAANNTCCNNG (where "N" is unspecified); the glucocorticoid response element (GRE) with a sequence of TGGTACAAATGTTCT, and the metal response element (MRE) with a sequence of CGNCCCGGNCNC. HSEs are located about 15 base pairs upstream from a transcription start site of a variety of genes whose expression dramatically changes in response to elevated temperatures. The response to steroid hormones depends on the presence in certain genes of a GRE positioned about 250 base pairs upstream from the transcription start point.

The complexity of these response elements can be seen in considering the *metallothionein* gene. Metallothionein is a metal binding protein produced by the *metallothionein* gene. It protects against heavy metal toxicity by removing excess amounts from the cell. Its concentration increases in response to the presence of heavy metals such as cadmium or in response to glucocorticoid hormones. The *metallothionein* gene promotion package consists of two general promoter elements, namely a TATA box and GC box; two basal level enhancers; four MREs; and one GRE. These elements function independently of one another with any one able to activate transcription of the gene to produce an increase in the metallothionein protein.

As expected, it is both the composition and the shape, which is driven by the composition, that are important. This shape is maintained through a combination of hydrophilic and hydrophobic interactions, cross-links, preferred bond angles, and inter- and intrachain interactions. It is a complex combination but one where we are beginning to understand some of the basics.

The age distribution for various genome sequences is done by comparing changes in similar sequences found in "older" species. This dating has several important assumptions. First, that the rate of sequence divergence is constant over time and between lineages. Second, that the "standard" older sequence is in fact a source of the sequence and that the date for this source is appropriate. These assumptions are at best appropriate so that results derived from such studies need to be considered in this light.

Our genomes contain a history of its development, including incorporation and infection by viruses and bacteria. Some of these "additions" form part of the so-called dead regions while others may allow desired activities to occur. Thus, it is possible that a foreign, incorporated bacterial sequence allows the encoding of monoamine oxidase that is an important degradative enzyme for the central nervous system. The presence of such apparently foreign information in our genome may mean that there is a dynamic nature to our genome that allows for the inclusion of new information into present genomes, and probably, the converse, the removal of segments of information from our genome.

Even so, the large majority of our genome is not borrowed, but rather developed on its own possibly through what are called *jumping genes* or *transposons* that caused them to be reproduced and inserted into the genomes. The *euchromatic* portion of the human genome has a higher density of transposables than that found for other species. Further, the human genome has more ancient transposables whereas other species have more recent transposables.

Most of these transposons move so that the new location is almost selected at random. Insertion of a transposon into an essential gene could kill the cell so that transposition is somewhat regulated and not frequent. Transposons are one of the simplest molecular parasites. In some cases they carry gene information that is of use to the host.

The most important group of transposons is believed to be the long interspersed element, or LINE groupings (there are three LINE families with only the LINE1 family active) that encodes instructions for whatever it needs, including copying its DNA into RNA, and copying the RNA back again into DNA, and finally moving out of and into the chromosome. LINEs are only about 6,000 base pairs in length (6 kb). Interestingly, most of these LINEs are found in the C and G rich or gene-poor regions of the genome. The LINEs have accompanying them other "parasites" called *Alu* elements that are only about 300 base-pairs long and these sequences are the most abundant sequences in our genome. (They are given the name Alu because their sequence generally includes one copy of the recognition sequence for the restriction endonuclease AluI.) While Alu elements cannot replicate on their own, they "borrow" the needed hardware from the LINE segments to reproduce. Because of their active nature, they can cause trouble. For instance, in the development of an egg or sperm cell, a replicating Alu segment can be inserted resulting in a child with a genetic disease. But Alu segments do perform positive functions. They become activated, helping modulate the body's response when the body is exposed to stresses such as sudden changes in temperature and light and exposure to alcohol. Alu segments are found only in the higher primates and are responsive to a large family of receptor proteins that allow cells to recognize potent hormones like estrogen, retinoic acid, and thyroid hormone. Their presence appears to allow the surrounding site to be more flexible and to slightly change in shape when exposed to these hormone and hormone-like chemical agents.

In humans, the LINE and Alu families account for about 60% of all interspersed repeat sequences, but there are not dominant families in the other species thus far studied. Alu segments compose about 1%–3 % of the total DNA and Alu and similar dispersed repeating sequences comprise about 5%–10 % of human DNA.

In humans, while less than 5% of the genome contains coding sequences, about 50% contain so-called repeat sequences. Such sequences are often included in the category of junk DNA yet they provide lots of information and some provide function. Such segments act, as noted above, as a kind of palaeontological record of past interactions with various bacteria and virus; they can act as passive markers for studying mutation and selection; and they can be active providing shape and function allowing the same sequence to behave in a different fashion because of the presence of these junk sequences.

Repeats can be divided into five classes. First, *transposon*-derived repeats, some of which have been briefly dealt with before. About 45% of our genome is derived from transposable elements. It is possible that some of the other "unique" DNA may also be derived from ancient transposable element copies that we have not yet recognized. Second, partially or inactive retroposed copies of cellular genes called *processed pseudogenes*. Third, short simple repeating sequences such as AAAAAA, CACACACACA, and so forth. Fourth, short segmental duplications that have been copied from one region of the genome into another region. These sequences are typically 10,000–300,000 base-pairs long (10–300 kb). Fifth, blocks of tandemly repeated sequences.

There are four types of so-called transposon-derived repeating sequences of which three transpose through RNA intermediates and one transposes directly as DNA (last one considered below). We have already identified the long interspersed elements, LINES. The second set are called *SINEs,* of which the Alu's are the only active members that exist in the human genome.

The third group is the LTR retrotransposons that are flanked by long-terminal direct repeats that contain all the transcriptional regulatory elements. LTR genes can encode a protease, reverse transcriptase, RNAse H, and integrase. Transposition occurs through a retroviral mechanism with reverse transcription occurring in a cytoplasmic virus-like particle. While a wide variety of LTRs exist it is believed that only the endogenous retroviruses (ERVs) are active in humans.

The last group of transposable elements is the DNA transposons that resemble bacterial transposons. They tend to have short life spans within a species. Humans have at least seven families of DNA transposons. Their replication is lessened by the presence of inactive copies so as the number of inactive copies accumulate, transposition becomes less efficient.

As part of the overall human genome, the LINEs, SINEs, LTR, and DNA retroposons make up 20%, 13%, 8%, and 3% (total of 44%) of the repeat sequences.

Such repeats are often included as "junk." Again, the so-called junk in our genome may not be junk but rather part of a complex of shape and electrical nature that forms the basis for the chemistries of the various polymeric molecules.

This massive amount of information should not be considered as insurmountable or only material to be marveled at but not understood. Much of the chemistry is already available to "mine" this information successfully. Much of it is understandable in somewhat simple terms, generally only after we have discovered the key to this simplicity. For instance, there is a marked decrease in the frequency of the dinucleotide CpG in some areas of the genome. The deficiency is believed to be due to the fact that most CpG nucleotides are methylated on the cytosine base and spontaneous deamination of the methyl-cytosine residue creates T residues. Thus, CpG dinucleotide sequences mutate to TpG dinucleotides. But there still remain some questions. There are certain regions or islands where the CpG sequences exist in a nonmethylated form and where the frequency of CpG occurs within the expected or normal rate. Why? These CpG islands are of particular interest because they are associated with the 5′ ends of genes.

Another broad finding in examining the human genome regards the rate of *recombination*. Recombination involves the cleavage and rejoining, insertion, of sequences of nucleic acids by enzymes. In fact, recombinant DNA is the result of such recombination. In general, the average recombination rate increases as the length of the chromosome arm decreases. Long chromosome arms have a recombination rate that is about half that of shorter arms. Second, the recombination rate is less near the *centromere* and greater in the more distance portions of the chromosomes. This effect is most pronounced for males. The centromere is an essential site for the equal and orderly distribution of chromosomal units during cell formation, meiosis.

Why these differences? A higher rate of recombination increases the likelihood of at least one crossover during meiosis of each chromosome arm. Such crossovers are necessary for normal meiotic disjunction of homologous chromosome pairs in eukaryote cells. Recombination occurs with the greatest frequency during meiosis, the process where diploid cells with two sets of chromosomes divide producing haploid gametes—sperm cells or ova—with each gamete having only one member of each chromosome pair.

The "crossing over" is not entirely random. Even so, in general, the frequency of homologous recombination in any region separating two points on a chromosome is proportional to the distance between the points. A homologous genetic recombination is simply the recombination between two DNAs of similar (not necessarily the same) sequence. Homologous recombination serves several functions. First, it contributes to the repair of certain types of DNA damage. Second, it provides a transient physical link between chromatids that encourages orderly segregation of chromosomes during the first meiotic cell division. Third, it enhances genetic diversity.

Since such crossover sequences are important, it is possible that they are present to an extent in each arm to insure that crossover occurs but the full answer is not currently known and since shorter arms are shorter, the density or frequency of them is greater.

We need to remember that the present knowledge of the human genome is a rough map without complete knowledge of the stop lights, detours, alternative routes, pot holes, and so on to use the metamorphic language relating a paper map to the actual physical terrain. Scaffolds are being built to fill these knowledge gaps with time allowing the scaffolding to become part of a solid building.

10.11 CHROMOSOMES

The preliminary investigation of chromosomes has resulted in several reoccurring themes being evident. One theme is that nature magnifies small differences, often the difference in only a single base pair can lead to marked differences in our overall predicted health, and so on. The second theme involves the interrelativeness of the genes with one another and with various proteins that are created by them. A third theme will not be dealt with to much extent. That theme concerns the fact that even though we talk about a common human genome, there are within this human genome sufficient differences to make each of us individuals with our own aspirations and dreams, tendencies toward particular foods and diseases, and so on. Even so, most of the human genome is the same with the small variances, including our outward environment, resulting in a divergence population of human beings.

We often think of the chromosomes as being flat with little or no geographical topology because the sheet of paper or screen we view them on is flat. They are not flat and it is that three-dimensional structure that assists the various genes to perform their function in designing needed proteins. The secondary structure of these features is more or less helical with the varying clefts shown in Figure 10.16 causing the DNA to have these varying structures. The transfer of information from the DNA template to protein, and less so RNA, synthesis is described in Section 10.4.

Our bodies have about 100 trillion cells. Inside each cell is the nucleus and, with the exception of egg and sperm cells, inside the nucleus are two copies of the human genome made from DNA with protein building genes contained within chromosomes that compose the human genome. There are a few exceptions to this, including the following. Not all genes are DNA, but some contain RNA. Not all genes code for proteins. Some are transcribed into RNA that becomes part of a ribosome or transfer RNA. While most gene-associated reactions are catalyzed by proteins, a few are catalyzed by RNA. Again, while many proteins are designed by a single gene, some are designed by several genes.

The human genome comes in 23 packages with each package being a chromosome. This number 23 is important. If we have more or less than 23 then we may well be in great trouble. Only

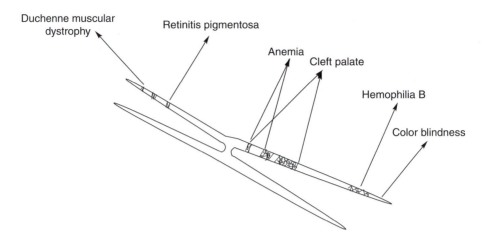

FIGURE 10.16   Selected mutations found in the X chromosome.

chromosome 21 can be present with more than one copy with the occupant having a healthy body, but unfortunately those with an extra chromosome 21 are not normal but rather have Down syndrome. The chromosomes are named in general order of size starting from the largest as chromosome 1 through 22 though it has recently been found that chromosome 21 is smaller than chromosome 22. We also have as the remaining chromosome the twined sex chromosomes with women having two large X chromosomes and men having one large X and a small Y. The X chromosome, in size, falls between chromosomes 7 and 8, while the Y chromosome is indeed the smallest. Note that even through the XY chromosome combinations are together one chromosome, researchers often refer to each part in terms of a chromosome though in truth they are only a part of the total twined sex chromosome. The variation from the smallest to the largest of the chromosomes is about 25 fold.

While 23 chromosomes is a relatively large number, it is by no means the largest number of chromosomes within a species. Normal chromosome numbers for selected organisms appears in Table 10.4.

All eukaryotic cells in our bodies contain the same 23 chromosomes with the same DNA base sequences. The lone differences are the mitochondria. The mitochondria in typical somatic cells contains less than 0.1% of the cell's DNA but in fertilized and dividing egg cells this number is greater. mtDNA is much smaller, often containing less than 20,000 base pairs. The value for humans is 16,569 base pairs. The mtDNA is a circular duplex. mDNA codes for the mitochondrial tRNAs and rRNAs but only a fraction of the mitochondrial proteins. More than 95% of the mitochondrial proteins are encoded by nuclear DNA. The mitochondria divides when the cell divides.

The association of a particular disease with a particular gene or group of genes is rapidly increasing. A spot check of www.ncbi.nlm.nih.gov/omim, the online version of Mendelian Inheritance in Man, OMIN, gives an ongoing updated progress report of this activity. Currently, about 1,500 disease-mutations have been entered. With the advent of the mapping comes a number of shifts in thinking and activity. Thus, we will move from so-called map-based gene discovery to looking at the particular activity of gene sequences; from association of particular gene-associated diseases to looking at tendency and susceptibility for given conditions and the variation with tendency/susceptibility between individuals; from looking at the activity of a single gene or gene location to investigating combined activities of several genes from varying locations; from so-called genomics or studying of genes themselves to proteomics and studying of the interaction between the genes and proteins; from gene action to gene regulation; and from specific mutations to the mechanisms and causes of such mutations. Much of this is a guessing game—hopefully an educated and educating guessing game and it is currently very costly. As new techniques and strategies are developed the cost should decrease.

TABLE 10.4
Number of Chromosome of Some Common Species

Organism	Chromosome Number*	Organism	Chromosome Number*
Bacteria	1	Mouse	20
Fruit fly	4	Rat	21
Pea	7	Rabbit	22
Frog	13	Human	23
Fox	17	Ape	24
Cat	19	Chicken	39

*The diploid chromosome number is double this number except for bacteria.

As noted above one of the important themes is that small, or seemingly small, changes in the genetic code can have profound effects. Many of the better known diseases contain the names of the discoveries or unfortunate victim(s). The Wolf–Hirschhorn disease and Huntington's chorea are such diseases. They are wholly genetically based or in their case, the result of a missing (Wolf–Hirschhorn) or mutated (Huntington's) gene. It is believed that the genetic origin of the disease is contained in chromosome 4. The coding that is responsible for Huntington's disease is the sequence CAG. This replication occurs at varying times. If this sequence occurs 35 times or less, then you will not develop the disease. Counter, if you have this sequence 39 time or more then we will develop this disease with the onset noticeably beginning in midlife with a slight lose in balance. There begins a decline in metal capacity, and an onset in the jerking of our limbs. This continues until death. It generally takes 15–30 years to run its course, but there is currently no cure. In general, the greater number of repeat sequences, the earlier the onset of the disease. If you have 40 repeat units, you will exhibit the disease at about 60; if 42 repeat units, you will exhibit the disease by 40; and at 50 you will be well along in the progress of the disease, and most probably dead, by 30. Thus, the frequency of this sequence in this particular chromosome determines one's outcome with respect to this disease. While both diseases are rare, it is the Huntington's disease that killed noted folk balladeer Woody Guthrie in 1967.

In the Huntington's disease, long repeat sequences of CAG appear. The CAG codon codes for glutamine so that areas that emphasize the buildup of this amino acid appear to offer a greater incidence of certain neurological diseases possibly whereever they are found. This may be due to a buildup of glutamine over a time with the glutamine-rich proteins being "sticky" and more apt to remain at the particular site rather than "moving on" to do what it was supposed to do. After some time this build up becomes great enough to block the healthy activity of the gene causing problems and possibly leading to death or noticeable loss of function of the cell. This buildup and death of a significant number of cells takes time so that is probably why these diseases take time to make themselves known. As an aside, it is interesting that some health potions emphasize the presence of glutamine but it must be remembered that while the presence of some glutamine is essential to healthy lives, it is not the presence of glutamine that may cause these diseases, but rather the coding on the DNA.

Other repeat sequences appear to also offer problems. Many sequences begin in C and end in G. Thus, we have large numbers of repeats of CCG and CGG (that code for proline and arginine) that are believed to give a disposition to certain nerve degeneration related diseases.

Another phenomenon is worth mentioning here. There appears to be a tendency for such repeats to become longer through each replication cycle. This is called *anticipation* and is believed to be related to a very slight tendency for the replication cycle to "lose count" of the number of repeats as the number of exact repeats becomes large, say 30 or more. Again, this is a reason why some diseases take time to manifest themselves since it takes time for the number of replications to increase to a "dangerous" length.

The Huntington scenario paints a sad picture for our ability to "cure" those with the disease. Are we to modify each of chromosome 4s in the billions of cells in our brain and if so how. It is not the sequence itself, but rather the length of the sequence that is the problem. All of us have some of these repeat sequences and they are necessary for other essential activities.

While certain behavioral and nonbehavioral diseases are believed to be monogenic, diseases such as the Huntington, cyctic fibrosis, Marfan, and Hirschsprung result in the specified disease, the outward appearance or result (phenotype) of the disease varies between individuals. For instance, for the Marfan syndrome, there is a level below which the mutant protein does not exhibit itself in an outward manner. Most of these diseases have modifier genes that cause modifications in the outward demonstration of the disease and play a key role in the clinical symptoms. Further, the particular metabolic pathways are often varied with several of the steps being important and the importance of each mechanistic pathway may vary with individual.

We are learning external ways to identify activities, actions, that may be related to our genome makeup. One of these observations involves changes in the capacity of individuals to "learn" with age. It appears that the ability to learn language, grammar precisely, decreases as we grow older and is most apparent in children. Thus, ability to learn a language appears to be gene related. There are genetic conditions that are related to our linguistic ability. One is the Williams syndrome where affected children have very low general intelligence, but have a vivid and loquacious ability to use language chattering on in long and elaborate sentences. Thus, they have a heightened ability to learn language. The Williams syndrome is caused by a change in a gene found on chromosome 11.

Another genetic-related disease is known as specific language impairment (SLI) where individuals with general intelligence have lowered linguistic ability. SLI is believed to be related to a gene found on chromosome 7.

Genes are related with all aspects of our lives. Someone has said we are what we eat. We can extend this to say that we become what our genes do with what we eat. There is a group of genes called *apolipoprotein* or *APO* genes. There are four basic types of APO genes interestingly know as A, B, C, and E (no explanation for what happened to D; known as APOA, APOB, etc.). Here we will focus on a specific gene that appears on chromosome 19 know as APOE. As we eat, the various food parts are digested, broken down. Both fats and cholesterol are brought through our blood stream by lipoproteins, some called *very low-density lipoproteins* (VLDLs). These fats and cholesterol are brought to various parts of the body to act as fuel and building blocks. As some of the triglycerides are delivered, the proteins now are called simply *low-density lipoprotein*, or LDL known to many of us as "bad cholesterol." After delivering the cholesterol, it becomes high-density lipoproteins (HDL) also know to us as good cholesterol and then returns to the liver to be replenished with cholesterol and fats. The APOE protein acts to affect the transfer between VLDL proteins and a receptor on a cell that needs some triglycerides. APOB serves a similar role except in delivering cholesterol. Thus, the presence and effectiveness of genes that code for the APO genes helps control our weight and health affecting such items as buildup on our arteries.

APOE is unusual in that it is polymorphic having several versions. The three most common are know as E2, E3, and E4. E3, deemed the best variety of APOE, is the most common in Europeans with about 80% having at least one copy and about 40% with two copies. But about 7% have two copies of the E4 gene, the worst variety, and they are at high risk of early heart disease. These trends of APOE are geographical and correlate with the frequency of heart disease. Thus, the frequency of E4 is about three times as high in Sweden and Finland as in Italy and the frequency of coronary heart disease is also about three times as high in Sweden and Finland in comparison to Italy. On a race basis, Orientals have the lowest frequency of E4 (ca 15%), with American blacks, Africans, and Polynesians all having higher values of E4, about 40%. Diet also contributes so that while New Guineans have a high frequency of E4, their diet is low in fats and they have a low incidence of heart disease. But when they change diet so that it is similar to many of ours, they become more susceptible to heart disease.

Our blood type is determined by a gene that is present on chromosome 9, near the end of the long arm. There are four general blood types, A, AB, B, and O. Some of these are "inter-mixable" while others are not. For instance, A blood from a person is compatible with A and AB; B with B and AB; and AB with only AB; and O blood is compatible with all of the blood types—a person with type O is then an universal donor. These compatibility scenarios are not race related. For all but the native Americans that have almost totally type O the rest of us have about 40% type O; another 40% type A; 15% type B; and 5% with type AB. (Some of the Eskimos are type AB or B and some Canadian tribe are type A.) A and B are codominant versions of the same gene and O is the "recessive" form of this gene.

The active codons of the blood-type gene are about one thousand base-pairs long and are divided into six short and one longer sequence of exons. The difference between the type A and type B gene is seven letters of which three do not make any difference in the amino acid coded for. The four truly different bases are positioned at sites 523, 700, 793, and 800 and are C, G, C, G for type A and G, A, A, C for type B blood. Type O people have just a single change from the type A people with a deletion in the type A base at base pair 258, omitting the G base. While this appears to be very minor, it is significant in that it causes a reading or frame-shift mutation. These seemingly minor changes are sufficient to cause the body to have an immune response to different types of blood. Even so, while this causes a different type of blood, it appears to have little or nothing to do with other parts of the overall human genome so that tendencies toward cancer, ageing, and so forth are not influenced by this change but it appears to have something to do with some general tendencies towards some diseases. For instance, those with AB blood are the most resistant toward cholera while those with type O blood are most susceptible. Those with two copies for the sickle-cell mutation generally contract sickle-cell anemia while those with one copy of the mutation are more susceptible to contracting sickle-cell anemia than the general public but they are more resistant to malaria. Some of these connections can be found in tracing the ancestry of individuals with the particular connections between blood type and susceptibility.

Recently, we have been working on Parkinson's disease. Parkinson's disease, and other similar diseases, are due to a depletion of dopamine in the corpus striatum. Direct addition of dopamine is not effective in the treatment presumably because it does not cross the blood–brain barrier. However, levodopa, the metabolic precursor of dopamine, does cross the blood–brain barrier and is believed to then be converted to dopamine in the basal ganglia.

On the short arm of chromosome 11 is a gene known as *D4DR* that manufactures the protein dopamine receptor. It is active in some parts of the brain and inactive in other parts. Dopamine is a neurotransmitter released from the end of neutrons by an electrical signal. When the dopamine receptor is exposed to dopamine, it also releases an electrical signal. In general, much of the brain activity is related to such stimulation of chemical reactions by electrical changes, and counter, electrical current calling for chemical reactions. Many of these exchanges occur essentially at the same time. Digressing for a moment, our brain is often compared to the operation of a computer. In some sense it is but in other senses it is not. In our brain "switches" are activated and closed, opened and shut, by not simply electrical charge, but rather by an electrical switch associated to a very sensitive, selective chemical site.

Brain sites that have an active D4DR are then part of the brain's dopamine-mediated system. A lower amount of dopamine causes this part of the brain's system to either shutdown or be less than fully active and in extreme cases resulting in Parkinson's and related diseases. Excess of dopamine may led to schizophrenia. Some of the hallucinogenic drugs act to increase the amount of dopamine. Thus, there is a tight balance between good health and health problems.

D4DR has a variable repeat sequence in it about 48 base pairs in length. Most of us have between 4 and 7 such sequences. The larger the number of repeat units the more ineffective is the dopamine receptor at capturing dopamine while a low number of such repeat sequences means the D4DR gene is highly responsive to dopamine. In some preliminary personality studies focusing on the number of these repeat sequence, it was found, in general, that those with only a few, like one or two, sequences appeared to be more adventuresome than those with a larger number of repeat sequences. Again, here we are looking at tendencies that are greatly shaped by our individual circumstances. Behavior tendencies are also implicated by other monoamines such as norepinephrine and serotonin.

Such wholly caused gene diseases are at one end of the spectrum. More likely, gene composition declares tendencies, some of these are somewhat random tendencies and others are related to our lifestyles—both voluntary and involuntary. Thus, single genes that dictate aggression, being good natured, intelligence, criminals, and so on are not present. Most of our tendencies are just that tendencies, and such tendencies are complex and involve the interaction of many genes and the associated proteins as well as external forces and opportunities. Genes in such multiple gene systems are called *quantitative trait loci* (QTIs) because they are apt to produce similar behaviors within different people.

Behavior-related illnesses and patterns, both so-called healthy and nonhealthy, are complex and involve many factors, including external factors, both learned and simply exposed to factors. As these studies continue two features are emerging. First, is a tendency (not certainty) for this trait to be inherited. This tendency is generally greater than is the tendency with respect to physical disease. Second, environment plays a role. Similar environments produce similar people and different environments for related people produce different people.

As noted before, there are probably few single gene-associated, monogenic, diseases and most involve a number of genes. This latter group of diseases is called *complex* or *multifactorial diseases*. As noted before, these complex diseases are called *quantitative trait locus* (OTL) disorders or diseases.

Similar animal studies are useful in studying such diseases where a so-called similar animal is available. The term similar animal simply means an animal that contracts the same disease because of a similarity in the disease-causing gene complex. Because of the similarity found between the genes between varying species, such similar animals should be available for many of the diseases.

Attention deficit hyperactivity disorder is believed to be related to genes associated with the dopamine system, namely DAT1, DRD4, DRD5. And schizophrenia has been reported linked to genes on chromosomes 1, 5, 6, 10, 13, 15, and 22. As the human genome map is better understood such combinations will become more evident. Finding such QTLs is the initial step. Next comes identifying the particular interactions between the QTLs, and between the various proteins produced by them and finally what, if anything, can or should be done to correct or modify the situation.

The Huntington-related problems, while deadly, are visually simple in relation to some other gene-related problems. Asthma is a disease that has multiple causes and symptoms and appears to be the consequence of groups of genes acting in multiple ways, some of which may be positive and others that cause asthma. Asthma, allergy, anaphylaxis, and eczema are all caused by mast cells altered and triggered by immunoglobulin-E molecules. I am allergic to certain foods and used to be to certain plants like rag weed. I outgrew much of the rag weed-like allergies but retain the food allergies. This is typical; allergies can come and go, are of varying severities, and can vary with age, sex, and race. While there is evidence to tie asthma to genes, the precise group of genes remains unknown and surely will be more complex than that of the Huntington-related diseases.

A brief review of the meiosis process is in order. In the first step, the chromosomes of a cell containing six chromosomes, three homologous pairs, are replicated and held together at their centromeres. Each replicated double-stranded DNA is called a *chromatid* or "sister chromatid." In the next step, the three homologous sets of chromatids align, forming tetrads that are held together by covalent bonding at homologous junctions called *chiasmata*. Crossovers, recombinations, occur such that the two tethered chromosomes segregate properly to opposite poles in the next step. This is followed by the homologous pairs separating and migrating toward opposite poles of the dividing cells. This first meiotic division gives two daughter cells, each with three pairs of chromatids. The homologous pairs again line up across the center or equator of the cell in preparation for separation of the chromatids, chromosomes. The second meiotic division produces four haploid daughter cells that can act as gametes. Each cell has three chromosomes, half the number of the diploid cell. The chromosomes have resorted and recombined.

We have just considered mitosis in general. We can take a simplistic look at the determination of whether a given embryo is a male or female. Females have two X chromosomes while males have one X and one Y so that the ability of the X chromosome to overwhelm the Y chromosomes and give only female embryos is favored on a statistical basis. On a size basis, the Y chromosome is the smallest of all the chromosomes while the X is among the largest. Further, the Y chromosome is largely composed of noncoding DNA giving few targets for the X chromosome to interact with. The gene on the Y chromosome that makes men men is called the *SRY* gene. The *SRY* gene interacts with the *DAX* gene on the X chromosome. In some sense, these two genes are antagonistic to one another where two *DAX* genes overcome the single *SRY* gene but one *SRY* gene overcomes one *DAX* gene so that depending upon the particular course of evens the outcome is a male or female. The *SRY* gene, when activated, ignites a whole cascade of events that leads to the maculation of the embryo. The *SKY* gene is peculiar in that it is remarkably consistent between men with essentially no variations in the coding regardless of race. Further, the human *SRY* gene is very different from those of other primates.

For many species this XY battle greatly favors one sex, generally the female, over the other. For instance in the butterfly *Acrea encedon*, 97% of the butterflies are female. But in humans the competition is such that the ratio of males to females is about, not exactly, one to one.

We will look at another example where the "equal" splitting of chromosome information is not entirely true. Some families have members that exhibit two related diseases. Those with the Prader–Willi syndrome are characterized with small hands and feet, underdeveloped sex organs, they are generally obese, and are also often mildly mentally retarded. Those with the Angelman's syndrome are taut, thin, insomnic, small-headed, move jerkily, have a happy disposition, always smiling, and are generally unable to speak and are mentally retarded. In both cases a section of chromosome 15 is missing. In the Prader–Willi syndrome the missing part is from the father's chromosome but in the Angelman's syndrome the missing part is from the mother's chromosome. Thus, the two diseases differ depending whether it is transmitted through the male or female.

Related to this are recent attempts to produce unisex children—that is embryos that are from male–male and those from female–female as sources of the chromosomal material. In both cases the attempt failed. The "two-mother-derived" chromosomes could not make a placenta and the "two-father-derived" chromosomes could not make a discernible head so that chromosomes from both sexes are required to give a healthy successful embryo.

We now move to what makes a single egg/sperm combination grow into a child. It is a combination of special events of which we will look at only one aspect. The machinery to construct a person is found about our chromosomes. One cluster of these developmental genes is found in the middle of chromosome 12. Within these genes is a grouping of homeotic genes that reside in the same general area. These genes are called the *Hox* genes and affect the parts of the body in the exact sequence that they appear in the fruit fly—mouth, face, top of head, neck, thorax, front half of abdomen, rear half of abdomen, and finally the other parts of the abdomen. Also, found in each of these *homeotic* genes is the same sequence of about 180 base-pairs long that is believed to act as a switch to turn on or off each gene referred to as the homeobox. Mice were examined and also found to have such *homeotic* genes and homeoboxes. Mice have 39 *Hox* genes in four clusters with some differences, but many similarities, with the fruit fly. We have the same Hox clusters as mice with one such cluster, Cluster C, on chromosome 12. A practical implication is that all the work done with other species, such as the fruit fly, may be useful as we look at our own genome with at least such developmental genes.

The similarity of the embryo genes between us and other species allows developmental scientists a filled table of data to sort over. It does not eliminate nor confirm a so-called master designer since one can argue about the simplicity and interrelatedness of the design as to origin and the ability through ordinary evolution to design such complex items as our human eye.

The imprinted region of chromosome 15 contains about 8 genes one of which is responsible, when broken, for Angelman syndrome—a gene called UBE$_3$A. Beside this gene are the two top candidates for the Prader–Willi syndrome when broken, one called the *SNRPN* gene and the other called *IPW*.

While these diseases normally occur because of mutation of these genes, it may also occur from a pair of parental chromosomes failing to separate, with the egg ending up with two copies of the parental chromosome. After fertilization with a sperm, the embryo has three copies of that chromosome, two from the mother and one from the father. While the embryo generally dies, in some cases it persists. If it persists and it is chromosome 21, then the result is a Down syndrome child. But normally the body detects the mistake and "kills" one of them. It does so nearly randomly so that there is about a one-third chance of eliminating the paternal-derived chromosome. Generally there is no problem but if the tripled chromosome is number 15 then there are two copies of UBE$_3$A, the maternally imprinted gene, and no copies of *SNRPN* (or *IPW*) exist resulting in a Prader–Willi syndrome child. Recently it has been found that UBE$_3$A is switched on in the brain. This leads us to another unfolding saga, that of imprinted genes being controlled "directly" by the brain. In mice, it is believed that much of the hypothalamus, found at the base of the brain, is built by imprinted genes derived from the father, while much of the forebrain is built by imprinted genes derived from the mother.

What is intelligence? There are different "kinds" of intelligence. The so-called intelligence quotient (IQ). Intelligence appears to be related to the ability to gain, understand, remember, and relate information and it is believed that about one half is in some ways related to our genes. So, unlike the Huntington diseases, while there is a genetic link it is not an absolute link. Several chromosomes and genes within these chromosomes are beginning to be identified as "smart" genes. One of these genes, called *IGF2R*, is found on the long arm of chromosome 6. This gene was first linked to liver cancer so it shows the variety of capability that may be present within a single gene. IGF2R is a large gene with the typical exons and introns. Among its varied activities is its involvement in the metabolism of sugar. One form of this gene is found in greater frequency in supposedly super-smart people than in so-called average intelligence people. Thus, there appears to be a relationship between the occurrence of one form of this gene and intelligence. This is circumstantial evidence at best but it is a start. An interesting side light is the observation that people with high IQs are, on the average, more effective at metabolizing sugar and that this gene is sometimes connected with insulin related proteins and the ability to metabolize sugar.

We again must not believe that IQ-related intelligence irrevocably binds us or propels us. Harvard did a study some time ago where they tried to relate intelligence to success (what ever success is) and came up with a relationship that success was related to intelligence times the *square* of effort.

Learning is related to intelligence. In general, the more apt we are to learn, retain, integrate, and use information the more intelligent we are. Since our brains are really limited in storage of information and for other reasons, intelligence requires an interesting mix of remembering and forgetting, a short range memory and a long range memory. Our input is also mixed. We have a kind of filtering system that filters out supposedly unwanted input, such as a constant ticking of a clock or constant hum of the fluorescent light.

One of the active agents is cyclic AMP. A protein called CREB (for CRE binding protein; CRE is simply a specific DNA unit, part of a gene, that is called the *cyclic AMP response element* or site) is activated altering the shape and functioning of the synapse in our brains when exposed to cyclic AMP or some related compound in our brains. Genes that are activated are called CRE genes with the name the initials of cyclic-AMP response elements. CREB, when phosphorylated, binds to the CREs near certain genes acting as a transcription factor and turning on or activating the genes. Animals without the CREB producing gene are able to learn but do not possess long-term memory. It is believed by some that the CREB related genes are in fact essential to our learning and memory and act as master switches in activating other genes necessary in our learning/memory process. The CREB gene is on chromosome 2. A related and essential gene that helps CREB perform is found on chromosome 16 and is given the name CREBBP.

Another essential "learning gene," related to alpha-integrin (integrins are proteins with two unlike units called alpha and beta that are anchored to the plasma membrane; they act as molecular adhesives and also as receptors and signal transducers), is also found on chromosome 16. This gene, called the *volado* (means in Chilean "forgetful") gene, appears to be a player in memory and it is not involved in the cyclic AMP sequence. The *volado* gene codes for one of the subunits of alpha-integrin. The *volado* gene appears to act to tighten connections between neurons as we learn.

We are aware of certain drugs that interfere with the activity of integrins by interfering with a process called long-term potentiation (LTP), which is an essential part of creating a memory. In the base of our brain is a part called the hippocampus (Greek for seahorse). A part of the hippocampus is called the *Ammon's horn* (named for the Egyptian god associated with the ram). The Ammon's horn has a large number of pyramidal-shaped neurons that assemble the inputs of secondary neurons. Single inputs appear not to "fire" these neurons but when two or more inputs arrive at the same time the neuron fires. Once fired, it is easier to again fire when one of the two original inputs arrive. Thus, in a real sense memories are made in the hippocampus. In relation to the human genome, the brain is much more complex, and it is controlled by a matrix of activities, including our genome. It operates on a three-dimensional network rather than a one or two dimension somewhat flat chromosome face. So artificial intelligence has a long way to go to mimic our brain.

It is then a group of genes found on several chromosomes that give us the ability to learn and retain information and we are just beginning to discover the genome elements responsible for this wonderful ability.

(10.12)

Cyclic AMP

Only vertebrates show an immune response. If foreign objects, called *antigens*, gain entry into our bloodstream, a molecular level protection system, called the *immune response*, goes into action. This response involves production of proteins capable of recognizing and destroying the antigen. It is normally mounted by certain white blood cells called the "B" and "T" cell lymphocytes and macrophages. B cells are called that because they mature in the bone marrow, and T cells mature in the thymus gland. Antibodies which "recognize" and bind antigens are immunoglobulin proteins secreted from B cells. Because the antigens can be quite varied from the protein inserted from an insect bite, to pollen, and so on the number of proteins that can "recognize" and bind this variety of invaders must be quite large. Nature's answer to creating such a great host of antibodies is found in the organization of the immunoglobulin genes that are scattered among multiple gene segments in germline cells (sperm and eggs). During our development and the formation of B lymphocytes, these segments are brought together and assembled by DNA rearrangement (genetic recombination) into complete genes. DNA rearrangement, or gene reorganization, provides a mechanism for creating a variety of protein isoforms from a limited number of genes. DNA rearrangement occurs in only a few genes, those encoding the antigen-binding proteins of the immune response—the T cell receptors and the immunoglobulins. The gene segments encoding the amino-terminated part of the immunoglobulin proteins are also quite susceptible to mutation. The result is a collective population of B cells within most of us with the ability of producing the needed vast number of antibodies. Thus, gene variety is produced near to the event of our conception.

As noted before, a single gene may play several roles, or at least the proteins derived from them may. Several genes have been associated with the early onset of Alzheimer's disease, two found on chromosome 14, one on chromosome 21, and interestingly one on chromosome 19, none other than APOE. It is not unexpected (many results are not unexpected after the fact) that a blood-lipid related gene is associated with a brain disease. It has been found for some time that those with Alzheimer's disease had high cholesterol levels. Again, the bad actor is the E4 variety. For those families that are prone to Alzheimer's disease, those with no E4 gene have about a 20% change of contracting the disease; those with one E4 almost 50% with a mean age of 75 for onset; and those with two E4 genes, the probability is more than 90% with the mean age of onset about 68 years of age. Other genes also affect the incidence of Alzheimer's disease. For instance the incidence of contracting the disease is much higher for whites with the same E4 amounts in comparison to blacks and Hispanics.

The difference between E4 and E3 is a signal base pair, the 334th base pair with the E4 having a G instead of an A.

The body is a marvelous "machine," growing and learning, and performing its own maintenance. Much of this maintenance is a sort of self preservation or self protection to maintain its own original molecular design. Involved with much of this are DNA repair enzymes that continuously monitor the genome to correct damaged nucleotides and nucleotide sequences that are damaged through self-inflected mutations or through environmentally related damage such as exposure to various chemical agents and radiation. There are currently about 150 known human DNA repair genes where the function (or at least one function) is know along with its location within the genome. The particular sequence is also known for most of these. The sequences for many of these have been know for several years but the specific location and proximity to other genes has only become known with the recent human genome project. These repair genes perform a number of functions. MSH2 and MSH3 found on chromosomes 2 and 5 are involved with mismatch and loop recognition repair; a group of genes known as fanconi anemia (FAN) genes, are found on chromosomes 3, 6, 9, 11, 16, and so on and are involved with repair of DNA cross-links; and so forth. The overall shape of these DNA repair proteins is rapidly being uncovered and active sites being identified.

Now let us look briefly at an aspect of aging. We are given, or so we are told in literature, our four score or 80 years. We are not able to describe why eighty, but can comment on why this number is not much larger. The human genome is much longer lived and its copying has occurred many times. Yet the cells in our body have only replicated a few times in comparison. Even the more active cells have replicated only several hundred times. Part of the answer resides on chromosome 14 in a gene called *TEPI*. This gene forms a protein that is part of the telomerase system. Lack of telomerase causes senescence. Addition of telomerase allows some cells a much longer lifetime. *Telomers*, produced by telomerase, occur at the end of chromosomes. These telomers are included in the so-called junk with a seemingly uncoded sequence, TTAGGG. This sequence is repeated many times. This sequence is the same for all mammals and is the same for most living species. Typically, each time the chromosome is reproduced, the number of "telomer sequences" decreases, at the average rate of about 30 base pairs a year, and may be partially responsible for the various cells "wearing out." By the time we reach our four score years, we have lost about 40% of the telomer sequences.

The telomerase contains RNA, which is used as the template for making telomeres, and a protein part that resembles reverse transcriptase, the enzyme responsible for the production of transposons and retroviruses. Telomerase acts to repair the ends of chromosomes relengthening the telomere ends. Thus, the lack of telomerase appears to cause the ageing and eventual death of at least some of our cells. The relation to aging is much less certain, and surely more complicated. Thus, those with Werner's syndrome, where rapid aging occurs, start out with the same average length of telomers, but the telomers shorten more rapidly so that at least cell aging involves not only the length of telomers, but also the rate at which they become shorter. Recently, it was found that certain genes on chromosome 6 appear with differing versions for long-lived males, and other versions for long-lived females.

In the laboratory, the immortal cell lines are those derived from cancer. The most famous is the HeLa cancer cells that many of us use as one of the cell lines tested against various anticancer agents. The HeLa cells are derived from Henrietta Lacks, a black woman, who died from cervical cancer. They are so strong, that they are know to invade other cell lines, both healthy and other cancer cell lines giving contaminated or mutated cell lines. HeLa cells have good telomerase levels. If antisense RNA is added to HeLa cells so that the RNA contains the opposite message to the ordinary RNA in the telomerase, the effect is to block the telomerase and the HeLa cells are no longer immortal and die after about 25 replications.

It is estimated that about 700 genes are probably involved in the overall aging process and *TEPI* is only one of these.

It can be argued as to whether all cancers occur as a direct result of our genes, but the relationship between cancer and genes is present. We are aware that many chemical agents and high-energy radiation that result in cancer do so through damaging DNA. Oncoviruses are also known to cause cancer and these oncoviruses are not viruses but are really genes. In general, cancer genes are genes that cause growth. Fortunately we have other genes that detect excessive growth and whose job it is to stop the growth. These genes are called *tumor-suppressor genes* as opposed to the oncogenes. The misfunctioning of either can result in cancer. If the oncogenes are not switched off then cancer occurs or if the tumor-suppressor genes are not permitted to work, then cancer occurs.

On the short arm of chromosome 17 is a gene called TP_{53}. This gene is a tumor-suppressor gene and it codes for the production of a protein called p53 that is being tested as an anticancer drug. TP_{53} is found to be broken in more than 50% of tested human cancers and is found broken in 95% of those with lung cancer. The most resistant cancers such as melanoma, lung, colorectal, and bladder cancers are ones where mutated TP_{53} are found. Further, where a patent initially responds to treatment, but then develops so-called resistant cells, again it is often found that the TP_{53} genes have mutated. (Thus, it should be possible to look at this gene to see if it has been mutated to see if additional chemo may be useful in the treatment of a particular cancer.) Those born with one of the two TP_{53} genes broken have a 95% chance of developing cancer, and generally early in their lives. We can look at the progress of colorectal cancer. Here, cancer begins with a mutation of the tumor-suppressor gene *APC*. If the developing polyp then undergoes a second mutation causing an oncogene to operate without restraint, the polyp becomes an adenoma. If the adenoma then undergoes a third mutation of a tumor-suppressor gene, then it continues to grow. If a fourth mutation occurs, now in the TP_{53} gene, it becomes a full blown carcinoma. Many other cancers follow a similar scenario often with TP_{53} as the final mutated tumor-suppressor gene. Thus, this gene appears to be important in the production of cancer and in the fight against it.

TP_{53} is about 1,179 base-pairs long and it encodes for the production of the protein p53. This protein is normally rapidly degraded by other enzymes with a half-life of only about 20 minutes. But when a certain signal occurs, protein production greatly increases and its degradation becomes less rapid. The signal appears to be caused by selective damage to DNA with the damaged parts calling for production of excess p53. The p53 protein then "takes over" the cell activating, essentially causing the cell, to either stop making DNA until repair is made, or signals to the cell to commit suicide. Another indicator for p53 is a shortage of cellular oxygen. Cancerous cells often outgrow their oxygen availability so that they send out new arteries to capture more oxygen for themselves. Some of the drugs being developed are aimed at preventing such adventurous artery formation.

Opposing forces operate in our bodies, often guided by our genes. These opposing forces or activities are both important but become dangerous when not held in check. Thus, *oncogenes* cause cell growth necessary for injury repair and cell replacement. They are held in check by tumor-suppressor genes. Interestingly, some oncogenes, such as the *MYC* gene, also hold in their code cell death. The death code is held in check by chemical signals know as survival signals. Thus, if the MYC cell begins to operate in a cancer mode, the cell is signaled to kill itself. There are in fact three different oncogenes, *MYC*, *RAS*, and *BCL-2* that appear to hold one another in balance. Normal cells can exist only if all three of these oncogenes are operating "correctly."

Mutations often occur but most have no long-term consequence. Some mutations are responsible for specific diseases and may cause tumor growth and cancer. Mutations are responsible for some of our diseases including genetic-related diseases. Figure 10.16 shows the general location on the X chromosome of the genes where selected disease-related mutations have been found to occur. There are a number of kinds of mutations. Many of them can be divided into two groups. Substitution mutations occur when one of the bases in a condon is changed from the intended base creating a flaw in the gene. Deletion mutations occur when a base is deleted often causing the other bases to shift altering the remaining codons. Knowing where such mutations occur allows scientists to better understand and treat such mutations.

When we take the sum total of humanity, it is not our human genomes, but rather a complex group of events, that include our genome that actually make us what we are. In some cases, the genome casts boundaries, but more likely, it tells us probabilities though some outcomes are so sure as to be nearly certainties. We are humans with moral, ethical, and gene-driven tendencies and freedoms. We have just begun another trip where moral, ethical, and gene-driven aspects all play a role. Good fortune and wisdom to us in this new adventure.

The secrets of the human genome are just beginning to be uncovered, discovered, understood, and finally utilized. It is an eventful, important trip to which we are both witness to and have an essential stake in, and some of us may take an active part in. New genes are continually being discovered. So while we have decoded the human genome, we have not unlocked most of its secrets.

New information is being found daily. Some of this information is maybe intuitive and consistent with information gathered from other sources. For instance, it is found that certain genes behave differently in men and women. While men and women have the same genes, the information expressed by the genes varies based on gender. The differences are particularly apparent in the liver where the number of copies of a particular gene varied by gender meaning that there is a general difference in how we metabolize drugs so that different doses of a drug is more suitable for one sex than the other. The difference is greater than previously suspected and is leading for different treatment strategies to be developed based on sex.

(Some ideas for the chromosome section were taken from the timely, lively, interesting, and easy to read book—Ridley, M. [1999]: *Genome*, HarperCollins, NY.)

10.12 SPLICEOSOMES

As noted before, the number of genes in the human genome is actually relatively small, in the range of 20,000–25,000 genes. Yet, these genes produce about 150,000 unique proteins so the notion that each gene is responsible for coding for a single protein is not true. The missing link is called the *spliceosome*, a large protein-RNA hybrid complex that is present in the nucleus of each of our cells. It splices and recombines the RNA transcribed from our DNA into different forms before translation into protein formation by the ribosome. The spliceosome cuts out introns and joins the remaining exons into unique combinations of mRNA. These combinations are responsible for the synthesis of the large number of proteins. The number of splice sites on our mRNA varies but is at least three. Some human mRNAs get spliced into thousands of different combinations.

The spliceosome is truly immense being about 3 megadaltons in size involving five RNAs and more than 150 proteins. It is present in only small amounts, less than one percentage of the dry weight of a cell. By comparison, ribosomes compose about 25% of a cell's dry weight. The spliceosome does not have a core structure but rather undergoes dramatic structural rearrangements as it operates. Ribosomes, by comparison have two RNAs and a few proteins adapting only several major conformational changes are it performs its functions. Further, spliceosomes morph into different compositions as they carry out their tasks. Thus, one team of RNA and protein may complex with an intron and then fall away with other members being added and deleted to the spliceosome as additional tasks are performed. Yet, the chemistry that is performed by the spliceosome is generally simple involving two transesterfication reactions cutting each end of the intron, and then forming one phosphodiester bond between the adjoining exons.

The reason for the complexity of the spliceosome is believed to be regulation. The spliceosome must respond to changes in the needs of a cell. It must not only perform many jobs but it must do them on a tight time scale.

10.13 PROTEOMICS

Identification of the protein target that interacts with a molecule is a bottleneck in drug discovery. Here we will deal with one branch of this important venture.

Probably for defensive reasons, because of the overwhelming amount of information, possibilities, and techniques required, subdivisions are present so that some groups focus on protein structures, others on DNA structures, others on the interaction between the two, others on target molecules/sites, and still others deal with specific diseases or biological responses. One such subdivision has been given the name "Proteomics" that deals with the interfacing or bridging between genomics or gene information and drug or target molecule activity. While we have uncovered more than 75,000 sequences of the human genome, we have just begun to look at this "raw" information for actual active or target sites. We need to remember that we are looking at the master template, the DNA genome, its interactions with various proteins at each of the real and potential sites of action, and finally target sites and molecules to effect specific biological responses. Let us remember that this is polymer science in its truest form.

Almost three quarters of the known proteins have no known cellular function but as we have learned in the past, nature seldom has true "junk" in its biological pile of macromolecules. In addition to identifying specific target and target molecules, other discoveries are important such as new enzymes, signaling molecules, pathways, and finally mechanistic behaviors and factors. Identification of such factors will allow better drug discovery and activity downstream.

Factors that will be needed in this hunt include the following: First, there is needed a more complete understanding of the behavior of particular proteins. This includes such seemingly pedestrian, but critical, activities as providing pure and structurally and chemically unaltered proteins. This "unaltered" form includes conformational as well as configurational aspects. It is known that proteins and protein fragments may be misfolded when reproduced employing a variety of reproductive techniques. These wrong structures can cause incorrect test results and when present as part of a large data bank, point the scientist in the wrong direction, or alternately, not point them in the right direction.

Second, while each of us have more than 100,000 proteins, only a fraction is expressed in any given cell type. Thus, it is a complex and puzzling problem to accurately match protein activity to some biological response. One emerging tool is to measure the relative abundance of mRNA in a cell because there is a not unexpected relationship between protein concentration and mRNA concentration. However, these correlations do not always follow, sometimes because the regulatory processes occur after transcription, so that caution must be exercised. Thus, the direct measure of the concentration of the particular protein is a better measure. Such protein determination is often achieved by traditional analytical tools such as coupling some chromatography technique such as electrophoresis with MALDI (mass spectrometry). Other biological techniques are employed and sometimes these are coupled with traditional analytical techniques.

Third, screens are being developed that look at the many potentially active sites on each of the genome fragments. These screens often evaluate selected catalytic activity of the particular sites. As we look at the large proteins and DNA fragments we need to be aware that the presence or absence of small possibly important molecules is critical to the site activity. Again, knowing the precise identity of the site in question is critical. We also need to be aware, that the particular site testing may be the wrong test and may not unlock the true site function or capabilities.

Fourth, simple language. Because of the need for the involvement of diverse groups, common language problems will develop. This is true even between seemingly like medically related specialities. Thus, it is important that we define our terms as we work with other colleagues in related fields.

Fifth, we need to recognize important driving or overriding factors. Structure is one of these. It is being found that structural similarities may be employed as one (notice only one) factor in determining site activity. Thus, it is important that precise three-dimensional site geometry is know including surrounding geometries. This is time and instrument intensive and short cuts are being developed, but as always caution must be exercised. Creation of shared structural data banks is occurring. Computer modeling efforts are particularly useful in helping solve such structural problems.

Sixth, if the vast array of sites and site-important molecules is not enough, protein–protein interactions are part of most cellular processes, including carbohydrate, lipid, protein, and nucleic acid metabolism, signal transduction, cellular architecture, and cell-cycle regulation. In fact, many of the major diseases are believed to involve a breakdown in such protein–protein interactions. These include some cancer, viral infections, and autoimmune disorders.

Techniques to discover the identity of such protein–protein interactions are evolving. One approach involves protein affinity chromatography. Here, the purified protein of interest is immobilized on a solid polymer support and proteins that associate with them are identified by electrophoresis and MALDI. There exists a wide number of modifications to the affinity chromatography approach. For instance, a number of proteins can be fixed to the support in such a manner to look at target molecule interactions as well as nucleic acid–protein interactions, and so on. You can also run through a variety of possible binders and select the ones that bind most strongly for further study. Thus, affinity chromatography is a powerful and versatile tool in this search.

Finally, a large sea of information is becoming available to us. How do we handle it? What sense does it make? Again, we will turn to two powerful tools, the computer and ourselves.

10.14 PROTEIN SITE ACTIVITY IDENTIFICATION

Scientists are developing a number of tools to look at the specific interactions that occur within proteins. As noted above, some deal with the interaction between proteins and genes while others are more general. Section 10.5 dealt with one such approach.

In general, protein target identification often employs genetic techniques such as expression cloning, expression profiling, screening of yeast mutations, and yeast three-hybrid assays. None of these techniques works for every situation.

Another approach in identifying the protein targets employs multicellular organisms. This approach is more aimed at identifying the targets of small molecules. The notion of using genetics to identify small-molecule targets is not new but work with worms and zebrafish embryos is new and represents the first multicellular approaches. In the worm approach, tens of thousands of genetically modified worms are exposed to the test molecules. The worms are watched for changes in their shape. In this genetic suppressor screening, the genes of the affected and unaffected worms are compared identifying the particular gene, and hopefully site on the gene, that is interacted with by the target molecule. By knowledge of the activity of the particular affected site, the potential for that molecule to be active in treating an illness associated with that site is obtained. As a test, the worms are treated with a drug that is specific for that illness and again, the site of activity identified.

10.15 SUMMARY

1. Physically there are little difference in the behavior, study, or testing of natural and synthetic polymers. The fundamental principles that underpin the behavior of macromolecules apply equally to both synthetic and natural polymers. Both groupings contain members that exhibit characteristics that are unique to that grouping. Even so, differences within even these groupings are by degree rather than kind with the fundamental laws continuing to be applicable.

2. Contributions from studying both natural and synthetic polymers are being used to forward the science of both sets of macromolecules.

3. Organic polymers are responsible for the very life—both plant and animal—that exists. Their complexity allows for the variety that is necessary for life to occur, to reproduce, and to adapt. Structures of largely linear natural and synthetic polymers can be divided into primary structures used to describe the particular sequence of (approximate) repeat units, secondary structure used to describe the molecular shape or conformation of the polymer, tertiary structure that describes the shaping or folding of macromolecules, and quaternary structure that gives the overall shape of groups of tertiary-structured macromolecules. The two basic secondary structures are those of the helix and sheets.

4. Proteins are composed of 20 different alpha-amino acids and contain peptide linkages similar to those present in polyamides. With the exception of glycine, all the amino acids contain a chiral carbon. The geometrical shape and behavior of the giant proteins is a product of the various preferred geometries that allow the molecules to balance factors such as preferred bond angle, secondary bonding forces, emphasizing hydrogen bonding, size, shape, hydrophobic/hydrophilic interactions, external/internal chemical environments, and cross-linking. Small chains of amino acids are referred to as peptides.

5. Secondary structures for proteins are generally fibrous and globular. Proteins such as keratines, collagen, and elastin are largely fibrous and have secondary structures of sheets and helixes. Many of the globular proteins are composed of protein chains present in secondary structures approximating helixes and sheets.

6. Enzymes are one important group of proteins. They serve as natural catalysts immobilizing various components that will be later joined or degraded.

7. The two major types of nucleic acids are DNA and RNA. Nucleic acids are polyphosphate esters containing the phosphate, sugar, and base moieties. Nucleic acids contain one of five purine/pyrimidine bases that are coupled within double-stranded helixes. DNA, which is an essential part of the chromosome of the cell, contains the information for the synthesis of protein molecules. For double-stranded nucleic acids, as the two strands separate, they act as a template for the construction of a complementary chain. The reproduction or duplication of the DNA chains is called *replication*. The DNA undergoes semiconservative replication where each of the two new strands contains one of the original strands.

8. The flow of biological genome knowledge is from DNA to RNA via transcription and from RNA to direct protein synthesis via translation.

9. Genetic engineering is based on chemical manipulations that are exactly analogous to those carried out by chemist in basic chemistry laboratories, but it involves the use of biological agents.

10. Much that is occurring with genes, proteins, mutations, chain folding, and other important molecular biology-related efforts is polymer chemistry applied to natural systems and we have much to offer to assist in these ventures.

GLOSSARY

α-helix: Right-handed helical conformation.

β arrangement: Pleated sheet-like conformation.

Active site-region of an enzyme or chromosome that binds the substrate molecule and catalytically transforms it.

Alpha helix: Helical conformation of a chain, usually with maximal interchain hydrogen
bonding; one of the most common "natural" structures.

AluI: One of a family of restriction endonucleases that are site-specific endodeoxyribonu-
cleases that cause cleavage of both strands of DNA within or near a specific site recognized
by the enzyme; its recognition sequence is AG/CT.

Antibody: Defense protein synthesized by the immune system of vertebrates.

Anticodon: Specific sequence of three nucleotides in tRNA, complementary to a codon for an
amino acid in mRNA.

Antigen: Molecules capable of eliciting the synthesis of a specific antibody.

Attenuator: RNA sequence involved in regulating the expression of certain genes; also functions as a transcription terminator.

Base pair: Two nucleotides in nucleic acid chains that are paired by hydrogen bonding of their
bases; like Gee-CAT representing the preferred pairing of G with C and A with T.

Chromosome: Single large DNA molecule and its associated proteins that contains many
genes. It stores and transmits genetic information.

Centromere: Specialized site within a chromosome that serves as the attachment point for the
mitotic or meiotic spindle during cell division that allows proteins to link to the chromosome. This attachment is essential for the equal and orderly distribution of chromosomes
sets to daughter cells. It is about 130 base pairs in length and is rich in A=T pairs.

Clones: Descendants of a single cell.

Cloning: Production of large numbers of identical DNA molecules, cells, or organisms from
a single ancestral DNA molecule, cell, or organism.

Codon: Sequence of three adjacent nucleotides in a nucleic acid that codes for a specific
amino acid.

Cofactor: A coenzyme or other cofactor required for enzyme activity.

Collagen: Protein present in connective tissue.

Complementary: Molecular surfaces with chemical groups arranges to interact specifically
with the chemical groupings on another molecular surface or molecule.

Denaturation: Change in conformation of a protein resulting from heat or chemicals.

Denatured: Partial or complete unfolding of the specific native conformation of a polypeptide,
protein, or nucleic acid.

Degenerate code: Code where a single element in one language is specified by more than one
element in a second language.

Deoxyribonucleic acid: Nucleic acid in which deoxyribose unit are present; compose the
human genome.

Diploid: Having two sets of genetic information—such as a cell having two chromosomes of
each type.

DNA profiling: Identification method based on variations between individual's DNA.

Domain: Distinct structural unit of a polypeptide.

Elastin: Protein that is the major material of arterial blood vessels and ligaments that is noted
for its flexibility.

Enhancers: DNA sequences that help the expression of a specific gene; may be located close
to or far from the particular gene.

Enzyme: Molecule, protein, or RNA that catalyzes a particular chemical reaction.

Engineering material: Material that can be machined, cut, drilled, sawed, and so on; must
have enough dimensional stability to allow these actions to be carried out on them.

Eukaryote: Unicellular or multicellular organism with cells having a membrane-bound
nucleus, several chromosomes, and internal organelles.

Exon: Segment of an eukaryotic gene that encodes a portion of the final product of the gene; part that remains after posttranscriptional processing and is transcribed into a protein or incorporated into the structure of an RNA. Also see intron.

Fibrillar protein: Hair like, insoluble, intermolecularly hydrogen-bonded protein.

Fibrils: Thread-like strands or bundles of fibers.

Gene: Chromosomal segment that codes for a single functional polypeptide or RNA chain.

Gene expression: Transcription, and in for proteins, translation, giving the product of a gene; a gene is expressed within its biological product is present and active.

Gene splicing: Enzymatic attachment of one gene or part of a gene to another.

Genetic code: Set of triplet code words in DNA or mRNA coding for the specific amino acids of proteins.

Genetic information: Information contained in a sequence of nucleotide bases in chromosomal DNA or RNA.

Genetic map: Diagram showing the relative sequence and position of specific genes within a chromosome.

Genome: All of the genetic information encoded in a cell or virus.

Genotype: Genetic makeup of an organism as distinct from its physical characteristics.

Globular proteins: Proteins with an overall globular structure formed from contributions of secondary structures, including sheets and helices.

Glycine: Simplest and only nonchiral alpha-amino acid.

Immune response: Ability of a vertebrate to create antibodies to an antigen, a molecule typically a macromolecule, foreign to the host.

Inducer: Signal molecule that when attached to a regulatory protein, produces an increase expression of a specific gene.

Intron: most genes are divided into two coding regions—one called *exons* or *coding regions* and the second called *intorns* or *noncoding regions*.

In vitro: Literally means "in glass," but today means outside the normal biological environment.

In vivo: Means "in life" and means within the normal biological environment.

Isoelectric point: pH at which an amino acid does not migrate to either the positive or negative pole in a cell.

Keratin: Fibrillar protein.

Latex: Stable dispersion of polymer particles in water.

Myosin: Protein present in muscle.

N-terminal amino acid: Amino acid with an amino end group.

Nucleoside: Contains a pentose, and base.

Nucleotide: Contains a phosphate, pentose, and base.

Oligosaccharide: Low molecular weight polysaccharide.

Phenotype: The observable, physical characteristics of an organism.

Plasmid: Extrachromosomal, independently replicating, small circular DNA molecule often used in genetic engineering.

Polynucleotide: Nucleic acid.

Polypeptide: Protein; often used for low molecular weight proteins.

Primer: Oligomeric molecule to which an enzyme adds additional monomeric subunits.

Probe: Labeled fragment of a nucleic acid containing a nucleotide sequence complementary to a gene or genomic sequence that one wants to detect.

Prokaryotic cell: Bacteria single-celled organism with a single chromosome, no nuclear envelope and no membrane-bounded organelles.

Primary structure: Term used to describe the primary configuration present in a protein chain.

Prosthetic group: Nonprotenicious group conjugated (connected) to a protein.

Protein: Polyamide in which the building blocks are alpha-amino acids joined by peptide (amide) linkages.

Purine base: Compounds consisting of two fused heterocyclic rings, namely a pyrimidine and an imidazole ring; essential part of nucleic acids.

Pyrimidine: A 1,3-diazine; essential part of nucleic acids.

Recombination: Enzymatic process whereby the linear arrangement of nucleic acid sequences in a chromosome is altered by cleavage and rejoining.

Recycling codes: designations that allow easy, quick identification of a number of plastics used in the manufacture of containers

Replication: Term used to describe duplication such as the duplication of DNA.

Repressor: Molecule, protein that binds to the regulatory sequence or operator for a gene blocking its transcription.

Ribosomal RNA, rRNA: Class of RNA that serve as components of ribosomes.

Ribosome: Very large complex of rRNAs and proteins—very large; site of protein synthesis.

Ribonucleic acid (RNA): Nucleic acid in which ribose units are present. Essential units of life and replication.

Secondary structure: Term used to describe the conformation of a protein molecule such as a helix.

Silent mutation: Mutation in a gene that causes no delectable change in the biological functioning of the gene.

Somatic cells: All body cells except germline cells; germline cells are cells that develop into gametes—that is, egg or sperm cells.

Spliceosome: Large protein-RNA complex that splices mRNA allowing for a great complexicity in synthesized proteins.

Sticky ends: Two DNA ends in the same or different DNA with short overhanging single-stranded segments that are complementary to one another helping the joining the two sites.

Structural gene: Gene coding for a protein or RNA as distinct from regulatory genes.

Template: Pattern or surface upon which mirror-image replication can occur.

Tertiary structure: Term used to describe the shape or folding of a protein.

Transcription: Enzymatic process where genetic information contained in one strand of DNA is employed to make a complementary sequence of bases in an mRNA.

Transfer RNA, tRNA: Class of RNS each of which combine covalently with a specific amino acid as the initial step in protein synthesis.

Translation: Process where the genetic information present in mRNA specifies the sequence of amino acids within a protein synthesis.

Transposons: Segments of DNA, found in almost all cells, that move or "hop" from one place on a chromosome to another on the same or different chromosome.

Transposition: Movement of one gene or set of genes form one site in the genome to another site.

Transcription: Term used to describe the transfer of information from DNA to RNA.

Translation: Term used to describe the transfer of information from RNA to protein synthesis.

Virus: Self-replicating nucleic acid–protein complex that requires an intact host cell for replication; its genome can be either DNA or RNA.

Zwitterion: Dipolar ion of an amino acid.

EXERCISES

1. Define a protein in polymer science language.
2. Which α-amino acid does not belong to the L-series of amino acids?

3. To which pole will an amino acid migrate at a pH above its isoelectric point?
4. Why is collagen stronger than albumin?
5. What are the requirements for a strong fiber?
6. Which protein would be more apt to be present in a helical conformation: (a) a linear polyamide with small pendant groups or (b) a linear polyamide with bulky pendant groups?
7. What is the difference between the molecular weight of (a) ribose and (b) deoxyribose?
8. What is the repeating unit in the polymer DNA?
9. Which is more acidic: (a) a nucleoside or (b) a nucleotide?
10. What base found in DNA is not present in RNA?
11. Why would you predict helical conformations for RNA and DNA?
12. If the sequence of one chain of a double helix of DNA is ATTACGTCAT, what is the sequence of the adjacent chain?
13. Why is it essential to have trinucleotides rather than dinucleotides as codons for directing protein synthesis?
14. Why is *E. Coli* most often used in gene splicing?
15. A protein that is rich in glutamic acid and aspartic acid will be rich in what kind of functional groups present as substituents on the alpha carbon? Will it be attracted, in electrophoresis, to the positive or negative side? What is this kind of protein polymer called?
16. The inclusion of which amino acids into proteins are responsible for the sulfur found in coal? What is the consequence of the presence of this sulfur?
17. Which of the amino acids contribute to the formation of nitrogen oxides? What is the consequence of the presence of this nitrogen to our atmospheric air?
18. What is the general shape of our enzymes? Why do they take this shape? What are they composed of?
19. What are essential amino acids?
20. Sulfur is contained in some amino acids. What else is sulfur utilized for in our bodies?
21. Since much of a protein is not active and is at times referred to as junk, why is it there?
22. Describe briefly the typical flow of biological information.
23. What role do "chaperonins" play in protein formation?
24. Why are the active sites of proteins found either as a cleft in the macromolecule or shallow depression on its surface and not thrust out where there is little hindrance for entry of the molecules?
25. Select a particular disease or illness and using the web find out what is known about its molecular biology including gene location.
26. What is the probability of a match if three tests were positive with the following probabilities: 1 in 100; 1 in 1,000, and 1 in 250?

ADDITIONAL READING

Bloomfield, V., Crothers, D., Tinoco, I. (2000): *Nucleic Acids*, University Science Books, Sausalito, CA.
Cheng, H. N., Gross, R. A. (2005): *Polymer Biocatalysis and Biomaterials*, Oxford University Press, NY.
Chiellini, E. (2001): *Biorelated Polymers*, Kluwer, NY.
Chiellini, E. (2001): *Biomedical Polymers and Polymer Therapeutics*, Kluwer, NY.
Dumitriu, S. (2001): *Polymeric Biomaterials*, Dekker, NY.
Ferre, F. (1997): *Gene Quantification*, Springer-Verlag, NY.
Gebelein, C. Carraher, C. (1995): *Industrial Biotechnological Polymers*, Technomic, Lancaster, PA.
Kaplan, D. (1994): *Silk Polymers*, ACS, Washington, DC.
Klok, H., Schlaad, H. (2006): *Peptide Hybrid Polymers*, Springer, NY.
Kobayashi, S., Kaplan, D., Ritter, H. (2006): *Enzyme-Catalyzed Synthesis of Polymers*, Springer, NY.
Malsten, M. (2003): *Biopolymers at Interfaces*, Dekker, NY.
McGrath, K., Kaplan, D. (1997): *Protein-Based Materials*, Springer-Verlag, NY.
Merrifield, R. B. (1975): Solid phase peptide synthesis, *Polymer Prep.*, 16:135.

Scholz, C., Gross, R. (2000): *Polymers from Renewable Resources*, ACS, Washington, DC.

Steinbuchel, A. (2003): *Biopolymers, Miscellaneous Biopolymers and Biodegradation of Synthetic Polymers*, Wiley, Hoboken.

Tsuji, H. (2008): *Degradation of Poly(Lactide)*, Nova, Hauppauge, NY.

Wuisman, P., Smit, T. (2009): *Degradable Polymers for Skeletal Implants*, Nova, Hauppauge, NY.

Yoon, J. S., Im, S. S., Chin, I. (2005): *Bio-Based Polymers-Recent Progress*, Wiley, Hoboken.

11 Organometallic and Inorganic–Organic Polymers

11.1 INTRODUCTION

Classical polymer chemistry emphasizes materials derived from about a dozen elements (including C, H, O, N, S, P, Cl, and F). The following two chapters deal with polymers containing additional elements. The present chapter focuses on inorganic and metal-containing polymers containing organic units.

Elements such as silicon, sulfur, and phosphorus catenate similar to the way carbon does, but such catenation generally does not lead to (homo) chains with high degrees of polymerization. Further, such products might be expected to offer lower thermal stabilities and possibly lower strengths than carbon-based polymers since their bond energies are generally lower (Table 11.1). The alternative of using heteroatomed backbones is attractive since the resultant products can exhibit greater bond energies (Table 11.1).

One common misconception concerns the type of bonding that can occur between inorganic and organic atoms. With the exception of the clearly ionic bonding, many of the inorganic–organic bonding is of the same general nature as that present in organic compounds. The percentage contribution of the organic–inorganic bonding due to covalent contributions is typically well within that found in organic acids, alcohols, and thio and nitro moieties (e.g., the usual limits are about 5% ionic character for the B–C bond to 55% ionic for the Sn–O and both are clearly directional bonding in character). Thus, the same spacial, geometrical rules apply to these polymers as to the more classical polymers such as PE, PS, nylons, polyesters, and PP. The exception is the ionomers where the metals are bonded through ionic bonding to the oxygen atoms.

The number of potential inorganic–organic polymers is great. The inorganic portions can exist as oxides, salts, in different oxidation states, different geometries, and so on. The importance of these inorganic–organic polymers can be appreciated by considering the following. First, photosynthesis, the conversion of carbon dioxide and water by sunlight to sugars is based on a metal-containing polymer—chlorophyll. Also, a number of critical enzymes, such as hemoglobin, contain a metal site as the key site for activity. Second, the inorganic–organic polymers produced thus far exhibit a wide range of properties not common to most organic polymers, including electrical conductivity, specific catalytic operations, wide operating temperatures, greater strengths, and greater thermal stabilities (Table 11.2). Third, inorganic–organic polymers form the basis for many insulators and building materials. Fourth, inorganic elements are present in high abundances in the Earth's crust (Table 11.3).

The topic of metal- and metalloid-containing polymers can be divided by many means. Here the topic will be divided according to the type of reaction employed to incorporate the inorganic atom into the polymer chain. While many other types of reactions have been employed to produce metal- and metalloid-containing polymers, including redox, coupling, ring-opening polymerizations, the present will focus on addition, condensation, and coordination reactions. Emphasis is given to unifying factors.

TABLE 11.1
General Magnitude of Bonds

Bond	General Bond Energy (kJ/mol)	Ionic Character* (%)	Bond	General Bond Energy (kJ/mol)	Ionic Character (%)
Al–O	560	60	P–O	400	40
B–C	360	5	P–S	320	5
B–N	440	20	S–S	240	0
B–O	460	45	Si–Si	220	0
Be–O	500	65	Si–C	300	10
C–C	340	0	Si–N	420	30
C–H	400	5	Si–O	440	50
C–N	300	5	Si–S	240	10
C–O	340	20	Sn–Sn	160	0
C–S	260	5	Sn–O	520	55
P–P	200	0	Ti–O	640	60
P–N	560	20			

*On the basis of Pauling electronegativity values. The percentage of ionic bonding should be less where pi-bonding occurs. Given to nearest 5%.

TABLE 11.2
Actual and Potential Applications for Organometallic and Metalloid Polymers

Biological	Anticancer, antiviral, treatment of arthritis, antibacterial, antifungal, antifouling, treatment of Cooley's anemia, algicides, molluscidides, contrast agents, radiology agents
Electrical/optical	Nonlinear optics, lithography, conductors, semiconductors, piezoelectronic, pyroelectronic, solar energy conversion, electrodes, computer chip circuitry
Analytical, catalytic, building	UV absorption, smart materials, nanocomposites, laser, sealants, paints, caulks, lubricants, gaskets

TABLE 11.3
Relative Abundances of Selected Elements in Earth's Upper (10 miles) Crust

Element	Weight (%)	Element	Weight (%)
Oxygen	50	Titanium	0.4
Silicon	26	Fluorine	0.3
Aluminum	7.3	Chlorine	0.2
Iron	4.2	Carbon	0.2
Calcium	3.2	Sulfur	0.1
Sodium	2.4	Phosphorus	0.1
Potassium	2.3	Barium	0.1
Magnesium	2.1	Manganese	0.1
Hydrogen	0.4		

11.2 INORGANIC REACTION MECHANISMS

Many of the polymerizations and monomer syntheses are simply extensions of known inorganic, organometallic, and organic reactions. The types and language used to describe inorganic–organic reaction mechanisms are more diversified than those employed by classical organic chemists.

The majority of inorganic reactions can be placed into one of two broad classes—oxidation/reduction (redox) reactions, including atom- and electron-transfer reactions and substitution reactions. Terms such as inner sphere, outer sphere, and photo-related reactions are employed to describe redox reactions. Such reactions are important in the synthesis of polymers and monomers and in the use of metal-containing polymers as catalysts and in applications involving transfer of heat, electricity, and light. They will not be dealt with any appreciable extent in this chapter.

Terms such as liability, inertness, ligand, associative, interchange, and dissociative are important when discussing substitution reactions. The ligand is simply (typically) the Lewis base that is substituted for and is also the agent of substitution. Thus, in the reaction between tetrachloroplatinate and diamines forming the anticancer and antiviral platinum II polyamines, the chloride is the leaving group or departing ligand, while the amine-functional group is the ligand that is the agent of substitution (Equation 11.1).

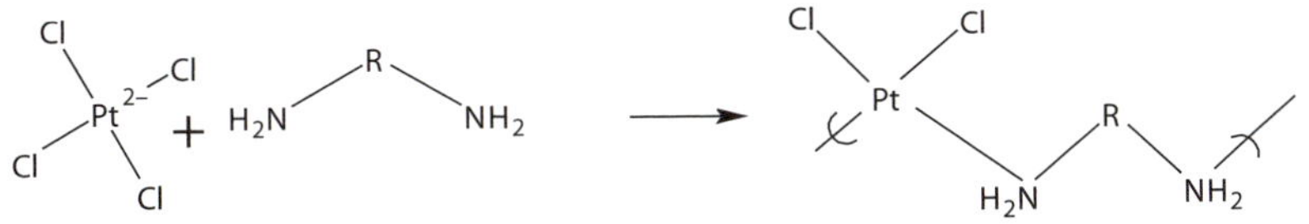

$$(11.1)$$

There is a difference between the thermodynamic terms stable and unstable and the kinetic terms labile and inert. Furthermore, the difference between the terms stable and unstable and the terms labile and inert are relative. Thus $Ni(CN)_4^{-2}$ and $Cr(CN)_6^{-3}$ are both thermodynamically stable in aqueous solution, yet, kinetically the rate of exchange of radiocarbon-labeled cyanide is quite different. The half-life for exchange is about 30 s for the nickel complex and 1 month for the chromium complex. Taube has suggested that those complexes that react completely within about 60 s at 25°C be considered labile while those that take a longer time be called inert. This rule of thumb is often given in texts but is not in general use in the literature. Actual rates and conditions are superior tools for the evaluation of the kinetic/thermodynamic stability of complexes.

The term "D mechanism" (dissociation) is loosely comparable to S_N1-type reaction mechanisms, but it does not imply an observed rate law. Here, a transient intermediate is assumed to live long enough to be able to differentiate between various ligands, including the one just lost, and between solvent molecules. Thus, the overall rate expression may be dependent on the nature of LL′, solvents, or some combination as pictured below where S = solvent, L is the leaving ligand, and L′ is the incoming ligand.

$$ML_4 \leftrightarrows ML_3 + L \qquad (11.2)$$

$$ML_3 + L' \rightarrow ML_3L' \qquad (11.3)$$

$$ML_3 + S \leftrightarrows ML_3S \qquad (11.4)$$

In the I_d mechanism, dissociative interchange, the transition state involves extensive elongation of the M⋯L bond, but not rupture.

$$ML_4 + L' \rightarrow [L^{..}ML_3^{..}L'] \rightarrow ML_3L' + L \qquad (11.5)$$

The ML_3L' species is often called an outer sphere complex, or, if ML_4 is a cation and L an anion, the couple is called an ion pair.

For the I_a mechanism, associative interchange, the interaction between M and L' is more advanced in the transition state than in the case of the I_d. The M···L' bonding is important in defining the activated complex. Both of these interchange mechanisms are loosely connected to the S_N 2-type mechanism.

For the A mechanism, associative, there is a fully formed intermediate complex ML_4L', which then dissociates, being roughly analogous to the E_1 type reaction mechanism.

It is important to remember that the same electronic, steric, mechanistic, kinetic, and thermodynamic factors that operate in regard to smaller molecules are in operation during a polymerization process.

11.3 CONDENSATION ORGANOMETALLIC POLYMERS

Condensation reactions exhibit several characteristics such as (typically) expulsion of a smaller molecule on reaction leading to a repeat unit containing fewer atoms than the sum of the two reactants, and most reactions can be considered in terms of polar (Lewis acid–base; nucleophilic–electrophilic) mechanisms. The reaction site can be at the metal atom (i.e., adjacent to the atom)

$$R_2MX_2 + H_2N\text{–}R\text{–}NH_2 \rightarrow \text{–(–}MR_2\text{–}NH\text{–}R\text{–}NH\text{–)–} + HX \quad (11.6)$$

or it can be somewhat removed from the metal site.

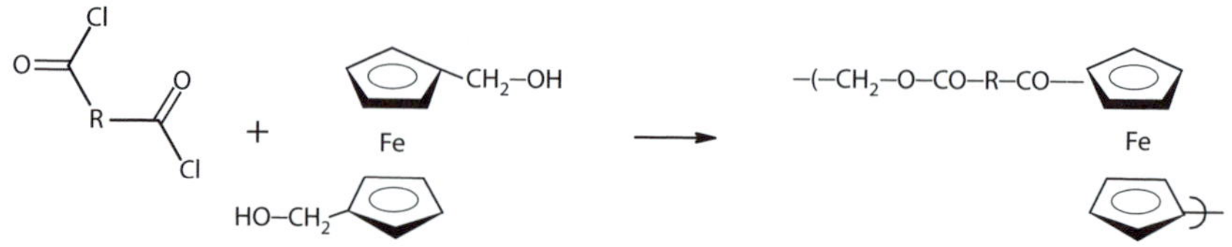

(11.7)

Research involving condensation organometallic polymers was catalyzed by the observation that many organometallic halides possess a high degree of covalent character within their composite structure and that they can behave as organic acid chlorides in many reactions, such as hydrolysis

$$R\text{–}\overset{\overset{\displaystyle O}{\|}}{C}\text{–}Cl + H_2O \rightarrow R\text{–}\overset{\overset{\displaystyle O}{\|}}{C}\text{–}OH + HCl \quad (11.8)$$

$$\text{–}\underset{|}{\overset{|}{M}}\text{–}Cl + H_2O \rightarrow \text{–}\underset{|}{\overset{|}{M}}\text{–}OH + HCl \quad (11.9)$$

and polyesterfication.

$$Cl\text{–}\overset{\overset{\displaystyle O}{\|}}{C}\text{–}R\text{–}\overset{\overset{\displaystyle O}{\|}}{C}\text{–}Cl + HO\text{–}R'\text{–}OH \rightarrow \text{–(–}\overset{\overset{\displaystyle O}{\|}}{C}\text{–}R\text{–}\overset{\overset{\displaystyle O}{\|}}{C}\text{–}O\text{–}R'\text{–}O\text{–)–} + HCl \quad (11.10)$$

$$Cl\text{–}\underset{|}{\overset{|}{M}}\text{–}Cl + HOR'\text{–}OH \rightarrow \text{–(–}\underset{|}{\overset{|}{M}}\text{–}O\text{–}R'\text{–}O\text{–)–} + HCl \quad (11.11)$$

Thus, many of the metal-containing polycondensations can be considered as extensions of organic polyesterfications, polyamination, etc. reactions.

11.3.1 POLYSILOXANES

The most important organometalloid polymers are the polysiloxanes based on the siloxane Si–O linkage found in glass and quartz (Chapter 12). The polysiolxanes were incorrectly named silicones by Kipping in the 1920s, but this name continues to be widely used. Originally it was wrongly believed to have a structure similar to a ketone, hence, silicone.

The production of silicate glass is believed to be a transcondensation of the siloxane linkages in silica. A comparable poly(silicic acid) is produced when silicon tetrachloride is hydrolyzed.

$$
SiCl_4 \longrightarrow Si(OH)_4 \longrightarrow -(-\underset{\underset{OH}{|}}{\overset{\overset{OH}{|}}{Si}}-O-)- \tag{11.12}
$$

Silicic acid   Poly(silicic acid)

The poly(silicic acid) further condenses producing a cross-linked gel. This cross-linking can be prevented by replacing the hydroxyl groups in silicic acid with alkyl groups. Ladenburg prepared the first silicone polymer in the nineteenth century by the hydrolysis of diethyldiethoxysilane. Kipping, in the early 1940s, recognized that these siloxanes could also be produced by the hydrolysis of dialkyldichlorosilanes giving a poly(silicic acid)-like structure where the hydroxyl groups are replaced by alkyl groups.

In 1945, Rochow discovered that a silicon–copper alloy reacted with organic chlorides forming a new class of compounds called organosilanes.

$$
CH_3Cl + Si(Cu) \longrightarrow (CH_3)_2SiCl_2 + Cu \tag{11.13}
$$

Dimethyldichlorosilane

These compounds react with water forming dihydroxylsilanes,

$$
(CH_3)_2SiCl_2 + H_2O \longrightarrow (CH_3)_2Si(OH)_2 + HCl \tag{11.14}
$$

and eventually dimeric, oligomeric, and finally polysiloxanes (Equation 11.15). Because of the toxicity of HCl, the chlorine groups are replaced by acetate groups leading to the familiar vinegar smell for many silicon caulks and sealants. Branching and cross-linking is introduced through the use of methyltrichlorosilane. Modern caulks and sealants are made using the tetrafunctional group tetraethoxysilane to introduce cross-linking into the resin.

Polysiloxanes, also called silicones, are characterized by combinations of chemical, mechanical, and electrical properties, which when taken together are not common to any other commercially available class of polymers. They exhibit relatively high thermal and oxidative stability, low power loss, high dielectric strength, and unique rheological properties, and are relatively inert to most ionic reagents. Almost all of the commercially utilized siloxanes are based on polydimethylsiloxane with trimethylsiloxy end groups. They have the widest use temperature range for commercial polymers suitable for outdoor applications from the winter of Nome, Alaska, to the summer of south Florida (about –80°F to 100°F; –60°C to 40°C). The first footprints on the moon were made with polysiloxane elastomeric boots.

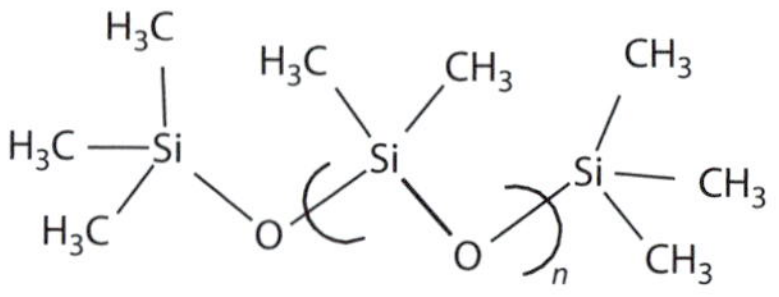

$$(11.15)$$

The reason for the low temperature flexibility is because of a very low T_g, about $-120°C$, which is the result of the methyl groups attached to the silicon atoms being free to rotate causing the oxygen and other surrounding atoms to "stay away" creating a flexible chain.

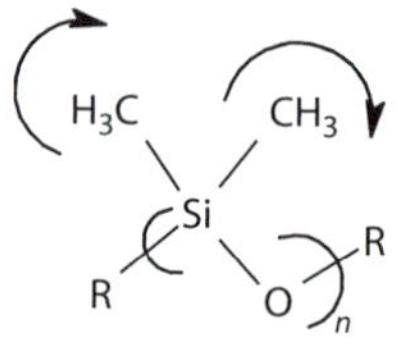

$$(11.16)$$

Polysiloxanes degrade by an unzipping mechanism forming six- and eight-membered rings. They also form these same rings when polymerized forming what is referred to as "wasted loops" because they must be removed before the polysiloxane is useful. This tendency to form six- and eight-membered rings (Equations 11.17 and 11.18) is based on the good stability of such siloxane rings.

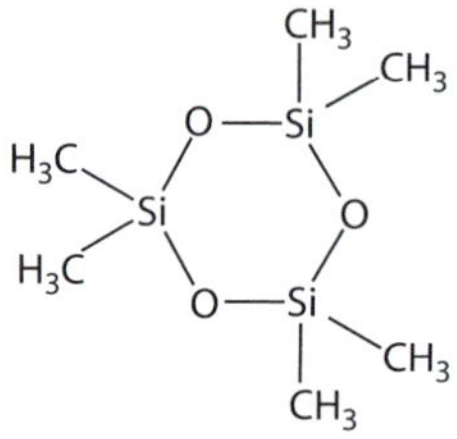

Six-membered loop (11.17)

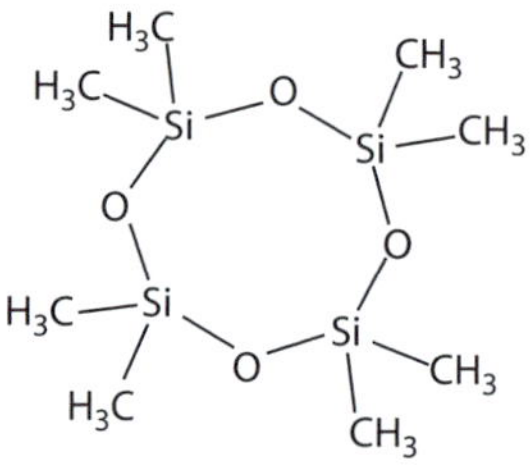

Eight-membered loop (11.18)

Polysiloxanes are used in a wide variety of applications. The viscosity or resistance to flow increases as the number of repeat units increases but physical properties such as surface tension and density remains about the same after a DP of about 25. The liquid surface tension is lower than the critical surface tension of wetting resulting in the polymer spreading over its own absorbed films. The forces of attraction between polysiloxane films is low resulting in the formation of porous films that allow oxygen and nitrogen to readily pass through but not water. Thus, semipermeable membranes, films, have been developed that allow divers to "breath air under water" for short periods.

As noted above, viscosity increases with DP allowing many of the uses to be grouped according to chain length. Low-viscosity fluids with DPs of 2–30 are used as antifoams and in the flow control of coatings applications. These applications are the direct consequence of the low attractions between polysiloxane chains, which, in turn, are responsible for their low surface tension. Thus, they encourage a coatings material to flow across the surface filling voids, corners, and crevices. Their good thermal conductivity and fluidicity at low temperatures allows their use as low-temperature heat exchangers and in low-temperature baths and thermostats.

Viscous fluids correspond to a DP range of about 50–400. These materials are employed as mold release agents for glass, plastic, and rubber parts. They are good lubricants for most metal to nonmetal contacts. They are used as dielectric fluids (liquids) in a variety of electrical applications, including transformers and capacitors; as hydraulic fluids in vacuum and hydraulic pumps; in delicate timing and photographic devices; as antifoam agents; components in protective hand creams; toners in photocopiers; in oil formulations when mixed with thickeners; and in inertial guidance systems. High-performance greases are formed by mixing the polysiloxane fluids with polytetrafluoroethylene or molybdenum disulfide. Brake fluids are formulated from polydimethylsiloxane fluids with DPs about 50. High-viscosity fluids with DPs about 700–6,000 are used as damping fluids for weighting meters at truck stops. They act as liquid springs in shock absorbers. The longer-chained fluids are used as impact modifiers for thermoplastic resins and as stationary phases in gas chromatography.

As with the alkanes, even longer chains form the basis for solid polysiloxanes that, according to design, can be classified as thermoplastics, engineering thermoplastics, elastomers, and when cross-linked as thermosets. Solid polysiloxanes are used in a variety of applications, including as sealants, thermostripping, caulking, dampening, O-rings, and window gaskets. Weather stripping on cooling units, trucks, and automobiles is often made of polysiloxanes.

Room temperature-vulcanizing (RTV) silicon rubbers make use of the room temperature reaction of certain groups that can be placed on polydimethylsiloxanes that react with water. When exposed to water, such as that normally present in the atmosphere, cross-links are formed creating an elastomeric product.

The first contact lenses were based on poly(methyl methacrylate). While they could be polished and machined, they did not permit gas exchange and were rigid. By early 1970s, these were replaced by soft contact lenses containing cross-linked poly(2-hydroxyethyl methacrylate) (HEMA). These so-called disposable lenses do permit gas exchange. More recently, Salamone and coworkers developed contact lenses based on the presence of siloxane units. Polysiloxanes have good gas permeability. These polymers are referred to as Tris materials and are generally copolymers containing units as shown below:

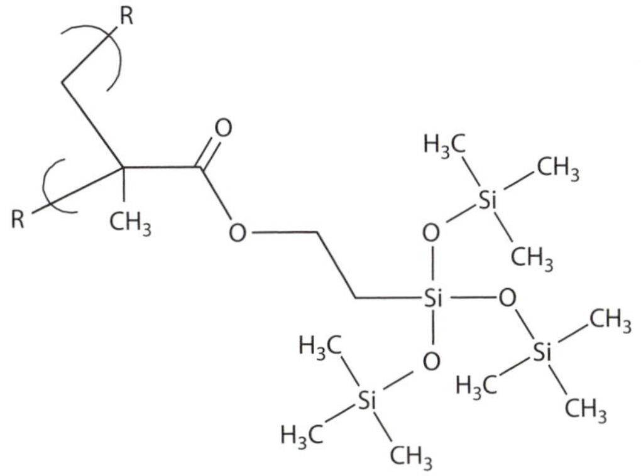

$$(11.19)$$

Polysiloxanes are widely employed as biomaterials. Artificial skin can be fabricated from a bilayer fabricated from a cross-linked mixture of bovine hide, collagen, and chondroitin sulfate derived from shark cartilage with a thin top layer of polysiloxane. The polysiloxane acts as a moisture and oxygen-permeable support to protect the lower layer from the "outer world." A number of drug delivery systems use polysiloxanes because of the flexibility and porous nature of the material.

The first Silly Putty (TM), also called Nutty Putty, was made more than 50 years ago from mixing together silicone oil with boric acid. The original formula has changed only a little, though colorants have been added giving the material brighter colors and some the ability to "glow-in-the-dark". Today, the formula contains about 70% dimethylsiloxane and boric acid, 17% quartz, 9% Thixatrol ST (a commercial rheology modifier that is a derivative of caster oil), and several other minor constituents. Silly Putty is a dilatant material (Chapter 13), meaning it has an inverse thixotropy (a thioxotropic liquid is one whose viscosity decreases with time). In essence, the resistance of flow increases faster than the increases in the rate of low. Thus, under short interaction times (Chapter 13), it behaves as a solid where the various molecular components resist ready movement, acting as a solid and under sharp impact like hitting it with a hammer or rapidly "snapping" it, it will act as a brittle material. Under a relatively long-interaction time the molecular chains are able to yield and the material acts as a liquid. Under moderate interaction times there is segmental movement and the material acts as a rubber.

Silly Putty is one of several materials whose discovery occurred at about the same time by different individuals. In this case, the two individuals are Earl Warrick, working for Dow Corning, and James Wright, a researcher for General Electric. The discoveries occurred in 1943 while searching for synthetic rubber during World War II. Initially no practical use was found. By 1949, it was found in a local toy store as a novelty item. Despite its good sales, the store dropped it after 1 year. The next year Peter Hodson began packaging it in the now familiar plastic egg. Today, it sells for about the same price it did in 1950. Silly Putty sells at a rate of about 6 million eggs, or 90 tons, yearly.

11.3.2 Organotin and Related Condensation Polymers

Carraher and coworkers have produced a wide variety of organometallic condensation polymers based on the Lewis acid–base concept. Polymers have been produced from Lewis bases containing amine, alcohol, acid, thiol, and related units, including a number of drugs, such as ciprofloxacin and acyclovir. Lewis acids containing such metals and metalloids as Ti, Zr, Hf, V, Nb, Si, Ge, Sn, Pb, As, Sb, Bi, Mn, Ru, P, Co, Fe, and S have been employed. These compounds have potential uses in the biomedical arena as antifungal, antibacterial, anticancer, Parkinson's treatment, and antiviral drugs. These polymers show promise in a wide variety of other areas, including electrical, catalytic, and solar energy conversion. Polymers referred to as polydyes, because of the presence of dye moieties in the polymer backbones, have impregnated paper products, plastics, rubber, fibers, coatings, and caulks, giving the impregnated material color, (often) added biological resistance, and special photo properties. Some of the polymers, including the metallocene-containing products, show the ability to control laser radiation. Depending on the range of radiation, laser energy can be focused allowing the material to be cut readily or it can be dispersed imparting to the material containing the polymer added stability toward the radiation.

A number of these polymers exhibit a phenomenon called "anomalous fiber formation," reminiscent of "metallic whiskers."

There are more organometallic compounds containing tin than for any other metal. Further, the volume of organotin compounds employed commercially is greater than for any other organometal. Worldwide, the production of organotin compounds, about 120 million pounds yearly, accounts for about 7% of the entire tin usage. Tin has been included into polymers for a variety of reasons. The two major reasons are its biological activities known for about 100 years, and the second reason involves its ability to help stabilize PVC. This second behavior accounts for about 70% by weight of the organotin compounds used commercially. These materials are employed to improve the heat stability of PVC as it is formed into piping. Today, because polymers leach much slower than small molecules, organotin polymers initially synthesized by Carraher and coworkers are being used as heat stabilizers in PVC piping.

The emphasis on organotin-containing polymers is the results of several factors. The first one involves the early discovery of their biological activity and more recently that these organotin compounds generally degrade to nontoxic inorganic compounds so that they are "environmentally friendly." The second reason concerns the commercial availability of a wide variety of organotin compounds that are suitable as monomers. Third, because of the biological activity of organotin compounds they were found to inhibit the growth of desirable species, in particular, in aquatic surroundings. This brought about federal laws that prohibited the use of leachable, monomeric-organotin compounds in a variety of coatings and protective applications. The result was a move to polymeric materials that did not suffer the same leachability and which were allowed by law.

Much of the recent activity with organotin compounds, including polymers, involves their use to inhibit a wide variety of microorganisms at low concentrations. These microorganisms include a variety of cancers, bacteria, yeasts, and viruses. Carraher, Roner, Barot, and coworkers have found that some simple organotin polyethers based on hydroxyl-terminated poly(ethylene glycols) inhibit a wide variety of cancers, including ones associated with bone, lung, prostate, breast, and colon cancers.

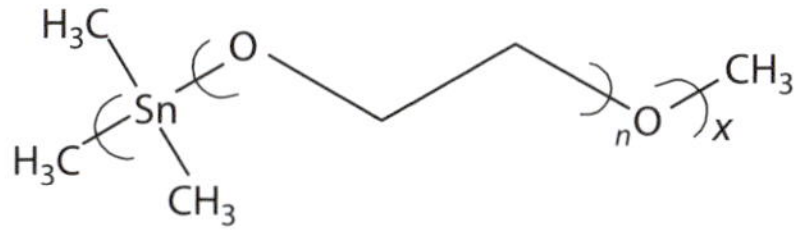

$$(11.20)$$

Two measures are typically employed to measure the ability of cancer cells to grow. The first is the concentration needed to inhibit 50% growth of the cell line, GI_{50}. Some of these polymers have GI_{50} values well below those of cisplatin, the most widely used anticancer drug. The second measure is the chemotherapeutic index, CI, which is the concentration of the compound that inhibits the growth of the healthy or normal cell by 50% divided by the concentration of the compound that inhibits the growth of the cancer or tumor cell by 50%. Larger values are desired since they indicate that a larger concentration is required to inhibit the healthy cells in comparison to the cancer cells, or, stated in another way, larger values indicate some preference for inhibiting the cancer cells in preference to the normal cells. Some of these polymers have CI values in the hundreds, allowing healthy cells to reproduce at concentrations where cancer cell lines are inhibited. Of additional interest is that some of these organotin polymers are water soluble, allowing for medical applications utilizing the material in simple pills.

A wide variety of organotin compounds developed by Carraher, Sabir, Roner, and others on the basis of known antiviral drugs such as acyclovir and known-antibacterial agents such as ciprofloxacin, norfloxacin, cephalexin (11.21), and ampicillin, inhibit a wide variety of viruses, including the ones responsible for many of the common colds, chicken pox, small pox, shingles, and herpes simplex.

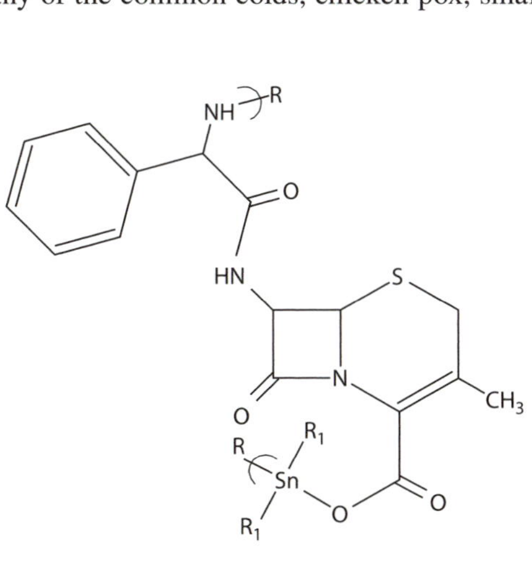

$$(11.21)$$

Some of the organotin-containing polymers selectively inhibit *Candias albicans*, the yeast responsible for yeast infections in men and women better than commercially available applications while leaving the normal flora unharmed. Others preferentially inhibit methicillin-resistant *Staphylococcus aureus* (MRSA), 11.22.

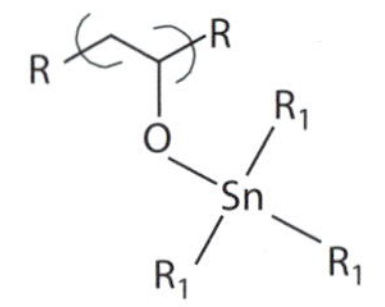

$$\text{(structure 11.22)} \tag{11.22}$$

Organotin polyamines containing the plant growth hormone kinetin (11.23) increases the germination of damaged seeds and thus it may help in providing food in third world countries. It also increases the germination rate of sawgrass seed from about 0% to more than 50% and may become an important agent in replenishing the "sea of grass" (actually sawgrass) in the Everglades.

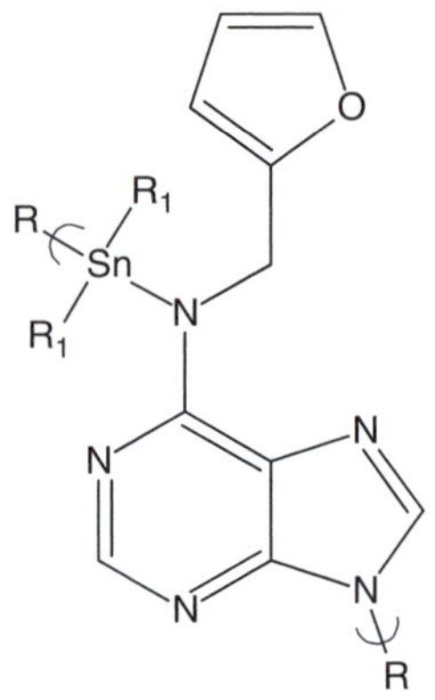

$$\text{(structure 11.23)} \tag{11.23}$$

Carraher and Battin recently reported the discovery of metal-containing polymers based on Group IVB metallocenes reacted with diamines where the conductivity was increased by as much as 10^7 through simple doping with iodine. Thus, such materials may have use as electrically conductive materials (Figure 11.1).

11.4   COORDINATION POLYMERS

Coordination polymers have served humankind since before recorded history. The tanning of leather and generation of selected colored pigments depend on the coordination of metal ions. A number of biological agents, including plants and animals, owe their existence to coordinated polymers such as hemoglobin. Many of these coordination polymers have unknown and/or irregular structures.

The drive for the synthesis and characterization of synthetic coordination polymers was catalyzed by work supported and conducted by the United States Air Force in a search for materials that exhibited high thermal stabilities. Attempts to prepare highly stable, tractable coordination polymers were disappointing. Typically, only oligomeric products were formed and the monomeric versions were often more stable than the polymeric versions.

Bailar listed a number of principles that can be considered in designing coordination polymers. Briefly these are as follows: (1) Little flexibility is imparted by the metal ion or within its immediate environment; thus, flexibility must arise from the organic moiety. Flexibility increases as the covalent nature of metal–ligand bond increases. (2) Metal ions only stabilize ligands in their immediate

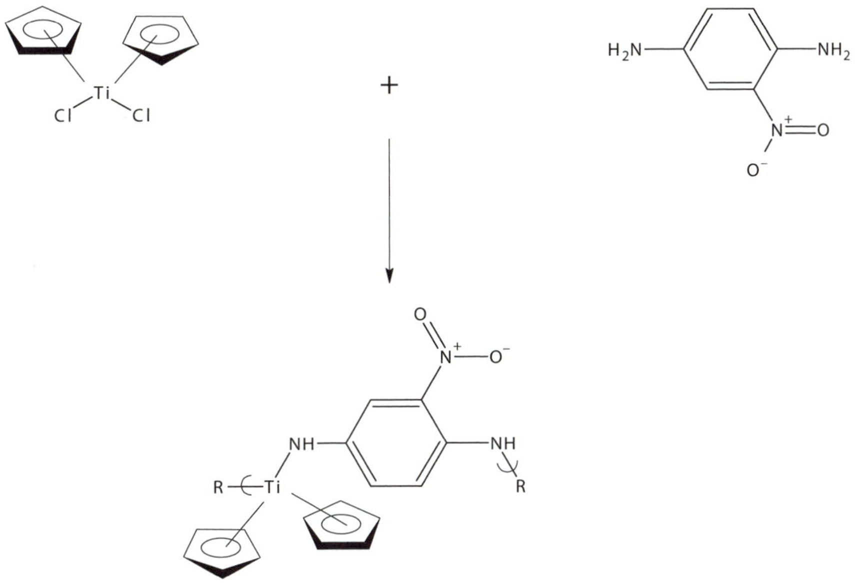

FIGURE 11.1 Synthesis of metallocene polyamines.

vicinity; thus, the chelates should be strong and close to the metal ions. (3) Thermal, oxidative, and hydrolytic stability are not directly related; polymers must be designed specifically for the properties desired. (4) Metal–ligand bonds have sufficient ionic character to permit them to rearrange more readily than typical "organic bonds." (5) Polymer structure (such as square planar, octahedral, linear, network) is dictated by the coordination number and stereochemistry of the metal ion or chelating agent. (6) Lastly, employed solvents should not form strong complexes with the metal or chelating agent or they will be incorporated into the polymer structure and/or prevent reaction from occurring.

Coordination polymers can be prepared by a number of routes, with the three most common being

(1) Preformed coordination metal complexes polymerized through functional groups where the actual polymer-forming step may be a condensation or addition reaction

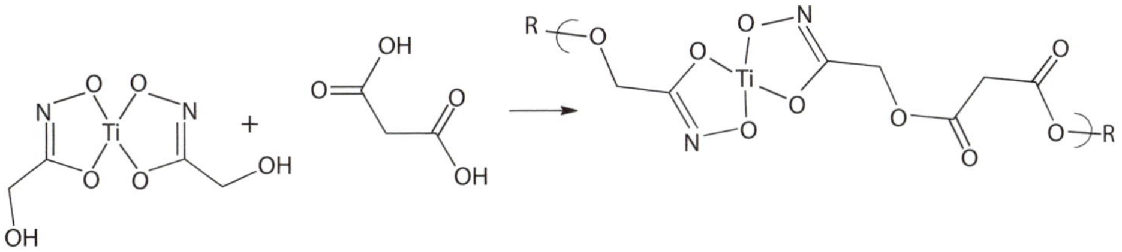

$$(11.24)$$

(2) Reaction with polymer-containing ligands; or

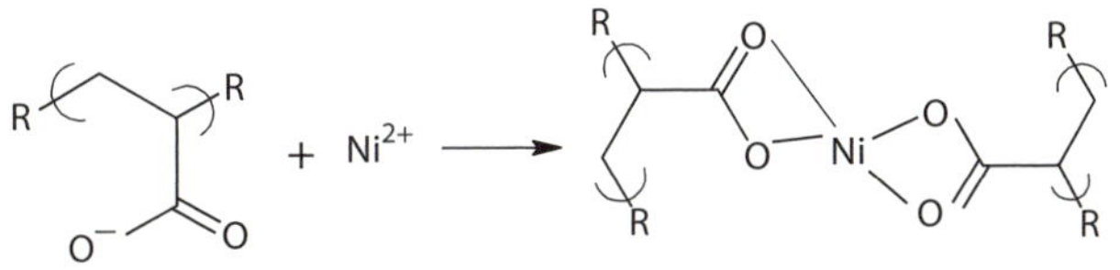

$$(11.25)$$

(3) Polymer formation through chelation

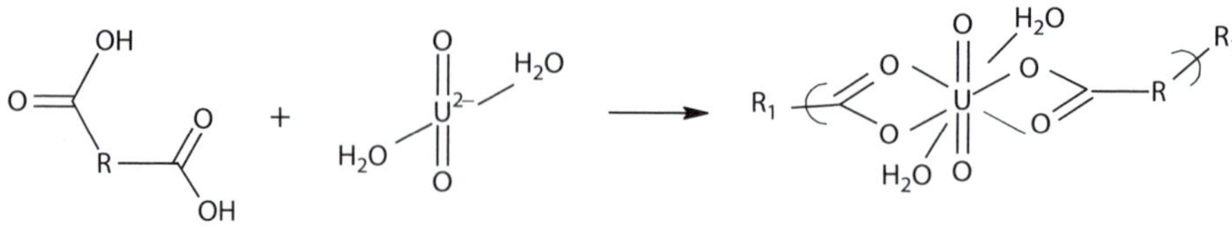

$$(11.26)$$

Carraher and coworkers employed the last two processes to recover the uranyl ion. The uranyl ion is the natural water-soluble form of uranium oxide. It is also toxic, acting as a heavy-metal toxin. Through the use of salts of dicarboxylic acids and poly(acrylic acid) the uranyl ion was removed to 10^{-5} M with the resulting product much less toxic and convertible to uranium oxide by heating.

Many of the organometallic polymers are semiconductors with bulk resistivities in the range of 10^3–10^{10} ohm cm suitable for specific semiconductor use. Further, some exhibit interesting photoproperties.

Simple chelation polymers are all about us, but they are not always recognized as such. During the winter, the north experiences freezing temperatures and the associated ice. Sodium chloride and calcium chloride are the most used freezing point lowering agents. They are inexpensive and readily available. On the negative side, they adversely affect surrounding plant life and must be reapplied generally after each rain–ice cycle.

Recently magnesium acetate has begun to be used for application in especially dangerous sites such as bridges. The acetate can be either internally bridged (11.27), where it is not polymeric,

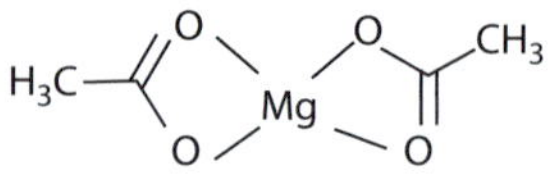

$$(11.27)$$

or it can be bridged forming a complex linear polymeric material (11.28).

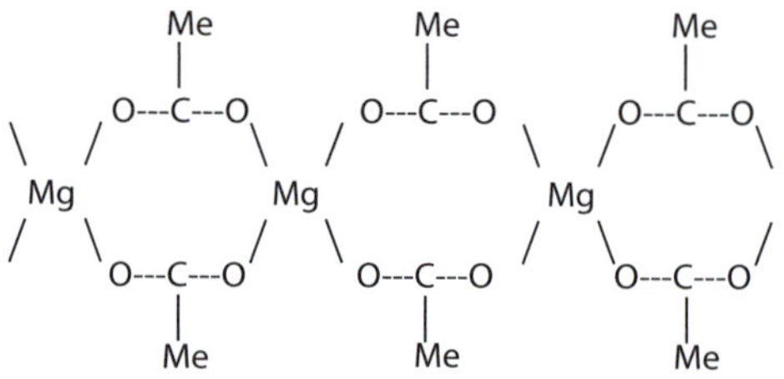

$$(11.28)$$

Hellmuth believes that the polymeric material forms as shown in Equation 11.28. (Note the C–O bonds contain double-bond character in Equation 11.28). If the magnesium acetate were simply the internally chelated material then it should be quickly washed away. The combination of calcium and magnesium acetates is known as CMA. CMA is more expensive than sodium chloride or calcium chloride, but it does not damage plant life and it has a much longer effective life. The polymer seeks the cracks and crevices in and around the pavement and remains until needed again to lower the freezing point of water. Often, one application of CMA is sufficient for a winter season.

11.4.1 Platinum-Containing Polymers

In 1964, Rosenberg and coworkers found that bacteria failed to divide but continued to grow. After much effort they found that the cause of this anomalous growth was a broken electrode and eventually identified the chemical as *cis*-dichlorodiamineplatinum II. This compound is now licensed under the name Platinol and is also known as cisplatin. Cisplatin is the most widely used anticancer drug. Carraher, Rosenberg, Allcock, Neuse, and others have reduced the toxic effects of cisplatin though placement of the platinum moiety into various coordination platinum polymers. Some of these polymers inhibit various cancer growths with much less toxic effects. Carraher and coworkers also found that many of these are also very active antiviral agents and some are able to prevent the onset of virally related juvenile diabetes in test animals. Recently, Carraher and Roner found that cisplatin derivatives formed from reaction of tetrachloroplatinum II and methotrexate inhibited a wide range of viruses in the nanograms/mL range. Similar results were found for the analogous product formed from reaction with tilorone (11.29).

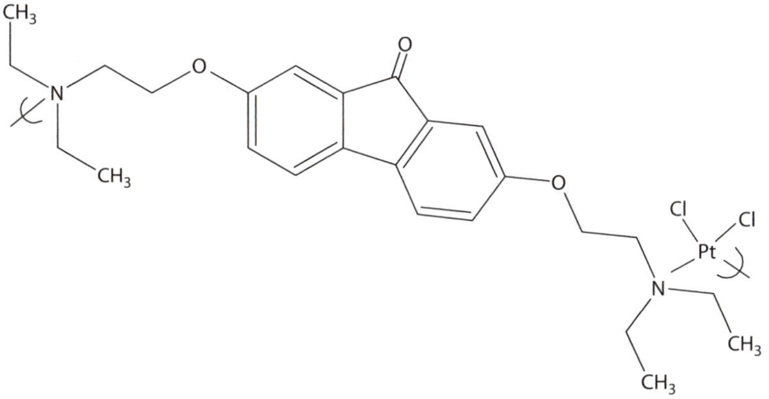

(11.29)

11.5 ADDITION POLYMERS

Sulfur nitride polymers (polythiazyls) [–(–S=N–)–], which have optical and electrical properties similar to those of metals, were first synthesized in 1910. These crystalline polymers, which are superconducive at 0.25 K, may be produced at room temperature using the solid-state polymerization of the dimer (S_2N_2). A dark blue–black amorphous paramagnetic form of poly(sulfur nitride) (Equation 11.30) is produced by quenching the gaseous tetramer in liquid nitrogen. The polymer is produced on heating the tetramer to about 300°C.

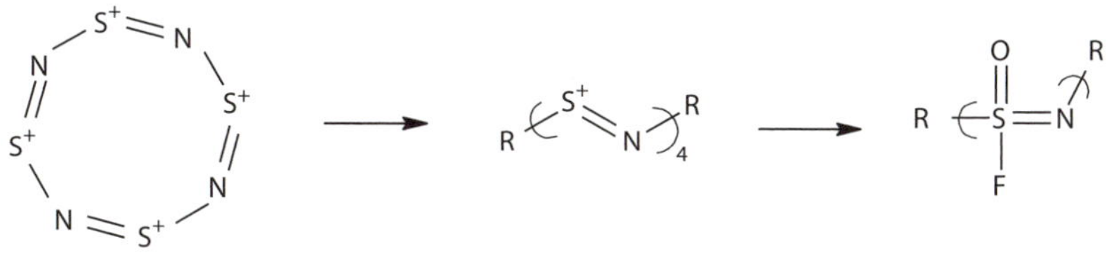

(11.30)

Rather than the simple up-and-down alternating structure for polythiazyls (11.31), the actual structure is more crankshaft like (11.32) caused by the uneven bond angles of SNS (120°) and NSN (103°).

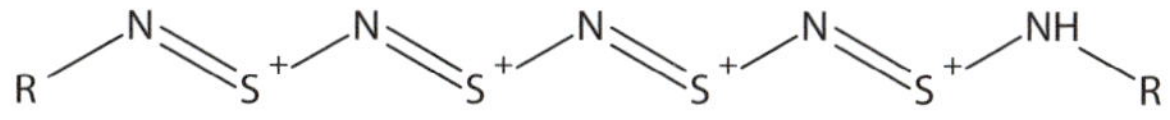

(11.31)

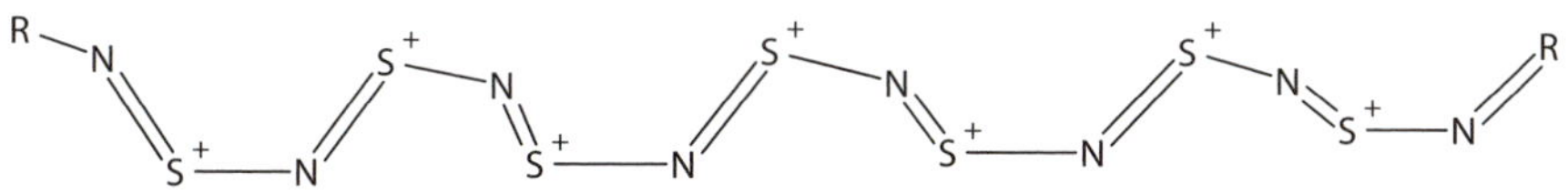

(11.32)

Much of the interest in the polysilanes, polygermanes, and polystannanes involves their sigma delocalization and their sigma-pi delocalization when coupled with arenes or acetylenes. This is not unexpected since silicon exists as a covalent network similar to diamond. In exhibiting electrical conductivity, germanium and tin show more typical "metallic" bonding. Some polystannanes have been referred to as "molecular metals."

Because of the interesting electronic and physical properties of polysilanes, a number of potential uses have been suggested, including precursors of beta-SiC fibers, impregnation of ceramics, polymerization initiators, photoconductors for electrophotography, contrast enhancement layers in photolithography, deep UV-sensitive photoresists, nonlinear optical materials, and self-developing by excimer laser for deep UV exposure. The unusual absorption spectra of polysilanes have indicated potential use in a number of conducting areas.

One area of active interest in ceramics is the formation of ceramics that may contain some fiber structure. Currently, ceramics, while very strong, are very brittle. Introduction of thermally stable fiber-like materials might allow the ceramics some flexibility before cleavage. Such materials might be considered as ceramic composites where the matrix is the ceramic portion and the fibers are the thermally stable fibers. Introduction of the fibers during the ceramic-forming step is a major obstacle that must be overcome. Carbon fibers have been investigated as have been other high-temperature materials such as the polysilanes. Polysilanes are formed from the six-membered ring through extended heating at 400°C (Equation 11.33).

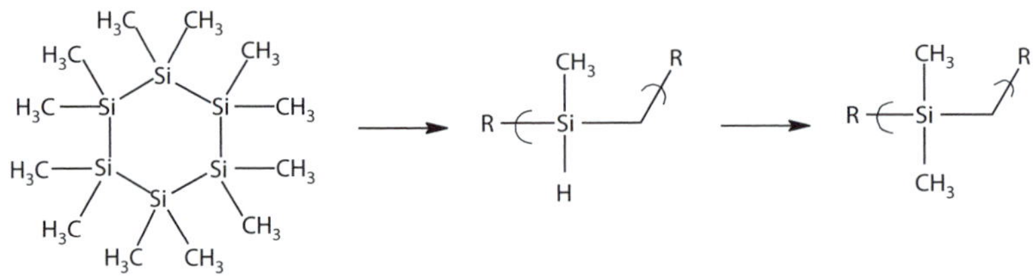

(11.33)

Further heating gives silicon carbide.

Table 11.4 lists a number of nonoxide ceramics that have been produced from the pyrolysis of polymers.

Boron carbonitride ceramics are formed from heating borazine with borazine derivatives (Equation 11.34).

Heating borazine yields a polymer connected by B–N bonds (Equation 11.35).

TABLE 11.4
Nonoxide Ceramics Produced from the Pyrolysis of Polymeric Materials

Polymer(s)	Resultant Ceramic
Poly(phosphonitric chlorides)	PN
Polysilanes, polycarbosilanes	SiC
Polyphenylborazole	BN
Polytitanocarbosilanes	Si–Ti–C
Polysilazanes	$Si_3 N_4$, Si–C–N

Borazine

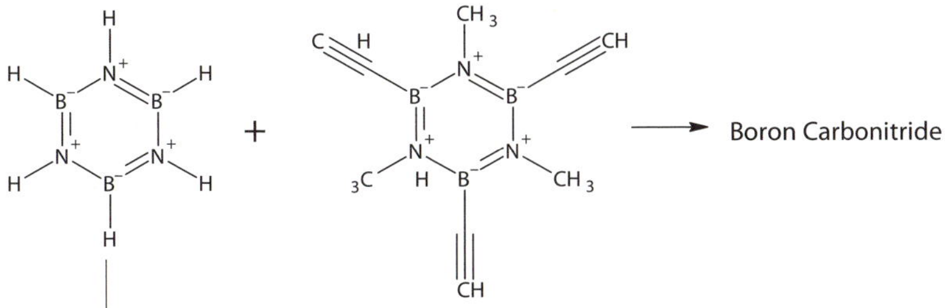

Boron Carbonitride   (11.34)

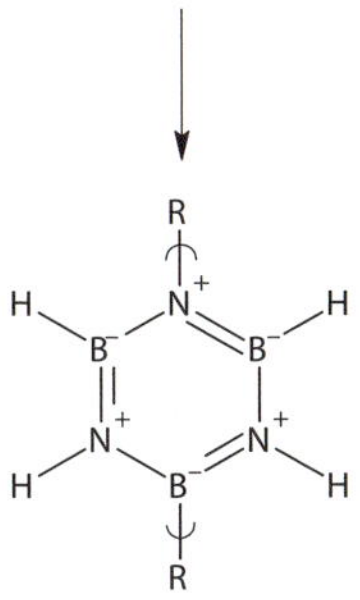

Polyborazylene   (11.35)

11.5.1 FERROCENE-CONTAINING AND RELATED POLYMERS

The landmark discovery of ferrocene by Kealy and Paulson in 1951 marked the beginning of modern organometallic chemistry. The first organometallic addition polymer was polyvinylferrocene synthesized by Arimoto and Haven in 1955. While polyvinylferrocene (11.36) had been synthesized, it was about another decade until the work of Pittman, Hayes and George, and Baldwin and Johnson allowed a launch of ferrocene-containing polymers.

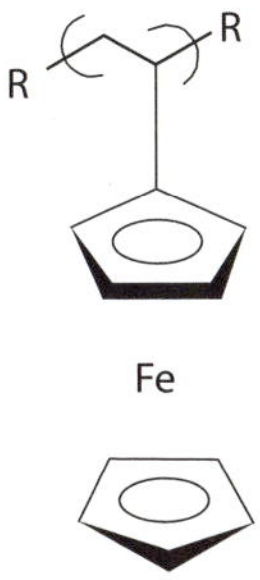

(11.36)

A large number of vinyl organometallic monomers have been prepared, homopolymerized, and copolymerized with classic vinyl monomers by Pittman and others. These include polymers containing Mo, W, Fe, Cr (11.37), Ir, Ru, Ti (11.38), Rh (11.39), and Co.

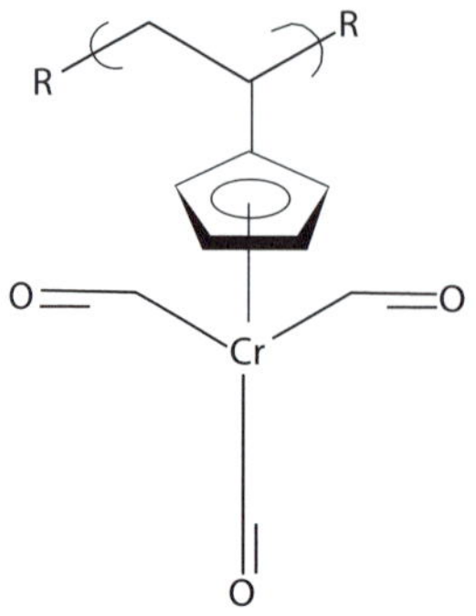

(11.37) 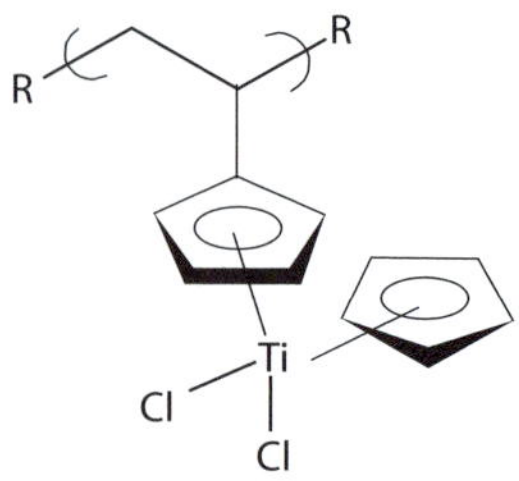   (11.38) 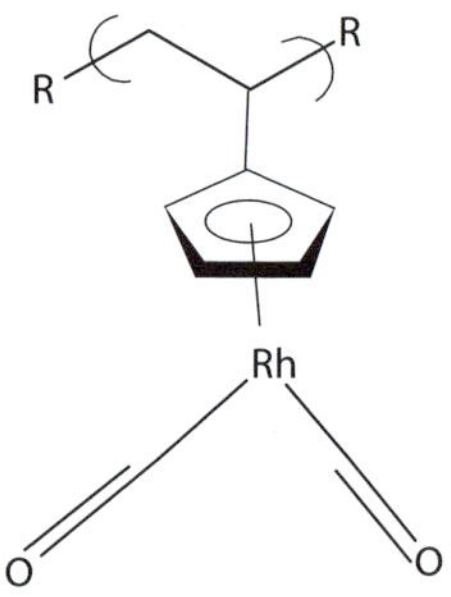   (11.39)

The effect that the presence of the organometallic function exerts in vinyl polymerizations is beginning to be fully understood. A transition metal may be expected, with its various readily available oxidation states and large steric bulk, to exert unusual electronic and steric effects during polymerization. The polymerization of vinyl ferrocene will be employed as an example. Its homopolymerization has been initiated by radical, cationic, coordination, and Ziegler–Natta initiators. Unlike the classic organic monomer styrene, vinylferrocene undergoes oxidation at the iron atom when peroxide initiators are employed. Thus, azo initiators (such as AIBN) are typically used. Here, we see one difference between an organic and an organometallic monomer in the presence of peroxide initiators. The stability of the ferricinium ion makes ferrocene readily oxidizable by peroxides, whereas styrene, for example, undergoes polymerization in their presence. Unlike most vinyl monomers, the molecular weight of polyvinylferrocene does not increase with a decrease in initiator concentration because of the unusually high chain-transfer constant for vinylferrocene. Finally, the rate law for vinylferrocene homopolymerization is first order in initiator in benzene. Thus, intramolecular termination occurs. Mossbauer studies support a mechanism involving electron transfer from iron to the growing chain radical giving a Zwitterion that terminates polymerization.

The high electron richness of vinylferrocene as a monomer is illustrated in its copolymerization with maleic anhydride, where 1:1 alternation copolymers are formed over a wide range of monomer feed ratios and $r_1r_2 = 0.003$. Subsequently, a large number of detailed copolymerization studies have been undertaken using metal-containing vinyl monomers.

Neuse acted as an early catalyst in the development of metal-containing polymers, including the use of ferrocene-containing polymers to fight cancer. One key feature of ferrocene is its ability to donate an electron from a nonbonding high-energy MO, resulting in the transformation of a neutral, diamagnetic site to the positive paramagnetic ferricenium ion radical (Equation 11.40). This formation occurs within a typical chemical environment and within selected biological environments.

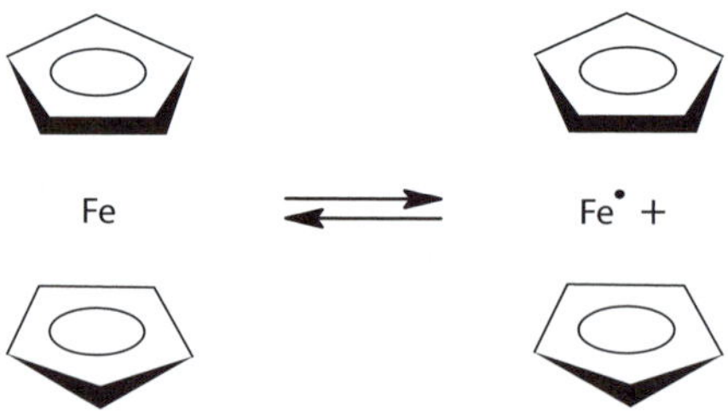

(11.40)

This ferricinium ion radical, as do other free radicals, readily reacts with other free radicals through recombination. Neuse and others have had good results at successfully inhibiting a wide variety of cancers using ferrocene-containing polymers. Often the employed compounds are elaborately designed with specially designed backbones and employing a ferrocene-containing unit as a tether dangling from the polymer backbone.

11.5.2 POLYPHOSPHAZENES AND RELATED POLYMERS

Other inorganic and metal-containing polymers have been formed using the addition approach. These include polyphosphazenes, polyphosphonitriles, and poly(sulfur nitride). Phosphonitrilic polymers (Equation 11.41) have been known for many years, but since they lacked resistance to water, they were not of interest as commercial polymers. However, when the pendant chlorine groups are replaced by fluorine atoms, amino, alkoxy, or phenoxy groups, these polymers are more resistant to hydrolysis. Allcock and coworkers have pioneered these efforts. Phosphonitrile fluoroelastomers are useful throughout a temperature range of –56°C to 180°C. Phosphazenes are produced by the thermal cleavage of a cyclic trimer obtained from the reaction of phosphorus pentachloride and ammonium chloride.

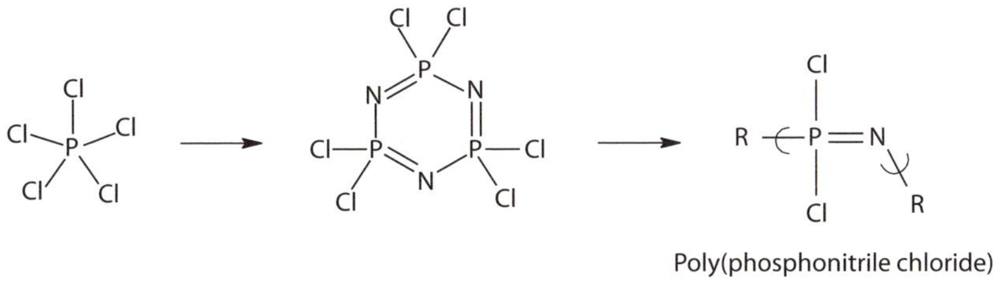

$$(11.41)$$

Poly(phosphonitrile chloride)

Amorphous elastomers are obtained when phosphazene is refluxed with nucleophiles, such as sodium trifluoroethoxide (Equation 11.42) or sodium cresylate, and secondary amines. Difunctional reactants such as dihydroxybenzenes (hydroquinone) produce cross-linked phosphazenes.

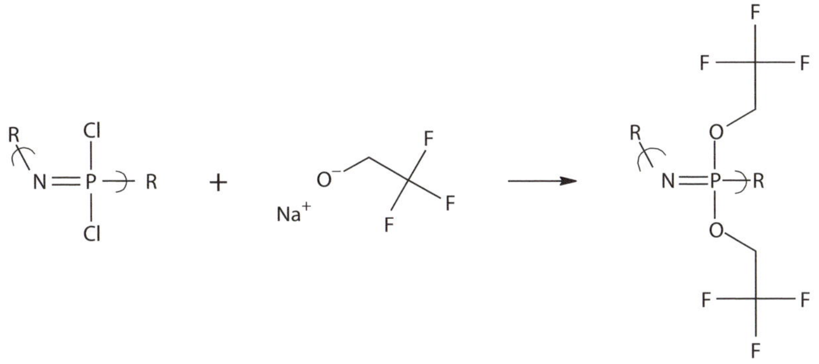

$$(11.42)$$

Polyphosphazenes generally exhibit very low T_g values consistent with the barriers to internal rotation being low and indicate the potential of these polymers for elastomer applications. In fact, theoretical calculations based on a rotational isomeric model assuming localized pi bonding predict the lowest (400 J/mol repeating unit) known polymer barrier to rotation for the skeletal bonds for polydifluorophosphazene. Temperature intervals between T_g and T_m are unusually small and generally fall outside the frequently cited relationship. This behavior may be related to complications in the first-order transition generally found for organo-substituted phosphazenes and not common to other semicrystalline polymers. Two first-order transitions are generally observed for organo-substituted phosphazenes with a temperature interval from about 150°C–200°C. The lower first-order transition can be detected using differential scanning calorimetry (DCS), differential thermal analysis (DTA), and thermomechanical analysis (TMA). Examination by optical micros-copy reveals that the crystalline structure is not entirely lost but persists throughout the extended temperature range to a higher temperature transition, which appears to be T_m, the true melting tem-perature. The nature of this transitional behavior resembles the transformation to a mesomorphic state similar to that observed in nematic liquid crystals. It appears from the relationship between the equilibrium melting temperature (heat and entropy of fusion; $T_m = H_m/S_m$) and the low value of H_m at T_m compared with the lower transition temperature that the upper transition, T_m, is character-ized by a very small entropy change. This may be due to an onset of chain motion between the two transitions leading to the small additional gain in conformational entropy at T_m. The lower transition is believed to correspond to the T_g.

Allcock and coworkers have employed polyphosphazenes in a variety of uses, including the broad areas of biomedical and electrical. From a practical point of view, polyphosphazenes are usually soft just above the lower transition so that compression molding of films can be carried out. This suggests that the lower transition temperature represents the upper temperature for most useful engineering applications of polyphosphazenes in unmodified forms.

11.5.3 BORON-CONTAINING POLYMERS

Organoboron has been incorporated into polymers employing a variety of techniques. Stock in the 1920s first created a boron hydride polymer during his work on boron hydrides. Much of the current interest in boron-containing polymers is a consequence of three factors. First, the presence of a "low-lying" (meaning low energy) vacant "p" orbital allows its use in moving electrons in a conjugated system. This is being taken advantage of through the synthesis of various pi-conjugated systems and their use in optical and sensing applications. This includes use in light-emitting diodes (LEDs), nonlinear optical systems, energy storage in batteries, and the construction of sensing devices. One such polymer structure is given in structure 11.43.

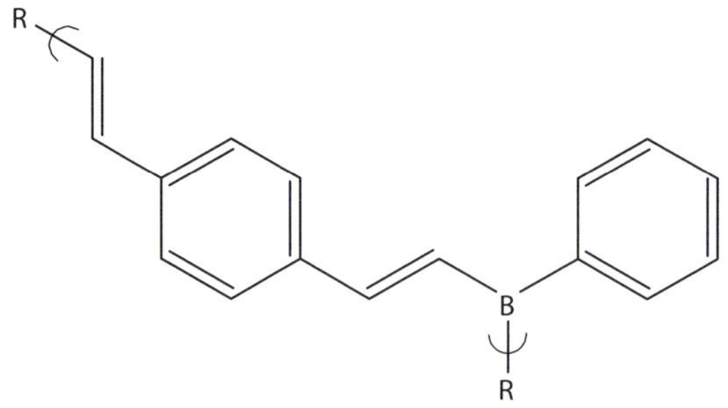

$$(11.43)$$

These polymers are mainly synthesized employing the hydroboration reaction which is simply the addition of a hydrogen from a boron hydride to a double or triple bond (Equation 11.44).

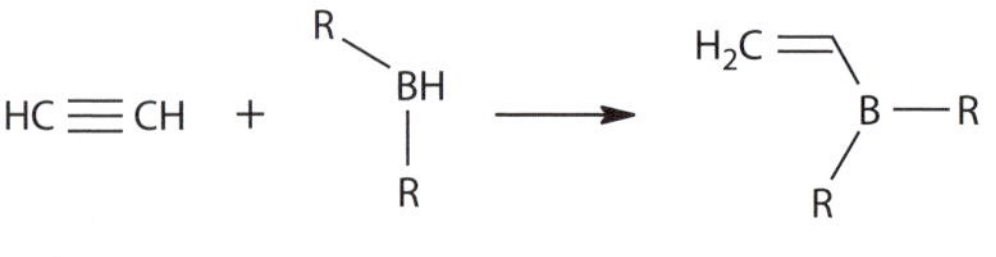

$$(11.44)$$

This low-lying vacant p orbital also allows boron-containing polymers to be luminescent again signaling potential optical applications. Many of these luminescent materials are NLO materials.

The second reason for interest in boron polymers involves their use as catalysts. While the boron atom can be used as the site of catalytic activity, more effort has involved the use of boron-containing materials as blocking and protecting agents and as cocatalysts. They are increasingly being used in catalytic asymmetric syntheses.

The third reason for interest in boron polymers involves the ability of many boron compounds to form cocoons about objects allowing a ready method for coating wires and fibers. Thus boron-containing polymers and monomers have been employed to form a surface layer of intumescent protective char that acts as a barrier to oxygen protecting the wires and fibers from ready oxidation. These coatings also provide flame retardancy to the coated materials. Recently, boron-containing units have been incorporated into polymers, resulting in materials that have added flame resistance through char formation. Somewhat related to this is the use of boron-containing polymers in forming high-strength fibers and whiskers for use in composites. Finally, boron has a high capture ability of neutrons so effort had gone into using this nuclear characteristic.

Today there exist a wide variety of boron-containing polymers, including ring systems such as borazines (11.45), boroxines (11.46), and triphosphatoborins (11.47),

Borazine   Boroxine   Triphosphatoborin

(11.45)   (11.46)   (11.47)

as well as metal, metalloid, ferrocene, and so on containing polymers each offering their own potential for exhibiting desired properties.

11.6   ION-EXCHANGE RESINS

Just as calcium ions form an insoluble compound through reaction with the carbonate ion (calcium carbonate) so also does it complex with a carbonyl functional group on a polymer. This concept forms the basis for many analyses, separations, and concentration techniques. Many of these techniques are based on organic resins, silicon dioxide intense compounds such as the zeolites, and on carbohydrate-related compounds such as dextran-based resins. The functional groups include typically fully charged sulfates, sulfonates, and acids and noncharges groups such as amines, imines, and hydroxyls. The functional groups are normally located on the surface of somewhat spherical beads. The use of negatively charged functional groups to preferentially capture cations is referred to as cation exchange. Amines are often protonated forming cations that attract anions and are called anion-exchange resins. The combinations of reactions generally are simple Lewis acid–base reactions.

While acid groups attract cations, the neutralized acid groups, salts, are more effective. The tendency of coordination is related to the size of the cation and the charge on the cation. In general, multiple-charged cations such as Ca^{+2}, Mg^{+2}, and $Fe^{+3,+2}$ are more strongly coordinated than single-charged cations as Na^{+1} and K^{+1}. This difference in tendency to coordinate is widely used in ion-exchange resin applications.

When two or more coordinations occur with a single cation, it is called chelation after the Greek word for the claw of a crab.

Applications of cation and anion resins are varied and include the purification of sugar, identification of drugs and biomacromolecules, concentration of uranium, calcium therapy to help increase the amount of calcium in our bones (that is, increase the bone density), and use as therapeutic agents for the control of bile acid and gastric acidity. In the latter use, a solid polyamide (Colestid) is diluted and taken with orange juice to help in the body's removal of bile acids. Removal of the bile acids

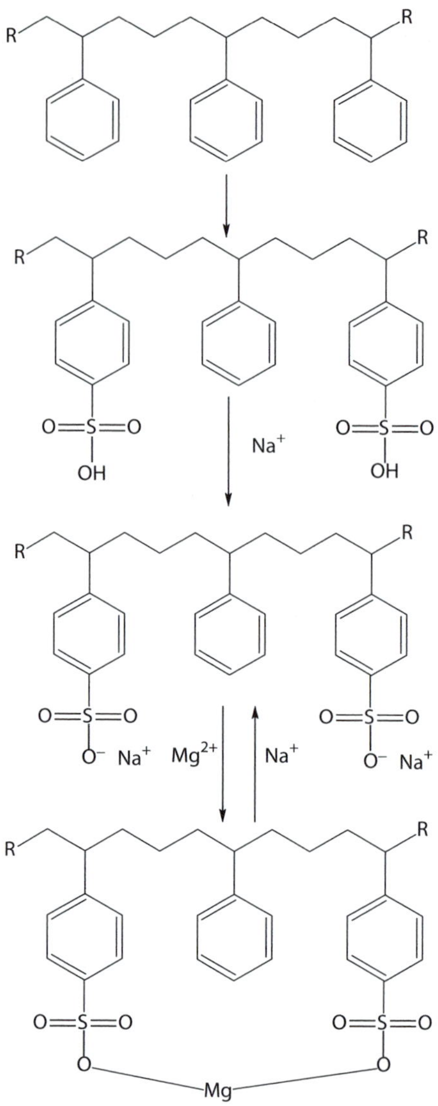

FIGURE 11.2   The ion-exchange cycle from top to bottom—sulfonation some of the phenyl rings; formation of the sodium form; metal chelation and regeneration.

causes the body to produce more bile acid from cholesterol, thus, effectively reducing the cholesterol level.

The Merrifield protein synthesis (employing chloromethylated polystyrene as the substrate) makes use of ion-exchange resins as do many of our industrial and home water purifiers. Water containing dissolved Ca^{+2}, Mg^{+2}, Fe^{+2}, and Fe^{+3} is called hard water. These ions act to reduce the effectiveness of detergents and soaps by coordinating with them producing solid scum. The precipitates may also be deposited in pipes and water heaters forming boiler scale. In bathtubs, they form the ring that must be scrubbed to remove it. The ions are generally from natural sources such as the passing of water over and through limestone ($CaCO_3$).

Most home water softeners are based on ion-exchange resins. The first ion-exchange materials used in softening water were naturally occurring polymeric aluminum silicates called zeolites.

Synthetic zeolites are also used today for this purpose. Today, most ion-exchange resins are based on styrene and divinylbenzene (vinylstyrene) resins that are then sulfonated. When the resin is ready for use, sodium ions generated from rock salt (simple sodium chloride) is passed through the resin bed replacing the hydrogen ion (protons). The sulfonate functional groups have a greater affinity for multiple-charged cations than for the single-charged sodium ion and the multivalent metal ions replace the sodium ions, resulting in the water having a lower concentration of the ions responsible for the water being hard. Eventually, the sulfonate sites on the resin become filled and the resin bed must be recharged by adding large amounts of dissolved sodium ions (derived from sodium chloride) which displace the more tightly bound but overwhelmingly outnumber "hard ions." After the system is flushed free of these "hard ions," the resin bed is again ready to give "soft water" for our use. This sequence is described in Figure 11.2.

11.7 SUMMARY

1. There is a wide variety of inorganic and metal-containing polymers. The potential uses are many and include the broad areas of biomedical, electrical, optical, analytical, catalytic, building, and photochemical applications.
2. The bond strength from any combination is higher than for many traditional polymers with many having superior thermal stabilities.
3. Metal and inorganic polymers can be formed through a variety of reaction types, including condensation, coordination, and addition reactions.
4. The majority of the condensation polymerizations can be considered extensions of typical Lewis acid–base reactions.
5. Polysiloxanes (silicons) offer a good combination of properties not found in organic polymers. Silicons are employed in a number of applications, including antifoaming agents, lubricants, caulks, sealants, gaskets, and as biomaterials.
6. Polyphosphazenes offer unique thermal properties and have shown a number of uses in the field of electronics and medicine.
7. The number and variety of organometallic polymers and potential applications for organometallic polymers is great. Because of the high cost of production of many of these materials, uses will often be limited to applications employing minute quantities of the polymers. This is not true for many polymers containing silicon, tin, and main-group materials since these are available in large quantities at reasonable cost.

GLOSSARY

Borazoles: Molecules composed of boron and nitrogen atoms.
Capping: Protecting end groups.
Carboranes: Molecules composed of carbon and boron atoms.
Coordination polymers: Polymers based on coordination complexes.
Metallocenes: Sandwich or distorted sandwich-like molecules generally containing two cyclopentadienes and a metal atom bonded to them.
Polyphosphonitrile: Polymer with a repeat unit of –P=N–
RTV: Room temperature vulcanization.
Siloxanes and Silicones: Polymers containing –Si–O– backbones.

EXERCISES

1. What is meant by "lost loops" in the production of silicones?
2. How could you produce a silicone with a low DP?

3. What would you estimate the solubility parameters of silicones to be?
4. Sodium silicate is water soluble (forming water glass), but silicones are water repellents. Explain the difference.
5. How might you polymerize an aqueous solution of sodium silicate?
6. How might you explain the good thermal stability of silicones?
7. Show the repeat unit for polydiethylsiloxane.
8. What are the reactants used to make phosphazenes?
9. Why would you predict that the chloro groups in phosphonitrilic polymers would be attacked by water?
10. Which phosphazene would be more flexible—one made by reaction of poly(phosphonitrilic chloride) with (a) sodium trifluoroethoxide or (b) sodium trifluorobutoxide?
11. Show the structure of borazole.
12. Since tin-containing organometallic polymers are used in marine antifouling coatings, what would by predict about their water resistance?
13. In addition to high cost, name another disadvantage of coordination polymers.
14. What is the ceiling temperature of sulfur nitride polymers?
15. What are the main uses of tin-containing polymers?
16. What is the basis for many of the electrical and optical applications of boron-containing polymers?
17. Why is there interest in the synthesis and study of metal-containing polymers?
18. What are some attractive features of polysiloxanes?
19. Name two biological polymers that contain metals and which are important to life.
20. What is a main difference between ionic and covalent bonds remembering that most bonds have both characteristics involved in them?

ADDITIONAL READING

Abd-El-Aziz, A., Carraher, C., Pittman, C., Sheats, J., Zeldin, M. (2003): *A Half Century of Metal- and Metalloid-Containing Polymers*, Wiley, Hoboken.

Abd-El-Aziz, A., Carraher, C., Pittman, C., Sheats, J., Zeldin, M. (2004): *Biomedical Applications of Metal-Containing Polymers*, Wiley, Hoboken.

Abd-El-Aziz, A., Carraher, C., Pittman, C., Sheats, J., Zeldin, M. (2004): *Ferrocene Polymers*, Wiley, Hoboken.

Abd-El-Aziz, A., Carraher, C., Pittman, C., Zeldin, M. (2005): *Coordination Polymers*, Wiley, Hoboken.

Abd-El-Aziz, A., Carraher, C., Pittman, C., Zeldin, M. (2005): *Group IVA-Containing Polymers*, Wiley, Hoboken.

Abd-El-Aziz, A., Carraher, C., Pittman, C., Zeldin, M. (2005): *Transition Metal Polymers*, Wiley, Hoboken.

Abd-El-Aziz, A., Carraher, C., Pittman, C., Zeldin, M. (2005): *Nanoscale Interactions of Metal-Containing Polymers*, Wiley, Hoboken.

Abd-El-Aziz, A., Carraher, C., Pittman, C., Zeldin, M. (2007): *Boron-Containing Polymers*, Wiley, Hoboken.

Abd-El-Aziz, A., Carraher, C., Pittman, C., Zeldin, M. (2007): *Polymers Containing Transition Metals*, Springer, NY.

Abd-El-Aziz, A., Carraher, C., Pittman, C., Sheats, J., Zeldin, M. (2006): *Nanoscale Interactions of Metal-Containing Polymers*, Wiley, Hoboken.

Abd-El-Aziz, A., Carraher, C., Pittman, C., Sheats, J., Zeldin, M. (2007): *Boron-Containing Polymers*, Wiley, Hoboken.

Abd-El-Aziz, A., Carraher, C., Pittman, C., Sheats, J., Zeldin, M. (2009): *Inorganic Supramolecular Assemblies*, Wiley, Hoboken.

Abd-El-Aziz, A., Carraher, C., Pittman, C., Sheats, J., Zeldin, M. (2010): *Photophysics and Photochemistry*, Wiley, Hoboken.

Abd-El-Aziz, A., Carraher, C., Pittman, C., Sheats, J., Zeldin, M. (2008): *Inorganic and Organometallic Macromolecules: Design and Application*, Springer, NY.

Abd-El-Aziz, A., Manners, I. (2007): *Frontiers in Transition Metal-Containing Polymers*, Wiley, Hoboken.

Allcock, H. R. (1972): *Phosphorus-Nitrogen Compounds*, Academic, NY.

Archer, R. (2004): *Inorganic and Organometallic Polymers*, Wiley, Hoboken.

Brinker, C., Scherer, D. (1990): *The Physics and Chemistry of Sol-Gel Processing*, Academic, Orlando, FL.

Brook, M. (1999): *Silicon in Organic, Organometallic, and Polymer Chemistry*, Wiley, NY.

Chandrasekhar, V. (2005): *Inorganic and Organometallic Polymers*, Springer, NY.

Chen, L., Hong, M. (2009): *Design and Construction of Coordination Polymers*, Wiley, Hoboken, NJ.

Dvornic, P., Owen, M. (2009): *Silicon-Containing Dendrite Polymers*, Springer, NY.

Ganachaud, F., Boileau, S., Boury, B. (2008): *Silicon Based Polymers: Advances in Synthesis and Supramolecular Organization*, Springer, NY.

Innocenzi, P., Zub, Y., Kessler, V. (2008): *Sol-Gel for Materials Processing: Focusing on Materials for Pollution Control, Water Purification, and Soil Remediation*, Springer, NY.

Jones, R. (2001): *Silicon-Containing Polymers*, Kluwer, NY.

Jones, R., Andeo, W., Chojnowski, J. (2000): *Silicon-Containing Polymers*, Kluwer, NY.

Kipping, F. S. (1927): Silicones, *J. Chem. Soc.*, 130:104.

Ladenburg, A. (1872): Silicones, *Ann. Chem.*, 164:300.

Manners, I. (2005): *Synthetic Metal-Containing Polymers*, Wiley, Hoboken.

Mark, J. E., Allcock, H. R., West, R. (2006): *Inorganic Polymers*, Oxford University Press, NY.

Marklund, S. (2008): *Organophosphorus Flame Retarders and Plasticizers*, VDM Verlag, NY.

Neuse, E. W., Rosenberg, H. (1970): *Metallocene Polymers*, Dekker, NY.

Pittman, C. Carraher, C., Zeldin, M., Culbertson, B., Sheats, J. (1996): *Metal-Containing Polymeric Materials*, Plenum, NY.

Rochow, E. G. (1951): *An Introduction to the Chemistry of Silicones*, Wiley, NY.

Schubert, U. S., Newkome, G. R., Manners, I. (2006): *Metal-Containing and Metallosupramolecular Polymers and Materials*, Oxford University Press, NY.

Stepnicka, P. (2008): *Ferrocenes: Ligands, Materials and Biomolecules*, Wiley, Hoboken, NJ.

Tsuchida, E. (2000): *Macromolecular-Metal Complexes*, Wiley, NY.

12 Inorganic Polymers

12.1 INTRODUCTION

Just as polymers abound in the world of organics, they also abound in the world of inorganics. Inorganic polymers are the major components of soil, mountains, and sand. Inorganic polymers are also extensively employed as abrasives and cutting materials (diamond, boron carbide, silicon carbide (carborundum), aluminum oxide), coatings, flame retardants, building and construction materials (window glass, stone, Portland cement, brick, tiles), and lubricants and catalysts (zinc oxide, nickel oxide, carbon black, graphite, silica gel, alumina, aluminum silicate, chromium oxides, clays).

The first somewhat man made, semisynthetic polymer was probably inorganic in nature. Alkaline silicate glass was used in the Badarian period in Egypt (about 12,000 BC) as a glaze, which was applied to steatite after it had been carved into various animal, and so forth, shapes. Frience, a composite containing powered quarts or steatite core covered with a layer of opaque glass, was employed from about 9,000 BC to make decorative objects. The earliest known piece of regular (modern day type) glass, dated to 3,000 BC, is a lion's amulet found at Thebes and now housed in the British Museum. This is a blue opaque glass partially covered with a dark green glass. Transparent glass appeared about 1,500 BC. Several fine pieces of glass jewelry were found in Tutankhamen's tomb (ca 1,300 BC), including two bird heads of light blue glass incorporated into the gold pectoral worn by the Pharaoh.

Because of the wide variety and great number of inorganic polymers, this chapter will focus on only a few of the more well-known inorganic polymers. Table 12.1 contains a partial listing of common inorganic polymers.

Along with the silicates that will be dealt within other sections, there are many other inorganic polymers based on other units. One of these is the hydroxylapatite or hydroxyapatite materials that have general formula of $Ca_5(PO_4)_3(OH)$, which is a member of the apatite group. Seventy percent of bone is composed of hydroxyapatite. Dental enamel is made from carbonated-calcium deficient hydroxylapatite nanotubes. It is believed that the initial step is the production of composite nanospheres of nanocrystallite apatite and amelogenin. These then aggregate forming nanorods about 50 nm in diameter and 250 nm long. These nanorods further assemble forming elongated crystals that compose our dental enamel.

12.2 PORTLAND CEMENT

Portland cement is the least expensive, most widely used synthetic inorganic polymer. It is employed as the basic nonmetallic, nonwoody material of construction. Concrete highways and streets span our country side and concrete skyscrapers silhouette the urban skyline. Less spectacular uses are found in everyday life as sidewalks, fence posts, and parking bumpers.

The name "Portland" is derived from the cement having the same color as the natural stone quarried on the Isle of Portland, a peninsula on the south of Great Britain. The word cement comes from the Latin word *caementum*, which means "pieces of rough, uncut stone." Concrete comes from the Latin word *concretus*, meaning "to grow together."

Common (dry) cement consists of anhydrous crystalline calcium silicates (the major ones being tricalcium silicate, Ca_3SiO_5, and β-dicalcium silicate, Ca_2SiO_4), lime (CaO, 60%), and alumina (a complex aluminum silicate, 5%). While cement is widely used and has been studied in good

TABLE 12.1
Important Inorganic Polymers

Agate	Chabazite	Glasses (many kinds)	Spodumene
Alumina	Chett	Graphite	Stilbite
Aluminum oxide	Chrysotile	Imogolite	Stishorite
Amphiboles	Concrete	Kaolinite	Sulfur nitride
Anthophylite	Cristobalite	Mesolite	Talc
Arsenic selenide	Crocidolite	Mica	Thomsonite
Arsenic sulfide	Diamond	Montmorillonite	Tremolite
Asbestos	Dickite	Muscovite	Tridymite
Berlinite	Epistilbite	Phosphorus oxynitride	Valentinite
Beryllium oxide	Feldspars	Polyphosphates (many)	Vermiculite
Boron nitride	Flint	Quartz	Wollastonite
Boron oxides	Fuller's earth	Rhodonite	Xonotlite
Boron phosphate	Garnet	Serpentine	Ziolites
Calcite	Germanium selenide	Silicon dioxides (many)	Zirconia
Carbon black	Gibbsite	Silicon carbide	

detail, yet its structure and the process whereby it is formed are not completely known. This is due to at least two factors. First, its three-dimensional arrangement of various atoms has a somewhat ordered array when a small (molecular level) portion is studied, but as larger portions are viewed, less order is observed giving only an average overall structure. This arrangement is referred to as short-range order and long-range disorder and is a good description of many three-dimensional, somewhat amorphous inorganic and organic polymers. Thus, there exists only an average structure for the cement that varies with amount of water and other components added, time after application (i.e., age of the cement), and source of concrete mix and location (surface or internal). Second, three-dimensional materials are insoluble in all liquids; therefore, tools of characterization and identification that require materials to be in solution cannot be employed to assist in the structural identification of cement.

When anhydrous cement mix is added to water, the silicates react, forming hydrates and calcium hydroxide. Hardened Portland cement contains about 70% cross-linked calcium silicate hydrate and 20% crystalline calcium hydroxide.

$$2Ca_3SiO_5 + 6H_2O \rightarrow Ca_3Si_2O_7 \cdot 3H_2O + 3Ca(OH)_2 \tag{12.1}$$

$$2Ca_2SiO_4 + 4H_2O \rightarrow Ca_3Si_2O_7 \cdot 3H_2O + Ca(OH)_2 \tag{12.2}$$

A typical cement paste contains about 60%–75% water by volume and only about 40%–25% solids. The hardening occurs through at least two major steps (Figure 12.1). First a gelatinous layer is formed on the surface of the calcium silicate particles. The layer consists mainly of water with some calcium hydroxide. After about 2 h, the gel layer sprouts fibrillar outgrowths that radiate from each calcium silicate particle. The fibrillar tentacles increase in number and length, becoming enmeshed and integrated. The lengthwise growth slows, with the fibrils now joining up sideways, forming striated sheets that contain tunnels and holes. During this time, calcium ions are washed away from the solid silicate polymeric structures by water molecules and react further, forming additional calcium hydroxide. As particular local sites become saturated with calcium hydroxide, calcium hydroxide itself begins to crystallize, occupying once vacant sites and carrying on the process of interconnecting about and with the silicate "jungle."

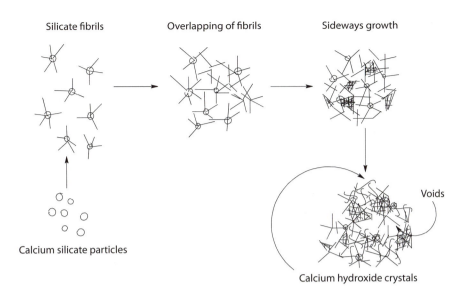

FIGURE 12.1 Steps in the hardening of Portland cement.

TABLE 12.2
Sample Concrete Mix

Material	Amount	
	By Volume	**By Weight**
Portland cement	90–100 lb (1 cubic foot)	90–100 lb (40–50 kg)
Water	5.5 gal	45 lb (20 kg)
Sand	2 cubic feet	200 lb (90 kg)
Gravel (small rocks)	3 cubic feet	250 lb (120 kg)

In spite of attempts by the silicate and calcium hydroxide to occupy all of the space, voids are formed, probably from the shrinkage of the calcium hydroxide as it forms a crystalline matrix. (Generally, crystalline materials have higher densities than amorphous materials; thus, a given amount will occupy less volume, leaving some unfilled sites.) Just as a chain is no stronger than its weakest link so is cement no stronger than its weakest sites, that is, its voids. Much current research concerns attempts to generate stronger cement with the focus on filling these voids. Interestingly, two of the more successful cement-void-fillers are also polymers–dextran, a polysaccharide, and polymeric sulfur.

Table 12.2 shows a typical concrete mix. The exact amounts may vary by as much as 50% depending on the intended use and preference of the concrete maker.

The manufacture of Portland concrete consists of three basic steps—crushing, burning, and finish grinding. As noted before, Portland cement contains about 60% lime, 25% silicates, and 5% alumina with the remainder being iron oxides and gypsum. Most cement plants are located near limestone ($CaCO_3$) quarries since this is the major source of lime. Lime may also come from oyster shells, chalk, and a type of clay called *marl*. The silicates and alumina are derived from clay, silicon sand, shale, and blast-furnace slag.

12.3 OTHER CEMENTS

There are a number of cements specially formulated for specific uses. *Air-entrained concrete* contains small air bubbles formed by the addition of soap-like resinous materials to the cement or to the concrete when it is mixed. The bubbles permit the concrete to expand and contract (as temperature changes) without breaking (since the resistance of air to changes in the concrete volumes is small). *Light-weight concrete* may be made through the use of light-weight fillers such as clays and pumice in place of sand and rocks or through the addition of chemical foaming agents that produce air pockets as the concrete hardens. These air pockets are typically much larger than those found in air-entrained concrete.

Reinforced concrete is made by casting concrete about steel bars or rods. Most large cement-intense structures such as bridges and skyscrapers employ reinforced concrete. *Prestressed concrete* is typically made by casting concrete about steel cables stretched by jacks. After the concrete hardens, the tension is released, resulting in the entrapped cables compressing the concrete. Steel is stronger when tensed, and concrete is stronger when compressed. Thus, prestressed concrete takes advantage of both of these factors. Archways and bridge connections are often made from prestressed concrete.

Concrete masonry is simply the name given to the cement building blocks employed in the construction of many homes, and it is simply a precast block of cement, usually with lots of voids. *Precast concrete* is concrete that is cast and hardened before it is taken to the site of construction. Concrete sewer piped, wall panels, beams, grinders, and spillways are all examples of precast cements.

The cements cited above are all typically derived from Portland cement. Following are non-Portland cements.

Calcium-aluminate cement has a much higher percentage of alumina than does Portland cement. Furthermore, the active ingredients are lime, CaO, and alumina. In Europe it is called *melted* or *fused cement*. In the United States it is manufactured under the trade name Lumnite. Its major advantage is its rapidity of hardening, developing high strength within a day or two.

Magnesia cement is largely composed of magnesium oxide (MgO). In practice, the MgO is mixed with fillers and rocks and an aqueous solution of magnesium chloride. This cement sets up (hardens) within 2–8 h and is employed for flooring in special circumstances.

Gypsum, or hydrated calcium sulfate ($CaSO_4 \cdot 2H_2O$), serves as the basis of a number of products, including plaster of Paris (also known as molding plaster, wall plaster, and finishing plaster). The ease with which plaster of Paris and other gypsum cements can be mixed and cast (applied) and the rapidity with which they harden contribute to their importance in the construction field as a major component for plaster wall boards. Plaster of Paris' lack of shrinkage in hardening accounts for its use in casts. Plaster of Paris is also employed as a dental plaster, pottery plaster, and as molds for decorative figures. Unlike Portland cement, plaster of Paris requires only about 20% water and dries to the touch in 30–60 min giving maximum strength after 2–3 days. Portland cement requires several weeks to reach maximum strength.

12.4 SILICATES

Silicon is the most abundant metal-like element in the earth's crust. It is seldom present in pure elemental form, but rather is present in a large number of polymers largely based on the polycondensation of the orthosilicate anion, SiO_4^{-4} as illustrated following:

$$SiO_4^{-4} \rightleftarrows Si_2O_7^{-6} + O^{-2} \qquad (12.3)$$
$$\phantom{SiO_4^{-4}}(1) (2)$$

$$Si_2O_7^{-6} + SiO_4^{-4} \rightleftarrows Si_3O_9^{-6} + 2O^{-2} \qquad (12.4)$$
$$(3)$$

$$2Si_2O_7^{-6} \rightleftarrows Si_4O_{10}^{-4} + 4O^{-2} \qquad (12.5)$$
$$(4)$$

$$2Si_2O_7^{-6} \rightleftarrows Si_4O_{11}^{-6} + 3O^{-2} \qquad (12.6)$$
$$(5)$$

$$2Si_2O_7^{-6} \rightleftarrows Si_4O_{12}^{-8} + 2O^{-2} \qquad (12.7)$$
$$(6)$$

$$3Si_2O_7^{-6} \rightleftarrows Si_6O_{18}^{-12} + 3O^{-2} \qquad (12.8)$$
$$(7)$$

$$SiO_4^{-4} \rightarrow\rightarrow\rightarrow SiO_2 \qquad (12.9)$$
$$(8)$$

The number listed with each product corresponds with the "Geometric ID Number" given in Table 12.3 and the "Structural Geometry" given in Table 12.3 and is depicted in Figure 12.2.

Each of these steps is based on a tetrahedral silicon atom attached to four oxygen atoms. The complexity and variety of naturally occurring silicates is due to two major factors. First, the ability of the tetrahedral SiO_4^{-4} unit to be linked together often giving polymeric structures. Second, the substitution of different metal atoms of the same approximate size as that of Si often occurs giving many different materials.

TABLE 12.3
Inorganic Polymeric Silicates as a Function of Common Geometry

Geometric ID Number (Text)	Basic Geometric Unit	Structural Geometry (Figure 12.2)	General Silicate Formula*	Examples*
1	Tetrahedran	A	SiO_4^{-4}	Granite Olivine—$(Mg, Fe)_2SiO_4$ Fosterite—Mg_2SiO_4 Topez
2	Double tetrahedran	B	$Si_2O_7^{-6}$	Akermanite—$Ca_2MgSi_2O_7$
3	Triple ring	C	$Si_3O_9^{-6}$	Wollastonite
4	Tetra ring	D	$Si_4O_{12}^{-8}$	Neptunite
5	Six ring	E	$Si_6O_{18}^{-12}$	Beryl—$Al_2Be_3Si_6O_{18}$
6a	Linear chain	F	$Si_4O_{12}^{-8}$	Augite, Enstatite-$MgSiO_3$ Diopside—$CaMg(SiO_3)_2$ Chrysotile—$Mg_6Si_4O_{11}(OH)_6$
6b	Double-stranded Ladder	G	$Si_4O_{11}^{-6}$	Hornblende
7	Parquet (layered)	H	$Si_4O_{10}^{-4}$	Talc—$Mg_3Si_4O_{10}(OH)_2$ Mica—$KAl_3Si_3O_{10}(OH)_2$ Kaolinite—$Al_2Si_2O_5(OH)_4$ (Condensed) silicic acid—$H_2Si_2O_5$
8	Network	I	SiO_2	Quartz, Feldspar (Orthoclase)—$KAlSi_3O_8$

*The formulas given are for the most part simplified.

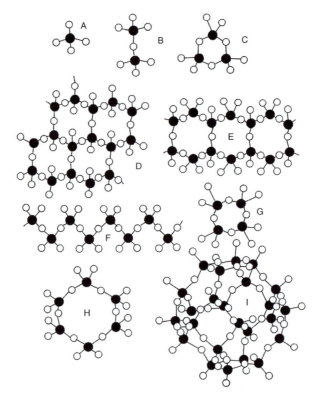

FIGURE 12.2 General silicate structures. From left to right the figures are (top row) A, B, C, (second row) D, E, (third row) F, G, (last row) H, and I with the letters corresponding to Figure 12.3 third column.

In the structures cited in Table 12.3, except for pure SiO_2, metal ions are required for overall electrical neutrality. These metal ions are positioned in tetrahedral, octahedral, and so on positions in the silicate-like lattice. Sometimes they replace the silicon atom. Kaolinite asbestos has aluminum substituted for silicon in the Gibbosite sheet. Further, sites for additional anions, such as the hydroxyl anion, are available. In ring, chain, and sheet structures neighboring rings, chains, and sheets are often bonded together by metal ions held between the rings. In vermiculite asbestos, the silicate sheets are held together by nonsilicon cations.

For sheet-layered compounds, the forces holding together the various sheets can be substantially less than the forces within the individual sheets. Similar to graphite, such structures may be easily cleaved parallel to the sheets. Examples of such materials are mica, kaolinite, and talc.

Bonding occurs through a combination of ionic and covalent contributions just as are present in organic polymers except that the ionic character is a little higher. "Back-bonding" from electrons associated with the oxygen to vacant orbitals in the silicon (or other tetrahedral metal atom) occurs giving the silicon–oxygen linkages some double or pi-bond character.

As noted before, cations other than silicon may occupy the tetrahedral centers. A major factor in predicting which cations will be found to substitute for silicon is ionic size. In general, cations whose size is about 0.03–0.1 nm are the best candidates. Si^{+4} has an ionic radius of about 0.041 nm. Cations such as Fe^{+2} (ionic radius = 0.07 nm), Al^{+3} (0.05 nm), Ca^{+2} (0.1 nm), and Mg^{+2} (0.065 nm) are most often found in silicate-like structures and meet this requirement.

Most silicate-like polymers can be divided into three major classes—the network structures based on a three-dimensional tetrahedral geometry (such as quartz), layered geometries with stronger bonding occurring within the "two-dimensional" layer (such as talc), and linear structures.

12.4.1 NETWORK

Quartz is an important network silicate (Section 12.10). A number of additional tetrahedral silicate-like materials possess some AlO_4 tetrahedra substituted for the SiO_4 tetrahedra. Such structures offer a little larger "hole" in comparison to the entirely SiO_4 structures, allowing alkali and alkaline-earth cations to be introduced. Feldspar (orthoclase) is such a mineral. The aluminosilicate networks are almost as hard as quartz. For feldspar and other tetrahedral networks, the number of oxygen atoms is twice the summation of silicon and other MO_4 cations.

The feldspars are widely distributed and comprise almost two-thirds of all igneous rocks. Orthoclase and albite, $NaAlSi_3O_8$, are feldspars, where one-fourth of the silicon atoms are replaced by aluminum and anorthite, $CaAl_2Si_2O_8$, has one-half of the silicon atoms replaced by aluminum. Because the ionic radius of Na^+ (0.095 nm) and Ca^{+2} (0.1 nm) are about the same, solid solutions are often formed between albite and anorthite. Good stones of albite and orthoclase are known as moonstones.

Some of the network structures exhibit a framework sufficiently "open" to permit ions to move in and out. The zeolite minerals used for softening water are of this type.

Ultramarines are three-dimensional cage-like structures. They differ from feldspars and zeolites because of the large spaces within the structures that can contain cations and anions but no water illustrating a natural "buckeyball-like" structure and cavity and a diversity of environment between the internal cage and external. Ultramarines can act as ion exchangers for both anions and cations. The blue color of ultramarines is due to the presence of the S_3^- ion although a yellow ion S_2^- also exists in the same structure.

12.4.2 LAYER

Layered structures typically conform to the approximate composition $Si_4O_{10}^{-4}$ or $Si_2O_5^{-2}$. For most of these, three of the oxygen atoms of each tetrahedron are shared by other tetrahedra, and the fourth oxygen is present on one side of the sheet.

In talc and kaolinite, the layers are neutral. Thus, the layers slide over one another easily imparting to these minerals a softness and ease in cleavage. In other minerals, the layers are charged and held together by cations. In mica, the aluminosilicate layers are negatively charged and cations, generally K^+, are present between the layers giving the entire system of layers electronic neutrality. The ionic attractive forces between the layers result in mica being much harder than talc and kaolinite. Even so, these intersheet bonding forces are less than the "within-the-sheet" bonding forces, permitting relatively easy and clean cleavage of mica. Mica is used as an insulator for furnaces and electric equipment. Montmorillonite is an important ingredient in soils and is employed industrially as a catalyst in the conversion of straight-chain hydrocarbons to more branched hydrocarbons and more recently as a sheet or clad material in the manufacture of tires.

Vermiculites are formed by the decomposition of mica. They contain layers of water and magnesium ions in place of the potassium ions. When heated to 800°C–1,100°C, vermiculite expands because of the conversion of the water to a gas. The expanded vermiculite has a low thermal conductivity and density and is used as a thermal and sound barrier and as an aggregate in light-weight concrete. It is also used as a moisture-retaining soil conditioner in planting.

A number of clays are layered silicate-like materials. Most clays contain finely divided quartzs, micas, and feldspars. Iron oxide-rich clays are employed to make pottery and terra cotta articles. Clays containing iron oxide and sand are used to make bricks and tiles. Clays rich in calcium and magnesium carbonate are known as marls and are used in the cement industry (Section 12.2).

Kaolinite is the main constituent in china clay used to make porcelain. The layers are largely held together by van der Waals' forces. Bentonite is used in cosmetics, as a filler for soaps, and as a plasticizer, and it is used in drilling muds as a suspension stabilizer. Bentonite and kaolinite clays are used, after treatment with sulfuric acid, to create acidic surface sites, as petroleum cracking catalysts. Asbestos also has a layered structure (Section 12.13).

12.4.3 CHAIN

Both single- and double-stranded chains are found. The most important members of single chains are the pyroxenes and include diopside. The most important double-chained minerals are the amphiboles. Some of these contain hydroxyl and fluoride ions, bonded directly to the metal cation and not to the silicon atom.

Jade, which has been valued in carving by eastern Asians for centuries, is generally one of two minerals—pyroxene or jadeite, $NaAl(SiO_3)_2$, and the amphibole nephrite, $Ca_2(Fe^{+2}$ and/or $Mg^{+2})_5$ $(Si_4O_{11})_2(OH)_2$. X-Ray diffraction has shown the presence of triple chains in nephrite.

Because the interchain bonding is weaker them the Si-O backbone bonding, these chain structures can generally be easily cleaved between the chains.

Several amphiboles are fibrous and fibers from them can be processed to give heat-insulating materials. Among these are tremolite and crocidolite. These minerals are also used as fibers in composites.

12.5 SILICON DIOXIDE (AMORPHOUS)

Silicon dioxide (SiO_2) is the major repeating general formula for the vast majority of rock, sand, and dirt about us and for the material we refer to as glass. The term glass can refer to many materials, but here we will use the ASTM definition that glass is an inorganic product of fusion that has been cooled to a rigid condition without crystallization. In this section, silicate glasses, the common glasses for electric light bulbs, window glass, drinking glasses, glass bottles, glass test tubes and beakers, and glass cookware, will be emphasized. Figure 12.3 contains a segment of an amorphous silicon dioxide structure. For some of the structure, there is a general distorted octahedral ring

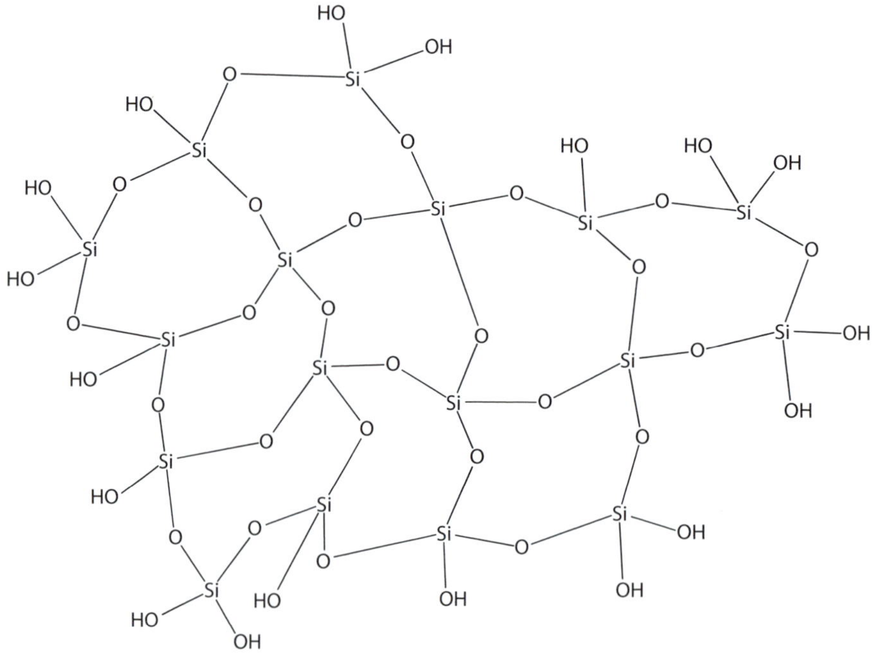

FIGURE 12.3 Segment of amorphous silicon dioxide structure.

TABLE 12.4
General Properties of Silicate Glasses

High transparency to light	Permanent (long-term) transparency
Hard	Scratch resistant
Chemically inert	Low thermal expansion coefficient
Good electrical insulator	High sparkle
Good luster	Low porosity
Good ease of reforming	Easily recyclable
(Relatively) Inexpensive	Available in large amounts

---

structure where each octahedral contains four silicon and four oxygen atoms. There is also a number of "dangling" hydroxyl groups. This amorphous structure allows the glass to be melted and some, but not much, flexibility.

Glass has many useful properties, as listed in Table 12.4. It ages (changes chemical composition and physical property) slowly, typically retaining its fine optical and hardness-related properties for centuries. Glass is referred to as a supercooled liquid or a very viscous liquid. Indeed, it is a slow-moving liquid as attested to by sensitive measurements carried out in some laboratories. Concurrent with this is the observation that the old stained glass windows adorning European cathedrals are a little thicker at the bottom of each small, individual piece than at the top of the piece. For most purposes though, glass can be treated as a brittle solid that shatters on sharp impact.

Glass is mainly silica sand (SiO_2) and is made by heating silica sand and powdered additives together in a specified manner and proportion much as a cake is baked, following a recipe that describes the items to be included, amounts, mixing procedure (including sequence), oven temperature, and heating time. The amounts, nature of additives, and so on, all affect the physical properties of the final glass.

Typically cullet, recycled or waste glass (5%–40%), is added along with the principle raw materials (mostly SiO_2). The mixture is thoroughly mixed and then added to a furnace where the mixture is heated to near 1,500°C to form a viscous, syrup-like liquid. The size and nature of the furnace corresponds to the glasses' intended uses. For small individual items, the mixture may be heated in a small clay (refractory) pot.

Most glass is melted in large (continuous) tanks that can melt 400–600 metric tons a day for production of different glass products. The process is continuous with the raw materials fed into one end as molten glass is removed from the other end. Once the process (called a *campaign*) is begun, it is continued indefinitely, night and day, often for several years until the demand is met or the furnace breaks down.

A typical window glass will contain 95%–99% silica sand with the remainder being soda ash (Na_2CO_3), limestone ($CaCO_3$), feldspar, and borax or boric acid along with the appropriate coloring and decolorizing agents, oxidizing agents, and so forth. As noted previously, 5%–40% by weight of crushed cullet is also added. The soda ash, limestone, feldspar, and borax or boric acid all form oxides as they are heated, which become integrated into the silicon structure:

$$Na_2CO_3 \rightarrow Na_2O + CO_2 \tag{12.10}$$

$$CaCO_3 \rightarrow CaO + CO_2 \tag{12.11}$$

$$R_2OAl_2O_3 \bullet 6H_2O \rightarrow R_2O + Al_2O_3 + 6SiO_2 \tag{12.12}$$

$$H_3BO_4 \rightarrow B_2O_3 \tag{12.13}$$

The exact structure varies according to the ingredients and actual processing conditions. As in the case of Portland cement, glass is a three-dimensional array that offers short-range order and long-range disorder—it is amorphous offering little or no areas of crystallinity. The structure is based on the silicon atoms existing in a tetrahedral geometry with each silicon atom attached to four oxygen atoms, generating a three-dimensional array of inexact tetrahedra. Thus, structural defects occur, due in part to the presence of impurities such as Al, B, and Ca, intentionally or unintentionally introduced. These impurities encourage the glass to cool to an amorphous structure since the different-sized impurity metal ions, and so on, disrupt the rigorous space requirement necessary to allow crystal formation.

Processing includes shaping and pretreatments of the glass. Since shaping may create undue sites of amorphous structure, most glass objects are again heated to near their melting point. This process is called *annealing*. Since many materials tend to form more ordered structures when heated and recooled slowly, the effect of annealing is to "heal" sites of major dissymmetry. It is important to heal these sites since they represent preferential locations for chemical and physical attack such as fracture.

Four main methods are employed for shaping glass. They are drawing, pressing, casting, and blowing. Drawing is employed for shaping flat glass, glass tubing, and for creating fibrous glass. Most flat glass is shaped by drawing a sheet of molten glass (heated so it can be shaped but not so it freely flows) onto a tank of molten tin. Since the glass literally floats on the tin, it is called "float glass." The temperature is carefully controlled. The glass from the float bath typically has both sides quite smooth with a brilliant finish that requires no polishing.

Glass tubing is made through drawing molten glass around a rotating cylinder of the appropriate shape and size. Air is blown through the cylinder creating the hollow tubing. Fibrous glass is made by forcing molten glass through tiny holes and drawing the resulting fibers helping to align the chain on a molecular level.

Pressing is accomplished by simply dropping molten glass into a form and then applying pressure to ensure the glass takes the form of the mold. Lenses, glass blocks, baking dishes, and ashtrays are examples of press-processed glass objects.

Casting involves filling molds with molten glass in much the same manner that cement and plaster of Paris molded objects are processes. Art glass objects are often made by casting.

Glass blowing is one of the oldest arts known to man. The objects are constructed or repaired by a skilled worker who blows into a pipe intruded into the molten glass. The glass temperature must be maintained to allow it to be moldable but not so it freely flows. Mass produced materials are manufactured employing mechanical blowers often employing a combination of glass blowing and molding to form the desired product.

As noted above, annealing encourages the removal of sites of stress and strain. Slow cooling results in a glass with more crystallinity that is stronger but more brittle. *Tempering* is the name given when the glass is rapidly cooled, resulting in an amorphous glass that is weaker but less brittle. The correlation between crystallinity, rate of cooling, and brittleness is demonstrated by noting that older window glass exposed to full sun for years, which is more brittle and can be more easily shattered since the sunlight raises the temperature sufficiently to permit small molecular movements (though even in full sunlight the glass is not near the temperature required for ready movement) and over the years gives a glass with small regions of greater order.

Silicon-based glasses account for almost all of the glasses manufactured. Silica is finely ground silica sand. Yet most sand is unsuitable for general glassmaking due to the presence of excessive impurities. Thus, while sand is plentiful, sand that is useful for the production of glass is much less common. In fact, the scarcity of large deposits of glass sand is one major reason for the need to recycle glass items. The second major reason is associated with the lowered energy requirements for glass to be made molten again for reshaping compared with a virgin glass mixture, that is, culled glass becomes molten at temperatures lower than virgin glass.

12.6   KINDS OF AMORPHOUS GLASS

The types and properties of glass can be readily varied by changing the relative amounts and nature of ingredients. *Soda-lime glass* is the most common of all glasses accounting for about 90% of glass made. Window glass, glass for bottles, and so forth, are all soda-lime glass. Soda-lime glass (75% silica, 15% soda (sodium oxide), 9% lime (calcium oxide), and the remaining 4% minor ingredients) has a relatively low softening temperature and low thermal shock resistance limiting its high-temperature applications.

Vycor, or 96% silicon glass, is made using silicon and boron oxide. Initially, the alkali-borosilicate mixture is melted and shaped using conventional procedures. The article is then heat-treated, resulting in the formation of two separate phases—one that is high in alkalis and boron oxide, and the other containing 96% silica and 3% boron oxide. The alkali–boron oxide phase is soluble in strong acids and is leached away by immersion in hot acid. The remaining silica–boron oxide phase is quite porous. The porous glass is again heated to about 1,200°C, resulting in a 14% shrinkage due to the remaining portions filling the porous voids. The best variety is "crystal" clear and called *fused quartz*. The 96% silica glasses are more stable and exhibit higher melting points (1,500°C) than soda-lime glass. Crucibles, ultraviolet filters, range burner plates, induction furnace linings, optically clear filters and cells, and super heat-resistant laboratory-ware are often 96% silicon glass.

Borosilicate glass contains about 80% silica, 13% boric oxide, 4% alkali, and 2% alumina. It is more heat-shock-resistant than most glasses due to its unusually small coefficient of thermal expansion (typically between 2 and 5×10^{-6} cm/cm/°C; for soda-lime glass it is $8–9 \times 10^{-6}$ cm/cm/°C). It is better known by such trade names as Kimax and Pyrex. Bakeware and glass pipelines are often borosilicate glass.

Lead glasses (often called *heavy glasses*) are made by replacing some or all of the calcium oxide by lead oxide (PbO). Very high amounts of lead oxide can be incorporated—up to 80%. Lead glasses are more expensive than soda-lime glasses, and they are easier to melt and work with. They are more easily cut and engraved, giving a product with high sparkle and luster (due to the high refractive indexes). Fine glass and tableware are often lead glass.

Silicon glass is made by fusing pure quartz crystals or glass sand (impure crystals), and it is typically about 99.8% SiO_2. It is high melting and difficult to fabricate.

Colored or *stained glass* has been made for thousands of years, first by Egyptians and later by Romans. Color is typically introduced by addition of transition metals and oxides. Table 12.5 contains selected inorganic colorants and the resulting colors. Because of the high clarity of glass, a small amount of coloring agent goes a long way. One part of cobalt oxide in 10,000 parts of glass gives an intense blue glass. The most well-known use for colored glass is the construction of stain-glass windows. In truth, there are many other uses such as industrial color filters and lenses.

Glazes are thin, transparent coatings (colored or colorless) fused on ceramic materials. *Vitreous enamels* are thin, normally opaque or semiopaque, colored coatings fused on metals, glasses, or

TABLE 12.5
Colorants for Stained Glass

Colorant	Color	Colorant	Color
Nickel (II) oxide	Yellow to purple	Calcium fluoride	Milky white
Cobalt (II) oxide	Blue	Iron (II) compounds	Green
Iron (III) compounds	Yellow	Copper (I) oxide	Red, blue, or green
Tin (IV) oxide	Opaque	Manganese (IV) oxide	Violet
Gold (III) oxide	Red		

TABLE 12.6
Leading U.S. Glass Companies

Owen-Illinois, Inc.	PPG Industries, Inc.
Corning Glass Works	Owens-Corning Fiberglass Corporation
Libbery-Owens-Ford Company	

ceramic materials. Both are special glasses but can contain little silica. They are typically low melting and often are not easily mixed in with more traditional glasses.

Optical fibers can be glass fibers that are coated with a highly refractive polymer coating such that light entering one end of the filer is transmitted through the fiber (even around corners as within a person's stomach), emerging from the other end with little loss of energy. These optical fibers can also be made to transmit sound and serve as the basis for transmission of television and telephone signals over great distances through cables. More about optical fibers in Section 12.12.

There are many silica-intensive fibers lumped together as *fibrous glass* or *fiberglass*. A general purpose fiberglass may contain silica (72%), calcium oxide (9.5%), MgO (3.5%), aluminum oxide (2%), and sodium oxide (13%). The fibers are produced by melting the "glass mixture" with the molten glass drawn through an orifice. The filaments are passed through a pan containing sizing solution onto a winding drum. The take-up rate of the filament is more rapid than the exit rate from the orifice acting to align the molecules and draw the fibers into thinner filaments. Thus, a fiber forced through a 0.1 cm orifice may result in filaments of 0.0005 cm diameter. This drawing increases the strength and flexibility of the fiberglass. Applications of fiberglass include insulation and use in composites.

Table 12.6 contains a listing of major glass-producing companies in the United States.

Optical glass for eyeglass lenses and camera lenses is typically soda-lime glass that is highly purified so that it is highly transmissive of light. Today, there exists many other special glass that are important in today's society as laser glasses, photosensitive glass, photochromic windows and eyeglass glass, invisible glasses, radiation absorbing glass, and so on. More about seeing lenses in Section 12.8.

12.7   SAFETY GLASS

Safety glass is defined as "glass" that diminishes the threat of injuries and robberies as a result of impacts, distortion, or fire.

In 1905, British inventor John C. Wood was working with cellulose and developed a method to adhere glass panes using celluloid as the adhesive. Wood's version of shatter-resistant glass was produced under the band name Triplex since it consisted of outer layers of glass with an inner layer of celluloid polymer.

About the same time, Edouard Benedictus, a French chemist, was climbing a ladder to get chemicals from a shelf and accidently (another discovery due to an accident) knocked a glass flask onto the floor. He heard the flask shatter but when he looked at the broken flask, the broken pieces hung together instead of breaking into many pieces and scattering over the floor. Benedictus learned from his assistant that the flask had recently held a solution of cellulose nitrate. The solution of cellulose nitrate dried to give a thin film that was transparent, so the flask was set aside as a "cleaned" flask. It was this thin film that coated the inside of the flask that held together the broken pieces. The film was formed on evaporation of cellulose nitrate prepared from cellulose and nitric acid.

Shortly after the laboratory accident, he read about a girl who had been badly cut from flying glass resulting from an automobile accident. Later he read about other people being cut by flying glass in automobile accidents. He remembered the flask that did not splinter into small pieces when broken because of the cellulose nitrate coating. He experimented with placing cellulose nitrate between sheets of glass and applying pressure to help adhere the glass with the cellulose nitrate. By 1909, Benedictus had patented the material.

Before its use in windshields, safety glass found its initial major application as the lenses for gas masks during World War I. Manufacturers found it easy to work with so that the technology and ability to manufacture safety glass windshields came easily.

As automobiles became more common, so did the hazards of mud, rocks, and so on, so that by 1904 windshields were introduced. These first windshields could be folded or moved if they became blocked by excessive mud. While the usefulness of the windshield was abundantly obvious, drivers found that they were dangerous during a wreck cutting passengers, drivers, and passer-byes alike. Because the drivers were believed to be primarily responsible for automobile safety, most manufactures were slow to adopt safety glass. In 1919, Henry Ford addressed the windshield problem by having safety glass windshields on his automobiles.

This safety glass turned yellow after several years of exposure to light. The bonding layer was replaced in 1933 by cellulose acetate, made from the reaction of cotton with acetic acid. By 1939, this was replaced by poly(vinyl butyral) (PVB) still in use today as the adhesive placed between sheets of glass to produce laminated safety glass. This is one of a very few modern-use materials that has retained the same basic materials for more than 60 years.

(12.14)

Poly(vinyl alcohol)/butyraldehyde   Poly(vinyl butyral)

Poly(vinyl butyral) is made from poly(vinyl alcohol) which itself is made from poly(vinyl acetate) because the monomer vinyl alcohol does not exist (Equation 12.14).

Today, safety glass is divided into three general categories, laminated safety glass, tempered safety glass, and armed glass. *Tempered safety glass* is made by heating the glass to its melting point, about 700°C, and then cooling it rapidly by blowing cold air onto its surfaces. The effect is similar to the production of stressed concrete where the concrete is allowed to harden under stress giving a stronger concrete. In the case of glass, when it is rapidly cooled, a structure is locked in that produces extra stress on the glass structure making it stronger. As the glass is cooled, the surfaces harden first locking in the overall glass volume. As the center cools, it forces the surfaces and edges into compression. With appropriate rapid cooling, the glass is not only stronger, but when shattered, produces granulates rather than sharp cutting shards. The typical force necessary to break tempered glass is about four times that required to shatter ordinary glass of the same thickness.

While the front "windshield" is made of safety glass, the remainder of the automotive glass windows are generally made from tempered glass. Tempered glass is also used for commercial building doors and windows, sidelights, patio-door assemblies, storm doors, shower and tub enclosures, refrigerator, oven, and stove shelves, and fireplace screens.

Armed glass is most commonly used as roofing on factory buildings. It is glass that has a built-in metal grill that strengthens the glass. The metal grill is often like chicken wire in appearance. The glass breaks similar to regular glass, but the wire mesh helps hold it into place. The visibility is reduced because of the presence of the metal mesh.

Laminated glass is used in automobiles and often used for added protection in windows, balconies, or sloping glass roofs. The laminated glass resists breakage in comparison to ordinary glass because the PVB inner layer(s) helps dampen sharp blows. Even so, laminated glass does crack and it is more easily cracked than tempered glass. But, it is harder to pierce than tempered glass because of the PVB inner layer(s). Another difference is that laminated glass can be cut, sawn, or drilled whereas tempered glass cannot. The PVB film also has ultraviolet-screening properties reducing discoloration of objects placed behind the safety glass. It also acts to dampen sound for additional soundproofing.

Laminated safety glass is available in different thicknesses depending on the number of PVB-glass layers. *Bullet-resistant glass* is one use for thick, multilayer laminated safety glass. The laminated construction allows the multilayer assembly to have some additional flexibility with the multiple layers yielding, allowing the PVB layers to absorb some of the energy of the bullet. The plastic layers help hold the shattered glass fragments together aiding in retaining a restraining barrier. Such glass is used in bank teller windows, and in windshields for aircraft, tanks, and special automobiles and trucks.

The ability to resist bullets and blasts is increased by increasing the number of layers of laminated safety glass. Such increases give an increased glass thickness and weight. So there is a trade-off between expected abuse and practicality.

12.8 LENSES

About 1,000 AD, reading stones started being used. These were segments of a glass sphere that were used to magnify letters. The reading stones were placed directly on the letters. By 1,350 AD, reading stones were hung with ribbons and strings near the eyes allowing the "wearer" to magnify objects within their "sight." In 1730, a London optician, Edward Scarlett, developed rigid sidepieces that rested on the ears of the user. These were the first somewhat modern reading and seeing glasses. The first bifocals were developed by Benjamin Franklin by cutting two lenses in half and placing one above the other. Until relatively recently, glasses fitting was largely trial-and-error with the person wearing the glasses responsible for determining the correct lenses. Even today, we can go to certain stores and select a pair of inexpensive glasses from a group of general eye glasses through this same process.

Below is a general menu between properties and type of reading glasses lenses that best demonstrate this property:

- Strongest (hardest to break)—polycarbonate
- Lightest—polycarbonate and plastic
- Greatest resistance to scratching—glass
- Clearest vision—glass and plastic
- Most responsive photochromic lens—glass
- Lightest photochromic lens—plastic
- Most resistant to heat and common household chemicals—glass
- Most resistant to flying objects—polycarbonate

In general, polycarbonate lenses are best for people

- Active in sports
- That use power tools

- With good vision in only one eye, and
- With refractive surgery

Plastic lenses are best when light weight and clear vision are important. They offer the widest range of bifocal designs.

Glass lenses are best for people

- Who work in a dusty environment
- Leave glasses in the car
- Wear sunglasses on the beach
- Want the best photochromic performance, and
- Who often take off and on their glasses

Reading glasses must pass a number of standard tests outlined in LCO993. Several of these are described following. The impact resistance test is described by the Federal Regulation on Impact Resistant Lenses (21 CFR 80.410) where a 5/8-in. diameter steel ball weighing 0.56 oz is dropped from a height of 50 in. onto the horizontal upper, convex surface of the lens. To pass, the lens must not fracture. The flammability test is passed when no evidence of ignition is found when the glasses assembly is placed in a preheated oven set to 200°C for 15 minutes.

The most common material for decent safety glasses is polycarbonate. Those glasses referred to as high impact must pass the ANSI test for high velocity impact. The requirement are that the lenses withstand an impact from a 6.35 mm (1/4 in.) diameter steel ball traveling at a velocity of 45.7 mps (150 fps). From a sample size of 20, none may shatter or the lenses fail the test.

Glass lenses can be made finished from a mold or the general shape made in a mold and latter worked providing the finished lens.

In the late 1960s, PPG introduced a new lens material that today is simply known as "plastic lenses." In comparison to crown glass, this material was lighter, produced thinner lenses, offered greater impact resistance, and was more flexible. While the new material is referred to by the sales people at eye glasses outlets as simply "plastic," its name is CR-39. The "CR" stands for Columbia Resin and the 39 was the batch or formula made by the Columbia Laboratories in Ohio. CR-39 is made from allyl diglycol carbonate (or diethylene glycol *bis*(allyl carbonate)) monomer. On heating, the two vinyl groups open up forming a cross-linked thermoset plastic that cannot be resoften on heating. By comparison, polycarbonate is a linear plastic that is a thermoplastic that can be resoften, and recycled, on addition of heat and pressure. The lenses are formed by melting polycarbonate pellets and injecting them into a mold. Lenses from CR-39 are made by casting the monomer into appropriate molds followed by polymerization creating the lens either as a finished product or further working the solid lens material creating the finished lens.

Initially, there was a problem casting lenses from CR-39 because there is typically shrinkage, with the polymer denser than the liquid monomer, when vinyl monomers are polymerized. In this case, there was 14% shrinkage. This was not a problem when casting flat lenses since the resulting lens dimensions were simply a little less. Further, the shrinkage was different creating optical distortions. Robert Graham, working with others for the Armorlite Company, overcame this problem by casting thick blanks where the back curve matched the finished front curve. The lens are then ground and polished to the required thickness and curve. CR-39 Monomer (top; 12.15) polymerized forming CR-39 Polymer (bottom; 12.16).

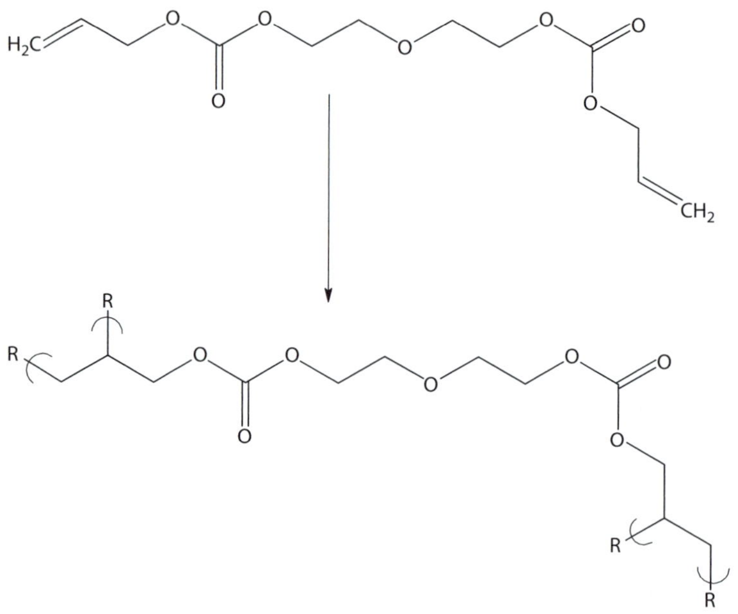

(12.15)

(12.16)

There still remained the problem of scratching. The CR-39 lenses are less scratch resistance than polycarbonate lenses. In the 1970s, the 3M (Minnesota Mining and Manufacturing Company) company had a research group that was expert in coatings, including lenses. One problem was the need, for some materials, to have them particularly free from dust to achieve mar-free coated surfaces. By the mid-1970s, they created a production facility that reduced airborne particles sufficiently to allow materials to be successfully coated. In 1979, 3M purchased the Armorlite Company and transferred their scratch-resistant coating technology to be used for the CR-39 lenses. American Optical introduced photolite photochromic lenses that change color with exposure to light in 1981. PPG, in 1983, discovered a new family of photochromics, the blue pyridobenzoxanines and by 1984 formed a joint venture with Intercast-Europe to manufacture and sell photochromic sunglasses called *Attiva sunglasses* made from CR-39. Many other companies have contributed to the presence of today's plastic lenses. The story illustrates how companies apply their interests and strengths to achieve the products we have today.

Today, there are new so-called high-index lenses made from a variety of polymers that are thinner than CR-39 and polycarbonate lenses.

12.9   SOL-GEL

In the sol-gel process, ceramic polymer precursors are formed in solution at ambient temperature; shaped by casting, film formation or fiber drawing; and then consolidated to furnish dense glasses or polycrystalline ceramics. The most common sol-gel procedures involve alkoxides of silicon, boron, titanium, and aluminum. In alcohol-water solution, the alkoxide groups are removed stepwise by hydrolysis under acidic or basic catalysis and replaced by hydroxyl groups, that then form –M–O–M– linkages. Branched polymer chains grow and interconnect forming a network that can span the entire solution volume. At this point, the gel point, the viscosity, and elastic modulus rapidly increase.

The gel is a viscoelastic material composed of interpenetrating liquid and solid phases. The network retards the escape of the liquid and prevents structural collapse. The shapes formed by casting, drawing of fibers, or film formation are locked in by the gel formation. Some gels are oriented by drawing or shearing. The gel is dried by evaporation forming a xerogel or by supercritical fluid extraction giving an aerogel. Consolidation to dense glasses or ceramics is carried out by thermal treatment and sintering.

Since both aerogels and xerogels have high surface areas and small pore diameters, they are used as ultrafiltration media, antireflective coatings, and catalysts supports. Final densification is carried out by viscous sintering.

The rate of silicate sol and gel formation is pH and water–alcohol sensitive as is the solubility of the amorphous silica that is formed. Silica networks are based on $(SiO_4)^{-4}$ tetrahedra modified by $(O_3 Si-O^-, M^+)$ units and often addition of boron oxide, aluminum oxide, titanium IV oxide, or zirconium IV oxide.

The nature of the reactants can be varied giving various silicate-like products. The formation of borosilicate glasses using the sol-gel approach is described below:

$$NaOR + B(OR)_3 + Si(OR)_4 + H_2O \rightarrow NaOH + B(OH)_3 + Si(OH)_4 + ROH \qquad (12.17)$$

$$NaOH + B(OH)_3 + Si(OH)_4 \rightarrow Na_2O \bullet B_2O_3 \bullet SiO_2 \bullet H_2O \rightarrow Na_2O \bullet B_2O_3 \bullet SiO_2 + H_2O \qquad (12.18)$$

The use of organically modified silicates (ceramers) gives a wide variety of products with a variety of structures and properties. Such ceramers have been used as adhesives for glass surfaces, protective coating for medieval stained glass, and as scratch-resistant coatings for plastic eyeglass lenses. They have also been used in the reinforcement of plastics and elastomers, and their nano-scale pores allow their use as porous supports, and as selective absorbents.

Sol-gel preparations of tetraethoxysilane can be spun into fibers once the appropriate viscosity has been achieved. These fibers are only slightly weaker than silica-glass fibers.

Hybrid materials have been made by incorporating end-capped poly(tetramethylene oxide) blocks to tetramethoxysilane sol-gel glasses. These materials have high extensibility with interdispersed organic and inorganic regions.

12.10 AEROGELS

Aerogels are highly porous materials where the pore sizes are truly on a molecular level, less than 50 nm in diameter. This gives a material with the highest known internal surface area per unit weight. One ounce can have a surface area equal to 10 football fields, more than 1,000 square meters in one gram.

Porous materials can be either open pored such as a common sponge, or closed pored such as the bubble-wrap packaging. Aerogels are open-pored materials such that unbonded material can move from one pore to another.

While in the gel state, the preaerogel has some flexibility, but as a solid aerogels behave as a fragile glass. It may be very strong in comparison to its weight, but remember it is very light. Aerogels are more durable when under compression. Compression can be simple such as sealing the aerogel sample in a typical food sealer packing. Aerogels are best cut using a diamond coated saw similar to that used by rock cutters to slice rocks.

When handled, aerogel samples will initially appear to exhibit some flexibility but then burst into millions of pieces. For large arrays of atoms, such as solid metals and polymers below their glass transition temperature, energy can be absorbed through bond flexing or bending. For polymers between their glass transition and melting points, kinetic energy can also be absorbed through segmental movement. Silica aerogels, being an inorganic polymer in the glassy state at room temperature, is a brittle material. As force is applied, there is very little bond flexing, so that the applied kinetic energy results in the collapse of the network with the force of impact spread over a large part of the aerogel and over a time because of the time required to transfer this energy from one cell to another within the aerogel matrix. Because the aerogel is open pored, gas contained within the solid is forced outward as collapse occurs. The frictional forces caused by the gas passing through a restricted opening are indirectly proportional to the square of the pore diameter. Because the pore sizes are so small, the rapidly moving gas also absorbs a lot of the energy. Thus, energy is absorbed by the aerogel through both collapse of the solid network structure and release of the gas within the aerogel.

Aerogels that are about 2–5 nm in diameter have large surface to volume ratios on the order of 10^9 meters^{-1} and high specific surface areas approaching 1,000 m^2/g. Such large surface to volume ratios makes the surface particularly active and potential materials as catalysts, absorbents, and catalyst substrates.

The precise chemical makeup of the surface depends on the materials used to make the aerogel and the method of processing. Typical aerogel sequences produce products whose surfaces are rich in hydroxyl groups. Because of the high surface area, -Si-OH groups act as weak acids and are reactive in typical Lewis acid–base reactions. As noted before, aerogels have many hydrogen-bonding hydroxyls at their surface making aerogels extremely hygroscopic. Dry aerogel materials will increase their weight by 20% through uptake of moisture from the air. This absorption is reversible and appears to have little or no effect on the aerogel. Water is removed through heating to 100°C–120°C.

While adsorption of water vapor has little effect on aerogels, contact with liquid water has devastating effects on aerogels. When water enters the nanometer-size pores, the surface tension of the water exerts capillary forces sufficient to fracture the silica backbone, resulting is a collapse of the complex matrix structure. This tendency to be attacked by water is overcome through conversion of the surface polar –OH groups to nonpolar –OR groups. The "R" is typically a trimethylsilyl group though any aliphatic group would work. Conversion can be accomplished within the wet stage (preaerogel) or after the supercritical drying. These treatments result in an aerogel that is called "hydrophobic" aerogel, which is stable in water.

The pore size of aerogels varies. The International Union of Pure and Applied Chemistry classifies materials with pore sizes of less that 2 nm as "micropores," 2–50 nm are called "mesopores," and those greater than 50 nm in diameter are called "macropores." While aerogels have some pores that fall within the micropore region, the majority of pores are in the mesopore region.

Most of the aerogels produced today are described as being transparent. While it might be assumed that since aerogels are made of the same material as window glass and quartz (SiO_2) that they would be transparent, this is not necessarily the case. Transparency requires a number of factors. Thus, so-called smokey and white quarts are colored because of the presence of impurities. Mixtures of amorphous and crystalline silicon dioxide can be made that are not transparent. The size and distribution of reflecting and refractive sites are important factors in determining if a material is transparent with the theme of "sameness" contributing to making a material transparent.

The majority of light we see is scattered light, that is, light that reaches our eyes in an indirect manner. The scattering phenomenon is what gives us blue skies, white to gray clouds, and poor visibility in fog. This scattering is not simply reflecting but results from the interaction of light with an inhomogeneous site. Light-scattering photometry is used to determine the size of polymers. Scattering is most effective when the scattering particle size is about that of the wave length of the light. For visible light, this occurs with scattering sites that are about 400–700 nm. Scattering centers that are much less in size than the incoming light wavelength are much less effective at scattering the light. Since the particle sizes in an aerogel are much smaller than the individual sites they are ineffective scattering sites. Similar to classical polymer chains, where the entire chain or segments of the entire chain act as a scattering site, clusters of individual sites within the aerogel act as scattering sites. Most of these scattering sites are again smaller than the wavelength of visible light, but some are within the range to scatter visible light so that a soft reflected light results. The different-sized scattering sites and variable wavelengths present in visible light cause a reddening of the transmitted light (red light has a longer wavelength and is scattered less by small clusters present in the aerogel), resulting in the blue appearance of the reflected light from the aerogel.

The good visible light transmission and good insulting power make aerogel materials of interest in window manufacturing. The visible transmission spectra of light shows little absorption in the range of about 300–2,700 nm giving aerogels a good visible light "window" making aerogels attractive for day-lighting applications. Aerogels provide about 40 fold more insulation than does fiberglass insulation. While such aerogels may eventually be used as the entire window component, for the present time they may act as the material sandwiched between two pains of clear plastic or glass. Thermoglass is generally simply glass sheets that are separated by a vacuum. The seals on such thermoglass often spring small leaks causing diminished insulation properties. Aerogel inner cores will not suffer from this problem. Currently, about 40%–50% of a house's heating bill literally goes out the window because of lost heat or cold through windows. A single one-inch thick glass pane of aerogel offers the insulation equivalent to more than 30 windowpanes of R-20 insulation rated glass.

Another commercial area that is being considered is the use of aerogels as nanocomposite materials. Approaches are varied. In one approach, material is added to the silica sol before gelation. The material can be inorganic, organic, polymeric, bulk fibers, woven cloths, and so forth. The additional material must be able to withstand the subsequent processing steps, including carbon dioxide drying. The added material must be present in a somewhat homogeneous manner throughout the system. Gentle agitation appears to be sufficient to give a product with decent homogeneity. Aerogels may be good materials for optical sensors. They have good visible transparency, high surface area, good temperature and chemical stabilities, and facile transport of gases through their pores.

12.11 SILICON DIOXIDE (CRYSTALLINE FORMS)—QUARTZ FORMS

Just as silicon dioxide forms the basis of glass, so also does it form the basis of many of the rocks, grains of sand, and dirt particles that compose the Earth's crust. Most rocks are inorganic polymers, but here we will deal with only a few of these containing significant amounts of silicon.

Silicon oxide crystallizes in mainly three forms—quarts, tridymite, and cristobalite. After the feldspars, quart is the most abundant material in the Earth's crust, being a major component of igneous rocks and one of the most common sedimentary materials, in the form of sandstone and sand. Quartz can occur as large (several pounds) single crystals, but normally is present as granular materials. The structure of quarts (Figure 12.4) is a three-dimensional network of six-membered

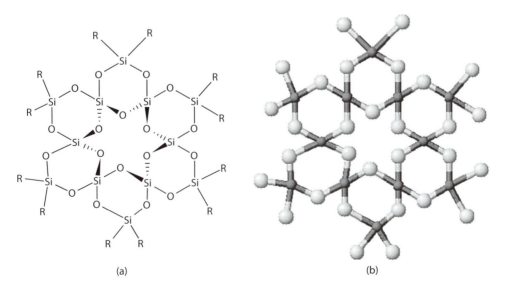

(a)   (b)

FIGURE 12.4 Structure of crystalline SiO$_2$ tetrahedra found in quartz (right structural formula and right ball-and-stick model of one unit cell).

Si-O rings (three SiO_4 tetrahedra) connected such that every six rings enclose a 12-membered Si-O (six SiO_4 tetrahedra) ring.

Quartz is found in several forms in all three major kinds of rocks—igneous, metamorphic, and sedimentary. It is one of the hardest minerals known. Geologist often divide quartz into two main groupings—course crystalline and cryptocrystalline quartz. Course crystalline quarts include six-sided quartz crystals and massive granular clumps. Some colored varieties of coarse crystalline quartz crystals, amethyst and citrine, are cut into gem stones. Others include pink (rose), purple, and milky quartz, but most coarse crystalline quartz is colorless and transparent. Sandstone is a ready example of granular quartz. Color is a result of the presence of small amounts of metal cations such as calcium, iron, magnesium, and aluminum.

Cryptocrystalline forms contain microscopic quartz crystals and include the chalcedony grouping of rocks such as chert, agate, jasper, and flint.

Quartz exhibits an important property that allows the *piezoelectric effect*. When pressure is applied to a slice of quartz, it develops a net positive charge on one side of the quartz slice and a negative charge on the other side. This phenomenon is the piezoelectric generation of a voltage difference across the two sides of the quartz crystal. Furthermore, the same effect is found when pressure is applied not mechanically, but through application of an alternating electrical field with only certain frequencies allowed to pass through the crystal. The frequencies allowed to pass vary with the crystal shape and thickness. Such crystals are used in radios, televisions, and radar. This effect also forms the basis for quarts watches and clocks. Voltage applied to a quartz crystal causes the crystal to expand and contract at a set rate, producing vibrations. The vibrations are then translated into a uniform measure of time.

While quartz crystals are suitable for the production of optical lenses, most lenses are manufactured from synthetically produced quartz due to the scarcity of good-grade large quartz crystals.

The feldspars are the most abundant minerals in the Earth's crust, accounting for about 60% of all igneous rocks. They are derivatives of silica, where about one-half or one-quarter of the silicon atoms have been replaced by aluminum. Feldspar is used in the manufacture of certain types of glass and pottery. Some feldspar crystals—such as moonstone (white perthilte), Amazon stone (green microcline), and multicolored labradorite, are used as gem stones and in architectural decorations. Some feldspar is used as a coating and filler in the production of paper.

Granite is a hard crystalline rock chiefly composed of quartz and feldspar. It is used in building bridges and building where great strength is needed. It is also employed in the construction of monuments and gravestones since it can be polished giving a lasting luster and because of its ability to withstand wear by the natural elements.

Sand is loose grains of minerals or rocks, larger than silt but smaller than gravel. Soil contains mineral (often in the form of small sand granules) and organic matter.

Micas are also composed of silicon–oxygen tetrahedra. The anionic charge on the silicate sheet is the result of the replacement of silicon by aluminum. Cations such as potassium are interspaced between these negatively charged sheets. Some mica is used in construction and electrical engineering applications. Synthetic mica is manufactured on a large scale for industrial consumption in coatings, as fillers, and so forth. Micas are one of the many layered silicon–oxygen-intense materials found in nature.

12.12 SILICON DIOXIDE IN ELECTRONIC CHIPS

Silicon dioxide plays a critical role in the electronics industry. The silicon used to produce silicon chips is derived from silicon dioxide. Semipure silicon dioxide (to about 99%) is prepared from the reaction of silicon dioxide with coke (a poor grade of graphite) using high temperature and an electronic arc.

$$SiO_2 + C \rightarrow Si + CO_2 \qquad (12.19)$$

Even so, this level of purity falls far short of the purity need to produce the chips used in computers. The purity required is about 99.9999996 or a level of impurity of about one part in a billion. This is achieved through multistep processes. One of these requires the silicon to be heated with HCl at high temperatures forming the desirable volatile trichlorosilane. The vapor is condensed then purified using distillation and absorption columns. The trichlorosilane is reacted with hydrogen gas at about 1,200°C depositing polycrystalline chip-grade silicon. The other product of this reaction is HCl, which can be again used to create more trichlorosilane, thus eliminating the production of unwanted byproducts.

$$Si + HCl \rightarrow HSiCl_3 \qquad (12.20)$$

$$HSiCl_3 + H_2 \rightarrow Si + HCl \qquad (12.21)$$

Silicon dioxide is also used to insulate regions of the integrated circuit. Here silicon dioxide is grown on the silicon surface by heating the surface to about 1,000°C in the presence of oxygen.

$$Si + O_2 \rightarrow SiO_2 \qquad (12.22)$$

An alternate approach employs heating gaseous tetraethoxysilane to form layers of silicon dioxide.

$$Si(OC_2H_5)_4 \rightarrow SiO_2 + 2H_2O + 4C_2H_4 \qquad (12.23)$$

12.13   SILICON DIOXIDE IN OPTICAL FIBERS

Today, almost all telecommunication occurs via optical fibers rather than metallic wires. Signal transmission with metallic wires was via electrons, while transmission through optical fibers is via photons. For a rough comparison, two small optical fibers can transmit the equivalent of more than 25,000 telephone calls simultaneously or in 1 second the information equivalent to about 2 h of TV shows. On a weight basis, 1 g of optical fiber has the transmission capability of about 300,000 grams of copper wire. The signal loss must be small. For a typical system, the loss more than a 10 mile distance is about that found for the transmission of light through an ordinary pane of window glass.

The optical fibers connect the two longest links of a typical optical fiber communications system. Briefly, an input signal enters an encoder generally in electrical form with the encoder transforming it into digitized bits of 1s and 0s. This electrical signal is then converted into an optical message in an electrical–optical converter. This converter is often a semiconductor laser that emits monochromatic coherent light. The message then travels through optical fibers to its target destination. Where that distance is large, such as between countries, repeater devices are employed that amplify the signal. Finally, at its destination, the message, in photonic form, is reconverted to an electric signal and then decoded.

The optical fiber is formed from a combination of polymeric materials. It typically consists of a core, cladding, and coating. The core does the actual transmission of the photons; the cladding constrains the light so it will travel within the core with little signal power loss and little pulse distortion; and the coating helps protect the inner material from damage and external pressures. There is a variety of materials that can be used in the construction of the optical fiber. On the basis of the core-cladding combination, there are three types of optical systems—the step-index, graded index, and single mode fibers.

Most of the "long-distance" systems use high-purity silica glass as the core material. These fibers are thin with a thickness of the order of 5–100 mm. Containment and retention of the signal is made possible because of the use of laser light and its total reflectance as it travels through the fiber. This containment is accomplished by using cladding and coating materials differing in refractive indexes from the core material. In the step-index approach, the index of refraction of the cladding is slightly less than that of the core. Here the output pulse is a little broader than the input because the light travels on slightly different paths as they travel through the fiber. This is overcome through the use of graded-index materials where impurities such as boron oxide or germanium dioxide are added to the silica glass so that the index of refraction varies parabolically across the cross section of the optical fiber core. Thus, the velocity of light varies according to where it is within the core, being greater at the periphery and less at the center. Thus, light that must travel a longer pathway through the outer periphery travel faster than those close to the center somewhat balancing themselves to minimize distortion.

Organic core optical materials are also in use. Such materials, in comparison to the silica cored materials, are lighter in weight, offer better ductility, have larger core diameters, and are less sensitive to vibrational stresses. They are often considered for shorter distances. For the step-index approach, where the refractive index of the cladding material should be a little less than that of the core material, a number of materials are possible. Two often used combinations are a polystyrene core with poly(methyl methacrylate) cladding and a poly(methyl methacrylate) core with fluorinated polymers as the cladding. Since the fiber core is to be optically clear, amorphous organic polymers are used. Also, to avoid unwanted scattering, the polymer must be of high purity since impurities will cause sites of differing refractive indexes and the associated difference in the speed of propagation of the signal. The organic fibers are generally made by melt spinning with the core-cladding structure formed by extrusion.

The types of optical loss for both silica-glass cored fibers and organic cored fibers are similar, but do differ in relative importance. In the visible wavelength region, there are both absorption and scattering losses. Absorption losses include higher harmonic molecular vibration modes in the infrared region and electronic transitional absorption in the ultraviolet region. Scattering losses include Rayleigh scattering and loss due to imperfections in the waveguide structure, mismatching of the core-cladding boundary interface, and birefringence due to fiber drawing as the fiber was formed.

12.14 ASBESTOS

Asbestos has been known and used for more than 2,000 years. Egyptians used asbestos cloth to prepare bodies for burial. The Romans called it *aminatus* and used it as a cremation cloth and for lamp wicks. Marco Polo described its use in the preparation of fire-resistant textiles in the thirteenth century. Asbestos is not a single mineral but rather a grouping of materials that give soft, thread-like fibers. These materials will serve as an example of two-dimensional sheet polymers containing two-dimensional silicate $(Si_4O_{10})^{-4}$ anions bound on either or both sides by a layer of aluminum hydroxide $(Al(OH)_3$; gibbsite) or magnesium hydroxide $(Mg(OH)_2$; brucite). Aluminum and magnesium are present as positively charged ions. These cations can also contain varying amounts of water molecules associated with them as hydrates. The spacing between silicate layers varies with the nature of the cation and amount of its hydration.

Due to its fibrous nature and ability to resist elevated temperatures (compared with most organic-based fabrics) it is used to make fabrics for the production of fire-resistant fabric, including laboratory glove-wear. Shorter fibers were used in electrical insulation, building insulation, and in automotive brake linings. Though asbestos has been known for thousands of years, it has only recently become known that asbestos can be dangerous. For instance, asbestos miners and manufacturing personnel who worked with it for 20 years or longer are 10 times more likely to contract asbestosis. Families of these workers and those living near the mines also have a greater than average chance of getting asbestosis. Asbestosis is a disease that blocks the lungs with thick fibrous tissue, causing shortness of breath and swollen fingers and toes. Bronchogenic cancer (cancer of the bronchial tubes) is prevalent among asbestos worker who also smoke. Asbestos also causes mesothelioma, a fatal cancer of the lining of the abdomen or chest. These diseases may lay dormant for many years after exposure.

It is believe that these diseases are caused by asbestos particles (whether asbestos or other sharp particles) about 5–20 μm in length corresponding to the approximate sizes of the mucous openings in the lungs. Thus, they become caught in the mucous openings. Because they are sharp, they cut the lining when people cough. Scar tissue and the repeated healing process cause scar tissue buildup and the opportunity for cancerous mutations to begin.

12.15 FLY ASH AND ALUMINOSILICATES

Aluminosilicates or aluminum silicates include industrial waste materials such as fly ash and ground granulated blast-furnace slag (GGBFS), and natural materials such as metakaolin, kaolin, microsilica, and volcanic ash. All are inorganic polymers derived from aluminum oxide and silicon oxide.

Fly ash is one residue created from the combustion of coal. Because there are different sources of coal, fly ash has a variable composition. Table 12.7 contains the general chemical composition as a function of the three major varieties of coal.

Fly ash consists of a variable mixture of somewhat spherical glasses ranging in size from 0.5–100 microns in diameter. It also contains minute amounts of many other elements, including arsenic, beryllium, boron, barium, copper, cadmium, chromium, thallium, vanadium, zinc, strontium, lead, and nickel. Fly ash is further divided by ASTM C618 standards into two general groupings, Class F and Class C. Class F typically comes from the combustion of older anthracite and bituminous coal that contains less than 10% lime (CaO). Pozzolanic materials combine with calcium hydroxide forming a cement-like material. Class F fly ash is pozzolanic with the glassy silica and alumina requiring a cementing agent such as Portland cement and water to produce a cement. Class C fly ash generally comes from younger lignite and subbituminous coal. Along with being pozzolanic, it is more self-cementing in comparison to Class F fly ash containing more lime, alkali, and sulfates.

Fly ash used to be simply land filled but recycling is increasing. Coal burning power plants produce about 80 million tons of fly ash yearly in the United States. About 30 million tons are currently recycled. Recycling of fly ash also reduces the need to quarry and the energy related to preparing concretes such as Portland cement. Fly ash is used with Portland cement allowing the amount of Portland cement to be reduced by up to 30% by mass. The resulting concrete is often greater in strength compared to employing only Portland cement. The use of fly ash to replace Portland cement is considered green-friendly since the production of 1 ton of Portland cement produces about 1 ton of carbon dioxide in comparison to zero carbon dioxide for fly ash.

Unlike soils, fly ash is more uniform in particle size so its addition to soil gives the mixture some interesting properties. Fly ash is used in the production of flowable fill or controlled low-strength material. It is used as a self-leveling, self-compacting backfill.

Asphalt concrete is a composite of asphalt and a mineral aggregate. Fly ash is used to fill voids between larger aggregates. Fly ash—containing asphalt concrete is stiffer resisting rutting.

Fly ash is a source of what is referred to as geopolymer. This term covers a group of inorganic synthetic aluminosilicate materials. Other major sources of geopolymer are volcanic materials and

TABLE 12.7
Percentage Average Chemical Composition of Fly Ash Derived from Different Coals

Composition	Bituminous	Subbituminous	Lignite
SiO_2	20–60	40–60	15–45
Al_2O_3	5–35	20–30	20–25
FeO, Fe_2O_3	10–40	5–10	5–15
CaO	1–12	5–30	15–40

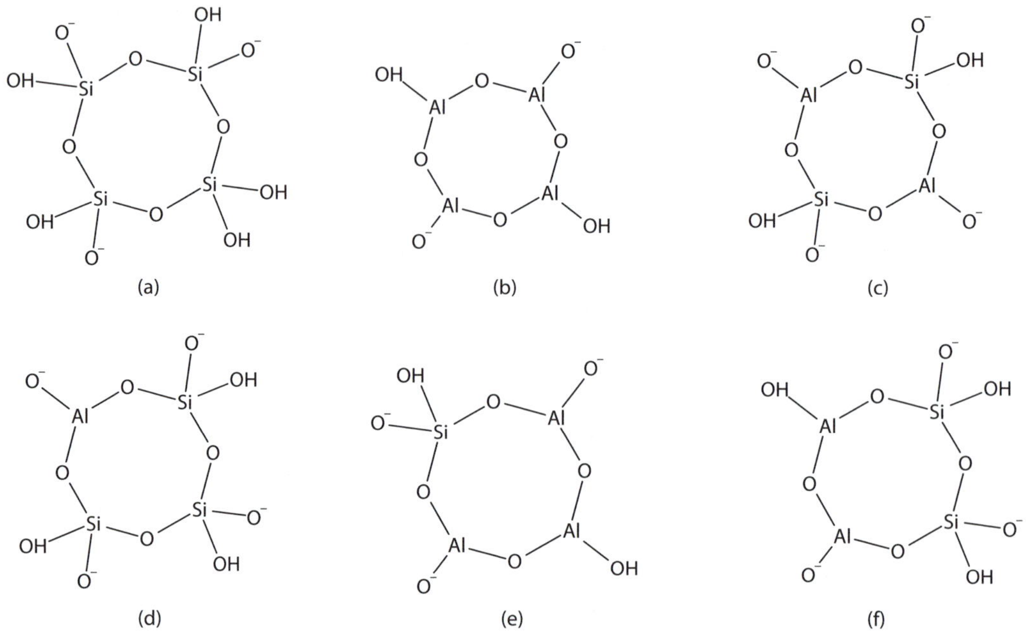

FIGURE 12.5 Examples of aluminosilicate rings, including those containing only silicone (a) and only aluminum (b).

the slag from metal smelting. Aluminosilicates are actually a mixture where much of the material exists as fused-ring clusters as shown in Figure 12.5.

Fly ash and aluminosilicate are environmentally conscious in that they are otherwise disposed in landfills and can substitute for materials with a large CO_2 footprint, but themselves they have no CO_2 footprint.

Each of these structures contain ionic oxygen "fingers" that can react with silicon oxide and vacant silicon sites on the sand and rocks of the concrete forming chemical bonds between the aluminosilicate concrete. Thus, aluminosilicates are being used as additives to Portland cement and as a building material themselves.

$$Al–O^- \text{ (from aluminosilicate)} + {}^+Si- \text{ (from rock)} \rightarrow Al–O–Si \qquad (12.24)$$

$$Al–O^- \text{ (from aluminosilicate)} + H–O–Si \text{ (from rock)} \rightarrow Al–O–Si + H_2O \qquad (12.25)$$

$$Si–O^- \text{ (from aluminosilicate)} + {}^+Si– \text{ (from rock)} \rightarrow Si–O–Si \qquad (12.26)$$

$$Si–O^- \text{ (from aluminosilicate)} + H–O–Si \text{ (from rock)} \rightarrow Si–O–Si + H_2O \qquad (12.27)$$

These bonds can also be formed through the calcium oxide portions of the concrete.

$$Si–O^- \text{ (from aluminosilicate)} + {}^+Ca- \text{ (from concrete)} \rightarrow Si–O–Ca \qquad (12.28)$$

$$Si–O^- \text{ (from aluminosilicate)} + H–O–Ca \text{ (from concrete)} \rightarrow Si–O–Ca + H_2O \qquad (12.29)$$

These inorganic polymers have a chemical composition similar to mica and zeolites but they are amorphous rather than crystalline.

The aluminosilicate geopolymers are generally formed from reaction of an aluminosilicate powder with an alkaline silicate solution or other activator.

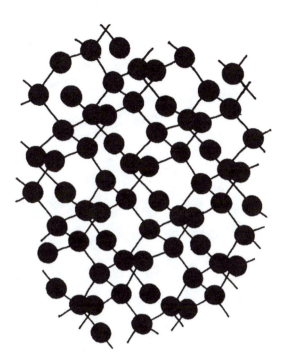

FIGURE 12.6 Representation of diamond, where each carbon is at the center of a tetrahedron composed of four other carbon atoms.

12.16 POLYMERIC CARBON—DIAMOND

Just as carbon serves as the basic building element for organic materials, so also does it form a building bock in the world of inorganic materials. Elemental carbon exists in many different forms, including the two longest known—diamond and graphite. Graphite is the more stable allotrope of carbon, with graphite readily formed from heating diamonds.

Natural diamonds (Figure 12.6) are believed to have been formed millions of years ago when concentrations of pure carbon were subjected by the Earth's mantle to great pressures and heat. They are the hardest known natural material. The majority of diamonds (nongem) are now man made. Most of the synthetic diamonds are no larger than a grain of common sand. The major use of synthetic diamonds is as industrial shaping and cutting agents to cut, grind, and bore (drill). By 1970, General Electric was manufacturing diamonds of gem quality and size through compressing pure carbon under extreme pressure and heat. It was found that addition of small amounts of boron to diamonds causes them to become semiconductors. Today, such doped diamonds are used to make transistors.

While diamonds can be cut, shaping is done by trained gem cutters striking the rough diamond on one of its cleavage plates. These cleavage plates are called *faces* and represent sites of preferential cleavage and reflection of light. This balance between strength and flexibility, crystalline and amorphous regions is demonstrated to one extreme by diamonds that are very crystalline, resulting in a strong, inflexible, and brittle material.

12.17 POLYMERIC CARBON—GRAPHITE

While diamond is the hardest naturally occurring material, the most common form of crystalline carbon is the much softer and flexible graphite. Graphite occurs as sheets of hexagonally fused benzene rings (Figure 12.7) or "hexachicken wire." The bonds holding the fused hexagons together

(a)

(b)

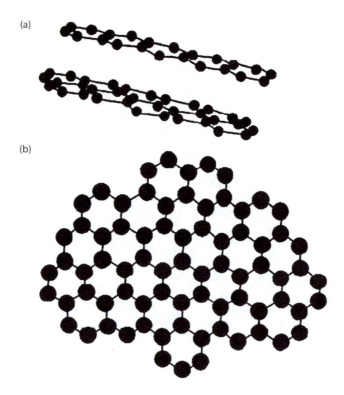

FIGURE 12.7 Representations of graphite emphasizing the layered (a) and sheet (b) nature of graphite.

are traditional covalent bonds. The bonds holding the sheets together are weaker than the bonding within the sheets consisting of a weak overlapping of pi-electron orbitals. Thus, graphite exhibits many properties that are dependent on the angle at which they are measured. They show some strength when measured along the sheet, but very little strength if the layers are allowed to slide past one another. This sliding allows the graphite its flexibility, much like the bending of bundles of proteins sliding past one another, allowing our hair flexibility. The fused hexagons are situated such that the atoms in each layer lie opposite to the centers of the six-membered rings in the next layer. This arrangement further weakens the overlapping of the pi electrons between layers such that the magnitude of layer-to-layer attraction is on the order of ordinary secondary van der Waals forces. The "slipperiness" of the layers accounts for graphite's ability to be a good lubricant.

The variance of property with angle of applied force, light, magnetism, and so on, is called *anisotropic behavior.* Calcite is anisotropic in its crystal structure, resulting in a dependency of its interaction with light with the angle of incidence of the light.

As with diamond, graphite's discovery and initial usage is lost in antiquity. It was long confused with other minerals such as molybdenite (MoS_2). At one time it was known as plumbago (like lead), crayon noir, silver lead, black lead, and *carbo mineralis.* Werner in 1789 first named it *graphit,* meaning (in Greek) "to write."

The Acheson process for graphite production begins by heating a mixture of charcoal, or coke, and sand. The silica is believed to be reduced to silicon that combines with carbon forming silicon carbide, which subsequently dissociates into carbon and silicon. The silicon vaporizes and the carbon condenses forming graphite. Graphite is also produced using other techniques.

Today, graphite is mixed with clay to form the "lead" in pencils. Graphite conducts electricity and is not easily burned so many industrial electrical contact points (electrodes) are made of graphite. Graphite is a good conductor of heat and is chemically inert, even at high temperatures. Thus, many crucibles for melting metals are graphite-lined. Graphite has good stability to even strong acids, thus it is employed to coat acid tanks. It is also effective at slowing down neutrons, and thus composite bricks and rods (often called *carbon rods*) are used in some nuclear generators to regulate the progress of the nuclear reaction. Its slipperiness allows its use as a lubricant for clocks, door locks, and handheld tools. Graphite is also the major starting material for the synthesis of synthetic diamonds. Graphite is sometimes used as a component of industrial coatings. Dry cells and some types of alkali-storage batteries also employ graphite. Graphite fibers are used for the reinforcement of certain composites.

12.18 INTERNAL CYCLIZATION—CARBON FIBERS AND RELATED MATERIALS

There are a number of important polymers that are formed through internal cyclization. In almost all cases, these are five- and six-membered rings with the vast majority being six membered. The tendency to form six-membered rings is related to a statistical feature. In studying the most probable distances from the beginning point using random statistics for units with a bond angle of about $109.5°$ (for a tetrahedral), the most probable distance for a chain of six units long is back at the starting point. The number of units required before the most probable distance is the starting point, which is dependant upon the bond angle and is called the *Kuhn element*. As note before, the Kuhn element for connected methylenes is six. The Kuhn element for sp^2 geometry is not six, but in this case, the driving force is the formation of six-membered rings with three alternating pi bonds—that is, the formation of the aromatic structure.

Often these internal cyclizations are incomplete giving products with mixed moieties. Even so, such internal cyclization is the source of a number of interesting and important polymers. A number of ladder-like structures have been synthesized from the internal cyclization of polymers. Following are several examples that illustrate this. The most important commercial products are those utilized to form the so-called carbon fibers.

Carbon fibers, and associated composite materials, are the result of internal cyclization. Polyacrylonitrile, when heated, undergoes internal addition forming condensed polycyclic material called "black orlon." Further heating to about $1,000°C$ removes the hydrogen atoms and most of the nitrogens giving a polyaromatic structure containing about 95% carbon (Figure 12.8). Further heating to about $2,800°C$ gives a product with almost 99% carbon. These products can be forced through tiny holes to form tiny fiber-like materials, fibrals, that are combined to give fibrous-like

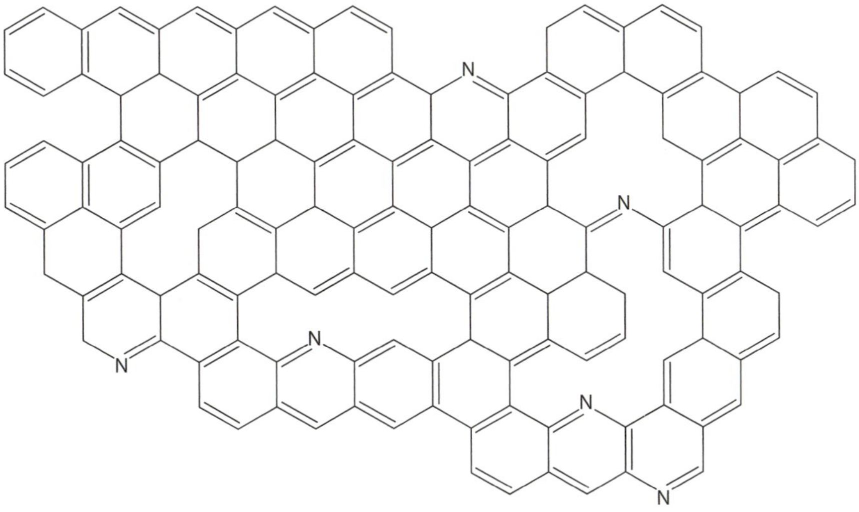

FIGURE 12.8 Idealized structure of "black orlon."

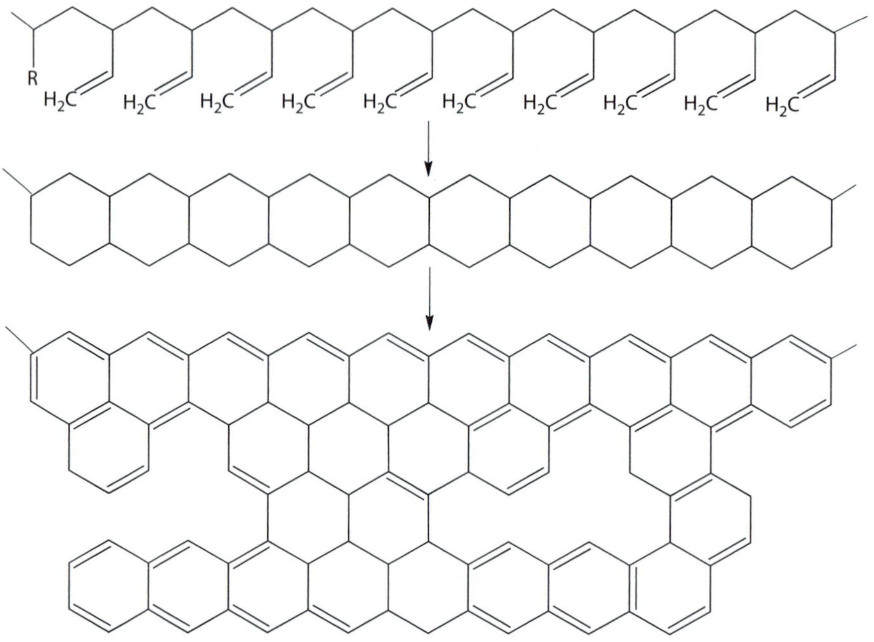

FIGURE 12.9   Idealized structures for the synthesis of carbon-like materials from poly(1,2-butadiene).

materials that can finally be woven together to give fabrics. The fibrals, fibers, and fabrics act as the fibrous portion of many high-strength composite materials. These fibers are light weight, very strong, chemically inert, and can form semiconductor and conductor materials.

Diene polymers undergo cyclization in the presence of cationic initiators such as sulfuric acid. 1,2- and 3,4-diene polymers undergo this cyclization forming extensive fused-ring groupings.

The polymerization of butadiene using certain catalytic systems such as butyl lithium and tetramethylenethylene diamine gives poly(1,2-polybutadiene). The polybutadiene, in turn, can undergo internal cyclization via cationic reactions forming a sort of linear saturated polycyclohexane. Further heating, resulting in dehydrogenation, with chloranil gives a fused-ring product similar to that of carbon fibers when heated to about 1,500°C (Figure 12.9). This material is sold under the trade name of Pluton by 3M. This sequence is described in Figure 12.9. While the final product is insoluble and infusible, prespinning is done on the soluble saturated intermediate.

12.19   CARBON NANOTUBES

There are several materials that form so-called "nanotubes," including boron–nitrogen compounds. Here we will focus on nanotubes derived from carbon. Carbon nanotubes (CNTs) have probably been made in small amounts since the first fires reduced trees and organic material to ashes. It has been found that certain ancient steel products may have possessed CNTs derived from the exposure of the processed steel to carbon sources. Thus, the Damascus steel used in making very strong weapons is believed to have profited from the presence of CNTs. It was not until recently, as part of the so-called nanorevolution, that we first recognized the existence of these nanotubes. In 1952, Radushkevich and Lukyanovich published pictures of tubes of carbon. This discovery was largely unnoticed. Others contributed to the early history of carbon in the form of tubes. In 1991, Sumio Iijima, NEC Fundamental Research Laboratory in Japan, first observed CNTs as byproducts of the arc-discharge synthesis of fullerenes and it is often accepted that this observation prompted the current activities with CNTs.

Carbon nanotubes are carbon allotropes that have attracted much attention. They have a diameter of about 1/50,000 that of a human hair. Some have suggested that CNTs will be one of the most important twenty-first century materials because of the exceptional properties and ready abundance of the feedstock, carbon. CNTs are generally classified into two groups. Multiwalled CNTs (MWCNTs) are composed of 2–30 concentric graphitic layers with diameters ranging from 10 to 50 nm with lengths that can exceed 10 μm. Single-walled CNTs (SWCNTs) have diameters ranging from 1.0 to 1.4 nm with lengths that can reach several micrometers.

An ideal CNT can be envisioned as a single sheet of fused hexagonal rings, that is, graphite, that has been rolled up forming a seamless cylinder with each end "capped" with half of a fullerene molecule. SWCNTs can be thought of as the fundamental cylindrical structure, with MWCNTs simply being concentric tubes. They can also be conceived of as being the fundamental building block of ordered arrays of single-walled nanotubes called *ropes*.

12.19.1 STRUCTURES

Carbon nanotubes are composed of carbon sp^2 bonded structures, similar to those of graphite. These bonds are stronger than typical carbon sp^3 bonds resulting in the strength of the CNTs. Unlike two-dimensional sheets of graphite, the CNTs align themselves into rope-like structures. Because of bonding similarity to graphite, these materials are often referred to as graphene.

Carbon nanotubes are composed of sp^2 bonds similar to those in graphite. They naturally form rope-like structures where the ropes are held together by van der Waals secondary forces.

Geometrically, CNTs can be described in terms of a two-dimensional graphene (graphite) sheet. A chiral vector is defined on the hexagonal lattice as

$$C_h = n\mathrm{x} + m\mathrm{y} \tag{12.30}$$

where **x** and **y** are unit vectors, and n and m are integers, also tube indices. The chiral angle is measured relative to the direction defined by $n\mathrm{x}$.

When the graphene sheet is rolled up forming a nanotube, the two ends of the chiral vector meet one another. The chiral vector thus forms the circumference of the CNTs circular cross section. Different values of n and m give different nanotube structures with different diameters (Figure 12.10).

There are three general types of CNT structure (Figure 12.11). The zigzag nanotubes correspond to $(n, 0)$ or $(0, m)$ and have a chiral angle of $0°$. The carbon–carbon position is parallel to the tube axis. Armchair nanotubes have (n, n) with a chiral angle of $30°$. The carbon–carbon positions are perpendicular to the tube axis. Chiral nanotubes have general (n, m) values and a chiral angle of between $0°$ and $30°$ and as the name implies, they are chiral.

In real life, nothing is perfect. As is the case with CNTs the defects are mainly inclusion of wrong-membered rings. Pentagonal defects, that is, the replacement of a hexagonal with a five-membered ring, results in a positive curvature causing the tube to curve inward like a horse shoe. The closure of an open cyclindrical surface necessarily involves topological defects—often formation of pentagons. Heptagonal defects result in a negative curvature with the lattice looking expanded around the defect.

The tendency to include pentagonal units can be seen by comparing the presence of pentagonal units in fullerene structures. The C_{60} structure contains 12 pentagons and 20 hexagons. The larger the fullerenes the smaller the ratio of pentagons to hexagons. This is consistent with the use of pentagons to "cause" sharper bends and greater curvature in comparison to hexagons. Interestingly, fullerene C_{60} is one of the most strained molecules known but it exhibits good kinetic stability. It begins to decompose at about 750°C. There are a number of higher numbered (carbon number) fullerenes, including C_{70}, C_{76}, C_{78} (two geometric isomers), C_{80}, ... Fullerenes can act as a source of the CNTs with the different-sized fullerenes producing different nanotubes. The three general structures of nanotubes can be produced using different fullerenes with C_{60} giving armchair

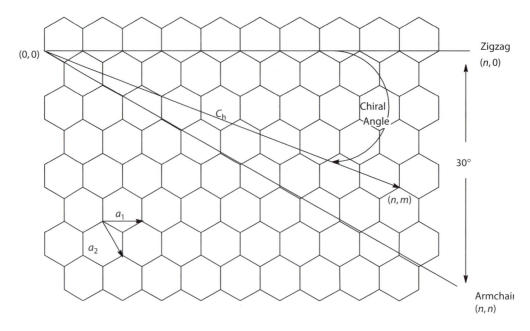

FIGURE 12.10 Graphite sheet representation showing the θ or chiral angle. For a chiral angle of $0°$, the structure is a zigzag; for $30°$ it is an armchair; and in between it is helical. The figure is drawn with a $30°$ angle between the zigzag and armchair.

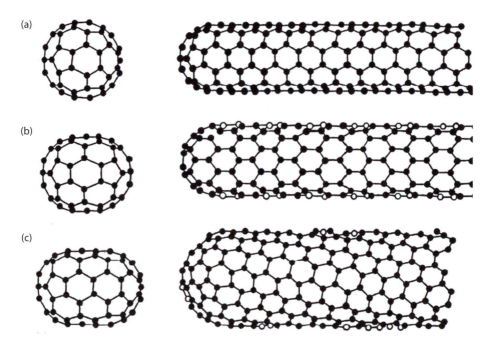

FIGURE 12.11 Representations of the three major structural forms of carbon nanotubes: armchair (a), zigzag (b), and helical (c) and of C_{60} (left, a), C_{70} (left, b), and C_{80} (left, c) fullerene structures.

nanotubes, C_{70} giving zigzag structures, and C_{80} giving helical forms of nanotubes. Figure 12.11 contains representations of these three fullerenes.

This difference in structure also influences the electrical conductivity with the armchair form being conductive or metal-like in its conductivity and most of the other forms act as semiconductors.

Two other forms of CNTs are recognized. The fullerite is a highly incompressible form with a diamond-like hardness. The torus is a doughnut-shaped nanotube. These circular nanotubes have extremely high magnetic moments and outstanding thermal stability. The particular properties are dependent on the radius of the tube.

One of the major reasons for the intense interest in CNT is their extreme and varied properties. CNTs are among the strongest and stiffest materials known. Tensile strengths to about 65 gigapascals (GPa) have been found. This translates to a cable of 1 mm² cross section capable of holding about 3,200 tons. A general density of CNTs is about 1.4 g/cc. Its specific strength is more than 300 times as great as steel. Under stress, tubes can undergo permanent deformation (plastic deformation). Interestingly, because the tubes are hollow, they are not nearly as strong as diamonds buckling under compression, torsion, and bending stress. Nanotubes conduct and transport along the lengthwise direction resulting in them often being referred to as being one dimensional. Table 12.8 contains a comparison between SWCNTs and competitive materials/techniques. Another reason for the intense interest is that CNTs are produced from readily available inexpensive materials, they are being considered for use as both bulk materials, such as in composites and clothing, and as components in computers, electrical devices, and so on.

The number of potential and real uses where an essential ingredient is CNTs is growing almost daily. We are able to control the length and kind of tubes formed. Because the starting material is so plentiful and inexpensive, this is an area where most of the limits are ones we impose on ourselves. Following is a discussion of some of the application areas for CNTs.

Electrical—Nanotubes can be metallic or semiconducting, depending on their diameters and helical arrangement. Armchair (n = m) tubes are metallic. For all other tubes (chiral and zigzag), when tube indices (n + m)/3 is a whole number integer the tubes are metallic, otherwise they are semiconducting. CNTs can in principle play the same role as silicon does in electronic circuits, but on a molecular scale where silicon and other standard semiconductors cease to work. Single

---

TABLE 12.8

Comparison between Selected Properties of Single-Walled CNTs and Competitive Materials/Techniques

Property	Single-Walled Carbon Nanotubes	Comparison
Size	0.6–1.8 nm in diameter	Electron beam lithography can create lines 50 nm wide and a few nm thick
Density	1.33–1.40 g/cc	Aluminum has a density of 2.7 g/cc and titanium has a density of 4.5 g/cc
Tensile strength	ca 45 billion Pascals	High-strength steel alloys break at about 2 billion Pascals
Resilience	Can be bent at large angles and restraightened without damage	Metals and carbon fibers fracture at grain boundaries
Current carrying capacity	Estimated at 1 billion A/cc	Copper wires burn out at about 1 million A/cc
Heat transmission	Predicted to be as high as 6,000 W/m·K	Diamond transmits 3,320 W/m·K
Field emission	Can activate phosphors at 1–3 V if electrodes are spaced 1 μm apart	Molybdenum tips require fields of 50–100 V/μm and have limited lifetimes
Temperature stability	Stable up to 2,800°C in vacuum, 750°C in air	Metal wires in microchips melt at 600°C–1,000°C

CNT bundles have been used to construct elementary computing circuits known as logic gates. Applications include composites and coatings for enclosures and gaskets and use as stealth material. They perform better than tungsten filaments in incandescent lamps. Metallic nanotubes can have electrical current densities on the order of one thousand times that of silver and copper.

Conductive Adhesives and Connectors—Because of their molecular-level characteristics, they have been used to connect electronic circuits and to bind together electrical units acting as a molecular solder.

Mechanical—CNTs have superior resilience and tensile strength. They can be bent and pressed over a large angle before they begin to ripple or buckle, finally developing kinks. Until the elastic limit is exceeded, the deformations are elastic with the deformation disappearing when the stress is removed. It is envisioned that buildings and bridges built from them may sway during an earthquake rather than fracturing and crumbling. Multiwalled nanotubes can easily slide past one another without friction resulting in the construction of rotating bearings used in rotational motors based on this observation.

Field Emission—When stood on end electrified CNTs act as a lightning rod concentrating the electrical field at their tips. While a lightning rod conducts an arc of electricity to a ground, a nanotube emits electrons from its tip at a rapid rate. Because the ends are so sharp, the nanotube emits electrons at lower voltages than do electrodes made from other materials and their strength allows nanotubes to operate for longer periods without damage. Field emission is important in several industrial areas, including lighting and displays. Commercial use of CNTs as field emitters has begun. Vacuum-tube lamps in six colors have been developed that are twice as bright as conventional light bulbs, longer-lived, and at least ten times more energy-efficient as conventional light bulbs.

Hydrogen and Ion Storage—While we can picture CNTs as being composed of hexagonal carbon atoms with lots of empty space between the carbons, atoms "thrown" against them generally just bounce off. Even helium atoms at an energy up to 5 eV do not readily penetrate the nanotube. Thus, the graphene sheet and CNTs are really membranes or fabrics that are one atom thick made of strong material that is also impenetrable (to a limit). Thus, CNTs can be used for hydrogen storage in their hollow centers with release being controlled allowing them to act as inexpensive and effective fuel cells.

Chemical and Genetic Probes—Nanotube-tipped atomic force microscopes can trace a strand of DNA and identify chemical markers that reveal DNA fine structure. A miniaturized sensor has been constructed based on coupling the electronic properties of nanotubes with the specific recognition properties of immobilized biomolecules through attaching organic molecules—handles—to these tubular nanostructures. In one study, the pi-electron network on the CNT is used to anchor a molecule that irreversibly adsorbs to the surface of the SWNT. The anchored molecules have a "tail" to which proteins, or a variety of other molecules, can be covalently attached. The result is that these molecules are immobilized on the sidewall of the nanotube with high specificity and efficiency. The molecule's tail is tipped with a succinimidyl ester group, which is readily displaced when an amine group attacks the ester function, forming an amide bond. Thus, the CNTs are used as both highly sensitive probes and highly selective immobilizing sites that allow specific reactions to occur.

Analytical Tools—SWCNTs are being used as tips of scanning probe microscopes. Because of their strength, stability, and controllable and reproducible size, the tup probes allow better image fidelity and longer tip lifetimes.

Solar Cells—Some nanotubes exhibit a photovoltaic effect and can replace material in solar cells acting as a transparent conductive film allowing light to pass to the active layers and generate photocurrent. It has also been shown that nanotubes can be used in conjunction with other materials to act as an ignition device that is triggered with a simple camera flash.

Filters—CNTs have been used to filter carbon dioxide from power plant emissions and so may be a future air pollution filter material. They have also been used to filter water removing the salt content by allowing water to pass through but capturing the chloride ions.

Catalyst Supports—CNTs have a very high surface area. Each carbon atom is exposed to the interior and exterior surface. Because of the chemist's ability to connect almost any moiety to their surface, a number of CNTs have been shown to act as outstanding catalyst supports when catalysts have been attached to them. Because of their high strength, they perform well to the rigors of being a catalysts support. Their ability to conduct electricity and heat suggests additional ways CNTs can be used to assist catalytic behavior.

Superconductors—CNTs offer unique electronic properties due to quantum confinement. According to quantum confinement, electrons can only move along the nanotube axis. Metallic CNTs are found to be high-temperature superconductors.

Fibers and Fabrics—Textiles spun from CNTs have been made into clothing such as slacks. Along with offering outstanding liquid shedding clothing made from them, the clothing possess outstanding wear resistance. Penetration-resistant body and vehicle armor can be made from composites containing CNTs fibers. They are also being studied for use in transmission line cables. Addition of nanotubes to concrete increases its tensile strength and stops crack propagation. Addition to polyethylene increases its elastic modulus. (For more fiber applications please see "Composites" below.)

Energy Storage—CNTs have a very high surface area (about 10^3 m²/g), good electrical conductivity and can be made very linear (straight). They have been used to make lithium batteries with the highest reversible capacity of any material and employed to make supercapacitor electrodes. CNTs are used in a variety of fuel cell applications where durability is important.

Biomedical Applications—CNTs allow cells to grown on and over them without adherence to the nanotubes and without toxic reaction. Potential applications range from their use within coatings and composites to be used within the body, for prosthetics, and in the construction of vascular stents and neuron growth and regulation.

Heat Conductivity—Their outstanding anisotropic thermal conductivity allows them to be used when heat must be moved from one location to another.

Composites—CNTs, because of their fiber nature on a molecular level makes them outstanding fibers for composites where strength, electrical and thermal conductivity, stiffness, and toughness are needed. Because CNTs contribute to high values of all of these properties, composites can be constructed where outstanding behavior in any one of these areas is needed. Sports wear such as lighter and stronger golf club shafts, bike parts, baseball bats, and tennis rackets have been produced from nanotubes-containing composites. Electronic motor brushes have been made from nanotubes composites that are better lubricated, cooler running, stronger, less brittle, and more accurately moldable in comparison to the traditional carbon black brushes.

Coatings—Because of their good strength, thermal and electrical conductivity, and slipperiness, CNTs offers potential interesting coatings applications. Thus, a building coated with a CNT-containing material that has great durability will remain free from dust, dirt, and organic growth, and can be used to capture energy via heat and electrical conductivity.

Films—Films have been formed from CNTs that are high strength, conductive, and transparent. They are being investigated for use in liquid crystal diodes (LCDs), photovoltaic devices, flexible displays (foldable TV screens), and touch screens and are intended to replace indium tin oxide (ITO) in many applications.

Nanomotors—Multiwalled nanotubes exhibit an interesting telescoping behavior where they can slide within the outer tube with essentially no friction allowing the creation of almost frictionless linear or rotating bearings. This behavior allowed the construction of the world's smallest rotational motor and nanorheostat.

Along with their high tensile strengths they also offer a high elastic modulus on the order of 1 TPa. It has a relatively low density of 1.3–1.4 g/mL. While they offer extremely high strengths when tension is applied, they are not as strong under compression because of their hollow structure. Another problem is defects. While we can synthesize almost defect-free tubes many of the tubes contain defects. Because these nanotubes are essentially one-dimensional structures, the defect acts as the weak link and can lower the tensile strength by 85%. The presence of defects also lowers the thermal and electronic conductivity properties. Most of the defects are of the Stone Wales variety, where a pentagon and heptagon pair is formed by rearrangement of the bonds.

As we are studying CNTs other interesting properties are being found. For instance, recently it was found that CNTs are like hot wheels tubes allowing water and gas to flow 100–10,000 times faster than as classical models predict. It is believed that just as Teflon sheds water because of its inert surface so also CNTs are slick because of its atomically smooth surface and lack of polarity differences within the tube allowing almost a zero friction situation. Under high pressure, CNTs can be forced together exchanging some of the sp^2 bonding for sp^3 bonding, giving the possibility for forming strong wiring. Electrically and thermally conductive carbon nanopaper can be made that is 250 times stronger than steel and much lighter that can be used as a heat sink for chipboards, blacklight for LCD screens, and to protect electronic devices.

Carbon nanotubes have been used as templates producing other nanotubes such as gold and zinc oxide nanotubes. These nanotubes are hydrophilic while CNTs are hydrophobic allowing alternative uses where the behavior of the tube can be varied according to its environment.

A major concern involves their possible toxicity. Research involving their potential and real toxic effects is just beginning. This is particularly important since these materials may become part of our everyday lives in many ways. Current results indicate that nanotubes can cross membrane barriers. If this is true, then access to our organs will occur. There is building biological evidence suggesting that under certain conditions nanotubes, of any origin, pose a real risk to our health.

Compounding the question of toxicity is the large variety of nanotubes potentially and really available each with its own toxicity profile. Further, the description of these various materials by name or other convention has yet to be established.

12.20   BITUMENS

The petroleum industry, including the commercial bitumen industry, was born in the United States in August 27, 1859 when Colonel Drake drilled a hole about 70 feet deep near Titusville, Pennsylvania to "bring in" the first producing well. By 1908, Henry Ford began to mass produce his Model "T" Ford creating an additional need for this petroleum in the form of gasoline. The distillation residue became more plentiful and a need for large-scale usage of bitumens increased.

Even so, the bitumens are a very old material. They were used in the water-proofing of the cradle that baby Moses was floated in. It was used by the ancient Egyptians in their mummification process. Bitumens were used in sand stabilization and for lighting the naval base by the Second Muslim Caliph, Omar ben Khattab, at Basra on Shattul-Arab on the West Coast of what is now Saudi Arabia around 640 AD.

Bitumens occur naturally or are formed as the residue in the distillation of coal tar, petroleum, and so on. Industrially, the two most important bitumens are asphalt and coal tar. Asphalt is a brown to black tar-like variety of bitumen that again occurs naturally or is the residue of distillation. Coal tar is the black, thick liquid obtained as the residue from the distillation of bituminous coal.

Bitumens are examples of materials that have only an approximate structure. Bitumens are carbon-intense small polymers with molecular weights from about 200 to 1,000 Da for coal tar with a calculated average number of carbons in a chain of about 15–70. Asphalt has a molecular weight averaging about 400–5,000 Da with a calculated average number of carbons in a chain of about

30 to about 400. Thus, they are generally oligomeric to short polymers. Asphalt has a C/H ratio of about 0.7 while coal tar has a *C/H* ratio of about 1.5 approaching that of a typical hydrocarbon where the *C/H* ratio is about 2.

As with most nonpolar hydrocarbon-intense polymers, bitumens exhibit good resistance to attack by inorganic salts and weak acids. They are dark, generally brown to black, with their color difficult to mask with pigments. They are thermoplastic materials with a narrow service temperature range unless modified with fibrous fillers and/or synthetic resins. They are abundant materials that are relatively inexpensive, thus their use in many bulk applications.

At temperatures above the T_g, bitumens generally show Newtonian behavior. Below the T_g bitumens have rheological properties similar to elastomers.

Bitumens are consumed at an annual rate in excess of 75 billion pounds in the United States. Bitumens are generally used in bulk such as pavements (about 75%), and in coatings for roofs (15%), driveways, adhesive applications, construction, metal protection, and so on where the bitumen acts as a weather barrier. Bituminous coatings are generally applied either hot or cold. Hot-applied coatings are generally either filled or nonfilled. Cold-applied coatings are generally either nonwater or water containing. In the hot-applied coatings, the solid is obtained through a combination of cooling and liquid evaporation while in the cold-applied coatings the solid material is arrived at through liquid evaporation. One often used coating employs aluminum pigments compounded along with solvents. These coatings are heat reflective and decrease the energy needs of building using them. The aluminum-metallic appearance is generally more desirable than black, and the reflective nature of the aluminum reflects light that may damage the bitumen coating allowing the coating a longer useful life. Today, many of the bitumen coatings contain epoxy resins, various rubbers, and urethane polymers.

12.21   CARBON BLACK

Carbon Black is another of the carbon-intensive materials. It is formed from the burning of gaseous or liquid hydrocarbons under conditions where the amount of air is limited. Such burning favors "soot" formation, that is, carbon black formation. It was produced by the Chinese more than 1,000 years ago. Today, it is produced in excess of 1.5 million tons annually in the United States. The most widely used carbon black is furnace carbon black. The particle size of this raw material is relatively large, about 0.08 mm. It is soft with a Mohs scale hardness of less than one.

In addition to carbon, carbon black also contains varying amounts of oxygen, hydrogen, nitrogen, and sulfur. A typical carbon black contains about 98% carbon, 0.3%–0.8% hydrogen, 0.3%–1.2% oxygen, 0.0%–0.3% nitrogen, and 0.3%–1% sulfur. The impurities in the water employed to quench the burning carbon is mainly responsible for the noncarbon and hydrogen content and the sulfur comes mainly from the feedstock.

While some describe the structure of carbon black as being small chain-like structures, recent information shows it as being somewhat graphite like in structure where the parallel planes are separated by about 0.35–0.38 nm, always greater than the interlayer distance for graphite of 0.335 nm (Figure 12.12). The microstructure of carbon black has been studied extensively employing varying spectroscopies, including ATR (attenuated total reflectance infrared), high-resolution transmission electron microscopy (TEM), and X-ray diffraction. Various models have been developed to describe the average structure in greater detail. These models are related to the structure of graphite.

As with other polymeric materials, the surface structure is different from the bulk and is related to the conditions under which the material was manufactured. Since the material is formed in air, the surface is rich in oxygen. These oxygen atoms play an important role in the resulting carbon black properties. The surface can be modified using a variety of treatments. Heat treatments above 800°C act to increase the amount of crystallinity in the overall structure. Under inert conditions surface groups are modified and the amount of oxygen decreased with heating from 200°C to 1,200°C. Plasma treatments are employed to modify the carbon black surface creating and destroying various functional groups in the presence of other reactants. Chemical oxidation is also employed to

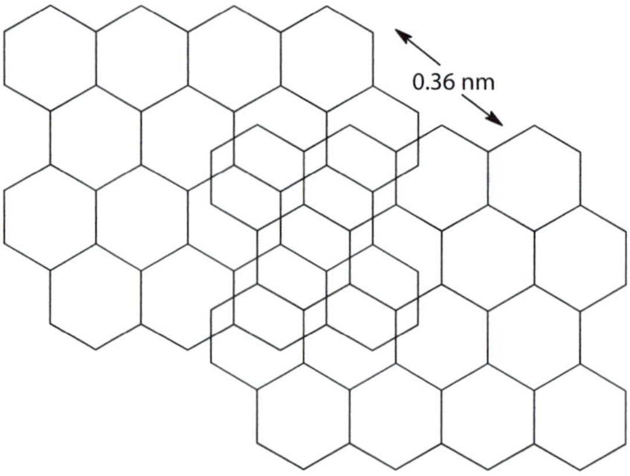

FIGURE 12.12 Sheets of carbon black.

modify carbon black employing oxidizing agents as air, nitric acid, and ozone. Surface grafting is also employed to provide desired surface functional groups.

The nature of the surface is especially important for carbon black since most applications employ the carbon black as an additive forming blends and alloys. These blends can be bound by simple physical contact or by chemical binding through formation of chemical bonding between the various phases.

A major use of carbon black is in rubbers. Incorporation of carbon black into a rubber matrix can result in an increase in strength-related properties such as abrasion resistance, viscosity, and modulus. Thus, carbon black is a reinforcing agent rather than simply an inexpensive additive. The reinforcing effect is dependent on the particle size, amount, aggregate structure, surface area, surface activity, and so on. The reinforcing effect mainly occurs because of the interaction between the interfacial surfaces of the rubber matrix and the carbon black. Carbon blacks can be divided roughly as to being reinforcing (<35 nm) or semireinforcing (>35 nm) depending on the particle size. The antiabrasion of carbon black increases with increase in particle size.

Carbon black is the most important additive to rubber composing of between 30% and 70% of the bulk rubber product. Tire goods consume about 65% of the carbon black, mechanical goods another 25%, with only about 10% employed for nonrubber applications.

A typical tire rubber formulation for tire tread will contain various rubbers, mainly styrene–butadiene (50%) and *cis*-polybutadiene (12%), various processing aids (2%), softeners (3%), vulcanizing agent (mainly sulfur; 1%), accelerators, and reinforcing filler (namely carbon black; 30%) so that by bulk, carbon black is the second most used material.

Of the 10% of carbon black used for nonrubber applications, about 35% is used for plastics, 30% for printing inks, 10% coating, and 5% for paper. In plastics, carbon black enhances a variety of properties, including UV shielding, electrical conductance, pigment, opacity, and mechanical properties. Plastics that use carbon black filler may contain only several percent to having over half of the weight being carbon black.

In xerography, carbon black is the most important pigment in printing or duplicating toner applications. Here, blackness, good dispersion, and needed electrical properties are provided by carbon black. Similarly, these properties are useful for applications in coatings, inks, and printing. Here though, rheological properties in liquid media are also important.

12.22 POLYSULFUR

Sulfur is present in the petrochemicals derived from once-living matter because of its presence in certain amino acids. Because of its removal from industrial waste, sulfur has been stockpiled and is available at a low price in large amounts. While the stable form of sulfur at room temperature is cycloocta-sulfur, S_8, linear polysulfur is formed on heating. Unfortunately, the thermodynamically stable form of sulfur is the cycloocta-sulfur monomer and the polymer undergoes depolymerization after awhile (Equation 12.31).

$$\tag{12.31}$$

Methods have been studied to inhibit this reversal process. Some have involved the addition of olefins such as limonene, myrcene, and cyclopentadiene to the ends to inhibit the depolymerization. Such stabilized polysulfur has been incorporated into concrete and asphalt mixes to strengthen them. Concrete blocks, posts, and parking tire restrainers containing polysulfur are now being produced.

12.23 CERAMICS

The term "ceramics" comes from the Greek word *keramos,* which means "potter's clay" or "burnt stuff." While traditional ceramics were often based on natural clays, today's ceramics are largely synthetic materials. Depending on which ceramic and which definition is to be applied, ceramics have been described as inorganic ionic materials and as inorganic covalent (polymeric) materials. In truth, many ceramics contain both covalent and ionic bonds and thus can be considered "to be or not to be" (shades of Shakespeare) polymeric materials. Many of the new ceramics, such as the boron nitriles and the silicon carbides, are polymeric without containing any ionic bonds.

Ceramics are typically brittle, strong; resistant to chemicals such as acids, bases, salts, and reducing agents; and they are high melting. They are largely composed of carbon, oxygen, and nitrogen and made from silicates such as clay, feldspar, bauxite, silica. But now ceramics contain other materials such as borides, carbides, silicides, nitrides.

Ceramics are generally made by two processes—sintering and fusing. In sintering, the starting material is reduced to a powder or granular form by a series of crushing, powdering, ball-milling, and so on. The ground preceramic material is then sized, separated according to particle size, using different-sized screens.

Ceramic material is generally shaped by pressing it into a form or through extruding, molding, jiggering, or slip casting. Slip casting uses a suspension of the preceramic material in water. The mixture must be dilute enough to allow it to be poured. Deflocculates are often added to assist in maintaining the suspension. The "slip" is poured into a plaster of Paris mold that absorbs water, leaving the finished shape. The preceramic material hardens next to the mold and surplus "slip" material poured off leaving a hollow item. At this point, the molded material is referred to as a "green body," which has little strength. Coffee pots and vases are formed using this technique.

In jiggering, machines press the preceramic material into a rotating mold of desired shape. Dinnerware products are often made using jiggering.

Abrasives and insulators are formed from simply pressing the preceramic material into a mold of desired shape. In extrusion, the preceramic material is forced through an opening in a "shaping" tool. Bricks and drainpipes are formed using extrusion.

After the product has dried, it is heated or fired in a furnace or kiln. Modern ceramics generally require certain heating schedules that include the rate and duration of heating and under what conditions such as in the presence or absence of air. This is similar to procedures used to produce carbon fibers, where the heating schedule is critical to the end product's properties.

In one approach, six-membered silicon-containing rings are pyrolyzed giving mixed carbosilane preceramic polymers through heating to 400°C, and subsequently forming silicon carbides or poly(carbosilanes) at 800°C (Equation 12.32).

$$
\begin{array}{c}
\text{Me}\quad\text{Me} \\
\backslash\ / \\
\text{Si} \\
/\ \backslash \\
(\text{Me})_2\text{Si}\quad\text{Si}(\text{Me})_2 \\
|\qquad\quad| \\
(\text{Me})_2\text{Si}\quad\text{Si}(\text{Me})_2 \\
\backslash\quad/ \\
\text{Si} \\
/\ \backslash \\
\text{Me}\quad\text{Me}
\end{array}
\rightarrow
\begin{array}{cc}
\text{Me} & \text{Me} \\
| & | \\
-(-\text{Si}-\text{HCH}-)- \text{ and } -(-\text{Si}-\text{HCH}-)- \rightarrow \text{SiC} \\
| & | \\
\text{H} & \text{Me}
\end{array}
\qquad (12.32)
$$

SiC fibers can be formed using dimethyl dichlorosilicon and diphenyl dichlorosilicon heated together (Equation 12.33).

$$
\begin{array}{cc}
\text{Me} & \text{Ph} \\
| & | \\
\text{Cl}-\text{Si}-\text{Cl} + \text{Cl}-\text{Si}-\text{Cl} \rightarrow \\
| & | \\
\text{Me} & \text{Ph}
\end{array}
\begin{array}{cc}
\text{Me} & \text{Ph} \\
| & | \\
-(-\text{Si}-)-(-\text{Si}-)- \rightarrow \text{SiC} \\
| & | \\
\text{Me} & \text{Ph}
\end{array}
\qquad (12.33)
$$

Such "ceramic" fibers offer uniquely strong and resistant inexpensive materials, including new ceramic composites that have great fracture toughness.

There are a number of other "nonoxygen" or nonoxide ceramics, including phosphonitric chlorides (PN backbone), boron nitriles (BN), aluminum nitriles (AlN), titanocarbosilanes (Si–Ti–C backbone), and silazanes (Si–C–N backbones).

Many ceramic products are coated with a glassy coating called a *glaze*. The glaze increases the resistance of the material to gas and solvent permeability and makes the surface smoother and in art objects used for decoration.

One group of advanced material ceramics are the zirconia ceramics. Most of these are based on zirconia (zirconium (IV) oxide) that contains small amounts of magnesia (MgO). They have bending strengths two to three times that of corundum and Alumina; high fracture toughness (about five times that of corundum ceramics); high resistance to wear and corrosion; and they have high density (5.8 gram/cc). They are used in the construction of shear blades in the textile industry, plungers in the food and drink industry, valves in the petroleum industry, and as milling balls in the materials industry.

Zirconia exists in three solid phases (Equation 12.34):

$$
\text{Monoclinic} \leftarrow 1{,}200°\text{C} \rightarrow \text{Tetragonal} \leftarrow 2{,}400°\text{C} \rightarrow \text{Cubic} \leftarrow 2{,}700°\text{C} \rightarrow \text{Liquid} \qquad (12.34)
$$

The transformations between the monoclinic and tetragonal phases involve large and abrupt volume changes introducing fractures in the material. This is minimized through the use of di- and tri-valent oxides of the cubic symmetry such as calcium oxide and MgO. This results in a lowering of the transition temperatures (M→T and T→C) and also lowers the expansion coefficient of the zirconia and the subsequent volume changes associated with the phase changes reducing ceramic fracturing. The addition of calcium oxide or MgO is said to "stabilize" the ceramic.

Strength, brittleness, and solvent permeability properties are limited because of lack of control of the ceramic composition on a macro- and microlevel. Even small particle sizes are large compared with the molecular level. There have been a number of attempts to produce uniform ceramic powders, including the sol-gel synthesis where processing involves a stable liquid medium, coprecipitation where two or more ions are precipitated simultaneously. More recently, Carraher and Xu have used the thermal degradation of metal containing polymers to deposit metal atoms and oxides on a molecular level.

12.24 HIGH-TEMPERATURE SUPERCONDUCTORS

12.24.1 DISCOVERY OF THE 123-COMPOUND

In early 1986, George Bedorz and K. Alex Muller reported a startling discovery—a ceramic material, La–Ba–Cu–O, lost its resistance to electrical current at about 30 K. This was the first report of a so-called high-T_c superconductor. Intensive efforts were then concentrated on substituting the component ions with similar elements on both the La and Ba sites. The first success was reported by Kishio et al. with an $(La_{1-x}Sr_x)_2CuO_4$ system that exhibited a higher T_c to about 37 K. Then the substitution on the La sites led Wu and coworkers to find another superconductor, the Y–Ba–Cu–O system with a T_c of 93 K in February 1987. This finally broke the technological barrier that would allow superconductivity at temperatures above liquid nitrogen. The superconducting phase was identified as $Y_1Ba_2Cu_3O_7$ (commonly referred to as the 123-compound). The 123-compound was the first of the 90 K plus superconductors to be discovered and it has been the most thoroughly studied.

12.24.2 STRUCTURE OF THE 123-COMPOUND

The structure of the 123-compound is related to that of an important class of minerals called *perovskites*. These minerals contain three oxygen atoms for every two metal atoms. It has six metal atoms in its unit cell and would be expected to have nine oxygens if it were an ideal perovskite. In fact, it has, in most samples, between 6.5 and 7 oxygens. In other words, by comparison to an ideal perovskite, about one-quarter of the oxygens are missing. The unit cell can be thought of as a pile of three cubes. Each cube has a metal atom at its center: barium in the bottom cube, yttrium in the middle one, and barium in the top one. At the corners of each cube a copper would be surrounded by six oxygens in an octahedral arrangement linked at each oxygen in an ideal perovskite. Each barium and yttrium would then be surrounded by 12 oxygens. But X-ray and neutron diffraction studies have shown that the unit cell does not conform to this simple picture because certain oxygen positions are vacant. All oxygen positions in the horizontal plane containing yttrium are vacant. The other vacancies are located in the top and bottom Cu-O planes.

The two copper oxide layers can be considered as polymeric since the covalent character is in the same range as for the carbon fluoride bond in Teflon. Thus, the 123-superconductors consist of two types of polymeric copper oxide layers held together by ionic bonding metals such as barium and yttrium. This theme of polymeric layers held together by ionic bonding to metals is common in the silicates and other minerals.

12.25 ZEOLITES

At least three major themes are helping drive polymer synthesis and use of polymers today. These involve synthesis and assembling on an individual scale (nanolevel); synthesis in confined spaces (selected inorganic zeolites and biological syntheses); and single-site catalysis (both selected biological and synthetic polymer synthesis). Superimposed on this is the applications aspects, including the human genome/biomedical, electronic/communications, and so on.

Zeolites are three-dimensional microporous crystalline solids. Zeolites include a whole group of aluminosilicates with an approximate formula of $SiAlO_4$. With respect to type of bonding, zeolites can be divided into three groups. The natrolite group (mesolite, thomsonite, edingtonite, natrolite) consist of structures build up from rings of four SiAlO$_4$ tetrahedral linked together into chains with fewer linkages between the chains so that cleavage along the chain direction is preferred. These materials generally have a fibrous character. In the heulandite group (stilbite, epistilbite, and heulandite) the SiAlO$_4$ tetrahedra form sheets of six-membered rings with few linkages between the sheets. These materials are mica-like in behavior. The third group, the so-called framework zeolites, have the density of bonding similar in all three directions. This group includes most of the zeolites mentioned below.

Framework zeolites can be described as aluminosilicates composed of tetrahedra linked by the sharing of oxygen atoms into rings and cages that can accommodate water molecules, metallic ions, as well as selected organic molecules. While there are a variety of structures, the framework zeolites can be briefly described as having an open arrangement of corner-sharing tetrahedra where the SiO$_4$ are partially replaced by AlO$_4$ units, where there are enough cations present to be neutral. There are well more than different synthetic and natural framework zeolites know today with more being found. Magic-angle nuclear magnetic resonance (NMR) indicates that there are five distinct zeolite groups where $n = 0$–4 for $Al(AlO)_n(SiO)_{4-n}$. The open structures give materials with lower densities (on the order of 2.0–2.2 g/mL), as expected, than similar materials with closed structures such as feldspar (density about 2.6–2.7 g/mL).

The three major applications of zeolites are absorption, catalysis, and ion exchange.

Molecular sieves was the name first given to framework zeolites dehydrated by heating in vacuum to about 350°C because of their ability to capture and remove water and certain other species. This sieve action is due to a regular pore structure such that materials are removed on a size exclusion basis. There exist zeolites with different ring or pore sizes so they can selectively remove selected materials. They are defined by the ring size of the opening. Thus, the term "8 ring" refers to a closed loop constructed from eight tetrahedral-coordinated silicon or aluminum atoms. Today, other materials, such as microporous silicas and aluminum phosphate, are also employed as molecular sieves.

Syntheses that occur within confined spaces typically give products with a specificity that is not available by other modes. The specificity may be general such as in the case of the synthesis of lignins in plants where synthesis between plant layers produces a largely two-dimensional material with only an average structure. It may be highly selective to not totally selective as in the case of zeolites and hollow nanofibers. Or it may be essentially totally selective as in the case of many of the biologically important proteins and nucleic acids where both spacial and electronic interactions act to give a highly "pre-ordained" structure.

Zeolites can accommodate a wide variety of cations that are loosely held and can be replaced by more tightly bound often "heavy metal" cations. Thus, they have been extensively employed as water softening materials removing calcium, iron, magnesium, and so on in "hard water." Smaller ringed zeolites have been used to remove water, carbon dioxide, and sulfur dioxide from low-grade natural gas streams. In medicine, they are used to extract oxygen from air for oxygen generation for patients that need to breathe a high-oxygen content atmosphere. Zeolites are also used in the laundry detergent market and in soil treatment, providing a source of slowly released nutrients that have been previously been added to the zeolite.

Zeolites and related ordered clay-associated materials have been suggested to be involved in the initial primeval synthesis of basic elements of life. They are also being involved in the synthesis of a number of polymeric and nonpolymeric materials. Zeolites come in a variety of forms with differing shapes and sizes with researchers associating the particular size and shape with a particular desired synthesis. This is somewhat akin to the considerations that are made in effectively employing crown phase transfer agents and related materials.

Zeolites have been employed in the preferential synthesis of optically active sites and in determining the particular products formed from certain reactions. In looking at the products formed from the decomposition and reformations involving an unsymmetrical ketone the major products are a combination of products listed below (Equation 12.35):

$$\begin{array}{cccc} & O & & O & & & O & \\ & || & & || & & & || & \\ Ph_3C\text{-}C\text{-}CMe_3 \rightarrow & Ph_3C\text{-}C\text{-}CPh_3 & + & Ph_3C\text{-}CPh_3 & + & Me_3C\text{-}C\text{-}C\text{-}Me_3 & + & Me_3C\text{-}CMe_3 \\ & A & & B & & & C & & D \end{array} \qquad (12.35)$$

Employing a single zeolite that selectively accommodates the Me_3C radical results in a preferential formation of D and little C because of preferential diffusion of the Me_3C radical as it is formed with the other radicals being "washed" past. The use of two zeolites in conjunction, one preferentially accepting the Me_3C radical and the second accepting the Ph_3C radical results in the preferential formation of B and D.

An additional consideration involves matching the so-called "hardness" or flexibility of the confined space. Zeolites offer hard or inflexible confined spaces whereas liquid crystals and some other polymer media offer softer confined spaces. While the zeolites offer a "safe haven" for selected species enabling the hosted species a relatively long existence, they do not allow for ready movement and mixing. By comparison, a more flexible container such as especially designed polyethylenes allow a more flexible environment that allows for variations in the product sizes and shapes and some assembling to occur within the more flexible confines. Thus, the reaction of 1-naphthyl esters was carried out in an appropriate zeolite and appropriate polyethylene. In solution, eight products are formed. In the zeolite, only one product, the result of a specific geminate-pair recombination, is formed. In the polyethylene, several products are formed primarily the isomeric products of geminate-pair recombinations.

12.26   SUMMARY

1. Inorganic polymers are widely employed in the construction and building businesses, as abrasives and cutting materials, as fibers in composites, as coatings and lubricants, and as catalysts. They also serve as the basis of rocks and soils.
2. Portland cement is the least expensive, most widely used synthetic polymer. It has a complex (short-range order and long-range disorder; average structure), three-dimensional structure.
3. There are many specialty cements, including reinforced concrete, light-weight concrete, prestressed concrete, gypsium, and heavy glass.
4. Silicates are among the most widely found materials on the face of the earth. They form the basis for much of the soil, sand, and rocks with most of this being of the crystalline variety of silicates. As amorphous materials they are found as the fiber material in fiberglass, as window glass, and a whole host of specialty glasses, including safety glass, borosilicate glasses, lead (or heavy) glasses, colored glasses, and glazes. Glass can be shaped by drawing, pressing, casting, and blowing. Most of these glasses have the approximate structure of SiO_2 or silicon dioxide. Many of the rocks are present in crystalline silicate forms as quartz and feldspar. Today, the sol-gel technique is important in making many materials, including aerogels, among the least-dense solids known today. In nature, silicates are found as polymeric sheets such as asbestos, three-dimensional materials such as the zeolites, and linear materials.
5. There are many widely used carbon polymers. These include the hardest known material, diamond, graphite a sheet material with little strength holding together the sheets, carbon fibers used in high-strength composites, CNTs one of the most important new materials for the twenty-first century, bitumens used in asphalt, and carbon black widely used as a filler material. Diamonds structurally have a carbon at the center of a tetrahedron composed of four other carbon atoms. They can be industrially synthesized and used for cutting and shaping. Graphite occurs as sheets of hexagonally fused benzene rings. The bonds holding the fused benzene rings together are covalent bonds while the bonding between the sheets results from the weaker overlapping of pi-electron orbitals. Thus, many of the properties of graphite are anisotropic. The weak forces holding the sheets together are responsible for its "slipperiness." Graphite is commercially made from charcoal or coke.

6. Many ceramics are partially polymeric in structure. These include the new superconductive materials that exist as polymeric sheets connected by metal ions similar to many of the silicate sheets.

GLOSSARY

Alumina: Aluminum oxide.

Annealing: Subjecting a material to near its melting point.

Asbestos: Group of silica-intensive materials containing aluminum and magnesium that gives soft, threadlike fibers.

Asbestosis: Disease that blocks the lung sacks with thick fibrous tissue.

Borosilicate glass: Relatively heat-shock-resistant glass with a small coefficient of thermal expansion, such as Kimax and Pyrex.

Calcium-aluminate cement: Contains more alumina than Portland cement.

Chrysotile: Most abundant type of asbestos.

Concrete: Combination of cement, water, and filler material such a rock and sand.

Diamonds: Polymeric carbon where the carbon atoms are at centers of tetrahedra composed of four other carbon atoms; hardest known natural material.

Feldspars: Derivatives of silica where one-half to one-quarter of the silicon atoms are replaced by aluminum atoms.

Fibrous glass (fiber glass): Fibers of drawn glass.

Float glass: Glass made by cooling sheets of molten glass in a tank of molten time: most common window glass is of this type.

Glass: Inorganic product of fusion that has been cooled to a rigid condition without crystallization; most glasses are amorphous silicon dioxide.

Glass sand: Impure quartz crystals.

Glaze: Thin, transparent coatings fused on ceramic materials.

Granite: Hard crystalline rock containing mainly quartz and feldspar.

Graphite: Polymeric carbon consisting of sheets of hexagonally fused rings where the sheets are held together by weak overlapping pi-electron orbitals; anisotropic in behavior.

Gypsum: Serves as the basis of plaster of Paris, Martin's cement, and so on; shrinks very little on hardening and rapid drying.

High-temperature superconductors: Polymeric copper oxide layers containing metal atoms that hold them together, which are superconductors above the boiling point of liquid nitrogen.

Inorganic polymers: Polymers containing no organic moieties.

Kaolinite: Important type of asbestos.

Lead glass (heavy glass): Glass where some or all of the calcium oxide is replaced by lead oxide.

Lime: Calcium carbonate from oyster shells, chalk, and marl.

Magnesia cement: Composed mainly of MgO; rapid hardening.

Optical fibers: Glass fibers coated with highly reflective polymeric coatings; allows light entering one end to pass though the fiber to the other end with little loss of energy.

Piezoelectric effect: Materials that develop net electronic charges when pressure is applied; sliced quart is piezoelectric.

Portland cement: Major three-dimensional inorganic construction polymer containing calcium silicates, lime, and alumina.

Precast concrete: Portland concrete cast and hardened before being taken to the site of use.

Prestressed concrete: Portland concrete cast about steel cables stretched by jacks.

Quartz: Crystalline forms of silicon dioxide: basic material of many sands, soils, and rocks.

Reinforced concrete: Portland concrete cast about steel rods or bars.

Safety glass: Laminated glass; sandwich containing alternate layers of soda-lime glass and PVB.

Sand: Loose grains of minerals or rocks larger than silt but smaller than gravel.

Sandstone: Granular quartz.

Silica: Based on SiO_2; finely ground sand.

Silicon glass: Made by fusing pure quartz crystals or glass sand; high melting.

Soda: Na_2O.

Soda ash: Na_2CO_3.

Soda-lime glass: Most common glass; based on silica, soda, and lime.

Soil: Contains mineral and organic particles; majority by weight is sand.

Tempered safety glass: Single piece of specially treated glass.

Tempering: Process of rapidly cooling glass resulting in an amorphous glass that is weaker but less brittle.

Vitreous enamels: Thin, normally somewhat opaque-colored inorganic coatings fused on materials.

Vycor: 96% silicon glass; made form silicon and boron oxides; best variety is called *fused quartz*.

EXERCISES

1. What properties of glass correspond to those of organic polymers?
2. Why is Portland cement an attractive large-bulk use building material?
3. Name five important synthetic inorganic polymers. Name five important natural inorganic polymers.
4. Describe what is meant by three-dimensional polymers. Name five important three-dimensional polymers. Name two general properties typical of three-dimensional polymers.
5. Why are specialty cements and concretes necessary?
6. What is intended by the comment that "glass is a supercooled liquid"?
7. What are the major techniques employed to shape glass?
8. Why are specialty glasses important in today's society?
9. Name two important inorganic fibers employed with resins to form useful materials.
10. Which would you predict to be more brittle-quartz, fibrous glass, or window glass?
11. Briefly describe the piezoelectric effect.
12. We do not live in a risk-free society. Discuss this statement in terms of asbestos.
13. What does anisotropic behavior mean? Why does graphite exhibit anisotropic behavior?
14. Briefly compare the structures of diamond and graphite.
15. Speculate as to why "nature" has emphasized polymers of carbon and silicon.
16. What are some advantages to having glass lenses?
17. What evidence do we have that glass is actually a very slow-moving liquid?
18. Why are glass and Portland cement so widely used in our society?
19. Why are inorganic materials employed rather than organic dyes to color or stain glass?
20. Are ceramics really polymeric? Why?
21. Why is the structure of Portland cement so difficult to study?
22. How is the silicon employed in optical fibers and computer chip manufacture related?
23. If asbestos of a size that was much larger than the size of the lung sacks was found, would it have a sizable market place in today's society?
24. Name some applications of CNTs.
25. Why is simple sand that is found at the beach not used in glass making?

ADDITIONAL READING

Andrady, A. (2008): *Science and Technology of Polymer Nanofibers*, Wiley, Hoboken, NJ.

Bruce, D., O'Hare, D. (1997): *Inorganic Materials*, 2nd Ed., Wiley, NY.

Bunsell, A., Berger, M-H. (1999): *Fine Ceramic Fibers*, Dekker, NY.

Bye, G. (1999): *Portland Cement*, Telford, London, UK.

Dresselhaus, M. S., Dresselhaus, G., Avouris, P. (2001): *Carbon Nanotubes: Synthesis, Structure, Properties, and Applications*, Vol. 80 (2001): Springer-Verlag, Heidelberg, Germany.

Fennessey, S. (2008): *Continuous Carbon Nanofibers*, VDM Verlag, NY.

Jaeger, R., Gleria, M. (2009): *Silicon-Based Inorganic Polymers*, Nova, Hauppauge, NY.

Mallada, R., Menendez, M. (2008): *Inorganic Membranes: Synthesis, Characterization and Applications*, Elsevier, NY.

Mark, J. (2005): *Inorganic Polymers*, Oxford University Press, Cary, NC.

Meyyappan, M. (2005): *Carbon Nanotubes: Science and Applications*, CRC Press, Boca Raton, FL.

Ohama, T., Kawakami, M., Fukuzawa, K. (1997): *Polymers in Concrete*, Routledge, NY.

Pinnavaia, T. J., Beall, G. W. (2000): *Polymer Clay Nanocomposites*, Wiley, Chichester, UK.

Poole, C. P., Owens, F. J. (2003): *Introduction to Nanotechnology*, Wiley, Hoboken.

Reich, S., Thomsen, C., Maultzsch, J. (2004): *Carbon Nanotubes*, Wiley-VCH, Darmstadt.

Schubert, U., Husing, N. (2000): *Inorganic Materials*, Wiley, NY.

Sirok, B., Blagoievic, B., Bullen, P. (2008): *Mineral Wool: Production and Properties*, CRC, Boca Raton, FL.

Weller, M. (1995): *Inorganic Materials Chemistry*, Oxford University Press, NY.

Wesche, R. (1999): *High-Temperature Superconductors*, Kluwer, NY.

13 Testing and Spectrometric Characterization of Polymers

Public acceptance of polymers is usually associated with an assurance of quality based on knowledge of successful long-term and reliable testing. In contrast, much of the dissatisfaction with synthetic polymers is related to failures that might have been prevented by proper testing, design, and quality control. The American Society for Testing and Materials (ASTM), through its various committees, has developed many standard tests, which may be referred to by all producers and consumers. There are also cooperating groups in many other technical societies: the American National Standards Institute (ANSDI), the International Standards Organization (ISO), and standards societies such as the British Standards Institution (BSI) in England, and Deutsche Normenausschuss (DNA) in Germany, and comparable groups in every nation with developed polymer technology throughout the world.

Testing is done by industry to satisfy product specifications and for public protection using standardized tests for stress–strain relationships, flex life, tensile strength, and so on. The U.S. tests are overseen by the ASTM through a committee arrangement. For instance, Committee D-1 oversees tests related to coatings, while Committee D-20 oversees tests on plastics. New tests are continuously being developed, submitted to the appropriate ASTM committee, and after adequate verification through "round robin" testing, finally accepted as standard tests. These tests are published by the ASTM. Each ASTM test is specified by a unique combination of letters and numbers, along with exacting specifications regarding data gathering, instrument design, and test conditions making it possible for laboratories throughout the world to reproduce the test and hopefully the test results if requested to do so. The Izod test, a common impact test, has the ASTM number D256–56(1961). The latter number, 1961, is the year the test was first accepted. The ASTM publication gives instructions for the Izod test specifying test material shape and size, exact specifications for the test equipment, detailed description of the test procedure, and how results should be reported. Most tests developed by one testing society have analogous tests or more often utilize the same tests so that they may have both ASTM, ISO, and other standardized society identification symbols.

A number of physical tests emphasizing stress–strain behavior will be covered in Chapter 14. Here we will concentrate on other areas of testing, emphasizing thermal and electrical properties and on the characterization of polymers by spectral means. Spectroscopic characterization generally concentrates on the structural identification of materials. Most of these techniques, and those given in Chapter 14, can also be directly applied to nonpolymeric materials such as small organic molecules, inorganic compounds, ceramics, and metals.

The testing of materials can be based on whether the tested material is chemically changed or is left unchanged. Nondestructive tests are those that result in no chemical change in the material, which may include many electrical property determinations, most spectroanalyses, simple-phase change tests (T_g and T_m), density, color, and most mechanical property determinations. Destructive tests result in a change in the chemical structure of at least a portion of the tested material. Examples include flammability and chemical resistance tests when the material is not resistant to the tested material.

13.1 SPECTRONIC CHARACTERIZATION OF POLYMERS

13.1.1 INFRARED SPECTROSCOPY

The infrared (IR) spectral range spans the region bound by the red end of the visible region to the microwave region at the lower frequencies. Molecular interactions that involve vibrational modes correspond to this energy region. IR is one of the most common spectronic techniques used today to identify polymer structure. Briefly, when the frequency of incident radiation of a specific vibration is equal to the frequency of a specific molecular vibration, the molecule absorbs the radiation. Today, most IR machines are rapid scan where the spectra are Fourier transformed. For the most part, IR band assignments for polymers are analogous to those made for small molecules.

In Fourier transform infrared spectroscopy (FTIR), the light is guided through an interferometer where the signal undergoes a mathematical Fourier transform giving a spectrum identical to the conventional dispersive IR.

With the advent of femtosecond infrared laser pulses, two-dimensional infrared correlative spectroscopy has become a new tool. Here, pump pulses are applied to the sample. After some time, that can be from zero to several picoseconds to allow the sample to relax, a second pulse is applied. The result is a two-dimensional plot of the frequency that resulted from the initial pump pulse and a second plot resulting from the relaxed state spectrum. This allows the coupling of various vibrational modes. In some ways, this is similar to two-dimensional nuclear magnetic resonance (NMR) spectroscopy in that the spectrum is spread out in two dimensions allowing certain "cross-peaks" to be observed.

Following are brief discussions of some of the more important techniques used specifically for polymer analysis.

Attenuated total reflectance IR (ATR-IR) is used to study films, coatings, threads, powders, interfaces, and solutions. (It also serves as the basis of much of the communications systems based on fiber optics.) ATR occurs when radiation enters from a more-dense material (i.e., a material with a higher refractive index) into a material that is less dense (i.e., with a lower refractive index). The fraction of the incident radiation reflected increases when the angle of incidence increases. The incident radiation is reflected at the interface when the angle of incidence is greater than the critical angle. The radiation penetrates a short depth into the interface before complete reflection occurs. This penetration is called the *evanescent wave*. Its intensity is reduced by the sample where the sample absorbs.

Specular reflectance IR involves a mirror-like reflection, producing reflection measurements of a reflective material or a reflection–absorption spectrum of a film on a reflective surface. This technique is used to look at thin (from nanometers to micrometers thick) films.

Diffuse reflectance infrared Fourier transform spectroscopy (DRIFTS) is used to obtain spectra of powders and rough polymeric surfaces such as textiles and paper. IR radiation is focused onto the surface of the sample in a cup resulting in both specular reflectance (which directly reflects off the surface having equal angles of incidence and reflectance) and diffuse reflectance (which penetrates into the sample subsequently scattering in all angles). Special mirrors allow the specular reflectance to be minimized.

Photoacoustic spectroscopy IR (PAS) is used for highly absorbing materials. In general, modulated IR radiation is focused onto a sample in a cup inside a chamber containing an IR-transparent gas such as nitrogen or helium. The IR radiation absorbed by the sample is converted into heat inside the sample. The heat travels to the sample surface and then into the surrounding gas causing expansion of the boundary layer of gas next to the sample surface. The modulated IR radiation thus produces intermittent thermal expansion of the boundary layer creating pressure waves that are detected as photoacoustic signals.

PAS spectra are similar to those obtained using ordinary FTIR except truncation of strong absorption bands occurs because photoacoustic signal saturation often occurs. PAS allows the structure to be studied at different thicknesses because the slower the frequency of modulation, the deeper the penetration of IR radiation.

Emission infrared spectroscopy is used for thin films and opaque polymers. The sample is heated so that energy is emitted. The sample acts as the radiation source and the emitted radiation is recorded giving spectra similar to those of classical FTIR. In some cases, IR frequencies vary because of differences in the structures at different depths and interactions between surface and interior emissions.

Infrared microscopy allows the characterization of minute amounts of a material or trace contaminants or additives. Samples as small as 10 μm can be studied using infrared microscopy. The microscope, often using fiber optics, allows IR radiation to be pin-pointed.

Today, there are many so-called hyphenated methods with IR. Hyphenated methods involving IR include gas chromatography-infrared spectroscopy (GC-IR) where the IR spectra are taken of materials as they are evolved through the column. Related to this is high-performance liquid chromatography-infrared spectroscopy (HPLC-IR), thermogravimetric infrared spectroscopy (TG-IR), and mass spectroscopy-infrared spectrometry (MS-IR).

13.1.2 RAMAN SPECTROSCOPY

Raman spectroscopy is similar to IR spectroscopy in that it investigates polymer structure focusing on the vibrational modes. Whereas IR is a result of energy being absorbed by a molecule from the ground state to an excited state, Raman spectroscopy is a scattering phenomenon where the energy of photons is much larger than the vibrational transition energies. Most of these photons are scattered without change (so-called Rayleigh scattering). Even so, some are scattered from molecular sites with less energy than they had before the interaction, resulting in Raman–Stokes lines. Another small fraction of photons have energies that are now greater than they originally had leading to the formation of anti–Stokes lines. Only the Raman–Stokes photons are important in Raman spectroscopy. While many chemical sites on a polymer are both IR and Raman active, that is, they both give rise to bands, some are less active or even nonactive because of the difference between groups that can absorb and those that scatter. These differences are generally described in terms of symmetry of vibration. Briefly, an IR absorption occurs only if there is a change in the dipole moment during the vibration, whereas a change in polarizability is required for Raman scattering to occur. Even so, the spectral are generally similar, and a comparison of the two allows for additional structural characterization beyond that obtained from either of the techniques alone.

Carraher and Williams showed that for many polymers, differences in symmetry and band production was similar for small molecules as they were for the same groups found in polymers. Thus, observations from the literature and for small "model-compound studies" are generally applicable to similar moieties present in polymeric systems for both Raman and IR spectral analyses.

As in the case with IR spectrometers, there exists a wide variety of specialty techniques especially applicable to polymer analysis.

In *surface-enhanced Raman spectroscopy* (SERS) samples are adsorbed onto microscopically roughened metal surfaces. Spectra are the intensities and frequencies of scattered radiation originating from a sample that has been irradiated with a monochromatic source such as a laser. SERS spectra are of molecules that are less than 50 Å from the surface.

13.1.3 NMR SPECTROSCOPY

Nuclear magnetic resonance spectroscopy is a powerful tool for polymer structure characterization. Certain isotopes have two or more energy states available when exposed to a magnetic field. The transitions between these energy states are the basis for NMR. These magnetically active nuclei have a property called *spin*. As a consequence of this spin, these nuclei have an angular momentum and magnetic moment. The ratio of these two properties is called the magnetogyric ratio. Each isotope has a distinct magnetogyric ratio that varies a little with the particular chemical environment in which they are placed.

While NMR has been a strong characterization tool for polymers for many years, it has increased in its usefulness because of continually improved instrumentation and techniques. When a nucleus is subjected to a magnetic field, two phenomenon are observed—Zeeman splitting and nuclear precession. Zeeman splitting creates $2I + 1$ magnetic energy states, where I is the spin quantum number. When the atomic mass and atomic number are even numbers, $I = 0$ so that these nuclei are unable to have multiple energy levels when exposed to a magnetic field. Thus, ^{12}C, which has both an even atomic number and atomic mass is NMR inactive, whereas ^{13}C, which has an uneven atomic mass, is NMR active. Nuclear precession is the motion of a spinning body whose axis of rotation changes orientation. The precessional frequency is equal to the magnetic field strength times the magnetogyric ratio.

In a magnetic field, NMR-active nuclei can be aligned with the magnetic field (low-energy state) or aligned against the field (high-energy state). At room temperature, there are slightly more nuclei in the lower-energy state than in the higher-energy state. As magnetic energy is supplied that corresponds to the energy gap, quantum level, between the low- and high-energy states, some nuclei in the low-energy state move to the high-energy state resulting in an absorption of energy, which is recorded as a NMR spectra. The difference between the two energy states is related to the strength of the external magnet. Better spectra are obtained when instruments with larger magnetic fields are employed.

Because of the small but consistent concentrations of ^{13}C present in all organic compounds, it is necessary to use more sophisticated NMR spectroscopy for determining the effect of neighboring electrons on these nuclei. However, ^{13}C-NMR spectroscopy is a valuable tool for investigating polymer structure.

Following is a short description of some of the current techniques.

Nuclear Overhauser Effect—The nuclear Overhauser effect (NOE) only occurs between nuclei that share a dipole coupling, that is, their nuclei are so close that their magnetic dipoles interact. Techniques that use NOE enhance ^{13}C spectra and allow spacial relationships of protons to be determined.

Two-Dimension NMR—Basically, the two-dimensional NMR techniques of nuclear Overhauser effect spectroscopy (NOESY) and correlation spectroscopy (COSY), depend on the observation that spins on different protons interact with one another. Protons that are attached to adjacent atoms can be directly spin-coupled and thus can be studied using the COSY method. This technique allows assignment of certain NMR frequencies by tracking from one atom to another. The NOESY approach is based on the observation that two protons closer than about 0.5 nm perturb one another's spins even if they are not closely coupled in the primary structure. This allows spacial geometry to be determined for certain molecules.

The use of actively shielded magnetic field gradients has made the use of pulsed field gradients possible. The use of pulsed field gradients reduces experiment time, minimizes artifacts, and allows for further solvent suppression.

In *pulsed NMR*, the magnetic field is turned on for the time necessary to rotate the magnetization vector into a plane called the *90° rotation* or *90° pulse*. The field is turned off and the magnetization vector rotates at a nuclear precession frequency relative to the coil. This induces a NMR signal that decays with time as the system returns to equilibrium. This signal is called the *free induction decay* (FID).

After a sample is excited, the spin loses excess energy through interactions with the surroundings eventually returning to its equilibrium state. This process is exponential and is called *spin-lattice relaxation*. The decay is characterized by an exponential time constant.

Two-dimensional experiments allow the more precise determination of coupling relationships. Such experiments are carried out by collecting a series of FID spectra. The time between the pulses is called the *evolution time*. The evolution time is systematically increased as each successive FID is obtained. Each new FID shows a continued change in the couplings in the polymer. The FID spectra are treated using Fourier transformation. A new series of FID spectra are now created by connecting points for each spectra and these new FIDs are again treated by Fourier transformation producing a two-dimensional spectra that are often presented as contour plots. Nuclei that share J-coupling produce a correlation peak. Such approaches allow better interpretation of dipole couplings, molecular diffusion, J-coupling, and chemical exchange.

Solids—Many polymers are either difficulty soluble or insoluble. NMR of solids generally gave broad lines because of the effects of dipolar coupling between nuclei and the effect of *chemical shift anisotropy* (CSA). Both of these effects are greatly reduced for polymers in solution and allow for decent spectra of soluble polymers in solution.

Chemical shift anisotropy effects are large for solids and are the result of the directional dependence of electronic shielding. CSA effects are overcome through rapidly spinning the sample at an angle to the magnetic field known as the *magic angle*. Solids probes use spinning rotors to hold and spin the sample. The sample is rotated at the magic angle and spun fast enough to remove CSA effects. High-power decouplers or multiple pulse line narrowing allow the decoupling between protons and carbon by using a series of pulses to average the dipolar interactions through spin reorienting. Cross-polarization uses dipolar coupling to increase the sensitivity of less sensitive nuclei. The combination of cross-polarization and magic angle spinning (CPMAS) allows good spectra to be obtained for solid polymers.

13.1.4 NMR APPLICATIONS

The various types of spectroscopes complement one another, sometimes giving "new" information and other times giving similar information. The amount of crystallinity and preference for syndiotactic, isotactic, and atactic structure has been determined employing a number of techniques, including X-ray. This information can be correlated with information that can be readily obtained from IR and NMR so that product control can be easily monitored. For instance, NMR spectra of poly(methyl methacrylate) (PMMA) made by different synthetic routes give different proton shifts that allow the determination of the amounts of isotactic, syndiotactic, and atactic material. Briefly, production of PMMA via so-called high-temperature free radical polymerization gives a largely atactic material. The proton chemical shifts for the α-methyl appear at about 8.8, 8.9, and 9.1 (chemical shifts are based on the tetramethylsilane peak having a value of 10.00 ppm). Largely isotactic PMMA produced using anionic polymerization shows an enhanced peak at about 8.8 that is assigned to the configuration where the α-methyl group in the PMMA repeat unit is flanked on both sides by units of the same configuration. It is the major peak in the 8.8–9.1 triad. PMMA produced employing low-temperature free radical polymerization gives largely syndiotactic product. The peak at 8.9 is assigned to the α-methyl present in a heteroatactic configuration so that the units on either side of it are unlike the central mer since it now becomes the major peak. The 9.1 peak is the largest for the higher-temperature product produced via free radical polymerization and is due to the triad central unit being flanked by a like and unlike mer.

Today's NMR capability allows the determination of additional structural features in solution and solid state, including the identification of end groups, branches, and defects. Also, sufficient experience of the positioning of bands has been gathered to computer programs to be developed that create spectra and assignments based on submitted structures. While such spectra can be created, they are no substitute for actually running the spectra but can be used to help in peak assignment.

Combinations of solid-state NMR and IR allow the molecular description of the affect of stress–strain exposure to various conditions, including chemical treatments, radiation, and heat on materials.

13.1.5 Electron Paramagnetic Resonance Spectroscopy

Electron paramagnetic resonance (EPR) spectroscopy, or electron spin resonance (ESR) spectroscopy is a valuable tool for measuring the relative abundance of unpaired electrons present in macromolecules. For example, macroradicals are formed by the homogeneous cleavage of nylon-66 chains when these filaments are broken, with the concentration of macroradicals increasing as the stress is increased.

13.1.6 X-Ray Spectroscopy

X-ray diffraction is a widely used tool for structural identification for almost all solids under the right conditions. X-ray diffractometers are generally either single-crystal or powder.

Single-crystal studies allow the absolute configurational determination of polymeric materials that have high degrees of crystallinity. Such determinations are costly with respect to time because of the complexity of polymeric materials.

Powder X-ray spectroscopy can employ smaller crystalline samples from 1 to several hundred nanometers. These crystallites have broadened peak profiles as a result of incomplete destructive interference at angles near the Bragg angle defined as

$$n\lambda = 2d \sin\theta \tag{13.1}$$

where n is the order of a reflection, λ is the wavelength, d is the distance between parallel lattice planes, and θ is the angle between the incident beam and a lattice plane known as the Bragg angle. This broadening allows determination of crystallite size and size distribution. (Note that this is not particle size.)

X-ray analysis of proteins and nucleic acids is especially important as the absolute structure is needed for many advances in the field of medicine and biochemistry.

13.2 SURFACE CHARACTERIZATION

Everything has a surface or an interface. These surfaces have their own kinetic and thermodynamic features that affect their formation and behavior. Sperling notes that for most polymers, the end groups reside perpendicular to the bulk of the polymers probably because the end is less hydrophobic to the bulk and the polymer surfaces generally are "faced" with an air atmosphere that is more hydrophilic. When a polymer solution is deposited onto a surface to "dry," the concentration has an influence on the orientation of the polymer chains at the surface in the dried solid. Thus, when the amount of polymer is small, the polymer chain lays parallel to the surface in a so-called "pancake" form (Figure 13.1). As the concentration increases, the surface is not able to accommodate the entire polymer chain and it begins to form an inner tangled chain with only the end and some of the chain segments facing the surface forming a "mushroom" shape. Finally, as the concentration of polymer increases, only the ends of the polymer chains occupy the surface with the polymer ends forming "brushes."

There is no exact, universally accepted structure of a surface. Here, the surface will be defined as the outermost atomic layers, including absorbed foreign atoms. The chemical and physical composition, orientation, and properties of surfaces typically differ from those of the bulk material.

Current surface characterization techniques fall into two broad categories—those that focus on the outermost few layers (to within the 10–20 atom layer boundary) and those whose focus includes components present to several thousand angstroms into the solid (hundred to several hundred layers).

Attenuated total reflectance typically employs special cells fitted onto traditional IR, FTIR, or ultraviolet (UV) instruments. While some outer surface aspects are gleaned from such techniques, information to several thousand angstroms is also present in the spectra from ATF.

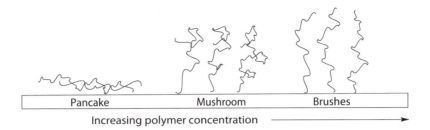

FIGURE 13.1 General surface structures as a function of polymer concentration.

Techniques that analyze the first few atomic layers generally involve low-energy electrons or ions since the incident radiation should penetrate only the top few layers. Normally a combination of techniques is employed to clearly define the composition of the outer layers. Special precautions are employed to minimize sample surface contamination.

Following is a brief presentation of some of these modern surface characterization techniques.

13.2.1 AUGER ELECTRON SPECTROSCOPY AND X-RAY PHOTOELECTRON SPECTROSCOPY

Auger electron spectroscopy (AES) and X-ray photoelectron spectroscopy (XPS) are two principle surface analysis techniques. They are used to identify the elemental composition, that is, the amount and nature of species present at the surface to a depth of about 1 nm.

In Auger transitions, incident electrons interact with the inner shell electrons (E_i) of the sample. The vacancy created by an ejected inner shell electron is filled by an outer shell electron (E_1), and a second outer shell electron (E_2) is ejected leaving the atom in a doubly ionized state. The electrons ejected from the outer shells are called *Auger electrons*, named after the Frenchman Pierre Auger, who discovered the effect. Thus, AES measures the energies of the Auger electrons (E_a) emitted from the first 10 angstroms of a sample surface. The energy equation is expressed as

$$E_a = E_1 - E_i + E_2 \tag{13.2}$$

The kinetic energies of these ejected electrons originating within the first 30 Å of the sample surface are measured by XPS. In XPS, a sample is bombarded by a beam of X-ray with energy $h\nu$ and core electrons are ejected with a kinetic energy E_k that overcomes the binding energy E_b and the work function (φ). These core electrons are called the *X-ray photoelectrons*. The energy equation is expressed as follows:

$$E_k = h\nu - E_b - \varphi \tag{13.3}$$

The energies of these photoelectrons are reflected in the kind and environment of the atoms present at the surface.

13.2.2 NEAR-FIELD SCANNING OPTICAL MICROSCOPY

Optical microscopes have one serious drawback, their resolution, resulting from the fundamental physics of lenses. Lord Rayleigh, more than 100 years ago, defined the presently accepted maximum optical lens resolution to be one-half the wavelength of the imaging radiation. In truth, conventional optical microscopy did not achieve this level of definition mainly because of out-of-focus light. This prevented the observation of atoms and single molecules.

Near-field scanning optical microscopy (NSOM) allows an extension of optical microscopy to near that of electron microscopy. The central feature is the optical element that is similar, and sometimes the same, to that employed in atomic force microscopy (AFM). Essentially, light is directed through the probe tip onto the sample from just immediately above the sample surface. The light emanating from the probe tip is smaller than the light's wavelength and spreads out over the surface. This results in the maximum influence occurring at the surface with little contribution from regions nearby (such as within 30 nm), resulting in little out-of-focus light. Depending upon the surface and sample thickness, the light is measured as absorption or fluorescence and collected and recorded electronically. NSOM can be fitted onto a regular optical microscope or coupled with scanning probe microscopy (SPM).

13.2.3 Electron Microscopy

The upper limit of magnification for optical microscopes is about 2,000 times. Thus, additional forms of microscopy have been developed that allow near to actual atomic observation of polymer surfaces. Electron microscopy utilizes an electron beam to act as the sensing radiation in place of light. High-energy electrons take on wave character as they approach the speed of light. The wavelength is inversely proportional to the electron speed or velocity. When accelerated over large voltages, electrons can be made to travel at speeds to permit wavelengths on the order of 0.003 nm. The electron beam is focused and the image is formed using magnetic lenses. The two most common forms of electron microscopy are transmission electron microscopy (TEM) and scanning electron microscopy (SEM).

In SEM, the surface of the polymeric surface is scanned using an electron beam with the reflected or back scattered beam of electrons collected and displayed on a cathode ray tube screen. The image represents the surface contour of the scanned material. Because the surface must be conductive, most polymer surfaces must be overlaid with a conductive coating. Magnifications up to about 50,000 are carried out using SEM.

Transmission electron microscopy utilizes an image formed by an electron beam that passes through the sample. This allows internal microstructures to be determined. Structural details of materials can be observed on an atomic level by looking at contrasts in the image caused by various concentrations of different elements. Very thin films are employed. Under good conditions, magnifications up to one million are possible employing TEM.

13.2.4 SPM

Scanning probe microscopy encompasses a group of surface-detection techniques that include AFM and scanning tunneling microscopy (STM) that allow the topographic profiling of surfaces. SPM techniques investigate only the outermost few atomic layers of the surface with nanometer resolutions and at times atomic level resolution.

Scanning tunneling microscopy is generally used with electrically conductive materials applied to polymeric materials giving overlays consisting of conducting material layered over the surface of the sample. STM experiments typically require extremely low pressures less than 1×10^{-10} mbar. By comparison, AFM can be run under room conditions and does not require the use of electrically conductive material. In with STM, the metallic tip is held close (about 0.5–1 nm) to the surface. A voltage, applied between the tip and sample surface, drives a tunneling current. The conductive surface reconstructs the atomic positions via minimizing the surface free energy. This gives topographic superstructures with specific electronic states which are recorded as surface contours or images.

Atomic force microscopy can be run under room conditions. AFM can be performed in either of two forms—a contact mode and a noncontact mode. It does not require the use of electrically conductive material since (in the contact mode) the tip actually "touches" the surface rather than residing immediately above it as is the case in STM. In both the contact and the noncontact mode, light is used as the sensing source rather than an applied voltage. In contact AFM, a cantilever with as sharp a point as possible is laid onto the sample surface with a small loading force in the range of 10^{-7}–10^{-10} N. Tips of differing size and shape are tailor made. Data is obtained optically by bouncing an incident laser beam onto the cantilever toward a quadrant detector on or in an interferometer. The AFM can work in two modes—either a contact mode or in a noncontact mode. In the noncontact mode, the attractive force is important and the experiment must be carried out under low pressures similar to those employed in STM.

In the contact mode, the tip acts as a low-load, high-resolution profiler. Along with structure determination, the AFM is also used to "move" atoms about allowing the construction of images at the atomic level. The AFM is also an important tool in the nanotechnology revolution.

Nanotubes are being used as points in some SPM units. The ends of these nanotubes can be closed or functionalized offering even "finer" tips and tips that interact with specific sites allowing manipulation on an atom-by-atom basis. These nanotubes are typically smaller than silicon tips and are generally more robust.

Atomic force microscopy is useful in identifying the nature and amount of surface objects. AFM, or any of its variations, also allow studies of polymer phase changes, especially thermal phase changes, and results of stress–strain experiments. In fact, any physical or chemical change that brings about a variation in the surface structure can, in theory, be examined and identified using AFM.

Today, there exist a wide variety of AFMs that are modifications or extensions of traditional AFM. Following is a brief summary of some of these techniques.

Contact mode AFM is the so-called traditional mode of AFM. Topography contours of solids can be obtained in air and fluids.

Tapping mode AFM measures contours by "tapping" the surface with an oscillating probe tip thereby minimizing shear forces that may damage soft surfaces. This allows increased surface resolution. This is currently the most widely employed AFM mode.

There are several modes that employ an expected difference in the adhesion and physical property (such as flexibility) as the chemical nature is varied. *Phase imaging* experiments can be carried out that rely on differences in surface adhesion and viscoelasticity. *Lateral force microscopy* (LFM) measures frictional forces between the probe tip and sample surface. *Force modulation* measures differences between the stiffness and/or elasticity of surface features. *Nanoindenting/scratching* measures mechanical properties by "nanoindenting" to study hardness, scratching, or wear, including film adhesion and coating durability. LFM identifies and maps relative differences in surface frictional characteristics. Polymer applications include identifying transitions between different components in polymer blends, composites, and other mixtures, identifying contaminants on surfaces, and looking as surface coatings.

Noncontact AFM measures the contour through sensing van der Waals attractive forces between the surface and the probe tip held above the sample. It provides less resolution than tapping mode AFM and contact mode AFM.

There are several modes that employ differences in a materials surface electronic and/or magnetic character as the chemical nature of the surface varies. *Magnetic force microscopy* (MFM) measures the force gradient distribution above the sample. *Electric force microscopy* (EFM) measures the electric field gradient distribution above a sample surface. EFM maps the gradient of the electric field between the tip and the sample surface. The field due to trapped charges, on or beneath the surface, is often sufficient to generate contrast in an EFM image. The voltage can be induced by applying a voltage between the tip and the surface. The voltage can be applied from the microscopes electronics under AFM control or from an external power supply. EFM is performed in one of three modes—phase detection, frequency modulation, FM, or amplitude detection. Three-dimensional plots are formed by plotting the cantilever's phase or amplitude as a function of surface location. *Surface potential microscopy* (SP) measures differences in the local surface potential across the sample surface. SP imaging is a nulling technique. As the tip travels above the surface the tip and the cantilever experiences a force whenever the surface potential differs from that of the tip. The force is nullified by varying the voltage of the tip so that the tip remains at the same potential as the immediate surface. The voltage applied to the tip to maintain this constant potential as the tip surveys the surface with the results plotted as a function of the surface coordinates creating a surface potential image. For best results, SP and EFM do the best job with conductive materials. *Force volume* measurements involve producing two-dimensional arrays of force-distance values allowing a mapping of the force variation and surface topology with individual force curves to be constructed.

Force-distance microscopy measures repulsive, attractive, and adhesion forces between the time and surface during approach, contact, and separation. This technique combines electrical with adhesion/physical property as a means to study sample surfaces.

Scanning thermal microscopy (SThM) measures two-dimensional temperature distributions across a sample surface. This is a special thermal technique.

In *Electrochemical microscopy* (ECSTM and ECAFM) the material is immersed in electrolyte solution and the surface and properties of conductive materials studied.

Information derived from several of these techniques go together to give a clearer idea of the nature of the surface.

Atomis force microscopy results can be utilized in conjunction with other techniques. While some techniques, such as SAXS and SANS, allow structural information to be inferred, AFM gives real space results. While some of the polymeric structural designs may not be unambiguously determined, many can be determined employing ATM. The major limitation concerns whether the structures observed at or near the surface are similar to those in the interior. We are well aware that surface composition differs from the interior composition. For instance, surfaces may be less organized being enriched in chain ends, loops, and switchboard chain segments. Further, for "sliced" samples does the "slicing" disturb the fine structure along the "cut" surface. For instance, the structure of linear polymers such as polyethylene has been suggested to consist of ordered or sharp folds, switchboard-like, loops with loose folds, buttressed loops and combinations of these features. Magonov and Godovsky and others recently investigated the surface structures of a number of polymers employing ATM. For single crystals of PE, ordered grains 10^{-12} nm in size are found. For melt-crystallized LLDPE spherulites of several microns are the major morphological features. Edge-on standing lamellae and lamellar sheets are found. Dark areas are assigned as amorphous regions. The lamellar edges are on the order of 25–40 nm while the strands are several microns in length. By comparison, melt-crystallized LDPE, which is only about 30% crystalline, shows only spherulitic patterns with ill-defined ring patterns. The grain sizes are about 15–25 nm with fibrillar structures visible. A sample of melt-crystalized ULDPE with low crystallinity (about 15%) gives largely an ill-defined surface consistent of the surface being largely amorphous. As higher force is applied to press through the surface layer, grains of 0–10 nm and finally 9–11 nm become visible with some grains up to about 100–150 nm visible

Other polymers have been studied. For instance isotactic-polypropylene, i-PP, shows well defined spherulites with grains (15–20 nm) embedded in an amorphous material. The grains are assembled in circles and in some cases, along the radial direction, an ordered texture exists. PVDF shows numerous spherulites with fibrils 12–15 nm in width. The granular nanostructure of spherulites has also been found for polyesters and polyurethanes. ATM and other studies (including WAXS and SAXS) suggest that the nanoscale grains are elementary building blocks of the crystalline architecture in most polymers. These grains or blocks can have more or less structure within them. The overall crystalline structure may be developed as a one-dimensional assembling of grains into fibrils and the two-dimensional structures an assembling of grains into lamella. A correlation between grain size and the size of molecular coils has not yet been answered using ATM.

Spin-cast films of poly(ethylene oxide) (PEO) show a flat crystalline morphology with lamellar sheets of different shapes. When melted and then cooled, PEO crystallizes with a similar morphology except the lamellar sheets are smaller. When it is again melted and cooled, crystallization proceeds more slowly and the PEO morphology is dominated by spiral crystallites formed via a screw dislocation mechanism. In all cases, the thickness of the lamellar sheets is about 12 nm indicating multiple folding of the PEO chains. The lamellar sheets disappear at about 60°C though the melting point is listed to be 70°C. On cooling, the lamellar structures reappear about 50°C.

The morphology of spin-cast film, thickness of 180 nm, from polycaprolactone shows many spherulitic structures with fibrillar nanostructures formed of lamellae lying edge on (about 10 nm thick) and areas with lamellar sheets lying flat on. Different crystalline structures are found when the sample is melted and crystallized as a function of temperature. These two studies reinforce the complex inner relationship between physical treatment and nanostructure.

While some structures show seemingly independent spherulitic structures on the surface we know from other studies that these structures are connected to one another and to the more amorphous regions overall giving a material with a characteristic flexibility and strength. In general, chains are shared with adjacent areas allowing a sharing of stress–strain factors.

Atomic force microscopy is important for biological as well as synthetic macromolecules. Several examples are given to illustrate applications. Collagen is an important natural protein that is present in many tissues, including bones, skin, tendons, and the cornea. It is also employed in medical devices such as artificial skin, tendons, cardiac valves, ligaments, hemostatic sponges, and blood vessels. There are at least 13 different types of collagen. ATF can image collagen molecules and fibers and their organization allowing identification of the different kinds of collagen and at least surface interactions.

Atomic force microscopy allows the study of cell membranes. The precise organization of such cell membranes is important since they play a role in cell communication, replication, and regulation. It is possible to study real-time interactions of such biologically important surfaces. Further, bilayers modeled or containing naturally produced bilayers are used as biosensors. Again, interactions of these biomembranes can be studied employing AFM. For instance, the degradation of bilayers by phospholipases, attachment of DNA, and so on can be studied on a molecular level. In another application, antibody–antigen interactions have been studied employing AFM. One application of this is the creation of biosensors to detect specific interactions between antigens and antibodies.

13.2.5 SECONDARY ION MASS SPECTROSCOPY

Secondary ion mass spectroscopy (SIMS) is a sensitive surface analysis tool. Here, the mass analysis of negative and positive ions sputtered from the polymer surface through ion bombardment is analyzed. The sputtering ion beam is called the *primary ion beam*. This beam causes erosion of the polymer surface removing atomic and molecular ions. Then these newly created ions, composing what is called the *secondary ion beam*, are analyzed as a function of mass and intensity. Depth of detection for SIMS is of the order of 20–50 Å. Because it is the ions in the secondary ion beam that are detected, the mass spectra obtained from SIMS is different from those obtained using simple electron impact methods. The extent of particular ion fragments observed is dependent on a number of factors, including the ionization efficiency of the particular atoms and molecules composing the polymer surface.

Secondary ion mass spectroscopy can detect species that are present on surfaces of the order of parts-per-million to parts-per-billion.

13.3 AMORPHOUS REGION DETERMINATIONS

Experimental tools that have been employed in an attempt to characterize amorphous regions are given in Table 13.1. Techniques such as birefringence and Raman scattering give information related to the short-range (< 20 Å) nature of the amorphous domains while techniques such as neutron scattering, electron diffraction, and electron microscopy gives information concerning the longer-range nature of these regions.

TABLE 13.1
Techniques Employed to Study the Amorphous Regions of Polymers

Short-range interactions	Long-range interactions
Magnetic birefringence	Electron diffraction
Raman scattering	Small-angle X-ray scattering
Depolarized light scattering	Electron microscopy
Rayleigh scattering	Density
Bruillouin scattering	Small-angle neutron scattering
NMR relaxation	
Wide-angle X-ray scattering	

Birefringence measures order in the axial, backbone direction. The birefringence of a sample can be defined as the difference between the refractive indices for light polarized in two directions 90° apart. Thus, a polymer sample containing polymer chains oriented in a preferential direction by stretching or some other method will exhibit a different refractive index along the direction of preferred chain alignment compared to that obtained at right angles. This change in birefringence gives information concerning the amount of order, thus, information about disorder.

Small-angle neutron scattering (SANS) results indicate that vinyl polymers exist in random coils in the amorphous state. Results from electron and X-ray diffraction studies show diffuse halos consistent with the nearest-neighbor spacings being somewhat irregular. It is possible that short-range order and long-range disorder exists within these amorphous regions.

13.4 MASS SPECTROMETRY

There are a number of mass spectrometry (MS) techniques available and directly applicable to polymers. Since high polymers have high molecular weights, determination of unbroken chains is not usual. Even so, determination of structures of ion fragments of segments of the polymers is straight forward and a valuable tool in determining the unit structure.

The exception to this is the application of matrix-assisted laser desorption/ionization mass spectrometry (MALDI MS). In 1981, Barber and Liu and coworkers independently introduced the concept of employing matrix-assisted desorption/ionization where the absorption of the matrix is chosen to coincide with the wavelength of the employed laser to assist in the volatilization of materials. In 1988, Tanaka, Hillenkamp and coworkers employed the laser as the energy source giving birth to MALDI MS.

Matrix-assisted laser desorption/ionization mass spectrometry was developed for the analysis of nonvolatile samples and was heralded as an exciting new MS technique for the identification of materials with special use in the identification of polymers. It has fulfilled this promise to only a limited extent. While it has become a well used and essential tool for biochemists in exploring mainly nucleic acids and proteins, it has been only sparsely employed by synthetic polymer chemists. This is because of lack of congruency between the requirements of MALDI MS and most synthetic polymers.

Classical MALDI MS requires that the material be soluble in a suitable solvent. A "suitable solvent" means a solvent that is sufficiently volatile to allow it to be evaporated before the procedure. Further, such a solvent should dissolve both the polymer and the matrix material. Finally, an ideal solvent will allow a decent level of polymer solubility, preferably a solubility of several percentage and greater. For most synthetic polymers, these qualifications are only approximately attainted. Thus, traditional MALDI MS has not achieved its possible position as a general use modern characterization tool for synthetic polymers. By comparison, MALDI MS is extremely useful for many biopolymers where the polymers are soluble in water. It is also useful in the identification of synthetic polymers, such as PEO where the solubility requirements are fulfilled. Thus, for PEO we have determined the molecular weight distribution of a series of compounds with the separations in ion fragment mass 44 Da corresponding to CH_2–CH_2–O units.

Recently Carraher and coworkers, especially Sabir and Cara Carraher, reported on the use of MALDI MS except focusing on the formation of fragments from the laser bombardment. This approach is named fragmentation matrix-assisted laser desorption/ionization mass spectrometry or simply FMALDI MS because it is the fragmentation fragments that are emphasized in the study. The technique should be applicable to any solid when the proper operating conditions are employed. While it is unable to give molecular weight distribution data it is able to generate identifiable and repeatable results to several thousand daltons. For lower molecular weight samples, entire chains are identifiable.

13.5 THERMAL ANALYSIS

Because polymeric materials are expected to perform under a variety of temperature conditions, thermal properties are important. Thermal property investigations can also allow better design of materials that meet the thermal requirements and may also give added structural data.

Major instrumentation involved with the generation of thermal property behavior of materials includes thermogravimetric analysis (TG, TGA), differential scanning calorimetry (DSC), differential thermal analysis (DTA), torsional braid analysis (TBA), thermomechanical analysis (TMA), thermogravimetric-mass spectrometry analysis (TG-MS) and pyrolysis gas chromatography (PGC). Most of these analysis techniques measure the polymer response as a function of time, atmosphere, and temperature.

One of the simplest techniques is PGC in which the gases, resulting from the pyrolysis of a polymer, are analyzed by gas chromatography. This technique may be used for qualitative and quantitative analysis. The latter requires calibration with known amounts of a standard polymer pyrolyzed under the same conditions as the unknown.

There are several different modes of thermal analysis described as DSC. DSC is a technique of nonequilibrium calorimetry in which the heat flow into or away from the polymer is measured as a function of temperature, atmosphere, and/or time. This differs from DTA, where the temperature difference between a reference and a sample is measured. Even so, the distinction is not always clear. DSC equipment measures the heat flow by maintaining a thermal balance between the reference and sample by changing a current passing through the heaters under the two chambers. For instance, the heating of a sample and a reference proceeds at a predetermined rate until heat is emitted or consumed by the sample. The circuitry is programmed to give a constant temperature between the reference and sample compartments. If an endothermic occurrence takes place, such as the melting of the sample, the temperature of the sample will be less than that of the reference. Greater current is then given to the sample to compensate for the loss in energy from melting. The amount of energy required to melt the sample is generally given as the area under the resulting curve and calculated by the associated computer program. The advantages of DSC and DTA over a good adiabatic calorimeter include speed, low cost, and the ability to use small samples.

The resultant plot of ΔT as a function of temperature is known as a thermogram. Figure 13.2 contains a thermogram emphasizing the T_g and T_m endothermic regions for a polymer sample that is approximately 50% crystalline. Several things should be noted. First, the magnitude of the area under the curve associated with T_m is greater than T_g because it requires more energy to gain entire chain mobility than to achieve segmental mobility. There is a small exothermic region after the T_g that reflects a tendency to form crystalline micelles when there is sufficient mobility to allow their formation.

Possible determinations from DSC/DTA measurements include (1) heat of transition, (2) heat of reaction, (3) sample purity, (4) phase diagram, (5) specific heat, (6) sample identification, (7) percentage incorporation of a substance, (8) reaction rate, (9) rate of crystallization or melting, (10) solvent

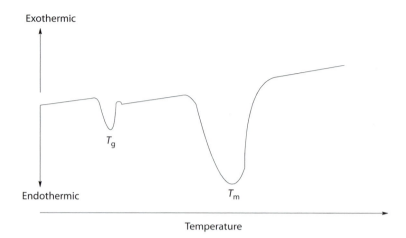

FIGURE 13.2 Representative DCS thermogram.

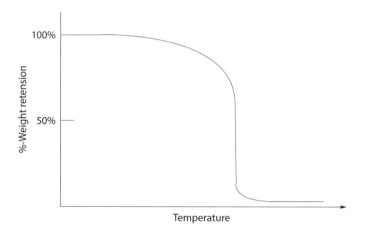

FIGURE 13.3 TGA thermogram for a typical vinyl polymer.

retention, and (11) activation energy. Thus, thermocalorimetric analysis can be a useful tool in describing the chemical and physical relationship of a polymer with respect to temperature.

In TGA, a sensitive balance is used to follow the weight change of a polymer as a function of time, atmosphere, or temperature. Figure 13.3 shows a typical TGA thermogram for a vinyl polymer such as PE. For such polymers where there is a similar thermal energy stability of the backbone, the TGA decrease generally occurs with one downward sweep beginning at about 200°C and being complete before 300°C. There is normally some residue that is high in carbon content remaining to more than 1,000°C.

Along with a determination of the thermal stability of the polymer, TGA can be used to determine the following: (1) sample purity, (2) identification, (3) solvent retention, (4) reaction rate, (5) activation energy, and (6) heat of reaction.

Thermomechanical analysis measures the mechanical response of a polymer looking at (1) expansion properties, including the coefficient of linear expansion, (2) tension properties such as measurement of shrinkage and expansion under tensile stress, that is, elastic modulus, (3) volumetric expansion, that is, specific volume, (4) single-fiber properties, and (5) compression properties such as measuring the softening or penetration under load.

In TBA, changes in polymer structure are measured. The name TBA is derived from the fact that measurements were initially made of fibers that were "braided" together to give the test samples connected between or onto vice-like attachments or hooks.

Reading and coworkers have pioneered in the adaptation of nanoassociated AFM for thermal analyses. The tip of the AFM is replaced with a miniature resistive heater that is used to heat or measure temperature differences on a nanoscale. This allows differentiation between phases and (at times) individual polymer chains and segments (such as block and graft segments of copolymers). T_g and T_m and melting are based on differences in thermal conductivity and diffusivity. This technique of using the microthermal sensor (MTDSC) is called calorimetric analysis with scanning microscopy (CASM). The probe can also be used to measure certain mechanical properties performing a microthermal mechanical analysis with scanning microscopy (MASH).

Newer techniques are being developed and modifications of existing techniques continue.

13.6 THERMAL PROPERTY TESTS

In general, thermal properties such as T_g, T_m, and softening temperatures are really ranges. Because of the large size of polymers, the structures are varied within them so that different amounts of heat are needed to cause disruption of the different sites when heating occurs. On cooling, time is needed to allow the chains to arrange themselves. Heating rates contribute to the particular range obtained. In general, the faster the heating rate, the wider the range. Some of these different conformational sites are shown in Figure 13.4.

13.6.1 SOFTENING RANGE

Softening ranges are dependent on the technique and procedure used to determine them. Thus, listings of softening ranges should be accompanied with how they were determined. Following are some of the techniques used to determine softening ranges. While DSC gives a more precise measurement of not only the T_g, it often approximates the softening range. The capillary method of determining the melting point of small molecules can also be used to gain some idea as to the softening point. Another related technique is to simply use a Fisher–John melting point apparatus and apply some pressure to a sample contained between two glass slides.

The viscat needle method (Figure 13.5) consists of determining the temperature at which a standard needle penetrates a sample. In the ring-and-ball method, the softening range is determined by noting the temperature at which the sample, held within a ring, can be forced through the ring by application of a standard force.

13.6.2 HEAT DEFLECTION TEMPERATURE

The heat deflection temperature (ASTM D-648) is determined by noting the temperature where a simple beam under a specific load deflects a specific amount (generally 0.01 inch; 0.25 mm;

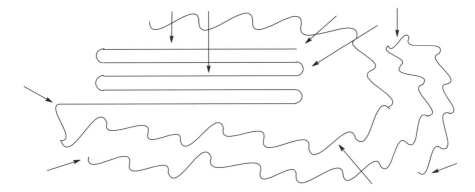

FIGURE 13.4 Simple chain arrangement for a linear polymer containing both ordered and unordered structures.

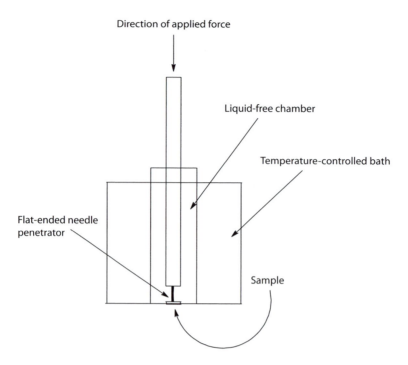

FIGURE 13.5   Vicat apparatus for determining softening points.

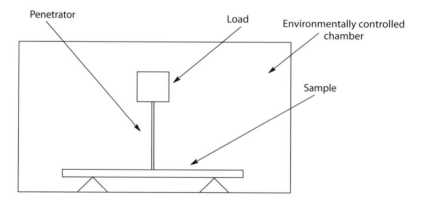

FIGURE 13.6   Deflection temperature test assembly.

Figure 13.6) within a specified environment. As in other ASTM, the experimental conditions are specified including sample size.

13.6.3   GLASS TRANSITION TEMPERATURES

Qualitatively, the T_g corresponds to the onset of short-range (typically 1–5 chain atoms) coordinated motion. Actually, many more (often 10–100) chain atoms may attain sufficient thermal energy sufficient to move in a coordinated manner at T_g. The T_g (ASTM D-3418) is the temperature at which there is an absorption or release of energy as the temperature is raised or lowered, respectively. The T_g can be measured by any means that allows changes in the conformation of the chains to be detected. Thus, most of the thermal techniques allow T_g to be determined. Also, most spectral techniques also allow T_g to be determined as well as techniques that measure volume changes. DMS is also used to measure chain changes by subjecting the sample to repeated small-amplitude stains in a cyclic fashion as a function of temperature. The polymer molecules store some of the imparted energy and dissipate a portion in the form of heat. Since the amount of energy stored and converted to heat is related to molecular motion, changes in the ratios of energy stored to energy converted to heat is used to measure T_g. Sperling compared literature reports of T_g values for some common polymers and found differences of several decades of degrees in the reported T_g values.

13.6.4 THERMAL CONDUCTIVITY

As energy in the form of heat, magnetic, or electric is applied to one side of a material, the energy is transmitted to other areas of the sample. Heat energy is largely transmitted through the increased amplitude of molecular vibrations. The heat flow Q from any point in a solid is related to the temperature gradient dt/dl with the thermal conductivity λ as follows:

$$Q = -\lambda \, (dt/dl) \tag{13.4}$$

Table 13.2 contains a listing of the thermal conductivities of selected materials. Notice the typical much smaller values for polymers compared with metals.

Most polymers have thermal conductivity values in the general range between 10^{-1} and 1 W/m-K. For polymers, transmission of thermal energy, heat, is favored by the presence of ordered crystalline lattices and covalently bonded atoms. Thus graphite, quartz, and diamond are relatively good thermal conductors. Crystalline polymers such as HDPE and i-PP exhibit somewhat higher thermal conductivities than amorphous polymers such as LDPE and a-PS. In general, thermal conductivity increases with increasing density and crystallinity for the same polymer. For amorphous polymers, where energy is transmitted through the polymer backbone, thermal conductivity increases as the chain length increases. Addition of small molecules, such as plasticizers, generally decreases thermal conductivity.

As long as a polymer does not undergo a phase change, thermal conductivity is not greatly affected by temperature changes. Aligning of polymers generally increases their thermal conductivities along the axis of elongation. For instance, the conductivity of HDPE increases 10-fold along the axis of elongation at 10% strain.

Foamed cellular materials have much lower thermal conductivities because the gas employed to create the foam is a poor conductor. Thus, foams are employed as commercial insulators in buildings, thermal jugs, and drinking mugs.

TABLE 13.2
Thermal Conductivities of Selected Materials

Material	Approximate Thermal Conductivity (W/m-K)	Material	Approximate Thermal Conductivity (W/m-K)
Copper	7,200	a-PS	0.16
Graphite	150	PS(foam)	0.04
Iron	90	PVC	0.16
Diamond	30	PVC(foam)	0.03
Quartz	10	Nylon 66	0.25
Glass	1	PET	0.14
PMMA	0.19	NR	0.18
HDPE	0.44	PU	0.31
LDPE	0.35	PU(foam)	0.03
i-PP	0.24	PTFE	0.27
PVC (35% plasticizer)	0.15		

13.6.5 Thermal Expansion

Coefficients of thermal expansion generally refer to differences in length, area, or volume as a function of a temperature unit. Relative to metals such as steel, polymers have large coefficients of thermal expansions. Polymers also have quite varied coefficients of thermal expansion. Both of these factors are troublesome when different materials are bound together, including composites, and exposed to wide temperature ranges. Such wide temperature ranges regularly occur in the aerospace industry (aircraft), within computer chips (and many other electrical devices), engines, motors, and so on. Thus, it is critical to match the coefficients of thermal expansions of materials that are to be bound through mechanical (such as screws and bolts) and chemical (polymer blends, alloys, adhered through use of an adhesive) means or stress will develop between the various components, resulting in fracture or separation.

For polymeric materials, factors that restrict gross movement, such as cross-linking, typically result in lowered coefficients of expansion. Thus, the typical range for coefficients of expansion for cross-linked thermosets is lower than the typical range found for thermoplastics. Further, materials such as glass, graphite, and concrete also exhibit low coefficients of expansion for the same reason.

13.7 FLAMMABILITY

Since many polymeric materials are used as clothing, household items, components of automobiles and aircraft, and so on flammability is an important consideration. Some polymers such as polytetrafluoroethylene and PVC are "naturally" flame resistant, but most common polymers such as PE and PP are not. Small-scale horizontal flame tests have been used to estimate the flammability of solid (ASTM D-635), cellular (ASTM D-1692–74), and foamed (ASTM D-1992) polymers, but these tests are useful for comparative purposes only. Large-scale tunnel tests (ASTM E-84) are more accurate, but they are also more expensive to run than ordinary laboratory tests cited before.

One of the most useful laboratory flammability tests is the oxygen index (OI) test (ASTM D-2043 and ASTM D-2863). In this test, the polymer is burned by a candle in controlled mixtures of oxygen and nitrogen. The minimum oxygen concentration that produces downward flame propagation is considered the OI or ignitability of the polymer.

It is interesting that we have the ability to make structures more flame resistant but in the past did not take advantage of these means. For instance, the steel inner structures of many high-rise buildings can be made to withstand most fires though coating the steel with flame resistant polymers. The practice of using such coatings was resisted because of added cost and the need to have workers trained to apply such coatings. Even so, because insurance companies lowered the premium for buildings possessing such coatings, today such coating is general practice.

13.8 ELECTRICAL PROPERTIES: THEORY

Polymers have served as important materials in the electronics industry. Generally, they have served as coating and containers because of their lack of conductivity; that is, they are nonconductors. More recently, polymers have become major materials as conductors such as polyacetylene and polypyrrole.

Some important dielectric behavior properties are dielectric loss, loss factor, dielectric constant, direct current (DC) conductivity, alternating current (AC) conductivity, and electric breakdown strength. The term "dielectric behavior" usually refers to the variation of these properties as a function of frequency, composition, voltage, pressure, and temperature.

The dielectric behavior is often studied by employing charging or polarization currents. Since polarization currents depend on the applied voltage and the dimensions of the condenser, it is customary to eliminate this dependence by dividing the charge, Q, by the voltage, V, to get a parameter called the *capacitance* (capacity; C):

$$C = Q/V \qquad (13.5)$$

and then using the dielectric constant ε, which is defined as

$$\varepsilon = C/C_o \qquad (13.6)$$

where C is the capacity of the condenser when the dielectric material is placed between its plates in a vacuum and C_o is the empty condenser capacity.

Dielectric polarization is the polarized condition in a dielectric resulting from an applied AC or DC field. The polarizability is the electric dipole moment per unit volume induced by an applied field or unit effective intensity. The molar polarizability is a measure of the polarizability per molar volume; thus, it is related to the polarizability of the individual molecules or polymer repeat unit.

Conductivity is a measure of the number of ions per unit volume and their average velocity in the direction of the applied field. Polarizability is a measure of the number of bound charged particles per cubic unit and their average displacement in the direction of the applied field.

There are two types of charging currents and condenser charges, which may be described as rapidly forming or instantaneous polarizations and slowly forming or absorptive polarizations. The total polarizability of the dielectric is the sum of contributions due to several types of charge displacement in the materials caused by the applied field. The relaxation time is the time required for polarization to form or disappear. The magnitude of the polarizability, k, of a dielectric is related to the dielectric constant, ε, as follows:

$$k = 3(\varepsilon - 1) / 4\pi (\varepsilon + 2) \qquad (13.7)$$

The terms "polarizability constant" and "dielectric constant" are often used interchangeably in a qualitative discussion of the magnitude of the dielectric constant. The k values obtained utilizing DC and low-frequency measurements are a summation of electronic (E), atomic (A), dipole (D), and interfacial (I) polarizations as shown in Figure 13.7. Only the contribution by electronic polarizations is evident at high frequencies. The contributions to dielectric constant at low frequencies are additive as shown in Figure 13.7.

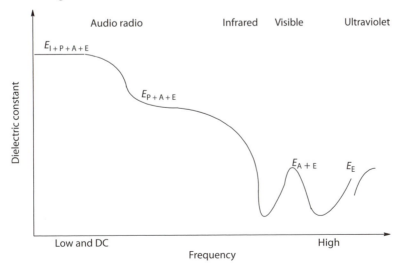

FIGURE 13.7 Relationship of dielectric constant with frequency emphasizing interfacial (I), dipole (P), atomic (A), and electronic (E) polarization contributions.

Instantaneous polarization occurs when rapid ($< 10^{-10}$ s) transitions occur, that is, at frequencies greater than 10^{10} Hz or at wavelengths less than 1 cm. Electronic polarization falls within this category and is due to the displacement of charges within the atoms. *Electronic polarization* is directly proportional to the number of bound electrons in a unit volume and inversely proportional to the forces binding these electrons to the nuclei of the atoms.

Electronic polarization occurs so rapidly that there is no observable effect of time or frequency on the dielectric constant until frequencies are reached that correspond to the visible and UV spectra. For convenience, the frequency range of the infrared through the UV region is called the *optical frequency range*, and the radio and the audio range is called the *electric frequency range*. Electronic polarization is an additive property dependent on the atomic bonds. Thus, the electronic polarizations and related properties are similar for both small molecules and polymers. Accordingly, values obtained for small molecules can be applied to analogous polymeric materials. This does not apply when the polymeric nature of the material plays an additional role in the conductance of electric charges, as in the case for whole-chain resonance or whole-chain delocalization of electrons.

Atomic polarization contributes to the relative motion of atoms in the molecule affected by perturbation by the applied field of the vibrations of atoms and ions having a characteristic resonance frequency in the infrared region. The atomic polarization is large in inorganic materials, which contain low-energy-conductive bonds and approaches zero for nonconductive polymers. The atomic polarization is rapid, and this, as well as the electronic polarization, constitutes the instantaneous polarization components.

The remaining types of polarization are absorptive types with characteristic relaxation times corresponding to relaxation frequencies. Debye, in 1912, suggested that the high dielectric constants of water, ethanol, and other highly polar molecules was due to the presence of permanent dipoles within each individual molecule and that there is a tendency for the molecules to align themselves with their dipole axes in the direction of the applied field. The major contributions to *dipole polarizations* are additive and are similar whether the moiety is within a small or large (macromolecule) molecule. Even so, the secondary contributions to the overall dipole polarization of a sample are dependent on both the chemical and physical environment of the specific dipole unit, its size, and its mobility. Thus, dipole contributions can be used to measure the T_g and T_m.

The polarizations noted above are the major types found in homogeneous materials. Other types of polarization, called *interfacial polarizations*, are the result of heterogeneity. Ceramics, polymers with additives and paper are considered to be electrically heterogeneous.

Table 13.3 contains often used electrical units.

TABLE 13.3
Selected Electrical Primary and Derived Units

Electrical Value	Symbol	SI Units	
		Primary	Derived
Capacitance	C	s^2-C^2/kg-m^2	farad
Conductivity	s	s-C^2/kg-m^3	1/ohm-meter
Dielectric constant	e, e_r	Simple ratio with no units	
Dielectric displacement	D	C/m^2	farad-volt/m^2
Electric charge	Q	C	coulomb
Electrical current	I	C/s	ampere
Electric polarization	P	C/m^2	farad-volt/m^2
Electric potential	V	kg-m^2/s^2-C	volt
Permittivity	e	s^2 C^2/kg-m^3	farad/meter
Resistance	R	kg-m^2/s-C^2	ohm
Resistivity	r	kg-m^3/s-C^2	ohm-meter

13.9 ELECTRIC MEASUREMENTS

Material response is typically studied using either direct (constant) applied voltage (DC) or alternating applied voltage (AC). The AC response as a function of frequency is characteristic of a material. In the future, such electric spectra may be used as a product identification tool, much like IR. Factors such as current strength, duration of measurement, specimen shape, temperature, and applied pressure affect the electric responses of materials. The response may be delayed because of a number of factors, including the interaction between polymer chains, the presence within the chain of specific molecular groupings, and effects related to interactions in the specific atoms themselves. A number of properties, such as relaxation time, power loss, dissipation factor, and power factor, are measures of this lag. The movement of dipoles (related to the dipole polarization, P) within a polymer can be divided into two types: an orientation polarization (P') and a dislocation or induced polarization.

The relaxation time required for the charge movement of electronic polarization E to reach equilibrium is extremely short (about 10^{-15} s) and this type of polarization is related to the square of the index of refraction. The relaxation time for atomic polarization A is about 10^{-3} s. The relaxation time for induced orientation polarization P' is dependent on molecular structure and it is temperature dependent.

The electric properties of polymers are also related to their mechanical behavior. The dielectric constant and dielectric loss factor are analogous to the elastic compliance and mechanical loss factor. Electric resistivity is analogous to viscosity. Polar polymers, such as ionomers, possess permanent dipole moments. These polar materials are capable of storing more electric energy than nonpolar materials. Nonpolar polymers are dependent almost entirely on induced dipoles for electric energy storage. Thus, orientation polarization is produced in addition to the induced polarization when the polar polymers are placed in an electric field. The induced dipole moment of a polymer in an electric field is proportional to the field strength, and the proportionality constant is related to the polarizability of the atoms in the polymer. The dielectric properties of polymers are affected adversely by the presence of moisture, and this effect is greater in hydrophilic than in hydrophobic polymers.

The Clausis–Mossotti equation 13.8 shows that the polarization of a polymer, P, in an electric field is related to the dielectric constant, e, the molecular weight, M, and the density, ρ.

$$P = (e - 1/e + 2) \, M/\rho \tag{13.8}$$

At low frequencies, the dipole moments of polymers are able to remain in phase with changes in a strong electric field resulting in low power losses. However, as the frequency increases the dipole reorientation may not occur sufficiently rapid to maintain the dipole in phase with the electric field and power losses occur.

13.9.1 DIELECTRIC CONSTANT

As previously note, the dielectric constant, e (ASTM D-150–74), is the ratio of the capacity of a condenser made with or containing the test material compared with the capacity of the same condenser with air as the dielectric. Polymers employed as insulators in electrical applications, should have low dielectric constants while those used as semiconductors or conductors should have high dielectric constants.

The dielectric constant is independency of electrical frequency at low to moderate frequencies but varies at higher frequencies. For most materials, the dielectric constant is approximately equal to the square of the index of refraction and to one-third the solubility parameter.

13.9.2 ELECTRICAL RESISTANCE

There are a number of electrical properties related to electrical resistance (ASTM D-257). These include insulation resistance, volume resistivity, surface resistivity, volume resistance, and surface resistance.

The bulk (or volume) specific resistance is one of the most useful general electrical properties. Specific resistance is a physical quantity that may vary more than 10^{23} in readily available materials. This unusually wide range of conductivity allows the wide variety of electrical applications. Conductive materials, such as copper, have specific resistance values of about 10^{-6} ohm cm, whereas good insulators such as polytetrafluoroethylene and low-density polyethylene have values of about 10^{17} ohm cm. Specific resistance is calculated from Equation 13.9, where R is the resistance in ohms, a is the pellet area in cm^2, t is the pellet thickness in cm, and p is the specific resistance in ohm cm.

$$P = R\,(a/t) \tag{13.9}$$

13.9.3 DISSIPATION FACTOR AND POWER LOSS

The dissipation factor (ASTM D-150) has been defined in several ways including the following:

Ratio of the real (in phase) power to the reactive (90° out of phase) power
Measure of the conversion of the reactive power to real power or heat
Tangent of the loss angle and the co-tangent of the phase angle; and
Ratio of the conductance of a capacitor (in which the material is the dielectric material) to its susceptibility.

Both the dielectric constant and dissipation factor are measured by comparison of results obtained with those obtained from a sample with know dissipation factor or dielectric constant values or substitution in an electrical bridge.

The power factor is the energy required for the rotation of the dipoles of a polymer in an applied electrostatic field of increasing frequency. Typical values vary from 1.5×10^4 for polystyrene to 5×10^{-2} for plasticized cellulose acetate. Values increase at T_g and T_m because of the increased chain mobility gained at T_g and T_m so that T_g and T_m have been measured using differences in the power factor as temperature is increased.

The loss factor is the product of the power factor and the dielectric constant, and is a measure of the total electric loss in a material.

13.9.4 ELECTRICAL CONDUCTIVITY AND DIELECTRIC STRENGTH

A steady current does not flow in a perfect insulator in a static electric field, but energy is "stored" in the sample as a result of dielectric polarization. Thus, the insulator acts as a device to story energy. In reality, some leakage of current occurs even for the best insulators.

The insulating property of materials breaks down in strong electrical fields. This breakdown strength, called the *electric* or *dielectric strength* (DS) is the voltage where material electrical failure occurs. The DS is often related to material thickness, L, as shown in Equation 13.10.

$$\text{DS is proportional to } L^{-0.4} \tag{13.10}$$

Breakdown may occur below the measured DS because of an accumulation of energy through inexact dissipation of the current. This leads to an increase in temperature and thermal breakdown. Breakdown means sudden passage of excessive current through the material that often is visible in the material.

The DS is high for many insulating polymers and may be high as 10^3 MV/m. The upper limit of the DS of a material is dependent on the ionization energy of the polymer. Electric or intrinsic decomposition (breakdown) occurs when electrons are removed from their associated nuclei, resulting in secondary ionization occurring and accelerated decomposition. The DS is reduced by mechanical loading of the sample and by increasing the temperature, both making the material more susceptible to degradation.

Dielectric strength (ASTM D-149) is an indication of the electrical strength of an insulating material and it is dependent on the particular test conditions.

Following are some relationships that are generally pertinent to describing the electrical properties of polymers.

For conductive polymers, and other materials, conductivity is defined by Ohm's law that says

$$U = R\,I \tag{13.11}$$

where I is the current in amperes through a resistor, U is the drop in potential in volts, and the relationship between I and U is called the *resistance* generally measured in ohms. Resistance, R, is measured by applying a known voltage across the material and measuring the current that passes though it. The reciprocal of resistance $(1/R)$ is called *conductance*. Ohm's law is an empirical law related in irreversible thermodynamics to the flow of the current I, as a result of a potential gradient that leads to energy being dissipated. (Not all materials obey Ohms law.) Gas discharges, semiconductors, and vacuum tubes are what are called *one-dimensional conductors* and generally deviate from Ohm's law.

For materials that obey Ohm's law, the resistance is proportional to the length, l, of a sample and inversely proportional to the material cross section, A.

$$R = \rho\, l/A \tag{13.12}$$

where ρ is the resistivity measured in ohm-meters.

The resistivity is the resistance per unit distance such as ohms/cm.

As noted above, the inverse of ρ is called *conductivity* and is generally given in Siemens/cm, S/cm (S = 1/ohms). The SI unit of conductivity is S/m but it is often reported as S/cm so attention should be paid as to which unit is being used. A graph of conductivity for many materials is given in Section 19.1.

Conductivity depends on a number of factors including the number density of charge carriers (number of electrons, n) and how rapidly they can move in the sample called *mobility* μ .

$$\text{Conductivity} = n\mu e_c \tag{13.13}$$

where e_c is the electron charge.

Conductivity also varies with temperature generally decreasing for "metallic" materials such as silver and copper but increasing as temperature is increased for semiconductive materials such as insulator, semiconductive and conductive polymers (Section 19.1).

The capacitance, C, relates the ability of a material to hold or store electrical charge and is defined as

$$C = Q/V \tag{13.14}$$

where Q is the charge I coulombs, C.

Alternating currents generally are displaced in time such that it is out of phase with the voltage by an angle δ. The tangent of this angle is the dissipation factor and is related to the power loss as follows:

$$W_{loss} = 2\pi f C_p (\tan \delta) V^2 \tag{13.15}$$

where W_{loss} is the work loss in watts, f is the frequency of the voltage, and C_p is the parallel capacitance of the sample. The relative dielectric constant, permittivity, is described as

$$\varepsilon = C_p/C_v \tag{13.16}$$

where C_v is the capacitance of vacuum. The dielectric loss constant, ε', also called the loss factor is described as follows:

$$W_{loss} \cong \varepsilon \tan \delta = \varepsilon' \tag{13.17}$$

13.10   OPTICAL PROPERTIES TESTS

Since polymers are often used as clear plastics or coatings and have many applications where transparency is an important property, knowledge of the optical properties of specific polymers is essential. The radiation scale, of course, includes microwave, infrared, UV, and visible regions. It is important to recognize the difference between refraction (associated with properties such as refractive index) and reflection (associated with properties such as haze). This difference is illustrated in Figure 13.8.

13.10.1   INDEX OF REFRACTION

Optical properties are related to both the degree of crystallinity and the actual polymer structure. Most polymers do not possess color site units so are colorless and transparent. But, some phenolic resins and polyacetylenes are colored, translucent, or opaque. Polymers that are transparent to visible light may be colored by addition of colorants, and some become opaque as the result of the presence of additives such as fillers, stabilizers, moisture, gases, and so on.

Many of the optical properties of a polymer are related to the refractive index n, which is a measure of the ability of the polymer to refract or bend light as it passes through the polymer. The

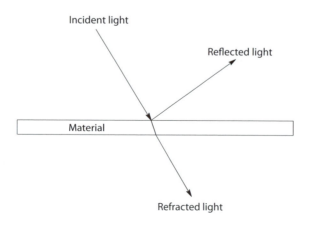

FIGURE 13.8   Refraction and reflection of incident light at the surface of a solid.

refractive index n is equal to the ratio of the sine of the angles of incidence, i, and refraction, r, of light passing into the polymer.

$$n = \sin i / \sin r \tag{13.18}$$

The magnitude of n is related to the density of the substance and varies from 1.000 and 1.3333 for air (actually in vacuum) and water, to about 1.5 for many polymers and 2.5 for white pigment titanium (IV) oxide (titanium dioxide). The value of n is often high for crystals and is dependent on the wavelength of the incident light and on the temperature. It is usually reported for the wavelength of the transparent sodium D line at 298 K. Typical refractive indices for polymers range from 1.35 for polytetrafluoroethylene to 1.67 for polyarylsulfone.

13.10.2 Optical Clarity

Optical clarity or the fraction of illumination transmitted through a material is related by the Beer-Lambert relationship:

$$\log I/I_o = -AL \quad \text{and} \quad I/I^o = e^{-AL} \tag{13.19}$$

where the fraction of illumination transmitted through a polymer (I/I_o) is dependent on the path length of the light (L) and the absorptivity of the polymer at that wavelength of light (A).

Clarity is typical for light passing through a homogeneous material, such as a crystalline ordered polymer or completely amorphous polymer. Interference occurs when the light beam passes through a heterogeneous material in which the polarizability of the individual units varies causing interference that disrupts optical clarity.

13.10.3 Absorption and Reflectance

Colorless materials range from being almost totally transparent to opaque. The opacity is related to the light-scattering process occurring within the material. Incident radiation passes through nonabsorbing, isotropic, optically homogeneous samples with essentially little loss in radiation intensity. Actually, all materials scatter some light. The angular distribution of the scattered light is complex because of the scattering due to micromolecular differences.

Transparency is defined as the state permitting perception of objects through a sample. *Transmission* is the light transmitted. In more specific terms, transparency is the amount of undeviated light, that is, the original intensity minus all light absorbed, scattered, or lost through other means. The ratio of reflected light intensity to the incident light intensity is called the *absorption coefficient*.

13.11 WEATHERABILITY

Polymers are used in almost every conceivable environment. They are tested for their ability to interact with radiation, weather, and microorganisms. Weathering includes the ability to resist attacks by freezing and heating cycles, resistance to frictional damage caused by rain and air, and influence of low and high temperatures as the polymeric materials are used. Many of the results are measured using chemical testing as described in the following section or "real time/condition" testing.

Moisture is an important factor for some polymers, especially those with noncarbon backbones, where hydrolysis, and subsequent degradation, can bring about drastic changes in chain length and consequently, polymer properties. Such attack can occur on the surface and is indicated, as can many chemical attacks, by a discoloration generally followed by crazing and cracking. Other attack can occur within the matrix with the polymer absorbing moisture.

Resistance to biological attack is important for many polymer applications, including almost all of the biomedical applications, food storage and protection, and coatings where microorganism destruction is important. Most synthetic polymers are "naturally" resistant to destruction by microorganisms. This is particularly true for nonpolar polymers, but less so for condensation polymers such as nylons and polyesters where microorganisms may recognize similarities to bonds they ordinarily hydrolyze. Various preservatives and antimicroorganism additives are added, when appropriate, to protect the material against microbial attack. Tests include destructive degradation and simple growth of the microorganism on the material.

13.12 CHEMICAL RESISTANCE

The classic test for chemical resistance (ASTM D-543) measures the percentage weight change of test samples after immersion in different liquid systems. Tests for chemical resistance have been extended to include changes in mechanical properties after immersion. Since chemical attack involves changes in chemical structure, it can be readily observed by many instrumental methods that measure chemical structure, in particular, surface structure.

Tables 13.4 and 13.5 contain a summary of typical stability values for a number of polymers and elastomers against typical chemical agents. As expected, condensation polymers generally exhibit good stability to nonpolar liquids while they are generally only (relatively) moderately or unstable toward polar agents and acids and bases. This is because of the polarity of the connective "condensation" linkages within the polymer backbone. By comparison, vinyl type of polymers exhibit moderate to good stability toward both polar and nonpolar liquids and acids and bases. This is because the carbon–carbon backbone is not particularly susceptible to attack by polar agents and nonpolar liquids, at best, will simply solubilize the polymer. All of the materials show good stability to water alone because all of the polymers have sufficient hydrophobic character to repeal the water.

TABLE 13.4
Stability of Various Polymers to Various Conditions

Polymer	Nonoxidizing Acid 20% Sulfuric	Oxidizing Acid 10% Nitric	Aqueous Salt Solution NaCl	Aqueous Base NaOH	Polar Liquids— Ethanol	Nonpolar Liquids— Benzene	Water
Nylon 6,6	U	U	S	S	M	S	S
Polytetrafluoroethylene	S	S	S	S	S	S	S
Polycarbonate	M	U	S	M	S	U	S
Polyester	M	M	S	M	M	U	S
Polyetheretherketone	S	S	S	S	S	S	S
LDPE	S	M	S	—	S	M	S
HDPE	S	S	S	—	S	S	S
Poly(phenylene oxide)	S	M	S	S	S	U	S
Polypropylene	S	M	S	S	S	M	S
Polystyrene	S	M	S	S	S	U	S
Polyurethane	M	U	S	M	U	M	S
Epoxy	S	U	S	S	S	S	S
Silicone	M	U	S	S	S	M	S

where S = satisfactory; M = moderately to poor; U = unsatisfactory.

TABLE 13.5
Stability of Selected Elastomeric Materials to Various Conditions

Polymers	Weather— Sunlight Aging	Oxidation	Ozone Cracking	NaOH— Diluted/ Concentrated	Acid— Diluted/ Concentrated	Degreasers Chlorinated Hydrocarbons	Aliphatic Hydrocarbons
Butadiene	P	G	B	F/F	F/F	P	P
Neoprene	G	G	G	G/G	G/G	P	F
Nitrile	P	G	F	G/G	G/G	G	G
Polyisoprene (Natural)	P	G	B	G/F	G/F	B	B
Polyisoprene (Synthetic)	B	G	B	F/F	F/F	B	B
Styrene–Butadiene	P	G	B	F/F	F/F	B	B
Silicone	G	G	G	G/G	G/F	B	F-P

where G = good; F = fair; P= poor; B = bad.

13.13 MEASUREMENT OF PARTICLE SIZE

Particle size is important in many polymer applications including coatings, creation of suspensions, and in quality control procedures such as the determination of contaminates.

Table 13.6 contains some of the analytical techniques that allow particle size determination. Before a technique is chosen, the relative particle size and type of size information needed should be determined. Simple microscopy measurements generally allow the determination of particle size and shape. Since some techniques require spherical samples for the best relationship between measured values and particle size, deviation of sample shapes from spherical introduces error. Thus, if the particle shape deviates scientifically from being spherical, another technique should be considered.

The type of measurements that are needed to accomplish the task should also be considered. Refractive index values are generally needed for measurements based on light scattering. Densities are often needed for techniques based on acoustics and sedimentation. Further, most approaches require the samples to be dissolved or suspended in a liquid. Thus, information related to how the liquid affects particle shape and association is also important.

Light obscuration (LO) is one of the major techniques used to determine particle size. LO is based on the observation that particles whose refractive index is different from the suspending liquid scattered light. This scattering is the same as that employed in molecular weight determination employing light-scattering photometry. In fact, light-scattering photometry can be employed to determine particle shape and size. Even so, LO instruments have been developed whose main function is particle size determination. Stirring is often required to maintain a somewhat homogeneous suspension. Wetting of the particles by the suspending liquid, often achieved by addition of a wetting agent, is also often required. Dispersion of the particles is assisted by sonication for situations where single particle size is important. Sonication is not recommended if particle aggregation sizes are important.

Sedimentation techniques are also utilized for particle size distribution for particles on the order of 0.1–50 microns. Capillary hydrodynamic chromatography (HDC) gives particle size distributions for particles of about 0.005–0.7 microns.

TABLE 13.6
Particle Size and Distribution Determination Techniques

Capillary hydrodynamic chromatography	Field flow fractionation
Fraunhofer diffraction	Light obscuration
Light-scattering photometry	Microscopy
Phase Doppler anemometry	Sedimentation
Ultrasonic spectroscopy	

13.14 MEASUREMENT OF ADHESION

Adhesion is the binding of two surfaces where they are held together by primary and/or secondary bonding forces. A discussion of the types of adhesives is given in Section 18.10. A discussion of surface analysis is given in Sections 13.2 and 13.3. Here we will look at the mechanical measurement of adhesion. As with all of polymer testing, the conditions of testing are important and experiments and experimental conditions should be chosen to reflect the particular conditions and operating conditions under which the material may be subjected to. During the fracture of adhered surfaces, the bonded surfaces and adhesive are both deformed.

Failure can be affected through a variety of mechanisms. The major mechanisms are through peeling, shearing, and detachment.

In peel separation, the adhesive simply "peels" away from the surface. Lap shear occurs when the adhered material is subjected to a force that is applied parallel to the bonding plane. Here the bond becomes deformed and stretched after initial rupture of some portion of the bond. It is a "sliding" type of failure. In tensile detachment, bond disruption occurs as force is applied at right angles to the bonding surface. Tensile detachment is a "ripping" type of bond disruption.

Scratch Testing—Often some type of "scratch" testing can be used to measure adhesion of thin films. Fingernail and pencil tip results are often used as a first measure of adhesion. Thus, a person presses their fingernail against the adhered surface watching to see if the adhesive is removed from the surface. Even in more complex scratch tests, such tests create both elastic and plastic deformations around the probe tip. The critical load for adhesion failure depends on the interfacial adhesion and on the mechanical properties of the adhesive. Today, there exist techniques that give better measures of adhesion. In the microimpact test, the test probe continuously impacts the adhesive along the wear track. This allows the effects of impacts adjacent to prior impacts to be studied, allowing the influence of large area damage near the impact to be studied.

Peel Testing—Peel angle is important with 180° and 90° peel angles being the most common. In general, a small amount of the adhesive is "peeled" away and the force necessary to continue the peel measured.

Tensile Detachment Testing—A simple description of tensile tests involves attaching two thick microscope slides to one another so the slides are at right angles to one another. After the bond is allowed to set up a desired amount, the ends of the top microscope are attached to the measuring device and the ends of the second, bottom, slide are attached to a load-bearing segment of the apparatus. Eventually, deformation and finally bond breakage occurs as the load is increased.

Lap Shear Testing—Shear can be applied in a number of ways—cyclic, intermittent, static (or constant), or increasing. A simple overlap shear test is described in ASTM-D-1002. This can be illustrated again using two strong microscope slides. Here, the microscope slides are adhered in parallel to one another except off-set. After the appropriate setup time, the top and bottom of the slide combination are attached to the shear tensile measuring device and the experiment carried out (Figure 13.9).

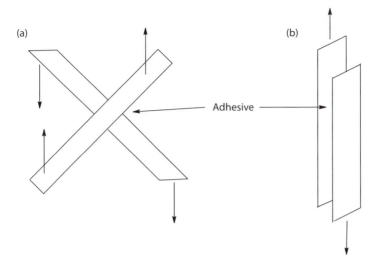

FIGURE 13.9 Descriptions of simple tensile detachment (a) and simple lap shear (b) assemblies for testing adhesion.

13.15 PERMEABILITY AND DIFFUSION

Permeability is a measure of the rate of passage of gas or liquid through a material. Diffusion is the movement of material from one chemical potential to another chemical potential. Thus, these two terms are similar. The diffusion of gases and vapors in polymers and the permeability of polymers to gases and vapors are of importance in packaging and coatings but also demonstrate the kinetic agitation of the diffused gas or vapor through the polymer matrix. The diffusion process, such as the passing of gas molecules through a polymer coating or membrane or the movement of plasticizer molecules within a polymer matrix, is generally described in terms of random jumps and hole filling by the smaller molecules.

The diffusion of larger organic vapor molecules is related to absorption. The rate of diffusion is dependent on the size and shape of the diffusate molecules, their interaction with the polymer molecules, and the size, shape, and stiffness of the polymer chains. The rate of diffusion is directly related to the polymer chain flexibility and inversely related to the size of the diffusate molecules.

Diffusion and permeability are inversely related to the density, degree of crystallinity, orientation, filler concentration, and cross-link density of a polymeric film. In general, the presence of smaller molecules, such as plasticizers, increases the rate of diffusion in polymers since they are more mobile and can create holes or vacancies within the polymer. The rate of diffusion or permeability is fairly independent of polymer chain length just as long as the polymer has a moderately high chain length.

Permeation of gases, liquids, and vapors through a polymeric film can be looked at as a three-step process as follows: (1) the rate of dissolution of the small molecules in the polymer; (2) the rate of diffusion of the small molecules in the polymer film in accordance with the concentration gradient, and (3) the energies of the smaller molecules on the opposite side of the polymer.

Permeation is dependent on the segmental motion of the polymer chains and the free volume within the polymer matrix. The free volume decreases and chain stiffness increases as the temperature of the polymer membrane or film is lowered toward the T_g. The free volume is predicted to be similar for all polymers at the T_g.

The diffusion of a liquid or gas through a membrane is similar to the diffusion in liquid systems. The mechanism for each case involves a transfer of a small molecule to a hole in the liquid or membrane. For diffusion, the jumping frequency from hole to hole is dependent on the activation energy,

E_D, for this jump, which is dependent of the size and shape of the diffusing molecule and the size of the holes in the membrane, that is, the free volume. For permeation, the activation energy is on the order of 20–40 kJ/mol. The values for E_D are somewhat higher.

The rate of diffusion, D, and the rate of permeability, P, increase exponentially as shown by the Arrhenius equation for diffusion Equation 13.20.

$$D = D_o \, e^{-E_D/RT} \tag{13.20}$$

where D_o is the rate of diffusion at some base temperature, R is the ideal gas constant, and T is the Kelvin temperature.

Diffusion can be described by Fick's law Equation 13.21,

$$F = -D\left[\frac{dc}{dx}\right] \tag{13.21}$$

where F is the weight of diffusate crossing a unit area per unit time and is proportional to the concentration gradient, dc/dx. The proportionality constant D is directly related to the pressure differential across the membrane and inversely related to the membrane thickness.

When the diffusion coefficient D is dependent on concentration, the diffusion process is said to be Fickian. In such cases, D is inversely related to solubility, S, and to permeability, P, as follows:

$$D \propto \frac{P}{S} \tag{13.22}$$

It has been suggested that linear alkanes diffuse through the holes of membranes by alignment with the segments of the organic polymer chains. Such alignments are more difficult for branched alkanes so that they diffuse more slowly.

The diffusion coefficient D is inversely related to the cross-link density of vulcanized rubbers. When D is extrapolated to zero concentration of the diffusing small molecules, it is related to the distance between the cross-links. Thus, as the cross-link density increases D becomes smaller, as expected. Further, the diffusion coefficient is less for crystalline polymers in comparison to the same polymer except in the amorphous state. In fact, this can be roughly stated as follows:

$$D_c = D_a\,(1 - x) \tag{13.23}$$

where D_c is the diffusion constant for the crystalline material, D_a is the diffusion constant for the amorphous material, and x is the extent of crystallinity.

The permeability coefficient P is related to the diffusion coefficient D and the solubility coefficient S as shown by Henry's law:

$$P = DS \tag{13.24}$$

Thus, the permeability values are high when the solubility parameter of the diffusion molecules are similar to that of the polymer film.

While some polymers exhibit non-Fickian diffusion below the T_g, many of these become Fickian as the temperature is raised to above the T_g.

13.16 SUMMARY

1. The American Society for Testing and Materials (ASTM) and comparable organizations throughout the world have established meaningful standards for the testing of polymers.
2. Spectroscopic techniques that are useful for small molecules are equally as important with macromolecules. These techniques give both structural data and data related to the morphology of the polymers.

3. Surface properties are important to the physical and chemical behavior of polymers. Similar to smaller molecules, polymer surface structures can be determined using a variety of techniques, including Auger electron spectroscopy, near-field optical microscopy, electron microcopy, SPM, SIMS, and certain IR and MS procedures.
4. Thermal analysis measurements allow the measure of polymer behavior as a function of temperature, time, and atmosphere. DSC/DTA measures changes in energy as temperature is changed and allows the determination of many valuable parameters, including T_g and T_m. TGA measures weight changes as a function of temperature.
5. The electrical properties of materials is important for many of the higher technology applications. Measurements can be made using alternating and/or direct current. The electrical properties are dependent on voltage and frequency. Important electrical properties include dielectric loss, loss factor, dielectric constant, conductivity, relaxation time, induced dipole moment, electrical resistance, power loss, dissipation factor, and electrical breakdown. Electrical properties are related to polymer structure. Most organic polymers are nonconductors, but some are conductors.
6. Important physical properties of polymers include weatherability, chemical resistance, and optical properties. Polymers generally show good to moderate chemical resistance when compared to metals and nonpolymers.

GLOSSARY

ASTM: American Society for Testing and Materials.

BSI: British Standards Institution.

Chemical shifts: Peaks in NMR spectroscopy.

Dielectric constant: Ratio of the capacitance of a polymer to that in a vacuum.

Dielectric strength: Maximum applied voltage that a polymer can withstand without structure change.

Differential scanning calorimetry (DSC): Measurement of the difference in changes in the enthalpy of a heated polymer and a reference standard based on power input.

Differential thermal analysis (DTA, DT): Measurement of the difference in the temperature of a polymer and a reference standard when heated.

Environmental stress cracking: Cracking of polymers.

Glass transition temperature: Temperature where the onset of local or segmental mobility begins.

Heat deflection temperature: Temperature at which a simple loaded beam undergoes a definite deflection.

Index of refraction: Ratio of the velocity of light in a vacuum to that in a transparent material or mixture.

Infrared spectroscopy (IR): Technique used for the characterization of polymers based on their molecular vibration and vibration-rotation spectra.

ISO: International Standards Organization.

Loss factor: Power factor multiplied by the dielectric constant.

Moh's scale: Hardness scale ranging from 1 for talc to 10 for diamond.

Nuclear magnetic resonance spectroscopy (NMR): Based on the absorption of magnetic energy by nuclei that have an uneven number of protons or neutrons; this absorption is dependent on the particular chemical structure and environment of the molecule.

Oxygen index (OI): Test for the minimum oxygen concentration in a mixture of oxygen and nitrogen that will support a candle-like flame of a burning polymer.

Power factor: Electrical energy required to rotate the dipoles in a polymer while in an electrostatic field.

Raman spectroscopy: Based on the interaction of vibrational modes of molecules with relative high-energy radiation where absorption is based on the chemical moieties that undergo changes in polarization.

Scanning probe microscopy (SPM): Group of surface-detection techniques that includes AFM and STM; measures surface depth differences.

Secondary ion mass spectroscopy (SIMS): Mass spectroscopy that looks at the ion fragments generated from the bombardment of surfaces with an ion beam.

Thermogravimetric analysis (TG, TGA): Measurement of the change in weight when a polymer is heated.

EXERCISES

1. Why are electrical tests for polymers important?
2. Which is the better insulator: a polymer with (a) a low or (b) high K factor?
3. Why are the specific heats of polymers higher than those of metals?
4. Which IR technique might give you good surface information?
5. Why is the use of the term flameproof plastics incorrect?
6. Which plastic should be more resistant when immersed in 25% sulfuric acid at room temperature: (a) HDPE, (b) PMMA, or (c) PVac?
7. Why is it important to know the nature and structure of the surface of materials?
8. Why is it important to know such things as the values of thermal expansion of materials?
9. How is NMR important in characterizing polymers?
10. What is the UV region of the spectrum?
11. Which would absorb in the UV region: (a) polystyrene, (b) hevea rubber, or (c) PVC?
12. What technique(s) would you use to determine crystallinity in a polymer?
13. What thermal instrumental technique would you use to determine T_g?
14. What is the difference between measuring physical properties and structural determinations?
15. Why is it important to have standard tests?
16. Are the measurements taken today any better than those taken 30 years ago?
17. In the TGA for most vinyl polymers, why is there a single somewhat smooth degradation plot of weight loss as a function of temperature?
18. A researcher finds using one testing method that the amount of amorphous character in a sample is 60% and using another testing method finds that the same sample is 70% crystalline. How can this occur?
19. Why are such terms as flame retardant or flame resistant used rather than flameproof?
20. What are some hurdles to adopting better protection in the building and automotive industries?

ADDITIONAL READING

Alfrey, T. (1948): *Mechanical Behavior of Polymers*, Interscience, NY.
Ando, I., Askakura, T. (1998): *Solid State NMR of Polymers*, Elsevier, NY.
Armer, M. (2009): *Raman Spectroscopy for Soft Matter Applications*, Wiley, Hoboken, NJ.
Askadskii, A. A. (1996): *Physical Properties of Polymers*, Gordon and Breach, NY.
Batteas, J., Michaels, C. A., Walker, G. C. (2005): *Applications of Scanned Probe Microscopy to Polymers*, American Chemical Society, Washington DC.
Blythe, A. (2005): *Electrical Properties of Polymers*, Cambridge University Press, Cambridge, UK.
Brandolini, A. Haney, D. (2000): *NMR Spectra of Plastics*, Dekker, NY.
Briggs, D., Clarke, D. R., Surest, S., Ward, I. M. (2005): *Surface Analysis of Polymers by XPS and Static SIMS*, Cambridge University Press, Cambridge, UK.
Carraher, C. E. (1977): Resistivity measurements, *J. Chem. Ed.*, 54:576.

Carswell, T. S., Nason, H. K. (1944): Classification of polymers, *Mod. Plastics*, 21:121.

Case, D. A., Zuiderweg, E. R. (2007): *NMR in Biophysical Chemistry*, Oxford University Press, NY.

Chalmers, J., Meier, R. (2008): *Molecular Characterization and Analysis of Polymers*, Elsevier, NY.

Cohen, S., Lightbody, M. (1999): *Atomic Force Microscopy/Scanning Tunneling Microscopy*, Kluwer, NY.

Craver, C. Carraher, C. E., (2000): *Applied Polymer Science*, Elsevier, NY.

Everall, N., Chalmers, J., Griffiths, P. (2007): *Vibrational Spectroscopy of Polymers: Principles and Practice*, Wiley, Hoboken, NJ.

Fawcett, A. H. (1996): *Polymer Spectroscopy*, Wiley, NY.

Friebolin, H. (1998): *Basic One- and Two-Dimensional NMR Spectroscopy*, Wiley, NY.

Furusho, Y., Ito, Y., Kihara, N., Osakada, K. (2004): *Polymer Synthesis and Polymer Analysis*, Springer, NY.

Gregoriou, V. (2004): *Polymer Spectroscopy*, Wiley, Hoboken.

Grellmann, W., Seidler, S. (2007): *Polymer Testing*, Hanser-Gardner, Cincinnati, OH.

Groenewoud, W. (2001): *Characterization of Polymer by Thermal Analysis*, Elsevier, NY.

Grulke, E. (2009): *Devolatilization and Purification of Polymers*, CRC, Boca Raton, FL.

Gupta, R. (2000): *Polymer and Composite Rheology*, Dekker, NY.

Hatada, K. (2004): *NMR Spectroscopy of Polymers*, Springer, NY.

Hatakeyama, T., Quinn, F. (1999): *Thermal Analysis*, Wiley, NY.

Koenig, J. (1999): *Spectroscopy of Polymers*, Elsevier, NY.

Kosmulski, M. (2001): *Chemical Properties of Material Surfaces*, Dekker, NY.

Lee, T. (1998): *A Beginners Guide to Mass Spectral Interpretations*, Wiley, NY.

Li, L. (2008): *Principles and Practice of Polymer Mass Spectrometry*, Wiley, Hoboken, NJ.

Maev, R. (2008): *Acoustic Microscopy: Fundamentals and Applications*, Wiley, Hoboken, NJ.

Marcomber, R. (1997): *A Completer Introduction to Modern NMR*, Wiley, NY.

Mark, J. (2004): *Physical Properties of Polymers*, Cambridge University Press, MA.

Mathias, L. (1991): *Solid State NMR of Polymers*, Plenum, NY.

McBrierty, V. J., Packer, K. (2006): *Nuclear Magnetic Resonance in Solid Polymers*, Cambridge University Press, Cambridge, UK.

Menard, K. (2008): *Dynamical Mechanical Analysis: A Practical Introduction*, CRC, Boca Raton, FL.

Menczel, J., Prime, R. (2008): *Thermal Analysis of Polymers: Fundamentals and Application*, Wiley, Hoboken, NJ.

Meyer, V. (1999): *Practical High-Performance Liquid Chromatography*, 3rd Ed., Wiley, NY.

Michler, M. (2008): *Electron Microscopy of Polymers*, Springer, NY.

Mirau, P. (2004): *A Practical Guide to Understanding the NMR of Polymers*, Wiley, Hoboken.

Moldoveanu, S. (2005): *Analytical Pyrolysis of Synthetic Organic Polymers*, Elsevier, NY.

Montaudo, G., Lattimer, R. P. (2001): *Mass Spectrometry of Polymers*, Taylor and Francis, Boca Raton, FL.

Oliver, R. W. A. (1998): *HPLC of Macromolecules*, 2nd Ed., Oxford, University Press, NY.

Petty, M. C., Bryce, M., Bloor, D. (1995): *An Introduction to Molecular Electronics*, Oxford University Press, NY.

Pham, Q. (2002): *Proton and Carbon NMR Spectra of Polymers*, Wiley, NY.

Rainer, M. (2008): *Mass Spectrometric Identification of Biomolecules- Employing Synthetic Polymers*, VDM Verlag, NY.

Riande, E. (2004): *Electrical Properties of Polymers*, Dekker, NY.

Roe, R. (2000): *Methods of X-Ray and Neutron Scattering in Polymer Science*, Oxford, NY.

Sanderson, M., Skelly, J. V. (2007): *Macromolecular Crystallography*, Oxford University Press, NY.

Sawyer, L., Grubb, D., Meyers, G. (2008): *Polymer Microscopy*, Springer, NY.

Scheirs, J. (2000): *Practical Polymer Analysis*, Wiley, NY.

Schmida, M., Antonietti, M., Coelfen, H., Koehler, W., Schaefer, R. (2000): *New Developments in Polymer Analysis*, Springer-Verlag, NY.

Seymour, R. B., Carraher, C. E. (1984): *Structure-Property Relationships in Polymers*, Plenum, NY.

Shah, V. (1998): *Handbook of Plastics Testing Technology*, Wiley, NY.

Smith A., Brockwell, D. (2006): *Handbook of Single Molecule Fluorescence Spectroscopy*, Oxford University Press, NY.

Solymar, L., Walsh, D. (1998): *Electrical Properties of Materials*, Oxford, NY.

Stuart, B. (2002): *Polymer Analysis*, Wiley, NY.

Tegenfeldt, J. (2008): *Solid-State NMR*, Wiley, Hoboken, NJ.

Turi, E. (1997): *Thermal Characterization of Polymeric Materials*, 2nd Ed., Academic, Orlando, FL.

Urban, M. (1996): *Attenuated Total Reflectance Spectroscopy of Polymers*, Oxford University Press, NY.

Wallace, G, Spinks, G., Kane-Maguire, L., Teasdale, P. (2008): *Conductive Electroactive Polymers: Intelligent Polymer Systems*, 3rd Ed., CRC, Boca Raton, FL.

Zaikiv, G. (2007): *Polymer and Biopolymer Analysis and Characterization*, Nova, Hauppauge, NY.

Zerbi, G. (1999): *Modern Polymer Spectroscopy*, Wiley, NY.

14 Rheology and Physical Tests

Rheology is the science of deformation and flow. Throughout this chapter, please remember that the main reasons for studying rheology (stress–strain relationships) are (1) to determine the suitability of materials to serve specific applications; and (2) to relate the results to polymer structure and form. Understanding structure–property relationships allow a better understanding of the observed results on a molecular level, resulting in a more knowledgeable approach to the design of materials. Look for these ideas as you study this chapter.

Polymers are *viscoelastic* materials meaning they can act as liquids, the "visco" portion, and as solids, the "elastic" portion. Descriptions of the viscoelastic properties of materials generally falls within the area called *rheology*. Determination of the viscoelastic behavior of materials generally occurs through stress–strain and related measurements. Whether a material behaves as a "viscous" or "elastic" material depends on temperature, the particular polymer and its prior treatment, polymer structure, and the particular measurement or conditions applied to the material. The particular property demonstrated by a material under given conditions allows polymers to act as solid or viscous liquids, as plastics, elastomers, or fibers, and so on. This chapter deals with the viscoelastic properties of polymers.

14.1 RHEOLOGY

The branch of science related to the study of deformation and flow of materials was given the name rheology by Bingham, whom some have called the father of modern rheology. The prefix *rheo* is derived from the Greek term *rheos*, meaning current or flow. The study of rheology includes two vastly different branches of mechanics called *fluid* and *solid mechanics*. The polymer scientist is usually concerned with viscoelastic materials that act as both solids and liquids.

The elastic component is dominant in solids, hence their mechanical properties may be described by *Hooke's law* (Equation 14.1), which states that the applied stress (s) is proportional to the resultant strain (γ) but is independent of the rate of this strain ($d\gamma/dt$).

$$s = E\gamma \tag{14.1}$$

Stress is equal to the force per unit area, and strain or elongation is the extension per unit length. For an isotopic solid, that is, one having the same properties regardless of direction, the strain is defined by *Poisson's ratio*, $V = \gamma_w/\gamma_l$, where γ_l is the change in length and γ_w is the change in thickness or lateral contraction.

When there is no volume change, as when an elastomer is stretched, Poisson's ratio is 0.5. This value decreases as the T_g of the polymer increases and approaches 0.3 for rigid solids such as poly(vinyl chloride) (PVC) and ebonite. For simplicity, the polymers dealt with here will be considered to be isotropic viscoelastic solids with a Poisson's ratio of 0.5, and only deformations in tension and shear will be considered. Thus, a shear modulus (G) will usually be used in place of Young's modulus of elasticity (E; Equation 14.2) where E is about 2.6G at temperatures below T_g. For comparison, the moduli (G) for steel, HDPE, and hevea rubber (NR) are 86, 0.087, and 0.0006 dynem², respectively.

$$ds = G\,d\gamma \quad \text{and} \quad s = G\gamma \tag{14.2}$$

The viscous component is dominant in liquids, hence their flow properties may be described by *Newton's law* (Equation 14.3) where η is the viscosity, which states that the applied stress s is proportional to the rate of strain $d\gamma/dt$, but is independent of the strain γ or applied velocity gradient.

$$S = \eta \frac{d\gamma}{dt} \qquad (14.3)$$

Both Hooke's and Newton's laws are valid for small changes in strain or rate of strain, and both are useful in studying the effect of stress on viscoelastic materials. The initial elongation of a stressed polymer below T_g is reversible elongation due to a stretching of covalent bonds and distortion of the bond angles. Some of the very early stages of elongation by disentanglement of chains may also be reversible. However, the rate of flow, which is related to slower disentanglement and slippage of polymer chains past one another, is irreversible and increases (and η decreases) as the temperature increases in accordance with the following form of the Arrhenius equation (Equation 14.4) in which E is the activation energy for viscous flow.

$$\eta = Ae^{E/RT} \qquad (14.4)$$

It is convenient to use a simple weightless *Hookean*, or ideal, *elastic spring* with a modulus G and a simple *Newtonian* (fluid) *dashpot* or shock absorber having a liquid with a viscosity of η as models to demonstrate the deformation of an elastic solid and an ideal liquid, respectively. The stress–strain curves for these models are shown in Figure 14.1.

In general terms, the Hookean spring represents bond flexing and slight separation of the atoms while the Newtonian dashpot represents chain and local segmental movement. It is customary to attempt to relate stress–strain behavior to combinations of dashpots and springs as indicators of the relative importance of bond flexing and segmental movement.

Again, in general terms, below their T_g polymers can be modeled as having a behavior where the spring portion is more important. Above their T_g where segmental mobility occurs, the dashpot portion is more important.

The relative importance of these two modeling parts, the spring and the dashpot, is also dependent on the rate at which an experiment is carried out. Rapid interaction, such as striking with a hammer, with a polymer is more apt to result in a behavior where bond flexibility is more important, while slow interactions are more apt to allow for segmental mobility to occur.

Since polymers are viscoelastic solids, combinations of these models are used to demonstrate the deformation resulting from the application of stress to an isotropic solid polymer. Maxwell joined the two models in series to explain the mechanical properties of pitch and tar (Figure 14.2a). He assumed that the

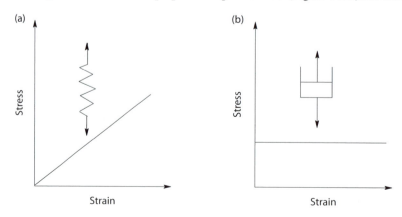

FIGURE 14.1 Stress–strain plots for a Hookean spring (a) where E is the slope (Equation 14.1), and a Newtonian dashpot (b) where s is a constant (Equation 14.3).

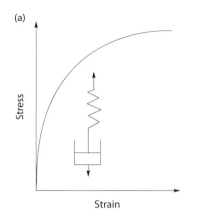
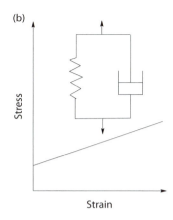

FIGURE 14.2   Stress–strain and strain–time plots for stress relaxation for the Maxwell model (a) and Voigt–Kelvin model (b).

contributions of both the spring and dashpot to strain were additive and that the application of stress would cause an instantaneous elongation of the spring, followed by a slow response of the piston in the dashpot. Thus, the relaxation time (τ), when the stress and elongation have reached equilibrium, is equal to η/G.

In the Maxwell model for viscoelastic deformation, it is assumed that the total strain is equal to the elastic strain plus the viscous strain. This is expressed in the two following differential equations from Equations 14.2 and 14.3.

$$\frac{d\gamma}{dt} = \frac{s}{\eta} + \frac{ds}{dt}\left(\frac{1}{G}\right)$$

(14.5)

The rate of strain $d\gamma/dt$ is equal to zero under conditions of constant stress (s), that is,

$$\frac{s}{\eta} + \frac{ds}{dt}\left(\frac{1}{G}\right) = 0$$

(14.6)

Then, assuming that $s = s_0$ at zero time, we get

$$s = s_0\, e^{-tG/n}$$

(14.7)

And, since the relaxation time $\tau = \eta/G$, then

$$s = s_0\, e^{-t/T}$$

(14.8)

Thus, according to Equation 14.8 for the *Maxwell model* or element, under conditions of constant strain, the stress will decrease exponentially with time and at the *relaxation time $t = \tau$*, s will equal $1/e$, or 0.37 of its original value, s_0.

The parallel dashpot–spring model was developed by Voigt and Kelvin and is known as the *Voigt–Kelvin model* (Figure 14.2b). In this model or element, the applied stress is shared between the spring and dashpot, and thus the elastic response is retarded by the viscous resistance of the liquid in the dashpot. In this model, the vertical movement of the spring is essentially equal to that of the piston in the dashpot. Thus, if G is much larger than η, the retardation time (η/G) or τ is small, and τ is large if η is much larger than G.

In the Voigt–Kelvin model for viscoelastic deformation, it is assumed that the total stress is equal to the sum of the viscous and elastic stress, $s = s_v + s_0$, so that

$$s = \eta\left(\frac{d\gamma}{dt}\right) + G\gamma$$

(14.9)

On integration one obtains

$$\gamma = \frac{s}{G\,(1 - e^{-tG/n})} = \frac{s}{G\,(1 - e^{-t/T})} \tag{14.10}$$

The retardation time τ is the time for the strain to decrease to $1 - (1/e)$ or $1 - (1/2.7) = 0.63$ of the original value. The viscoelastic flow of polymers is explained by approximate combinations of the dashpot and spring. The plots of the real data are compared with those predicted by various models. The relative importance of the various components of the model that fits the experimental data, dashpot and spring combinations, indicate the importance that the types of chain movement represented by the dashpot and spring have for that particular polymer under the particular experimental conditions.

14.1.1 RHEOLOGY AND PHYSICAL TESTS

While polymer melts and noncross-linked elastomers flow readily when stress is applied, structural plastics must resist irreversible deformation and behave as elastic solids when relatively small stresses are applied. These plastics are called ideal or Bingham plastics with their behavior described mathematically by

$$s - s_{0} = \eta \left(\frac{d\gamma}{dt} \right) \tag{14.11}$$

Figure 14.3 contains the stress–strain plots for a number of important flow behaviors. A Bingham plastic exhibits Newtonian flow above the stress yield or stress value (s_{0}). Liquids that undergo a decrease in viscosity with time are called *thixotropic* or false bodied. Those that undergo an increase in viscosity with time are called *rheopectic* (shear thickening). The term *creep* is used to describe the slow slippage of polymer chains over a long period of time. The Herschel–Buckley equation (Equation 14.12) is a general equation that reduces to the Bingham equation when $\eta = 1$ and to the Newtonian equation when $\eta = 1$ and $s_{0} = 0$ and where φ is related to viscosity.

$$(s - s_{0})\, \eta = \varphi \left(\frac{d\gamma}{dt} \right) \tag{14.12}$$

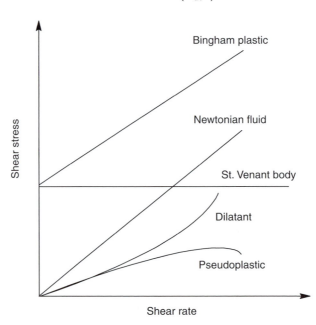

FIGURE 14.3   Various types of polymer flow.

Eyring explained liquid flow using a random hole-filling model in which the holes (vacancies or free volume) account for about 15% of the total volume at room temperature. The size of these holes is similar to that for small molecules. The number of holes increases as the temperature increases, and thus flow or hole filling is temperature dependent. Small molecules jump into the holes leaving empty holes when their energy exceeds the activation energy. The activation energy for jumping or moving into a hole is smaller, per individual unit, for linear molecules that fill the holes by successive correlated jumps of chain segments along the polymer chain. The jump frequency (ϕ) is governed by a segmental factor with both values related to molecular structure and temperature.

For convenience and simplicity, polymers have generally been considered to be isotropic in which the principle force is shear stress. While such assumptions are acceptable for polymers at low shear rates, they fail to account for stresses perpendicular to the plane of the shear stress, which are encountered at high shear rates. For example, an extrudate such as the formation of a pipe or filament, expands when it emerges from the die in what is called the *Barus* or *Weissenberg effect* or *die swell*. This behavior is not explained by simple flow theories.

Viscoelastic behavior can be divided into five subclassifications (Figure 14.4). From 1 to 2, Figure 14.4, the material behaves as a viscous glass or Hookean elastic or glass, where chain segmental motion is quite restricted and involves mainly only bond bending and bond angle deformation. The material behaves as a glass such as window glass.

At 2, the material is undergoing a glassy transition into the rubbery region, 3 to 4, which is often referred to as the viscoelastic region where polymer deformation is reversible but time dependent and associated with both side chain and main chain rotation. In the rubbery region, local segmental mobility occurs but total chain flow is restricted by a physical and/or chemical network. At 5, rubbery flow or viscous flow occurs, where irreversible bulk deformation and slippage of chains past one another occurs. Each of these viscoelastic regions is time dependent. Thus, given a short interaction time, window glass acts as a Hookean glass or like a solid, yet observation of glass over many years would permit the visual observation of flow, with the window glass giving a viscous flow response, thus acting as a fluid. In fact, most polymers give a response as noted in Figure 14.5 for the response of a ball dropped on to the polymeric material at different rates onto the polymeric material that is heated to different temperatures. Commercial Silly Putty or Nutty Putty easily illustrates three of these regions. When struck rapidly it shatters as a solid (glass region), when dropped

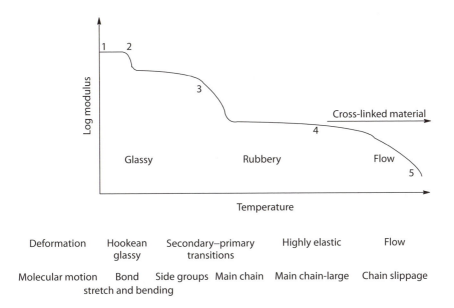

FIGURE 14.4 Characteristic conformational changes as temperature is changed.

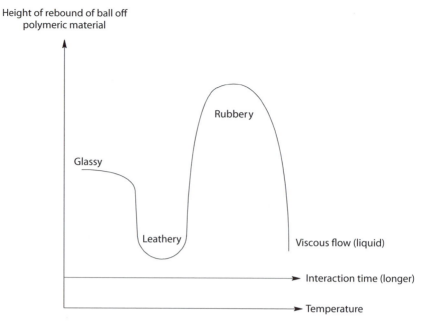

FIGURE 14.5 Regions of material response as a function of interaction (reaction) time and temperature for a typical polymeric material.

at a moderate rate it acts as a rubber (rubbery region), and when allowed to reside in its container for some time, it flows to occupy the container contour, acting as a liquid. The exact response (Figure 14.5) varies between materials.

The leathery region deserves special comment. This region corresponds to the T_g of a material where the impact energy is absorbed, just a energy is absorbed when a material melts, resulting in some segmental mobility rather than simply being released, resulting in a smaller rebound. The extent of energy absorption is dependent on a number of factors, including the proportion of amorphous area.

There have been many attempts to predict long-range behavior from short-range behavior. This is a dangerous practice if the end prediction is material failure because, for most polymeric materials, failure is catastrophic and not "well-behaved." But, for other properties, application of the Boltzman time–temperature superposition master curves can be used to some advantage, particularly pertaining to predicting behavior as depicted in Figure 14.4. This transposition for amorphous polymers is accomplished using a shift factor (a_T) calculated relative to a reference temperature T_r, which may be equal to T_g. The relationship of the shift factor to the reference temperature and some other temperature T, which is between T_g and $T_g + 50$ K, may be approximated by the Arrhenius-like equation (14.13).

$$\log a_T = -\left(\frac{b}{2.3 T T_g} \right) \left(T - T_g \right)$$

(14.13)

According to the more widely used *Williams, Landel, and Ferry* (WLF) *equations*, all linear, amorphous polymers have similar viscoelastic properties at T_g. At specific temperatures above T_g, such as $T_g + 25$ K, where the constants C_1 and C_2 are related to holes or free volume, the following relationship holds:

$$\log a_T = -\frac{C_1 \left(T - T_g \right)}{\left[C_2 + \left(T - T_g \right) \right]}$$

(14.14)

14.1.2 Response Time

We can get a first approximation of the physical nature of a material from its response time. For a Maxwell element, the relaxation time is the time required for the stress in a stress–strain experiment to decay to $1/e$ or 0.37 of its initial value. A material with a low relaxation time flows easily so it shows relatively rapid stress decay. Thus, whether a viscoelastic material behaves as a solid or fluid is indicated by its response time and the experimental time scale or observation time. This observation was first made by Marcus Reiner who defined the ratio of the material response time to the experimental time scale as the *Deborah Number*, D_n. Presumably, the name was derived by Reiner from the Biblical quote in Judges 5, Song of Deborah where it says "The mountains flowed before the Lord."

$$D_n = \frac{\text{Response time}}{\text{Experimental time scale}} \tag{14.15}$$

A high Deborah Number designates the solid behavior of a viscoelastic material while a low Deborah Number corresponds to a viscoelastic material behaving as a fluid. Thus, window glass has a high relaxation time at room temperature. Application of a stress to produce a little strain and looking for it to return to its approximate prestressed state will take a long time as we count our observation time in hours, days, and weeks. Thus, it would have a relatively high Deborah Number under this observation time scale and be acting as a solid. Yet, if we were to have as our observation scale millions of years, the return to the prestressed state would be rapid with the glass acting as a viscous liquid and having a low Deborah Number. Again, this represents only a first approximation measure of the behavior of a material.

14.2 TYPICAL STRESS–STRAIN BEHAVIOR

For perspective, Figure 14.6 contains general ranges for the three major polymer groupings with respect to simple stress–strain behavior.

Most physical tests involve nondestructive evaluations. For our purposes, three types of mechanical stress measures (described in Figure 14.7) will be considered. The ratio of stress to strain is called *Young's modulus*. This ratio is also called the *modulus of elasticity* and *tensile modulus*. It is calculated by dividing the stress by the strain.

$$\text{Young's modulus} = \frac{\text{Stress (Pa)}}{\text{Strain (mm/mm)}} \tag{14.16}$$

Large values of Young's modulus indicate that the material is rigid and resistant to elongation and stretching. Many synthetic organic polymers have Young's modulus values in the general range of about 10^5–10^{10} Pa for stiff fibers. Polystyrene has a Young's modulus of about 10^9 Pa and soft rubber a value of about 10^6 Pa. For comparison, fused quartz has a Young's modulus of about 10^{10}; cast iron, tungsten, and copper have values of about 10^{11}; and diamond has a value of about 10^{12} Pa. Thus, PS represents a glassy polymer at room temperature and is about 100 times as soft as copper, but soft rubber, such as present in rubber bands, is about 1,000 times softer than PS.

Carswell and Nason assigned five classifications to polymers (Figure 14.8). It must be remembered that the ultimate strength of each of these is the total area under the curve before breaking. The soft and weak class, such as polyisobutylene, is characterized by a low modulus of elasticity, low yield (stress) point, and moderate time-dependent elongation. The Poisson ratio, that is, ratio of contraction to elongation, for soft and weak polymers is about 0.5, which is similar to that found for many liquids.

In contrast, the Poisson ratio for the hard and brittle class of polymers, such as polystyrene, approaches 0.3. These polymers are characterized by a high modulus of elasticity, a poorly defined yield point, and little elongation before failure. However, soft and tough polymers, such as plasticized PVC, have a low modulus of elasticity, high elongation, a Poisson ratio of about 0.5–0.6, and a well-defined yield point. Soft and tough polymers stretch after the yield point with the area under the entire curve, toughness, or ultimate strength, greater than for the hard and brittle polymers.

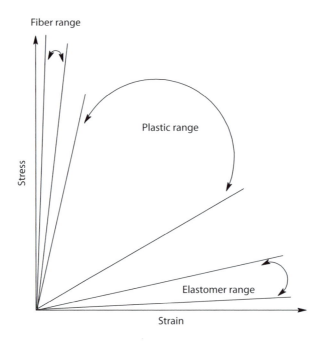

FIGURE 14.6   Typical stress–strain behavior for fibers, plastics, and elastomers.

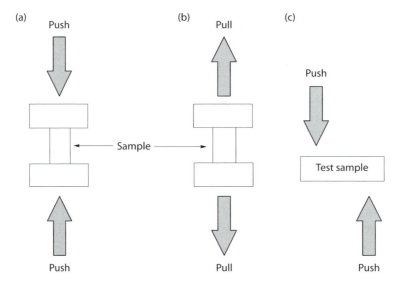

FIGURE 14.7   Major types of stress tests: compressive (a); pulling stress or tensile strength (b); and shear stress (c).

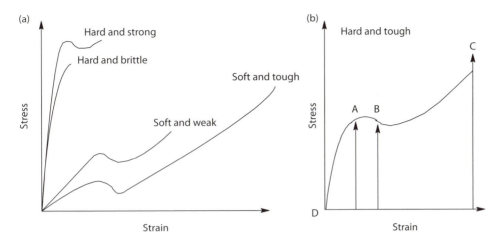

FIGURE 14.8 Typical stress–strain curves for plastics where (b) A is the elongation at yield point, B is the yield stress, C is the elongation at break, and the area under the curve is the ultimate strength.

Rigid PVC is representative of hard and soft polymers. These polymers have a high modulus of elasticity and high yield strength. The curve for hard and tough polymers, such as acrylonitril–butadiene–styrene (ABS) copolymers, shows moderate elongation before the yield point followed by nonrecoverable elongation.

In general, the behavior of all classes of polymer behavior is Hookean before the yield point. The reversible recoverable elongation before the yield point, called the *elastic range*, is primarily the result of bending and stretching of covalent bonds in the polymer backbone. This useful portion of the stress–strain curve may also include some recoverable uncoiling of polymer chains. Irreversible slippage of polymer chains is the predominant mechanism after the yield point.

Since these properties are time dependent, the soft and weak polymers may resemble the hard and strong polymers if the stress is rapidly applied, and vice versa. These properties are also temperature-dependent. Hence, the properties of soft and tough polymers may resemble hard and brittle when temperature is decreased. The effects of temperature and the mechanisms of elongation are summarized in Figure 14.4.

As noted in Figure 14.7, the major modes for applying stress are axial (compression or tension), flexural (bending or shear), and torsional (twisting; not shown). Superimposed on these can be any number of cyclic arrangements. Several common fluctuating stress-time modes are typically employed. In the regular and sinusoidal time-dependent mode, the stress is applied, as both compressional and torsional, in a regular manner with respect to both time and amount. In a second cyclic arrangement stress, again both compressional and torsional, is applied in an uneven manner with respect to amount and time so that the maxima and minima are asymmetrical relative to the zero stress level.

Associated with such cyclic application of stress is the term *fatigue* to describe the failure that occurs after repeated applications of stress. The fatigue values are almost always less than measurements, such as tensile strength, obtained under static load. Thus, it is important that both static and cyclic measurements be made.

Data is often plotted as stress versus the logarithm of the number of cycles, N, to failure for each sample. Generally, tests are taken using about two-thirds of the static parameter value. There is some limiting stress level below which fatigue does not occur (over some reasonably extended time). This value is called the *fatigue* or *endurance limit*. The *fatigue strength* is defined as the stress at which failure will occur for some specified number of cycles. *Fatigue life* is the number of cycles that are needed to cause failure.

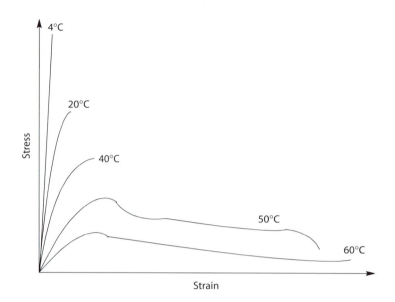

FIGURE 14.9 Influence of temperature on the stress–strain behavior of a sample of PMMA. (Modeled after Carswell, T. S., Nason, H. K. (1944): Effects of Environmental Conditions on the Mechanical Properties of Organic Plastics, Symposium on Plastics, American Society for Testing Materials, Philadelphia, PA; Copyright, ASTM, 1916 Race Street, Philadelphia, PA, 19103; Used by permission.)

Polymeric materials are dissimilar to metals in a number of respects. For instance, the range of strain–strain behavior is more complex and varied for polymers. Highly elastic polymers can exhibit a modulus as low as 7 MPa or as high as 100 GPa for stiff polymeric materials. The modulus values for metals are higher (50–400 GPa) but not as varied. Tensile strengths for polymers are on the order of 100 MPa for polymers and for some metal alloys 4,000 MPa. By comparison, polymers are more deformable with elastic polymers having elongations in excess of 500% while even "moldable" metals having elongations not exceeding 100%. Further, stress–strain behavior for polymers is more temperature dependant in comparison to metals. For instance, Figure 14.9 contains the stress–strain behavior for a poly(methyl methacrylate) (PMMA), material. Below the T_g the material behaves as a stiff, brittle solid similar to the behavior given for hard and brittle polymers (Figure 14.9). As the temperature increases, the PMMA gains enough thermal energy to allow for some segmental mobility. The T_g varies greatly for PMMA depending upon the tacticity with isotactic PMMA having a T_g of about 45°C and for syndiotactic the T_g is about 130°C. The material in Figure 14.9 is largely isotactic with a T_g of about 50°C–60°C. By 50°C, the material behaves as a soft and tough material in Figure 14.9.

14.3 STRESS–STRAIN RELATIONSHIPS

Mechanical testing involves a complex of measurements, including creep and shear strength, impact strengths, and so forth. Stress–strain testing is typically carried out using holders where one member is movable contained within a load frame. Studies typically vary with either the stress or strain fixed and the result response measured. In a variable stress experiment, a sample of given geometry is connected to the grips. Stress, load, is applied, generally by movement of the grip heads either toward one another (compression) or away from one another (elongation). This causes deformation, strain, of the sample. The deformation is recorded as the force per unit area necessary to achieve this deformation. The applied load versus deformation is recorded and translated into stress–strain curves such as those given in Figures 14.3, 14.8, and 14.10.

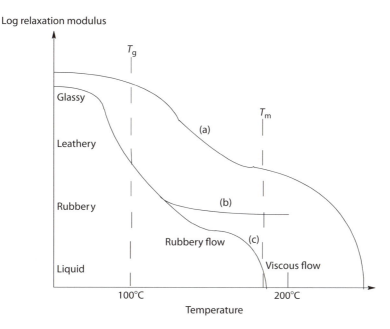

FIGURE 14.10 Logarithm of the relaxation modulus as a function of temperature for three polymer samples. Sample (a) is (largely) crystalline vinyl polymer; sample (b) is an amorphous vinyl polymer that contains light cross-linking; and sample (c) is an amorphous vinyl polymer. The T_g for the amorphous polymer is about 100°C and the T_m for the crystalline polymer is about 180°C.

For many materials, between their T_g and T_m, recovery is not complete. This incomplete recovery is called *creep*. Creep is time dependent and is described as follows:

$$\text{Compliance} = \frac{\text{Stress}}{\text{Strain}} \tag{14.17}$$

where the compliance and strain are time dependent.

Creep behavior is similar to viscous flow. The behavior shown in Equation 14.17 shows that compliance and strain are linearly related and inversely related to stress. This linear behavior is typical for most amorphous polymers for small strains over short periods of time. Further, the overall effect of a number of such imposed stresses are additive. Noncreep-related recovery occurs when the applied stress is relieved. Thus, amorphous polymers and polymers containing substantial amounts of amorphous regions act as both elastic solids and liquids above their T_g.

Even materials with low creep under the limited times tests are often performed may undergo microlevel creep that eventually results in reorientation of polymer chains. For many materials, this results in amorphous regions becoming more crystalline, resulting in a material that is less flexible and, while stronger, more brittle.

Phase changes, phase transitions, refer to changes of state. Most low molecular weight materials exist as solids, liquids, and gasses. The transitions that separate these states are melting (or freezing) and boiling (or condensing). By contrast, polymers do not vaporize "intact" to a gas, nor do they boil. The state of a polymer, that is, its physical response character, depends on the time allotted for interaction and the temperature as well as the molecular organization (crystalline, amorphous-mix). The term *relaxation* refers to the time required for response to a change in pressure, temperature, or other applied parameter. The term *dispersion* refers to the absorption or emission of energy at a transition. In practice, the terms relaxation and dispersion are often used interchangeably.

Stress–strain plots for many polymers, that are amorphous or contain a mixture of amorphous/crystalline regions, can be given for various conditions, resulting in the general behavior shown in Figure 14.4. As noted before, the viscoelastic behavior of polymers is both temperature and time dependant. One measure of the tendency to deform and rebound is to look at the stress relaxation of materials. Stress relaxation measurements are carried out by rapidly straining a material by tension to a predetermined but low strain level and measuring the stress needed to maintain this strain as a function of time. The stress needed to maintain the strain decreases with time due to a molecular relaxation processes with the various bond flexing and chain and segmental movements influencing this time. Figure 14.10 (similar to Figure 14.4) contains an idealized plot of the logarithm of the relaxation modulus as a function of temperature for a typical vinyl (largely) crystalline polymer, curve a; lightly cross-linked vinyl amorphous polymer, curve b; and an amorphous vinyl polymer, line c. The T_g for the polymer is about 100°C and the T_m is about 180°C for this illustration.

Four general behavior patterns emerge and correspond to the same regions given in Figure 14.5.

The first region is called the *glassy region*. The Young's modulus for glassy polymers is somewhat constant over a wide range of vinyl polymers having a value of about 3×10^9 Pa. Here the molecular motions are restricted to vibrational and short-range rotational movement, often the former sometimes referred to as bond flexing. Two diverse approaches are often used to describe molecular motion in this range. The cohesive energy density approach measures the energy theoretically required to move a detached segment into the vapor state per unit volume. Using polystyrene as an example, a value of about 3×10^9 Pa is calculated for the modulus. For largely hydrocarbon polymers, the cohesive energy density is typically similar to that of polystyrene and similar theoretical modulus values are found. The second approach is based on the carbon–carbon bonding forces as seen using vibrational frequencies. Again, values on the order of 3×10^9 Pa are found. Thus, in the glassy or solid region, below the T_g, only bond flexing and short-range rotational movement occurs, and the response time is fast for all three samples.

After the precipitous drop that occurs near the T_g, the modulus again is about constant, often on the order of 2×10^6 Pa, until the melting occurs. Over this rubbery range moderate segmental chain mobility is possible when stress is applied, but wholesale mobility does not occur. Thus, somewhat elastic behavior occurs where segmental mobility allows the material to be stretched often to beyond 100% with an approximate return to the prestretched structure after release of the stress. The rubbery region is particularly "interaction-time" sensitive. For instance, if applied stress is maintained, chain slippage will occur for noncross-linked materials, limiting return to the original molecular conformation after release of the stress. For the same polymer, the width of this plateau is somewhat molecular weight dependant such that the longer the polymer chains the broader the plateau.

As the temperature is increased there is sufficient energy available to melt the crystalline polymer, the T_m, and also for the amorphous polymer so that in both cases ready wholesale movement of polymer chains occurs. The entire polymer now behaves as a viscous liquid such as molasses. For the cross-linked material wholesale mobility is not possible so it remains in the rubbery region until the temperature is sufficient to degrade the material.

The crystalline portion of the material does not exhibit a T_g and only a T_m so the leathery and rubbery areas are muted since only bond flexing is largely possible until the T_m is reached. The crystalline polymer is not totally crystalline so it exhibits "amorphous" polymer behavior corresponding to the amount of amorphous regions present. The lightly cross-linked material behaves similarly to the amorphous material through the first three regions but cross-linking prevents ready wholesale movement of the polymer chains preventing it to flow freely. Finally, the amorphous material, curve c, shows all four behaviors. Here, above the T_g, segmental movement occurs and the time required for the chains to return to a relaxed state is considerably greater than that required for simple return of bond flexing. Finally, as the temperature approaches the T_m wholesale movement of the polymer chains becomes possible and the material behaves as a viscous liquid.

14.4 SPECIFIC PHYSICAL TESTS

14.4.1 TENSILE STRENGTH

Tensile strength can be determined by applying force to the material until it breaks as shown in Equation 14.18 (including typical units).

FIGURE 14.11 Typical "dumbbell" or "dog-bone" shaped sample used for stress–strain measurements.

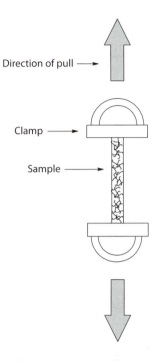

FIGURE 14.12 Typical tensile test assembly where both heads are movable.

$$\text{Tensile strength (Pascals)} = \frac{\text{Force required to break sample (N)}}{\text{cross-sectional area (m}^2\text{)}} \qquad (14.18)$$

Tensile strength, which is a measure of the ability of a polymer to withstand pulling stresses, is usually measured using a dumbbell shaped specimen (ASTM D-638–72; Figures 14.11 and 14.12). These test specimens are conditioned under standard conditions of humidity and temperature before testing.

Both heads may be movable as in Figure 14.12 or more often only one head is moving and the other held constant.

The *elastic modulus* (also called *tensile modulus* or *modulus of elasticity*) is the ratio of the applied stress to the strain it produces within the region where the relationship between stress and strain is linear. The *ultimate tensile strength* is equal to the force required to cause failure divided by the minimum cross-sectional area of the test sample.

When rubbery elasticity is required for sample performance, a high ultimate elongation is desirable. When rigidity is required, it is desirable that the material exhibit a lower ultimate elongation. Some elongation is desirable since it allows the material to absorb rapid impact and shock. The total area under the stress–strain curve is indicative of overall toughness.

Pulling stress is the deformation of a test sample caused by application of specific loads. It is specifically the change in length of the test sample divided by the original length. Recoverable strain or elongation is called *elastic strain*. Here, stressed molecules return to their original relative locations after release of the applied force. Elongation may also be a consequence of wholesale movement of chains past one another, that is, creep or plastic strain. Many samples undergo both reversible and irreversible strain.

Flexural strength, or cross-breaking strength, is a measure of the bending strength or stiffness or a test bar specimen used as a simple beam in accordance with ASTM D-790. The specimen is placed on supports (Figure 14.13). A load is applied to its center at a specified rate with the loading at failure called the *flexural strength*. However, because many materials do not break even after being greatly deformed, by agreement, the modulus at 5% strain is used as the flexural strength for these samples.

14.4.2 TENSILE STRENGTH OF INORGANIC AND METALLIC FIBERS AND WHISKERS

The tensile strength of materials is dependent on the treatment and form of the material. Thus, the tensile strength of isotropic bulk nylon-66 is less than that of anisotropic oriented nylon-66 fiber. Inorganics and metals also form fibers and whiskers with varying tensile strengths (Table 14.1). Fibers are generally less crystalline and larger than whiskers.

Many of these inorganic fibers and whiskers are polymeric, including many of the oxides (including the so-called ceramic fibers), carbon and graphite materials, and silicon carbide. Carbon and graphite materials are similar but differ in the starting materials and the percentage of carbon. Carbon fibers derived from polyacrylonitrile are about 95% carbon, while graphite fibers are formed at higher temperatures giving a material with 99% carbon.

These specialty fibers and whiskers exhibit some of the highest tensile strengths recorded (Tables 14.1 and 14.2) and they are employed in applications where light weight and high strength are

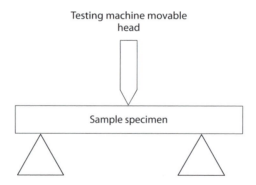

FIGURE 14.13   Typical flexural strength test.

TABLE 14.1
Tensile Strengths of Inorganic and Metallic Materials as a Function of Form

Material	Form	Tensile Strength (MPa)
Graphite	Bulk	1,000
	Fiber	2,800
	Whisker	15,000
Glass	Bulk	1,000
	Fiber	4,000
Steel	Bulk	2,000
	Fiber	400
	Whisker	10,000

TABLE 14. 2
Ultimate Tensile Strength of Representative Organic, Inorganic, and Metallic Fibers

Material	Tensile Strength (MPa)	Tensile Strength/Density
Aluminum silica	4,100	1,060
Aramid	280	200
Beryllium carbide	1,000	400
Beryllium oxide	500	170
Boron–tungsten boride	3,450	1,500
Carbon	2,800	1,800
Graphite	2,800	1,800
UHMW Polyethylene	380	3,800
Poly(ethylene terephthate)	690	500
Quartz	900	400
Steel	4,000	500
Titanium	1,900	400
Tungsten	4,300	220

required. Organics offer weight advantages, typically being less dense than most inorganic and metallic fibers. Uses include in dental fillings, the aircraft industry, production of lightweight fishing poles, automotive antennas, light weight strong bicycles, turbine blades, heat-resistant reentry vessels, golf club shafts, and so on. Many are also used as reinforcing agents in composites.

14.4.3 COMPRESSIVE STRENGTH

Compressive strength is defined as the pressure required to crush a material defined as

$$\text{Compressive strength (Pa)} = \frac{\text{Force (Newtons)}}{\text{cross-sectional area (m}^2)} \tag{14.19}$$

Compressive strength, or ability of a specimen to resist a crushing force, is measured by crushing a cylindrical specimen (ASTM-D-695) as shown in Figure 14.14. Here a sample of specified dimensions is placed between two heads, one movable and one set. Force is applied to the movable head moving the head at a constant rate. The ultimate compression strength is equal to the load that causes failure divided by the minimal cross-sectional area. Since many materials do not fail in compression, strength reflective of specific deformation is often reported.

14.4.4 IMPACT STRENGTH

Impact strength is a measure of the energy needed to break a sample—it is not a measure of the stress needed to break a sample. The term *toughness* is typically used in describing the impact strength of a material but does not have an universally accepted definition but is often described as the area under stress–strain curves.

There are a number of impact-related tests. Impact tests fall into two main categories: (1) falling-mass tests and (2) pendulum tests. Figure 14.15 shows an assembly suitable for determining impact strength of a solid sample by dropping a material of specified shape and weight at a given distance.

Impact resistance is related to impact strength. Two of the most utilized tests are the Izod (ASTM D-256) and Charpy (ASTM D-256) tests illustrated in Figure 14.16.

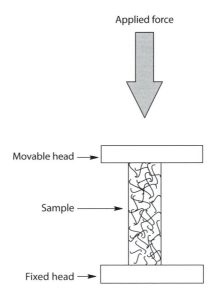

FIGURE 14.14   Representation of test apparatus for measurement of compression-related properties.

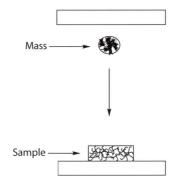

FIGURE 14.15   Assembly used to measure the impact strength of solid samples.

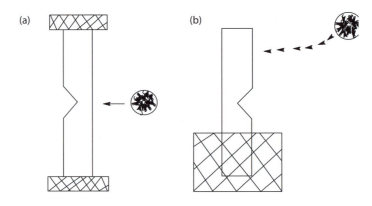

FIGURE 14.16   Pictorial description of Charpy (a) and Izod (b) pendulum impact tests.

The Izod impact test measures the energy required to break a notched sample under standard conditions. A sample is clamped in the base of a pendulum testing apparatus so that it is cantilevered upward with the notch facing the direction of impact. The pendulum is released from a specified height. The procedure is repeated until the sample is broken. The Izod value is useful when comparing samples of the same polymer but is not a reliable indicator of toughness, impact strength, or abrasive resistance. The Izod test may indicate the need to avoid sharp corners in some manufactured products. Thus, nylon is a notch-sensitive material and gives a relatively low Izod impact value, but it is considered a tough material.

Unnotched or oppositely notched specimens are employed in the Charpy test.

Overall, impact strength tests are measures of toughness or the ability of a sample to withstand sharp blows, such as being dropped.

14.4.5 Hardness

The term hardness is a relative term. *Hardness* is the resistance to local deformation that is often measured as the ease or difficulty for a material to be scratched, indented, marred, cut, drilled, or abraded. It involves a number of interrelated properties such as yield strength and elastic modulus. Because polymers present such a range of behavior, they are viscoelastic materials, the test conditions must be carefully described. For instance, elastomeric materials can be easily deformed, but this deformation may be elastic with the indentation disappearing once the force is removed. While many polymeric materials deform in a truly elastic manner returning to the initial state once the load is removed, the range of total elasticity is often small resulting in limited plastic or permanent deformation. Thus, care must be taken in measuring and in drawing conclusions from results of hardness measurements.

Static indention is the most widely employed measurement of hardness. Here, permanent deformation is measured. Hardness is often computed by dividing the peak contact load by the projected area of impression. The indentation stresses, while concentrated within the area of impact, are generally more widely distributed to surrounding areas. Because of the presence of a combination of elastic and plastic or permanent deformation, the amount of recovery is also measured. The combination of plastic (permanent deformation) and elastic (noncreep associated reversible changes) deformation are determined by noting the recovery of the material after impact. The indention is then a measure of a complex of structure dependent on chain length, orientation, amounts and distributions of amorphous and crystalline combinations, and so on.

Following is a description of some of the most used hardness tests.

14.4.5.1 Brinell Hardness

Here, a steel ball is pushed against the flat surface of the test specimen. The standard test uses a ball of specific size pushed against the test surface with a given force.

14.4.5.2 Rockwell Hardness

Rockwell hardness tests [ASTM-D785–65 (1970)], depicted in Figure 14.17, measures hardness in progressive numbers on different scales corresponding to the size of the ball indenter and force of the load. Thus, varying loads are applied. Depth AB corresponds to the depth that the indicator penetrates with a heavy load. The difference between A and B, that is B–A, is used to calculate the Rockwell hardness values. While Rockwell hardness allows differentiation between materials, factors such as creep and elastic recovery are involved in determining the overall Rockwell hardness. Rockwell hardness may not be a good measure of wear qualities or abrasion resistance. For instance, polystyrene has a relatively high Rockwell hardness value, yet it is easily scratched.

In the most used Rockwell tests, the hardness number does not measure total indentation but only the irreversible portion after a heavy load is applied for a given time and reduced to a minor load for a given time. In Table 14.3, the "*M*" value is for a major load of 980 N (100 kgf) and a minor load

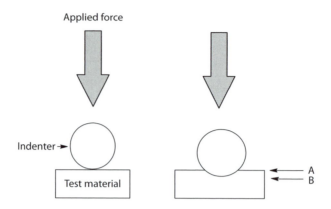

FIGURE14.17   Description of the Rockwell hardness test.

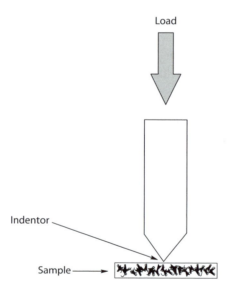

FIGURE 14.18   Barcol instrument employing a sharp indentor.

of 98 N (10 kgf) using an indenter with a 6.25 mm diameter. The values given in the R column are for a lower initial load of 588 N (60 kgf) and a minor load of 98 N employing a larger indenter with a diameter of 12.5 mm.

Hardness may also be measured by the number of bounces of a ball or the extent of rocking by a Sward Hardness Rocker. Abrasion resistance may be measured by the loss in weight caused by the rubbing of the wheels of a Tabor-abrader (ASTM-D-1044).

The shore durometer is a simple instrument used to measure the resistance of a material to the penetration of a blunt needle. In the Barcol approach, a sharp indentor is used to measure the ability of a sample to resist penetration by the indentor (Figure 14.18). The values given in Table 14.3 are for one specific set of conditions and needle area for the Barcol and Brinell hardness tests.

Table 14.3 contains comparative hardness values for five hardness scales including the classical Mohs scale, which ranges from the force necessary to indent talc given a value of 1 to that needed to scratch diamond given a Mohs value of 10. In the field, a number of relative tests have been developed to measure relative hardness. The easiest test for scratch hardness is to simply see how hard you have to push your fingernail into a material to indent it. A more reliable approach involves scratching the material with pencils of specified hardness (ASTM-D-3363) and noting the pencil hardness necessary to indent the material.

TABLE 14.3
Comparative Hardness Scales

		Hardness Scale				
		Rockwell				Type of
Mohs	Brindell	M	R	Shore	Barcol	Material
2 (Gypsum)	25	100			55	Hard
	16	80			40	Plastics
	12	70	100	90	30	
	10	65	97	86	20	
	8	60	93	80		
	6	54	88	74		
1 (Talc)	5	50	85	70		
	2	32	ca50	89		Soft Plastics
	1	23		42		
	0.8	20		38		Rubbery
	0.5	15		30		

14.4.6 SHEAR STRENGTH

The measure of shear strength typically utilizes a punch-type shear fixture (Figure 14.19). The shear strength is equal to the load divided by the area. Thus, the sample is mounted in a punch-type shear fixture and punch—pushed down at a specified rate until shear occurs. This test is important for sheets and films but is not typically employed for extruded or molded products.

14.4.7 ABRASION RESISTANCE

Abrasion is the wearing away of a materials surface by friction. The most widely used tests to measure abrasion resistance employs Williams, Lamborn, and Tabor abraders (ASTM D-1044). In each test the abrader is rubbed on the material's surface and the material loss noted.

14.4.8 FAILURE

The failure of materials can be associated with a number of parameters. Two major causes of failure are creep and fracture. The tensile strength is the nominal stress at the failure of a material. Toughness is related to ductility. For a material to be tough, it often takes a material having a good balance of stiffness and give.

Calculations have been made to determine the theoretical upper limits with respect to the strength of polymers. Real materials show behaviors near to those predicted by the theoretical calculations during the initial stress–strain determination, but vary greatly near the failure of the material. It is believed that the major reasons for the actual tensile strength at failure being smaller than calculated are related to imperfections, the nonhomogeneity of the polymeric structure. These flaws, molecular irregularities, act as the weak-link in the polymer's behavior. These irregularities can be dislocations, voids, physical crack, and energy concentrations. Even with these imperfections, they have high strength-to-mass ratios (Table 14.2) with the highest tensile strength/density value for UHMW polyethylene.

The fracture strengths of polymers are generally lower than those of metals and ceramics. The mode of failure for thermosets is generally referred to as the materials being brittle. Cracks, related to bond breakage, occur at points of excess stress. These create weak spots and may lead to fracture if the applied stress, appropriate to create bond breakage, continues.

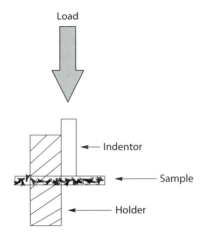

FIGURE 14.19   Assembly employed to measure shear strength.

Fracture of thermoplastics can occur by either a ductile or brittle fashion, or some combination. Rapid application of stress (short reaction time), lower temperatures, application of the stress by a pointed or sharpened object all increase the chances that failure will occur by a brittle mechanism as expected. For instance, below the T_g thermoplastics behave as solids, and above the T_g polymer fracture increasingly moves to a ductile mode. Above the T_g they display some plastic deformation before fracture. For some thermoplastics, crazing occurs before fracture. These crazes are a form of defect and are associated with some yielding or void formation either through migration or elimination of some part of the polymer mix or because of chain migration through chain realignment or other mechanism. Fibrillar connections, from aligned chains, often form between void sides. If sufficient stress is continued, these fibrillar bridges are broken and the microvoids enlarge eventually leading to crack formation. Crazes can begin from scratches, flaws, molecular inhomogeneities, and molecular-level voids formed from chain migration (such as through crystallization) or small molecule migration from a site as well as application of stress.

14.5   SUMMARY

1. Rheology is the study of deformation and flow of materials. Polymers are viscoelastic materials meaning they can act as liquids and as solids.
2. Many of the rheological properties of materials are determined using stress–strain measurements.
3. Models are used to describe the behavior of materials. The fluid or liquid part of the behavior is described in terms of a Newtonian dashpot or shock absorber while the elastic or solid part of the behavior is described in terms of a Hookean or ideal elastic spring. The Hookean spring represents bond flexing while the Newtonian dashpot represents chain and local segmental movement. In elastomers, the spring also represents chain elongation.
4. In general, the spring-like behavior is more important for polymers below their T_g, whereas the dashpot behavior is more important for polymers above their T_g.
5. Mathematical relationships have been developed to relate models composed to varying degrees of these two aspects of behavior to real materials.
6. Stress–strain behavior is closely dependent upon the particular conditions under which the experiment is carried out.
7. There are a number of specific tests that are used to measure various physical properties of materials. Many of these tests are described in detail by the various governing bodies that specify testing procedures and conditions.

GLOSSARY

Barcol impressor: Instrument used to measure the resistance of a polymer to penetration or indentation.

Biaxially stretching: Stretching in two directions perpendicular to each other.

Bingham, E. C.: Father of rheology.

Bingham plastic: Plastic that does not flow until the external stress exceeds a critical threshold value.

Coefficient of expansion: Change in dimensions per degree of temperature.

Compression strength: Resistance to crushing forces.

Creep: Permanent flow of polymer chains.

Dashpot: Model for Newtonian fluids consisting of a piston and a cylinder containing a viscous liquid.

Dilatant: Shear thickening agent; describes a system where the shear rate does not increase as rapidly as the applied stress.

Elastic range: Area under a stress–strain curve to the yield point.

Flexural strength: Resistance to bending.

Hooke's law: Stress is proportional to strain.

Impact strength: Measure of toughness.

Isotropic: Having similar properties in all directions.

Maxwell element or model: Model in which an ideal spring and dashpot are connected in series; used to study the stress relaxation of polymers.

Modulus: Stress per unit strain; measure of the stiffness of a polymer.

Newtonian fluid: Fluid whose viscosity is proportional to the applied viscosity gradient.

Newton's law: Stress is proportional to flow.

Poisson's ratio: Ratio of the percentage change in length of a sample under tension to its percentage change in width.

Pseudoplastic: Shear thinning agent: system where the shear rate increases faster than the applied stress.

Relaxation time: Time for stress of a polymer under constant strain to decrease to $1/e$ or 0.37 of its original value.

Retardation time: Time for the stress in a deformed polymer to decrease to 63% of the original value.

Rheology: Science of flow.

Rheopectic: Liquid whose viscosity increases with time.

Shear: Stress caused by planes sliding by each other.

Shear strength: Resistance to shearing forces.

Stress (s): Force per unit area.

Stress relaxation: Relaxation of a stressed sample with time after the load is removed.

Tensile strength: Resistance to pulling stresses.

Thixotropic: Liquid whose viscosity decreases with time.

Velocity gradient: Flow rate.

Viscoelastic: Having the properties of a liquid and a solid.

Voigt–Kelvin model or element: Model consisting of an ideal spring and dashpot in parallel in which the elastic response is retarded by viscous resistance of the fluid in the dashpot.

Williams, Landel, and Ferry equation (WLF): Used for predicting viscoelastic properties at temperatures above T_g when these properties are known for one specific temperature.

Yield point: Point on a stress–strain curve below which there is reversible recovery.

EXERCISES

1. What is the difference between morphology and rheology?
2. Which of the following is/are viscoelastic: (a) steel, (b) polystyrene, (c) diamond, or (d) neoprene?
3. Define G in Hooke's law.
4. Which would be isotropic: (a) a nylon filament, (b) an extruded pipe, or (c) ebonite?
5. What are the major types of stress tests?
6. Describe in molecular terms what a dashpot represents.
7. Which will have the higher relaxation time: (a) unvulcanized rubber or (b) ebonite?
8. Describe in molecular terms what a spring represents.
9. Define what compliance is.
10. What does viscoelastic mean?
11. Below the glass transition temperature a material should behave in what manner?
12. At what temperature would the properties of polystyrene resemble those of hevea rubber at 35 K above its T_g?
13. Define Young's modulus. Should it be high or low for a fiber?
14. Define the proportionality constant in Newton's law.
15. What term is used to describe the decrease of stress at constant length with time?
16. In which element or model for a viscoelastic body, (a) Maxwell or (b) Voigt–Kelvin, will the elastic response be retarded by viscous resistance?
17. What is the most important standards organization in the Unites States?
18. For most polymeric materials, is failure abrupt or continuous?
19. Is it safe to predict long-range properties from short time period results for plastics?
20. Describe the importance of relating models to stress–strain results.
21. If a sample of polypropylene measuring 5 cm elongates to 12 cm, what is the percentage of elongation?
22. If the tensile strength is 705 kg/cm^2 and the elongation is 0.026 cm, what is the tensile modulus?
23. Define creep.
24. Briefly describe a Voigt–Kelvin model.
25. What changes occur in a polymer under stress before the yield point?
26. What changes occur in a stress polymer after the yield point?
27. How can you estimate relative toughness of polymer samples from stress–strain curves?
28. What effect will a decrease in testing temperature have on tensile strength?
29. What effect will an increase in the time of testing have on tensile strength?
30. Describe a Maxwell model.

ADDITIONAL READING

Alfrey, T. (1948): *Mechanical Behavior of High Polymers*, Interscience, NY.

Boyd, R., Smith, G. (2007): *Relaxations in Polymers*, Cambridge University, NY.

Bicerano, J. (2002): *Prediction of Polymer Properties*, 3rd Ed., Dekker, NY.

Brinson, H., Brinson, L. (2007): *Polymer Engineering Science and Viscoelasticity: An Introduction*, Springer, NY.

Brostow, W., D'Souza, N., Menesses, V., Hess, M. (2000): *Polymer Characterization*, Wiley, NY.

Darinskii, A. A. (2006): *Molecular Mobility and Order in Polymer Systems*, Wiley, Hoboken.

Dealy, J. M., Larson, R. G. (2006): *Molecular Structure and Rheology of Molten Polymers*, Hanser-Gardner, Cincinnati, OH.

Greenhalgh, E., Hiley, M., Meeks, C. (2009): *Failure Analysis and Fractography of Polymer Compositions*, CRC, Boca Raton, FL.

Gupta, R. K. (2000): *Polymer and Composite Rheology*, 2nd Ed., Dekker, NY.

Han, C. (2007): *Polymer Rheology*, Oxford University, NY.

Hu, J. (2008): *Design Automation*, CRC, Boca Raton, FL.

Koppmans, R. (2009): *Polymer Melt Fracture*, CRC, Boca Raton, FL.

Mark, J. (2006): *Physical Properties of Polymers Handbook*, Springer, NY.

Mark, J., Erman, B., Bokobza, L. (2007): *Rubberlike Elasticity*, Cambridge University Press, Cambridge, UK.

Moore, D., Pavan, A., Williams, J. (2001): *Fracture Mechanics Testing Methods for Polymers, Adhesives, and Composites*, Elsevier, NY.

Morrison, F. A. (2001): *Understanding Rheology*, Oxford University Press, Oxford, UK.

Nielsen, L. E. (1974): *Mechanical Properties of Polymers*, Dekker, NY.

Riande, E. (1999): *Polymer Viscoelasticity*, Dekker, NY.

Rubinstein, M., Colby, R. H. (2003): *Polymer Physics*, Oxford University Press, Oxford, UK.

Scheirs, J. (2000): *Practical Polymer Analysis*, Wiley, NY.

Shaw, M., MacKnight W. (2008): *Introduction to Polymer Viscoelasticity*, Wiley, Hoboken, NJ.

Strobl, G. (2007): *The Physics of Polymers*, Springer, NY.

Urban, M. W. (2005): *Stimuli-Response Polymeric Films and Coatings*, Oxford University Press, NY.

Ward, I., Sweeney, J. (2005): *An Introduction to the Mechanical Properties of Solid Polymers*, Wiley, Hoboken.

Wineman, A., Rajagopal, K. (2000): *Mechanical Response of Polymers*, Cambridge University Press, Cambridge, UK.

Xu, Z. (2009): *Surface Engineering of Polymer Membranes*, Springer, NY.

Zaikov, G., Kozlowski, R. (2005): *Chemical and Physical Properties of Polymers*, Nova, Commack, NY.

15 Additives

Most polymeric materials are not wholly a single polymer, but they contain chemicals added to them to modify some physical, biological, and/or chemical behavior. These added chemicals, or additives, are generally added to modify properties, assist in processing, and introduce new properties to a material. Coloring agents, colorants, are added giving a product or component with a particular color often for identification or ecstatic purposes. Antibacterial agents are added to protect the material from certain microbial attack. Composites contain a fiber and continuous polymer phase, resulting in a material that has a greater flexibility and strength than either of the two components. Some of these additives are polymeric while others are not. They may be added as gases, liquids, or solids. Often some combination of additives is present. A typical tire tread recipe has a processing aid, activator, antioxidant, reinforcing filler, finishing aid, retarder, vulcanizing agent, and accelerator as additives (Table 15.1). A typical water-based paint has titanium dioxide as the white pigment, China clay as an extender, a fungicide, a defoaming agent, a coalescing liquid, a surfactant-dispersing agent, and calcium carbonate as another extender. In total, additives are essential materials that allow the polymeric portion(s) to perform as needed. Some of these additives are present in minute amounts and others are major amounts of the overall composition. Some additives are expensive and others are added to simply give bulk and thus are inexpensive. While the identity, amount, and action of many additives were developed on less than a scientific basis, today adequate scientific knowledge is known so that there exists a rational for the use of essentially all additives.

Typical additives include the following:

Antiblocking agents	Lubricants
Antifoaming agents	Mold release agents
Antifogging agents	Odorants or fragrances
Antimicrobial agents	Plasticizers
Antioxidants	Preservatives
Antistatic agents	Reinforcements
Blowing agents	Slip agents
Coloring agents	Stabilizers, including
Coupling agents	Radiation (UV, VIS)
Curing agents	Heat
Fillers	Viscosity modifiers
Flame retardants	Flow enhancers
Foaming agent	Thickening agents
Impact modifiers	Antisag agents
Low-profile materials	

In this chapter, we will look at a variety of additives.

15.1 PLASTICIZERS

Flexibilizing of polymers can be achieved through internal and external plasticization. Internal plasticization can be produced through copolymerization giving a more flexible polymer backbone or by grafting another polymer onto a given polymer backbone. Thus, poly(vinylchloride-covinyl

TABLE 15.1
Typical Contents of a Modern Tire Trea

Material	% (by Weight)	Purpose
Natural rubber	30	Elastomer
Styrene–butadiene rubber	30	Elastomer
Carbon black	27	Reinforcing filler
Aromatic oil	5	Extender
Stearic acid	2	Accelerator
Aryl diamine	2	Antioxidant
Zinc oxide	2	Accelerator
Sulfur	1	Vulcanizing agent
Antiozonate	0.5	Antioxidant
Parafin wax	0.5	Processing aid
N,N-Diphenyl guanidine	0.1	Delayed accelerator

acetate) is internally plasticized because of the increased flexibility brought about by the change in structure of the polymer chain. The presence of bulky groups on the polymer chain increases segmental motion and placement of such groups through grafting acts as an internal plasticizer. Internal plasticization achieves its end goal at least in part through discouraging association between polymer chains. However, grafted linear groups with more than 10 carbon atoms can reduce flexibility because of side-chain crystallization when the groups are regularly spaced.

External plasticization is achieved through incorporation of a plasticizing agent into a polymer through mixing and/or heating. The remainder of this section focuses on external plasticization.

Plasticizers should be relatively nonvolatile, nonmobile, inert, inexpensive, nontoxic, and compatible with the system to be plasticized. They can be divided on the basis of their solvating power and compatibility. Primary plasticizers are used as either the sole plasticizer or the major plasticizer with the effect of being compatible with some solvating nature. Secondary plasticizers are materials that are generally blended with a primary plasticizer to improve some performance such as flame resistance, mildew resistance, or to reduce cost. The division between primary and secondary plasticizers is at times arbitrary. Here we will deal with primary plasticizers.

According to the ASTM D-883 definition, a plasticizer is a material incorporated into a plastic to increase its workability and flexibility or dispensability. The addition of a plasticizer may lower the melt viscosity, elastic modulus, and T_g.

Waldo Semon patented the use of tricresyl phosphate as a plasticizer for poly(vinyl chloride) (PVC) in 1933. This was later replaced by the less toxic di-2-ethylhexyl phthalate (DOP), which is now the most widely used plasticizer. The worldwide production of plasticizer is on the order of 3.2 million tons annually. Volume wise, about 90% of the plasticizers are used with PVC and PVC-containing systems.

The effect of plasticizers has been explained by the lubricity, gel, and free volume theories. The lubricity theory states that the plasticizer acts as an internal lubricant and permits the polymers to slip past one another. The gel theory, which is applicable to amorphous polymers, assumes that a polymer, such as PVC, has many intermolecular attractions that are weakened by the presence of a plasticizer. In free volume theories, it is assumed that the addition of a plasticizer increases the free volume of a polymer and that the free volume is identical for polymers at T_g. There may be some truth in most of these theories. It is believed that a good plasticizer solubilizes segments allowing them some degree of mobility creating free volume through Brownian movement. In turn, this lowers the temperature where segmental mobility can occur making the material more flexible.

Most plasticizers are classified as to being general purpose, performance, or specialty plasticizers. General purpose plasticizers are those that offer good performance inexpensively. Most plasticizers are of this grouping. Performance plasticizers offer added performance over general purpose plasticizers generally with added cost. Performance plasticizers include fast solvating materials such as butyl benzyl phthalate and dihexyl phthalate; low temperature plasticizers such as di-*n*-undecyl phthalate and di-2-ethylhexyl adipate; and so-called permanent plasticizers such as tri-2-ethylhexyl trimellitate (TOTM), triisononyl trimellitate, and diisodecyl phthalate. Specialty plasticizers include materials that provide important properties such as reduced migration, improved stress–strain behavior, flame resistance, and increased stabilization.

The three main chemical groups of plasticizers are phthalate esters, trimellitate esters, and adipate esters. In all three cases, performance is varied through the introduction of different alcohols into the final plasticizer product. There is a balance between compatibility and migration. Generally, the larger the ester grouping the less the migration up to a point where compatibility becomes a problem and where compatibility now becomes the limiting factor.

The most used phthalate ester is DOP (15.3), made from the reaction of 2-ethyl hexanol (15.1) with phthalic anhydride (15.2).

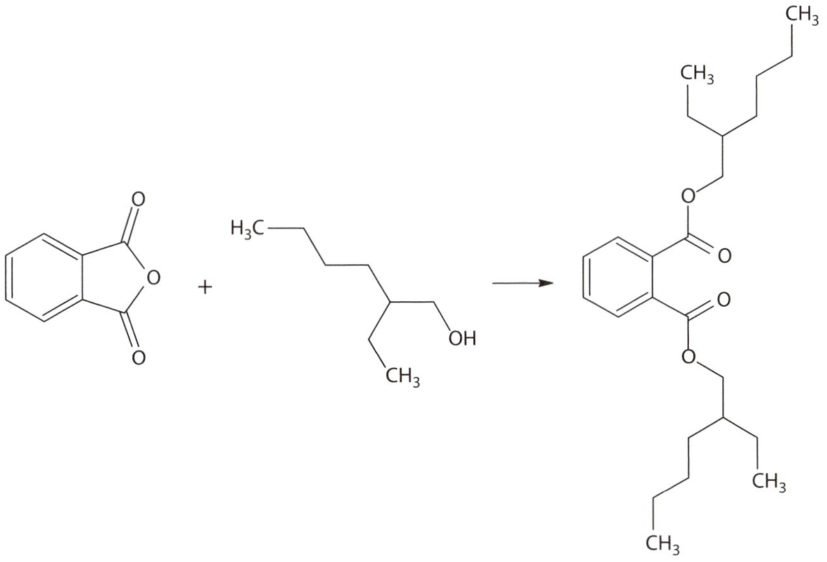

| Phthalic anhydride | 1-Ethyl hexanol | DOP |
| (15.1) | (15.2) | (15.3) |

The most important adipate ester is also the di-2-ethylhexyl ester, di-2-ethylhexyl adipate, DOA (15.4).

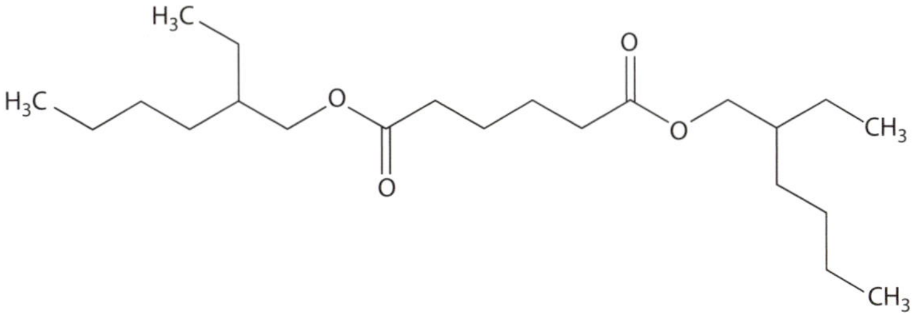

(15.4)

DOA

There are several widely used trimellitate esters, including TOTM (15.5).

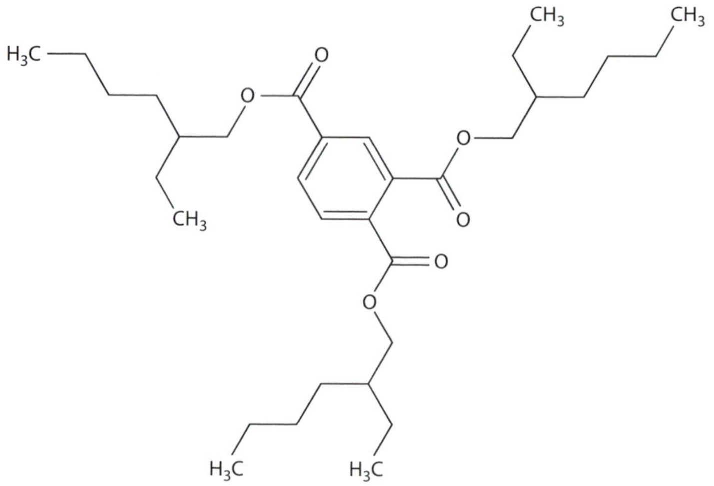

(15.5)

TOTM

Compatibility is the ability to mix together without forming different phases. A good approach to measure likelyhood of compatibility is solubility parameters (Section 3.2). Plasticizers with solubility parameters and type of bonding similar to those of the polymer are more apt to be compatible than when the solubility parameters are different. The solubility parameter for PVC of 9.66 is near that of good plasticizers for PVC such as DOP with a solubility parameter of 8.85. Table 15.2 contains solubility parameters for typical plasticizers.

The development of plasticizers has been plagued with toxicity problems. Thus, the use of highly toxic polychlorinated biphenyls (PCBs) has been discontinued. Phthalic acid esters, such as DOP, may be extracted from blood stored in plasticized PVC blood bags and tubing. These aromatic esters are also distilled from PVC upholstery in closed automobiles in hot weather. These problems have been solved by using oligomeric polyesters as nonmigrating plasticizers in place of DOP for appropriate situations. Recently, some limited tests have indicated a relationship of prenatal exposure to phthalates and reproductive abnormalities of male babies. The chemical industry is monitoring these results taking appropriate steps to limit or eliminate exposure to dangerous materials.

TABLE 15.2
Solubility Parameters of Typical Plasticizers

Plasticizer	Solubility Parameter (H)
Paraffinic oils	7.5
Dioctyl phthalate	7.9
Dibutoxyethyl phthalate	8.0
Tricresyl phosphate	8.4
Dioctyl sebacate	8.6
Triphenyl phosphate	8.6
Dihexyl phthalate	8.9
Hydrogenated terphenyl	9.0
Dibutyl sebacate	9.2
Dibutyl phthalate	9.3
Dipropyl phthalate	9.7
Diethyl phthalate	10.0
Glycerol	16.5

Experimental investigations and assessments show that the average person takes in about 2 grams a year of external plasticizers. Most of this is from traces of DOA migrating from food packaging. The so-called no observed effect level (NOEL) for DOA in rodents is about 40 mg/kg of body weight per day. Extrapolation for a person equates to 1,000 g of plasticizer for a "safety factor" of about 500. Even so, increased efforts to evaluate the safety of plasticizers continue.

Plasticizers extend the lower temperature range for use of materials since they discourage polymer chain associative behavior and encourage segmental flexibility increasing the rotational freedom effectively decreasing the material's typical T_g.

When present in small amounts, plasticizers can act as *antiplasticizers* increasing the hardness and decreasing the elongation of the material partly due to their ability to fill voids. Inefficient plasticizers require larger amounts of plasticizers to overcome the initial antiplasticization. However, good plasticizers such as DOP change from being antiplasticizers to plasticizers when less than 10% of the plasticizer is added to PVC.

External plasticizers are not permanent. Plasticizer molecules associate with one another eventually creating "preferred" migration routes to the material's surface where the plasticizer is rubbed or washed away. The preferential association of plasticizers also leaves some sites less flexible and creates variations in the material's stress–strain and expansion–contraction behaviors.

As noted elsewhere in this text, the health issues related to plasticizers is an issue particularly with respect to their use in materials that come in contact with our food. While the leachability of most plasticizers is small, it is real and ppm to ppt amounts of some plasticizers have been found to leach to the surfaces of food packaging materials. Thus, efforts are aimed at lowering to near zero the migration of plasticizers to surfaces as well as using safe plasticizers so that even when migration occurs, the health effects are near zero.

15.2 ANTIOXIDANTS

Antioxidants retard oxidative degradation. Heat, mechanical shear, and ultraviolet (UV) radiation can be responsible for the formation of free radicals, which in turn, can act to shorten polymer chains and increase cross-linking, both leading to deterioration in material properties. Free radical production often begins a chain reaction. Primary antioxidants donate active hydrogen atoms to free-radical sites thereby quenching or stopping the chain reaction. Secondary antioxidants or synergists act to decompose free radicals to more stable products.

Polymers such as polypropylene (PP) are not usable outdoors without appropriate stabilizers because of the presence of readily removable hydrogen atoms on the tertiary carbon atoms. PP and many other polymers (R–H) are attacked during processing or outdoor use in the absence of stabilizers because of chain degradation reactions as shown in the following:

By initiation:

$$R\text{–}H \quad \longrightarrow \quad R^{\bullet} + H^{\bullet}$$

Polymer $\longrightarrow$ Free radicals

$$(15.6)$$

By propagation:

$$R^{\bullet} \ + \ O_2 \quad \longrightarrow \quad ROO^{\bullet}$$

Free   Oxygen   Peroxy free
radical   radical

$$(15.7)$$

$$ROO^{\bullet} \ + \ R\text{–}H \quad \longrightarrow \quad ROOH \quad + \quad R^{\bullet}$$

Peroxy free Polymer   Dead polymer   Free
radical   radical

$$(15.8)$$

By termination:

$$R^\bullet + R^\bullet \rightarrow R\text{-}R$$
$$\text{Dead polymer}$$

(15.9)

$$R^\bullet + ROO^\bullet \rightarrow ROOR$$
$$\text{Dead polymer}$$

(15.10)

$$ROO^\bullet + ROO^\bullet \rightarrow ROOR + O_2$$
$$\text{Dead polymer} \quad \text{Oxygen}$$

(15.11)

These degradation reactions are retarded by the presence of small amounts of antioxidants, but they are accelerated by the presence of heavy metals such as cobalt(II) ions as follows:

$$ROOH + Co^{+2} \rightarrow Co^{+3} + RO^\bullet + OH^{-1}$$

(15.12)

$$ROOH + Co^{+3} \rightarrow Co^{+2} + H^{+1} + ROO^\bullet$$

(15.13)

Naturally occurring antioxidants are present in many plants and trees such as the hevea rubber trees. The first synthetic antioxidants were synthesized independently by Caldwell and by Winkelman and Gray by the condensation of aromatic amines with aliphatic aldehydes.

Many naturally occurring antioxidants are derivatives of phenol and hindered phenols, such as di-*tert*-butyl paracresol (Equation 15.14). It has the ability to act as a chain transfer agent forming a stable free radical that does not initiate chain radical degradation. However, the phenoxy free radical may react with other free radicals producing quinone derivatives.

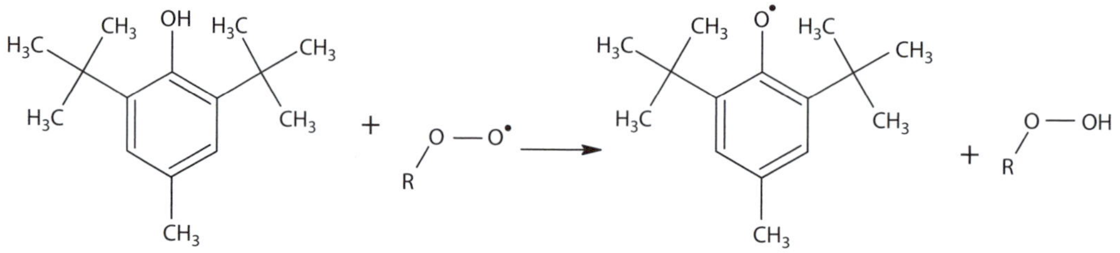

(15.14)

Since carbon black has many stable free radicals, it may be added to polymers such as polyolefins to retard free radical by attracting and absorbing other free radicals. It is customary to add small amounts of other antioxidants to enhance the stabilization by a "synergistic effect" whereby many antioxidant combinations are more stable than using only one antioxidant.

15.3 HEAT STABILIZERS

Heat stabilizers are added to materials to impart protection against heat-induced decomposition. Such stabilizers are needed to protect a material when it is subjected to a thermal-intense process (such as melt extrusion) or when the material is employed under conditions where increased heat stability is needed.

In addition to the free radical chain degradation described for polyolefins, other types of degradation occur, including dehydrohalogenation that occurs for PVC. When heated, PVC may lose hydrogen chloride forming chromophoric conjugated polyene structures eventually yielding fused aromatic ring systems.

This type of degradation is accelerated in the presence of iron salts, oxygen, and hydrogen chloride. Toxic lead and barium and cadmium salts act as scavengers for hydrogen chloride and may be used for some applications such as wire coating. Mixtures of magnesium and calcium stearates are less toxic and effective. Among the most widely used heat stabilizers for PVC are the organotin products. Many of these are based on organotin polymers discovered by Carraher.

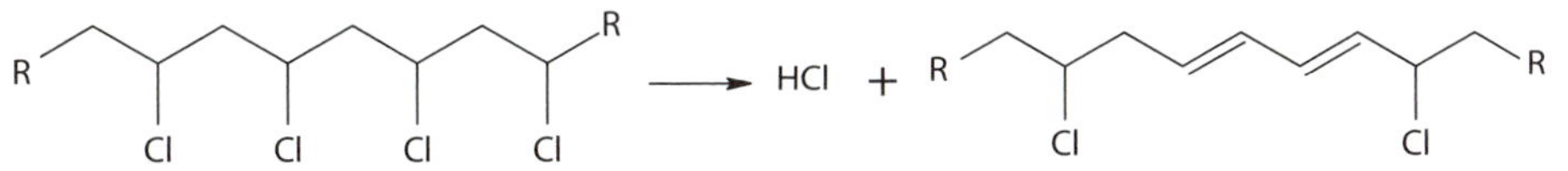

$$\text{(15.15)}$$

15.4 UV STABILIZERS

Ultraviolet stabilizers act to quench UV radiation. While much of the sun's high-energy radiation is absorbed by the atmosphere, some radiation in the 280–400 nm (UV) range reaches the earth's surface. Since the energy of this radiation is in the order of 70–100 kcal/mol (280–400 kJ/mol), UV radiation is sufficient to break many chemical bonds, and thus cause polymer chains to break. This breakage generally does one of two things: (1) it can increase the amount of cross-linking through subsequent formation of bonds after the initial bond breakage; or (2) bond breakage can result in a decreased chain length. Either of these results in a decreased overall strength of the polymeric material, and often the yellowing of many organic polymers. Cross-linking also results in the embrittlement of these polymers.

Poly(ethylene) PE, PVC, polystyrene (PS), poly(ethylene terephthalate) (PET), and PP are degraded at wavelengths of 300, 310, 319, 325, and 370 nm, respectively. The bond energy required to cleave the tertiary carbon hydrogen bond in PP is only 320 kJ/mol. This corresponds to a wavelength of 318 nm.

Since the effect of UV radiation on synthetic polymers is similar to its effect on the human skin, it is not surprising that UV stabilizers such as phenyl salicylate have been used for commercial polymers and in suntanning lotions for many years. Phenyl salicylate rearranges in the presence of UV radiation forming 2,2'-dihydroxybenzophenone. The latter and other 2-hydroxybenozophenones act as energy transfer agents, that is, they absorb energy forming chelates that release energy at longer wavelengths by the formation of quinone derivatives.

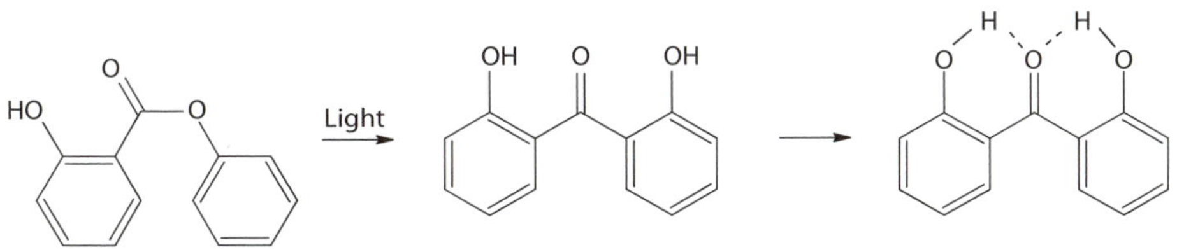

$$\text{(15.16)}$$

More than 100,000 tons of UV stabilizers are used by the U.S. polymer industry annually.

15.5 FLAME RETARDANTS

Flame retardants inhibit or resist the ignition and spread of fire. There are several general types of flame retardants that have been added to synthetic polymers. These include synthetic materials generally halocarbons and phosphorus-containing materials. Many of these present health problems. This includes most of the halocarbons such as PCBs.

Another class of flame retardants is minerals such as aluminum hydroxide, boron compounds, magnesium hydroxide, and antimony trioxide.

In general, flame retardants resist fire by several mechanisms. Some, such as magnesium hydroxide and aluminum hydroxide decompose when exposed to fire through endothermic means removing heat from the fire. Inert fillers such as calcium carbonate and talc dilute the portion of material that can burn. Thus, a home made of concrete block is less apt to burn completely than a home that is made of wood since concrete block reduces the amount of flammable material.

While some polymers such as PVC are not readily ignited, most organic polymers, like hydrocarbons, will burn. Some will support combustion, such as polyolefins, styrene–butadiene rubber (SBR), wood, and paper, when lit with a match or some other source of flame. The major products for much of this combustion are carbon dioxide (or carbon monoxide if insufficient oxygen is present) and water.

Since many polymers are used as shelter and clothing and in household furnishing, it is essential that they have good flame resistance. Combustion is a chain reaction that may be initiated and propagated by free radicals. Since halides and phosphorus radicals couple with free radicals produced in combustion terminating the reaction, many flame retardant are halide (halogen) or phosphorus-containing compounds. These may be additives; external retardants sprayed onto the material such as antimony oxide and organic bromides; or internal retardants such as tetrabromophthalic anhydride that are introduced during the polymerization process so that they are part of the polymer chain.

Fuel, oxygen, and high temperature are essential for the combustion process. Thus, polyfluorocarbons, phosphazenes, and some composites are flame resistant because they are not good fuels. Fillers such as alumina trihydrate (ATH) release water when heated and hence reduce the temperature of the combustion process. Compounds such sodium carbonate, which releases carbon dioxide when heated, shield the reactants from oxygen. Char, formed in some combustion processes also shield the reactants from a ready source of oxygen and retard the outward diffusion of volatile combustible products. Aromatic polymers, such as PS, tend to char and some phosphorus and boron compounds catalyze char formation aiding in controlling the combustion process.

Synergistic flame retardants such as a mixture of antimony trioxide and an organic bromo compound are more effective than single flame retardants. Thus, mixtures are often employed to protect materials and people.

Since combustion is subject to many variables, tests for flame retardancy may not correctly predict flame resistance under unusual conditions. Thus, a disclaimer stating that flame retardancy tests do not predict performance in an actual fire must accompany all flame-retardant products. Flame retardants, like many organic compounds, may be toxic or may produce toxic gases when burned. Hence, care must be exercised with using fabrics or other polymers treated with flame retardants.

Combustion is a chain reaction that can be initiated and propagated by free radicals such as the hydroxyl free radical. The hydroxyl radical may be produced by reaction of oxygen with macroalkyl radicals as shown below:

$$R-CH_2^{\bullet} \quad + \quad O_2 \quad \rightarrow \quad R-CHO \quad + \quad HO^{\bullet}$$

| Macroalkyl radical | Oxygen | Dead polymer | Hydroxyl radical | (15.17) |

$$HO^{\bullet} \quad + \quad R-CH_3 \quad \rightarrow \quad R-CH_2^{\bullet} \quad + \quad H_2O$$

| Hydroxyl | Polymer | Macroradical | Water | (15.18) |

Halide-containing compounds have often been used to suppress flame propagation through reduction of free radical concentration as follows:

$$HX + HO^\bullet \rightarrow H_2O + X^\bullet \tag{15.19}$$

Hydrogen   Hydroxyl   Water   Halogen
halide   radical   radical

$$X^\bullet + R{-}CH_2^\bullet \rightarrow R{-}CH_2X \tag{15.20}$$

Halogen   Macroradical   Dead polymer
radical

15.6 COLORANTS

Color is a subjective phenomenon whose esthetic value has been recognized for centuries. Since it is dependent on the light source, the object, and the observer, color is not subject to direct measurement, though instruments can measure a color for reproducibility. Colorants that provide color in polymers may be soluble dyes or comminuted pigments.

Some polymeric objects, such as rubber tires, are black because of the presence of high proportions of carbon black filler. Many other products, including some paints, are white because of the presence of titanium dioxide (titanium (IV) oxide), the most widely used inorganic pigment. More than 50,000 tons of colorants are used annually by the polymer industry.

Pigments are classified as organic or inorganic. Organic pigments are brighter, less dense, and smaller in particle size than the more widely used, more opaque, inorganic colorants. Iron oxides or ochers are available as yellow, red, black, brown, and tan.

Carbon black is the most widely used organic pigment, but phthalocyanine blues and greens are available in many shades and are also widely used.

15.7 CURING AGENTS

The use of curing agents began with the serendipitous discovery of vulcanization of hevea rubber with sulfur by Charles Goodyear in 1838. The conversion of an A- or B-stage phenolic novolac resin with hexamethalenetetramine in the early 1900s was another relatively early use of curing (cross-linking) agents. Organic accelerators, or catalysts, for the sulfur vulcanization of rubber were discovered by Oenslager in 1912. While these accelerators are not completely innocuous, they are less toxic than aniline, used before the discovery of accelerators. Other widely used accelerators are thiocarbanilide and 2-mercaptobenzothiazole (Captax).

Captax (15.21) is used to the extent of 1% with heval rubber and accounts for the major part of the more than 30,000 tons of accelerators used annually in the United States. Other widely used accelerators include 2-mercaptobenzothiazole sulfonamide (Santocure; 15.22), used for the vulcanization of SBR; dithiocarbamates; and thiuram disulfides. Thiuram disulfide (15.23) is a member of a group called ultraaccelerators that allow the curing of rubber at moderate temperatures and may be used in the absence of sulfur.

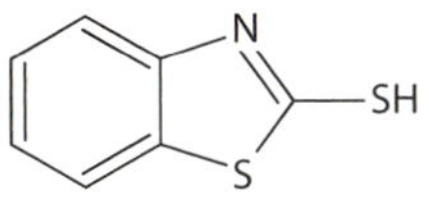

2-Mercaptobenzothiazole
(15.21)

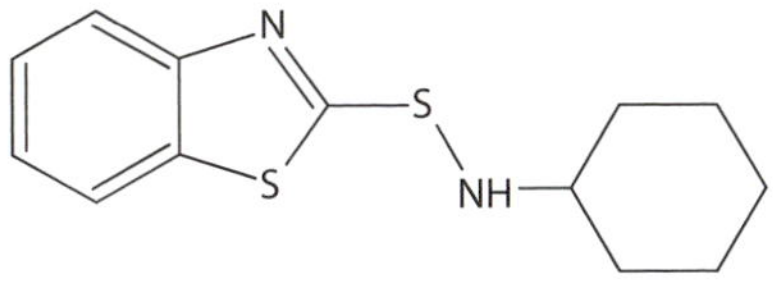

2-Mercaptobenzothiazole
sulfenamide (15.22)

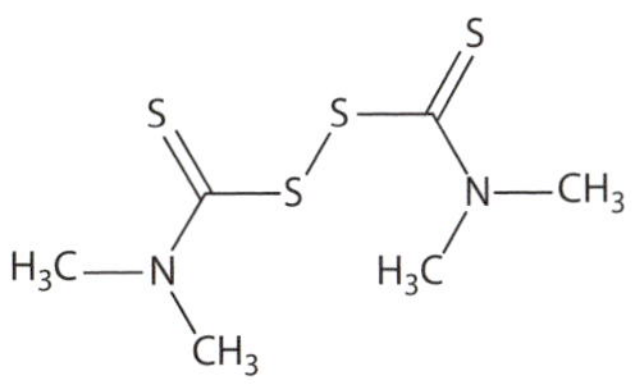

Tetramethyl thiuram
disulfide (15.23)

Initiators such as benzoyl peroxide are used not only for the initiation of chain-reaction poly-merization, but also for the curing of polyesters and ethylene–propylene copolymers, and for the grafting of styrene on elastomeric polymer chains.

Unsaturated polymers such as alkyd resins can be cured or "dried" in the presence of oxygen, a heavy metal, and an organic acid called a *drier*. The most common organic acids are linoleic, abi-etic, naphthenic, octoic, and tall oil fatty acids.

15.8 ANTISTATIC AGENTS—ANTISTATS

Antistatic agents (antistats) dissipate static electrical charges. Insulating materials, including most organic plastics, fibers, films, and elastomers, can build up electrical charge. Because these largely organic materials are insulators, they are not able to dissipate the charge. Such charge buildup is particularly noticeable in cold, dry climates and lead to dust attraction and sparking.

Antistatic agents can be either internal or external. External antistats are applied to the surface by wiping, spraying, and so on. These surface treatments act to prevent static charge buildup. Internal antistats are added during the processing and become an integral part of the bulk composition of the material. Because surface treatments are often worn away through washing, waxing, and handling, the external antistats must be replenished. Internal antistats are added to allow the antistats to come ("bleed") to the surface over a long time, giving the material long-term protection. Many antistatic agents are long-chain aliphatic amines and amides, esters of phosphoric acid, poly(ethylene gly-col) esters (PEGs), and quaternary ammonium salts such as docosyltrimethylammonium chloride (15.24) or cocamidopropyl betaine (15.25).

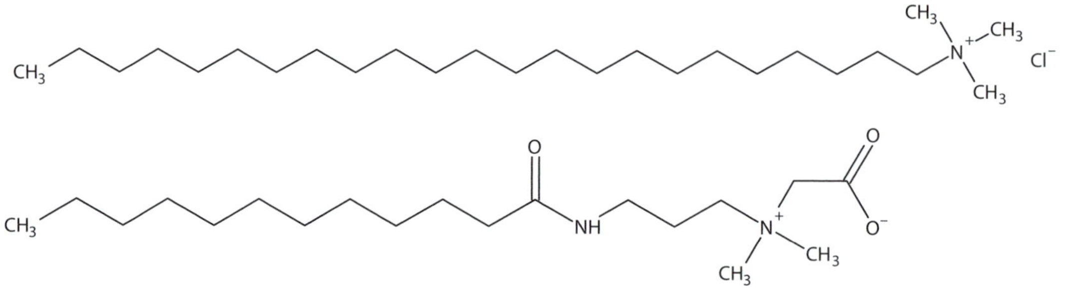

Docosyltrimethylammonium chloride (top) (15.24)

Cocamidopropyl betaine (bottom) (15.25)

Antistatic agents often have both a hydrophobic (above the hydrocarbon chain; 15.24, 15.25) and a hydrophilic (salts, quaternary amine, amide; 15.24, 15.25) portion. The hydrophobic portion is attracted to the hydrophobic part of most vinyl polymers while the hydrophilic portion interacts with moisture in the air.

15.9 CHEMICAL BLOWING AGENTS

Chemical blowing agents (CBAs) are employed to create lighter weight material through formation of foam. Physical CBAs are volatile liquids and gases that expand, volatilize during processing through control of the pressure and temperature.

Cellular polymers not only provide insulation and resiliency, but are usually stronger on a weight basis than solid polymers. Fluid polymers may be formed by the addition of low-boiling liquids such as pentane or fluorocarbons by blowing with compressed nitrogen gas, by mechanical heating, and by the addition of foaming agents. While some carbon dioxide is produced when polyurethanes are produced in the presence of moisture, auxiliary propellants are also added to the prepolymer mixture to give the desired amount of foaming. The most widely used foaming agents are nitrogen-producing compounds such as azobisformamide. Other foaming agents that decompose at various temperatures are available.

15.10 COMPATIBILIZERS

Compatibilizers are compounds that provide miscibility or compatibility to materials that are otherwise immiscible or only partially miscible yielding a homogeneous product that does not separate into its components. Typically, compatibilizers act to reduce the interfacial tension and are concentrated at phase boundaries. Reactive compatibilizers chemically react with the materials they are to make compatible. Nonreactive compatibilizers perform their task by physically making the various component materials compatible.

15.11 IMPACT MODIFIERS

Impact modifiers improve the resistance of materials to stress. Most impact modifiers are elastomers such as ABS, butadiene–styrene (BS), methacrylate–butadiene–styrene, acrylic, ethylene-vinyl acetate, and chlorinated polyethylene.

15.12 PROCESSING AIDS

Processing aids are added to improve the processing characteristics of a material. They may increase the rheological and mechanical properties of a melted material. Acrylate copolymers are often utilized as processing aids.

15.13 LUBRICANTS

Lubricants are added to improve the flow characteristics of a material during its processing. They operate by reducing the melt viscosity or by decreasing adhesion between the metallic surfaces of the processing equipment and the material being processed. Internal lubricants reduce molecular friction, consequently decreasing the material's melt viscosity and allowing it to flow more easily. External lubricants act by increasing the flow of the material by decreasing the friction of the melted material as it comes in contact with surrounding surfaces. In reality, lubricants such as waxes, amides, esters, acids, and metallic stearates act as both external and internal lubricants.

15.14 MICROORGANISM INHIBITORS

While most synthetic polymers are not directly attacked by microorganisms such as fungi, yeast, and bacteria, they often allow growth on their surfaces. Further, naturally occurring polymeric materials such as cellulosics, starch, protein, and vegetable oil-based coatings are often subject to microbiologic deterioration. Finally, some synthetics that contain linkages that can be "recognized" by enzymes contained with the microorganism (such as amide and ester linkages) may also be susceptible to attack.

One major antimicrobial grouping once was organotin-containing compounds. These monomeric organotin-containing compounds are now outlawed because of the high "leaching" rates of these material affecting surrounding areas. Even so, polymeric versions are acceptable and can be considered "non-leaching" or slowly leaching.

Organic fungistatic and bacteriostatic additives are currently employed, but in all cases, formation of resistant strains and the toxicity of the bioactive additive must be considered.

15.15 SUMMARY

1. Most polymers contain materials added to modify some chemical and/or chemical property that allows them to better fulfill their intended use. These added materials are called *additives*.
2. Plasticizers allow polymer chains to move past one another allowing wholesale flexibility. They can enhance flexibility above and below the glass transition temperature of a polymer. Plasticization can occur through addition of an external chemical agent or may be incorporated within the polymer itself.
3. Antioxidants retard oxidative degradation.
4. Heat stabilizers allow some protection to heat-induced decomposition.
5. UV stabilizers act to quench UV radiation. UV radiation is strong enough to break chemical bonds, and thus cause polymer chains to break either increasing the amount of cross-linking through subsequent formation of bonds or can result in a decreased chain length. Either of these result in a decreased overall strength of the polymeric material.
6. Flame retardants impart to the polymers some ability to resist ready combustion. Since fuel, oxygen, and high temperature are essential for the combustion of polymers, the removal of any of these prerequisites retards combustion. Flame retardants act through a variety of mechanisms, including char formation, combination with free radical species that promote further combustion, through release of water, and so on.
7. A variety of inorganic and organic compounds are used to color polymers.
8. The rate of cross-linking of polymers, such as natural rubber, is increased through the use of accelerators, often misnamed as catalysts.
9. Antistats reduce the electrostatic charge on the surface of polymers.
10. Gas-producing additives are essential for the formation of cellular products such as foam cushions.
11. Biocides are used to prevent or retard attack on polymers by microorganisms.
12. Lubricants server as processing aids that discourage the sticking of polymers to metal surfaces during processing.
13. Other important additives are coloring agents, curing agents, antistatic agents, chemical blowing agents, and microorganism inhibitors.
14. Each of these additives performs a critical role in allowing polymers to be processed and utilized giving the variety of useful products we have today.

GLOSSARY

Accelerator: Catalyst for the vulcanization of rubber.
Acicular: Needle shaped.
Antioxidant: Additive that retards polymer degradation by oxidative modes.
Antiplasticization: Hardening and stiffening effect observed when small amounts of a plasticizer are added to a polymer.
Antistat: Additive that reduces static charges on polymers.
Aspect radio: Ratio of length to diameter.
Biocide: Additive that retards attack by microorganisms.
Blocking: Sticking of sheets of film to one another.
Bulk molding compound (BMC): Resin-impregnated bundles of fibers.
Bound rubber: Rubber adsorbed on carbon black and that is insoluble in benzene.
Carbon black: Finely divided carbon made by the incomplete combustion of hydrocarbons.
Cellular polymers: Foams.
Chemical blowing agent: Volatile liquids and gasses that expand and/or volatilize during processing of a material creating pockets leading to lighter weight materials.
Comminuted: Finely divided.

Colorant: Color causing material; usually a dye or pigment.

Compatibilizers: Chemicals that provide miscibility or compatibility to materials that are otherwise immiscible or only partially miscible, giving a more homogeneous material.

Coupling agents: Products that improve the interfacial bond between the filler and resin.

Curing agent: Additive that causes cross-linking.

Drier: Catalyst that aids the reaction of polymers with oxygen.

Drying: Cross-linking of an unsaturated polymer chain.

Energy transfer agent: Molecule that absorbs high energy and reradiates it in the form of lower energy.

Extender: Term sometimes applied to an inexpensive filler.

Filler: Usually a relatively inert material used as the discontinuous additive; generally inexpensive.

Flame retardant: Additive that increases the flame resistance of a polymer.

Foaming agent: Gas producer.

Free volume: Volume not occupied by polymer chains.

Gel theory: Theory that assumes that in the presence of a pseudo-three-dimensional structure, intermolecular attractions are weakened by the presence of a plasticizer.

Heat stabilizers: Additives that retard the decomposition of polymers at elevated temperatures.

Impact modifiers: Materials that improve the resistance of materials to stress.

Low-profile resins: Finely divided incompatible resins that made a rough surface smooth.

Lubricants: Materials that improve the flow characteristics of materials, generally during processing.

Lubricity theory: Theory that plasticization occurs because of increased polymer chain slippage.

Microballoons: Hollow glass spheres.

Mold release agent: Lubricant that prevents polymers from sticking to mold cavities.

Novaculite: Finely ground quartzite rock.

Plasticizer: Material that increases the flexibility of a material.

Plastisol: Suspension of finely divided polymer in a liquid plasticizer with the plasticizer penetrating and plasticizing the polymer when heated.

Promoter: Coupling agent.

Side-chain crystallization: Stiffening effect noted when long, regularly spaced pendent groups are present on a polymer chain.

Synergistic effect: Enhanced effect beyond that expected by simply an additive effect.

Ultraaccelerator: Catalyst that cures rubber at low temperatures.

UV stabilizer: Additive that retards degradation caused by UV radiation.

Vulcanization: Cross-linking with heat and sulfur.

EXERCISES

1. Which would be stronger: (a) a chair made from Polypropylene or (b) one of equal weight made from cellular Polypropylene?
2. What advantage would a barium ferrite-filled PVC strip have over an iron magnet?
3. Providing that the volumes of the fibers are similar, which would give the stronger composite: fibers with (a) small or (b) large cross-sections?
4. How would you make abrasive foam from polyurethane?
5. Why is a good interfacial bond between the filler surface and the resin important?
6. Cellulose nitrate explodes when softened by heat. What would you add to permit processing at lower temperatures?
7. PVC was produced in the 1830s but no used until the 1930s. Why?

8. What was the source of "fog" on the inside of the windshield until the 1930s.
9. Can you propose a mechanism for antiplasticization?
10. What naturally occurring fiber is more resistant to microbiological attack than nylon?
11. The T_g decreases progressively as the size of the alkoxy group increases from methyl to decyl in polyalkyl methacrylates, but then increases. Explain.
12. PP is now used in indoor/outdoor carpets. However, the first PP products deteriorated rapidly when subjected to sunlight because of the presence of tertiary hydrogen atoms present. Explain.
13. Which is more resistant to attack by microorganisms: (a) PVC or (b) plasticized PVC?
14. Lead stearate is an effective thermal stabilizer for PVC, yet its use in PVC pipe is not permitted. Why?
15. When PVC sheeting fails when exposed for a long time to sunlight it goes through a series of color changes before becoming black. Explain.
16. What is the advantage of using epoxidized soybean oil as a stabilizer for PVC.
17. Why do PVC films deteriorate more rapidly when used outdoors in comparison to indoors?
18. Which of the following is more resistant to UV light: (a) PP, (b) PE, or (c) PVC?
19. Sometimes molded plastic have opaque material called "bloom" on the surface. Explain.
20. Why is the presence of static charges on polymer surfaces not desired?
21. Many ingredient combinations are called recipes. Why?
22. How were recipes developed?
23. Certain live animal testing has shown that certain additives may be harmful. If you headed a chemical company what would you do if one of the additives was suspected of being harmful?
24. Why is there an overlap between materials found in sun screens and those employed to help protect polymers from degradation from light?
25. Why are there so many additives that are used in polymers?

ADDITIONAL READING

Bart, J. (2005): *Additives in Polymers*, Wiley, Hoboken.
Bolgar, M., Groger, J., Hubball, J., Meronek, S. (2007): *Handbook for the Chemical Analysis of Plastics and Additives*, CRC, Boca Raton, FL.
Bukharov, S., Nuqumanova, G., Kochner, A. (2009): *Polyfunctional Stabilizers of Polymers*, Nova, Hauppauge, NY.
Craver, C., Carraher, C. E. (2000): *Applied Polymer Science*, Elsevier, NY.
Datta, S., Lohse, D. (1996): *Polymeric Compatibilizers*, Hanser Gardner, Cincinnati, OH.
Lutz, J. Grossman, R. (2000): *Polymer Modifiers and Additives*, Dekker, NY.
Marklund Sundkvist, A. (2008): *Organophosphorus Flame Retardants and Plasticizers*, VDM Verlag, NY.
Merhari, L. (2009): *Hybrid Nanocomposites for Nanotechnology: Electronic, Optical, Magnetic, and Biomedical Applications*, Springer, NY.
Mikitaev, A., Kozlov, G., Zaikov, G. (2009): *Polymer Nanocomposites: Variety of Structural Forms and Applications*, Nova, Hauppauge, NY.
Okdmsn, K., Sain, M. (2008): *Wood-Polymer Composites*, CRC, Boca Raton, FL.
Pethrick, R., Zaikov, G., Horak, D. (2007): *Polymers and Composites: Synthesis, Properties, and Applications*, Nova, Hauppauge, NY.
Powell, G. M., Brister, J. E. (1900): *Plastigels*, US Patent 2,427,507.
Wilkie, C., Morgan, A. (2009): *Fire Retardancy of Polymeric Material*, 2nd Ed., CRC, Boca Raton, FL.

16 Reactions on Polymers

Reactions occurring on polymers in nature and industry is widespread. In nature, reactions on polymers serve as the basis for information transfer, synthesis of needed biomaterials, degradation of biomaterials, and in fact reactions of polymers are at the core of life itself. Nature also reacts with synthetic polymers as they age, degrade, cross-link, and so on. Synthetic polymers also serve as the basic material for the production of many important fibers, elastomers, and plastics. The nature of these reactions is govern similarly whether the source and site of polymer is nature or made-made. There are some differences especially with respect to the precision of the interaction and predicted outcome. Some polymer interactions, such as those that transfer information, must occur with a very high degree of precision for each and every incident. By comparison, naturally induced degradation of natural and synthetic polymers through weathering occurs through general steps that can be described in some statistical manner.

Reactions on synthetic polymers often mimic similar reactions involving small molecules where size is the main difference. There are some exceptions such as where near groups may hinder or assist in the reaction where differences occur. Here, the main differences are often kinetic, though some geometrical differences are found for specific cases. Reactions where the rate of reaction in polymers is enhanced by the presence of neighboring groups are called *anchimeric assistance* reactions.

While there are many possible routes, most degradations can be described as occurring through two general routes—unzipping and random scission. Polymers such as polysiloxanes and polysulfur undergo unzipping reactions forming preferred internal cyclic products unless the end group is capped in such a manner as to discourage unzipping. In unzipping, one end begins to "unzip" and this process continues down the chain until it is completed. Random scission is the normal degradation pathway for most natural (such as polycarbonates [PC]) and synthetic polymers. Here, a long chain is attacked at some site, normally one that is exposed and of the exposed sites, one that is stressed. Thus, while random scission implies a random statistical manner of chain breakage, superimposed on this are more complex considerations of exposure and stress and the likelihood that a particular site is susceptible to that particular type of bond breakage.

This chapter describes many of the important reactions of polymers. Synthesis and curing (cross-linking) of polymers and telomerization are chemical reactions of polymers that have been discussed in previous chapters.

16.1 REACTIONS WITH POLYOLEFINES AND POLYENES

As with other carbon, hydrogen, and oxygen-containing materials, the main products of combustion in the presence of oxygen are water and carbon dioxide. Such polymers can be reacted with various reactants giving products analogous to those obtained from small alkanes and alkane-intense compounds. The moderating conditions between reactions of small molecules and polymers involve contact between the polymer segments and the reactants.

Polyolefines, like simple alkanes, can be chlorinated by chlorine giving hydrogen chloride and chlorinated products such as Tyrin, used as plasticizers and flame retardants, and poly(vinyl dichloride), that has better heat resistance than PVC and is used for hot water piping.

Reactions with polyenes are similar to the reactions of alkenes. Thus, Hermann Staudinger found that polyenes such as *Hevea brasiliensis* could be hydrogenated, halogenated, hydrohalogenated, and cyclized. This classic work was done in the early 1900s. In fact, Berthelot hydrogenated *H. brasiliensis* in 1869. Chlorinated rubber (Tornesit and Parlon) is produced by the chlorination of rubber giving products with varying amounts of chlorine and is used for the coating of concrete. Hydrohalogenation of *H. brasiliensis* gives a product (such as Pliofilm) that is a packaging film.

The product obtained by the partial hydrogenation of polybutadiene (Hydropol) has been used as a wire coating and a saturated ABA copolymer (Kraton) is produced by the hydrogenation of the ABA block copolymer of styrene and butadiene.

16.2 REACTIONS OF AROMATIC AND ALIPHATIC PENDANT GROUPS

Polymers with aromatic pendant groups, such as polystyrene (PS), undergo all of the characteristic reactions of benzene, including alkylation, halogenation, nitration, and sulfonation. PS has been sulfonated by fuming sulfuric acid with cross-linked products used as ion-exchange resins. Living polymers (cation propagation) are used to form a variety of products with hydroxyl and dihydroxyl products forming telechelic and macromers. Polyaminostyrene is formed from the nitration of PS. These polyaminostyrene products can be diazotized giving polymeric dyes.

Esters such as poly(vinyl acetate) (PVAc) (Equation 16.1), may be hydrolyzed producing alcohols such as poly(vinyl alcohol) (PVA), which will have the same degree of polymerization (DP) as the ester. In truth, PVAc does not totally hydrolyze, but with reasonable effort the extent of hydrolysis is greater than 90%. Since PVAc is not water soluble, but PVA is, the extent of water solubility is dependent on the extent of hydrolysis.

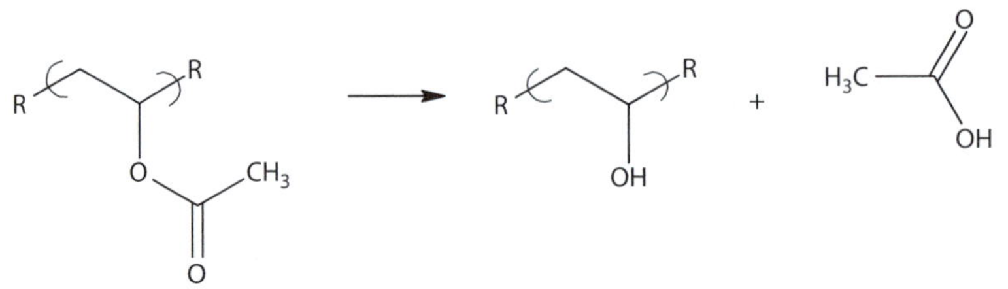

$$(16.1)$$

Esters of poly(carboxylic acids), nitrites, and amides may be hydrolyzed to produce poly(carboxylic acids). Thus, polyacrylonitrile, polyacrylamide, or poly(methyl acrylate) may be hydrolyzed producing poly(acrylic acid) (Equation 16.2).

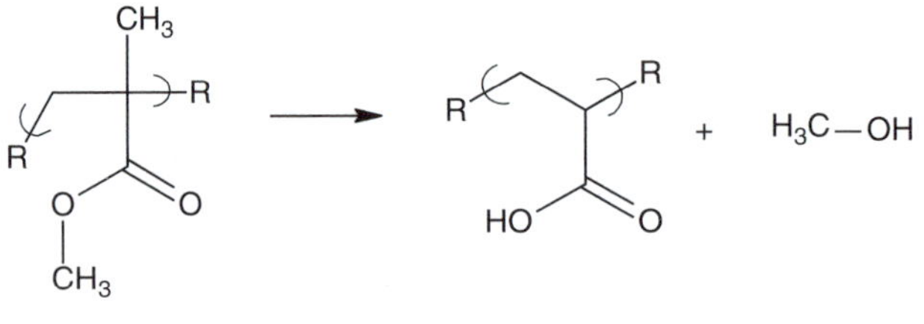

$$(16.2)$$

Poly(acrylic acid) and partially hydrolyzed polyacrylamide are used for the prevention of scale in water used for boilers and as flocculating agents in water purification.

Neutralization of ionic polymers, such as poly(acrylic acid), causes the now fully negatively charged carboxylate groups to repel one another, resulting in the chain changing conformation from that of a free-draining random shape to that approaching a rigid rod. This increase in length results in an increase in polymer viscosity with the extent of viscosity increasing related to the extend of ionic groups present. Addition of salts, such as sodium chloride, allow the negatively charged chains to return to a more random conformation and subsequent decrease in viscosity.

Some of the largest polymers, with molecular weights of the order of 10^{-15} million Da, contain sodium acrylate and acrylamide groups. Simple stirring can result in chain breakage for these extremely high molecular weight chains. Copolymers containing sodium acrylate and acrylamide groups are used in tertiary oil recovery and in water purification.

16.3   DEGRADATION

Here, the term degradation includes any change, decrease, in polymer property because of the impact of environmental factors, namely light, heat, mechanical, and chemicals. Seven polymers represent the majority of the synthetic polymers. These are the various polyethylenes (PE), polypropylene (PP), nylons, poly(ethylene terephthalate) (PET), PS, poly(vinyl chloride) (PVC), and PC. Each of these has their own particular mode of degradation. Even so, there are some common generalities for the condensation (PET, PC, nylons) polymers that contain a noncarbon in the backbone and vinyl (PE, PP, PS, PVC) polymers that contain only a carbon backbone.

Some reactions on polymers are intended and give a material with different desired properties. The (positive) modification of polymers is an area of vigorous activity. Other reactions on polymers are unintended and generally result in a material with unfavorable properties. Included in the latter are a whole host of polymer degradation reactions. Some of these degradation reactions are covered elsewhere. Here we will focus on some general concepts. Degradation can be promoted by many means and any combination of means. The major means of polymer degradation are given in Table 16.1.

Backbone chain scission degradation can be divided as occurring via depolymerization, random chain breakage, weak-link or preferential site degradation, or some combination of these general routes. In depolymerization, monomer is split off from an activated end group. This is the opposite of the addition polymerization and is often referred to as "unzipping."

$$\text{P–MMMMMMMM} \xrightarrow{-M} \text{P–MMMMMMM} \xrightarrow{-M} \text{P–MMMMMMM} \xrightarrow{-M} \text{P–MMMMM} \xrightarrow{-M} \text{etc.} \quad (16.3)$$

Chain scission is similar to the opposite of stepwise polycondensation where units are split apart in a random manner.

$$P_m - P_n \rightarrow P_m + P_n \quad (16.4)$$

Depolymerization can result in backbone degradation and/or in the formation of cyclic or other products. The thermal degradation of PVA and PVC occurs with the splitting-out of water or HCl

---

TABLE 16.1
Major Synthetic Polymer Degradative Agents

Degradation Agent	(Most) Susceptible Polymer Types	Examples
Acids and bases	Heterochain polymers	Polyesters, polyurethanes
Moisture	Heterochain polymers	Polyesters, nylons, polyurethanes
High-energy radiation	Aliphatic polymers with quaternary carbons	Polypropylene, LDPE, PMMA, poly(alpha-methylstyrene)
Ozone	Unsaturated polymers	Polybutadienes, polyisoprene
Organic liquids/vapors	Amorphous polymers	
Biodegradation	Heterochain polymers	Polyesters, nylons, polyurethanes
Heat	Vinyl polymers	PVC, poly(alpha-methylstyrene)
Mechanical (applied stresses)	Polymers below T_g	

followed by a combination of further chain degradation to give finally small products and formation of complex cyclic products. Elimination of HCl further accelerates additional HCl elimination and increased property loss. PVC degradation is decreased by addition of agents that impede degradation such as those that neutralize HCl, trap free radicals, and/or that react with the forming double bonds to impede further depolymerization. Commercial PVC often contains organotin or antimony mercaptide compounds that act as stabilizers.

In general, for vinyl polymers thermal degradation in air (combustion) produces the expected products of water, carbon dioxide, and char along with numerous hydrocarbon products. Thermally, simple combustion of polymeric materials gives a complex of compounds that varies according to the particular reaction conditions. Application of heat under controlled conditions can result in true depolymerization generally occurring via an unzipping. Such depolymerization may be related to the ceiling temperature of the particular polymer. Polymers such as poly(methyl methacrylate) PMMA and poly(alpha-methylstyrene) depolymerize to give large amounts of monomer when heated under the appropriate conditions. Thermal depolymerization generally results in some char and formation of smaller molecules, including water, methanol, and carbon dioxide.

Most polymers are susceptible to degradation under natural radiation, sunlight, and high temperatures even in the presence of antioxidants. Thus, low-density polyethylene (LDPE) sheets, impregnated with carbon black, become brittle after exposure to 1 year's elements in South Florida. High-density polyethylene (HDPE), while more costly, does stand up better to these elements, but again after several seasons, the elements win and the HDPE sheets become brittle and break. Long-term degradation is often indicated in clear polymers by a yellowing and a decrease in mechanical properties.

Most polymers are subject to oxidative degradation, particularly in the presence of other "enticers" such as heat, a good supply of air, various catalysts, high-energy radiation, including ultraviolet (UV) and higher energy visible light, and mechanical stressing that not only exposes additional polymer to the "elements" but also brings about the actual breakage of bonds subsequently leading to additional breakdown.

While polymers that contain sites of unsaturation, such as polyisoprene and the polybutadienes, are most susceptible to oxygen and ozone oxidation, most other polymers also show some susceptibility to such degradation, including natural rubber, PS, PP, nylons, PEs, and most natural and naturally derived polymers.

Because of the prevalence of degradation by oxidation, antioxidants are generally added. These antioxidants are generally compounds that readily react with free radicals or those that may act to lessen the effects of "enticers" such as UV radiation.

Mechanical degradation, while applied on a macrolevel, can result in not only chain rearrangement but also in chain degradation. Such forces may be repetitive or abrupt and may act on the polymer while it is in solution, melt, elastic, or below its T_g. Passage of polymer melts through a tiny orifice for fabrication purposes can result in both chain alignment and chain breakage. In the case of rubber, mastication of the elastomer, breaking polymer chains, is intentional allowing easier deformation and processability. While shearing itself can result in chain breakage, chain breakage is often associated with localized heat buildup that is a consequence of chains "rubbing" together, and so on (molecular friction).

Most heterochained polymers, including condensation polymers, are susceptible to aqueous-associated acid or base degradation. This mode of degradation is referred to as hydrolysis. This susceptibility is due to a combination of the chemical reactivity of heteroatom sites and to the materials being at least wetted by the aqueous solution allowing contact between the proton or hydroxide ion to occur. Both of these factors are related to the difference in the electronegatives of the two different atoms resulting in the formation of a dipole that acts as a site for nucleophilic/electrophilic chemical attack and that allows polar materials to come in contact with it. Such polymers can be partially protected by application of a thin film of hydrocarbon polymer that acts to repel the aqueous solutions.

Enzymatic degradation is complex and not totally agreed upon. Microbes have enzymes, some of which are capable of breaking selected bonds such as those that appear naturally—including amide, ester, and ether linkages—and including natural, naturally derived and synthetic materials. While a purpose of these enzymes is to digest nutrients for the host, when polymers with susceptible linkages come in contact with a microbe that contains appropriate enzymes, polymer degradation can occur. While often similar to acid- and base-associated degradations, enzymatic degradations are more specific bringing about only specific reactions. Even so, it is often difficult to differentiate between the two and both may occur together.

Some polymer deterioration reactions occur without loss in molecular weight. These include a wide variety of reactions where free radicals (most typical) or ions are formed and cross-linking or other nonchain session reaction occurs. Cross-linking discourages chain and segmental chain movement. At times this cross-link is desired such as in permanent press fabric and in elastomeric materials. Often the cross-links bring about an increased brittleness beyond that desired.

Some degradation reactions occur without an increase in cross-linking or a lessening in chain length. Thus, minute amounts of HCl, water, ester, and so on elimination can occur with vinyl polymers giving localized sites of double bond formation. Because such sites are less flexible and because such sites are more susceptible to further degradation, these reactions are generally considered as unwanted.

Cross-linking reactions can give products with desired increased strength, memory retention, and so on but accompanying such cross-inking can be unwanted increases in brittleness. Throughout the text, cross-linking is an important reaction that allows the introduction of desirable properties. For instance, cross-linking is the basis for many of the elastomeric materials.

Stabilizers are often added to prevent some forms of degradation. Hindered-amine light stabilizers are added to scavenge free radicals that are light produced in many polymers. UV stabilizers are added to absorb UV radiation converting some of it into heat. Antioxidants are added to stop free radical reactions caused by UV radiation.

16.4 CROSS-LINKING

Cross-linking reactions are common for both natural and synthetic and vinyl and condensation polymers. These cross-links can act to lock in "memory" preventing free-chain movement. Cross-linking can be chemical or physical. Physical cross-linking occurs in two major modes. First, chain entanglement acts to cause the tangled chains to act as a whole. Second, crystalline formations, large scale or small scale, act to lock in particular structures. This crystalline formation typically increases the strength of a material as well as acting to reduce wholesale chain movement.

Chemical cross-linking often occurs through use of double bonds that are exploited to be the sites of cross-linking. Cross-linking can be effected either through use of such preferential sites as double bonds, or through the use of other especially susceptible sites such as tertiary hydrogens. It can occur without the addition of an external chemical agent or, as in the case of vulcanization, an external agent, a cross-linking agent as sulfur, is added. Cross-linking can be effected through application of heat, mechanically, though exposure to ionizing radiation and nonionizing (such as microwave) radiation, through exposure to active chemical agents, or though any combination of these.

Cross-linking can be positive or negative depending on the extent and the intended result. Chemical cross-linking generally renders the material insoluble. It often increases the strength of the cross-linked material but decreases its flexibility and increases its brittleness. Most chemical cross-linking is not easily reversible. The progress of formation of a network polymer has been described in a variety of ways. As the extent of cross-linking increases, there is a steady increase in the viscosity of the melt. At some point, there is a rapid increase in viscosity and the mixture becomes elastic and begins to feel like a rubber. At this point, the mixture is said to be "gelled." Beyond this point the polymer is insoluble. The extent of cross-linking can continue beyond this gel point.

Flory and others derived expressions describing the extent of reaction at the gel point using statistical methods. In general, the $\overline{M}_w$ at an extent reaction p can be described by the expression

$$\overline{M}_{w,p} = \frac{\overline{M}_{w,o}\,(1+p)}{(1-p)\,(f-1)} \tag{16.5}$$

where $\overline{M}_{w,o}$ is the weight-average molecular weight at no reaction, and $\overline{M}_{w,p}$ is the weight-average molecular weight at an extent of reaction p and f is the functionality. At the gel point, molecular weight becomes quite large approaching infinity, giving eventually

$$p_{gel} = \frac{1}{f-1} \tag{16.6}$$

For a system containing two types of functional groups

$$(p_{gel})^2 = \frac{1}{r\,(f'-1)\,(f''-1)} \tag{16.7}$$

where r is the ratio of the two types of functionality f' and f'.

This equation is valid when the ratio of the two types of reactant groups is about equal and where the reactivity of members within each of the two types is about the same, and finally, when the reactivity of the two types of sites does not change during the process.

For the case where cross-linking occurs with the cross-linking sites contained in already formed polymer chains, average functionalities can be calculated based in terms of the weigh-average or number-average molecular weight. For the weight-average molecular weight, the average degree of polymerization, $\overline{DP}_w$, can be described, where $\overline{M}_c$ is the weight-average molecular weight between cross-linkable sites and p is the fraction of repeat units reacted, that is, the degree of conversion.

$$\overline{DP}_w = \frac{\overline{M}_w}{\overline{M}_c} = \frac{f(1+p)}{(1-p)\,(1-f)} \tag{16.8}$$

If $\overline{DP}_{w,o}$ is the weight-average degree of polymerization of the initial mixture of long chains then

$$\overline{DP}_w = \frac{\overline{DP}_{w,o}\,(1+p)}{(1-p)\,(\overline{DP}_{w,o}-1)} \tag{16.9}$$

and at the gel point the degree of polymer approaches infinity giving

$$p_{gel} = \frac{1}{\overline{DP}_{w,o}-1} \tag{16.10}$$

Equating the two equations for p_{gel} gives

$$\overline{DP}_{w,o}-1 = f-1 \tag{16.11}$$

Similar expressions can be derived based on the number-average molecular weight so that at the gel point

$$P_{gel} = \frac{1}{\overline{DP}_{n,o}}$$

(16.12)

The degree of cross-linking can be expressed in terms of cross-links per gram or per unit volume. If C is the moles of cross-links per unit volume, n is the number of network chains per unit volume, d is the density of cross-linked polymer, and $\overline{M}_c$ is the number-average molecular weight of the polymer segments between cross-links, then

$$C = \frac{n}{2} = \frac{d}{2\overline{M}_c}$$

(16.13)

such that the number of moles of cross-links per gram of network polymer is $1/2\,\overline{M}_c$.

Swelling and mechanical measurements are generally employed to experimentally determine the degree of cross-linking.

16.5 REACTIVITIES OF END GROUPS

Polymers containing one reactive end group are often referred to as macromolecular monomers or *macromers*. They can be represented as

M–M–M–M–M–M–M–M–M–M–M–M–M–M–M–M–M–M–M–M–Reactive group   (16.14)

They are used in the formation of a number of products, including star polymers and some graft polymers.

Polymers that contain two active end groups are referred to as *telechelic* polymers. These are then of the form

Reactive group–M–M–M–M–M–M–M–M–M–M–M–M–M–M–M–Reactive group   (16.15)

Telechelic polymers are also used in the synthesis of many products, including segmented polyurethane and polyester products.

Macromers are then "short" polymers that contain an active end group. This end group can be a site of unsaturation, heterocycle, or other group that can further react. Macromers are usually designed as intermediates in the complete synthesis of a polymeric material. These macromers can be introduced as side chains (grafts) or they may serve as the backbones (comonomer) of polymers. The macromers can also act as separate phases.

Macromers have been used to produce thermoplastic elastomers. Generally, the backbone serves as the elastomeric phase while the branches serve as the hard phases. These structures are often referred to as "comb"-shaped because of the similarity between the rigid part of the comb and its teeth and the structure of these graft polymers.

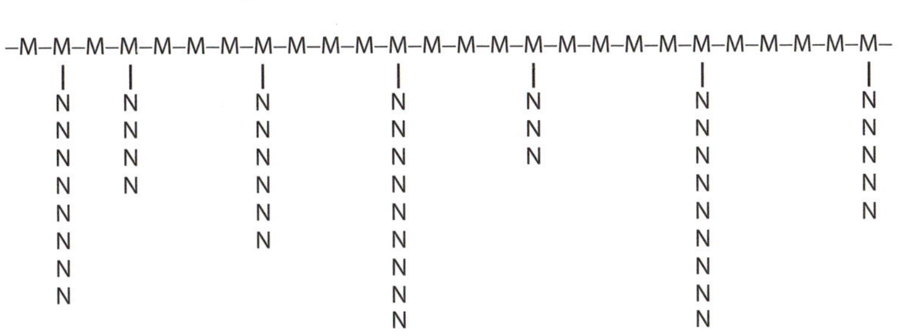

```
–M–M–M–M–M–M–M–M–M–M–M–M–M–M–M–M–M–M–M–M–M–M–M–M–M–M–
   |     |           |           |           |                 |           |
   N     N           N           N           N                 N           N
   N     N           N           N           N                 N           N
   N     N           N           N           N                 N           N
   N     N           N           N                             N           N
   N                 N           N                             N           N
   N                 N           N                             N
   N                             N                             N
   N                             N                             N
                                 N                             N
```

(16.16)

Block copolymers are also produced using such macromers. Again, the various units of the polymer can be designed to act as needed with hard/soft or hydrophilic/hydrophobic or other combinations. Also, these macromers generally have two active functional groups that allow polymerization to occur.

Telechelic polymers can also be used as cross-linkers between already formed polymers because both ends are active. These functional groups could be either two vinyl ends or two Lewis acids or bases, such as two hydroxyl or amine groups. Interpenetrating polymer networks (IPNs) and related structures such as dendrites and stars can also be formed using macromers.

The use of placing specific end groups onto polymers and oligomers is widespread and will be illustrated here for dimethylsiloxanes, the most widely used of the siloxanes. These end groups are given various names depending on their intended use. Some names for these end groups are capping agents, blocking agents, chain-length modifiers, and coupling agents. Typical siloxane end groups include trimethylsiloxyls, acyloxyls, amines, oximes, alcohols, and alkoxyls. These reactive end groups are reacted forming a wide variety of useful materials. Most silicone room temperature vulcanizing (RTV) adhesives, sealants, and caulks, are moisture curing, that is, they contain a hydroxyl-capped siloxyls that are reacted with acyloxyls, amines, oximes, or alkoxyls moisture-sensitive compounds.

16.6 SUPRAMOLECULES AND SELF-ASSEMBLY

The terms self-assembly, self-organization, and self-synthesis are closely related and sometimes used interchangeably. *Self-assembly* involves the aggregation of molecules, including macromolecules, into thermodynamically stable structures that are held together often using secondary bonding, including hydrogen bonding, Van der Waal's and electrostatic forces, pi–pi interactions and hydrophobic and hydrophilic interactions. The term *self-organization* is used for situations where the secondary bonding interactions are more specific and directional, giving a higher degree of order to the self-assembled molecules. Finally, the term *self-synthesis* includes self-assembly and self-organization, but also includes situations where self-replication and template-type synthesis occurs.

Self-assembly is the spontaneous organization of molecules into stable, well-defined structures with the driving forces being noncovalent associations. The final structure is normally near or at the thermodynamic equilibrium arrangement allowing it to readily spontaneously form. Such formations can be done under conditions where defects are either minimized or eliminated. In nature, self-assembly is common such as the folding of proteins, formation of the DNA double helix, and so on.

Self-assembled monolayers (SAMs) are the most widely studied nonnatural self-assembly systems. They are generally spontaneously formed from chemisorption and self-organization of organic molecules onto appropriate surfaces.

Natural polymers utilize a combination of primary and secondary forces and bond angles and distances to form polymers with both long-range (multimacromolecular) and short-range structures with both structures essential for the "proper" functioning of the macromolecular structure. While most synthetic polymer chemists have focused on what is referred to as primary and secondary structures (short-order structure control), work is just beginning on developing the appropriate structure control to allow tertiary and quaternary structural control (long-range control). While the "backbones" of these structures are held together with primary bonds, the secondary, tertiary, and quaternary structures are generally "driven" by secondary forces with the resulting tertiary and quaternary structures fixed in place through a combination of these secondary forces and small amounts of ionic and covalent cross-linking.

As noted above, self-assembled structures normally are the result of what are often referred to as weaker force interactions that include Van der Waal's forces, pi–pi interactions, capillary interactions, and so on. Structures that successfully form ordered structures often take advantage of these short-range attractive forces but also of longer-range repulsive forces. Almost all of these structures have some thermodynamic stability with the overall structure more stable than the unassembled parts. These somewhat overall weak-energy driving forces make the structures especially susceptible to small variations in the conditions leading to ready reversibility.

Several strategies are being developed that allow this long-range control, including the use of secondary forces to "hold" in place monomers that subsequently will be polymerized "in place." In another approach, already molecular "architectured " templates are employed to hold the polymer, prepolymer, or monomers in the desired shape with subsequent reactions and interactions enacted producing a material with a somewhat "robust" tertiary and quaternary structure. Some of these "molecular molds" are being produced using nanotechnology.

Self-assembly tendencies are apparent in the simple crystallization of inorganics and organics. Structure, size, and chemical tendencies (such as "like-liking-like" and "unlikes" repelling, secondary and primary bonding tendencies) are all involved. Proteins "self-assemble" in a much more diverse manner than so-called "simple" crystallization of common organics and inorganics. As noted above, we are just beginning to understand the nuances involved with the self-assembling, formation of giant macromolecules, including organizations such as those present in the cells of our bodies. We are beginning to understand the major factors involved in making the cell membranes and are starting to mimic these features to form synthetic biological-like membranes. We are using self-assembling concepts and approaches to develop a large number of interesting and potentially useful macromolecular materials.

One of the applications of molecular self-assembling is the formation of ultrathin films using both synthetic and natural surfaces as two-dimensional templates. As noted above, the same chemical and physical factors that we recognize in other areas are at work here. We will begin considering the formation of a simple bilayer membrane such as that present in natural cell membranes. Using the concept of like-liking-like and unlikes rejecting one another, the orientation of molecules with two different polar environments will vary depending upon the particular environment in which they are placed. For a common soap molecule with hydrophilic and hydrophobic ends, the like ends will congregate together and will reside either internally together or externally together. This is exactly the same concept as given in most general chemistry texts when considering the formation of micelles in commercial detergents. In the presence of water the hydrophilic ends will face outward, and in a nonpolar organic solvent the polar ends will face inward. Researchers have extended these simple concepts to include specially designed molecules that contain not only the heads and tails, but also spacers, conductors, and to vary the flexibility of the various parts of the molecule, spacing and number of heads and tails, and so on.

An important concept in the creation of some of these structures is that a primary driving force is the solute–solvent immiscibilities (energy; enthalphic). Thus, the magnitude of the cohesive energy of the solute may be a secondary factor in determining these supermolecular or supramolecular structures for such systems.

The self-assembling character of bilayer membranes is demonstrated by the formation of free-standing cast films from aqueous dispersions of synthetic bilayer membranes. The tendencies for association are sufficiently strong as to allow the addition of "guest" molecules (nanoparticles, proteins, and various small molecules) to these films where the connective forces are secondary in nature and not primary. Synthetic polymer chemists have made use of these self-assembling tendencies to synthesize monolayer films. Essentially, a monomer that contains both reactive groups and hydrophobic and hydrophilic areas is "cast" onto an appropriate template that "self-assemblies" the monomer, holding it for subsequent polymerization. Thus, a bilayer structure is formed by

$$
\begin{array}{c}
\text{O} \qquad\quad \text{O} \qquad\quad \text{Me} \\
\| \qquad\quad \| \qquad\quad | \\
H_3C\text{-}(\text{-}CH_2\text{-})_{15}\text{-}O\text{-}C\text{-}CH\text{-}NH\text{-}C\text{-}(\text{-}CH_2\text{-})_{10}\text{-}N\text{-}Me, Br \\
| \qquad\qquad\qquad\qquad\qquad | \\
H_3C\text{-}(\text{-}CH_2\text{-})_{15}\text{-}O\text{-}C\text{-}CH_2 \qquad\qquad Me \\
\| \\
O
\end{array}
\qquad (16.17)
$$

The *bis*-acrylate monomer (Equation 16.18)

$$
\begin{array}{c}
\text{O} \qquad\qquad\qquad\qquad \text{O} \\
\| \qquad\qquad\qquad\qquad \| \\
H_2C\text{=}CH\text{-}C\text{-}O\text{-}(\text{-}CH_2\text{-}CH_2\text{-}O\text{-})_{14}\text{-}O\text{-}C\text{-}CH\text{=}CH_2
\end{array}
\qquad (16.18)
$$

is accommodated by this bilayer. The *bis*-acrylate is photopolymerized. While there is some change in the particular bond lengths, the bilayer still holds on and a coherent film is formed. The bilayer template is subsequently washed away by addition of methanol leaving a flexible, self-supporting film. Polymerization can also result in the creation of different bond lengths that can act to "release" the newly formed monolayer. Multilayered structures can be made by simply allowing the outer surface of one monolayer to act as the template for the next monolayer.

Inorganics can also be synthesized and used as templates. Thus, controlled siloxane networks were formed when dispersions of alkoxysilanes (such as $[MeO]_3SiMe$) are mixed with the suitable template matrixes. Ultrafine particles of metal oxides can be used as starting materials for the formation of metal oxide films. For instance, a mixture of a double-chained ammonium amphiphile and an aqueous solution of aluminum oxide particles (diameter about 10–100 nm) gives a multilayered aluminum oxide film when calcinated at over 300°C.

Metal–ligand structures, such as porphyrin-based structures, are able to "control" the geometry of forming superstructures. The interactions between the various internally chelated porphyrin metal and a combination of the planarity and pi–pi interactions between the porphyrin rings drive the resulting structures. Connecting groups can act as spacers or act as additional geometry-determining features. Smaller metal-based groups have been used to self-assemble structures such as nanocages that can have within them molecules that give the combination cage and captive molecule unique chemical and electrical environments. A nanocage has been formed using six *cis*-protected palladinum II nitrate molecules and four tridendate tripyridyl molecules. These structures, unlike other self-assembly molecules considered here, have primary bonds as major contributors to the self-assembly process. Many of these excursions are the exercise of combining organometallic chemistry with structural knowledge and purpose.

Graphite and carbon fibers have been used as templates. Thus, nylon 6 has been polymerized on a graphite matrix. Such syntheses of polymers in the presence of a solid template where the solid acts as a template have been described as polymerized-induced epitaxy (PIE). The monomer and resulting film is adsorbed on the template surface through only van der Waal's forces. After polymerization, the polymer is washed from the template. The recovered polymer retains "special" structural features introduced by the template.

Block copolymers with amphilic groups have been used to give molecules with several "levels" of molecular architecture. Thus, a block copolymer of styrene and 4-vinylpyridine and long-tailed alkylphenols contains polar groups in the backbone that associate with the alkylphenol chains forming hydrogen bonds, resulting in a bottle-brush-like structure. If the alkyl tails on the polymer are strongly repulsed by the amphiphilic portions, microphase separation between the tails and the rest of the copolymer results. Here, the alkylphenol portion, which is hydrogen bonded to the vinylpyridine blocks, separates as a microphase inside the poly(4-vinylpyridine) block domains. If the copolymer is heated, the bottle-brush structure undergoes an order–disorder transition around 100°C and a second-order transition about 150°C as the amphiphilic alkylphenol chains diffuse into the PS-rich domains and cylindrical structures are formed.

While much of the emphasis is on inter- and intramolecular interactions, secondary bonding and forces associated with association and dissociation involve attractive forces, we are finding that phobic effects are also important and for some systems are actually the major factors. Briefly, this can be described by such sayings as "the enemy of my enemy is my friend," or "given the choice between bad and worse, bad wins out." Formation of many self-assemblies is due in large measure to such phobic factors.

As we learn more about what drives molecular shapes, we are finding more applications utilizing this information in designing, on a molecular level, oligomeric and polymeric chains. It is also becoming more apparent to us that we have been forming organized structures without knowing it. While simple layered and linear structures are generally employed to describe the concept, the self-assembling approach holds for any two- and three-dimensional structure under the appropriate conditions. As in many of the areas of research, the potential is only limited by our imagination.

Considering only Lewis acid–base or donor–acceptor interactions we can envision a hydrogen bond donor site such as an alcohol, acid, thiol, or amine and an acceptor site such as a carbonyl oxygen on another molecule or part of the molecule. These components will bind with one another, acting to bind either the molecules containing the two differing bonding sites or if the two sites are on the same molecule attempt to contort, twist the molecule allowing the preferred bonding to occur.

Synthetic shapes are generally limited to sheets and polyhedral structures. Yet nature produces a much wider variety of shapes, including curves, spirals, ripples, bowls, pores, tunnels, spheres, and circles. We are beginning to master such shapes. We are beginning to make these shapes based on especially "grown" shapes that act as templates for further growth. For instance, Geofreey Ozin and coworkers mixed together alumina, phosphoric, and decylamine in an aqueous solution of tetraethylene glocol. After a few days, millimeter-sized aluminophosphate solid spheres and hollow shells were formed with the surfaces sculpted into patterns of pores, meshes, ripples, bowls, and so on. A decylammonium dihydrogenphosphate liquid–crystal phase was formed and this surfactant, along with the glycol, was forming bilayer vesicles. The vesicles acted in different ways with some fusing to one another, others splitting apart or collapsing giving a variety of structures. Thus, appropriate conditions can be selected that favor certain template structures producing an array of geometric structures. Further, the templates themselves can be used to make selective separations. In a related study, the group employed a silica precursor, tetraethyl orthosilicate. Here the orthosilicate units assembled together forming micelles that in turn acted as liquid-seed crystals growing other assemblies with varying shapes. Rapid growth in the axial direction produces rope-like structures that can be made to form circles and loops through application of external forces. Other structures included egg shapes, disks, spirals, knots, and spheres.

Metal coordination is another important bonding opportunity with respect to self-assembly. This is important in many natural molecules such as hemoglobin and chlorophyll where the metal atom acts as both the site of activity and as a "centralizing" agent with respect to shape and thus acts as a nucleating agent for self-assembly.

Numerous metal chelating designs can be envisioned. Structure 16.19 is one made by Daniel Funeriu and coworkers. The end structure is dependent upon the nature of the metal. For instance, a wreath-shaped double-helical complex is formed when $FeCl_2$ is added with each wreath containing five iron ions with each iron having three bonding sites. Further, the wreath size is such that it will selectively bind the chloride or other similarly sized ions because the source of the iron is the iron chloride. The ratio of reactants is also important and by varying the ratio different structures can be formed, including wire and tape-like structures. Again, it is up to the researcher to utilize information at hand to construct these self-assemblies.

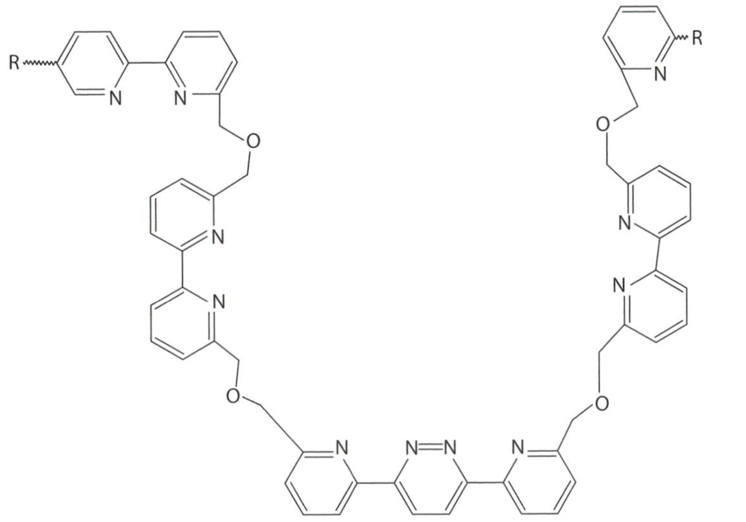

(16.19)

There are many potential and real applications of self-assembly.

Pharmaceutical chiral drug sales top $100 billion yearly worldwide. More than half of the drugs on the market are asymmetric molecules with about 90% administered as racemates. In general, one optical center of a drug will have the desired activity while the other can produce negative side effects. Because of this the Food and Drug Administration (FDA) in 1992, issued a statement saying that for every new racemic drug, the two enantiomers must be treated as separate substances that are required to undergo pharmocokinetic and topological studies.

One direct approach to the separation of chiral compounds is called *molecular imprint polymers* (MIPs) that involves the formation of a three-dimensional cavity with the shape and electronic features that are complementary to the imprinted or target molecule.

While MIPs are part of the current nanorevolution, it's roots are found in the antibody formation theory of Pauling's. While the particulars were wrong, the general concept is good.

In the formation of MIPs, the target drug is added to a solvent along with selected polymers. It is important that the liquid, self-assembling polymer(s), and template molecule complement one another. The specific bonding can be a combination of covalent and noncovalent bonding. Here we will look at an instance involving only noncovalent bonding. The main secondary bonding interactions include metal–ligand complexations, hydrogen bonding, and ionic, dipolar, hydrophobic, and pi–pi interactions.

Because most drugs have both polar and nonpolar regions, solvents and vinyl monomers that contain both polar and nonpolar regions are often employed. Where aromatic rings are present, polymers such as 4-vinylpyridine and styrene are often utilized because of their ability to "fit" such structures, bond through overlap of pi systems, and be readily polymerized via a variety of methods (such as UV, heat, and use of free radical initiators). Hydrogen bonding solvents are generally discouraged because of their tendency to form strong bonds with the template molecule and after evacuation of the template site, with polar portions of the template site. Often dipolar aprotic solvents are employed that offer both polar and nonpolar sites.

In general, the sequence is

mixing together of the template, polymer, and solvent → self-assembly about the template → polymerization → extraction of template molecule → grinding, sieving, and column packing.

A number of drugs have been successfully separated using MIPs. Naproxen(TM), (S)-6-methosy-α-methyl-2-naphthaleneacetic acid, is a nonsteroidal antiinflammatory drug (NSAID) that is administered as the "S" enantiomer. Naproxen (16.20) will be used to illustrate the MIP sequence. Naproxen has both polar and nonpolar domains. It also has a fused-ring aromatic site.

(16.20)

Solvents and self-assembling polymer(s) are chosen that have both polar and nonpolar portions. The polymers and solvents then self-assemble about the target molecule. This arrangement is then exposed to UV radiation, heat, and/or catalysts that effectively form a polymeric "cocoon" about the target molecule. After polymerization and the formation of the "cocoon" about the target molecule, the solvent molecules and target molecule are removed exposing a partially completed cavity with both structural (both shape and spatial configuration) and electronic characteristics complementing the target molecule. For naproxen the solvent is tetrahydrofurane (THF) and the monomer is 4-vinylpyridine.

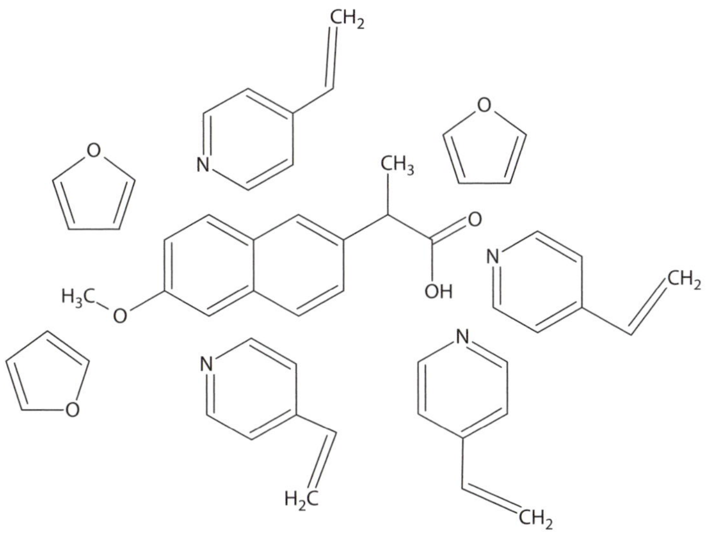

(16.21)

An illustration of a possible self-assembly arrangement involving only the vinylpyridine and THF is given in 16.21. About the acid group there are two vinylpyridines both with the nitrogens pointed toward the naproxen template molecule, one making use of hydrogen bonding and one bonded through dipolar interactions to the electron deficient "acid" carbon. This gives an electron-rich region. About the naphthalene-ring portion are also located vinylpyridine molecules except they have the nitrogens "pointed" away creating a nonpolar region. These self-assemblies vary from template site to template site and this variation in specific site structure is at least partially responsibility for the broadening of the chromatography peak associated with the template molecule. Assembly also occurs above and below the naproxen again with appropriate secondary bonding occurring, including pi-bonding interactions between the aromatic sites on the naproxen and the furane and vinylpyridine.

The ratio of THF and 4-vinylpyridine is important since both assemble about the naproxen. If there are too many solvent molecules present, the "cocoon" cavity will not be sufficient to retain the template molecule imprint. If the concentration of vinylpyridine is too high, the "cocoon" structure will be too complete and prevent both the exit of the target molecule and entrance of other naproxen molecules during the separation procedure.

Added along with the "imprinting" vinylpyridine are cross-linkers, spacers or "porogens." These cross-linkers or spacers should be miscible with the other ingredients but have shapes that are dissimilar so that they do not also become an integral part of the assembly about the template molecule. For naproxen the cross-linker is ethylene glycol dimethylacrylate (EDMA) (16.22).

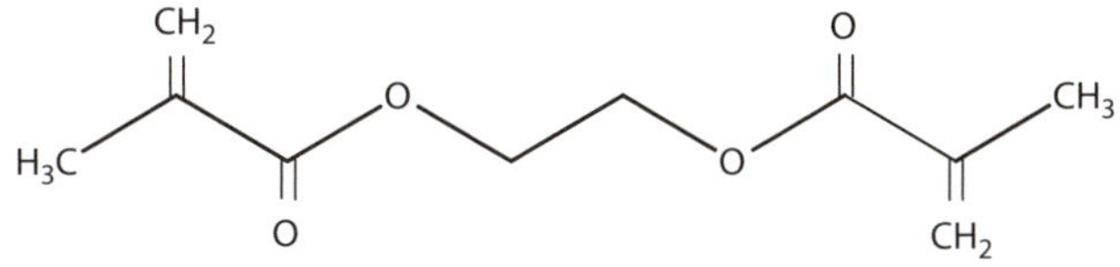

(16.22)

Again, the conditions, amounts, and identity of reactants are carefully selected to allow ready entrance and acceptance of the target molecules.

Eventually, a matrix is formed containing the molecularly imprinted sites locked into a matrix such as illustrated in 16.23.

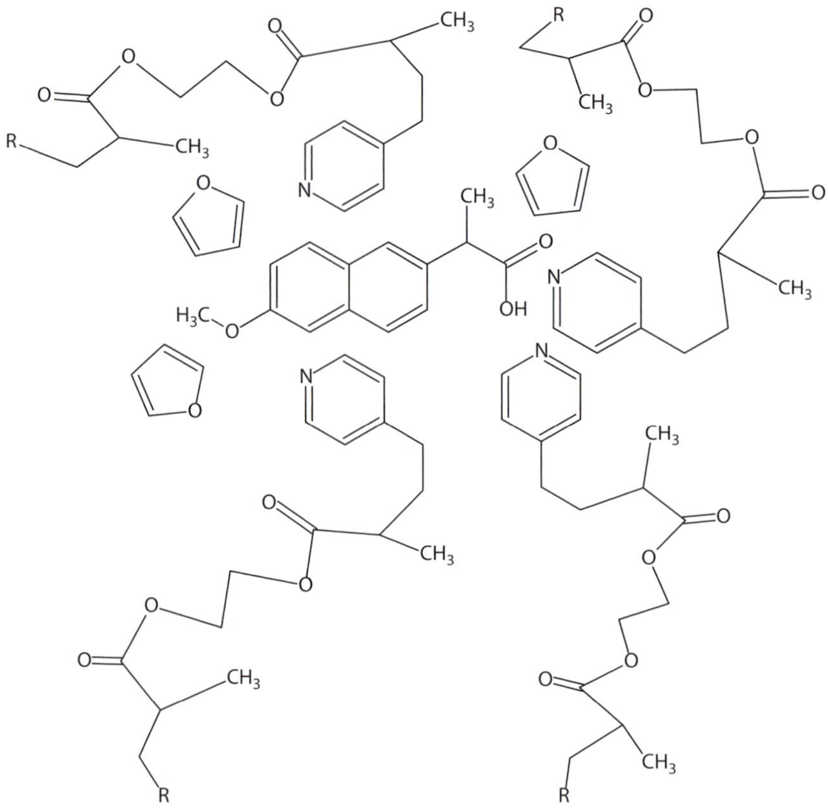

(16.23)

After polymerization, the MIP or functionalized polymer matrix is dried, ground, sieved, and packed.

On the column, the eluding liquid is important. The liquid must dissolve the desired compound but it should not be too good of a solvent or it will inhibit the release of the desired molecule allowing it to interact with the template cavity. In the case of naproxen, there is a further consideration. Naproxen has an acid function that was "templated" in the protonated form. Thus, acetic acid, along with THF and heptane, was added to insure that the naproxen was present in the needed correct geometry.

As with most processes, MIPs have both positive and negative aspects. On the positive side, with the correct choice of original polymer and solvent, almost any molecule can be employed as a template, including much larger molecules. (This approach is not realistic for synthetic polymers where the short-range geometry is varied and where the long-range geometry is not fixed. It might be feasible for natural polymers where the structures are fixed.) Further, since the desired compound is the one that is preferentially being attracted by the template site, it is the last to be eluted from the column so that additional work is not needed to identify where the target molecules are. Increased temperatures can also be employed, presumably to the range of 180°C–200°C so that high-temperature isolations can be effected. In general, with the possible exception of the target compound, the reagents are inexpensive so that such selective molecular sieves are relatively inexpensive.

On the negative side, there is excessive broadening of the elution band caused by a number of features, including the heterogeneity of the imprint as noted before.

One problem with nanomolecules is ready alignment. On a molecular level, atomic force microscopy (AFM) and related tools can be used to align such individual or nanomolecules but this is impractical on a large scale. Thus, strategies are being developed to accomplish this. One approach for single-walled carbon nanotubes (SWNTs) is to physically or chemically anchor molecules onto tubes and to have these anchors direct the nanomolecules into desired arrays or organizations.

While simple layered and linear structures are generally used to descript the concept, the self-assembling approach holds for any two- and three-dimensional structures under the appropriate conditions. As in many areas of research, the potential is only limited by our imagination fueled by our understanding and use of available and newly created tools.

16.7 TRANSFER AND RETENTION OF OXYGEN

Today, the polymer scientist should be aware of synthetic, inorganic, and biological macromolecules. The field of biological macromolecules is large and is one of the most rapidly expanding areas of knowledge. It involves gene splicing and other related biological aspects, including biological engineering, neurobiology, medicine, drugs, and so on; the very elements of life and death, of thought and caution, pain and health, of biological transference, of energy, and of biological matter. The polymer scientist can learn from these advances but must also contribute to their understanding on a molecular and chain aggregate level. The investigation of these biological macromolecules is done using state-of-the-art instrumentation and techniques and the use of scientific intuition. The world of the natural macromolecule is yielding information allowing an understanding on a molecular level. A striking example involves oxygen transfer and retention in mammals.

Oxygen retention and transfer involves the iron-containing organometallic planer porphyrin-containing structure called *heme* (Figure 16.1). The iron is bonded through classical coordination. The ferrous or iron (II) has six coordination sites. Four of these coordination sites are occupied by the nitrogen atoms of the four pyrrole-related rings of the porphyrin. A fifth site is occupied by one of the nitrogens of an imidazole side chain found as part of the protein structure and located just opposite the planer porphyrin moiety. The sixth site acts to bind oxygen. The ion remains in the +2 oxidation state, whether oxygen is being bound or the site is vacant. An additional histidine is present, residing in the protein chain opposite the sixth site of the iron atom. This second histidine does not bind iron but serves to stabilize the binding site for oxygen. Experimentally, heme does not bind oxygen. Instead, a complex protein wrapping is necessary to both assist binding and protect the binding site from foreign competitor molecules that could render the heme site inactive, either through structural change, change in iron oxidation state, or through occupation of this site, thus preventing oxygen access to the active binding site.

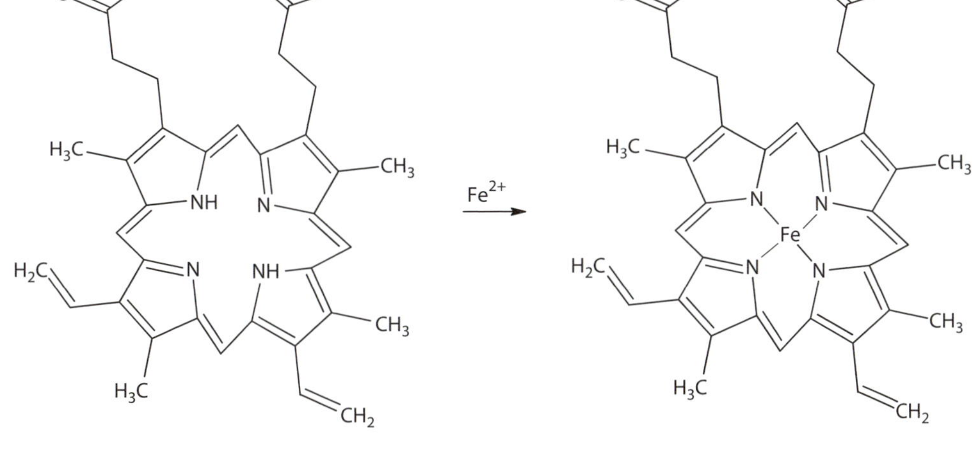

FIGURE 16.1 Porphyrin structures that serve as the basis of heme (a). Upon addition of iron, this porphyrin, which is called *protoporphyrin IX*, forms the heme group (b).

The precise electronic environment of iron deserves special comment. In *deoxyhemoglobin*, the iron atom has four unpaired electrons, but in *oxyhemoglobin* iron has no unpaired electrons. The iron in the oxygen-free deoxyhemoglobin is referred to as "high spin" iron, whereas iron in oxyhemoglobin is called "low-spin" iron Hund's rule of maximum multiplicity calls for the most energy-favored, lowest energy form to be the structure containing the highest number of unpaired electrons. The binding of oxygen, itself with two unpaired electrons, is probably the result of a favored energy of binding brought about through the coupling of the two sets of unpaired electrons—the favorable energy allowing the violation of Hund's rule.

There are two major protein-heme-binding macromolecules. These are myoglobin (Figure 16.2), which is used as an oxygen storage molecule in mammalian muscle, and hemoglobin, which is used in oxygen transport. *Myoglobin* is single stranded with one heme site per chain, whereas *hemoglobin* is composed of four protein chains, each one containing a single heme site (Figure 16.4). In hemoglobin, there are two sets of equivalent chains composing its quaternary structure. These two chain types are referred to as α and β chains. The α chains contain 141 amino acid units and the β chains contain 146 units. The myoglobin contains 153 amino acid units. Each of these chains are similar and each forms the necessary environment to allow the heme site to bind oxygen in a reversible manner. The protein segments are described as being loosely helical with about 60%–80% of the structure helical.

While the chains are similar in overall structure, there exist somewhat subtle differences. For instance, the quaternary structure of hemoglobin permits interaction between the four chains. Structural movement brought about through binding of oxygen at one of the four heme sites acts to make it easier for subsequent oxygen addition at the other heme sites. Such cooperative binding of oxygen is not possible in the single-chained myoglobin. The consequence of this cooperative binding is clearly seen in a comparison of oxygen binding by both myoglobin and hemoglobin as a function of oxygen pressure (Figure 16.3). Oxygen binding occurs by myoglobin at even low oxygen pressures. This is referred to as hyperbolic behavior. By comparison, hemoglobin binding increases more slowly as the oxygen pressure increases, occurring in a more sigmoidal fashion. Thus, initial binding of oxygen at a heme site is relatively difficult, becoming increasingly easier as the number of heme sites bonded increases.

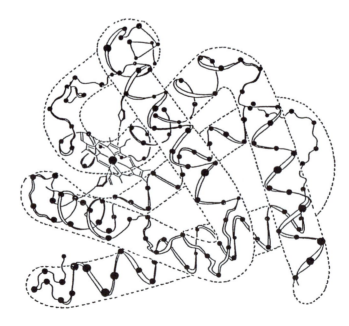

FIGURE 16.2 Generalized myoglobin structure showing some amino acid units as open circles to illustrate the "folded" tertiary structure.

The differences in oxygen binding characteristics are related to the differing roles of hemoglobin and myoglobin. Thus, myoglobin is employed for the storage of oxygen in muscle. Binding must occur even at low oxygen contents. Hemoglobin is used for the transport of oxygen and becomes saturated only at higher oxygen concentrations. The oxygen content in the alveoli portion of our lungs is on the order of 100 torr (1 atm of pressure = 760 torr). Here, almost total saturation of the heme binding sites in hemoglobin occurs. By comparison, the oxygen level in the capillaries of active muscles is on the order of only 20 torr, allowing the hemoglobin to deliver about 75% of its oxygen and for myoglobin to almost reach saturation in oxygen binding.

Conformational changes accompany the binding and release of oxygen. These changes are clearly seen by superimposing the oxygen-containing form of hemoglobin—oxyhemoglobin—over the nonoxygen-containing form of hemoglobin—deoxyhemoglobin (Figure 16.4). It is interesting to note that most enzymes have their active sites on the surface but enzymes, such as myoglobin, that are dealing with "small molecules" such as oxygen, have their active sites "hidden away" in the internal portion so that other unwanted molecules cannot easily access them with the protein "tunnel" acting to restrain unwanted suitors.

Another metal-containing important enzyme, *chlorophyll* (16.24) is structurally related to myoglobin and it is responsible for photosynthesis, the conversion of water and carbon dioxide, with the aid of solar energy, to carbohydrates.

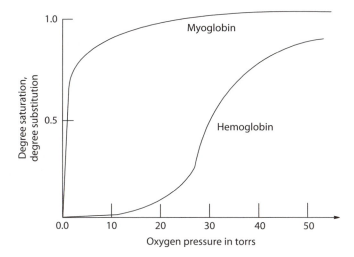

FIGURE 16.3 Degree of saturation as a function of oxygen pressure.

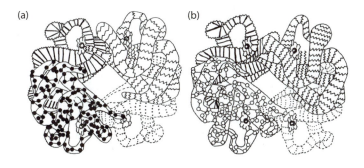

FIGURE 16.4 Space-filling models of deoxyhemoglobin (a), and oxyhemoglobin (b). Notice the small shifts in the overall geometry of the various protein chains and the decreased size of the inner core.

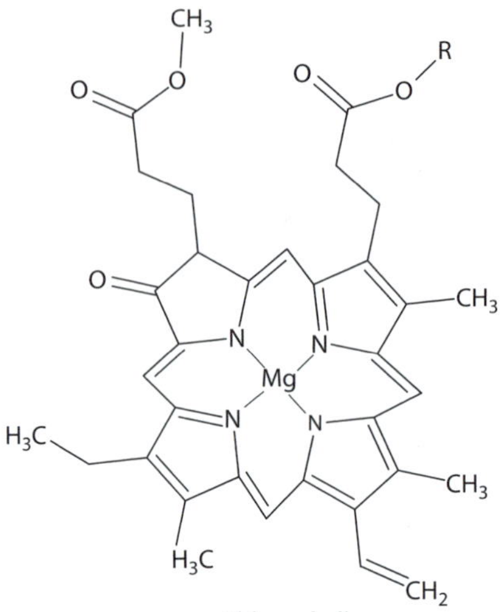

Chlorophyll

(16.24)

16.8 NATURE'S MACROMOLECULAR CATALYSTS

Probably the most important reactions occurring on polymers involve the catalytic activity of a class of proteins called *enzymes*. The catalytic action is a result of a lowering of the activation energy for the rate-determining step in the reaction. In general terms, the catalytic action results from the formation of a complex between the enzyme and the molecule undergoing reaction. The decreased activation energy is a result of the reacting molecules being held by the enzyme in such a manner as to favor the appropriate reaction occurring. The two classical models employed to describe the formation of the complex between the reacting molecules(s) and the enzyme are the lock-and-key model and the induced-fit model. Briefly, the *lock-and-key model* calls for an exact or highly similar complementary fit between the enzyme and the reacting molecule(s) (Figure 16.5). Geometry plays an essential role in permitting the electronic (polar, electrostatic, and so on; attractions/repulsions) interactions to form the necessary complex with the correct geometry. Release is encouraged by the new geometry and electronic distribution of the resulting products of the reaction being sufficiently dissimilar. The *induced-fit model* is similar except that the enzyme originally does not fit the required shape (geometrical and electronic cavity). The required shape is achieved upon binding—the binding causing needed "assisting factors" or proximity and orientation to effect a decrease in the energy of the transition state. Figure 16.5 shows how the enzyme sucrase breaks sucrose into its basic units of glucose and fructose. The sucrose is held by the enzyme in such a fashion so that water easily and efficiency breaks the ether bond forming two hydroxyl units that are released because they no longer have the required shape, size, and electronic features of the enzyme cavity.

Enzyme reactions generally follow one of two kinetic behaviors. Briefly, the oxygen binding curve for myoglobin is hyperbolic, whereas that for hemoglobin it is sigmoidal (Figure 16.6). In general, it is found that similar enzymes such as myoglobin follow a similar hyperbolic relationship between reaction extent and reaction time. More complex enzymes such as hemoglobin follow a sigmoidal relationship between reaction extent and reaction time. The primary difference involves the ability of different portions of the overall hemoglobin structure, removed reaction sites, to affect each other. Molecules in which various removed sites affect the reactivity of other removed sites are called *allosteric enzymes*.

The *Michaelis–Menten model* is commonly employed in describing *nonallosteric enzyme* reactions. The overall model can be pictured as follows (16.25), where E represents the enzyme, M the

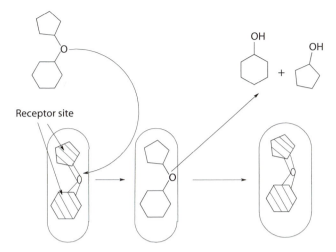

FIGURE 16.5   There are two major models for the binding of reactant molecules to the active sites of nonallosteric enzymes. The sequence describes the essential steps in the lock-and-key model, where the reactant(s) is attracted to the active site on the enzyme where the active site is a cavity of the same general size, shape, and (complementary) electronic features. Binding occurs and the appropriate reaction(s) occurs resulting in a change in the geometry and electronic configuration of the product, causing its release. The second model (not shown) is the induced-fit model where the individual steps are similar to the lock-and-key except the reactants "induce" a change in the conformation of the active site on the enzyme, allowing it to accept the reactant(s).

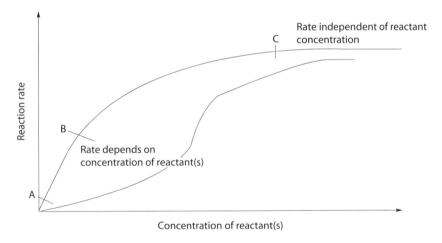

FIGURE 16.6   Dependence of reaction showing sigmoidal (bottom) and hyperbolic (top) behavior. The top plot also shows the initial rate of reaction as a function of reactant concentration when the concentration of enzyme remains constant.

reacting molecule(s), $E + M \rightarrow EM$ is associated with k_1, and the reverse reaction associated with k_{-1} and $EM \rightarrow E + P$ associated with k_2.

$$E+M \underset{k_{-1}}{\overset{k_1}{\rightleftarrows}} EM \overset{k_2}{\rightleftarrows} E+P \tag{16.25}$$

Here EM represents the enzyme complex and P the product(s). The rate of complex formation is described as

$$\text{Rate of complex formation} = \frac{d[EM]}{dt} = k_1[E][M] \tag{16.26}$$

The complex then either returns to form the initial reactants or forms the product(s) and the free enzyme, E. In kinetic terms, the change, or rate of breakdown of the complex is described as

$$\text{Rate of complex change} = -\frac{d[EM]}{dt} = k_{-1}[EM] + k_2[EM] \tag{16.27}$$

The negative sign associated with the equation means that the terms are describing the rate of decrease in complex concentration. The rate of complex formation is rapid, and fairly soon the rate at which the complex is formed is equal to the rate at which it breaks down. This situation is called a *steady state*. Mathematically, this is described by

$$\frac{d[EM]}{dt} = -\frac{d[EM]}{dt} \quad \text{and} \tag{16.28}$$

$$k_1[E][M] = k_{-1}[EM] + k_2[EM] \tag{16.29}$$

The initial reaction rate, A to B, varies until the number of reactants clearly outnumbers the number of reaction sites on the enzyme, at which time, C, the rate becomes zero order, independent of the reactant concentration. Often it is difficult to directly measure the concentration of E as the reaction progresses. Thus, the concentration of E is generally substituted for using the relationship

$$[E] = [E_o] - [EM] \tag{16.30}$$

where $[E_o]$ is the initial enzyme concentration.

Substitution of this description for [E] into Equation 16.29 gives

$$k_1 ([E_o][M] - [EM][M]) = k_{-1}[EM] + k_2[EM] \tag{16.31}$$

Separating out [EM] on the right-hand side gives

$$k_1 ([E_o][M] - [EM][M]) = (k_{-1} + k_2)[EM] \tag{16.32}$$

Bringing together all of the rate constants form both sides of the equation gives

$$[E_o][M] - [EM][M] = k'[EM] \tag{16.33}$$

Moving the [EM] containing terms to the right side gives

$$[E_o][M] = k'[EM] + [EM][M] \tag{16.34}$$

and separating out [EM] from the right side gives

$$[E_o][M] = [EM] (k' + [M]) \tag{16.35}$$

Now division of both sides by k' + [M] gives

$$[EM] = \frac{[E_0][M]}{k' + [M]} \tag{16.36}$$

The initial rate of product formation, R_i, for the Michaelis–Menten model depends only on the rate of complex breakdown, that is,

$$R_i = k_2[EM] \tag{16.37}$$

Substitution from Equation 16.36 into Equation 16.37 gives

$$R_i = \frac{k_2[E_0][M]}{k' + [M]} \tag{16.38}$$

This expression is dependent on the concentration of M and describes the initial part of the plot given in Figure 16.6.

Generally, the concentration of M far exceeds that of the enzyme sites such that essentially all of the enzyme sites are complexed, that is, $[EM] = [E_o]$. (This is similar to a situation that occurs regularly in south Florida where four- and six-lane roads are funneled into a two-lane section of road because of a wreck or road construction.) Thus, the rate of product formation is maximized under these conditions. This maximum rate, R_m, allows us to substitute $[E_o]$ for $[EM]$ in Equation 16.37 giving

$$R_m = k_2[E_o] \tag{16.39}$$

Since the enzyme concentration is constant, the rate of product formation under these conditions is independent of [M] and is said to be zero order (Figure 11.6).

The maximum rate is directly related to the rate at which the enzyme "processes" or permits conversion of the reactant molecule(s). The number of moles of reactants processed per mole of enzyme per second is called the *turnover number*. Turnover numbers vary widely. Some are high, such as for the scavenging of harmful free radicals by catalase, with a turnover number of about 40 million. Others are small such as the hydrolysis of bacterial cell walls by the enzyme lysozyme, with a turnover number of about one-half.

The Michaelis–Menten approach does not describe the behavior of *allosteric enzymes*, such as hemoglobin, where rate curves are sigmoidal rather than hyperbolic. A more complex mode is called for to account for the biofeedback that occurs with allosteroid enzymes. Such affects may be positive such as those associated with hemoglobin, where binding by one site changes the geometry and electronic environment of the other remaining sites, allowing these addition sites to bind oxygen under more favorable conditions. The effects may also be negative, such as that of cytidine triphosphate, which inhibits ATCase and catalyzes the condensation of aspartate and carbamoly phosphate-forming carbamoyl aspartate.

Two major models are typically used to describe these situations: the concerted model and the sequential model. In the *concerted model*, the enzyme has two major conformations—a relaxed form that can bind the appropriate reactant molecule(s) and a tight form that is unable to tightly bind the reactant molecule(s). In this model, all subunits containing reactive sites change at the same time (Figure 16.7). An equilibrium exists between the active and inactive structures. Binding at one of the sites shifts the equilibrium to favor the active relaxed form.

The major feature in the *sequential model* is the induction of a conformational change from the inactive tight form to the active relaxed form as the reacting molecule(s) is bound at one of the sites. This change from an unfavorable to a favorable structure is signaled to other potentially reactive sites bringing about a change to the more favored structural arrangement in these other sites (Figure 16.8).

Structural changes can be brought about through simple electrostatic and steric events caused by the presence of the reacting molecule(s). Structural changes also result as cross-linking and other primary bonding changes occur.

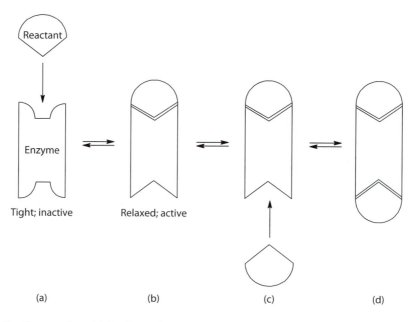

FIGURE 16.7 Concerted model for allosteric enzymes. The major steps are (a) and (b). An equilibrium exists between the tight (a) and relaxed (b) forms of the allosteric enzyme. The reactant molecule(s) approaches the reactive site of one of the enzyme sites present in the relaxed form (c). Binding occurs, shifting the equilibrium to the relaxed form(s). The second site is bound (d).

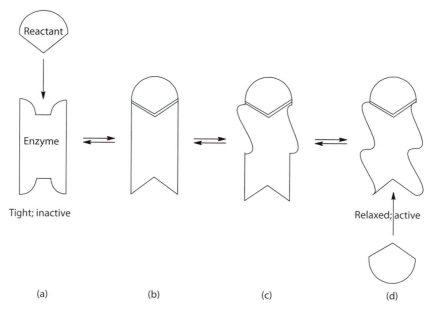

FIGURE 16.8 Sequential model for allosteric enzymes. The individual steps are as follows: the reactive molecule(s) approach the reactive site (a), which typically has a cavity similar to, but not the same as, the reactant molecule(s); a conformational change is effected so that the reactive molecule(s) can be bound (b); the bound portion of the enzyme changes shape; this shape change is transmitted to the other unit(s) containing active sites (c); the remaining enzyme portion containing active sites undergoes a conformational change that makes binding easier.

16.9 PHOTOSYNTHESIS

The recent environmental issues related to the green house effect and atmospheric contamination heightens the importance of obtaining energy from clean sources such as photosynthesis. Photosynthesis also acts as a model for the creation of synthetic light-harvesting systems that might mimic chlorophyll in its ability to convert sunlight into usable energy. The basis of natural photosynthesis was discovered by Melvin Calvin, one of my academic grandfathers. Using ^{14}C as a tracer, Calvin and his team found the pathway that carbon follows in a plant during photosynthesis. They showed that sunlight supplies the energy through the chlorophyll site, allowing the synthesis of carbon-containing units, saccharides or carbohydrates. Chlorophyll is a metal embedded in a protein polymer matrix and illustrates the importance of metals in the field of photochemistry and photophysics. A brief description of the activity of chlorophyll in creating energy from the sun follows.

The maximum solar power density reaching Earth is approximately 1,350 W/m^2. When this energy enters the Earth's atmosphere, the magnitude reaching the surface drops approximately to 1,000 W/m^2 due to atmospheric absorption. The amount that is used by plants in photosynthesis is about seven times the total energy used by all humans at any given time, thus it is a huge energy source. On Earth, organisms convert about 10^{11} tons of carbon into biomass yearly.

Solar energy is clean and economical but it must be converted into useful forms of energy. For example, solar energy can be used as a source of excitation to induce a variety of chemical reactions.

Plants and algae are natural examples of the conversion of light energy and this energy is used to synthesize organic sugar-type compounds through photosynthesis. This process has a great importance for survival of all life on our planet because light provides the energy and reduces the carbon dioxide as well as produces molecular oxygen necessary for oxygen consuming organisms. In fact, this process can be considered as a vital link between material and energy cycling in the biosphere.

Photoenergy transfer can be considered to occur through two primary mechanisms. The first is the Förster mechanism, which is also known as the columbic mechanism or dipole-induced dipole interaction. Here, the emission band of one molecule (donor) overlaps with the absorption band of another molecule (acceptor). In this case, a rapid energy transfer may occur without a photon emission. This mechanism involves the migration of energy by the resonant coupling of electrical dipoles from an excited molecule (donor) to an acceptor molecule. On the basis of the nature of interactions present between the donor and acceptor this process can occur over a long distances (30–100 Å). The Dexter mechanism is a nonradiative energy transfer process, which involves a double electron exchange between the donor and the acceptor. Although the double electron exchange is involved in this mechanism, no charge-separated state is formed. For this double electron exchange process to operate, there should be a molecular orbital overlap between the excited donor and the acceptor molecular orbital. For a bimolecular process, intermolecular collisions are required as well. This mechanism involves short-range interactions (~6–20 Å and shorter).

In photosynthesis, green plants and some bacteria harvest the light coming from the sun by means of their photosynthetic antenna systems. The light harvesting starts with light gathering by antenna systems, which consist of pigment molecules, including chlorophylls, carotenoids, and their derivatives. The absorbed photons are used to generate excitons, which travel via Förster energy transfers toward the reaction centers (RCs). This overall series of processes is represented in Figure 16.9. The series can be remembered by the initials ARC, where A represents antenna pigments; R represents the reaction centers; and C represents chlorophylls and carbohydrates.

In reaction centers, this energy drives an electron-transfer reaction, which in turn initiates a series of slower chemical reactions and the energy is saved as redox energy inducing a charge separation in a chlorophyll dimer called the *special pair* (chlorophyll)$_2$. Charge separation, which forms the basis for photosynthetic energy transfer, is achieved inside these reaction centers (16.40).

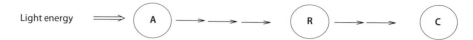

FIGURE 16.9 Light is absorbed by the antenna pigments, A, transferring the energy until it reaches the reaction center, R, where it is used as the driving force for electron-transfer reactions where specialized chlorophylls use it to form carbohydrates.

$$(Chlorophyll)_2 + energy \rightarrow (Chlorophyll)_2^+ + e^- \tag{16.40}$$

Specialized reaction center proteins are the final destination for the transferred energy. Here, it is converted into chemical energy through electron-transfer reactions. These proteins consist of a mixture of polypeptides, chlorophylls (plus the special pair), and other redox-active cofactors. In the RCs, a series of downhill electron transfers occur, resulting in the formation of a charge-separated state. On the basis of the nature of the electron acceptors, two types of RCs can be described. The first RC type (photosystem I) contains iron–sulfur clusters (Fe_4S_4) as their electron acceptors and relays, whereas the second type (photosystem II) features quinones as their electron acceptors. Both types of RCs are present in plants, algae, and cyanobacteria, whereas the purple photosynthetic bacteria contain only photosystem II and the green sulfur bacteria contain a photosystem I. To gain a better understanding of these two types of RCs each will be further discussed.

16.9.1 PURPLE PHOTOSYNTHETIC BACTERIA

In the mid-1980s, Deisenhofer reported his model for the structure of photosystem II for two species of purple photosynthetic bacteria (*Rhodopseudomonas viridis* and *Rhodobacter*) based on X-ray crystallography of the light-harvesting device II. Photosynthetic centers in purple bacteria are similar, but not identical models for the green plants. Since they are simpler and better understood, they will be described here. The photosynthetic membrane of purple photosynthetic bacteria is composed of many phospholipid-filled ring systems (LH II) and several larger dissymmetric rings (LH I) stacked almost like a honeycomb. Inside the LH I is a protein called the RC as illustrated in Figure 16.10.

The light-harvesting antenna complex LH II is composed of two *bacteriochlorophyll a* (BCHl) molecules that can be classified into two categories. The first one is a set of 18 molecules arranged in a slipped face-to-face arrangement and is located close to the membrane surface perpendicularly to these molecules. The second ring is composed of nine BCHl in the middle of the bilayer. The first 18 BCHl have an absorption maximum at 850 nm and are collectively called *B850*, while the second (nine BCHl) have an absorption maximum at 800 nm and are called *B800*. These structures are contained within the walls of protein cylinders with radii of 1.8 and 3.4 nm. Once the LH II complex antenna absorbs light, a series of very complex nonradiative photophysical processes are triggered. First the excitation energy migrates via energy transfers involving the hopping of excitation energy within almost isoenergetic subunits of a single complex. This is followed by a fast energy transfer to a lower energy complex with minimal losses (Figure 16.11). These ultrafast events occur in the singlet state (S_1) of the BCHl pigments and are believed to occur by a Förster mechanism occurring over relatively long distances (30–100 Å).

The energy collected by the LH II antenna is transferred to another antenna complex known as LH I, which surrounds the RC. The photosynthetic reaction centers of bacteria consist mainly of a protein, which is embedded in and spans a lipid bilayer membrane. In the RC, a series of electron-transfer reactions are driven by the captured solar energy. As a result of these electron-transfer reactions, the captured solar energy is converted to chemical energy in the form of a charge separation process across the bilayer. The photosynthetic reaction center is where the harvested solar energy

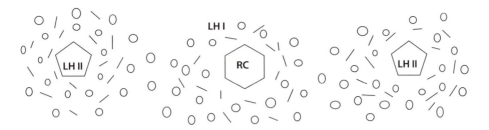

FIGURE 16.10 Simplified drawing showing two light harvesting II assemblies next to one light harvesting I unit. The circles are polypeptides and the lines represent rings of interacting bacteriochlorophylls *a* (called *B850*). In the middle of LH I, there is a protein called the *reaction center* (RC) where the primary photo-induced electron transfer takes place from the special pair of bacteriochlorophylls *b*.

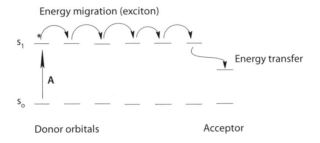

FIGURE 16.11 The exciton and energy transfer processes.

is converted to chemical energy via charge separation across the bilayer by means of an electron-transfer reaction.

A special BCHl (P870) pair is excited either by the absorption of a photon or by acquiring this excitation energy from an energy transfer from the peripheral antenna BCHl (not shown for simplicity) triggering a photo-induced electron transfer inside the RC (36). Two photo-induced electrons are transferred to a plastoquinone located inside the photosynthesis membrane. This plastoquinone acts as an electron acceptor and is consequently reduced to a semiquinone and finally to a hydroquinone. This reduction involves the uptake of two protons from water on the internal cytoplasmic side of the membrane. This hydroquinone then diffuses to the next component of the apparatus, a proton pump called the *cytochrome bc1 complex.*

The next step involves the oxidation of the hydroquinone back to a quinone and the energy released is used for the translocation of the protons across the membrane. This establishes a proton concentration and charge imbalance (proton motive force [pmf]). Thus, the oxidation process takes place via a series of redox reactions triggered by the oxidized special pair BCHl (P870) which at the end is reduced to its initial state. The oxidation process is ultimately driven, via various cytochrome redox relays, by the oxidized P870. Oxidized P870 becomes reduced to its initial state in this sequence. Finally, the enzyme adenosine triphosphate (ATP) synthase allows protons to flow back down across the membrane driven by the thermodynamic gradient, leading to the release of ATP formed from adenosine diphosphate and inorganic phosphate (Pi). The ATP fills the majority of the energy needs of the bacterium.

16.9.2 GREEN SULFUR BACTERIA

The observation of a photosynthetic reaction center in green sulfur bacteria dates back to 1963. Green sulfur bacteria reaction centers are of the type I or the Fe-S-type (photosystem I). Here the electron acceptor is not the quinine; instead chlorophyll molecules (BChl 663, 8¹-OH-Chl a, or Chl a) serve as primary electron acceptors and three Fe_4S_4 centers (ferredoxins) as secondary acceptors. A quinone molecule may or may not serve as an intermediate carrier between primary electron acceptor (Chl) and secondary acceptor (Fe–S centers).

A large number of chlorophyll antennas are used to harvest the solar energy, which in turn are used to excite the special pair P_{700}. The P_{700} donor will in turn transfer an electron to a primary acceptor (A_0, phyophytin) and in less than 100 ps to a secondary acceptor (A_1, a phylloquinone). The electron received by A_1 is in turn transferred to an iron–sulfur cluster and then to the terminal iron–sulfur acceptor.

16.10   MECHANISMS OF PHYSICAL ENERGY ABSORPTION

Let us consider a force, stress, acting on a material producing a deformation. The action of this force can be described in terms of modeling components—a Hookean spring and a Newtonian dashpot. In the Hookean spring, the energy of deformation is stored in the spring and may be recovered by allowing the spring to act on another load or through release of the stress; in either case, the site is returned to zero strain. A Newtonian dashpot is pictorially a frictionless piston and is used to describe chains flowing past one another. The energy of deformation is converted to heat. In actuality, the deformation of most plastics results in some combination of Hookean and Newtonian behavior. The Newtonian behavior results in net energy adsorption by the stressed material, some of this energy producing the work of moving chains in an irreversible manner while some of the energy is converted to heat.

There are three major mechanisms of energy absorption: shear yielding, crazing, and cracking. The latter two are often dealt with together and called *normal stress yielding*.

We can distinguish between a crack and a craze. When stresses are applied to polymeric materials, the initial deformation involves shear flow of the macromolecules past one another if it is above T_g, or bond bending, stretching, or breaking for glassy polymers. Eventually, a crack will begin to form, presumably at a microscopic flaw, which will then propagate at high speed, often causing catastrophic failure. The applied stress results in a realigning of the polymer chains. This results in greater order, but decreased volume occupied by the polymer chains, that is, an increase in free volume. This unoccupied volume often acts as the site for opportunistic smaller molecules to attack, leading to cracking and crazing and eventually property failure.

A *crack* is an open fissure, whereas a *craze* is spanned top to bottom by fibrils that act to resist entrance of opportunistic molecules such as water vapor. Even here, some smaller molecular interactions can occur within the void space, and eventually the specimen is weakened.

Crazing and cracking can be induced by stress or combined stress and solvent action. Most typical polymers show similar features. To the naked eye, crazing and cracking appear to be a fine, microscopic network of cracks generally advancing in a direction at right angles to the maximum principle stress. Such stress yielding can occur at low stress levels under long-term loading. Suppression of stress yielding has been observed for some polymers by imposition of high pressure.

In shear yielding, oriented regions are formed at 45° angles to the stress. No void space is produced in shear yielding. Crazing often occurs before and in front of a crack tip. As noted before, the craze portion contains both fibrils and small voids that can be exploited after the stress is released or if the stress is maintained. Materials that are somewhat elastic are better at preventing small stress related crazing and cracks. Most plastics are not ideal elastomers and additional microscopic voids occur each time a material is stressed.

All three mechanisms result in a difference in the optical properties of the polymeric material because of the preferential reorientation, with realignment of the polymer chains resulting in a change in optical properties such as refractive index, allowing detection through various optical methods including visual examination, microscopy, and infrared spectroscopy of films and sheets. Crazed and cracked sites of optically clear materials appear opaque, whereas shear-yielded sites may appear to be "wavy" when properly viewed by the naked eye employing refracted light.

It is important to emphasize that the surface layers of most polymeric materials are different from the bulk material and are often more susceptible to environmental attack. Thus, special surface treatments are often employed in an attempt to protect the surface molecules.

Directly related to energy absorption is energy dissipation. Generally, the better a material can dissipate or share applied energy the more apt it is to retain its needed properties subsequent to the applied energy. Polymers dissipate applied energies through a variety of mechanisms, including rotational, vibrational, electronic, and translational modes.

One area that illustrates aspects important to energy dissipation is the fabrication of protective armor. Such armor includes helmets, vests, vehicle exteriors and interiors, riot shields, bomb blankets, explosive containment boxes (aircraft cargo), and bus and taxi shields. In each case energy dissipation is a critical element in the desired behavior of the device. To illustrate this let us look at body armor.

Most of the so-called bulletproof vests were made of polycarbonates. More recently, layers of PE were found to have similar "stopping power" for a lesser weight of material.

Today, most body armor is a complex of polymeric materials. Rapid dissipation of energy is critical, allowing the impact energy to be spread into a wide area. Materials should be strong enough so as not to immediately break when impacted and they need to have enough contact with other parts of the body armor to allow ready transfer of some of the impact energy. If the material can adsorb some of the energy through bond breakage or heating, then additional energy can be absorbed at the site of impact. Along with high strength, the material should have some ability to stretch, to move allowing the material to transfer some of the energy to surrounding material. If the connective forces between the components are too strong, total energy dissipation is reduced because a strong bond discourages another way of reducing the impact energy, that is, allowing the various materials to slide past one another. Thus, a balance is needed between material strength, strength of bonding holding the components together, and the ability to readily dissipate the impact energy.

Recently, it was found that some sequences of layered materials are more effective at energy dissipation than others. One of the better combinations is obtained when aramid layers are adjacent to UHMWPE.

Another factor is breaking up the projectile. This is again done using polymeric materials—here composites such as boron carbide ceramics in combination with aramids, UHMWPE, or fibrous glass.

One of the most recent approaches employed to increase the effectiveness of body armor involves the use of *shear thickening fluids* (STFs) or dilatants. These STF combinations are generally referred to as "liquid armor." As force is inflected on a STF, it resists flow different to many liquids such as water. STFs viscosity increases as the rate of shear increases. It is a non-Newtonian material. The STF effect occurs when closely packed materials are combined with sufficient liquid to fill the gaps between them. At low velocities, such as simple motion, the liquid easily flows. At higher velocities, such as the intrusion of a rapid moving projectile, the liquid is not able to move rapidly enough to fill the gaps created between the molecules. This results in an increased friction from the now nonsolvent protected portion of the mixture rubbing against one another ultimately resulting in increased viscosity.

Shear thickening fluid material is placed between the Kevlar sheets and performs several functions. Because it is a liquid under normal conditions, it adds to the flexibility of the body armor. Second, it is more effective at preventing penetration of a bullet so allows less material to be used to achieve the same "stopping power," allowing the armor to weight less. Finally, it helps blunt the projectile through its action as a STF. Most STFs are polymeric. The liquid is generally low molecular weight poly(ethylene glycol) (PEG) with silica particles dispersed in it. The PEG wets the Kevlar sheets assisting the Kevlar–STF combination to spread the impact dissipating the effect of the projectile.

In the future, body armor may be flexible. Experimentation is underway with shear thickening material using PEG with nanoparticles that remains flexible until rapidly struck such as with a bullet whereupon it acts as a solid protecting the wearer from the major impact. Kevlar is also being employed to protect space craft and space men from discarded space junk and small meteors. Thus experimentation in body armor is being applied to additional areas where impact protection is essential.

16.11   BREAKAGE OF POLYMERIC MATERIALS

When a plastic is broken by a sharp blow or cut, are polymer chains broken? The important factors include the nature of the polymer, chain length, and arrangement of the chains.

Sperling and coworkers looked at the question of how many chains are broken and the defining factors related to this breakage when a polymeric material is cut or broken. They used various chain lengths of PS and employed a dental drill as the cutting implement.

Pictorially, the problem can be described as looking at a robin pulling a worm out of a hole. Does the robin get the entire worm or some fraction of the worm? The factors are similar and deal with the length of the worm and how far into the hole it is. If it is largely within the hole then it can grasp the dirt, roots, and so on about it to "hold on for dear life." If not, then the entire worm is a meal for the "early bird."

It turns out that the question dealt with here is related to determining the critical length of fibers that are to be used in a composite. When determining the optimum fiber length of a fiber in a matrix, measurements are made using fibers of differing lengths. If a fiber can be removed from the matrix unbroken, then it is too short, and if the fiber breaks before it can be removed, then the fiber is too long. Thus, fiber lengths should be such that the fiber just begins to be broken rather than allowing it to be removed in tact. In a composite, the worm is the fiber and the soil is the matrix. For the plastic, the worm is the individual chain and the soil is the remainder of the plastic. For the composite, the fiber contains many individual polymer chains, while for the situation dealt with here individual polymer chains will be examined.

The length of fiber or chain that can be removed without breaking is related to the frictional and attractive energies between the fiber and the matrix or other polymer chains holding onto the chain. Thus, if the strength holding together the polymer backbone is greater than the frictional energy holding the chain in place, the polymer chain will be removed unbroken. In general, what was found through calculations was that PS chains to 300 units in length are capable of being removed in tact without breakage. This is in rough agreement with what Sperling found experimentally. Thus, individual PS chains up to about 300 units in length are removed from the plastic without chain breakage.

The relationship between chain length and chain breakage was found to be directly related to the typical length of chain necessary to produce physical cross-links, that is, chain entanglements. (This is probably due to the fact that chain entanglements greatly increase the "apparent" chain length and frictional energy needed to overcome to move a chain.) Typically, at least one chain entanglement is needed to guarantee some chain breakage. For many vinyl polymers, including PS, one chain entanglement occurs for every 300 units. Experimentally it was found that as the length of the PS chain increases so does the number of chain entanglements so that with a chain length of about 2,000 (or an average of seven chain entanglements), 50% of the chains are broken and when the chain length is about 4,000 (or an average of 13 entanglements), approximately 100% of the chains break.

The production of chain entanglements is statically directly related to polymer length for linear chains, and almost independent of the nature of the vinyl unit for many polymers.

Chain length and entanglement are also related to the strength of the polymeric material. As chain length increases, the number of entanglements increases as does the strength. At about eight entanglements, the relationship between number of entanglements (and chain length) and polymer strength levels off with only small changes in polymer strength occurring as the chain length and number of entanglements further increases as shown in Figure 16.12.

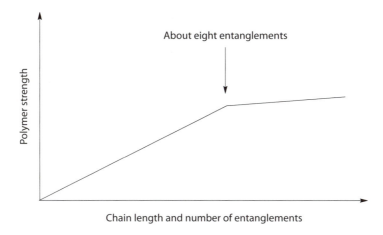

FIGURE 16.12 Idealized relationship between chain length and number of entanglements and polymer strength.

Because the distance between each entanglement is about 300 units, to achieve nearly maximum strength eight entanglements are needed for a chain of about 2,400 (300 units/entanglement times eight entanglements) units is needed. Calculations relating chain strength (related to the energy necessary to break a chain) and frictional force to hold a chain of varying length in place, that is

$$\text{Chain strength } \alpha \text{ Frictional force as a function of chain length} \qquad (16.41)$$

have been carried out with chain lengths of about 2,000–3,000 required before the frictional force necessary to break the chain occur. This is in rough agreement with the chain length of 2,400 calculated above using chain entanglement as the limiting factor.

As the polymer chain length increases so does the viscosity of the melted polymer, requiring more energy to process the polymers using any technique that requires the polymer to flow. This includes most of the molding processing typically used to process vinyl polymers. For many applications, the maximum strength is not needed so that industry looks to a balance between chain length and necessary strength. This "balance" is often chosen such that the chain length is sufficient to give seven entanglements rather than the eight required to insure about 100% chain breakage and maximum polymer strength, as noted above.

There are exceptions to this. One exception involves ultrahigh molecular weight linear PE (UHMWPE) which has few chain entanglements so it is easier to pull long chains from a PE matrix without chain breakage. The equivalent energy necessary to pull a UHMWPE chain from a PE matrix occurs at about a chain length of 100,000 (or a DP of about 3,300), much larger than that for PS.

For condensation polymers, the attractive forces between chains and chain units is greater so that physical chain entanglement is not necessarily the limiting factor, but rather other factors, including localized crystallization becomes important.

16.12 SUMMARY

1. Polymers undergo reactions that are analogous to smaller molecules. Variation generally involves the need for reactants to have contact with the active site. This is more difficult for polymers.
2. Polymer degradation typically occurs via random chain scission, depolymerization, or both, resulting in a loss of chain length and properties associated with polymer length.

3. Among the most important reactions on polymers are those that occur with biomacromole-cules such as involved in the transfer of oxygen and the activity of enzymes. Shape, size, electronic configuration are all essential factors in the transfer of oxygen and activity of enzymes. The transfer of oxygen can be mathematically described using the classical Michaelis–Menten approach. The two major models describing the activity of nonallosteric enzymes such as myoglobin are referred to as the lock-and-key model and the induced-fit model. Hemoglobin is an example of an allosteric enzyme where the two most popular models describing this behavior are the concerted and sequential models.

GLOSSARY

Anchimeric reactions: Reactions enhanced by the presence of a neighboring group.

Chain scission: Breaking of a polymer chain.

Curing: Cross-linking producing a polymer network.

Cyclized rubber: Isomerized rubber containing cyclohexane rings.

Heme: Iron-containing active site for hemoglobin and myoglobin.

Hemoglobin: Allosteric enzyme responsible for the transport of oxygen in our bodies; contains four myoglobin-like units.

Hydrogenation: Addition of hydrogen to an unsaturated compound.

Induced-fit model: One of two basic models employed to describe enzymatic behavior of nonallosteric molecules. Here, the steps are similar to the lock-and-key model except the reactants "induce" a change in the conformation ini the conformation of the active site allowing the active site to bind with the reactant.

Kraton: Trade name for ABA block copolymer of styrene (A) and butadiene (B).

Kuhn element: Number of repeat units needed so that the most probable distance between the first and the growing end is back at the first unit again.

Ladder polymer: Double-stranded polymer for added stability.

Lock-and key model: One of two basic models to describe the selectivity and catalytic nature of nonallosteric enzymes. In this model the reactant(s) is attracted to the active site on the enzyme, which is of the same general size, shape, and complementary electronic nature as the reactant.

Macromolecular monomers (macromers): Polymers containing one reactive end group.

Myoglobin: Nonallosteric enzyme responsible for the storage of oxygen in our bodies. The activity follows the kinetic scheme described by Michaelis–Menten.

Ozonolysis: Reaction of an unsaturated organic compound with ozone.

Photosynthesis: Metabolic pathway for the conversion of carbon dioxide into organic compounds, mainly carbohydrates, using sunlight as the energy source.

Telechelic polymers: Polymers containing two active ends.

Telomerization: Abstraction of an atom by a macroradical.

Topochemical reactions: Reactions on the surface.

EXERCISES

1. In general, which metal ions are better chelated by polymers such as the salt of poly(acrylic acid)?
2. What is the general mechanism for the curing of step-reaction polymers?
3. Write the formula for perdeuterated PE.
4. What is the major difference between reactions occurring on benzene and polystyrene.
5. When would you expect a polymer to undergo degradation via random scission?
6. How might you cross-link a PE coating after it is applied to a wire?
7. How might you prepare a block copolymer of styrene and an alternating copolymer of ethylene and propylene?

8. What is the similarity between completely hydrogenated *Hevea brasiliensis* and completely hydrogenated gutta percha?

9. What part does the porphyrin play in transmitting oxygen?

10. Would you expect PP chains with a DP of 200 to break when a PP plastic is broken into two pieces?

11. What are the most important structural factors involved in natural catalysis?

12. What product would be produced from the ozonolysis of polybutadiene?

13. Write the structural formula for the polymeric hydrolytic products from (a) PVAc and (b) poly(methyl methacrylate).

14. Why is commercial methylcellulose more soluble in water than native cellulose?

15. Why is CMC used in detergent formulations?

16. Would you expect the addition of chlorine or hydrogen to a double bond in a polymer to occur 100%?

17. What is the DS of cellulose nitrate when it is used as an explosive?

18. Why is the DS of cellulose triacetate only 2.8 and not 3.0?

19. Which is more polar: (a) cellulose triacetate or (b) cellulose diacetate?

20. Why is poly(vinyl butyral) not 100% vinyl butyral units?

21. If you have PVA and you want to make it less hydrophilic using a condensation reaction, what might you use?

22. What products would you expect from the combustion of most vinyl polymers?

23. What monomer would be obtained by the decomposition of PVA?

24. Which would be more resistant to nitric acid: (a) polystyrene or (b) perfluoropolystyrene?

25. What ions would be removed form water by sulfonated polystyrene: (a) cations or (b) anions?

26. What reaction occurs when tannic acid is added to proteins such as those present in cowhide?

27. Which of the following should be most susceptible to degradation by acids and bases; (a) PS, (b) PVC, (c) PET, (d) Nylon 6,6, (e) proteins, (f) DNA.

28. Propose a procedure for recovering monomeric methyl methacrylate from scrap PMMA.

29. Why is PVC so thermally unstable?

30. Why are synthetic polymer scientists looking at results obtained from the biomolecular scientists?

31. Why is the active portion of myoglobin "hidden?"

32. What role does the porphyrin structure play in the activity hemoglobin?

33. Outline the three general steps in operation for photosynthesis.

ADDITIONAL READING

Adler, H. P. (2005): *Reactive Polymers*, Wiley, NY.

Cheng, S., Gross, R. (2008): *Polymer Biocatalysis and Biomaterials*, Oxford University, Ithaca, NY.

Chiellini, E. (2003): *Biodegradable Polymers and Plastics*, Kluwer, NY.

Ciferri, A. (2005): *Supramolecular Polymers*, 2nd Ed., Taylor and Francis, Boca Raton, FL.

Kausch, H. H. (2005): *Intrinsic Molecular Mobility and Toughness of Polymers*, Springer, NY.

Kharitonov, A. (2008): *Direct Fluorination of Polymers*, Nova, Hauppauge, NY.

Lee, Y. (2008). *Self-Assembly and Nanotechnology*: A Force Balance Approach, Wiley, Hoboken, NJ.

Lyland, R., Browning, I. (2008): *RNA Interference Research Progress*, Nova, Hauppauge, NY.

Malmsten, M. (2003): *Biopolymers at Interfaces*, 2nd Ed., Taylor and Francis, Boca Raton, FL.

Moeller, H. (2007): *Progress in Polymer Degradation and Stability Research*, Nova, Hauppauge, NY.

Rotello, V., Rotello, K. V., Thayumanavan, S. (2008): *Molecular Recognition and Polymers: Control of Polymer Structure and Self-Assembly*, Wiley, Hoboken, NJ.

Rudnik, E. (2008): *Compostable Polymer Materials*, Elsevier, NY.

Scott, G. (2003): *Degradable Polymers*, Kluwer, NY.

van Leeuwen, P. (2008): *Supramolecular Catalysis*, Wiley, Hoboken, NJ. Witten, T. A. (2004): *Structured Fluids*, Oxford University Press, NY.

Yan, M., Ramstrom, O. (2005): *Molecularly Imprinted Materials*, Taylor and Francis, Boca Raton, FL.

17 Synthesis of Reactants and Intermediates for Polymers

Many of the reactants used for the production of polymers are standard organic chemicals. However, because of the high purity requirements and large amounts needed, special conditions have been developed that allow large amounts of high purity reactants to be made in high yield. The first section of this chapter deals with the availability of the general feedstocks. The remaining sections deal with the synthesis of particular polymer reactants.

There is an industrial turn toward green materials. Much of this is found in developing processes that allow the creation of monomers from natural resources. You will see the move toward the creation of synthetic polymers from natural materials as you view Sections 17.2 and 17.3. The emphasis on green chemistry is also found when reactions produce "environmentally friendly" byproducts."

17.1 MONOMER SYNTHESIS FROM BASIC FEEDSTOCKS

Most of the monomers widely employed for both vinyl and condensation polymers are derived indirectly from simple feedstock molecules. This synthesis of monomers is a lesson in inventiveness. The application of the saying that "necessity is the mother of invention" has led to the sequence of chemical reactions where little is wasted and byproducts from one reaction are employed as integral materials in another. Following is a brief look at some of these pathways traced from basic feedstock materials. It must be remembered that often many years of effort were involved in discovering the conditions of pressure, temperature, catalysts, and so on that must be present as one goes from the starting materials to the products.

Fossil fuels refer to materials formed from the decomposition of once living matter. Because these once living materials contain sulfur and heavy metals such as iron and cobalt, they must be removed either prior or subsequent to use.

The major fossil fuels are coal and petroleum. Marine organisms were typically deposited in mud and under water, where anaerobic decay occurred. The major decomposition products are hydrocarbons, carbon dioxide, water, and ammonium. These deposits form much of the basis for our petroleum resources. Many of these deposits are situated so that the evaporation of the more volatile products such as water and ammonia occurred, giving petroleum resources with little nitrogen- or oxygen-containing products. By comparison, coal is formed from plant material that has decayed to graphite carbon and methane.

Only about 5% of the fossil fuels consumed today are used as feedstocks for the production of today's synthetic carbon-based products. This includes the products produced by the chemical and drug industries with a major portion acting as the feedstocks for plastics, elastomers, coatings, fibers, and so on.

The major petroleum resources contain linear, saturated hydrocarbons (alkanes), cyclic alkanes, and aromatics. For the most part, this material is considered to have low free-energy content.

Raw or crude petroleum materials are separated into groups of compounds with similar boiling points by a process called *fractionation*. Table 15.1 contains a brief listing of typical fraction-separated materials. Accompanying or subsequent to this fractionation occurs a process called "cracking" whereby the hydrocarbon molecules are heated over catalysts that allow the hydrocarbons molecules to break up and then reform into structures that contain more branching that allow for appropriate

TABLE 17.1
Typical Straight Chain Hydrocarbon Fractions Obtained from Distillation of Petroleum Resources

Boiling Range (°C)	Average Number of Carbon Atoms	Name	Uses
<30	1–4	Gas	Heating
30–180	5–10	Gasoline	Automotive fuel
180–230	11, 12	Kerosene	Jet fuel, heating
230–300	13–17	Light gas oil	Diesel fuel, heating
300–400	18–25	Heavy gas oil	Heating

combustion in our automobiles and trucks. Under other conditions, the cracking allows the formation of other desired feedstock molecules, including methane, ethane, ethylene, propylene, benzene, and so on, that eventually become our plastics, fibers, elastomers, sealants, coatings, composites, and so on.

In 1925, Phillips Petroleum Company was only one of dozens of small oil companies in Oklahoma. The only distinction was the large amount of natural gasoline (or naphtha), the lightest liquid fraction (Table 17.1), found in its crude oil. As was customary, Phillips, and most of the other oil companies of the time, employed a distillation process to isolate the butane and propane. Even so, they were sued for the use of this distillation process probably because they were small, had no real research capacity of their own, and no real legal defense team. Frank Phillips elected to fight supposedly including an argument that the ancient Egyptians had used a similar process to create an alcoholic equivalent to an Egyptian alcoholic drink. Phillips won the suit but became convinced that if the company was to remain successful it would need to have a research effort.

During the early dust bowl years, 1935, they established the oil industry's first research team in Bartlesville, OK. George Oberfell, hired by Phillips to fight the lawsuit, planned the initial research efforts that involved three main initiatives. First, develop technology to use light hydrocarbons in new ways as motor fuels. Second, develop markets for butane and propane. Finally, find new uses for the light hydrocarbons outside the fuel market. All three objectives were achieved.

Frederick Frey and Walter Shultze were instrumental early researchers. Frey was among the first to dehydrogenate paraffins catalytically to olefins and then the olefins to diolefins that serve as feedstocks to the production of many of today's polymers. In competition with Bakelite, he discovered the preparation of polysulfone polymers made from the reaction of sulfur dioxide and olefins creating a hard Bakelite-like material. Frey and Schultz also developed a process that allowed the production of 1,3-butadiene from butane that allowed the synthesis of synthetic rubber.

Probably Frey's most important invention involved the use of hydrogen fluoride to convert light olefins, produced as byproducts of a catalytic cracker, into high octane motor and aviation fuels. This process is still widely used. It came at a critical time for America's World War II efforts allowing fuel production for the Allied forces. This fuel allowed aircraft faster liftoffs, more power, and higher efficiency.

The major one carbon feedstock is methane and it serves as the feedstock to a number of important monomers, including hexamethylene tetramine and melamine, used in the synthesis of a number of cross-linked thermosets as well as vinyl acetate, ethylene, ethylene glycol, and methyl methacrylate (Table 17.1).

Formaldehyde, produced in the methane stream, serves as the basis for the formaldehyde-intensive resins, namely phenol-formaldehyde, urea-formaldehyde (UF), and melamine-formaldehyde resins, as noted above. Formaldehyde is also involved in the synthesis of ethylene glycol, one of the two comonomers used in the production of polyethylene terephthalate (PET). Formaldehyde also serves as the basic feedstock for the synthesis of polyacetals.

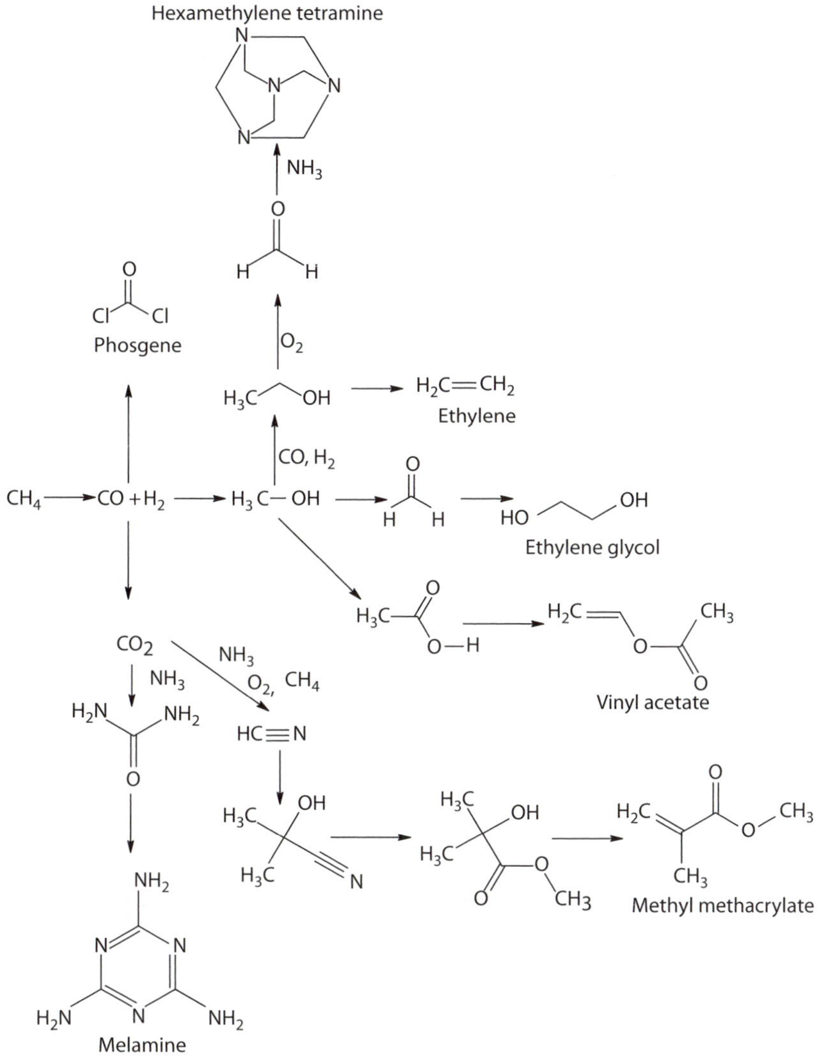

FIGURE 17.1 Monomer synthesis chemical flow diagram based on methane feedstock.

Another important use for methane is its conversion to synthesis gas (or syn-gas), a mixture of hydrogen gas and carbon monoxide as shown in Figure 17.1. Synthesis gas can also be derived from coal. When this occurs, it is called *water gas*. Interestingly, the reaction of methane giving carbon monoxide and hydrogen can be reversed so that methane can be produced from coal through this route.

One major two-carbon feedstock is ethylene. From Figure 17.2 you can see that a number of the monomers are directly synthesized from ethylene. Again, while the "react" arrow goes directly from ethylene to the product, as noted above, it often took years to develop an economical procedure to obtain the product in essentially 100% yield. Here, depending on the reactions conditions, a wide variety of intermediates and products are formed that allow the synthesis of a number of polymers, including poly(acrylic acid), poly(vinyl chloride), polystyrene, poly(vinyl acetate), polyesters (in particular PET), and poly(methyl acrylate). Of course, ethylene is itself part of the polymer feedstock pool being the feedstock for all of the polyethylenes.

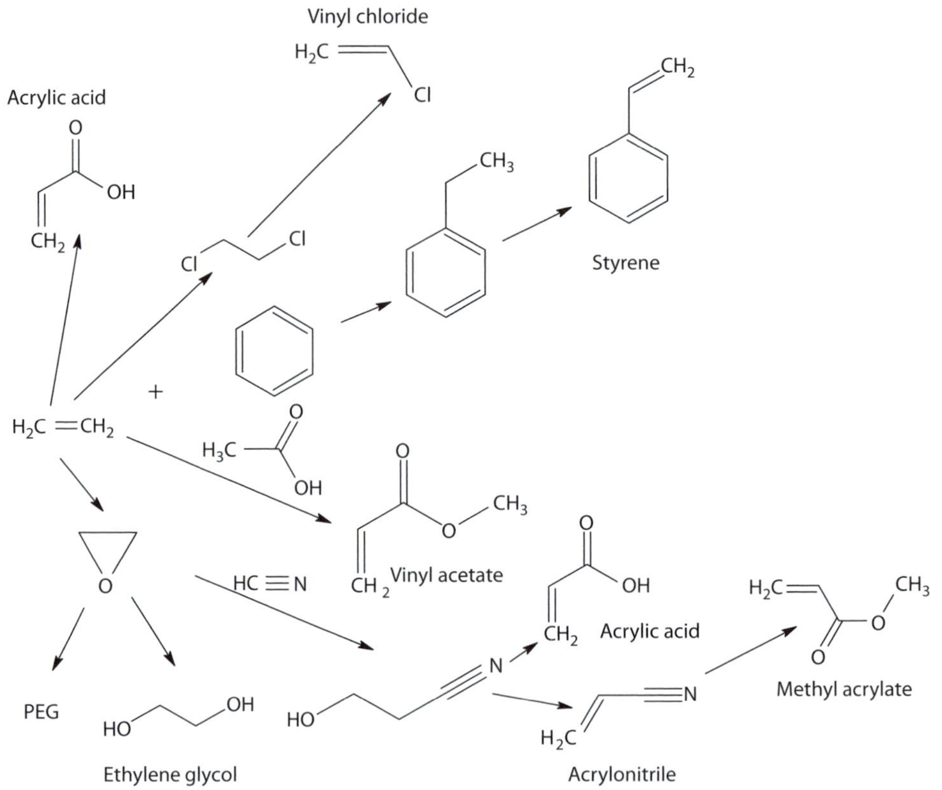

FIGURE 17.2 Monomer synthesis chemical flow diagram based on ethylene feedstock.

Another two-carbon feedstock is acetylene. Acetylene is typically obtained from coal by converting coke calcium carbide and then treating the calcium carbide with water. As shown in Figure 17.3, a number of important monomers can be made from acetylene. Even so, because of the abundance of other feedstocks from petroleum reserves, only some of the routes shown in Figure 17.3 are widely used.

Propylene is the basic three-carbon building block (Figure 17.4). Again, its polymerization gives polypropylene. The ingeniousness of some of the synthetic routes is shown in the conversion of benzene, through reaction with propylene, to cumene and the consequent oxidation forming phenol and acetone that is subsequently converted to bisphenol A, a basic building block for certain polyesters. Phenol is involved in the synthesis of the phenol-formaldehyde resins, adipic acid, and 1,6-hexamethylenediamine. Acetone, in turn, is also involved in numerous important synthetic steps either as a reactant or solvent. It is involved in the synthesis of methyl methacrylate and isoprene.

The major four-carbon feedstock molecules are 1,3-butadiene and isobutylene, both involved in the synthesis of many monomers and intermediates. Butadiene is copolymerized with styrene to form styrene–butadiene rubber (SBR) and with acrylonitrile to form acrylonitrile–butadiene–styrene (ABS) rubbers.

Benzene forms the basis for a number of monomers (Figure 17.5), including those that retain their aromatic character like styrene, and those that do not, like adipic acid.

In summary, monomer synthesis from basic, readily available inexpensive feedstocks based on fossil fuels is both an art and a science developed over the past half century or so. It represents a delicate balance and interrelationship between feedstocks and so-called byproducts from one reaction that become critical reactants in another reaction. Monomer and polymer synthesis continues to undergo change and improvement as the natural environment and societal and worker health

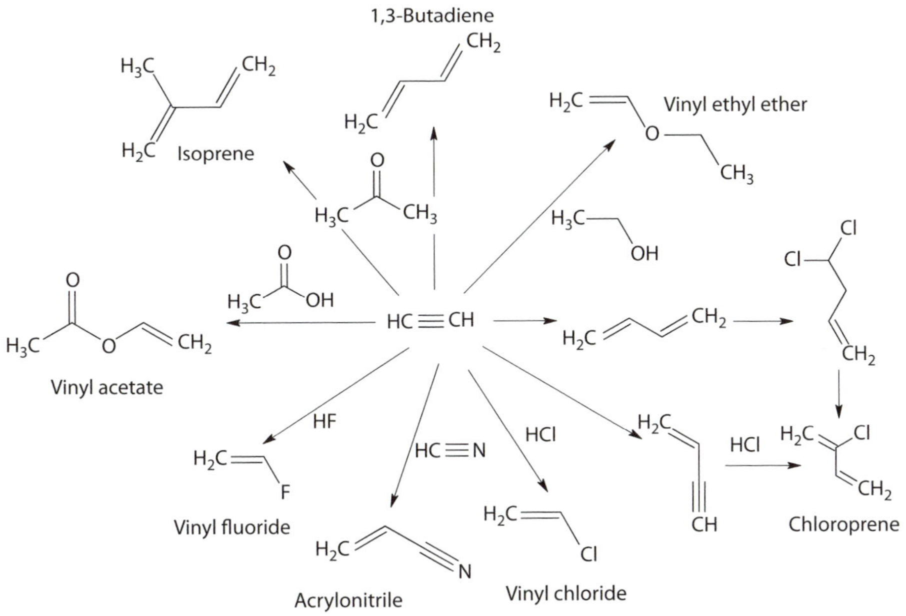

FIGURE 17.3 Monomer synthesis chemical flow diagram based on acetylene feedstock.

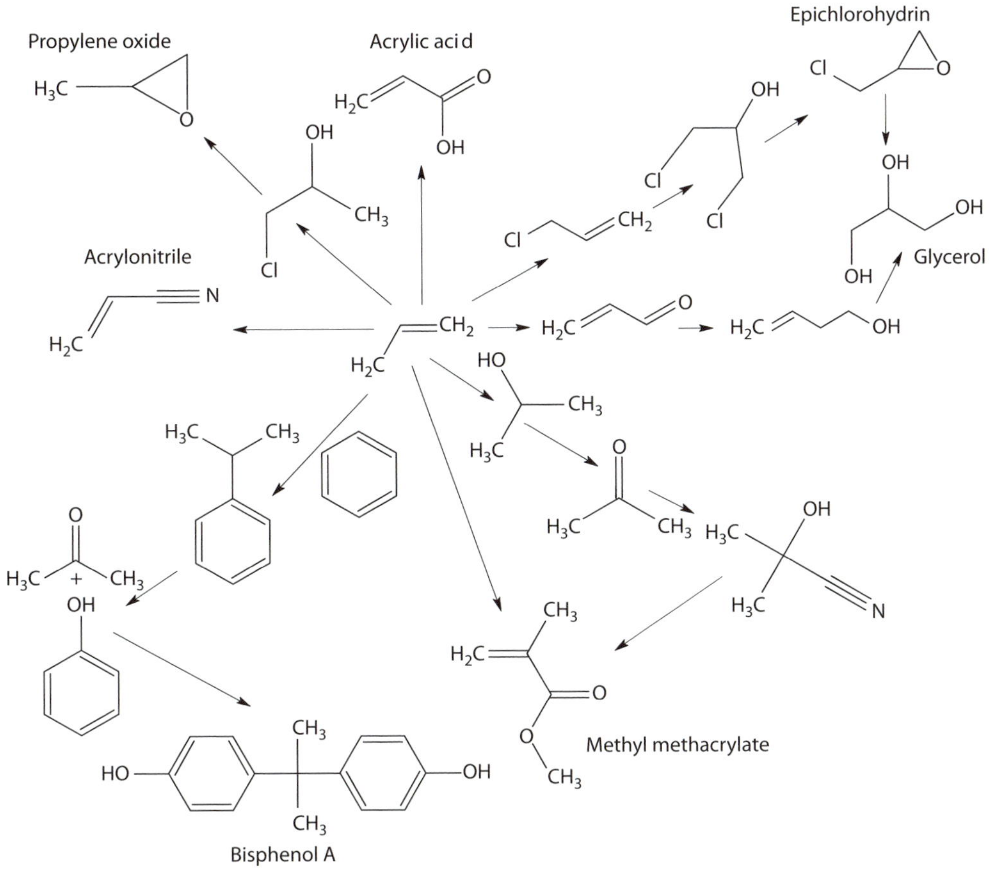

FIGURE 17.4 Monomer synthesis chemical flow diagram based on propylene feedstock.

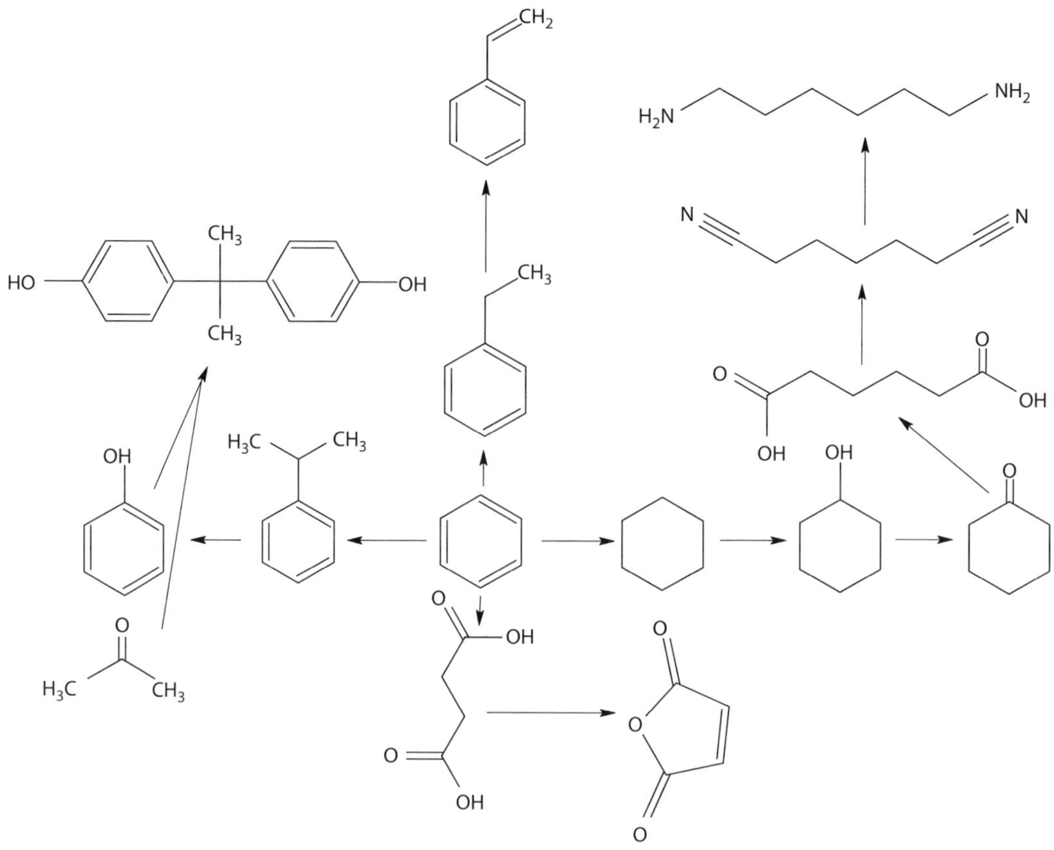

FIGURE 17.5 Monomer synthesis chemical flow diagram based on benzene feedstock.

continue to be dominant factors. Many of these and other monomers are now being produced by various microbes. Some of these efforts are described here and in Sections 19.15 through 19.17.

17.2 REACTANTS FOR STEP-REACTION POLYMERIZATION

Adipic acid (1,4-butanedicarboxylic acid) is used for the production of nylon-66 and may be produced from the oxidation of cyclohexane as shown below. Cyclohexane is obtained by the Raney nickel-catalytic hydrogenation of benzene. Both the cyclohexanol and cyclohexanone are oxidized to adipic acid by heating with nitric acid.

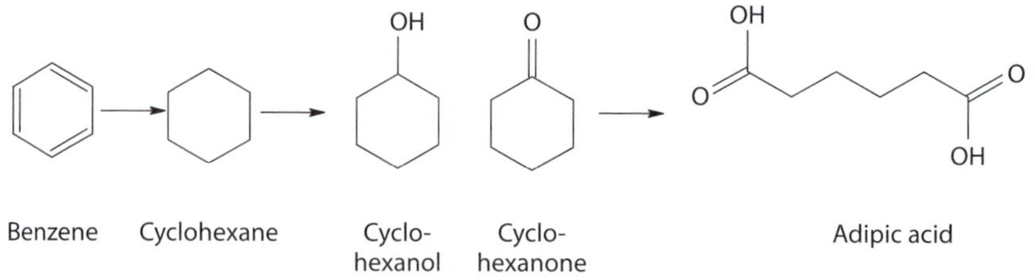

| Benzene | Cyclohexane | Cyclo-hexanol | Cyclo-hexanone | Adipic acid | (17.1) |

A new method that emphases green chemistry principles involves oxidizing cyclohexene with hydrogen peroxide through a tungsten catalyst and a phase-transfer catalyst, producing adipic acid and water as the byproduct.

Adipic acid can also be made from tetrahydrofurane (THF), obtained from furfural a naturally derived material (Equation 17.2). It is carbonylated in the presence of nickel carbonyl-nickel iodide catalyst. Furfural is a chemurgic product obtained by the steam-acid digestion of corn cobs, oat hulls, bagasse, or rice hulls.

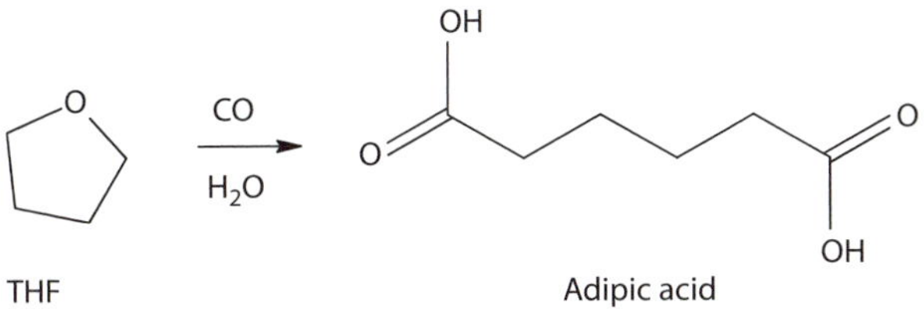

(17.2)

THF   Adipic acid

Adiponitrile may be produced from the hydrodimerization of acrylonitrile or from 1,3-butadiene via 1,4-dicyanobutene-2. Adiponitrile is then hydrogenated forming 1,6-hexanediamine.

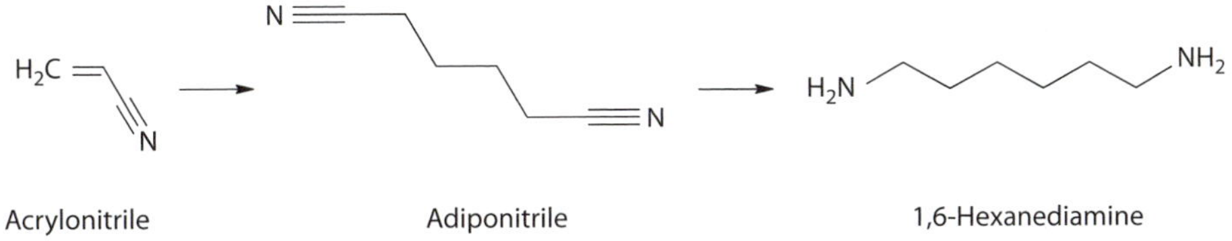

(17.3)

Acrylonitrile   Adiponitrile   1,6-Hexanediamine

1,6-Hexanediamine can also be made by the liquid-phase catalytic hydrogenation of adiponitrile or adipamide, which is made from adipic acid.

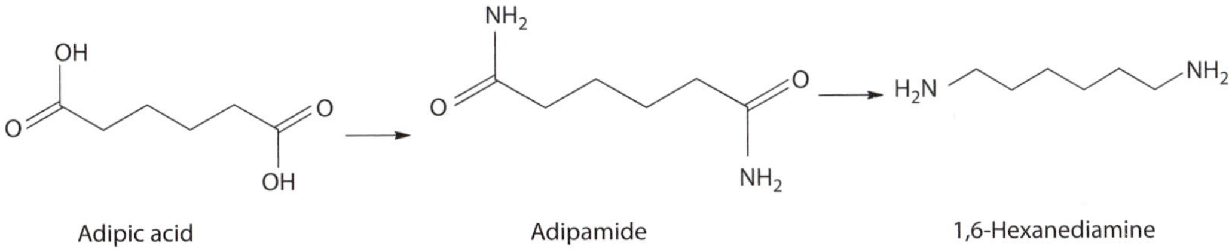

Adipic acid   Adipamide   1,6-Hexanediamine

(17.4)

Sebacic acid (1,8-octane dicarboxylic acid), which is used to make nylon-610, has been produced from 1,3-butadiene and by the dry distillation of caster oil (ricinolein). The cleavage of ricinoleic acid gives 2-octanol and the salt of sebacic acid.

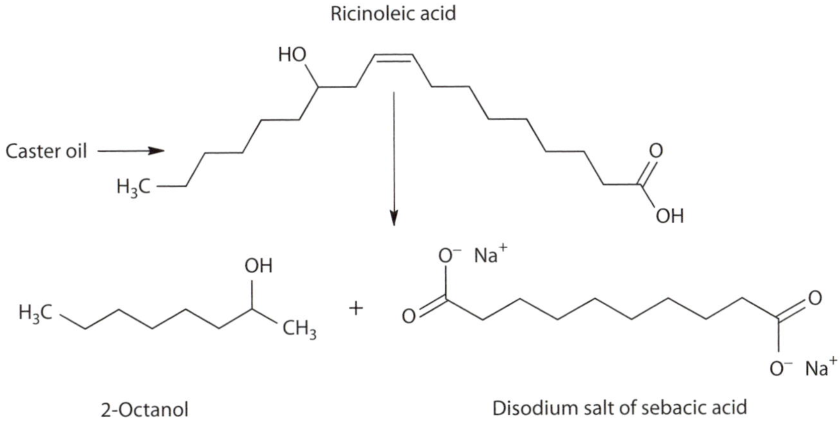

Ricinoleic acid

Caster oil →

(17.5)

2-Octanol   Disodium salt of sebacic acid

Phthalic acid (1,2-benzene dicarboxylic acid), isophthalic acid (1,3-benzene dicarboxylic acid), and terephthalic acid (1,4-benzene dicarboxylic acid) are made by the selective oxidation of the corresponding xylenes. Terephthalic acid may also be produced from the oxidation of naphthalene and by the hydrolysis of terephthalonitrile. The oxidation of p-xylene by oxygen from the air is generally done using acetic acid as a solvent in the presence of a catalyst such as cobalt–manganese. The yield is close to 100%.

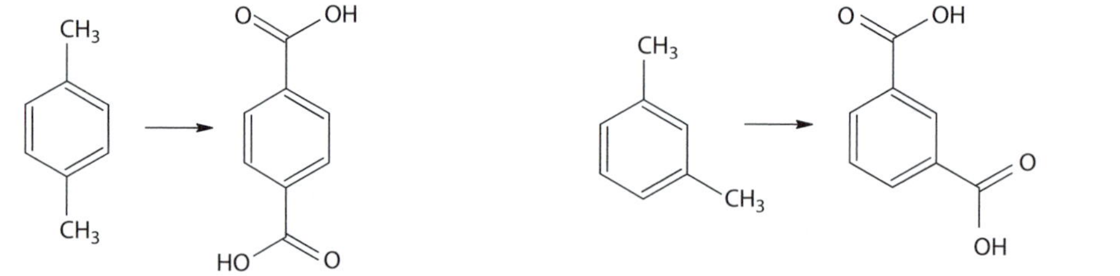

p-Xylene   Terephthalic acid (17.6)   m-Xylene   Isophthalic acid (17.7)

Malic anhydride (2,5-furandione) is made as a byproduct in the production of phthalic anhydride, and by the vapor phase oxidation of butylene or crotonaldehyde. It is also obtained by the dehydration of maleic acid and by the oxidation of benzene. Maleic anhydride is used for the production of unsaturated polyester resin. This reactant, like many reactants, is fairly toxic and should be treated as such.

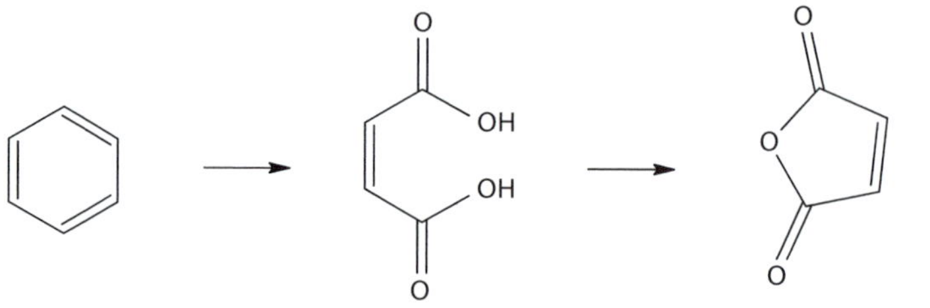

(17.8)

Benzene   Maleic acid   Maleic anhydride

2-Pyrrolidone is a lactone used for the production of nylon-4. This reactant may be produced by the reduction ammoniation of maleic anhydride. ε-Caprolactam, used in the production of nylon-6, may be produced by the Beckman rearrangement of cyclohexanone oxime (Equation 17.11). The oxime may be produced by the catalytic hydrogenation of nitrobenzene, the photolytic nitrosylation of cyclohexane (Equation 17.9), or the reaction of cyclohexanone and hydroxylamine (Equation 17.10). Nearly one-half of the production of caprolactam is derived from phenol.

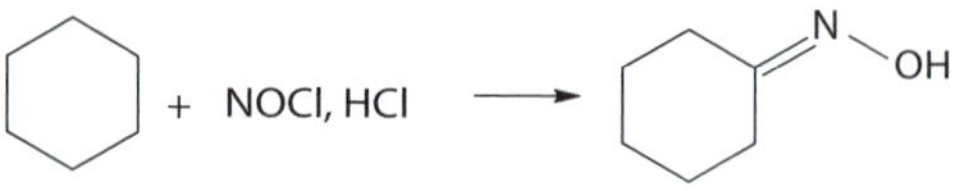

+ NOCl, HCl   (17.9)

+ O_2 (17.10)

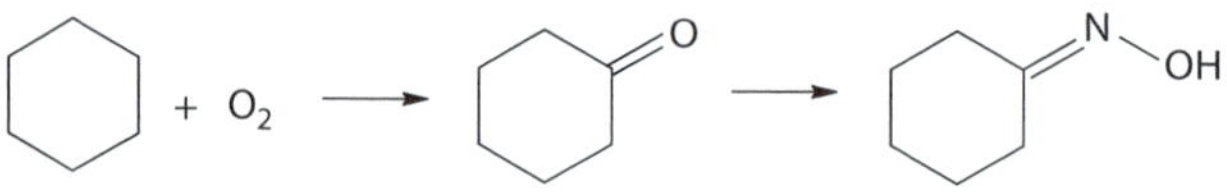

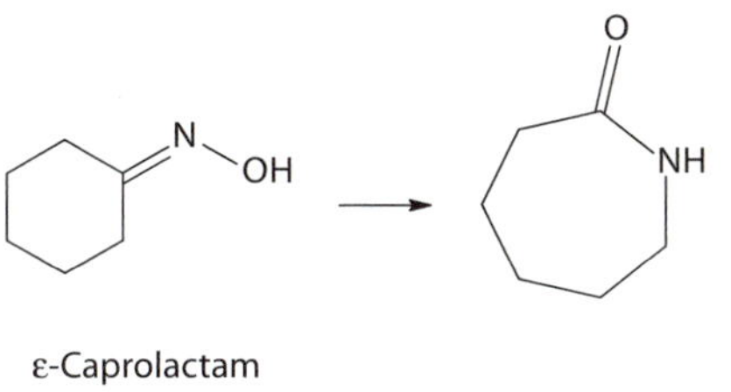

(17.11)

ε-Caprolactam

Ethylene oxide, used for the production of ethylene glycol and poly(ethylene oxide) (PEO), is obtained by the catalytic oxidation of ethylene. Ethylene glycol, used in the production of PET, is produced by the hydrolysis of ethylene oxide.

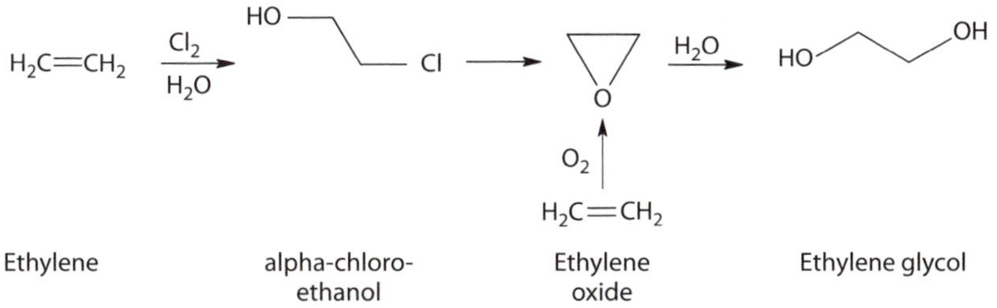

(17.12)

| Ethylene | alpha-chloro-ethanol | Ethylene oxide | Ethylene glycol |

Glycerol, used for the production of alkyds, is produced by the catalytic hydroxylation or the hypochlorination of allyl alcohol. Allyl alcohol is produced by the reduction of acrolein, that is in turn, obtained by the oxidation of propylene.

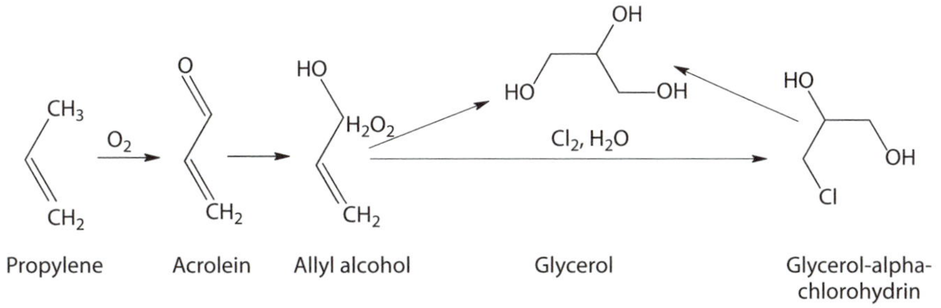

(17.13)

| Propylene | Acrolein | Allyl alcohol | Glycerol | Glycerol-alpha-chlorohydrin |

Until recently, glycerol was obtained mainly from epichlorohydrin. But because glycerol forms the backbone of fats, it is produced as the fats are degraded giving the fatty acid and glycerol. It is also a byproduct of the production of biodiesel production through the transesterification of vegetable oils. Thus, glycerol is almost solely obtained today from these sources.

Pentaerythritol, used in the production of alkyds, is produced by a crossed Cannizzaro reaction of the aldol conensation product of formaldehyde and acetaldehyde. The byproduct formate salt is a major source of formic acid.

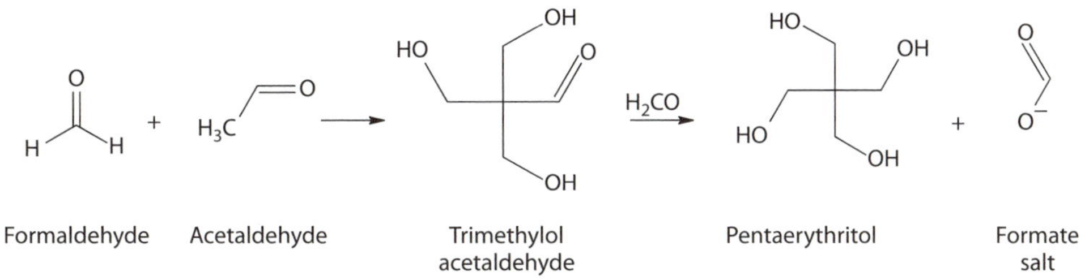

(17.14)

| Formaldehyde | Acetaldehyde | Trimethylol acetaldehyde | Pentaerythritol | Formate salt |

2,4-Toluene diisocyanate (TDI), used for the production of polyurethanes and polyureas, is obtained by the phosgenation of 2,4-toluenediamine. Phosgene is obtained by the reaction of chlorine and carbon monoxide.

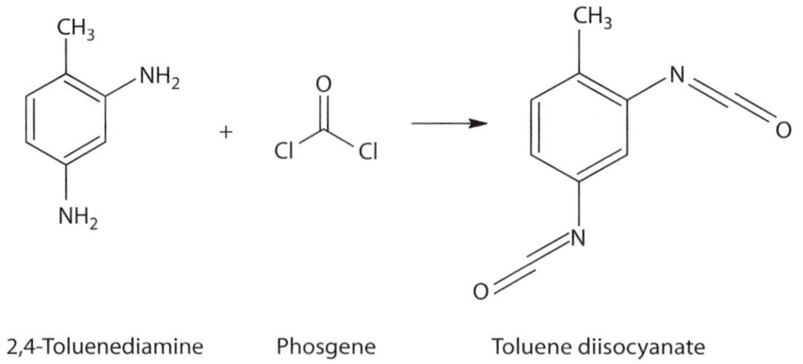

| 2,4-Toluenediamine | Phosgene | Toluene diisocyanate |

$$(17.15)$$

Formaldehyde is employed as a basic unit for many industrial adhesives such as the phenolic plastics formed from reaction of phenol and formaldehyde. It also serves as one of the reactants in the formation of amino plastics in the production of UF resins. Formaldehyde can self condense forming the cyclic trimer trioxane and the polymer paraformaldehyde. It is industrially produced from the catalytic oxidation of methanol. In turn, hexamethylenetetramine is produced by the condensation of ammonia and 30% aqueous formaldehyde (formalin).

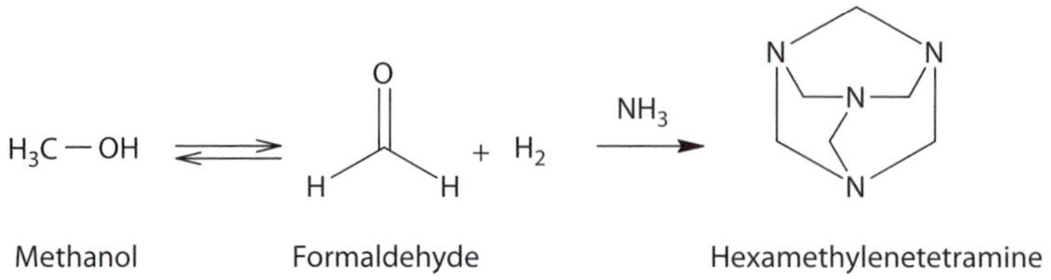

Methanol   Formaldehyde   Hexamethylenetetramine

$$(17.16)$$

While some phenol is produced by the nucleophilic substitution of chlorine in chlorobenzene by the hydroxyl group (Equation 17.17), most is produced by the acidic decomposition of cumene hydroperoxide (Equation 17.18) that also gives acetone along with the phenol. Some of the new processes for synthesizing phenol are the dehydrogenation of cyclohexanol, the decarboxylation of benzoic acid, and the hydrogen peroxide hydroxylation of benzene.

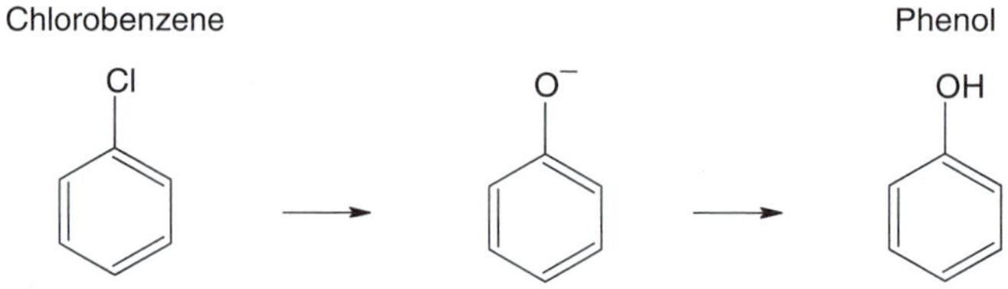

Chlorobenzene   Phenol

$$(17.17)$$

Benzene   Cumene   Cumene hydroperoxide   Phenol   Acetone

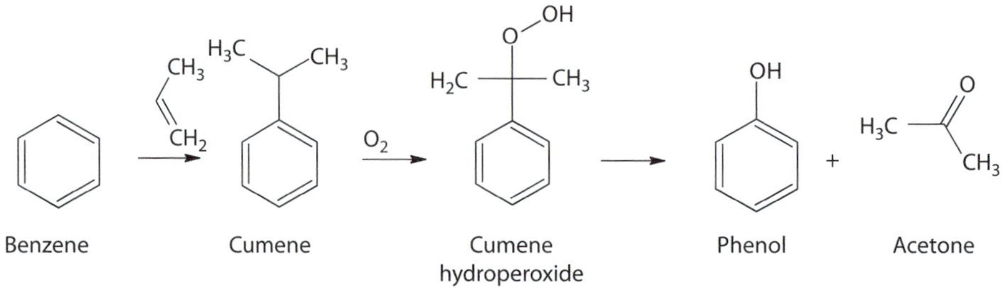

$$(17.18)$$

Urea is highly water soluble and offers an efficient avenue for the human body to expel excess nitrogen. The individual atoms come from water, carbon dioxide, aspartate, and ammonia and are involved in the urea cycle metabolic pathway. Urea, which is used for the production of UF resins, is made by the in situ decomposition of ammonium carbamate, which is made by the condensation of ammonia and carbon dioxide at 200°C and 200 atm.

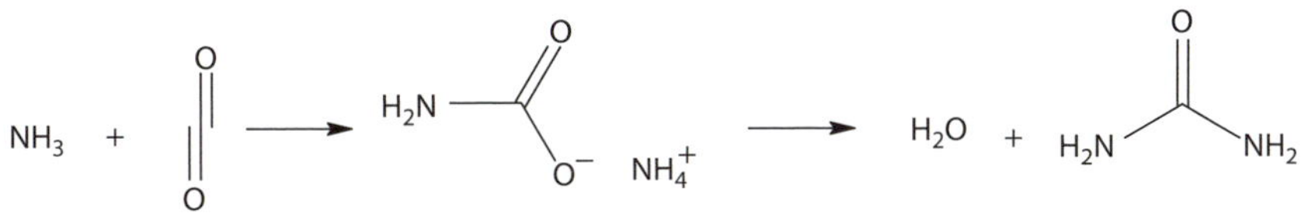

$$(17.19)$$

Ammonia Carbon dioxide   Ammonium carbamate   Urea

Melamine (cyanuramide), used in the production of melamine-formaldehyde resins, is obtained by heating dicyanodiamide (Equation 17.20), which is obtained by heating cyanamide. Today, most melamine is produced by heating urea (Equation 17.21).

Calcium   Cyanamide   Dicyanodiamide   Melamine
cyanamide

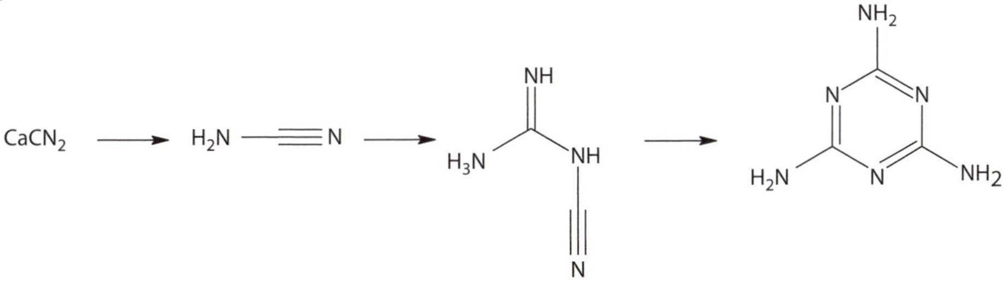

$$(17.20)$$

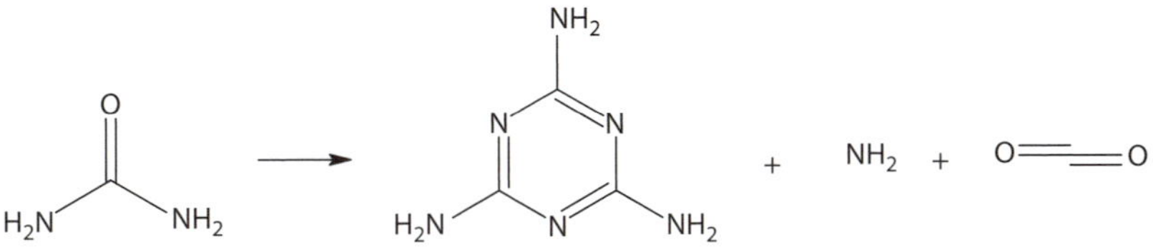

Urea   Melamine   Ammonia   Carbon dioxide (17.21)

Bisphenol A ([*bis*-4-hydroxphenol]dimethylmethane), used for the production of epoxy resins and polycarbonates, is obtained by the acidic condensation of phenol and acetone. Here, the carbonium ion produced by the protonation of acetone attacks the phenol molecule at the para position producing a quinoidal oxonium ion that loses water and rearranges to a p-isopropylphenol carbonium ion. The water attacks another phenol molecule, also in the para position, giving another quinoidal structure that rearranges to bisphenol A. It has been found that bisphenol A may be involved in one of the endocrine systems. The consequences of this are still being determined.

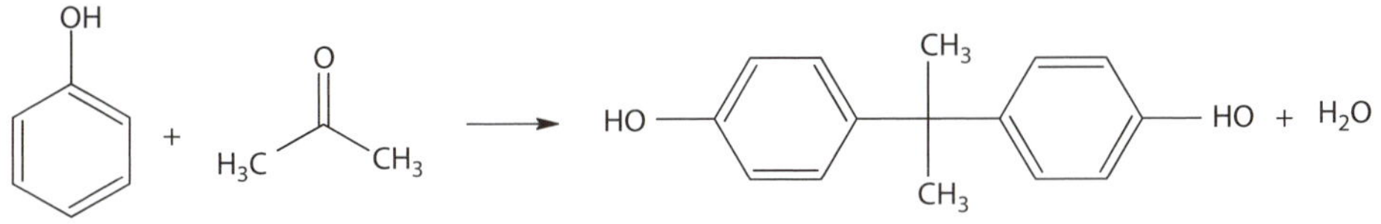

$$(17.22)$$

Phenol   Acetone   Bisphenol A

Epichlorohydrin (chloropropylene oxide) is used for the production of epoxy resins. It is produced by the dehydrochlorination of 2,3-dichloro-1-propanol (Equation 17.23). The hydrin is produced by the chlorohydrination of allyl chloride.

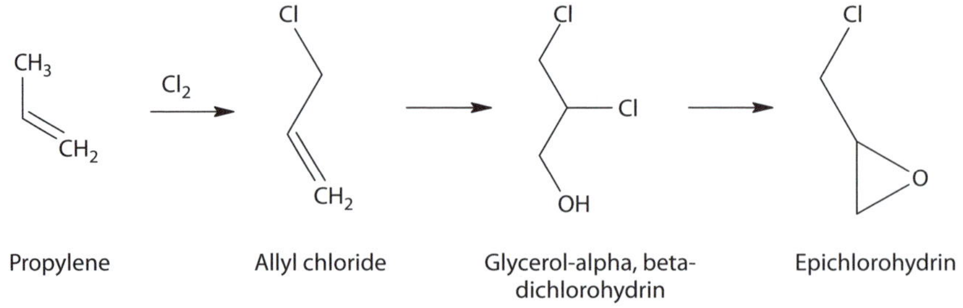

(17.23)

| Propylene | Allyl chloride | Glycerol-alpha, beta-dichlorohydrin | Epichlorohydrin |

Methyltrichlorosilane is produced by the Grignard reaction of silicon tetrachloride and methylmagnesium chloride (Equation 17.28). Dimethyldichlorosilane, used in the synthesis of polydimethylsiolxane, is obtained by the reaction of methylmagnesium chloride and methyltrichlorosilane (Equation 17.25).

$$H_3CMgCl + SiCl_4 \rightarrow H_2CSiCl_3 + MgCl_2 \qquad (17.24)$$

$$H_3CMgCl + H_3CSiCl_3 \rightarrow (H_3C)_2SiCl_2 + MgCl_2 \qquad (17.25)$$

17.3 SYNTHESIS OF VINYL MONOMERS

Styrene is generally produced by the catalytic vapor-phase dehydrogenation of ethylbenzene (Equation 17.26). Ethylbenzene is made by the Friedel-Crafts condensation of ethylene and benzene. Ethylbenzene, in the vapor phase, is passed over a solid catalyst bed. Most catalysts are based on iron(III) oxide containing potassium oxide or potassium carbonate. Steam serves several roles acting as an energy source and it removes coke that forms on the iron oxide catalyst. Typically, several reactors are used in series with each "pass" of the vapor ethylbenzene resulting in increased styrene production. The main byproducts are benzene and toluene. Styrene is also produced by the palladium acetate-catalyzed condensation of ethylene and benzene and by the dehydration of methylphenylcarbinol obtained by the propylation of ethylbenzene. Because of the toxicity of styrene, its concentration in the atmosphere must be severely limited.

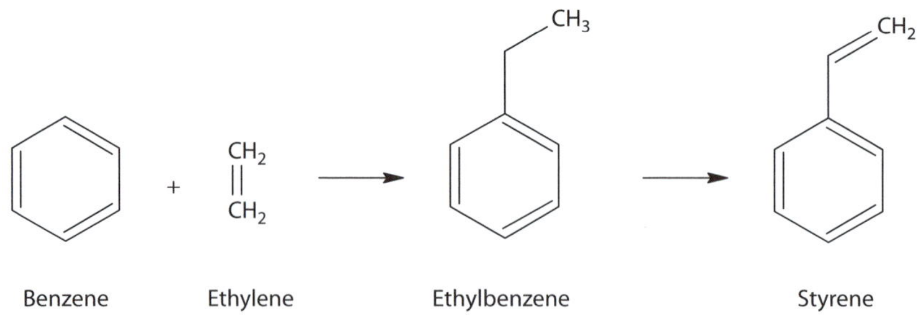

(17.26)

| Benzene | Ethylene | Ethylbenzene | Styrene |

A process to produce styrene monomer and propylene oxide simultaneously was introduced in 1969 and it is also employed to produce styrene industrially (Equation 17.27).

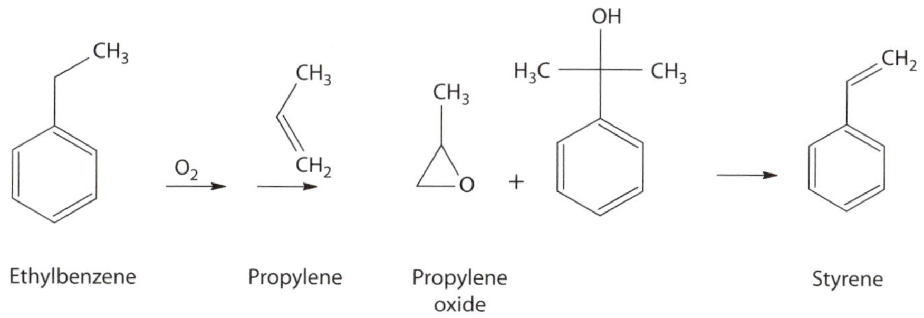

Ethylbenzene   Propylene   Propylene oxide   Styrene   (17.27)

Vinyl chloride, formerly obtained from acetylene, is now produced by the transcatalytic process where chlorination of ethylene, oxychlorination of the byproduct hydrogen chloride, and dehydro-chlorination occur in a single reactor.

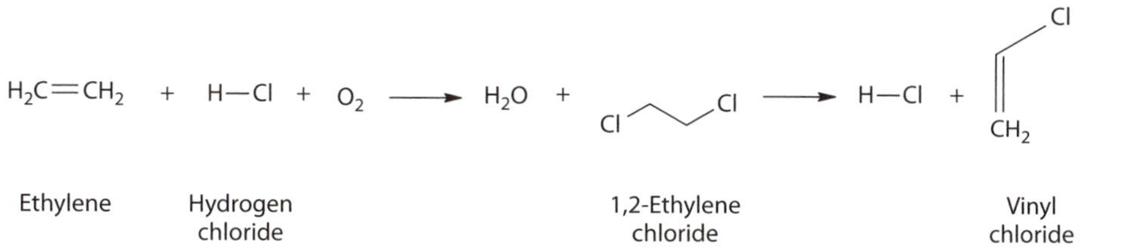

Ethylene   Hydrogen chloride   1,2-Ethylene chloride   Vinyl chloride   (17.28)

Vinyl chloride is also produced by the direct chlorination of ethylene and the reaction of acetylene and hydrogen chloride (Equation 17.29). The hydrogen chloride generated in the chlorination of ethylene can be employed in reaction with acetylene, allowing a useful coupling of these two reactions (Equation 17.30). Today, most vinyl chloride is produced from reaction of ethylene and chlorine forming ethylene dichloride. The ethylene dichloride is heated under pressure resulting in its decomposition to vinyl chloride and hydrogen chloride (Equation 17.29).

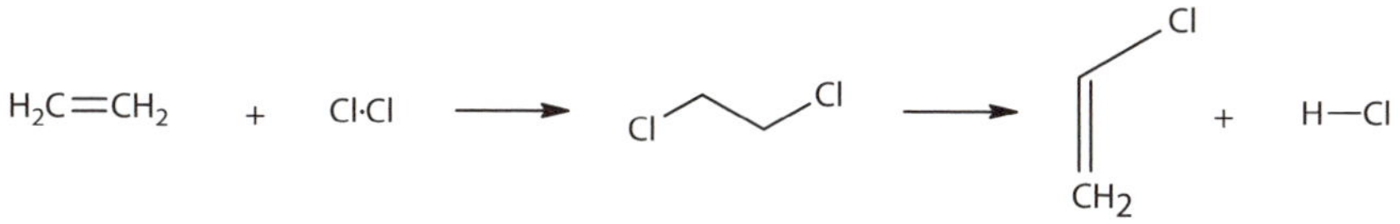

(17.29)

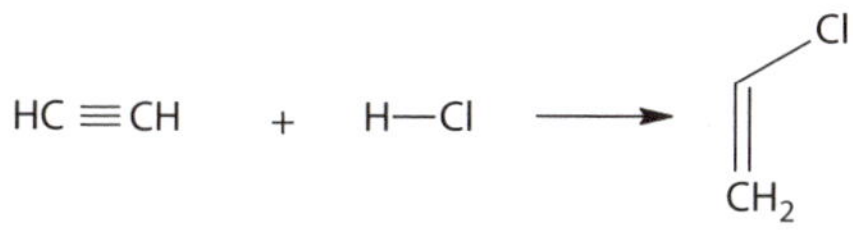

(17.30)

Vinylidene chloride, or 1,1-dichloroethene, is produced by the pyrolysis of 1,1,2-trichloroethane at 400°C in the presence of lime or base (Equation 17.31). Since both vinylidene chloride and vinyl chloride are carcinogenic, their concentrations must be kept low. It was widely used to form polymer, mainly poly(vinylidene chloride), which formed the basis for a cling wrap called *Saran warp*. Research suggested that Saran wrap may pose a health danger by leaching when microwaved. Because of this, most cling wrap formulations have changed to polythene.

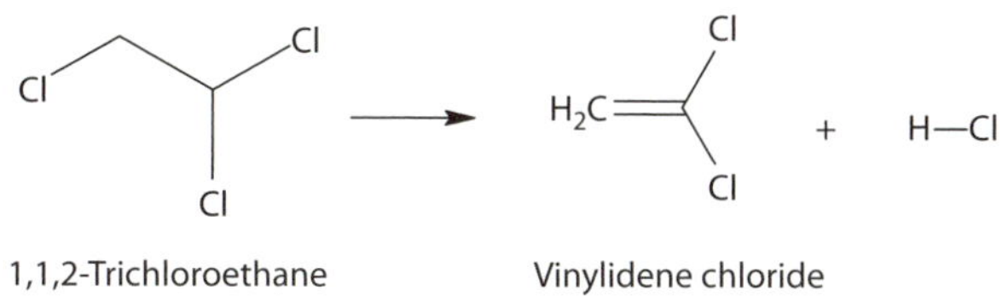

1,1,2-Trichloroethane   Vinylidene chloride   (17.31)

Vinyl acetate was produced by the catalytic acetylation of acetylene, but this monomer is now produced by the catalytic oxidative condensation of acetic acid and ethylene (Equation 17.32). Other vinyl esters can be produced by the transesterification of vinyl acetate with higher boiling carboxylic acids.

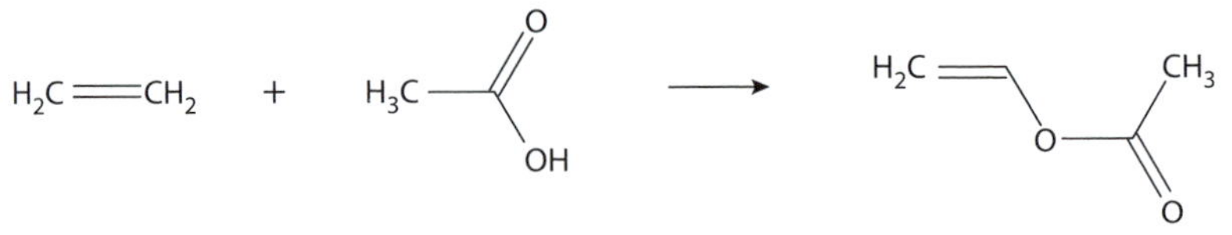

$$
\text{H}_2\text{C}=\text{CH}_2 \quad + \quad \text{H}_3\text{C}-\overset{\text{O}}{\underset{\text{OH}}{\text{C}}} \quad \longrightarrow \quad \text{H}_2\text{C}=\text{vinyl acetate structure} \tag{17.32}
$$

Ethylene   Acetic acid   Vinyl acetate

Acrylonitrile (vinyl cyanide) is produced by the Sohio process involving the ammoxidation of propylene. Again, since this monomer is carcinogenic, care must be taken to minimize exposure to it.

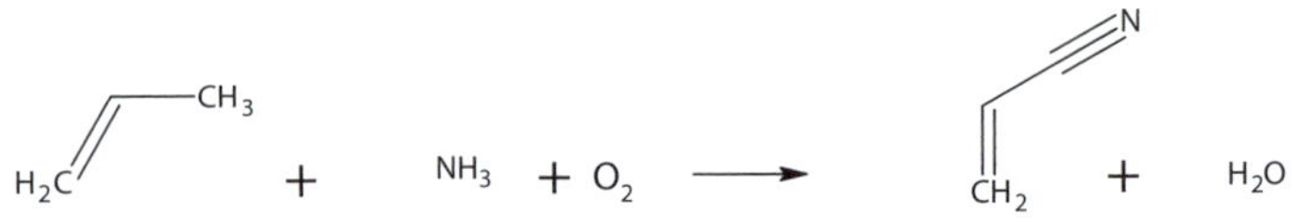

$$
\text{H}_2\text{C}=\text{CH—CH}_3 \quad + \quad \text{NH}_3 + \text{O}_2 \quad \longrightarrow \quad \text{acrylonitrile} \quad + \quad \text{H}_2\text{O} \tag{17.33}
$$

Propylene   Ammonia   Acrylonitrile

Tetrafluoroethylene is produced from the thermal dehydrochlorination of chlorodifluoromethane (Equation 17.35), which, in turn, is produced from chloroform and hydrogen flouride (HF) (Equation 17.34).

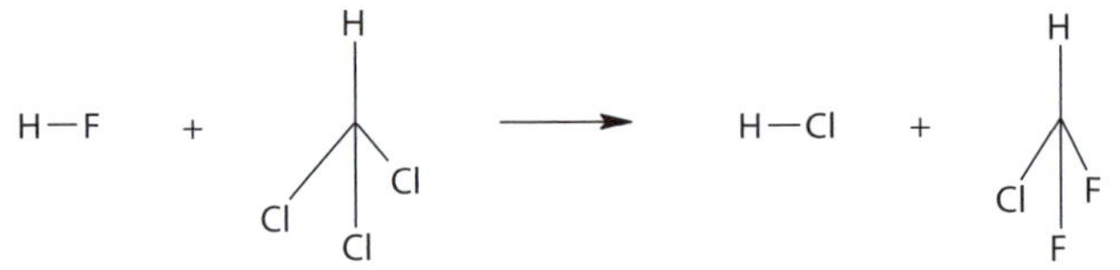

Hydrogen   Chloroform   Hydrogen   Chlorodifluoro-
fluoride   chloride   methane

$$
\text{H—F} \quad + \quad \text{CHCl}_3 \quad \longrightarrow \quad \text{H—Cl} \quad + \quad \text{CHClF}_2 \tag{17.34}
$$

$$
\text{CHClF}_2 \quad \longrightarrow \quad \text{H—Cl} \quad + \quad \text{CF}_2=\text{CF}_2 \tag{17.35}
$$

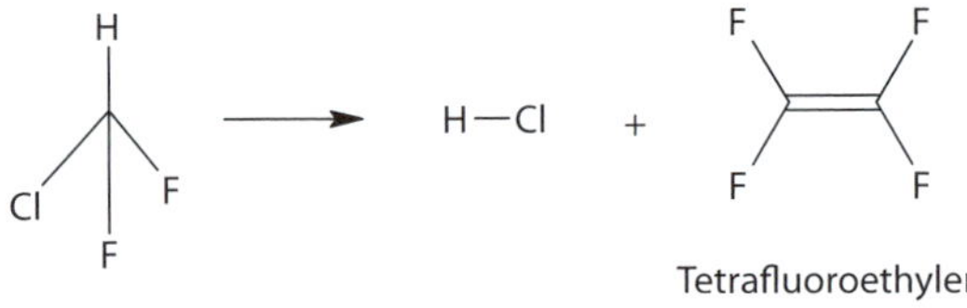

Tetrafluoroethylene

Trifluoromonochloroethylene is obtained from the zinc metal dechlorination of trichlorotrifluoroethane. The latter is produced by the fluorination of hexachloroethane.

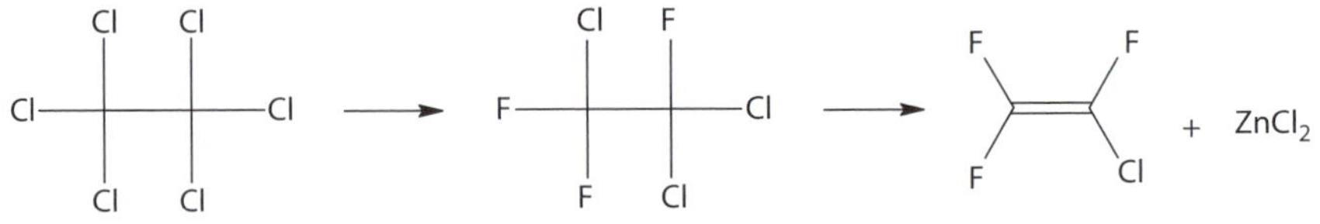

$$
\text{C}_2\text{Cl}_6 \quad \longrightarrow \quad \text{CCl}_2\text{F—CClF}_2 \quad \longrightarrow \quad \text{CF}_2=\text{CClF} \quad + \quad \text{ZnCl}_2 \tag{17.36}
$$

Hexachloroethane   Trichlorotrifluoro-   Trifluoromonochloro
                          ethane   ethylene

Vinylidene fluoride is produced by the thermal dehydrochlorination of 1-monochloro-1,1,-difluoroethane.

1-Monochloro-1,1-di-
fluoroethane

Vinylidene fluoride

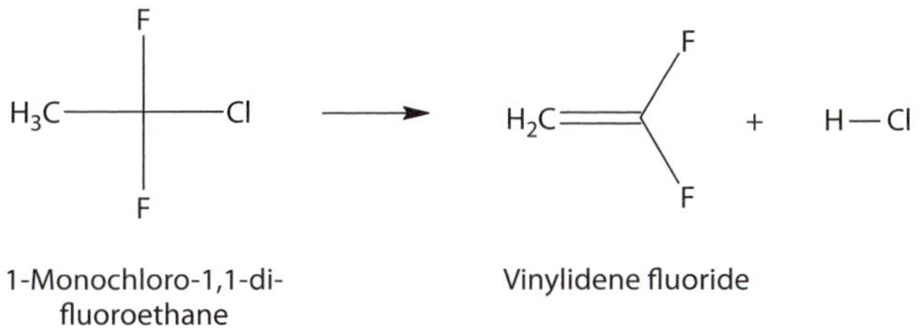

(17.37)

Vinyl fluoride may be obtained by the catalytic hydrofluorination of acetylene (Equation 17.38).

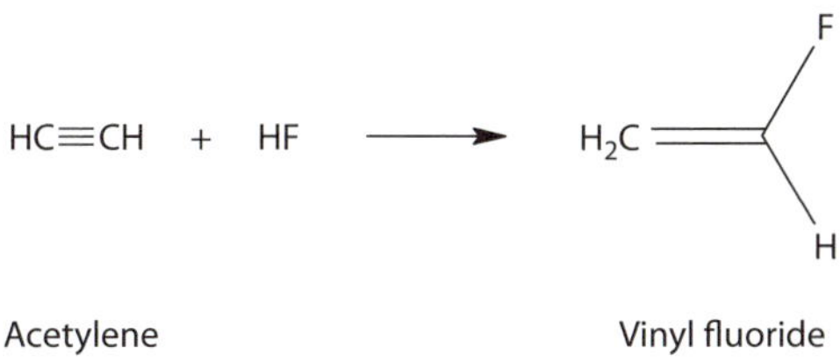

$HC \equiv CH$ + HF

Acetylene

Vinyl fluoride

(17.38)

Vinyl ethyl ether is obtained by the ethanolysis of acetylene in the presence of potassium ethoxide.

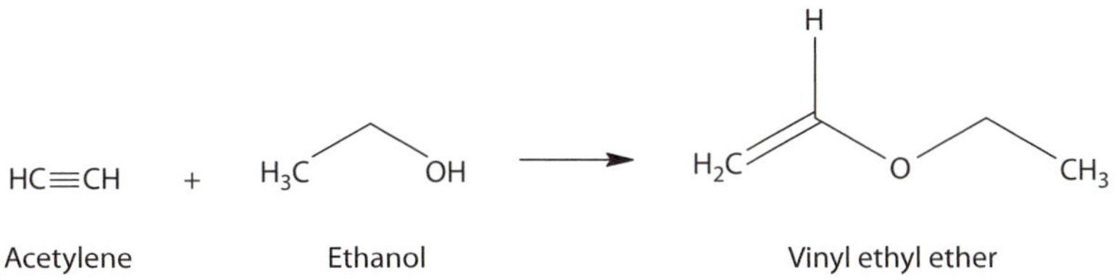

$HC \equiv CH$ +

Acetylene

Ethanol

Vinyl ethyl ether

(17.39)

1,3-Butadiene, used for the production of elastomers, is produced by the catalytic thermal cracking of butane and as a byproduct of other cracking reactions.

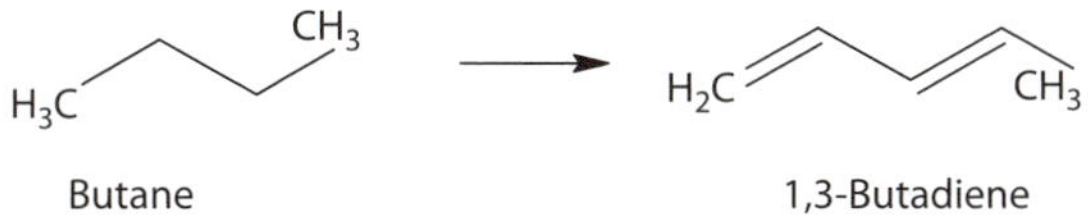

Butane

1,3-Butadiene

(17.40)

While butadiene is produced as a byproduct of the steam cracking process used in the production of ethylene and other olefins in the United States, Europe, and Japan, it is produced in other parts of the world from ethanol, a green material. In the single-step process, ethanol is converted to butadiene, hydrogen, and water through passing the ethanol over metal oxide catalysts. In a two-step process, ethanol is oxidized to acetaldehyde, which then reacts with additional ethanol-producing butadiene.

The isoprene monomer is not readily available from direct cracking processes. Several routes are employed for its synthesis. One route begins with the extraction of isoamylene fractions from catalytically cracked gasoline streams. Isoprene is produced by subsequent catalytic dehydrogenation.

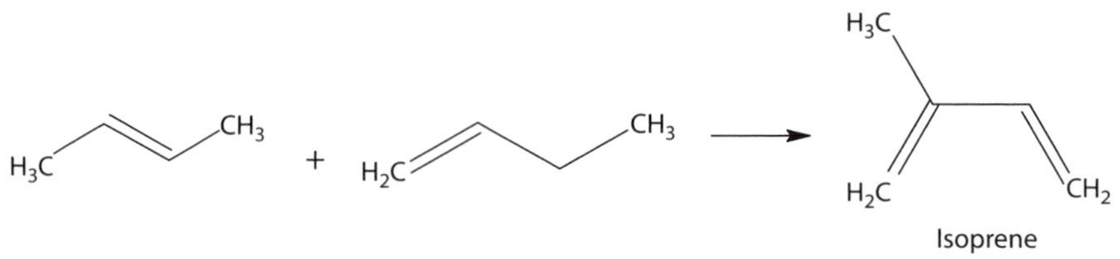

$$(17.41)$$

Isoprene

Dimerization of propylene is also used to produce isoprene. Several steps are involved. Initially, dimerization of propylene to 2-methyl-1-pentene occurs. Then isomerization to 2-methyl-2-pentene is effected. Finally, the 2-methyl-2-pentene is pyrolyzed to isoprene and methane. Another multistep synthesis starts with acetylene and acetone. Perhaps the most attractive route involves formaldehyde and isobutylene (Equation 17.42).

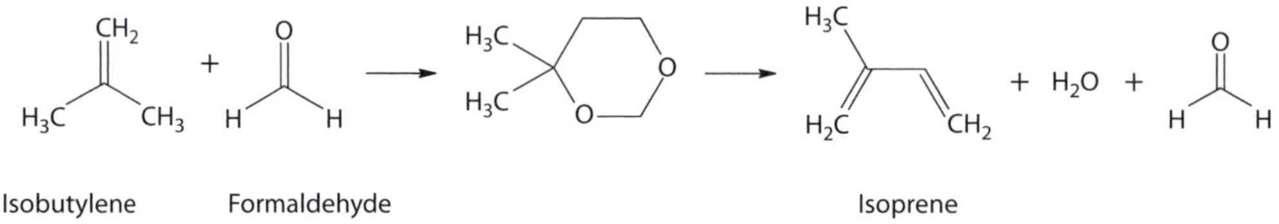

Isobutylene   Formaldehyde   Isoprene

$$(17.42)$$

Chloroprene, used for the production of neoprene rubber, is obtained by the dehydrochlorination of dichlrobutene. The latter is produced by the chlorination of 1,3-butadiene, which, in turn, is synthesized from acetylene.

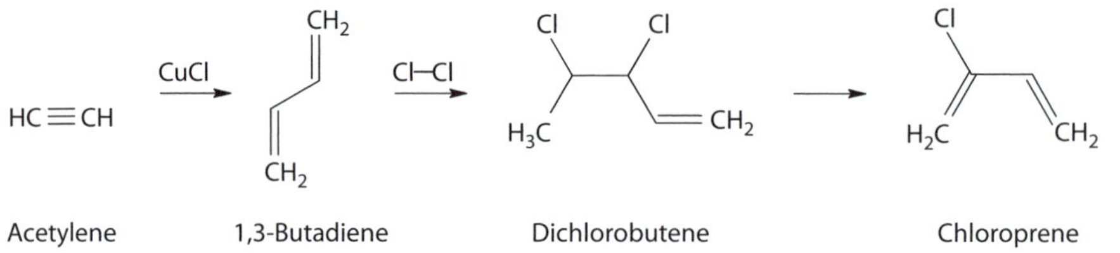

Acetylene   1,3-Butadiene   Dichlorobutene   Chloroprene $\quad(17.43)$

Acrylic acid can be prepared by the catalytic oxidative carbonylation of ethylene or by heating formaldehyde and acetic acid in the presence of KOH.

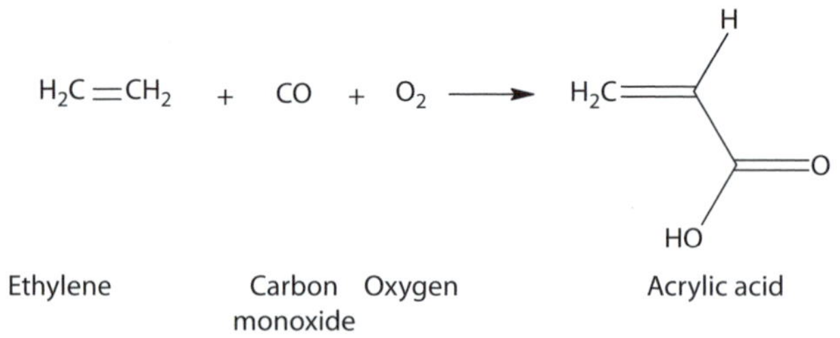

$$H_2C=CH_2 \;+\; CO \;+\; O_2 \longrightarrow$$

$$(17.44)$$

Ethylene   Carbon Oxygen   Acrylic acid
                 monoxide

Methyl acrylate may be obtained by the addition of methanol to the reactants in the previous synthesis (Equation 17.44) for acrylic acid or by the methanolysis of acrylonitrile (Equation 17.45).

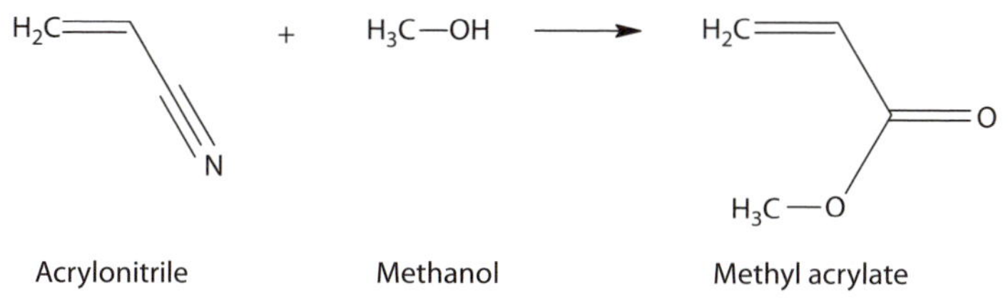

$$\text{(17.45)}$$

| Acrylonitrile | Methanol | Methyl acrylate |

Methyl methacrylate may be prepared by the catalytic oxidative carbonylation of propylene in the presence of methanol.

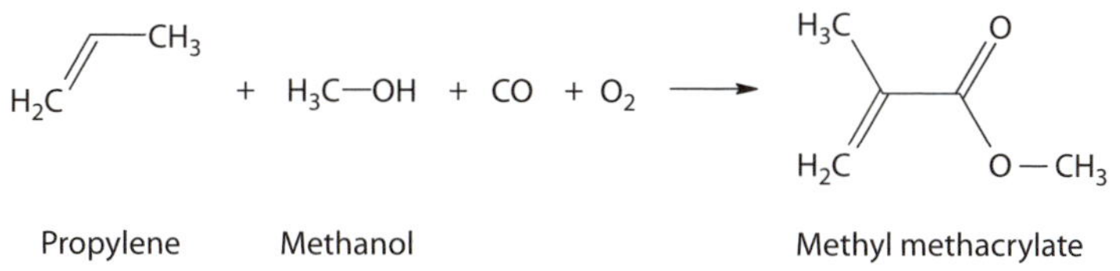

$$\text{(17.46)}$$

| Propylene | Methanol | Methyl methacrylate |

Most production employs acetone and hydrogen cyanide as the beginning materials. The intermediate cyanohydrin is converted through reaction with sulfuric acid to give a sulfate ester of the methacrylamide. This sulfate ester is subsequently hydrolyzed giving ammonium bisulfate and methyl methacrylate.

A newer process employs ethylene, carbon monoxide, and methanol passed over a catalyst to produce methyl propionate. The methyl propionate is combined with formaldehyde producing (green chemistry) water and methyl methacrylate.

17.4 SYNTHESIS OF FREE RADICAL INITIATORS

Free radical initiators are compounds containing covalent bonds that readily undergo hemolytic cleavage producing free radicals. The most widely used organic free radical initiators are peroxides and azo compounds. Here we will briefly describe the synthesis of the more widely employed free radical initiators.

Benzoyl peroxide is produced when benzoly chloride and sodium peroxide are stirred in water.

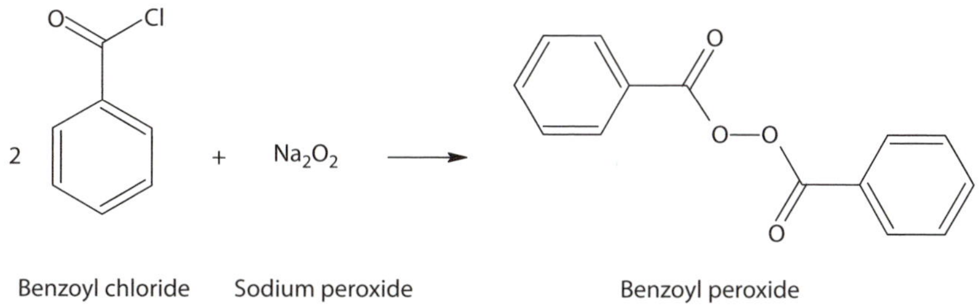

$$\text{(17.47)}$$

| Benzoyl chloride | Sodium peroxide | Benzoyl peroxide |

tert-Butyl hydroperoxide is produced by the acid-catalyzed addition of hydrogen peroxide to isobutylene.

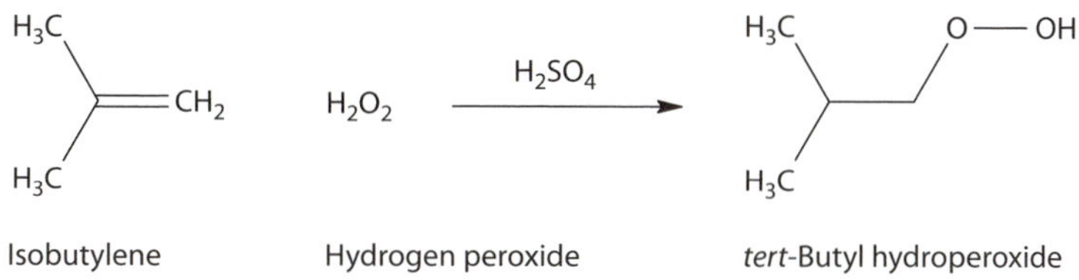

$$\text{(17.48)}$$

| Isobutylene | Hydrogen peroxide | *tert*-Butyl hydroperoxide |

tert-Butyl peroxide is produced with *tert*-Butyl hydroperoxide is added to isobutylene.

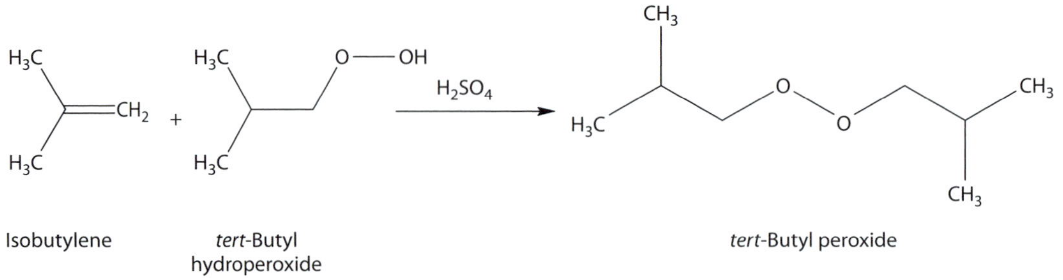

Isobutylene   *tert*-Butyl
                     hydroperoxide   *tert*-Butyl peroxide

$$(17.49)$$

Dicumyl peroxide is produced by the air oxidation of cumene.

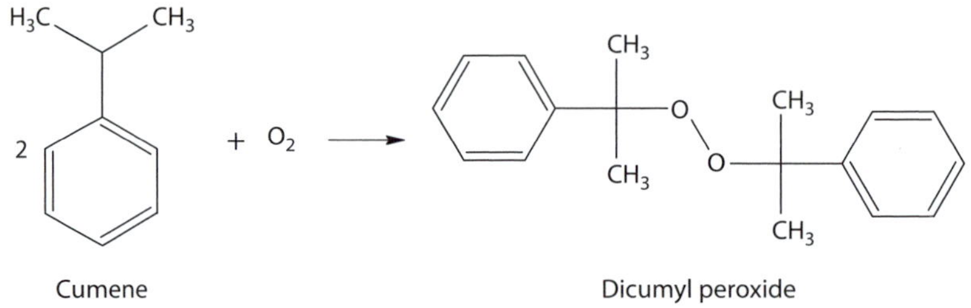

Cumene   Dicumyl peroxide

$$(17.50)$$

All initiators are potentially explosive compounds and must be stored and handled with care.

2,2′-Azobisisobutyronitrile (AIBN) is obtained from the reaction of acetone with potassium cyanide and hydrazine hydrochloride. As shown in Equation 17.51, the reaction produces hydrogen cyanide and hydrazine. The later reacts with acetone forming acetone dihydrazone that reacts with HCN producing a substituted hydrazone. This hydrazone is then oxidized to AIBN by addition of sodium hypochlorite.

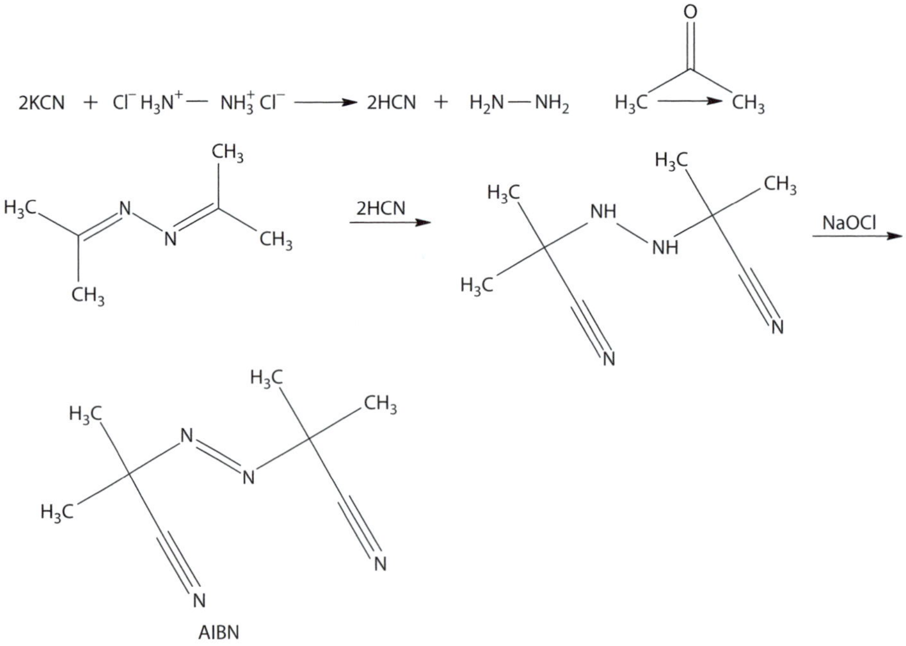

$$(17.51)$$

AIBN

When methyl ethyl ketone is used in place of acetone, AIBN is produced.

17.5 SUMMARY

1. Feedstocks for the synthesis of monomers of basic polymeric materials must be readily available and inexpensive because they are utilized in polymer synthesis in large quantities, allowing the polymeric materials to be inexpensive. Basic feedstocks are petrochemical and are coal based.
2. Monomer synthesis is both an art and science developed by major and ongoing research efforts, allowing the inexpensive and safe availability of the starting materials upon which the polymer industry is based. Commercial monomer synthesis is based on both the availability of inexpensive materials and on a "interconnectiveness" between products and synthetic byproducts that are essential to the synthesis of other essential materials.
3. The precise conditions of synthesis are continually being refined. They are based on "high" science.
4. There is an increased emphasis on green chemistry in the production of monomers.

GLOSSARY

Cannizzaro reaction: An internal oxidation–reduction reaction of aldehydes.
Carbamide: Urea.
Carcinogenic: Cancer causing.
Chemurgic Compound; compound made from a plant source.
Friedel-Crafts condensation: Condensation that takes place in the presence of a Lewis acid such as aluminum chloride.
Grignard reagent: RMgX.
Raney nickel: A porous nickel catalyst produced from a nickel–aluminum alloy.

EXERCISES

1. Why are there so many methods for the preparation of adipic acid?
2. Write equations for the industrial synthesis of the following:
   - a. Adipic acid
   - b. Hexamethylenediamine
   - c. Sebacic acid
   - d. Terephthalic acid
   - e. Maleic anhydride
   - f. ε-Caprolactam
   - g. Ethylene glycol
   - h. Glycerol
   - i. Pentaerythritol
   - j. TDI
   - k. Hexamethylenetetramine
   - l. Phenol
   - m. Urea
   - n. Melamine
   - o. Bisphenol A
   - p. Epichlorohydrin
   - q. Methyltrichlorosilane
   - r. Styrene
   - s. Vinyl chloride
   - t. Vinyl acetate
   - u. Acrylonitrile
   - v. Vinyl ethyl ether
   - w. Methyl methacrylate
3. Name a reactant or monomer produced by the following:
   - a. Grignard reaction
   - b. Friedel–Crafts reaction
   - c. Beckman rearrangement
   - d. A chemurgic process
   - e. A crossed Cannizzaro reaction
4. Why are catalysts so important in the synthesis of monomers?
5. Name three monomers that are derived from methane.
6. Why is it important to be able to make many different monomers from the same starting feedstock?
7. What is one basic feedstock for the synthesis of bisphenol A?
8. Why are most of the major monomer producers oil companies?
9. In the synthesis of peroxides as initiators why is added care wise?
10. Why is it important to save at least some of our petroleum for future generations?
11. Name three monomers whose synthesis illustrates green chemistry.

ADDITIONAL READING

Belgacem, M., Gandini, A. (2008): *Monomers, Polymers, and Composites from Renewable Resources*, Elsevier, NY.

Boundy, R. H., Boyer, R. F., Stroesser, S. M. (1965): *Styrene*, Hafner, NY.

D'Amore, A. (2006): *Monomers and Polymers: Reactions and Properties*, Nova, Commack, NY.

Leonard, E. C. (1970): *Vinyl and Diene Monomers*, Wiley, NY.

Pethrick, R., Zaikov, G., Pielichowski, J. (2008): *Progress in Monomers, Oligomers, Polymers, Composites and Nanocomposites*, Nova, Hauppauge, NY.

Summers, J., Zaikov, G. E. (2006): *Basic Research in Polymer and Monomer Chemistry*, Nova, Commack, NY.

Yokum, R. H., Nyquist, E. B. (1974): *Functional Monomers, Their Preparation, Polymerization and Application*, Dekker, NY.

18 Polymer Technology

Today, nearly 10,000 American companies are active in the general area of synthetic polymers. Following is a brief description of these companies divided according to their function.

Manufacturers: There are more than 200 major manufacturers of general purpose polymers and numerous other manufacturers of specialty polymers.

Processors: Some companies manufacture their own polymeric materials for subsequent processing, but the majority purchases the necessary polymeric materials from other companies. Processors may specialize in the use of selected polymers, such as nylons and polycarbonates, or focus on particular techniques of processing, such as coatings, films, sheets, laminates, and bulk-molded and reinforced plastics.

Fabricators and Finishers: The majority of companies are involved in the fabrication and finishing of polymers, that is, production of the end products for industrial and general public consumption. Fabrication can be divided into three broad areas: machining, forming, and fashioning. Machining includes grinding, sawing, screwing, and other techniques. Forming includes molding and other methods of shaping and joining by welding, gluing, screwing, and other techniques. Fashioning includes cutting, sewing, sheeting, and sealing. Fabrication sequences vary with the polymeric material and desired end product.

While much classic polymer technology was developed without the benefit of science, modern polymer technology and polymer science are closely associated. The technology of fibers, elastomers, coatings, composites, drug delivery, and plastics is discussed in this chapter.

Chemistry is moving center stage in many areas of medicine, biology, engineering, environmental science, and physics. While solid-state physics is traditionally based on silicon, polymers offer a much wider vista of opportunities for application and fine tuning those applications. Some areas are based on single crystals that may be small in our sight, but are large when compared to individual molecules. Even single silicon wafers with a minimum pattern dimension of 200 nm are on the order of ten times the size of individual molecules. Eventually, electronic, photonic, and stress–strain behavior individuality can be placed into single giant chains creating chains that behave as entire assemblies behave today.

18.1 POLYMER PROCESSING

Polymer processing can be defined as the process whereby raw materials are converted into products of desired shape and properties. Thermoplastic resins are generally supplied as pellets, marbles, or chips of varying sizes and they may contain some or all of the desired additives. When heated above their T_g, thermoplastic materials soften and flow as viscous liquids that can be shaped using a variety of techniques and then cooled to "lock" in the micro and gross structure.

Thermosetting feedstocks are normally supplied as meltable and/or flowable prepolymer, oligomers, or lightly or noncross-linked polymers that are subsequently cross-linked forming the thermoset article.

The processing operation can be divided into three general steps—preshaping, shaping, and postshaping. In preshaping, the intent is to produce a material that can be shaped by application of heat and/or pressure. Important considerations include the following:

- Handling of solids and liquids, including mixing, low, compaction, and packing
- Softening through application of heat and/or pressure
- Addition and mixing/dispersion of added materials

- Movement of the resin to the shaping apparatus through application of heat and/or pressure and other flow aiding processes, and
- Removal and recycling of unwanted solvent, unreacted monomer(s), byproducts, and waste (flash)

The shaping step may include any single or combination of the following:

- Die forming (including sheet and film formation, tube and pipe formation, fiber formation, coating, and extrusion)
- Molding and casting
- Secondary shaping (such as film and blow molding, thermoforming), and
- Surface treatments (coating and calendering)

Postshaping processes include welding, bonding, fastening, decorating, cutting, milling, drilling, dying, and gluing.

Polymer processing operations can be divided into five broad categories:

- Spinning (generally for fibers)
- Calendering
- Coating
- Molding, and
- Injection

Table 18.1 lists some of the major shapes produced by each of these processing groups.

TABLE 18.1
Major Forms of Polymer Processing Groupings

Process	Typical Form of Product
Calendering	Films, sheets
Coating	Film
Injection	Solid
Reaction injected	
Reciprocating screw	
Two-stage	
Molding	
Blow	Hollow
Displacement	
Extrusion	
Injection/transfer	
Stretch	
Cold Solid	
Compression	Solid, hollow
Rotational	Solid, hollow
Thermoforming	Hollow
Transfer	Solid
Spinning	Fibers
Dry	
Gel	
Melt	
Reaction	
Wet	

Essentially, all of the various processing types utilize computer-assisted design (CAD) and computer-assisted manufacture (CAM). CAD allows the design of a part and incorporates operating conditions to predict behavior of the pieces before real operation. CAD also transfers particular designs and design specifications to other computer-operated systems (CAMs) that allow the actual construction of the part or total apparatus. CAM systems operate most modern processing systems many allowing feedback to influence machine operation.

Processing and performance are interrelated to one another and to additional factors. Jaffe relates these major groups of factors in an interactive diamond given below (Figure 18.1). Understanding these factors and their interrelationships becomes increasingly important as the specific performance requirements become more specific. Performance is related to the chemical and physical structure and to the particular processing performed on the material during its lifetime. The physical structure is a reflection of both the chemical structure and the total history of the synthesis and subsequent exposure of the material to additional forces. These "additional" forces are included under the broad idea of processing and include any influence that contributes to the secondary (and greater) structure—stress–strain, light, chemical, and so on. The portion of the diamond relating processing to physical structure encompasses the study of structure–property relationships. A single material may be processed using only a single process somewhat unique to that material (such as liquid crystals) or by a variety of processes (such as polyethylene) where the particular technique is dictated by such factors as end use and cost.

18.2   SECONDARY STRUCTURES—MESOPHASES

The primary and secondary structures greatly influence possible processing scenarios. Here, the secondary structure is generally the same as the physical structure and the primary structure is generally the same as the chemical structure. The end properties and uses are governed by intrinsic properties that in turn are related to the primary and secondary structures—the chemical and physical structures.

The term "meso" will be used to describe local chain organizations that occur within the nano- and microscale regions (Figure 18.2). While the terms mesophase and mesoregions have been employed in describing order within liquid crystals, the definition will be broaden to include other ordered regions within a material's physical or secondary structure. We generally describe polymer secondary structure in simple terms of ordered or disordered and crystalline or amorphous. The ordered regions can be further described in terms of mesoregions or mesophases according to their permanency and ability to influence changes within and about these regions.

Jaffe describes four mesophase classifications. *Permanent mesogens* are materials whose microstructures are highly fixed such as in liquid crystalline polymers (LCs). LCs are characterized by highly ordered structures in the quiescent state. They exhibit relatively low viscosities in uniaxial flows and can

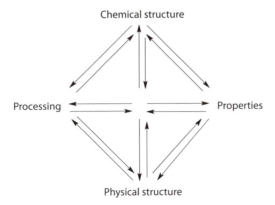

FIGURE 18.1   Relationships that influence the important interrelationships that exist for polymeric materials with respect to processing and end-product properties.

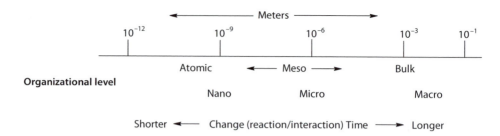

FIGURE 18.2   Relationship between organizational level and size.

be easily realigned through application of processing forces such as extrusion. To be processed, such polymers must be in the mesogenic state below their decomposition temperature. This can be achieved through the use of a specific solvent or the introduction of special comonomer units that allow them to melt (or soften), but that are introduced in such a manner as to preserve their LC character. As solids, such materials exhibit high molecular orientation, high tensile moduli (near to theoretical), poor compression (i.e., little unoccupied volume), poor shear behavior, and high tensile strengths (on the order of 4 GPa). Such materials are anisotropic conductors and generally offer good liquid and gas barrier properties. Properties are controlled by the inherent chemical structure, molecular orientation, defect occurrences, and stress-transfer mechanisms. Defects often act as the "weak-links" in a chain limiting mechanical properties so that defect detection, and elimination/curtailment are important and can be dependant on the processing conditions. Only certain processing techniques are suitable.

Accessible mesogens are formed from polymers that are thermotropic (i.e., polymers that have a phase organization that is temperature dependent) but have an accessible isotrophic phase below their decomposition temperature. Such polymers can be processed either when the material is in its mesogen or ordered state using LC-type processing forming strong well-ordered products, or at temperatures where the ordered mesogen structure is absent. In temperature-assisted systems, the material is rapidly cooled, quenched, preventing the mesogen structure formation producing a metastable isotropic glass or rubber. The metastable material can be processed employing less energy and force followed by a simple annealing and slower cooling that allows the formation of the ordered mesogenic structures along with the appearance of associated properties. Examples of *assembled mesogens* include groups of polymer coils and polymers with side chains that can form such mesogens. In the former case, tertiary-mesophase structures can be formed when the bundles of coiled chains come together.

Transient mesogens are regions present in flexible, random coil polymers often caused by application of external forces, including simple flowing/shearing. These regions occur through local segmental chain movements that happen within the chain network at points of minimum chain entropy such as sites of entanglements. They are fibrillar-like and appear to be the nucleating phase and key to the row and shish-kabob-like structures in oriented polymer crystals.

These latter groups include many of the so-called crystalites and crystalline regions of common polymers.

Understanding the factors that govern the formation of mesogens will assist in determining the processing conditions for the production of materials with specified amounts, sizes, and distribution of such crystalline microstructures. Mesophases can be local or permeate the entire structure. They can be large or small, and present in a random or more ordered arrangement.

18.3   FIBERS

18.3.1   Polymer Processing—Spinning and Fiber Production

18.3.1.1   Introduction

Most polymeric materials are controlled by the Federal Trade Commission (FTC) with respect to the relationship between the name and the content, including fibers. While the FTC controls industry in the United States, the international standards are generally determined by the International Organization for Standardization (ISO). Table 18.2 contains a brief listing of some of the ISO and FTC names for some of the most utilized fibers.

Fiber production continues to increase for most general groupings. Table 18.3 contains approximate fiber production by fiber type.

The dimensions of a filament or yarn are expressed in terms of a unit called the "tex," which is a measure of the fineness or linear density. One tex is 1 g/1,000 m or 10^{-6} kg/m. The tex has replaced denier as a measure of the density of the fiber. One denier is 1 g/9,000 m, so 1 denier = 0.1111 tex.

While some natural polymers produced "natural" fibers, fibers from synthetic and regenerated natural polymers are generally produced using one of the spinning processing techniques. Three spinning processes are generally employed in the large-scale commercial production of fibers. The first produces fiber from the melted polymer-melt spinning. The other two techniques form fibers from concentrated polymer solutions—dry and wet spinning. Figure 18.3 illustrates the essentials of these three spinning techniques. Table 18.4 is a listing of the most common polymers made into fibers by these three processes.

18.3.2 MELT SPINNING

Melt spinning was developed in the 1930s. In melt spinning, the polymer is melted or extruded, clarified by filtration, and pumped through a die having one or more small holes. The die is called a *spinneret*. The number, shape, and size of the hole can vary considerably. The number of holes ranges from several holes to several thousand holes.

TABLE 18.2
Generic Names for Synthetic Fibers According to the ISO and FTC

ISO	FTC
Acetate	Acetate
Acrylic	Acrylic
Aramid	Aramid
Chlorofiber	Vinyon/Saran
Cupro/viscose/modal/deacetylated acetate	Cupra/rayon
Elastane	Spandex
Glass	Glass
Modacrylic	Modacrylic
Nylon/polyamide	Nylon/polyamide
Polyester	Polyester
Polyethylene/polypropylene–polyolefin olefin	Vinylal/vinal

TABLE 18.3
Global Production of Fibers by Fiber Type (2005)

Fiber Type	Global Production Million Tons	Global Production Percentage
Cellulosic	2.0	6
Acrylic	2.7	8
Nylon	3.9	11
Olefin	5.9	17
Polyester	21	58

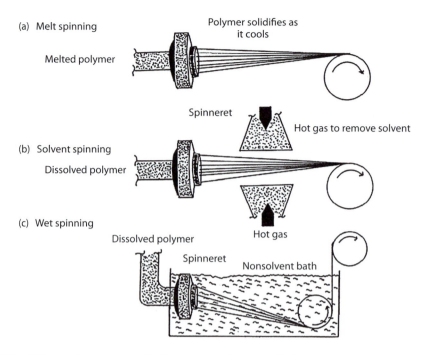

FIGURE 18.3   Fiber production using the three major spinning techniques.

TABLE 18.4
(Preferred) Spinning Processes

Melt Spinning	Dry Spinning	Wet Spinning Coagulation	Regeneration
Nylon	Acetate	Acrylic	Viscose
Polyester	Triacetate	Aramid	Cupro
Polyethylene	Acrylic	Elastane/Spandex	
Polypropylene	Elastine/Spandex	Poly(vinyl chloride)	
Poly(vinylene chloride)	Poly(vinyl chloride)		
	Aramid		
	Modacrylic		
	Vinyal		

For continuous filament formation, the number of holes is on the order of 10–100. The exit hole is usually circular giving round fibers. Other shaped holes are also employed that produce fibers with varying shapes.

The extruded fiber is then often uniaxially stretched by take-up rollers rotating at different speeds. The fiber stretching encourages the polymer chains to align on a molecular level producing increased strength in the direction of the pull.

To produce the melted polymer, the polymer chips, rods, marbles, or sheets are heated forming a melted pool of material. To minimize oxidation, the melted polymer is blanketed by an inert gas such as nitrogen or argon. The fluidity (inverse of viscosity) of the melt increases with increased temperature as does the cost to provide the necessary energy and tendency for unwanted reactions to occur. Thus, the polymer melt is generally assisted to and through the spinneret by means of an extruder that may also be used to supply some or all of the heating.

TABLE 18.5
Typical Spinning Temperatures for Selected Polymers

(Typical) Polymer	Melting Point (°C)	Spinning Temperature (°C)
Nylon-6	220	280
Nylon-6,6	260	290
Poly(ethylene terephthalate)	260	290
Poly(vinylidene chloride), copolymers	120–140	180
Poly(p-phenylene sulfide)	290	300
Polyethylene	130	220–300
Polypropylene	170	250–300

Typical melt spinning temperatures are given in Table 18.5.

Many nylon and polyester assemblies are configured so that there is a continuous progression from the melt formation of the polymer, and without hardening, the melted polymer is melt spun into fibers.

Monofilament is produced at a lower spinning speed, in comparison with chopped filament, because of the problem of heat buildup within the monofilament. The monofilament is generally cooled by passing it through cold water or by winding it on to a cold quench roll.

18.3.3 DRY SPINNING

Polymer concentrations in the order of 20%–40% are employed in dry and wet spinning. In the dry spinning process, the solution is filtered and then forced through a spinneret into a spinning cabinet through which heated air is passed to dry the filament. For economical reasons, the gas is usually air, but inert gasses such as nitrogen and superheated water are sometimes used.

Volatile solvents are used to assist in the drying. Water has been used for some systems, such as poly(vinyl alcohol), where the polymer is water soluble. Solvent removal and recycling is important. Spinning is usually carried out using either low (about 1%–2%) or high (10%–50%) solvent in the filament. The amount of solvent influences the drawing process. In high solvent cases, the filaments are plasticized allowing greater extension of the filament and greater alignment of the polymer chains to occur at lower temperatures and lower stresses. The extra solvent is removed just before, during, or subsequent to stretching. Just before extrusion, the polymer solution is heated to just above its boiling point, increasing the likelihood for ready removal of the solvent.

Dry spinning produced fibers have lower-void concentrations in comparison to melt spun fibers because the presence of solvent molecules cause voids that are often "remembered" by the polymer. This is reflected by greater densities and lower dyeability for the dry spun fibers.

Fibrous glass is the most important inorganic fiber. It is produced by melt spinning in both a continuous filament and staple form. The molten glass is fed directly from the furnace, or melted from rods or marbles, to the spinneret. As the fibrous glass emerges it is attenuated, quenched, lubricated, and wound forming a yarn or continuous filament. The temperature for spinning is on the order of 1,200°C–1,500°C. This temperature is important since it controls the output, and in conjunction with the removal speed, helps control the properties of the resultant fiber, including thickness and density.

18.3.4 WET SPINNING

Wet spinning is similar to dry spinning except that fiber formation results from the coagulation of the polymer solution as it is introduced into a nonsolvent bath. Since the coagulation process is relatively long, the linear velocity of wet spinning is less than for either melt or dry spinning. Wet spinning allows the placement of holes in the spinneret face to be closer together allowing productivity to be increased. Even so, it remains the slowest of the traditional spinning processes. The equipment used for wet spinning is similar to that used in dry spinning thought it is not necessary to heat the polymer solution to a high temperature. The spinnerets are immersed in tanks containing the nonsolvent. Wet spinning is the most complex of the three spinning processes, typically including washing, stretching, drying, crimping, finish application, and controlled relaxation to form tow material. Spinning of natural-derived materials generally include additional steps including ageing or ripening to achieve the desired viscosity and chain length.

Fibers made from wet spinning generally have high-void contents in comparison to all of the other processes giving them increased dyeability. The surface is rougher with longitudinal serrations and from a round die hole it has an approximately circular to bean-shaped diameter.

Hollow fibers for gas and liquid separation are prepared through passing air through the material just prior to entrance into the nonsolvent bath.

18.3.5   Other Spinning Processes

There are a number of lesser used, but still important, spinning processes. Following is a summary of some of these. In *reaction spinning,* a prepolymer is generally used that is further reacted upon by a material that is may be in solution in a bath. Further treatments may included cross-linking of the fiber. The most important example is the production of selected segmented polyurethane (PU) elastomeric fibers. Here the prepolymer is the soft segment generally a low molecular weight polyether or polyester. Reaction with an aromatic diisocyanate converts the end groups, generally hydroxyls, into isocyanate groups. The bath contains a diamine such as 1,2-diaminoethane. The reaction between the amine and isocyanate forms the hard urea linkages. Some segmented PUs, such as Lycra (TM), are formed using conventional dry spinning.

Fibers can be formed from intractable materials such as ceramics and polytetrafluoroethylene through extrusion of a suspension of fine particles in a solution of a matrix polymer. The matrix polymer-intractable material is coagulated embedding and aligning the intractable material in the matrix polymer. The filament is then heated decomposing the matrix polymer. During this process, the material is sintered and drawn giving small, often with little flexibility, fibrils.

Gel spinning is used to produce high strength and modulus fibers. High molecular weight (such as 10^6 Da for polyethylene) polymer is dissolved in a high-temperature solvent at low concentration (1%). The hot solution is extruded into a cooling zone such as a liquid nonsolvent. The resulting gel-like filament contains polymer with lots of entrapped liquid. This gel-like filament can be easily highly drawn. The drawing can be carried out even though liquid is removed before drawing. What occurs is that the low density of polymer chains in the gel allows a decreased chain entanglement allowing greater linear chain conformations to occur as the fiber is drawn.

Some low orientation polymers exhibit what is referred to as necking. In necking, a filament extends preferentially at only selected sites known as necks or necking sites. This behavior occurs with many thermoplastic materials near their T_g. At lower temperatures brittle fracture may occur at high tensions instead of necking. At higher temperatures filament extension occurs uniformly without preferred necking sites. Commercially, filament extension is carried out at sufficient temperatures to avoid necking. For multifilament yarns, filament elongation is generally carried out by first heating the filament with subsequent application of the stress necessary to stretch the filament. For monofilaments and tows, the heating and application of the stress occur together.

Along with centrifugal spinning, there are several additional fiber forming techniques that are employed in fiber formation that do not employ a spinneret. In *electrostatic spinning,* a high voltage, generally >5,000 V, is applied to a viscous solution of the polymer dissolved in a volatile solvent with a high dielectric constant but low conductivity contained in a fine capillary tube. A stream of filaments emerge from the capillary. These filaments are collected on a suitable surface.

18.4 NONSPINNING FIBER PRODUCTION

Fibers can also be made using specific conditions employing blow molding of a melt. They can also be mechanically made by machining. Thus, polytetrafluoroethylene fibers are made by machining a thin film from a block of the polymer and then drawing the film at 300°C.

Fibers can be made from directly pulling some of the polymer from the melt. Similarly, fibers can be made using the interfacial process with fibers being formed as reaction of the two core-actants occurs at or near the interface. Neither instance has been used in industrial-scale fiber formation.

Fibers are commercially made from uniaxially drawn film. The film is extruded, slit into tape-like strips, drawn, fibrillated, and wound. As in the case of spinning, the drawing produces preferred alignment of the polymer chains along the axis of pull. If the drawing precedes slitting, the fiber gives some cross-orientation and is less apt to split.

Fibrillation can be achieved mechanically by drawing and pulling thin sheets of polymer. This is compounded if twisting is also involved. A rough idea of this process can be demonstrated by cutting several ribbons of film from a trash bag. Take one and pull. It will elongate and eventually form a somewhat thick filament-like material. Do the same to another strip except also twist it.

Film can be heated and/or stretched and cut eventually giving filament-like materials. Unfibrillated slit-film materials are used in weaving sacks and other packaging. Randomly fibrillated slit-film material is used to make twins and ropes while controlled fibrillated material is used to make yarns for use in carpet backings and furnishing fabrics.

Whiskers can be made of some metals from simple scraping of the metal to from filament-like whiskers of high strength. Carraher and coworkers have produced a number of metal-containing polymers, generally rigid-rod like, that spontaneously form fiber-like structures from the reaction solution or when mechanically agitated, fiber-like organizations form.

18.4.1 Natural Fibers

Most plant and animal materials contain natural fibers that have been concerted into useful fibers for thousands of years, including ropes, building materials, brushes, textiles, and brushes (Table 18.6). Animal protein fibers such as wool and silk are no longer competitive with synthetic fibers with respect to cost but are still often utilized in the production of high-end rugs. Some of these rugs are hundreds of years old yet retaining their color and physical properties.

TABLE 18.6
Common Natural Sources of Fibers

Animal	Vegetable
Alpaca	Cotton
Angora	Hemp
Camel	Jute
Cashmere	Linen
Mohair	Ramie
Silk	Sisal
Vicuna	
Wool	
Mineral	**Derived from Plants**
Asbestos	Paper
Fiber glass	Rayon (and related materials)
Metal-intense whickers, fibers	Modal

Plant fibers such as cotton, abaca, agave, flax, hemp, kapok, jute, kenaf, and ramie are still in use but even cotton is no longer "king."

Regenerated proteins from casein (lanital), peanuts (ardil), soybeans (aralac), and zine (vicara) are used as specialty fibers. Regenerated and modified cellulose products, including acetate, are still widely used today and the production of fibers is similar to that described above for synthetic fiber production. Most regenerated cellulose (rayon) is produced by the viscose process where an aqueous solution of the sodium salt of cellulose xanthate is precipitated in an acid bath. The relatively weak fibers produced by this wet spinning process are stretched to produce strong rayon.

18.5 ELASTOMERS

Before World War II, hevea rubber accounted for more than 99% of all elastomers used, but synthetic elastomers account for more than 70% of all rubber used today. Natural rubber and many synthetic elastomers are available in latex form. The latex may be used, as such, for adhering carpet fibers or for dipped articles, such as gloves, but most of the latex is coagulated and the dried coagulant used for the production of tires and mechanical goods.

More than 5.5 billion pounds of synthetic rubber is produced annually in the United States. The principle elastomer is the copolymer of butadiene (75%) and styrene styrene–butadiene–rubber (SBR) produced at an annual rate of more than 1 million tons by the emulsion polymerization of butadiene and styrene. The copolymer of butadiene and acrylonitrile (Buna-H, NBR) is also produced by the emulsion process at an annual rate of about 200 million pounds. Likewise, neoprene is produced by the emulsion polymerization of chloroprene at an annual rate of more than 125,000 tons. Butyl rubber is produced by the low-temperature cationic copolymerization of isobutylene (90%) and isoprene (10%) at an annual rate of about 150,000 tons. Polybutadiene (BR), polyisoprene, and ethylene–propylene copolymer rubber (EPDM) are produced by the anionic polymerization of about 600,000, 100,000, and 350,000 tons, respectively. Many other elastomers are also produced.

18.5.1 ELASTOMER PROCESSING

The processing of elastomeric (rubbery) material is quite varied dependent on the end use, form of the material (that is dry or in solution) and material processed. Latex forms of rubber can be properly mixed with additives using simple (or more complex) stirring and agitation. The mixing/agitation should be such as to not cause a separation or breakdown of the latex or foam formation. Straight, coagulant, and heat-assisted dipping processes are commonly used to produce a variety of tubes, gloves, and so on. Latexes are also used to make thread for the garment industry, and adhesives for shoes, carpets, and tape formation.

Following we will look at the processing involving bulk rubber. The manufacture of rubber products from this material can be divided into four steps:

1. Mastication
2. Incorporation or compounding
3. Shaping, and
4. Vulcanization

The shaping and vulcanization steps are combined in a number of processes such as transfer and injection molding or may be separated as in the extrusion and subsequent vulcanization sequence. An outline of these steps is given in Figure 18.4.

We will look at the processing of natural dry rubber first since its processing is similar to other elastomers and because of its historical importance. Natural rubber is a dispersion of rubber particles in water. Unvulcanized raw rubber obtained by coagulation and drying has large chains with chain lengths in the order of 10^5 carbons.

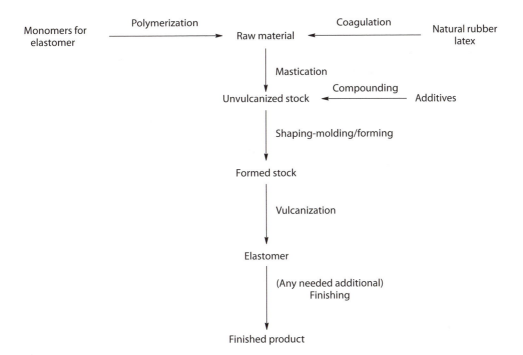

FIGURE 18.4 Outline of steps involved in the processing to form elastomeric materials.

Mastication is intended to bring the material to the necessary consistency to accept the compounding ingredients. Mastication results in a lowering of chain size to an average chain length in the order of 10^4 carbons. Two basic types of internal mixers are in use. The Banbury has rotors rotating at different speeds creating a "kneading" action such as that employed in handling bread dough. A shearing action between the rotors and the walls of the mixer is also achieved. The Shaw Intermix employs rotors that turn at the same speed and closely intermesh causing an intracompound friction for mixing thus closely resembling a mill's mixing action.

The next step is the incorporation of various additives—*compounding*. Typical additives include filler, processing aids, activators, processing aids, age resistors, sulfur, antioxidants and antiozone compounds, extenders, plasticizers, blowing agents, pigments, and accelerators.

An important aspect in the compounding is the amount of crystallization of the rubber. If the rubber is in a highly crystalline state, it will mix poorly if at all. Thus, partially crystallized rubber must be heated before it will yield to mixing.

Stabilizers are materials that help the rubber withstand oxidative ageing and ozone attack. They act by intercepting the active free radicals breaking the free-radical-associated degradation process. Amines and phenols are generally employed. Reinforcing fillers, of which carbon black is the most important, are added to improve the mechanical properties such as hardness, abrasion resistance, modulus, and tear resistance. It is believed that the rubber adheres to the carbon surface. Carbon black also helps in retarding ultraviolet (UV) degradation and increases the electrical conductivity, reducing triboelectric charging and acting as an antistatic material.

Natural rubber can be compounded without fillers to give a vulcanized material with high elongation (to about 800%) and high tensile strength (about 28 MPa).

The internal mixers fragment the large rubber molecules by high-shear forces. Depending on the particular assembly and ingredients, the created free radicals can combine to give larger molecules or may form smaller chains. Breakdown is often assisted by the use of chemical peptizers such as thiophenols, mixtures of salts of saturated fatty acids, and aromatic disulfides. The fatty acids mainly generally act as dispersing agents and processing aids.

The viscous prerubber is now *shaped* by addition to a mold of the desired shape. Addition can be achieved by simply pouring the material into the mold but usually the material is added to the mold employing the usual molding addition (extrusion, compression, and transfer) techniques. The material can also be treated using most of the other "thermoplastic" processing techniques such as calendering, coating, and extrusion.

The material is now heated to cure, set, or *vulcanize* (all terms are appropriate) the material into the (typically) finished shape. Between 1% and 5% of sulfur (by weight) is added in typical black rubber mixes, giving a vulcanized material with an average of about 500 carbon atoms between cross-links. Larger amounts of sulfur will give a tougher material eventually giving a somewhat brittle, but quite strong, ebonite as the amount of sulfur is increased to about 40%. Sometimes additional finishing may be desirable including painting, machining, grinding, and cutting.

These steps are typical for most of the synthetic elastomers. The use of sulfur for vulcanization is common for the production of most elastomers. Magnesium and zinc oxides are often used for the cross-linking of polychloroprene, CR. Saturated materials such as ethylene–propylene (EPM) and fluoroelastomers are cross-linked using typical organic cross-linking agents such as peroxides.

Carbon black is widely used as a reinforcing agent for most synthetic elastomers. Carbon black is especially important for synthetic elastomers such as SBR nitrile rubber (NBR), and BR that do not crystallize at high strains. Thus, noncarbon filled SBR has a tensile strength of about 2 MPa and with addition of carbon black this increases to about 20 MPa.

The above processing applies to the processing of typical bulk carbon backbone-intensive elastomers. Other important classes of elastomers are also available. PUs represent a broad range of elastomeric materials. Most PUs are either hydroxyl or isocyanate-terminated. Three groups of urethane elastomers are commercially produced. Millabile elastomers are produced from the curing of the isocyanate group using trifunctional glycols. These elastomers are made from high polymers made by the chain extension of the PU through reaction of the terminal isocyanate groups with a polyether or polyester. Low molecular weight isocyanate-terminated PUs are cured through a combination of chain extension by reaction with a hydroxyl-terminated polyether or polyester and trifunctional glycols giving cast elastomers. Thermoplastic elastomers are block copolymers formed from the reaction of isocyanate-terminated PUs with hydroxyl-terminated polyethers or polyesters. These are generally processed as thermoplastic materials as are the thermoplastic elastomers. Many of these materials have little or no chemical cross-linking. The elastomeric behavior is due to the presence of physical hard domains that act as cross-links. Thus, SBR consists of soft butadiene blocks sandwiched between polystyrene (PS) hard blocks. These hard blocks also act as a well dispersed fine-particle reinforcing material increasing the tensile strength and modulus. The effectiveness of these hard blocks greatly decreases above the T_g (about 100°C) of PS.

Polysiloxanes (silicons) form another group of important elastomers. Again, processing typically does not involve either carbon black or sulfur.

18.6   FILMS AND SHEETS

Films, such as regenerated cellulose (cellophane), are produced by precipitating a polymeric solution after it has passed through a slit die. Other films, such as cellulose acetate, are cast from a solution of the polymers, but most films are produced by the extrusion process. Some relatively thick films and coextruded films are extruded through a flat slit die, but most thermoplastic films, such as polyethylene (PE) film, are produced by air blowing of a warm extruded tube as it emerges from a circular die (Figure 18.5). Films and sheets are also produced employing calendering. Calendering is also used to apply coatings to textiles or other supporting material.

The most widely used films are low-density polyethylene (LDPE), cellophane, polyethylene terephthalate (PET) poly(vinyl chloride) (PVC) cellulose acetate, polyfulorocarbons, nylons, polypropylene (PP), PS, and linear low-density polyfluorocarbons (LLDPE). The strength of many films is improved by biaxial orientation, stretching. Most of the thermoplastics used as films may also be

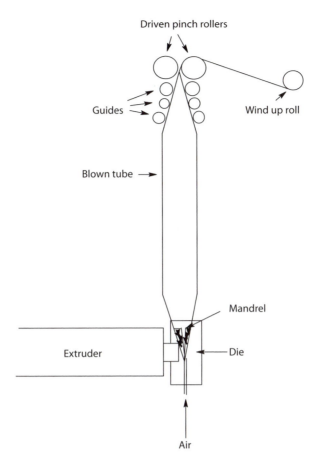

FIGURE 18.5 Film formation employing extrusion.

extruded as relatively thick sheets. These sheets may also be produced by pressing a stack of film at elevated temperature (laminating) or by the calendering process.

Wire is coated by being passed through a plastic extruder, but most materials are coated from solutions, emulsions, or hot powders. The classic brushing process has been replaced by roll coating, spraying, and hot powder coating. The application of polymers from water dispersions to large objects, such as automobile frames, has been improved by electrodeposition of the polymer onto the metal surface.

Printing inks are highly filled solutions of resins. The classic printing inks were drying oil-based systems but the trend in this almost billion dollar business is toward solvent-free inks.

18.6.1 CALENDERING

Calendering is simply the squeezing or extruding of a material between pairs of corotating, parallel rollers to form film and sheets. It can also be used to mix and impregnate such as in the case of embedding fiber into slightly melted matrix material to form impregnated composite tapes. It can also be used to combine sheets of material such as sheets of impregnated paper and fiber woven and nonwoven mats to form laminar composite materials. It is also used in processing certain rubber material and textiles. Calendering is also employed in conjunction with other processing techniques such as extrusion in the formation of films from extruded material. It is also used to coat, seal, laminate, sandwich, finish, and emboss.

The major bulk-processed thermoplastic processed using calendering is PVC sheets and film, including blends and copolymers. A sample recipe to produce PVC sheet might include a plasticizer such as a dialkyl phthalate, pigment, filler, lubricant, and stabilizer.

Because of the variation in flexibility, the terms film and sheet vary with materials. For PVC, films have a thickness of 6 mil (0.15 mm) and less while sheets are thicker than this. While PVC is relatively rigid with a tensile modulus greater than about 690 MPa (105 psi), thin films are easily folded. Films are generally shipped as rolls with the PVC rolled about a central rigid core. Sheets are generally shipped as flat layered sheets.

The major bulk-processed thermoplastic processed using calendering is PVC sheets and film, including blends and copolymers. A sample recipe to produce PVC sheet might include a plasticizer such as a dialkyl phthalate, pigment, filler, lubricant, and stabilizer. A partial flow chart illustrates the particular features of the production of a PVC sheet as follows:

Raw materials → Mixing of raw materials in the specified amounts (with heating) → Cooling → Milling-powdering/chipping of mixed material → Feeding of stock into first calender nip → Calendering → Stripping and drawdown (if needed) → Embossing (if needed) → Relaxing and tempering → Cooling → Trimming (if needed) → Stacking (or continued processing) → Quality control → Fabrication into final product → Sales.

In addition to controlling the recipe materials and PVC properties, including molecular weight and molecular weight distribution, the major processing considerations are:

- Calender speed
- Temperature
- Thickness/gauge
- Orientation
- Finish, and
- Embossing

Films and thin sheets are typically drawn to impart additional unidimensional strength. For films, both unidirectional and bidirectional drawing is used.

18.7 POLYMERIC FOAMS

Before 1920, the only flexible foam available was the natural sponge, but chemically foamed rubber and mechanically foamed rubber latex were introduced before World War II. These foams may consist of discrete unit cells (unicellular, closed cell), or they may be composed of interconnecting cells (multicellular, open cells) depending on the viscosity of the system at the time the blowing agent is introduced. More than 1.5 million tons of foamed plastic is produced annually in the United States.

Unicellular foams are used for insulation, buoyancy, and flotation applications while multicellular foams are used for upholstery, carpet backing, and laminated textiles. Expanded PS, which is produced by the extrusion of PS beads containing a volatile liquid, is used to produce low-density moldings such as foamed drinking cups and insulation board. Foamed products are also produced from PVC, LDPE, urea resins, ABS, and PU. PU foams are versatile materials that range from hard (rigid) to soft (flexible). These are produced by the reaction of a polyols and a diisocyanate.

18.8 REINFORCED PLASTICS (COMPOSITES) AND LAMINATES

18.8.1 COMPOSITES

Theoretical and material considerations for composites are dealt within Chapter 8. Here we will focus more on processing considerations.

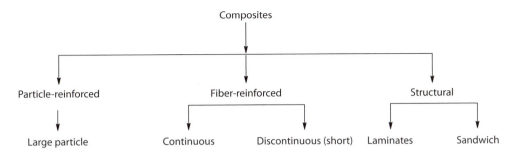

FIGURE 18.6   Classification of polymer-intense composites.

TABLE 18.7
Materials Used for Heat Protection in the Space Shuttle Orbiter

Identifier	Maximum Operating Temperature (°C)	Description of Material
Advanced flexible reusable surface insulation	810	Quartz batting sandwiched between AFRSI quartz and fibrous glass fabric
Felt-reusable surface insulation	400	Nylon felt with a silicone rubber FRSI coating
High-temperature reusable surface insulation	1,250	Silica tiles, borosilicate glass coating insulation HRSI with silicon boride added
Low-temperature reusable surface	650	Silica tiles with a borosilicate coating insulation—LRSI
Reinforced carbon–carbon–RCC	1,650	Pyrolyzed carbon–carbon coated with silicon carbide, SiC

The locations are given in Figure 18.7.

Composites are generally composed of two phases, one called the *continuous* or *matrix phase* that surrounds the second called the *discontinuous* or *dispersed phase*. There are a variety of polymer-intense composites that can be classified as shown in Figure 18.6. Many of these composite groups are used in combination with other materials, including different types of composites and like types of composites except differing in orientation.

Many naturally occurring materials such as wood are reinforced composites consisting of a resinous continuous phase and a discontinuous fibrous reinforcing phase.

Composites are also used extensively where light but very strong materials are needed such as in the construction of the new Boeing 767 where composites play a critical role in the construction of the exterior. They are also used where excessive high heat stability is needed such as in the reusable space vehicle (Table 18.7).

Here we will briefly look at each of the main groupings of composites.

18.8.1.1 Particle-Reinforced Composites-Large-Particle Composites

Some materials to which fillers have been added can be considered as composites. These include a number of the so-called cements, including concrete (Section 12.2). As long as the added particles are relatively small, of roughly the same size, and evenly distributed throughout the mixture there can be a reinforcing effect. The major materials in Portland cement concrete are the Portland cement, a fine aggregate (sand), course aggregate (gravel and small rocks), and water. The aggregate particles act as inexpensive fillers. The water is also inexpensive. The relatively expensive material is the Portland cement. Good strength is gained by having a mixture of these such that there is a

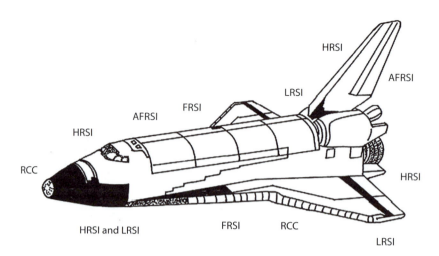

FIGURE 18.7 Location of various advanced materials, including composites, employed for heat protection in the Space Shuttle Orbiter. The descriptions of the materials are given in Table 18.7.

dense packing of the aggregates and good interfacial contact, both achieved by having a mixture of aggregate sizes—thus the use of large gravel and small sand. The sand helps fill the voids between the various larger gravel particles. Mixing and contact is achieved with the correct amount of water. Enough water must be present to allow a wetting of the surfaces to occur along with providing some of the reactants for the setting up of the cement. Too much water creates large voids and weakens the concrete.

18.8.1.2 Fiber-Reinforced Composites

Mathematically, the critical fiber length necessary for effective strengthening and stiffening can be described as follows:

> Critical fiber length = (Ultimate or tensile strength times fiber diameter/2) times the fiber-matrix bond strength OR the shear yield strength of the matrix—which ever is smaller.

Fibers where the fiber length is greater than this critical fiber length are called *continuous fibers* while those that are less than this critical length are called *discontinuous* or short *fibers*. Little transference of stress and thus little reinforcement is achieved for short fibers. Thus, fibers whose lengths exceed the critical fiber length are used.

Fibers can be divided according to their diameters. Whiskers are very thin single crystals that have large length to diameter ratios. They have a high degree of crystalline perfection and are essentially flaw free. They are some of the strongest materials know. Whisker materials include graphite, silicon carbide, aluminum oxide, and silicon nitride. Fine wires of tungsten, steel, and molybdenum are also used but here, even though they are fine relatively to other metal wires, they have large diameters. The most used fibers are "organic fibers," which are either crystalline or amorphous or semicrystalline with small diameters.

18.8.1.3 Processing of Fiber-Reinforced Composites

These exists a wide variety of particular operations but briefly they can be described in terms of filament winding, preimpregnation of the fiber with the partially cured resin, and pultrusion. Pultrusion is used to produce rods, tubes, beams, and so on with continuous fibers that have a constant cross-sectional shape. The fiber (as a continuous fiber bundle, weave or tow) is impregnated with a thermosetting resin and pulled through a die that shapes and establishes the fiber to resin ratio. This stock is then pulled though a curing die that can machine or cut producing the final shape such as filled and hollow tubes and sheets.

The term used for continuous fiber reinforcement *preimpregnation* with a polymer resin that is only partially cured is "prepreg." Prepreg material is generally delivered to the customer in the form of a tape. The customer than molds and forms the tape material into the desired shape finally curing the material without having to add any additional resin. Preparation of the prepreg can be carried out using a calendering process. Briefly, fiber from many spools are sandwiched and pressed between sheets of heated resin with the resin heated to allow impregnation but not so high as to be very fluid.

Thus, the fiber is impregnated in the partially cured resin. Depending upon the assembly the fiber is usually unidirectonal, but can be made so that the fibers are bidirectional or some other combination. The process of fitting the prepreg into, generally onto, the mold is called "lay-up". Generally a number of layers of prepreg are used. The lay up may be done by hand, called *hand lay up*, or done automatically, or some combination of automatic and hand lay up. As expected, hand lay up is more costly but is needed where one-of-a-kind products are produced.

In *filament winding,* the fiber is wound to form a desired pattern, usually but not necessarily hollow and cylindrical. The fiber is passed through the resin and then spun onto a mandrel. After the desired number of layers of fiber is added, it is cured. Prepregs can be filament wound. With the advent of new machinery, complex shapes and designs of the filament can be readily wound.

18.8.1.4 Structural Composites

Structural composites can be combinations of homogeneous and composite materials. Laminar composites are composed of two-dimensional sheets that generally have a preferred high-strength direction. The layers are stacked so that the preferred high-strength directions are different, generally at right angles to one another. The composition is held together by a resin. This resin can be applied as simply an adhesive to the various surfaces of the individual sheets or the sheet can be soaked in the resin before laying the sheets together. In either case, the bonding is usually of a physical type. Plywood is an example of a laminar composite. Laminar fibrous glass sheets are included as part of the modern ski construction. These fibrous glass sheets are fiber-reinforced composites used together as laminar composites.

Laminar materials are produced by a variety of techniques. Coextrusion blow molding produces a number of common food containers that consist of multilayers such as layers consisting of PP/adhesive/poly(vinyl alcohol)/adhesive/adhesive/PP.

Sandwich composites are combinations where a central core(s) is surrounded generally by stronger outer layers. Sandwich composites are present in the modern ski and as high-temperature stable materials used in the space program. Some cores are very light acting something like a filler with respect to high strength, with the strength provided by the outer panels. Simple corrugated cardboard is an example of a honeycomb core sandwich structure except that the outer paper-intense layers are not particularly strong. Even in the case of similar PE and PP corrugated structures, the outer layers are not appreciatively stronger than the inner layer. In these cases, the combination acts to give a light weight somewhat strong combination, but they are not truly composites but simply exploit a common construction.

18.8.2 LAMINATING

Laminating is a simple binding together of different layers of materials. The binding materials are often thermosetting plastics and resins. The materials to be bound together can be paper, cloth, wood, or fibrous glass. These are often referred to as the reinforcing materials. Typically sheets, impregnated by a binding material, are stacked between highly polished metal plates, subjected to high pressure and heat in a hydraulic press producing a bonded product, which may be subsequently treated, depending on its final use (Figure 18.8 (a)). The end product may be flat, rod-shaped, tubular, rounded, or some other formed shape.

Reinforced plastics differ from high-pressure laminates in that little or no pressure is employed. For instance, in making formed shapes, impregnated reinforcing material is cut to a desired

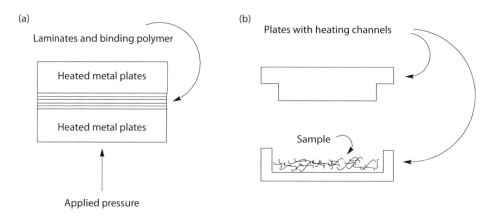

FIGURE 18.8   Assembly employed for (a) the fabrication of laminates and (b) reinforced plastics.

shape, the various layers are added to a mold, and the molding is completed by heating the mold (Figure 18.8 (b)). This process is favored over the high-pressure process because of the use of a simpler, lower-cost mold, and production of strain-free products.

18.9   MOLDING

Molding is a general technique that can be used with plastics and thermosetting materials when employing mobile prepolymer. Molding is used to produce sheet-like, foamed, hollow, or solid materials from very small to very large objects. Here we will look at various molding processes.

18.9.1   Injection Molding

The most widely employed processing techniques for thermoplastics are extrusion and injection molding. Injection molding is also used to produce some thermoset products utilizing fluid prepolymer.

Injection molding involves forcing, injecting, a molten polymer into a mold where is cools becoming solid. The mold separates allowing the molded material to be released. The mold parts are again joined and the process begins again.

Injection molding allows the rapid, economical production of small to large parts. It provides close tolerances and the same machine can be used to mold many different articles. Parts can be molded combining the polymer with other polymers and with any number of additives. Further, it can be run so that various parts can be easily married as part of an entire or combined-parts production assembly of an article. The ability to easily modify the operating conditions of the injection molding machine is important because of the variety of articles that may be needed, variety of material employed to produce the same (general) article, variety of materials to produce different injection molded articles, and the variability of supposedly the same polymer material from batch to batch.

Injection molding is not new. A patent was issued in 1872 for an injection molding machine for camphor-plasticized cellulose nitrate, celluloid. Almost all of the machines used today are reciprocating or two-stage screw types. Both types employ a reciprocating Archimedean-like screw similar to that of a screw extruder. A few are of the plunger type.

A traditional injection apparatus consists of a hopper that feeds the molding powder to a heated cylinder where the polymer is melted and forced forward by a reciprocating plunger or screw. The cooled part is ejected when the mold opens and then the cycle is repeated. The molten material passes from the nozzle through a tapered sprue, a channel or runner, and a small gate into the cooled mold cavity. The polymer in the mold is easily broken off at the gate site and the materials in the sprue, runner, and gate are ground and remolded. An illustration of such an injection molding press is given in Figure 18.9. The hopper (a) feeds the molding powder to a heated cylinder (b) where the polymer is melted and forced forward by a reciprocating plunger (c) (or screw). The molten material advances toward a spreader or torpedo into a cool, closed, (here) two-piece mold (d). The cooled part is ejected when the mold opens and then the cycle is repeated. The molten plastic is passed from the nozzle through a tapered sprue, runner, and a small gate into the cooled mold cavity. The plastic in the narrow gate section is easily broken off and excess material remaining within the sprue, runner, and gate ground and remolded.

In a reciprocating screw machine, the material is collected in front of the screw that continues to move backward as additional material is melted. The area where the melted material is collected corresponds to the heating chamber or pot in a two-stage system. The material is melted by the internally generated heat caused by the friction of the polymer segments and chains rubbing against one another. The screw is also good at mixing so that additive introduction and mixing of different polymers can be achieved in the same step with overall polymer melting.

As the size of the molded product becomes larger it is more difficult to control uniformity and to maintain a sufficient clamping force to keep the mold closed during filling. Reaction injection molding (RIM) overcomes these problems by largely carrying out the polymerization reaction in the mold. The most widely used RIM materials are PU and PU-reinforced elastomeric materials Most of the automotive interior panels (such as dashboards) are produced using RIM.

On a molecular level, partially crystalline to amorphous polymers are normally used. As the material is heated Brownian motion occurs resulting in a more random chain arrangement. When an unidirectional force is applied to a resting polymer melt the chains tend to move away from the applied force. If the applied force is slow enough to allow the Brownian movement to continue to keep the polymers in a somewhat random conformation then the movement of the polymer melt is proportional to the applied stress; that is, the flow is Newtonian.

As the rate of movement increases, chain alignment occurs along the direction of flow with movement too fast for the Brownian factors to return the system to a somewhat random state and flow is then non-Newtonian. Most systems are operated, at least at the injection stage, under non-Newtonian conditions so that some polymer alignment occurs. If the polymer melt flow rate continues to increase, polymer chains align parallel to the flow plane and eventually reach a point where it again becomes Newtonian. Even so, the polymer chains have been aligned as the flow processes moved through the non-Newtonian flow range.

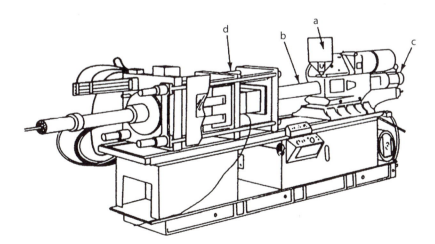

FIGURE 18.9   Cross-section of an injection molding press. (From Seymour, R. (1975): *Modern Plastics Technology*, Reston Publishing Co., Reston, Virginia.)

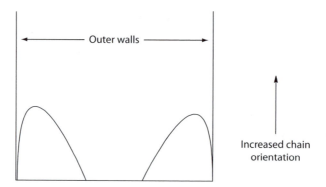

FIGURE 18.10   Idealized relationships between the distance from the outer wall and chain alignment.

As the molten polymer is injected into the cold mold it rapidly solidifies locking in at least some of the "orientated" chain conformations. As the material enters the cold mold, the flow turbulence occurring with the outermost layers is generally sufficient to result in a more randomized, more amorphous outer structure. As the outermost chains cool, they "drag" the next chains effectively aligning them giving a more ordered structure. Finally, the cooling of the inner material is slowed because of the heat uptake of the outer layers allowing Brownian movement to again somewhat randomize these chains. Thus, the structure of the molded part is varied and can be further varied by controlling the flow rate, cooling rate, and flow and cooling temperatures for a specific injected-produced material. Figure 18.10 contains an idealized relationship between the distance from the outer wall of the wall of a tube and the amount of chain aligning.

18.9.2   BLOW MOLDING

Most molded material, as well as most processed material, will have a different surface or skin composition compared with the bulk or core material. Take a look at a common disposable PS foam plate. The surface or skin is smooth. Break it and look at the core and it is different being more cellular. This difference is greater than having simply a difference in appearance. There also exist different fine molecular-level differences. Molecular structure and associated bulk properties are controlled in part by the particular processing and processing particulars.

Blow molding has been used for many years in the creation of glass bottles. In about 1872, the blow molding of thermoplastic objects began by the clamping of two sheets of cellulose nitrate between two mold cavities. Steam was injected between the two sheets softening the sheets and pushing the material against the mold cavities. But, it was not until the late 1950s that large-scale use of blow molding began with the introduction of blow-molded high-density polyethylene (HDPE) articles.

Figure 18.11 contains a sketch of an extrusion blow-molding scheme. Here a heat-softened hollow plastic tube, or parison, is forced again the walls of the mold by air pressure. The sequence of material introduction into the mold and subsequent rejection of the material from the mold is generally rapid and automated. Approximately one million tons of thermoplastics are produced by this technique annually.

While there is a wide variety of blow-molding techniques, there are three main blow-molding procedures:

1. Injection blow molding that employs injection molded "test-tube" shaped preforms or parisons
2. Extrusion blow molding that uses an extruded tube preform or parison, and
3. Stretch blow molding that employs an injection molded, extrusion-blow molded preform, or extruded tube preform

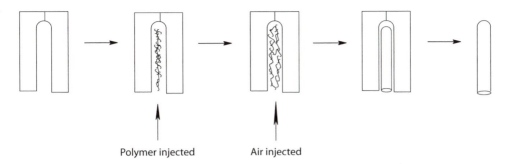

Polymer injected   Air injected

FIGURE 18.11 Steps employed in simple extrusion blow molding of a test tube. From left to right: mold closed, soften material introduced, air or other gas injected forcing the softened thermoplastic against the walls of the mold, and after suitable cooling, the mold is opened giving the molded plastic test tube.

The major difference between injection and extrusion blow molding is the way the soft hollow tube (called a *preform* or *pairson*) is made. In injection blow molding two different molds are used. One mold forms the preform and the other mold is used in the actual blow-molding operation to give the final shaped article. In the molding process the soften material preform from the preform mold is introduced into the blowing mold and blow molded to fit the cavity of the second "finished" blow mold. This process is sometimes also called *transfer blow molding* because the injected preform is transferred from the preform mold to the final blow mold. This allows better control of the product wall thickness and the thickness of the various curved locations.

Injection blow molding is typically used to produce smaller articles, generally with a total volume of 500 mL or less. Because two molds are used, there is little waste material that must be recycled and there is no bottom weld joint. It allows the production of small articles that at times are very difficult to manufacture in any other way.

Extrusion blow molding is the most common process used to produce hollow articles larger than 250 mL up to about 10,000 L. In extrusion blow molding the soften material is extruded continuously or intermittently. The preform is introduced, the mold halves close, and air or other gas is introduced forcing the preform material against the mold surfaces. After cooling, the mold is opened and the formed article rejected. Articles with handles and off-set necks can be manufactured using extrusion blow molding. Unlike injection blow molding, waste that must be cut away and recycled is produced as the two halves of the mold are pressed together.

In continuous extrusion blow molding the preform is continuously produced at the same rate as the article is molded, cooled, and released. To avoid interference with the preform formation, the mold-clamping step must be rapid to capture the preform and move it to the blow mold station. There are various modifications of this that allow essentially continuous operation.

The stretching is best done just above the materials T_g allowing a balance between good alignment because of ease in chain movement, and a decreased tendency to form crystalline areas in the melt allowing ready flow of material. A diagram illustrating this in found in Figure 18.12 for a typical polymer.

In the one-step process, preform production, stretching, and blowing all occur in the same machine. In the two-step sequence, the preform is produced in a separate step. The preform can be stretched before blowing in either the one-step or two-step process. In the one-step process, the preform is simply stretched just before, during, or just after the air is blown into the preform forcing it against the cavity walls.

Multilayered articles can be made by coinjection blow-molding or coextrusion methods. A three-layer system generally contains a barrier layer sandwiched between two "exterior" layers. These are actually laminar products. In the coextrusion sequence, several extruders can be used to place the material into the mold. The multilayer container is then produced from blowing air into the preform.

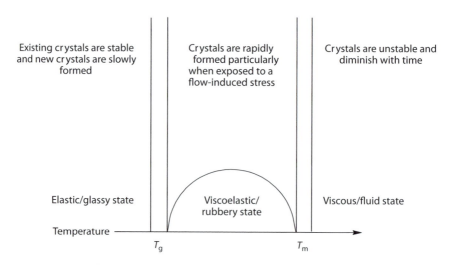

FIGURE 18.12 Idealized formation of order, crystals, as a function of temperature. The raised curve between the T_g and T_m ranges illustrate an idealized rate of crystal formation. This figure is sometimes referred to as the "Molders' Diagram of Crystallization."

18.9.3 ROTATIONAL MOLDING

In rotational molding, also known as rotomolding, the mold (or cavity) is filled with material, either as a solid power or as a liquid. The mold is closed, placed in a heated oven, and then rotated biaxially. The mold is then cooled, opened, and the article recovered. Powders of about 35 mesh (500 μm) are typical though different sizes are also employed. The distribution of particles and additives is determined by mixing and rotation ratio.

Almost any mold design can be incorporated into rotational molding. Tanks used for agricultural, chemical, and recreational vehicle industries are made using rotational molding as are containers used for packaging and material handling, battery cases, portable toilets, vacuum cleaner housings, light globes, and garbage containers. Rotational molding produces little waste and produces a material with uniform wall thickness as well as strong corner sections.

18.9.4 COMPRESSION AND TRANSFER MOLDING

While there are a number of molding processes, compression and transfer molding are the main techniques for molding articles from thermosetting materials. In compression or transfer molding, the material, thermoplastic or prethermoset material, is heated sufficiently to soften or plasticize the material to allow it to enter the mold cavity. The soften material is held against the mold by pressure. For thermoplastics it is then cooled below the T_g thus locking in the shape. For thermosets it is held until the cross-linking occurs thereby locking in the article shape.

The most widely employed molding process is compression molding where the material is placed in the bottom half of an open-heated mold. The second half of the mold is closed and brings heat and pressure against the material softening the material further and eventually allowing it to cross-link, if it is a thermoset. When completed, the pressure is released and article removed from the mold. Generally excess material, or flash, is produced. Figure 18.13 contains a representation of a simple compression molding assembly.

Compression molding is one of the oldest materials handling processes. Ancient Chinese employed compression molding to form articles from paper mache. Rubber articles were made in the early nineteenth century from composites of woody fibers and gum shellac. Baekeland used compression molding to make many of his early phenol–formaldehyde products.

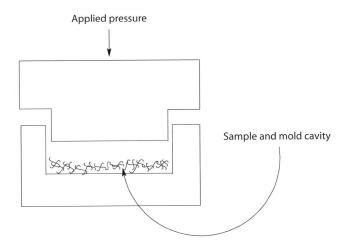

FIGURE 18.13   Representation of a compression molding assembly.

Transfer (or plunger) molding introduces the material to be molded after the mold is closed through a small opening or gate. This process can be used when additional materials, such as a glass globe or other designed object, are placed in the mold before closing the mold.

In true transfer or pot-type molding, the mold is closed and placed in a press. The soften material is introduced into an open port at the top of the mold. The plunger is placed into the pot and the press closed. As the press closes it pushes against the plunger forcing the molding material into the mold cavity. Excess molding compound is used to ensure that there is sufficient material to fill the mold. After the material is cured and/or cooled the plunger is removed and the part removed from the mold. In plunger molding, the plunger is part of the press rather than part of the mold. Because of this it can be smaller than the pot-type plunger. The clamping action of the press keeps the mold closed. Here there is less material waste compared with the pot-type molding.

In cold molding, the compound is compacted in a mold at around room temperature. The compressing operation is similar to that employed in the production of KBr pellets from powdered KBr. The compound generally contains a lot of filler and binder. The compacted material is removed from the mold and placed in an oven where it becomes cross-linked. Ceramic materials are often produced using cold molding.

While most molding involves thermosetting materials such as phenol, urea, and melamine–formaldehyde prepolymers, many elastomeric and thermoplastic materials are also molded. These include unsaturated polyesters, alkyd resins, epoxys, PVC, silicones, synthetic and natural elastomers, and diallyl phthalate polymers where the molded end products are also generally thermosets. Because all of these materials are good heat-insulating materials, the charge is generally preheated before it is introduced into the mold. With thermosets, slow heating may give cross-linking before the desired time, so that rapid heating is preferred followed by fast introduction into the mold. In compression molding, the rapid closing of the mold causes some frictional heating and in transfer molding, frictional heating is produced from the rapid and forced flow of the material through small gates into the mold cavity.

Some thermosets are postheated to finish curing allowing better control of the final amount of cross-linking and thus properties. Without postcuring, the product may continue to undergo cross-linking over the next months giving a material with varying properties over this period of time. Some of the thermosets, particularly the formaldehyde resins, give off some gas during polymerization. These gasses can be retained within the mold, increasing the pressure, or be released during the process. If the gasses are retained during the process the part can be recovered without noticeable effects, but if it is removed too soon blisters and ruptures may occur as a result of these gasses. However, often the gas is released periodically during the molding process. The time of opening is called the *dwell*. The step is referred to as breathe and dwell. Timing and duration of the breath and dwell steps are important.

While curing reactions occur at room pressures, it is important for good mold contact to employ high pressures, generally on the order of 20–70 MPa though some molding processes can get by with low pressure (0.7–7 MPa). Under pressure, the molecules behave as non-Newtonian fluids and some ordering occurs. Further, forced flow of polymers and prepolymers into the mold causes some aligning of the molecules. Thus, the fine structure, and associated properties, can be somewhat controlled by flow rates, heating/cooling, curing rate and extent, and pressure with speed associated with many of these factors. For instance, if a homogeneous, isotropic behaving material is wanted the flow rate into the mold should be slow and flow pathway short. Further, for thermosetting materials, the time that the material is preheated should be low and the rate rapid. Thus, there exist many balances where the end result is reached, not surprisingly, through a mix of science and practice (trial and error). Each machine, mold, and material will present a new opportunity for determining the optimum set of conditions.

In solvent molding a mold is immersed in a solution and withdrawn, or a mold is filled with a polymer and evaporation or cooling occurs producing an article such as a bathing cap. Solvent molding and casting are closely related.

18.9.5 THERMOFORMING

Thermoforming involves heating a sheet or thick film just above its T_g or T_m, stretching it against a rigid mold, cooling, and trimming the formed part. Inexpensive aluminum, wood, epoxy, and steel molds are often employed. This allows the construction of inexpensive molds that allow the production of low-volume articles. All thermoplastic materials that can be formed into sheets can be thermoformed provided that the heating does not exceed the ability of the sheet to support itself.

Thermoforming is employed to convert extruded sheets into smaller items such as packaging containers, plates, trays, bath tubs, pick-up truck liners, freezer liners, cabinetry, and cups. The skin packaging that involves a flexible plastic skin drawn tightly over an article on a card backing is made by thermoforming. Thermoforming permits the production of small to large articles, including those with thin walls such as drinking cups. Thus, thermoforming is employed to produce articles with a relatively high surface to thickness ratio. Figure 18.14 illustrates the operation of a simple plug-assisted vacuum thermoforming assembly.

Multilayered materials can be readily formed using thermoforming, including food packaging that may involve inclusion of layers of ethylene–vinyl alcohol copolymers, PS, polyolefins, and/or copolymers of vinylene dichloride and vinyl chloride. Microwavable food trays from (crystallized) poly(ethylene terephthalate) are manufactured using thermoforming.

Polystyrene is the most widely used resin material for thermoforming. High-impact PS (HIPS) is the most widely used being employed largely in the packaging areas, including disposables

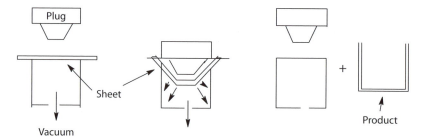

FIGURE 18.14 Steps in plug-assisted vacuum thermoforming. Initially vacuum is applied and the plug is pulled into the sheet pushing it into the mold. Further vacuum pulls the sheet against the walls of the mold. Finally, vacuum is turned off and the plume and formed product removed.

(foam drinking cups, lids, lunch trays, and food service containers), packaging for medical devices, and food packaging (meat and poultry trays, egg cartons, dairy and delicatessen containers, and barrier packages). Acrylonitrile–butadiene–styrene (ABS) is thermoformed to produce more durable articles such as refrigerator inner door liners, recreational vehicle and boat parts, automotive panels, picnic coolers, and luggage exteriors. HDPE is used in the manufacture of pick-up truck liners, golf cart tops, and sleds. Crystallized PET is used in the thermoforming of food trays that can be heated in a microwave or regular oven. Frozen-food oven-friendly trays are made from coextruded polycarbonate–polyetherimide. Polycarbonates and poly(methyl methacrylate) (PMMA) are thermoformed to produce skylights, windows, tub and shower stalls, and outdoor display signs. PVC is used in the production of blister packaging of pharmaceuticals, foods, cosmetics, and hardware.

Thermoforming is easily carried out for materials where the "sag" temperature is broad. Amorphous resins such as foamed, oriented, and HIPS, acrylics, and PVC are primary resins employed in thermoforming. Semicrystalline thermoformable resins include both HDPE and LDPE. Control of conditions is more important where the "sag" temperature range is narrower such as for semicrystalline resins like PP, some polyesters, and fluoropolymers. The range for effective sag can be increased through the introduction of appropriate branching, copolymerization, and addition of selected additives, including crystallization modifiers.

While some molding techniques are adaptable to the production of thermoset materials, thermoforming is carried out using thermoplastic materials.

There are a wide variety of thermoforming techniques in use today.

Vacuum can be used in a process called *basic vacuum forming*. The sheet is fixed to a frame, heated, and vacuum applied that implodes the sheet to conform to the mold contour. The vacuum site is generally at the base of the male mold. It cools as it comes into contact with the cold mold. For thick sheets, extra cooling is supplied by means of forced air- or mist-sprayed water. Articles formed using vacuum forming typically have thinner walls the further the sheet must travel to the mold location. The excess plastic material is trimmed and reused. At times vacuum forming is run in line with a sheeting extruder. It is similar to the plug-assisted system described in Figure 18.14 except no plug is used.

In drape forming the thermoplastic sheet is clamped and heated and the assembly then sealed over a male mold. The mold may be forced into the sheet or the sheet may be pulled into the mold by introduction of a vacuum between the sealed sheet and the mold. By draping the sheet over the mold, the part of the sheet touching the mold remains close to the original thickness. Foamed PS and polyolefins are generally used in this procedure.

In pressure forming, positive pressure is employed to assist the sheet-contents into the mold. The major advantage is a decreased cooling time for pressure forming. Crystalline PS, HDPE, and oriented PP are used in this procedure.

After the heated sheet is sealed across the mold, a shaped plug is pushed into the sheet, stretching it as it enters into the mold cavity in plug-assisted forming. The plug is generally of such a size and shape as to assist in the formation of the final mold shape and generally occupies about 90% of the mold volume. As it gets near the bottom of the mold, full vacuum is applied.

A variety of the plug-assisted process are the prestretching-bubble techniques. In pressure-bubble plug-assisted forming a heated sheet is sealed across a female cavity and pressure is blown through the cavity forcing the sheet from the mold. An "assist-plug" is then forced against the blown bubble with the heated sheet beginning to form about the plug as it forces the sheet against the female mold. As it nears contact with the female mold bottom, vacuum is applied through the mold causing the material to collapse onto to mold. Alternately, positive pressure can be applied on the "plug" side forcing the sheet against the mold walls. In the reverse of the pressure-bubble technique, called *vacuum snapback* forming, the heated sheet is sealed against a vacuum female cavity and controlled vacuum draws the concave shaped sheet away from the entering male mold. The male mold is then pressed against the sheet and vacuum applied through the male mold and/or pressure applied from the female cavity side forcing the material against the male mold. Luggage and automotive parts are made using this technique.

In pressure-bubble vacuum snapback the heated sheet is clamped and sealed against a pressure box. Air is forced through the female pressure box forcing the sheet to push outward from the pressure box. A male mold is then pressed against the bubble and as it pushes into the pressure box, excess air is forced from the pressure box forcing the heated sheet to take the shape of the male mold. The major difference between the vacuum snapback and pressure-bubble vacuum snapback is that in the vacuum snapback process vacuum from a female pressure box distorts the sheet away from the male mold and into the female pressure box, while in the pressure-bubble vacuum snapback excess pressure from the female pressure box forces the heated sheet toward the male mold and away from the female pressure box.

Similar to pressure-bubble vacuum snapback forming, air slip forming seals a heated sheet to the surface of a pressure chamber employing a male mold. It differs in the way the bubble is produced. Here, the heated sheet is clamped above a male mold. Pressure against the mold is created by the upward motion of the male mold toward the sheet causing it to bubble away from the oncoming mold. At the right time, a vacuum is applied through the male mold that causes the sheet material bubble to collapse and form about the male mold.

Trapped-sheet contact heat pressure forming utilizes a heating plate that contains many small vacuum and air pressure holes. A sheet is placed between the heating plate and the female mold. Initially, excess air pressure from the mold pushes the sheet into contact with the heating plate. The heating plate then heats the sheet and after desired heating, vacuum is applied from the female and/ or pressure applied through the heating plate pulls/pushes the heated sheet material into the female mold. Additional pressure can be used for trimming the article. Candy and cookie box liners and some medical packaging is made using this process.

These bubble-associated processes are aimed at prestretching the heated sheet to allow more even walls and bottoms to be formed.

In matched-mold forming a heated sheet is placed between a matched female and a male mold parts. As the two mold halves close, they distort the sheet to their shape. The air between the mold halves is removed. The article walls are more uniform than for many of the thermoforming techniques. This technique is used for the production of foamed PS and foamed polyolefins food containers.

A number of other techniques have been developed either to handle special materials or to create specific articles.

18.10   CASTING

Casting is employed in making special shapes, sheets, films, tubes, and rods from both thermoplastic and thermoset materials. The essential difference between most molding processes and casting is that no added pressure is employed in casting. In casting, the polymer or prepolymer is heated to a fluid, poured into a mold, cured at a specific temperature, and removed for the mold. Casting of films and sheets can be done on a wheel or belt or by precipitation. In the case of a wheel or belt, the polymer is spread to the desired thickness onto a moving belt as the temperature is increased. The film is dried and then stripped off. "Drying" may occur though solvent evaporation, polymerization, or cross-linking.

18.11   EXTRUSION

Extrusion involves a number of processing operations and is widely used. We will look at extrusion as it is involved in several of these processes. These processing operations are used together or separately. A representative extruder is shown in Figure 18.15. The extruder accepts granulated thermoplastic in a hopper (c), and forces it from the feed throat (d) through a die (f). The die may

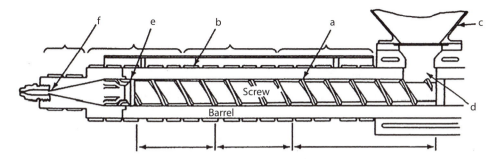

FIGURE 18.15 Sketch details of screw and extruded zones. (From Seymour, R. (1975): *Modern Plastics Technology*, Reston Publishing Co., Reston, Virginia.)

be circular for the production of a rod or pipe, or flat for the production of a sheet, or it may have any desired profile for the continuous production of almost any uniformly shaped product. The screw (a) advances the polymer through a heated cylinder (barrel) (b) to a breaker plate and a protective screen pack (e) before it enters the die (f). The extrusion process may be divided into a feed or transport zone, a compression or transition zone, and a metering zone. More than one million tons of extruded pipe are produced annually in the United States.

In extrusion, a fluid material, generally rendered a fluid material through heating is forced through a shaping device. Since there is a need for quickness and because the preshaped material is quite viscous, extrusion requires high pressure to drive or force the melt through a die. The melts can be extruded as pipers, sheets, films, or into molds.

Along with moving and shaping the molten material, extruders also act to disperse additives and are often the agent for creating heat thus enabling the material to become molten.

18.12 COATINGS

The fundamental purpose for painting is decorating, whereas the purpose for coating is for protection. In truth, we often do not differentiate between the two terms. Government edicts concerning air, water, solid particulates, and worker conditions are having real effects on the coatings industry with the generation of new coating techniques. Paint solvents, in particular, are being looked at in view of increased environmental standards. The volatile organic compound (VOC) regulations under Titles I and VI of the Clean Air Act specify the phasing out of ozone-depleting chemicals—namely chlorinated solvents. Baseline solvent emission are to be decreased. These, and related regulations, affect the emission of all organic volatiles, whether in coatings or other volatile-containing materials.

A major driving force in coatings continues to be a move toward water-based coatings. Another is to eliminate the "odor" of the coating. Most waterborne coatings actually contain about 8%–10% nonaqueous solvent. The odor we get as the coating is drying is mainly due to this solvent evaporating. Work continues to develop the right balance of properties and materials that allow the latex particles to flow together and coalesce into suitable films without the need of nonaqueous liquids.

Another area of active research is the development of paints that dry under extreme or unusual conditions, including under water and on cool substrates. The latter allows the painting season for exterior coating to be extended, particularly in the northern states.

Work continues on making more durable exterior paints. Remember that there is a difference in requirements of exterior and interior paints. For instance interior paints are generally required to be faster drying and more durable against scraps and punctures since it is the inside of the house that generally experiences such traumatic events. By comparison, exterior paints need to remain flexible and adhered under a wide variety of humidity and temperature. A more durable exterior coating should allow it a longer lifetime because it can better withstand stress caused by the pounding of the rain, sticks, and human-afflicted dings and dents. Binders or coatings resins are critical to the performance of coatings. They bind the components together. Since the primary cost of most commercial application of coatings is labor, the market will allow price increases for products that give added positive properties. Table 1.7 contains a listing of the production of paints for the year 2005.

Paint manufactures in the United States sell about 1,450 million gallons of coating material annually or about 5 gal for every man, woman, and child. Paint is typically a mixture of a liquid and one or more colorants (pigments). The liquid is called a *vehicle* or *binder* (adhesive) and may include a solvent or thinner along with the coating agent. The colored powders are called *pigments,* which may be prime or inert. Prime pigments give the paint its color. These may be inorganic, such as titanium dioxide (titanium (IV) oxide; the most widely used pigment by far) for white (but contained in many colored paints as well), iron oxides for browns, yellows, and reds, or organic compounds such as phthalocyanine for greens and blues. Inert pigments such as clay, talc, calcium carbonate, and magnesium silicate make the paint last longer and may contribute to the protective coating as do mica chips in some latex paints that actually form a clad on drying. The paint may also contain special additives and catalysts. Thus, many paints in wet areas contain an agent to fight fungus, rot, and mold.

Vehicles include liquids such as oils (both natural and modified natural) and resins and water. A latex vehicle is made by suspending synthetic resins, such as poly(ethyl acrylate), in water. This suspension is called an *emulsion*, and paints using such vehicles are called *latex*, waterborne, or emulsion paints. When the vehicle comes in contact with air, it dries or evaporates, leaving behind a solid coating. For latexes, the water evaporates, leaving behind a film of the resin.

Paints are specially formulated for specific purposes and locations. Following is a brief description of the most popular paint types.

Oil paints—Oil paints consist of a suspension of pigments in a drying oil, such as linseed oil. The film is formed by a reaction involving atmospheric oxygen that polymerizes and cross-links the drying oil. Catalysts may be added to promote the cross-linking reaction. Oil paints, once dried, are no longer soluble, although they can be removed through polymer degradation using the appropriate paint stripper.

Oil varnishes—Varnish coatings consist of a polymer, either natural or synthetic, dissolved in a drying oil together with appropriate additives as catalysts to promote cross-linking with oxygen. When dried, they produce a clear, tough film. The drying oil is generally incorporated, along with the dissolved polymer, into the coating.

Enamels—Classical enamel is an oil varnish with a pigment added. The added polymer is typically selected to provide a harder, glossier coating that the oil varnish mixture. Today, there are latex enamels that are similar to the oil enamels except no natural oil is present.

Lacquers—Lacquers consist of polymer solutions to which pigments have been added. The film is formed through simple evaporation of the solvent leaving the polymer film as the coating. These coatings are formed without subsequent cross-linking, thus the surface exhibits poor resistance to some organic solvents.

Latex paints—Latex paints today account for more than one-half of the commercial paint sold. They are characterized by quick drying (generally several minutes to several hours), little order, and easy cleanup (with water). Latex paints are polymer latexes to which pigments have been added. The film is formed by coalescence of the polymer particles on evaporation of the water. The polymer itself is not water-soluble, though these paints are called *waterborne coatings* because the polymer emulsion is "carried" by water.

The composition of latex paints is variable depending on the end use, manufacturer, and intended cost. Ethyl acrylate (18.1) and butyl acrylate (18.2) are frequently employed as the synthetic resin monomer. Most synthetic resins for coatings are copolymers of the acrylics. Vinyl acetate or other monomers are sometimes incorporated.

Ethyl acrylate derived unit (18.1)   Butyl acrylate derived unit (18.2)

The T_g of the copolymer must be below the application temperature to allow diffusion of water from the latex to occur as it is drying forming the protective film.

18.12.1   PROCESSING

Many different process are used to apply a thin layer of liquid-melted polymer or polymer solution or dispersion) including rollers, spraying, calendering, and brushing. Here we will look at the industrial application of coatings onto film and sheet-like materials. Figure 18.16 contains examples of general coating processes employed to achieve this. The moving sheet is called a *web*. In *roll coating*, Figure 18.16(a), the lower roller picks up the coatings material that then transfers it to the second roller and finally to one surface of the mat. Spacing of the rollers, viscosity of the polymer solution in the dip tank, and roller speed and size control the thickness of the applied coating.

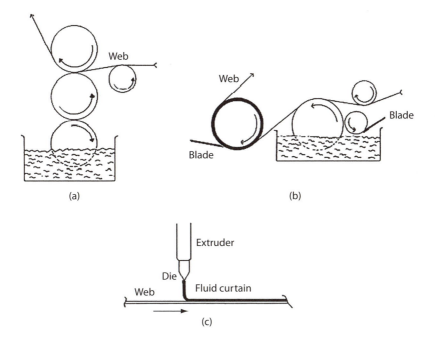

FIGURE 18.16 Three principle industrial coatings processes for films and thin sheets—(a) roll coating, (b) blade coating, and (c) curtain coating.

In *blade coating*, Figure 18.16(b) flexible blade helps control the coating thickness. The blade may be located after the bath or within the bath as part of the roller system. The blade is flexible and adjustable so that the amount of coating material can be controlled by application of force on the blade and/or distance the blade is from the web or roller. Both of these processes are forms of calendering. The third main coating process involves direct application of the coating material forming a so-called *fluid curtain*, Figure 18.16(c). This process, *curtain coating*, is usually used in conjunction with a curing process. The coating thickness is controlled by web speed, polymer concentration, and rate of application from the extruder.

Coatings can be divided according to formulation, drying mechanism or other system of categorizing. Table 18.8 contains a summary of coatings according to drying mechanism.

A coating is normally a mixture of various components. For instance, the label on a latex emulsion-type paint might have as major components poly(ethyl acrylate), titanium dioxide as the white pigment, and water. It could also have china clay and calcium carbonate as extenders, carboxymethylcellulose as a colloid thickener, a defoaming agent, a plasticizer, a surfactant dispersing aid, additional coloring agents, and an added fungicide.

The hiding power is a measure of the ability of the coating to achieve a specified degree of "hiding" or obliteration. Industrially, it is often tested by comparing the reflectance of the coated surface overpainting a black surface (that is, the tested paint applied over a black surface) with white panels. The ability to cover or hide is related to the scattering of incident light hitting the surface and returning to the observer or light meter. As the film surface increases, the ability of light to penetrate the surface coating and be scattered from the (black for tests) undercoating lessens. For a simple white latex paint, no absorption occurs and we can consider the scattering occurring at the interfaces of the transparent polymer matrix and the dispersed pigment particles. The scattering coefficient of polymers can be obtained from reflectance measurements. While the refractive indices for most polymers do not widely vary (generally about 1.5), the scattering coefficients can vary widely. For good scattering, the refractive index of the polymer should differ from that of the pigment. For

---

TABLE 18.8
Major Coatings Systems as a Function of Drying Mechanism

A. Film formation through chemical reaction
   (a) Through reaction of unsaturated double-bond sites with oxygen
       Alkyds
       Drying oils—fatty acid and related
   (b) Cold curing and thermosetting by reaction with vehicle components
       Epoxies
       Polyurethanes
       Unsaturated polyesters
       Urea and melamine–formaldehyde
B. Film formation through evaporation
   (a) Solution types
       Bituminous coatings
       Cellulose derived-acetate, acetate butyrate, ethyl, nitrate
       Chlorinated rubbers
       Poly(acrylic esters) such as poly(methyl methacrylate)
       Poly(vinyl chloride) copolymers
   (b) Dispersion types
       Poly(acrylic esters) such as poly(methyl methacrylate)
       Poly(vinyl acetate) and copolymers
       Poly(vinyl chloride) organosols
       Poly(styrene–cobutadiene)

---

instance, while calcium carbonate, with a refractive index of about 1.6, is often used as a pigment in paints, it has a much lower hiding power than titanium dioxide (rutile titanium (IV) oxide), with a refractive index of about 2.8.

Scattering efficiency increases as the pigment surface area becomes larger, thus smaller particles aid in increasing the scattering to a lower limit determined by the wavelength of light, thereafter reduced size produces a rapid loss in scattering efficiency. For good scattering, and good hiding power, the particles should be dispersed in a homogeneous manner so that dispersing agents are commonly used.

18.12.1.1 Rainwear

When it rains we grab the umbrella and put on a raincoat. Charles Macintosh is credited with the discovery of waterproof coats, aka raincoats. The first raincoats predated Macintosh's concept by hundreds of years. As early as the thirteenth century, South American Indians coated cloth with natural rubber latex making waterproof caps and footwear. The latex is natural rubber, a nonpolar material and as such it repels polar water molecules. As this latex was shipped to Europe bacteria attacked the natural rubber making it useless. In 1748, Francois Fresneau developed a process of dissolving natural rubber in nonpolar turpentine forming a rubber solution that was poured onto fabric rendering it "rain" proof. A contemporary of Macintosh, James Syme found that coal tar naphtha, again a nonpolar liquid, dissolved rubber and could be used to make waterproof fabric. He turned to over ventures rather than commercializing his process.

Coal tar naphtha is a waste byproduct from the conversion of coal into gas that was used as the fuel for streetlamps. Macintosh was under contract of the Glasgow Gas Light Company to work with the waste byproducts. He was extracting ammonia from the byproducts for use in his father's dye business. He was left with unusable coal tar naphtha. So, he looked for uses for this waste solvent. Shortly, he found that it dissolved rubber, which could be poured onto fabric creating water-proofed material. The material remained sticky and had an unpleasant odor. He conquered the problem of tackiness by sandwiching the treated fabric between two other fabric layers. The smell remained. Macintosh patented the idea in 1823 but his rainproof material was not well received, presumable because of the unpleasant odor. Fortunately for him, the military purchased waterproof fabric from him, "bad smell and all."

Improvements continued making the material lighter, more flexible, and less smelly. Use of cold-cure vulcanization where the rubber was treated with sulfur compounds eliminated the "sticky" problem allowing a single layer of impregnated fabric to be adequate.

Today, there exists a variety of polymeric materials that offer rainproof umbrellas and clothing, including "breathable" fabrics. Gore-Tex derived materials offer a combination of layers, including a breathable polymer membrane made of microporous polytetrafluoroethylene under a layer of PU. The polytetrafluoroethylene pores provide a layer of air. The PU has a high diffusivity for water vapor taking away any water that is on the skin, such as water vapor or sweat. The temperature inside the clothing is generally greater than the external temperature and acts as a "driving force" to "push" the water through the polytetrafluoroethylene pores and away from the skin.

18.13 ADHESIVES

In contrast to coatings, which must adhere to one surface only, adhesives are used to join two surfaces together. Resinous adhesives were used by the Egyptians at least 6,000 years ago for bonding ceramic vessels. Other adhesives, such as casein from milk, starch, sugar, and glues from animals and fish, were first used at least 3,500 years ago.

Adhesion occurs generally through one or more of the following mechanisms. Mechanical adhesion with interlocking occurs when the adhesive mixture flows about and into two rough substrate faces. This can be likened to a hook and eye, where the stiff plastic hooks get caught in the fuzz-like maze of more flexible fibers. Chemical adhesion is the bonding of primary chemical groups. Specific or secondary adhesion occurs when hydrogen bonding or polar (dipolar) bonding occurs.

Viscosity adhesion occurs when movement is restricted simply due to the viscous nature of the adhesive material.

Adhesives can be divided according to the type of delivery of the adhesive or by type of polymer employed in the adhesive. Following are short summaries of adhesives divided according to these two factors.

Solvent-Based Adhesives—Solvent-based adhesion occurs through action of the adhesive on the substrate. Solidification occurs on evaporation of the solvent. Bonding is assisted if the solvent partially interact, or, in the case of model airplane glues and PVC piping glues, actually dissolves some of the plastic (the adherent). Thus, model airplane glues and PVC glues often contain volatile solvents that dissolve the plastic, forming what is called a *solvent weld*. Some of the PVC glues actually contain dissolved PVC to assist in forming a good weld. A major difference between home-use PVC adhesive and industrial PVC adhesive is added color for the commercial adhesive so building inspectors can rapidly identify that appropriate solvent welding has been accomplished.

Latex Adhesives—Latex adhesives are based on polymer latexes and require that the polymers be near their T_g so that they can flow and provide good surface contact when the water evaporates. It is not surprising that the same polymers that are useful as latex paints are also useful as latex adhesives (such as PMMA). Latex adhesives are widely employed for bonding pile to carpet backings.

Pressure-Sensitive Adhesives—Pressure-sensitive adhesions are actually viscous polymer melts at room temperature. The polymers must be applied at temperatures above their T_g to permit rapid flow. The adhesive is caused to flow by application of pressure. When the pressure is removed, the viscosity of the polymer is sufficient to hold and adhere to the surface. Many tapes are of this type where the back is smooth and coated with a nonpolar coating so as not to bond with the sticky surface. The two adhering surfaces can be separated, but only with some difficulty.

Hot-Melt Adhesives—Hot-melt adhesives are thermoplastics that form good adhesives simply by melting, followed by subsequent cooling after the plastic has filled surface voids. Nylons are frequently employed in this manner. Electric glue guns typically operate on this principle.

Reactive Adhesives—Reactive adhesives are either low molecular weight polymers or monomers that solidify by polymerization and/or cross-linking reactions after application. Cyanoacrylates, phenolics, silicon rubbers, and epoxies are examples of this type of adhesive. Plywood is formed from impregnation of thin sheets of wood with resin with the impregnation occurring after the resin is placed between the wooden sheets.

Thermosets—A number of thermosets have been used as adhesives. Phenolic resins were used as adhesives by Leo Baekeland in the early 1900s. Phenolic resins are still used to bind together thin sheets of wood to make plywood. Urea resins have been used since 1930 as binders for wood chips in the manufacture of particle board. Unsaturated polyester resins are used for body repair and PUs are used to bond polyester cord to rubber in tires, bond vinyl film to particle board, and to function as industrial sealants. Epoxy resins are used in the construction of automobiles and aircraft and as a component of plastic cement.

Elastomers—Solutions of natural rubber have been used for laminating textiles for over a century. The Macintosh raincoat, invented in 1825, consisted of two sheets of cotton adhered by an inner layer of natural rubber. SBR is used as an adhesive in carpet backing and packaging. Neoprene (polychloroprene) may be blended with a terpene or phenolic resin and used as a contact adhesive for shoes and furniture.

Pressure-Sensitive Tape—Pressure-sensitive tape, consists of a coating of a solution of a blend of natural rubber and an ester of glycerol and abietic acid (rosin) on cellophane, was developed over half a century ago. More recently, natural rubber latex and synthetic rubber have been used in place of the natural rubber solution. The requirement for pressure-sensitive adhesives is that the elastomers have a T_g below room temperature. Today, there are many other formulations used in the production of pressure-sensitive tapes.

Contact Adhesives—Contact adhesives are usually applied to both surfaces, which are then pressed together. Liquid copolymers of butadiene and acrylonitrile with carboxyl end groups are used as contact adhesives in the automotive industry.

Thermoplastics—A number of thermoplastics have been used as adhesives. Polyamides and copolymers of ethylene and vinyl acetate (EVA) are used as melt adhesives. Copolymers of methyl methacrylate and other monomers are used as adhesives in the textile industry. Poly(vinyl acetate) is often used in school glues.

Anaerobic Adhesives—Anaerobic adhesives consist of mixtures of dimethacrylates and hydroperoxides (initiators) that polymerize in the absence of oxygen. They are used for anchoring bolts.

Cyanoacrylates—One of the most interesting and strongly bonded adhesives are cyanoacrylates (Super Glue; Krazy Glue). These monomers, such as butyl-alpha-cyanoacrylate (18.3), polymerize spontaneously in the presence of moist air, producing excellent adhesives. These adhesives, which have both the cyano and ester polar groups, are used for household adhesive problems as well as in surgery, mechanical assemblies, and as a "fast fix" for athletic cuts such as in boxing.

(18.3)

Many seemingly simple adhesive applications are actually complex. The labels on commercial dry cell batteries can contain over a dozen layers each present for a specific purpose. While price is a major consideration, ease of application is another. Thus, while many naturally derived adhesives are less expensive, synthetic materials may be chosen because of ready application and consistency of the end product.

One common use of adhesives is as the "working" ingredient of tapes. Numerous tapes are available, many with interesting stories. The ingredients vary with the intended use. Table 18.9 contains the main ingredients of some important tapes. But, knowing the ingredients is only the start. There is much science involved. For instance, important factors involved with pressure-sensitive adhesion are a balance between allowing molecular interaction between the adhesive and adherent (often referred to as "wetting") and the dynamic modulus of the adhesive mixture. This also involves a balance between "pull-off-rate" and "wetting rate." Mechanical adhesion with interlocking and diffusion factors is less important than for permanent adhesion. Pressure-sensitive adhesives, such as present in "pull-off" tabs such as Post-It Notes, contain components similar to those present in more permanent Scotch tape, except that particles of emulsified glass polymer are added to reduce the contact area between the adhesive and the substrate. Some polymers, such as PE, might appear to be decent adhesive materials, but even in its melt, it is not exceptionally tacky. This is believed a result of the high degree of chain entanglement. Since the dynamic modulus increases with increasing chain entanglement, PE is not "tacky" (does not easily contact and wet a substrate), and is not a useful pressure-sensitive adhesive.

Super glue was initially discovered in 1942 as part of the war effort in a search to make clear plastic gun sights. Super glue was discovered but it stuck to everything so was discarded. In 1951, Harry Coover and Fred Joyner, Eastman Kodak researchers, rediscovered it. It was first sold as a commercial product in 1958.

TABLE 18.9
General Composition of Typical "Scotch" Tapes (By Weight)

Paint masking tape
    Paper backing 55%–74%
    Natural rubber adhesive 26%–45%

Transparent tape
    Acrylic adhesive about 36%
    Polypropylene film (backing) about 64%

Post-it notes
    Acrylic adhesive <0.5%
    Inks and dyes <0.1%
    Paper >99.5%

Vinyl electrical tape
    Poly(vinyl chloride) backing 48%–54%
    Polyester adipate backing 21%–27%
    Hydrotreated light naphthal 4%–8%
    Epoxidized soybean oil 2%–4%
    Antimony trioxide (Sb_2O_3) flame control 2%–4%
    Piperylene-2-methyl-2-butene polymer 2%–4%
    Branched alkyl phthalate plasticizer 2%–4%
    Fillers and processing aids 2%–4%

It polymerizes rapidly, generally within a minute or so continuing to harden for about 24 h. Acetone, from many finger nail polishes, can be used to soften super glue.

Super glue is based on cyanoacrylates, generally methyl-2-cyanoacrylate, ethyl-2-cyanoacrylate (sold under trade names such as Super Glue and Krazy Glue), n-butyl cyanoacrylate (used in veterinary glues such as Vetbond and LiquiVet and skin glues such as Indermil and Histoacryl), and octyl-2-cyanoacrylate (used as a medical glue with trade names such as LiquiBand, FloraSeal, Dermabond, SurgiSeal, and Nexaband). Cyanoacrylate glue has a low shearing strength that allows its use as a temporary adhesive where it is employed to hold an object in place and then sheared off later. It is used to assemble prototype electronics and to hold nuts and bolts in place and in building model aircraft. It is used as a forensic tool to capture fingerprints on glass and metals. The glue was first used in 1966 as a spray to retard bleeding in wounded soldiers on the battlefield until they could be properly cared for in a hospital. It has been used to stop cuts in professional fighters and to repair bones in animals. Most glues are a mixture of the cyanoacrylate but with about 10% of other material such as PMMA, hydroquinone, and small amounts of organic sulfonic acid.

Cyanoacrylates are acrylic resins that polymerize rapidly in the presence of water via attack of a nucleophile such as the hydroxide ion. This is shown below for methyl-2-cyanoacrylate.

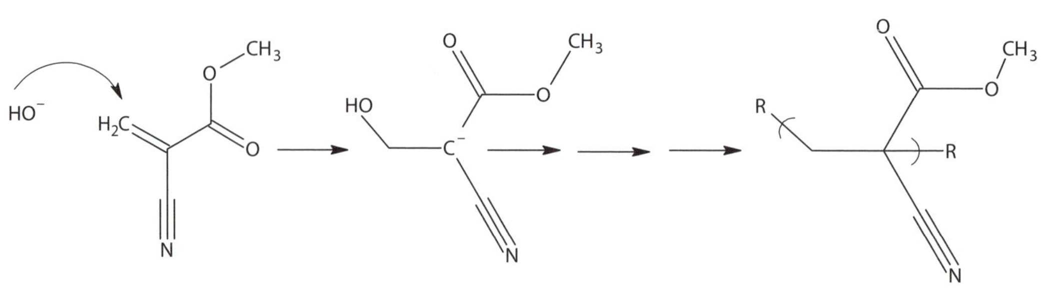

$$(18.4)$$

The 3M company has been instrumental in the discovery and improvement of many tapes. The first pressure-sensitive adhesive tape was discovered in 1925. During the depression an increasing number of items were taped together rather than discarded. Double-sided tape was used to hold together layers of metal skins to aircraft frames. But what we now know as Scotch tape was discovered by a young 3M engineer Richard Drew in 1930. Then 3M was a small sand paper manufacturing company. Two toned automobiles were becoming popular. Painters would mask one section while painting the second color. Parts of the first color were often ripped off when the masking material, typically simply newsprint, was removed. Drew noticed this and worked for the next 2 years developing pressure-sensitive glue that he applied to the edges of one side of some crepe paper. While it could be easily removed by the painters, it also kept falling off. The story goes that one of the painters told a 3M representative to tell his "Scotch" bosses (an indication that 3M was cheap) to put adhesive all over the tape and not just on the edges. They did and the name stuck as did the tape. Today, 3M has hundreds of tapes sold under the "Scotch Tape" name.

Drew was then asked to develop a waterproof tape that could seal insulation panels for refrigerated railroad cars. He and fellow workers developed the clear cellophane-backed tape that is familiar today but today with a variety of different formulations.

Post-It Notes is another of the host of 3M tapes. It was discovered by Art Fry using an adhesive developed by a fellow coworker Spenser Silver with the help of Jesse Kopes in 1968. The adhesive was simply an acrylic adhesive similar to that used for many of the more permanent tape adhesives, but diluted with nonadhesive material. Silver promoted his "low-tack" reusable pressure-sensitive adhesive for 5 years within the 3M origination but with little success. Fry attended one of the sales pitches by Silver and used the formula to anchor his bookmark in his hymnbook. 3M commercialized the product in 1977 but customers would not try this new product. After giving away free samples to the residents of Boise, Idaho they found that people liked it so a new promotion was tried and the rest is history. By 2003, 3M came out with "super" Post-It Notes that were simply the old formula with less nonactive filler giving the Post-In Notes better adhesion.

Self-adhesive tapes generally consist of four thin layers as pictured in Figure 18.17. The flexible backing is often about 50–100 μm thick and can be composed of a variety of materials, including polyethylene, PP, paper, cloth, foil, vinyl, or foam.

Backings are coated on both sides. The release coating is often a poly(vinyl carbamate) (18.5) and it prevents the tape from adhering with itself. The other side of the backing is coated with a primer that attaches the backing and adhesive material. The major component of the adhesive is polymeric. For instance, styrene block copolymers are often employed as the adhesive for packaging and double-sided tapes while office tapes often employ water-based acrylics similar to those employed in some paints. Silicon adhesives are often used for medical tapes.

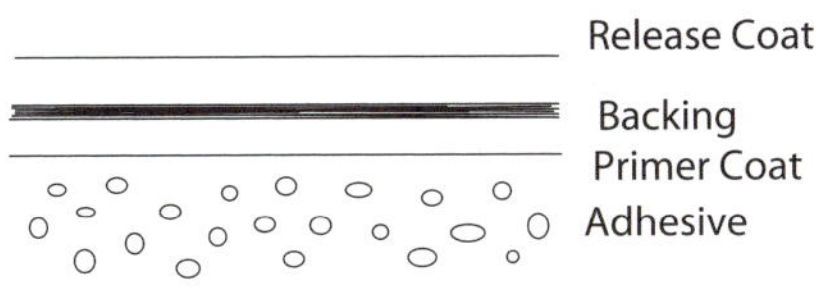

FIGURE 18.17   Components of a typical pressure-sensitive tape.

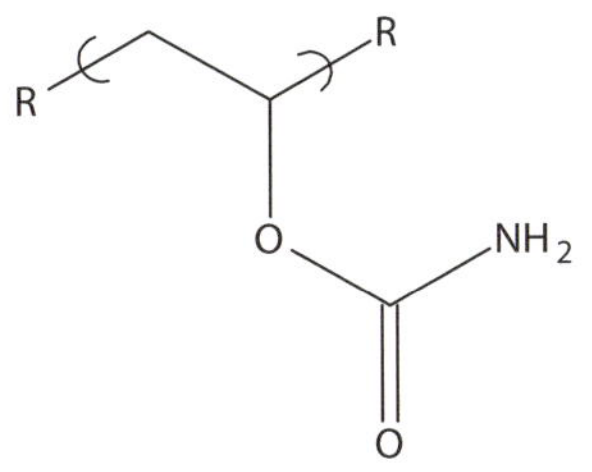

(18.5)

The polymers themselves are often not sticky so agents are added to promote adhesion. These additives are varied. One major additive is referred to as the tackifying agent that may be present in amounts that are in range of the polymer.

Duct tape, like other self-adhesive tapes, generally owe their adhesion properties to pressure-sensitive combinations that are soft, solid polymer blends that adhere to the surface through van der Wall's forces when slight pressure is applied. Thus, no chemical reaction occurs but adhesion depends on cumulative second forces. The polymer component to the adhesive in duct tape is usually natural rubber that is not itself sticky. A polyterprene resin is added that partially dissolves the natural rubber but it does not break up the cross-links in the natural rubber. This makes the natural rubber softer and allows some flow.

The adhesive layer is generally about 20 μm thick for most office tapes and 50 μm for more heavy duty tapes such as duct tape.

18.14 SUMMARY

1. The processing of polymeric materials is well advanced in science, engineering, and technology and continues to advance. The properties of the end product are dependent on the chemical structure, processing, and physical structure. The properties depend on the meso-structure of the particular polymer.
2. Fibers are processed using mainly one of three processes—melt spinning where the polymer is melted and forced through spinnerets, dry spinning where the dissolved polymer is forced through a spinneret and the solvent removed, and wet spinning where the dissolved polymer is forced through a spinneret into a nonsolvent bath. Fibers are usually stretched as they are formed allowing the chains to further align giving added strength in the direction of the pull.
3. Elastomers are processed using four basic steps—mastication, incorporation or compounding, shaping, and vulcanization.
4. Film and sheets are made from precipitation and/or regeneration from a polymer solution or melt. Calendering is also used where the polymer is passed between a series of counterrotating rollers.
5. Unicellular (closed cell, discrete unit cells) and multicellular (open cells, interconnecting cells) foams are made through the chemical and physical introduction of gasses.
6. A wide variety of composites are made. Composites have two phases, the continuous or matrix phase that surrounds the discontinuous or dispersed phase, usually a fiber material. Composites are strong with many uses.
7. Molding employs a mobile prepolymer that may be thermoset or using a thermoplastic polymer. The polymer can be injection molded (often for solid objects), blow molded (for hollow objects such as bottles), rotational molded, compression molded, transfer molded, and thermoformed. Casting is closely related to molding except pressure is typically not used.
8. In extrusion, a fluid material, is forced through a shaping device. The melts are extruded as pipes, sheets, films, or into molds.
9. Coatings and paints are used for decoration and protection. There are many different kinds of coatings with the majority being latex-based where the polymer is present in a suspension that forms a film when the water evaporates. Industrially, coatings are applied using roll coating, blade coating, and curtain coating processes.
10. The difference between coatings and adhesives is that coatings must adhere on only a single surface, but adhesives must bind together two surfaces. There is a wide variety of adhesives that make use of the various polymer properties. Adhesion can occur through physical locking together, through chemical (primary and secondary bond formation) joining, or through simply highly viscous material holding together materials.
11. Polymers are major materials in the nanotechnology revolution, including as conductive (photo and electronic) materials. Delocalization of electrons throughout a polymer chain or matrix

is important for electronic conductance. This is often accomplished through doping that encourages flow of electrons.

12. Polymers are also major materials in the biomedical areas as materials and in the delivery of drugs.

13. For new polymeric materials to enter the marketplace, many things must be in place, including a "flagship" property that meets a particular market need, ready availability, and money. It takes about $1 billion to introduce and establish a new polymer into the marketplace.

GLOSSARY

Abaca: Hemp-like fiber.

Acrilan: Polyacrylonitrile-based fibers.

Acrylic fibers: Polyacrylonitrile-based fibers.

Adhesive: Material that binds, holding together two surfaces.

Biaxial orientation: Process where a material, normally a film, is stretched in two directions at right angles to each other.

Buna-N: Acrylonitrile–butadiene copolymer.

Calender: Machine for making polymeric sheets using counterrotating rolls.

Casting: Production of film by evaporation of a polymeric solution.

Cellophane: Regenerated cellulosic film.

Charge: Amount of polymer used in each molding or processing cycle.

Coextruded film: Film produced by the simultaneous extrusion of two or more polymers.

Dacron: Trade name for PET fiber.

Draw: Depth of mold cavity.

Drying oils: Liquids employed in coatings that will be cured, cross-linked.

Dry spinning: Process for obtaining fiber by forcing a solution of a polymer through holes in a spinneret and evaporating the solvent from the extruded material (extrudate).

Elastomer: Rubber.

Electrodeposition: Use of an electric charge to deposit polymer film or aqueous dispersions onto a metal substrate.

Engineering material: Material that can be machined, cut, drilled, sawed, and so on; must have sufficient dimensional stability to allow these actions to occur.

Extrusion: Fabrication process in which a heat-softened polymer is continually forced by a screw through a die.

Fibrillation: Process for producing fiber by heating and pulling twisted film strips.

Filament: Continuous thread.

Filament winding: Process in which filament are dipped in a prepolymer, wound on a mandrel, and cured.

Gate: Thin sections of runner at the entrance of a mold cavity.

Green materials: Includes materials that do not have a negative impact on the environment; biomass-derived materials

Hemp: Fiber from plants of the nettle family.

Hevea rubber: Natural rubber; *Hevea brasiliensis*.

Hycar: Trade name for Buna-N elastomer.

Jute: Plant fiber used for making burlap.

Kodel: Trade name for PET fiber.

Lacquers: Polymer solutions to which pigments have been added.

Lamination: Plying up of sheets.

Latex: Stable dispersion of a polymer in water.

Mechanical goods: Generally industrial rubber products like belts.

Melt spinning: Process of obtaining fibers by forcing molten polymer through holes in a spinneret and cooling the filaments.

Molding powder or compound: Premix of resin and other additives used as a molding resin.

Multicellular: Open celled.

Neoprene: Trade name for polychloroprene.

Nonwoven textiles: Sheet produced by binding fibers with a heated thermoplastic.

Oil paints: Suspension of pigments in a drying oil.

Oil varnish: Polymer dissolved in a drying oil.

Parison: Short plastic tube that is heated and expanded by air in the blow-molding process.

Perlon: Trade name for some PU fibers.

Photoconductive: Material that is conductive when exposed to light.

Photoresponsive: Material whose properties change when exposed to light.

Pigment: Coloring material; colorant.

Polyacetylene: Polymer whose conductivity increases when doped to be a conductor.

Printing ink: Highly pigmented coatings used in printing.

Pultrusion: Process in which filaments are dipped in a prepolymer, passed through a die, and cured.

Rotational molding: Polymer added to a warm, rotating mold; centrifugal force distributes the polymer evenly.

Rovings: Multiple untwisted strands of filaments.

Runner: Channel between the spruce and the mold cavity.

Screen pack: Metal screen that prevents foreign material form reaching the die in an extruder.

Specific strength: Strength based on mass rather than area.

Spinneret: Metal plate between the nozzle and runner.

Sprue: Tapered orifice between nozzle and runner. Term also used to apply to plastic material in the sprue.

Structural foams: Polymeric foamed article with a dense surface.

Styrofoam: Trade name for foamed PS.

Technology: Applied science.

Tenacity: Fiber strength.

Thermoforming: Shaping of a hot thermoplastic sheet.

Transfer molding: Process in which a preheated briquette or preform is forced through an orifice into a heated mold cavity.

Vehicle: Liquid in a coating.

Wet spinning: Obtaining fibers by precipitation of polymeric solutions.

EXERCISES

1. Which is more important: (a) polymer science or (b) polymer technology?
2. Name three important natural fibers.
3. Name an important regenerated fiber.
4. Why is secondary cellulose acetate more widely used than the tertiary cellulose acetate?
5. What is the difference between rayon and cellophane?
6. Name three important synthetic fibers.
7. Name an elastomer produced by (a) cationic, (b) anionic, (c) free radical, and (d) step-reaction polymerization techniques.
8. How is LDPE film produced?
9. Why is there a trend toward the use of less solvent in polymeric coatings?
10. What is meant by trade sales?
11. What are the general steps needed before a new drug comes to market.

12. How would you produce a unicellular foam?
13. Which foam is preferable for upholstery: (a) unicellular or (b) multicellular?
14. Why do nonflame-retardant foams burn readily?
15. Why reinforced plastic has been used as an automobile body?
16. Why is graphite-reinforced epoxy resin a good candidate for parts in future automobiles?
17. Why are molded thermoplastics used more than molded thermosets.
18. Which of the following might you expect to increase their conductivity when doped: (a) PS, (b) PPO, (c) nylon 66, or (d) aramids?
19. Why are structural foams used for complex furniture?
20. What are some of the advantages of a blow-molded PET bottle?
21. Why would an article be thermoformed instead of molded?
22. What is the limit to the length of an extrudate such as PVC pipe?
23. Name three popular laminates.
24. Why are the terms painting and coating often used interchangeably?
25. Why have latex waterborn coatings been popular with the general public?
26. Differentiate between oil paints, oil varnishes, latexes, enamels, and lacquers.
27. Briefly discuss the popular mechanisms related to adhesion and the general types of adhesives.
28. How might the mesophase structure be changed?
29. What is a disadvantage of employing spinning of polymers to form fibers using a nonsolvent bath assembly?
30. Are PU foams closed or opened celled?
31. Name a common laminate found in many kitchens.
32. Why are there so many different kinds of adhesives?
33. What is an advantage to using cyanoacrylates at a crash site?
34. There is a black room that you wish to lightened up by painting it a lighter color. What should you do?

ADDITIONAL READING

Aldissi, M. (1992): *Intrinsically Conducting Polymers*, Kluwer, Dordrecht.
Bachmann, K. J. (1995): *Materials Science of Microelectronics*, VCH, NY.
Baird, D., Collias, D. (1998): *Polymer Processing Principles and Design*, Wiley, NY.
Belcher, S. (1999): *Practical Extrusion Blow Molding*, Dekker, NY.
Benedek, I., Feldstein, M. (2009): *Handbook of Pressure-Sensitive Adhesives and Products*, Taylor and Francis, Boca Raton, FL.
Bieleman, J. (2000): *Additives for Coatings*, Wiley, NY.
Cahn, R., Haasen, P., Kramer, E. (1996): *Processing of Polymers*, VCH, NY.
Carraher, S., Carraher, C. (1994–1997). ISO process and practices in a series of articles appearing in Polymer News 19: 373; 20:147; 20:278; 21:21; 21:167: 22:16.
Chanda, M., Roy, S. (2008): *Plastics Fundamentals, Properties, and Testing*, CRC, Boca Raton, FL.
Clemitson, I. R. (2008): *Castable Polyurethane Elastomers*, Taylor and Francis, Boca Raton, FL.
Contescu, C., Putyera, K. (2009): *Encyclopedia of Nanoscience and Nanotechnology*, Taylor and Francis, Boca Raton, FL.
Dadmun, M. (2008): *Modification of Interfaces with Polymers*, Wiley, Hoboken, NJ.
Ericksen, J. (2008): *Orienting Polymers*, Springer, NY.
Fried, J. R. (1995): *Polymer Science and Technology*, Prentice Hall, Upper Saddle River, NJ.
Groza, J., Shackelford, J., Lavernia, E., Powers, M. (2007): *Materials Processing Handbook*, Taylor and Francis, Boca Raton, FL.
Grulke, E. (1993): *Introduction to Polymer Processing*, Prentice Hall, Englewood Cliffs, NH.
Hatziliriakos S. G., Migler, K. B. (2005): *Polymer Processing Instabilities*, Taylor and Francis, Boca Raton, FL.
Hearle, J., Morton, W. (2008): *Physical Properties of Textile Fibres*, CRC, Boca Raton, FL.
Jaques, R. (2000): *Plastics and Technology*, Cambridge, NY.

Jimenez, A., Zaikov, G. (2008): *Recent Advances in Research on Biodegradable Polymers and Sustainable Composites*, Nova, Hauppauge, NY.

Johnson, P. (2001): *Rubber Processing*, Hanser, Gardner, Cincinnati, OH.

Kestleman, V (2001): *Adhesion of Polymers*, McGraw-Hill, NY.

Klingender. H. (2008): *Handbook of Specialty Elastomers*, Taylor and Francis, Boca Raton, FL.

Lagaly, G., Richtering, W. (2006): *Mesophases, Polymers, and Particles*, Springer, NY.

Martin, J., Dickie, R., Ryntz, R., Chin, J. (2008): *Service Life Prediction of Polymeric Materials: Global Perspectives*, Springer, NY.

O'Brian, K. T. (1999): *Applications of Computer Modeling for Extrusion and Other Continuous Polymer Processes*, Oxford, Oxford, England.

Olmsted, B. (2001): *Practical Injection Molding*, Dekker, NY.

Piringer, O., Baner, A. (2008): *Plastic Packaging Materials: For Food and Pharmaceuticals*, Wiley, Hoboken, NJ.

Provder, T. (2006): *Film Formation, Process, and Morphology*, Oxford University Press, NY.

Reyne, M. (2008): *Plastic Forming Processes*, Wiley, Hoboken, NJ.

Rudin, A. (1998): *Elements of Polymer Science and Technology*, 2nd Ed., Academic, Orlando, FL.

Schwartz, M. (2009): *Smart Materials*, Taylor and Francis, Boca Raton, FL.

Shibaev, V., Lam, L. (1993): *Liquid Crystalline and Mesomorphic Polymers*, Springer-Verlag, NY.

Sinha, R. (2002): *Outlines of Polymer Technology: Manufacture of Polymers and Processing Polymers*, Prentice Hall, Englewood Clifts, NJ.

Sudarshan, T., Dahorte, N. (1999): *High-Temperature Coatings*, Dekker, NY.

Wallace, G., Spinks, G. (1996): *Conductive Electroactive Polymers*, Technomic, Lancaster, PA.

Wicks, Z., Jones, F., Papas, S. P. (1999): *Organic Coatings Science and Technology*, Wiley, NY.

Williams, P. (2007): *Handbook of Industrial Water Soluble Polymers*, Wiley, Hoboken, NJ.

19 Selected Topics

The previous chapters allowed the weaving of polymer science fundamentals and applications into a carpet necessary for those involved in the sciences, engineering, biology, and the biomedicines to understand as they ply their trade. The present chapter enhances the areas where an understanding of basic polymer principles is important as polymers are employed in many areas in today's society that is becoming increasingly dependent on materials that are polymeric.

19.1 CONDUCTIVE POLYMERIC MATERIALS

Conductance behavior is dependent on the material and what is conducted. For instance, polymeric materials are considered poor conductors of sound, heat, electricity, and applied forces in comparison to metals. Typical polymers have the ability to transfer and "mute" these factors. For instance, as a force is applied, a polymer network transfers the forces between neighboring parts of the polymer chain and between neighboring chains. Because the polymer matrix is seldom as closely packed as a metal, the various polymer units are able to absorb (mute; absorption through simple translation or movement of polymer atoms, vibrational and rotational changes) as well as transfer (share) this energy. Similar explanations can be given for the relatively poor conductance of other physical forces.

Even so, polymers can be designed that compete with metals and other nonpolymer materials in the area of conductance. Some of these are described subsequently. Covered elsewhere are other materials that act similar to metals in the conductance of specific phenomena. For instance, force transference of ceramics is similar in some ways to that of metals because of a number of factors, including the inability of the tightly packed ceramics to mute applied forces and their ability to directly "pass along" the results of such applied forces or stresses.

19.1.1 Photoconductive and Photonic Polymers

Some polymeric materials become conductive when illuminated with light. For instance, poly(*N*-vinylcarbazole) is an insulator in the dark, but when exposed to ultraviolet (UV) radiation it becomes conductive. Addition of electron acceptors and sensitizing dyes allows the photoconductive response to be extended into the visible and near-infrared (NIR) regions. In general, such photoconductivity is dependant on the materials ability to create free-charge carriers, electron holes, through absorption of light, and to "move" these carriers when a current is applied.

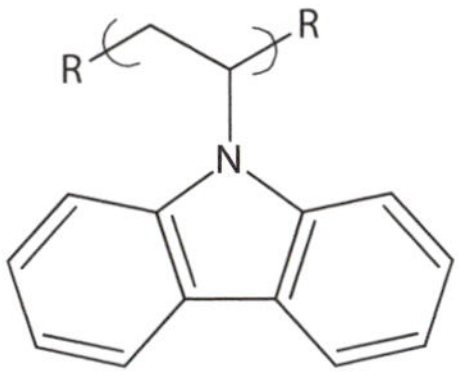

Poly(*N*-vinylcarbazole)

(19.1)

Related to this are materials whose response to applied light varies according to the intensity of the applied light. This kind of behavior is referred to as nonlinear behavior. In general, polymers with whole-chain delocalization or large-area delocalization where electrons are optically excited may exhibit such nonlinear optical behavior.

A photoresponsive sunglass whose color or tint varies with the intensity of the sunlight is an example of nonliner optical material. Some of the so-called "smart" windows are also composed of polymeric materials whose tint varies according to the incident light. Currently much material is stored using electronic means but optical storage is becoming common place with the use of CD-ROM and WORM devices. Such storage has the advantages of rapid retrieval and increased knowledge density (i.e., more information stored in a smaller space).

Since the discovery that doped polyacetylene becomes electrically conductive, a range of polymer-intense semiconductor devices has been studied, including normal transistors and field-effect transistors (FETs), and photodiodes and light-emitting diodes (LEDs). Like conductive polymers, these materials obtain their properties because of their electronic nature, specifically the presence of conjugated pi-bonding systems.

In electrochemical light-emitting cells, the semiconductive polymer can be surrounded asymmetrically with a hole-injecting material on one side and a low work function electron-injecting metal (such as magnesium, calcium, or aluminum) on the other side. The emission of light results from a radiative charge carrier recombining in the polymer as electrons from one side and holes from the other recombine.

Poly(p-phenylene vinylene) (PPV) was the first reported (1990) polymer to exhibit electroluminescence. PPV is employed as a semiconductor layer. The PPV layer was sandwiched between a hole-injecting electrode and electron-injecting metal on the other. PPV has an energy gap of about 2.5 eV and thus produces a yellow green luminescence. Today, other materials are available that give a variety of colors.

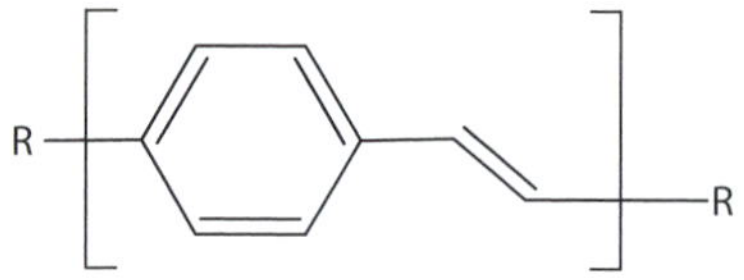

Poly(p-phenylene vinylene)

(19.2)

A number of PPV derivatives have been prepared. Attachment of electron-donating substituents such as dimethoxy groups (19.3) act to stabilize the doped cationic form and thus lower the ionization potential. These polymers exhibit both solvatochromism (color change as solvent is changed) and thermochromism (color is temperature dependent).

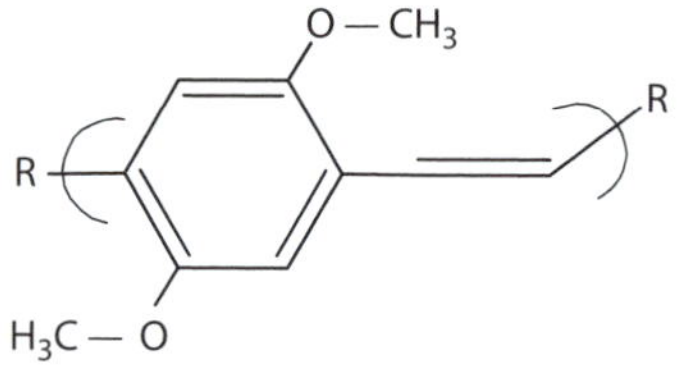

Poly(2,5-dimethoxy-p-phenylene vinylene)

(19.3)

19.1.2 ELECTRICALLY CONDUCTIVE POLYMERS

The search for flexible, noncorrosive inexpensive conductive materials has recently focused on polymeric materials. This search has increased to include, for some applications, nanosized fibrils and tubes. The conductivity for general materials is noted in Figure 19.1. As seen, most polymers are nonconductive and in fact, are employed in the electronics industry as insulators. This includes polyethylene and poly(vinyl chloride). The idea that polymers can become conductive is not new and is now one of the most active areas in polymer science. The advantages of polymeric conductors

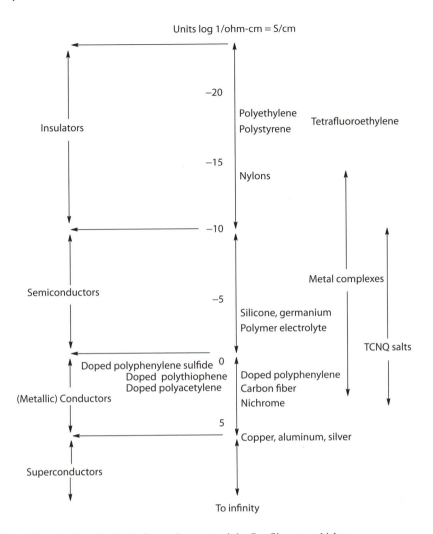

FIGURE 19.1 Electrical conductivity for various materials. S = Sieman = 1/ohm.

includes lack of corrosion, low weight, ability to "lay" wires on almost a molecular level, and ability to run polymeric conductive wires in very intricate and complex designs. The topic of conductive carbon nanotubes has already been covered (Section 12.17).

The Nobel Prize in Chemistry for 2000 was given to Alan MacDiarmid, Alan Heeger, and Hideki Shirakawa for the discovery and development of electrically conductive polymers. In 1975, MacDiarmid and Heeger began studying the metallic properties of inorganic poly(sulfur nitride) but shifted their efforts to polyacetylene after visiting with Shirakawa. While the synthesis of polyacetylene was known for years, Shirakawa and coworkers, using a Ziegler–Natta catalyst, prepared it as a silvery film in 1974. But, in spite of its metallic appearance, it was not a conductor. In 1977, using techniques MacDiarmid and Heeger developed for poly(sulfur nitride), Shirakawa, MacDiarmid, and Heeger were able to increase the conductivity of *trans*-polyacetylene samples, after doping, to a conductivity of about 10^3 S/m or 1 S/cm. They found that oxidation with chlorine, bromine, or iodine vapor made polyacetylene film 10^9 times more conductive than the nontreated film. This treatment with a halogen was called "doping" by analogy with the doping employed with semiconductors. Other oxidizing "doping" agents have been used, including arsenic pentafluoride. Reducing agents such as metallic sodium have also been successfully used. This chemical doping transforms the polyacetylene from an insulator or semiconductor to a conductor.

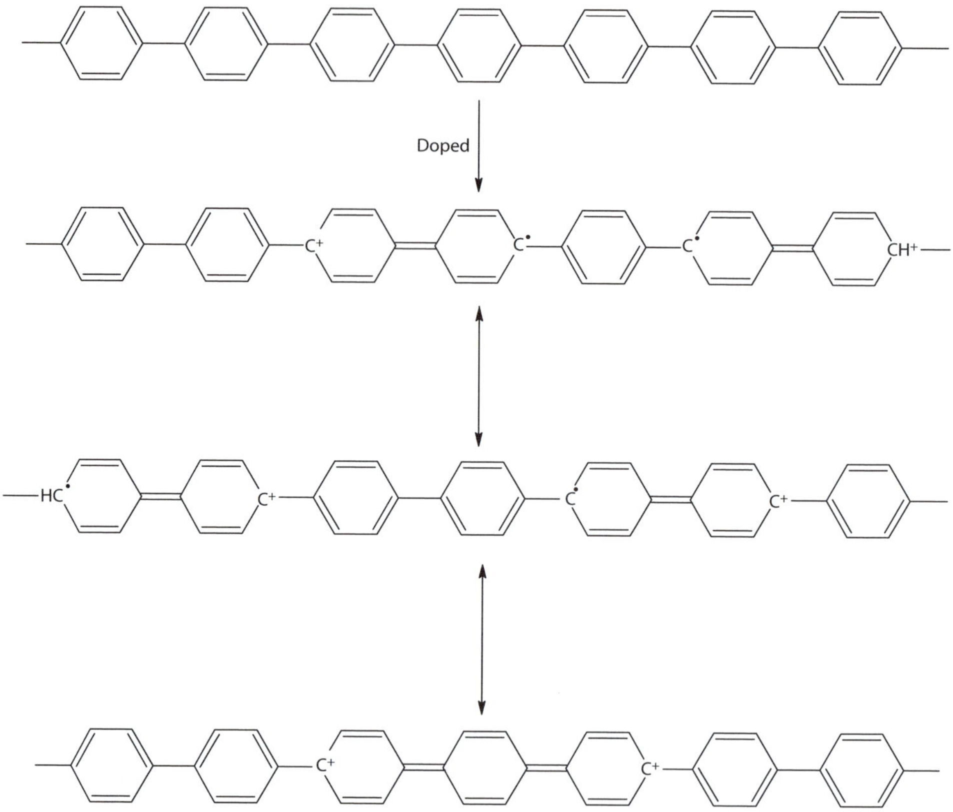

FIGURE 19.2 Poly-*p*-phenylene resonance structures after doping.

A critical structural feature for conductive polymers is the presence of conjugated double bonds. For polyacetylene, every backbone carbon is connected by a stronger localized sigma bond. They are also connected by a network of alternating less localized and weaker pi bonds. While conjugation is present, it is localized enough to prevent ready delocalization of the pi-bond electrons. The dopants cause the introduction of sites of increased or deficiency of electrons (Figure 19.2). When there is a deficiency of electrons, or holes, created electrons flow to fill this hole with the newly created hole causing other electrons to flow to fill the new hole, and so on allowing charge to migrate within and between the polyacetylene chains.

In general, the Huckel theory predicts that pi electrons can be delocalized over an entire chain containing adjacent pi bonds with the activation energy for delocalization decreasing as the chain length increases. The basic mechanism of electrical conductance is often explained in terms of the band theory borrowed from more classical semiconductors. As noted before, for essentially all conductive polymers, pi bonds are critical for conductance with the overlapping of pi clouds, allowing conductivity of the electric current to occur. Highly delocalized electronic structures with conjugated pi electrons along the backbones are the norm for good conductivity. The ionization potential to move from the valence band to the conduction band should be small.

Doping provides a ready mechanism for delocalization to occur. Doping is the mode of creating electron sinks or deficiencies and electron excesses that are necessary to breach, or get to, the conductive band. The list of often used oxidant dopants includes I_2 (probably the most widely used because of the ease of using it), AsF_5, $AlCl_3$, Br_2, and O_2. Anionic counterions include $Sb_2F_{11}^-$, I_3^-, and AsF_6^-. These dopants create defects that can be viewed as either a radical cation for oxidation dopants (the doping agents acting to remove electrons), or radical anion for reduction doping (the doping agents acting to increase the number of electrons). This is referred to as *polaron* with the oxidation or reduction resulting in the formation of a *bipolaron*. The formation of polarons is shown in Figure 19.2 where resonance structures illustrate the movement of electrons. The bottom resonance structure shows the formation of a bipolaron from the combination of two polarons. While conductance within a single chain is classically used to illustrate the concept, conductance actually occurs between chains as well as within chains. Further, for conductance to be effective, transference between chains is necessary.

Ordinary polyacetylene is composed of small fibers (fibrils) that are randomly oriented. Conductivity is decreased because of the contacts between the various random fibrils. Two approaches have been taken to align the polyacetylene fibrils. The first approach is to employ a liquid crystal solvent for the acetylene polymerization and to form the polymer under external perturbation. The second approach is to mechanically stretch the polyacetylene material causing the fibrils to align. The conductivity of polyacetylene is about 100 greater in the direction of the "stretch" in comparison to that perpendicular to the stretch direction. Thus, the conductivity is isotropic. By comparison, the conductivity of metals such as copper and silver is anisotropic. Of interest is the nonconductivity of diamond, which has only ordered sigma-bonds and thus no "movable" electrons and the isotropic behavior of graphite. Graphite, similar to polyacetylene, has a series of alternating pi bonds (Section 12.16) where the conductivity in the plane of the graphite rings is about 10^6 times that at right angles to this plane.

Polyacetylene has been produced by several methods, many utilizing the Zeigler–Natta polymerization systems. Both *cis* and *trans* isomers exist (19.4 and 19.5). The *cis*-polyacetylene is copper colored with films having a conductivity of about 10^{-8} S/m. By comparison, the *trans*-polyacetylene is silver colored with films having a much greater conductivity on the order of 10^{-3} S/m. The *cis* isomer is changed into the thermodynamically more stable *trans* isomer by heating. As noted above, conductivity is greatly increased when the *trans*-polyacetylene is doped (to about 10^2–10^4 S/cm). Conductivity is dependent on the micro or fine structure of the fibrils, doping agent, extent, and technique, and aging of the sample.

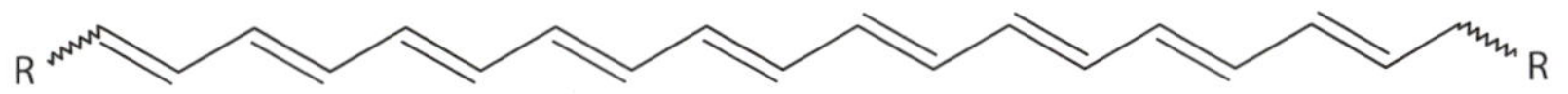

$$(19.4)$$

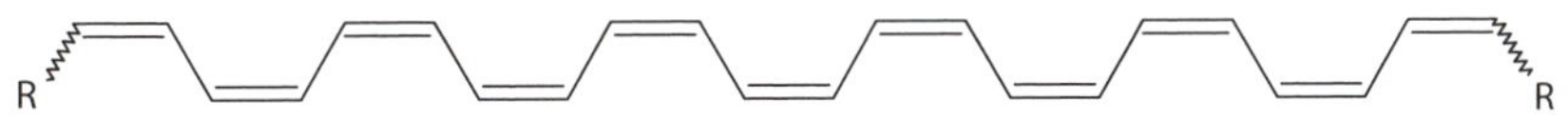

$$(19.5)$$

Polyacetylene was initially produced using Ziegler-Natta systems producing what have become known as Shirakawa polyacetylene. These materials are not easily processable and are mainly fibrillar. Recently other approaches have been taken. In the Durham route, the metathesis polymerization of 7,8-*bis*(trifluoromethyl)tricyclo[4.2.2.0]deca-3,7,9-triene gives a high molecular weight soluble precursor polymer that is thermally converted to polyacetylene (Equation 19.6). The precursor polymer is soluble in common organic liquids and is easily purified by reprecipitation. The end product can be aligned giving a more compact material with bulk densities on the order of 1.05–1.1 g/cc.

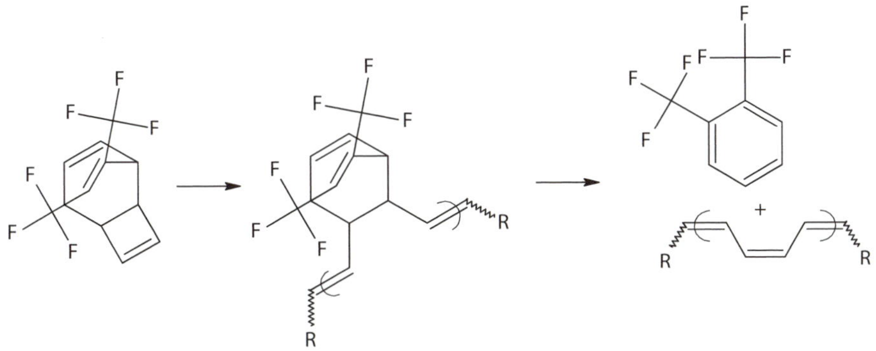

$$(19.6)$$

Grubbs and others have used the ring-opening metathesis polymerization to produce thick films of polyacetylene and polyacetylene derivatives (Equation 19.7).

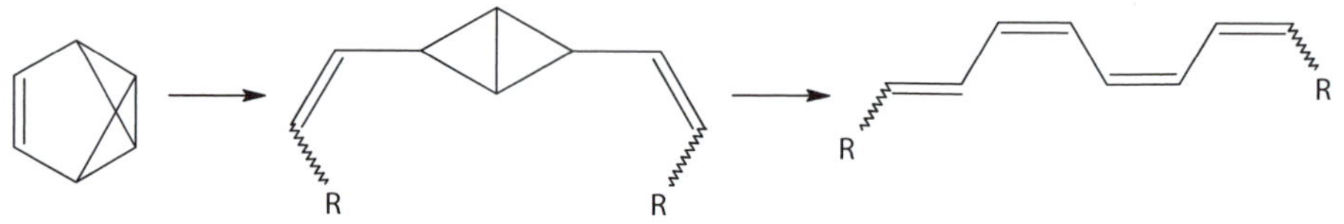

$$(19.7)$$

Polyacetylene has good inert atmospheric thermal stability but oxidizes easily in the presence of air. The doped samples are even more susceptible to air. Polyacetylene films have a lustrous, silvery appearance, and some flexibility. Other polymers have been found to be conductive. These include poly(p-phenylene) prepared by the Freidel–Crafts polymerization of benzene, polythiophene and derivatives, PPV, polypyrrole, and polyaniline. The first polymers commercialized as conductive polymers were polypyrrole and polythiophene because of their greater stability to air and the ability to directly produce these polymers in a doped form. While their conductivities (often on the order of 10^4 S/m) are lower than that of polyacetylene, this is sufficient for many applications.

Doped polyaniline is employed as a conductor and as an electromagnetic shielding for electronic circuits. Poly(ethylenedioxythiophene) (PEDOT) doped with polystyrenesulfonic acid is used as an antistatic coating material to prevent electrical discharge exposure on photographic emulsions and is also used as a hole-injecting electrode material in polymer light-emitting devices. Organic soluble substituted polythiophenes with good conductivities have been prepared. Poly(3-hexylthiophene) has a room temperature conductivity of about 100 S/cm; poly(3-metnylthiophene) has a conductivity of 500 S/cm; and a poly(3-alkylether)thiphene with a conductivity of about 1,000 S/cm reported. The unsubstituted polythiophene has a conductivity in the range of 50–100 S/cm. The fact that all of these substituted polythiophenes have similar conductivities indicates that there is little twisting of the backbone conjugation as alkyl substituents are added.

Polythiophene derivatives are being used in field-effect transistors. Polypyrrole is being used as microwave-absorbing "stealth" screen coatings and in sensing devices. PPV derivatives are being used in the production of electroluminescent displays.

Following are the structures of some of the more common conjugated polymers, along with poly(acetylene), that can be made conductive through doping. As noted before, doping causes and electrical imbalance that allows electrons to flow when an electrical potential is applied. The band gap is the energy needed to promote an election from the valence band to the empty energy or conductive band. Metals have zero band gaps while insulators such as polyethylene have large band gaps meaning that a lot of energy is needed to promote an electron to an empty band. Semiconductors have small band gaps where valence electrons can be moved into the conductance band through application of relatively small potential energies.

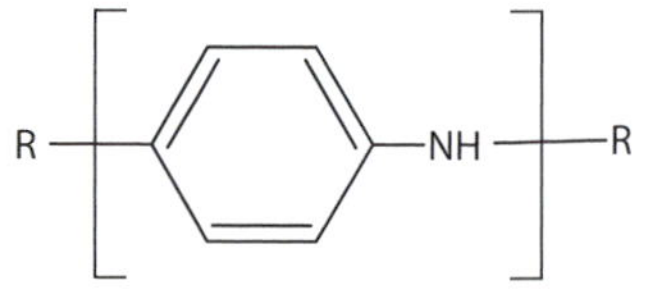

Polyalanine (19.8) 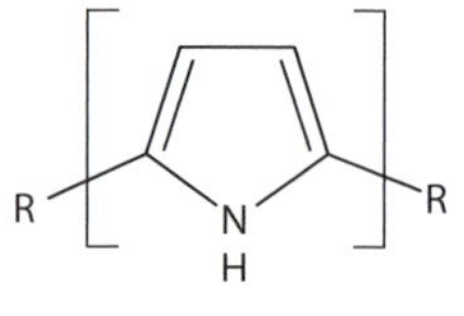　　　　Polypyrrole (19.9) 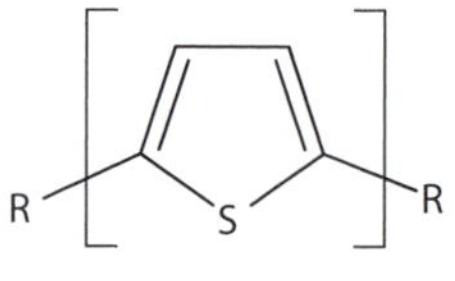　　　　Polythiophene (19.10)

The optical behavior can be used to help in the understanding of these conductive materials. Pi conjugation often occurs in the visible region so that most, if not all, of these conductive polymers are colored. Color changes are an important probe in assisting the characterizing the effects of various doping agents. Changes in the spectra allow various mechanisms to be studied. Because many of the polythiophene derivatives are organic soluble, spectral changes can be more easily studied. The electronic absorption spectra of a variety of polythiophene derivatives indicate that the band edge for conductivity begins about 2 eV.

Water-soluble derivatives of polythiophene have been made allowing counterions bound to the polymer backbone to "self-dope" with the protons (such as lithium and sodium ions) injecting electrons into the pi system. Thus, combinations of sodium salts and so-called "proton salts" (such as prepared from poly-3-(2-ethanesulfonate)thiophene) have been prepared that are both water soluble and conducting.

Another area of activity involves the synthesis of material with small band gaps that would allow activation to occur at room temperature without doping. Polyisothianaphthene has been produced with a band gap of about 1 eV. More recently, polymers with alternating donor and acceptor units with band gaps of about 0.5 eV and less have been developed.

While the amount of electricity that can be conducted by polymer films and "wires" is limited, on a weight basis the conductivity is comparable to that of copper. These polymeric conductors are lighter, some more flexible, and they can be "laid down" in "wires" that approach being one atom thick. They are being used as cathodes and solid electrolytes in batteries, and potential uses include in fuel cells, "smart" windows, nonlinear optical materials, light-emitting diodes, conductive coatings, sensors, electronic displays, and in electromagnetic shielding.

There is a large potential for conducting polymers as corrosion inhibiting coatings. For instance the corrosion protection ability of polyaniline is pH dependent. At lower pHs polyaniline-coated steel corrodes about 100 more slowly than noncoated steel. By comparison, at a pH of about 7 the corrosion protection time is only twice for polyaniline-coated steel. Another area of application involves creation of solid-state recharagable batteries and electrochromic cells. Polyheterocycles have been cycled thousands of times with retention of more than 50% of the electrochromic activity for some materials after 10,000 cycles. Infrared polarizers based on polyaniline have been shown to be as good as metal wire polarizers.

They will also find uses in nonlinear optical devices such as in optoelectronics, which is for signal processing and optical communication. Some of the new conducting polymers offer such benefits as flexibility, high damage threshold, ultrafast response in the subpicosecond range, and good mechanical strength. Polyheterocyclic conducting polymers have shown a wide variation in color as they are electronically converted between oxidized and reduced forms. The instability of some of the polymers is being used to monitor moisture, radiation, mechanical, and chemical destruction.

Work is also being done with polymers that "naturally" contain an excess or deficiency of electrons. For instance, boron has a valence electronic configuration of $2s^2 2p^1$ and on bond formation forms a sp^2 trigonal planar molecular geometry with vacant "p" orbitals giving such compounds a "natural" hole through which delocalized electrons can flow.

By comparison, phosphorus has five valence electrons plus available vacant d orbitals. These five valence electrons typically form three (molecular geometry of trigonal pyramid; sp^3) and five (sp^3 with one pi bonds such as occurring in nucleic acids and polyphosphate and phosphonate esters and sp^3d trigonal bypyramid) bonds. The focus with respect to electrical conductivity is on the three-bonded phosphorus polymers because they contain an unbonded electron pair. The unbonded electron pair is capable of bridging the electron gap between conjugated units automatically creating increased electrons that can promote conductivity. Structures of some of these central units are shown below.

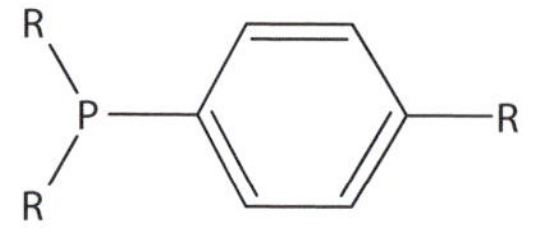

Arylphosphane
(19.11)

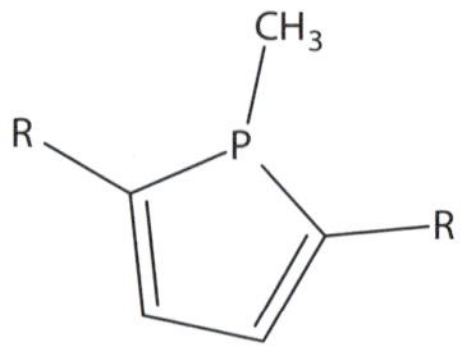

Phosphole
(19.12)

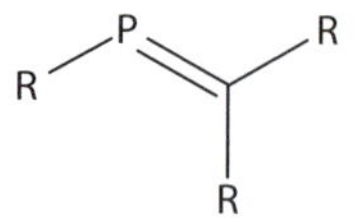

Phosphalkene
(19.13)

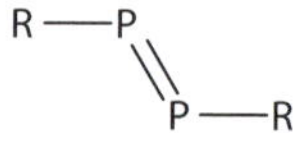

Diphosphene
(19.14)

Some of the arylphosphanes have exhibited nonlinear optical (NOL) behavior. They show potential application in constructing organic light-emitting diodes (OLEDs). Efforts are underway to employ unbonded electron phosphorus-containing units as endgroups for organic-conducting polymers such as polythiophenes, 19.15, with the intent to use these unbonded electron sites as metal-chelating units, giving materials with possible conductivity, luminescence, and interesting optical and magnetic properties.

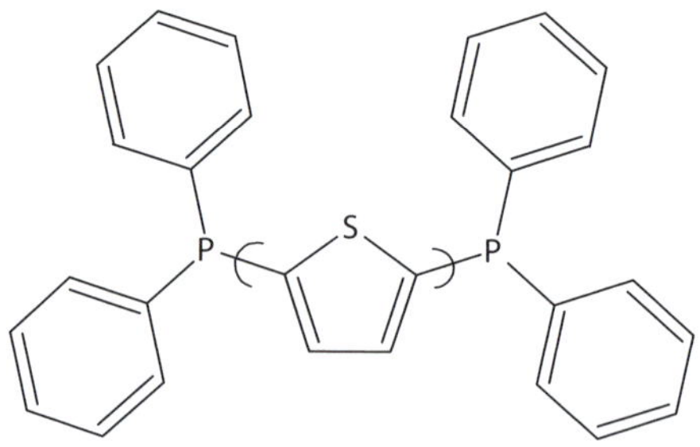

(19.15)

Complexation of Lewis bases with a wide variety of metals and organometallics has occurred and show promise in a variety of application areas, including light-induced conductors. This is an area of great promise in a wide variety of areas, including communications, electronics, solar energy conversion, and catalysis.

19.1.3 NANOWIRES

Part of the nanorevolution involves the electronics industry and the synthesis of so-called nano or molecular wires for electronic applications. Basically, molecular wires contain a series of double or double and triple bonds that have what can be referred to as "whole-chain" resonance. Polyacetylene is an example of a molecule that exhibits whole-chain resonance or delocalization of electrons where the activation energy for delocalization is relatively low. The search is on for nanowires that are more flexible. Often this flexibility is achieved through a balance of the conductive inner core backbone that is characteristically rigid with a flexible outer core composed of covalently bonded groups that act to decrease the tendency to form highly crystalline structure and which encourage flexibility as illustrated below. As in many of the new areas creativity is important. Often, the molecular wires are created one step at a time with the eventual joining of chains or simply the use of the single chain.

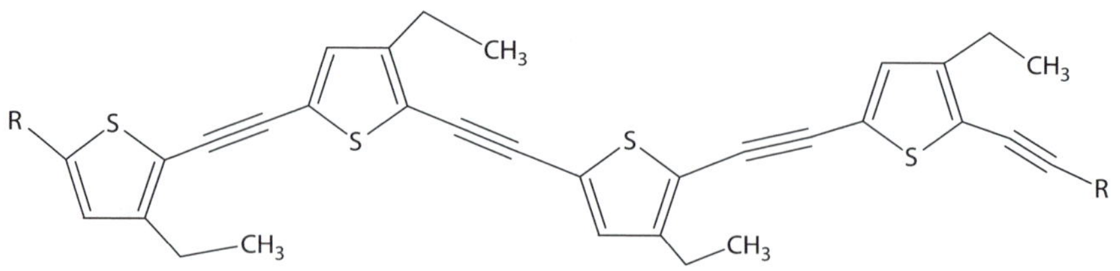

(19.16)

Poly(thiophene ethynylene)

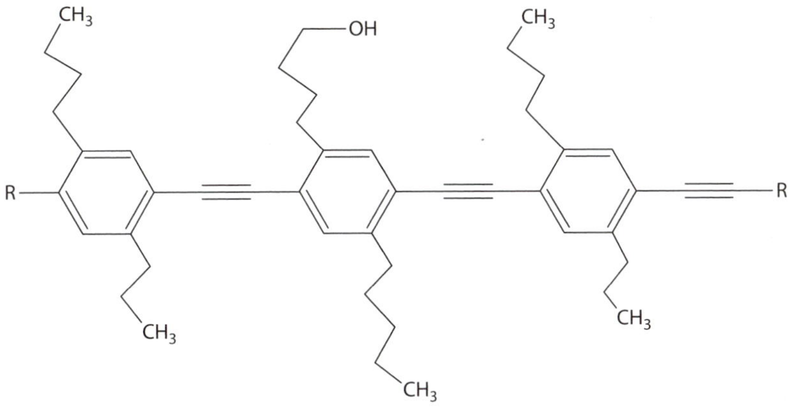

(19.17)

Phenylene–acetylene backbone copolymer

While we call these materials polymeric, in truth they are oligomeric, often intentionally made to be oligomers with specific lengths intended to connect other molecules or molecular devices with the distance between the "gaps" corresponding to the distance of the molecular wires. Thus, we can design molecules with particular lengths, conductivity, and the ability to molecularly connect to other molecules. When needed, nonconductive molecules, such as methylene units, can be added to decrease electronic conductance. These nonconductive moieties are referred to as barriers. Below, 19.18, is a phenylene–acetylene grouping containing a barrier methylene (noted by the arrow).

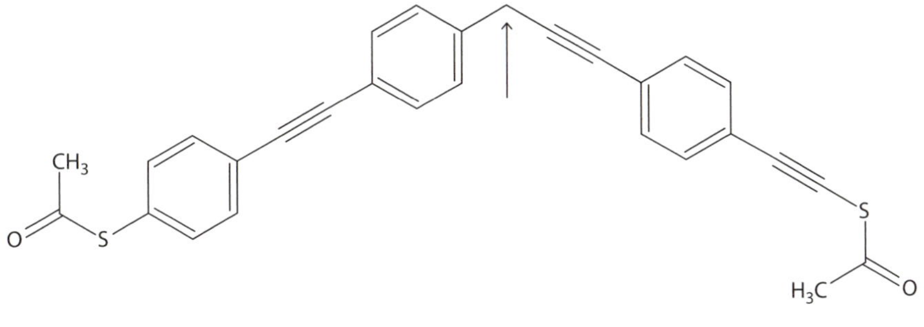

(19.18)

Chain ends are often functionalized with groups that can react with specific sites on another molecule. These functionalized sites are referred to as molecular alligator clips. The chain shown above has two such functionalized groups or alligator clips, one on each end. These molecular alligator clips are normally designed to attach or anchor to a metal surface. Thus, this molecule is a candidate to connect two electrodes, acting as a molecular wire.

Of importance to their use in electronics, these chains are able to conduct in the microamp region, the same region that most computing instruments operate and individual chains can conduct in the tenth of a microamp range.

19.2 NONLINEAR OPTICAL BEHAVIOR

Nonlinear optics (NLO) involves the interaction of light with materials, resulting in a change in the frequency, phase, or other characteristics of the light. Second-order NLO behavior includes second harmonic generation of light that involves the doubling of the frequency of the incident light; mixing of frequencies where the frequency of two light beams is either added or subtracted; and electro-optic effects occurring that results in frequency and amplitude change and rotation of polarization. NOL behavior has been found in inorganic and organic compounds and in polymers. The structural requirement is the absence of an inversion center requiring the presence of asymmetric centers and/or poling. Poling is the application of a high-voltage field to a material that orients some or all of the molecules generally in the direction of the field. The most effective poling is found when the polymers are poled above the T_g (that allows a better movement of chain segments) and then cooled to lock in the "polled" structure. Similar results are found for polymers that contain side chains that are easily poled. Again, cooling helps lock in the polled structure. At times, cross-linking is also employed to help lock in the polled structure.

Third-order NLO behavior generally involves three photons resulting in affects similar to those obtained for second-order NLO behavior. Third-order NLO behavior does not require the presence of asymmetrical structures.

Polymers that have already been found to offer NLO behavior include polydiacetylenes and a number of polymers with liquid crystal side chains. Polymers are also employed as carriers of materials that themselves are NLO materials. Applications include communication devices, routing components, and optical switches.

19.3   PHOTOPHYSICS AND PHOTOCHEMISTRY—BASICS

Photophysics involves the absorption, transfer, movement, and emission of electromagnetic, light, energy without chemical reactions. By comparison, photochemistry involves the interaction of electromagnetic energy that results in chemical reactions. Let us briefly review the two major types of spectroscopy with respect to light. In absorption the detector is placed along the direction of the incoming light and the transmitted light measured. In emission studies, the detector is placed at some angle, generally 90°, away from the incoming light. Remember that energy, $E = h\nu$.

When absorption of light occurs, the resulting polymer, P*, contains excess energy and is said to be excited.

$$P + h\nu \rightarrow P^* \tag{19.19}$$

The light can be simply reemitted.

$$P^* \rightarrow h\nu + P \tag{19.20}$$

Of much greater interest is the migration, either along the polymer backbone or to another chain, of the light. This migration allows the energy to move to a site of interest. Thus, for plants, the site of interest is chlorophyll. These "light-gathering" sites are referred to as antennas. Natural antennas include chlorophyll, carotenoids, and special pigment-containing proteins. These antenna sites harvest the light by absorbing the light photon and storing it in the form of an electron that is promoted to an excited singlet energy state by the absorbed light.

Bimolecular occurrences can occur leading to an electronic relaxation called *quenching*. In this approach, P* finds another molecule or part of the same chain, A, transferring the energy to A.

$$P^* + A \rightarrow P + A^* \tag{19.21}$$

Generally, the quenching molecule or site is in its ground state.

Eliminating chemical rearrangements, the end results for quenching are most likely electronic energy transfer, complex formation, or increased nonradioactive decay. Electronic energy transfer involves an exothermic process where part of the energy is absorbed as heat energy, and part is emitted as fluorescence or phosphorescence radiation. Polarized light is taken on in fluorescence depolarization also known as *luminescence anisotropy*. Thus, if the chain segments are moving at about the same rate as the reemission, part of the light is depolarized. The extent of depolarization is then a measure of the segmental motions.

Complex formation is important in photophysics. Two terms need to be described here. First, an *exciplex* is an excited state complex formed between two different kinds of molecules, one that is excited and the other that is in its grown state. The second term, *excimer*, is similar except the complex is formed between like molecules. Here we will focus on excimer complexes that form between two like polymer chains or within the same polymer chain. Such complexes are often formed between two aromatic structures. Resonance interactions between aromatic structures, such as two phenyl rings in polystyrene (PS), give a weak intermolecular force formed from attractions between the pi electrons of the two aromatic entities. Excimers involving such aromatic structures give strong fluorescence.

Excimer formation can be described as follows where [PP]* is the excimer:

$$P* + P \rightarrow [PP]* \tag{19.22}$$

The excimer decays giving two ground state aromatic sites and emission of fluorescence.

$$[PP]* \rightarrow hv + 2P \tag{19.23}$$

As always, the energy of the light emitted is less than that originally taken on. By studying the amount and energy of the fluorescence radiation decay rates, depolarization effects, excimer stability, and structure can be determined.

Light has dualistic properties of both waves and particles. For the particle properties we can consider light as being composed of particles known as photons, each of which has the energy of Planck's quantum hc/λ; where h is the Plank's constant, c is the velocity of light, and λ is the wavelength of the radiation. Ejection of electrons from an atom as a result of light bombardment is due to the particle behavior while the observed light diffraction at gratings is attributed to the wave properties. The different processes related to light interactions with molecule can be represented by Figure 19.3.

The absorption of light by materials produces physical and chemical changes. On the negative side, such absorption can lead to discoloration generally as a response to unwanted changes in the material's structure. Absorption also can lead to a loss in physical properties such as strength. In the biological world it is responsible for a multitude of problems, including skin cancer. It is

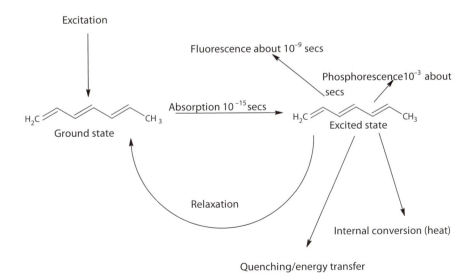

FIGURE 19.3 Processes associated with light absorption by a molecule.

one of the chief modes of weathering by materials. Our focus here will be on positive changes effected by the absorption of light. For many years, absorption of light has intentionally resulted in cross-linking and associated insolubilization. This forms the basis for coatings and negative-lithographic resists. Light-induced chain breakage is the basis for positive-lithographic resists. Photoconductivity forms the basis for photocopying and photovoltaic effects and is the basis for solar cells being developed to harvest light energy.

It is important to remember that the basic laws governing small and large molecules are the same.

The Grotthus–Draper law states that photophysical/photochemical reactions only occur when a photon of light is absorbed. This forms the basis for the First Law of Photochemistry, that is, only light that is absorbed can have a photophysical/photochemical effect.

We can write this as follows:

$$P + light \rightarrow P* \tag{19.24}$$

where P* is P after it has taken on some light energy that it has acquired energy during a photo-chemical reaction. The asterisk is used to show that M is now in an excited state.

There are two kinds of spectra, namely excitation (or emission) and absorption. The absorption and excitation spectra are distinct but usually overlap, sometimes to the extent that they are nearly indistinguishable. The excitation spectrum is the spectrum of emitted light by the material as a function of the excitation wavelength. The absorption spectrum is the spectrum of absorbed light by the material as a function of wavelength. The origin of the occasional discrepancies between the excitation and absorption spectra are due to the differences in structures between the ground and excited states, or the presence of photo reactions, or the presence of nonradiative processes that relax the molecule to the ground state without passing through the luminescent states.

Visible color is normally a result of changes in the electron states. Molecules that reside in the lowest energy level are said to be in the ground state or unexcited state. We will restrict our attention to the electrons that are in the highest occupied molecular orbital (HOMO) and the lowest unoccu-pied molecular orbital (LUMO). These orbitals are often referred to as the frontier orbitals.

Excitation of photons results in the movement of electrons from the HOMO to the LUMO. This is pictured in Figure 19.4.

Photon energies can vary. Only one photon can be accepted at a time by an orbital. This is stated in the Stark-Einstein Law also known as the Second Law of Photochemistry—if a species absorbs radiation, then one particle (molecule, ion, atom, etc.) is excited for each quantum of radiation (pho-ton) that is absorbed.

Remember that a powerful lamp will have a greater photon flux than a weaker lamp. Further, photons enter a system one photon at a time. Thus, every photon absorbed does not result in bond

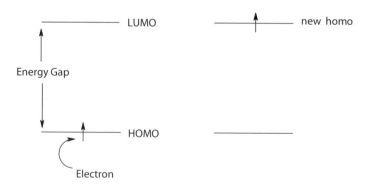

FIGURE 19.4 Representation of a photon being absorbed by a single molecule of chromophore.

breakage or other possible measurable effect. The quantum yield, φ, is a measure of the effectiveness for effecting the desired outcome, possibly bond cleavage and formation of free radicals.

$$\phi = \frac{\text{number of molecules of reactant consumed}}{\text{number of photons consumed}} \qquad (19.25)$$

Quantum yields can provide information about the excited electronic states such as the rates of radiative and nonradiative processes. Moreover, they can also find applications in the determination of chemical structures and sample purity. The emission quantum yield can be defined as the fraction of molecules that emits a photon after direct excitation by a light source. So emission quantum yield is also a measure of the relative probability for radiative relaxation of the electronically excited molecules.

Quantum yields vary greatly from the photons being very ineffective (10^{-6}) to being very effective (10^6). Values greater than 1 indicate that some chain reaction, such as in a polymerization, occurred.

We often differentiate between the primary quantum yield that focuses on only the first event (here the quantum yield cannot be greater than 1), and secondary quantum yield that focuses on the total number of molecules formed via secondary reactions and here the quantum yield can be high.

Luminescence is a form of cold body radiation. Older TV screens operated on the principle of luminescence, where the emission of light occurs when they are relatively cool. Luminescence includes phosphorescence and fluorescence. In the TV, electrons are accelerated by a large electron gun sitting behind the screen. They are accelerated by a large voltage. In the black and white sets, the electrons slam into the screen surface that is coated with a phosphor that emits light when hit with an electron. Only the phosphor that is hit with these electrons gave off light. The same principle operates in the old generation color TVs except the inside of the screen is coated with thousands of groups of dots, each group consisting of three dots—red, green, and blue.

The kinetic energy of the electrons is absorbed by the phosphor and reemitted as visible light to be seen by us.

Fluorescence involves the molecular absorption of a photon that triggers the emission of a photon of longer wavelength (less energy; Figure 19.3). The energy difference ends up as rotational, vibrational, or heat energy losses.

Here excitation is described as

$$P_o + h\nu_{ex} \rightarrow P_1 \qquad (19.26)$$

and emission as

$$P_1 \rightarrow h\nu_{em} + P_o \qquad (19.27)$$

where P_o is the ground state, P_1 is the first excited state.

The excited state molecule can relax by a number of different, generally competing pathways. One of these pathways is conversion to a triplet state that can subsequently relax through phosphorescence or some secondary nonradiative step. Relaxation of the excited state can also occur through fluorescence quenching. Molecular oxygen is a particularly efficient quenching molecule because of its unusual triplet ground state.

Watch hands that can be seen in the dark allow us to read the time in the dark. These watch hands typically are painted with phosphorescent paint. Like fluorescence, phosphorescence is the emission of light by a material previously hit by electromagnetic radiation. Unlike fluorescence, phosphorescence emission persists as an afterglow for some time after the radiation has stopped. The shorter end of the duration for continued light emission is 10^{-3} s but the process can persist for hours and days.

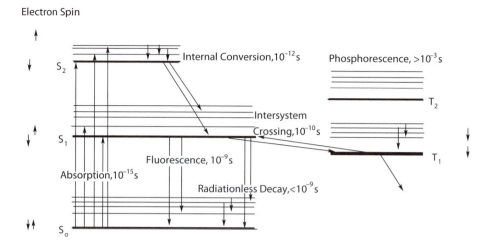

FIGURE 19.5 Jablonski diagram showing various processes associated with light absorption and emission and their time scale. The arrows on the far right and left describe the relative spin states of the "paired" electrons.

The energy level diagram representing the different states and transitions is called a *Jablonski diagram* or a *state diagram* (Figure 19.5). Here S_0 represents the electronic ground state while S_1 and S_2 represent the first and second singlet excited states, respectively. The first and second triplet states are denoted T_1 and T_2. For singlet states, all electron spins are paired and the multiplicity of this state is 1. On the other hand, in the triplet states two electrons are no longer antiparallel and the multiplicity is 3. The triplet state is more stable than the singlet counterpart (S) and the source for this energy difference is created by the difference in the Coulomb repulsion energies between the two electrons in the singlet verses the triplet states and the increase in degrees of freedom of the magnetic spins. Because the electrons in the singlet excited state are confined in the same orbital, the Coulomb repulsive energy between them is higher than in the triplet excited state where the electrons are now in separate orbitals. The splitting between these two states (S–T) is also dependent on the nature of the orbital. Let's consider a case where the two orbitals involved in a transition are similar (i.e., two p orbitals of an atom, or two π orbitals of an aromatic hydrocarbon). For this situation the overlap between them may be high and the two electrons will be forced to be close to each other resulting in the S–T splitting being large. The other situation is the case where the two orbitals are different (i.e., n→π* or d→π transitions) resulting in a small overlap. Since the overlap is small, the two electrons will have their own region of space to spread in resulting in a minimization of the repulsive interactions between them and hence the S–T splitting will be small.

Absorption occurs on a time scale of 10^{-15} s. When inducing the promotion of an electron from the HOMO to the LUMO, the molecule passes from an electronic ground singlet state S_0 (for diamagnetic molecules) to a vibrational level of an upper singlet or triplet excited state S_n or T_n, respectively. The energy of the absorbed photon determines which excited state is accessible. After a while, the excited molecule relaxes to the ground state via either radiative (with emission of light) or nonradiative processes (without emission of light). The radiative processes (for diamagnetic molecules) include either the spin-allowed fluorescence or spin-forbidden phosphorescence. Nonradiative processes include intersystem crossings (ISC), a process allowing a molecule to relax from the S_n to the T_n manifolds, and internal conversions (IC and IP), a stepwise (vibrational) energy loss process relaxing molecules from upper excited states to any other state without or with a change in state multiplicity, respectively.

An IC is observed when a molecule lying in the excited state relaxes to a lower excited state. This is a radiationless transition between two different electronic states of the same multiplicity and is possible when there is a good overlap of the vibrational wave functions (or probabilities) that are involved between the two states (beginning and final). IC occurs on a time scale of 10^{-12} s, which is a time scale associated with molecular vibrations. A similar process occurs for an IP when it is accompanied by a change in multiplicity (e.g., triplet T_1 to S_0). Upon nonradiative relaxation heat is released. This heat is transferred to the media by collision with neighboring molecules.

Fluorescence is a radiative process in a diamagnetic molecule involving two states (excited and ground states) of the same multiplicity (i.e., $S_1 \rightarrow S_0$ and $S_2 \rightarrow S_0$). Fluorescence spectra show the intensity of the emitted light versus the wavelength. A fluorescence spectrum is obtained by initial irradiation of the sample normally at a single wavelength, where the molecule absorbs light. The lifetime of fluorescence is typically on the order of 10^{-8}–10^{-9} s (i.e., ns time scale) for organic molecules and faster for metal-containing compounds (10^{-10} s or shorter).

In general, the fluorescence band, typically $S_1 \rightarrow S_0$, is a mirror image of the absorption band ($S_0 \rightarrow S_1$). This is particularly true for rigid molecules such as aromatics. Once again, the Franck–Condon principle is applicable and hence the presence of vibronic bands is expected in the fluorescence band. However, there are numerous exceptions to this rule, particularly when the molecule changes its geometry in the excited states. Another observation is that the emission is usually red shifted in comparison with absorption because the vibronic energy levels involved are lower for the fluorescence and higher for the absorption processes. The difference in wavelength between the 0–0 absorption and emission band is usually known as the Stokes shift. The magnitude of the Stokes shift gives an indication of the extent of geometry difference between the ground and excited states of a molecule as well as the solvent–solute reorganization.

Another nonradiative process that can take place is known as intersystem crossing from a singlet to a triplet or triplet to a singlet state. This process is very rapid for metal-containing compounds. This process can take place on a time scale of $\sim 10^{-6}$–10^{-8} s for an organic molecule while for organometallics it is $\sim 10^{-11}$ s. This rate enhancement is due to spin-orbit coupling present in the metal-containing systems, that is, an interaction between the spin angular momentum and the orbital angular momentum, which allows mixing of the spin angular momentum with the orbital angular momentum of S_n and T_n states. Thus, these singlet and triplet states are no longer "pure" singlets and triplets and the transition from one state to the other is "less forbidden" by multiplicity rules. A rate increase in intersystem crossing can also be achieved by the "heavy atom effect" arising from an increased mixing of spin and orbital quantum number with increased atomic number. This is accomplished either through introduction of heavy atoms into the molecule via chemical bonding (internal heavy atom effect) or with the solvent (external heavy atom effect). The spin-orbit interaction energy for atoms grows with the fourth power of the atomic number Z. In addition to the increase in the intersystem crossing rate, heavy atoms exert more effects that can be summarized as follows. Their presence acts to (1) decrease the phosphorescence lifetime due to increase in the nonradiative rates; (2) decrease the fluorescence lifetime; and (3) increase the phosphorescence quantum yield. The presence of a heavy atom not only affects the rate for intersystem crossing but also the energy gap between the singlet and the triplet state, where the rate for the intersystem crossing increases as the energy gap between S_1 and T_1 decreases. Moreover, the nature of the excited state exerts an important effect on the intersystem crossing. For example, the $S_1(n, \pi^*) \rightarrow T_2(\pi, \pi^*)$ (e.g., as in benzophenone) transition occurs almost three orders of magnitude faster than the $S_1(\pi, \pi^*) \rightarrow T_2(\pi, \pi^*)$ transition (such as in anthracene).

Relaxation of triplet state molecules to the ground state can be achieved by either internal conversion (nonradiative IP) or phosphorescence (radiative). Emissions from triplet states (i.e., phosphorescence) exhibit longer lifetimes than fluorescence. These long-lived emissions occur on time scale of 10^{-3} s for organic samples and 10^{-5}–10^{-7} s for metal-containing species. This difference between the fluorescence and phosphorescence is associated with the fact that it involves a spin-forbidden electronic transition. Moreover, the phosphorescence bands are always red shifted in comparison with their fluorescence counterpart because of the relative stability of the triplet state compared to the singlet manifold. Nonradiative processes in the triplet states increase exponentially with a decrease in triplet energies (energy gap law). Hence, phosphorescence is more difficult to observe when the triplet states are present in very low energy levels. It is also often easier to observe phosphorescence at lower temperatures where the thermal decay is further inhibited.

19.4   DRUG DESIGN AND ACTIVITY

There are a number of steps that should be completed before a drug is brought to market. Generally, first a compound is found to be active in treating some illness. This may be done with cell and bacterial studies followed up with live animal tests. This drug is then modified in an attempt to increase its activity and minimize negative side effects. When possible, the target is identified. The target is typically one that is identified with the illness. The target is validated by determining the target's function and determining that the target activity is modified by association with the drug.

A series of studies are undertaken to evaluate the effectiveness and toxicity of the drug compound. All along this process, large decreases in the number of compounds that make it through each hurdle occur. Generally, extensive animal testing occurs. If the drug is believed to be effective in the treatment of the illness then additional steps occur. Initially, a drug company files an Investigational New Drug Application (INDA) with the Food and Drug Administration to get permission to begin testing the drug in humans. The potential drug then begins a series of clinical trials with humans. *Phase I* clinical trials involve testing the drug in a small number (20–100) of healthy individuals to test the drug's safety, tolerance, length of activity, effective dosage, and dosing schedule. *Phase II* clinical trials involve testing the drug in larger numbers of individuals (100–500) that possess the particular illness to gain additional information of efficacy, side effects, safety and appropriate dosage, and schedule. *Phase III* clinical tests involve testing the drug in larger numbers of volunteers (100s–1,000s) to again gain more information of the efficacy, side effects, and safety. In general, only one or two dosage levels and schedules are studied in Phase III studies. If results are consistent with the drug successfully treating the illness with only minor side effects application is made to the Food and Drug Administration to market it. Even after the drug has come to market, ongoing monitoring occurs for safety and side effects. To accomplish such a gauntlet of testing it can cost from $200 to $800 million.

Most drugs are smaller molecules. Polymer-containing drug formulations are common. Here, the drug is contained within some polymeric matrix that assists in controlling the release either through control of drug diffusion through the chains or by erosion, degradation, or solubility, of the polymer network.

But there are some drugs that are polymeric offering advantages over smaller molecules. Along with polymers being simply depositories of the drugs, polymers can perform active roles in drug therapy. Polymeric drugs can act in two divergent ways—as a control release agent and as a drug itself. The polymeric drug may act in a control release manner. A polymer that contains a known therapeutic portion can degrade releasing this portion over some time. Thus, a polymer that contains the drug L-dopa (19.28) used to treat Parkinson's disease degrades over time releasing L-dopa over a period of time.

(19.28)

Polymer-containing L-dopa

The polymer may act as a drug itself or the polymer drug may also act in both roles. Following is a brief discussion of advantages of polymeric drugs that may apply to either mode of control release or/and as a drug itself. Because of their size, polymers with chain lengths of about 100 units and greater typically are unable to easily move through biological membranes, thus movement is restricted. This can result in limiting negative side effects, such as damage to the kidneys, because the polymer can reside in only certain body sites. Also, this limited mobility can assist in directing the polymer drug. This also allows the use of lower and more level drug concentrations reducing renal, kidney, and so forth damage. In some cases, multiple attachments by the same polymer chain can be beneficial. Thus, many cancer drugs act to "tie up" the DNA chains inhibiting cell replication. Here, multiple attachments to the DNA, possible for polymer drugs that contain numerous binding units within each single chain, might be beneficial in decreasing the cells effectiveness in overcoming single DNA "chemical knots" (chemical bonding together of the two DNA strands).

The polymeric nature may inhibit premature drug deactivation. Thus, cisplatin (19.29), the most widely used anticancer drug, is converted into numerous inactive, but more toxic, platinum-containing compounds before it arrives at the targeted cancer cells. Placement of the active platinum-containing platinum moiety into a polymer (19.30) decreases this tendency to hydrolyze into these unwanted cisplatin compounds because of the greater hydrophobic character of the polymeric drug.

Cisplatin (19.29)   Polymer containing a cisplatin-like unit (19.30)

Polymers may also evade the microorganism's defense. Recently, it was found that many cells become drug resistant after repeated treatment. Cells have groups of molecules that protect it from outside invasion. The small molecule drugs are considered "outside" invaders. Thus, drugs introduced into their environment are considered as "outside" invaders and the cell manufactures greater numbers of these "house-keepers" to rid the cell of the invaders. These "house-keeping" agents are often not very specific and remove other similar drugs that intrude their cell. These "house-keeping" proteins may not be as effective at eliminating polymer drugs.

Polymer drugs are also known to remain within human hosts longer than smaller molecules. Depending on the illness and treatment, this preferential retention may be simply due to physical retention through entanglement with biological outer-layer materials or through some specific additional interaction. Tumor-associated cells are frequently hyperpermeable to plasma proteins and other macromolecules. These "leaky" vasculatures and limited lymphatic drainage, typical of tumor and missing in normal tissue, result in the accumulation of polymers. Thus, such polymers reside in the interstitial space of these cells. This results in *enhanced permeability and retention* (EPR) of large chains.

Drug design, today, typically aims at certain specific biological activities. Thus, cancer treatments focus on controlling cell growth (Figure 19.6). Cell growth can be considered as occurring in four dependent steps. Drugs are designed to control cell growth at any one of these steps directly or indirectly. The indirect intervention of cell growth has as a target any one of the many steps that influence the cell growth cycle. Essentially all chemo therapies are based on the continual growth of cancer cells compared with healthy cells generally being in some rest state.

Drugs aiming to control the same problem may target different sites. For instance, in the treatment of breast cancer, tamoxifen acts to control cell growth by blocking estrogen receptors on the cancer cell while arimidex acts to reduce the body's estrogen production.

Cancer cells have three main characteristics that are different from healthy cells. First, they are immortal, able to replicate themselves hundreds of times while healthy cells generally replicate themselves less than two dozen times over our lifetime. Second, cancer cells are not contact inhibited meaning that they will continue to grow forming tumors. By comparison, healthy cells replicate until they

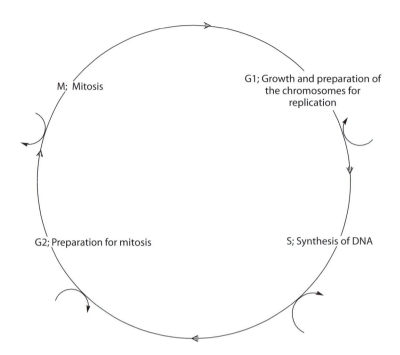

FIGURE 19.6   Cell growth cycle.

touch another cell and then stop reproduction. This is called *contact inhibition*. Finally, cancer cells are normally in a growth mode whereas healthy cells are generally in a rest mode. For most situations, drug action takes advantage of the cancer cells being in a growth mode needing an intake of various materials to keep growing. Thus, any cell that is growing at the time a chemo drug is administered will get a dose of the chemo drug, most likely a lethal dose. Since the cancer cells are almost always in the growth mode, they will be most affected. Recently, we have found a family of organotin polymers that inhibit a variety of cancer cells but do not affect the growth cycle of healthy cells. This points out another possible avenue to create cancer cell specific drugs. Healthy cells have their autoimmune system in tack, whereas cancer cells have damaged autoimmune systems. These organotin polymers may be warded off by the healthy cells, whereas the cancer cells, are inhibited by them.

19.5   SYNTHETIC BIOMEDICAL POLYMERS

Synthetic polymers have been studied for some time for their use in the general field of medicine. Initial results were average. More recently, because of a better understanding of the importance of surface, avoidance of contaminants, control of molecular weight and molecular weight distribution the use of synthetic polymers in the biomedical sciences is increasing. Polymers are viewed as important biomedical materials for a number of reasons, some of which appear contradictory—but only because different uses require different properties. Some of the important properties are the ability to tailor-make structures, surface control, strength, flexibility, rigidity, inertness/reactivity, light weight, ease of fabrication, ability to achieve a high degree of purity, lack of and because of their water solubility/compatibility, bioerodability, and the ability of some of them to withstand long-term exposure to the human body—a truly hostile environment. Fighting against some of the biomaterials are their limited (by volume) use—thus, researchers are often limited to using commercial materials made for other applications, but as the use of these important materials increases, manufactures will become more interested in tailormaking materials if for no other reason than the high cost per unit volume.

Long-term inertness without loss of strength, flexibility, or other necessary physical property is needed for use in artificial organs, prostheses, skeletal joints, and so on. Bioerodability is needed when the polymer is used as a carrier such as in controlled release of drugs, removal of unwanted materials, or where the materials purpose is short lived such as in their use as sutures and as frames for natural growth.

While the nature of the material is important, the surface of the material is also often critical. The human body wants to wrap around or connect to bodies within its domain. In some cases, the desired situation is little or no buildup on the polymer. Here, surface "slickness" is needed. Siloxanes and flurinated materials such as polytetraflouroethylene (PTFE) are generally slick materials but other materials can be made slick through surface treatments that presents to the body few voids and irregularities at the atomic level. In other cases, bodily buildup is desired and surfaces and materials that assist this growth are desired. Surface hydrophobicity/hydrophilicity, presence/absence of ionic groups, chemical and physical (solid or gel) surface are all important considerations as one designs a material for a specific application.

Ability to function long term is an ongoing problem. In general, polyurethanes degrade after about 1.5 years, nylons lose much of their mechanical strength after about 3 years, and polyacrylonitrile loses about 25% strength after 2 years. On the long side, PTFE loses less than 10% strength after about a year and siloxanes retain most of their properties after 1.5 years.

Following is a brief look at some of the varied uses, real and projected, for synthetic polymers. These are intended to be illustrative only. The field of biomedical materials is rapidly growing and it is extensive.

There has been a lot of effort to construct artificial hearts for human use. These hearts are largely composed of polymeric materials. Even with the problems associated with organ transplant, the use of artificial hearts may never be wide spread. This is because of several trends. First, many biological solutions to biological problems resides with the use of biological materials to solve biological problems. With the advent of increased effectiveness of cloning and related advances suitable biological replacements may be grown from a person's cells thus ensuring compatibility. Second, related to the first, regeneration of essential parts of vital organs is becoming more practical. Third, because of nanotechnology and related electrical and optical advances, surgery to locate and repair imperfections is improving. Forth, our autoimmune system is divided into two main systems—one guards against "small" invading organisms such as viruses, bacteria, and pollen. The second group acts to reject whole organisms such as the heart through rejection of foreign body tissue. Purine nucleoside phosphorylase (PNP), is a human enzyme that serves at least two major functions. First, PNP acts to degrade unleashed molecules, including foreign nucleic acids. PNP is necessary to our immune system as it fights disease. Some anticancer drugs are synthetic nucleosides and nucleotides such as polyIpolyC that are employed to directly attack selected cancerous tumors. PNP degrades such nucleic acid-related materials before they reach the tumor. Neutralization of PNP just before administration of the synthetic nucleic acid-related material would allow lower dosages to be used. Second, PNP is an integral part of the body's T-cell immunity system that rejects foreign tissue. Effort is currently underway to effectively inhibit only the PNP action allowing the first autoimmune system to work while allowing organ transplantation without rejection. Recently, through the use of zero gravity conditions, crystals of PNP have been grown of sufficient size to allow structural determination. With this structure determined, efforts are underway to detect sites of activity and drugs that would allow only these sites to be neutralized when needed.

Even so, synthetic polymers have been important in replacing parts of our essential organs. Thus, silicon balls are used in the construction of mechanical heart valves. Many of these fail after some time and they are being replaced by a flap valve made from pyrolytic carbon or polyoxymethylene.

Aneurisms can be repaired through reinforcement of the artery wall with a tube of woven PTFE or polyester (PET). Replacement of sections of the artery can be done using a tube of porous PTFE. One remaining problem is the difference in elasticity between the woven and porous materials and the arteries themselves.

Carbon-fiber composites are replacing screws for bone fracture repair and joint replacements. These fiber composites are equally as strong and are chemically inert. By comparison, the metals they replace are often alloys that may contain metals that the patient may be allergic to.

Polymers are also used as sutures. Fighters and other athletics have used poly(alpha-cyanoacry-lates), super glues, to quickly stop blood flow in surface cuts. Today, super glue is also used for, in place of, or along with more traditional polymeric suture threads for selected surface wounds, internal surgery, and in retinal and corneal surgery. The alpha-cyanoacrylates monomers (19.31) undergo anionic polymerization in the presence of water forming polycyanoacrylates. More about sutures is detailed in Section 19.4.

$$H_2O + H_2C \qquad\qquad\longrightarrow\qquad\qquad \tag{19.31}$$

Poly(alpha-cyanoarylate)

Siloxanes are the most extensively used synthetic biomaterial. They are used for a number of reasons, including flexibility, chemical and biological inertness, low capacity to bring about blood clotting, overall low degree of biological toxicity, and good stability within biological environments.

Artificial skin had been made from a bilayer fabricated from a cross-linked mixture of bovine hide, collagen, and chondroitin-b-sulfate derived from shark cartilage with a thin top layer of siloxane. The siloxane layer acts as a moisture- and oxygen-permeable support and to protect the lower layer from the "outer world" allowing skin formation to occur in conjunction with the lower layer. Poly(amino acid) films have also been used as an "artificial" skin. Research continues in search of a skin that can be effectively used to cover extensive wounds and for burn patients.

Elastomeric siloxanes have also been used in encapsulating drugs, implant devices, and in maxillofacial applications to replace facial portions lost through surgery or trauma. Transcutaneous nerve simulators are made from "conductive" siloxanes. These are employed in the treatment of chronic severe pain through application of low-level voltage to the nerves disrupting transmission of pain impulses to the brain. Siloxanes are also used in extracorporeal blood oxygenation employed in the heart–lung assist machine that is routinely used in open heart surgery. The "heart" of the apparatus is the membrane that must allow ready transport of oxygen and carbon dioxide yet retain moisture and the blood cells. The siloxane membranes can be made using a PE or PTFE screen in an organic dispersion of silicon rubber. When dried, thin films are obtained that are used in the heart–lung assist device.

Siloxane-containing devices have been also used as contact lenses, tracheostomy vents, tracheal stents, antireflux cuffs, extracorporeal dialysis, ureteral stents, tibial cups, synovial fluids, toe joints, testes penile prosthesis, gluteal pads, hip implants, pacemakers, intraaortic balloon pumps, heart valves, eustachian tubes, wrist joints, ear frames, finger joints, and in construction of brain membranes. Almost all of the siloxane polymers are based on various polydimethylsiloxanes.

The kidney removes waste material from our blood. Because of partial or total kidney failure, many persons are on hematolysis. The first hemadialysis units were large and by today's standards not very effective and the semipermeable tubes, made mainly of cellophane, had very limited lifetimes. Initially, dialysis treatment was expected to be for only terminal patients but as the life expectancy of dialysis patients increased as did the demand for dialysis, smaller, more efficient dialysis machines have emerged. At the heart of these advances is the filtering material. Today bundles of microhollow fibers are used in the construction of hemadialysis cells. The fibers are "heparinized" to discourage blood clotting. The fibers are mainly made of polyacrylonitrile. Polycarbonate, cellulose acetate, and rayon fibers are also being used.

The most widely used hip joint replacement is largely a polished cobalt–chromium alloy that moves against a specially designed ultrahigh molecular weight polyethylene (UHMWPE). This

UHMWPE is highly crystalline and highly cross-linked through gamma radiation. Tests have shown that the UHMWPE wears out at an average of 0.1 mm each year. Since most assemblies employ about a 10 mm-thick layer of UHMWPE, the lifetime of the hip joint replacement based on only the wear of the UHMWPE is about 100 years.

Controlled release of drugs can be envisioned as occurring via three major routes. One approach utilizes diffusion-controlled release through membranes or matrices. Here the rate of release is controlled by the permeability of the membrane or matrix. In the second approach, the drug is captured within a matrix that undergoes degradation, usually through aqueous-assisted solubilization or degradation (including hydrolysis). Here the rate of drug release is dependent on the break up of the typically polymeric matrix. For the second approach, a number of polymers have been used including poly(glycolic acid) (PGA) and polyanhydrides. The third approach involves simple degradation of a drug-containing polymer where the drug moiety is present as part of the polymer backbone or as side chains. Degradation of the polymer results in the release of the drug in some fashion.

Controlled release of drugs using polymer intensive materials is becoming more common place. The release "pack" can be attached externally such as many of the "nicotine patches" that deliver controlled amounts of nicotine transdermally. The release "pack" can also be introduced beneath the skin or within the body as is the case with many diabetes treatment assemblies.

A number of siloxane-containing controlled release packs have been devised and are being used. Glaucoma, motion sickness, and diabetics have been treated using drugs dispersed in a silicon matrix. This kind of pack needs to be placed near the site of intended activity for greatest effectiveness.

Implant materials can be divided into two general categories dependent on the time requirement. Those that are present for release of a drug or to hold a broken bond in place until sufficient healing occurs are termed *short-term implant materials*. The second group includes materials that are to function over a longer time such as for the life of the patient. In the first case, degradation is generally required while for the longer-term material inertness and long-term stability are typically required. There are times when this is not true. For instance, some of the newer biomaterials act as scaffolds that promote tissue growth by providing a three-dimensional framework with properties that encourage favorable cell growth. This material may be designed to be either short or long term. One approach to designing scaffolding material involves placing certain amino acid-containing units on the polymeric scaffold that encourage cell growth.

Another aspect related to control release of drugs concerns the type of structures that currently appear to be working. Not unexpectedly, because of compatibility and degradation purposes, most of the effort on the control release formulations includes polymers that have both a hydrophobic and hydrophilic portion with the material necessarily containing atoms in addition to carbon. Another concern is that the products of degradation are not toxic or do not go on to form toxic materials. It has also been found that amorphous materials appear to be better since they are more flexible and permit more ready entrance of potential degradative compounds.

Another area of activity involves the synthesis of supermolecular layers that are connected through cross-linking giving essentially one molecule thick micelles. Depending on the particular template and solvents employed, these monolayers can be designed to have almost any combination of hydrophilic and hydrophobic sites. Again, specific control of release rates, degradation times and routes, biocompatibility or incompatibility is possible. Many of these micelle-based delivery systems are based on a poly(ethylene oxide)-b-poly(propylene oxide)-b-poly(ethylene oxide) triblock or on a polypeptide and poly(ethylene oxide) combination. Drug delivery has also been achieved using conducting electroactive polymers formed through controlled ionic transport of counterions (dopants) in and out of membranes.

Hydrogels have been used that shepherd drugs though the stomach and into the more alkaline intestine. Hydrogels are cross-linked, hydrophilic polymer networks that allow the smaller drugs access to their interior and that can be designed to inflate, swell at the desired site, to deliver the drug. These hydrogels have largely been formed from materials with a poly(acrylic acid) backbone. More about hydrogels in Section 19.13.

Also included in the general grouping of biomaterials are the new electronics that are being developed, including nanotechnology. These electronic biomaterials will need to be encapsulated sufficiently to protect them from the body and to protect the body from hostile actions against the presence of the foreign object or they will need to be made with a coatings material that will act as an encapsulating material to the electronics device. Special care will need to be taken with respect to adhesive materials that connect the electrical device to the particular site for activity and/or interactions between the electromagnetic messages and surrounding tissue. Again, adhesive, encapsulating, and shielding materials will be largely polymeric.

19.5.1 DENTISTRY

Polymers are used extensively in dentistry. Somewhat permanent dental materials exist in a hostile environment, the mouth, and often must resist fracture and wear under extreme forces. They must perform without fatiguing. They must also be compatible with the biological environment.

The original modern day fillings or amalgams contained various metals and alloys. More recently, composite resins often referred to as plastic or white fillings are being employed. They often consist of a mixture of powered glass and polymer resin. They are strong, compatible, resistant, and cosmetically approach the appearance of natural teeth. Even crowns are generally metallic with a ceramic coating to approach the appearance of natural teeth.

Dental restorative composites generally consist of di-and tri-functional monomer systems that provide the cross-linking necessary to form a strong matrix once polymerized. Reinforcing fillers such as silanized quartzes, ceramics, hybrid filler such as prepolymerized resins on fumed or pyrogenic silica, and ceramics act as the dispersed phase of the composite. The diluent or continuous phase is the mono and difunctional monomer that decreases the viscosity of di- and tri-functional monomers that finally act to form the stable composite matrix. Dental sealants are similar except the filler material is either not present or in lesser concentration. Many of the restorative composites are based on modified dimethacrylates the most common being 2,2-*bis*[*p*-(2'-2-hydroxy-3'-methacryloxypropoxy) phenyl]-propane, *bis*-GMA (19.32), and triethyleneglycol dimethacrylate (TEGDMA) (19.33).

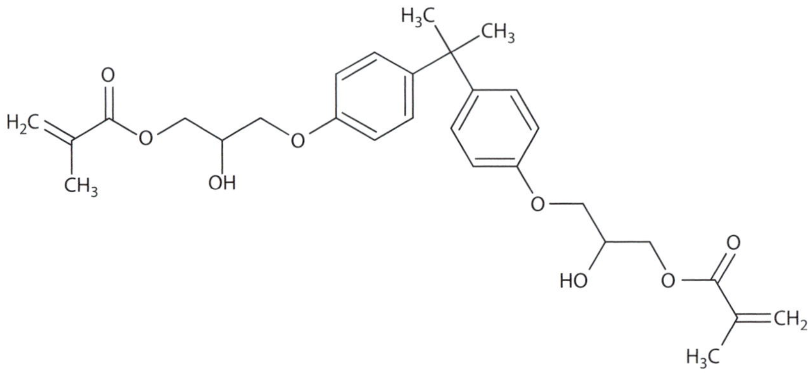

(19.32)

bis-GMA

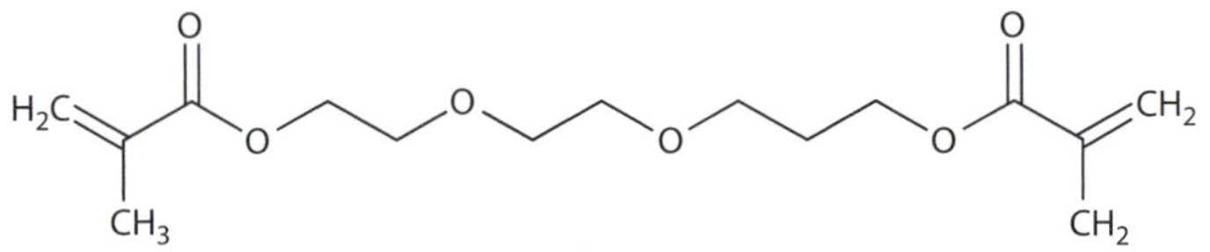

(19.33)

TEGDMA

These two are used either separately or as a mixture. Because of the presence of the somewhat flexible ethylene oxide and related units and use of appropriate fillers, these materials give composite fillings with lower polymerization shrinkage, enhanced mechanical properties, lower solubility and water adsorption, better thermal expansion characteristics, good biocompatibility, with aesthetic properties closely matching those of the tooth itself.

Some other fillings employ urethanedimethacrylate (1,6-*bis*(methacryloxy-2-ethoxycarbonylamino)-2,4,4-trimethylhexane) (UDMA) in place of *bis*-GMA. This is an active area of research with new monomer systems being introduced in an ongoing manner.

Cavity varnishes are used to seal the exposed dentinal tubules and protect the pulp from the irritation of chemicals in the filling materials. They are generally largely natural rubber or a synthetic polymeric resin such as 2-hydroxyethyl methacrylate (HEMA).

Fillings should be viewed as nonpermanent with a lifetime often of a decade. Many materials are employed in fillings, including metals, alloys, amalgams, composites, and glass ionomer cements. Glass ionomer cements are a mixture of glass and an ionomer referred to as a polyalkonic acid, which is simply an ionomer with carboxylic salt groups as the chelating moieties. The ionometric polymer chelates both the tooth and glass binding them together into the filling. Today's glass ionomer cements are especially designed for dental application. For instance, many contain releasable fluoride that prevents carious lesions with the fluoride content recharged through use of fluoride-containing toothpaste. Newer formulations containing light-cured resins do not release fluoride as readily.

Almost all denture bases are made of methacrylic (acrylic) resins that give good fit and a natural appearance. A compression molding process is used where the monomer-polymer dough or slurry contain poly(methyl methacrylate) or poly(methyl acrylate). Often there is a change in the contour of the soft tissue and a liner is fitted onto the denture base. Silicon reliners are often used for this purpose.

Plastic acrylic denture teeth are made by injection or transfer molding. Acrylic teeth have a higher strength than porcelain teeth and break less readily. However, they cold flow, have a greater water absorption, and they have a higher wear rate than porcelain teeth.

Many of the dental polymer cements are glass-ionomer combinations made from ionomer-forming polymers that contain acid groups such as poly(acrylic acid), poly(itaconic acid) (19.34), poly(maleic acid) (19.35), and poly(2-butene-1,2,3-tricarboxylic acid) (19.36). These polyalkenoate cements are set up through reaction with an alumino-silicate-fluoride glass with the poly-acids hydrolyzing the glass network releasing the aluminum and calcium ions and forming a silaceous hydrogel. The acid groups chelate with the released metal cations forming a cross-linked matrix.

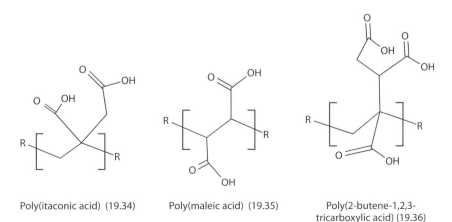

Poly(itaconic acid) (19.34)   Poly(maleic acid) (19.35)   Poly(2-butene-1,2,3-tricarboxylic acid) (19.36)

19.6 SUTURES

Since surgery begun, about 5,000 BC, tying together the surgical site required a stitching material we now call sutures. Surgery, combat, accidents, and hunting could result in wounds requiring closure with sutures.

As time progressed, various suture materials were used, essentially all polymeric, and initially all natural. Early materials included linen strips, grasses, mandibles of pincher ants, silk, animal hair, and parts of animals, including muscles, tendons, and intestines. The earliest use of so-called cat gut was by the ancient Greek physician Galen. The eighteenth century brought along the use of animal hide and silver wire. By the twentieth century, synthetic sutures began to dominate as sutures of choice. The introduction of polyglactin structures in the 1970s started the transition from natural to more synthetic adsorbable sutures.

Sutures are the largest group of devices implanted into humans. There is a wide variety of materials today available, each with known advantages and limitations.

Today, sutures come in a variety of kinds and sizes. The standard sizes of sutures (with diameters in inches) follow a system set by the United States Pharmacopeia. The scale originally was 0–3 with zero being the smallest. As the ability to make strong thin sutures increased, smaller suture diameters were indicated by additional zeros, thus a 00 was smaller than 0. Today, suture size is given by a number representing the diameter ranging in general order from 10 to 1 and then 1/0–12/0 with 1/0 being the largest and 12/0 being the smallest with a diameter smaller than a human hair strain. A 4/0 corresponds to a 0000-sized suture and a 6/0 corresponds to a 000000 suture. Thus, in the x/0 series, the x corresponds to the number of zeros that reflect suture diameter.

Table 19.1 contains a listing of suture sizes and typical uses.

As expected, smaller sutures of the same material generally exhibit lower tensile strengths and can break more easily.

Suture size has some correspondence to the tissue being bound together with thinner tissues like the face requiring thinner sutures. Thinner sutures are also required for facial surgery to limit scar formation.

A surgeon considers many factors in choosing which suture to use. These considerations include (a) knot security, (b) tensile strength and tensile strength lifetime, (c) minimal tissue drag and inflammation, (d) handling, (e) size, (f) inhibition of infection, and (g) potential of allergenic reaction.

Other less obvious factors come into play when choosing a suture. For instance, the age and health of the patient should be considered. An older person in poor health generally requires a longer time to heal, thus for absorbate sutures, the required degradation time needs to be increased. Conversely, for an active child, while the degradation time may be less, the suture may be exposed to more frequent sudden stresses and the body tissues are generally thinner. The presence of immunodeficiency is also an important factor since they are more susceptible to infection. Patients who have taken steroids, undergone chemotherapy, infected with human immunovirus (HIV) and so on

TABLE 19.1
Suture Sizes and Typical Uses

Size	Typical Uses
0 and larger	Abdominal wall closure, drain sites, arterial lines, fascia, orthopaedic uses
2/0	Blood vessels, viscera, fascia
3/0	Trunk, limbs, gut
4/0	Neck, hands, tendons, mucosa
5/0	Face, neck, blood vessels
6/0	Face
7/0 and smaller	Microsurgery, ophthalmology

where the immunosystem is stressed are included in this grouping. In some way, all of us fit into this category, to some extent, because the surgery or action that has inflected the need for the suture compromises our immune system. Thus, where possible, a health picture or history should be taken before scheduled surgery.

Sutures can be a single strand called a *monofilament*, or many filaments twisted or braided together. The importance of specific factors that may not be initially important may actually be of greater importance. For instance, braided sutures offer advantages of being stronger, more flexible, and offer a more secure knot formation, but they offer disadvantages such as the presence of crevices, where infection can find a home and they may inflect a sawing action on the tissue, cutting the tissue with each movement. Further, braided sutures, because of their rough morphology, cause tissues to swell more. While knotting or tying off are generally required to maintain suture location and ability to hold the tissues together, the stresses on the suture caused by the formation of the knot result in the knotted material typically being the least resistant to hydrolysis and are generally the sites where the initial break occurs.

Sutures are sold and packaged with specific needles already threaded and the sutures already cut top length. They can be colored (dyed) or clear and coated or uncoated. Black colored sutures are typical in instances where ready identification of the suture is important, such as internal surgery. Other coatings can also be applied that make the surface smoother or give the surface some therapeutic activity.

The United States Pharmacopeia is the agency that determines standard procedures and parameters for suture testing. In Europe, it is the British Pharmacopeia (for the United Kingdom) and European Pharmacopoeia that specifies the test procedures and parameters for suture testing. Product testing involves both biological, such as biocompatibility, and physical examination. Unlike many materials areas, most of the testing is done on suture material already being used. A number of factors are responsible for this trend. A major factor involves the lack of understanding between the test results and materials because of the complex hostile environment where the sutures operate. Researchers believe that it is important to understand these relationships using known materials before looking at newer candidates. This does not mean that the area of sutures is static. New suture materials are continually being introduced but most of these advances involve modification of existing materials.

Biocompatibility is an important consideration. Again, many unintended consequences are often found, unfortunately often after the fact. In the area of synthetic hip and knee replacement, use of alloys is often called for, yet some patients are found to have long-term allergies to certain metals present as minute amounts that eventually require replacement of the joint material. Thus, extensive testing is required before a new suture becomes commercially available.

Much of the testing involves looking at real-time biodegradation results on physical properties. There is an effort to develop standardized tests that can be done outside a patient with this effort now being done by comparing the tests done outside the patient with results found with the patient.

Stress–strain properties, often described in terms of tensile strength, are usually used to describe the ability of a suture to withstand various surgical (such as pulling together the materials to be held together) and postsurgical stresses. Postsurgical abrupt stresses, such as a cough, are important considerations. Long-term failure of a suture, caused by even small leakages, can result in formation of an edema or hemorrhage. Thus, both long-term and abrupt behavior needs to be considered in choosing a suture. For example, a suture used in a lung would need to have a high elasticity, slow degradation, and high tensile strength.

Diameters are measured using a dead-weight gauge with a specific foot pressure of weight applied with diameters measured to within 0.02 mm. The diameter is measured at three points on the suture, about one-fourth, one-half, and three-fourth of the strand length. Since the knot strength is so important, a knot pull strength is measured. Here, the suture is tied about a flexible rubber tubing of 6.5 mm inside diameter and 1.6 mm wall thickness. The suture is attached to the testing machine and tested at a specified rate of elongation. A similar needle attachment test is done.

Suture manufacture is specialized. In the production from raw materials or isolation from natural materials, purification is especially important because small amounts of byproducts and contaminants can have a serious effect of the properties of the suture and the behavior within the patient. Processes for suture fabrication include melt spinning, extrusion, and braiding.

With the exception of sutures made from steel, all sutures are polymeric. Sutures are generally divided into two broad categories. Absorbable sutures are absorbed by the healing tissue through proteolysis or hydrolysis. Nonabsorbable sutures remain in the tissue, often for the lifetime of the patient.

In general, absorbent sutures are composed of materials that are natural to mammals, such as catgut, and to materials that are either quite susceptible to hydrolysis and/or polymers derived from natural materials such as polyglactin, which is a copolymer of lactic and glycolic acid. Nonabsorbent sutures can be made from natural materials such as cotton that is a plant material and not animal material, polymers that range from being hydrophobic to hydrophilic, and finally steel.

There is a difference between what is called permanent or nonabsorbent and the actual degradation rate of the material. Silk, PE, nylon, and cotton sutures generally degrade with time, while steel sutures remain reasonably intact throughout the patient's lifetime. Further, there is a difference between such physical properties as tensile strength and absorbent and nonabsorbent sutures. As just noted, silk, PE, nylon, cotton, and in fact most nonabsorbent sutures lose tensile strength with time, that time being a month or year or longer.

Because only a little suture is generally employed and is important in the overall surgical procedure, cost is generally not a major consideration. Following are brief discussions of the major kinds of suture material.

Cotton and linen are not widely used today as sutures. Cotton is usually used as a twisted monotifilament, such as thread. It is used where rapid healing is expected because of its short-term resistance to degradation. Atramat surgical cotton, unlike "ordinary" cotton, has good strength and a uniform diameter but it is weaker than silk. It offers good handling and good knot security. *Linen* is obtained from linen fibers and is similar to surgical cotton in most ways.

Catgut—Catgut sutures are seldom actually catgut. The name is derived from past practices. Today, so-called catgut is derived from the submucosa of sheep jejunum and the ileum or serosa of beef intestine that is cut into longitudinal ribbons.

For *plain catgut*, the ribbons are treated with dilute formaldehyde to increase tensile strength and resistance to enzymatic lysis. They are sterilized using cobalt 60 irradiation. Plain catgut is, in fact, largely highly purified collagen. It is generally used only with wounds that heal rapidly because of its rapid degradation with the necessary tensile strength remaining for only a week. It can also give adverse tissue reaction because it is in reality a protein. It is seldom used today because of the adverse tissue reactions and rapid degradation.

Chromic catgut is created by treatment of plain catgut with basic chromium salts that result in the catgut being more resistant to absorption and stronger. Chromic catgut retains a reasonable tensile strength for 2–3 weeks. It is often the suture of choice for Pomeroy tubal ligation procedures because it dissolves quickly decreasing the opportunity for a fistula to form between the two ends of the bisected tube.

Both varieties of catgut are not employed for facial or surface use because of their tendency for scar formation and inflammatory response.

Silk, used for sutures, is obtained from the cocoon of the *Bombyx mori* silk worm. Tension force is gradually lost until tissue encapsulation occurs. Tissue reactivity may be moderate because silk is a protein and its interaction with the body is not benign. It is classified as nonabsorbent because it retails much of its strength for more than 2 months and 50% to half a year, but loses most of its strength after 2 years. While stronger than cotton, surgical silk is weaker than the synthetic nonabsorbable sutures. Much of the silk is coated with silicon, allowing a smooth travel through tissue and avoiding a saw-cutting effect. It is widely used in general surgery, gastroenterology, gynecology, obstetrics, ophthalmology, and so forth.

While natural sutures can undergo enzyme-induced degradation, synthetic sutures generally undergo mainly only physical hydrolysis as the major path of degradation.

The ring-opening polymerization (ROP) of *p*-dioxanone (19.37) gives *polydioxanone* (19.38). Polydioxanone sutures are generally offered as a monofilament in a variety of sizes. They offer a minimum of tissue reactivity with absorption occurring over about a 200 day process. The original tensile resistance holds to about 70% after 4 weeks and 50% after 6 weeks. It is often employed in tissue coaptation that heals slowly. It offers adequate tissue support and good knot security. While it offers good initial tissue support, it is not used where long-term tissue support is necessary. It is used in general surgery, gastroenterology, gynecology, urology, as well as in subdermal plastic surgery procedures. As with many sutures, it is colored by application of a dye for easy location during the surgical procedure.

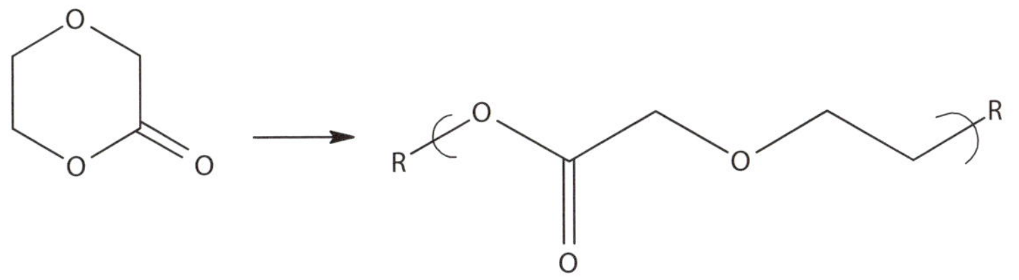

p-Dioxanone (19.37)   Polydioxanone (19.38)

Polyglactin (structurally the same as *polyglyconate*) is a structural copolymer of lactic and glycolic acid synthesized from glycolide (19.39) and 3,6-dimetnyl-1,4-dioxane (19.40). Polyglactin (19.41) was the second synthetic suture offered commercially, beginning in 1975. It is usually sold as a braided material offered in a wide variety of sizes. A variety of copolymer compositions are offered under varying trade names. For instance, Vicryl is a 90/10 composition of poly(glycolide-co-lactide), while Panacryl is a 5/95 copolymer composition. Vicryl Rapide is a new polyglactin suture introduced into usage in the United States, but it has been used for sometime in Europe. It retains 50% strength for 5 days and is essentially absorbed in 40 days. Coated Vicryl Rapide is used for skin closure in small superficial areas, and in the vulvar and episitomy areas because it falls off in about 1 week to 10 days, eliminating the necessity for suture removal. It is also being used in gynecological surgery to tie bleeders in areas close to vital structures because of its fast absorption and minimal tissue reaction decreasing the opportunity for kinking and obstruction formation.

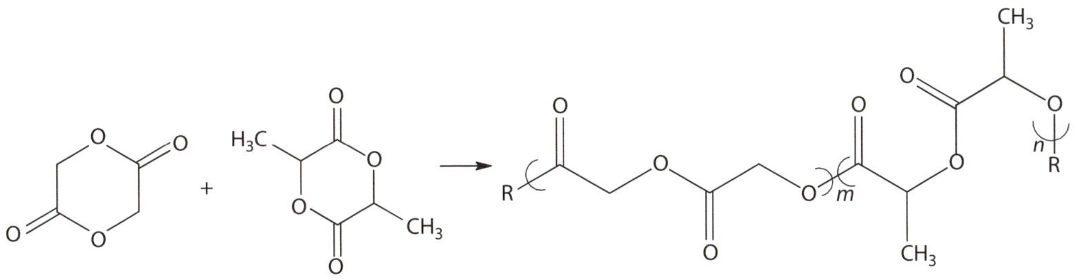

1,4-dioxane-2,5-dione; glycolide (19.39)   3,6-dimethyl-1,4-dioxane (19.40)   Polyglactin (19.41)

Polyglycolic acid is a homopolymer of glycolide and was the initial synthetic suture offered for sale under the trade name of Dexon. It is formed into monofilaments from the melt with the filaments stretched as they are formed to assist in aligning the polymer chain thus increasing the filament strength. Polyglycolic acid (Equation 19.42) is also braided into sutures.

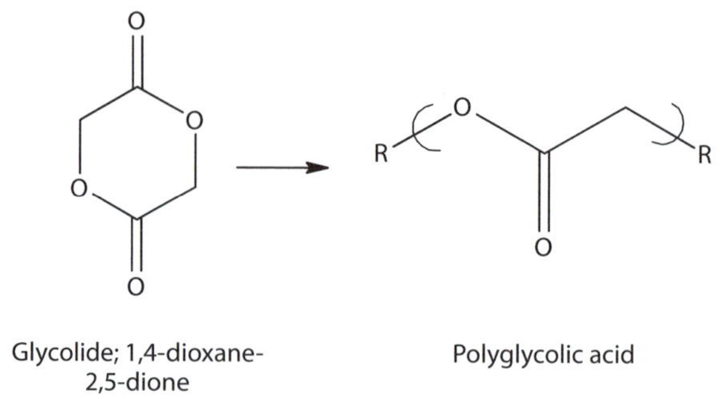

Glycolide; 1,4-dioxane-   Polyglycolic acid
2,5-dione

(19.42)

The *polyamide* (Nylon) material that is employed as a suture material is the standard nylon 6,6. It is usually used as a monofilament. Tissue reaction is minimal and, in time, it is encapsulated by connective tissue. Hydrolysis decreases the chain length and strength by about 10% a year. It is used in neurology, ophthalmic, and plastic surgery. It has good tensile strength and little elongation change. It offers good knot security and low tissue drag. It is the suture material of choice for chain-saw leg wounds since it offers good strength and flexibility needed for knee and leg action.

Polyester sutures are made from PET. They show better handling than nylon offering greater initial strength than nylon, polypropylene (PP), silk, with only stainless steel being stronger. It offers less skin reaction than silk and is equally manageable, offering good knot security and ease in tying the knot. It is often employed as a braided material. Uncoated polyester sutures, such as Dacron, Mersilene, and Surgidac, offer the best knot security. Teflon coated sutures, such as Tevdek, Polydek, and Ethiflex, handle better.

Polyethylene sutures are used for general surgery offering a smooth surface area that produces little tissue drag. They are inert.

Polytetrafluoroethylene, Teflon is often impregnated into sutures and used as a coating promoting reduction of tissue drag and biological inertness. Gore-Tex sutures are a monofilament PTFE and part of a grouping of PTFE biological implants called *ePTFE*. Gore-Tex is 50% air by volume. It has good tensile strength, inertness, slipperiness, ties like silk, and offers good knot security. Suture-related bleeding is reduced because they can be swaged to needles that closely approximate the thread diameter. It is especially used for vaginal reconstructive surgery for perivesical and sacrospinous ligament support.

Polybutester (19.43) is a segmented block copolymer of butylene terepthalate units and tetramethylene ether glycol units. The terephthalate portion is rigid giving the butylene terephthalate blocks a semirigid or hard property. The tetramethylene ether glycol or tetramethylene ether oxide units are flexible and contribute to the "softness" of the polymer. Thus, polybutester makes use of the hard/soft concept often employed in polymer science to give a material with some strength that acts flexible or soft. It is used with soft tissue applications, including opthalmic and cardiovascular surgery, but not microsurgery and neural tissue surgery. It offers minimal acute inflammatory reaction and is gradually encapsulated. It offers long-term strength retention. Polybutester-coated Ethibond sutures also handle well causing less tissue reaction in comparison to the other polyester sutures.

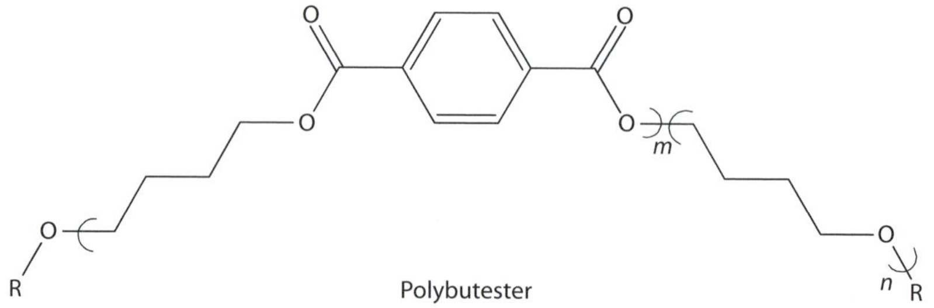

(19.43)

Polybutester

Isotactic *PP* is employed as a suture material. It is extruded as a monofilament. Tissue reaction is minimal and, in vivo, it offers outstanding stability making it a good candidate when permanent support is needed. It resists repeated bending over a long time period, retains good strength, offers minimal rejection and good controlled elongation making it a suture of choice for cardiovascular and cuticular applications. It also offers minimal resistance or drag as it is moved through tissues creating a minimal of tear and tissue dragging. It provides good knot security. It is also used in plastic reconstruction, gynecology and obstetrics, orthopedics, and general surgery.

Polyhexafluoropropylene (19.44) is a recently available suture material used for soft tissue neurological, cardiovascular, vascular, and ophthalmic surgeries. It offers the inertness of PTFE. It is available as a monofilament.

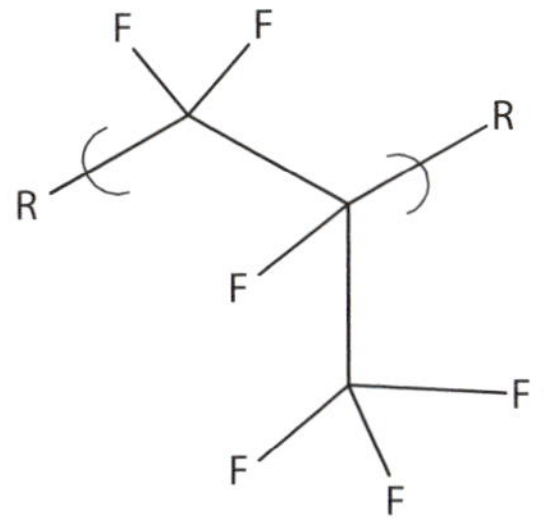

(19.44)

Polyhexafluoropropylene

Stainless steel wire offers the greatest strength and knot security and it seldom promotes tissue reaction. But, it is difficult to handle with a tendency of puncturing gloves and tissues, kinks, deforms, and undergoes fatigue. Thus, it is not widely used and, with the availability of the new high-strength synthetic sutures, continues to be replaced.

In summary, there exists a variety of suture materials with only one that is not polymeric. They serve as essential materials in surgery.

19.7 GEOTEXTILES

Because of the recent rash of hurricanes like Katrina and tsunamis we have become more aware of the need for protection against their violence. Geotextiles play a major role in this protection. Reinforced soil was used by Babylonians 3,000 years ago in the construction of their pyramid-like towers, ziggurats. One of these famous towers, the Tower of Bable, collapsed. For thousands of years, the Chinese used wood, straw, and bamboo for soil reinforcement, including the construction of the Great Wall. In fact, the Chinese symbol for civil engineering can be translated as "earth and wood." The Dutch have made extensive use of natural fibrous materials in their age-old battle with the sea. The Romans employed wood and reed for foundation reinforcement. By the 1920s, cotton fabrics were tested as a means for strengthening road pavements in the United States but these field trials were not followed up by application.

The modern materials for geotextiles are those produced by the textile industry, some since the early 1900s. In the 1950s the original technology for manufacturing plastic nets was developed by the packaging industry. The 1960s saw manufacturing capabilities for nonwoven fabrics made from continuous filaments being developed. The stage was set for the birth of the geotextiles industry. In 1957, as part of the Netherlands project to reclaim and protect its ocean-side, nylon-woven fabric sandbags were used. In 1958 PVC woven fabric was used for coastal erosion in Florida. In 1958 and 1959, sandbags made from synthetic polymers were used in West Germany and Japan for soil erosion control.

Geotextiles come as mats, textiles, webs, nets, grids, and sheets. When retention of the contained material is desired synthetic polymers such as PP, polyesters, nylons, polyethylenes, and poly(vinyl chloride) are used because they resist rapid degradation. When only short-term retention is needed natural materials such as cotton are used. Geotextiles are not always made from fibers but include film material such as polyethylene and PP sheets used to retain moisture but retard weed growth in gardens.

Geotextiles perform a number of functions. They help control fluid transmission such as helping to direct the flow of rain water to an outlet. They are used to separate materials for containment and to stop mixing such as highway embankment retention. It can be used for filtration control that allows for liquid and small particulates to pass through but to retain larger materials such as rocks. In sandbags for flood control, geotextiles simply hold together a material such as sand for breach control.

19.8 SMART MATERIALS

We have had smart materials as materials for a long time though the term is relatively new. Some of the first smart materials were piezoelectric materials, including poly(vinylene fluoride), that emit an electric current when pressure is applied and change volume when a current is passed through it. Most smart materials are polymeric or have a critical portion of the smart system that is polymeric.

Today research involves not only the synthesis of new smart materials, but also on the application of already existing smart materials. Much of the applications of smart materials involve the assemblage of smart materials and envisioning uses for these smart materials. Thus, it is possible that since application of pressure to a piezoelectric material causes a discharge of current, a portion of a wing could be constructed such that apparent "warpage" of a wing would result form an "electronic feedback" mechanism employing a computer coupled with a complex system of electronic sensing devices. Almost instantaneous, self-correcting changes in the overall wing shape would act to allow safer and more fuel-efficient air flight. Piezoelectric sensors could also be used to measure application of "loads" through reaction of the piezoelectric sensors to stress–strain.

Smart materials are materials that react to applied force—electrical, stress–strain (including pressure), thermal, light, and magnetic. A smart material is not smart simply because it responds to external stimuli, but it becomes smart when the interaction is used to achieve a defined engineering or scientific goal. Thus, most materials, including ceramics, alloys, and polymers, undergo volume changes as they undergo phase changes. While the best known phase changes involve changes in state such as melting/freezing, many materials, specifically polymers, offer more subtle phase changes. For polymers, the best known subtle phase change is associated with the glass transition, T_g, where local segmental mobility occurs. Volume changes associated with T_g are well known and used as a measure of the amount of crystallinity present in a polymer. Thus, when this volume change is used to effect some desired change, such as switching on and off an electric circuit, the polymer becomes a smart material. Multiple switching devices can be constructed using polymers with varying T_g values.

The use of smart materials as sensing devices and shape-changing materials has been enhanced because of the increased emphasis on composite materials that allow the introduction of smart materials as components.

A smart material assembly might contain the following:

- Sensor components that contain smart materials that monitor changes in some parameters such as temperature, light, magnetic field, and/or current
- Communications networks that relay changes detected by the sensor components through fiberoptics or conductive "wire"
- Actuator parts that react to the external stimuli such as changes in temperature, current, and so on

The actuator part may also be a smart material such as a piezoelectric bar placed in a wind foil that changes orientation according to a current imputed by the computer center allowing a machine such as an automobile to handle better and be more fuel efficient.

Muscles contract and expand in response to electrical, thermal, and chemical stimuli. Certain polymers, including synthetic polypeptides, are known to change shape on application of electric current, temperature, and chemical environment. For instance, selected bioelastic smart materials expand in salt solutions and may be used in desalination efforts and as salt concentration sensors. Polypeptides and other polymeric materials are being studied in tissue reconstruction, as adhesive barriers to prevent adhesion growth between surgically operated tissues, and in controlled drug release where the material is designed to behave in a predetermined matter according to a specific chemical environment.

Most current efforts include three general types of smart materials: piezoelectric, magnetostrictive (materials that change their dimension when exposed to a magnetic field) and shape memory alloys (materials that change shape and/or volume as they undergo phase changes). Conductive polymers and liquid crystalline polymers can also be used as smart materials since many of them undergo relatively large dimensional changes when exposed to the appropriate stimulus such as an electric field.

New technology is being combined with smart materials called *micromachines*, machines that are smaller than the width of a human hair. Pressure and flow meter sensors are being investigated and commercially manufactured.

As with so many areas of polymers smart materials have our imagination as the limit.

19.9 HIGH-PERFORMANCE THERMOPLASTICS

Engineering plastics are also referred to as high-performance thermoplastics or advanced thermoplastics. An *engineering plastic* is simply one that can be cut, sawed, drilled, or similarly worked with. Along with the ability to be worked with, high-performance thermoplastics generally also can be used at temperatures exceeding 200°C. These materials are also referred to as high-temperature thermoplastics. As the advantages of polymeric materials become evident in new areas, the property requirements, including thermal stability, will increase causing the polymer chemist to seek new materials or "old" materials produced in new ways to meet these demands.

Table 19.2 contains some of the new advanced thermoplastics that are currently available.

TABLE 19.2
Advanced High-Temperature Thermoplastics and Applications

Material	Heat Deflection Temperature, °C	Properties
Poly(arylene carbonates)	—	Leaves no degradation residue
Polyamide-imides	280	Good wear and good friction and solvent resistance
Polyanilines	70	Electrical conductor
Polyarylates (Aromatic polyesters)	175	Good toughness, UV stability, flame retarder
Polybenzimidazoles	440	Good hydrolytic, dimensional, and compressive stability
Polyetherimides	220	Good chemical, creep, and dimensional stability
Polyethersulfones	200	Good chemical resistance and stability to hydrolysis
Polyimides	360	Good toughness
Polyketones	330	Good chemical resistance, strength, and stiffness
Poly(phenylene ether)	170	Often alloyed with polystyrene
Poly(phenylene sulfide)	260	Good dimensional stability and chemical resistance
Polyphenylenesulfone	260	Good chemical resistance
Polyphthalamide	290	Good mechanical properties
Polysulfone	175	Good rigidity

Many of these polymers are being utilized as light-weight replacements for metal because of their strength, high dimensional stability and resistance to chemicals and weathering.

Nylon-66 was the first engineering thermoplastic and in 1953 represented the entire annual engineering thermoplastic sales. Plastic nylon-66 is tough, rigid, and does not need to be lubricated. It has a relatively high-use temperature (to about 270°C) and is used in the manufacture of many items from hair brush handles to automotive gears.

Nylon-46 was developed by DSM Engineering Plastics in 1990 and sold under the trade name Stanyl giving a nylon that has a higher heat and chemical resistance for the automotive industry and in electrical applications. It has a T_m of 295°C and can be made more crystalline than nylon-66. A number of other nylons, including the aromatic nylons, aramids, are strong and can operate at high temperatures, and they have good flame-resistant properties.

The next general grouping of polymers to enter the engineering thermoplastic market was the polyacetals derived from formaldehyde and known as polyoxymethylenes (POMs). While formaldehyde can be easily polymerized, it also depolymerizes easily through an unzipping mechanism. Most industrial polyacetals are capped preventing ready unzipping. POMs are highly crystalline, creep resistant, rigid, fatigue resistant, mechanically strong and tough, solvent resistant, and self lubricating. Unfortunately, even when capped, care must be taken since they have a tendency to unzip at high temperatures. Polyacetals are used in the manufacture of rollers, housings, bearings, gears, and conveyor chains.

While a number of polyesters have been considered as engineering thermoplastics, only several have become widely used. While PET is widely used, it does not crystallize rapidly discouraging its use in rapid molding processes. Polyester engineering thermoplastics require a high degree of crystallinity. Thus, PET is widely used as a fiber but not as widely employed as an engineering thermoplastic.

By comparison poly(butyl terephthalate) (PBT) is widely used as an engineering thermoplastic since it crystallizes rapidly allowing rapid fabrication. It has low moisture absorption, good self-lubrication, good retention of mechanical properties at elevated temperatures, and offers good solvent resistance and fatigue resistance. It is used in numerous "under-the-hood" automotive applications, including ignition systems. It is also used in athletic goods, small- and large-appliance components, power tools, and electrical applications.

Polycarbonates have been made using a variety of bisphenols but the most widely produced materials are derived from bisphenol A (BPA). Polycarbonates exhibit good impact strength, are heat-resistant, have good electrical properties, and high dimensional stabilities. They also offer good transparency that allows their use in replacing glass where good transparency is needed such as in safety glazing, automotive headlamps and taillights, CDs, and in ophthalmic applications. They also offer inherent flame resistance due in part to the presence of the carbonate moiety that is already largely oxidized so as not to act as a fuel.

Poly(alkylene carbonates) leave no residue when decomposed. Thus, they are used as binders for holding ceramic and metal powders together long enough for them to be made into the desired products. These polymers also give good moisture and oxygen barriers and they are abrasion resistant, offer good clarity and give tough films and are being considered in food and medical packaging applications.

Poly(phenylene ether) (PPE) or poly(phenylene oxide) (PPO) shows total compatibility with PS, allowing a number of different combinations to be formed, giving rise to the Noryl family of blends and alloys. These two combine each offering needed properties. The PPO brings fire retardancy and contributes to a high heat distortion temperature above 100°C allowing the products to be cleaned and used in boiling water. PS contributes ease of fabrication to the combination. Both contribute to the blends outstanding electrical properties and good water resistance. Furthermore, the blends give materials with a lower density than most engineering materials.

Polyarylsulfones offer materials with good thermal-oxidative stability, solvent resistance, creep resistance, and good hydrolytic stability. Their low flammability and smoke evolution encourages their use in aircraft and transportation applications. They hold up to repeated steam sterilization cycles and are used in a wide variety of medical applications such as life support parts, autoclavable tray systems, and surgical and laboratory equipment. Blow-molded products include suction bottles, surgical hollow shapes, and tissue culture bottles. Poly(phenylene sulfide) has a number of automotive uses, including as an injection-molded fuel line connector and as part of the fuel filter system.

The aromatic polyketones offer good thermal stability, good environmental resistance, good mechanical properties, resistance to chemicals at high temperatures, inherent flame retardancy, good friction and wear resistance, and good impact resistance. Poly(ether ether ketone) (PEEK) became commercialized in 1980 under the name Victrex. It is used in the chemical process industry as compressor plates, pump impellers, bearing cages, valve seats, and thrust washers; in the aerospace industry as aircraft fairings and fuel valves and in the electrical industry as semiconductor wafer carriers and in wire coatings.

Polyetherimide (PEI) was first announced by General Electric scientists in 1982. It offers good stiffness, impact strength, transparency, low smoke generation, broad chemical resistance, good heat resistance, good electrical properties, and good flame resistance. It also offers good processability, and a good resin can be made form it, allowing for easy molding. They are used in internal components of microwave ovens, electrical applications, and in transportation.

With the advent of the soluble stereoregulating catalysts, so-called older polymers have been synthesized with additional control over the structure giving products with enhanced strength and dimensional stability. Amorphous PS is relatively brittle, requiring a plasticizer to allow it to be flexible. The use of soluble stereoregulating catalysts allowed the synthesis of syndiotactic PS with a T_m of about 270°C and a T_g of about 100°C with good solvent and chemical resistance. DOW commercialized s-PS under the trade name Questra in 1997. It is used in specialty electrical and under-the-hood automotive applications.

19.10   CONSTRUCTION AND BUILDING

The use of polymeric materials as basic structural materials is widespread and of ancient origin. In past times, the building materials were largely rocks and plant materials, both largely polymeric. Today, the major structural building materials are concrete, and in many places wood, again, both largely polymeric. The proportion of concrete to wood varies on the type of building and location. In general, the structural materials are concrete and steel for large buildings. For home dwellings, it may be almost only concrete such as in the Middle East where wood is not plentiful with more wood used for framing in the United States.

For home building, the first material is concrete for the slab, here in Florida, or some form of concrete for the basement. Frames consisting of wooden boards are generally the next major addition. This is followed by a plywood roof covered by roofing material, again the major choices of fiber glass shingles, wood shakes, and ceramic and concrete tiles are all polymeric. Shortly, the structure is enclosed with the addition of a wooden door and glass windows. Table 19.3 contains the proportion of materials used in building and construction and is an average of homes and large buildings for the United States for 2000. Of these basic materials, over three quarters are polymer intensive. Table 19.4 contains general use categories for polymers in home construction.

The use of synthetic polymers in building and construction is also increasing at a rapid rate. Flooring is a mixture of wood, synthetic and clad wood, carpet, and tile, all polymeric. While carpets were once be derived from natural materials such as cotton and wool, today almost all of them are derived from synthetic polymers and include nylon, polyester, olefins, and polyacrylonitrile.

Uses of plastics are given in Table 19.5.

As lighter and stronger polymeric materials become available, their impact on the building and construction industry and on other industries will increase. Further, as materials that perform

TABLE 19.3
Use of Materials in Building and Construction

Material	Weight	Percentage by Weight (Billions of Pounds)
Concrete*	250	50
Lumber*	60	12
Ceramic*	50	10
Wood panels*	20	4
Iron and steel	15	3
Plastics*	10	2
Other†	95	19

*Indicates a polymeric material.
†Some of these are polymeric, such as paints, sealants, and coatings.

TABLE 19.4
Common Applications of Polymers in Home Construction

Application	Typical Polymers Used
Foundation	Cement (concrete)
Framing	Wood
Thermal insulation	Foamed polystyrene, polyurethane, Fiberglas®
Vapor barrier	Polyethylene
Siding	Wood, polyvinyl chloride
Paints	Acrylics, polyurethanes
Electrical insulation	Polyisoprene (rubber), chlorinated polyethylene
Flooring	Wood, complex silicates (cement/ceramic tile), carpet (nylon, polyester, polyolefin)
Roofing	Plywood (sheathing), Fiberglas® (shingles), complex silicates (cement, ceramic tile)

TABLE 19.5
Major Plastic Applications in Building and Construction

Plastic	Use(s)
ASA	Window frames
Acrylics	Lighting fixtures and glazing
PVC and chlorinated PVC	Hot and cold water pipe; moldings, siding, window frames, floor tiles
Melamine and urea formaldehyde	Laminating for counter tops, adhesives for wood, plywood, and particle board
Phenol formaldehyde	Electrical devices and plywood adhesive
P(ethylene terephthalate)	Counter tops and sinks
Polycarbonates	Window and skylight glazing
Polyethylene	Pipes, wire and cable coverings, plastic lumber, vapor barriers
Poly(ethylene oxide)	Roofing panels
Polystyrene	Insulation and sheathing
Polyurethane	Insulation and roofing systems
Polypropylene	Vapor barrier sheeting, pallets, brushes

specific tasks become available, they too will become integrated into the building and construction industry. This includes devices for gathering and storage of solar energy and "smart" materials including "smart windows."

19.11   FLAME-RESISTANT TEXTILES

Combustion generally occurs through a burning process where there is a fuel, here a polymer, and oxygen. Because most organic polymers have high-hydrocarbon content and/or contain only carbon, hydrogen, and oxygen, the major products of combustion are carbon dioxide and water, or when there is an insufficient amount of oxygen, carbon monoxide, and water. For some polymers such as PS and PVC, formation of fused-ring compounds result in formation of char that acts as a barrier to oxygen acting to impede the combustion process. In truth, the burning process is more complex occurring in at least three environments—the surface where oxidation is predominant; the inner surface where the amount of oxygen is depleted yet where evolution of created gases occurs; and the bulk where there is a lack of oxygen and produced gases are trapped. Even so, as combustion continues, more of the polymer sample is exposed becoming surfaces as the "old" surface is burnt away. These new surfaces generally react with incoming oxygen forming mainly the typical combustion products.

Most combustion is exothermic feeding on itself. Table 19.6 contains the heats of combustion for selected polymers. Those polymers containing large amounts of oxygen typically have lower heats of combustion because they are already partially oxidized. Thus, POM and PET have low heats of combustion.

Heats of combustion are not indicators of the tendency to burn or the rate of burning. Other indicators are employed as markers with respect to their tendency to burn. The limiting oxygen index (LOI) is one of these markers. Briefly, the sample, in a predescribed standard form, is set afire in an upward-flowing oxygen–nitrogen gas mixture and a stable flame is established. The ratio of oxygen to nitrogen is reduced until the sample flame becomes unstable and is extinguished. The minimum oxygen content that supports combustion is the LOI of the sample. Such tests are quantitative but must be considered first approximations.

Table 19.7 lists the LOIs of selected polymers. Material with LOI values above about 0.25 are considered to be self-extinguishing under normal atmospheric conditions. Addition of flame retardants increases the LOI. While most polymers decompose mainly to carbon dioxide and water, some give off harmful gases. Thus, PVC splits off hydrogen chloride when burning. Some compounds such as polyacrylonitrile might be suspected to give off hydrogen cyanide, but in truth they form complex fused-ring systems incorporating the cyanide moiety into the system resulting in the emission of little or no hydrogen cyanide.

Asbestos was the first material used as a flame-resistant material but because of its negative health effects, it has been replaced by other materials. Today, the most widely used flame-resistant textile is Nomex™ (19.45). It is inherently flame resistant not requiring an additional application or need to worry that the coating may be removed through cleaning. Nomex is an aramid or aromatic

TABLE 19.6
Heat of Combustion of Selected Polymers

Polymer	ΔH (kJ/g)	Polymer	ΔH (kJ/g)
POM	17	Nylon-66	32
Cellulose	18	PS	42
PVC	20	NR	45
PET	22	PP	46
PMMA	26	PE	47

TABLE 19.7
Limiting Oxygen Index of Selected Polymers

Polymer	LOI
PP	0.17
PE	0.17
PMMA	0.17
PS	0.18
PC	0.27
PVC	0.48
PTFE	0.95

nylon material. Along with being flame resistant, it is very strong and so is also used in tire cord. Because of its stiffness and high cost, it is seldom used as a general textile material. Here we will focus on fiber-intensive uses of such aromatic nylons.

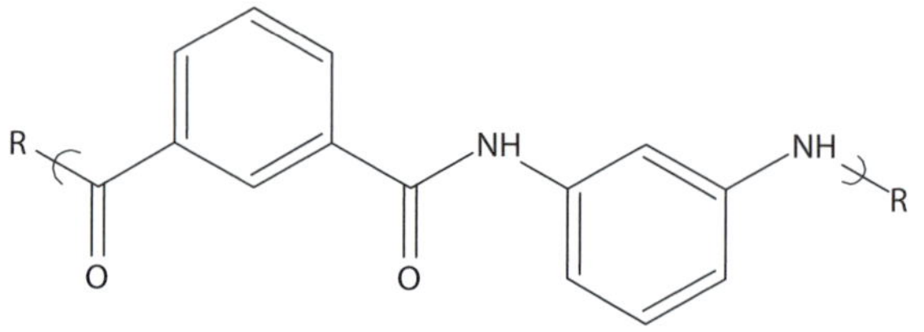

$$(19.45)$$

Cost of a material depends on many factors. Some polymers retain unwanted materials, such as unreacted monomer, that must be removed. Removal is generally done by dissolving the polymer and reprecipitating the polymer leaving the unwanted material in solution. Aramid polymers are difficult to dissolve so did not become commercially available until the 1980s. Paul Morgan and Charles Carraher noticed that some materials stay dissolved when produced in rapidly stirred systems. Morgan turned this observation into use when he noticed that aromatic nylons, when made in rapidly stirred systems, stayed dissolved long enough so that the solution can be run into a nonsolvent that turned the aromatic nylons to solids as they passed through spinnerets into the nonsolvent. Thus, the aramid fibers were formed without needing to redissolve them. This allowed aromatic nylons to be available at a competitive cost.

Aramids resist melting, dripping, supporting combustion in air, or burning. These are all positive characteristics. It is employed by firefighters for both routine fire-fighting materials but also as common station wear. Clothing made from aramids is durable and made to be comfortable. The "fire-fighting" outerwear is thicker than that employed as innerwear by the firemen. Racers also take advantage of the durability and flame-resistant characteristics of aramids by wearing clothing made of it. In a pile up it provides the valuable seconds that allows the drivers to be pulled safely from the burning wreckage.

19.12 WATER-SOLUBLE POLYMERS

The solubility rules governing polymers are similar to those governing water solubility of smaller organic molecules except the extent of polymer solubility and range of polymeric structures are more limited. Figure 19.7 contains structures of some of the commercially available water-soluble polymers. As with smaller molecules, the presence of highly electronegative atoms that can participate in hydrogen bonding is normal for water-soluble polymers. Such groups include alcohols, amines, imines, ethers, sulfates, carboxylic acids and associated salts, and to a lesser extent, thiols. For charged species, such as sulfates, pH and the formation of charged species are also important

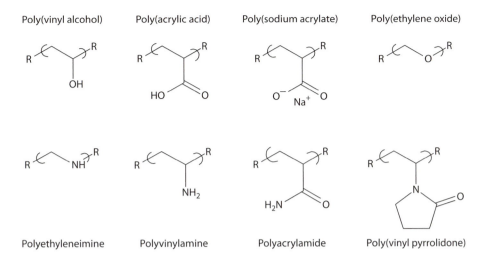

Poly(vinyl alcohol) Poly(acrylic acid) Poly(sodium acrylate) Poly(ethylene oxide)

Polyethyleneimine Polyvinylamine Polyacrylamide Poly(vinyl pyrrolidone)

FIGURE 19.7 Structures of commercially available water-soluble polymers.

factors. Thus, the copolymer derived from vinylamine and vinyl sulfonate is not water soluble, but the corresponding sodium salt of this copolymer is water soluble.

The amount and rate of water solubility is decreased by the presence of cross-linking and the substitution of nonpolar units for polar units within the polymer. Most water-soluble polymers posses both hydrophobic and hydrophilic portions and this combination affects the shape of the polymer chains in solution. Many water-soluble polymers exist in random or partially helical chains in water solution where the chains are partially extended allowing further hydrogen bonding with water to occur. Since polar groups present in similar configurations and conformations repeal one another, many water-soluble polymers approach rigid rods. Such structures are somewhat favored because the organic backbones tend to reside within the interior of the polymer chain with the polar groups thrust outward toward the water.

There also exist a number of water-soluble copolymers. Thus, copolymers containing to about 80% ethylene units with units of sodium acrylate are water soluble and used in the formation of many materials, including ionomers.

19.13 ANAEROBIC ADHESIVES

Almost everything we use daily has adhesives connected with them. Our soda bottles use adhesives to attach the label; our tables—wooden, glass, metal, or a mixture—are connected by adhesives; our shoes are held together by adhesives; our automobiles, trucks, airplanes, trains—all connected by adhesives; our tires, stoves, washers, all contain adhesives as essential materials, and so on.

Ancient adhesives involved and generally required oxygen to affect the adhesion. Today, there are many application areas where adhesion is required but in the absence of oxygen. These adhesives are called *anaerobic adhesives*. Most anaerobic adhesives are based on dimethyl acrylates. Thus, adhesive is applied to the threads of a bolt. As long as the screw surface remains in contact with oxygen, curing does not occur. On screwing the bolt into the channel or nut, the adhesive's contact with oxygen is cut off and its contact with a metal, typically iron or copper, begins the adhesion process.

An often employed adhesive material for anaerobic adhesives is tetraethyleneglycol dimethacrylate (TEGMA). Reaction begins through the free radical polymerization of TEGMA. This reaction is inhibited by the presence of oxygen as shown in Figure 19.8. Thus, as long as oxygen is present, any TEGMA molecules that become unintentionally active react either with another TEGMA molecule (not preferred) or with oxygen rendering it largely inactive. The second reaction can occur at any point leading to passive oligomers. In fact, anaerobic adhesives are shipped with a blanket of oxygen layered over the adhesive to be sure that a ready supply of oxygen is present to prevent premature polymerization from occurring. Even on application, the monomer is exposed to oxygen up until it is married in such a manner as to cut off the supply of oxygen. At this point a hardener comes into play.

The hardener systems are complex. The most employed hardener system contains a three part system—the actual radical-producing molecule, here a cumene hydroperoxide, an accelerator, here *N,N*-dimethyl-*p*-toluidine, and saccharin, which acts as a metal complexing material and a reducing agent for metal ion, here copper (Figure 19.9). The reaction between the saccharin and *N,N*-dimethyl-*p*-toluidine consumes any remaining oxygen (Figure 19.9, top left). An aminal is produced that dissolves surface metal ions reducing them to a lower oxidation number, here

$$Cu^{+2} \longrightarrow Cu^{+1} + e^- \qquad (19.46)$$

The Cu^{+1} catalyzes the formation of radicals from the cumene hydroperoxide, which then begins the polymerization of the TEGMA and TEGMA molecules with only one reactive end (Figure 19.10). Those with two active ends result in the formation of cross-linked materials. A similar reaction occurs with iron and several metals such as zinc, gold, silver, cadmium, magnesium, titanium, and alloys that contain any of these metals.

The "aminal" reaction is cyclic ensuring the formation of an abundance of radicals that effect the polymerization of TEGMA.

Because the vast majority of the monomers react at both ends, these adhesives are thermosets and are brittle. Thus, such adhesives are not suitable for areas requiring flexibility. Because they are thermosets, they are typically resistant to oils, water, and solvents. Thus, the adhesive plays two roles—bonding and protection. These properties make anaerobic adhesives suitable for automotive and truck motor mounts and electric engines.

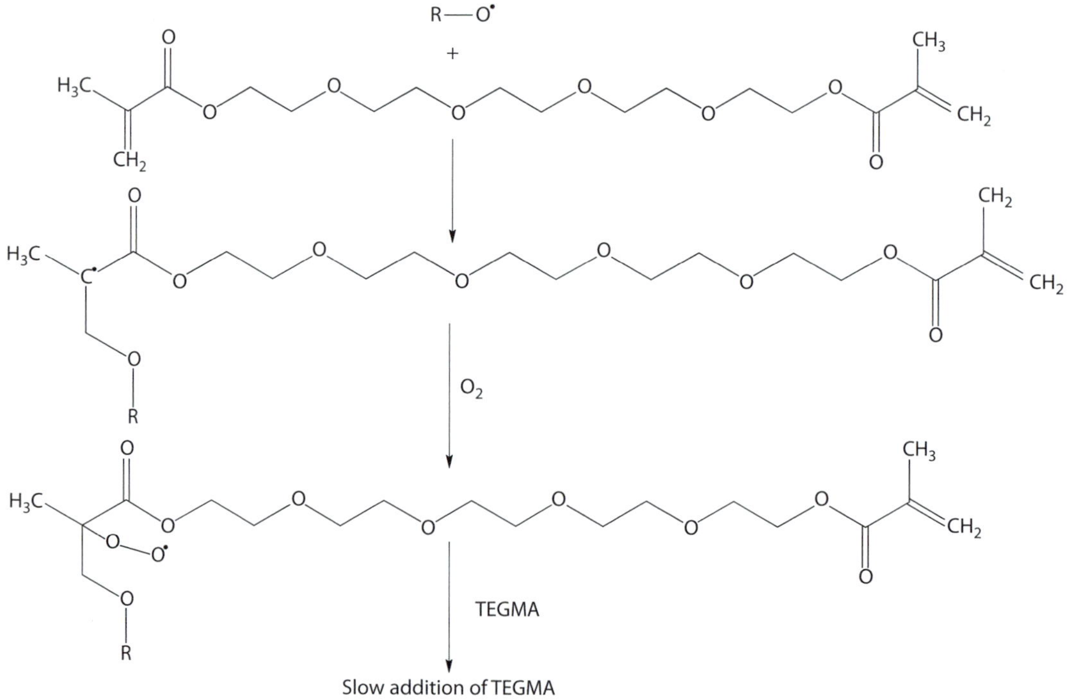

FIGURE 19.8   Formation of passive monomer-derived molecules.

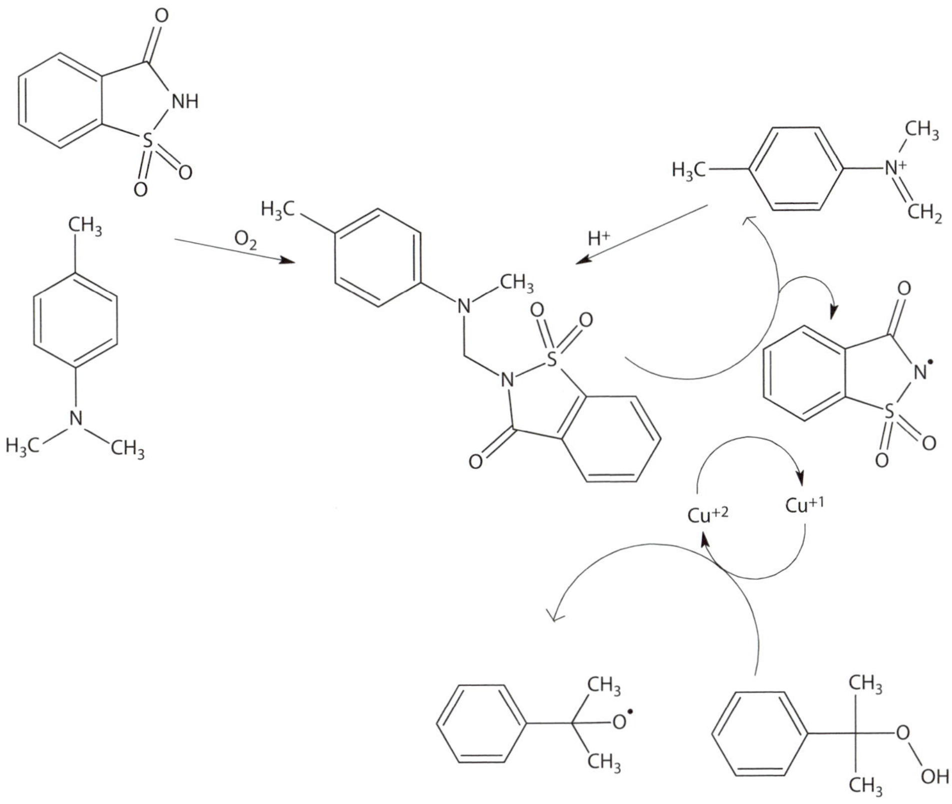

FIGURE 19.9 Reactions involved in the hardening process.

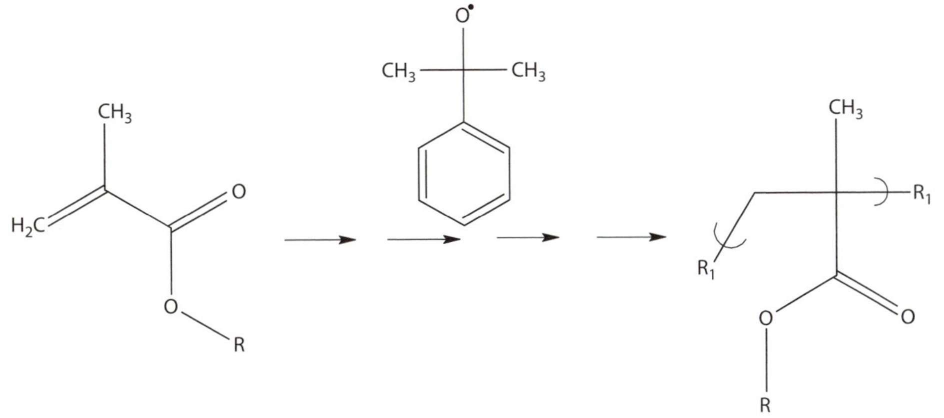

FIGURE 19.10 Polymerization of one end of the TEGMA.

Along with their use with bolts, they are also employed in securing screws in various mechanical and electrical applications.

Bolts are generally disconnected by heating to about 400°C thereby thermally degrading the adhesive matrix allowing the bolt to become disengaged.

Different mixtures of anaerobic adhesives are available that offer a variety of curing times, strengths, and compatibilities. A typical cure time is 20 min for cure to begin, vibration resistance in about an hour. Some fast cure systems will have about 20% cure in a few minutes and are almost cured in about 2 h. Some instant cure systems are ready for use within seconds.

19.14   HYDROGELS

Hydrogels are simply water-filled gels. They are characterized by being hydrophilic yet not completely soluble in water. Those hydrogels that are able to absorb large amounts of water are referred to as superwater adsorbents.

Hydrogel structures may contain the hydrophilic units in the polymer backbone or as side chains. Those polymers that contain the hydrophilic units in the backbone include acrylic, poly(vinyl alcohol), N-vinyl-2-pyrrolidine, and acrylamide-containing materials, derivatives of poly(ethylene oxide), and ionomers and glycopolymers. Polymers that contain such functional groups as –OH, –COOH, –SO$_3$H, –CONH$_2$ can act as foundations for hydrogels. These polymers generally contain some cross-linking that locks in a three-dimensional structure that prevents water solubility and also helps retain water. The cross-links can be chemical or physical.

Applications of hydrogels include highly absorbent diapers based on poly(sodium acrylate), contact lenses based on poly(2-hydroxyethyl methacrylate) (polyHEMA) and switches based on variations of swelling of the hydrogels. A number of drug delivery systems have also been based on hydrogels.

19.15   EMERGING POLYMERS

A number of small-scale polymeric materials will continue to enter the marketplace on a regular basis. These include biomaterials and electronics materials where the cost per pound is high and the poundage is low, generally well less than a 100 tons a year. These are materials that fulfill specific needs.

The number of new larger-scale giant molecules that enter the marketplace will be small. It has been estimated that it takes about $1 billion to introduce and establish a new material. It is a daunting task with no guarantee of success. In the past, new giant molecules could be introduced that offered improvements in a number of areas and thus would attract a market share in a number of application areas. Today, there are already a wide range of materials for most large-scale application areas that compete for that particular market share so that it is difficult for any material to significantly break into any market area. A new material needs a "flagship" property that a particular market needs.

DuPont and Shell have developed a new polyester, poly(trimethylene terephthalate) (PTT; 19.47) that is structurally similar to PET, except that 1,3-propanediol (PDO) is used in place of ethylene glycol. The extra carbon in Sorona allows the fiber to be more easily colored giving a textile material that is softer to the touch with greater stretch. Further, it offers good wear and stain resistance for carpet use. The ready availability of the monomer PDO is a major consideration with efforts underway to create PDO from the fermentation of sugar through the use of biocatalysts for this conversion. Sorona and Lycra blends have already been successfully marketed. Sorona is also targeted for use as a resin and film.

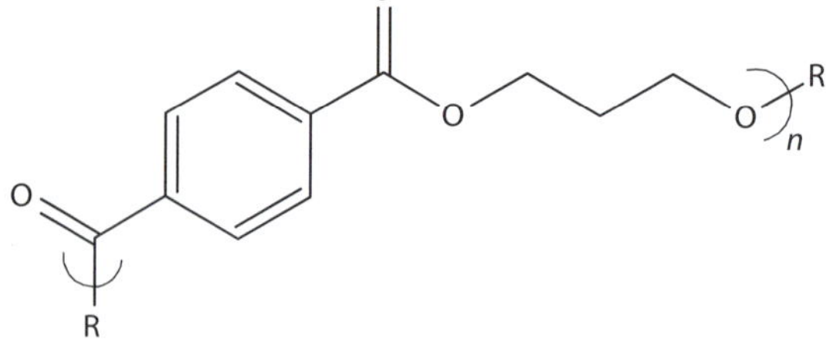

(19.47)

Poly(trimethylene terephthalate)

DMS introduced in 1990 nylon 4,6 called *Stanyl* (19.48), based on the reaction between adipic acid and 1,4-diaminobutane. Stanyl can withstand temperature of about 310°C, allowing it to create a niche between conventional nylons and high-performance materials. It was not able to break into the film market and has only now begun to be accepted for tire cord applications. About 22 million pounds of Stanyl was produced in 2001.

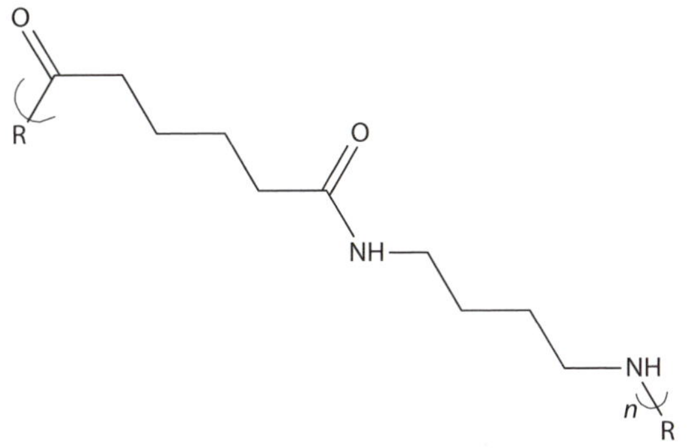

(19.48)

Nylon-4,6

In 2007, DSM announced the development of PA4T nylon (19.49). It is intended to address new needs for miniaturization and the convergence of electronic devices such as cell phones and computers. It offers good dimensional stability, compatibility with lead free soldering, high melting point, good mechanical strength at elevated temperatures, high stiffness, and good processability. Specific applications include central processing unit (CPU) sockets, high-temperature bobbins, notebook computer memory module connectors, and memory card connectors.

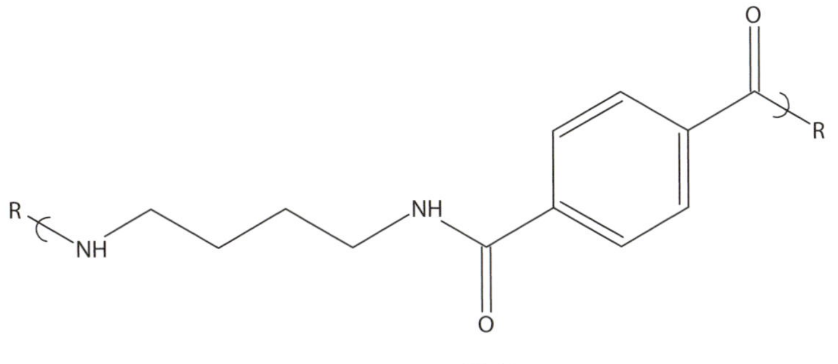

(19.49)

PA4T

The development of PA4T nylon illustrates the use of polymers in the intermediate synthesis scale, larger than specialty chemical use and smaller than bulk chemical use. The development of polymers that fit into this category will greatly increase as special properties are needed and the price per weight can be high.

In 1997, Dow introduced syndiotactic PS under the trade name Questra. The technology for the production of Questra is based on relatively new technology and science involving soluble stereo-regulating catalysts that produce PS that has a fixed and repeating geometry as each new styrene monomer unit is added to the growing PS chain. Targeted areas include medical, automotive, and electronic applications.

Several other produces have been developed based on the relatively new soluble stereoregulating catalysis systems. Index, an ethylene-styrene interpolymer, was introduced in 1998 and is intended to compete with block copolymers such as styrene–butadiene–styrene, flexible PVC, polyurethanes, and polyolefins. It is being used as a modifier for PS and polyethylene. Dow is also developing soundproofing and packaging foam applications for Index. Hoechst Celanese (now Ticona) developed Topas, a cyclo-olefin copolymer, in the 1980s and in 2000 began commercial production of it. Topas has high moisture-barrier properties and is being considered for use in blister packaging for pharmaceuticals. It is also being used in resealable packages where it provides stiffness to the sealing strip. It is also being used in toner resin applications and is being blended with linear low-density and low-density polyethylene providing stiffness and to improve sealing properties.

A number of new materials are looking toward being involved in the upcoming move toward blue-light CDs. For any of these to become important materials in this area they will need to improve on the present polycarbonate-based materials.

GE introduced in 2000 a new polyester carbonate based on resorcinol arylates it called *W-4*. It is now marketed as Sollx. Sollx does not need to be painted; it offers good weather, chip, scratch, and chemical resistance and is being used as the fenders for the new Segway Human Transporter. It is also aimed at automotive uses, including as body panels. Sollx is coextruded into two layers, one clear and one colored, to simulate automotive paint. It is then thermoformed and molded into the finished product.

A number of new materials have been developed because of the health fears associated with the monomer BPA, which is the comonomer for most polycarbonates (PCs). The replacement should possess similar properties to PC and also be available in large quantity and inexpensive. One material that has become available is a copolymer polyester developed by Eastman. The precise structure for this material, Tritan, is proprietary, but is believed to be based on the diol tetramethylcyclobutandiol, 19.50. The ring system contributes the necessary stiffness and the methylenes supply the flexibility.

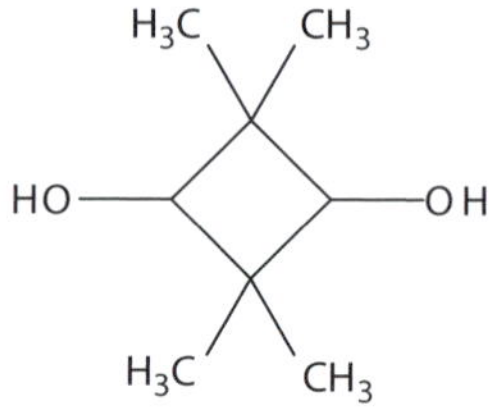

(19.50)

Several new ventures are based on using natural, renewable materials as the starting materials instead of petrochemicals. These products are know as "green" products since they are made from renewable resources and they can be composted. Along with the microbial production of PDO by Shell and DuPont to produce nylon 4,6, Cargill Dow is making PLA beginning with corn-derived dextrose. The polylactide (PLA) (19.51) is made from corn-derived dextrose, which is fermented making lactic acid. The lactic acid is converted into lactide, a ring compound, that is polymerized through ring opening.

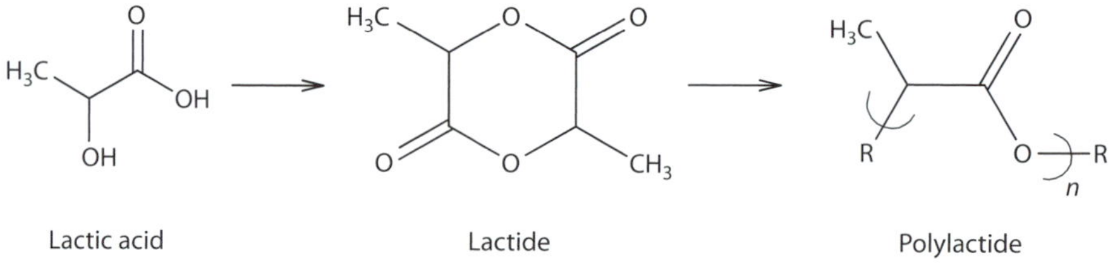

|　Lactic acid　|　Lactide　|　Polylactide　|　(19.51)

Polylactide looks and processes like PS. It has the stiffness and tensile strength of PET and offers good odor barrier and resists fats, oils, and greases. PLA is being considered for use in fibers and in packaging. As a film, PLA has good dead fold properties, that is, it has the ability to be folded and to stay folded. It is being used as a fiber for apparel and carpeting applications. It is being sold as a bridge between synthetic and natural fibers in that it processes like synthetic fibers but has the touch, comfort, and moisture management of natural fibers.

As noted above, new materials must "fight-it-out" with existing materials for existing and developing areas. This is particularly true in the automotive area. Thus, polyethylene-polyethylene-copolypropylene panels and bumpers have replaced "rubber" HIPS as automotive bumpers. Other materials were available with greater toughness and scratch resistance, but they were also more expensive. Here, classical monomers were employed to develop nonclassical materials.

In the automotive industry, the rule of thumb is that every 10% reduction of vehicle weight gives about a 5% increase in fuel economy. Weight reduction is only one of the driving forces for the replacement of metals with polymeric materials. It is worth mentioning other driving forces because these are also involved in the selection of certain polymers over other polymers. Reduction in cost is also a factor. Reduction in cost involves a number of factors, including, but not exclusive of, simple pound per pound cost. Polymers are out performing metals because of the ease with which complex structures can be made. Some polymers lend themselves to easy fabrication and this represents both a cost savings and enhances the opportunity for the material to become an automotive material. Interestingly, the number of parts also is involved in deciding what material to be used. For instance, if the number of a part to be used is less than about 100,000 annually, then injection-molded plastic parts are less expensive than ones stamped from steel. Injection-molded parts also provide aesthetics that are hard to obtain from steel and glass. In fact, many of the plastics can be made to resemble chromes and gold so are used to enhance the look of some automobiles. Polycarbonate tail lenses are favored over glass for a number of reasons, including ability to make more complex and attractive shapes. Circular taillights are no longer the only option.

Flexibility of design as well as material flexibility favors polymers. Again, plastics can be made into almost any shape and when necessary even bent a little to fit them into almost impossible spaces within the automobile's interior.

Along with traditional plastics, composites are also becoming more widely employed. GM has traditionally employed fiber-glass composites for the Corvette's body panels (now using carbon-fiber composites for some of these panels), but is now looking to use composites for the body panels of other automobiles.

Performance is also a consideration and in some instances favors polymers over steel. DuPont developed a nylon water jacket spacer for the 2006 Toyota Crown and Lexus GS-300. It did not reduce weight, but did increase the heat-transfer efficiency between the engine coolant system and the cylinders lowering the fuel consumption by 1% or an equivalence of more than 50 pounds.

Another indirect way that polymers can assist fuel economy is by allowing alternative fuel uses. Plastics can "create" free space needed to store bulky batteries encouraging their use as alternative fuel sources.

Along with the more classical polymers some new ones have been developed for automotive use. Thus, GE developed Noryl GTX resins that are being used on the fenders of the some cars, including the Volkswagen New Beetle and Hummer H3. The Noryl GTX resins are alloys of PPO and nylon-66.

The automotive industry is continuing looking for new materials and new uses for more classical materials.

Many of the so-called new materials are actually "old materials" arranged in different settings. Following is a brief description of three of these materials. Trgris, developed by Milliken, is a PP thermoplastic composite. It is made from coextruded PP tape yarn that has a highly drawn core creating additional strength for the PP yarn. The pulling encourages better alignment of the PP chains. The tape yarn is then woven into a fabric and layered and finally heated with nonyarn PP, with pressure applied forming rigid sheets that are used for panel applications or premold material. Thus, the yarn acts as the discontinuous phase with the nonoriented PP used as the continuous phase forming a PP–PP composite. In some fashion, Trgris is similar to plywood in that various layers are stacked, generally with the sheets placed at different angles to one another. The entire PP composite is strong and it is lighter than most composites. It also has good impact resistance since the entire composite is more intimately mixed, because of the sameness of the PP that is both the discontinuous and continuous phases, allowing more efficient transfer of impact energies. Also, because it is one single material, it is easy to recycle lending to it being a more environmentally friendly material. It is being used in automotive applications, including at the race track. Because of its outstanding impact resistance, it is used as the race car splitter and to reinforce other parts of the race cars. If it does fail, unlike other composite materials that often shatter leaving unwanted debris on the track, Trgris materials retain their entirety. Further, because of the layering, fractures do not propagate throughout the entire Trgris part. Other uses are in transportation as liners and floors; in water sports in the construction of small boats and watercraft; in construction of blast resistance materials; and it is being used in the construction of playground and outdoor objects.

A second material recently introduced as Impaxx is an energy absorbing foam developed by Dow Automotive. It reduces weight, space, cost, and improves safety. It is composed of foamed PS so is similar to disposable coffee cups. Impaxx is a closed foam material. On impact, the closed foam cells compress and buckle absorbing energy and when excessive impact energy is applied the cells rupture again absorbing energy. Impaxx is used in more than 2 million automobiles, including the Honda Pilot, Chevy Malibu, and Ford Crown Victoria as well as a number of racing cars. It is used as headliners and pillar trims, steering column and lower instrument panels, under carpet, front and rear doors, and in bumper systems. Again, it is advertised as being green friendly since it is all PS and can be readily and easily recycled.

CarbonX is made by Chapman Thermal Products and is in competition with Nomex and similar materials as a flame-resistant material. CarbonX is made by a controlled heating of polyacrylonitrile forming a fully oxidized fused polycyclic composition similar to that described in Chapter 10. The material can be heated to 500, 1,000°C or higher giving a material that is then stable to that temperature since material that is unstable has already been driven off. The oxidized polyacrylonitrile (OPAN) is forced through small holes resulting in the formation of tiny fiber-like materials, which are not particularly flexible. Thus, the OPAN is often woven with other fibers, including steel fibers, giving the material additional strength and flexibility.

Even if some burning occurs, the OPAN fabric expands removing oxygen. CarbonX materials are replacing some of the fabrics previously made by Nomex. Nomex and CarbonX garments are made for use by fire fighters, steel mill workers, military, race track drivers and other personal, and police and include undergarments, socks, suits, head gear, shoelaces, and escape blankets. CarbonX comes in your choice of colors as long as you want black. Other colors are achieved through blending colored fibers in with the CarbonX fibers.

Safety glass is also undergoing changes. Along with the poly(vinyl butrylate) (PVB) innerlayers, other materials are being used, including PET, EVA, and ionoplast as innerlayers. DuPont, Solutia, and PPG have pioneered in lots of these efforts so that the applications are not just for the automotive windshields but for many architectural uses. The U-shaped Grand Canyon Skywalk that extends 70 feet from the rim and 4,000 feet above the Colorado river employs laminated glass. The balustrades have laminated bent glass of two 3/8 in. layers with a 0.06 in. PVB interlayer. The 2-in. thick glass floor has five layers of glass alternating with four layers of SentryGlas Plus. The SentryGlas Plus has an ionoplast resin composition that is five times stronger and with 100-fold rigidity compared with the traditional PVB innerlayers. The walkway can support more than 70 million pounds; can withstand sustained winds in excess of 100 miles h; and an 8.0 magnitude earthquake.

Head-up displays (HUDs) allow drivers to view important information as Global Positioning Systems (GPSs) on the windshield. These systems use laminated glass combinations as Wedge, which is a Butacite PVB interlayer that converts the windshield into a transparent liquid crystal display. Wedge is also employed as outer windows for large-scale buildings with the ability to color the glass in an assortment of colors and hues.

Layered glass also contributes to savings of heat and our health. Tempered glass absorbs about 60 UV radiation, while laminated glass blocks more than 90% of the UV radiation equivalent to a sun protection factor (SPF) of 50.

19.16   GREEN MATERIALS

There is an increasing emphasis in so-called "green materials." There are other terms often associated with the term green materials. These other terms include renewable resource materials and natural materials. The emphasis is the replacement of nongreen materials by these green materials. Each of these terms has slightly different meanings. The term "renewable materials" is generally employed for rapidly renewable materials such that are replenished within a short time such as a year. The term "natural materials" emphasizes materials that are derived from nature. Along with these descriptions, a green material also encompasses the energy requirements, processing procedures, and recycling ability of the material

The shift from petroleum-based feedstock toward renewable, green, materials is rapidly accelerating with much of this emphasis based on developing monomers from these green sources. In 2008, the world market for biobased chemicals, excluding biofuels, was $1.6 billion and this is projected to be about $5 billion by 2015. Concrete measures of this emphasis can be seen by the construction of plants capable of producing 100 million pounds per year of such green monomers as PDO and polymers such as polyhydroxyalkanoate, and polylactic (PLA).

Natural materials include materials that are "naturally" found about us. This includes oil, coal, and natural gas. These materials are not considered green materials because they are not readily replenished. Other natural materials include silicon dioxide-intense materials such as mica that is utilized in some coatings producing clad-like coverings and diatomaceous earth that is being employed in the construction of some tires to retain tire pressure. It also includes many renewable materials such as cellulose, chitosan, vegetable oils, and lignin. These renewable materials are readily available in the billions of tons yearly, renewable on a regular yearly cycle, and are greatly underused materials. Carraher and others have worked with chitosan to produce anticancer drugs and lignin to produce materials with structural integrity as alternative industrial materials. Sperling and coworkers have employed naturally derived oils producing various rubbers.

Here we will describe other green materials. We emphasize polymers that are commercially available and that are really derived from natural sources, not simply said to be possibly derived from natural sources. Nylon-66 was initially advertised to be derivable from various natural sources, and it is, but today it is manufactured from monomers that are part of the stream of petroleum derived feedstocks. Thus, nylon-66 is not included in this discussion as a green material though it may become part of the green material revolution as natural, biomass-derived, sources for the reactants become available on a large scale. This may change as industry seeks ways to produce already employed monomers from biological feedstocks. For example, Coca-Cola will employ beverage bottles that are traditionally produced from synthetic PET employing ethylene glycol derived from sugar and molasses rather than ethylene glycol produced from petroleum and natural gas.

The movement from a petroleum-based material to a green material is not straightforward and involves many considerations. It is a journey that we need to be careful of but a journey we must begin. We also need to be aware that the large bulk of petroleum-based use is not as materials but rather as fuel.

There are some concerns that need to be considered in the production of products from green materials. A major emphasis in the production of materials from synthetic polymers is that they are inert, not offering unwanted biological activity and decomposition. The idea that green materials should be naturally recyclable runs against this concept. Another concern involves the actual net energy and resources necessary to produce the monomers/polymer/product. An analysis needs to be made concerning the various components such as energy and resource (including water) requirements before a green material is accepted as a replacement. Another consideration is the ready availability of the particular green material. A material may be green but if it is not readily available or can be seen as readily available in sufficient amount then it may not be a strong candidate to be employed on a large-scale commercial basis. Also, the behavior or properties of the green material should as least approach the materials that may be replaced.

As our bioengineering revolution matures it should be possible to employ various biomasses as resources that can be converted into classical monomers such as ethylene, propylene, styrene, tere- phthalic acid, and so on from reactions caused by genetically designed microbes. In fact, it may be possible for combinations of microbes to produce finished polymers that may have better tacticity and chain orientation producing polymers with superior physical properties.

The latest major emphasis on a major green material is PLA discussed in the previous section. The use of rayon and other material derived from cellulose has also been discussed.

There are many vegetable oils that are currently available on an industrial scale. These include palm, soybean, cotton seed, castor, and rapeseed oils. While these have been employed in the pro- duction of many commercial materials such as coatings, pharmaceuticals, plasticizers, and building materials, the particular fatty acids are being used as polymeric materials. Three of the five most common fatty acid substituents, oleic acid, linoleic acid, and linolenic acid, are unsaturated with these sites of unsaturation available for cross-linking. These are pictured in Figure 19.11. These fatty acids have been directly polymerized through the double bond(s).

Castor oil produces ricinoleic acid on its hydrolysis. Ricinoleic acid has a single olefinic site as well as both an acid and alcohol group (Figure 19.11). Thus, it can be polymerized by reaction through its double bond or by condensation, as shown by Carraher and workers, with a Lewis acid. Polyesters have also been produced through formation of the lactone followed by ROP.

One area of activity involves the production of polyurethanes employing fatty acids as well as the oils themselves. Often the oil is functionalized with hydroxyl groups through reaction at the unsaturated sites. The reaction is analogous to that employed to produce a number of soft-hard block polyurethanes.

Carbon dioxide is readily available and renewable. It is employed in the production of PC but can also be utilized in the production of various polymers through reaction with different heterocycles such as aziridines, episulfides, and epoxides. The production of PP carbonate (PPC) (19.52) has been known for about 40 years but it has yet to be commercially produced.

$$(19.52)$$

There are a number of biodegradable polyesters, including PGA (19.53), PLA, poly-3-hydroxybu- tyrate; PHB (19.54), and polycaprolactone (PCL; 19.55). PGA, PLA, and PCL are synthesized from the acid-catalyzed ROP of the internal ester. PHB is made from microorganisms such as *Alcaligenes eutro- phus* and *Bacillus megaterium* from natural materials such as starch and glucose. PHB has properties similar to those of PP. It is stiff and brittle, has a relatively high melting point of about 180°C, and unlike PP it is biodegradable. However, PP is much tougher. PHB is commercially available from Metabolix. It is an example of a larger grouping of synthetic polyesters called *polyhydroxyalkanoates* (PHAs).

PGA (19.53)   PHB (19.54)   PCL (19.55)

This theme of employing bioengineers, namely bacteria, to produce polymers is illustrated in the syn- thesis of a group of more than 150 linear polyesters called, PHAs (19.56). More than 150 species of both gram-negative and gram-positive bacteria have been found to produce a variety of PHAs. For instance,

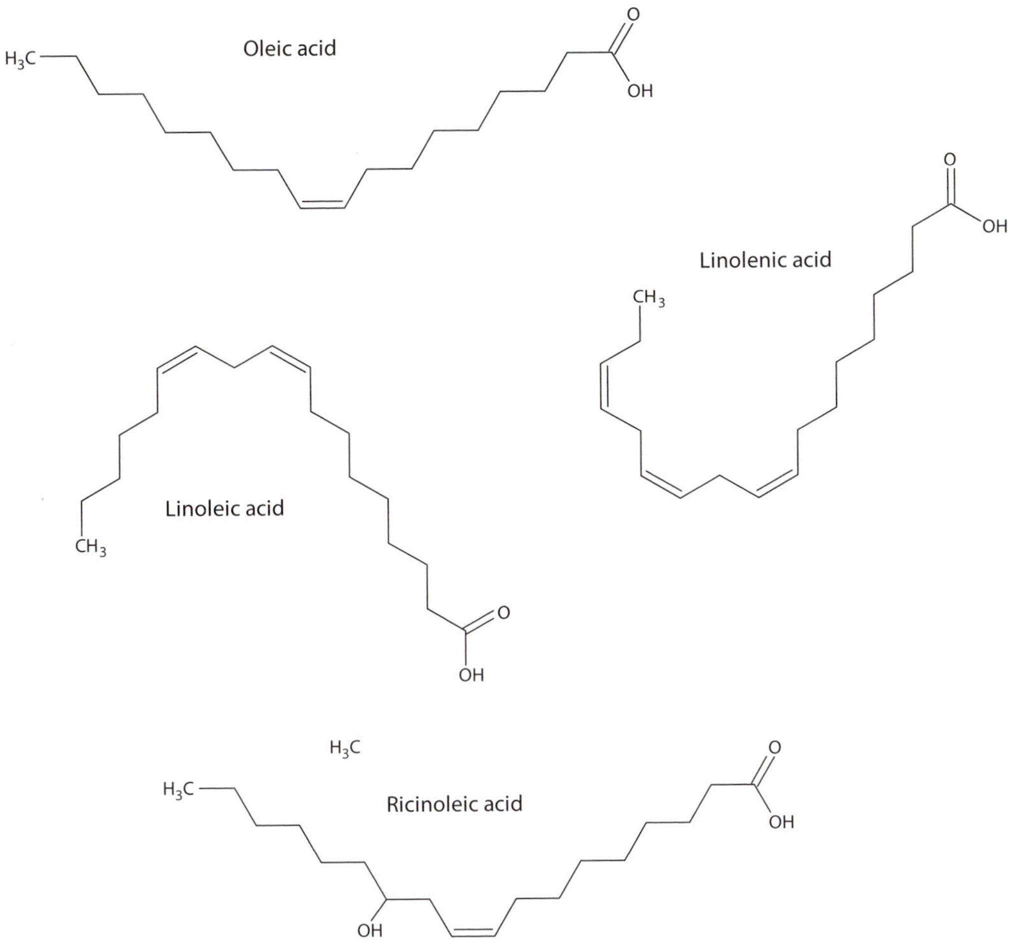

FIGURE 19.11 Vegetable oil derived fatty acids.

Alcaligenes eutrophus is placed in a medium that allows it to rapidly reproduce. After a sufficient population is obtained, the conditions are changed resulting in the production of a range of PHBs with general structures as shown below. *A. eutrophus* typically produces so-called short chain PHAs where $n = 3-5$. By comparison, *Pseudomonas oleovorans* produce longer chained PHAs, where $n = 6-14$.

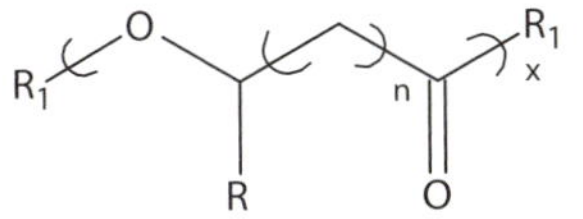

(19.56)

PHA General structure

PHB has $n = 1$ and R = methyl. PHB is biocompatible and is a metabolite normally present in blood. Thus, a wide range of biomaterial uses can be envisioned for it. The copolymer of PHB and PHV (where R = ethyl and $n = 1$), poly(-3-hydroxybutyrate-co-3-hydroxyvalerate) is being used as a packaging material. Compared to PHB, the copolymer is tougher, less rigid, and easier to process.

The general properties can be predicted knowing the values of R and n. As expected, the smaller the value of n the stiffer the backbone and the larger the R the greater the flexibility because the ability of the chains to coalesce is less because of the greater steric requirements. Values of "x" (19.56) generally range from 100 to more than 30,000.

Polyhydroxyalkanoates are viewed by microorganisms as an attainable energy source so are biodegradable. The end products under aerobic conditions are carbon dioxide and water, while methane is produced in anaerobic environments. They can also degrade under simple physical conditions with the ester moiety attacked by water, acid, and base.

Both nylon 9 and nylon 11 can be synthesized from biomaterials. Nylon 11 (19.57) is commercially available and synthesized from monomers produced from vegetable oils, particularly from the castor plant.

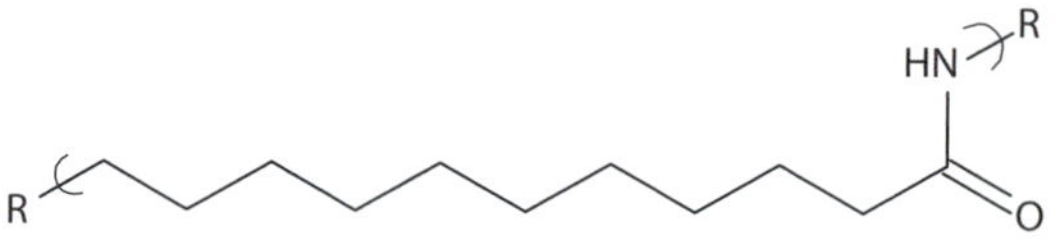

Nylon 11

$$(19.57)$$

DuPont has developed several polymers derived from monomers created from renewable resources. Sorona, 3GT, is a thermoplastic polyester fiber based on the reaction between PDO, where the PDO is created from corn, and terephthalic acid creating the PTT, 3GT (19.58). The corn is converted to sugar and the sugar fermented in the presence of certain bacteria that convert it to PDO.

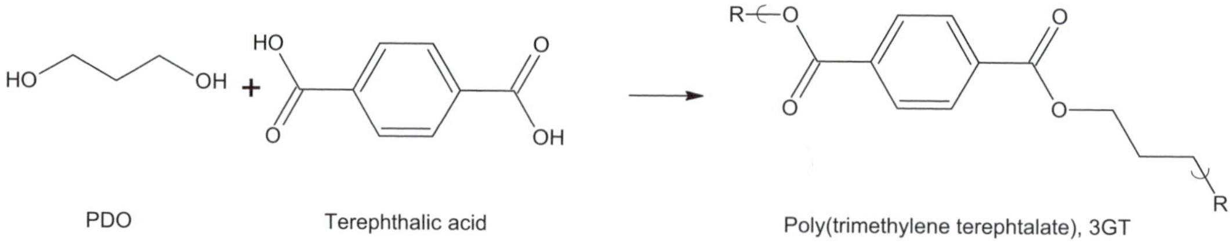

PDO   Terephthalic acid   Poly(trimethylene terephtalate), 3GT

$$(19.58)$$

Genomatica is producing commercial quantities of the similar diol 1,4-butanediol (BDO) using microbes. They continue to work on developing microbes that will produce BDO in larger yields and from more diverse natural feedstocks.

Hytrel is a thermoplastic polyester and elastomer (TPC-ET) containing 20%–60% polyol, Cerenol, derived from corn. Cerenol is synthesized from the self-condensation of PDO with itself and includes a variety of polyols of different chain lengths with molecular weights from about 500–2,000. Below, 19.59 shows the formation of a trimethylene ether glycol from reaction with three molecules of PDO. This diol is then reacted in the usual fashion giving polyesters with the usual hard (terephthalic derived moiety) and soft (polyol derived moiety) segments.

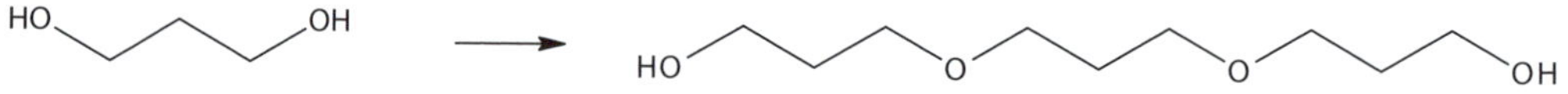

PDO   Trimethylene ether glycol

$$(19.59)$$

There are a variety of blends that are using various biomaterials, including celluloses and proteins that are also being investigated and employed.

Succinic acid is an intermediary in the energy-producing Krebs cycle so its production by microbes has been investigated. After several years of effort, several international groups are producing it on a commercial basis. It can be used in polymer synthesis as well as in the production of a variety of plasticizers.

Recently, some effort has focused on the synthesis of monomers employing enzymes. Currently, the focus is on investigating enzymes that can carry out specific steps to convert various carbon-containing materials into other carbon-containing materials with the aim at developing a series of enzymes that will convert natural-occurring feedstocks into the desired monomers. Eventually, enzymes will be modified that will allow several synthetic steps to be carried out employing a single enzyme. One of the initial efforts is the production of BioIsoprene™. BioIsoprene was developed by Genencor who is working with Goodyear Tire & Rubber Co. to produce the monomer isoprene for synthetic rubber tire product, as well as potential use in a number of other areas, including golf balls, adhesives, and surgical gloves.

Thus, there exist a number of possible green polymeric materials that are possible for commercialization. More are being developed.

19.17 NEW MATERIALS—ADDITIONAL ASPECTS

In considering new materials there are concerns that are not often obvious. Sustainability and availability are important. Are the starting materials available and will they remain available.

Another issue is health. As we are seeing with PC, we often are not aware of the particular health ramifications of the components of a particular polymer and their degradation products. It must be emphasized that synthetic polymers have not been found to offer health concerns. But, some monomers are known to offer health concerns. If the polymer does not offer health concerns, then why is there a concern over PCs? The answer is that polymers are not entirely only polymer. Most polymeric materials contain other materials, including intended additives and some unreacted monomer. It is the presence of small, often parts per billion, amounts of unreacted BPA monomer that is of concern because some studies have linked low-level exposure to BPA (19.60) with increased rates of breast and prostate cancers, early onset of puberty in girls, and reproductive problems.

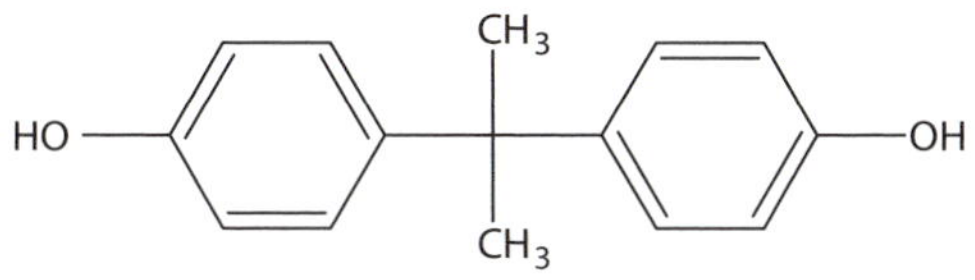

(19.60)

After the fact, it is interesting to note the structural similarity of BPA with some other synthetic hormones some of which are known to offer health problems. One of these is diethylstilbestrol (4,4′[(1E)-1,2-ethenediyl]bisphenol) (DES) (19.61). DES is a synthetic estrogen that mimics estrogen, one of the primary ovarian hormones.

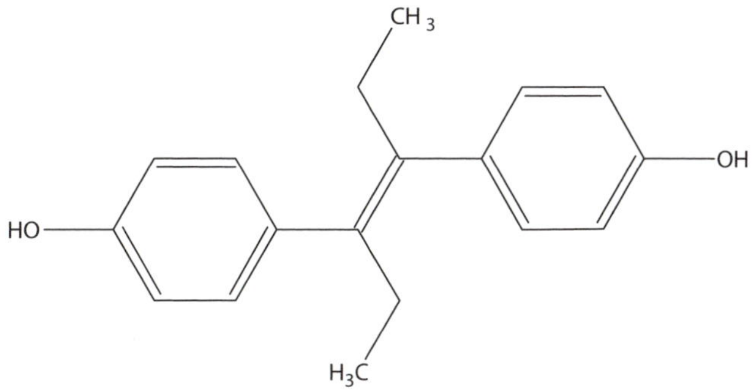

DES

(19.61)

Diethylstilbestrol (4,4′[(1E)-1,2-ethenediyl]bisphenol) was first used in 1938 for women in an effort to prevent miscarriage or premature deliveries. In 1953, a double-blind study showed that DES did little to improve premature deliveries or miscarriage. Even so, it was still widely marketed until the early 1970s for this use. By 1971, it was estimated that 5–10 million people were exposed to DES. In 1971, the Food and Drug Administration issued a Drug Bulletin advising physicians to halt prescribing DES. DES was linked to a rare vaginal cancer in female offspring. Further research has shown that DES is a teratogen that can cause malformation of an embryo or fetus.

As with many chemicals, there is a counter to the health problems. DES is currently used with animals. Its primary use is to treat urinary incontinence in spayed female cats and dogs. It has also been used to prevent unwanted pregnancy in dogs and cats. DES has been used to treat breast and prostate cancer in humans. DES is effective against estrogen receptor positive (ER+) tumors. The use of estrogens as potent antiandrogens in hormonal therapy of metastatic prostate cancer has also been described. Thus, there exist several studies that indicate the potential usefulness of DES as a positive drug in the treatment of specific cancers.

Even with the structural similarities, there was no reason to believe that BPA itself was a health concern since many other compounds have similar structures but are currently not suspected to offer health concerns. (Many other compounds will probably be found to offer health concerns when more intently investigated so simply because a compound is not presently known to offer health issues is no guarantee that future investigations will continue to give the compound a pass on health issues.)

Degradation products can also be a concern. For instance, poly(vinyl chloride) degrades when burnt giving hydrogen chloride and a number of fused-ring hydrocarbons that may be cancer causing. Under normal use conditions this is not a concern but when PVC is incinerated this is of concern and additional precautions should be taken.

Let us now look at some additional aspects with respect to BPA. Because of the heat resistance, superior optical properties, and toughness PCs are widely used in the manufacture of automotive parts, sports safety equipment, and CDs and DVDs. The exposure levels from there materials are currently not believed to be of concern. But, because of the properties of PCs, they are also used in the manufacture of a number of products where there is a concern. PCs are used to make baby bottles, food-storage containers, reusable water bottles, and table wear. Further, BPA is also employed in the manufacture of epoxy resins often applied to the insiders of food and beverage cans forming a protective barrier between the metal and the food product.

Several approaches to the problem of BPA presence are being taken. First, reduce the amount of BPA in the PCs. This has been partially successful but the amount of monomer is not zero. Second, is to find a replacement for bisphenyl A. Much effort had gone into this. There are a number of hurdles that must be overcome for the replacement. One hurdle is that the replacement should be nontoxic. Another aspect is it should be available through some inexpensive synthetic route in large amount. Another requirement is that the replacement should not significantly reduce the physical properties of the resultant PC. This requires a structure that balances the stiffness supplied by the phenylenes with the flexibility supplied by the propylene unit. One such monomer that is being investigated is tetramethylcyclobutanediol (19.50). The ring system contributes the stiffness and the methylenes supply the flexibility.

Third, find a replacement for PCs. A replacement material needs to supply the properties that allow PCs to be successful in the marketplace. One of the critical properties is transparency. PCs are one of the few materials that show crystal-clear transparency. While other materials appear transparent, most of these are not sufficiently clear. Thus, PP and polyethylene can be made to be clear but in comparison, they appear cloudy. Crystal-clear transparency allows many applications such as detecting whether a solution contained within a PC container is itself really clear. Even so, there are a number of materials that can be made clear. These include styrene–acrylonitrile copolymers, poly(methyl methacrylate) (acrylics), and PS. But, in comparison to PCs, they are not tough, another essential property if the material is to replace PCs. This lack of adequate toughness excludes these materials from such applications as high-impact sports equipment, including drinking bottles. Another important property is a high T_g so that they can be repeatedly heated in hot water for sterilization without distortion. Stereoregular PP and linear PE can be heated in boiling water for sterilization purposes, but there is some distortion. Thus, while the relatively low T_g for PP and PE is necessary for them to be flexible, it does not contribute to the lack of distortion necessary for many PC products. Some polymers such as polyimides and polysulfones have the necessary high T_g but they are more expensive than PC.

Several substitute materials are coming onto the marketplace that have similar properties to PC. Tritan, a copolyester made by Eastman, is one such material. While the precise structure of Tritan is not currently public, it is based on tetramethylcyclobutanediol and a derivative of cyclobutanediol. Tritan has good heat tolerance and resistance to hydrolysis and chemical decomposition in comparison to PC. Another material is based on adjusting the ratio of tetramethylcyclobutanediol and DPO in dimethyl terephthalate-based copolyesters. The material is touted as being green since it can use bacterially produced PDO and because no solvent is needed for their synthesis creating less waste.

19.18 SUMMARY

1. Polymers are major materials in the nanotechnology revolution, including as conductive (photo and electronic) materials. Delocalization of electrons throughout a polymer chain or matrix is important for electronic conductance. This is often accomplished through doping that encourages flow of electrons. Doping provides a ready mechanism for delocalization to occur. Doping is the mode of creating electron sinks or deficiencies and electron excesses that are necessary to breach, or get to, the conductive band.

2. Polymers are also major materials in the biomedical areas as materials and in the delivery of drugs. Polymeric drugs can act as control agents to deliver specific biologically active agents or can act as drugs themselves. Polymeric drugs have advantages over smaller drugs because of their size, which have begun to be taken advantage of in the medical area.

3. Polymers are used extensively in dentistry as composites, fillings, dental bases, teeth, cements, and as adhesives.

4. Sutures are the largest group of devices implanted into humans. Sutures are employed to hold together parts of the body generally through the use of fibers. There is a wide variety of materials available today, each with known advantages and limitations and essentially all are polymeric.

5. Smart materials are materials that react to applied force—electrical, stress–strain (including pressure), thermal, light, and magnetic. A smart material is not smart simply because it responds to external stimuli, but it becomes smart when the interaction is used to achieve a defined engineering or scientific goal.

6. Engineering plastics are also referred to as high-performance thermoplastics or advanced thermoplastics. An engineering plastic is simply one that can be cut, sawed, drilled, or similarly worked with. Many of those today must have a relatively high use temperature generally in excess of 200°C. Many of these polymers are being utilized as light-weight replacements for metal because of their strength, high-dimensional stability and resistance to chemicals and weathering. Illustrations of high-performance thermoplastics are a number of nylons and polyesters, PC, polyimides, PPO, and polysulfones.

7. The use of polymeric materials as basic structural materials is widespread and of ancient origin. These materials include concrete, wood, glass, and a wide variety of plastics and elastomers. In fact, with the exception of steel, most of a house is polymeric.

8. Polymers form the basis for fire resistant textiles. For instance, many of the firefighters and race car drivers wear clothing made from aromatic nylons because these materials resist melting, dripping, supporting combustion in air, or burning.

9. For new polymeric materials to enter the marketplace, many things must be in place, including a "flagship" property that meets a particular market need, ready availability, and money. It takes about $1 billion to introduce and establish a new polymer into the marketplace.

10. There is an increased awareness and much activity involving the development of green materials and green practices. This awareness involves public awareness, increased petroleum cost, recognition that the petroleum reserves are limited, and increased ability to employ microorganisms to

synthesize needed materials. This awareness also involves realizing that we need to take care of our environment with us being responsible care takers of the world in which we reside.

GLOSSARY

Anaerobic adhesives: Adhesives that cure or set in the absence of oxygen.

Biological compatibility: No negative biological interactions.

Bioerodability: Material degradation caused by biological means.

Cisplatin (*cis*-diaminedichloroplatinum II): Most widely used anticancer drug.

Doping: Addition of materials that create electron sinks or deficiencies and electron excesses that are necessary to create electrically conductive polymers.

Engineering plastics: Also referred to as high-performance thermoplastics or advanced thermoplastics. An engineering plastic is simply one that can be cut, sawed, drilled, or similarly worked with.

Excimer: Similar to an exciplex except the complex is formed between like molecules.

Exciplex: An excited state complex formed between two different kinds of molecules, one that is excited and the other that is in its grown state.

Green materials: Includes materials that do not have a negative impact on the environment; biomass-derived materials.

Hydrogels: Cross-linked, hydrophilic polymer networks that allow smaller drugs access to their interior and that can be designed to inflate, swell at the desired site, to deliver a drug.

Limiting oxygen index (LOI): Minimum oxygen level where burning is sustained.

Magnetostrictive material: Materials that change their dimension when exposed to a magnetic field.

Nanowires: Oligomeric molecules that contain an electrical conducting core with chemically functional groups on each end.

Nonlinear optics (NLO): Involves the interaction of light with materials resulting in a change in the frequency, phase, or other characteristics of the light.

Photochemistry: Area of study involving the interaction of electromagnetic energy that results in chemical reactions.

Photoconductive: Material that is conductive when exposed to light.

Photophysics: Area of study involving the absorption, transfer, movement, and emission of electromagnetic, light, energy without chemical reactions.

Photoresponsive: Material whose properties change when exposed to light.

Piezoelectric material: Materials that emit a current when pressure is applied. The pressure may be applied by simply exposing the material to an electromotive force.

Shape memory alloys: Materials that change shape and/or volume as they undergo phase changes.

Smart materials: Materials that react to applied force—electrical, stress–strain (including pressure), thermal, light, and magnetic.

Transdermal: Across the skin.

EXERCISES

1. Compare and contrast cancer and healthy cells.
2. Describe a smart material.
3. What are properties that make aromatic fibers good material for fire resistant materials?
4. Which of the following would you expect might become electrically conductive if doped and why? Polyethylene, poly(vinyl chloride), PS, poly-*p*-phenylene, and PPV.
5. What advantageous might conductive carbon nanotubes have over polyacetylene?

6. What disadvantageous do conductive carbon nanotubes have in comparison with some of the newly developed conductive polymers?
7. Describe briefly how doping works.
8. Give one advantage polymeric drugs might have over small molecule drugs.
9. Polymer hearts have been devised and tested for some time. What are several reasons why this area of research may not produce widespread use of synthetic hearts?
10. Most suture material is colorless. Why is color added?
11. What are some important consideration in developing a biomaterial?
12. What are denture bases made of?
13. You cut yourself using a chain saw. Which of the following sutures might be used and why? Cutgut, cotton thread, or nylon filament.
14. Which of the following polymers might be water soluble? Polyethylene, nylon-66, poly(ethylene oxide), copolymer containing about 50% ethylene and about 50% acrylic acid units, and poly(vinyl alcohol), and PET.
15. Which of the following is not an emerging thermoplastic? Nylon-66, PET, PPO, or PP.
16. Since PVC has a much higher LOI than PC, why is it not employed to make fire retardant fabric?
17. Would you predict poly(ethylene oxide), $-(CH_2-CH_2-O-)-$, to be water soluble?
18. Why is PFTE not employed as a general high bulk used in construction material?
19. Why is there such a wide variety of polymers employed in biomedical applications?
20. Why is there a number of fluorine-containing polymers employed in industry and in biomedicine?
21. Why is there an industrial move toward so-called green material?
22. Why is there such a broad definition as to what green materials mean?
23. Name three polymeric green materials.

ADDITIONAL READING

Aldissi, M. (1992): *Intrinsically Conducting Polymers*, Kluwer, Dordrecht.

Bachmann, K. J. (1995): *Materials Science of Microelectronics*, VCH, NY.

Barford, W. (2009): *Electronic and Optical Properties of Conjugated Polymers*, Oxford University, Ithaca, NY.

Blythe, T., Bloor, D. (2005): *Electrical Properties of Polymers*, Cambridge University Press, Cambridge, UK.

Boyd, R. (2008): *Nonlinear Optics*, Elsevier, NY.

Bundy, K. (2007): *Fundamentals and Biomaterials: Science and Applications*, Springer, NY.

Celina, M., Assink, R. (2007): *Polymer Durability and Radiation Effects*, Oxford University, Ithaca, NY.

Chanda, M., Roy, S. (2008): *Plastics Fundamentals, Properties, and Testing*, CRC, Boca Raton, FL.

Chiellini, E., Gil, H., Braunegg, G., Buchert, J., Gatenholm, P., vander Aee, M. (2001): *Biorelated Polymers*, Kluwer, NY.

Craig, R. G., Powers, J. M. (2002): *Restorative Dental Materials*, Mosby, St. Louis.

Dadmun, M. (2008): *Modification of Interfaces with Polymers*, Wiley, Hoboken, NJ.

Donald, A. M., Windle, A. H., Hanna, S. (2006): *Liquid Crystalline Polymers*, Cambridge University Press, Cambridge, UK.

Dumitriu, S. (2002): *Polymeric Biomaterials*, 2nd Ed. Dekker, NY.

Dutton, H. (2000): *Understanding Optical Communication*, Prentice Hall, Englewood Cliffs, NJ.

Freund, M. (2007): *Self-Doped Conducting Polymers*, Wiley, Hoboken, NJ.

Galaev, I., Mattiasson, B. (2007): *Smart Polymers: Applications in Biotechnology and Biomedicine*, CRC, Boca Raton, FL.

Grand, A., Wilkie, C. A. (2000): *Fire Retardancy of Polymeric Materials*, Dekker, NY.

Hadziioannou, G., Malliaras, G. G. (2007): *Semiconducting Polymers: Chemistry, Physics and Engineering*, Wiley, Hoboken.

Inzelt, G. (2008): *Conducting Polymers*, Springer, NY.

Jenkins, M. (2007): *Biomedical Polymers*, CRC, Boca Raton, FL.

Jimenez, A., Zaikov, G. (2008): *Recent Advances in Research on Biodegradable Polymers and Sustainable Composites*, Nova, Hauppauge, NY.

Kausch, H. (2006): *Radiation Effects on Polymers for Biological Use*, Springer, NY.

King, E. (2008): *Shape-Memory Polymers: Industrial Applications of Smart Materials*, Wiley, Hoboken, NJ.

Lee, K. S. (2006): *Polymers for Photonics Applications*, Springer, NY.

Lippert, T. K. (2006): *Polymers and Light*, Springer, NY.

MacDiarmid, A. G. (2001): *Rev. Mod. Phys.*, 73, 701 (Nobel lecture).

Mahapatro, A. (2006): *Polymers for Biomedical Applications*, Oxford University Press, Oxford, UK.

Marder, S., Lee, K. (2008): *Photoresponsive Polymers*, Springer, NY.

Martin, J., Dickie, R., Ryntz, R., Chin, J. (2008): *Service Life Prediction of Polymeric Materials: Global Perspectives*, Springer, NY.

Moliton, A. (2005): *Optoelectronics of Molecules and Polymers*, Springer, NY.

Morgan, A. B., Wilkie, C. A. (2007): *Flame Retardant Polymer Nanocomposites*, Wiley, Hoboken.

Nelson, G., Wilkie, C. A. (2001): *Fire and Polymers*, Oxford University Press, NY.

Piringer, O., Baner, A. (2008): *Plastic Packaging Materials: for Food and Pharmaceuticals*, Wiley, Hoboken, NJ.

Prasad, P. N., Williams, D. J. (1991): *Nonlinear Optical Effects in Molecules and Polymers*, Wiley, NY.

Ravve, A. (2006): *Light-Associated Reactions of Synthetic Polymers*, Sprionger, NY.

Rockett, A. (2008): *The Materials Science of Semiconductors*, Springer, NY.

Roth, S., Carroll, D. (2006): *One-Dimensional Metals: Conjugated Polymers, Organic Crystals, Carbon Nanotubes*, Wiley, Hoboken.

Satchi-Fainaro, R., Ducan, R. (2006): *Polymer Therapeutics*, Springer, NY.

Scharf, T. (2007): *Polarized Light in Liquid Crystals and Polymers*, Wiley, Hoboken.

Schnabel, W. (2007): *Polymers and Light: Fundamentals and Technical Applications*, Wiley, Hoboken, NJ.

Shibaev, V., Lam, L. (1993): *Liquid Crystalline and Mesomorphic Polymers*, Springer-Verlag, NY.

Svenson, S. (2006): *Polymeric Drug delivery I and II*, Oxford University Press, NY.

Uchegbu, I., Schatzlein, A. G. (2006): *Polymers in Drug Delivery*, Taylor and Francis, Boca Raton, FL. Wallace, G., Spinks, G. (1996): *Conductive Electroactive Polymers*, Technomic, Lancaster, PA.

Wallace, G., Spinks, G., Kane-Maguire, L., Teasdale, P. (2003): *Conductive Electroactive Polymers: Intelligent Materials Systems*, CRC Press, Boca Raton, FL.

Werner, C. (2006): *Polymers for Regenerative Medicine*, Springer, NY.

Yui, N., Mrsny, R., Park, K. (2004): *Reflexive Polymers and Hydrogels: Understanding and Designing Fast Response Polymeric Systems*, CRC Press, Boca Raton, FL.

20 Solutions

CHAPTER 1

1. Paper, wood, coatings, window, meat, bread, hair, and toothbrush.
2. Because polymers are important items in today's society and are all about us present in essentially all of our daily activities from food to sleeping, to clothing, and so on.
3. Because polymer science is relatively new, so many of the major contributors are still alive.
4. All are mainly polymeric except for water.
5. Clays, bricks, sand, glass, mica, and diamonds.
6. Polyethylene, polypropylene, polystyrene, polyesters, nycons, and so on.
7. a. low-density polyethylene, b. polystyrene, and c. polyepropylene.
8. It takes time to optimize conditions of synthesis, scale up producing the material for commercial usage, further characterization of the material, developing a market, and so on.
9. Lighter, do not rust, less expensive, and so on.
10. Much of what we wear, eat, and handle are polymeric. We are largely water and polymers with our skin and hair being proteins and the driving force for our living include nucleic acids and proteins.
11. There are different reasons such as the presence of a trained workforce and the proximity of the raw materials, favorable local tax structure, and so on.
12. Many items, such as sneakers and tires, are a complex of materials rather being made from one or two materials. Thus, simple codes would not be able to appropriately describe the composition of such complex products.
13. By becoming better informed; taking an active role in being involved in the many possible ways, including informed recycling and influencing legislation.

CHAPTER 2

1. –MMMMMMMMMMMMMM–,

–CCCCCCCCCCCCCC–,
| | | | | | |
X X X X X X X

–CCCCCCCCC–
| |
M M
M M

(a)

(b)

(c)

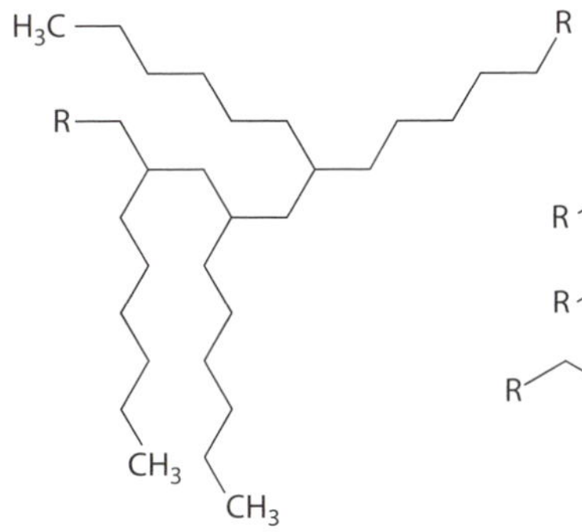

(d)

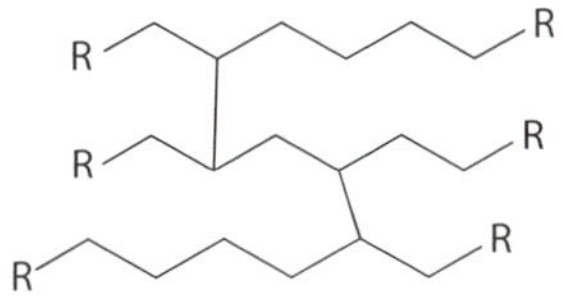

(e)

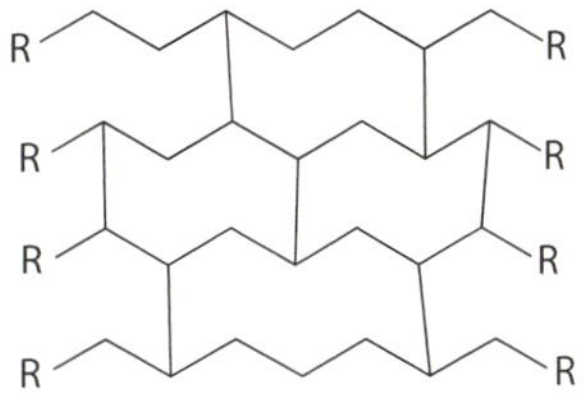

(f)

2. (a) LDPE and (b) LDPE.

3. (a) about 109.5; (b) about 109.5; zigzag chains characteristic of alkanes.

4. Contour length are both about the same since the backbone for each is composed entirely of carbon atoms. Given a C–C bond length of 0.126 nm this means the effective length for each unit is 2 × 0.126 nm = 0.252 nm. Thus, the contour length is 0.252 nm times 2,000 units = 504 nm.

5. (d), (g), and (i).

6. 1,000.

7. (a) $-CH_2-CH(CH_3)-$, (b) $-CH_2-CHCl-$, and (c) $-CH_2-CH(CH_3)=CH-CH_2-$

8. (c).

9. (d), (e), (f), and (g).

10. (a).

11. (a).

12. (a).

13. (a).

14. (a) $-CH_2-CH(OH)-CH_2-CH(OH)-$, (b) $-CH_2-CH(OH)-CH(OH)-CH_2-$

15.

(a) Syndiotactic–polypropylene or simply sPP.

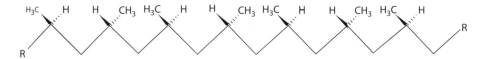

(b) Isotactic–polypropylene or simply iPP.

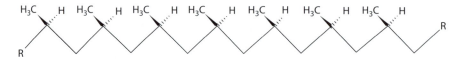

16. See Figure 2.7; simply extend the end methyl groups making them methylene groups.

17. (a) HDPE, LDPE, hevea rubber, and so on, (b) PVC, and so on, (c) nylon-66, cellulose, silk, and so on.

18. (b).

19. (a).

20. (a).

21. (b).

22. Low or no flow; slow cooling rate; linear polymers.

23. 378 nm.

24. (a).

25. (a) because of a more regular structure.

26. (b).

27. Intramolecular hydrogen bonds.

28. (a).

29. Being transparent depends of having a homogeneous structure so (a) is the least homogeneous and thus has varying refractive indexes causing it to appear hazy.

30. (a).

31. (a) and (c).

32. (a).

33. (b).

34. Vacant space and sufficient energy.

35. Chemical (formation of primary bonds) and physical (chain entanglement and crystalline formation).

CHAPTER 3

1. (b), (c), (d), and e.
2. 50,000.
3. $\overline{M}_n = 1.57 \times 10^6$; $\overline{M}_w = 1.62 \times 10^6$.
4. (a) 1 and (b) 2.
5. $\overline{M}_n < \overline{M}_v < \overline{M}_w < \overline{M}_z$.
6. (b), (c), and (d).
7. LVN = KM^a.
8. GPC, ultracentrifugation.
9. (d), (e), and (f).
10. Weight-average molecular weight.
11. 0.5.
12. 2 (end groups).
13. 2.0.
14. 2.25×10^8.
15. (c).
16. A type of double extrapolation for the determination of the weight-average molecular weight for high polymers in which both the concentration and the angle of the incident beam are extrapolated to zero.
17. Number-average molecular weight.
18. Because of higher vapor pressure, the solvent evaporates faster from a pure solvent than from a solution; an application of Raoult's law.
19. The melt viscosity is proportional to the molecular weight to the 3.4 power.
20. The extremely high molecular weight polymer is much tougher.
21. For a monodisperse system.
22. Entropy.
23. Many, such as membrane osmometry, end-group analysis. Number-average molecular weight.
24. Inexpensive equipment and relatively fast.
25. (b).
26. Yes.
27. All of them.
28. B is a constant related to the interaction of the solvent and polymer.
29. One in which London forces are the predominant intermolecular attractions.
30. (b).
31. Cohesive energy density is equal to the strength of the intermolecular forces between molecules, which is equal to the molar energy of vaporization per unit volume.
32. (b).
33. Higher.
34. $\Delta S = (\Delta H - \Delta G/T)$ at constant T.
35. A good solvent.
36. 0.
37. Theta temperature.
38. (b).
39. 548 cal/cc.
40. 0.
41. 1,000,000 g.

42. Wetting or polymer coming into contact with the solvent, invasion of the polymer matrix by the solvent (swelling) and finally going into solution.
43. Because there is a relationship between chain length and property.
44. The contribution of the polar group becomes less significant as the alkyl portion increases.
45. The answer depends on what "better" means. From a toxicity standpoint, it is probably *n*-pentane since it is the least toxic. From a solubility standpoint, it is b.
46. a. Benzene is a better solvent, therefore the value of "*a*" is greater.
47. About 0.5–0.8
48. When $\alpha = 1$.
49. Root mean square end-to-end distance.
50. As shown in the Arrhenius equation, log viscosity = log A + $E/2.3RT$, viscosity is inversely related to *T*.
51. No, it can be used with both synthetic and natural polymers.
52. Ethanol because it has the OH group that can hydrogen bond to the water and because it is the only polar molecule.
53. It is an effort to try to find solvents that will minimize the enthalpy term.
54. MALDI MS requires that the polymer be soluble is a volatile solvent such as water and most polymers are not soluble in such solvents.

CHAPTER 4

1. (b), (c), and (e).
2. No. The polymer chains would probably be too flexible.
3. (a). It is less flexible.
4. 100,000.
5. Log $k = (-E_a/2.3RT) + \log A$. Therefore, the value of log k increases as the first term in the equation decreases because of an increased temperature.
6. The dimer.
7. A trimer or a tetramer, since two dimer molecules may react or a dimer may react with a molecule of the reactant.
8. Stretch the filament permitting formation of more hydrogen bonds.
9. Poly(tetramethylene adipamide) or nylon 46.
10. (a). A stable six-membered ring could also be formed.
11. (a).
12. 1 h.
13. (b).
14. 99.
15. It would be a stiff linear high molecular weight polymer with lots of hydrogen bonds.
16. Phenolic and amino plastics, Bakelite.
17. $1/1 - p$, where p = extent of reaction. Allows $\overline{DP}$ to be predicted as the extent of reaction varies.
18. Poly(methylene oxide). No, it must be capped.
19. Polyurethane. Coatings, elastomers, foams, sometimes fibers, and plastics.
20. (a). It would have a more stable geometry.
21. It would be weak at the angle perpendicular to the stretch.
22. (b).
23. (a). There would be more surface available for attack.
24. Phosgene and a diol such as bisphenol A.
25. Increase the number of methylene groups like nylon-9,9.
26. Reduce the number of methylene groups so that it would absorb more moisture.

27. (a). The amide is a stiffening group.
28. Because of the presence of bulky pendant groups.
29. Bakelite is a thermal set material and not moldable once it is formed. Thus prepolymer is used to make molded products.
30. Not particularly, since the polyamine reacts producing a polyurea. However, other propellants are used commercially in foam production.
31. Use an excess of the glycol.
32. $\overline{DP}_n = 1/1 - p$.
33. DP steadily increases.
34. Because of the presence of strong polar groups.
35. The precursor, furfural, is produced from waste corn cobs.
36. Some group that can be cross-linked generally a carbon–carbon double bond.
37. No: resorcinol is trifunctional and very reactive.
38. Phenolic resins were developed before there was much information on polymer science. Those involved in an art have a tendency to create new terms to describe what they have observed.
39. Most people prefer to use light-colored dinnerware.
40. All three.
41. (a). (b) would be a branched chain.
42. Microfibers have an extremely small diameter that allows a fabric to be woven that is light weight and strong. Microfibers can be tightly woven so that wind, rain, and cold do not easily penetrate but allow perspiration to pass through them. Microfibers are also very flexible because the small fibers can easily slide back and forth on one another.
43. Polyurethanes are mainly employed as foam and elastomer materials.
44. Open-celled foams allow air to readily pass through them, allowing the material to be more flexible. Closed-celled foams retain the air trapped within them, resulting in retention of heat/cold so they are good insulators.
45. LCs are composed of materials that reorient themselves on application of heat, pressure, or some other external change.
46. LCs are used in watches, computer screens, and in TV screens.
47. PEGs are water soluble and impart water solubility when attached to otherwise water-insoluble drugs. Further, PEG is generally biologically benign.
48. Aramides is the name given to polyamides or nylons that have aromatic groups, generally phenylenes, in their backbone.
49. Stiff, strong, and high melting.

CHAPTER 5

1. (a) A small amount of reactants, dimer, trimer, and oligomers, plus low molecular weight polymers, (b) high molecular weight polyisobutylene plus monomer.
2. A Lewis base–cocatalyst complex. Actually, the proton is the initiator here.
3. A carbonium ion.
4. A macrocarbonium ion.
5. A gegenion.
6. (a).
7. (b).
8. The macroions have similar charges.
9. (a).
10. (b).
11. (a) HDPE or PP, (b) IR, EP, or butyl rubber (IIR), and (c) PP.
12. (a).

13. (b).
14. (a) and (b).
15. (d), (e), and (f).
16. (b).
17. v = average DP.
18. Polyisobutylene.
19. $R_i = kC[M]$, where C is the catalyst–cocatalyst complex.
20. Increase the rate.
21. No observable change.
22. (a).
23. (c) and (d).
24. A cation.
25. They are equal.
26. Average DP = R_p/R_t.

27.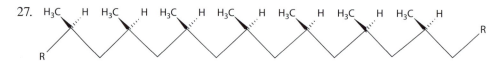

28. Zero percentage.
29. The degree of polymerization is independent of [C].
30. No, they are thermally unstable.
31. Cap the ends by esterification or copolymerize with something like a dioxolane.
32. Because formaldehyde is produced in the decomposition, which is the inverse of the propagation.
33. Narrower molecular weight range, will polymerize more monomer, better stereoregularity, no need for clean up.
34. Monomer is inexpensive. Depending on the reaction conditions, lots of different polyethylenes can be made. Lots of varied uses requiring PE with differing properties and cost.
35. Add a water-soluble linear polymer such as poly(ethylene oxide).
36. Because of its regular structure.
37. Low cost of monomers, monomers available in large amount, useful in many different applications.
38. Polyesters.
39. Convert them to solid polymers by cationic chain polymerization, using sulfonic acid as the initiator.
40. A carbanion.
41. Macrocarbanions.
42. Because they are stable macrocarbanions capable of further polymerization.
43. (c).
44. Nylon-6.
45. –C(O)–(–CH$_2$–)–NH–
46. TiCl$_3$ and (Et)$_2$AlCl.
47. *cis*-1,4; *trans*-1,4; and 1,2.
48. The propagating species in anionic chain polymerization is a butyl-terminated macroanion. Propagation species in complex coordination catalyst polymerization is an active center with an alkyl group form the cocatalyst as the terminal group. Propagation takes place by the insertion of the monomer between the titanium atom and the carbon atom by way of a pi complex.

49. Increased crystallinity, higher T_g, less permeable, higher stress–strain ratio.

50.

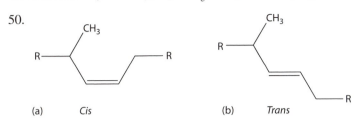

(a)   *Cis*   (b)   *Trans*

51.

(a)   (b)

52. A chromia catalyst supported on silica (Phillips catalyst).
53. *cis*-Polyisoprene.
54. Ethylene–propylene copolymer.
55. Entrenchment of catalysts that must be removed; often broad molecular weight distribution; and, in comparison with soluble stereoregulating catalysts, not as exacting control of tacticity of resulting products.
56. Soluble steroregulating catalysts overcome the problems by having a greater control of the tacticity of the resulting polymers; little or no entrapment of catalysts; and narrower molecular weight distributions.
57. Overall cost including shipping; how strong the bags need to be as well as their ability to withstand punctures; ability to withstand natural elements of water, fertilizer, sunlight, and heat/cold.
58. Mixing the HDPE when melted and rapidly cooling, locking in some amorphous character.
59. Many properties are drastically different depending on the particular form of the material. The one material with the lower elongation may be largely crystalline whereas the material with the high elongation may be an amorphous form. This points out the need that must be taken when comparing values from different tables.
60. Yes. Butadiene polymers typically have C=C within them that can be exploited to form cross-links.
61. While PE is a simple structure, the variations in structure offer a wide range of properties and application opportunities. Also, the ethylene monomer is readily available on a large scale and it is inexpensive.

CHAPTER 6

1. (a) H/:BF$_3$OH; (b) Na/:NH$_2$; and (c) (H$_3$C)$_2$–C(CN)–N/N–C(CN)–(CH$_3$)$_2$
2. Free radical chain polymerization.
3. 12.5%.
4. Coupling.
5. Neither since both will reduce the average DP.
6. LDPE, PVC, PS, and so on.
7. None, provided the system is a homogeneous solution.

8.

(a) (b) (c)

9. Since the rate of termination has decreased, more monomer may be added gradually to produce high molecular weight polymers.
10. Initiation, which is governed by the rate of production of free radicals.
11. (a).
12. Monomeric styrene.
13. Ln 2 = 0.693.
14. Measure the volume of N_2 produced (given off).
15. About 30–50 kcal/mol.
16. One may polymerize at low temperatures and also regulate the rate of production of free radicals.
17. PVC degrades when heated in a furnace giving hydrogen chloride that can damage the metal in the furnace. With respect to the ozone layer, PVC is also a minor source of chlorine. Counter, PVC piping is very important and its use allows the "non-use" of many metals whose production has potentially a greater negative influence on the environment.
18. (a), (b), and (c) = 1×10^{-11} mol/L, assuming monomer is still present.
19. 5 ± 3 kcal/mol.
20. R_p is proportional to $[I]^{1/2}$.
21. The average DP decreases.
22. Steady-state number of polymerizing or growing chains; rate of termination is equal to rate of initiation; steady state for initiator radical.
23. Vinyl acetate is obtained by hydrolysis of poly(vinyl acetate).
24. Advantage—simple equipment; disadvantage—bad heat control, broad molecular weight, may require further purification.
25. Neither; it remains unchanged.
26. Ethylene has no electron-donating groups, and is less polar than isobutylene.
27. Disproportionation.
28. The termination mechanism at higher temperatures is disproportionation.
29. 0–6 kcal/mol.
30. $T_c = 61°C$. Therefore, poly(α-methylstyrene) would decompose when heated above 61°C.
31. Teflon is the trade name and was widely advertized.
32. (b). The bond strength for C–Cl is less.
33. The $H_3C(CH_2)_{11}S\cdot$ produced by chain transfer is an active free radical that initiates new chains.
34. No, the heat from the exothermic reaction might cause an explosion.
35. They are the same.
36. No, but one should avoid contact with the skin and ascertain that there is a minimal amount of monomer, which is done industrially before it is sold, residue in the polymer.
37. Good slippery behavior, good thermal and corrosive resistance, good inertness.
38. The droplets are relatively few in number, and the chances of an oligoradical entering them is small.
39. The primary free radicals are water soluble.
40. The initial polymerization rate would be retarded.
41. Styrofoam packing and cups, bottles, disposable egg cartons.

42. The structure of PTFE is regular. The large pendant group in FEP destroys this regularity.
43. Both the average DP and R_p are proportional to $N/2$, where $N/2$ is the number of active micelles.
44. (b). The additional chlorine on the repeating unit increases the specific gravity.
45. Probably because they cost less and often are easier to synthesize.
46. Would still probably go by a chain-type mechanism but the reaction would be much faster at a given temperature so would probably be carried out at lower temperatures.
47. Probably for a number of reasons, including the great potential for use and profit. In both cases, it is good to understand the system so that you can optimize it and tailor it as needed.
48. An engineering plastic is a plastic that can be cut, sawed, drilled, and so on. Teflon can be cut, sawed, and drilled, but it often undergoes distortion when these operations occur so it is often called a "near" engineering plastic.
49. A plastisol is made by heating finely divided PVC with a plasticizer.
50. A comonomer that can accept dye molecules is added to the polymer.
51. PMMA scratches more easily; about as clear; higher impact strength; less environmentally stable; less resistant to solvents; lighter (less dense).
52. In live animal tests, generally with mice, some plasticizers have shown negative health effects. While the amounts and reactions in humans may not be the same, these results signal a concern that care should be taken and further examination of the use of these specific plasticizers be monitored.

CHAPTER 7

1. Let S = styrene and A = acrylonitrile units.
   (a) –(–S–)– and –(A–)–;
   (b) –SASASASASA–; (c) –AASASSASAAASSASSASASSSSASAA–;
   (d) –AAAAAAAASSSSSSSSSAAAAAAAAASSSS–;
   (e) –SSSSSSSSSSSSSS–

2. Equimolar.
3. $r_1 = k_{11}/k_{12}$ and $r_2 = k_{22}/k_{21}$.
4. No, they are constant.
5. They vary in accordance with the Arrhenius equation.
6. Vinylidene chloride, about 76%; vinyl chloride about 24%.
7. Styrene and so on.
8. (a) cationic, (b) anionic or free radical, and (c) Ziegler–Natta.
9. Alternating.
10. $(M_1)_n(M_2)_m(M_1)_n$
11. $r_1 r_2 = 1$
12. (b).
13. Most elastomeric material is cross-linked and thus are thermoset and not easily processable.
14. (a) 50% of each, (b) 91% S and 9% MMA, and (c) 94% MMA and 6% S.
15. Isobutylene, 71%; isoprene, 29.6%.
16. Isobutylene, 96%.
17. Add a mixture of monomers to maintain the original ratio of reactants.
18. B can also form crystalline segments themselves.
19. Vinyl chloride, 68%.
20. The presence of vinyl acetate mers in the chain breaks up the regularity characteristic of PVC so that the copolymer has a lower T_g, is more soluble in organic solvents, more flexible, and is more readily processed.

21. The methacrylic acid mers present decrease the crystallinity and improve the toughness and adhesive properties. The salts act like cross-links at ordinary temperatures.
22. They are all essentially the same copolymers of butadiene and styrene.
23. Chemically or through having materials where one end is miscible in one phase and the other end miscible in the other phase.
24. (b).
25. For immiscible blends when the phase separated portions are able to operate independently.
26. Depends. Molecular weight wise they are low polymers but some can be described in terms repeat units.
27. An alternating copolymer is rapidly produced until the maleic anhydride is consumed.
28. They are cross-linked, yet can be molded through use of pressure and heat.
29. Allows prediction of conditions to get copolymers with certain composition and thus properties. Also, allows the basis for control of copolymer composition.
30. Graft is with a lyophilic monomer such as styrene.
31. A carboxylic acid group.
32. Ionomers are generally flame resistant. With respect to recycling, while they are thermosets, they can be formed through a combination of heat and pressure so are similar to thermoplastics in this respect.
33. If the functionality is only two then only linear products are formed. Dendrites occur when the structures are somewhat three-dimensional so functionalities greater than two are required for their production.
34. Specialization allows a savings on equipment and accumulation of an in-depth knowledge base. Further, it maximizes research and selling efforts as well as identifying to others that they are "experts" within these areas.
35. The properties of the simple random copolymer will generally be some blending of the properties of the two monomers while the block copolymer may give properties that are characteristic of homopolymers of each individual monomer.
36. Possibly one of the most important properties is memory or rebound that allows a classical elastomer to return to its original shape after deformation.
37. One possible area of use is the observation that dendrites often act to separate units from one another, so catalyst sites might be embedded into the dendrite structure with the various dendrite "wings" acting to keep them apart, allowing them to act independently.

CHAPTER 8

1. For example, PMMA, rubber bands, cellophane, PE, rayon, nylon-66.
2. Lignin.
3. Wood flour.
4. Formica table tops, plywood, fibrous glass-reinforced plastic boats, graphite-reinforced gold shafts, and so on.
5. Treat the surface with a silane or melt it and convert it to a fiber.
6. According to the Einstein equation, the viscosities will be equal.
7. (a).
8. The intermolecular attractions impede motion.
9. T_g increases.
10. (b).
11. Advantages—more fibrous, less discoloration; Disadvantages—higher cost, more difficult to process because of bulk factor.
12. Wood flour below the surface and alpha-cellulose and paper near and at the surface.
13. Inexpensive, biodegradable, not dependent on oil prices.
14. Can cut you and if breathed could cut parts of your body.

15. Convenience, uniformity, speed, reproducibility, less exposure of workers to volatile monomers.
16. No. Before 1910, it was patented by Goodyear and used as a colorant. A compounder made a mistake and used a 100-fold excess and found that it was a good filler. Another accidental discovery.
17. You would expect that on the basis of size, $O_2 > N_2$, that nitrogen would diffuse faster. Diffusion can occur in both directions, into the tire and out of the tire. The tire can be considered as a membrane where there is a drive for the concentration of components on both sides to be equal. Since the concentration of nitrogen in air (78%) is greater than that of oxygen (21%), the driving force for removal of nitrogen from the tire is less. Thus, use of certain nanocomposites such as those that contain the silicates allow for the formation of a barrier that helps prevent movement of gas into and out of the tire.
18. Nonsystematic testing is dangerous but since that is your task you might take several different lengths of the new fiber and place them in some resins and note if they adhere to the resin and to test their general strength. If there are promising results then further testing is called for. Again, remember such unorganized testing may miss combinations that offer truly outstanding performance.
19. Asbestos (health concerns), peanut shells, palm prongs (leaves), and corn husks.
20. In phase separation, the fiber material and resin material are not mixed together on a molecular level, thus you do not get the reinforcing effect of having the strong fiber material dispersed throughout the composite.
21. Since the fiber material in metal composites weights more, the density of these materials is generally greater than for regular organic composites.
22. Metal matrix composites, in comparison to polymer-intense composites, offer no moisture absorption, greater fire resistance, higher use temperatures, greater radiation resistance, greater stiffness and strength, and higher thermal and electrical conductivities.

CHAPTER 9

1. Enzymes are present in the human digestive system for the hydrolysis of the alpha linkages but not for the beta linkages in polysaccharides such as cellulose and starch.
2. Cellobiose is a beta-glucoside, whereas maltose is an alpha-glucoside.
3. Because of the presence of intermolecular hydrogen bonds.
4. It is the same since there is an equilibrium between the alpha and beta conformations of the D-glucose.
5. Cellulose mostly taken from wood pulp with the lignin removed.
6. Three.
7. Secondary cellulose acetate.
8. It forms an amine salt.
9. (a).
10. It is branched, therefore discourages the formation of hydrogen bonding.
11. 0.5.
12. Melanin
13. Increases.
14. Because of the formation of ordered regions.
15. An elastomer is an amorphous polymer with a low T_g and low intermolecular forces that increases in entropy, order when stretched.
16. Guayule grows on arid soil in temperate areas and has a high content of natural rubber.
17. (b). Isoprene is the monomer in the synthetic process, and isopentenyl pyrophosphate is the precursor in natural rubber.
18. That NR is a member of the terpene family.
19. Because of the formation of crystals.
20. Change in entropy, $-T(dS/dl)$.

21. NR. S, accelerator, antioxidant, carbon black, ZnO, and processing aids.
22. It warms up.
23. They often cost more, are contaminated, and are inferior to synthetic products.
24. A polar solvent, since it is a polyol.
25. Much of it is formed between polysaccharide fibers and thus takes on a somewhat flat or two-dimensional structure.
26. (b).
27. At least 60 million tons.
28. Yes, casein is no longer used as a molding resin and, hence the article may have some value as a collector's item.
29. There are a number of problems. Essentially, all of them can be overcome. Problems include public acceptance; fuel delivery; motors that can run on the synthetic fuels that are not so costly in comparison to gas-fueled engines; reliable and inexpensive fuel production; reliable supply of the feedstock; and so on.
30. The impediments to the use of any new material are similar to those described in answer 29, including public acceptance, manufacture costs, reliable feedstock source, and so on.
31. There are not correct or incorrect answers here. Considerations might include searching the literature and, as Carothers did, look at reactions that occur with phenols and reactions on aromatic units for a start and see if any of these might give potentially useful properties.
32. In truth, there is one human race but we choose to differentiate based on a number of characteristics with color being one. Since the major factor in skin color difference is melanin, then it is a factor when color is used to differentiate humans. Since melanin is not associated with any other factor than color, it is important to remember that differentiation based solely on color is only skin deep.
33. Titanium dioxide (titanium (IV) oxide) is the most widely used paint pigment.

CHAPTER 10

1. It is a polyamide.
2. Glycine, because it has no chiral carbon atom.
3. To the positive pole.
4. Collagen has strong intermolecular hydrogen bonds.
5. It should be a linear crystalline polymer with strong intermolecular hydrogen bonds.
6. (b).
7. DNA is usually greater.

8. 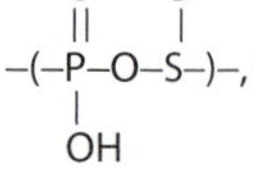
$$-(-P-O-S-)-$$, where B = base, S = deoxyribose sugar.

with O double-bonded to P, B bonded to the second position, and OH below P.

9. (b).
10. Thymine.
11. The presence of bulky pendant groups and hydrogen bonding favors the helical arrangement.
12. TAATGCAGTA.
13. The maximum of 16 dinucleotide combinations is not sufficient to direct the initiation, termination, and insertion of 20 amino acids in a protein chain; 64 trinucleotide combinations is more than ample. Further, 2 codons is more easily mixed up, whereas 3 codons is less easily mixed up.
14. *E. coli* is relatively harmless, much is known about it, it is easily work with, inexpensive, rapidly growing, lots of experience working with it.

15. Acid or carboxylic acid groups. Positive side under most conditions. Polyelectrolyte.
16. Methionione and cystein. Formation of sulfur oxides and eventually acid rain if not collected before it becomes part of our atmospheric air.
17. All amino acids contain the amine or nitrogen-containing functional group. Formation of nitrogen oxides and eventually acid rain if not collected before it becoming part of our atmospheric air.
18. Globular or spherical. To allow the hydrophobic areas to cluster internally and to place the hydrophilic portions to the exterior coming into contact with the water-rich exterior. Composed of amino acids or proteins.
19. Ones that must be obtained from sources other than ourselves such as in vitamin pills and, more preferred, our food.
20. Sulfur is also used as a cross-linking agent such as for our hair to lock in the overall structure.
21. We are finding out more and more that what we consider as "junk" is essential for the bioactive molecule to perform its function. If nothing else, it forms the template that presents the active sites in the needed structure.
22. The flow of biological information can be mapped as follows:
    DNA —> RNA —> Protein —> Cell structure and Function
23. As the name implies, in general, chaperonins accompany and help proteins fold as they are formed.
24. The clefts and indentions provide necessary protection to the active site such that unwanted molecules will not occupy, distort, or destroy the active site.
25. Students/teachers call as to what disease/illness is selected.
26. One in $0.01 \times 0.001 \times 0.004 = 4 \times 10^{-9}$ or a good match. Or 1 in $2.5 \times 10^8 = 250,000,000$ or 1 in 250 million, near the population of the United States, which is near 300 million.

CHAPTER 11

1. The formation of cyclic compounds instead of linear polymers.
2. By using a mixture of monochloroaldylsilanes and dichloroaldylsilanes.
3. Like the alkyl pendant groups, 6–7 H.
4. The silanol groups in silicic acid are polar, but the alkyl groups in silicones are much less polar.
5. Acidify it. The silicic acid will spontaneously polymerize.
6. Failure of polymers at elevated temperatures is usually due to the breaking of covalent bonds in the main polymer chain. The $-O-Si-O-$ bonds are very strong.
7. $-O-Si(Et)_2-)-$
8. Phosphorus pentachloride and ammonium chloride.
9. $PCl_5 + 4H_2O \rightarrow H_3PO_4 + 5HCl$, therefore

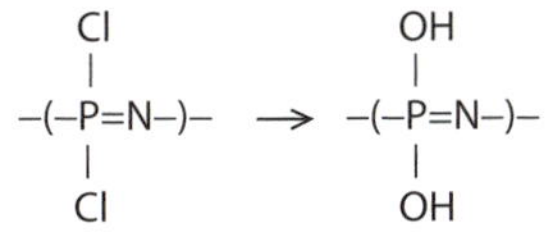

10. (b).

11.

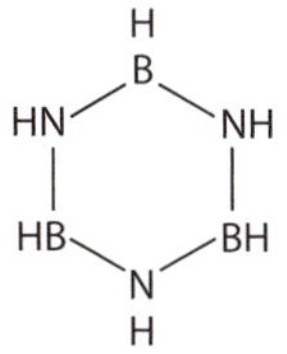

12. They decompose slowly in water.
13. Often undefined with combinations of structures.
14. Evidently below 145°C.
15. As heat stabilizers for PVC processing and a variety of uses to inhibit unwanted microorganisms.
16. The presence of a vacant low-lying *p* orbital.
17. Because of the potential properties that are not readily available in more traditional polymers.
18. Readily available starting materials, wide use temperatures, and variability in properties from more traditional polymers.
19. Chlorophyll and hemoglobin. There are others.
20. Directional bonding that is present in bonds with some degree of covalent character.

CHAPTER 12

1. Held together by directional covalent bonds; acts physically like many organic polymers that are above their T_g; very viscous, acts as a solid on rapid impact but as a liquid on a much elongated time scale.
2. Readily available on a large scale; inexpensive; relatively nontoxic; stands up well to most natural elements such as rain, cold, heat, mild acids and bases, and light; strong.
3. Fiberglass, colored glass, concrete, and so on.
4. Polymers where the structure cannot be adequately described on the basis of a single flat plane. Diamond, Portland cement, sand, topez, beryl, and so on. (see Table 10.1). Insoluble; many exhibit long-range disorder; difficult to characterize structurally.
5. Wide variety of applications and conditions of application with some materials more appropriate than others for specific applications.
6. Flows like a liquid, but the flow rate is very low.
7. Four main methods are employed for shaping glass. They are drawing, pressing, casting, and blowing. Drawing is employed for shaping flat glass, glass tubing, and for creating fibrous glass. Most flat glass is shaped by drawing a sheet of molten glass (heated so it can be shaped but not so much that it freely flows) onto a tank of molten tin. Since the glass literally floats on the tin, it is called "float glass."
8. Wide variety of applications that require materials that may possess glass-like properties— good resistance to natural elements, easily shaped, polished, and cut, many transmit light and can be colored.
9. Fiberglass and asbestos; fibers generally impart greater strength and flexibility.
10. Quartz—highly crystalline.
11. When pressure is applied to a slice of quartz, it develops a net positive charge on one side of the quartz slice and a negative charge on the other side. This phenomenon is the piezoelectric generation of a voltage difference across the two sides of the quartz crystal. Also, the same effect is found when pressure is applied not mechanically, but through application of an alternating electrical field with only certain frequencies allowed to pass through the crystal.
12. Most of what we do and use have risks involved: taking a sun bath and getting skin cancer; taking a water bath and slipping on the soap; walking across a street and being run over; and so on. Some risks may be necessary but others are not. We have many additives in our foods that prevent spoilage, but if certain additives are found to be cancer causing we should omit the use of those additives and find safe substitutes. We have found that the widespread use of asbestos as insulation is unacceptable and are currently using fiberglass as an appropriate alternative.
13. Anisotropic means that properties vary with direction. Since graphite is a sheet material, properties such as strength vary depending on whether you measure the strength along the flat sheet (it is relatively high) or between the sheets (it is relatively weak).

14. Quartz and graphite are both made from carbon and both are highly ordered in structure. Quartz has tetrahedral carbons while graphite has trigonal planer carbons, which are layered one upon the other, and so on.

15. It may be related to such factors as abundance and the ability of both silicon and carbon to form a large variety of structures. It is also probably related to the ability of carbon to connect with itself and for silicon to react with oxygen that, in turn, react with other Si-O containing units forming the silicates.

16. Relatively to the alternative materials employed to manufacture optical lenses, glass holds up better to high temperatures; it is also clear, transparent, and resistant to abrasion and scratching and is easily cleaned. It is superior in its resistance to the elements and to common chemicals.

17. Supposedly, the individual plates of the old glass windows found in Europe are a little thicker at the bottom than at the top.

18. Readily available in large quantities, inexpensive, and they possess the properties that allow them to perform needed applications.

19. Most organic dyes decompose at relatively high temperatures necessary to incorporate the coloring material into the glass.

20. In truth, it depends on the particular ceramic and who is doing the evaluation. Many ceramics contain polymeric sheets, cluster, and/or chains that are connected to one another through ionic linkages, thus they are hybrids between ionic compounds and polymers.

21. The structure is varied; the material is not soluble; there are lots of "kinds" or "types" of Portland cement; structure varies with amounts of materials added and with age. It does not have a precise structure but only an average structure.

22. Both require very pure silicon dioxide.

23. Probably not because of the number of better alternatives and it is difficult to live down its negative past history. Further, there is no reason to believe that the somewhat brittle asbestos compositions might not eventually break down into sizes that would negatively affect the lungs. Also, the problem with asbestos is not only its size but also its sharpness. It would be reasonable to believe that the new asbestos might also be sharp creating health problems when ingested.

24. Carbon nanotubes are being used/studied as tips in atomic force microscopy, electrical wires and switches, manufacture of super slick materials, and so on.

25. It might be but it probably is not sufficiently free from impurities to give a good clear final product. Beach sand also is composed of minor amounts of other materials that do not behave as sand. In any case, beach sand is generally a mixture of a variety of silicon-intense materials and not a good source of "glass-making" silicon dioxide.

CHAPTER 13

1. Because many polymers, often being poor conductors, are so widely used in electrical applications.

2. (a).

3. Because they are poor conductors of heat.

4. ATF.

5. Almost every organic compound will burn if the conditions are good for combustion. Tests such as the oxygen index are useful as comparative tests only.

6. (a).

7. The nature of the surface is important to many of the overall properties of materials.

8. It is particularly important when two or more materials are placed together because their thermal expansions need to be similar or they may break apart.

9. NMR tools can be used to determine structural units, but at times assist in describing interactions between chains and within chains such as chain folding.

10. About 190–800 μm.

11. (a).
12. Preferably X-ray diffraction; NMR may also be used, and so on.
13. DSC, DTA, TML, TBA, and so on.
14. Structural characterization focuses on determining the structure, generally repeat unit or secondary structure, of a material. Such techniques may also give information on the physical properties of a material. In fact, physical properties are directly related to overall polymer structure.
15. Standard tests allow tests to be carried out in laboratories over the world and results to be compared in some reliable fashion. They also assist the researcher in carrying out the tests since such ASTM tests are described in great detail.
16. The answer depends on what "better" means. Because of modern instrumentation, we have both "better" instruments capable of obtaining useful information not available to scientists 30 years ago. Further, many modern instruments are set up to take duplicate measurements quickly and perform statistical treatments of the data, allowing a greater confidence in results.
17. Vinyl polymers have carbon–carbon backbones that have very similar thermal stabilities so that when one degrades, releasing a fragment resulting in weight loss, other carbon–carbon bond scissions occur that results in the production of some statistical average of fragmentation occurring dependent on a number of factors, including heating rate and polymer structure.
18. Most tests that determine the amount of crystallinity and amorphous character a sample has, give results dependent on the particular test. Since the polymer has a variety of molecular-level structures that can be classified as either depending on the particular test, it is not surprising that the end result does not add up to 100%.
19. Few materials are truly flame proof but only resist burning to some extent. Thus, even the suits used by firemen will burn under the right set of conditions. Further, these terms help guard the manufacturer against law suites when their material catches fire under extreme conditions and are intended to discourage people to assuming they will not be hurt if they are in a fire because of the presence of the particular flame-resistant material.
20. The hurdles vary but include increased cost and the need to train people to carry out such safety applications/practices. In the automotive industry, added weight may also be a hurdle. Overall, change typically must overcome hurdles to occur. Even so, industry is sensitive to issues of safety issues.

CHAPTER 14

1. Morphology is the study of shape, while rheology is the study of flow and deformation.
2. (b) and (d).
3. G is the shear modulus.
4. (c).
5. Compressive, pulling, and shear.
6. Movement of part or all of the chain.
7. (a) has a low modulus and high viscosity, therefore the viscosity/G will be large.
8. Bond flexing.
9. Compliance is a ratio of strain divided by stress.
10. Having both flow and glassy or solid behavior.
11. As a solid or a glassy material.
12. At a temperature of 35°C above the T_g of the polystyrene, that is about 140°C.
13. Young's modulus is a ratio of stress to strain. High.
14. The coefficient of viscosity is equal to the ratio of the applied stress to the applied velocity gradient.
15. Stress relaxation.
16. (b).
17. ASTM.

18. Abrupt.
19. No. Failure is abrupt.
20. Models give some idea of the importance of different kinds of behavior such as the importance of bond flexing and chain mobility.
21. The percentage change of elongation = (change in length/length) × 100 = (12 − 5/5) × 100 = 140%.
22. $E = TS/EI = 27{,}100 \ kg/cm^2$.
23. Permanent flow of polymer chains.
24. Spring and dashpot in series.
25. Primarily reversible bond stretching and distortion of bond angles in the polymer chain.
26. Irreversible uncoiling and slippage of chains.
27. By relative areas under the curves.
28. Increase.
29. Decrease.
30. Spring and dashpot in parallel.

CHAPTER 15

1. (b).
2. Light weight and flexibility.
3. According to the Kelly–Tyson equation, the two composites should have equal strength.
4. Fill it with alumina, corrundum, or silicon carbide.
5. To allow the stress on the softer resin to be transferred to the stronger filler.
6. Add a plasticizer such as camphor or TCP.
7. PVC decomposes at processing temperatures; therefore, it is not useful until it was plasticized by TCP and later by DOP; and with the use of organotin polymers even later.
8. Silicon rubber.
9. The stiffness of polymers such as PVC is due to intermolecular attractions. A small amount of plasticizer permits chain orientation, which results in increased attractions.
10. None.
11. As the size of the bulky groups increases, T_g decreases. However, long pendant groups are attracted to one another forming ordered structures, resulting in increased T_g values.
12. Appropriate antioxidants are added.
13. (a).
14. The water in lead-stabilized pipes might be contaminated with lead ions.
15. Degradation often includes the formation of fused ring aromatic compounds that are colored.
16. It is a natural product, inexpensive, nontoxic, and serves as a plasticizer.
17. Because most high-energy UV radiation is filtered out indoors.
18. (b).
19. It is believed that the "bloom" is the result of lubricants to the surface.
20. Gathers dust, can shock you, and can start fires.
21. It is a reference that predates much of the science we have today but simply means that if we follow the directions (recipe) then we will get a specific outcome. Also, by following such a "recipe" the outcome can be roughly repeated.
22. Just as in cooking, recipes were often developed by trial and error based on the outcome. As the desired outcome changes new recipes, often simple modifications of the procedure of amounts in the recipe, change.
23. Animal testing is not always a good indicator of health hazards. For instance, some tests for cancer agents are done on mice that have a tendency to form tumors. Even so, until tests can be performed that show that the additive is safe it should be replaced with another additive that has not been shown to possibly have the bad tendency and which data shows is safe. It is better to error on the side of caution.

24. Both sun tan lotion and additives added to plastics form a similar role and act to protect in a similar fashion so it is not unexpected that the additives are often the same or similar.

25. Additives are added to enhance or modify a particular property. Such there are so many polymers used in such a variety of ways there are many properties that need to be modified. Thus, there are many materials that are used to help achieve the desired outcome.

CHAPTER 16

1. So-called polyvalent ions; that is cations with more than a univalent charge such as Ca^{+2} and Fe^{+3}.
2. Cross-linking through the double bond.
3. $-(-CD_2-CD_2-)-$
4. Time. Also, yield will be generally lower for PS.
5. When unzipping is not likely to occur. For most polymers.
6. Expose it to high-energy radiation.
7. Hydrogenate the block copolymer of styrene and isoprene.
8. They are both alternating copolymers of propylene and ethylene.
9. It controls the approach of oxygen and other molecules by size and electronic character.
10. No. It is below the entanglement length.
11. Size, shape, and electronic nature.
12. Succinic aldehyde.
13. PVA and PAA.
14. The introduction of a few methoxyl groups reduces the hydrogen bonding forces so that the polymer with the remaining hydroxyl groups will dissolve in water.
15. It forms a protective film around dirt particles and prevents their redeposition.
16. No. Both steric and accessibly prevent 100% reaction.
17. The maximum, that is, about 2.8.
18. All of the hydroxyl groups are inaccessible in the hetergeneous acetylation reaction.
19. (b) has more free hydroxyl groups.
20. There are residual acetyl groups in the reactant PVA, and the acetal formation requires two 1,3-hydroxyl groups. Also, on a statistical basis some of the –OH groups become isolated.
21. Can use a fatty acid and employ esterification.
22. Formation of carbon dioxide and water.
23. None. Any vinyl alcohol would rearrange to acetaldehyde.
24. (a) Polystyrene
25. (a) Cations
26. Helps remove fat and other unwanted materials leaving a protein-rich material that will last longer.
27. (c), (d), (e), and (f).
28. Heat polymer and condense the monomer produced by thermal depolymerization.
29. It readily forms hydrogen chloride and aromatic fused ring systems.
30. Material is material. Responses are structure and geometrical driven whether the material is biological or synthetic. Often, synthetic chemists would like materials that had similar behavior to biomolecules, so they look to see what the key structural features are to give the particular biomolecule response and attempt to make use of these findings to achieve the desired outcome except with synthetic materials. Scientists dealing with most biomolecules are also polymer scientists, applying their efforts on biomaterial.
31. A better term is "protected" since part of the structure is present to hold the iron in an active geometry, while much of the remaining structure protects the active iron site from unwanted intruders that can damage its natural action.

32. The porphyrin acts to hold the iron site in the correct geometrical arrangement allowing it to function as needed.
33. Light is absorbed by the antenna pigments, A, transferring the energy until it reaches the reaction center, R, where it is used as the driving force for electron transfer reactions where specialized chlorophylls use it to form carbohydrates.

CHAPTER 17

1. Because chemical engineers are seeking better, safer, and more economical sources of this compound that is used on a large scale.
2. See the appropriate equations in the chapter. (a) 17.1a or 17.2; (b) 17.3 or 17.4; (c) 17.5; (d) 17.6; (e) 17.8; (f) 17.11; (g) 17.12; (h) 17.13; (i) 17.14; (j) 17.15; (k) 17.16; (l) 17.17 or 17.18; (m) 17.19; (n) 17.20 or 17.21; (o) 17.22; (p) 17.23; (q) 17.24 or 17.25; (r) 17.26; (s) 17.28; (t) 17.32; (u) 17.33; (v) 17.39; (w) 17.46.

(a)

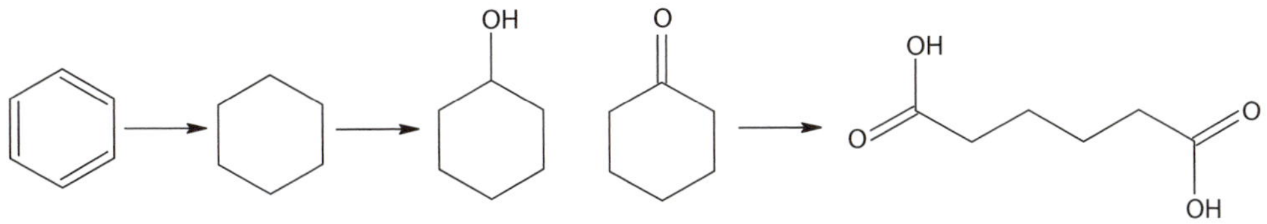

(b)

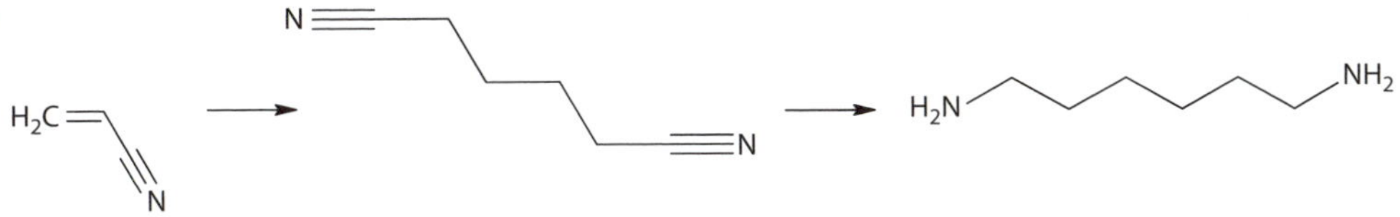

(c)

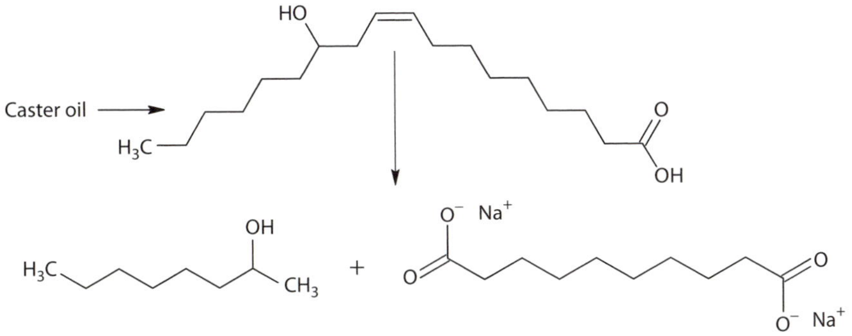

(d)

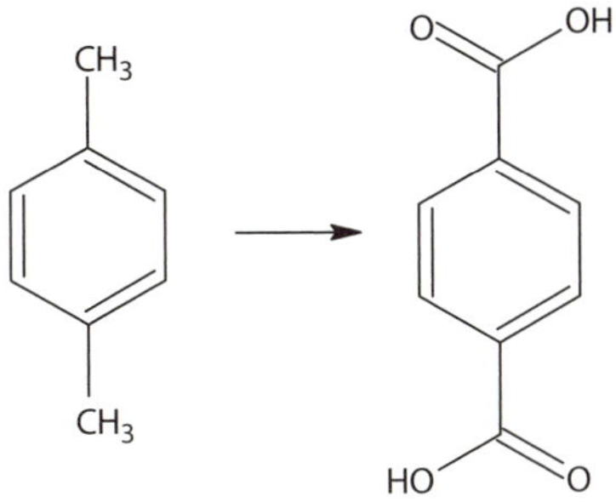

(e)

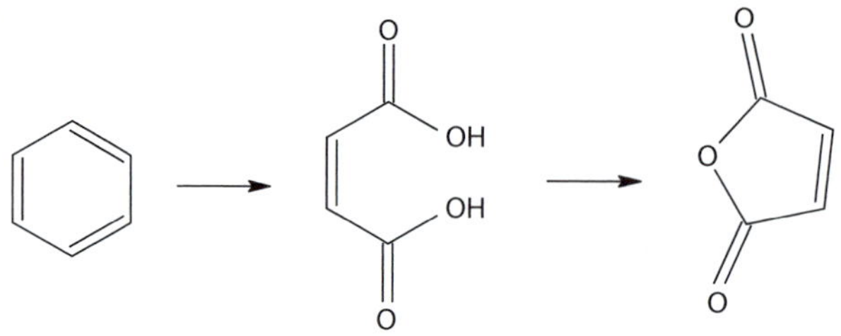

(f)

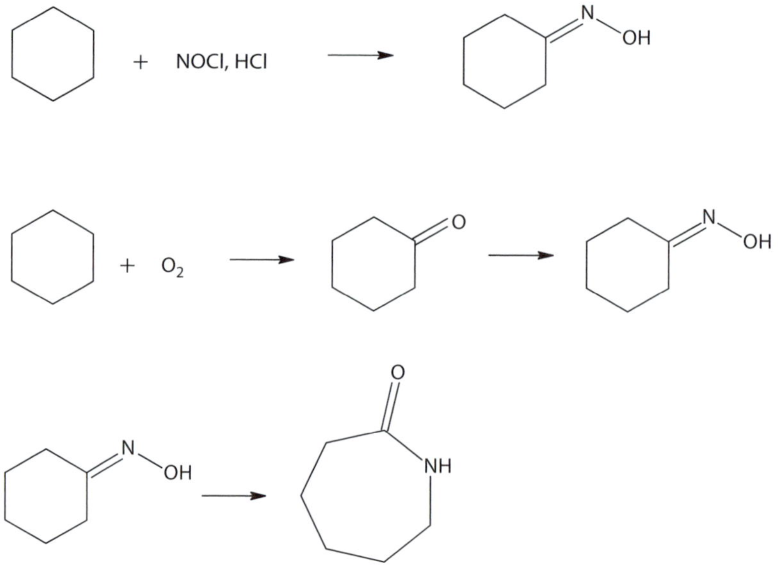

(g)

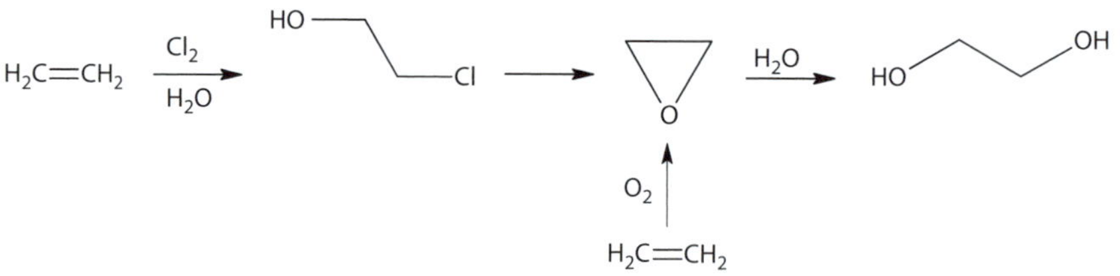

(h)

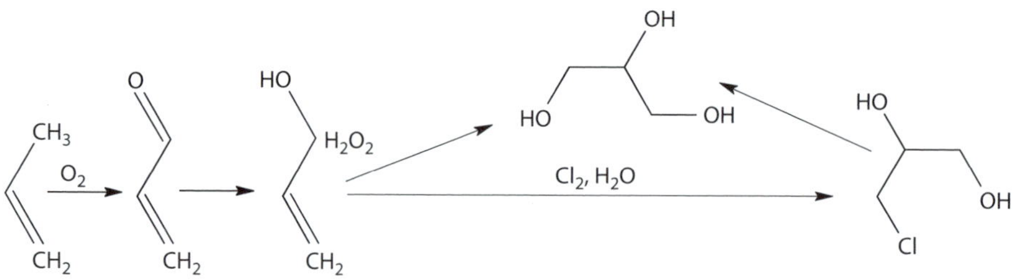

(i)

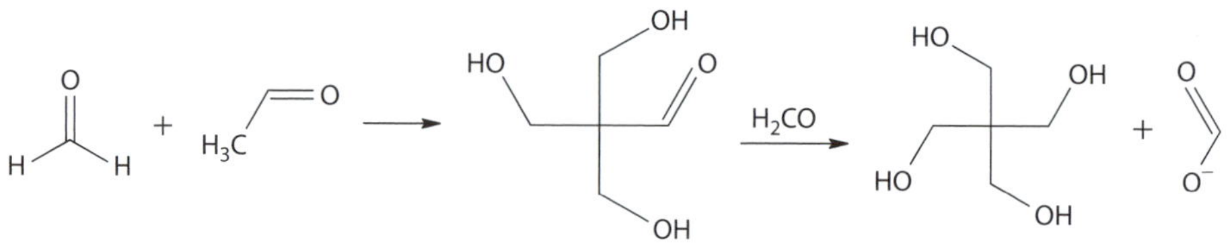

(j)

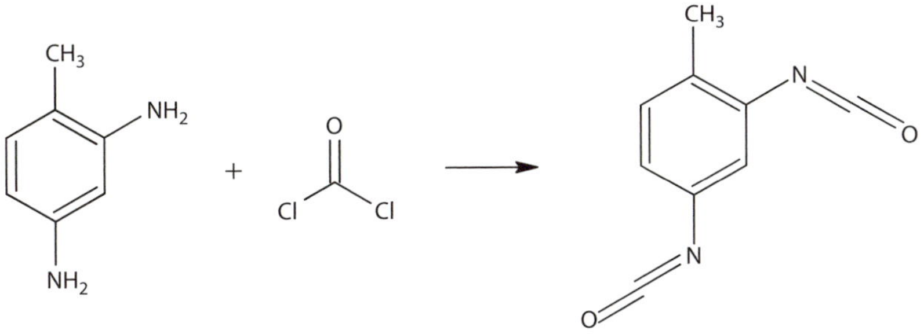

(k)

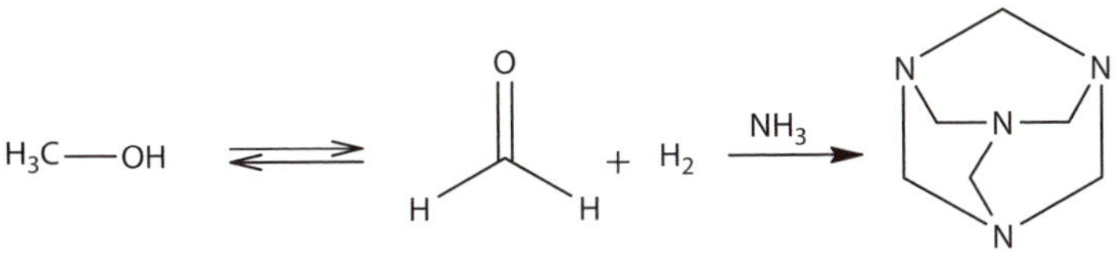

(l)

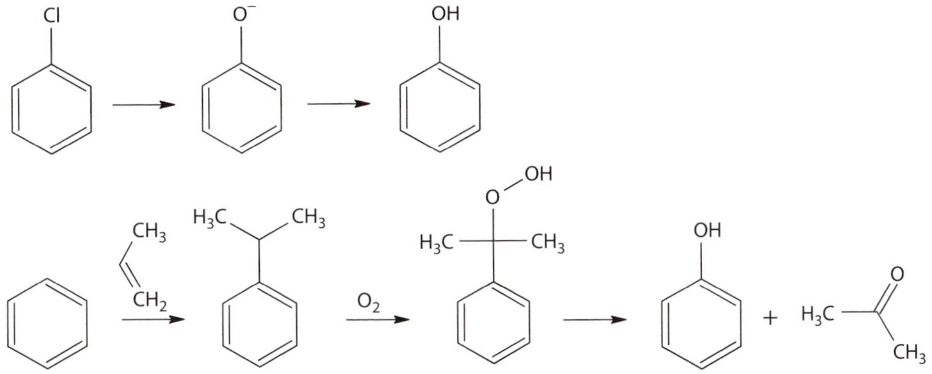

(m)

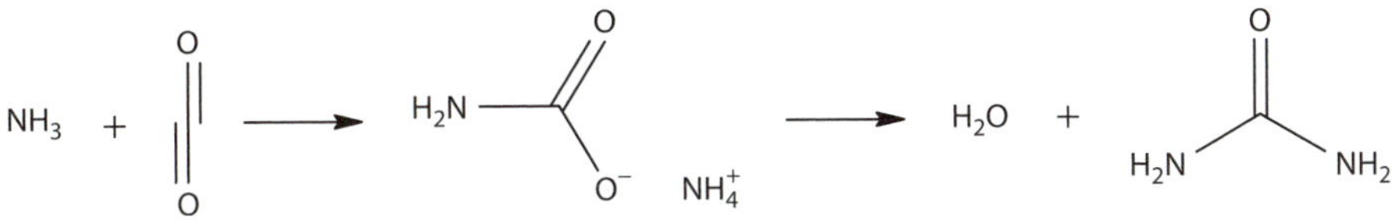

(n)

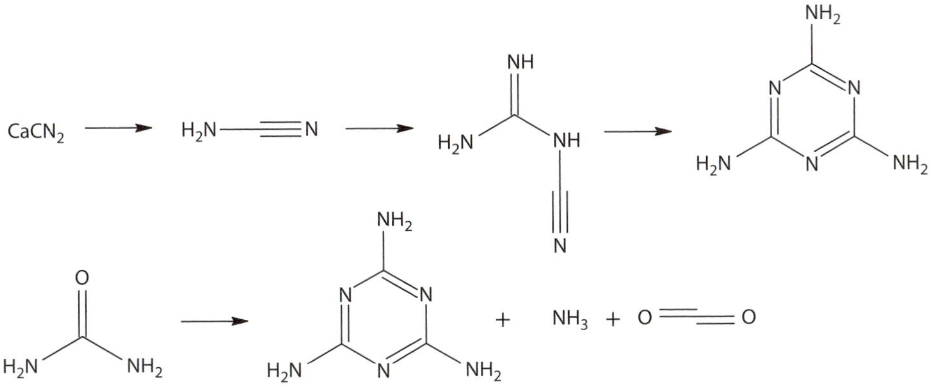

(o)

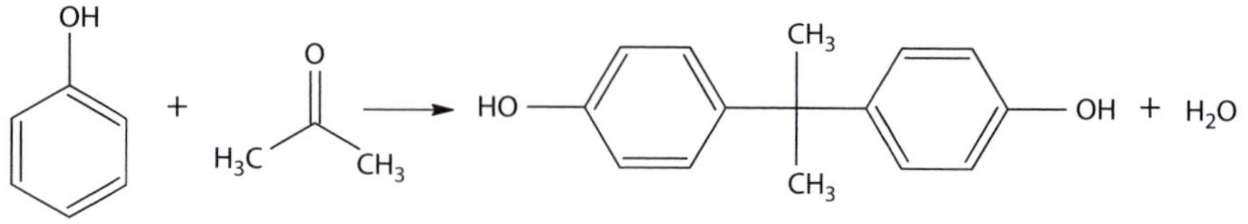

(p)

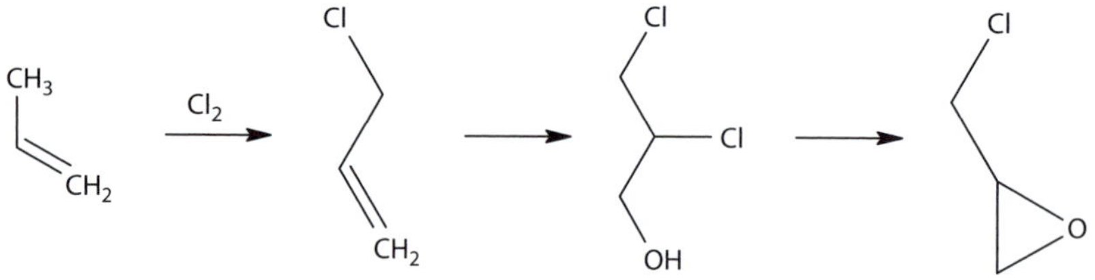

(q) $H_3CMgCl + SiCl_4 \longrightarrow H_2CSiCl_3 + MgCl_2$

(r)

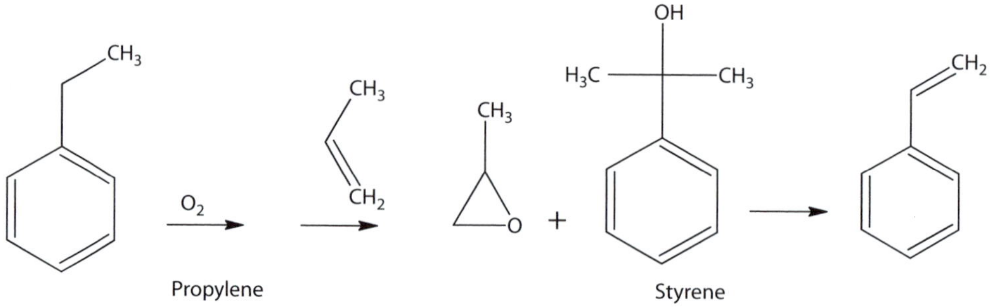

Propylene   Styrene

(s)

$H_2C{=}CH_2 + H{-}Cl + O_2 \longrightarrow H_2O +$ $\longrightarrow H{-}Cl +$

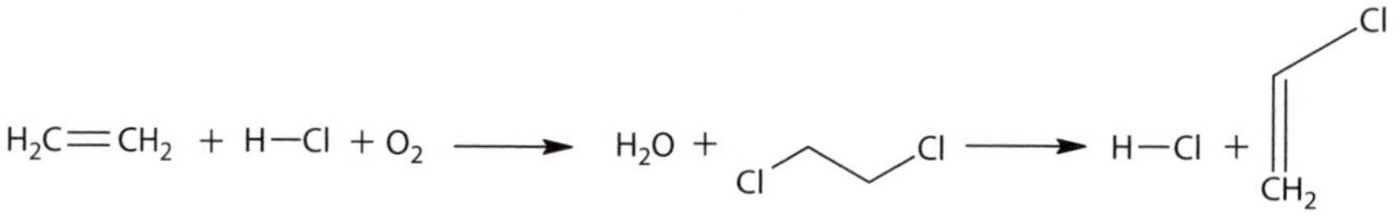

(t)

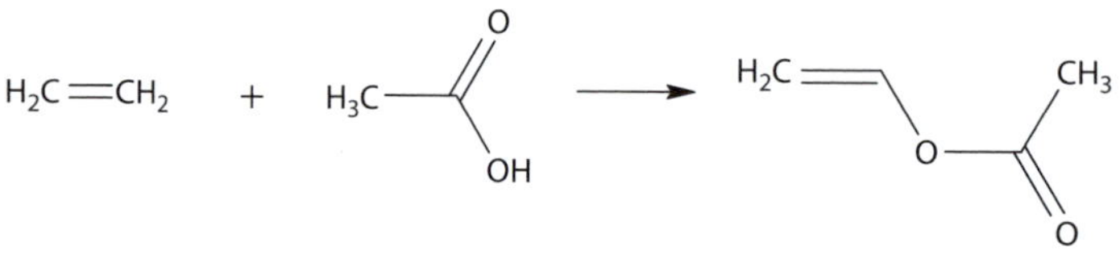

(u)

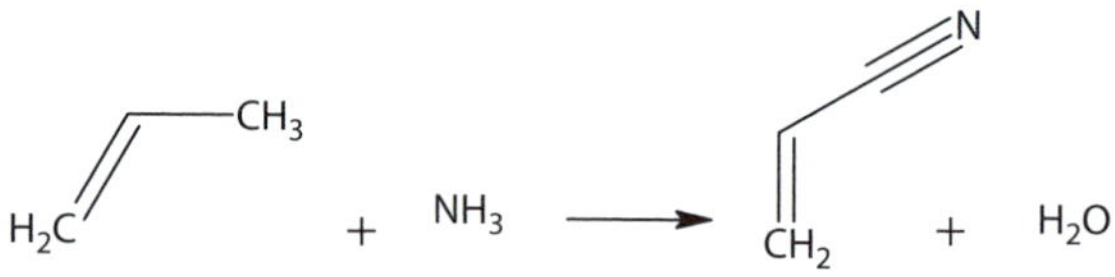

(v)

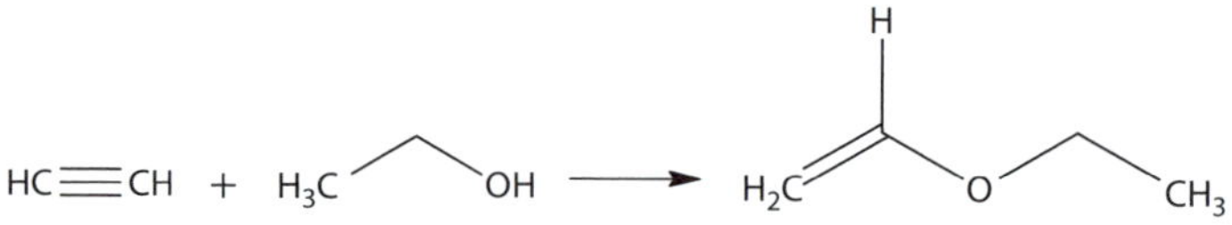

(w)

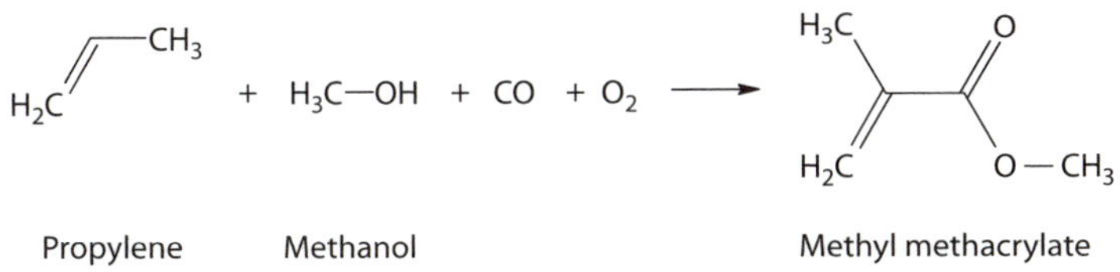

Propylene   Methanol   Methyl methacrylate

3. (a) silanes, (b) styrene, (c) ε-caprolactam, (d) adipic acid or sebacic acid, (e) pentaerythritol.
4. Catalysts perform at least two important functions. These functions are speed and directing the synthesis to give the desired product rather than a broad range of products.
5. Vinyl acetate, methyl methacrylate, ethylene.
6. This allows needed flexibility. The need for certain monomers changes so that the particular monomer stream can be varied as to need. Also, there are only a limited number of basic feedstocks that are available in large quantities, inexpensive, and where we have the ready knowledge and equipment to perform the needed syntheses.
7. Propylene.
8. Petrochemicals serve as the feedstocks for most monomer synthesis. It is natural for a company that produces the basic feedstock to also produce the derived monomers.
9. Peroxides are employed as initiators because they decompose when heated to temperatures that are near to those employed in their synthesis. Further, the decomposition reaction is highly exothermic and can cause explosions to occur.
10. Petroleum serves as not only a fuel, but in the long term more importantly, they serve as many of the basic materials for the synthesis of polymers, drugs, and so on.
11. Glycerol, adipic acid, butadiene, and so on.

CHAPTER 18

1. They both are important and dependent on each other.
2. Cotton, wool, silk, hemp, jute, and so on.
3. Rayon, and so on.
4. It is soluble in less-expensive solvents such as acetone.
5. They are both regenerated cellulose.
6. Polyester, nylon, acrylic fiber, polyurethane, polyolefins.
7. (a) butyl rubber, (b) polybutadiene, (c) SBR, neoprene, (d) silicone, polyurethane elastomer, Thiokol.
8. By air blowing an extruded tube.
9. For prevention of environmental pollution.
10. Over-the-counter.
11. Lots of testing from preliminary cell inhibition, to animal, and finally to human testing if it is found to offer distinct advantages over what is currently available and if it is safe.
12. Delay the blowing step until a viscous, strong, high molecular weight polymer is present.
13. (b).
14. Because of the availability of large surface areas.
15. Fibrous glass-reinforced polyester resin in the Corvette.
16. It has a high specific modulus and low coefficient of expansion.
17. Most molded thermosets are molded by high-cost compression molding.
18. (b) and (d) offer whole chain resonance so may benefit from doping.
19. They have high specific strength, and once an intricate design has been incorporated in the die, the intricate design, such as that resembling hard wood carving, is reproduced at low cost.
20. Low cost, light weight, recyclable, nontoxic, and less hazardous than glass.

21. The molds are relatively inexpensive, and large articles can be easily produced by the thermo-forming process.
22. The length is limited by the ability to store and transport the extrudate. Actually, pipe can be made in continuous lengths by extruding on the job site. Flexible pipe such as LDPE can be coiled allowing for very long continuous pipes to be made and used.
23. Formica, safety glass, plywood, Macintosh raincoat.
24. Most paints coat and most coatings contain a pigment that allows it to "look nice." Relatively nontoxic, easy clean up.
25. Relatively nontoxic with easy clean up.
26. *Oil Paints*—Oil paints consist of a suspension of pigments in a drying oil, such as linseed oil. The film is formed by a reaction involving atmospheric oxygen, which polymerizes and cross-links the drying oil. *Oil Varnishes*—Varnish coatings consist of a polymer, either nat-ural or synthetic, dissolved in a drying oil together with appropriate additives as catalysts to promote cross-linking with oxygen. *Enamels*—Classical enamel is an oil varnish with a pig-ment added. *Lacquers*—Lacquers consist of polymer solutions to which pigments have been added. *Latex Paints*—Latex paints are polymer latexes to which pigments have been added.
27. Adhesion occurs generally through one or more of the following mechanisms. Mechanical adhesion with interlocking occurs when the adhesive mixture flows about and into two rough substrate faces. This can be likened to a hook and eye, where the stiff plastic hooks get caught in the fuzz-like maze of more flexible fibers. Chemical adhesion is the bonding of primary chem-ical groups. Specific or secondary adhesion occurs when hydrogen bonding or polar (dipolar) bonding occurs. Viscosity adhesion occurs when movement is restricted simply due to the vis-cous nature of the adhesive material.
    Solvent-Based Adhesives—Solvent-based adhesion occurs through action of the adhesive on the substrate. Solidification occurs on evaporation of the solvent. *Latex Adhesives*—Latex adhesives are based on polymer latexes and require that the polymers be near their T_g so that they can flow and provide good surface contact when the water evaporates. *Pressure-Sensitive Adhesives*—Pressure-sensitive adhesions are actually viscous polymer melts at room temper-ature. The polymers must be applied at temperatures above their T_g to permit rapid flow. The adhesive is caused to flow by application of pressure. When the pressure is removed, the viscos-ity of the polymer is sufficient to hold and adhere to the surface. *Hot-Melt Adhesives*—Hot-melt adhesives are thermoplastics that form good adhesives simply by melting, followed by subse-quent cooling after the plastic has filled surface voids. *Reactive Adhesives*—Reactive adhesives are either low molecular weight polymers or monomers that solidify by polymerization and/or cross-linking reactions after application. *Thermosets*: A number of thermosets have been used as adhesives. Phenolic resins are still used to bind together thin sheets of wood to make ply-wood. *Elastomers*—Solutions of natural rubber have been used for laminating textiles for over a century. *Pressure-Sensitive Tape*—Pressure-sensitive tape, consists of a coating of a solution of a blend of natural rubber and an ester of glycerol and abietic acid (rosin) on cellophane, was developed over half a century ago. More recently, natural rubber latex and synthetic rubber have been used in place of the natural rubber solution. The requirement for pressure-sensitive adhe-sives is that the elastomers have a T_g below room temperature. *Contact Adhesives*—Contact adhesives are usually applied to both surfaces, which are then pressed together. Liquid copo-lymers of butadiene and acrylonitrile with carboxyl end groups are used as contact adhesives in the automotive industry. *Thermoplastics*—A number of thermoplastics have been used as adhesives. Polyamides and copolymers of ethylene and vinyl acetate (EVA) are used as melt adhesives. Copolymers of methyl methacrylate and other monomers are used as adhesives in the textile industry. Poly(vinyl acetate) is often used in school glues. *Anaerobic Adhesives*—Anaerobic adhesives consist of mixtures of dimethacrylates and hydroperoxides (initiators) that polymerize in the absence of oxygen. *Cyanoacrylates*—Cyanoactylates polymerize spontane-ously in the presence of moist air, producing excellent adhesives.

28. One method is simply to heat the material to above its T_g or T_m.
29. Solvent removal. Solvent recovery.
30. Both depending on the end use. Pillows are open celled, allowing a "softer" more flexible behavior while insulation foam is more rigid and closed celled retaining heat/cold better.
31. Countertops; plywood.
32. Adhesives hold things together. There are lots of things we want to hold together under a variety of conditions.
33. Dries fast so can minimize bleeding; clean. Works to adhere many different surfaces including skin.
34. Probably the easiest thing is to "hide" the black painted wall using specially designed coatings with a high "hiding" ability. Then you can paint it.

CHAPTER 19

1. Cancer cells have three main characteristics that are different from healthy cells. First, they are immortal, are able to replicate themselves hundreds of times, while healthy cells generally replicate themselves less than two dozen times over our lifetime. Second, cancer cells are not contact inhibited meaning that they will continue to grow forming tumors. By comparison, healthy cells replicate until they touch another cell and then stop reproduction. Finally, cancer cells are normally in a growth mode, whereas healthy cells are generally in a rest mode.
2. Smart materials are materials that react to applied force—electrical, stress–strain (including pressure), thermal, light, and magnetic. A smart material is not smart simply because it responds to external stimuli, but it becomes smart when the interaction is used to achieve a defined engineering or scientific goal.
3. Aramides do not melt, drip, support combustion in air, or burn.
4. Poly-p-phenylene and poly(p-phenylene vinylene) because they offer whole chain resonance in their backbones.
5. Carbon nanotubes do not have to be doped and are more flexible and stronger.
6. Carbon nanotubes are not soluble, whereas some of the newly developed conductive polymers are soluble in organic and others in water allowing them to be "laid out" more easily without the need of creating the nanotubes on site.
7. The dopants cause the introduction of sites of increased or deficiency of electrons. When there is a deficiency of electrons, or hole created, electrons flow to fill this hole with the newly created hole causing other electrons to flow to fill the new hole, and so on, allowing charge to migrate within and between the polyacetylene chains.
8. There are many. Because of their size, polymers with chain lengths of about 100 units and greater typically are unable to easily move through biological membranes, thus movement is restricted. This can result in limiting negative side effects, such as damage to the kidneys, because the polymer can reside in only certain body sites. Also, this limited mobility can assist in directing the polymer drug.
9. This is because of several trends. First, many biological solutions to biological problems resides with the use of biological materials to solve biological problems. With the advent of increased effectiveness of cloning and related advances, suitable biological replacements may be grown from a person's cells thus ensuring compatibility. Second, related to the first, regeneration of essential parts of vital organs is becoming more practical. Third, because of nanotechnology and related electrical and optical advances, surgery to locate and repair imperfections is improving. Forth, our autoimmune systems are being studied with the hope that rejection of substitute hearts may be more easily overcome.
10. For quick identification of used and unused suture's location. It is often difficult to see when operating.

11. Some of the important properties are the ability to tailor-make structures, surface control, strength, flexibility, rigidity, inertness/reactivity, light weight, ease of fabrication, ability to achieve a high degree of purity, lack of and because of their water solubility/compatability, bioerodablity, and the ability of some of them to withstand long-term exposure to the human body—a truly hostile environment.

12. Almost all denture bases are made of methacrylic (acrylic) resins that give good fit and a natural appearance.

13. Nylon, because tissue reaction is minimal. Nylon is strong and cuts on the leg often are exposed to high stresses as the person heals. It has good tensile strength and little elongation change. It offers good knot security and low tissue drag.

14. Poly(ethylene oxide), copolymer containing about 50% ethylene and about 50% acrylic acid units, and poly(vinyl alcohol).

15. Only PP.

16. PVC decomposed giving the toxic gas hydrogen chloride when burned.

17. Yes. There is a sufficient proportion of the polymer that hydrogen bond to water, allowing it to be water soluble.

18. Its cost is too high and it does not have good dimensional stability meaning it is not easily cut, drilled, or sawed.

19. Polymers are light weight, can be designed to be permanent or short lived in the body, can be designed to become part of the bodie's complex structure, high strength, varying requirements needed in biomaterials that can be incorporated into a polymer structure, and so on.

20. Probably a number of reasons can be cited, including general inertness and a tendency toward being slick.

21. There are both political and environmental reasons. Green materials are in some ways relatively rapidly regenerating lowering our dependence on less-rapidly regenerated petroleum resources. They should also more easily be returned to the earth renewing the nutrients needed to sustain the earth. Further, they should require less nonsolar energy and should offer less pollution. Increase in petroleum cost and recognition that the petroleum reserves are limited, and so on.

22. Again, there are political and environmental reasons. Many things drive this move. For instance, industry wants to capture popular press to have their products included under the umbrella of green materials. But, also industry recognizes the need to make products that do not have a negative impact on the environment and is working to make materials that perform well and are economically and environmentally better, and so on.

23. PLA, PHB, and ricinoleic acid.

Appendix A
Symbols

A	Arrhenius constant
ABA	Acrylonitrile-butadiene acrylate
ABS	Copolymer of acrylonitrile, butadiene, and styrene
ACS	Acrylonitrile-chlorinated polyethylene styrene terpolymer
AIBN	$2,2'$-Azobisisobutyronitrile
AMA	Acrylate maleic anhydride terpolymer
AMMA	Acrylate-methyl methacrylate copolymer
AN	Acrylonitrile
ANSI	American National Standards Institute
AP	Ethylene–propylene copolymers
APO	Amorphous polyolefin
AS	Acrylonitrile styrene copolymer
ASA	Acrylonitrile–styrene–acrylonitrile block
ASTM	American Society for Testing Materials
ATR	Attenuated total reflectance spectroscopy
AU	Polyurethane
BMC	Bulk molding compound
BPO	Benzoyl peroxide
BSI	British Standards Institute
bp	Boiling point
CA	Cellulose acetate
CAB	Cellulose acetate butyrate
CAR	Carbon fiber
CED	Cohesive energy density
CFRP	Carbon reinforced plastics
CMC	Carboxymethylcellulose
CN	Cellulose nitrate
COC	Cycloolefin copolymer
CPE	Chloronated polyethylene
CPVC	Chlorinated poly(vinyl chloride)
CR	Neoprene
CTA	Cellulose triacetate
CTFE	Chlorotrifluoroethylene
C_p	Specific heat
C_s	Chain-transfer constant
DAIP	Diallyl isophthalate plasticizer
DAP	Diallyl phthalate plasticizer

DNA	Deoxyribonucleic acid
DP	Degree of polymerization
DRS	Dynamic reflectance spectroscopy
DS	Degree of substitution
DSC	Differential scanning calorimetry
DTA	Differential thermal analysis
E_a	Activation energy or energy of activation
E	Young's modulus
EAA	Ethylene acrylic acid copolymer
EC	Ethyl cellulose
ECTFE	Ethylene–chlorotrifluoroethylene copolymer
EEA	Ethylene–ethyl acetate copolymer
EGG	Einstein-Guth-Gold equation
EMAC	Ethylene–methyl acrylate copolymer
EP	Epoxy resin
EPDM	Poly(ethylene-copropylene) cross-linked
EPM	Poly(ethylene-copropylene)
EPM	Ethylene–propylene copolymer
EPR	Ethylene propylene rubber
EPR	Electron paramagnetic resonance spectroscopy
EPS	Expanded polystyrene
ESR	Electron spin resonance spectroscopy
ET	Thiokol
ETA	Electrothermal analysis
ETFE	Ethylene tetrafluoroethylene polymer
EU	Polyether polyurethane
EVA	Ethylene–vinyl acetate copolymer
EVOH	Ethylene–vinyl alcohol copolymer
f	Aspect ratio
FEP	Fluorinated ethylene propylene
FRP	Fibrous glass reinforced polyester; fiber reinforced plastic
G	Gibbs free energy; modulus; molar attraction constant
GF	Glass reinforced
GPC	Gel permeation chromatography
GRS	Poly(butadiene-co-styrene)
HDPE	High-density polyethylene
HIPS	High-impact polystyrene
HMC	High-strength molding compound
I	Ionomer
IIR	Butyl rubber
IPN	Interpenetrating polymer network
IR	Infrared
ISO	International Standards Organization
IUPAC	International Union of Pure and Applied Chemistry
K	Constant in Mark–Houwink equation; Kelvin
LC	Liquid crystal
LCP	Liquid crystal polymer

LDPE	Low-density polyethylene
LLDPE	Linear low-density polyethylene
LPE	Linear polyethylene
MABS	Methyl methacrylate ABS copolymer
MBS	Methyl methacrylate butadiene styrene terpolymer
MDPE	Medium-density polyethylene
MDI	Methylene diphenylisocyanate
MF	Melamine-formaldehyde resin
MP	Melamine phenolic
MWD	Molecular weight distribution
$\overline{M}_n$	Number-average molecular weight
$\overline{M}_v$	Viscosity-average molecular weight
$\overline{M}_w$	Weight-average molecular weight
$\overline{M}_z$	Z-average molecular weight
NBR	Poly(butadiene-co-acrylonitrile); nitrile butadiene rubber
NBS	National Bureau of Standards
NMR	Nuclear magnetic resonance spectroscopy
NR	Natural rubber
OI	Oxygen index
P	Phenolic
PA	Polyamide; nylon
PAA	Poly(acrylic acid)
PAEK	Polyaryletherketone
PAK	Polyester alkyd
PAI	Polyamide-imide
PAL	Polyanaline
PAN	Polyacrylonitrile
PARA	Polyaryl amide
PAS	Polyarylsulfone
PB	Polybutylene
PBAN	Polybutylene–acrylonitrile copolymer
PBI	Polybenzimidazole
PBS	Polybutadiene-styrene copolymer
PBT	Poly(butylene terephthalate)
PC	Polycarbonate
PCB	Polychlorinated biphenyls
PCL	Polycaprolactone
PCT	Poly(cyclohexylene terephthalate)
PCTFE	Polymonochlorotrifluoroethylene
PE	Polyethylene
PEEK	Poly(ether ether ketone)
PEG	Poly(ethylene glycol)
PEI	Polyetherimide
PEK	Polyetherketone
PEO	Poly(ethylene oxide)
PES	Polyethersulfone

PET, PETE	Poly(ethylene terephthalate)
PEX	Cross-linked polyethylene
PF	Phenol-formaldehyde resin
PGC	Pyrolysis gas chromatography
PI	Polyimide
PIB	Polyisobutylene
PMMA	Poly(methyl methacrylate)
PMP	Polymethylpentene
PMR	Proton magnetic resonance spectroscopy
PMS	Polymethylstyrene
PNF	Poly(phosphonitrilic fluorides)
PO	Polyolefin
PolyEd	Polymer Education Committee
POM	Polyoxymethylene, polyformaldehyde, acetals
PP	Polypropylene
PPC	Chlorinated polypropylene
PPE	Poly(phenylene ether)
POM	Polyoxymethylene
PPO	Poly(phenylene oxide)
PPOX	Polypropylene oxide
PPS	Poly(phenylene sulfide)
PPSU	Poly(phenylene sulfone)
PPT	Poly(propylene terephthalate)
PS	Polystyrene
PCTFE	Polytrifluoromonochloroethylene
PTFE	Polytetrafluoroethylene, Teflon
PTME	Poly(tetramethylene terephthalate)
PU	Polyurethane
PUR	Polyurethane rubber
PVA	Poly(vinyl alcohol); sometimes poly(vinyl acetate)
PVAc	Poly(vinyl acetate)
PVAI	Poly(vinyl alcohol)
PVB	Poly(vinyl butyral)
PVC	Poly(vinyl chloride)
PVDC	Poly(vinylidene chloride)
PVDF	Poly(vinylidene fluoride)
PVF	Poly(vinyl fluoride)
PVK	Poly(vinyl carbazole)
PVOH	Poly(vinyl alcohol)
PVP	Poly(vinyl pyrrolidone)
RIM	Reaction injection molding
RNA	Ribonucleic acid
ROMP	Ring-opening metathesis polymerization
ROP	Ring-opening polymerization
S	Entropy; radius of gyration
SAN	Poly(styrene-coacrylonitrile)
SB	Styrene butadiene copolymer

SBR	Poly(butadiene-costyrene) elastomer
SBS	Styrene butadiene styrene block copolymer
SEBS	Styrene ethylene butylene styrene block copolymer
SEM	Scanning electron microscopy
SI	Silicon
SIS	Styrene isoprene styrene block copolymer
SMA	Poly(styrene-co-maleic anhydride)
SMC	Sheet-molding compound
SMMA	Styrene methyl methacrylate copolymer
SN	Sulfur nitride
SPE	Society of Plastics Engineering
SPI	Society of the Plastics Industry
SR	Synthetic rubber
SRP	Styrene–rubber plastics
TDI	Toluenediisocyanate
TEO	Thermoplastic elastic olefin
TGA, TG	Thermal gravimetric analysis
TMC	Thick molding compound
TMMV	Threshold molecular weight value
TPE	Thermoplastic elastomer
TPE	Rubber-toughened nylons
TPU	Thermoplastic urethane
TPX	Poly-4-methylpentene
T_c	Ceiling temperature; cloud point temperature
T_g	Glass transition temperature
T_m	Melting point
UF	Urea–formaldehyde resin
UHMWPE	Ultrahigh molecular weight polyethylene
ULDPE	Ultralow-density polyethylene
ULPE	Ultralinear polyethylene
UV	Ultraviolet
VA	Vinyl acetate
VLDPE	Very low-density polyethylene
WLF	Williams-Landel-Ferry equation
WS	Polyurethane
XLPE	Cross-linked polyethylene
XPS	Expandable polystyrene

Appendix B
Trade Names

Trade or Brand Name	Product	Manufacturer
Abafil	Reinforced ABS	Rexall
Abalyn	Abietic acid derivative	Hercules
Absafil	ABS	Fiberfil
Absinol	ABS	Dart
Abson	ABS	Goodrich
Accepta	Polyester	KoSa
Acctuf	PP copolymer	Amoco Polymers
Acelon	Cellulose acetate	May & Baker
Acetophane	Cellulose acetate films	UCB-Sidac
Aclar	Polyfluorocarbon film	Allied
Aclon	Fluoropolymer	Allied Signal
ACP	PVC	Alpha Gary
Acralen	Styrene–butadiene latex	Farbenfabriken Bayer AG
Acrilan	Acrylic fibers	Solutia
Acronal	Polyalkyl vinyl ether	General Aniline Film
Acrylacon	Fibrous glass-reinforced polymers	Rexall
Acrylafil	Reinforced polymers	Rexall
Acrylaglas	Fibrous glass-reinforced styrene–acrylonitrile	Dart
Acrylicomb	Acrylic honeycomb	Dimensional Plastics
Acrylite	Acrylic	Cyro Inds.
Acrylux	Acrylic Plastics	Westlake
Acryrex	Acrylic	Chi Mei Ind.
Acrilan	Polyacrylonitrile	Chemstrand
Acrylan-Rubber	Butyl acrylate-acrylonitrile copolymer	Monomer Corp.
Acrylite	PMMA	American Cyanamide
Acrysol	Thickeners	Rhome & Haas
Actol	Polyethers	Allied
Adell	Thermoplastic resin	Adell
Adipol	lasticizer	FMC
Adiprene	Urethane elastomers	DuPont
Adpro	PP	Huntsman
Aerodux	Resorcinol–formaldehyde resin	Ciba
Aerpflex	PE	Anchor
Aeron	Plastic coated nylon	Flexfilm
Ffcolene	PS and SAV copolymers	Pechiney–Saint-Gobain
Afcoryl	ABS	Pechiney–Saint-Gobain
Affinity	PE copolymers with poly(alpha-olefins (plastomers)	Dow
Agro	Rayon fibers	Beaunit Mills
Akulon	Nylon-6, -66	AKU, DSM

continued

Trade or Brand Name	Product	Manufacturer
Akuloy	Nylon-6,-66 alloys	DSM
Aim	PS	Dow
Alathon	HDPE and copolymers	DuPont/Lyondell Polymers
Albertols	Phenolic resins	Chemische Werke
Albis	Nylon-6, -66	Albis Canada
Alcryn	TP elastomer	DuPont
Aldocryl	Acetal resin	Shell
Alfane	Epoxy resin cement	Atlas Mineral
Algil	Styrene copolymer monofilament	Shawinigan; Polymer Corp.
Algoflon	Polytetrafluoroethylene	Montecatini/Auismont
Alkathene	PE	Imperial
Alkon	Acetal copolymers	Imperial
Alkor	Furan resin cement	Atlas Minerals
Alloprene	Chlorinated natural rubber	Imperial
Alphatec	TP elastomer	Alpha Gary
Alphalux	Poly(phenylene oxide)	Marbon
Alsibronz	Muscovite mica	Franklin Mineral
Alsilate	Clays	Freeport Kaolin
Alsynite	Reinforced plastic panels	Reichhold
Amberlac	Modified alkyl resins	Rhom & Haas
Amberol	Phenolic resins	Rhom & Haas
Amberlite	Ion-exchange resins	Rhom & Haas
Amco	Olefin fiber	Sampson Rope Tech.
Ameripol	Polyisoprene	Firestone
Amer-Plate	PVC sheets	Ameron Corrosion Control
American	Olefin fiber	Sampson Rope Tech.
Amerith	Cellulose nitrate	Celanese
Amilan	Nylon	Tojo Rayon/Toray Ind.
Amoco	TP resin	Amoco
Amodel	Polyphthalamide, PPA	Amoco Polymers
Ampcoflex	Rigid PVC	Atlas Mineral
Ancorex	ABS extrusions	Anchor
Angel Hair	Olefin fiber	Wayn-Tex
Anso (group of trade names)	Nylon-6	Honeywell Nylon
Antron (and related trade names)	Nylon-66 fiber	DuPont/INVISTA
Anvyl	Vinyl extrusions	Anchor
Apec	High-temperature PC	Bayer
Apex N	PVC blend with nitrile rubber	Teknor Apex
API	PS	American Polymers
Apogen	Epoxy resins	Apogee
Aqualoy	Nylon-6/12,-66, PP	ComAlloy
Aquathene	PE	Quantum
Araclor	Polychlorinated polyphenyls	Monsanto
Aralac	Protein fiber	Imperial
Araldite	Epoxy resins	Ciba
Ardel	Polyarylate	Amoco Polymers
Ardil	Protein fiber	Imperial
Armite	Vulcanized fiber	Spaulding Fibre
Armorite	Vinyl coating	Armitage
Arnel	Cellulose triacetate	Celanese
Arnite	PET	Algemene Kuntstzijde, DSM

Trade or Brand Name	Product	Manufacturer
Arnitel	Polyester block copolymer with polyester	DSM
Arochem	Modified phenolic resins	Ashland
Aroclor	Chlorinated polyphenyls	Monsanto
Arodure	Urea resins	Ashland
Arofene	Phenolic resins	Ashland
Aroplaz	Alkyd resins	Ashland
Aropol	Polyester resins	Ashland
Aroset	Acrylic resins	Ashland
Arpak	Expandable PP bead	JSP
Arpro	Expandable PP bead	JSP
Arothane	Polyester resins	Ashland
Artfoam	Rigid urethane foam	Strux
Arylon, Arylon T	Polyaryl ethers	Uniroyal
Asaparene	Linear block styrene copolymer with butadiene	Asahi
Ashlene	Nylon-6, -66, -6/12	Ashley Polymers
Astrel	Polyarylsulfone	3M
Astryn	PP alloy, cohomopolymers, TPO	Montell
Ascot	Polyolefin sheet-coated spunbonded	Appleton Coated Paper
Astralit	Vinyl copolymer sheets	Dynamit Nobel
Astroturf	Synthetic turf–nylon and PE	Monsanto
Astyr	Butyl rubber	Montecatini
Atlac	Polyester cast resin	ICI
Attane	ULDPE	Dow Plastic
Aurum	TP polyimide	Mitsui Toatsu
Avisco	PVC film	FMC
Avistar	Polyester film	FMC
Avisun	PP	Avisum
Avora (and related trade names)	Polyester	KoSa
Avron	Rayon fiber	American Viscose
Azdel	Fibrous glass-reinforced ABS copolymer sheet	Azdel
Bakelite	Phenol–formaldehyde	Union Carbide
Bapolene	PE	Bamberger
Barden	Kaolin clay	Huber
Barex	Barrier resin/Acrylonitrile copolymers	Vistron/BP Chemicals
Barricaut	Polyester	Honeywell
Basofil	Melamine fibers and materials	Basofil Fibers, LLC
Baydur	Structural foam PUR RIM	Bayer
Bayblend	PC/ABS	Bayer
Baygal	Polyester casting resin	Farbenfabriken Bayer AG
Baylon	Polycarbonate	Farbenfabriken Bayer AG
Baypren	Polychloroprene	Farbenfabriken Bayer AG
Beckacite	Modified phenolic resin	Reinhhold; Beck, Koller
Beckamine	Urea–formaldehyde resin	Reichhold; Beck, Koller
Beckopox	Epoxy resins	Reichhold
Beckosol	Alkyl resins	Reichhold
Beetle	Urea–formaldehyde resin	American Cyanamid/Cytec Ind.
Beltec	Polyester	Honeywell
Bemberg	Rayon fiber	Beaunit Mills
Benvic	PVC	Solvay
Bexloy	Ionomer	DuPont

continued

Trade or Brand Name	Product	Manufacturer
Bexone F	Poly(vinyl formal)	British Xylonite
Benvic	PVC	Solvay & Cie
Bexphane	PP	Bakelite Xylonite
Biobarrier	Olefin	Reemay
BioFresh	Acrylic fibers	Sterling Fibers
Blanex	Cross-linked PE	Richhold
Blapol	PE	Richhold
Blendex	ABS resin	Borg-Warner
Bolta Flex	Vinyl sheeting and film	General Tire & Rubber
Bolta Thene	Rigid oldfin sheets	General Tire & Rubber
Boltaron	Plastic sheets	General Tire & Rubber
Bondstrand	Filament wound fiberglass reinforced plastics	American Corrosion Control
Bondtie	Olefin	Sampson Rope Tech.
Borofil	Boron filaments	Texaco
Boronol	Polyolefins with boron	Allied Resinous
Bostik	Epoxy and polyurethane adhesives	Bostik-Finch
Bounce-Back	Acrylic fibers	Solutia
Bronco	Supported vinyl or pyroxyline	General Tire & Rubber
Budene	cis-1,4-Polybutadiene	Goodyear
Butacite	Poly(vinyl acetal) resins	DuPont
Bukaton	Butadiene copolymers	Imperial
Butaprene	Styrene–butadiene elastomers	Firestone
Butarez CTL	Telechelic butadiene polymers	Phillips
Buton	Butadiene–styrene resin	Enjay
Bu-Tuf	Polybutene	Petrotex
Butvar	Poly(vinyl butyral) resin	Shawinigan
BXL	Polysulfone	Union Carbide
Cab-O-Sil	Colloidal silica	Cabot
Cabot	TP resin	Cabot
Cadco	Plastic rod	Cadillac
Cadon	Nylon filament	Chemstrand
Calprene	Linear and branched styrene copolymers with butadiene	Repsol
Capran	Nylon-6	Allied Signal
Carbaglas	Fiberglass-reinforced polycarbonate	Dart
Cordura	Nylon-66 fibers	INVISTA
Carillon	Aliphatic PK	Shell
Caromastic	Epoxyl coal tar coating	Carboline
Carbopol	Water-soluble resins	Goodrich
Carboset	Acrylic resins	Goodrich
Carbospheres	Hollow carbon spheres	Versar
Carbowax	Poly(ethylene glycols)	Union Carbide
Cariflex I	cis-1,4-Polyisoprene	Shell
Cariflex TR	Linear and branched styrene block copolymers	Shell
Caroma	PVC	Shell
Carinex	PS	Shell
Carolux	Flexible polyurethane foam	North Carolina Foam
Castear	Cast polyolefin films	Exxon-Mobile
Castethane	Castable poluurethanes	Upjohn
Catalac	Phenol–formaldehyde	Catalin
Cefor	PP	Shell

Trade or Brand Name	Product	Manufacturer
Celanex	Thermoplastic foam and sheeting	Celanese
Celatron	PS	Celanese
Celbond	Polyester	KoSa
Celcon	Acetal copolymers	Celanese
Celgard	Microporous PP film	Celanese
Cellasto	Microcellular urethane elastomer	North American Urethane
Cellofoam	PS foam board	United States Mineral Products
Cellonex	Cellulose acetate	Dynamit Nobel
Cellon	Cellulose acetate	Dynamit Nobel
Cellosize	Hydroxyethyl cellulose	Union Carbide
Celluliner	Expanded PS foam	Gilmon Brothers
Cellulite	Expanded PS foam	Gilman Brothers
Celluloid	Plasticized cellulose nitrate	Celanese
Celpak	Rigid polyurethane foam	Dacar
Celstar	Acetate fibers	Celanese Acetate
Celramic	Glass nodules	Pittsburgh Corning
Centrex	ASA, ASA + AES	Bayer
Cerex	Styrene copolymers	Monsanto
Cevian	ASA, ASA + PBT, SAN	Daicel
C-Flex	Linear block styrene copolymer with ethylene–butylene	Consolidated Polymer Tech.
Chemigum	Urethane elastomer	Goodyear
Chemlon	Nylon-6,-66	Chem Polymer
Chem-o-sol	PVC plastisol	Chemical Products
Chempro	Ion-exchange resin	Freeman
Celthane	Rigid polyurethane foam	Dacar
Chemfluor	Fluorocarbon plastics	Chemplast
Chemglaze	Polyurethane-based coating	Lord
Chemgrip	Epoxy adhesives for TFE	Chemplast
Chlorowax	Chlorinated paraffins	Diamond Alkali
Chromspun	Acetate fibers	Voridian
Cibanite	Aniline–formaldehyde resin	Ciba
Cimglas	Fiberglass-reinforced polyester	Cincinnati Milacron
cis-4	cis-1,4-Polybutadiene	Phillips
Claradex	ABS	Shin-A
Clocel	Rigid polyurethane foam	Baxenden
Clopane	PVC tubing and film	Clopay
Cloudfoam	Polyurethane foam	International foam
Co-Rexyn	Polyester resins, coatings, pastes	Interplastic
Cobex	PVC	Bakelite Xylonite
Cobocell	Cellulose acetate butyrate tubing	Cobon
Coboflon	Teflon tubing	Cobon
Cobothane	Ethylene–vinyl acetate tubing	Cobon
Collodion	Cellulose nitrate solution	Generic name
Colovin	Calendered vinyl sheeting	Columbus Coated Fabrics
Comforel	Polyester	INVISTA
ComFortrel (and related trade names)	Polyester	Wellman, Inc.
Comshield	PP	ComAlloy
Comtuf	Impact resistant resins	ComAlloy
Conathane	Polyurethane compounds	Conap
Conductrol	Acrylic fibers	Sterling Fibers

continued

Trade or Brand Name	Product	Manufacturer
Conolite	Polyester laminate	Woodall
Coolmax	Polyester fiber	INVISTA
Coperba	Linear styrene block copolymers with butadiene	Petroflex
Coral rubber	cis-Polyisoprene	Firestone
Cordo	PVC foam and film	Ferro Corp.
Cordoflex	Poly(vinylidene fluoride) solutions	Ferro Corp.
Cordura	Regenerated cellulose	DuPont
Corfam	Poromeric film	DuPont
Corlite	Reinforced foam	Snark
Coro-Foam	Urethane foam	Cook Paint and Varnish
Corval	Rayon fiber	Courtaulds
Corvel	Plastic coating powders	Polymer Corp.
Corvic	Vinyl polymers	Imperial
Courlene	PE fiber	Courtaulds
Coverlight HTV	Vinyl-coated nylon fabric	Reeves Brothers
Covol	Poly(vinyl alcohol)	Corn Products
Crastin	PBT	DuPont
Creslan	Acrylonitrile–acrylic ester	American Cyanamid
Creslite	Acrylic fibers	Sterling Fibers
Cresloft	Acrylic fibers	Sterling Fibers
Cronar	PE	Dupont
Crowelon	Olefin fiber	Crowe Rope Inds.
Crown Fiber	Olefin fiber	Nexcel Synthetics
Cryowrap	Thermoplastic sheets and films	W.R. Grace
Cryovac	PP film	W.R. Grace
Crystalex	Acrylic resin	Rhom & Haas
Crystalon	Rayon fiber	American Enka
Crystalor	Polymethylpentene	Phillips Chemical
Crystic	Polyester resins	Scott Bader
CTI	Nylon-66	M.A. Hanna
Cumar	Coumarone–indene resins	Allied Chemical
Curithane	Polyaniline polyamines	Upjohn
Curon	Polyurethane foam	Reeves Brothers
Cyanaprene	Polyurethane	American Cyanamid
Cyanolit	Cyanoacrylate adhesive	Leader, Denis, Ltd.
Cycloset	Cellulose acetate fiber	DuPont
Cycolac	Acrylonitrile–butadiene–styrene copolymer	Borg-Warner
Cycogel	ABS	Nova Polymers
Cycolac	ABS, ABS + PBT	GE Plastics
Cycolin	ABS/PBT	GE Plastics
Cycoloy	PC/ABS blend	GE Plastics
Cycopoac	ABS and nitrile barrier	Borg-Warner
Cyglas	TS polyester	Cytec Inds.
Cymac	Thermoplastic molding materials	American Cyanamid
Cymel	Melamine molding compound	Cytec Inds.
Cyovin	Self-extinguishing ABS graft polymer blends	Borg-Warner
Cyrex	Acrylic/PC alloy	Cyro Inds.
Cyrolite	Acrylic	Cyro Inds.
Dacovin	Rigid PVC	Diamond Alkali
Dacron	Polyester fiber	DuPont/DAK Americas, LLC
Dapon	Diallyl phthalate prepolymer	FMC

Trade or Brand Name	Product	Manufacturer
Daponite	Dapon-fabric laminates	FMC
Daran	PVC emulsion coatings	W.R. Grace
Daratak	PVC emulsions	W.C. Grace
Darex	Styrene copolymer resin	W.R. Grace
Darlon	Polyacrylonitrile fiber	Farbenfabriken Bayer AG
Darvan	Poly(vinylidene cyanide)	Celanese
Darvic	PVC	Imperial
Davon	Tetrafluoroethylene polymer	Davies Nitrate
Degalan	PMMA	Degussa
Delcron	Polyester fiber	DAK Americas, LLC
Delrin	Acetal polymers	DuPont
Densite	Urethane foam	Tenneco Chemical
Derakane	Polyester resin	Dow
Desmopan	Polyurethanes	Farbenfabriken Bayer AG
Desmophen	Polyesters and polyethers for polyurethanes	Farbenfabriken Bayer AG
Devran	Epoxy resins	Devoe & Reyaolds
Dexel	Cellulose acetate	British Celanese
Dexon	PP acrylic	Exxon-Mobile
Dexplex	TPO	D & S Polymers
Dexsil	Polycarboranesiloxane	Olin Mathieson
Diakon	PMMA	Imperial
Diaron	Melamine resins	Richhold
Dielux	Acetals	Westlake Plastics
Diene	Polybutadiene	Firestone
Dimension	Nylon-6 alloy	Allied Signal
Diolen	PET	ENKA-Glazstoff
Dion	Polyester resins	Diamond Alkali
Dolphon	Epoxy and polyester resins	John C. Dolph
Dorlastan	Spandex fiber	Dorlastan Fibers LLC
Dorvon	PS foam	Dow
Doryl	Poly(diphenyl oxide)	Westinghouse Electric
Dow Corning	Silicons	Dow Corning
Dowex	Ion-Exchange resinis	Dow
Dowlex	HDPE, LLDPE	Dow Plastics
Drexflex	TP elastomer	D & S Plastics
Dri-Lite	Expanded PS	Poly Foam
Dry Step	Nylon-6	Honeywell Nylon
DSDN	Nylon-66 fibers	INVISTA
DSP	Polyester fiber	Honeywell Int.
Duco	Cellulose nitrate lacquers	DuPont
Dulac	Lacquers	Sun Chemical
Dulux	Polymeric enamels	DuPont
Duolite	Ion-exchange resins	Diamond Alkali
Duracel	Lacquers	Mass & Waldstein
Duracon	Acetal copolymers	Polyplastics
Duraflex	Polybutylene	Shell
Dural	Acrylic modified PVC	Alpha Gary
Duralon	Furan molding resins	U. S. Stoneware
Durane	Polyurethanes	Raffi & Swanson
Duramac	Alkyd resins	Commercial Solvents Corp.
Duraplex	Alkyd resins	Rohm & Haas
Duraspan	Spandex fibers	Carr-Fulflex

continued

Trade or Brand Name	Product	Manufacturer
Durel	Polyarylate	Hoechst-Celanese
Durelene	PVC tubing	Plastic Warehousing
Durethan	Nylon-6	Farbenfabriken Bayer AG
Durethan U	Polyurethanes	Farbenfabriken Bayer AG
Durethene	PE film	Sinclair-Koppers
Durez	Phenolic resins	Occidental
Durathon	Polybutylene resins	Witco
Durite	Phenolic resins	Borden
Duron	Phenolics	Firestone Foam
Duron	Olefin fibers	Drake Extrusion
Dural	Ethylene–propylene copolymers	Montecatini
Duraspun	Acrylic fibers	Solutia
Dyal	Alkyd resins	Sherwin-Williams
Dyalon	Urethane elastomers	Thombert
DyeNAMIX	Nylon-66 fiber	Solitia
Dylark	SMA copolymer	Nova Chemicals
Dylene	PS	Nova Chemicals
Dylite	Expandable PS	Nova Chemicals
Dynaflex	SBS, SEBS	GLS Plastics
Dyneema	Olefin fiber	DSM High Performance Fibers
Dyroam	Expanded PS	W.C. Grace
Dylan	PE resins	Sinclair-Koppers
Dynaflexl	Linear block styrene copolymers	GLS
Dynel	Vinyl chloride-acrylonitrile copolymers	Union Carbide
Dylel	ABS copolymers	Sinclair-Koppers
Dylene	PS resins	ARCO Polymer
Dylite	Expandable PS	Sinclair-Koppers
Dynafilm	PP film	U. S. Industrial Chemical & National Distillers & Chemical Corp.
Dynel	Modacrylic fiber	Union Carbide
Dyphene	Phenol-formaldehyde resins	Sherwin-Williams
E-Foam	Epoxies	Allied
Eastabond	PET	Eastman Chemical
Eastalloy	PC + polyester	Eastman Chemical
Eastapak	PET	Eastman Chemical
Eastar	Polyesters	Eastman Chemical
Eastman	Thermoplastic resin	Eastman
Easypoxy	Epoxy adhesive kits	Conap
Ebolan	TEF materials	Chicago Gasket
Ecavyl	PVC	Kuhlmann
Eccosil	Silicon resins	Emerson & Cummings
Ecdel	Polyester block copolymer with polyether	Eastman Kodak
Ecoprene	TP elastomer	Rubber & Plastics Solutions
Edistir	PS	Enichem
Ektar	PET, PBT, PCT polyesters	Eastman
Elastalloy	TP elastomer	Eastman
Elastolit	Urethane engineering thermoplastic	North American Urethanes
Elastollan	Polyurethane block copolymer with polyether/polyester	BASF
Elastollyx	Urethane engineering thermoplastic	North American Urethanes
Elastolur	Urethane coating	BASF

Trade or Brand Name	Product	Manufacturer
Elastonate	Urethane isocyanate prepolymers	BASF
Elastonol	Urethane polyester polyols	North American Urethanes
Elastopel	Urethane engineering thermoplastic	North American Urethanes
Elastothane	Polyurethane elastomer	Thiokol
Electrafil	Electrically conductive polymers	DSM
Electroglas	Cast acrylic	Glasflex
Elexar	Linear styrene block copolymer with ethylene–butylene	Teknor Apex
El Rexene	Polyolefin resins	Rexall Chemical
El Rey	LDPE	Rexall Chemical
Eltex	HDPE	Teknor Apex
Elustra	Olefin fibers	FiberVisions Products
Elvace	Acetate-ethylene copolymer	DuPont
Elvacet	Poly(vinyl acetate) emulsion	DuPont
Elvacite	Acrylic resins	DuPont
Elvamide	Nylon resins	DuPont
Elvanol	Poly(vinyl alcohol) resins	DuPont
Elvax	Poly(ethylene-covinyl acetate)	DuPont
Elvic	PVC	Solvay
Emac	EMA copolymer	Chevron Chemical
Empee	PE, PP	Monmouth
Enathene	Ethylene butyl acrylate polymer	Quantum
Engage	Polyethylene copolymer with Poly(alpha-olefins)	Dow
Enka	Nylon-66 fiber	Polyamide Ind. Fibers
Enkalure	Nylon fiber	American Enka
Enrad	Preirradiated PE	Enflo
Enrup	Thermosetting resin	United States Rubber
Ensocote	PVC lacquer coatings	Uniroyal
Ensolex	Cellular plastic sheets	Uniroyal
Ensolite	Cellular plastic sheets	Uniroyal
Epibond	Epoxy adhesive resin	Furane Plastics
Epikote	Epoxy resins	Shell Chemical
Epilox	Epoxy resins	Leuna
Epi-Rez	Epoxy cast resins	Celanese
Epi-Tex	Epoxy–ester resins	Hoechst-Celanese
Epikote	Epoxy resins	Shell Chemical
Epocap	Two-part epoxy compounds	Hardman
Epocryl	Epoxy acrylate resins	Shell Chemical
Epodite	Epoxy resins	Showa Highpolymer
Epolast	Two-part epoxy compounds	Hardman
Epolene	Low-melt PE	Eastman Chemical
Epolite	Epoxy compounds	Hexcel
Epomarine	Two-part epoxy compounds	Hardman
Epon	Epoxy resins	Shell Chemical
Eponol	Linear polyether resins	Shell Chemical
Eposet	Two-part epoxy compounds	Hardman
Epotuf	Epoxy resins	Reichhold
Epoxylite	Epoxy resins	Epoxylite Corp.
Eref	PA/PP alloy	Solvay
Escalloy	Crack resistant PP	ComAlloy
Escon	PP	Enjay

continued

Trade or Brand Name	Product	Manufacturer
Escor	Acid terpolymer	ExxonMobile
Escorene	PP	ExxonMobile
ESP	Polyester fiber	KoSa
Essera	Olefin fibers	American Fibers & Yarns
Estane	Polyurethane resins	B. F. Goodrich
Estron	Cellulose acetate filament	Eastman Chemical
Ethafoam	PE foam	Dow
Ethocel	Ethyl cellulose	Dow
Ethofil	Fiberglass-reinforced PE	Dart
Ethoglas	Fiberglass-reinforced PE	Dart
Ethosar	Fiberglass-reinforced PE	Dart
Ethron	PE	Dow
Ethylux	PE	Westlake Plastics
Europrene Sol T	Linear and branched styrene copolymers	Enichem
Evalca	EVA copolymer	Eval
Evenglo	PS	Sinclair-Koppers
Everflex	PVA copolymer emulsion	W. C. Grace
Everlon	Polyurethane foam	Stauffer Chemical
Evolutia	Acrylic fibers	Solutia
Exact	Polyethylene copolymer with poly(alpha-olefins)	ExxonMobil
Excelite	PE tubing	Thermoplastic Processes
Exon	PVC	Firestone Plastics
Extane	Polyurethane tubing	Pipeline Service
Extrel	Plastic film	Exxon-Mobil
Extren	Fiberglass-reinforced polyester	Morrison Molded Fiber Glass
Fabrikoid	Pyroxylin-coated fabrics	DuPont
Facilon	Reinforced PVC fabric	Sun Chemical
Fassgard	Vinyl-coated nylon	M. S. Fassler
Fasslon	Vinyl coating	M. S. Fassler
Felor	Nylon filaments	DuPont
Ferrene	PE	Ferro
Ferrex	PP	Ferro
Ferroflex	PP blend with EPDM or EPR	Ferro
Ferrocon	Polyolefin	Ferro
Ferroflo	PS	Ferro
Ferropak	PP/PE alloy	Ferro
Fertene	PE	Montecatini
Fibercast	Reinforced plastic pipe	Fibercast
Fiberfil	Fiber reinforced material	DSM
Fiberglas	Fibrous glass	Owens-Corning Fiberglass
FiberLoc	Nylon-66 fiber	INVISTA
Fibermesh	Olefin fibers	SI Corp.
Fiberite	Phenolic molding compounds	Fiberite
Fibro	Rayon	Courtaulds NA
Fina	Polyolefin	Fina Oil
Finaclear	PS, SBS	Fina Oil
Filabond	Unsaturated polyester	Reichhold
Finaclear	PS, SBS	Fina Oil
Fillwell (and related trade names)	Polyester	Wellman, Inc.

Trade or Brand Name	Product	Manufacturer
Finaprene	Linear styrene block copolymers with butadiene	Fina Oil
Flexalloy	PVC	Teknor Apex
Flexomer	ULDPE	Union Carbide
Flexprene	TP elastomer	Teknor Apex
Floterrope	Olefin fiber	Sampson Rope Tech.
Flakeglas	Glass flakes for reinforcements	Owens-Corning Fiberglas
Flexane	Polyurethanes	Devcon
Flesocel	Polyurethane foams	Baxenden Chemical
Flexomer	PE copolymer with poly(alpha-olefins)	Union Carbide
Flexothene	PP blend with EPDM or EPR	quistar
Flexprene	Linear block styrene copolymer with butadiene	Teknor Apex
Floranier	Cellulose	Rayonier
Flovic	Poly(vinyl acetate)	Imperial Chemical
Fluokem	Teflon spray	Bel-Art Products
Fluon	Polytetrafluoroethylene and powders	Imperial Chemical
Fluorel	Poly(vinylidene fluoride)	3M
Fluorglas	PTEF-impregnated materials	Dodge Industries
Fluorobestos	Asbestos-Teflon composite	Raybestos Manhattan
Fluorocomp	Reinforced fluoropolymer	LNP
Fluorocord	Fluorocarbon materials	Raybestos Manhattan
Fluorofilm	Cast Teflon fioms	Dilectrix Corp.
Fluoron	Polychlorotrifluoroethylene	Stokes Molded Products
Fluoroplast	PTFE	U.S. Gasket
Fluororay	Filled fluorocarbon	Raybestos Manhattan
Fluorored	TFE compounds	John L. Dore Co.
Fluorosint	TFE-fluorocarbon composites	Polymer Corp.
Foamex	Poly(vinyl formal)	General Electric
Foamthane	Polyurethane foam	Pittsburgh Corning
Foamspan	TP foam	ComAlloy
Foraflon	PVDF	Atochem
Formadall	Polyester premix	Woodall Industries
Formaldafil	Fiberglass-reinforced acetals	Dart Industries
Formaldaglass	Fiberglass-reinforced acetals	Dart Industries
Formaldasar	Fiberglass-reinforced acetals	Dart Industries
Formex	Poly(vinyl acetal)	General Electric
Formica	Thermosetting laminates	Formica Corp.
Formion	Ionomer	A. Schulman
Formrezel	Liquid resins for urethane elastomers	Witco Chemical
Formvar	Poly(vinyl formal)	Shawinigan Resins Corp.
Forticel	Cellulose propionate	Celanese
Fortiflex	PE	Solvay
Fortisan	Saponified cellulose acetate	Celanese
Fortilene	PP	Solvay
Fortrel	(and related trade names) Polyester fiber	Fiber Ind./Wellman, Inc.
Fortron	PPS	Hoechst-Celanese
Fostacryl	Poly(styrene-coacrylonitrile)	Foster Grant
Fostalene	Plastic	Foster Grant
Fosta-Net	PS foam mesh	Foster Grant
Fosta Tuf-Flex	High-impact PS	Foster Grant

continued

Trade or Brand Name	Product	Manufacturer
Fostafoam	Expandable PS beads	Foster Grant
Fostalite	Light stabilized PS	Foster Grant
Fostarene	PS	Foster Grant
FPC	PVC resins	Firestone
FR-PC	PC	Lucky
FTPE	Fluorelastomer	3M
Furname	Epoxy and furan resins	Atlas Mineral Products
Futron	Polyester	Fusion Rubbermaid
Galalith	Casein plastics	Generic name
Gantrez	Poly(vinyl ether-comaleic anhydride)	General Aniline & Film
Gantrez	Fibrous glass roving	Johns-Manville
Gapex	Nylon	Ferro
Garan Finish	Sizing for glass fibers	Johns-Manville
Gedamine	Unsaturated polyester	Charbonnages
Geloy	ASA, ASA + PC, ASA + PVC	GE Plastics
Gelva	Poly(vinyl acetate)	Shawinigan Resins
Gelvatex	Poly(vinyl acetate) emulsions	Shawinigan Resins
Gelvatol	Poly(vinyl alcohol)	Shawinigan Resins
Geolast	TP elastomer	Advanced Elastomer Sys.
Geon	PVC	Geon
Genaire	Poromeric film	General Tire & Rubber
Genal	Thermosets	General Electric
Genthane	Polyurethane elastomers	General Tire & Rubber
Genetron	Fluorinated hydrocarbon monomers and polymers	Allied Chemical
Gentro	Butadiene copolymer	General Tire & Rubber
Geon	PVC	B.F. Goodrich
Gil-Fold	PE sheets	Gilman Brothers
Ginny	Acrylic fibers	Solutia
Glaskyd	Glass-reinforced alkyd resins	American Cyanamid/CYTEC
Glastic	Thermoset resins	Glastic
Glospan	Spandex fiber	RadiciSpandex Corp.
Glyptal	Alkyd coating	General Electric
Goldrex	Acrylic	Hanyang Chemical
Gracon	PVC	W.C. Grace
GravoFLEX	ABS sheets	Hermes Plastics
GravoPLY	Acrylic sheets	Hermes Plastics
Grex	PE	W.R. Grace
Grilamid	Polyamide copolymer with polyether/ polyester; nylon-12	EMS America Grilon
Grilon	Nylon-6,-66	EMS America Grilon
Grivory	Nylon	EMS America Grilon
Halar	Polyfluorocarbons	Allied/Ausimont
Halex	Polyfluorocarbons	Allied Chemical
Halon	Polyfluorocarbons	Ausimont
Hanalac	ABS	Miwon
Haysite	Polyester laminates	Synthane-Taylor/Haysite
Hercocel	Cellulose acetate	Hercules Powder
Hercose	Cellulose acetate-propionate	Hercules Powder
Herculoid	Cellulose nitrate	Hercules Powder
Herculon	PP	Hercules Powder
Hercuprene	Thermoset resins	J-Von

Trade or Brand Name	Product	Manufacturer
Herox	Nylon	DuPont
Heterofoam	Fire-retardant urethane foam	Hooker Chemical
Hetron	Fire-retardant polyester resins	Ashland
Heaveaplus	Copolymer of methyl methacrylate and rubber	Generic name
Hex-One	HDPE	Gulf Oil
H-film	Polyamide film from pyromellitic anhydride and 4,4-diaminodiphenyl ether	DuPont
Hi-Blen	ABS polymers	Japanese Geon
Hi-fax	HDPE	FMC; Hercules Powder
Hifax	PP blend with EPDM or EPR	Himont
HiGlass	Glass filled PP	Himont
Hipack	PE	Showa Highpolymer
Hi-Sil	Amorphous silica	PPG
Hi-Styrolux	High-impact PS	Westlake Plastics
Hitalex	PE	Hitachi Chemical
Hitanol	Phenol–formaldehyde resins	Hitachi Chemical
HiWal	HDPE	General Polymers
Hivalloy	PP alloy	Montell
Hollofiber	Polyester fiber	Wellman, Inc.
Hostacom	Reinforced PP	Hoechst-Cellanese
Hostadur	PET	Farbwerke Hoechst AG
Hostaflon C2	Polychlorotrifluorethylene	Farbwerke Hoechst AG
Hostaflon TF	PTFE	Farbwerke Hoechst AG
Hostaform	Acetal copolymer	Hoechst-Cellanese
Hostalen	PE	Farbwerke Hoechst AG
Hostalen GC	HDPE/LDPE	Hoechst
Hostalloy	Polyolefin alloy	Hoechst-Cellanese
Hostyren	PS	Hoechst
Huntsman	Thermoplastic	Huntsman
Hurcuprene	Linear block styrene copolymer with butadiene or ethylene–butylene	J-Von
Hy (and related trade names)	Olefin fibers	FiberVision Prods.
Hycar	Butadiene acrylonitrile copolymer	B.F. Goodrich
Hydrepoxy	Water-based epoxys	Allied Products
Hydrin	Epichlorohydrin rubber	Goodrich-Hercules
Hydro Foam	Expanded phenol-formaldehyde	Smithers
Hydropol	Hydrogenated polybutadiene	Phillips Petroleum
Hydrotec	Polyester fibers	DAK Americas, LLC
Hyflon	Fluoropolymer	Auismont
Hylar	PVDF	Auismont
Hypalon	Chlorosulfonated PE	DuPont
Hytrel	Polyester block copolymer with polyether/polyester	DuPont
Imbue	Polyester fibers	KoSa
Impet	PET	Hoechst-Cellanese
Implex	Acrylic resins	Rohm and Haas
Impressa	Olefin fibers	American Fibers & Yarns
Innova	Olefin fibers	American Fibers & Yarns
Insurok	Phenol–formaldehyde molding compounds	Richardson Company
Intamix	Rigid PVC	Diamond Shamrock
Iotek	Ionomer	ExxonMobel

continued

Trade or Brand Name	Product	Manufacturer
Interpol	Polyurethane	Cook Composites
Ionac	Ion-exchange resins	Permutit Company
Irrathene	Irradiated PE	General Electric
Irvinil	PVC resins	Great American Chemical
Isoderm	Urethane integral-skinning foam	Upjohn
Isofoam	Polyurethane foam resins	Isocyanate Products
Isomid	Polyester–polyamide film magnet wire	Schenectady Chemicals
Isoplast	TPU	Dow
Isoteraglas	Isocyanate elastomer-coated Dacron glass fabric	Natvar Corp.
Isothane	Polyurethane foam	Bernel Foam Products
Ipuiace	PPO/PPE	Mitsubishi
Iupilon	Polycarbonate	Mitsubishi Edogawa
Iupital	Acetal	Mitsubishi
Ixan	PVDF	Solvay
Ixef	Polyacrylamide	Solvay Polymers
Jetfoam	Polyurethane foam	International Foam
Jet-Kote	Furane resin coatings	Furane Plastics
J-Plast	TP elastomer	J-Von
Kadel	PAEK	Amoco Polymers
Kalex	Urethane resin	Di-Acro Kaufman
Kalspray	Rigid urethane foam	Baxenden Chemical
Kamax	Acrylic copolymer	AtoHaas
Kapton	Polyamide	DuPont
Kardel	PS film	Union Carbide
Kaurit	Phenol–formaldehyde resins	Badische Anilin & Coda
Kelburon	PP/EP	DSM
Kel-F	trifluorochloroethylene resins	3M
Keltrol	Copolymers	Textron
Kematal	Acetal copolymers	Imperial
Kemcor	LDPE, HDPE	Kemcor Australia
Kematal	Acetal copolymer	Hoechst-Celanese
Kenflex	Hydrocarbon resins	Kenrich Petrochemicals
Ken-U-Thane	Polyurethane	Kenrich Petrochemicals
Ketac	Ketone-aldehyde resins	American Cyanamid
Kevlar	Aramid materials	DuPont
Kibisan	SAN	Chi Mei Ind.
Kibiton	SBS	Chi Mei Ind.
Koblend	Polycarbonate/ABS	EniChem America
Kodacel	Cellulose acetate film	Eastman Chemical
Kodapak	PET	Eastman
Kodar	Copolyesters	Eastman Chemical
Kohinor	Vinyl	Rimet
Kollidon	Poly(vinyl pyrrolidone)	General Aniline & Film
Kolorbon	Rayon fiber	American Enka
Kopa	Nylon-6,-66	Kolan America
Kopox	Epoxy resins	Koppers
Korad	Acrylic films	Rhom & Haas
Korez	Phenolic resin cements	Atlas Mineral Products
Koroseal	PVC	B.F. Goodrich
Kosmos	Carbon black	United Carbon Company
Kotol	Resin solutions	Uniroyal

Trade or Brand Name	Product	Manufacturer
Kralac	ABS resins	Uniroyal
Kralastic	ABS	Uniroyal
Kralon	HIPS and ABS resins	Uniroyal
Kraton	Butadiene block copolymers and linear and branched styrene block copolymers	Shell Chemical
Kraton D	SBS or SIS terpolymers	Shell Chemical
Kraton IPN	SEBS-Polyester	Shell Chemical
Krene	Plasticized vinyl film	Union Carbide
K-Resin	Butadiene-styrene copolymer	Phillips Petroleum
Kriston	Allyl ester casting resins	B.F. Goodrich
Krystal	PVC sheet	Allied Chemical
Krystaltite	PVC shrink film	Allied Chemical
Kydene	Acrylic-PVC powder	Rhom & Haas
Kydex	Acrylic-PVC sheets	Rhom & Haas
Kylan	Chitan	Generic name
Kynar	Poly(vinylidene fluoride)	Atochem
Laden	PS	SABIC
Lamabond	Reinforced PE	Columbia Carbon Co.
Lamar	Mylar vinyl laminates	Morgan Adhesives
Laminac	Polyester resins	American Cyanamid
Lanital	Fiber from milk protein	Shia Viscosa
Laguval	Polyester resins	Farbenfabriken Bayer AG
Last-A-Foam	Plastic foam	General Plastics Mfg.
Lekutherm	Epoxy resins	Farbenfabriken Bayer AG
Lemac	Poly(vinyl acetate)	Borden Chemical
Lemol	Poly(vinyl alcohol)	Borden Chemical
Levapren	Ethylene-vinyl acetate copolymers	Farbenfabriken Bayer AG
Lexan	Polycarbonate resin	General Electric
Lock Foam	Polyurethane foam	Nopco Chemical Co.
Lofguard	Polyester fiber	KoSa
Lomod	Polyester block copolymer with polyether	General Electric
Lubriloy	Nylon-6,-66, 6/12, PBT	Comalloy
Lucel	Acetal copolymer	Lucky
Lucet	Acetal copolymer	Lucky
Lucite	PMMA and copolymers	DuPont
Ludox	Colloidal silica	DuPont
Lumarith	Cellulose acetate	Celanese
Lumasite	Acrylic sheet	American Acrylic Corp.
Lumax	PBT alloy	Lucky
Lumite	Saran filaments	Chicopee Manufacturing Co.
Lupan	SAN	Lucky
Lupol	Polyolefin	Lucky
Lupon	Nylon066	Lucky
Lupos	ABS	Lucky
Lupox	PBT	Lucky
Lupoy	Polycarbonate/ABS	LG Chemical
Luran	SAN, ASA	BASF
Lusep	PPS	Lucky
Lustran	Molding and extrusion resins, ABS	Bayer
Lustrex	PS	Monsanto
Lutonal	Poly(vinyl ethers)	Badische Anilin & Soda-Fabrik AG

continued

Trade or Brand Name	Product	Manufacturer
Lutrex	Poly(vinyl acetate)	Foster Grant
Luvican	Poly(vinyl carbazole)	Badische Anilin & Soda-Fabrik AG
Luxis	Nylon-6	Westover
Lycra	Spandix fibers	DuPont/INVISTA
Macal	Cast vinyl films	Morgan Adhesive Co.
Madurik	Melamine–formaldehyde resins	Casella Farbwerke Mainkur
Magnacomp	Nylon 6; 6,10, PP	LPN
Magnum	ABS	Dow Plastics
Makrofol	Polycarbonate film	Naftone, Inc.
Makrolon	Polycarbonate	Farbenfabriken Bayer AG
Malecca	Copolymers with styrene	Denki Kagaku
Maranyl	Nylon	ICI Americas
Marqesa	Olefin fibers	American Fibers & Yarns
Marquesa Lana	Olefin fibers	Shaw Inds.
Marafoam	Polyurethane foam	Marblette Co.
Maraglas	Epoxy resin	Marblette Co.
Maranyl	Nylons	Imperial
Maraset	Epoxy resin	Marblette Co.
Marathane	Urethane materials	Allied Products
Maraweld	Epoxy resin	Marblette Co.
Marbon	PS and copolymers	Borg-Warner
Marlex	Polyolefin resins	Phillips Chemical
Marvibond	Metal-plastic laminates	Uniroyal
Marvinol	PVC	Uniroyal
Mater-Bi	Biodegradable polymers	Novamont
Melan	Melamine resins	Hitachi Chemical
Meldin	Polyimides	Dixon Corp.
Melinex	PET	Imperial
Melit	Melamine–formaldehyde resins	Societa Italiana Pesine
Melmac	Melamine molding materials	American Cyanamid
Melolam	Melamine resin	Ciba Geigy
Melurac	Melamine-urea resins	American Cyanamid
Meraklon	PP	Montecatini
Merlon	Polycarbonate	Mobay Chemical
Meryl (and related trade nemes)	Nylon-6 and 66 fibers	Nylstar
Metallex	Cast acrylic sheets	Hermes Plastics
Methocel	Methylcellulose	Dow
Meticone	Silicone rubber	Hermes Plastics
Metre-Set	Epoxy adhesives	Metachem Resins Corp.
Micarta	Thermosetting laminates	Westinghouse Electric Corp.
Microblocker	Olefin fibers	SI Corp.
Microdenier Sensure	Polyester fiber	Willman, Inc.
Microlux	Polyester fiber	KoSa
Micro-Matte	Extruded acrylic sheet with matte finish	Extrudaline, Inc.
Micronex	Carbon black	Columbian Carbon Co.
Micropel	Powdered nylon	Nypel, Inc.
Microsol	Vinyl plastisol	Michigan Chrome & Chemical
Microthene	Powdered PE	Quantum
MicroSafe	Acetate fibers	Celanese Acetate
MicroSupreme	Acrylif fibers	Sterling Fibers

Trade or Brand Name	Product	Manufacturer
Milastomer	Thermoplastic elastomer	Mitsui
Milmar	Polyester	Morgan Adhesives
Mindel	PSU, PSU alloy	Amoco Polymers
Mini-Vaps	Expanded PE	Malge Co.
Minit Grip	Epoxy adhesives	High Strength Plastics Corp.
Minit Man	Epoxy adhesives	Kristal Draft, Inc.
Minlon	Reinforced nylon	Dupont
Mipolam	PVC	Dynamit Nobel
Mipoplast	PVC sheets	Dynamit Nobel
Mirafi	Olefin fibers	TenCate Geosynthetics
Mirasol	Alkyd resins	C.S. Osborn Chemicals
Mirbane	Amino resins	Showa Highpolymer Co.
Mirrex	Calendered PVC	Tenneco Chemicals
Mista Foam	Urethane foam	M.R. Plastics & Coatings
Modulene	PE resins	Muehlstein & Co.
Mogal	Carbon black	Cabot Corp.
Molplen	PP	Novamont Corp.
Moltopren	Polyurethane foam	Farbenfabriken Bayer AG
Molycor	Fiberglass-reinforced epoxy tubing	A.O. Smith, Inland Inc.
Monocast	Nylon	Polymer Corp.
Montac	Polyamide copolymer with polyether/ polyester	Monsanto
Montrek	Polyethyleneimine	Dow
Moplen	PP	Montecatini
Morthane	Polyurethane block copolymer with polyether/polyester	Morton Int.
Mowlith	Poly(vinyl acetate)	Farbwerke Hoechst AG
Mowiol	Poly(vinyl alcohol)	Farbwerke Hoechst AG
Mowital	Poly(vinyl butyral)	Farbwerke Hoechst AG
Multibase	ABS	Multibase
Multi-Flex	Linear block styrene copolymers with ethylene–propylene	Multibase
Multi-Hips	PS	Multibase
Multi-Pro	PP	Multibase
Multi-San	SAN	Multibase
Multrathane	Urethane elastomer	Mobay Chemical
Multron	Polyesters	Mobay Chemical
Mycalex	Inorganic molded plastic	Mycalex Corp. America
Mylar	Polyester film	DuPont
Napryl	PP	Pechiney-Saint-Gobain
NAS	SMMA acrylic	Nova Chemicals
Natene	PE	Pechiney-Saint-Gobain
Natsyn	cis-1,4-Polyisoprene	Goodyear
Naugahyde	Vinyl-coated fabric	U.S. Rubber Co.
NeoCryl	Acrylic resins and emulsions	Polyvinyl Chemicals
Neoprene	Polychloroprene	Dupont
NeoRez	Styrene emulsions and urethane solutions	Polyvinyl Chemicals
NeoVac	PVA emulsions	Polyvinyl Chemicals
Nepoxide	Epoxy resin coating	Atlas Minerals & Chemicals
Nestrite	Phenolic and urea–formaldehyde resins	James Ferguson & Sons
Nevidene	Coumarone–indene resin	Neville Chemical Co.
Nevillac	Modified coumarone–indene resin	Neville Chemical Co.

continued

Trade or Brand Name	Product	Manufacturer
Niax	Polyol polyesters	Union Carbide
Nimbus	Polyurethane foam	General Tire & Rubber
Nipeon	PVC	Japanese Geon Co.
Nipoflex	Ethylene-vinyl acetate copolymer	Toyo Soda Mfg.
Nipolon	PE	Toyo Soda Mfg.
Nitrocol	Nitrocellulose-based pigment dispersions	J. C. Osburn Chemicals
Noan	Styrene–methyl methacrylate copolymer	Richardson Corp.
Nob-Lock	PVC sheets	Ameron Corrosion Control
Nomex	Aramid Nylon	DuPont
Nopcofoam	Polyurethane foam	Nopco Chemical Co.
Norchem	LDPE resin	Northern Petrochemical Co.
Nordel	Ethylene-propylene	DuPont
Norsophen	Phenolic resins	Norold Composites
Nortuff	HDPE, PP	Polymerland
Noryl	Poly(phenylene oxide), PPO alloy	General Electric
Novalast	Thermoplastic elastomers	Nova Polymers
Novalene	Thermoplastic elastomers	Nova Polymers
Novamid	Nylon	Mitsubishi
Novapol	LLDPE, LDPE, HDPE	Nova Chemicals
Novatemp	PVC	Novatec
Novelle	Olefin fibers	FiberVisions
Novodur	ABS polymers	Farbenfabriken Bayer AG
Novon	Starch-based polymer	Novon
NSC	Nylon, PS	Thermofil
Nuclon	Polycarbonate	Pittsburgh Plate Glass Co.
Nucrel	EMMA copolymer	DuPont
Nukem	Acid-resistant resin cements	Amercoat Corp.
Numa	Spandex fibers	American Cyanamid
Nupol	Thermosetting acrylic resin	Freeman Chemical
Nybex	Nylon 6,12	Nova Chemicals
Nydur	Nylon 6	Durethan
Nyglathane	Glass-filled polyurethane	Nypel, Inc.
Nylafil	Reinforced nylon	DSM
Nylaglas	Fiberglass-reinforced nylon	Dart
Nylamid	Nylon	Polymer Service
Nylasar	Fiberglass-reinforced nylon	Dart
Nylasint	Sintered nylon parts	Polymer Corp.
Nylast	Thermoplastic elastomer	Allied Signal
Nylatron	Filled nylons	DSM
Nylene	Nylon	Custom Resins
Nylind	Nylon 66	DuPont
Nylon	Polyamides	DuPont
Nylo-Seal	Nylon-11 tubing	Imperial-Eastman
Nyloy	Nylon 66, PC, PP	Nytex Composites
Nylux	Nylons	Westlake Plastics
Nypel	Nylon 6	Allied Signal
Nyplube	TFE-filled nylons	Nypel, Inc.
Nyreg	Glass-reinforced nylons	Nyper, Inc.
Nytron	Nylon 66	Nytex Composites
Oasis	Expanded phenol–formaldehyde	Smithers Co.
Olefane	PP film	Avisum Corp.
Olefil	Filled PP resin	Amoco Chemicals

Trade or Brand Name	Product	Manufacturer
Oleflo	PP resin	Amoco Chemicals
Olehard	Filled PP	Chiso America
Olemer	Propylene copolymer	Avisum Corp.
Oletac	Amorphous PP	Amoco Chemicals
Ontex	Thermoplastic elastomer	D & S Plastics
Opalon	PVC	Monsanto
Oppanol B	Polyisobutylene	Badische Anilin & Soda-Fabrik AG
Oppanol C	Poly(vinyl isobutylether)	Badische Anilin & Soda-Fabrik AG
Optema	EMA copolymer	Exxon
Optix	Acrylic	Plaskolite
Orel	Polyester fiber	DuPont
Orevac	Polyamide copolymer with polyether/ polyester	Atochem
Orgalacqe	Epoxy and PVC powders	Aquitaine-Organico
Orgamide R	Nylon-6	Aquitaine-Organico
Orlon	Acrylic fibers	DuPont
Ortix	Poromeric film	Celanese
Oxy	Vinyl	Occidental
Oxyblend	Vinyl	Occidental
Oxyclear	PVC	Occidental
Panda	Vinyl and urethane-coated fabrics	Pandel-Bradford
Panelyte	Laminates	Thiokol
Panlite	PC	Teijin Chemical
Papi	Polymethylene, poly(phenyl isocyanate)	Upjohn
Paracon	Polyester rubber	Bell Telephone Labs.
Paracryl	Butadiene–acrylonitrile copolymers	U.S. Rubber Co.
Paradene	Coumarone–indene resins	Neville Chemical Co.
Paralac	Polyester resin	ICI
Parfe	Rayon fiber	Beaunit Mills Corp.
Parlon	Chlorinated rubber	Hercules Corp.
Parylen C	Polymonochloro-p-xylene	Union Carbide
Parylen N	Polyxylene	Union Carbide
Paxon	HDPE	Paxon
Pearlon	PE film	Visking Corp.
Pebax	Polyamide block copolymer with polyether/polyester	Atochem
Pee Vee Cee	PVC	ESB Corp.
Pelaspan	Expandable PS	Dow
Pellethene	Thermoplastic urethane	Dow Plastics
Pentalyn	Abietic acid derivatives	Hercules
Pentec	Polyester fiber	Honeywell Int.
Penton	Chlorinated polyether resins	Hercules
Perbunan N	Butadiene–acrylonitrile copolymers	Farbenfabriken Bayer AG
Perlon	Polyurethane filament	Farbenfabriken Bayer AG
PermaRex	Cast epoxy	Permali
Permasoft	Nylon-6	Beaulieu
Peremelite	Melamine resin	Melamine Plastics
Permutit	Ion-exchange resin	Permutit Co.
Perspex	Acrylic resins	ICI
Petlon	PBT	Albis
Petra	Polyester sheets	Allied Chemical
Petrothene	PE	Quantum

continued

Trade or Brand Name	Product	Manufacturer
Pevalon	Poly(vinyl alcohol)	May & Baker Ltd.
Phenoweld	Phenolic adhesive	Hardman, Inc.
Philjo	Polyolefin film	Phillips-Joana Co.
Philprene	Sytrene–butadiene rubber	Phillips Petroleum
Pibiter	Poly(butylene terephthalate)	EniChem
Picco	Hydrocarbon resins	Hercules
Piccocumaron	Hydrocarbon resins	Hercules
Piccoflex	Acrylonitrile–styrene resins	Pennsylvania Industrial Chemical
Piccolastic	PS resin	Pennsylvania Industrial Chemical
Piccolyte	Terpene polymer resins	Hercules
Piccotex	Vinyl-toluene copolymers	Pennsylvania Industrial Chemical
Piccoumaron	Coumarone-indene resins	Pennsylvania Industrial Chemical
Piccovar	Alkyl-aromatic resins	Pennsylvania Industrial Chemical
Pienco	Polyester resins	American Petrochemical
Pil-Trol	Acrylic fibers	Solutia
Plaskon	Amino resins	Allied Chemical
Plastacel	Cellose acetate flake	DuPont
Plastylene	PE	Pechiney-Saint-Gobain
Plenco	Phenolic resins	Plastics Engineering Co.
Pleogen	Polyester resins and gels	Whittaker Corp.
Plexiglas	Acrylic sheets	Rohm & Haas
Plexigum	Acrylate and methacrylate resins	Rohm & Haas
Plicose	PE	Diamond Shamrock
Pliobond	Adhesive	Goodyear
Pliofilm	Rubber hydrochloride	Goodyear
Plioflex	PVC	Goodyear
Pliolite	Cyclized rubber	Goodyear
Pliothene	PE rubber blends	Ametek/Westchester
Pliovic	PVC	Goodyear
Pluracol	Polyethers	Wyandotte Chemicals
Pluragard	Urethane foams	BASF Wyandotte
Pluronic	Polyethers	BASF Wyandotte
Plyocite	Phenol-impregnated materials	Reichhold Chemicals
Plyophen	Phenolic resins	Reichhold Chemicals
Pluronics	Block polyether diols	Wyandotte Corp.
PMC	Melamine formaldehyde	Sun Coast
Pocan	Poly(butylene terephthalate)	Albis
Polarguard (and related trade names)	Polyester fibers	KoSa
Polex	Oriented acrylics	Southwestern Plastics
Pollopas	Urethane–formaldehyde materials	Dynamit Nobel
Polvonite	Cellular plastic materials	Voplex Corp.
Polyallomer	Ethylene block copolymers	Eastman Chemicals
Poly-Dap	Diallyl phthalate resins	U.S. Polymeric
Polycarbafil	Fiberglass-reinforced polycarbonates	Dart
Polycure	Cross-linked PE	Crooke Color & Chemical
Poly-eth	PE	Gulf Oil
Poly-eze	Ethylene copolymers	Gulf Oil
Polyflon	Fluoropolymers	Daikin
Polyfoam	Polyurethane foam	General Tire & Rubber
Poly-Gard	Solventless epoxies	Richhold Chemicals
Polyimidal	Polyimide thermoplastics	Raychem Corp.

Trade or Brand Name	Product	Manufacturer
Polylasting	Olefin fibers	Blue Mountains Inds.
Polylite	Polyester resins	Reichold Chemicals
Polyloom	Olefin fibers	TC Thiolon
Polylumy	PP	Kohjin Co.
Polyman	ABS Alloy	Schulman
Polymet	Plastic-filled sintered metal	Polymer Corp.
Polymin	Polyethyleneimine	Badische Anilin & Soda-Fabrik AG
Polymul	PE emulsions	Diamond Shamrock
Poly-pro	PP	Gulf Oil
Polyox	Water soluble resins	Union Carbide
Polysizer	Poly(vinyl alcohol)	Showa Highpolymer Co.
Polystar	Olefin fiber	Nexcel Synthetics
Polyteraglas	Polyester-coated Dacron glass fabric	Natvar Corp.
Polytron	PVC alloy	Geon
Polytrope	PP blend with EPDM or EPR	Schulmam
Poly Tying	Olefin fibers	Blue Mountain Inds.
Polyvin	PVC	Schulman
Polyviol	Poly(vinyl alcohol)	Wacker Chemie GmbH
Powminco	Asbestos fibers	Powhatan Mining Co.
PPO	Poly(phenylene oxide)	Hercules
Premi-glas	Glass reinforced SMC	Premix
Prevail	ABS/Polyurethane	Dow Plastics
Prevex	PE	GE Plastics
Primef	PS	Solvay
Prism	RIM Polyurethane	Bayer
Pro-fax	PP resins	Hercules Powder Co.
Profil	Fiberglass-reinforced PP	Dart
Proglas	Fiberglass-reinforced PP	Dart
Prohi	HDPE	Protective Lining Corp.
Prolan	Olefin fiber	Ronile
Propathene	PP	Imperial
Propax	PP	PolyPacific
Propiofan	Poly(vinyl propionate)	BASF
Propylsar	Fiberglass-reinforced PP	Dart
Propylus	PP	Westlake Plastics
Protectolite	PE film	Protective Lining Corp.
Protron	Ultrahigh-strength PE	Protective Lining Corp.
Pulse	Polycarbonate/ABS	Dow Plastics
Purilon	Rayon	FMC Corp.
PYR-ML	Polyimide	DuPont
Quadrol	Poly(hydroxy amine)	Wyandotte Chemicals
QualiFlo	Polyester fiber	Reemay
Quelflam	Polyurethanes	Baxenden Chemical
Quintac	Linear styrene block copolymer with isoprene	Nippon Zeon
Radel	Poly(ether sulfone)	Amoco
Radilon	Nylon 6	Radicinovacips
Radipol	Nylon 66	Radicinovacips
Ravinil	PVC	ANIC, SPA
Raybrite	Alpha-cellulose filler	Rayonier, Inc.
Rayflex	Rayon	FMC Corp.
Regalite	Press-polished PVC	Tenneco Advanced Materials

continued

Trade or Brand Name	Product	Manufacturer
Ren-Flex	PP blend with EPDM or EPR	D & S
REN-Shape	Epoxy materials	Ren Plastics
Ren-Thane	Urethane elastomers	Ren Plastics
Reny	Nylon 66	Mitsubishi
Replay	PS	Huntsman
Reprean	Ethylene copolymer	Discas
Resiglas	Polyester resins	Kristal Draft, Inc.
Resimene	Urea and melamine resins	Monsanto
Resinol	Polyolefins	Allied Resinous Products
Resinox	Phenolic resins	Monsanto
Resistoflex	Poly(vinyl alcohol)	Resistoflex Corp.
Resollm	Melamine resins	Monsanto
Resolite	Urea–formaldehyde resins	Ciba Geigy
Restfoam	Urethane foam	Stauffer Chemical
Restirolo	PS	Societa Italiana Resine
Retain	PE	Dsow Plastics
Retrieve	Polyester fiber	Marglen Inds.
Rexolene	Cross-linked polyolefin	Brand-Rex Co.
Rexolite	PS	Brand-Rex Co.
Reynolon	Plastic films	Reynolds Metals Co.
Reynosol	Urethane, PVC	Hoover Ball & Bearing Co.
Reflex	PP	Resene
Reximac	Alkyds	Commercial Solvents Corp.
Rezyl	Alkyd varnishes	Sinclair-Koppers
Rhodiod	Cellulose acetate	M & B Plastics
Rhonite	Resins for textile finishes	Rohm & Haas
Rhoplex	Acrylic emulsions	Rohm & Haas
Riblene	PE	ANIC, SPA
Richfoam	Polyurethane foam	E.R. Carpenter Co.
Rigidex	PE	BP Chemicals
Rigidite	Acrylic and polyester resins	American Cyanamid
Rigidsol	Rigid plastisol	Watson-Standard Co.
Rigolac	Polyester resins	Showa Highpolymer Co.
Rilsan	Nylon-11	Aquitaine-Organico
Rimplast	Blends of TPEs with silicone rubbers	Petrarch Systems
Riteflex	Polyester block copolymer with polyether	Hoechst
Rolox	Two part epoxies	Hardman
Ronfalin	ABS	DSM
Roskydal	Urea-formaldehyde resins	Farbenfabriken Bayer Ag
Royalbrite	Nylon-6	Royal-American
Royalex	Cellular thermoplastic sheets	Uniroyal
Royalite	Thermoplastic sheet materials	Uniroyal
Roylar	Polyurethanes	Uniroyal
Rucoam	Vinyl materials	Hooker Chemical
Rucon	PVC	Hooker Chemical
Rucothane	Polyurethanes	Hooker Chemical
Rulan	Flame-retardant plastic	DuPont
Rynite	Polyester (PET, PBT)	DuPont
Ryton	Poly(phenylene sulfide)	Phillips Petroleum
Sabre	PC + PET	Dow Plastics
Saflex	Poly(vinyl butyral)	Monsanto

Trade or Brand Name	Product	Manufacturer
Safom	Polyurethane foam	Monsanto
Salus	Olefin fiber	FFT
Santoprene	PP dynamic vulcanizate with nitrile rubber	AES
Saran	PVC and poly(vinylidene chloride) copolymerDow Plastics	
Sarlink 1000	PP dynamic vulcanizate with nitrile rubber	DSM
Sarlink 3000 and 4000	PP dynamic vulcanizate with EPDM	DSM
Satinflex	PVC	Alpha Grey
Satin Foam	Extruded PS foam	Dow
Scotch	Adhesives	3M
Scotchcast	Epoxy resins	3M
Scotchpak	Polyester film	3M
Scotchpar	Polyester film	3M
Scotchweld	Adhesives	3M
Schulamid	Nylon 6; Nylon 66	Schulman
Schulink	Cross-linkable HDPE	Schulman
Sclair	PE	Nova Chemicals
SEF	Modacrylic fibers	Solutia
Seilon	Thermoplastic sheets	Seiberling Rubber Co.
Selar	Nylon, PET	DuPont
Selectron	Polyester resins	PPG
Selecttrofoam	Polyurethane foam	PPG
Sensura	Polyester fiber	Wellman, Inc.
Serelle	Polyester fiber	KoSa
Serene	Polyester	KoSa
Shareen	Nylon	Courtaulds
Shell	Polyolefins	Shell
Shimmereen	Nylon-6	Honeywell Nylon
Shinite	PBT	Shinkong
Shuvin	Vinyl molding materials	Reichhold Chemicals
Silastic	Silicone materials	Dow Corning
Silkey Touch	Nylon-6 fibers	Honeywell Nylon
Silon-TSR	PDMS/PTFE	BioMed Sciences
Sipon	Alkyl and aryl resins	Alcoloa, Inc.
Silastomer	Silicones	Midland Silicones
Silbon	Rayon paper	Kohjin Co.
Silocet	Silicon rubber	ICI
Sinite	PBT	EniChem
Sinvet	PC	EniChem
Sirfen	Phenol-formaldehyde resins	Societa Italiana Resine
Sir-pel	Poromeric film	Georgia Bonded Fibers
Sirtene	PE	Societa Italiana Resine
Skinwich	Polyurethane integral-skinning foam	Upjohn
Smart Yarns	Acrylic fibers	Solutia
Soarnol	EVA copolymer	Nichimen
Softlite	Ionomer foam	Gilman Brother
Solarflex	Chlorinated PE	Pantascote Co.
Solef	PVDF	Solvay
Solithane	Urethane prepolymers	Thiokol
Solprene	Branched block copolymer of styrene and butadiene	Phillips Petroleum

continued

Trade or Brand Name	Product	Manufacturer
Solvic	PVC	Solvay
Sonite	Epoxy resin	Smooth-On, Inc.
Sovar	Poly(vinyl acetate)	Shawinigan Resins Corp.
Solvic	PVC	Solvay & Cie
Soreflon	PTFE	Rhone-Poulenc
Spandal	Polyurethane laminates	Baxenden Chemical
Spandex	Polyurethane copolymers filaments	DuPont
Spandofoam	Polyurethane foam	Baxenden Chemical
Spandoplast	Expanded PS	Baxenden Chemical
Spectar	Polyester copolymers	Eastman
Spectra (and related trade names)	Olefin fibers and materials	Honeywell Int.
Spectran	Polyester	Monsanto Textiles
S-polymers	Butadiene–styrene copolymers	Esso Labs.
Spraythane	Urethane resin	Thiokol
Spunnaire	Polyester fiber	Wellman, Inc.
Stainmaster (and related trade names)	Nylon-66 fibers	INVISTA
Standlite	Phenol–formaldehyde resins	Hitachi Chemical Co.
Stanyl	Nylon 46	DSM
Starex	Poly(vinyl acetate)	International Latex & Chemical Corp.
Statex	Carbon black	Columbian Carbon Co.
Stay Gard	Nylon-6 fibers	Honeywell Nylon
Stearon	Linear block styrene–butadiene copolymers	Firestone
Stepton	Linear styrene block copolymers	Kurary
Stereon	Styrene–butadiene block copolymer	Firestone
Steripur	Polyester fibers	DAK Americas, LLC
Stretch-aire	Polyester fibers	KoSa
Structo-Foam	Foamed PS slab	Stauffer Chemical Co.
Strux	Cellular cellulose	Aircraft Specialities
Stylafoam	Coated PS sheets	Gilman Brothers
Stymer	Styrene copolymer	Monsanto
Stypol	Urea-formaldehyde resins	Freeman
Styrafil	Fiberglass-reinforced PS	Dart
Styraglas	Fiberglass-reinforced PS	Dart
Styrex	Resin	Dow
Styrocel	Espandable PS	Styrene Products Ltd.
Styroflex	Biaxially oriented PS film	Natvar Corp.
Styrofoam	Extruded expanded PS	Dow
Styrolus	PS	Westlake Plastics
Styron	PS	Dow
Styronol	PS	Allied Resinour Prods.
Styropor	PS	BASF
Substraight	Polyester fiber	Honeywell Int.
Sulfasar	Fiberglass-reinforced polysulfone	Dart
Sulfil	Fiberglass-reinforced polysulfone	Dart
Sullvac	Acrylonitrile–butadiene–styrene terpolymer	O'Sullivan Rubber Corp.
Sumiplex	Acrylics	Sumitomo
Sunlon	Nylon resins	Sun Chemical Corp.
Sunprene	PVC elastomer	Schulman
Suntra	PPS	Sunkyong Industries

Trade or Brand Name	Product	Manufacturer
Supec	PPS	GE Plastics
Superkleen	PVC	Alpha Gary
Super Aeroflex	Linear PE	Anchor Plastic Co.
Super Coilife	Epoxy potting resin	Westinghouse Electric
Super Dylan	HDPE	Arco Polymer Co.
Superflex	Grafted high-impact PS	Gordon Chemical
Superflow	PS	Gordon Chemical
Supplex	Nylon-66 fibers	INVISTA
Suprel	ABS-PVC	Vista Chemicals
Surflex	Ionomer film	Flex-O-Glass, Inc.
Surlyn	Ionomer resin	DuPont
Syn-U-Tex	Ureathane–formaldehyde and melamine–formaldehyde	Celanese Coatings Co.
Swedcast	Acrylic sheet	Swedlow, Inc.
Sylgard	Silicon casting resins	Dow Corning
Sylplast	Urea-formaldehyde resins	Sylvan Plastics, Inc.
Syntex	Alkyd resins	Celanese Corp.
Synthane	Laminated plastic products	Synthane Corp.
Syretex	Styrenated alkyd resins	Celanese Coatings Co.
Tactasse	Nylon-66 fiber	INVISTA
Taipol	Linear and branched styrene block copolymers	Taiwan Synthetic Rubber Co.
Tairilin	Polyester fiber	Nan Ya Plastics Corp.
TanClas	Spray or dip plastisol	Tamite Inds.
Technyl	Nylon 66	Rhone-Poulenc
Tecoflex	PUR	Thermidics
Tedlar	Polyvinyl fluorocarbon resins	DuPont
Tedur	PPS	Albis
Teflon	Polytetrafluoroethylene and related materials	DuPont
Teflon FEP	TFE copolymer	DuPont
Teflon TFE	PTFE	DuPont
Tefzel	PE-TFE fluoropolymers	DuPont
Teglac	Alkyl coatings	American Cyanamid
Tego	Phenolic resins	Rohm & Haas
Tekon	Linear block styrene copolymer with ethylene–butadiene	Teknor Apex
Tekton	Olefin fiber	Reemay
Telar	Olefin fiber	FFT
Telcar	PP blend with EPDM or EPR	Teknor Apex
Terluran	ABS	BASF
Tempra	Rayon fiber	American Enka Corp.
Tempreg	Low-pressure laminates	U.S. Plywood Corp.
Tencel	Lyocell fibers	Tencel
Tenite	Cellulose derivatives	Eastman Kodak
Tenn Foam	Polyurethane foam	Morristown Foam Corp.
Tensylon	Olefin fibers	SI Corp.
Teracol	Poly(oxytetramethylene glycol)	DuPont
Tere-Cast	Polyester casting resins	Reichhold Chemicals
Terluran	ABS polymers	Badisch Anilin & Soda-Fabrik AG
Terucello	Carboxymethyl cellulose	Showa Highpolymer Co.
Terylem	PET	ICI

continued

Trade or Brand Name	Product	Manufacturer
Terylene	Polyester fiber	ICI
Tetra-Phen	Phenolic resins	Georgia-Pacific
Tetra-Ria	Amino resins	Georgia-Pacific
Tetraloy	Filled TFE molding resins	Whitford Chemical
Tetronic	Polyethers	Wyandotte Chemical Corp.
Texalon	Nylon	Texapol
Texicote	Poly(vinyl acetate)	Scott Bader Co.
Texileather	Pyroxylin-leather cloth	General tire & Rubber
Texin	Urethane elastomer	Mobay Chemical
Textolite	Laminated plastic	General Electric
Thermaflow	Reinforced polyesters	Atlas Powder Co.
Thermalux	Polysulfones	Westlake Plastics
Thermasol	Vinyl plastisols and organosols	Lakeside Plastics
Thermax	Carbon black	Commercial Solvents Corp.
Thermco	Expanded PS	Holland Plastics
Thiokol	Poly(ethylene sulfide)	Thiokol
Thornel	Graphite filaments	Union Carbide
Thurane	Polyurethane foam	Dow
T-Lock	PVC sheets	Amercoat Corp.
Topas	Cycloolefin copolymer	Hoechst-Celanese
Topel	Rayon fiber	Courtaulds
Topex	PBT	Tong Yang Nylon
Toplex	PC/ABS	Multibase
Toray	PBT	Toray Industries
Torlon	Polyamide-imide	Amoco Polymers
Toyolac	ABS, ABS/PC	Toray Industries
TPX	Poly-4-methylpentane-1	Imperial
Trace	Olefin fibers	American Fibers & Yarns
Trans-4	*trans*-1,4-Polybutadiene	Phillips Petroleum
Trans-Stay	Polyester film	Transiwrap Co.
Trevarno	Resin-impregnated cloth	Coast Mfg. & Supply Corp.
Triax	PC/ABS, ANS/Nylon	Bayer
Tribit	PBT	Sam Yang
Trilene	Nylon-6 fiber	Berkley
Tri-Foil	TFE-coated aluminum foil	Tri-Point Inds.
Trilon	TFE	Dynamit Nobel
Triocel	Rayon acetate	Celanese Fibers
Trirex	PC	Sam Yang
Trithene	TFE	Union Carbide
Trolen	PE	Dynamit Nobel
Trolitan	Phenol–formaldehyde	Dynamit Nobel
Trosifol	Poly(vinyl butyral) film	Dynamit Nobel
Trosiplast	PVC	Dynamit Nobel
Trubyte	Acrylic-based multicomponent dental system	Dentsply
Trulon	PVC resin	Olin Corp.
Tuffak	Polycarbonate	Rohm & Haas
Tufrex	ABS	Bayer
Tuftane	Polyurethane	B. F. Goodrich
Trusite	Olefin fibers	Nexcel Synthetics
Tusson	Rayon fiber	Beaunit Mills Corp.

Trade or Brand Name	Product	Manufacturer
Tybrene	ABS polymers	Dow
Tygon	Vinyl copolymer	U.S. Stoneware Co.
Tylose	Cellulose ethers	Farbwerke Hoechst AG
Tynex	Nylon bristles and filaments	DupPont
Typar	Olefin fibers	Reemay
Typelle	Olefin fibers	Reemay
Tyril	Styrene-acrylonitrile copolymer	Dow
Tyrilfoam	Styrene/acrylonitrile foam	Dow
Tyrin	Chlorinated PE	Dow
Tyrite	Olefin fiber	Nexcel Synthetics
Tyvec	Olefin fibers	DuPont
Udel	PSO	Amoco
Uformite	Urea resins	Rohm & Haas
Ultem	Polyetherimide	GE Plastics
Ultradur	PBT	BASF
Ultraform	Acetal	BASF
Ultraline	Olefin fibers	Samson Rope Tech.
Ultramid	Nylon	BASF
Ultrason-E	Poly(ether sulfone)	BASF
Ultrason-S	Polysulfone (PSO)	BASF
Ultrastyr	ABS	Enichem America
Ultrathene	EVA Copolymers	Quantum
Ultrapas	Melamine-formaldehyde resins	Dynamit Nobel
Ultron (and related trade names)	Nylon-66 fibers	Solutia
UltraFlo	Polyester fiber	Reemay
Ultra Touch	Nylon-6 fibers	Honeywell Nylon
Ultryl	PVC	Phillips Petroleum
Ultura	Polyester fiber	Wellman, Inc.
Unichem	PVC	Colorite Plastics
Unifoam	Polyurethane foam	William T. Burnett & Co.
Unipoxy	Epoxy resins and adhesives	Kristal Kraft Co.
Unival	PE	Union Carbide
Unox	Epoxies	Union Carbide
Urac	Urea–formaldehyde resins	American Cyanamid
Urafil	Fiberglass-reinforced polyurethane	Dart
Uraglas	Fiberglass-reinforced polyurethane	Dart
Uralite	Polyurethanes	Hexcel Corp.
Urapol	Polyurethane elastomeric coatings	Gordon Chemicals
Urapac	Rigid polyurethanes	North American Urethanes
Urapol	Polyurethane elastomeric coatings	Poly Resins
Urecoll	Urea–formaldehyde resins	BASF
Uscolite	ABS copolymer	U.S. Rubber Co.
U-Thane	Rigid insulation polyurethane	Upjohn
Uvex	Cellulose acetate butyrate	Eastman Kodak
Valox	Polyesters (PBT, PCT, PET)	General Electric
Valsof	PE emulsions	United Merchants & Manfs.
Valtra	PS	Chevron Chemical
Vandar	Polyester alloy	Hoechst-Celanese
Varcum	Phenolic resins	Reichhold Chemicals
Varex	Polyester	McCloskey Varnish Co.

continued

Trade or Brand Name	Product	Manufacturer
Varkyd	Alkyd and modified alkyd resins	McCloskey Varnish Co.
Varsil	Silicon-coated fiberglass	New Jersey Wood Finishing
V del	Polysulfone resins	Union Carbide
Vector	Linear styrene block copolymers with butadiene/isoprene	Dexco
Vectra	PP fibers	Exxon-Mobile
Velene	PS-foam laminates	Scott Paper Co.
Velon	PVC	Firestone Tire & Rubber
Verel	Modacrylic staple fibers	Eastman Chemicals
Versamid	Polyamide resins	General Mills Inc.
Versel	Polyester thermoplastic	Allied Chemical Corp.
Versi-Ply	Coextruded film	Pearson Inds.
Vespel	Polymelitimide	DuPont
Vestamid	Nylon-12	Chemische Werke Huls AG
Vestolit	PVC	Chemische Werke Huls AG
Vestyron	PS	Chemische Werke Huls AG
Vibrin	Polyester resins	Uniroyal
Vibrin-Mat	Polyester-glass molding material	W. R. Grace
Vibro-Flo	Epoxy and polyester coating powders	Armstrong Products
Vicara	Protein fiber	Virginia-Caroline Chem. Co.
Viclan	PVC	Imperial
Victrex	PEEK	ICI
Videne	Polyester film	Goodyear Tire & Rubber
Vinac	Poly(vinyl acetate) emulsions	Air Reduction Co.
Vinapas	Poly(vinyl acetate)	Wacker Chemie GmbH
Vinidur	PVC	BASF
Vinoflex	PVC	BASF
Vinol	Poly(vinyl alcohol)	Air Reduction Co.
Vinsil	Rosin derivative	Hercules
Vinylite	Poly(vinyl chloride-covinyl acetate)	Union Carbide
Vinyon	Poly(vinyl chloride-coacrylonitrile)	Union Carbide
Vipla	PVC	Montecatini Edison SPA
Viscalon	Rayon fiber	American Enka
Viskon	Nonwoven fabrics	Union Carbide
Vista	PVC	Vista Chemical
Vistanex	Polyisobutylene	Enjay
Vitel	Polyester resins	Goodyear
Vithane	Polyurethanes	Goodyear
Viton	Copolymer of vinylidene fluoride and hexafluoropropylene	DuPont
Vituf	Polyester resins	Goodyear
Volara	Closed-cell LDPE foam	Voltek, Inc.
Volaron	Closed-cell LDPE foam	Voltek, Inc.
Volasta	Closed-cell medium density PE foam	Voltek, Inc.
Voranol	Polyurethane foam	Dow
Vulcaprene	Polyurethane	Imperial
Vulkollan	Urethane elastomer	Mobay Chemical Co.
Vult-Acet	Poly(vinyl alcohol) latexes	General Latex & Chemical
Vultafoam	Polyurethane foam	General Latex & Chemical
Vultathane	Polyurethane coatings	General Latex & Chemical
Vybak	PVC	Bakelite Xylonite Ltd.

Trade or Brand Name	Product	Manufacturer
Vybex	Polyester	Ferro
Vycron	Polyester fiber	DuPont
Vydyne	Nylon resins	Monsanto
Vygen	PVC	General Tire & Rubber
Vynaclor	PVC emulsions	B. F. Goodrich Chemical
Vynaloy	Vinyl sheets	B. F. Goodrich Chemical
Vynex	Rigid vinyl sheeting	Nixon-Baldwin Chemicals
Vyram	PP dynamic vulcanizate with NR	AES
Vyrene	Spandex fiber	US Rubber Co.
Vythene	PVC + PUR	Alpha Gary
Wear-Dated (and related trade names)	Acrylic and Nylon-66 fibers	Solutia
WeatherBloc	Acrylic fiber	Sterling fibers
Webril	Nonwoven fabric	Kendall Co.
Weldfast	Epoxy and polyester adhesives	Fibercast Co.
Wellamid	Nylon-66 and -6 molding resins	Wellman, Inc.
Welltite	Olefin fibers	Wellington
Wellon	Nylon-66 and -6 fibers	Wellman, Inc.
Wellstrand	Nylon-66 and -6 fibers	Wellman, Inc.
Welvic	PVC	Imperial
Whirlclad	Plastic coatings	Polymer Corp.
Whirlsint	Powdered polymers	Polymer Group
Whitcon	Fluoroplastic lubricants	Whitford Chemical Corp.
Wicaloid	Styrene/butadiene emulsions	Ott Chemical Co.
Wicaset	PVC emulsions	Ott Chemical Co.
Wilfex	Vinyl plastisols	Flexible Products Co.
Xenoy	PC/Polyester	GE Plastics
XT Polymer	Acrylics	American Cynamid
Xydar	Liuqid crystal polymers	Amoco
Xylon	Nylon-66 and nylon-6	Dart
Xylonite	Cellulose nitrate	B. X. Plastics, Inc.
Zantrel	Rayon fiber	American Enka
Zee	PE wrap	Crown Zellerback
Zefran	Acrylic fiber	Dow
Zefsport	Nylon-6 fibers	Honeywell Nylon
Zeftron (line of trade names)	Nylon-6 fibers	Honeywell Nylon
Zelux	PE films	Union Carbide
Zemid	PE, HDPE	DuPont Canada
Zendel	PE	Union Carbide
Zeonex	Polymethylpentene (PMP)	Nippon Zeon
Zerlon	Acrylic/styrene copolymer	Dow
Zerok	Protective coatings	Atlas Minerals & Chemicals
Zetafax	Poly(ethylene-coacrylic acid)	Dow
Zetafin	Poly(ethylene-coethyl acrylate)	Dow
Zylar	Acrylic copolymer	Novacor
Zytel series	Nylons	DuPont

Appendix C
Syllabus

Three interrelated questions can be addressed when considering the construction of a course. These questions are (1) topics to be covered, (2) order in which these topics should be covered, and (3) proportion of time to be spent on each topic. Just as in any other area of science and engineering, there exists a healthy variety of topics, extent of coverage of each topic, and order of covering the material. There is no "right" answer and this text is developed so that chapters can be dealt with in essentially any order with the most important topics dealt with near the beginning of most chapters and (possibly) less important material covered later in the chapter.

The study of polymers is expanding at a rapid rate with too much fundamental material to be handed in a single introductory course, yet the basic elements can be included in such a course and are included in this text. Some topics that are today considered to be fundamental were not known a decade ago. Each of the fundamental topics are placed into perspective in the current text building upon the foundational courses of chemistry—organic, physical, inorganic, and analytical chemistry.

One assumption agreed upon by most academic and industrial polymer scientists and engineers and associated education committees is that there should be both a core of material common to introductory courses and a portion that reflects individual interests and training of teachers, student bodies, and local preferences and circumstances. Thus, not every topic needs to be covered to present a meaningful introductory polymer course. Some years ago, PolyEd, the education arm of Polymer Chemistry and Polymeric Materials: Science and Engineering, developed, with the help of polymer scientists and engineers, a listing of basic topics and preferred coverage. The results are given in Table C.1.

Basically, the committee proposed that all lecture courses include portions of the first seven topics with the level and extent of coverage guided by such factors as available class time, additional topics covered, interest of instructor, student interests, class composition, and so forth. It must be emphasized that the "optional topics" listed should not be considered limiting and that additional topics can be introduced.

TABLE C.1
Preferred Topics in Introductory Polymers

Topic	Amount of course time (%)	Chapter
Introduction	5	1
Morphology	10	2
Stereochemistry		
Molecular interactions		
Crystallinity/amorphousity		
Molecular weights	10	3
General types		
Solubility		
Testing and characterization	10	13, 14
Structure/property relationships		
Physical tests		
Spectral identification		
Stepwise polymerizations and condensation polymers	10	4
Chain-reaction polymerizations and addition polymers	10	5, 6
Ionic and free radical kinetics of polymerization		
Polymers produced by chain-reaction polymerization		
Copolymerization	10	7
Kinetics		
Types of copolymers		
Blends		
Principle copolymers		
Optional topics	35	
Natural and biomedical polymers		9, 10
Organometallic polymers		11
Inorganic polymers		12
Reactions of polymers		16
Rheology (flow properties, viscoelasticity)		14
Additives		15
Synthesis of polymer reactants		17
Polymer technology		18

Appendix D
Polymer Core Course
Committees

In 1979/80, the American Chemical Society Committee on Professional Training noted the following:

> In light of the current importance of inorganic chemistry, biochemistry, and polymer chemistry, advanced courses in these areas are especially recommended and students should be strongly encouraged to take one or more of these. Furthermore, the basic aspects of these three important areas should be included at some place in the core material.

In light with this directive, the Polymer Education Committee formed committees that focused on the integration of polymer topics in the foundational courses. Committees, hereafter called the Core Course Committees, were formed to develop avenues where polymers would be included in the foundational courses to enhance these courses. Polymer topics, principles, and illustrations were identified that would help and enhance these courses. The reports of these committee deliberations were published in the *Journal of Chemical Education* as follows:

- Introduction (describing the overall project) 60(11):971 (1983)
- General Chemistry 60(11):973 (1983)
- Inorganic 61:230 (1984)
- Physical 62:780 (1985) and 62:1030 (1985)
- Chemical Engineering 62:1079 (1985)

These reports act as a starting point for those teaching the specific foundational courses to introduce polymers in these foundational courses.

The newest guidelines call for the intergradation of polymer topics within all of the foundational courses. With this in mind, new committees are being formed to assist teachers to fulfill this.

Appendix E
Structures of Common Polymers

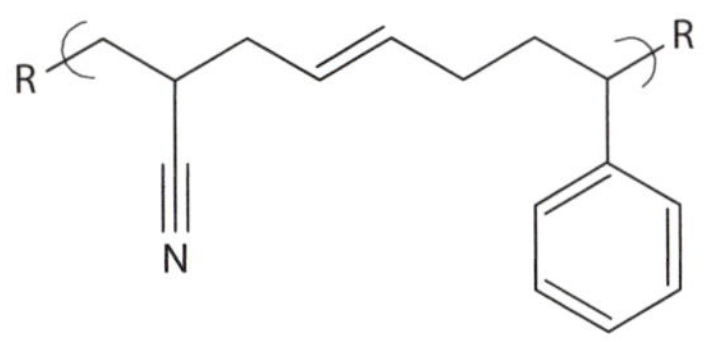

Acrylonitrile–butadiene–styrene (ABS)

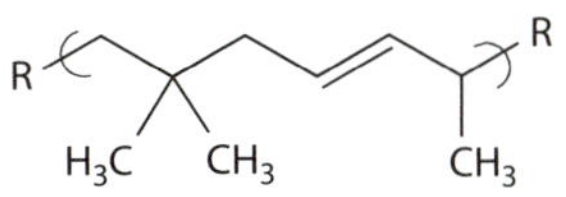

Butyl rubber

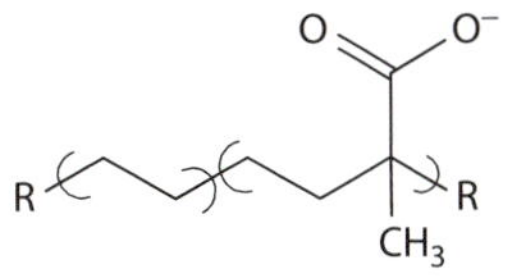

Ethylene–methacrylic acid copolymer (Ionomer)

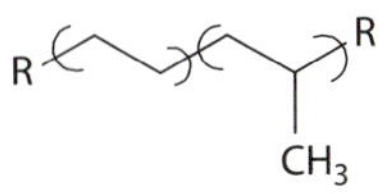

Ethylene–propylene elastomer

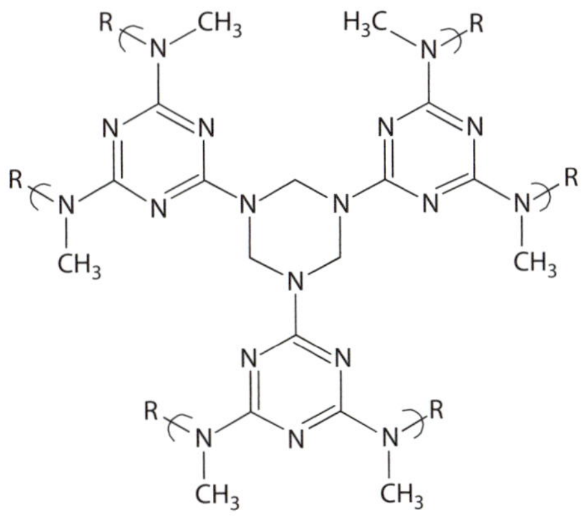

Melamine–formaldehyde resin (MF)

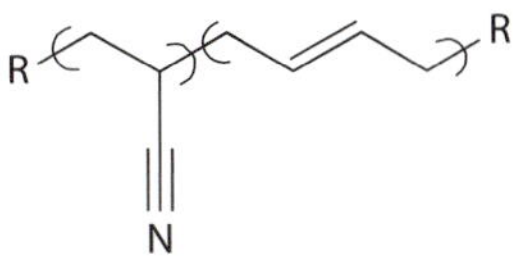

Nitrile rubber (NRB)

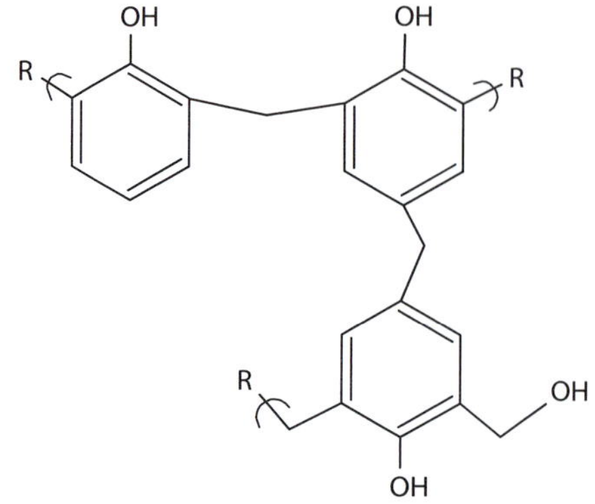

Phenol–formaldehyde resin (PF)

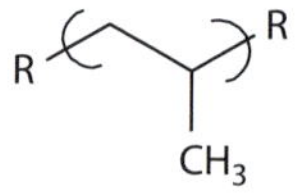

Polyacetaldehyde

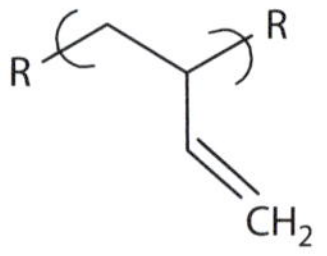

Polyacrolein

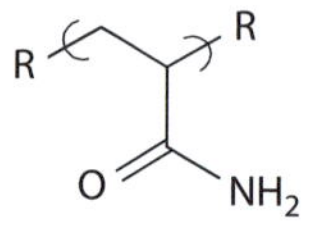

Polyacrylamide

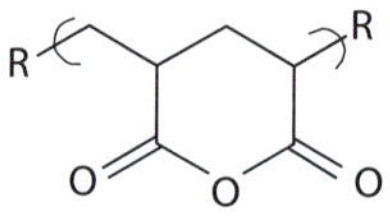

Poly(acrylic anhydride)

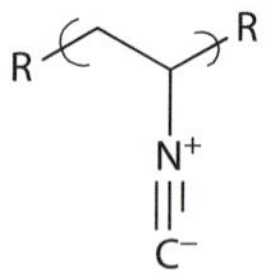

Polyacrylonitrile

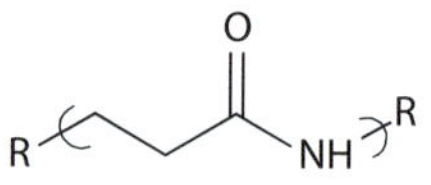

Poly(β-alanine);
nylon-3

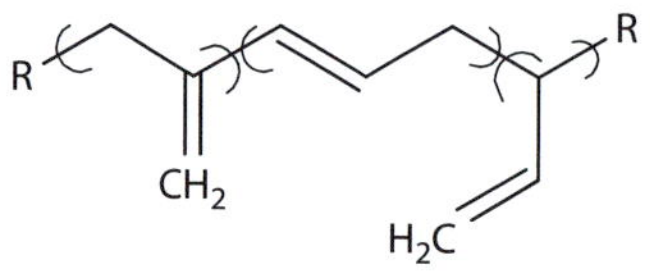

Polyallene

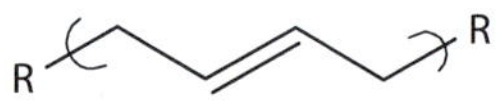

Polybutadiene, butadiene
rubber (BR)

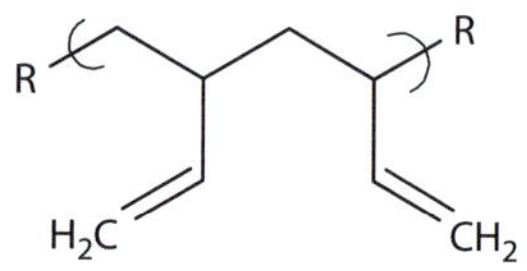

1,2-Polybutadiene

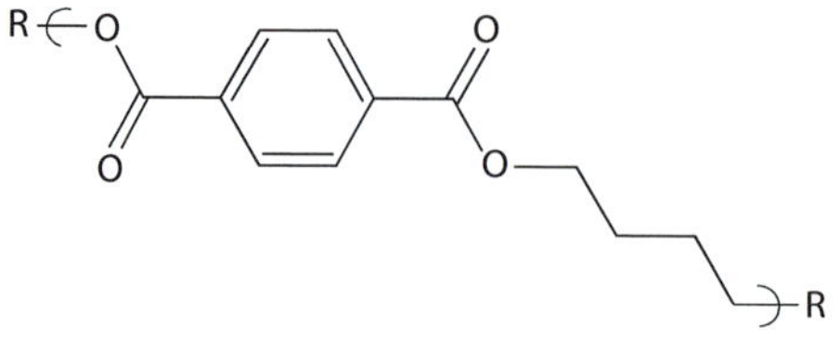

Poly(butylene terephthalate) (PBT)

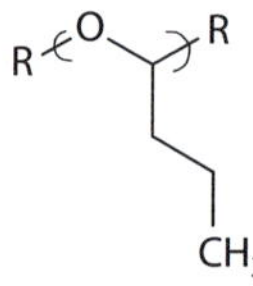

Poly(n-butyraldehyde)

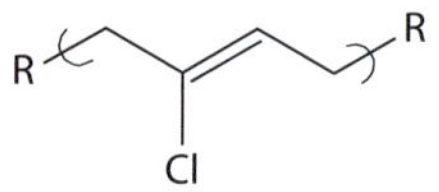

Polychloroprene

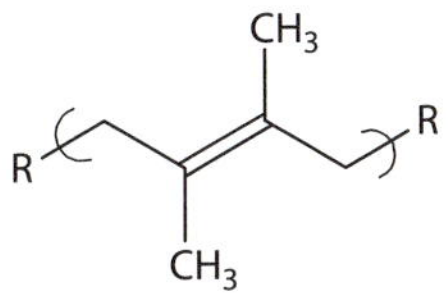

Poly(2,3-dimethylbutadiene)

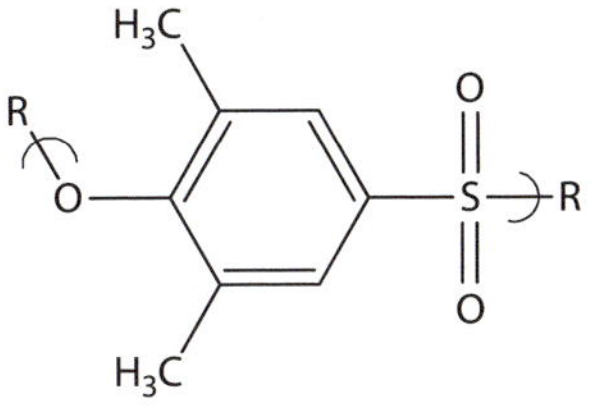

Poly(3,5-dimethyl-1,4-phenylene sulfonate)

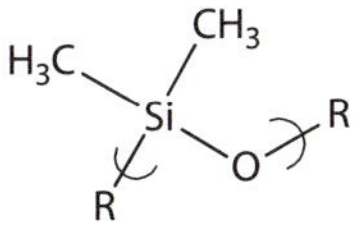

Polydimethylsiloxane

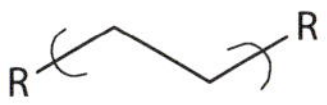

Polyethylene (PE)

Poly(ethylene glycol) (PEG)

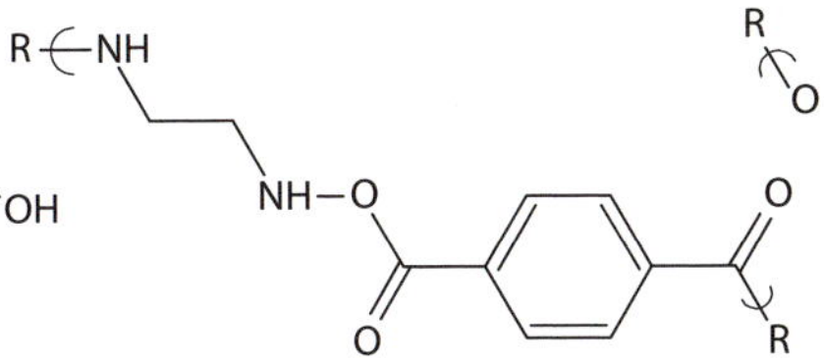

Poly(ethylene terephthalamide)

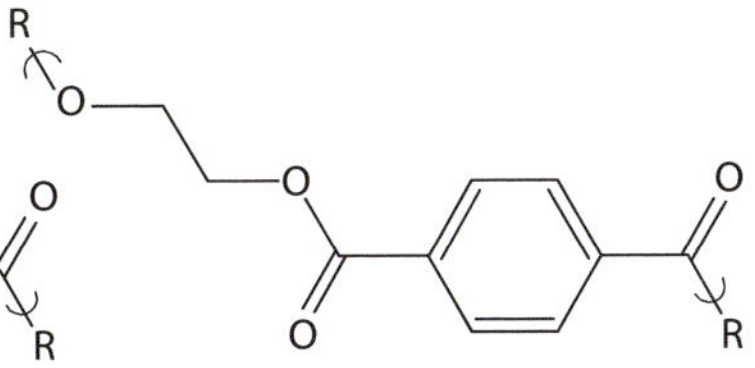

Poly(ethylene terephthalate) (PET)

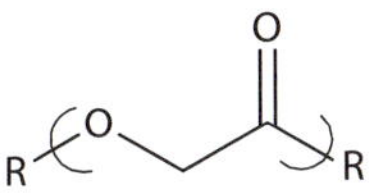

Poly(glycolic ester)

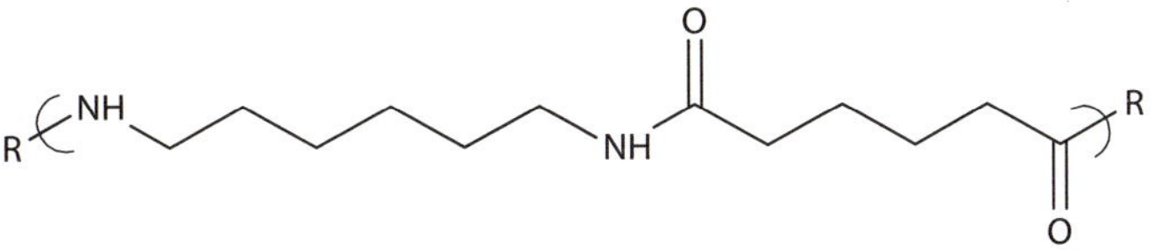

Poly(hexamethylene adipamide) (nylon-66)

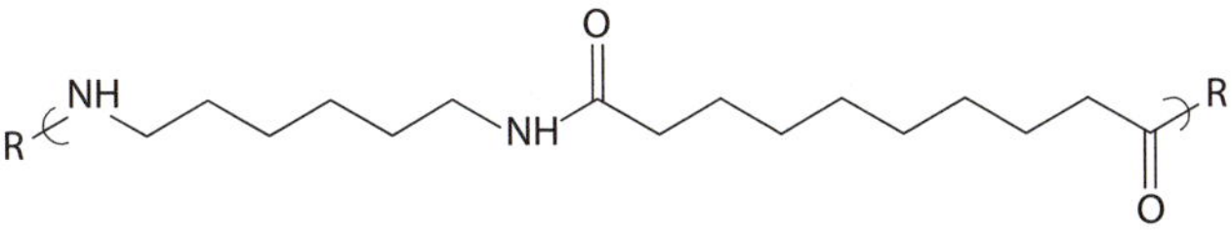

Poly(hexamethylene sebacamide) (nylon-610)

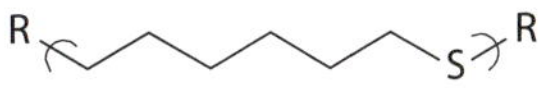

Poly(hexamethylene thioether)

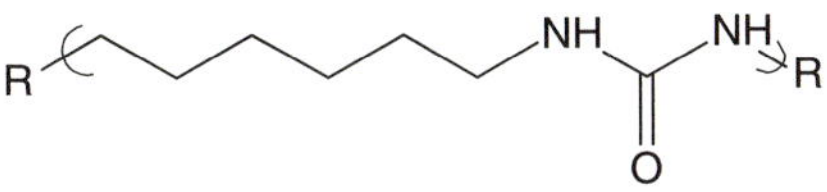

Poly(hexamethylene urea)

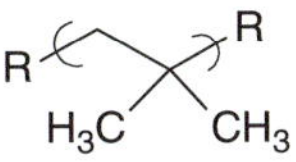

Polyisobutylene (PIB)

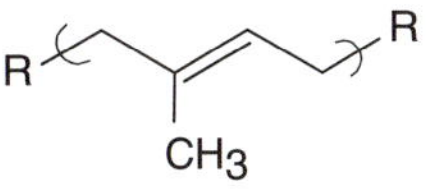

Polyisoprene

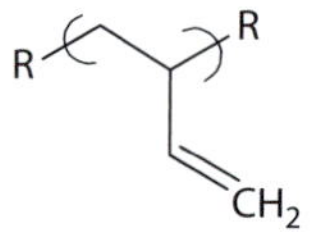

Poly-3,4-isoprene

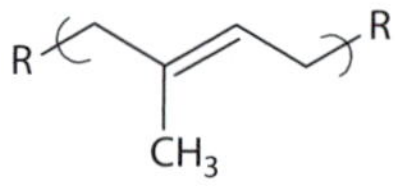

Poly-*trans*-1,4-isoprene

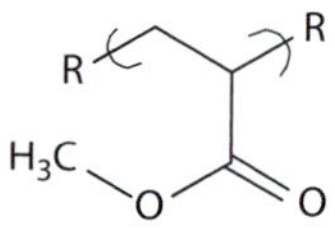

Poly(methyl acrylate)

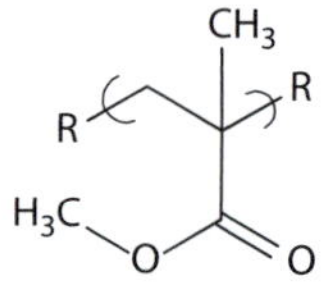

Poly(methylmethacrylate)

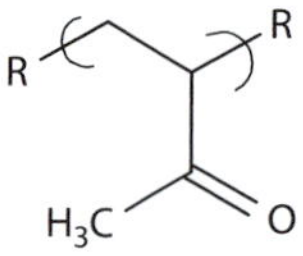

Poly(methyl vinyl ketone)

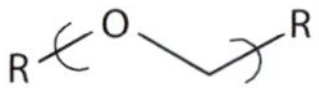

Polyoxymethylene polyacetal

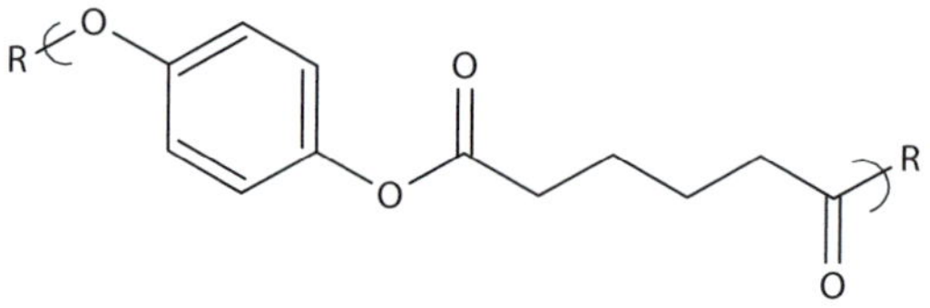

Poly(1,4-phenylene adipate)

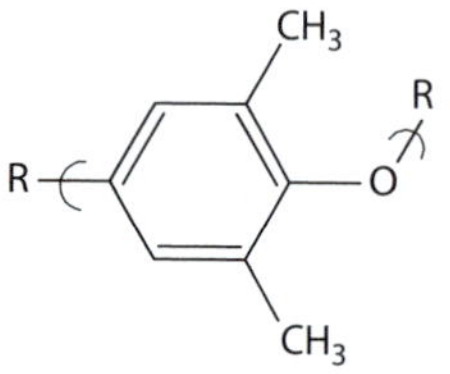

Poly(phenylene oxide) (PPO) Poly(2,6-dimethyl-p-phenylene ether)

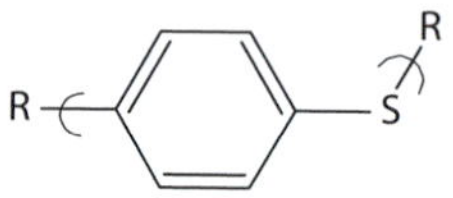

Poly(phenylene sulfide) (PPS)

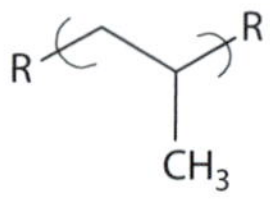

Polypropylene (PP)

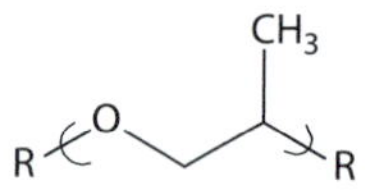

Poly(propylene glycol) (PPG)

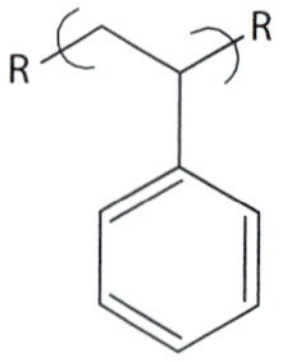

Polystyrene

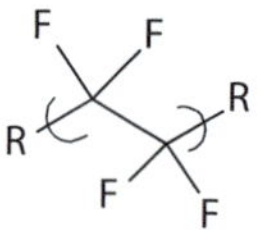

Polytetrafluoroethylene (PTFE)

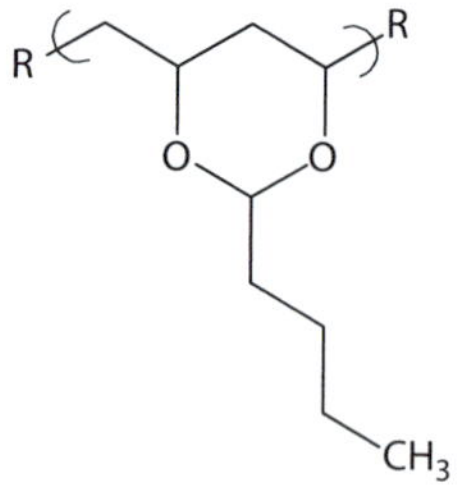

Poly(vinyl butyral)

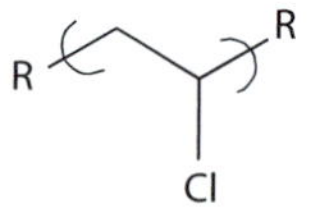

Poly(vinyl chloride) (PVC)

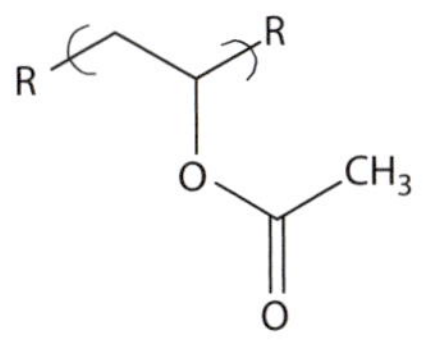

Poly(vinyl acetate)
(PVAc)

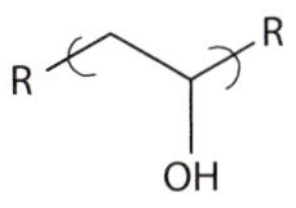

Poly(vinyl alcohol)
(PVA)

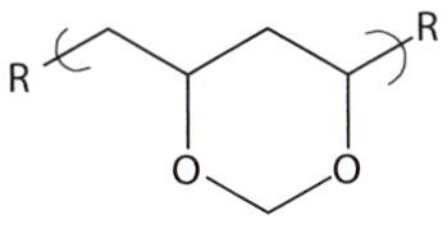

Poly(vinyl formal)
(PVF)

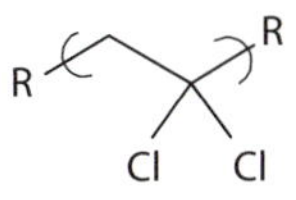

Poly(vinylidene
chloride)

Poly(vinyl pyridine)

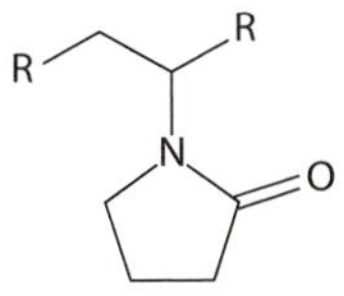

Poly(vinyl pyrrolidone)

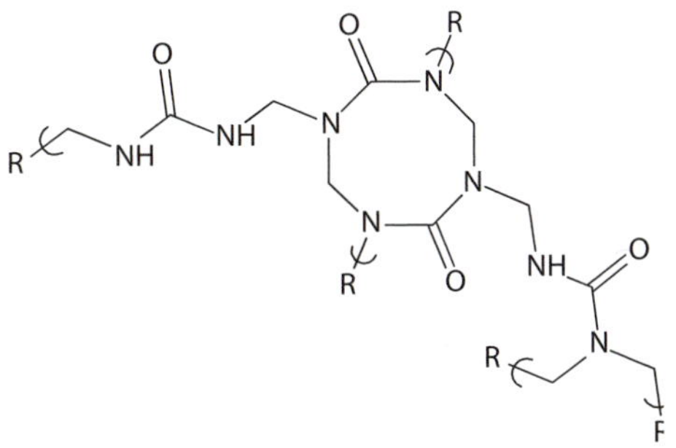

Urea–formaldehyde resin (UF)

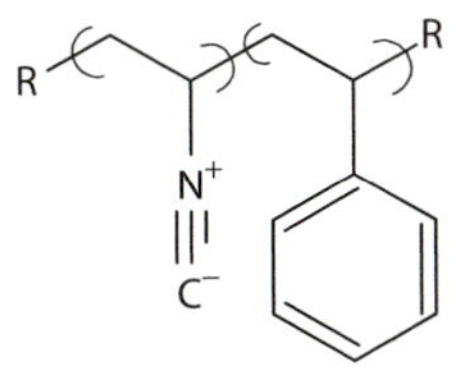

Styrene–acrylonitrile (SAN)

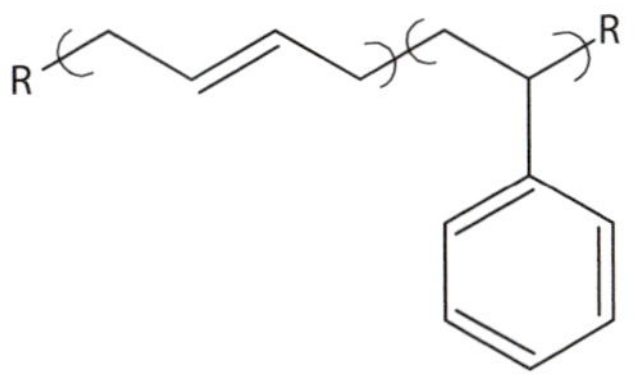

Styrene–butadiene rubber (SBR)

Appendix F
Mathematical Values and Units

TABLE F.1
Prefixes for Multiples and Submultiples

Multiple/Submultiple	Prefix	SI Symbol
10^{12}	tetra	T
10^{9}	giga	G
10^{6}	mega	M
10^{3}	kilo	k, K
10^{2}	hecto	h
10^{1}	deka	da
10^{0}		
10^{-1}	deci	d
10^{-2}	centi	c
10^{-3}	milli	m
10^{-6}	micro	μ
10^{-9}	nano	n
10^{-15}	femto	f
10^{-18}	atto	a

TABLE F.2
Units of Measure

Quality	Unit	SI symbol	Formula
Acceleration			m/s^2
Amount of substance	mole	mol	
Bulk modulus			N/m^2
Chemical potential	joule	J	N m
Compressibility			1/Pa
Density			kg/m^3
Electrical charge	coulomb	C	A s
Electrical capacitance	farad	F	A s/V
Electrical conductivity	siemens	S	A/V
Electrical current	ampere	A	
Electrical field strength			V/m
Electrical inductance	henry	H	V s/A
Electrical resistance	ohm	Ω	V/A

continued

TABLE F.2 (continued)
Units of Measure

Quality	Unit	SI symbol	Formula
Electromotive force	volt	V	W/A
Energy	joule	J	N m
Enthalpy	joule	J	N m
Entropy			J/K
Force	newton	n	kg m/s^2
Frequency	hertz	Hz	cycles/s^2
Gibbs free energy	joule	J	N m
Heat capacity			J/K
Heat flow			J/s m^2
Length	meter	m	
Illuminance	lux	lx	l m/m^2
Luminance			cd/m^2
Luminous flux	lumen	lm	cd sr
Luminous intensity	candela	cd	
Magnetic field strength			A/m
Magnetic flux	weber	Wb	V s
Magnetic flux density	tesla	T	Wb/m^2
Magnetic permeability			H/m
Magnetic permittivity			F/m
Mass	kilogram	kg	
Power	watt	W	J/s
Pressure	pascal	Pa	N/m^2
Resistivity			Ω m
Shear modulus			N/m^2
Surface tension			N/m
Temperature	kelvin	K	
Thermal conductivity			W/mk
Thermal expansion			1/K
Time	second	s	
Velocity			m/s
Viscosity (dynamic)			Ns/m^2
Viscosity (kinematic)			m^2/s
Voltage	volt	V	W/A
Volume			m^3
Wavelength			1/m
Work			N/m
Young's modulus			N/m^2

TABLE F.3
Physical Constants

	Symbol	SI	cgs
Acceleration (due to gravity at earth's surface at Equator)	g	9.7805 m/s^2	9.7805×10^2 cm/s^2
Avogadro's constant	N$_a$	6.02252×10^{23} 1/mol	6.02252×10^{23} 1/mol
Boltzmann's constant	k	1.3806×10^{-23} J/K	1.3806×10^{-16} erg/K
Electron charge		1.602×10^{-19} C	1.602×10^{-20} emu
Faraday's constant	F	9.6487×10^4 1/mol	9.6487×10^3 emu/mol
Gas constant	R	8.314 J/mol K	1.987 cal/mol K
Gradational constant	G	6.67×10^{-11} N m^2/kg^2	6.67×10^{-8} dyne cm^2/g^2
Permittivity of a vacuum	go	8.84×10^{-12} F/m^2	1.0 dyne cm^2/statcoul^{-2}
Permeability of a vacuum	μ	1.25×10^{-6} H/m	
Planck's constant	h	6.626×10^{-34} J s	6.626×10^{-27} erg s
Velocity of light in a vacuum	c	2.9979×10^8 m/s	2.9979×10^{10} cm/s

Appendix G
Comments on Health

Most polymers are nontoxic under the normal and intended use. (Some biopolymers, such as snake venom, should not be dealt with except under very controlled conditions.) Most of the additives employed are also relatively nontoxic. Even so, care should be exercised when dealing with many of the monomers of synthetic polymers and when dealing with polymeric materials under extreme conditions such as in commercial and domestic fires.

G.1 FIRE

Fire hazards involve not only burning (most deaths occur from the ingestion of volatiles produced by the fire). Carbon monoxide, the major cause of death, causes unconsciousness in less than 3 minutes due to its preferential attack on hemoglobin.

Interestingly, one important observation concerning burning in general is where colored smoke is produced. Some materials burn producing lots of dark-colored smoke. Some of this colored smoke may be due to the production of aromatic substances containing fused-ring systems that may contain harmful chemicals, including respiratory toxins and cancer-causing agents.

G.2 MEASURES OF TOXICITY

Toxicity involves the affect of various materials on living objects, including bacteria, plants, mice, fish, and humans. Tests to determine the toxicity of materials are typically done in a number of ways, including inhalation, simple skin contact, and injection under the skin.

While mainly concerned with the affects of various agents on humans, most standard tests are carried out on animals, often a suitable test animal that is believed to be a good model for transferring results from the animal tests to humans. Table G.1 contains some of the toxicity values found in today's literature.

While commercially available, synthetic polymers are relatively nontoxic, the monomers vary greatly in toxicity. This points out the need for monomers and other potentially toxic chemicals to be removed from the polymers. Table G.2 contains the time–weight average (TWA) for some monomers as cited by the U.S. Occupational Standards. For comparison, entries for some well-known toxic materials have been added.

G.3 CUMULATIVE EFFECTS

While exposure of the general public to toxins is to be avoided, exposure of people that deal with commercial chemicals on a daily basis is even more important. Such people must take special care to avoid exposure since the toxicities of many of these chemicals are accumulative in our bodies slowly building to levels that may be unhealthy.

Most of the toxic, environmentally unwanted chemicals of a decade ago have been eliminated from the common workplace. This includes halogenated hydrocarbons such as carbon tetrachloride and aromatic hydrocarbons such as benzene and toluene. Further, chemicals that are known to be potentially toxic, such as some monomers, are being eliminated from the polymeric materials to within the limits of detection.

TABLE G.1
Description of Toxic Measures

TDLo/Tpxic Dose Low—The lowest dose introduced by any route other than inhalation over any period of time that produces any toxic effect in humans or to produce carcinogenic, teratogenic, mutagenic, or neoplastic effects in humans and animals

TCLo/Toxic Concentration Low—Any concentration in air that causes any toxic effect in humans or produces a carcinogenic, teratogenic, mutagenic, or neoplastigenic toxic effect in humans or animals

LDLo/Lethal Dose Low—The lowest dose introduced by any route other than by inhalation over a time to have caused death in humans or the lowest single dose to have caused death in animals

LD$_{50}$/Lethal Dose Fifty—A calculated dose expected to cause the death of 50% of a tested population from exposure by any route other than inhalation

LCLo/Lethal Concentration Low—The lowest concentration in air to have caused death in a human or animal when exposed for 24 h or less

LC$_{50}$/Lethal Concentration Fifty—A calculated concentration of a substance in air that would cause death in 50% of a test population from exposure for 24 h or less

EEGL/Emergency Exposure Guideline Level—Exposure limits for very short exposure

WEEL/Workplace Environmental Exposure Level—Exposure limits for healthy workers exposed repeatedly without adverse health effects

OEL/Occupational Exposure Limits—Worker exposure guide

PREL/Permissible Exposure Limits—Worker exposure limits for no ill effect

NOAEL/No Observable Adverse Effect Level—Safe usage level

LOAEL/Lowest Observable Adverse Effect Level—Safe usage level

TABLE G.2
TWA Values for Selected Monomers and Additional Recognized Toxins

Chemical	TWA (ppm)	Chemical	TWA (ppm)
Acetic anhydride	5	Ethylene oxide	50
Acrylonitrile	20	Formaldehyde	3
Benzene	10	Hydrazine	1
1,3-Butadiene	1,000	Hydrogen cyanide	10
Carbon monoxide	50	Phenol	5
Chloroprene	25	Styrene	100
1,2-Ethylenediamine	10	Vinyl chloride	500

Industrial recognition of customer and employee safety is a major factor and included in such international programs as ISO 9000 and ISO 14000. Further, a number of national and state agencies and associations deal with aspects of the environmental and personal safety issues on an ongoing basis.

G.4 ENVIRONMENT

Today, industry and business recognize that part of doing business is taking care of the environment. Along with various agencies, business and industry are concluding that good environmental practices are good business. Advances continue with respect to lowering potentially harmful emissions into our water and air shared by all of us. Chemical industries are taking the lead in this clean-up process. A combination of watchful vigilance and trust is needed to continue this effort.

Appendix H
ISO 9000 and 14000

The International Organization for Standardization (ISO) is an international organization with members in about 100 countries working to develop common global standards.

The ISO 9000 series encompasses the product development sequence from strategic planning to customer service. Currently, it is a series of five quality system standards with two of the standards focusing on guidance and three contractual standards.

ISO 9000 certification is often obtained to promote a company's perceived quality level, for supplier control, and to promote certain management practices, often total quality management (TQM) practices. It is acting as a global standardizing "tool" with respect to business and industry in its broadest sense, including banking, volunteer organizations, and most aspects of the chemical (including the polymer) industry.

ISO 9000 requires what is called a "third party" assessment but involves developing "first" and "second" party strategies. "First party" refers to the supplier company that requests ISO 9000 certification. "Second party" refers to the customer whose "needs" have been met by the "first party" through the use of quality management procedures achieved through ISO 9000 compliance. "Third party" refers to an outside reviewer that "certifies" that the "first party" has satisfied ISO 9000 procedures.

While ISO 9000 is a management tool, it affects the way "industry does business" and deals with quality control issues such as how machinery and parts manufactured by a company are monitored for quality. It focuses on satisfying the "customer," the "purchaser" of the raw materials, manufactured parts or assembled items, and includes the eventual "end-customer"—the general public. It is an attempt to assure quality goods.

ISO 14000 is a series of standards intended to assist in managing the impact of manufactured materials, including finished products and original "feedstocks." It addresses the need to have one internationally accepted environmental management system that involves "cradle to grave" responsibility for manufactured materials emphasizing the impact of products, operations, and services on the environment.

Appendix I
Electronic Education Web Sites

The information concerning polymers on the World Wide Web is rapidly expanding. This is a valuable source for information giving both applied and fundamental information on a wide range of polymer-related topics. As you search, please be aware that there exist specific pooled information sites on many topics, including those considered "hot" topics such as the human genome. There are also web clusters that deal with special topics such as nanomaterials and electrically conductive materials. Have fun "surfing the web." It is an important source of information about polymers.

Sites that you should consider visiting and that are not obvious because of their names are as follows:

- www.polyed.org, which is a general Web site for PolyEd, the joint polymer education committee of the American Chemical Society divisions of Polymer Chemistry and Polymeric Materials: Science and Engineering. It contains many connections to other important Web sites dealing with polymers.
- www.uwsp.edu/chemistry/ipec, which is a general Web site for the Intersociety Polymer Educational Council which is a joint society venture that focuses on K-12 science education employing polymers as the connective vehicle. The Polymer Ambassadors have their own site at www.polymerambassadors.org.
- www.pslc.ws/macrog/index.htm or simply type in "macrogalleria" and you will be taken to a fun and educational Web site that focuses on the relationship between everyday items and their polymeric nature and the fundamentals that underlie their use in these materials.

Appendix J
Stereogeometry of Polymers

The precise stereogeometry of molecules is important in determining the physical properties of a material and it is critical in determining the biological properties of materials. Most synthetic and nonspecific natural polymers are a mix of stereoshapes with numerous stereocenters along the polymer chain. For polypropylene, every other backbone carbon is most likely a stereocenter. Even polyethylene has stereochemical sites whenever there is branching. The imprecise structures of most natural nonspecific polymers such as the polyisoprenes and polysaccharides have stereocenters at each branch.

For stereospecific natural polymers the entire geometry is critical to the proper behavior and activity of the material.

Thus, stereogeometry is important and a brief review is in order.

We have two general types of isomers—constitutional isomers that have the same number and kind of atoms but connected in a different order, such as cis and trans arrangements, and stereoisomers. We have two types of steroisomers—diastereomers that are not mirror images of one another and *enantiomers* that are mirror image stereoisomers. Just as our hands cannot be superimposed on its mirror image, enantiomers are nonsuperimposible on one another. *Chiral* objects or sites are enantiomeric. In fact, the term "chiral" comes from the Greek word for hand, *kheir*.

By compairson, an *achiral* object, like a simple nail, ball, basket, white pocket-less T-shirt, are all superimposible on their mirror image.

A collection containing only one enantiomeric form of a chiral molecule is referred to by several names, including enantiopure, enantiomerically pure, or *optically pure*. A sample containing predominantly one enantiomer is called *enantiomerically enriched* or *enantioenriched*. A collection containing equal amounts of two enantiomeric forms of a chiral molecule is called a *racemic mixture* or *racemate*.

Unlike other stereoisomers, enantiomers have essentially identical physical properties and consequently are difficult to separate. A process where enantiomers are separated is called *resolution*.

Diastereomers are nonenantiomeric isomers that result when more than one stereocenter is present in a molecule. The distinction between diastereomers and enantiomers is not always clear but in general, enantiomers have mirror images whereas diastereomers are not mirror images of one another. As such diastereomers have different physical properties, different boiling and melting points, solubilities, and so on.

The total number of stereoisomers due to tetrahedral stereocenters does not exceed 2^n, where n is the number of tetrahedral stereocenters. For a compound with two stereocenters n = 2, giving a total of four (maximum) steroisomers.

Following are four formula isomers first drawn in the more conventional straight-chain manner and below in more conformationally correct forms drawn for a compound, 2,3-dichlorohexane, with two stereocenters.

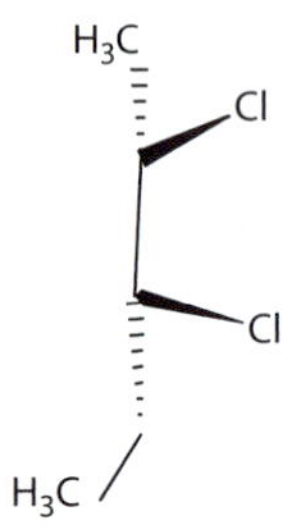

1. 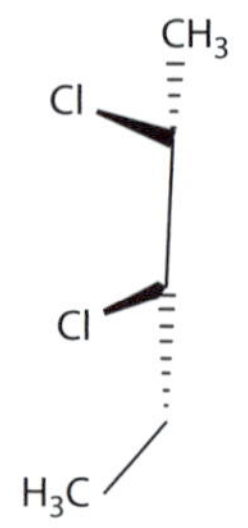   2. 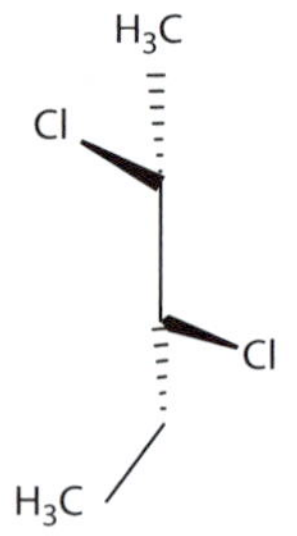   3. 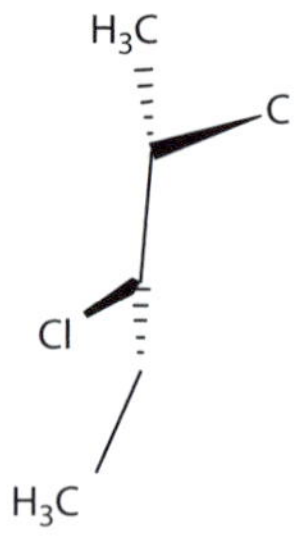   4.

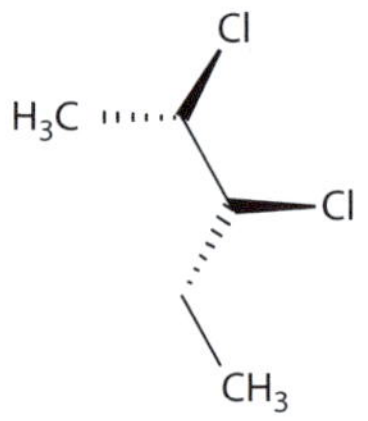

1. 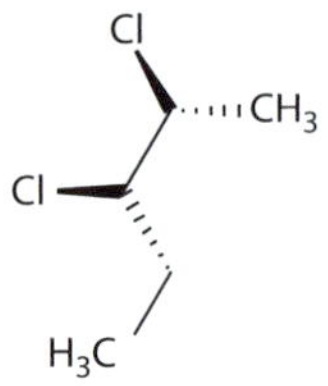   2. 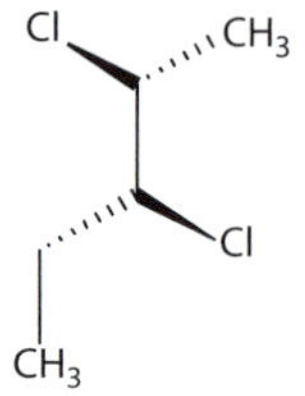   3. 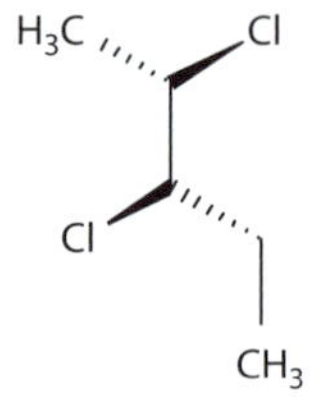   4.

Structures 1 and 3, as well as 2 and 4 and structures 1 and 4, and so on are steroisomers but they are not mirror images of one another, thus they are diastereomers. Structures 1 and 2, as well as 3 and 4, are mirror images of one another and so we have a pair of compounds that are not superimposable and they are enantiomers to one another, that is, 1 and 2 are enantiomers and 3 and 4 are enantiomers.

Enantiomers can rotate the plane of polarized light. If the rotation is positive then the enantiomer is given the symbol "+" or "d" and it is referred to as the dextrorotatory enantiomer. Counter, if the compound causes the light to be rotated in the negative direction, the compound is given the symbol "−" or "l" and referred to as the levorotatory enantiomer. An equal mixture of two enantiomers, racemates, does not rotate plane-polarized light because the rotation due to one enantiomer is canceled by that of the other. There is no relationship between the absolute configuration, S or R, and the direction of rotation of plane-polarized light (+ or −).

The absolute configuration about each stereogeometrical site is most often determined using the Cahn–Ingold–Prelog sequence rules. (These rules are found in most organic texts.) For tetrahedral carbons containing four different groups we determine the group with the highest priority and assign it the number 1 and the group with the lowest priority the number 4. We then view the molecule having at its center the particular stereocarbon in question and arrange at the top the number 1 group. In doing so we can see if the direction of going from 1 —> 2 —>3 is clockwise, R, or counterclockwise, S.

For structure 1 given above, looking only at the top stereocarbon, the arrangement is counterclockwise and so this particular site is designated as S. For structure 2 given above, again looking only at the top stereocarbon, the arrangement is clockwise and so that carbon is designated as R.

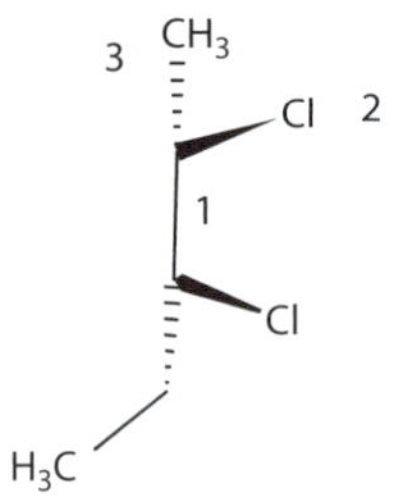

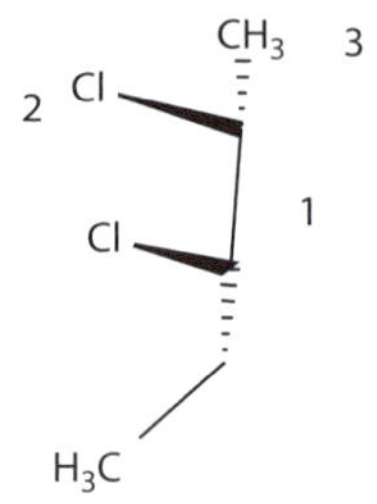

Now looking at only the second carbon, we have for the first compound a clockwise arrangement meaning it is R; and for the second compound, the arrangement is counterclockwise so it is designated as S.

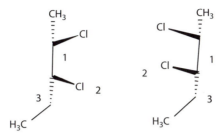

Thus, the two stereocarbons in the first compound would be designated as S, R or 2S, 3R and compound two as R, S or 2R, 3S with the numbers indicating the position of the carbon atoms.

Often, you will find that both the stereogeometry and rotation of light are given.

We can develop a concept map that describes the possible geometrical isomers as follows.

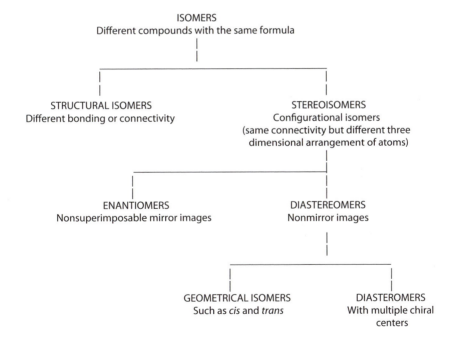

With the exception of alanine, all of the naturally occurring amino acids contain a chiral carbon adjacent to the amino acid grouping. All of these amino acids are of the l or L form meaning they rotate light in a negative direction. The rules governing specifying the absolute configuration are such that you can get both S and R forms of the amino acids. Thus, l-phenylalanine is an S enantiomer while l-cysteine is a R enantiomer.

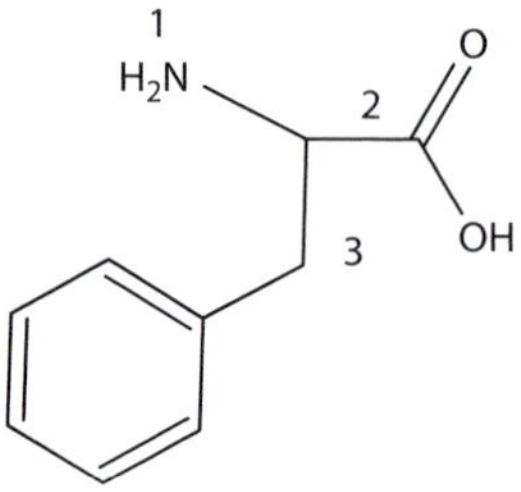

L-phenylalanine or
s-phenylalanine

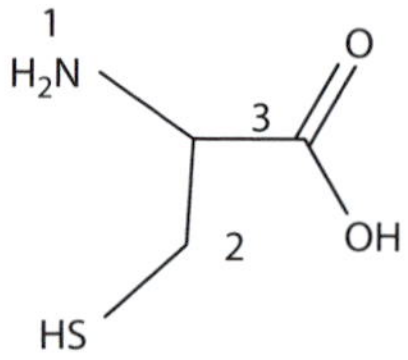

L-cysteine or
R-cysteine

As noted above, with the exception of alanine, the addition of amino acids to form polypeptides allows for a large number of sterochemical isomers to be formed, even considering that all are of the L form. But nature does not allow for this diversity but rather selects only one configuration for a sequence to occur in its synthesis of structural-specific proteins such as those employed as enzymes. Even those employed for other activities, such as muscle, have a specific geochemistry. In fact, the cell produces only geometry-specific polypeptides.

Nature is also selective in the geometry involved in nucleic acid synthesis. This specificity involves both the base order and the particular sugar employed. For DNA, the employed sugar is β-2-deoxy-D-ribose, deoxyribose (below left). Deoxyribose has three chiral centers but only one of them is employed in the synthesis of nucleic acids. Ribose, the sugar employed in the synthesis of ribonucleic acid (RNA), has four geometric sites (below right).

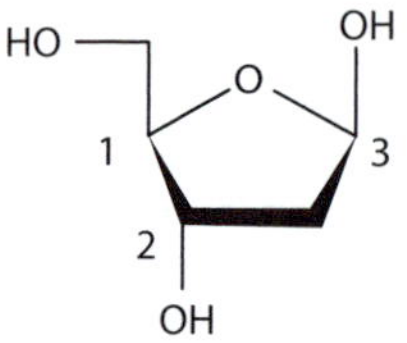

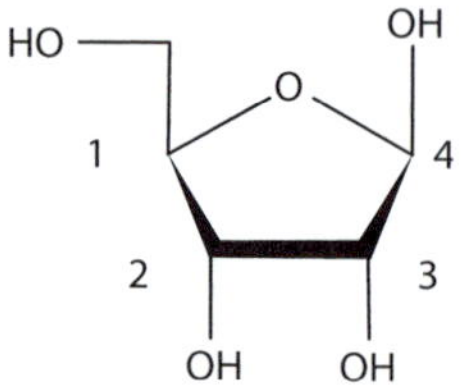

Now let us examine simple, only one site of substitution per repeat unit, vinyl polymers. When we look at a polymer chain we will focus only on combinations of diads or couples. For our discussion, we will use segments of poly(vinyl chloride). The geometries can be divided into three general groups. One where the substitutes, or here chloride atoms, are all identical with adjoining neighbors compose meso diads. Polymers or sections of polymers that contain meso diads are referred to as isotactic.

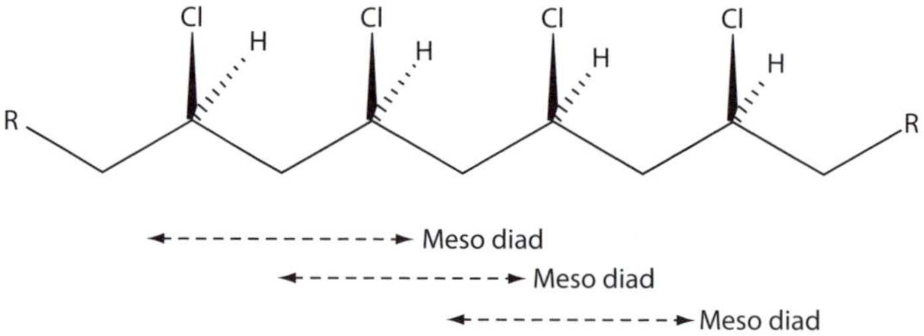

Meso diad

Meso diad

Meso diad

Isotactic poly(vinyl chloride) segment

The second grouping is where the geometry of the substitutes alternate on the chiral carbons that contain the chloride atoms. Here each diad is racemic. Such segments are referred to as syndiotactic segments.

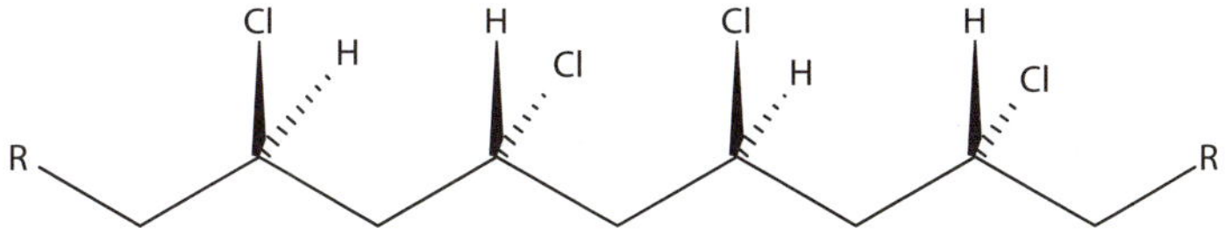

Racemic diad
Racemic diad
Racemic diad

Syndiotactic poly(vinyl chloride) segment

The third group consists of mixtures of racemic and meso diads. These sequences are given the name atactic or "having nothing to do with tacticity or orderly arrangement."

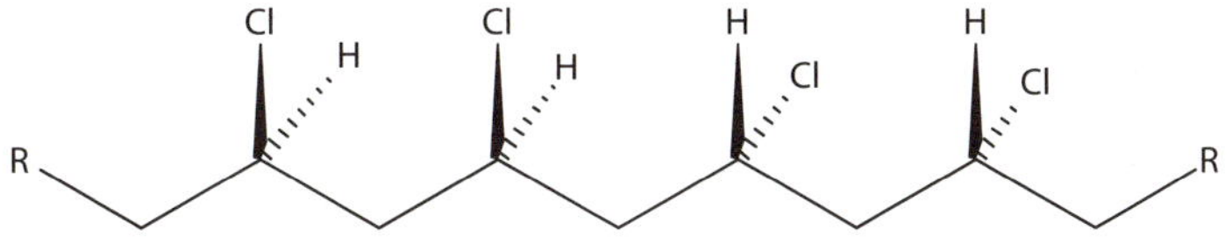

Meso diad
Racemic diad
Meso diad

Atactic poly(vinyl chloride) segment

Stereoregular polymers are those that contain large segments of ordered segments. In truth, even stereoregular polymers contain some atactic regions. Even so, polymers that contain large fractions of ordered segments exhibit a greater tendency to form crystalline regions and to exhibit, relative to those containing large amounts of atactic regions, greater stress/strain values, greater resistance to gas flow, greater resistance to chemical degradation, lower solubilities, and so on.

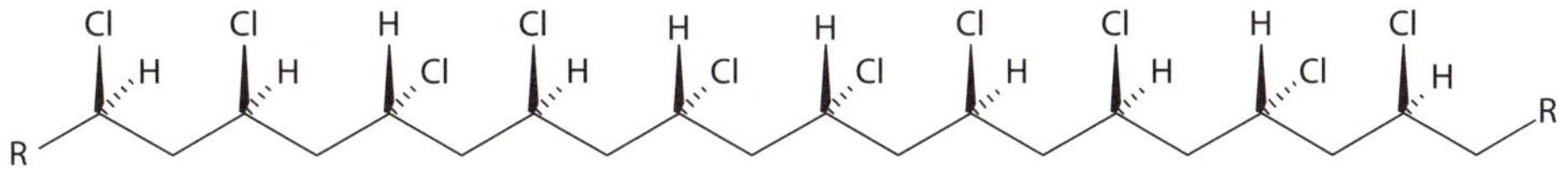

Atactic poly(vinyl chloride) segment

While the situation with respect to simple vinyl polymers is straight forward, the tacticity and geometrical arguments are more complicated for more complex polymers. Here we will only briefly consider this situation. Before we move to an illustration of this, let us view two related chloride-containing materials, pictured below. We notice that by inserting a methylene between the two chlorine-containing carbons the description of the structure changes from being racemic to meso. Thus, there exists difficulty between the historical connection of meso with isotactic and racemic with syndiotactic.

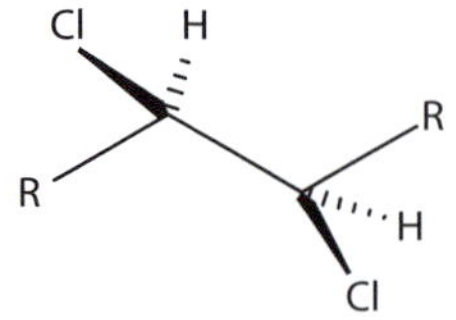

2R, 3R-dichloro diad
Racemic

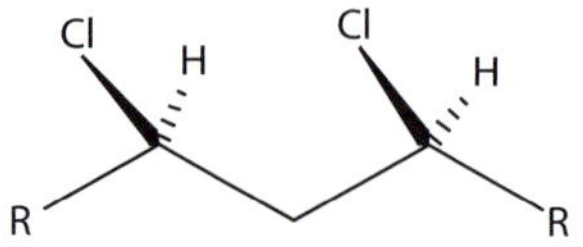

2R, 4S-dichloro diad plus inserted methylene
Meso

Let us now move to the insertion of another methylene forming the following segments. The first set contains racemic pairs.

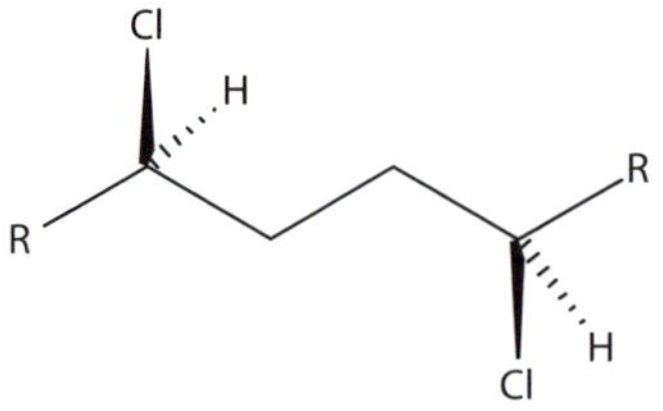

2R, 5R-dichloro segment
Racemic pairs

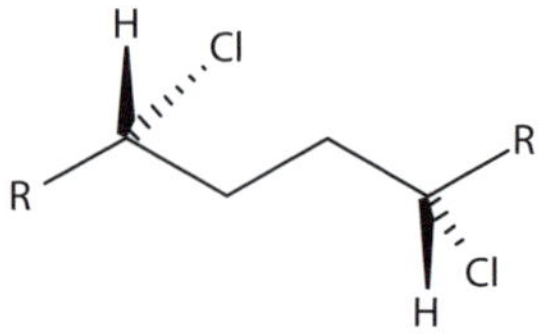

2S, 5S-dichloro segment

The second pair contain a meso pairing.

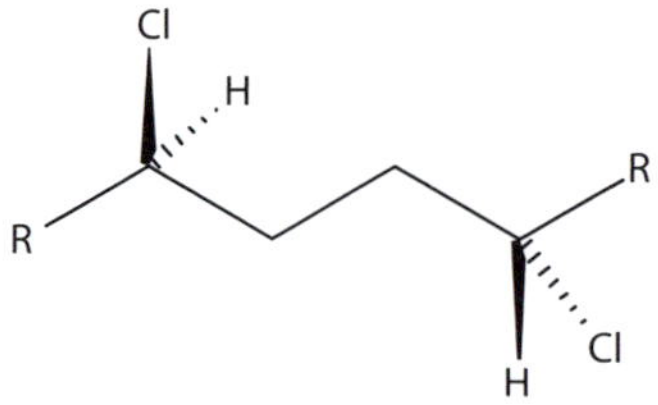

2R,5S-dichloro segment

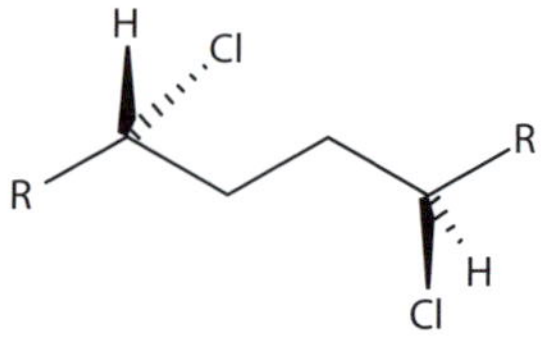

2S,5R-dichloro segment

Meso pair

Now let us look at triad segments of our poly(vinyl chloride). The first one had meso adjacent units and is isotactic by definition of the meso, racemic argument but the adjacent chlorides are not on one side of the plane.

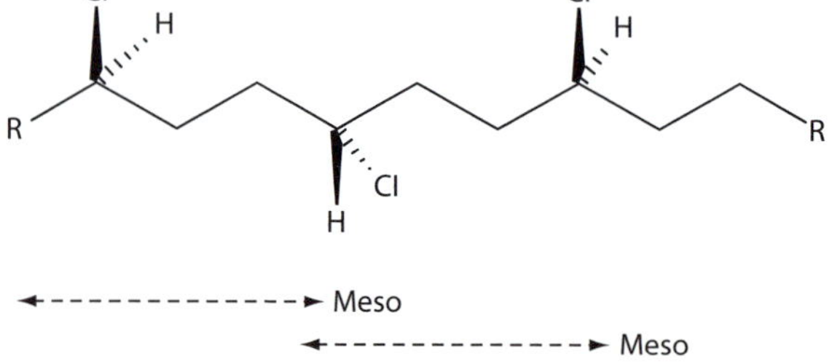

The second pair contains racemic diads and is syndiotactic by the meso, racemic argument but with the chloride atoms on the same side.

Syndiotactic

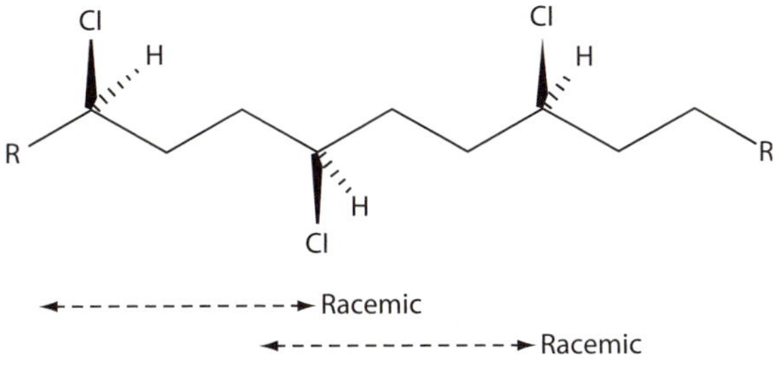

We will now consider segments of a polymer derived from the polymerization of propylene oxide. Here the simplest approach is to simply consider this an extension of the case immediately above except where the chloride atoms are substituted by methyl radicals and the next methylene is now an oxygen atom. Thus, we can make the same assignments based on the meso, racemic considerations.

Syndiotactic

Isotactic

Atactic

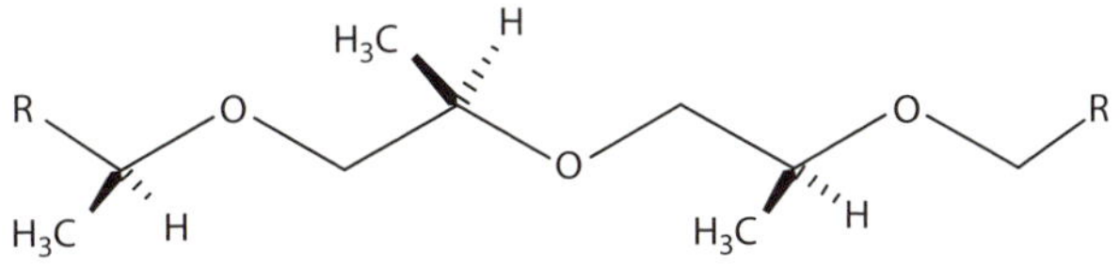

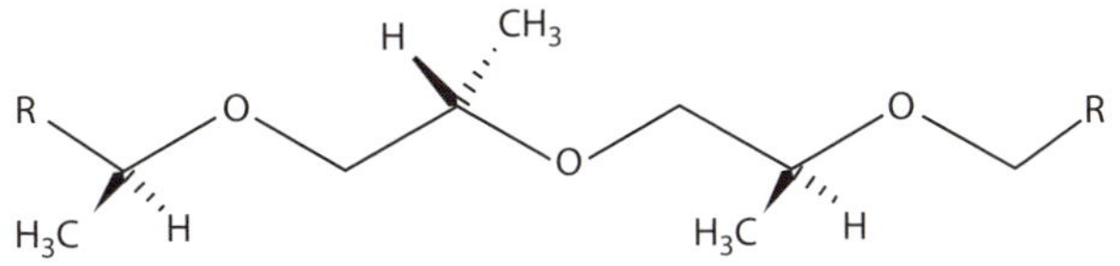

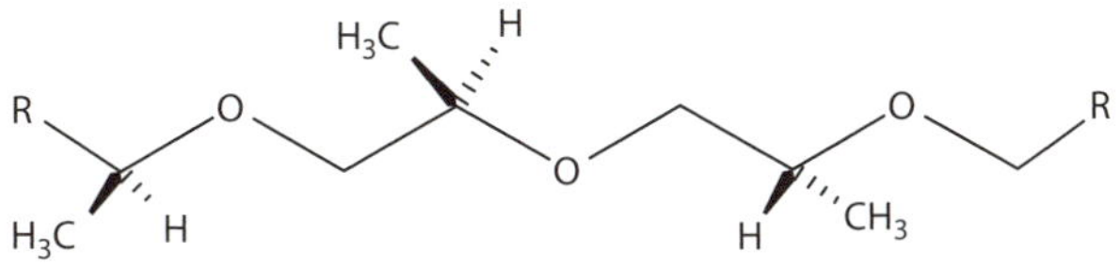

Similarity, we can make assignments for poly(lactic acid) except considering that the carbon next to the chloride-containing carbon has a methyl group, the next following methylene is now a carbonyl, and the next following methylene is an oxygen.

Syndiotactic

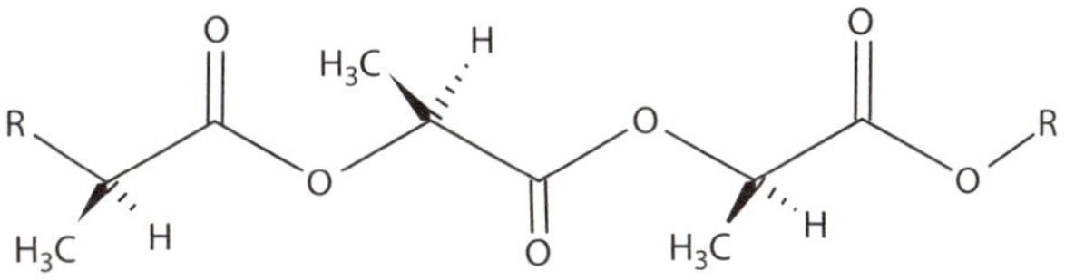

Isotactic

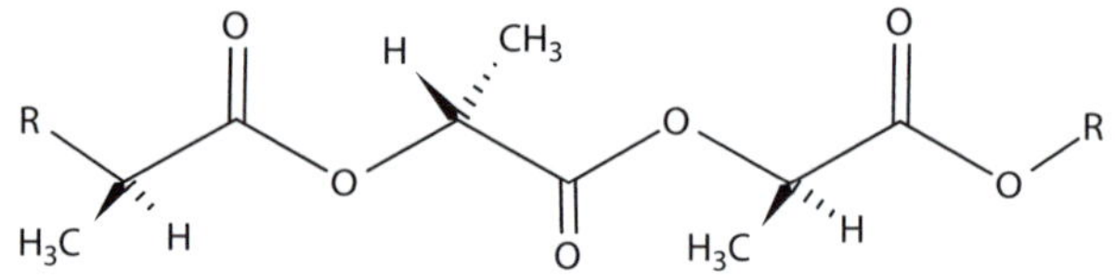

Atactic

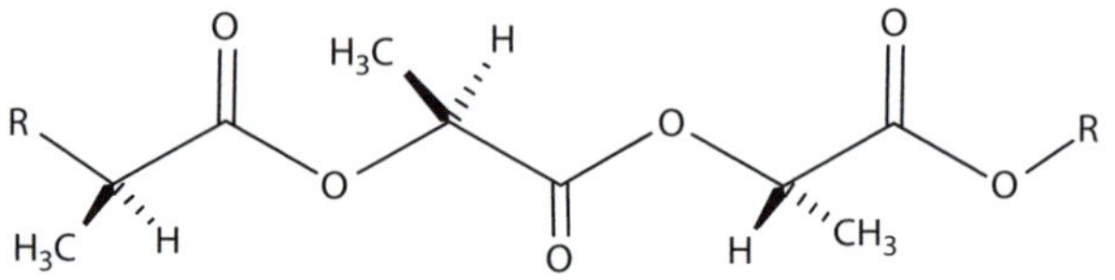

Appendix K
Statistical Treatment
of Measurements

In research and product development and control, there exists a variability in the particular value measured such as percentage yield, melting point, tensile strength, and electrical conductivity as you repeat the measurement. Accuracy concerns how close to the true value your measured value is. Unless there is an established value for a particular material, the values you obtain may well contribute to the "true" value. Precision deals with the closeness of a group of measurement to one another. Today, most modern instruments making spectral measurements collect many spectra in a short time and those measurements undergo some type of statistical treatment, such as FT-IR, so that the statistical treatment of these results has already been completed. In comparison many measurements are done more or less singularly. This is true for most physical testing measurements. Thus, to evaluate the tensile strength of a polycarbonate plastic sheeting sample "dog-bones" are cut from several sheets picked at random and tested under an appropriate set of conditions. These results are then statistically treated and the reported value given along with the variability. Following is a brief summary of one of the more common statistically treatments for such measurements.

The first step involves calculation of the average value, A, which is simply the summation of the individual values, A_i, divided by the number of measurements or observations, n. This is described mathematically as $A = (\sum A_i)/n$, where the summation is for all of the "n" values.

The most common statistical measure of the variability, dispersion, or scatter is the standard deviation, s, defined as

$$s = [(\sum (A_i - A)^2)/n - 1]^{1/2}$$

The smaller the value of "s," the greater the precision of the measurements. Some testing call for the precision to be within some "s" value such as one "s" value or two "s" or three "s," and so on.

Appendix L
Combinatorial Chemistry

Langer and coworkers synthesized a series of copolymers containing various amounts of diacrylate and amine monomers investigating copolymer composition with the ability to act as transport DNA into cells. They screened 140 copolymers as synthetic gene-delivery vectors. Of these 56 were able to bind to DNA. These polymers were then screened for their ability to facilitate the transfer of plasmid DNA into a common monkey cancer cell line. Two of the copolymers with quite varied compositions showed good activity—one expected and the other unexpected. The expected copolymer composition would have been a selected composition in a typical search and the other would have been omitted. Thus, combinatorial-like approaches can offer unexpected results to problems.

Appendix M
Polymerization Reactors

Polymerization can occur within glass ampules, large-scale batch reactors, within laboratory beakers, flow-through systems, and so forth. The processes used for small preparation in the research laboratory can be similar or dissimilar to that employed for the industrial-scale preparation of pound and larger quantities. While the kind or polymerization influences molecular weight and molecular weight distribution, polymer structure, and composition as well as some of the physical characteristics, the kind of reactor also influences these factors. The reactor must allow adequate temperature control, mix of reactants and, if needed, catalysts (and at times a number of additives), reactant homogeneity, blending/mixing, and so on. It must also allow for the economical "mass production" of the material. While there exists a wide variety of commercial reactors we will look at only three of the most used styles—batch, plug flow, and continuous stirred tank reactors (CSTR).

M.1 BATCH

In batch reactions, the reactants are added (charged) to the reactor, mixed for a specific time and temperature, and then removed (discharged). Batch reactors are generally simple and can vary from being relatively small (such as gallon size) to large (several hundred gallon size) with the reaction occurring under varying conditions throughout the reaction vessel with time giving products that vary with time and secondarily, location within the vessel. This second condition is referred to as the polymerization occurring under nonsteady state or unsteady state conditions.

The general material balance can be described as follows:

$$\begin{matrix} \text{Rate of monomer} \\ \text{flow into reactor} \end{matrix} = \begin{matrix} \text{Rate of monomer} \\ \text{flow from reactor} \end{matrix} + \begin{matrix} \text{Rate of monomer loss} \\ \text{through reaction} \end{matrix} + \begin{matrix} \text{Rate of polymer} \\ \text{accumulation} \\ \text{in reactor} \end{matrix}$$

In a batch system, the first two terms are zero since monomer is only added once and leaves only once, after the reaction is completed.

Thus,

$$0 = \text{Rate of monomer loss through reaction} + \text{Rate of polymer accumulation}$$

$$0 = d[M]/dt + R_p$$

or

$$-d[M]/dt = R_p$$

For free radical polymerization we have

$$R_p = k'[M][I]^{1/2} = k''[M] \quad \text{or}$$

$$dt = d[M]/k''[M]$$

Integration gives

$$\log([M]/[M_0]) = -k''t \text{ and}$$

$$[M] = [M_0]e^{-k''t} \quad \text{and}$$

$$\% \text{ Conversion} = 100 \,([M_0] - [M]/[M_0]) = 100(1 - e^{-k''t})$$

This was derived assuming uniform concentration so that good mixing is important for this relationship to hold. It also assumes a constant temperature. Both these assumptions are only approached in most batch systems. Further, stirring becomes more difficult as conversion increases so that both control of localized temperature and concentration becomes more difficult. In reality, this relationship holds for only a few percentage points of conversion. Overall, temperature is a major concern for vinyl polymerizations because they are relatively quite exothermic. This is particularly important for bulk polymerizations. This, coupled with the general rapid increase in viscosity leads to the Trommsdorff-like effects.

M.2   PLUG FLOW (TUBULAR)

A plug flow or tubular flow reactor is tubular in shape with a high length to diameter l/d, ratio. In an ideal case (as in most ideal cases such as an ideal gas, this only approached reality), flow is orderly with no axial diffusion and no difference in velocity of any members in the tube. Thus, the time a particular material remains within the tube is the same as that of any other material. We can derive relationships for such an ideal situation for a first-order reaction. One that relates extent of conversion with mean residence time, t, for free radical polymerizations is

$$[M] = [M_0]\, e^{-k''t} \quad \text{and} \quad k'' = -(1/\tau)\, \ln\,([M]/[M_0])$$

Again, while such relationships are important, they are approximate at best. For vinyl polymerizations, temperature control is again difficult with temperature increasing from the cooling reactor wall to the center of the tube, and along with high and different viscosities led to broad molecular weight distributions. Further, these factors contribute to differences in initiator and monomer concentrations again leading to even greater molecular weight distributions.

M.3   CONTINUOUS STIRRED TANK REACTOR

In the continuous stirred tank reactor, instant mixing to achieve a homogeneous reaction mixture is assumed so that the composition throughout the reactor is uniform. During the reaction, monomer is fed into the system at the same rate as polymer is withdrawn. The "heat" problem is somewhat diminished because of the constant removal of heated products and the addition of nonheated reactants.

In a CSTR, each reaction mixture component has an equal change of being removed at any time regardless of the time it has been in the reactor. Thus, in a CSTR, unlike the tubular and batch systems, the residence time is variable. The residential times can take the exponential form

$$R(t) = e^{-t/\tau}$$

where $R(t)$ is the residence time distribution, t is the time, and τ is the mean residence time, which is a ratio of the reactor volume to the volumetric flow rate. The residence time distribution influences the mixing effectiveness which in turn determines the uniformity of the composition and temperature of the reactants in the reactor and ultimately the primary and secondary polymer structure.

Table M.1 contains a listing of selected polymerization processes and most industrially employed reactor types.

TABLE M.1

Listing of Selected Polymerization Processes and the Most Industrially Employed Reactor Types

Polymerization Reaction	Polymerization Process	Batch	Plug Flow	CSTR
Step wise	Solution	X	X	
Chain-free radical	Bulk	X		X
	Solution	X	X	X
	Suspension	X		
	Emulsion	X		X
	Precipitation	X		X
Chain-ionic	Solution	X		
	Precipitation	X	X	X

Appendix N
Material Selection Charts

In the selection of a material for a specific application many considerations are involved. Today, for the most part charts and other relationships are computerized. Here we will look at their use but only by employing a graphical chart for illustration. Let us consider making a shaft for a blade that mixes salt water with nonsalt water for controlled saline irrigation. The shaft material must be strong and light weight and must be able to absorb twisting shear. While strength and weight are only two important considerations, we will focus on only these two. A mathematical relationship between weigh or mass and strength for a cylindrical shaft can be made such that

mass is proportional to [density/(shear stress)$^{2/3}$] times some safety factors.

This tells us that the best light-weight material to make our shaft out is a material with a low density/(shear stress)$^{2/3}$ ratio.

Often, the inverse of this ratio is employed and given the name performance index P. (There are performance indexes for many different relationships between various physical behaviors.) Here then

$$P = (\text{shear stress})^{2/3}/\text{density}$$

Taking the log of both sides gives

$$\log P = 2/3\log \text{ shear stress} - \text{density}$$

Rearrangement gives

$$\log \text{ shear strength} = 3/2 \log + 3/2 \log P$$

This expression tells us that a plot of the log of the shear strength versus log density will give a family of straight and parallel lines each with a slop of 3/2 with each straight light corresponding to a different performance index, P. These lines are called *design guidelines*. Figure N.1 contains a general plot of log shear strength versus density for a number of materials grouped together under common headings. For instance, polytetrafluoroethylene exists in the midrange, extreme right on the "Polymers" circle and so has a relatively high density and strength while polypropylene exists in the upper left corner of the "Polymers" circle and has a relatively low density and relatively good strength. Such charts allow the quick focusing in on the general type of material that exhibits needed characteristics. Today, most of this is done by computer.

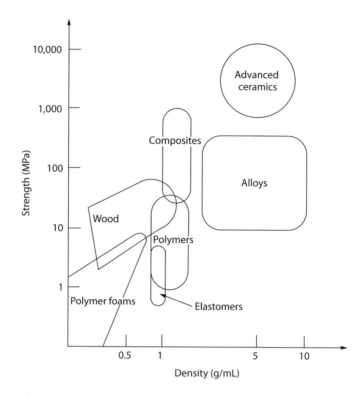

FIGURE N.1   Materials selection chart for a material's strength as a function of density.

Index

Note: *Italicized* page references denote figures and tables.

A

ABA block copolymer, 158, 230, 522
Abaca, 582
Abietic acid, 310, 604
Abrasion resistance, 169, 244, 267, 500
    of ionomers, 247
    measurement of, 499
Absolute molecular weight methods, 65, 86
Absolute time in pregroove (ATIP), 113
Absorption coefficient, 473
Accelerators, 515
Accessable mesogens, 576
Acetal resin, 700
Acetylene, 556, *557*, 565, 566, 567, 568
Achras sapota, 308
Acrea encedon, 367
Acrilan, 216
Acrylic acid, 555, 568
Acrylic fibers, 217, 597
Acrylics, 214, 216
Acrylonitrile, 556, 559, 566, 568
    anionic polymerization of, 156
Acrylonitrile–Butadiene–Styrene (ABS), *5*, 242
    copolymers, 176
    terpolymers, 242
Acrylonitrile copolymer, 176, 662
Active regions, 358
Addition polymerization, 150, 187
Addition polymers, 93, 393–399
    boron-containing polymers, 398–399
    ferrocene-containing, 395–397
    polyphosphazenes as, 397–398
Additives, in polymeric materials, 42, 156, 233, 413, 428, 467, 507
    antioxidant, 511–512
    antistatic agents, 516
    chain transfer, 198
    chemical blowing agents, 516–517
    colorants and curing agents, 515
    compatibilizer, 517
    curing agents, 515–516
    dielectric current as, 467, 468
    flame retardants, 513–514
    gasoline, 172
    heat stabilizers, 512–513
    impact modifiers & microorganism inhibitors, 517
    lubricant, 517
    oil, 178
    plasticizer, 507–511
    processing aid, 517
    PVC and, 210
    ultraviolet stabilizers, 513

Adhesion, 129, 218, 261, 603–604, 643
    measurement of, 476
Adhesives, 4, 122, 129, 156, 200, 213
    anaerobic, 605, 649–652, 690
    PVC, 604
Adipic acid, 556, 558–559
Aerogels, 421–423
Affinity chromatography, 66, 337, 374
Agar, 300, 301
Agarose, 300
Agave, 582
Aging, 315
Air-entrained concrete, 408
Alcaligenes eutrophus, 659
Alfrey-Price equation, 228
Alginic acid, 300
Alkali, 283
Alkyds, 107, 561
Alligator balloon, 311
Allophanate, 124
Allosteric enzymes, 538
    concerted model, 541
    sequential model, 541
Alloy, 126, 244–245, 273–274, 440
    metallic, 261
    silver, 113
Alpha-amylase, 295
α-cellulose, 282
α chains, 536
Alpha helix, 327, *327*, 328
Alpha-linkage, 283
Alternation copolymer, 223
Alternative splicing, 357
AluI, 359
Alumina, 263, 407, 408
Aluminosilicates and fly ash, 427–429
Amber, 311
American Chemical Society Committee on Professional Training, 731
American National Standards Institute (ANSI), 419, 449
American Society for Testing and Materials (ASTM), 257, 449
Aminatus, 426
Amino acid, 67, 321, 322–325, 536
    protein separation, 324
    structure of, *323*
Amino plastics, 133–135
Amino resin, 144, 715
Ammon's horn, 369
Amonton's Laws, 205
Amorphous, 24, 25, 35, 38, 42
    chain arrangements, 11
    and crystalline structures, 38